Edited by
Wolfgang Knoll and
Rigoberto C. Advincula

Functional Polymer Films

Related Titles

Ariga, K. (ed.)

Organized Organic Ultrathin Films

Fundamentals and Applications

2012

ISBN: 978-3-527-32733-1

Barner-Kowollik, C., Gründling, T., Falkenhagen, J., Weidner, S. (eds.)

Mass Spectrometry in Polymer Chemistry

2011

ISBN: 978-3-527-32924-3

Friedbacher, G., Bubert, H. (eds.)

Surface and Thin Film Analysis

A Compendium of Principles, Instrumentation, and Applications

2011

ISBN: 978-3-527-32047-9

Chujo, Y. (ed.)

Conjugated Polymer Synthesis

Methods and Reactions

2011

ISBN: 978-3-527-32267-1

Mathers, Robert T./Michael A. R. Meier (eds.)

Green Polymerization Methods

Renewable Starting Materials, Catalysis and Waste Reduction

2011

ISBN: 978-3-527-32625-9

Leclerc, M., Morin, J.-F. (eds.)

Design and Synthesis of Conjugated Polymers

2010

ISBN: 978-3-527-32474-3

Kumar, C. S. S. R. (ed.)

Nanostructured Thin Films and Surfaces

2010

ISBN: 978-3-527-32155-1

Advinaia, R., et. al.

Polymer Brushes

ISBN: 978-3-527-31033-3

Matyjaszewski, K., Advincula, R. C., Saldivar-Guerra, E., Luna-Barcenas, G., Gonzalez-Nunez, R. (eds.)

New Trends in Polymer Sciences

2009

ISBN: 978-3-527-32735-5

Matyjaszewski, K., Müller, A. H. E. (eds.)

Controlled and Living Polymerizations

From Mechanisms to Applications

2009

ISBN: 978-3-527-32492-7

Elias, H.-G.

Macromolecules

2009

ISBN: 978-3-527-31171-2

Matyjaszewski, K., Davis, T. P.

Handbook of Radical Polymerization

E-Book

ISBN: 978-0-470-35609-8

Edited by Wolfgang Knoll and Rigoberto C. Advincula

Functional Polymer Films

Volume 1
Preparation and Patterning

WILEY-VCH

WILEY-VCH Verlag GmbH & Co. KGaA

The Editors

Prof. Dr. Wolfgang Knoll
AIT Austrian Institute
of Technology GmbH
Donau-City-Straße 1
1220 Vienna
Austria

Prof. Dr. Rigoberto Advincula
University of Houston
Department of Chemistry
136 Fleming Bldg.
Houston, TX 77204-5003
USA

■ All books published by **Wiley-VCH** are
carefully produced. Nevertheless, authors,
editors, and publisher do not warrant the
information contained in these books,
including this book, to be free of errors.
Readers are advised to keep in mind that
statements, data, illustrations, procedural
details or other items may inadvertently be
inaccurate.

Library of Congress Card No.: applied for

**British Library Cataloguing-in-Publication
Data**
A catalogue record for this book is available
from the British Library.

**Bibliographic information published by the
Deutsche Nationalbibliothek**
The Deutsche Nationalbibliothek
lists this publication in the Deutsche
Nationalbibliografie; detailed bibliographic
data are available on the Internet at
<http://dnb.d-nb.de>.

© 2011 Wiley-VCH Verlag & Co. KGaA,
Boschstr. 12, 69469 Weinheim,
Germany

Composition   Laserwords Private Ltd.,
Chennai
Printing and Binding   betz-druck GmbH,
Darmstadt
Cover Design   Schulz Grafik-Design,
Fußgönheim

Printed in the Federal Republic of Germany
Printed on acid-free paper

ISBN: 978-3-527-32190-2

Contents of Volume 1

Contents of Volume 2

Preface

Thick films, thin films, even ultrathin ones, or just industrial coatings? The appropriate classification is a question of what constitutes a barrier layer or a coating material and what the ultimate function is. For sure, there are still many unsolved technological problems and fundamental scientific interest in understanding the phenomena related to the behavior and the properties of a polymeric or organic film on a solid surface. In a truly industrial application, one thinks of a polymer film as a paint coating applied to a wall or a packaging product. On the other hand, we look at these films as a system to be probed with the latest surface analytical tools or to be imbued with unique properties and functions. Smart or stimuli-responsive coatings, nanostructured thin films or devices all evoke different responses to what an organic or polymeric film does. In an inevitably interdisciplinary approach, everyone has something to bring to the table: A physicist or a surface scientist is interested in the phenomena of adsorption, relaxation, mesophase separation; a materials chemist is interested in developing new synthetic methodologies or in studying polymerization in the confined dimensions of a thin film; a chemical engineer is interested in understanding the transport phenomena and barrier properties of new coatings; a biologist or biophysicist is interested in biomimetic systems that enable the replication of *in-vivo* conditions for quantitative measurements on surfaces; and finally a nanotechnology-oriented scientist or engineer thinks about the unique nanoscale dimension by which structure–property relationships can be derived. The relationship and findings with colloidal and core-shell particle structures is not remote and can be extrapolated. While this book addresses mostly some of the fundamentals and state-of-the-art of organic and polymer materials, there is room and interest for everyone to contribute and learn. A curious mind will always come out with a fresh new insight and an inspiration to learn, understand, and even invent. *This book is dedicated to such an individual.*

The collection is divided into two volumes: *Volume 1.* Preparation and Patterning; *Volume 2.* Characterization and Applications. The volumes are meant to address the main thrusts of this body of literature. The idea is to equip the reader with the basic knowledge and at the same time with the state-of-the-art for each theme, almost tracing the development of the field and the interdisciplinary nature of the endeavor. Taken together as a whole, the volumes should thoroughly provide the reader with a good understanding of what is collectively known as the *field of organic polymer thin films*. The editors and authors do not claim that this collection represents the most important works in the field, nor can it replace some of the revered classics in the fields. There are more up-to-date review articles that will always supplant the information in these chapters. The editors would also like to extend their apologies to any author or part of work that has not been invited to contribute or has not been cited. It is not intentional – rather there is always room for another project. Lastly, the editors heartily thank all the authors and researchers who have contributed to this book. We hope you will like what we have done here and you will be proud of your chapter.

Rigoberto Advincula and Wolfgang Knoll

List of Contributors

Rigoberto C. Advincula
University of Houston
Department of Chemistry
Department of Chemical and
Biomolecular Engineering
136 Fleming Bldg.
Houston, TX 77204-5003
USA

Gregory L. Baker
Michigan State University
Department of Chemistry
East Lansing, MI 48824
USA

Débora T. Balogh
Universidade de São Paulo (USP)
Instituto de Física de São Carlos
(IFSC)
Av. Trabalhador São-carlense, 400
CP 369, 13560-970
São Carlos
São Paulo
Brazil

Zhiyi Bao
Michigan State University
Department of Chemistry
East Lansing, MI 48824
USA

Merlin L. Bruening
Michigan State University
Department of Chemistry
East Lansing, MI 48824
USA

Tao Chen
TU Dresden
Makromolekulare Chemie
Helmholtzstraße 10
01069 Dresden
Germany

Pawilai Chinwangso
University of Houston
Department of Chemistry
136 Fleming Bldg.
Houston, TX 77204-5003
USA

Anuja De Silva
Cornell University
Department of Chemistry
and Chemical Biology
116 Baker Laboratory
Ithaca, NY 14850
USA

Gero Decher
Centre National de la Recherche
Scientifique (CNRS UPR022)
Institute Charles Sadron
23 rue du Loess
67034 Strasbourg-Cedex
France

and

Université Louis Pasteur (ULP)
1 rue Blaise Pascal
67008 Strasbourg-Cedex
France

Robert Ducker
Duke University
Center for Biologically Inspired
Materials and Materials Systems
and Department of Mechanical
Engineering and Materials
Science
144 Hudson Hall
Box 90300
Durham, NC 27708
USA

Rita J. El-khouri
Centre National de la Recherche
Scientifique (CNRS UPR022)
Institute Charles Sadron
23 rue du Loess
67034 Strasbourg-Cedex
France

and

Université Louis Pasteur (ULP)
1 rue Blaise Pascal
67008 Strasbourg-Cedex
France

Olivier Felix
Centre National de la Recherche
Scientifique (CNRS UPR022)
Institute Charles Sadron
23 rue du Loess
67034 Strasbourg-Cedex
France

Marystela Ferreira
Rodovia João Leme dos Santos
Km 110 - SP - 264
Bairro do Itinga
Sorocaba, São Paulo,
CEP 18.052-780
Brazil

Renate Förch
Austrian Institute of
Technology (AIT)
Donau-City-Str. 1
1220 Vienna
Austria

Andres Garcia
Liquidia Technologies Inc.
P.O. Box 110085
Research Triangle Park
NC 27709
USA

Patrick Guenoun
UMR 3299 CEA/CNRS SIS2M
Laboratoire Interdisciplinaire sur
l'Organisation Nanométrique et
Supramoléculaire (LIONS) C.E.A.
Saclay
Bât. 125, pièce 243
91191 Gif sur Yvette Cedex
France

Tsunemi Hiramatsu
Osaka University
Department of Applied Physics
Yamada-oka 2-1
Osaka
Suita 565-0871
Japan

Hidekazu Ishitobi
RIKEN
Nanophotonics Laboratory
2-1 Hirosawa
Saitama
Wako 351-0198
Japan

Andrew C. Jamison
University of Houston
Department of Chemistry
136 Fleming Bldg.
Houston, TX 77204-5003
USA

Satoshi Kawata
RIKEN
Nanophotonics Laboratory
2-1 Hirosawa
Saitama
Wako 351-0198
Japan

and

Osaka University
Department of Applied Physics
Yamada-oka 2-1
Osaka
Suita 565-0871
Japan

Eunkyoung Kim
Yonsei University
Department of Chemical and
Biomolecular Engineering
262 Seongsanno
Seodaemun-gu
Seoul, 120–749
Republic of Korea

Jeonghun Kim
Yonsei University
Department of Chemical and
Biomolecular Engineering
262 Seongsanno
Seodaemun-gu
Seoul, 120–749
Republic of Korea

Yuna Kim
Yonsei University
Department of Chemical and
Biomolecular Engineering
262 Seongsanno
Seodaemun-gu
Seoul, 120–749
Republic of Korea

Wolfgang Knoll
Austrian Institute of
Technology (AIT)
Donau-City-Str. 1
1220 Vienna
Austria

T. Randall Lee
University of Houston
Department of Chemistry
136 Fleming Bldg.
Houston, TX 77204-5003
USA

Jason Locklin
University of Georgia
Department of Chemistry
Faculty of Engineering
and Center for Nanoscale
Science and Engineering
Athens, GA 30602
USA

Nicholas Marshall
University of Georgia
Department of Chemistry
Faculty of Engineering
and Center for Nanoscale
Science and Engineering
Athens, GA 30602
USA

Christopher K. Ober
Cornell University
Department of Materials
Science and Engineering
310 Bard Hall
Ithaca, NY 14850
USA

Osvaldo N. Oliveira
Universidade de São Paulo (USP)
Instituto de Física de São Carlos
(IFSC)
Av. Trabalhador São-carlense, 400
CP 369, 13560-970
São Carlos
São Paulo
Brazil

Mi-Kyoung Park
University of Houston
Department of Chemistry
Department of Chemical and
Biomolecular Engineering
136 Fleming Bldg.
Houston, TX 77204-5003
USA

Soojin Park
Ulsan National Institute of
Science and Technology (UNIST)
Interdisciplinary School of
Green Energy
Ulsan 689–798
Korea

and

University of Massachusetts
Department of Polymer Science
and Engineering
120 Governors Drive
Amherst, MA 01003
USA

Thomas P. Russell
University of Massachusetts
Department of Polymer Science
and Engineering
120 Governors Drive
Amherst, MA 01003
USA

Zouheir Sekkat
MAScIR
INANOTECH
ENSET
Optics and Photonics Laboratory
Rabat
Morocco

and

Hassan II Academy of
Science and Technology
Rabat
Morocco

and

RIKEN
Nanophotonics Laboratory
2-1 Hirosawa
Saitama
Wako 351-0198
Japan

and

Osaka University
Department of Applied Physics
Yamada-oka 2-1
Osaka
Suita 565-0871
Japan

Yulia Sergeeva
Centre National de la Recherche
Scientifique (CNRS UPR022)
Institute Charles Sadron
23 rue du Loess
67034, Strasbourg-Cedex
France

and

Université Louis Pasteur (ULP)
1 rue Blaise Pascal
67008 Strasbourg-Cedex
France

S. Kyle Sontag
University of Georgia
Department of Chemistry
Faculty of Engineering
and Center for Nanoscale
Science and Engineering
Athens, GA 30602
USA

Rafael Szamocki
Centre National de la Recherche
Scientifique (CNRS UPR022)
Institute Charles Sadron
23 rue du Loess
67034 Strasbourg-Cedex
France

and

Université Louis Pasteur (ULP)
1 rue Blaise Pascal
67008 Strasbourg-Cedex
France

Mamoru Tanabe
Osaka University
Department of Applied Physics
Yamada-oka 2-1
Osaka
Suita 565-0871
Japan

Hiroaki Usui
Tokyo University of Agriculture
and Technology
Department of Organic and
Polymer Materials Chemistry
Nakacho, Koganei
Tokyo 184-8588
Japan

Jungmok You
Yonsei University
Department of Chemical and
Biomolecular Engineering
262 Seongsanno
Seodaemun-gu
Seoul, 120–749
Republic of Korea

Stefan Zauscher
Duke University
Center for Biologically Inspired
Materials and Materials Systems
and Department of Mechanical
Engineering and Materials
Science
144 Hudson Hall
Box 903 00
Durham, NC 27708
USA

Jianming Zhang
Duke University
Center for Biologically Inspired
Materials and Materials Systems
and Department of Mechanical
Engineering and Materials
Science
144 Hudson Hall
Box 90300
Durham, NC 27708
USA

Ying Zheng
Michigan State University
Department of Chemistry
East Lansing, MI 48824
USA

Part I
Preparation

1
A Perspective and Introduction to Organic and Polymer Ultrathin Films: Deposition, Nanostructuring, Biological Function, and Surface Analytical Methods

Rigoberto C. Advincula and Wolfgang Knoll

Monolayer and multilayer ultrathin films of organic, polymeric, and/or hybrid materials have gained much attention over the last several decades owing to their fundamental importance in understanding materials' properties in confined geometries and their potential applications as smart and/or stimuli-responsive coatings, in microelectronics, electro-optics, sensors, nanotechnology, and biotechnology, to mention but a few. Ultrathin films are defined at a scale that is smaller than what can be industrially accessed by spin-casting or roll-to-roll transfer methods for the deposition of coatings. "Ultrathin" usually refers to thicknesses in the submicrometer or even sub-100-nm scale for a coating on a relatively flat solid support or surface. By going to the nanoscale, the main advantage is the ability to control nanostructured architectures in which self-assembly or directed assembly of organic materials are formed as ultrathin films on the substrates. These monolayers may in fact be ordered assemblies themselves that display long-range ordering including crystallinity, liquid crystallinity, nanoscopic, and/or mesoscopic structures. Subsequently, multilayers can be produced by repetitive or alternating "layer-by-layer" deposition of individual monolayers. Thereby, it is possible to control the molecular orientation and organization at the nanoscale thus precisely tuning the macroscopic properties of the organic and polymer thin films. This stacking can be in the form of oriented or isotropic "random" stacking, which can be replicated by self- or directed assembly.

The oldest technique for fabricating multilayer ultrathin films (and the most extensively studied prior to the 1990s) is the formation by the sequential deposition of individual monolayers known as the Langmuir–Blodgett (LB) or Langmuir–Blodgett–Kuhn (LBK) technique. The monolayers are equilibrated at the air/water interface and then transferred onto a solid substrate either by dipping the substrate in a vertical deposition step or by horizontal transfer (Langmuir–Schaefer technique). Multilayers can be realized by repetitive dipping. The LB technique, indeed, provided scientists with the practical capability to construct ordered monomolecular assemblies that can be probed with surface sensitive analytical techniques. However, as many have realized, the LB technique requires special equipment and has severe limitations with respect to substrate size and topology as well as film quality and stability. Another approach to assembling

Functional Polymer Films, First Edition. Edited by Wolfgang Knoll and Rigoberto C. Advincula.

layered ultrathin film structures based on chemisorption was also reported widely in the early 1980s. These are called self-assembled monolayers (SAMs) and are usually based on the adsorption of amphiphilic or reactive molecules on specific surfaces thus forming monolayers with a certain degree of thermal stability. Also in the 1980s, multilayer films were reported to be prepared by a two-step sequential reaction protocol of deposition followed by chemical reaction of the end-groups involving, for example, protected–deprotection schemes with silanes and hydroxylated surfaces employing the conversion of a nonpolar terminal group to a hydroxyl group. Once a subsequent monolayer is adsorbed on the "activated" monolayer, multilayer films may be built by repetition of this process. This has also been a route toward inorganic multilayer films, for example, sequential complexation of Zr^{4+} and α,ω-bisphosphonic acid. However, self-assembled films based on covalent or coordination chemistry are restricted to certain classes of organics, and high-quality homogeneous multilayer films or large-area films cannot be reliably obtained because of the high steric demand of covalent chemistry and the requirement for 100% reactivity in each step.

The layer-by-layer deposition of oppositely charged molecules and polyelectrolytes was first reported in the early 1990s. This was a revival of a previous method reported in the 1960s relying mostly on electrostatic attraction. It was the ability to form multilayers with precise control over the total thickness, that is, in the range of a few tenths of nanometers up to micrometers, without the use of the more expensive LB deposition equipment that has allowed this fabrication concept to become the most popular ultrathin film-preparation method to date. This is due to its characteristic repetitive deposition steps, using essentially a "beaker technique" that even a high-school student can play around with. A number of parameters can be used to control the resulting ultrathin film structure and thickness: concentration, ionic strength, dipping time, and so on. These films can also be prepared with a nearly unlimited range of functional groups incorporated within the structure of the film. Additionally, it has significant advantages over other techniques; for example, this process is independent of the substrate size and topology, the assembly is based on spontaneous adsorptions, and no stoichiometric control is necessary to maintain surface functionality. In the late 1990s, the layer-by-layer deposition of polyelectrolytes has been extended onto charged micrometer-sized particles. It has been shown that polyelectrolyte hollow capsules could thus be prepared by removing the core after the deposition to form "hollow-shell" particles. Since then, the technique has demonstrated an enormous variety of assembly mechanisms, using different substrate surfaces, shape of templates, and transformation in protocols, with the term layer-by-layer becoming synonymous with nanostructured materials and assembly of thin-film coatings, shell, and hollow-shell or tube formation. From the mechanistic side, other than simple electrostatic attraction of oppositely charged species, this technique now includes the use of other noncovalent interactions, for example, hydrogen bonding, stereocomplexation, dipole interactions, and so on. Covalent coupling protocols include click chemistry, thiol-ene-chemistry, and other chemical mechanisms. The list of species and objects that have been assembled includes a wide variety

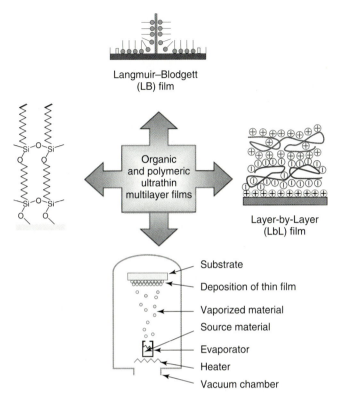

Figure 1.1 Ultrathin organic molecular and macromolecular assembles including vapor-deposition methods based on monolayer and ultimately multilayer deposited systems.

of molecules, macromolecules, dendrimers, block-copolymer micelles, graphene, carbon nanotubes, nanoparticles, biological objects, and so on. For the substrates employed this has gone from simple glass slides and Si wafers to large-area substrates and nanoparticle surfaces. For the protocols employed, the list includes spin coating, spray assembly, large-area dipping, and roll-to-roll processes. A schematic diagram of these ultrathin molecular and macromolecular assembles including vapor deposition methods are shown in Figure 1.1. While the material of interest is mainly organic in nature, inorganic and organic–inorganic hybrid materials can be utilized in such protocols.

Of high interest in the past decade is the grafting of polymers by directly growing them from surface-tethered initiators otherwise known as surface-initiated polymerization or SIP to form polymer brushes (Figure 1.2). The "grafting-from" method is a departure from previous methods of grafting polymers by physical adsorption, for example, diblock-copolymer or amphiphilic polyelectrolytes including polymers with "anchoring groups." Other than physical adsorption, grafting by chemical adsorption of a preformed polymer, for example, one with an anchoring group at the end is also another route. This is almost an extension of

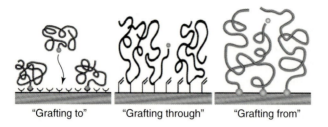

"Grafting to"   "Grafting through"   "Grafting from"

Figure 1.2 Different methods of grafting polymers on surfaces based on methods of attachment of a preformed polymer or direct growth of the polymer from or through a surface based on specific polymerization mechanisms.

the SAM technique as applied to macromolecules and is sometimes referred to as a *"grafting-to"* approach. Another member of this class of polymer film formation is a "grafting-through" approach where the polymerization stitches through surface-tethered monomers. The reason the SIP has garnered so much popularity is that it has enabled the formation of polymers brushes with a high grafting density in which the polymer main chain is extended away from the surface. It has also enabled the polymerization in confined environments by well-known addition polymerization mechanisms including free-radical, anionic, cationic, metathesis, ring-opening, and living free-radical polymerization. The latter being the most reproducible and well controlled of all the polymerization processes so far reported.

Another aspect of thin polymer films concerns the more recent developments in the field of block-copolymer mesophase and nanophase structuring (Figure 1.3). While this aspect has been well studied in the bulk for the last three decades, new methods for templating, patterning, and applications take advantage of new polymer synthesis and microscopic and scattering characterization methods. The well-known phase separation between polymers of dissimilar χ-interaction parameters has taken on new meaning when applied to ultrathin films and in the presence of externally applied fields, for example, of static electric, magnetic, electromagnetic, and so on,. nature. It is of particular interest to introduce ternary compositions, for example, adding surfactants or nanoparticles and how this third

Figure 1.3 Different mesoscopic and nanoscopic structures based on phase separation induced by composition or field effects in thin films.

component affects the various traditionally reported mesophases of cylindrical, lamellar, or bicontinuous nature. On the mesoscopic scale one should emphasize the various aspects and methods for patterning. These include: photolithography, electron-beam lithography, soft lithography based on the popular microcontact printing technique, imprint lithography, and nanolithography in conjunction with surface probe microscopy.

It is well beyond the scope of this book to include here the most recent developments in plasma polymerization, physical and chemical vapor depositions, organic molecular beam epitaxy, hybrid materials layering, sol-gel and polymer coatings, electrodeposition, and electroplating. Even developments in traditional spin coating, roll-to-roll printing or a combination of the above techniques for new hybrid or multilayer films have been reported and are considered state-of-the-art.

There are many aspects of new materials that have been described in the literature that are based on new synthesis strategies or rely on the use of functional polymers. Other than the traditional polymers used in bulk and as thin-film coatings such as those based on vinyl polymers, methacrylates, urethanes, and polyesters, the most popular if not the ones with the highest potential for thin-film devices are in the area of electro-optical systems based on conducting or π-conjugated polymers. Conductive polymers can be accessed by electropolymerization or chemical oxidative methods. It is also possible to deposit preformed polymers that can be prepared by metal-mediated coupling reactions or by metathesis reactions. In this case there is a high interest in their crystallinity, orientation, charge carrier mobility, electrochromic properties, and so on. Perhaps in light of the recent Nobel Prize in Physics for 2010, graphene ultrathin films are expected to take center stage in terms of electronic and electro-optical applications.

Some of the most important developments in recent years refer to films for biological applications, made from building blocks of biological origin or are of a biomimetic nature. Simply speaking, one can use any of the above techniques and apply the resulting films for biological problems or conversely, looking for a materials solution by using well-known biological models. For example, by bringing together materials science and the biological world, it is possible to introduce tethering of the ubiquitous polyethylene glycol (PEG) for new biological architectures and interactions. PEG brushes can be made to be highly resistant to biofouling. Stimuli-responsive polymers with hydrogel structures including poly(N-isopropyl acryl amide) or PNIPAM polymers are equally important for drug release. Yet another development concerns architectures based on layer-by-layer systems or polymer brushes that can be used to control drug release or cell chemotaxis, ion gates, DNA capture, or to control cell proliferation on surfaces. Other examples are lipid bilayer membranes that can be artificially tethered to a solid surface using polymer-tethered lipids. An important aspect of biological systems as thin films is their applications in sensing (Figure 1.4), where it is important to enable specificity coupled with transduction. This concept allows for studies that enable quantification of phenomena normally observed only in *in-vivo*

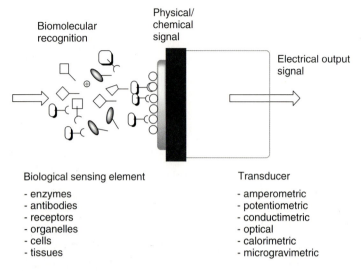

Figure 1.4 Biomimetic systems can be coupled as thin-film materials on a variety of flat substrate surface connected with a transduction mechanism.

conditions and is a very good connection of new materials toward biophysics- and bioengineering-related projects.

The last but not the least aspect of organic polymer ultrathin films is developments related to surface analysis and characterization. A traditional division is the use of optical/spectroscopic, microscopic, and scattering methods. However, one should add other techniques like piezoelectric (acoustic) or electrochemical (including impedance) into this menu. Traditional optical/spectroscopic methods include transmission, absorption, and fluorescence spectroscopies. New variations include single molecule fluorescence and fluorescence photocorrelation spectroscopies. Microscopy techniques include optical, surface probe microscopies (SPMs), as well as electron microscopies including transmission electron microscopy (TEM) and scanning electron microscopy (SEM). SPM methods, which includes atomic force microscopy (AFM) and scanning tunneling microscopy (STM), have truly been of wide utility to complement the TEM and SEM methods. Scattering methods include X-ray and neutron diffraction, scattering, reflectometry, and so on. There are many techniques that are very specific to the geometry of the measurement. Optical methods deserve their own classification since they can include many hyphenated techniques and functions dealing with spectroscopy and microscopy. Well-known methods include waveguides, ellipsometry, interferometry, surface plasmon resonance, dielectric spectroscopy, Raman scattering, and so on Surface plasmon resonance (SPR) spectroscopy together with optical waveguiding spectroscopy (OWS) in particular, has been a method of choice not only for optical and dielectric measurements but also for their versatility toward hyphenated techniques–a recent combination with AFM and electrochemistry is an example (Figure 1.5). This has enabled applications in materials science and in the investigation of biophysical

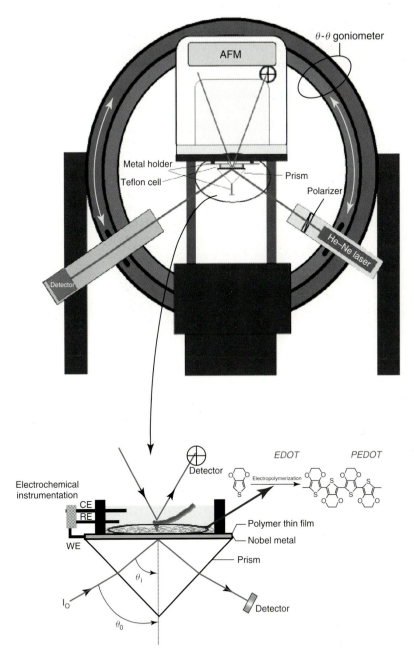

Figure 1.5 A combined electrochemical–atomic force microscopy–surface plasmon resonance spectroscopy (EC-SPR-AFM) instrumentation for the simultaneous and real-time detection of optical and electrochemical phenomena coupled with morphology investigation in films – including electropolymerization of conducting polymers.

phenomena based on coupling the films of interest on a gold- or silver-coated glass/prism surface. The open environment on one side means that the experiment can be done *in situ* or in real time with simultaneous parameters monitored as a function of reflectivity, angle, time, and wavelength. Applications with microscopy and fluorescence further extend its utility as a characterization tool or for sensing.

In the area of electrochemical analysis, it is possible to use this to probe electron transfer and redox properties of polymer species. In impedance analysis it is possible to use model circuits to explain the behavior of ion and/or charge transport and film structure although for certain systems it is not always easy to derive a generally accepted model. Acoustic methods such as the quartz crystal microbalance (QCM) and surface acoustic waves (SAWs) methods are useful for probing not only the deposition of mass but also the mechanical properties of thin films. There are many more unconventional techniques that can be mentioned. These include, the surface force apparatus (SFA), streaming potential measurements, Kelvin probe methods, photobleaching experiments, and so on.

This chapter has covered—in a very comprehensive but succinct way—our perspective of what constitutes the field of organic and polymer thin films.

2
Multifunctional Layer-by-Layer Architectures for Biological Applications

Rita J. El-khouri, Rafael Szamocki, Yulia Sergeeva, Olivier Felix, and Gero Decher

2.1
Introduction

Layer-by-layer (LbL) deposition technique was introduced in 1991 [1–4] and since then, has been utilized in numerous facets of research [5–14]. In the simplest description LbL films are patterns of nanoscopic layers in the z-direction, in which each layer carriers a single complementary feature from its nearest neighbors giving rise to a seamlessly balanced final construct. Complex living systems require hierarchical organization and compartmentalization on different length scales ranging from molecular/supermolecular to organs and macroscopic life forms. Inspired by nature, the intrinsic patterns that evolve through LbL assembly can also integrate multiple functionalities and additionally be used to form barriers between layers, Figure 2.1. LbL films are generally prepared under mild aqueous conditions, making them an attractive alternative to other film-deposition methods. Furthermore, LbL film deposition is nondiscriminatory toward template size, shape, or material. Enormous potential has been recognized for the biosciences field [12, 14–16]. Specifically, LbL has provided a means for templating biomolecules without sacrificing structure or bioactivity. The most far-reaching example is the controlled cell adhesion and even the inclusion of layers containing living cells [17–19]. In addition to the incorporation of biomolecules, LbL films are versatile in that film architecture can be manipulated to include various functionalities, provide different topological features, as well as different physical properties. One of the most interesting and recently exploited architectural features of LbL assembly is the tunable construction of films toward the controlled degradation of the layers. The ability to manipulate the described parameters displays the aptitude of LbL assembly. It is for these reasons that LbL has become an alternative to conventional technologies used to prepare drug-delivery and gene-therapeutic platforms. Herein, we report current progress in the development and application of biofunctional LbL films.

Functional Polymer Films, First Edition. Edited by Wolfgang Knoll and Rigoberto C. Advincula.
© 2011 Wiley-VCH Verlag GmbH & Co. KGaA. Published 2011 by Wiley-VCH Verlag GmbH & Co. KGaA.

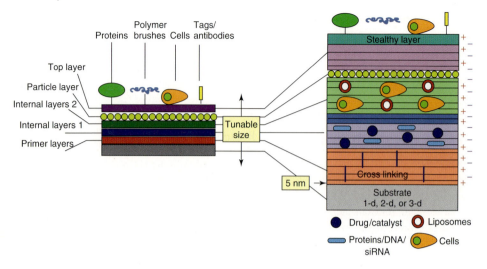

Figure 2.1 Illustration of the many ways layer-by-layer (LbL) is used to both template biomolecules, and biofunctionalize films.

2.1.1
LbL Polyelectrolyte Multilayer (PEM) Formation

LbL assembly is an easy method for the preparation of multifunctional films. In a few words, on a substrate carrying a net charge, an alternately charged polyelectrolyte (P1) is deposited, Figure 2.2. The substrate is rinsed to remove unadsorbed P1s. It should be noted that the substrate used could be any solvent accessible surface (any shape, chemical composition, dimension). To the first P1 layer a polyelectrolyte carrying an alternate charge (P2) is deposited forming the first "layer pair," the substrate is rinsed to remove unbound P2s. Subsequently,

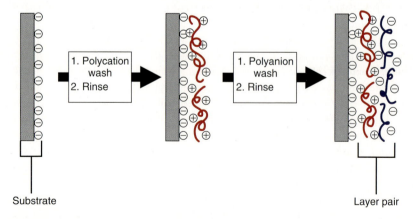

Figure 2.2 Illustration of LbL deposition method.

additional layer pairs can be deposited, giving rise to the polyelectrolyte multilayer (PEM) formation and yielding intrinsically patterned surfaces along the layer normal. The growth of these layer pairs has been described to occur in different regimes including stagnation/sublinear growth, linear growth, and superlinear growth [20, 21]. Typically, linearly growing films are formed from materials that are kinetically trapped in their positions, while in superlinear growing films parts of the constituents can diffuse almost freely within the construct. It should be noted that aside from the electrostatic interactions in the described films, there is an additional entropic gain due to the release of counterions [22, 23]. Film thickness can be readily tuned via number of layer pairs deposited. Although in the described example films are assembled via electrostatic interactions, LbL film formation is not limited to these interactions, but a rather wide variety of assemblies have been demonstrated, for example, hydrogen bonding, charge transfer, covalent bonding, and biological recognition have recently been reviewed [24].

2.1.2
LbL Nomenclature

Although there is variation within the literature in the nomenclature used to describe PEM-coated substrates, in this article, layer pairs will be denoted as P1/P2, where P1 and P2 describe the abbreviated name for each respective polyelectrolyte. Templates coated with multiple types of layer pairs or superstructures (supramolecular assemblies) will be described as *(P1/P2)-(P3/P4)*, in which the order, that is, described begins with the bottom most layer pair type and moves upward in film assembly. Additionally, the layer pairs are within brackets that denote multilayer pairs and the subscript outside the bracket denotes the number of layer pairs deposited on a template; hence in $(P1/P2)_5$ there are five layer pairs of P1/P2. When describing films that are deposited on a particle or particular planar support this will be denoted as, surface type-$(P1/P2)_3$. Finally, frequently the terminating layer is not composed of a full layer pair; therefore in this instance it will be denoted as $(P1/P2)_3$-$(P3/P4)_6$-P3, where P3 is the topmost layer.

2.2
Drug Delivery and LbL

There have been many new approaches in the field of drug delivery [25] and the LbL assembly method has gained importance in improving traditional therapies. The demonstration of using LbL to deposit sensitive biomolecules [26–32] in a controlled and biofriendly environment encouraged researchers to investigate LbL deposition for preparing drug-delivery systems. Unlike conventional methods used to prepare drug-delivery systems, the LbL technique offers templates that can be fine-tuned to the nanometer regime, while the drug concentration can be dosed in a wide range [1, 12]. One of the most important factors toward building a drug-delivery system is the controlled spatiotemporal release of the encased drug or bioactive materials [6, 33]. There have been numerous accounts of LbL assemblies that degrade under different external stimuli as well as via self-degradation, and such systems have been recently reviewed [12, 15, 34].

Since LbL assembly is nonsubstrate discriminatory LbL-based drug-delivery systems have been prepared both on planar surfaces as well as micro and nanoparticles. In the former case, one of the driving forces to build drug-delivery systems on a "planar" substrate initially originated from applications requiring functional implantable materials, such as coronary stents. Some of the problems with implants arise from inflammatory and immune reactions leading to rejection of the foreign material. LbL assembly has been used to coat implantable materials in order to prevent rejection *in vivo*. More recently, there has been an effort to assemble slow degrading films with encased drugs to yield sustained therapeutic delivery and prevent infection over longer time scales, for example, stents. Some of these advances will be highlighted below.

The development of particle-based drug-delivery systems has also benefited from LbL assembly method. Initially micrometer-sized LbL coated particles (corona or shell) and hollow capsules were prepared with the intention of ultimately being used as an alternative to the conventional drug-delivery systems [35–37]. In short, a micrometer-sized particle is LbL coated and the drug is deposited within the layers (corona) or could be initially imbedded in the particle "core" [37]. Upon completion of LbL assembly the core can be further removed by dissolution, leaving a hollow shell with an encased drug in the layers or resting inside the shell. Such micrometer-sized forms provide drug-delivery systems with controlled drug loading, and degradation rates, through parameters such as particle size, and for certain applications composition, especially of the surface layers [33, 38]. Submicrometer to nanosized systems provide controlled cellular uptake and subsequent delivery [39, 40]. Additionally, any particle should be stealthy in order to prevent premature clearance from the blood stream and promote targeting through enhanced circulation times.

2.2.1
Trends in Drug Release from Planar LbL Films

2.2.1.1   Progress in Degradable LbL Films toward Drug Delivery
The controlled degradation of a drug-enriched LbL film is an important topic for the development of viable delivery platforms [41]. One current interest is the utilization of hydrolytically controlled degradable templates [42]. Vazques *et al.* [43] first demonstrated LbL films composed of poly(β-amino esters) (Scheme 2.1, **Poly-1 (Poly-1)**) that could be gradually hydrolytically eroded at physiological conditions. Unlike other degradable LbL films, the benefit to working with hydrolytically degradable films is that it enables continuous elution of drugs without the need of enzymatic or cellular interactions. In 2005 Wood *et al.* [44] prepared films composed of **Poly-1** and loaded with either polysaccharide therapeutics, heparin (HEP), or chondroitin sulfate (CS). In all cases, deposition of the multilayer proceeded by superlinear growth permitting quick access to films with a thickness superior to

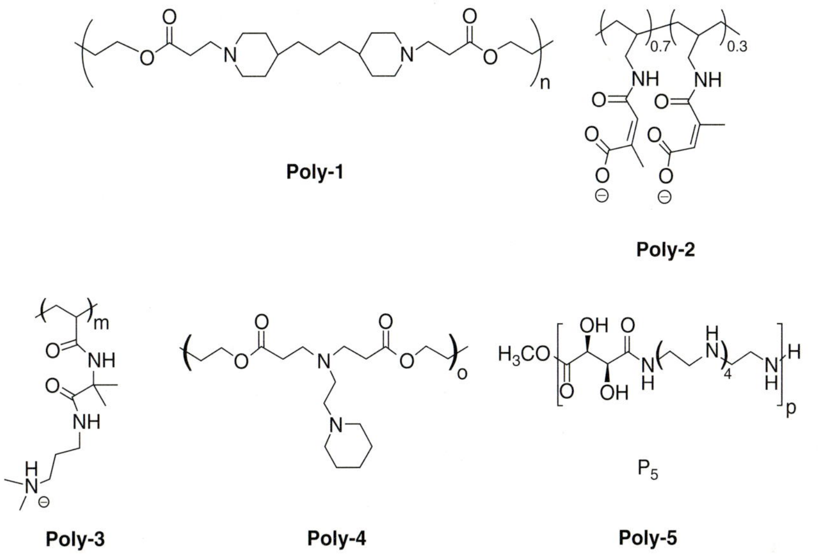

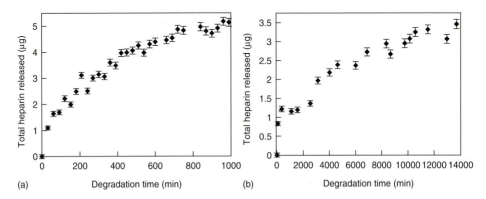

Scheme 2.1 Chemical structures of degradable polymers.

linear growing films. Films were exposed to two different pH environments of 6.2 and 7.4 in order to monitor degradation of **Poly-1**, and release of embedded HEP or CS in LbL film. Figure 2.3 shows the release rates of HEP samples over time of exposure under the two different pH environments; (a) at pH 7.4 and (b) at pH 6.2. **Poly-1** has been cited to degrade more rapidly under neutral conditions, and in this study degradation rates went from 10 days to 24 h when switching from pH 6.2–7.4. Owing to the tunable nature of degradation of **Poly-1**, similar poly(β-amino esters) have been explored for the controlled delivery of a number of small molecule drug compounds in LbL films.

Figure 2.3 Heparin release from degradable (polymer 1/heparin)$_{20}$ thin films at (a) pH 7.4 and (b) pH 6.2. (Adapted from Ref. [39].)

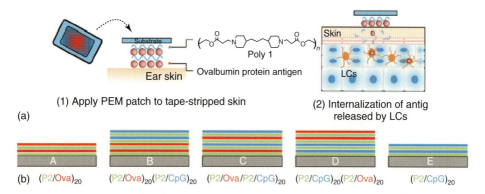

(1) Apply PEM patch to tape-stripped skin

(2) Internalization of antig released by LCs

(a)

(b) (P2/Ova)$_{20}$ (P2/Ova)$_{20}$(P2/CpG)$_{20}$ (P2/Ova/P2/CpG)$_{20}$ (P2/CpG)$_{20}$(P2/Ova)$_{20}$ (P2/CpG)$_{20}$

Figure 2.4 (a) Schematic illustration of the events taking place following the application of PEM patch onto skin. (b) Schematic architectures of antigen (ova) and adjuvant (CpG DNA) codelivery films tested. (Adapted from Ref. [41].)

Liu and colleagues have described the preparation of LbL films composed of a degradable anionic polymer, Scheme 2.1, **Poly-2** [45]. Such templates provide a means for tunable delivery of cationic drugs. In this study, layers of **Poly-2** and cationic poly(allylamine hydrochloride) (PAH) were found to grow superlinearly. When these films were placed under slightly acidic conditions, full degradation of ~10 nm film occurred within 48 h.

Some applications of hydrolytically degradable polymers may benefit the delivery of multiple drugs from a single substrate. Su *et al.* [46] prepared films on flexible Poly(dimethylsiloxane) (PDMS) surfaces of poly(b-amino ester)s with two different drug components (Figure 2.4b). The focus of the study was to prepare surfaces that could eventually be used as a method of vaccination through skin contact, therefore, protein antigen – ovalbumin (ova) and an adjuvant cargo-CpG DNAs (deoxyribonucleic acids) were incorporated within the LbL layers (Figure 2.4a, only ova depicted). In order to understand and regulate the release of ova and CpG DNAs the authors explored a number of different multilayer architectures (Figure 2.4b). Dried films were found to degrade upon contact with skin. After degradation the embedded drugs were found to be uptake by skin cells once in the tissue environment.

2.2.1.2 Micelle Encased Drugs in LbL Films

Some drugs that need to be eluted are hydrophobic in nature. Since LbL assembly takes place under mild aqueous conditions, micelles have been used in order to encase such hydrophobic drugs. Moreover, there has been substantial focus on the integration of these drug-loaded micelles into LbL films that ultimately undergo controlled degradation and drug release [47–49]. Qi *et al.* [50] have prepared multilayers using two different polymeric micelles that have either a polycationic or polyanionic corona. Each micelle type was impregnated with dye molecules serving as model compounds. Release of the dye molecules was explored in the presence and absences of micelles in solution. It was found that under both conditions the dye molecules were released from the film after 30 min of exposure. The LbL samples that were immersed into micelle-rich solutions actually ejected the dye molecules more rapidly then in the case of micelle deficient solutions. This points to the fact that the release rates, for hydrophobic molecules, not only depend on the degradability of the LbL films but also on the solubility of the drug in the surrounding solution.

Another interesting example of LbL assembly with drug-incorporated micelles was reported by Kim and colleagues [51] in which multilayers were assembled via hydrogen-bonding rather electrostatic interactions. In this case, the micelles were composed of poly(ethylene oxide)-block-poly(ε-caprolactone) (PEO-b-PCL) and were loaded with an antibacterial drug, triclosan. The counter polymer layer used caused strong hydrogen bonding to occur between the poly(acrylic acid) (PAA) and simple poly(ethylene oxide) (PEO), Figure 2.5a. The construction of PAA/PEO multilayers and their degradation was first explored by Sukhishvili and Granick

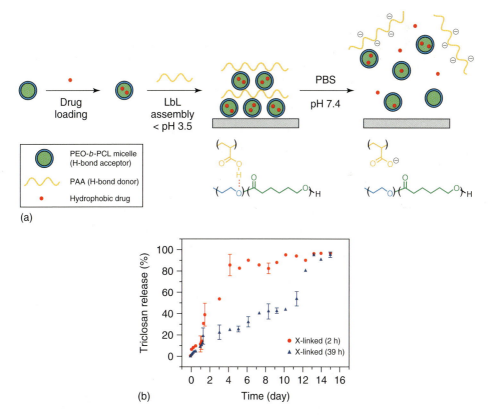

Figure 2.5 (a) Schematic representation of hydrogen-bonding LbL assembly of block copolymer micelles for hydrophobic drug delivery vehicles from surfaces. (b) Release profile of triclosan from (PEO-b-PCL/PAA)$_{30}$ film in phosphate buffer (PBS) at pH 7.4 from cross-linked film with different degrees of linking. (Adapted from Ref. [46].)

[52] and were later used as sacrificial layers by Ono and Decher [53]. In Kim's study, the hydrogen-bonded film was shown to grow linearly with respect to the number of micelle/polymer layer pairs deposited. The micelle layer thickness was very close to the diameter of the micelle in solution. In order to better understand the full potential of the surface degradation and ultimate drug release, the films were subsequently thermally cross-linked. As depicted in Figure 2.5b, triclosan was fully released from the 2 h cross-linked film over a 4-day period while the sample cross-linked for 39 h took 13 days to fully liberate the triclosan. More recently, the same research team has demonstrated the preparation of advanced templates encasing different hydrophobic drug compounds [54].

Stents are medical devices that once implanted are used to prevent or counteract constrictions in tube-like tissues. The new generation of stents possess active surface coatings for enhanced performance, one example being drug-eluting stents (DESs). Voegel and coworkers [55] demonstrated control of cell growth on the inner lumen of stents by simple LbL in the past. Common complications with conventional bare metal stents are restenosis or reoccurrence of narrowing of blood vessels, which may lead further to thrombosis. This is usually induced by the abnormal proliferation and migration of vascular smooth muscle cells (VSMCs) and subsequent thickening in the arterial intima. Currently, there are DESs that use biopolymers such as hyaluronic acid (HA) and HEP and that show promise in resisting thrombosis and decreasing restenosis in preclinical trials. One of the main difficulties in fabricating DESs that are multitherapeutic is loading hydrophobic antiproliferative drugs within hydrophilic surface layers. Kim and colleagues have recently demonstrated the use of HA-γ-poly(lactic-co-glycolic acid) based micelles for embedding hydrophobic paclitaxel (PTX) in LbL films of HEP/poly-L-lysine (PLL) on metallic stents [56]. It was found that when the LbL-coated stent (with the loaded micelle encased PTX) was exposed to coronary-artery smooth-muscle cells (CASMCs) over a five-day incubation period, there was a significant reduction in cellular proliferation in comparison to the conventional bare-metal stent, Figure 2.6. It should be noted that this study demonstrates a further extension of LbL coated stents toward multitherapeutic systems.

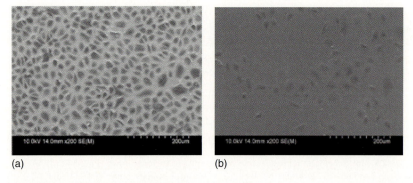

(a)　　　　　　　　　　　　　　　　(b)

Figure 2.6 SEM images of CASMC under (a) control and (b) Hep/PTX multilayer. (Adapted from Ref. [49].)

2.2.2
Trends in Direct Drug Delivery Using Nanoparticles

2.2.2.1 LbL Coated Drug Particles

Finding galenic formulations for poorly soluble drugs is a longstanding issue in the pharmaceutical sciences [57, 58]. Intravenous administration of hydrophobic drugs is often difficult and cumbersome because such compounds tend to aggregate in aqueous media. One of the most used techniques that have been employed to solubilize hydrophobic drugs is to encase the drug inside a micelle [59–62]. Although micelles have been found to prevent the problem of drug aggregation and increased the amount of actual drug delivered intravenously, there are still many limitations to these methods such as low threshold-loading efficacy, and difficulty to control the release rates. It was tempting and a consequent to explore the versatility of LbL film architecture of nanosized carrier systems for delivery of poorly soluble bioactive compounds. Lvov *et al.* [63] have previously described a method for preparing microencapsulated urease within LbL assemblies of PAH/poly(styrene sulfonate) (PSS). Baladushevitch and coworkers [64, 65] discussed the encapsulation of the protein α-Chymotrypsin and factors that regulated the protein release from a LbL microaggregate. Based on this idea of encapsulating specific drug compounds within a stable LbL shell, a number of research teams have applied this concept toward the preparation of nanosized encapsulation of hydrophobic drug compounds [66–68]. Fan *et al.* [66] have reported a method for encapsulation of insulin model drug in an LbL shell. As illustrated in Figure 2.7, first micrometer-sized aggregates of insulin were prepared via a salting-out method and coated with an initial poly(α, β,-L-malic acid) (PMA) layer. Water-soluble chitosan (CHI) was used as complementary cationic layer, and an LbL film was deposited using CHI and PMA. Upon completion of the PEM film assembly, the particles were exposed to ultrasound for a few minutes yielding nanosized polyelectrolyte coated aggregates, (an example scanning electron microscopy (SEM) of these particles is depicted in Figure 2.7). These coated drug nanoparticles are stabilized via the PEM shells and cannot be prepared otherwise.

A similar method of nanosized drug particle preparation is described by Agarwal *et al.* [69]. Poorly soluble and potent anticancer drugs tamoxifen (TMF) and paclitaxel (PXT) were prepared into nanosized stable particles using poly(dimethyldiallylamide ammonium chloride) (PDDA) and PSS LbL films off the surface of the drug particles preventing aggregation. Figure 2.8 is the TMF release curves for drug crystals without LbL (1), nanoparticles of sonicated noncoated TMF (2), PDDA coated TMF nanoparticles (3), and PDDA/PSS coated TMF nanoparticles (4). Sample (4) had the slowest drug release rate out of all four samples. Unlike the pure drug crystal, which took 2 h for complete TMF release, LbL-coated nanoparticles reached complete release after 10 h, similar results were obtained for PXT coated nanoparticles. It was also demonstrated that the LbL-coated drug nanoparticles could be labeled with tumor-specific antibody tags and used toward targeted drug delivery.

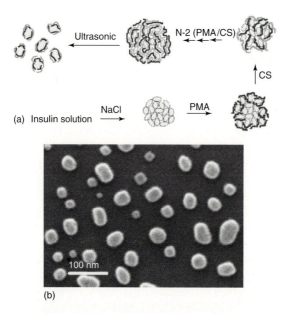

Figure 2.7 (a) Scheme of the LbL adsorption of negatively (black) and positively (gray) charged polyelectrolytes on protein particles. (b) The SEM photograph of the insulin–polyelectrolyte nanoparticles with six polyelectrolyte adsorption cycles after ultrasonic treating for 2 min at 10 °C. A scale bar represents 100 nm. (Adapted from Ref. [59].)

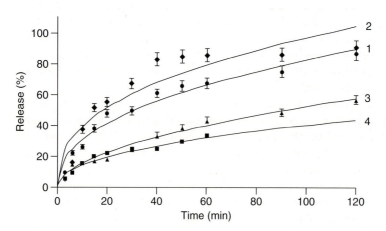

Figure 2.8 Controlled drug release from the LbL nanocolloidal particles. Dissolution rate of free tamoxifen (as drug crystals without sonication – 1, and nanoparticles of sonicated noncoated drug – 2) and tamoxifen release form the circa 125 nm LbL nanocolloidal particles with different coating composition: PDDA coating – 3, and (PDDA/PSS)$_3$ coating – 4. (Adapted from Ref. [62].)

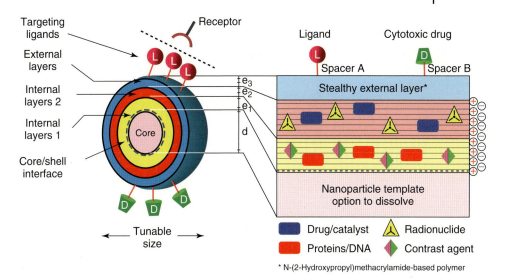

Figure 2.9 Schematic depiction of nanoparticles coated with multilayer shells as new drug-delivery system. (Adapted from Ref. [63].)

2.2.2.2 Multifunctioning Nanocarriers for Localized Drug Delivery and Tracking Abilities

One of the most valuable features of LbL deposition method is that each layer is built in a modular fashion. This provides a means to incorporate various functionalities within a single particle, Figure 2.9. Recently, our research team has described a model system of a multifunctional nanoparticle (MFNP) [70]. A colloidal core composed of a gold nanoparticle (AuNP) was coated with primer layers of (PAH/PSS)$_5$-PAH creating a stable and defined polyamine surface. The AuNP-(PAH/PSS)$_5$-PAH was capped with a functional terpolymer (F-HPMA) using covalent LbL assembly. The F-HPMA terpolymer was mainly composed of N-(2-hydroxypropyl)methacarylamide), providing a stealthy corona layer and steric stabilization. The two minority monomer repeat units were N-methacryloyl-glycyl-glycyl thiazolidine-2-thione to allow for covalent LbL coupling, and pharmacologically active monomer N-methyloyl-glycyl-DL-phenylalanyl-leucyl-glycyl doxorubicin (Ma-Y-Dox) containing an ezymatically cleavable oligopeptide spacer (Y) and a known chemotherapeutic drug (Dox), Figure 2.10a,b. The coated AuNPs showed only negligible aggregation in buffer media. The release rates of Dox from AuNP-(PAH/PSS)$_5$-PAH-(F-HPMA) was shown to depend on the presence of cathepsin B, an enzyme, that is, specific for cleaving at Gly–Phe–Leu–Gly Y oligopeptide sequences. Stealthing by the F-HPMA corona was visualized with the help of the plasmon absorption of the gold cores. Macrophages exposed to particles without F-HMPA stealthing turned dark pink/purple after less then 6 h, while macrophages exposed to F-HMPA-coated particles remained pale in color

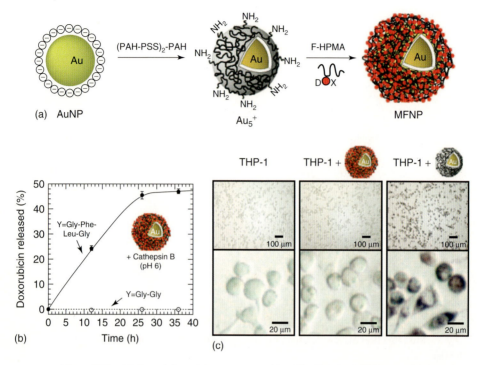

Figure 2.10 (a) From left to right: 13-nm-sized gold nanoparticles (AuNPs) as obtained after synthesis, stabilized by adsorbed sodium citrate; AuNPs coated with five primer layers of PAH and PSS and further coated with a external layer of F-HPMA (yielding MFNP). The red circles represent doxorubicin moieties (Dox) and are to scale with respect to the size of Dox molecules and the density of Dox moieties on the nanoparticle surface. (b) Cathepsin B induced release of doxorubicin from MFNPs by specific cleavage of the tetrapeptide

(Y = Gly–Phe–Leu–Gly) spacer between doxorubicin and the F-HPMA terpolymer backbone. The control with a Y = Gly–Gly spacer is not cleaved. (c) Optical micrographs of TPA differentiated THP-1 leukemia monocytes after 72 h of incubation without nanoparticles (images on the left "THP-1" depicts the control), incubated with MFNPs (center images demonstrate the stealthiness), and incubated with Au_5^+ (images on the right show the strong uptake of particles without F-HPMA layer). (Adapted from Ref. [63].)

even after 72 h. When color change occurred this was interpreted, as the particles were taken up by the macrophages.

Majewski and coworkers [71] reported the development of a similar multifunctional nanocarrier using LbL as modular method to build in specific functionalities. They describe the preparation of Maghemite cores coated with primer layers of $(PAA/PAH)_n$-PAA. In this case, a poly(ethyleneimine)-poly(ethylene glycol) (PEI-PEG) copolymer was placed as the outermost layer to provide stealthiness and increase colloidal stability. The anticancer drug cisplatin was loaded directly into the $(PAH/PAA)_n$-PAA-(PEI-PEG) shell via direct substitution of the cisplatin chlorine ligands with free carboxyl and amine groups. When Jurkat human adult T-leukemia cells were exposed to these particles, cellular death occurred at a lower half-maximum concentration (41.9 ng ml^{-1} at 72 h) then assays using free cisplatin (1.6 µg ml^{-1} at 72 h).

2.3
Interaction of Cells with LbL Films: Adhesion, Proliferation, Stimulation, and Differentiation

LbL films are also being exploited in the development of surfaces intended for cellular adhesion and growth [72]. In addition to preparing such templates, it has been demonstrated that LbL-coated substrates could be used to stimulate cell differentiation in a controlled manner. Contributions toward building and understanding LbL films for the purpose of controlled cellular adhesion and processes plays a large role in the development of any implantable templates. Therefore, it is critical to explore the current progress that has been made in studying and preparing such templates. The focus of this section is some of the most promising developments that have been made toward preparing LbL-deposition-based substrates that cells adhere to, grow, and differentiate/stimulated on in a controlled fashion.

2.3.1
Cell Interactions with Pure PEM Films

A number of research teams have found that simple LbL films, composed only of polyelectrolytes, could be easily tuned to promote cellular adhesion by adjusting electrostatic interactions. Initially, it was demonstrated that by placing a simple LbL film on a glass substrates, cellular adhesion and proliferation improved [73–77]. Generally cells have a preference for surfaces with a last layer carrying an opposite charge to itself [78, 79]. Salloum *et al.* [80] studied the adhesion of smooth muscle cells (SMCs) as a function of surface charge and changes in hydrophobicity. Interestingly, the pH value at the time of film assembly has been established to have a long-term effect in cell adhesion and growth studies [81]. Mendelsohn *et al.* [82] have reported on the assembly of LbL films using (PAA/PAH) at two different pH values; acidic 2, and close to neutral 6.5. When NR6WT fibroblasts are incubated on the corresponding substrates, it was found that cells grew only on the films prepared at neutral pH. The acidic films were quite bioinert with little to no cellular adhesion. The films composed under acidic conditions yielded high net positive charges within the substrate, which in turn resulted in cell destruction. However, the more neutral preparation yielded films that were less densely charged, and were more attractive to cells and facilitate growth. The same research team recently reported the preparation of (PAA/PAH) films at pH 6 that were further exposed to a pH 2 environment following assembly [83]. They demonstrated that these samples are antimicrobial and attribute these properties to the swelling of the film and the high charge density. Such films can simply be "switched" to cytophilic when post-treated with pH 6 media.

As noted before [78] the last layer of an LbL film can be used to tailor control for cellular adhesion. In 2009 Saravia and Toca-Herrara [84] observed that cells spread more efficiently when films are terminated with a positively charged PEI layer, rather then samples with a negatively charged PSS top layer. Similar effects were documented by Hernadez-Lopez *et al.* [85] when assembling films composed of negative and positively charged N,N-disubstituted hydrazine phosphorus-containing dendrimers. Fetal cortical rat neurons attached faster on top of films with a last layer of cationic dendrimers then on films terminated with negatively charged layers. The chemical functionalities of the last surface layer can also dictate the fate of cells to adhere to the substrate. An interesting example is the work by Kidambi *et al.* [86] in which hepatocytes only adhere to surfaces with PSS as the last layer in (PDDA/PSS) multilayers. This preferential adhesion is due for a high affinity of the cells to the sulfonate groups in the PSS layer.

2.3.2
Importance of Mechanical Properties

Aside from chemical constitution, surface rigidity has been demonstrated to play an important role for controlling interfacial cellular adhesion. LbL assembly offers a means to assemble interfacial films in which chemical composition and mechanical properties are independently controlled. Mechanical properties are typically controlled by assembly pH [87], cross-link density [77, 88–90], and ionic strength [91–94] during the assembly process. In 2004 Richert *et al.* [88, 89] described the preparation of $(PLL/HA)_n$ films on glass that were additionally cross-linked using 1-ethyl-3-(3-dimethylaminopropyl)carbodiimide/N-hydroxysuccinimide (EDC/NHS) in an effort to increase surface rigidity. The elastic Young's modulus of the cross-linked films was six- to eight-fold higher than the noncross-linked films. Note that the measured modulus in the cross-linked films is comparable to the modulus of tissue composed of SMCs. Both film types were coated with collagen and it was found that cellular adhesion/spreading was strongly enhanced on the cross-linked systems. Similar behavior has also been demonstrated in polyacrylamine hydrogel films, which happen to have a comparable Young's modulus as the cross-linked films [89, 95]. EDC/sulfonated-NHS cross-linked (PLL/HA) films coated with collagen have also been shown to increase cellular adhesion and growth [96]. Unlike in the previously described section, the terminating layer seems to have no effect on the adhesion and cellular spreading. Another method of introducing rigidity within LbL films to increase cell adhesion, was recently described by Vazquez and colleagues [97]. This team prepared (PLL/HA) multilayer films containing photocross-linkable vinylbenzyl units. The investigators reported an increase in adhesion of mouse myoblast cells with increasing film rigidity. Such systems offer great potential for photolithographic techniques toward tissue engineering.

Recently, Mousallem *et al.* [98] prepared LbL films that not only facilitate cellular adhesion but also modulate the cellular phenotype based on film rigidity. In this work (PAH/PAA), PEMs were assembled using two different pH solutions of PAA. Under acidic conditions, PAA formed thicker and more flexible films then

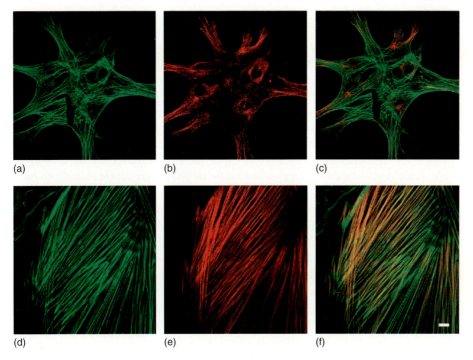

(a)　　　　　　　　(b)　　　　　　　　(c)

(d)　　　　　　　　(e)　　　　　　　　(f)

Figure 2.11 Localization of total actin and smooth muscle R-actin in A7r5 cells cultured on native and cross-linked PEMs. Cells were grown for three days on native (a–c) and cross-linked (d–f) (PAH/PAA)-4-PAH-coated coverslips. Actin filaments are stained with Phalloidin-Alexa 488 (green) and smooth muscle R-actin is labeled with a specific anti-R-actin antibody and Alexa 546-secondary antibody (b and e). Overlaid dual-labeled images (c) and (f); scale bar = 10 μm). (Adapted from Ref. [91].)

when neutral solutions were used. Both film types were additionally thermally cross-linked where the films prepared under acidic conditions showed faster cross-linking kinetics then films prepared using neutral PAA solutions. In addition, films immobilized at more acidic pHs showed enhanced rigidity then the neutral films. Following the assembly and cross-linking steps, rat aortic A7r5 SMC cells were cultured on both film types, as well as noncross-linked PEMs. Figure 2.11 is an example in which images (a–c) are of cells cultured on non cross-linked PEMs, while (d–f) depict cells cultured on cross-linked PEMs. The actin filaments are stained with Phalloidin-Alexa 488 (green, a and d) and the smooth muscle α-actin is labeled Alexa 546-secondary antibody (red, b and e). On the PEMs that were not cross-linked cells show the morphology of motile synthetic cells and expressed "synthetic" phenotype markers. In contrast the cross-linked PEMs induced expression of "contractile" phenotype marker proteins. Upon placement of Ca^{2+} on cross-linked PEMs cell cultures, it was found that the cells were stimulated and contracted demonstrating cell viability on such interfaces.

2.3.3
Importance of Surface Topology

Surface topography and roughness contributes significantly to the ultimate adhesion and proliferation of cultured cells. Kommireddy and colleagues implemented this idea by incorporating 21-nm nanoparticles of TiO_2 into films of (PDDA/PSS) in an effort to increase surface roughness [99]. Mouse stem cells were cultured on the LbL treated, rough substrates and it was found that the cells attached and proliferated. The authors investigated cell adhesion and spreading based on the layer number in the film. It was found that with increased layer number the surface roughness grows linearly (up to 140 nm for six TiO_2 layers), cell spreading occurred more rapidly. The findings in this study support that with rough samples, cells are able to attach more readily and hence proliferate. Lu *et al.* [100] found similar results in templates of (PAA/PAH) where micrometric line patterns were prepared using a so-called "room-temperature imprinting technique" in order to introduce roughness within the film. These line structures can be varied in lateral sizes and vertical height. Samples with 6.5 μm broad and 1.29 μm high lines separated by 3.5 μm were found to be cytophobic, while templates with coarser line structures (69 μm broad, 107 μm high, separated by 43 μm) were cytophilic. A number of articles demonstrate the use of nanoporous LbL films to increase surface roughness and achieve higher cellular adhesion rates. One example was described by Hajicharalambous and coworkers [101] in which nanoporous (PAA/PAH)$_n$ films were used for cellular adhesion and migration studies. After LbL assembly, the films were immersed in an acidic solution in order to initiate pH-induced phase separation, yielding nanopores. The size of the pores could be controlled via the pH of the acidic solution used in the formation process; 100-nm pores when pH 2 is used and 600-nm pores when pH 2.3 is used. In this study human corneal epithelial cells (HCECs) were cultured on substrates with various pore sizes. It was found that adhesion and proliferation increased on all nanoporous templates. Cell migration was most rapid on substrates containing the smallest pore sizes of 100 nm. Actin within the cellular cytoskeleton was studied closely in this work, and it was found that in larger pore size PEM as well as nonporous PEMs, the actin fibers localized primarily along the cell periphery and showed diffuse and undefined structure, Figure 2.12. Alternatively, on the templates with smaller pore sizes the actin structure remained well defined and the fibers transversed the entire cell cross-section, which is depicted in Figure 2.12.

2.3.4
Introduction of Chemical Functionality into LbL Film

It has already been documented that surface charge, stiffness, and topology governs the fate of cultured cells on such LbL-based templates. Chemically functionalized LbL interfaces have also furnished a method for modifying the surface in order to enhance cellular adhesion and proliferation.

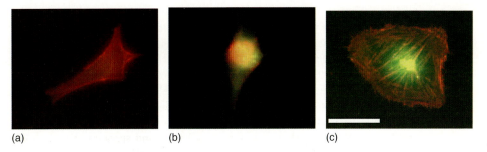

(a)   (b)   (c)

Figure 2.12 Immunofluorescence staining for actin structure and vinculin focal adhesion for an HCEC adhered on (a) nonporous, (b) submicrometer, and (c) nanoporous surface. Scale bar = 50 μm. (Adapted from Ref. [94].)

2.3.4.1 Adsorption of Adhesive Proteins on Multilayers toward Assisted Cell Adhesion

The inclusion of adhesive proteins within LbL films has been one approach toward the preparation of functional surfaces for enhanced cellular adhesion. Kirchhof *et al.* explored (CHI/HEP) multilayers on glass with a terminal layer of adhesive protein plasma fibronectin (pFN) [102]. Osteoblast MG-63 cells were cultured on untreated glass, PEM coated substrates with and without a terminal pFN layer. It was shown that the best cellular adhesion and spreading occurred on pFN PEM templates that were assembled under slightly basic conditions. Additionally, the layer just beneath the pFN layer also helped to control the success of cell attachment. More specifically, samples with HEP as the layer resting under the terminal pFN layer yielded less cell adhesion, while the corresponding CHI samples demonstrated better cell attachment. In a similar study, Wittmer *et al.* [103] showed the use of (PLL/dextran sulfate (DS)) films with pFN as the terminal layer, and found that these templates also promoted good adhesion and proliferation of human umbilical vein endothelial cells (HUVECs). Other research teams have also prepared similar surfaces that yielded cell adhesion and proliferation [104, 105].

2.3.4.2 Covalent Modification of Polyelectrolyte Films

Prefunctionalized Polymers for LbL Formation   Chemically tailored LbL films using covalently functionalized polyelectrolytes have been exploited toward controlling cell/surface interactions. Prior to multilayer assembly, polymers are covalently modified in order to incorporate specific functionalities.

After the polymers are modified, then LbL assembly is used in order to build films of the customized polymers. The benefit toward preparation of LbL films using previously functionalized polymers is that a large degree of control over the purity, and degree of functionality of the deposited material is possible. Moreover, because most of the harsh chemical modifications are conducted prior to film formation, LbL assembly generally occurs under mild conditions. In addition, the deposition conditions are similar or identical for different kinds of tailored polymers, making it simpler to adapt even to an industrial process. The described template preparation method has been used toward the development of cell-adhesive substrates. It was already demonstrated that the preparation of LbL films with a terminal layer of PLL to which arginine-glycine-aspartic acid (RGD) polypeptide (a sequence of cell-adhesion extracellular matrix proteins) was attached covalently enhances considerably osteroblast adhesion [106, 107]. In a similar effort, Swierczewska *et al.* [108] recently reported the modification of PEI and PAA by attaching an elastine-like polypeptide (ELP). Following the covalent attachment of ELP to both polyelectrolytes, the investigators further prepared PEMs using the customized polymers. It was established that these templates demonstrated enhanced cellular adhesion and proliferation. Other peptides have also been explored in a comparable fashion [109]. In a similar approach Chluba and coworkers [110–112] prefunctionalized PLL and poly(L-glutamic acid) (PGA) with α-melanocyte-stimulating hormone (α-MSH). Melanocytes show melanogenesis when grown on such films and an anti-inflammatory effect was found for monocytes [111]. The mechanism of drug delivery to the cells is mainly due to local degradation of the multilayer.

Postfunctionalized LbL Films A different approach toward the preparation of functionalized LbL films, is the postfunctionalization of the assembly. A few selected examples will be depicted in this section. In 2009 Buck *et al.* [113] reported the preparation of postfunctionalized LbL films formed by covalent amide formation of a polyamine and poly(2-vinyl-4,4′-dimethylazlactone) (PVDMA). The terminal layer in these films was PVDMA giving an excess of azlactone on the surface, which was further modified with amine-functionalized bioactive molecules. Some of the substrates were patterned with decylamine (a cell-adhesion and growth-promoting molecule), and D-glucamine (that prevents cell attachment). When CV-1 in Origin, and carrying the SV40 genetic material (COS-7) mammalian cells were cultured on these patterned templates, cellular growth occurred regioselectively where cells only grew on the decylamine portions. Moreover, the same trend was exhibited for the bacterial pathogen *Pseudomonas aeruginosa* cells.

Analogous to Buck's work, Kinnane and coworkers [114] developed a new approach for tuning surface–cell interactions using click chemistry. In this work simple Cu(II)-catalyzed cycloadditions of alkynes with azides yield a very elegant way for preparing reactive LbL films that can be postfunctionalized. In order to build "click chemistry" ready films, azide or alkyne prefunctionlized PEG were used in LbL assembly. In addition to the antifouling nature of PEG, other surface modifications could be made via the covalent attachment of a range of biomolecules such as carbohydrates, antibodies, DNA, and peptides, to the unreacted azides and alkynes, Figure 2.13. As an example, cell adhesion promoting RGD peptide was attached to PEG multilayers by click chemistry with the remaining excess alkyne groups on the film. Monkey kidney epithelial cells were seeded on RGD-functionalized PEG films and showed specific adhesion and growth. The combination of two desirable characteristics, namely, antifouling and the possibility for simple functionalization makes such templates a very promising

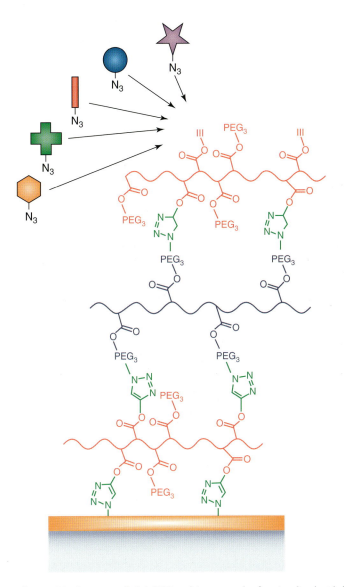

Figure 2.13 Structure of click PEG multilayers. Primer layers of PEI and PAA-azide were electrostatically adsorbed to enable the buildup of PEG layers. Covalent linking between layers of adsorbed click polymers occurs through the copper(I)-catalyzed cycloaddition of alkyne and azides. Using free click groups available at the surface, films can be functionalized with biomolecules such as carbohydrates, antibodies, or peptides, modified with click groups. When such biomolecules are attached to a low-biofouling surface, these functionalized materials are capable of promoting specific interactions with cells. (Adapted from Ref. [13].)

method for biomaterial engineering. Click chemistry has the advantage that it is highly selective, thus mostly preventing undesired side reactions from occurring.

Wischerhoff and coworkers [115] have recently described another unique application of LbL assembly toward the tunable adhesion of cells. The main focus of this work was to develop a new method of controlling the thickness of polymer brushes on solid substrates. The authors argue that in order for controlled cell adhesion to be achieved, it is critical to have templates with fine-tuned thicknesses. Furthermore, a new approach was demonstrated toward building polymeric brushes with controlled thickness by taking advantage of the nanolevel control of LbL assembly method. Templates were first coated with exponentially growing PEMs, with a macroinitiator as the terminal layer. Following LbL assembly, polymers were grafted from the substrate using atom-transfer radical polymerization (ATRP). The key to this approach is the dependence of brush thickness is on the number of deposited layer pairs, rather then polymerization conditions (time, deactivator in reaction). Thus, it is easy to fine tune the final thickness of the polymeric brush, solely based on layer-pair number alone. In order to demonstrate the robustness under biological conditions, cellular-adhesion studies were performed. In these particular studies thermoresponsive polymers were grafted from the templates. These polymers have a lower critical solution temperature (LCST) (at physiological temperature), which can modulate cell adhesion by temperature changes, which trigger conformational changes and swelling. At 37 °C the polymer brush collapses, allowing fibroblast cells to adhere onto the surface. However, at lower temperatures the polymer chains expand and cells tended to minimize contact with the more hydrated surface. When cells were cultured on these templates at 37 °C they grew and spread. Interestingly, upon lowering the temperature to 22 °C (same surface) the cells rounded up to minimize their contact with the surface, indicating low adhesion. This phenomenon was found to be reversible for several cycles.

2.3.5
Implantable Materials

A requirement for implantable materials is biocompatibility. Depending on the application, such templates need to be selective for cell adhesion, show antifouling properties and/or require antimicrobial properties. Recently there have been a number of interesting examples of such templates, some of which will be highlighted here [88, 116–127]. Rubner and coworkers [16] described a more detailed overview of this field in a recent review.

It has been previously established that biocompatibility and antimicrobial properties can be introduced to a surface via PEM assembly of (HA/CHI). Adding a cell-adhesion promoter to the surface yields interfaces that can comprise both, antimicrobial and cell adhesive properties. Recently, Chua *et al.* [128] modified Ti substrates with (HA/CHI) multilayers for the preparation of implantable templates. While osteoblast adsorption is fully inhibited on (HA/CHI) multilayers on Ti they presented a new method for introducing cell-adhesion functionality by EDC/NHS coupling of RGD peptide to the film [128]. It was successfully demonstrated that osteoblast adsorption and proliferation was increased compared to bare Ti and that bacterial adhesion was reduced by 80%. Li *et al.* [129] revealed almost complete platelet adhesion inhibition on Ti substrates coated with (collagen/sulfated-CHI) multilayers. Choi *et al.* [130] deposited covalently cross-linked multilayers on Ti surfaces composed of poly(vinyl alcohol) (PVA) and a water-soluble phosphorylcholine-functionalized polymer with phenylboronic acid moiety (poly(2-methacryloyloxyethyl phosphorylcholine-co-n -butylmethacrylate-co-p-vinylphenylboronic acid) (PMBV)) forming a biocompatible hydrogel. Cells were found to successfully adhere and proliferate on (PVA/PMBV) hydrogels.

Füredi-Milhofer and coworkers [131] engineered a very interesting approach toward the preparation of bone mimicking coatings. They deposited films containing amorphous calcium phosphate (ACP) nanoparticles embedded between PGA/PLL multilayers on Ti surfaces. By dipping those coated Ti substrates in a metastable calcifying solution, the ACP template was transformed into calcium octaphosphate and/or apatite. Cell adhesion did not take place when the coatings were topped with calcium phosphate, however, when the same surface was capped with another PEM, the coated Ti showed excellent biocompatibility both *in vitro* and *in vivo*.

LbL-coated biomaterial based templates have also been explored toward use as implantable substrates. Kerdjoudj and coworkers [132] demonstrated LbL coating of cryopreserved arteries from umbilical cords toward the preparation of potential vascular implants. Some of the current limitations with defrosted blood vessels are the change in structural integrity of the vascular walls and alteration of their biomechanical properties. These physical faults can ultimately lead to structural failures and ruptures after implantation. Additionally, the loss of endothelium is provoked by the cryopreservation and can lead to thrombosis and/or restenosis when blood comes in direct contact with the extracellular matrix. Kerdjoudj *et al.* [132] explored coating these vessels with LbL films in order to remedy the current physical and mechanical problems with defrosted vessels. The investigators coated the implants with (PAH/PSS) multilayers, which resulted in an increase in the mechanical stability in comparison to nontreated arteries. It was also documented that the multilayer coating enhanced the cell adhesion and spreading properties allowing the re-endothelialization. As captured in Figure 2.14, the cell cultures on untreated substrates led to transformation into round-shaped nonadhesive endothelial cells whereas the adhesive elongated cell morphology of fresh arteries was conserved on (PAH/PSS) films. On the LbL-coated arteries high Von Willebrand factor expression displayed phenotype preservation and indicated that the surfaces show optimal compliance to endothelial cells. This study shows that LbL-modified cryopreserved blood vessels had an increase in mechanical strength, while maintaining structure and functionality, that is, similar to fresh arteries. Furthermore, the modified vessels facilitated endothelial cell adhesion and growth. The same team has explored the *in vivo* behavior of these LbL-coated arteries and found that they remained patent longer then uncoated arteries [133]. The described progress alone displays the promise for the preparation of novel biobased graft materials.

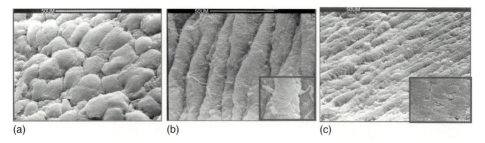

(a)   (b)   (c)

Figure 2.14 Scanning electron microscopy (SEM) images of untreated (a), with polyelectrolyte multilayer treated (b) cryopreserved umbilical arteries seeded with endothelial cells, and fresh umbilical arteries (c) (Inset: endothelial cells under a different observation direction). (Original magnification Å ~ 1000 for all images except insert of B with magnification Å ~ 2000). (Adapted from Ref. [124].)

2.3.6
Cell Stimulation from LbL Films

Films that are capable of not only allowing cell adhesion but also govern cell behavior has become a popular forefront of LbL research. In the following section we will describe some of the LbL-based surfaces that supported cells are also chemically or electrically stimulated.

2.3.6.1 LbL Films and Chemical Stimulation

The formation of artificial tissues by LbL deposition of living cells with polyelectrolytes has already been reported on several occasions [19, 134]. The Strasbourg team recently established a completely new technique for the development of templates that could be used toward tissue engineering based on sprayed PEMs with embedded cells [18, 134]. These templates have been used toward the stimulation of cells via incorporation of hormones within the film. Substrates were initially prepared using (HA/PLL) multilayers, followed by the formation of a thin calcium-alginate/cell gel layer. This cell layer was deposited via spraying a solution of alginate with dispersed fibroblast followed by spraying a Ca^{2+} solution, and resulting in surface gel formation. Precise control of gel thickness could be achieved via spray time. Cell-encased templates were prepared using exponentially growing films of (HA/PLL) or (PGA/PLL). Each film type yields different film structure and porosity that could further be manipulated. In initial investigations, fibroblasts were introduced into the alginate gel (see Figure 2.15) to test cell proliferation. It was found that after eight days 80% of the cells survived. In a second series of tests we studied the cell response when incorporating a hormone within the (HA/PLL) films. Specifically, melanocytes in calcium-alginate were immobilized in between multilayers of hormone rich α-MSH linked PGA and PLL. It was found that all cells distributed in the alginate gel responded to the biological stimulus. This response was not just limited to the cells that were in close proximity with the α-MSH-PGA containing multilayer, furnishing films with additional diffusion of the hormone. This result is rather important as it allows an efficient delivery

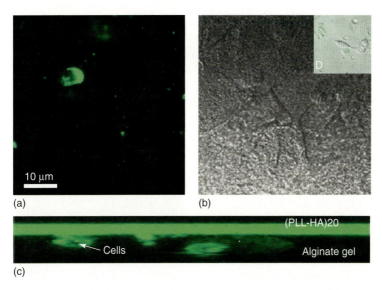

(a)

(b)

(PLL-HA)20

Cells

Alginate gel

(c)

Figure 2.15 Confocal microscopy observation after 2 h of alginate and fibroblastic cells spraying: (PLL-HA)20-PLLFITC-GEL/cells. The inset is the combination of (a) and (b), with the green channel superimposed with the conventional transmission channel. (c) Confocal Z section of the same image. The fact that the fibroblastic cells appear in green proves that they are in contact with the PLL-FITC of the underlying multilayer. (Adapted from Ref. [128].)

and cell response in gel/cell architectures that are much thicker than the characteristic cell size. However, when using linear growing (PAH/PSS) intermediate layers between the (α-MSH-PGA/PLL) multilayer and gel/cell layers, it was found that hormone delivery was strongly reduced or even completely block (based on layer-pair numbers of PAH/PSS). Such template modifications demonstrate the ability to easily tune hormone delivery within a cell rich multilayer. Studies using poly(D-glutamic acid) (PDGA) and poly-D-lysine (PDL), the D form equivalents of PGA and PLL, showed that cell response was dramatically decreased. These results depict that cells come in contact with α-MSH through partial degradation of the film by melanocytes. Finally, it was noted that the described degradation mechanism is in direct competition with the diffusion mechanism, and ultimately the results highly depend on the nature of the bioactive component.

As in the case of drug-delivery from surfaces, cell growth and even morphology can also be controlled by slow release of a growth factor from a LbL film. Crouzier *et al.* [135] cross-linked (PLL/HA) multilayers and loaded them with human recombinant bone morphogenetic protein 2 (rhBMP-2), a bone growth factor. The loaded amount was tuned by varying solution concentration and film thickness. The concentration of adsorbed rhBMP-2 was determined by microfluorimetry and it was found that the amount of rhBMP-2 increased up to 500-fold in the film compared to the original adsorption solution due to strong interactions between the protein and the multilayer. C2C12 myoblast cells were cultured on the treated templates. On films containing rhBMP-2 the myoblasts differentiated into osteoblasts, whereas control films without rhBMP-2 myotubes were produced. This behavior was found to be dose dependent, for example, films loaded with rhBMP-2 from a solution with concentrations higher then $1\,g\,ml^{-1}$ led to osteoblast differentiation, whereas at lower concentrations myotube phenotypes were found. The rhBMP-2-doped films maintained bioactivity for up to 10 days. This was attributed to the strong interaction between the trapped growth factor and the slow release of rhBMP-2 from the film. Furthermore, the growth factor within the multilayer film is protected against degradation by proteases.

Dierich *et al.* [136] described the preparation of active multilayers, which were found to induce stem-cell differentiation to bone cells. The PEMs were composed of PLL and PGA with embedded transforming growth factor β_1 (TGFβ_1) and bone morphogenetic proteins (BMPs). Embryonic stem cells were cultured on these degradable templates. Interestingly, both TGFβ_1 and BMPs are needed to be present within the film in order to induce bone formation. More recently, the same research team demonstrated chemically similar microcapsules used toward injectable transplantation of embryonic stem cells for a treatment of degenerative cartilage and bone disease [137, 138].

2.3.6.2 LbL Films and Electrical Stimulation

The Kotov research team recently described the preparation of LbL-based interfaces that could be used toward electrical neuron stimulation [139–141]. Specifically, this research team has demonstrated the preparation of PEMs using single-walled carbon nanotubes (SWNTs). The use of SWNT multilayers with PEM is motivated by a number of important features that the nanotubes provide to the interface. Such materials have a high potential for biomedical applications as they show good mechanical properties, corrosion resistance, and biocompatibility coupled with high conductivity. In this regard, one of the first SWNT-encased templates that has been described by this research team was composed of positively charged SWNT and PAA. To these films, NG108-15 cells were cultured, and the cells differentiated. It was also cited that the templated cells grew neurons and branches, which actually increased the contact area to the conductive SWNT film. The same research team has also demonstrated the preparation of PEMs composed of negatively charged SWNT and PEI [142]. Neural stem cells (NSCs) were found to differentiate to neurons on (SWNT/PEI) films. Recently, an extension of this work was described in which the investigators prepared films composed of SWNT and laminine, an extracellular matrix protein [140]. The motivation for using laminine as the alternate layer was to mimic the surrounding cell media of the *in vivo* environment. Similar to the PEI films, NSCs differentiated to neurons successfully on these templates. In addition to cellular differentiation, these templates have also been explored for electrical cellular stimulation of neuron-like NG108-15 cells. Two methods of electrical stimulation were investigated, (i) by inward excitation via a voltage-clamp pipette electrode and (ii) by extrinsic stimulation through the underlaying SWNT film. In both methods, multilayer supported neural cells were found to become stimulated via the measured activation potential from these films. This activation potential only occurs when the ion channels within the neural cell membranes open, and ion exchange is allowed to occur. In comparison with standard substrates for NSC culturing and differentiation, such as poly-L-ornithine (PLO), SWNT multilayers showed comparable growth, proliferation, and even expression of specific neural markers. All these results prove that the cell differentiation occurred in a similar process as the PLO samples. In terms of electrical interaction with neural cells, SWNT multilayers out performed state-of-the-art neural-interface materials; iridium oxide (IrOx) and poly(3,4-ethylenedioxythiophene) (PEDOT) [141]. SWNT multilayers showed reduced impedance, increased cathodic charge capacity, and facilitated charge transfer. In summary, these improvements demonstrate that SWNT based films actually show a high potential toward improved templates for neural cell differentiation and stimulation.

2.3.7
Patterned LbL Films for Cell Templating

2.3.7.1   Two-Dimensional Patterns
Selective interactions of cells to a surface toward pattern formation could potentially benefit a number of growing technologies such as tissue engineering, cell arrays, and biosensors. Upon preparing PEMs using LbL assembly, intrinsic z-directional architecture patterns arise, and could be further manipulated. One of the first examples of patterned PEMs was reported by Hammond and coworkers [143, 144]. Such templates are envisioned toward patterning cells and could ultimately be used in the preparation of positional multiplex arrays. Prior to the LbL assembly method, 2D cell patterns were usually prepared by first patterning cell-inert regions on a cell-adhesive background or vise versa. Carter [145] prepared one of the first cell-patterned surfaces in the late 1960s in which fibroblasts were oriented on islands of Pd. Thirty years later, Ingber and Whitesides and coworkers [146, 147] used microcontact printing to pattern gold surfaces with self-assembled monolayers (SAMs) of adhesive fibroblast regions on an antifouling ethylene glycol background. One of the drawbacks with the SAM method to prepare templates is only a handful of surface types can be employed, such as gold (Au) and some metal oxides. On the contrary, LbL assembly technique is nonsubstrate discriminatory making it a broader approach toward building patterned templates. Moreover, a number of research teams have exploited LbL assembly in order to pattern cells on both 2D and 3D templates.

Yang and coworkers [148, 149] described an early example of surface cell patterning by LbL technique in 2003. In this work PEM films were formed using cell-inert (PAA/PAH). Following film formation the investigators used ink-jet printing and photolithography in order to form cell-adhesive/cell-inert patterns throughout the PEM. The same research team later communicated a simple method of forming such patterns using stamping technique [150]. After PEM film formation, patterns of RGD are stamped to the surface interface yielding a cell-adhesive pattern. These surfaces showed a regioselective cell adhesion for WT NR6 fibroblasts in which cells only adhered on RGD treated surface regions. The density of RGD in the cell-adhesive regions could be controlled by the stamping conditions, which had a strong effect on cell attachment, morphology, and cytoskeleton protein organization.

Lee *et al.* [151] used a similar stamp based technique for surface cell patterning via LbL. In this work a PDMS stamp is placed on top of a (PAH/PSS) multilayers, which was further exposed to a solution of cell-inert diblock-copolymer poly(ethylene glycol)-poly(D,L-lactide) (PEG-PLA). After the PEG-PLA layer was deposited, the stamp was removed leaving a cell-inert pattern on a cell-adhesive film. This technique of inducing pattern formation was coined "micromolding in capillaries" (MIMIC) [152]. Fibroblasts mammalian cells, bacteria, and spores were all respectively cultured on these patterned templates and only adhered to the (PAH/PSS) interfacial regions. Figure 2.16 depicts the high contrast in cell adhesion between the different regions. The patterns appeared very sharp and different spot and line shapes could be achieved depending on the PDMS stamp used. Other research teams have also described similar stamp-based patterning techniques [153–157].

Mohamed *et al.* [158] described the preparation of patterned cell templates using a combination of photolithography and LbL assembly. This work demonstrates the preparation of interdigitated micropatterns on a single template, in which different cell types could be confined. A template is first coated with a photoresist, and using UV lithography portions of the photoresist are removed. To this template, LbL films of (PSS/PLL) are deposited and the photoresist is then lifted-off leaving behind a patterned surface. Subsequent photoresist deposition and UV lithography could be conducted on the same template in order to deposit an additional pattern of another LbL film composition. The templates were used to selectively confine a number of different cell types within the same sample template. In a similar approach, there has also been some work in the preparation of heterotypic cell patterns, that is, coculturing of cells [154, 159].

2.3.7.2 Three-Dimensional Scaffolds

Three-dimensional inverted colloidal crystal (ICC) hydrogel scaffolds have been explored as a template to culture cells. The preparation of such substrates could potentially be used as a tissue-engineering scaffold. Specifically, in the case of stem cells, the rate and direction of cell growth is strongly dictated by the cell–cell interaction and the 3D microenvironments. Lee *et al.* [160] described the preparation of LbL modified polyacrylamide ICCs toward cell growth applications. LbL films of (nanoclay particles/PDDA) were deposited within the ICC matrix. It was noted that the nanoclay particles actually reinforced the fragile hydrogel matrix, as well as yielding a surface with more biocompatible properties. In addition, the materials themselves are highly transparent making it easy to investigate via optical techniques such as confocal microscopy. Thymus epithelial cells and human premyelote monocytes were cocultured within the 3D scaffold in order to study the behavior of coexisting cells in such an environment. As can be seen in Figure 2.17, epithelial cells (green) covered the entire cavity surface, while the monocytes (red) went around the porous structure. The thymus epithelial cells block the outer pore layers and entrap the monocytes inside the 3D scaffold. Recently, the same

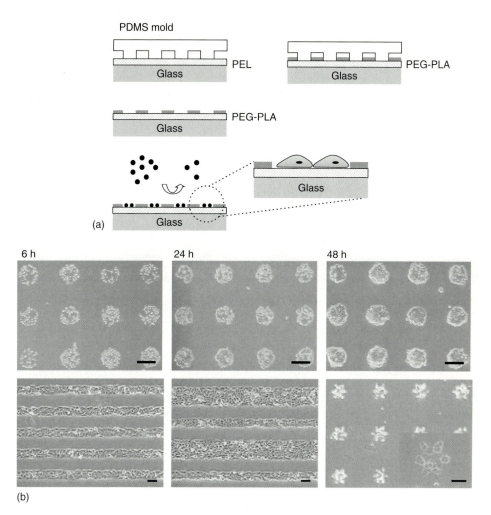

Figure 2.16 (a) Schematic diagram of preparation of orthogonal surfaces using micromolding in capillaries (MIMIC): surface modification with polyelectrolytes (PELs) (PAH/PSS); Polydimethylsiloxane (PDMS) micromold placement and filling the micromold with a poly(ethylene glycol)–poly(D,L-lactide) diblock-copolymer (PEG–PLA) by capillary action; removing the PDMS micromold; loading of biomolecules (proteins or cells) onto the fabricated surface. (b) Control of cell adhesion onto the various types of micropatterns. Optical images of cell proliferation on micropatterns selectively performed on 200-m circles, line cell patterns with identical width (100 m), and line cell patterns with alternative widths (200 and 100 m). The cells precisely recognized PEL regions so that boundary lines were clearly differentiated. The scale bar indicates 100 m. Star-shaped patterns of fibroblast cells; after one day of culturing, cells were spread out. The cell patterns could be stable for at least two weeks if suitable culture conditions were maintained. (Adapted from Ref. [146].)

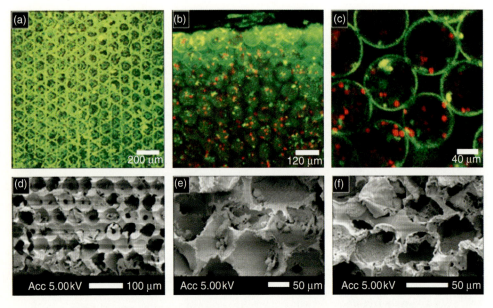

Figure 2.17 Confocal images (a–c) of co-cultured ICC hydrogel scaffolds; thymic epithelial cells (green) and monocytes (red). (a) Bottom area image shows the surface of the scaffold was densely covered with thymic epithelial cells. Most of the monocytes around the edge of the scaffold were released. (b) A cross-sectional image after cutting the cocultured ICC scaffold with a razor blade shows decreasing thymic epithelial cell density moving into the inside of the ICC scaffold. Monocytes were distributed through the whole ICC scaffold and a similar number of cells were entrapped at each pore. (c) A lateral section image of 80 mm in depth. SEM images (d–f) of cocultured hydrogel scaffolds. (d) Cross-sectional image of the scaffold's interior. (e) Entrapped monocytes. (f) Thymic epithelial cells covering pores and channels. (Adapted from Ref. [154].)

investigators published the culturing of human bone stromal cells and CD34+ hematopoietic stem cells (HSCs) on the surface of (nanoclay/PDDA) coated 3D scaffold [161]. By putting stromal cells in the porous ICC construct interior, the natural *in vivo* 3D microenvironment of bone marrow was mimicked. As a result of this imitated environment, HSC expansion occurred as well as B-lymphocyte differentiation. In comparison to standard 2D culturing plates the cell growth and differentiation was much faster demonstrating the importance of the 3D environment for HSC growth and differentiation.

2.4
Plasmid DNA (pDNA)-Based Gene Therapy and LbL Films

Similar to protein and small molecule drug-delivery LbL-assembly-based systems, there has been an effort to build substrates that are capable of releasing plasmid deoxyribonucleic acid (pDNA) in a controlled fashion. As demonstrated earlier in this review there are a number of different possibilities in achieving substrates that can degrade in a timely manner. At present, the most promising films are composed of a polymer that is cleavable under physiological conditions via enzymatic hydrolysis [162–164], hydrolysis [165, 166], or under reductive conditions [167, 168]. In addition, understanding the fate of the delivered pDNA is important, and the mechanism of cellular uptake plays an important role in achieving successful transfection. In solution-based gene therapy pDNA complexes with linear poly(ethyleneimine) (LPEI) to form dense particles called *"polyplexes."* It has been demonstrated that polyplexes have high gene transfection rates. Other types of polyplexes using different polycations have also been studied and found to yield similar transfection rates. Moreover, the preparation of LbL films composed of both pDNA/polycation, and polyplexes have been explored [169–171]. In addition, viral-based LbL-assembled substrates for gene therapy have also been described [172–174].

2.4.1
LbL Films and Plasmid DNA (pDNA) for Gene Therapy

2.4.1.1 Nonviral Surface-Based Transfection
The development of pDNA rich films with controlled degradation has yielded substrates that could be used toward surface-mediated cellular transfection [6, 175]. Moreover, these films are envisioned to be used toward localized gene therapy. As a nonviral gene-delivery system, DNA-loaded LbL films exhibit low toxicity and high stability upon storage. As opposed to viral vectors, they do not induce side effects such as mutations in host chromosome or immune reaction that can lead to a fatal outcome for the patient.

This section will describe some of the progress that has been made in studying pDNA-rich LbL films and advancements regarding *in vitro* and *in vivo* cellular transfection.

While the first incorporation of DNA and other polynucleotides was carried out in our team [176–178], the Lynn research team was one of the first groups to present pDNA transfection experiments conducted with LbL films [179, 180]. One of the earliest developments from this research group uses **Poly-1** (Scheme 2.1) and pDNA (pEGFP-N1) (is encoded with gene that activates green fluorescence's protein (GFP)) films on quartz slides (PSS/PAH were used as primer layers). As established earlier in this review, **Poly-1**/HEP LbL films actually erode over the course of 24 h under physiological conditions. The $(PSS/PAH)_n$-**(Poly-1**/pDNA$)_n$-coated slides were incubated on top of COS-7 cells in a "forward-mediated transfection" approach, and the collected images are depicted in Figure 2.18. The majority of the cells that expressed GFP were located directly under the coated slides. The investigators suggested that films composed of **Poly-1** degraded in a manner that results in condensed pDNA, which in turn could improve the internalization of the pDNA. The same research team has also described coating stainless steel intravascular stents with (**Poly-1**/pDNA) assemblies. The LbL films remained preserved under balloon expansion; a mechanical method to simulate *in vivo* challenges. Similar to

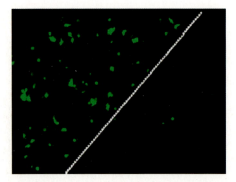

Figure 2.18 Fluorescence microscopy images showing the expression of EGFP and the localization of transfection in COS-7 cells. Direct transfection was mediated using a quartz substrate functionalized on a single side with a thin multilayered film composed of eight layers of polymer 1 and pEGFP plasmid. The image was recorded after 48 h of contact with the film-coated substrate. This image was recorded through a 4× objective showing the extent to which transfection was localized to cells growing under the film-coated substrate. The dotted line in this image indicates the edge of the film-coated substrate. (Adapted from Ref. [173].)

the original study, the coated stent was exposed to COS-7 cells in the presence and absence of additional transfection agents [181]. In both cases it was found that cells did express GFP after exposure to the coated stent. The Reineke research team has described the use of poly(glycoamidoamine)s (PGAA) (**Poly-5**) in the formation of pDNA/PGAA PEMs that could also be used toward cellular transfection [182].

Zhang *et al.* [183, 184] have recently explored the intracellular DNA pathway of pDNA delivered from an LbL film. In this work, PLL was tagged with β-cyclodextrin (CD); a group of cyclic oligosaccharides that have been shown to improve the bioavailability of drug compounds by forming inclusion complexes [185, 186]. Primarily, CDs have demonstrated an ability to bind to nucleotides, and even enhance their transfection efficiencies *in vivo*. First expressed by Lvov *et al.* [176], condensed pDNA polyplexes were used within the LbL film in order to promote transfection. Specifically, the PEMs were composed of (PLL/HA) primer multilayers and (PLL-CD/pDNA)$_n$ on top. Transfection experiments in the presence of serum and Hela cells were conducted on multilayers composed of PLL, CD, PLL-CD, jetPEI (LPEI), and bare pDNA. Comparative experiments were also conducted using polyplex complexes in solution composed of the respective materials. The pDNA used in this study was pCMV-Luc, which is encoded for the luciferase gene. Figures 2.19a,b represent the cellular uptake of the pCMV-Luc as well as the documented toxicity data. JetPEI complexes yielded the highest transfection in comparison to the other solution complexes. When comparing the multilayer systems, the highest transfection rates were demonstrated in the PLL-CD samples. Cellular viability was enhanced for the PLL-CD PEMs, in comparison to the solution complexes. It was speculated that the enhanced viability could be due to increased transfection. The surface-mediated transfection versus solution methods

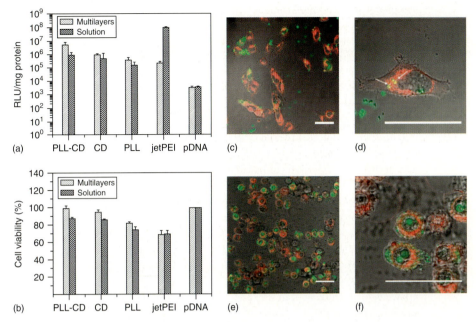

Figure 2.19 (a) Transfection efficiency and (b) cytotoxicity of the complexes of pDNA with PLL, CD, and PLL-CD at a N/P ratio of 3 for *HeLa* cells grown on (PLL-HA)$_5$-(complexes)-(PLL-HA)$_5$ films or in the absence of multilayers for 24 h. Cells were lysed for luciferase activity quantification. Gene expression determined from the luciferase assay was expressed as RLU/mg of protein. JetPEI/pDNA complex and naked pDNA were used as controls. Internalization of the complexes of pDNA with PLL-CD in *HeLa* cells grown on the surface of (PLL-HA)$_5$-PLL-CD/pDNA complexes-(PLL-HA)$_5$ multilayered films (e,f) or in the absence of multilayers (c,d). The concentration of pDNA was 60 μM, and the N/P ratio was 3. The images correspond to the overlay of light transmission and fluorescence confocal images and were taken at 24 h post transfection. Yellow spots (arrows) correspond to areas where the complexes and FM4-64 are colocalized. Red and green spots correspond to areas containing only FM4-64 and YOYO-1 labeled complexes, respectively. Images are representative of more than 90% of the observed cells (*n* = 3 specimens). Scale bars = 40 μm. (Adapted from Ref. [169].)

were additionally compared by confocal microscopy. In these sets of experiments pDNA was stained with YOYO-1, green fluorescent label. In addition they used a marker of membrane endocytosis FM4-64 (red fluorescent label) was used in order to differentiate between actual cellular transfection and intracellular trafficking of the pDNA. In the instances of cocolonization of the two respective tags yellow fluorescence occurs. An example of the data obtained is depicted in Figure 2.19. Figure 2.19c,d are Hela cells exposed to PLL-CD solution complexes, while Figure 2.19e,f are Hela cells exposed to PEM of (PLL/HA)$_5$-PLL-CD/pDNA. Cells that were exposed to solution-based agents showed only a limited number of complexes that entered the cell, and the few that did yielded yellow spots. This suggests that these complexes enter the cell through endocytosis. With the PEM samples numerous green spots within the cellular cytoplasm were found and no yellow spots were observed. It was concluded that the pDNA from the PEM samples was internalized by the cell via a nonendocytotic pathway. This could explain why transfection efficiencies on the PEM samples were much higher than in the solution complexes.

Lin *et al.* [187] have described the preparation of PEM composed of the HA and complexed chitosan-pDNA polyplexes. The same team previously prepared similar templates using pDNA–LPEI polyplexes and PGA, that failed to product interesting transfection results due to cytotoxicity [188]. When increasing the layer-pair number from one to five higher transfection was observed. This increased transfection was due to the greater amount of possible pDNA that could be delivered within the film.

Reductively degradable films based on poly(amido amine) have been used in the preparation of pDNA PEMs. These surfaces degrade when exposed to reductive environments such as the intracellular and certain extracellular milieu via cleavage of the disulfide bonds [167]. Blacklock *et al.* [189] also described the use of reducible hyperbranched poly(amido amine) (RHB) (Figure 2.20a) and pDNA to form a PEM on a flexible stainless steel substrate. This work is one of the only studies, to our knowledge, in which both *in vitro* and *in vivo* animal experiments were explored. It was expected that films composed of (RHB/pDNA) multilayers would begin to erode when in contact with the reducing microenvironment of the plasma membrane. In this regard, there would be a controlled disassembly of pDNA-containing PEMs. The used pDNA was encoded for secreted alkaline phosphate (SEAP) and GFP. It was found in *in vitro* transfection studies that (RHB/pDNA)$_n$ films actually showed higher and longer-lasting activity than control solution phase pDNA-LPEI polyplexes. *In vivo* transfection activity was evaluated by implanting the coated substrate and measuring SEAP secreted into the blood circulation of rats, Figures 2.20b–e. The implant took eight days to completely heal, and no adverse reactions occurred. Figure 2.20f shows that SEAP secretion peaked on day 5 after implantation for (RHB/pDNA)-coated substrates. The sharp decline of SEAP levels after day 5 was suspected to be due to fibrous capsules that formed around the LbL-coated substrates. Although there remain many questions that are still unanswered in regards to using these templates, such preliminary *in vivo* studies [190] shows promise for the future of LbL-assembly-based transfection templates.

2.4.1.2 Multiple Plasmid Delivery from LbL Films

LbL films containing multiple plasmids and able to deliver each respective plasmid in a time-controlled fashion present the most advanced platforms that have been developed to date. Jessel *et al.* [191] first demonstrated films that contained two different plasmids that could be released in a time-controlled manner and yield transfection of the two separate plasmids. The PEMs were composed of PLL, PGA, and the two different pDNA as well as cationic CD pyridylamino-β-cyclodextrin (pCD) as a transfection agent. Based on the architectural sequence of the PEMs, biological activity can be induced successively via the layer of pDNA, that is, released

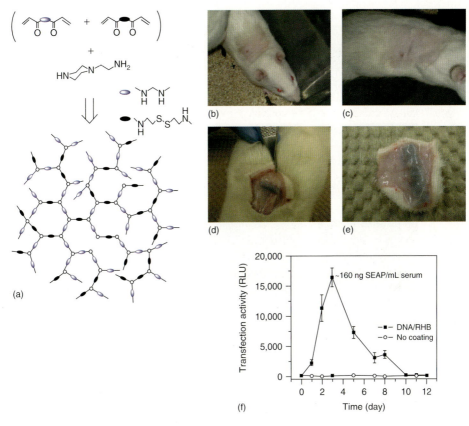

Figure 2.20 (a) Synthesis of reducible hyperbranched poly(amido amine) (RHB). Implant excision. Healed wound observed eight days after implantation of steel mesh coated with (DNA/RHB)₁₅ films (b,c). Implant excision on day 12 (d,e). In vivo transfection activity of (DNA/RHB)₁₅ films. Stainless steel mesh coated with (DNA/RHB)15 (−) and control noncoated mesh were implanted subcutaneously in rats and levels of SEAP expression in blood plasma were measured (f). (Adapted from Ref. [182].)

from the film. *In vitro* studies were conducted on three separate cell types; Chinese hamster ovary cells (CHO), macrophages, and COS. Two different plasmids were used, SPT7pTL, which expresses human SPT7 nuclear transcription factor, and pEGFP expressing GFP as cytoplasmic protein. Figure 2.21 shows the results of cells incubated on top of multilayered-coated plates with film architectures of (PLL-PLGA)₅-pCD-(PLGA-PLL)₅-PLGA-pCD-(PLGA-PLL)₅ (Figure 2.21a,b) and/or (PLL-PLGA)₅-pCD-pEGFP-pCD-(PLGA-PLL)₅-PLGA-pCD-hSPT7pTL-pCD-(PLGA-PLL)₅ (Figure 2.21c–h). Within the first 2 h (Figures 2.21c,d) the expression of SPT7 occurs and persisted over the next few hours. After 4 h the GFP began to express within the cytoplasm and increased over the next 8 h (Figures 2.21e–h). Interestingly, reversing the order of the two plasmids furnished the expression of the respective released vectors. These results demonstrate the capacity of LbL

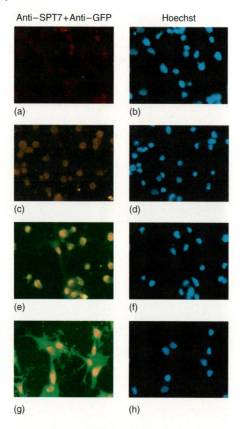

Figure 2.21 Expression of SPT7 and EGFP in COS cells grown on the surface of (PLL-PLGA)₅-pCD-(PLGA-PLL)₅-PLGA-pCD-(PLGA-PLL)₅ multilayered films (a and b) or (PLL-PLGA)₅-pCD-pEGFP-pCD-(PLGA-PLL)₅-PLGA-pCD-hSPT7pTL-pCD-(PLGA-PLL)₅ multilayered films for 2 h (c and d), 4 h (e and f), and 8 h (g and h). The expression of SPT7 (red) and GFP (green) was detected by using mouse monoclonal anti-SPT7 and rabbit polyclonal anti-GFP as primary antibodies and Cy3-conjugated goat antimouse and Alexa Fluor 488 goat antirabbit as secondary antibodies (a, c, e, and g). Nuclei were visualized by Hoechst 33258 staining (b, d, f, and h). (Adapted from Ref. [184], Copyright (2006) National Academy of Sciences, U.S.A.)

assembly for depositing specific nanoscale architectures that yield controlled release of multiple transfection agents from a single substrate.

More recently, Zhang *et al.* [192] have demonstrated the preparation of multiple plasmid PEMs using poly(β-amino esters). Wood *et al.* [193] first demonstrated the preparation, and degradation of films composed of **Poly-1** and anionic polysaccharides, and found that intermediate layers of cross-linked polyelectrolytes could be used to control the rates of released polysaccharides. Based on these findings, Zhang and coworkers [194] prepared LbL films composed of **Poly-4** and three different plasmid DNAs; pDsRed-Cy5, Pegfp-Cy3, and pLuc (used as spacer layers, not fluorescently labeled). They selected the

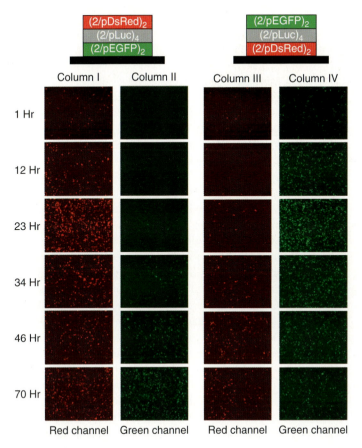

Figure 2.22 Representative low-magnification fluorescence microscopy images (4×) showing relative levels of enhanced green fluorescent protein (EGFP; green channel) and red fluorescent protein (RFP; red channel) expressed in COS-7 cells. Cells were transfected with samples of DNA released from films having the structures (**2**/pEGFP)2(**2**/pLuc)4(**2**/pDsRed)2 (columns I and II) and (**2**/pDsRed)2(**2**/pLuc)4(**2**/pEGFP)2 (columns III and IV), as indicated schematically. The relative levels of EGFP and RFP observed correspond qualitatively to relative levels of each plasmid released and collected over each of the following time periods: 0–1, 1–12, 12–23, 23–34, 34–46, and 46–70 h. (Adapted from Ref. [187].)

pLuc plasmid to inhibit displacement of the other plasmids within the film. As can be seen in Figure 2.22, preliminary results show that depending on the film architecture different plasmids are expressed at different times. The same research team has made additional contributions toward the development of multiplasmid-rich LbL films using an array of hydrolytically degradable polymers.

It is worth noting the developments conducted by Meyer *et al.* [195] in which they describe the preparation of coatings that combine simultaneous interfacial gene delivery and peptide-based cell signaling. LbL films composed of condensed pDNA-LPEI polyplexes and PGA tagged peptide with PAH and PSS serving at intermediate layers. Peptide activity in the embedded PEM remained, while preliminary data revealed effective transfection rates.

2.4.1.3 Current Progress in Transfection and LbL-Coated Nanoparticles

Microcapsules used toward delivering pDNA in a targeted and controlled release for cellular transfection from LbL films have played a substantial role in recent research [196–198]. Furthermore, Jewel and Lynn [6] has already provided a review that highlights the current progress in the field of micrometer-sized LbL-assembly-based transfection from particles. Nanosized LbL-coated particles for cellular transfection have recently become of high interest. As mentioned earlier, polyplexes are polycation condensed pDNA nanoparticles [199–202]. Cellular transfection rates of nonviral pDNA polyplexes with additional LbL layers have shown improved gene transfer compared to bare polyplexes. Described as *"recharging"* of the polyplexes, these particles offer more opportunity toward the development of targeted gene transfection [203, 204]. Trubetskoy *et al.* [205, 206] demonstrated the adsorption of anionic succinylated poly-L-lysine (SPLL) onto polyplexes composed of PLL. Subsequent deposition of additional layers using cationic PLL yielded PEMs coated polyplexes [206]. The same research team also demonstrated that pDNA-LPEI-$(PAA/PEI)_n$ polyplexes actually had increased levels of cell transfection and gene expression [205]. More recently, Saul *et al.* [207] described the same pDNA-LPEI-$(PAA/PEI)_n$ coated polyplexes and detailed transfection studies where the coated polyplexes were purified with size exclusion chromatography between layer deposition, Figure 2.23. The purpose of the additional purification steps was to remove any random subpopulations and obtain a pure homogeneous sample. Although highly enhanced transfection efficiencies were demonstrated in this study, the importance of purifying the polyplexes between each layer deposition was also revealed.

2.4.1.4 siRNA and LbL Assembly on Planar Surfaces and Nanoparticles toward Gene Therapy

Currently, some efforts toward incorporating short interferring RNA (siRNA) into LbL films for the preparation of gene therapeutic templates have been explored. The function of siRNA is it silences specific gene expression at the post-transcriptional level via mediating degradation of complementary messenger ribonucleic acid (mRNA). SiRNAs provide an alternative approach toward degrading viral RNA and could possibly be used for viral infection treatments [208]. The current issues in the field of siRNA research are: the efficiency, safety, and duration of delivery. LbL assembly has been demonstrated in a number of recently published papers to provide a possible remedy to the challenges in siRNA delivery and the elimination of premature degradation.

Recksiedler *et al.* [209] described the preparation and controlled release of RNA-rich PEMs in 2006. Since then, a number of interesting studies have been conducted using siRNA-rich PEMs in which additional cell studies were demonstrated. Dimitrova *et al.* [210] constructed films using $(PLL/PGA)_n$ as primer layers,

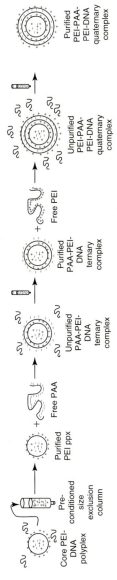

Figure 2.23 Schematic representation of multilayer system for the formation of polyplexes. Core polyplexes were formed at N:P ratio of 5 and purified over size exclusion chromatography (SEC) to remove free PEI. PAA was then added at a PAA:DNA ratio of 8 and the resulting ternary complexes were purified over SEC to remove free PAA. A second layer of PEI was then added to form quaternary complexes using N:P ratio of 40 and the resulting complexes were purified over SEC. (Adapted from Ref. [200].)

followed by siRNA-LPEI complexes, and then (HA/CHI) multilayers were used in the intermediate and top portions of the assembly. In order to understand the feasibility of these coated surfaces, hepatitis C virus (HCV) was used as an infection model. Infected cells were placed on top of the siRNA (siHCV)-rich films and inhibition/replication studies were reported. Films composed of five layers of siRNA-PEI (compared to one, and three layers) actually revealed the most sustained release of the siRNA and a decrease in HCV replication of the infected cells. These results demonstrated that the siRNAs that are deep within the architecture are efficiently available for the target cells. In addition, cellular viability showed that the toxicity of this approach was lower then with solution phase siRNA-LPEI complexes.

LbL-coated nanoparticles for the delivery of siRNA has also been recently explored by Elbakry *et al.* [211]. AuNPs were first coated with mercaptoundecanoic acid (MUA) and then LbL films were formed using (PEI/siRNA)$_n$, Figure 2.24a. As described in a previous section, AuNPs provide a simple 3D template that could be used to deliver drugs, as well as provides a means to monitor the delivery route. Following the preparation of the siRNA PEM coated AuNPs, the particles were then incubated with CHO-K1 cells, which successfully took up the particles. In gene-silencing experiments it was found that when CHO-KI/EGFP (enhanced green fluorescent protein) cells were exposed to LbL coated AuNPs with siRNA against EGFP, the GFP fluorescence was reduced, Figure 2.24b. In addition cellular viability remained unharmed when incubated with the LbL-coated AuNPs. It was found that the architecture of the LbL films plays a large role in the ultimate success of gene silencing.

Mehrotra *et al.* [212] described the preparation of LbL-assembled degradable multilayer films for region specific delivery of siRNA via the forward mediated transfection approach. Figures 2.25a–f illustrates the protocol required for the preparation of these films using soft lithography. First, (PAA/PEG)$_n$ PEMs were prepared under acidic conditions. Then siRNA-LPEI condensed nanoparticles were deposited on a PDMS stamp, and stamped across the (PAA/PEG)$_n$ film. Following this step, PEM substrates containing the patterned siRNA-LPEI were incubated on top of cells at physiological pH, similar to the method described by the Lynn research team [179]. This approach toward transfection was coined "multilayer mediated forward transfection" or MFT method. Unlike the majority of the transfection methods described above where cell adhesion plays an important role in transfection efficiencies, MFT methods demonstrate a new class of templates that merely depend on the film erosion and release of pDNA or siRNA. Figure 2.25g depicts fluorescent microscopy images of patterned MFT samples. As seen in the microscopy image, cells that were exposed to the patterned siRNA-LPEI regions fluorescently illuminated. Interestingly, the cells remained healthy throughout the MFT process. In addition to the described work, siRNA particle sizes, and layer-pair numbers were investigated. Finally, gene silencing was observed in the MFT samples. This described work is an interesting approach toward the development of cellular microarrays.

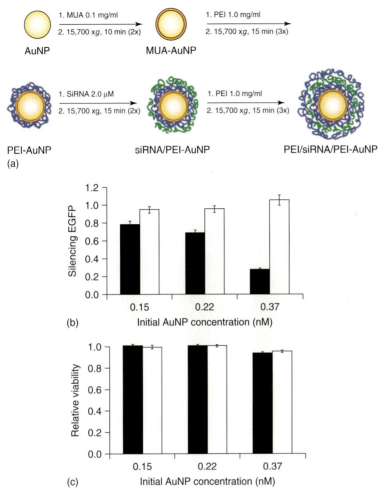

Figure 2.24 (a) Flowchart illustrating the LbL deposition applied to AuNPs. (b) Gene silencing of EGFP in CHO-K1 cells stably expressing EGFP and (c) relative cell viability after addition of PEI/siRNA/PEI-AuNPs at various initial concentrations. (Adapted from Ref. [204].)

2.5
Exotic Applications of LbL Inspired by Biology

2.5.1
Digitally Encoded Cell Carriers

Multiplexed methods of assaying many different cells in a robust manner still remains a growing field of research. Covered in the previous section one method of obtaining multiplexed templates is via the preparation of positional arrays [213].

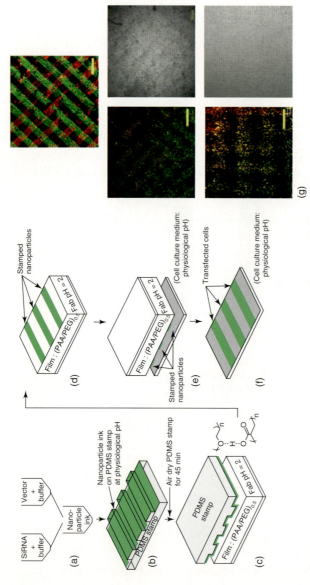

Figure 2.25 (a–f) Diagram illustrating the multilayer mediated forward transfection (MFT) of cationic vector complexed siRNA for patterned delivery. (g) Fluorescent images demonstrating patterned siRNA delivery to HeLa cells with multilayer mediated forward transfection (MFT) using (PAA/PEG) 6.5 multilayer assembly, fluorescent dsRNA oligomers (100 pmol), and Lipofectamine 2000 (LF2k, 5 µg). Nanoparticles and HeLa cell patterns transfected with Fluorescein and Alexa Fluor 555-labeled oligomers (overlaid images). Top panel: confocal laser scanning microscopy (CLSM) images of LF2k-fluorescent oligomer nanoparticles arrayed onto multilayer. Middle and bottom panels: HeLa cell patterns transfected with fluorescent oligomers and their corresponding phase-contrast images acquired using CLSM (middle panel) and conventional fluorescence microscopy (bottom panel). Scale bar represents 500 µm. (Adapted from Ref. [205].)

As described by Fayazpour *et al.* [214], positional arrays are simple templates that carry multiple wells in which many different cell types can be templated and are further exposed to a drug compound. The identity of the cell type can be located via the *x,y*-position within the array. Although these templates are frequently used, there is always the issue of cross-contamination. An alternative to positional arrays is nonpositional arrays [214]. Such arrays do not depend on location of a cell on the templates to identify cell type, but rather the identity of the cell is made via a bar code written on a microparticle [214]. These particles offer an alternative to the positional arrays, and alleviate the concerns of cross-contamination. Recently Fayazpour *et al.* [214] have described the LbL coating of the encoded microparticles and identified that three different cell types were able to grow on the surface of the particles, Figures 2.26a–c. In addition, the investigators embedded ferromagnetic nanoparticles within the LbL films on the surfaces of the microparticle allowing positioning of the microcarriers in a magnetic field. Interestingly, they immobilized adenoviral vectors (bearing genetic code for GFP and red fluorescent protein (RFP)) on the surface of the LbL coated microcarriers, and found the encoded bead could transduce the cells grown on the surface. Figures 2.26d,f depicts microcarriers that have been coated with an additional adenoviral layer and nonviral coated microcarriers (Figure 2.26g). The microcarriers with the additional adenoviral layer actually appear to transduce the attached cells making them fluoresce while nonviral-coated surfaces did not fluoresce. Only the cells that were in contact with the surface of the viral-coated particle actually become transduced. The encoding on the inside of the microcarriers still remained visible throughout the entire LbL-coating process for all mixtures.

2.5.2
"Artificial Cells" via LbL Assembly

Recently, Picart and Discher [215] revisited earlier work that was conduced by Kreft *et al.* [216] on the preparation of multicompartment microcapsules toward compartmentalized reactions. In the original work by Kreft *et al.* the strategy toward building such multicompartment capsules was an alternative to other approaches used to build similar templates. All of these methods require the use of harsh organic solvents, which could be detrimental to biomolecules. In this respect, Kreft *et al.* described a milder method of preparing mutlicompartment microcapsules using LbL assembly. Figure 2.27a illustrate the preparation of these hierarchal constructed templates. First, the authors began with immobilizing proteins and nanoparticles into calcium carbonate microspheres. Then, a PEM is deposited on the surface of the microsphere followed by growth of a calcium carbonate shell with a second protein encased in the layer. An additional PEM was deposited, and then the calcium carbonate was washed away with ethylenediaminetetraacetic acid (EDTA). This resulted in a multicomponent shell within shell microcapsule with compartmentalized contained proteins. Picart and Discher compared this unique construct to a cell, describing the internal shell as the cellular nucleus encasing a different protein then the protein in the exterior regions or the "cytoplasm."

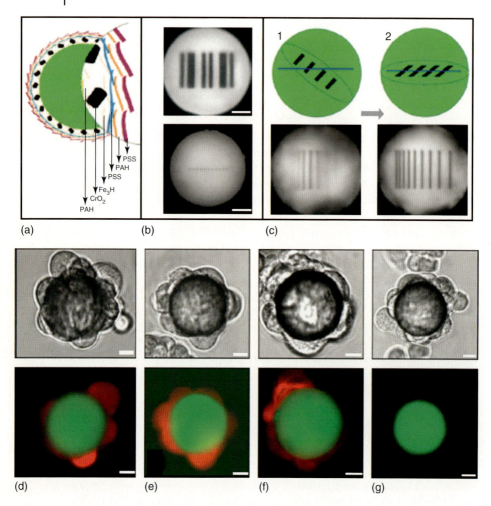

Figure 2.26 (a) Schematic representation of the LbL coating of green fluorescent polystyrene microspheres. The polycation PAH and the polyanion PSS are alternately adsorbed to the surface of the microcarrier. Magnetic CrO_2 NPs are immobilized in the LbL coating. (b,c) Confocal images of the central plane of fluorescent microspheres encoded with, respectively, a bar code (b) and a dot code (c). The scale bar represents 10 mm. (d–g) Transmission (top) and merged red/green fluorescence (bottom; lex1/4567 and 488 nm) images of Ad-RFP-coated microcarriers loaded with Vero-1 cells. In (g) microcarriers were used that did not contain viral particles in their coating (negative control). Note that nonencoded microcarriers were used. The scale bar represents 10 mm. (Adapted from Ref. [207].)

The investigators used these microcapsules as a reaction vessel in order to run the bioenzymatic oxidation of glucose with glucose oxidase (GOD) in the presence of peroxidase (POD). The same reaction was demonstrated in planar LbL films by Onda *et al.* [30]. In this reaction oxidation of glucose yields H_2O_2. Amplex Red was selected as an electron donor, and in the presence of POD becomes converted to

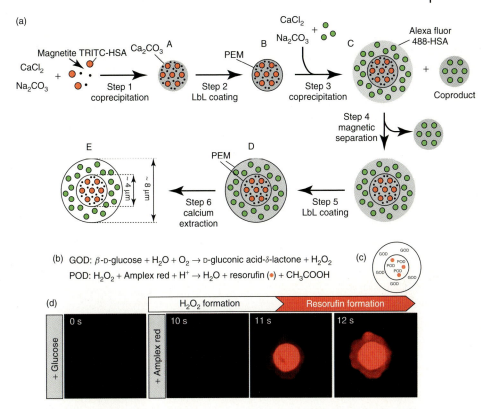

Figure 2.27 (a) General route for the synthesis of shell-in-shell microcapsules. A: Initial core; B: core-shell particle; C: ball-in-ball particle (type I); D: ball-in-ball particle (type II); and E: shell-in-shell microcapsule. (b) Reaction schemes of coupled enzymatic test by using GOD and POD inside shell-in-shell capsules. (c) Localization of GOD and POD within shell-in-shell capsules; (d) CLSM imaging *in situ* of resorufin formation. (Adapted from Ref. [209].)

fluorescent resorufin. In the original construct POD is encased in the inner shell, while GOD was placed in the outer compartment, Figure 2.27b. The glucose was first added to the bulk solution followed by Amplex Red. Initially, the GOD sitting in the outer compartment should oxidize the glucose, and H_2O_2 was produced. Upon addition of the Amplex Red, resorufin is initially formed only in the inner shell and then begins to migrate to the outer compartment, Figure 2.27d. It should be noted that only small compounds could diffuse through the PEM layers while larger enzymes remain separated. The demonstrations of multicompartment microcapsules that are prepared under mild conditions have provided a means to incorporate solvent sensitive biomolecules with a protective shell. The example provided in this section only demonstrates that these systems could be used to run sensitive bioenzymatic reactions in a controlled fashion. On a higher level, such multicompartment capsules mimic the hierarchal order seen in biology with cells [217]. Continuous developments are being made in the preparation of more

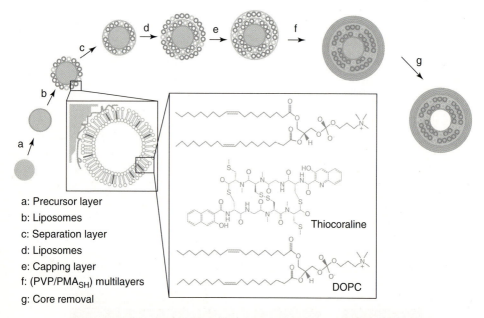

a: Precursor layer
b: Liposomes
c: Separation layer
d: Liposomes
e: Capping layer
f: (PVP/PMA$_{SH}$) multilayers
g: Core removal

Figure 2.28 Schematic illustration of the assembly of capsosomes: encapsulation of a double layer of liposomes containing a small hydrophobic peptide (antitumor compound, TC) in the lipid membrane within a polymer carrier capsule. A colloidal template is coated with a PLL precursor layer (a), followed by the deposition of the first layer of liposomes (Lx or Lzw) (b), a polymer separation layer (PLL, PMAc, or both) (c), a second layer of liposomes (Lx or Lzw) (d), and a PMAc capping layer (e). The carrier capsule is assembled by the alternate deposition of PVP and PMA$_{SH}$ (f) and capsosomes are obtained upon removal of the core (g). (Adapted from Ref. [213].)

complex multicompartment constructs, and as Picart and Discher mention there still remain many challenges in developing true synthetic systems mimicking living cells.

More recently, Stadler *et al.* [218] described the preparation of capsosomes; polymer-carrier capsules containing multilayers of liposomes on the surface. Figure 2.28 illustrates the preparation of such templates [219, 220]. First, a particle core was coated with a precursor PLL layer, followed by deposition of zwitterionic liposomes. Then, a separation layer was deposited on top of the liposome layer, which was composed of (PLL or poly-(methacrylic acid)-co-(cholesteryl methacrylate) (PMAc)), followed by another liposome layer. A capping layer of PMAc was deposited followed by a final multilayer of poly(N-vinylpyrrolidone) (PVP) and thiol-modified poly-(methacrylic acid) (PMA$_{SH}$). The particle core is then dissolved, leaving a hollow center surrounded by layers of small compartments. It is stated that the polymer separation layer is essential in order for the liposomes to adsorb. Alternatively, Loew *et al.* [221] described the preparation of a similar multicompartment template using LbL assembly of vesicles without the use of additional spacer layers. Rather, in this work the authors bind each vesicle layer via DNA hybridization. Primer layers were deposited on a silica bead in which the last layer

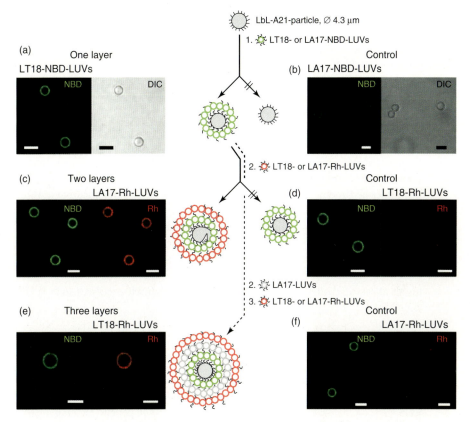

Figure 2.29 Coating of the LbL-A21 particles with oligonucleotide-containing vesicles: a scheme (not to scale) of the sequential layer-on-layer coating with 100-nm large unilamellar vesicles (LUVs) and images of the 4.3-μm particles obtained by differential interference contrast (DIC) and confocal fluorescence microscopy. The fluorescence originates from the lipid analogs *N*-NBD-PE (NBD, green) and *N*-Rh-PE (Rh, red) incorporated into LUVs that are attached to the LbL-A21 particles. Bars correspond to 5 μm. (a) One layer of LT18-NBD-LUVs. (b) Control for a nonspecific binding: addition of noncomplementary LA17-NBD-LUVs: no binding. (c) Two layers (the first layer: LT18-NBD-LUVs, the second layer: LA17-Rh-LUVs). (d) Control addition of noncomplementary LT18-Rh-LUVs: almost no binding. (e) Three layers (the first layer: LT18-NBD-LUVs, the second layer: unlabeled LA17-LUVs, the third layer: LT18-Rh-LUVs). (f) Control addition of noncomplementary LA17-Rh-LUVs: some binding is observed. Note that the presentation of bound single oligonucleotides as strands radiating out from the particle and vesicles surfaces is intended only to indicate that oligonucleotides are available for formation of dsDNA strands. (Adapted from Ref. [214].)

was poly(methacrylic acid) (PMAA) with a covalently linked oligonucleotide. Then, layers of vesicles were attached using vesicles containing complementary lipophilic oligonucleotides in the outer membrane leaflet. Figure 2.29 is the illustration of the deposition process with fluorescent microscopy results from the deposited fluorescently tagged vesicle layers. In the work by Stadler *et al.*, the capsosomes were loaded with liposomes impregnated with hydrophobic model peptide thiocoraline (TC), which is also an anticancer agent. The preparation of TC-doped capsosomes would provide insight into the incorporation of transmembrane proteins in future capsosomes. Preliminary investigations demonstrate that TC-doped capsosomes actually reduced cellular viability when exposed to colon cancer cells, revealing that capsosomes could be used in the future as drug-delivery systems. The ultimate goal of this work was to provide building blocks toward the development of artificial cells. The preparation of a microcapsule with millions of nanocompartments is a major first step toward this goal.

Together, LbL assembly used in the preparation of 3D compartmentalized microcapsules has been demonstrated using a number of different techniques. The described methods most benefit from the use of LbL for incorporation of biomolecules under mild conditions. In this regard, the LbL deposition method has allowed for progress toward the preparation of the so-called "artificial cells", mainly because it provides a mild method of preparing barriers without harming encased sensitive biomolecules. Although it may seem in the distant future, duplicating an increasing number of the functionalities and process occurring in living cells is highly attractive challenge for multistep nanoscale assembly processes such as LbL deposition.

Abbreviations

CASMC	Coronary-artery smooth muscle cell
CHO	Chinese hamster ovary cells
COS	**C**V-1 in **O**rigin, and carrying the **S**V40 genetic material
HCEC	Human corneal epithelial cell
HCV	Hepatitis C virus
HSC	Hematopoietic stem cell
HUVEC	Human umbilical vein endothelial cell
NSC	Neuron stem cell
pFN	Plasma fibronectin
SMC	Smooth muscle cell
VSMC	Vascular smooth muscle cell

Substances and materials

α-MSH	α-melanocyte stimulating hormone
ACP	Amorphous calcium phosphate
AuNP	Gold nanoparticle
CD	β-cyclodextrin
CLSM	confocal laser scanning microscopy
CS	chondroitin sulfate
DNA	deoxyribonucleic acid

Dox	Doxorubicin
DS	Dextran sulfate
EDC	1-Ethyl-3-(3-dimethylaminopropyl)carbodiimide
EDTA	Ethylenediaminetetraacetic acid
ELP	Elastine-like polypeptide
GFP	Green fluorescent protein
GOD	Glucose oxidase
HA	Hyaluronic acid
HEP	Heparin
ICC	Inverted colloidal crystal
IrOx	Iridium oxide
LUVs	Large unilamellar vesicles
LPEI	Linear poly-ethyleneimine
Ma-Y-Dox	N-methyloyl-glycyl-DL-phenylalanyl-leucyl-glycyl doxorubicin
MFNP	Multifunctional nanoparticle
mRNA	Messenger ribonucleic acid
MUA	Mercaptoundecanoic acid
NHS	N-hydroxysuccinimide
P	Polymer
PAA	Poly(acrylic acid)
PAAm	Poly-allylamine
PAH	Poly(allylamine hydrochloride)
pCD	Pyridylamino-β-cyclodextrin
PDDA	Poly(diallyldimethylammonium chloride)
PDGA	Poly(D-glutamic acid)
PDL	Poly-D-lysine
PDMS	Poly(dimethylsiloxane)
pDNA	Plasmid deoxyribonucleic acid
PEDOT	Poly(3,4-ethylenedioxythiophene)
PEG	Poly(ethylene glycol)
PEI	Poly-ethyleneimine
PEM	Polyelectrolyte multilayer
PEO	Poly(ethylene oxide)
PEO-b-PCL	Poly(ethylene oxide)-block-poly(ε-caprolactone)
PGA	Poly(L-glutamic acid)
PBS	Phosphate Buffer
PLA	Poly-D,L-lactide
PLL	Poly-L-lysine
PLO	Poly-L-ornithine
PMA	Poly(α, β, L-malic acid)
PMA$_{SH}$	Thiol-modified poly(methacrylic acid)
PMAA	Poly(methacrylic acid)
PMAc	Poly(methacrylic acid)-co-(cholesteryl methacrylate)
PMBV	Poly(2-methacryloyloxyethyl phosphorylcholine-co-n-butyl methacrylate-co-p-vinylphenylboronic acid)

POD	Peroxidase
PSS	Poly(styrene sulfonate)
PTX	Paclitaxel
PVA	Poly(vinyl alcohol)
PVDMA	Poly(2-vinyl-4,4'-dimethylazlactone)
PVP	Poly(N-vinylpyrrolidone)
PXT	paclitaxel
RFP	Red fluorescent protein
RGD	Arginine-glycine-aspartic acid
RHB	Reducible hyperbranched poly(amido amine)
rhBMP-2	Recombinant bone morphogenetic protein 2
SEC	size exclusion chromatography
SAM	Self-assembled monolayer
SEAP	Secreted alkaline phosphate
siRNA	Short interferring ribonucleic acid
SPLL	Succinylated poly-L-lysine
SWNT	Single-walled carbon nanotube
TC	Thiocoraline
TMF	tamoxifen
VB	Vinylbenzyl

Methods and concepts

ATRP	Atom-transfer radical polymerization
DES	Drug-eluting stent
LbL	Layer-by-layer
LCST	Lower critical solution temperature
MFT	Multilayer mediated forward transfection
MIMIC	Micromolding in capillaries
SEM	Scanning electron microscopy

References

1. Decher, G. (1997) Fuzzy nanoassemblies: toward layered polymeric multicomposites. *Science*, **277** (5330), 1232–1237.
2. Decher, G. and Hong, J.D. (1991) Buildup of ultrathin multilayer films by a self-assembly process. 1. Consecutive adsorption of anionic and cationic bipolar amphiphiles on charged surfaces. *Makromol. Chem. Macromol. Symp.*, **46**, 321–327.
3. Decher, G. and Hong, J.D. (1991) Buildup of ultrathin multilayer films by a self-assembly process. 2. Consecutive adsorption of anionic and cationic bipolar amphiphiles and polyelectrolytes on charged surfaces. *Ber. Bunsen. Phys. Chem.*, **95** (11), 1430–1434.
4. Decher, G., Hong, J.D., and Schmitt, J. (1992) Buildup of ultrathin multilayer films by a self-assembly process. 3. Consecutively alternating adsorption of

anionic and cationic polyelectrolytes on charged surfaces. *Thin Solid Films*, **210** (1–2), 831–835.

5. Ariga, K., Hill, J.P., and Ji, Q.M. (2007) Layer-by-layer assembly as a versatile bottom-up nanofabrication technique for exploratory research and realistic application. *Phys. Chem. Chem. Phys.*, **9** (19), 2319–2340.

6. Jewell, C.M. and Lynn, D.M. (2008) Multilayered polyelectrolyte assemblies as platforms for the delivery of DNA and other nucleic acid-based therapeutics. *Adv. Drug. Deliver. Rev.*, **60** (9), 979–999.

7. Decher, G. (1996) *Layered Nanoarchitectures via Directed Assembly of Anionic and Cationic Molecules; in Comprehensive Supramolecular Chemistry*, vol. 9, Pergamon Press, Oxford.

8. Decher, G., Eckle, M., Schmitt, J., and Struth, B. (1998) Layer-by-Layer assembled multicomposite films. *Curr. Opinion Coll. Interf. Sci.*, **3**, 32–39.

9. Bertrand, P., Jonas, A.M., Laschewsky, A., and Legras, R. (2000) Ultrathin polymer coatings by complexation of polyelectrolytes at interfaces: suitable materials, structures and properties. *Macromol. Rapid. Commun.*, **21**, 319–348.

10. Hammond, P.T. (2000) Recent explorations in electrostatic multilayer thin film assembly. *Curr. Opinion Coll. Interf. Sci.*, **4**, 430–442.

11. Decher, G. and Schlenoff, J.B. (2003) *Multilayer Thin Films: Sequential Assembly of Nanocomposite Materials*, Wiley-VCH, Weinheim, p. 524.

12. Tang, Z.Y., Wang, Y., Podsiadlo, P., and Kotov, N.A. (2006) Biomedical applications of layer-by-layer assembly: from biomimetics to tissue engineering. *Adv. Mater.*, **18** (24), 3203–3224.

13. Johnston, A.P.R., Cortez, C., Angelatos, A.S., and Caruso, F. (2006) layer-by-layer engineered capsules and their applications. *Curr. Opinion Coll. Interf. Sci.*, **11**, 203–209.

14. Jessel, N., Lavalle, P., Ball, V., Ogier, J., Senger, B., Picart, C., Schaaf, P., Voegel, J., and Decher, G. (2007) Polyelectrolyte multilayer films in *Elements of Macromolecular Structural Control*, vol. 2 (eds Y. Gnanou, L. Leibler, and K. Matyiaszewski), Wiley-VCH, Weinheim, pp. 1249–1306.

15. Picart, C. (2008) Polyelectrolyte multilayer films: from physico-chemical properties to the control of cellular processes. *Curr. Med. Chem.*, **15** (7), 685–697.

16. Lichter, J.A., Van Vliet, K.J., and Rubner, M.F. (2009) Design of antibacterial surfaces and interfaces: polyelectrolyte multilayers as a multifunctional platform. *Macromolecules*, **42** (22), 8573–8586.

17. Decher, G., Felix, O., Saulnier, B., Izquierdo, A., Schaaf, P., Voegel, J., Jessel, N., and Ball, V. (2006) Méthode de construction de matériaux vivants fonctionnels, matériaux obtenus et applications Inventors. FR290114.

18. Grossin, L., Cortial, D., Saulnier, B., Felix, O., Chassepot, A., Decher, G., Netter, P., Schaaf, P., Gillet, P., Mainard, D., Voegel, J.C., and Benkirane-Jessel, N. (2009) Step-by-step build-up of biologically active cell-containing stratified films aimed at tissue engineering. *Adv. Mater.*, **21** (6), 650–655.

19. Matsusaki, M., Kadowaki, K., Nakahara, Y., and Akashi, M. (2007) Fabrication of cellular multilayers with nanometer-sized extracellular matrix films. *Angew. Chem. Int. Ed.*, **46** (25), 4689–4692.

20. Picart, C., Mutterer, J., Richert, L., Luo, Y., Prestwich, G.D., Schaaf, P., Voegel, J.C., and Lavalle, P. (2002) Molecular basis for the explanation of the exponential growth of polyelectrolyte multilayers. *Proc. Natl. Acad. Sci. USA*, **99** (20), 12531–12535.

21. Losche, M., Schmitt, J., Decher, G., Bouwman, W.G., and Kjaer, K. (1998) Detailed structure of molecularly thin polyelectrolyte multilayer films on solid substrates as revealed by neutron reflectometry. *Macromolecules*, **31** (25), 8893–8906.

22. Dubas, S.T. and Schlenoff, J.B. (1999) Factors controlling the growth of polyelectrolyte multilayers. *Macromolecules*, **32** (24), 8153–8160.

23. Schlenoff, J.B. (2003) Charge balance and transport in polyelectrolyte multilayers in *Multilayer Thin Films: Sequential Assembly of Nanocomposite Materials* (eds G. Decher and J.B. Schlenoff), Wiley-VCH, Weinheim, pp. 99–132.

24. Decher, G. (2003) Polyelectrolyte multilayers, an overview in *Multilayer Thin Films: Sequential Assembly of Nanocomposite Materials* (eds G. Decher and J.B. Schlenoff), Wiley-VCH, Weinheim, pp. 1–46.

25. Langer, R. (1998) Drug delivery and targeting. *Nature*, **392** (6679), 5–10.

26. Lvov, Y., Ariga, K., Ichinose, I., and Kunitake, T. (1995) Assembly of multicomponent protein films by means of electrostatic layer-by-layer adsorption. *J. Am. Chem. Soc.*, **117** (22), 6117–6123.

27. Qiu, X.P., Leporatti, S., Donath, E., and Mohwald, H. (2001) Studies on the drug release properties of polysaccharide multilayers encapsulated ibuprofen microparticles. *Langmuir*, **17** (17), 5375–5380.

28. Thierry, B., Winnik, F.M., Merhi, Y., and Tabrizian, M. (2003) Nanocoatings onto arteries via layer-by-layer deposition: toward the in vivo repair of damaged blood vessels. *J. Am. Chem. Soc.*, **125** (25), 7494–7495.

29. Richert, L., Lavalle, P., Payan, E., Shu, X.Z., Prestwich, G.D., Stoltz, J.F., Schaaf, P., Voegel, J.C., and Picart, C. (2004) Layer by layer buildup of polysaccharide films: physical chemistry and cellular adhesion aspects. *Langmuir*, **20** (2), 448–458.

30. Onda, M., Lvov, Y., Ariga, K., and Kunitake, T. (1996) Sequential actions of glucose oxidase and peroxidase in molecular films assembled by layer-by-layer alternate adsorption. *Biotechnol. Bioeng.*, **51** (2), 163–167.

31. Volodkin, D.V., Madaboosi, N., Blacklock, J., Skirtach, A.G., and Mohwald, H. (2009) Surface-supported multilayers decorated with bio-active material aimed at light-triggered drug delivery. *Langmuir*, **25** (24), 14037–14043.

32. Vodouhe, C., Le Guen, E., Garza, J.M., Francius, G., Dejugnat, C., Ogier, J., Schaaf, P., Voegel, J.C., and Lavalle, P. (2006) Control of drug accessibility on functional polyelectrolyte multilayer films. *Biomaterials*, **27** (22), 4149–4156.

33. Sukhorukov, G.B. and Mohwald, H. (2007) Multifunctional cargo systems for biotechnology. *Trends Biotechnol.*, **25** (3), 93–98.

34. Liu, S.Q. and Tang, Z.Y. (2010) Nanoparticle assemblies for biological and chemical sensing. *J. Mater. Chem.*, **20** (1), 24–35.

35. Caruso, F., Caruso, R.A., and Mohwald, H. (1998) Nanoengineering of inorganic and hybrid hollow spheres by colloidal templating. *Science*, **282** (5391), 1111–1114.

36. Tong, W.J. and Gao, C.Y. (2008) Multilayer microcapsules with tailored structures for bio-related applications. *J. Mater. Chem.*, **18** (32), 3799–3812.

37. Donath, E., Sukhorukov, G.B., Caruso, F., Davis, S.A., and Mohwald, H. (1998) Novel hollow polymer shells by colloid-templated assembly of polyelectrolytes. *Angew. Chem. Int. Ed.*, **37** (16), 2202–2205.

38. De Geest, B.G., Sanders, N.N., Sukhorukov, G.B., Demeester, J., and De Smedt, S.C. (2007) Release mechanisms for polyelectrolyte capsules. *Chem. Soc. Rev.*, **36** (4), 636–649.

39. Ferrari, M. (2005) Cancer nanotechnology: opportunities and challenges. *Nature Rev. Cancer*, **5** (3), 161–171.

40. Peer, D., Karp, J.M., Hong, S., Farokhzad, O.C., Margalit, R., and Langer, R. (2007) Nanocarriers as an emerging platform for cancer therapy. *Nature Nanotechnol.*, **2** (12), 751–760.

41. Choi, J., Konno, T., Takai, M., and Ishihara, K. (2009) Smart controlled preparation of multilayered hydrogel for releasing bioactive molecules. *Curr. Appl. Phys.*, **9** (4), E259–E262.

42. Lynn, D.M. (2006) Layers of opportunity: nanostructured polymer

assemblies for the delivery of macro-molecular therapeutics. *Soft Matter.*, **2** (4), 269–273.

43. Vazquez, E., Dewitt, D.M., Hammond, P.T., and Lynn, D.M. (2002) Construction of hydrolytically-degradable thin films via layer-by-layer deposition of degradable polyelectrolytes. *J. Am. Chem. Soc.*, **124** (47), 13992–13993.

44. Wood, K.C., Boedicker, J.Q., Lynn, D.M., and Hammon, P.T. (2005) Tunable drug release from hydrolytically degradable layer-by-layer thin films. *Langmuir*, **21** (4), 1603–1609.

45. Liu, X.H., Zhang, J.T., and Lynn, D.M. (2008) Polyelectrolyte multilayers fabricated from 'charge-shifting' anionic polymers: a new approach to controlled film disruption and the release of cationic agents from surfaces. *Soft Matter*, **4** (8), 1688–1695.

46. Su, X.F., Kim, B.S., Kim, S.R., Hammond, P.T., and Irvine, D.J. (2009) Layer-by-layer-assembled multilayer films for transcutaneous drug and vaccine delivery. *ACS Nano*, **3** (11), 3719–3729.

47. Ma, N., Zhang, H.Y., Song, B., Wang, Z.Q., and Zhang, X. (2005) Polymer micelles as building blocks for layer-by-layer assembly: an approach for incorporation and controlled release of water-insoluble dyes. *Chem. Mater.*, **17** (20), 5065–5069.

48. Nguyen, P.M., Zacharia, N.S., Verploegen, E., and Hammond, P.T. (2007) Extended release antibacterial layer-by-layer films incorporating linear-dendritic block copolymer micelles. *Chem. Mater.*, **19** (23), 5524–5530.

49. Hu, X.F. and Ji, J. (2010) Construction of multifunctional coatings via layer-by-layer assembly of sulfonated hyperbranched polyether and chitosan. *Langmuir*, **26** (4), 2624–2629.

50. Qi, B., Tong, X., and Zhao, Y. (2006) Layer-by-layer assembly of two different polymer micelles with polycation and polyanion coronas. *Macromolecules*, **39** (17), 5714–5719.

51. Kim, B.S., Park, S.W., and Hammond, P.T. (2008) Hydrogen-bonding layer-by-layer assembled biodegradable polymeric micelles as drug delivery vehicles from surfaces. *ACS Nano*, **2** (2), 386–392.

52. Sukhishvili, S.A. and Granick, S. (2002) Layered, erasable polymer multilayers formed by hydrogen-bonded sequential self-assembly. *Macromolecules*, **35** (1), 301–310.

53. Ono, S.S. and Decher, G. (2006) Preparation of ultrathin self-standing polyelectrolyte multilayer membranes at physiological conditions using pH-responsive film segments as sacrificial layers. *Nano Lett.*, **6** (4), 592–598.

54. Kim, B.S., Smith, R.C., Poon, Z., and Hammond, P.T. (2009) MAD (multi-agent delivery) nanolayer: delivering multiple therapeutics from hierarchically assembled surface coatings. *Langmuir*, **25** (24), 14086–14092.

55. Schultz, P., Vautier, D., Richert, L., Jessel, N., Haikel, Y., Schaaf, P., Voegel, J.C., Ogier, J., and Debry, C. (2005) Polyelectrolyte multilayers functionalized by a synthetic analogue of an anti-inflammatory peptide, alpha-MSH, for coating a tracheal prosthesis. *Biomaterials*, **26** (15), 2621–2630.

56. Kim, T.G., Lee, H., Jang, Y., and Park, T.G. (2009) Controlled release of paclitaxel from heparinized metal stent fabricated by layer-by-layer assembly of polylysine and hyaluronic acid-g-poly(lactic-co-glycolic acid) micelles encapsulating paclitaxel. *Biomacromolecules*, **10** (6), 1532–1539.

57. Hageluken, A., Grunbaum, L., Nurnberg, B., Harhammer, R., Schunack, W., and Seifert, R. (1994) Lipophilic beta-adrenoceptor antagonists and local-anesthetics are effective direct activators of G-proteins. *Biochem. Pharmacol.*, **47** (10), 1789–1795.

58. Yokogawa, K., Nakashima, E., Ishizaki, J., Maeda, H., Nagano, T., and Ichimura, F. (1990) Relationships in the structure tissue distribution of basic drugs in the rabbit. *Pharm. Res.*, **7** (7), 691–696.

59. Jones, M.C. and Leroux, J.C. (1999) Polymeric micelles - a new generation of colloidal drug carriers. *Eur. J. Pharm. Biopharm.*, **48** (2), 101–111.

60. Lasic, D.D. (1992) Mixed micelles in drug delivery. *Nature*, **355** (6357), 279–280.

61. Torchilin, V.P. (2007) Micellar nanocarriers: pharmaceutical perspectives. *Pharm. Res.*, **24** (1), 1–16.

62. Muranishi, S. (1990) Absorption enhancers. *Crit. Rev. Ther. Drug Carrier Syst.*, **7** (1), 1–33.

63. Lvov, Y., Antipov, A.A., Mamedov, A., Mohwald, H., and Sukhorukov, G.B. (2001) Urease encapsulation in nanoorganized microshells. *Nano Lett.*, **1** (3), 125–128.

64. Balabushevitch, N.G., Sukhorukov, G.B., Moroz, N.A., Volodkin, D.V., Larionova, N.I., Donath, E., and Mohwald, H. (2001) Encapsulation of proteins by layer-by-layer adsorption of polyelectrolytes onto protein aggregates: factors regulating the protein release. *Biotechnol. Bioeng.*, **76** (3), 207–213.

65. Dai, Z.F., Heilig, A., Zastrow, H., Donath, E., and Mohwald, H. (2004) Novel formulations of vitamins and insulin by nanoengineering of polyelectrolyte multilayers around microcrystals. *Chem. Eur. J.*, **10** (24), 6369–6374.

66. Fan, Y.F., Wang, Y.N., Fan, Y.G., and Ma, J.B. (2006) Preparation of insulin nanoparticles and their encapsulation with biodegradable polyelectrolytes via the layer-by-layer adsorption. *Int. J. Pharm.*, **324** (2), 158–167.

67. Zahr, A.S. and Pishko, M.V. (2007) Encapsulation of paclitaxel in macromolecular nanoshells. *Biomacromolecules*, **8** (6), 2004–2010.

68. Zahr, A.S., de Villiers, M., and Pishko, M.V. (2005) Encapsulation of drug nanoparticles in self-assembled macromolecular nanoshells. *Langmuir*, **21** (1), 403–410.

69. Agarwal, A., Lvov, Y., Sawant, R., and Torchilin, V. (2008) Stable nanocolloids of poorly soluble drugs with high drug content prepared using the combination of sonication and layer-by-layer technology. *J. Control. Release*, **128** (3), 255–260.

70. Schneider, G.F., Subr, V., Ulbrich, K., and Decher, G. (2009) Multifunctional cytotoxic stealth nanoparticles. a model approach with potential for cancer therapy. *Nano Lett.*, **9** (2), 636–642.

71. Thierry, B., Al-Ejeh, F., Khatri, A., Yuan, Z., Russell, P.J., Ping, S., Brown, M.P., and Majewski, P. (2009) Multifunctional core-shell magnetic cisplatin nanocarriers. *Chem. Commun.*, (47), 7348–7350.

72. Boudou, T., Crouzier, T., Ren, K.F., Blin, G., and Picart, C. (2010) Multiple functionalities of polyelectrolyte multilayer films: new biomedical applications. *Adv. Mater.*, **22** (4), 441–467.

73. Boura, C., Menu, P., Payan, E., Picart, C., Voegel, J.C., Muller, S., and Stoltz, J.F. (2003) Endothelial cells grown on thin polyelectrolyte mutlilayered films: an evaluation of a new versatile surface modification. *Biomaterials*, **24** (20), 3521–3530.

74. Mhamdi, L., Picart, C., Lagneau, C., Othmane, A., Grosgogeat, B., Jaffrezic-Renault, N., and Ponsonnet, L. (2006) Study of the polyelectrolyte multilayer thin films' properties and correlation with the behavior of the human gingival fibroblasts. *Mater. Sci. Eng. C-Bio. S.*, **26** (2–3), 273–281.

75. Brunot, C., Grosgogeat, B., Picart, C., Lagneau, C., Jaffrezic-Renault, N., and Ponsonnet, L. (2008) Response of fibroblast activity and polyelectrolyte multilayer films coating titanium. *Dent Mater.*, **24** (8), 1025–1035.

76. Tryoen-Toth, P., Vautier, D., Haikel, Y., Voegel, J.C., Schaaf, P., Chluba, J., and Ogier, J. (2002) Viability, adhesion, and bone phenotype of osteoblast-like cells on polyelectrolyte multilayer films. *J Biomed. Mater. Res.*, **60** (4), 657–667.

77. Wittmer, C.R., Phelps, J.A., Lepus, C.M., Saltzman, W.M., Harding, M.J., and Van Tassel, P.R. (2008) Multilayer nanofilms as substrates for hepatocellular applications. *Biomaterials*, **29** (30), 4082–4090.

78. Richert, L., Lavalle, P., Vautier, D., Senger, B., Stoltz, J.F., Schaaf, P., Voegel, J.C., and Picart, C. (2002) Cell interactions with polyelectrolyte multilayer films. *Biomacromolecules*, **3** (6), 1170–1178.

79. Olenych, S.G., Moussallem, M.D., Salloum, D.S., Schlenoff, J.B., and Keller, T.C.S. (2005) Fibronectin and cell attachment to cell and protein resistant polyelectrolyte surfaces. *Biomacromolecules*, **6** (6), 3252–3258.

80. Salloum, D.S., Olenych, S.G., Keller, T.C.S., and Schlenoff, J.B. (2005) Vascular smooth muscle cells on polyelectrolyte multilayers: hydrophobicity-directed adhesion and growth. *Biomacromolecules*, **6** (1), 161–167.

81. Niepel, M.S., Peschel, D., Sisquella, X., Planell, J.A., and Groth, T. (2009) pH-dependent modulation of fibroblast adhesion on multilayers composed of poly(ethylene imine) and heparin. *Biomaterials*, **30** (28), 4939–4947.

82. Mendelsohn, J.D., Yang, S.Y., Hiller, J., Hochbaum, A.I., and Rubner, M.F. (2003) Rational design of cytophilic and cytophobic polyelectrolyte multilayer thin films. *Biomacromolecules*, **4** (1), 96–106.

83. Lichter, J.A. and Rubner, M.F. (2009) Polyelectrolyte multilayers with intrinsic antimicrobial functionality: the importance of mobile polycations. *Langmuir*, **25** (13), 7686–7694.

84. Saravia, V. and Toca-Herrera, J.L. (2009) Substrate influence on cell shape and cell mechanics: HepG2 cells spread on positively charged surfaces. *Microsc. Res. Tech.*, **72** (12), 957–964.

85. Hernandez-Lopez, J.L., Khor, H.L., Caminade, A.M., Majoral, J.P., Mittler, S., Knoll, W., and Kim, D.H. (2008) Bioactive multilayer thin films of charged N,N-disubstituted hydrazine phosphorus dendrimers fabricated by layer-by-layer self-assembly. *Thin Solid Films*, **516** (6), 1256–1264.

86. Kidambi, S., Lee, I., and Chan, C. (2004) Controlling primary hepatocyte adhesion and spreading on protein-free polyelectrolyte multilayer films. *J. Am. Chem. Soc.*, **126** (50), 16286–16287.

87. Thompson, M.T., Berg, M.C., Tobias, I.S., Rubner, M.F., and Van Vliet, K.J. (2005) Tuning compliance of nanoscale polyelectrolyte multilayers to modulate cell adhesion. *Biomaterials*, **26** (34), 6836–6845.

88. Richert, L., Boulmedais, F., Lavalle, P., Mutterer, J., Ferreux, E., Decher, G., Schaaf, P., Voegel, J.C., and Picart, C. (2004) Improvement of stability and cell adhesion properties of polyelectrolyte multilayer films by chemical cross-linking. *Biomacromolecules*, **5** (2), 284–294.

89. Richert, L., Engler, A.J., Discher, D.E., and Picart, C. (2004) Elasticity of native and cross-linked polyelectrolyte multilayer films. *Biomacromolecules*, **5** (5), 1908–1916.

90. Francius, G., Hemmerle, J., Ohayon, J., Schaaf, P., Voegel, J.C., Picart, C., and Senger, B. (2006) Effect of crosslinking on the elasticity of polyelectrolyte multilayer films measured by colloidal probe AFM. *Microsc. Res. Tech.*, **69** (2), 84–92.

91. Fery, A., Scholer, B., Cassagneau, T., and Caruso, F. (2001) Nanoporous thin films formed by salt-induced structural changes in multilayers of poly(acrylic acid) and poly(allylamine). *Langmuir*, **17** (13), 3779–3783.

92. Heuvingh, J., Zappa, M., and Fery, A. (2005) Salt softening of polyelectrolyte multilayer capsules. *Langmuir*, **21** (7), 3165–3171.

93. Dubas, S.T. and Schlenoff, J.B. (2001) Swelling and smoothing of polyelectrolyte multilayers by salt. *Langmuir*, **17** (25), 7725–7727.

94. Jaber, J.A. and Schlenoff, J.B. (2006) Mechanical properties of reversibly cross-linked ultrathin polyelectrolyte complexes. *J. Am. Chem. Soc.*, **128** (9), 2940–2947.

95. Engler, A.J., Richert, L., Wong, J.Y., Picart, C., and Discher, D.E. (2004) Surface probe measurements of the elasticity of sectioned tissue, thin gels and polyelectrolyte multilayer films: correlations between substrate stiffness and cell adhesion. *Surf. Sci.*, **570** (1–2), 142–154.

96. Schneider, A., Francius, G., Obeid, R., Schwinte, P., Hemmerle, J., Frisch, B., Schaaf, P., Voegel, J.C., Senger, B., and Picart, C. (2006) Polyelectrolyte multilayers with a tunable Young's modulus: Influence of film stiffness

on cell adhesion. *Langmuir*, **22** (3), 1193–1200.

97. Vazquez, C.P., Boudou, T., Dulong, V., Nicolas, C., Picart, C., and Glinel, K. (2009) Variation of polyelectrolyte film stiffness by photocross-linking: a new way to control cell adhesion. *Langmuir*, **25** (6), 3556–3563.

98. Moussallem, M.D., Olenych, S.G., Scott, S.L., Keller, T.C.S., and Schlenoff, J.B. (2009) Smooth muscle cell phenotype modulation and contraction on native and cross-linked polyelectrolyte multilayers. *Biomacromolecules*, **10** (11), 3062–3068.

99. Kommireddy, D.S., Sriram, S.M., Lvov, Y.M., and Mills, D.K. (2006) Stem cell attachment to layer-by-layer assembled TiO2 nanoparticle thin films. *Biomaterials*, **27** (24), 4296–4303.

100. Lu, Y.X., Sun, J.Q., and Shen, J.C. (2008) Cell adhesion properties of patterned poly(acrylic acid)/poly(allylamine hydrochloride) multilayer films created by room-temperature imprinting technique. *Langmuir*, **24** (15), 8050–8055.

101. Hajicharalambous, C.S., Lichter, J., Hix, W.T., Swierczewska, M., Rubner, M.F., and Rajagopalan, P. (2009) Nano- and sub-micron porous polyelectrolyte multilayer assemblies: biomimetic surfaces for human corneal epithelial cells. *Biomaterials*, **30** (23–24), 4029–4036.

102. Kirchhof, K., Hristova, K., Krasteva, N., Altankov, G., and Groth, T. (2009) Multilayer coatings on biomaterials for control of MG-63 osteoblast adhesion and growth. *J. Mater. Sci.: Mater. Med.*, **20** (4), 897–907.

103. Wittmer, C.R., Phelps, J.A., Saltzman, W.M., and Van Tassel, P.R. (2007) Fibronectin terminated multilayer films: protein adsorption and cell attachment studies. *Biomaterials*, **28** (5), 851–860.

104. Semenov, O., Malek, A., Voros, J., and Zisch, A. (2009) Human stem cell sheets engineered on polyelectrolyte multilayer coatings for prospective transplantation. *Tissue Eng. Part A*, **15** (3), 714–715.

105. Li, M.Y., Mills, D.K., Cui, T.H., and McShane, M.J. (2005) Cellular response to gelatin- and fibronectin-coated multilayer polyelectrolyte nanofilms. *IEEE Trans. Nanobiosci.*, **4** (2), 170–179.

106. Vivet, V., Mesini, P., Cuisinier, F.J.G., Gergelly, C., and Decher, G. (2002) Polyelectrolyte multilayer films containing poly(L-lysine) grafted with a RGD-tripeptide. *Abstr. Pap. Am. Chem. Soc.*, **223**, U410–U410.

107. Picart, C., Elkaim, R., Richert, L., Audoin, T., Arntz, Y., Cardoso, M.D., Schaaf, P., Voegel, J.C., and Frisch, B. (2005) Primary cell adhesion on RGD-functionalized and covalently crosslinked thin polyelectrolyte multilayer films. *Adv. Funct. Mater.*, **15** (1), 83–94.

108. Swierczewska, M., Hajicharalambous, C.S., Janorkar, A.V., Megeed, Z., Yarmush, M.L., and Rajagopalan, P. (2008) Cellular response to nanoscale elastin-like polypeptide polyelectrolyte multilayers. *Acta Biomater.*, **4** (4), 827–837.

109. Werner, S., Huck, O., Frisch, B., Vautier, D., Elkaim, R., Voegel, J.C., Brunel, G., and Tenenbaum, H. (2009) The effect of microstructured surfaces and laminin-derived peptide coatings on soft tissue interactions with titanium dental implants. *Biomaterials*, **30** (12), 2291–2301.

110. Benkirane-Jessel, N., Lavalle, P., Meyer, F., Audouin, F., Frisch, B., Schaaf, P., Ogier, J., Decher, G., and Voegel, J.C. (2004) Control of monocyte morphology on and response to model surfaces for implants equipped with anti-inflammatory agents. *Adv. Mater.*, **16** (17), 1507–1150.

111. Chluba, J., Voegel, J.C., Decher, G., Erbacher, P., Schaaf, P., and Ogier, J. (2001) Peptide hormone covalently bound to polyelectrolytes and embedded into multilayer architectures conserving full biological activity. *Biomacromolecules*, **2** (3), 800–805.

112. Jessel, N., Atalar, F., Lavalle, P., Mutterer, J., Decher, G., Schaaf, P., Voegel, J.C., and Ogier, J. (2003) Bioactive coatings based on a polyelectrolyte multilayer architecture functionalized by embedded proteins. *Adv. Mater.*, **15** (9), 692–695.

113. Buck, M.E., Breitbach, A.S., Belgrade, S.K., Blackwell, H.E., and Lynn, D.M. (2009) Chemical modification of reactive multilayered films fabricated from poly(2-alkenyl azlactone)s: design of surfaces that prevent or promote mammalian cell adhesion and bacterial biofilm growth. *Biomacromolecules*, **10** (6), 1564–1574.

114. Kinnane, C.R., Wark, K., Such, G.K., Johnston, A.P.R., and Caruso, F. (2009) Peptide-functionalized, low-biofouling click multilayers for promoting cell adhesion and growth. *Small*, **5** (4), 444–448.

115. Wischerhoff, E., Glatzel, S., Uhlig, K., Lankenau, A., Lutz, J.F., and Laschewsky, A. (2009) Tuning the thickness of polymer brushes grafted from nonlinearly growing multilayer assemblies. *Langmuir*, **25** (10), 5949–5956.

116. Thierry, B., Winnik, F.M., Merhi, Y., Silver, J., and Tabrizian, M. (2003) Bioactive coatings of endovascular stents based on polyelectrolyte multilayers. *Biomacromolecules*, **4** (6), 1564–1571.

117. Tan, Q.G., Ji, J., Zhao, F., Fan, D.Z., Sun, F.Y., and Shen, J.C. (2005) Fabrication of thromboresistant multilayer thin film on plasma treated poly (vinyl chloride) surface. *J. Mater. Sci.: Mater. Med.*, **16** (7), 687–692.

118. Serizawa, T., Yamaguchi, M., and Akashi, M. (2002) Alternating bioactivity of polymeric layer-by-layer assemblies: anticoagulation vs procoagulation of human blood. *Biomacromolecules*, **3** (4), 724–731.

119. Houska, M., Brynda, E., Solovyev, A., Brouckova, A., Krizova, P., Vanickova, M., and Dyr, J.E. (2008) Hemocompatible albumin-heparin coatings prepared by the layer-by-layer technique. The effect of layer ordering on thrombin inhibition and Platelet adhesion. *J. Biomed. Mater. Res. A*, **86A** (3), 769–778.

120. Bratskaya, S., Marinin, D., Simon, F., Synytska, A., Zschoche, S., Busscher, H.J., Jager, D., and van der Mei, H.C. (2007) Adhesion and viability of two enterococcal strains on covalently grafted chitosan and chitosan/kappa-carrageenan multilayers. *Biomacromolecules*, **8** (9), 2960–2968.

121. Fu, J.H., Ji, J., Yuan, W.Y., and Shen, J.C. (2005) Construction of anti-adhesive and antibacterial multilayer films via layer-by-layer assembly of heparin and chitosan. *Biomaterials*, **26** (33), 6684–6692.

122. Rudra, J.S., Dave, K., and Haynie, D.T. (2006) Antimicrobial polypeptide multilayer nanocoatings. *J. Biomater. Sci.: Polym. Ed.*, **17** (11), 1301–1315.

123. Etienne, O., Gasnier, C., Taddei, C., Voegel, J.C., Aunis, D., Schaaf, P., Metz-Boutigue, M.H., Bolcato-Bellemin, A.L., and Egles, C. (2005) Antifungal coating by biofunctionalized polyelectrolyte multilayered films. *Biomaterials*, **26** (33), 6704–6712.

124. Guyomard, A., De, E., Jouenne, T., Malandain, J.J., Muller, G., and Glinel, K. (2008) Incorporation of a hydrophobic antibacterial peptide into amphiphilic polyelectrolyte multilayers: a bioinspired approach to prepare biocidal thin coatings. *Adv. Funct. Mater.*, **18** (5), 758–765.

125. Lee, D., Cohen, R.E., and Rubner, M.F. (2005) Antibacterial properties of Ag nanoparticle loaded multilayers and formation of magnetically directed antibacterial microparticles. *Langmuir*, **21** (21), 9651–9659.

126. Li, Z., Lee, D., Sheng, X.X., Cohen, R.E., and Rubner, M.F. (2006) Two-level antibacterial coating with both release-killing and contact-killing capabilities. *Langmuir*, **22** (24), 9820–9823.

127. Malcher, M., Volodkin, D., Heurtault, B., Andre, P., Schaaf, P., Mohwald, H., Voegel, J.C., Sokolowski, A., Ball, V., Boulmedais, F., and Frisch, B. (2008) Embedded silver ions-containing liposomes in polyelectrolyte multilayers: cargos films for antibacterial agents. *Langmuir*, **24** (18), 10209–10215.

128. Chua, P.H., Neoh, K.G., Kang, E.T., and Wang, W. (2008) Surface functionalization of titanium with hyaluronic acid/chitosan polyelectrolyte multilayers and RGD for promoting osteoblast

functions and inhibiting bacterial adhesion. *Biomaterials*, **29** (10), 1412–1421.

129. Li, Q.L., Huang, N., Chen, J.L., Wan, G.J., Zhao, A.S., Chen, J.Y., Wang, J., Yang, P., and Leng, Y.X. (2009) Anticoagulant surface modification of titanium via layer-by-layer assembly of collagen and sulfated chitosan multilayers. *J. Biomed. Mater. Res. A*, **89A** (3), 575–584.

130. Choi, J.Y., Konno, T., Matsuno, R., Takai, M., and Ishihara, K. (2008) Surface immobilization of biocompatible phospholipid polymer multilayered hydrogel on titanium alloy. *Colloid Surf. B*, **67** (2), 216–223.

131. Sikiric, M.D., Gergely, C., Elkaim, R., Wachtel, E., Cuisinier, F.J.G., and Furedi-Milhofer, H. (2009) Biomimetic organic-inorganic nanocomposite coatings for titanium implants. *J. Biomed. Mater. Res. A*, **89A** (3), 759–771.

132. Kerdjoudj, H., Boura, C., Moby, V., Montagne, K., Schaaf, P., Voegel, J.C., Stoltz, J.F., and Menu, P. (2007) Re-endothelialization of human umbilical arteries treated with polyelectrolyte multilayers: a tool for damaged vessel replacement. *Adv. Funct. Mater.*, **17** (15), 2667–2673.

133. Kerdjoudj, H., Berthelemy, N., Rinckenbach, S., Kearney-Schwartz, A., Montagne, K., Schaaf, P., Lacolley, P., Stoltz, J.F., Voegel, J.C., and Menu, P. (2008) Small vessel replacement by human umbilical arteries with polyelectrolyte film-treated arteries in vivo behavior. *J. Am. Coll. Cardiol.*, **52** (19), 1589–1597.

134. Decher, G., Felix, O., Saulnier, B., Izquierdo, A., Voegel, J., Schaaf, P., Jessel, N., and Ball, V. (2007) Method for Construction Functional Living Materials, Resulting Materials and Uses Thereof. W02007132099, EP2018194.

135. Crouzier, T., Ren, K., Nicolas, C., Roy, C., and Picart, C. (2009) Layer-by-layer films as a biomimetic reservoir for rhBMP-2 Delivery: controlled differentiation of myoblasts to osteoblasts. *Small*, **5** (5), 598–608.

136. Dierich, A., Le Guen, E., Messaddeq, N., Stoltz, J.F., Netter, P., Schaaf, P., Voegel, J.C., and Benkirane-Jessel, N.

(2007) Bone formation mediated by synergy-acting growth factors embedded in a polyelectrolyte multilayer film. *Adv. Mater.*, **19** (5), 693–69+.

137. Facca, S., Cortezb, C., Mendoza-Palomaresa, C., Messadeqc, N., Dierichc, A., Johnstonb, A.P.R., Mainardd, D., Voegel, J.C., Caruso, F., and Benkirane-Jessel, N. (2010) Active multilayered capsules for in vivo bone formation. *Proc. Natl. Acad. Sci. USA*, **107**, 3406–3411.

138. Nadiri, A., Kuchler-Bopp, S., Mjahed, H., Hu, B., Haikel, Y., Schaaf, P., Voegel, J.C., and Benkirane-Jessel, N. (2007) Cell apoptosis control using BMP4 and noggin embedded in a polyelectrolyte multilayer film. *Small*, **3** (9), 1577–1583.

139. Gheith, M.K., Pappas, T.C., Liopo, A.V., Sinani, V.A., Shim, B.S., Motamedi, M., Wicksted, J.R., and Kotov, N.A. (2006) Stimulation of neural cells by lateral layer-by-layer films of single-walled currents in conductive carbon nanotubes. *Adv. Mater.*, **18** (22), 2975–297+.

140. Kam, N.W.S., Jan, E., and Kotov, N.A. (2009) Electrical stimulation of neural stem cells mediated by humanized carbon nanotube composite made with extracellular matrix protein. *Nano Lett.*, **9** (1), 273–278.

141. Jan, E., Hendricks, J.L., Husaini, V., Richardson-Burns, S.M., Sereno, A., Martin, D.C., and Kotov, N.A. (2009) Layered carbon nanotube-polyelectrolyte electrodes outperform traditional neural interface materials. *Nano Lett.*, **9** (12), 4012–4018.

142. Jan, E. and Kotov, N.A. (2007) Successful differentiation of mouse neural stem cells on layer-by-layer assembled single-walled carbon nanotube composite. *Nano Lett.*, **7** (5), 1123–1128.

143. Jiang, X.P. and Hammond, P.T. (2000) Selective deposition in layer-by-layer assembly: functional graft copolymers as molecular templates. *Langmuir*, **16** (22), 8501–8509.

144. Hammond, P.T. and Whitesides, G.M. (1995) Formation of polymer microstructures by selective deposition of polyion multilayers using

patterned self-assembled monolayers as a template. *Macromolecules*, **28** (22), 7569–7571.

145. Carter, S.B. (1967) Haptotactic islands: a method of confining single cells to study individual cell reactions and clone formation. *Exp. Cell. Res.*, **48** (1), 189–193.

146. Singhvi, R., Kumar, A., Lopez, G.P., Stephanopoulos, G.N., Wang, D.I.C., Whitesides, G.M., and Ingber, D.E. (1994) Engineering cell-shape and function. *Science*, **264** (5159), 696–698.

147. Chen, C.S., Mrksich, M., Huang, S., Whitesides, G.M., and Ingber, D.E. (1997) Geometric control of cell life and death. *Science*, **276** (5317), 1425–1428.

148. Yang, S.Y., Mendelsohn, J.D., and Rubner, M.F. (2003) New class of ultra-thin, highly cell-adhesion-resistant polyelectrolyte multilayers with micropatterning capabilities. *Biomacro-molecules*, **4** (4), 987–994.

149. Yang, S.Y. and Rubner, M.F. (2002) Micropatterning of polymer thin films with pH-sensitive and cross-linkable hydrogen-bonded polyelectrolyte mul-tilayers. *J. Am. Chem. Soc.*, **124** (10), 2100–2101.

150. Berg, M.C., Yang, S.Y., Hammond, P.T., and Rubner, M.F. (2004) Con-trolling mammalian cell interactions on patterned polyelectrolyte multilayer surfaces. *Langmuir*, **20** (4), 1362–1368.

151. Lee, J.H., Kim, H.E., Im, J.H., Bae, Y.M., Choi, J.S., Huh, K.M., and Lee, C.S. (2008) Preparation of orthogonally functionalized surface using micro-molding in capillaries technique for the control of cellular adhesion. *Colloid Surf. B*, **64** (1), 126–134.

152. Shim, H.W., Lee, J.H., Kim, B.Y., Son, Y.A., and Lee, C.S. (2009) Facile preparation of biopatternable surface for selective immobilization from bac-teria to mammalian cells. *J. Nanosci. Nanotechnol.*, **9** (2), 1204–1209.

153. Forry, S.P., Reyes, D.R., Gaitan, M., and Locascio, L.E. (2006) Facilitat-ing the culture of mammalian nerve cells with polyelectrolyte multilayers. *Langmuir*, **22** (13), 5770–5775.

154. Fukuda, J., Khademhosseini, A., Yeh, J., Eng, G., Cheng, J.J., Farokhzad, O.C., and Langer, R. (2006) Mi-cropatterned cell co-cultures using layer-by-layer deposition of extracellular matrix components. *Biomaterials*, **27** (8), 1479–1486.

155. Reyes, D.R., Perruccio, E.M., Becerra, S.P., Locascio, L.E., and Gaitan, M. (2004) Micropatterning neuronal cells on polyelectrolyte multilayers. *Langmuir*, **20** (20), 8805–8811.

156. Kidambi, S., Sheng, L.F., Yarmush, M.L., Toner, M., Lee, I., and Chan, C. (2007) Patterned co-culture of pri-mary hepatocytes and fibroblasts using polyelectrolyte multilayer templates. *Macromol. Biosci.*, **7** (3), 344–353.

157. Kidambi, S., Lee, I., and Chan, C. (2008) Primary neuron/astrocyte co-culture on polyelectrolyte multi-layer films: a template for studying astrocyte-mediated oxidative stress in neurons. *Adv. Funct. Mater.*, **18** (2), 294–301.

158. Mohammed, J.S., DeCoster, M.A., and McShane, M.J. (2006) Fabrica-tion of interdigitated micropatterns of self-assembled polymer nanofilms containing cell-adhesive materials. *Langmuir*, **22** (6), 2738–2746.

159. Khademhosseini, A., Suh, K.Y., Yang, J.M., Eng, G., Yeh, J., Levenberg, S., and Langer, R. (2004) Layer-by-layer deposition of hyaluronic acid and poly-L-lysine for patterned cell co-cultures. *Biomaterials*, **25** (17), 3583–3592.

160. Lee, J., Shanbhag, S., and Kotov, N.A. (2006) Inverted colloidal crystals as three-dimensional microenvironments for cellular co-cultures. *J. Mater. Chem.*, **16** (35), 3558–3564.

161. Nicholsa, J.E., Cortiella, J.Q., Lee, J., Niles, J.A., Cuddihy, M., Wang, S.P., Bielitzki, J., Cantu, A., Mlcak, R., Valdivia, E., Yancy, R., McClure, M.L., and Kotov, N.A. (2009) In vitro analog of human bone marrow from 3D scaf-folds with biomimetic inverted colloidal crystal geometry. *Biomaterials*, **30** (6), 1071–1079.

162. Lu, Z.S., Li, C.M., Zhou, Q., Bao, Q.L., and Cui, X.Q. (2007) Covalently linked

DNA/protein multilayered film for controlled DNA release. *J. Colloid Interface Sci.*, **314** (1), 80–88.

163. Ren, K.F., Ji, J., and Shen, J.C. (2006) Construction and enzymatic degradation of multilayered poly-L-lysine/DNA films. *Biomaterials*, **27** (7), 1152–1159.

164. Itoh, Y., Matsusaki, M., Kida, T., and Akashi, M. (2006) Enzyme-responsive release of encapsulated proteins from biodegradable hollow capsules. *Biomacromolecules*, **7** (10), 2715–2718.

165. Lu, Z.Z., Wu, J., Sun, T.M., Ji, J., Yan, L.F., and Wang, J. (2008) Biodegradable polycation and plasmid DNA multilayer film for prolonged gene delivery to mouse osteoblasts. *Biomaterials*, **29** (6), 733–741.

166. Cai, K.Y., Hu, Y., Wang, Y.L., and Yang, L. (2008) Build up of multilayered thin films with chitosan/DNA pairs on poly(D,L-lactic acid) films: physical chemistry and sustained release behavior. *J. Biomed. Mater. Res. A*, **84A** (2), 516–522.

167. Blacklock, J., Handa, H., Manickam, D.S., Mao, G.Z., Mukhopadhyay, A., and Oupicky, D. (2007) Disassembly of layer-by-layer films of plasmid DNA and reducible TAT polypeptide. *Biomaterials*, **28** (1), 117–124.

168. Chen, J., Huang, S.W., Lin, W.H., and Zhuo, R.X. (2007) Tunable film degradation and sustained release of plasmid DNA from cleavable polycation/plasmid DNA multilayers under reductive conditions. *Small*, **3** (4), 636–643.

169. Bengali, Z., Rea, J.C., Gibly, R.F., and Shea, L.D. (2009) Efficacy of immobilized polyplexes and lipoplexes for substrate-mediated gene delivery. *Biotechnol. Bioeng.*, **102** (6), 1679–1691.

170. Meyer, F., Ball, V., Schaaf, P., Voegel, J.C., and Ogier, J. (2006) Polyplex-embedding in polyelectrolyte multilayers for gene delivery. *Biochim. Biophys. Acta -Biomembr.*, **1758** (3), 419–422.

171. Cai, K.Y., Hu, Y., Luo, Z., Kong, T., Lai, M., Sui, X.J., Wang, Y.L., Yang, L., and Deng, L.H. (2008) Cell-specific gene transfection from a gene-functionalized poly(D,L-lactic

acid) substrate fabricated by the layer-by-layer assembly technique. *Angew. Chem. Int. Ed.*, **47** (39), 7479–7481.

172. Dimitrova, M., Arntz, Y., Lavalle, P., Meyer, F., Wolf, M., Schuster, C., Haikel, Y., Voegel, J.C., and Ogier, J. (2007) Adenoviral gene delivery from multilayered polyelectrolyte architectures. *Adv. Funct. Mater.*, **17** (2), 233–245.

173. Suci, P.A., Klem, M.T., Arce, F.T., Douglas, T., and Young, M. (2006) Assembly of multilayer films incorporating a viral protein cage architecture. *Langmuir*, **22** (21), 8891–8896.

174. Chong, S.F., Sexton, A., De Rose, R., Kent, S.J., Zelikin, A.N., and Caruso, F. (2009) A paradigm for peptide vaccine delivery using viral epitopes encapsulated in degradable polymer hydrogel capsules. *Biomaterials*, **30** (28), 5178–5186.

175. Shen, H., Tan, J., and Saltzman, W.M. (2004) Surface-mediated gene transfer from nanocomposites of controlled texture. *Nature Mater.*, **3** (8), 569–574.

176. Lvov, Y., Decher, G., and Sukhorukov, G. (1993) Assembly of thin-films by means of successive deposition of alternate layers of DNA and poly(Allylamine). *Macromolecules*, **26** (20), 5396–5399.

177. Sukhorukov, G.B., Mohwald, H., Decher, G., and Lvov, Y.M. (1996) Assembly of polyelectrolyte multilayer films by consecutively alternating adsorption of polynucleotides and polycations. *Thin Solid Films*, **285**, 220–223.

178. Decher, G., Lehr, B., Lowack, K., Lvov, Y., and Schmitt, J. (1994) New nanocomposite films for biosensors – layer-by-layer adsorbed films of polyelectrolytes, proteins or DNA. *Biosens. Bioelectron.*, **9** (9–10), 677–684.

179. Zhang, J.T., Chua, L.S., and Lynn, D.M. (2004) Multilayered thin films that sustain the release of functional DNA under physiological conditions. *Langmuir*, **20** (19), 8015–8021.

180. Jewell, C.M., Zhang, J.T., Fredin, N.J., and Lynn, D.M. (2005) Multilayered polyelectrolyte films promote the direct

and localized delivery of DNA to cells. *J. Control. Release*, **106** (1–2), 214–223.

181. Jewell, C.M., Zhang, J.T., Fredin, N.J., Wolff, M.R., Hacker, T.A., and Lynn, D.M. (2006) Release of plasmid DNA from intravascular stents coated with ultrathin multilayered polyelectrolyte films. *Biomacromolecules*, **7** (9), 2483–2491.

182. Taori, V.P., Liu, Y.M., and Reineke, T.M. (2009) DNA delivery in vitro via surface release from multilayer assemblies with poly(glycoamidoamine)s. *Acta Biomater.*, **5** (3), 925–933.

183. Zhang, X., Sharma, K.K., Boeglin, M., Ogier, J., Mainard, D., Voegel, J.C., Mely, Y., and Benkirane-Jessel, N. (2008) Transfection ability and intracellular DNA pathway of nanostructured gene-delivery systems. *Nano Lett.*, **8** (8), 2432–2436.

184. Zhang, X., Voegel, J.C., Mely, Y., and Benlirane-Jessel, N. (2008) Transfection ability and intracellular DNA pathway of nano-structured gene delivery systems. *Hum. Gene Ther.*, **19** (10), 1153–1153.

185. Gonzalez, H., Hwang, S.J., and Davis, M.E. (1999) New class of polymers for the delivery of macromolecular therapeutics. *Bioconjug. Chem.*, **10** (6), 1068–1074.

186. Arima, H., Kihara, F., Hirayama, F., and Uekama, K. (2001) Enhancement of gene expression by polyamidoamine dendrimer conjugates with alpha-, beta-, and gamma-cyclodextrins. *Bioconjug. Chem.*, **12** (4), 476–484.

187. Lin, Q.K., Ren, K.F., and Ji, J. (2009) Hyaluronic acid and chitosan-DNA complex multilayered thin film as surface-mediated nonviral gene delivery system. *Colloid Surf. B*, **74** (1), 298–303.

188. Ren, K.F., Ji, J., and Shen, J.C. (2005) Construction of polycation-based non-viral DNA nanoparticles and polyanion multilayers via layer-by-layer self-assembly. *Macromol. Rapid Commun.*, **26** (20), 1633–1638.

189. Blacklock, J., You, Y.Z., Zhou, Q.H., Mao, G.Z., and Oupicky, D. (2009) Gene delivery in vitro and in vivo from bioreducible multilayered polyelectrolyte films of plasmid DNA. *Biomaterials*, **30** (5), 939–950.

190. Aviles, M.O., Lin, C.H., Zelivyanskaya, M., Graham, J.G., Boehler, R.M., Messersmith, P.B., and Shea, L.D. (2010) The contribution of plasmid design and release to in vivo gene expression following delivery from cationic polymer modified scaffolds. *Biomaterials*, **31** (6), 1140–1147.

191. Jessel, N., Oulad-Abdeighani, M., Meyer, F., Lavalle, P., Haikel, Y., Schaaf, P., and Voegel, J.C. (2006) Multiple and time-scheduled in situ DNA delivery mediated by beta-cyclodextrin embedded in a polyelectrolyte multilayer. *Proc. Natl. Acad. Sci. USA*, **103** (23), 8618–8621.

192. Zhang, J.T., Montanez, S.I., Jewell, C.M., and Lynn, D.M. (2007) Multilayered films fabricated from plasmid DNA and a side-chain functionalized poly(beta-amino ester): surface-type erosion and sequential release of multiple plasmid constructs-from surfaces. *Langmuir*, **23** (22), 11139–11146.

193. Wood, K.C., Chuang, H.F., Batten, R.D., Lynn, D.M., and Hammond, P.T. (2006) Controlling interlayer diffusion to achieve sustained, multiagent delivery from layer-by-layer thin films. *Proc. Natl. Acad. Sci. USA*, **103** (27), 10207–10212.

194. Liu, X.H., Zhang, J.T., and Lynn, D.M. (2008) Ultrathin multilayered films that promote the release of two DNA constructs with separate and distinct release profiles. *Adv. Mater.*, **20** (21), 4148–414+.

195. Meyer, F., Dimitrova, M., Jedrzejenska, J., Arntz, Y., Schaaf, P., Frisch, B., Voegel, J.C., and Ogier, J. (2008) Relevance of bi-functionalized polyelectrolyte multilayers for cell transfection. *Biomaterials*, **29** (5), 618–624.

196. De Koker, S., De Geest, B.G., Cuvelier, C., Ferdinande, L., Deckers, W., Hennink, W.E., De Smedt, S., and Mertens, N. (2007) In vivo cellular uptake, degradation, and biocompatibility of polyelectrolyte microcapsules. *Adv. Funct. Mater.*, **17** (18), 3754–3763.

197. Saurer, E.M., Jewell, C.M., Kuchenreuther, J.M., and Lynn, D.M. (2009) Assembly of erodable, DNA-containing thin films on the surfaces of polymer microparticles: toward a layer-by-layer approach to the delivery of DNA to antigen-presenting cells. *Acta Biomater.*, **5** (3), 913–924.

198. Kakade, S., Manickam, D.S., Handa, H., Mao, G.Z., and Oupicky, D. (2009) Transfection activity of layer-by-layer plasmid DNA/poly(ethylenimine) films deposited on PLGA microparticles. *Int. J. Pharm.*, **365** (1–2), 44–52.

199. Elouahabi, A. and Ruysschaert, J.M. (2005) Formation and intracellular trafficking of lipoplexes and polyplexes. *Mol. Ther.*, **11** (3), 336–347.

200. Boussif, O., Lezoualch, F., Zanta, M.A., Mergny, M.D., Scherman, D., Demeneix, B., and Behr, J.P. (1995) A versatile vector for gene and oligonucleotide transfer into cells in culture and in-vivo – polyethylenimine. *Proc. Natl. Acad. Sci. USA*, **92** (16), 7297–7301.

201. Fischer, D., Bieber, T., Li, Y.X., Elsasser, H.P., and Kissel, T. (1999) A novel non-viral vector for DNA delivery based on low molecular weight, branched polyethylenimine: effect of molecular weight on transfection efficiency and cytotoxicity. *Pharm. Res.*, **16** (8), 1273–1279.

202. Zhang, S.B., Xu, Y.M., Wang, B., Qiao, W.H., Liu, D.L., and Li, Z.S. (2004) Cationic compounds used in lipoplexes and polyplexes for gene delivery. *J. Control. Release*, **100** (2), 165–180.

203. Kasturi, S.P., Sachaphibulkij, K., and Roy, K. (2005) Covalent conjugation of polyethyleneimine on biodegradable microparticles for delivery of plasmid DNA vaccines. *Biomaterials*, **26** (32), 6375–6385.

204. Trimaille, T., Pichot, C., and Delair, T. (2003) Surface functionalization of poly(D,L-lactic acid) nanoparticles with poly(ethylenimine) and plasmid DNA by the layer-by-layer approach. *Colloid Surf. A*, **221** (1–3), 39–48.

205. Trubetskoy, V.S., Wong, S.C., Subbotin, V., Budker, V.G., Loomis, A., Hagstrom, J.E., and Wolff, J.A. (2003) Recharging cationic DNA complexes with highly charged polyanions for in vitro and in vivo gene delivery. *Gene Ther.*, **10** (3), 261–271.

206. Trubetskoy, V.S., Loomis, A., Hagstrom, J.E., Budker, V.G., and Wolff, J.A. (1999) Layer-by-layer deposition of oppositely charged polyelectrolytes on the surface of condensed DNA particles. *Nucleic Acids Res.*, **27** (15), 3090–3095.

207. Saul, J.M., Wang, C.H.K., Ng, C.P., and Pun, S.H. (2008) Multilayer nanocomplexes of polymer and DNA exhibit enhanced gene delivery. *Adv. Mater.*, **20** (1), 19–25.

208. Ryther, R.C.C., Flynt, A.S., Phillips, J.A., and Patton, J.G. (2005) siRNA therapeutics: big potential from small RNAs. *Gene Ther.*, **12** (1), 5–11.

209. Recksiedler, C.L., Deore, B.A., and Freund, M.S. (2006) A novel layer-by-layer approach for the fabrication of conducting polymer/RNA multilayer films for controlled release. *Langmuir*, **22** (6), 2811–2815.

210. Dimitrova, M., Affolter, C., Meyer, F., Nguyen, I., Richard, D.G., Schuster, C., Bartenschlager, R., Voegel, J.C., Ogier, J., and Baumert, T.F. (2008) Sustained delivery of siRNAs targeting viral infection by cell-degradable multilayered polyelectrolyte films. *Proc. Natl. Acad. Sci. USA*, **105** (42), 16320–16325.

211. Elbakry, A., Zaky, A., Liebkl, R., Rachel, R., Goepferich, A., and Breunig, M. (2009) Layer-by-layer assembled gold nanoparticles for siRNA delivery. *Nano Lett.*, **9** (5), 2059–2064.

212. Mehrotra, S., Lee, I., and Chan, C. (2009) Multilayer mediated forward and patterned siRNA transfection using linear-PEI at extended N/P ratios. *Acta Biomater.*, **5** (5), 1474–1488.

213. Seo, J., Lee, H., Jeon, J., Jang, Y., Kim, R., Char, K., and Nam, J.M. (2009) Tunable layer-by-layer polyelectrolyte platforms for comparative cell assays. *Biomacromolecules*, **10** (8), 2254–2260.

214. Fayazpour, F., Lucas, B., Vandenbroucke, R.E., Derveaux, S., Tavernier, J., Lievens, S., Demeester, J., and De Smedt, S.C. (2008) Evaluation of digitally encoded layer-by-layer

coated microparticles as cell carriers. *Adv. Funct. Mater.*, **18** (18), 2716–2723.

215. Picart, C. and Discher, D.E. (2007) Materials science – embedded shells decalcified. *Nature*, **448** (7156), 879–880.

216. Kreft, O., Prevot, M., Mohwald, H., and Sukhorukov, G.B. (2007) Shell-in-shell microcapsules: a novel tool for integrated, spatially confined enzymatic reactions. *Angew. Chem. Int. Ed.*, **46** (29), 5605–5608.

217. Jiang, Y.J., Sun, Q.Y., Zhang, L., and Jiang, Z.Y. (2009) Capsules-in-bead scaffold: a rational architecture for spatially separated multienzyme cascade system. *J. Mater. Chem.*, **19** (47), 9068–9074.

218. Stadler, B., Chandrawati, R., Goldie, K., and Caruso, F. (2009) Capsosomes: subcompartmentalizing polyelectrolyte capsules using liposomes. *Langmuir*, **25** (12), 6725–6732.

219. Hosta-Rigau, L., Stadler, B., Yan, Y., Nice, E.C., Heath, J.K., Albericio, F., and Caruso, F. (2010) Capsosomes with multilayered subcompartments: assembly and loading with hydrophobic cargo. *Adv. Funct. Mater.*, **20** (1), 59–66.

220. Stadler, B., Chandrawati, R., Price, A.D., Chong, S.F., Breheney, K., Postma, A., Connal, L.A., Zelikin, A.N., and Caruso, F. (2009) A microreactor with thousands of subcompartments: enzyme-loaded liposomes within polymer capsules. *Angew. Chem. Int. Ed.*, **48** (24), 4359–4362.

221. Loew, M., Kang, L., Dahne, L., Hendus-Altenburger, R., Kaczmarek, O., Liebscher, J., Huster, D., Ludwig, K., Bottcher, C., Herrmann, A., and Arbuzova, A. (2009) Controlled assembly of vesicle-based nanocontainers on layer-by-layer particles via DNA hybridization. *Small*, **5** (3), 320–323.

3
The Layer-by-Layer Assemblies of Polyelectrolytes and Nanomaterials as Films and Particle Coatings

Mi-Kyoung Park and Rigoberto C. Advincula

Ultrathin multilayer films of organic, polymeric, and hybrid materials have gained much attention over the last three decades due to their applications in microelectronics, electro-optics, sensors, nanotechnology, and biotechnology [1, 2]. The main advantages of these systems originate from their nanostructured architecture in which the self-assembled or ordered monolayers are formed on the substrates. These monolayers may in fact be ordered assemblies themselves that display long-range ordering including liquid crystallinity or mesoscopic structures. Subsequently, multilayers are produced by the repetitive "layer-by-layer" (LbL) deposition of these monolayers. Therefore, it is possible to control molecular orientation and organization on the nanoscale that precisely tunes the macroscopic properties of the organic and polymer thin films [3, 4].

Perhaps one of the oldest technique for fabricating multilayer ultrathin films (and the most extensively studied prior to the 1990s) from individual monolayers is the Langmuir–Blodgett (LB) technique. In this technique, monolayers are formed at the air/water interface and then transferred onto a solid substrate by dipping the substrate in a vertical deposition method [5, 6]. Multilayers are realized by repetitive dipping. Kuhn and coworkers reported, in the early 1970s, that different layers of LB films of functionalized cyanine dyes have distance-dependent Förster energy transfer on the nanoscale, which is not exhibited by the individual monolayers [7]. LB was, indeed, the first technique to provide scientists with the practical capability to construct ordered molecular assemblies. These films are highly ordered and have controlled uniform thickness [8]. However, as many have realized, the LB technique requires special equipment and has severe limitations with respect to substrate size and topology as well as film quality and stability.

Another approach to assemble layered ultrathin film structures based on chemisorption was also reported in the early 1980s. These are usually based on two-step sequential reactions between two chemical species, for example, silane and hydroxylated surface with conversion of a nonpolar terminal group to a hydroxyl group [9, 10]. Once a subsequent monolayer is adsorbed on the "activated" monolayer, multilayer films may be built by repetition of this process [11]. In 1988, Mallouk and coworkers showed that multilayer films could be prepared simply by sequential complexation of Zr^{4+} and α, ω-bisphosphonic acid [12]. However,

Functional Polymer Films, First Edition. Edited by Wolfgang Knoll and Rigoberto C. Advincula.
© 2011 Wiley-VCH Verlag GmbH & Co. KGaA. Published 2011 by Wiley-VCH Verlag GmbH & Co. KGaA.

self-assembled films based on covalent or coordination chemistry are restricted to certain classes of organics, and high-quality homogeneous multilayer films cannot be reliably obtained because of the high steric demand of covalent chemistry and for a 100% reactivity in each step [13].

The LbL self-assembly, which is based on the alternate physisorption of mutually attracted species of molecules, macromolecules (polymers), and nanoparticles (nano-objects) have gained much popularity in scientific research and practical technological applications. The trend in publications starting in the early 1990s has reached up to 10,500 publications to date, just by referring to the topic "layer-by-layer" at the ISI Web of Knowledge search engine.

The LbL deposition of oppositely charged bolaform amphiphiles and polyelectrolytes, was first reported by Decher and coworkers in the early 1990s [14, 15]. It was the ability to form multilayers with precise control over the total thickness, that is, in the range of a few angstroms up to micrometers, without the use of the more expensive LB technique that first garnered great attention. This is due to its characteristic repetitive deposition steps, using essentially a "beaker technique." Furthermore, highly tailored polymer ultrathin films can be prepared with a nearly unlimited range of functional groups incorporated within the structure of the film. Additionally, it had significant advantages over other techniques; this process is independent of the substrate size and topology, the assembly is based on spontaneous adsorptions, and no stoichiometric control is necessary to maintain surface functionality [13, 15, 16]. In the late 1990s, the Möhwald group reported the LbL deposition of polyelectrolytes onto charged polystyrene latex particles, and have also shown that polyelectrolyte hollow capsules can be prepared by removing the core after the deposition to form "hollow shell" particles [16a–c].

Since then, the technique has undergone a great variety of assembly mechanism, substrate-surface, shape of templates, and transformation in protocols, with the term LbL becoming synonymous with nanostructured materials and assembly of thin film coatings, shell, and hollow-shell formation. For the mechanistic side, other than simple electrostatic attraction of oppositely charged species, this has now included the use of other noncovalent; hydrogen bonding, stereocomplexation, dipole interactions, and covalent; click chemistry, thiol-ene-chemistry, and other mechanisms. For the species and objects that have been assembled, this has now included a wide variety of molecules, macromolecules, nanoparticles, and biological species; dendrimers, block-copolymer micelles, bacteria, graphene, carbon nanotubes, and so on. For the substrates employed this has gone from simple glass slides, Si wafers, to large-area substrates, and nanoparticle surfaces. For the protocols employed, this has now included spin coating, spray assembly, large area dipping, and roll-to-roll processes. The reader is referred to the most recent review articles in the last three years on this topic as summarized in Table 3.1 [17].

Table 3.1 gives the reader the height, depth, and breadth of how the field has been transformed and the variety of protocols and applications to which the LbL process has been referred to. Note that Table 3.1 is not meant to be exhaustive and doubtless there are many other recent articles and reviews that will be of higher relevance to the individual reader. One of the earlier collection of fundamentals and applications

Table 3.1 Recent reviews in layer-by-layer assemblies on flat surfaces and colloidal particles.

Title	Year	Authors	References
Recent developments in supramolecular approach for nanocomposites	2010	Ariga, K. *et al.*	[17a]
Multiple functionalities of polyelectrolyte multilayer films: new biomedical applications.	2010	Picart, C. *et al.*	[17b]
LbL multilayer capsules: recent progress and future outlook for their use in life sciences	2010	Parak, W. *et al.*	[17c]
Polymer/clay and polymer/carbon nanotube hybrid organic–inorganic multilayered composites made by sequential layering of nanometer scale films	2010	Kotov, N. *et al.*	[17d]
Nanofabrication by self-assembly	2009	Ozin, G. *et al.*	[17e]
The pros and cons of polyelectrolyte capsules in drug delivery	2009	Mohwald, H. *et al.*	[17f]
Self-assembly of composite nanotubes and their applications	2009	Li, J. *et al.*	[17g]
Design of antibacterial surfaces and interfaces: polyelectrolyte multilayers as a multifunctional platform	2009	Rubner, M. *et al.*	[17h]
Functional multilayered capsules for targeting and local drug delivery	2009	Akashi, M. *et al.*	[17i]
Multilayer-derived, ultrathin, stimuli-responsive hydrogels	2009	Sukhishvili, S. *et al.*	[17j]
Molecular-engineered stimuli-responsive thin polymer film: a platform for the development of integrated multifunctional intelligent materials	2009	Minko, S. *et al.*	[17k]
Comparative electrochemical study of myoglobin loaded in different polyelectrolyte layer-by-layer films assembled by spin-coating	2009	Hu, *et al.*	[17l]
Modulating the release kinetics through the control of the permeability of the layer-by-layer assembly: a review	2009	Tabrizian, M. *et al.*	[17m]
Designer polymer-quantum dot architectures	2009	Vancso, G. *et al.*	[17n]
Coupling of soft technology (layer-by-layer assembly) with hard materials (mesoporous solids) to give hierarchic functional structures	2009	Ariga, K. *et al.*	[17o]

(continued overleaf)

Table 3.1 *(continued).*

Title	Year	Authors	References
Layer-by-layer assembly on stimuli-responsive microgels	2008	Richtering, W. *et al.*	[17p]
Composite layer-by-layer (LbL) assembly with inorganic nanoparticles and nanowires	2008	Kotov, N. *et al.*	[17q]
Monolayers and multilayers of conjugated polymers as nanosized electronic components	2008	Zotti, G. *et al.*	[17r]
Template synthesis of nanostructured materials via layer-by-layer assembly	2008	Caruso, F. *et al.*	[17s]
Biomaterials and biofunctionality in layered macromolecular assemblies	2008	Ariga, K. *et al.*	[17t]
Assembly of nanoparticles on patterned surfaces by noncovalent interactions	2008	Reinhoudt, D. *et al.*	[17u]
Polyelectrolyte multilayer films: from physicochemical properties to the control of cellular processes	2008	Picart, C. *et al.*	[17v]
Next-generation, sequentially assembled ultrathin films: beyond electrostatics	2007	Caruso, F. *et al.*	[17w]
Electrochemically enabled polyelectrolyte multilayer devices: from fuel cells to sensors	2007	Hammond, P. *et al.*	[17x]
Layer-by-layer assembly as a versatile bottom-up nanofabrication technique for exploratory research and realistic application	2007	Ariga, K. *et al.*	[17y]
Recent developments in the properties and applications of polyelectrolyte multilayers	2006	Schlenoff, J. *et al.*	[17z]

in this field was published back in 2003 by Decher and Schlenoff [18]. The rest of this review looks at the origins and fundamental assumptions of the LbL process. Then, the author summarizes the work in his group and their contributions to the field.

3.1
Layer-by-Layer (LbL) Self-Assembly Technique

The experiment of building multilayers by the "LbL" method or alternate "electrostatic self-assembly" (ESA), was actually first reported by Iler in 1966 [19]. He demonstrated the deposition of alternate layers of positively charged alumina particles with negatively charged silica on glass substrate. However, the importance of LbL method was not realized until a quarter of a century later. In 1991 and 1992,

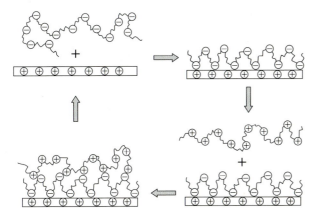

Scheme 3.1 Scheme of the electrostatic layer-by-layer self-assembly.

Decher and Hong reported a series of experiments of multilayer film build up by a self-assembly process with bipolar amphiphiles [14], bipolar amphiphiles with polyelectrolytes, and polyelectrolytes [18] on charged surfaces. They showed that the multilayer ultrathin organic films can be prepared by the consecutive deposition of oppositely charged polyelectrolytes from dilute aqueous solution onto charged solid substrates (Scheme 3.1).

This technique utilizes the electrostatic attraction and complex formation between polyanions and polycations to form supramolecular multilayer assemblies of polyelectrolytes. In principle, the adsorption of molecules carrying more than one equal charge allows charge reversal on the surface, which has two important consequences: (i) repulsion of equally charged molecules and thus self-regulation of the adsorption and restriction to a single layer, and (ii) the ability of an oppositely charged molecule to be adsorbed in a second step on top of the first one. There seems to be no limitation to the maximum number of layers that can be deposited, and films with up to 1000 layers have been prepared [20]. This technique is characterized by several advantages: (i) the preparative procedure is simple and an elaborate apparatus is not required; (ii) a large variety of water soluble polyions are used; (iii) the individual layer has a defined molecular thickness; (iv) any charged surface is employable. Furthermore, the process can be easily automated, and is *a priori* environmentally friendly, since water is usually used as a solvent [3, 13, 16, 21].

Since the first report by Decher, this technique has experienced an explosion of growth in both applications and fundamental theoretical and experimental advances. As an extension of this technique, a variety of materials (e.g., biological macromolecules, surfactants, phospholipids, nanoparticles, inorganic crystals, and multivalent dyes as shown in Table 3.2) have been successfully incorporated into polyelectrolyte films by replacing one of the polyelectrolytes with another species of the same charge to fabricate composite polyelectrolyte multilayer films.

Many types of charged molecules and nano-objects seem to be suitable for deposition by the LbL method, but for the most part polyelectrolytes have been

Table 3.2 Charged molecules, macromolecules, and nano-objects that have been used for electrostatic self-assembly.

Polyelectrolytes

Conjugated polymer; poly(phenylene vinylene) precursor [22], poly(*p*-phenylene) [23], polyaniline [24], sulfonated polyaniline [25], polythiophenes [26]

Dendrimers [27], liquid-crystalline polyelectrolytes [28], diazo-resins [29], azo-polymers [30]

Bio-organic materials

Proteins [31], virus [32], lipids [33], albumin [34], DNA [35] polypeptides [36], enzymes [37], avidin [38], bacteriorhodopsin [39]

Polysaccharides [40]; chitosan [41], dextan sulfate [42], cellulose sulfate [43]

Inorganic materials

Charged nano-objects; silica [44], metal oxides [45], semiconductor nanoparticles (CdS [46], TiO_2 [47], CdSe [48], CdTe [49]), metal colloids (Au [50], Pt [51]), charged latex spheres [52], microcrystallites [53], metallo-supramolecular complexes [54]

Clay platelets; Montmorillonte [55], hectorite [56], saponite [57], α-zirconium phosphate [58], graphite oxide [59], MoS_2 [60]

Small organic materials

Bolaamphiphiles [61], phthalocyanine [62], azobenzene dyes [63], cyanine dyes [64]

---

employed. Therefore, in the rest of this chapter, the properties of polyelectrolytes and adsorption of polyelectrolytes on substrate will be addressed first, followed by a review on the mechanism of formation of multilayers, multilayer structure, and controlling factors.

3.2
The Properties of Polyelectrolytes

Polyelectrolytes are charged macromolecules containing a large number of ionizable or ionic groups. In solution under appropriate conditions, the ionizable groups in a polyelectrolyte dissociate into polyions and a number of small ions. These small ions are oppositely charged and are referred to as the *counterions* [65]. Polyelectrolytes are usually classified as polycations, and polyanions depending on the ionic groups. Examples of typical polycations and polyanions used in self-assembled LbL assembly are presented in Figure 3.1. Also, there is a special case of polyelectrolytes, the "polyampholytes," carrying both anionic and cationic groups covalently bound to the macromolecules, which are represented in nature by an abundant number of proteins.

Depending on the degree of dissociation of polyelectrolytes in aqueous media, they are classified as being weak or strong. Strong polyelectrolytes such as poly(sodium 4-styrene sulfonate) (PSS) and poly(diallydimethylammonium chloride) (PDADMAC) fully dissociate in aqueous solutions in the total pH range

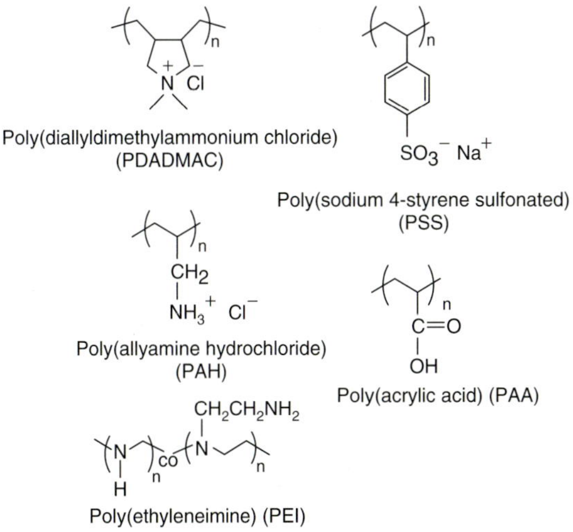

Poly(diallyldimethylammonium chloride)
(PDADMAC)

Poly(sodium 4-styrene sulfonated)
(PSS)

Poly(allyamine hydrochloride)
(PAH)

Poly(acrylic acid) (PAA)

Poly(ethyleneimine) (PEI)

Figure 3.1 Typical polycations and polyanions used in self-assembled layer-by-layer assembly.

between 0 and 14 due to the strong acid and base groups attached to the polymer repeat units. In contrast, the amount of ionization of weak polyelectrolytes (e.g., poly(acrylic acid) (PAA) and poly(allyamine hydrochloride) (PAH)) fully depends on the pH of the solution. They form a polyion–counterion system only in a limited pH range, and remain as an undissociated polyacid in the acid range or an undissociated polybase in the alkaline range, respectively.

The pH value of a weak polyanion is given by the equation [66].

$$pH = pK_a + \log\left(\frac{\alpha}{1-\alpha}\right)$$
$$= pK_0 + \Delta pK + \log\left(\frac{\alpha}{1-\alpha}\right) \qquad (3.1)$$

where α is the fraction of dissociated acidic groups (degree of ionization), pK_a is the negative logarithm of the effective (α-dependent) dissociation constant, and pK_0 is the negative logarithm of the intrinsic dissociation constant. ΔpK represents the contribution of the polyelectrolyte field to the standard free energy of ionization of a single group. In other words, ΔpK represents the shift in the dissociation constant that is due to the change in the electrostatic free energy (G_e) of a polyanion upon variation in the number, n, of negatively charged groups,

$$\Delta pK = \frac{0.4343}{kT}\left(\frac{\delta G_e}{\delta n}\right)_{\kappa} \qquad (3.2)$$

Generally, polyanions have a higher pK_a than the monomer acid, and polycations have a lower pK_a than the monomer base, and the titration curves become much broader than those of small molar-mass analogs. For example, acetic acid has a pK_a of 4.8 while PAA has an apparent $pK_a \sim 5.8$ or higher depending on the ionic strength of the solution.

Because of the restriction of charged groups along the polymer backbone by chemical bonds and the long-range nature of electrostatic interaction, polyelectrolytes can exhibit totally unexpected physical properties. In solution, a repelling force between these charges results in charges along the polymer chain, either on the main chain or in the side groups. The overall effect of the force is to stretch the whole polymer molecule in the solution. The addition of low molecular weight salts increases the conductivity in the solution, resulting in a lowered Debye screening length. This screens the repelling interaction between the charges, resulting in a less-extended, and more coiled conformation of the polymer.

According to the Debye–Hückel theory, coulombic fields are screened in the electrolyte with a *Debye screening length* (κ^{-1}) given by [67].

$$\kappa^{-1} = \left(\frac{\varepsilon k_B T}{4\pi e^2 n} \right)^{0.5} \tag{3.3}$$

where ε is the dielectric constant, k_B is the Boltzmann constant, T is the temperature, e is the electron charge, and n is the concentration of ions. At low ionic strength, κ^{-1} is large and the polyelectrolyte chain is in a much more extended conformation than the random coil of a neutral polymer. Whereas, at high ionic strength, κ^{-1} is small, the electrostatic repulsion between like-charged units on the polyelectrolyte chain is reduced and the polyelectrolyte chain progresses to a random coil conformation.

Besides the solution pH and ionic strength that were mentioned above, the charge density, nature of the ions, molecular weight, and the noncoulombic interactions such as van der Waals interaction, hydrogen bonding, and other molecular interactions play a very important role in determining the conformation of the macromolecule. The solution behavior of polyelectrolytes can affect their adsorption on a surface.

3.3
Adsorption of Polyelectrolytes

The physical properties of charged macromolecules in contact with a solid surface are fundamentally different from those of similar layers consisting of uncharged polymers. In contrast to those of neutral polymer films, the structure and properties of polyelectrolyte layers are almost exclusively dominated by electrostatic interactions. Mutual repulsion between the charged polymer segments and electrostatic forces between the polyelectrolyte molecules and the surface markedly influence the strength of interaction with the substrate and the physical properties of the layers. The conformations of a polyelectrolyte (and polymer in general) adsorbed on a surface can be described in terms of three types of subchains: *trains*, which have all their segments in contact with the substrate, *loops*, which have no contact with the surface and connect two trains, and *tails*, which are nonadsorbed chain ends [68].

Several theoretical approaches including self-consistent-field theory [69–71], Monte Carlo simulations [72–74], and scaling theory [75] have been applied to deal with polyelectrolyte adsorption on a charged surface. Despite these intensive investigations, the full picture of polyelectrolyte adsorption on charged surfaces is still far from complete, due to the delicate interplay between chain interconnectivity and the long-range nature of electrostatic interaction as well as the existence of interactions at several length scales.

Although electrostatic attraction between polyelectrolytes and an oppositely charged surface plays a key role in adsorption, the primary driving force is presumably entropy, not enthalpy [3]. The complexation of the polyions with an oppositely charged surface liberates undissociated low-molar mass counterions, therefore the entropy of the system increases. An additional entropic gain can be obtained from the release of solvent molecules from the solvation shell of the polymer-bound ionic groups. Since surface–counterion and polyelectrolyte–counterion interactions are replaced by surface-polyelectrolyte and anion–cation interactions, the enthalpic change should be small.

Because of the importance of the electrostatic interactions, polyelectrolyte adsorption depends strongly on electrostatic parameters such as the surface charge and the polymer charge, which can depend on both the pH and the ionic strength. Polyelectrolytes adsorbing on an oppositely charged surface experience both an electrostatic attraction to the surface (favoring adsorption) and an electrostatic repulsion within the layer (counteracting adsorption). When strong polyelectrolytes are adsorbed onto an oppositely charged surface, they are adsorbed rather flat and form thin layers due to the strong repulsion between segments [70]. If this repulsion is screened, for example, by adding salt, the adsorbed amount increases and the adsorbed layer becomes thicker. Böhmer *et al.* [69] have calculated the fraction of segments in trains, loops, and tails as a function of the square root of salt concentration based on self-consistent-field theory [71]. They have shown that at low salt concentrations, the polyelectrolyte adsorbs very flat conformations, and around 90% of the segments of adsorbed chains are in contact with the surface. The fraction of segments in trains becomes lower with rising ionic strength, whereas the fraction of segments in loops and tails become larger.

For weak polyelectrolytes, the adsorption shows richer behavior with respect to variation of pH change, addition of salt, molecular weight, and the presence of nonelectrostatic interactions [76]. In the case of adsorption of polyanions on a surface at high pH, where the polymer is fully charged, the situation is similar to that of strong polyelectrolytes, and consequently, the adsorbed amount is low. As the pH decreases, the charge on the polyelectrolyte reduces. Thus, the segment–segment repulsion becomes less important and the adsorbed amount increases (Figure 3.2).

If the surface and the polyanion are oppositely charged and the interaction between segments and the surface is not too high, a maximum adsorption occurs in the adsorbed amount around 1–1.5 pH units below pK. If the pH is too low, the charge of the polyelectrolyte will be decreased [71]. The attractive electrostatic interaction between polyelectrolyte and the surface vanishes, and therefore the adsorbed amount decreases. The effect of the polyelectrolyte length, linear charge density of the polyelectrolyte, volume fraction of the polyelectrolyte, nonelectrostatic polyelectrolyte–surface interaction, and surface charge density on adsorption

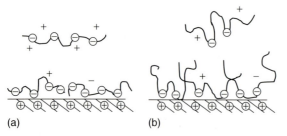

Figure 3.2 The conformation of weak polyelectrolytes adsorbed on an oppositely charged surface (a) at high charge density (high pH) and (b) at low charge density (low pH).

of weak polyelectrolytes on an oppositely charged surface has been studied by Linse using the mean-field lattice theory [76]. He suggested that as the salt concentration increases, the adsorbed amount is reduced while the thickness of the adsorbed amount increases. At low salt concentration and at constant surface charge density, the amount adsorbed is governed by the surface charge density through a polyelectrolyte–surface charge matching.

As mentioned above, the adsorption of polyelectrolytes can increase (strong polyelectrolyte) or decrease (weak polyelectrolyte with low charge density) with increasing salt concentration. Recently, van de Steeg *et al.* [77] made extensive numerical investigations of the adsorption of polyelectrolytes at oppositely charged surfaces by using the mean-field lattice theory. They introduced the concepts of "screening-enhanced adsorption" and "screening-reduced adsorption" regime. These regimes apply to the cases where the adsorbed amount increases and decreases, respectively, upon the addition of salt. The former is more typical for a highly charged polyelectrolyte, that is, in the case where adsorption is limited by electrostatic repulsion between the segments of the polyelectrolyte. Screening of this repulsion leads to an increase in polyion adsorptivity, provided the nonelectrostatic polymer–surface interaction is strong enough to keep the screened polyelectrolyte adsorbed (Figure 3.3). However, salt screens not only

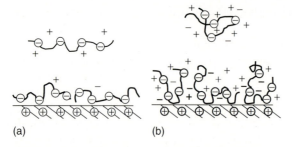

Figure 3.3 The conformation of polyelectrolytes adsorbed on an oppositely charge surface (a) without salt and (b) with salt in "screening-enhanced adsorption" regime.

repulsion between charged segments of polyelectrolyte, but also the electrostatic attraction between the segments and the surface. Thus, in polyelectrolyte adsorption added salt has two antagonistic effects. It depends on the balance between electrostatic and nonelectrostatic attraction, whether or not increasing the salt concentration leads to an increase or a decrease in adsorption. In the case where the adsorption is dominated by this electrostatic attraction (relatively high surface charge, low polyelectrolyte charge, and weak nonelectrostatic contribution) an increase of electrolyte concentration results in decreased adsorption.

The polyelectrolyte can be displaced from the surface at a certain critical salt concentration, c_{sc}, which depends on the segment charge τ, and the surface charge density σ_0, according to $c_{sc} \propto (\sigma\tau)^{10/11}$. Now, the screening-reduced adsorption regime occurs only for a very low segment charge and sufficiently high surface charge density. The majority of the cases constitute the screening-enhanced adsorption regime. Naturally, a full range of situations intermediate between the above two regimes is feasible, and this often leads to the adsorbed amount as a function of salt concentration displaying a maximum. Kinetically, the adsorption mechanism can be thought of as a two-step process; (i) an initial, fast process during which the polymer chains or aggregates diffuse from solution to bare substrate, and thereby a monolayer is formed; (ii) a subsequent slow buildup process where the chains penetrate through the existing monolayer in conjunction with conformational rearrangement of the chains. Whereas the first step usually takes place within minutes, the second step can be as long as several hours or days, due to reconformation of the adsorbed layer. The latter process is often much slower than that of uncharged polymer adsorption due to the electrostatic repulsion by the polymers that are already adsorbed [3, 78]. During the initial stages of the adsorption, the transport process is rate limiting, unless the adsorption energy becomes too low or when there is an electrostatic potential barrier, for example, for a polyelectrolyte adsorbing on a surface with a like charge. As a consequence, a variation of the ionic strength or the pH influences the adsorption rate mainly through its effect on the diffusion coefficient [79]. This diffusion coefficient is low for a highly charged polyelectrolyte in a solution with a low salt concentration, because the polyelectrolyte coil is strongly swollen due to the internal repulsion between the charged segments. This repulsion is reduced, and thus the transport rate enhanced, by either increasing the ionic strength or decreasing the segment charge.

One specific aspect of polyelectrolyte adsorption is *overcompensation*. The charge of a planar surface can be inverted by adsorption of an oppositely charged polyelectrolyte. Even in the case of a highly charged polyelectrolyte that has a rather flat conformation, slight charge overcompensation can still occur on the adsorbed film surface since not all the charged segments will be paired with oppositely charged surface sites due to entropic and steric constraints. Joanny [80] reported on theoretical modeling of charge inversion using the classical self-consistent mean field theory. At low ionic strength, the overcompensated charge per unit area is proportional to the inverse screening length κ and the thickness of the adsorbed layer is of the order of the thickness of a single adsorbed chain. At higher ionic strength, the electrostatic interaction is strongly screened and is equivalent to an effective excluded volume. The overcompensated charge is then proportional to the bare surface charge. However, the theoretical approach for overcompensation of polyelectrolyte adsorption by self-consistent mean field theory has limitations since it assumes that the adsorbed polyelectrolyte layer has a thermodynamic equilibrium structure. Further detail of overcompensation will be discussed in the following section.

3.4
Polyelectrolyte Multilayer Formation on Flat Surfaces

The procedure for the build-up of polyelectrolyte multilayers on charged surfaces is extremely simple, as shown in Figure 3.4. If one starts from a positively charged substrate to grow a film, the first layer is deposited through polymer adsorption by dipping the substrate into a polyanion solution. Typically, a solid support with a charged surface is exposed to the solution of a polyion of opposite charge for a short time. The method is relatively rapid, as adsorption steps last typically between 1 min and 1 h. Under proper conditions, polymeric material with more than the stoichiometric number of charges (relative to the substrate) is adsorbed, so that the sign of the surface charge is reversed. In consequence, when the substrate is exposed to a second solution containing polycation, an additional polyelectrolyte layer is adsorbed. This again reverses the sign of the surface charge again. Consecutive cycles with alternating adsorption of polyanions and polycations result in the stepwise growth of polymer films. The samples are rinsed after each immersion in a polyelectrolyte solution to remove excess polymers that are not

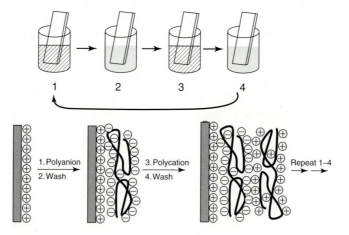

Figure 3.4 Schematic of the layer-by-layer deposition process. Steps 1 and 3 represent the adsorption of a polyanion and polycation respectively, and steps 2 and 4 are washing steps.

tightly bound. This ensures that no free polyanions interact in solution with other components that will come in contact with the film. Drying is sometimes applied after each rinsing step [81].

3.4.1
Mechanism

Although the basic idea of the polyelectrolyte multilayer is fairly simple, the theoretical description is quite complex because of the long range of the Coulombic interaction attracting the layers to each other. The simplest adsorption theory of polyelectrolytes may be called the *ion-exchange model* [82]. Due to entropic gain upon the adsorption, the adsorbed polyelectrolyte successfully competes with small ions and compensates the surface charges, as shown in Equation 3.4 Pol^- and Pol^+ are charged polymer segments, Pol^-Pol^+ is an ion pair, and M^+ and A^- are salt counterions. The subscript "m" refers to the multilayer surface [83].

$$Pol^-M^+_{(m)} + Pol^+A^-_{(aq)} \Longleftrightarrow Pol^-Pol^+_{(m)} + M^+_{(aq)} + A^-_{(aq)} \qquad (3.4)$$

If the salt ions are radiolabeled it is possible to observe, *in situ*, the displacement of these ions by polyelectrolytes. The ion-exchange model is based on thermodynamic equilibrium. Thus, the adsorbed amount of polyelectrolytes depends on the charge density of polyelectrolytes, surface charge density, ionic strength, and change of pH. All these factors can be explained by the model, and experimentally it has been qualitatively verified. However, there are some features that cannot be understood in the framework of the simple ion-exchange theory, such as irreversibility and overcompensation. Therefore, the model of *kinetically hindered equilibrium* has been proposed [84]. As mentioned in the previous section, adsorption often proceeds in two steps; initial transport-adsorption and reconformation. If reconformation is slower than adsorption, charge overcompensation is expected. During adsorption, polyelectrolyte chains anchor by only some of their charged groups to the surface, and these latter sites are occupied by other polyelectrolytes chains before they have time to reconform in order to occupy the neighboring charged sites of the surface. The attached polymer molecules progressively create a surplus charge that leads to an electrostatic barrier that prevents the attachment of other polyelectrolyte chains by repelling them. This phenomenon is self-regulating. The polyelectrolyte molecules adsorb only until they are electrostatically repelled from the surface, but not until the binding sites are saturated, which corresponds to "true equilibrium."

The first indication of irreversibility can be found during the buildup of polyelectrolyte multilayers; even though a deposition cycle terminates with a rinse with water. Typically multilayers remain unchanged (i.e. irreversibility upon dilution) but only unbounded polymers are removed. It has been shown that the thickness of multilayer films immersed for 10 days in pure water or exposed to water vapor for two months presents unchanged thickness [84]. Several additional experiments were designed to prove the extent of irreversibility. For example, a radiolabeled positive layer (14-labeled PM2VP), a polyvinylpyridine was adsorbed on top of a

10-bilayer-pair PSS/PDADMAC multilayer. After rinsing, the modified multilayers were immersed in 10 mM PDADMAC to allow for the exchange of labeled and unlabeled positive polyelectrolytes. No evidence of desorption was found on the time scale of hours, and after several weeks, only 30–40% had exchanged [85]. According to the model of kinetically hindered equilibrium, since it is far away from "true equilibrium," it is impossible to break the few bonds by simple washing [84]. This leads to the observed difficulty/impossibility in removal of adsorbed polyelectrolytes from a surface on polyelectrolyte dilution, decrease of salt concentration, or change of pH. Another fact that should be mentioned here, is that even though the electrostatic binding energy of polyelectrolyte–surface or polyelectrolyte–polyelectrolyte is very weak, about kT (k is the Boltzmann constant), the irreversibility is observed due to the additive effect of many ion pairs [84, 85].

The key to polyelectrolyte multilayer buildup is *overcompensation*. The surface charge should be inverted on each deposition step for the next immersion in oppositely charged polymers. Otherwise, one could not add as many polycation/polyanion layers as one may wish. The charge reversal during buildup can be monitored by measurements of the ζ-potential. ζ-potential values for the multilayer of poly(vinyl sulfonate) (PVS) with an ionene on a planar surface have shown the reversal in sign upon the alternating deposition of positively charged ionene and negatively charged PVS [85]. Electrophoretic measurements have been routinely conducted to determine the ζ-potential of multilayer-coated colloidal latex particles. Alternatively positive and negative potentials are observed, when polycation and polyanion formed the outmost layer, respectively [17, 17b, 86]. The LbL deposition of polyelectrolytes on colloidal particles will be discussed in detail in the following section. Schlenoff *et al.* [87] have used radiolabeled counterions to determine surface excess charge in PSS/PDADMAC multilayers deposited from 0.1 M NaCl. They found 45 and 46%, respectively, of charges within the top positive or negative layer remained uncompensated.

As mentioned in the previous section, the polyelectrolytes adsorbed at the liquid/solid interface form not only trains, which are in contact with the surface but also dangling loops and tails into solution since all the chains are flexible and hydrated. This behavior is suggested by the range of the bridging forces between two polycation monolayers observed in the surface force measurements. It has been found that mobile chains extended as far as 2.5–4 nm from each surface [84]. Also, the order of the multilayer perpendicular to the surface has been investigated with neutron and X-ray scattering methods. It has been shown that no short-range order exists between adjacent polycation/polyanion bilayers; there is no Bragg peak when every anion is deuterated [13]. Obviously, there is interpenetration that is the positively and negatively charged chains mingling. These loops and tails are responsible for the charge overcompensation that, in turn, determines the amount adsorbed in the next step. In other words, the growth is controlled by the loop size or the degree of surface charge overcompensation. Arys *et al.* have demonstrated that the thickness of xAyAxAy multilayers (where A denotes a polyanion, x and y denote polycations) was significantly different from the thickness of yAxAyAx multilayers although they both contain the same number of layers of A, x, and y [15].

The possibilities for loops and tails formation are different in the vicinity of a solid surface or in the loose outer layer of polyelectrolyte multilayers. Therefore, the first deposited layer on substrate is different from the last deposited layers due to the effect of the substrate. Since the substrate is harder and smoother than polyelectrolyte multilayers chain intermingling is not yet possible; therefore a different amount of polyelectrolyte is adsorbed. This often causes a nonlinear growth of the multilayer thickness for the first few layers versus the number of dipping cycles. Such initial nonlinear growth have been observed in many cases. A few layers need to be adsorbed until the thickness increase per polycation/polyanion pair is constant [88].

Schlenoff *et al.* have proposed a mechanism for distributing excess polymer surface to model the growth of multilayers of strong polyelectrolytes [87]. Two parameters are required for the semiempirical analysis: the surface, or unrestricted, charge overcompensation level, ϕ, which is assumed to decrease exponentially from the film surface to bulk, and the characteristic length for this decay, l_{cp}, which is termed the *charge penetration length*. The overall thickness of a film that includes a transition from nonlinear to linear growth is modeled as the sum of thickness increments, t_n,

$$t = \Sigma t_n = t_1 + \Sigma t_n^+ + \Sigma t_n^- \tag{3.5}$$

The anomalous first layer, thickness t_1, has been separated from the increments due to positive and negative layers. For positive and negative layers, respectively,

$$t_n^+ = 2l_{cp}\left(\frac{\phi - 1}{\phi}\right)f^+(1 - e^{-t/l_{cp}}) \tag{3.6}$$

$$t_n^- = 2l_{cp}\left(\frac{\phi - 1}{\phi}\right)f^-(1 - e^{-t/l_{cp}}) \tag{3.7}$$

where $f^+(f^-)$ is the bulk volume fraction of positive (negative) polymer. These equations were evaluated using an iterative computer program to fit experimental data of the film thicknesses for PDADMAC/PSS multilayers adsorbed from 0 or 0.1 NaCl aqueous solution. Modeling of data reveals that only modest levels of polymer charge overcompensation are required to account for large increments in polymer thickness, realized at high salt concentration, since the excess charge is distributed over several layers. Experimentally, ϕ appears to be roughly independent of salt concentration. The thickness increment is primarily controlled by l_{cp}, which is about 2.5 nominal layers for the system studied.

3.4.2
Structure of Polyelectrolyte Multilayers

The structure of multilayers obtained by LbL self-assembly deposition has been investigated by scattering techniques, in particular neutron and X-ray reflectometry [21, 92–94]. These methods are extremely useful for the structural characterization of thin films. The reflectivity spectra of a neutron or X-ray beam contain information on the scattering length density profile perpendicular to the surface as well as surface roughness. X-rays are well suited for *ex situ* studies of multilayer build-up and characterization of the film surface at the solid/air interface. This information can also be gathered from neutron reflectivity. The main advantage of neutrons over X-rays is the possibility of examining the internal film structure, buried interfaces, and films at solid/liquid interfaces upon isotopic labeling [95].

In X-ray reflectivity experiments on multilayer films composed of polyanion/polycation pairs, X-ray reflectograms have exhibited Kiessig fringes, which result from the interference of X-ray beams reflected at the substrate/film and film/air interface. Kiessig fringes can provide a good measure of the film thickness, and are very sensitive to the roughness of the interfaces. It has been shown that the polyelectrolyte multilayer films yield well-defined Kiessig fringes indicating good flatness (or smoothness) of the films [21, 89]. However, Bragg reflections corresponding to the internal layer structure of the film are usually not observed for multilayer films with $(AB)_n$ architecture (A is a polycation, B is a polyanion, and n is the number of deposition cycles). Absence of Bragg peaks can be caused by (i) the electron densities of two consecutive layers being too close to yield enough contrast, and (ii) the interpenetration of polymer segments of adjacent layers leading to a smearing of the electron density profile [96]. Bragg peaks are still not found in X-ray, and neutron reflectivity spectra even when contrast is enhanced by selectively deuterating every second layer $((AB_d)_n$, where B_d is a perdeuterated polyanion). However, when one starts to deuterate specific layer positions in a multilayer film, Bragg peaks are observed by neutron reflectometry [93]. It has been shown that for $((BA)_n(B_dA))$ architecture, the Bragg peaks started appearing when n is larger than 3, which clearly demonstrated the stratification of an internal layer [90]. Also, stratification was proven for multilayers containing more than two types of polyelectrolytes. $(ABCB)_n$, $((AB)_n(AC))_m$, $((AB)_n(AC)(AB)_m(AC))_p$ films have been assembled that give rise to Bragg peaks detected by X-ray and neutron reflectivity [91]. Thus, these experiments indicate that the absence of Bragg peaks for normal $(A/B)_n$ type films does not arise from small density differences between different layers, but rather from the interpenetration between neighboring layers. Schmitt *et al.* [92] obtained the roughness from three different interfaces for a 48-layer sample where one out of six layers was deuterated by neutron reflectivity. The substrate/film interface is smooth, $\sigma^{s/f} \sim 4$ Å, the film/air interface has a much higher roughness but is still almost molecularly smooth, $\sigma^{f/a} \sim 13$–15 Å, and the internal interfaces between the individual layers show the largest roughness, $\sigma^{int} \sim 19$ Å. They concluded that this may be attributed to chain–chain interdigitation between molecules from adjacent layers. The length scale of chain–chain interdigitation was calculated to be ~ 12 Å, which is less than the layer thickness 20–30 Å.

This type of interdigitated, yet stratificated structure of polyelectrolyte multilayer is called "*fuzzy layered assemblies,*" which was first named by Decher [13]. On the basis of these results from reflectivity measurements, together with the inability to detect significant amounts of small counterions ($\sim 7\%$), polyelectrolyte multilayers should have a 1 : 1 stoichiometry of anionic and cationic groups in the case of strong polyelectrolytes. Decher suggested the polyelectrolyte multilayer film model: each layer is represented by an arbitrarily chosen sinusoidal concentration profile. The

50% overlap of layers of equal charge has the consequence that at any point inside the film (the substrate/film and film/air interfaces are different), the sum of the concentration of equal ionic groups is unity in both the cationic and anionic case. This model agrees with the observation of a single Bragg peak for the architecture $(ABAB_d)_n$ where the deuterium labels are placed in every fourth layer.

Interpenetration can be reduced by using more rigid ionic blocks for the multilayer assembly. Arys *et al.* reported the investigation on the polyelectrolyte self-assemblies made from a lyotropic ionene and a strong polyelectrolyte. They showed that highly ordered polyelectrolyte films can be obtained, consisting of a regular lamellar nanostructure extending over considerable distances in the films, with preferential orientation of chain fragments occurring in the films [15].

Ladam *et al.* [93] have investigated the buildup of the first layers of PSS/PAH multilayers by streaming potential measurements (SPMs) and scanning angle reflectometry. They suggested the basic structure of a multilayer film as subdivided into three zones: a precursor zone (I), a core zone (II), and an outer zone (III). The first layers being deposited close to the substrate, they will possess a slight smaller thickness than that in zone II. In zone II film properties are constant. It is shown by SAR that a regular build-up regime, in which the thickness increment per layer is constant, can be reached after the deposition of the first six polyelectrolyte layers. This also gives an indication of the extension of zone I. When approaching the outer part of the multilayer, the local properties should again vary because of the solution environment. The outer region over which the local film properties are no longer similar to those in zone II constitutes the outer zone of the film, or zone III. When the film is fabricated, zone I is completed first. As more layers are added zones I and III will preserve their respective thickness, while zone II will grow in thickness. There is exact charge compensation in zone II, however, an excess charge is entirely located in the outer zone III. The layers in zone III should not be charge compensated and thus show classic polyion-like behavior. This means that they should swell in pure water and collapse in the salt solutions owing to the screening of the electrostatic repulsion between the like charges.

3.4.3
Controlling Factors

One special feature of these molecular films is that the macroscopic properties can be controlled by the microscopic structure. The thickness of the self-assembled films is in general increased by the addition of salt to the aqueous polyion dipping solutions [94]. As mentioned in the previous section, the chain configuration of polyelectrolytes in solution is drastically affected by the presence of salt. When no salt is added to the polyelectrolyte solution, the like charges within a single polyelectrolyte chain repel each other, such that the chain exists in an almost fully extended, rodlike configuration. In the presence of additional salt, counterions screen some of the charges allowing the polyelectrolyte chain to fold into a random coil configuration. If the chain retains its solution configuration when it adsorbs to the substrate, then the structural details of the multilayer film will be strongly influenced by the salt concentration. The dependence of total thickness of PAH/PSS multilayer films has been examined using neutron reflectivity by Lösche *et al.* [92]. They found that both the layer pair thickness and the internal layer roughness values scale fairly linearly between ionic strengths $I = 0.5$ and $3\,M$ NaCl. At low ionic concentration, an increase of the layer thickness with increasing I is no longer linear, but the thickness shows a square-root dependence. The thickness of 10 layer pairs of PSS/PDADMAC also shows almost linear dependence on salt concentration between 10^{-2} and $2\,M$ NaCl. Recently, McAloney *et al.* [97] investigated the salt effects on PSS/PDADMAC multilayer film morphology by atomic force microscopy (AFM). Ten-bilayer films that were deposited with less than $0.3\,M$ NaCl were flat and featureless, with similar characteristics to the underlying silica substrate. When the multilayers formed at and above this salt concentration, a wormlike or vermiculate morphology was observed. At low salt concentrations, in particular, when the Debye screen length is larger than the Bjerrum radius (the distance between charges where the electrostatic energy is equal to the thermal energy, $7.2\,\text{Å}$ in the case of monovalent ions in water at room temperature), the polyelectrolyte forms an extended conformation. When the ionic concentration increases such that the Debye screen length is smaller than the Bjerrum radius, the charges are screened by the counterions therefore the polyelectrolyte starts forming coils. In water, the transition will occur at $I = 0.2\,M$ at room temperature. This is consistent with AFM results, where a vermiculate morphology can be observed when $I \geq 0.3\,M$. Whereas an increase in thickness with increasing salt concentration occurs due to a change in chain conformation, it also results in decreased electrostatic attraction between film surface and polyelectrolyte. This counteracts the net increase in the film thickness due to the coiling effects. Therefore, one would expect a plateau or even a decrease in film thickness at a certain ionic strength. However, no plateau or decrease in thickness for PAH/PSS multilayers is observed up to the highest experimental salt concentration. In order for polyelectrolytes to accumulate at the surface with added salt, an additional chemical interaction of polymer with surface must be invoked; essentially, these are additional noncovalent interaction contributions (e.g., hydrophobicity, dipolar interactions).

The growth of polyelectrolyte multilayer thin films is also greatly influenced by the charge density of the polyelectrolytes. It has been shown that there is a critical charge density below which no multilayer growth is observed. Several studies have investigated the role of charge density on the multilayer growth of an alternately deposited strong polyelectrolyte (PSS) with copolymers of charged and uncharged monomer units [81, 98]. It has been found that above a critical charge density, the adsorbed amounts are nonmonotonic with charge density and show a maximum at intermediate values. However, with decreasing charge density more molecules adsorb at the surface to compensate the surface charge. Below a critical charge density, two situations can be distinguished: (i) at a very low polymer charge, the attraction between the polymer molecules is too weak to cause complexation at the surface, (ii) at a higher polymer charge, the complexation occurs but the individual copolymer chains are very weakly bound. In the subsequent adsorption steps, exposure to the oppositely charged polyelectrolyte results in removal of the weakly bounded copolymer chains. These probably form complexes with the oppositely charged polyelectrolyte in solution where desorption of the complex can also take place. When the enthalpic gain from electrostatic interactions is small, as in the case of weakly charged materials, the formation of complexes is favored by entropic considerations, that is, polyelectrolytes in complexes have more degrees of freedom than in a multilayer on a surface.

The critical charge density can be found at very low values when nonelectrostatic interactions such as H-bonding, hydrophobic, and charge transfer are present between polyelectrolytes. Schoeler *et al.* [99] have shown that the multilayer formation of PAA and a lowly charged copolymer that contained acrylamide is possible at charge densities of just 8% due to secondary interactions (H-bonding). Furthermore, it is even possible to form multilayers with uncharged polymers solely using H-bonding interactions. Nonetheless, it has been found that for the formation of multilayers by alternate adsorption of PSS and a statistical copolymer of diallydimethyl ammonium chloride (DADMAC) and N-methyl-N-vinylacetamide (NMVA) (where no secondary interaction exists) a minimum degree of charge density between 75 and 53% is needed for the formation of stable films [100].

The molecular organization, structure, and thickness of weak polyelectrolyte multilayers are extremely sensitive to pH of the dipping solution. This is primarily due to the fact that the linear charge density of a weak polyelectrolyte can vary with changes in pH when operating near the pK_a of the polymer [101]. Therefore, it is possible to systematically vary the linear charge density of weak polyelectrolytes, such as PAA and PAH, via simple adjustment of the pH of the dipping solution. For example, when PAA (pKa $\sim$ 5) is deposited at low pH (<4), the partially ionized PAA adsorbs in loop-rich conformations. In contrast, when PAA is deposited at high pH (>6), fully charged molecules form thin, flat layers. Furthermore, the pH of the dipping solution can affect not only the degree of ionization of an adsorbing polymer but also that of an already adsorbed polymer on the surface. This provides a rich but complex parameter space within which molecular organization can be manipulated. Shiratori and Rubner [102] showed that the thickness of an adsorbed polycation or polyanion layer can be varied from 5 to 80 Å simply by controlling the pH. Very thick layers were obtained when a fully charged chain was combined with a nearly fully charged chain of opposite charge. It has also been shown that composition of the layers, degree of layer interpenetration, and surface wettability can be controlled for the PAA/PAH system by adjusting the pH of solutions [103]. Furthermore, it has been reported that nanoporous films can be obtained from multilayers composed of PAA/PAH by pH-induced [103] or salt-induced [104] structural changes of preformed PAA/PAH multilayers.

3.4.4
Adsorption Kinetics

Understanding the adsorption kinetics of polyelectrolytes is a key for the optimization of LbL deposition process. Several techniques including quartz crystal microbalance (QCM) [105–107], surface plasmon spectroscopy (SPR) [108, 109], ellipsometry [110], UV-vis spectroscopy [111], and second-harmonic generation (SHG) [112] have been utilized for investigating adsorption kinetics. Typically, the alternate adsorption of polyions onto oppositely charged surfaces displays the pseudo-first-order kinetics, and is saturated within 10–20 min. Lvov *et al.* [113] reported the reciprocal of the pseudo-first-order rate constant, τ, for PSS and PAH which is about 2.1 ± 0.2 min, using *in-situ* QCM technique. Thus, the adsorption is essentially completed in 8–10 min since 3τ corresponds to 95 of saturation. Baba *et al.* [114] studied the adsorption kinetics for three kinds of polyelectrolyte combinations; PSS as polyanion and PDADMAC, PEI, or PAH as polycation. It was found that 70% saturation is achieved within the first 15 min and PSS has different τ values for each pair, roughly coinciding with the trend PDADMAC > PAH > PEI. Kurth and Osterhout [106] have reported the adsorption study of met-allosuprmolecular coordination-polyelectrolyte on negatively charged surfaced by SPS. It was found that adsorption kinetics shows two distinct regions. Adsorption is a diffusion-controlled transport process obeying a square root t dependence at the beginning of adsorption. Once a certain amount is adsorbed, adsorption can be described by a first-order rate law. The adsorption kinetics of a weakly charged polycation, PAH, and a polyanion containing an azobenzene chromophore, P-Azo, over the pH range of 3–11 was investigated using UV-vis spectroscopy and ellipsometry. Films assembled with the pH of the dipping solution near that of pK_a value of the polymer produced both the thickest films and displayed remarkably rapid adsorption isotherms (saturation achieved in 60 s) due to their low charge density. In some PAH/P-Azo films, a significantly large thickness was achieved in less than 5 s, which was more than 2 orders of magnitude faster than what is usually observed in the case of strongly charged polyelectrolyte [107].

3.5
Polyelectrolyte Multilayer Formation on Colloidal Particles

The LbL self-assembly technique has also been applied to fabricate core-shell type polyelectrolyte multilayer-coated colloidal particles. Oppositely charged polyelectrolytes can be consecutively adsorbed onto charged colloidal templates [17a–c]. Multilayer shells of polyelectrolytes [17, 17b], proteins [109], and a variety of nanoparticles [111] have been deposited onto particle templates, giving rise to novel colloidal entities. An interesting extension of these tailor-made core-shell particles has been the subsequent removal of the sacrificial core, resulting in hollow capsules of polymer, inorganic, or polymer–inorganic composites [115]. A variety of colloidal particles, melamine resin latexes [116], biological cells [117], and organic [118] and inorganic crystals [119] have already been used as templates for the capsule preparation. Figure 3.5 provides a scheme for the fabrication of hollow polyelectrolyte shells. The core-shell particles and hollow capsules are of increasing interest in various fields, including materials science, catalysis, pharmaceutics, and medicine.

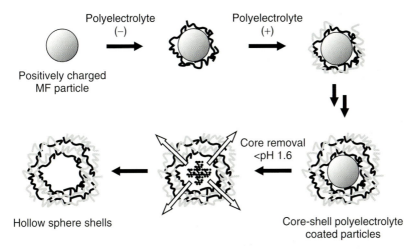

Positively charged
MF particle

Polyelectrolyte
(−)

Polyelectrolyte
(+)

Core removal
<pH 1.6

Hollow sphere shells

Core-shell polyelectrolyte
coated particles

Figure 3.5 Schematic illustration of the polyelectrolyte deposition process and of subsequent core decomposition. The initial steps involve stepwise film formation by repeated exposure of the colloids to polyelectrolytes of alternating charges. The excess polyelectrolyte is removed by cycles of centrifugation and washing before the next layer is deposited. After the desired number of polyelectrolyte layers is deposited, the coated particles are exposed to 100 mm HCl (pH < 1.6).

These novel three-dimensional hollow polymer shells combine the advantages of thin organic films of polyelectrolytes with the easy handling and applications of colloids. The most important novel features of the fabricated polyelectrolyte shells are (i) the thickness of the capsule wall depends on the number of assembled polyelectrolyte layers, which can be adjusted in the nanometer range, (ii) they can be fabricated with controlled physical and chemical properties, and (iii) they offer novel structures for micro- and nanocompartmentalization of materials [17]. In addition, the fabricated shells are readily permeable to small (circa 1–2 nm in diameter) polar molecules [120] and are extremely stable against chemical and physical influences. It is envisioned that different species can be incorporated into the shell structure giving them unique tailored properties, and making them attractive for a wide range of applications. For example, the incorporation of catalytic materials inside or onto the inner surface of the shells, together with the easy diffusion of reaction products and substrates, is a promising way to both control and increase the efficiency of catalytic processes. The hollow polyelectrolyte shells may also have potential as micro- and nanocarriers for molecules and nanoparticles, as well as biological species for the controlled release and targeting of drugs. They can also serve as cages for chemical reactions in restricted volumes. The control of the chemical composition of the inner surface (or likewise the outer surface) can be used to produce nucleation centers for subsequent crystal growth in constrained environments, which is also a novel way to produce nanocomposites [17, 121]. More recent reviews are mentioned in Table 3.1 and their cited references [19].

3.5.1
Formation of Core-Shell Particles

In order to fabricate colloidal core-multilayer shell particles, a colloidal solution (typically a few wt%) is exposed to an excess amount of an oppositely charged polyelectrolyte solution to cause saturation adsorption. The excess polyelectrolyte is then removed by centrifugation or filtration [122] with washing in water before adding the oppositely charged polyelectrolyte. The formation of a multilayer film can be verified using measurement of the electrophoretic mobility. Knowing the particle size, electric field, and solvent viscosity, the data can be converted into a zeta (ζ)-potential. The oppositely charged polyelectrolyte layer on colloidal particles results in reversal of the surface charge, showing a reversal in the sign of the ζ-potential. Subsequent consecutive adsorption of an oppositely charged polymer produces particles that exhibit alternating ζ-potential signs (+ or −). Such alternating values qualitatively demonstrate a successful recharging of the particle surface, and are characteristic of stepwise growth of multilayer films on colloids [17, 17b, 88].

The polymer coating can also be followed by single-particle light scattering (SPLS), a technique that enables determination of the thickness of adsorbed layers as well as the state of the coated colloid particles with respect to aggregation. The SPLS measurements typically yield the distribution of scattering intensity per particle, shifting to higher values depending on the number of coating processes. The average thickness can be calculated using the Rayleigh–Debye–Gans theory knowing the molar refractive index of the polymer [123]. For example, the thickness of a PAH/PSS layer pair formed in 0.5 M NaCl was found to be 2–3 nm [124b]. Recently, polyelectrolyte layers absorbed on colloidal particles have been investigated by means of small-angle neutron scattering (SANS). Unlike SPLS, where the thickness is calculated based on the refractive index of the layer that in turn is a function of the water content, the thickness of the polyelectrolyte coating on colloidal particles can be directly determined in an aqueous environment using SANS. It was found that a thickness of a system consisting of PSS and PAH is 16.6 Å per single layer [125]. Direct visualization of the morphology of the core-shell particles has been made using AFM, scanning electron microscopy (SEM), and transmission electron microscopy (TEM) and has conformed the regular assembly of polyelectrolytes on colloidal particles.

One of the most important features on the fabrication of core-shell particles via the LbL process is that various charged species can be deposited on the templates, for example, pure polymer or nanocomposite inorganic–organic multilayers can be assembled. For example, semiconductor nanocrystals such as thiol-capped CdTe and HgTe have been incorporated in multilayer thin films on latex particles [124]. Also, core-shell particles consisting of a PS latex colloidal core and Fe (II) metallosupramolecular polyelectrolyte (Fe(II)-MEPE)/PSS multilayer shells have been fabricated [126]. Furthermore, the LbL procedure can be extended to colloidal templates with diameters in the nanometer regime. Gittins and Caruso [127] have shown that sub-100-nm sized gold nanoparticles (between 15 and 50 nm in diameter) modified with an anionic alkanethiol can be coated with alternate layers of oppositely charged polyelectrolytes, PSS, and PDADMAC. In addition to using polymer latex particles as the colloidal template, biocolloidals with diameters in the micrometer range can also be successfully templated. The encapsulation of protein aggregates, ibuprofen (nonsteroidal anti-inflammatory drug) [128], and single living yeast cells [129] with the alternate layer of polyelectrolytes have been reported. More recent reviews on bio- and biomedical applications are mentioned in Table 3.1 and their cited references [19].

3.5.2
Production of Hollow Capsules

After the assembly of core-shell particles, the templated core can be removed either chemically or thermally. If the shell has sufficiently high permeability for removal of the decomposed core constituents, then the shell can be preserved and hollow shells obtained. One of the most common sacrificial cores is a weakly cross-linked melamine formaldehyde (MF) particle that exhibits a positively charged surface. The MF core can be decomposed by exposure of the coated particles to an acidic solution of pH $<$ 1.6 without destroying the coating [118]. Low molecular weight fragments then permeate through the wall leaving a hollow volume inside. For microsized particles, this process can be visualized by fluorescence microscopy and are shown to occur within 10 s. Erythrocytes obtained from human blood cells are also often used as cores and can be decomposed by deproteinizing (oxidizing) agents [130]. Basically, one can choose solvent for the removal of the core that decomposed the templated core but leaves the polymer films. For example, tetrahydrofuran (THF) is used for the removal of PS cores [131], and an acidic solution (pH $=$ 3) for $CdCO_3$ [132]. Inorganic capsules can by obtained by calcination of the core coated with nanoparticles/polymer multilayer. The calcination process removes the organic matter (the colloidal core and bridging polymer) during heating to 450 $^\circ$C. This also causes condensation of the silica nanoparticles, hence providing structural integrity for the hollow spheres [116]. Inorganic macroporous TiO_2 and TiO_2/SiO_2 materials have been prepared using assemblies of close-packed coated colloidal spheres as templates, infiltrating titanium (IV) isopropoxide into the close-packed spheres and thereafter removing the colloidal cores and polyelectrolytes by calcination [133].

One promising feature of these capsules is that by varying the composition and thickness of the walls, their properties can be tuned over a wide range. Interestingly, the multilayers composed of polyanion, PSS with two different combinations of polycations, PAH, or PDADMAC, exhibit very different properties such as elasticity, mechanical stability, and swelling and shrinking upon annealing and exposure to electrolyte due to the different chemical nature of PAH and PDADMAC.

The elasticity and mechanical stability of the capsules are definitely important for a variety of practical applications. For example, deformation and rupture of polyelectrolyte multilayer capsules under shear stress may apparently limit the use of these novel structures [134]. Stability and elasticity of hollow capsules have been investigated by the osmotic pressure method. The polyelectrolyte multilayer capsule is permeable to water and small molecules but impermeable to polymers. Therefore, when the capsules are suspended into a solution of PSS, intact capsules deform as a result of the osmotic pressure difference created. The number of deformed capsules as a function of the PSS concentration was counted to obtain the elasticity modulus of multilayers using confocal laser scanning microscopy (CLSM). The elasticity modulus of PSS/PAH multilayers was found to be 500–750 MPa, depending on the molecular weight of PAH [135]. The multilayer formed by means of electrostatic interactions of strong polyelectrolyte resembles a polymer glass rather than a rubber material. On the other hand, the elasticity modulus of PSS/PDADMAC multilayers was 136 MPa, which is considerably smaller than that of the PSS/PAH multilayer [136]. The apparent difference of the stability and elasticity between the PSS/PAH capsules and PSS/PDADMAC capsules can be explained as a result of the different chemical natures of PAH and PDADMAC, which is responsible for an overall weaker interaction between PDADMAC and PSS compared with the PAH/PSS interaction.

Heating-induced morphological changes of the hollow spheres composed of PSS/PAH were investigated by CLSM and scanning force microscopy (SFM). During heating of the capsule solution at 70 °C for 2 h, the diameter of the capsules is reduced (~10%) accompanied by a parallel layer thickness increase [135]. It was suggested that higher temperature facilitates the transient breaking of electrostatic bonds between oppositely charged polyelectrolytes and provides the energy necessary for overcoming the electrostatic attraction. As a consequence, the polymer molecules become capable of changing their largely two-dimensional arrangement corresponding to a low-entropy state toward a more coiled arrangement. The shrinking process will stop when the polyion molecules have reached a coiled conformation, thus the system keeps a hollow capsule structure. Interestingly, such a behavior is in opposition to that of the hollow capsule composed of PSS/PDADMAC. When the capsules were incubated in H_2O at 40 °C for 2 h the capsule size increased by 40%. In parallel with the capsule volume increase, the capsule wall become thinner [137]. On the other hand, pronounced diameter reduction was found when PSS/PDADMAC capsules were exposed to a concentrated electrolyte solution, that is, 0.5 N NaCl at room temperature [138]. This reduction in volume was supposedly mainly caused by the compression of the capsule wall due to the ionic screening from the electrolyte. The highly porous microstructure of the multilayers and loosely bound PSS/PDADMAC complex is thought to be responsible for the structure of the PSS/PDADMAC capsules being easily modulated upon annealing and salt exposure. The structure of the PSS/PDADMAC capsule wall can be assumed to be a highly coiling layer, in which irregular pores are distributed through the entire multilayer film. The multilayer is held together by randomly distributed electrostatic bonds. A number of fixed charges are not able to form ion pairs likely because of topological restrictions brought about by the two bulks and stiff polymers. The elevated temperature generally facilitates the thermal motion of the polymer segments, and the probability for bond breaking increases. During the annealing process, the unbounded segments should move in the direction of increased entropy. This would correspond to a more three-dimensional structure of the layer constituents resulting in an increase in the layer thickness together with a decrease in the diameter of the capsule.

One of the significant properties of these capsules is their semipermeability. It has been shown that they are permeable for small molecules such as dyes and ions, while they exclude compounds with a higher molecular weight [123]. The permeability of the capsules for a fluorescently labeled polymer and fluorescent small dye was studied by means of CLSM. No fluorescence was observed in the capsule interior when the polymer was added. However, when a small dye was added into the microcapsule solution, immediate (less than 1 min) recovering of fluorescence was observed after bleaching. The impermeability for large molecules can be used to create pH gradients across the capsule wall that can be used to perform specific chemistry inside. Sukhorukov *et al.* [139] have investigated this pH difference between the bulk and internal solution by adding a polymeric acid (polystyrene sulfonic acid). When a polymeric acid is dissolved outside, it dissociates protons that only partly penetrate into the interior, because the negatively charged polymer chain remains outside. This results in a Donnan equilibrium between the inside and outside of the capsule wall. It was found that the pH in the capsule interior is more basic. The difference in pH is larger than a unit for capsules of 3.3 μm in diameter.

The possibility to change and control the capsule permeability would be desirable for many applications, such as medicine, biotechnology, catalysis, cosmetics, and nutrition, where selective encapsulation and release is required. Recently, several studies have been devoted to PSS/PAH microcapsules in an effort to control their permeability properties. Antipov and coworkers [138] were able to obtain sustained release of fluorescent materials by adjusting the number of PAH/PSS multilayers. At least 8–10 layers of polyelectrolytes were needed to sustain fluorescein release. Also, increasing the number of layers prolonged the release for several minutes. Later, it was found that these PAH/PSS microcapsules are closed or open to high molecular weight polymers depending on the pH of the solution. The capsules are closed at a pH value of 8 or higher, but open at a pH lower than 6. They suggested that this is attributed to the charging effects on their walls creating local defects [134].

3.6
Applications of Layer-by-Layer Films and Particles

Due to its simplicity and versatility, the LbL deposition technique has led to numerous investigations on possible applications of the multilayer films and particle coatings. More recent reviews on their applications are mentioned in Table 3.1 and their cited references [19]. Multilayer thin films of conjugated polymer, semiconductor nanoparticles, and dyes have found applications in electronic and photonic devices. Light-emitting diodes (LEDs) have been fabricated from multilayer thin films of polyions with conjugated polymers (e.g. PPV precursor [24], PPP [25]) or semiconductor nanoparticles (thiol-capped CdSe [140], CdTe [141]). Recently, it has been demonstrated that field effect transistors (FETs) can be made with multi-layer of cationic and anionic phthalocyanine derivatives [142]. Also, a photovoltaic device, a dye-sensitized solar cell, has been made by LbL assembly of TiO_2 and a polyelectrolyte [143]. Multilayers of azo-polymers and small dye molecules have applications in optical storages [144], surface-relief gratings [145], and nonlinear optical (NLO) devices [146]. Most proteins are water-soluble and amphoteric, thus electrostatic adsorption is quite useful for the construction of various protein organization and hierarchical assemblies. Several applications for such films have been reported such as biosensors [147] and enzyme immobilization [148]. The insertion of inorganic nanoparticles or platelets with defined chemical and physical properties at precise locations in a multilayer is of interest for fundamental research, as well as for a number of potential applications. Clay platelets (montmorillonite [57] and hectorite [58]) have been incorporated into inorganic–organic composites. Hybrid polymer–clay assemblies have been fabricated and were found to have unique mechanical, electrical, and gas-permeation properties. As a result, they have been used in gas-permeation membranes [57a]. The Rubner group reported the PAH/PAA multilayers can be utilized as nanoreactor for both metallic (Ag) and semiconductor nanoparticles (PbS) [149]. They took advantage of the weak polyelectrolyte, PAA, in which free carboxylic acid binding groups can be tuned by adjusting the pH of dipping solution. Furthermore, metallodielectric photonic structures were fabricated by alternating deposition of a block of PAH/PSS and a block of PAH/PAA where Ag was selectively synthesized [150]. Other potential applications are separation membranes [151], chromatography columns [152], high charge density batteries [153], and antireflection surfaces [154].

Polyelectrolyte hollow capsules have potential applications in drug delivery and microreactors. It has been shown that small drug molecules or proteins can be loaded or encapsulated in the hollow capsules and then can be released under proper conditions (pH, salt, and oxidation) [132, 134, 155].

3.7
Advincula Group Research on Layer-by-Layer Films and Particle Coatings

The Advincula Research Group at the University of Houston has had a number of projects and applications related to the LbL approach for the last 10 years [156]. Like many other groups, the attraction to the technique is that of the simplicity of the process and the amenability toward nanostructuring in materials. This has resulted in the use of a variety of polymer and nanomaterials including the application of sol-gel coatings (not reviewed in this chapter). An important aspect of the LbL technique is that one can apply innovative surface-sensitive spectroscopic, microscopic, and other field-dependent analytical techniques. This enables the discovery of phenomena of fundamental properties that is correlated with versatility and control in the build-up of such layers. Therefore, meaningful structure–property relationships can be determined. Lastly, the ability to demonstrate the concept to a beginning student (from high school) up to a seasoned researcher (post-doctoral) is one of the best attractions for engaging in this field.

One of the earliest studies involved the use of the QCM technique or QCM to monitor the stoichiometric and nonstoichiometric adsorption in polyelectrolytes [156a]. It was found that the nonlinearity of the deposition is related to some of the desorption phenomena but otherwise, most adsorptions followed a Langmuir isotherm, concentration-dependent process. To demonstrate the formation of hybrid materials, ultrathin film self-assembly of organic–inorganic zirconium metal coordination polymers was investigated and the deposition monitored by surface plasmon resonance spectroscopy [156b]. Many charged small molecule dyes can be deposited by LbL. It was important to demonstrated the order, aggregation, and orientation of low molecular weight azobenzene dyes and polycations by alternate multilayer films [156c]. As it turned out, these types of systems have relevance for photoalignment and applications in liquid-crystalline materials displays and surface-relief gratings (Figure 3.6) [156f,g,i,k,l,m,n,o,q,r,v].

In collaboration with the Caruso group, the first demonstration of cross-linked, luminescent spherical colloidal and hollow-shell particles was demonstrated [156d]. This has enabled chemistry to be done locally on the shell of the particle by converting it to visible fluorescent species from the corresponding polyfluorene

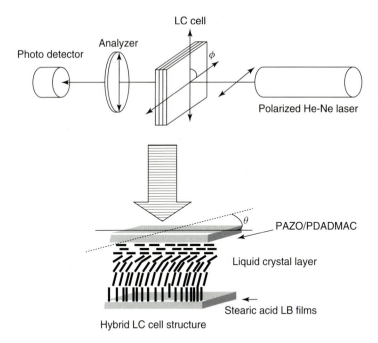

Figure 3.6 Experimental setup for measurement of the photoinduced birefringence of the cell. Hybrid LC cell structure and actinic light exposure conditions at defined angles. (The *cell axis* is defined as the direction of the longer sides of the rectangular cell, and the polarization plane angle (θ) contained by the cell axis and the polarization plane of actinic light.) Ref. [156i] (with permission Am. Chem. Soc.)

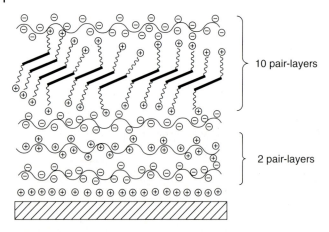

Figure 3.7 Schematic view of the sexithiophene/polystyrene sulfonated (6TN/PSS) alternating multilayer buildup on solid support substrates. Ref. [156e] (with permission Am. Chem. Soc.)

precursors. An important goal in the field of organic field effect transistor (OFET) devices is to utilize processable oligothiophene derivatives. The group was one of the first to report on nanostructured ultrathin films of water-soluble sexithiophene bolaform amphiphiles prepared using LbL self-assembly (Figure 3.7) [156e]. This included the intercalation of clay platelets in the nanostructuring of sexithiophenes [156h].

SPR spectroscopy has been used as an effective tool for investigating the LbL deposition process both by *in-situ* and *ex-situ* methods (Figure 3.8) [156i]. An important development is the use of a combined electrochemical-SPR or electrochemical surface plasmon resonance (EC-SPR) set-up to also monitor the changes in the redox properties *in situ* with changes in the optical properties. Another important tool was the use nanoindentation measurements to characterize the nanomechanical properties of LbL polymer–clay nanocomposites [156p]. This enabled the measurement of hardness and modulus for these materials showing that intercalated LbL properties of clay and polyelectrolyte films are closer to copper films. An interesting development was the report of an ambipolar OFET behavior in phthalocyanine dyes fabricated using the LbL approach including gas-sensing properties [156s,x].

The first report on the preparation of electrostatically adsorbed clay platelets and surface initiated polymerization of adsorbed initiators on planar films was reported in 2003 [156t]. In this case, the clay platelets anchored the cationic free-radical initiators demonstrating the possibility of monitoring the surface-initiated polymerization (SIP) on nanoparticle surfaces directly. The monitoring of electroactivity of polyaniline multilayer films in neutral solution and the electrocatalyzed oxidation of beta-nicotinamide adenine dinucleotide was reported utilizing both SPR and QCM [156u]. The self-assembly and characterization of polyaniline and sulfonated polystyrene multilayer-coated colloidal particles and hollow shells was first reported

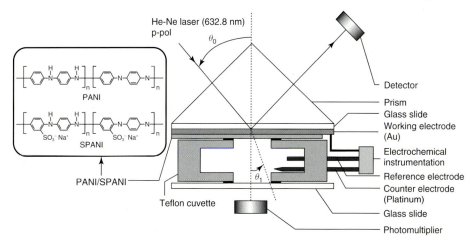

Figure 3.8 SPR setup for the excitation of surface plasmons in the Kretschmann geometry (a thin metal film is evaporated onto the base of a glass prism and acts as a resonator driven by the photon field) and structural formulas of the employed PANI and SPANI. Ref. [156i] (with permission Am. Chem. Soc.)

in 2003 [156w]. In this case, a high oxidative stability on the conducting polymer was observed. The formation of nanocomposite hydrogen-bonded multilayer ultrathin films facilitated the simultaneous sexithiophene and Au nanoparticle formation through a complementary redox reaction [156y]. This unusual coupling proved that redox chemistry inside the LbL firm can drive the formation of dendronic shaped nanoparticles. The formation of a pH-sensitive bipolar ion-permselective ultrathin film was reported based on the use of photocross-linking benzophenone modified polyelectrolytes and their cross-linking [156z]. Effectively, the same benzophenone system was used to demonstrate sustained release control via photocross-linking of polyelectrolyte LbL hollow capsules (Figure 3.9) [155aa].

In the area of biomaterials, the use of the surface sol-gel nanostructuring resulted in very good osteoblast adhesion and matrix mineralization on sol-gel-derived titanium oxide [156ab,af]. This is an important step toward control of bone growth in bioimplant devices. An important development in the area of nanostructured layers is their use for electronanopatterning. The first report on phthalocyanine and sexithiophene dyes made use of their LbL films to demonstrate possible applications as memory devices [156ac]. The same azobenzene dyes and polyelectrolytes used in previous publications, have also been utilized for the pattern formation of subvisible wavelength surface-relief gratings using current-sensing atomic force microscopy (CS-AFM) (Figure 3.10) [156ad]. This was primarily driven by the thermal backrelaxation of the dyes in the presence of joule heating. Fuzzy ternary particle systems by surface-initiated atom-transfer radical polymerization (ATRP) from LbL colloidal core-shell macroinitiator particles was demonstrated [156ae], including polymerization from flat surface substrates and demonstration

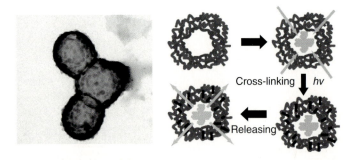

Figure 3.9 Shown in this figure is the formation (TEM images) and permeability of polyelectrolyte multilayer hollow-shell capsules by photocross-linking and controlled-release (fluorescence) studies. The hollow shells were prepared by alternate layer-by-layer (LbL) adsorption of photocross-linkable benzophenone-modified poly(allylamine hydrochloride) and poly(sodium 4-styrenesulfonate). A model drug, rhodamine B (RB), was successfully loaded into the polyelectrolyte hollow capsules. The release kinetics of RB was investigated using fluorescence spectroscopy. The permeability of RB through the hollow shells was effectively controlled based on UV irradiation time or degree of cross-linking. Ref. [156aa] (with permission Am. Chem. Soc.)

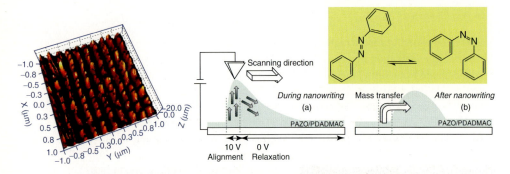

Figure 3.10 AFM image of the formation of subvisible wavelength surface-relief gratings using current-sensing atomic force microscopy on LbL films of azobenzene containing polyelectrolytes. Scheme shows mass transport as influenced by joule heating of the AFM tip and the corresponding thermal backrelaxation of the azobenzene dyes. Ref. [156ad] (with permission Am. Chem. Soc.)

of hydrogel effects (Figure 3.11) [156ak]. These papers first demonstrated the possibility of combining the LbL approach with SIP to form polymer brushes using polyelectrolyte macroinitiators.

Most recently, the LbL technique has been effectively utilized for the electronanopatterning of electrochemically active and nanostructured polyethylenedioxythiophene (PEDOT)/PSS, again demonstrating possible applications for memory devices [156ag]. The PEDOT/PSS is a popular conducting polymer material not only for its commercial availability but also its good electron–hole-transporting and electrochromic properties. This technique holds promise toward controlling these properties at the nanoscale level. The reactivity or electropolymerizability

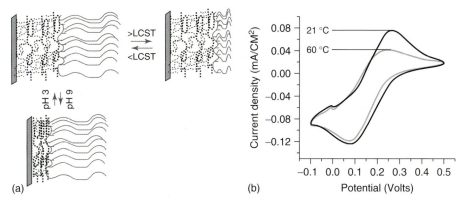

Figure 3.11 Substrate modification using the combined LbL and surface-initiated polymerization (SIP) methods showing: (a) pH and temperature dependence on swelling and contraction. (b) Also shown in the molecular probe ion transport response on two temperatures based on cyclic voltammetry. Ref. [156ak] (with permission Am. Chem. Soc.)

of precursor monomers bearing electroactive groups have been demonstrated in the electrocopolymerization of LbL-deposited polythiophene and polycarbazole precursor ultrathin films [156ah], including LB films of carbazole functionalized poly(p-phenylenes) [156ai]. This precursor-polymer approach enables the use of electrochemistry to control the degree of cross-linking in conjugated polymers. Signal enhancement and tuning of surface plasmon resonance was demonstrated in Au nanoparticle/polyelectrolyte LbL ultrathin films [156aj]. This demonstrated that it is possible to exert further control on the local surface plasmon field enhancement by adding other plasmonic structures on the dielectric layer. Nanostructured ultrathin

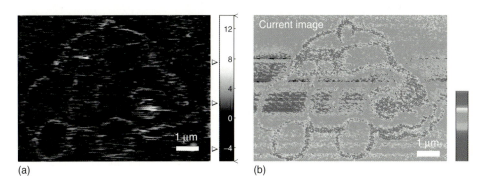

Figure 3.12 CS-AFM nanopatterning of a "nanocar" on a 10-bilayer P4VPCBZ/PAA film at 10 V with a writing speed of 0.8 μm/s (a) topographic image and (b) current image. The current image was obtained by scanning at 1 V after the patterning. Color bar range is 0–14.7 pA. Ref. [156al] (with Permission Am. Chem. Soc.)

films of alternating sexithiophenes and electropolymerizable polycarbazole precursor layers were investigated by electrochemical surface plasmon resonance (EC-SPR) spectroscopy, as in previous studies [156am]. However, what is interesting is that the electrochemical cross-linking can be monitored more quantitatively. The ability to combine electrochemical cross-linking and patterning of nanostructured polyelectrolyte–carbazole precursor ultrathin films garnered a cover page with the journal, *Macromolecules* (Figure 3.12) [156al]. This is of high interest because controlled thickness, layer order, polymer architecture, and type of electropolymerizable monomer can be coupled with the parameters for electronanopatterning

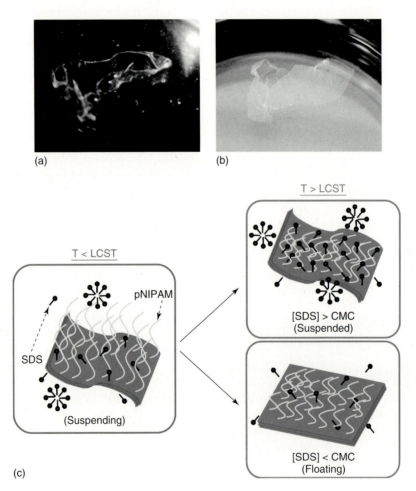

Figure 3.13 (a,b) Hydrodynamic effect of the anionic surfactant SDS on the pNIPAM nanosheet. The pNIPAM nanosheet (1 cm × 2 cm) (a) was suspended in water at 25 °C and (b) floated up to the air/water interface at 40 °C. (c) Scheme explaining the effect of anionic surfactants. Ref. [156an] (with Permission Am. Chem. Soc.)

using the CS-AFM technique. Lastly and recently, it was demonstrated that the hydrodynamic transformation of a freestanding polymer nanosheet formed by LbL can be induced by a thermoresponsive surface (Figure 3.13) [156an]. The technique holds promise for the preparation of bioadhesives and drug-delivery systems through nanosheets that can be fabricated and incorporated in medical devices or therapeutic systems.

Acknowledgments

The authors acknowledge the contribution of the past and present members of the Advincula Research Group. The authors would like to acknowledge funding from NSF DMR-0602896, DMR-1006776, CBET-0854979, CHE 10-41300, and the Robert A Welch Foundation, E-1551 and Texas NHARP 01896. We would also like to thank KSV Instruments (Biolin), Viscotek (Malvern Instruments), Agilent Technologies, and Optrel for their technical support.

References

1. (a) Ariga, K. (2009) *Organized Organic Ultrathin Films: Fundamentals and Applications*, John Wiley & Sons, Inc.; (b) Frank, C. (1998) *Organic Thin Films: Structure and Applications*, ACS Symposium Series 695, American Chemical Society; (c) Tsukruk, V. (1997) *Prog. Polym. Sci.*, **22**(2) 247–311.

2. Berlin, A. and Zotti, G. (2000) *Macromol. Rapid Commun.*, **21**, 301.

3. Bertrand, P., Jonas, A., Laschewsky, A., and Legras, R. (2000) *Macromol. Rapid Commun.*, **21**, 319.

4. Fendler, J.H. (1996) *Chem. Mater.*, **8**, 1616.

5. Langmuir, I. (1920) *Trans. Faraday Soc.*, **15**, 62.

6. Blodgett, K.B. (1935) *J. Am. Chem. Soc.*, **57**, 1007.

7. Kuhn, H. and Möbius, D. (1971) *Angew. Chem. Int. Ed. Eng.*, **10**, 620.

8. Ulman, A. (1991) *An Introduction to Ultrathin Organic Films from Langmuir–Blodgett to Self-Assembly*, Academic Press, London.

9. Netzer, L. and Sagiv, J. (1983) *J. Am. Chem. Soc.*, **105**, 674.

10. Tillman, N., Ulman, A., and Penner, T.L. (1989) *Langmuir*, **5**, 101.

11. Ulman, A. (1996) *Chem. Rev.*, **96**, 1533.

12. Lee, H., Kepley, L.J., Hong, H.-G., and Mallouk, T.E. (1988) *J. Am. Chem. Soc.*, **110**, 618.

13. Decher, G. (1997) *Science*, **277**, 1232.

14. (a) Decher, G. and Hong, J.-D. (1991) *Macromol. Chem. Macromol. Symp.*, **46**, 321; (b) Decher, G. and Hong, J.-D. (1991) *Ber. Bunsen-Ges. Phys. Chem.*, **95**, 1430; (c) Decher, G., Hong, J.-D., and Schmitt, J. (1992) *Thin Solid Films*, **210/211**, 831.

15. Arys, X., Jonas, A.M., Laschewsky, A., and Legras, R. (2000) Supramolecular polyelectrolyte assemblies, in *Supramolecular Polymers* (ed. A. Ciferri), Marcel Dekker, New York pp. 505–564.

16. (a) Caruso, F., Donath, E., and Möhwald, H. (1998) *J. Phys. Chem. B*, **102**, 2011; (b) Caruso, F., Lichtenfeld, H., Giersig, M., and Möhwald, H. (1998) *J. Am. Chem. Soc.*, **120**, 8523; (c) Donath, E., Sukhorukov, G.B., Caruso, F., Davis, S.A., and Möhwald, H. (1998) *Angew. Chem. Int. Ed. Engl.*, **37**, 2202.

17. (a) Mandal, S., Lee, M., Hill, J., Vinu, A., and Ariga, K. (2010) *J. Nanosci. Nanotechnol.*, **10**(1), 21–33; (b) Boudou, T., Crouzier, T., Ren, K., Blin, G., and Picart, C. (2010)

Adv. Mater., **22**(4), 441–467; (c) del Mercato, L., Rivera-Gil, P., Abbasi, A., Ochs, M., Ganas, C., Zins, I., Sonnichsen, C., and Parak, W. (2010) *Nanoscale*, **2**(4), 458–467; (d) Podsiadlo, P., Shim, B., and Kotov, N. (2009) *Coord. Chem. Rev.*, **253**, 2835–2851; (e) Ozin, G., Hou, K., Lotsch, B., Cademartiri, L., Puzzo, D., Scotognella, F., Ghadimi, A., and Thomson, J. (2009) *Mater. To-day*, **12**(5), 12–23; (f) De Geest, B., Sukhorukov, G., and Mohwald, H. (2009) *Expert Opin. Drug Deliv.*, **6**(6), 613–624; (g) He, Q., Cui, Y., Ai, S., Tian, Y., and Li, J. (2009) *Curr. Opin. Colloid Interface Sci.*, **14**(2), 115–125; (h) Lichter, J., Van Vlie, K., and Rubner, M. (2009) *Macromolecules*, **42**(22), 8573–8586; (i) Matsusaki, M. and Akashi, M. (2009) *Expert Opin. Drug Deliv.*, **6**(11), 1207–1217; (j) Kozlovskaya, V., Kharlampieva, E., Erel, I., and Sukhishvili, S. (2009) *Soft Matter*, **5**(21), 4077–4087; (k) Tokarev, I., Motornov, M., and Minko, S. (2009) *J. Mater. Chem.*, **19**(38), 6932–6948; (l) Song, S., Liu, H., Guo, X., and Hu, N. (2009) *Electrochim. Acta*, **54**(24), 5851–5857; (m) Mansouri, S., Winnik, F., and Tabrizian, M. (2009) *Expert. Opin. Drug. Deliv.*, **6**(6), 585–597; (n) Tomczak, N., Janczewski, D., Han, M., and Vancso, G. (2009) *Prog. Polym. Sci.*, **34**(5), 393–430; (o) Ariga, K., Ji, Q., Hill, J., and Vinu, A. (2009) *Soft Matter*, **5**(19), 3562–3571; (p) Wong, J. and Richtering, W. (2008) *Curr. Opin. Colloid Interface Sci.*, **13**(6), 403–412; (q) Srivastava, S. and Kotov, N. (2008) *Acc. Chem. Res.*, **41**(12), 1831–1841; (r) Zotti, G., Vercelli, B., and Berlin, A. (2008) *Acc. Chem. Res.*, **41**(9), 1098–1109; (s) Wang, Y., Angelatos, A., and Caruso, F. (2008) *Chem. Mater.*, **20**(3), 848–858; (t) Ariga, K., Hill, J., and Ji, Q. (2008) *Macromol. Biosci.*, **8**(11), 981–990; (u) Maury, P., Reinhoudt, D., and Huskens, J. (2008) *Curr. Opin. Colloid Interface Sci.*, **13**, 74–80; (v) Picart, C. (2008) *Curr. Med. Chem.*, **15**(7), 685–697; (w) Quinn, J., Johnston, A., Such, G., Zelikin, A., and Caruso, F. (2007) *Chem. Soc. Rev.*,

36(5), 707–718; (x) Lutkenhaus, J. and Hammond, P. (2007) *Soft Matter*, **3**(7), 804–816; (y) Ariga, K., Hill, J., and Ji, Q. (2007) *Phys. Chem. Chem. Phys*, **9**(19), 2319–2340; (z) Jaber, J. and Schlenoff, J. (2006) *Curr. Opin. Colloid Interface Sci.*, **11**(6), 324–329.

18. Decher, G. and Schlenoff, J. (2003) *Multilayer Thin Films*, Wiley-VCH.

19. Iler, R.K. (1966) *J. Colloid Interface Sci.*, **21**, 569.

20. Lenahan, K.M., Wang, Y.X., Liu, Y., Claus, R.O., Heflin, J.R., Marciu, D., and Figura, C. (1998) *Adv. Mater*, **10**, 853.

21. Hammond, P.T. (2000) *Curr. Opin. Colloid Interface Sci.*, **4**, 430.

22. Ho, P.K.H., Granstrom, M., Friend, R.H., and Greenham, N.C. (1998) *Adv. Mater.*, **10**, 769.

23. Baur, J.W., Kim, S., Balanda, P.B., Reynolds, J.R., and Rubner, M.F. (1998) *Adv. Mater.*, **10**, 1452.

24. (a) Ram, M.K., Salerno, M., Adami, M., Faraci, P., and Nicolini, C. (1999) *Langmuir*, **15**, 1252; (b) Cheung, J.H., Stockton, W.B., and Rubner, M.F. (1997) *Macromolecules*, **30**, 2712.

25. Onoda, M. and Yoshino, K. (1995) *Jpn. J. Appl. Phys.*, **34**, L260.

26. (a) Lukkari, J., Salomaki, M., Viinikanoja, A., Aaritalo, T., Paukkunen, J., Kocharova, N., and Kankare, J. (2001) *J. Am. Chem. Soc.*, **123**, 6083; (b) Zotti, G., Schiavon, G., Zecchin, S., Berlin, A., and Giro, G. (2001) *Synth. Met.*, **121**, 1381.

27. Watanabe, S. and Regen, S.L. (1994) *J. Am. Chem. Soc.*, **116**, 8855.

28. (a) Gao, M., Kong, X., Zhang, X., and Shen, J. (1994) *Thin Solid Films*, **244**, 815; (b) Cochin, D., Paßmann, M., Wilbert, G., Zentel, R., Wischerhoff, E., and Laschewsky, A. (1997) *Macromolecules*, **30**, 4775.

29. Sun, J., Wu, T., Liu, F., Wang, Z., Zhang, X., and Shen, J. (2000) *Langmuir*, **16**, 4620.

30. (a) Yang, K., Balasubramanian, S., Wang, X., Kumar, J., and Tripathy, S. (1998) *Appl. Phys. Lett.*, **73**, 3345; (b) Dante, S., Advincula, R., Frank, C.W., and Stroeve, P. (1999) *Langmuir*, **15**, 193.

31. (a) Caruso, F., Furlong, D.N., Ariga, K., Ichinose, I., and Kunitake, T. (1998) *Langmuir*, **14**, 4559; (b) Ma, H., Hu, N., and Rusling, J.F. (2000) *Langmuir*, **16**, 4969; (c) Cassier, T., Lowack, K., and Decher, G. (1998) *Supramol. Sci.*, **5**(3–4), 309–315.

32. Lvov, Y., Haas, H., Decher, G., Möhwald, H., Michailov, A., Mtchedlishvily, B., Morgunova, E., and Vainshtain, B. (1994) *Langmuir*, **10**, 4232.

33. Wang, L.Y., Schonhoff, M., and Möhwald, H. (2002) *J. Phys. Chem. B*, **106**, 9135.

34. Ball, V., Winterhalter, M., Schwinte, P., Lavalle, P., Voegel, J.C., and Schaaf, P. (2002) *J. Phys. Chem. B*, **106**, 2357.

35. (a) Lang, J. and Liu, M. (1999) *J. Phys. Chem. B*, **103**, 11393; (b) Sukhorukov, G.B., Montrel, M.M., Petrov, A.I., Shabarchina, L.I., and Sukhorukov, B. (1996) *Biosens. Bioelectron.*, **11**, 913; (c) Pei, R., Cui, X., Yang, X., and Wang, E. (2001) *Biomacromolecules*, **2**, 463.

36. (a) Rmaile, H.H. and Schlenoff, J.B. (2003) *J. Am. Chem. Soc.*, **125**, 6602; (b) Cooper, T.M., Campbell, A.L., and Crane, R.L. (1995) *Langmuir*, **11**, 2713.

37. Onta, M., Lvov, Y., Ariga, K., and Kunitake, T. (1996) *Biotechnol. Bioeng.*, **51**, 163.

38. Anzai, J., Kobayashi, Y., Suzuki, Y., Takeshita, H., Chen, Q., Osa, T., Hoshi, T., Du, and X. (1998) *Sens. Actuators, B*, **52**, 3.

39. He, J.A., Samuelson, L., Li, L., Kumar, J., and Tripathy, S.K. (1998) *Langmuir*, **14**, 1674.

40. Lvov, Y., Onda, M., Ariga, K., and Kunitake, T. (1998) *J. Biomater. Sci. Polym. Ed.*, **9**, 345.

41. (a) Voigt, A., Lichtenfeld, H., Sukhorukov, G.B., Zastrow, H., Donath, E., Baumler, H., and Möhwald, H. (1999) *Ind. Eng. Chem. Res.*, **38**, 4037; (b) Saremi, F. and Tieke, B. (1998) *Adv. Mater.*, **10**, 388.

42. Decher, G., Lehr, B., Lowack, K., Lvov, Y., and Schmitt, J. (1994) *Biosens. Bioelectron.*, **9**, 677.

43. (a) Lehr, B., Seufert, M., Wenz, G., and Decher, G. (1995) *Supramol. Sci.*, **2**, 199; (b) Müller, M., Brissova, M., Rieser, T., Powers, A.C., and Lunkwitz, K. (1999) *Mater. Sci. Eng. C*, **8/9**, 163.

44. (a) Lvov, Y., Ariga, K., Onda, M., Ichinose, I., and Kunitake, T. (1997) *Langmuir*, **13**, 6195; (b) Ariga, K., Lvov, Y., Onda, M., Ichinose, I., and Kunitake, T. (1997) *Chem. Lett.*, 125.

45. (a) Fendler, J.H. (1996) *Chem. Mater*, **8**, 1616; (b) Rosidian, A., Liu, Y., and Claus, R.O. (1998) *Adv. Mater.*, **10**, 1087.

46. Halaoui, L.I. (2001) *Langmuir*, **17**, 7130.

47. Kotov, N., Dekany, I., and Fendler, J.H. (1995) *J. Phys. Chem.*, **99**, 13065.

48. Gao, M., Richter, B., Kirstein, S., and Mohwald, H. (1998) *J. Phys. Chem. B.*, **102**, 4096.

49. Lesser, C., Gao, M., and Kirstein, S. (1999) *Mater. Sci. Eng. C*, **8/9**, 159.

50. Schrof, W., Rozouvan, S., Van Keuren, E., Horn, D., Schmitt, J., and Decher, G. (1998) *Adv. Mater.*, **10**, 339.

51. Liu, Y., Wang, Y., Lu, H., and Claus, R.O. (1999) *J. Phys. Chem. B*, **103**, 2035.

52. Serizawa, T., Kamimura, S., and Akashi, M. (2000) *Colloids Surf., A*, **164**, 237.

53. He, J.A., Yang, K., Kumar, J., Tripathy, S.K., Samuelson, L.A., Oshikiri, T., Katagi, H., Kasai, H., Okada, Sh., Oikawa, H., and Nakanishi, H. (1999) *J. Phys. Chem. B*, **103**, 11050.

54. (a) Schütte, M., Kurth, D.G., Linford, M.R., Cölfen, H., and Möhwald, H. (1998) *Angew. Chem. Int. Ed. Eng.*, **37**, 2891; (b) Caruso, F., Schüler, M., and Kurth, D.G. (1999) *Chem. Mater.*, **11**, 3394.

55. (a) Kotov, N.A., Magonov, S., and Tropsha, E. (1998) *Chem. Mater.*, **10**, 886; (b) Fan, X., Park, M.-K., Xia, C., and Advincula, R. (2002) *J. Mater. Res.*, **17**, 1622.

56. Kleinfeld, E.R. and Ferguson, G.S. (1994) *Science*, **265**, 370.

57. van Duffel, B., Schoonheydt, R.A., Grim, C.P.M., and De Schryver, D. (1999) *Langmuir*, **15**, 7520.

58. Keller, S.W., Kim, H.-N., and Mallouk, T.E. (1994) *J. Am. Chem. Soc.*, **116**, 8817.

59. Cassagneau, T., Guerin, F., and Fendler, J.H. (2000) *Langmuir*, **16**, 7318.

60. Ollivier, P.J., Kovtyukhova, N.I., Keller, S.W., and Mallouk, T.E. (1998) *J. Chem. Soc., Chem. Commun.*, 1563.

61. (a) Toutianoush, A., Saremi, F., and Tieke, B. (1999) *Mater. Sci. Eng. C*, **8-9**, 343; (b) Saremi, F. and Tieke, B. (1998) *Adv. Mater.*, **10**, 388.

62. Zhang, X., Gao, M., Kong, X., Sun, Y., and Shen, J. (1994) *J. Chem. Soc., Chem. Commun.*, 1055.

63. Advincula, R., Park, M.-K., Baba, A., and Kaneko, F. (2003) *Langmuir*, **19**, 654.

64. Kometani, N., Nakajima, H., Asami, K., Yonezawa, Y., and Kajimoto, O. (2000) *J. Phys. Chem. B*, **104**, 9630.

65. Dautzenberg, H., Jaeger, W., Kötz, J., Philipp, B., Seidel, Ch., and Stscherbina, D. (1994) *Polyelectrolytes*, Hanser Publishers, Munich, pp. 1–4.

66. Wandrey, C. and Hunkeler, D. (2002) Study of polyion counterion interaction by electrochemical methods, in *Handbook of Polyelectrolytes and Their Applications* (eds S.K. Tripathy, J. Kumar, and H.S. Nalwa), American Scientific Publishers, Stevenson Ranch pp. 147–172.

67. Cohidar, H.B. (2002) Characterization of polyelectrolytes by dynamic laser light scattering, in *Handbook of Polyelectrolytes and Their Applications* (eds S.K. Tripathy, J. Kumar, and H.S. Nalwa), American Scientific Publishers, Stevenson Ranch.

68. Fleer, G.J., Cohen Stuart, M.A., Scheutijens, J.M.H.M., Cosgrove, T., and Vincent, B. (1993) *Polymers at Interfaces*, Chapman & Hall, London, pp. 30–32.

69. Böhmer, M.R., Evers, O.A., and Scheutjens, J.M.H.M. (1990) *Macromolecules*, **23**, 2288.

70. Shubin, V. and Linse, P. (1997) *Macromolecules*, **30**, 5944.

71. Vermeer, A.W.P., Leermakers, F.A.M., and Koopal, L.K. (1997) *Langmuir*, **13**, 4413.

72. Åkesson, T., Woodward, C. and Jönsson, B. (1989) *J. Chem. Phys.*, **91**, 2461.

73. Beltrán, S., Hooper, H.H., Blanch, H.W., and Prausnitz, J.M. (1991) *Macromolecules*, **24**, 3178.

74. Granfeldt, M.K., Jönsson, B., and Woodward, C.E. (1991) *J. Phys. Chem.*, **95**, 4819.

75. Dobrynin, A.V., Deshkovski, A., and Rubinstein, M. (2001) *Macromolecules*, **34**, 3421.

76. Linse, P. (1996) *Macromolecules*, **29**, 326.

77. van de Steeg, H.G.M., Cohen Stuart, M.A., Keizer, A., and Bijsterbosch, B.H. (1992) *Langmuir*, **8**, 2538.

78. Schwarz, S., Eichhorn, K.J., Wischerhoff, E., and Laschewsky, A. (1999) *Colloids Surf., A*, **159**, 491.

79. Hoogeveen, N.G., Cohen Stuart, M.A., and Fleer, G.J. (1996) *J. Colloid Interface Sci.*, **182**, 133.

80. Joanny, J.F. (1999) *Eur. Phys. J. B.*, **9**, 117.

81. Castelnovo, M. and Joanny, J.-F. (2000) *Langmuir*, **16**, 7524.

82. Lowack, K. and Helm, C.A. (1998) *Macromolecules*, **31**, 823.

83. Dubas, S.T. and Schlenoff, J.B. (1999) *Macromolecules*, **32**, 8153.

84. Kayushina, R., Lvov, Y., Stepina, N., Belyaev, V., and Khurgin, Y. (1996) *Thin Solid Films*, **284–285**, 246.

85. Laschewsky, A., Mayer, B., Wischerhoff, E., Arys, X., Bertrand, P., Delcorte, A., and Jonas, A. (1996) *Thin Solid Films*, **284–285**, 334.

86. Caruso, F. and Möhwald, H. (1999) *J. Am. Chem. Soc.*, **121**, 6039.

87. Schlenoff, J.B., Ly, H., and Li, M. (1998) *J. Am. Chem. Soc.*, **120**, 7626.

88. Schlenoff, J.B. and Dubas, S.T. (2001) *Macromolecules*, **34**, 592.

89. (a) Decher, G. and Schmitt, J. (1992) *Prog. Colloid Polym. Sci.*, **89**, 160; (b) Lvov, Y., Decher, G., and Möhwald, H. (1993) *Langmuir*, **9**, 481.

90. Lösche, M., Schmitt, J., Decher, G., Bouwman, W.G., and Kjaer, K. (1998) *Macromolecules*, **31**, 8893.

91. Tarabia, M., Hong, H., Davidov, D., Kirstein, S., Steitz, R., Neumann, R., and Avny, Y. (1998) *J. Appl. Phys.*, **83**, 725.

92. Schmitt, J., Grunewald, T., Decher, G., Pershan, P.S., Kjaer, K., and Lösche, M. (1993) *Macromolecules*, **26**, 7058.

93. Ladam, G., Schaad, P., Voegel, J.C., Schaaf, P., Decher, G., and Cuisinier, F. (2000) *Langmuir*, **16**, 1249.

94. (a) Lehr, B., Seufert, M., Wenz, G., and Decher, G. (1995) *Supramol. Sci.*, **2**, 199; (b) Hong, H., Tarabia, M., Chayet, H., Davidov, D., Faraggi, F.Z., Avny, Y., Neumann, R., and Kirstein, S. (1996) *J. Appl. Phys.*, **79**, 3082; (c) Kolarik, L., Furlong, D.F., Joy, H., Struijk, C., and Rowe, R. (1999) *Langmuir*, **15**, 8265.

95. Von Klitzing, R. and Steitz, R. (2002) Internal structure of polyelectrolyte multilayers, in *Handbook of Polyelectrolytes and Their Applications* (eds S.K. Tripathy, J. Kumar, and H.S. Nalwa), American Scientific Publishers, Stevenson Ranch.

96. Decher, G., Lvov, Y., and Schmitt, J. (1994) *Thin Solid Films*, **244**, 772.

97. McAloney, R.A., Sinyor, M., Dudnik, V., and Goh, M.C. (2001) *Langmuir*, **17**, 6655.

98. Hoogeveen, N.G., Cohen Stuart, M.A., Fleer, G.J., and Bohmer, M.R. (1996) *Langmuir*, **12**, 3675.

99. Schoeler, B., Poptoshev, E., and Caruso, F. (2003) *Macromolecules*, **36**, 5258.

100. (a) Schoeler, B., Kumarswamy, G., and Caruso, F. (2002) *Macromolecules*, **35**, 889; (b) Voigt, U., Jaeger, W., Findenegg, G.H., and Klitzing, R. (2003) *J. Phys. Chem. B.*, **107**, 5273.

101. Yoo, D., Shiratori, S.S., and Rubner, M.F. (1998) *Macromolecules*, **31**, 4309.

102. Shiratori, S.S. and Rubner, M.F. (2000) *Macromolecules*, **33**, 4213.

103. Mendelsohn, J.D., Barrett, C.J., Chan, V.V., Pal, A.J., Mayes, A.M., and Rubner, M.F. (2000) *Langmuir*, **16**, 5017.

104. Fery, A., Scholer, B., Cassagneau, T., and Caruso, F. (2001) *Langmuir*, **17**, 3779.

105. Ariga, K., Lvov, Y., and Kunitake, T. (1997) *J. Am. Chem. Soc.*, **119**, 2224.

106. Kurth, D.G. and Osterhout, R. (1999) *Langmuir*, **15**, 4842.

107. Mermut, O. and Barrett, C.J. (2003) *J. Phys. Chem. B*, **107**, 2525.

108. Advincula, R., Aust, E., Meyer, W., and Knoll, W. (1996) *Langmuir*, **12**, 3536.

109. (a) Schüler, C. and Caruso, F. (2000) *Macromol. Rapid Commun.*, **21**, 750; (b) Caruso, F. and Schüler, C. (2000) *Langmuir*, **16**, 9595; (c) Lvov, Y. and Caruso, F. (2001) *Anal. Chem.*, **73**, 4212.

110. Seenrfors, T., Bogdanovic, G., and Tiberg, F. (2002) *Langmuir*, **18**, 6410.

111. Caruso, F., Caruso, R.A., and Möhwald, H. (1998) *Science*, **282**, 1111.

112. Breit, M., Gao, M., von Plessen, G., Lemmer, U., Feldmann, J., and Cundiff, S.T. (2002) *J. Chem. Phys.*, **117**, 3956.

113. Lvov, Y., Ariga, K., Onda, M., Ichinose, I., and Kunitake, T. (1999) *Colloids Surf., A*, **146**, 337.

114. Baba, A., Kaneko, F., and Advincula, R. (2000) *Colloids Surf., A*, **173**, 39.

115. Caruso, F. (2000) *Chem. Eur. J.*, **6**, 413.

116. Gao, C., Moya, S., Donath, E., and Möhwald, H. (2002) *Macromol. Chem. Phys.*, **203**, 953.

117. (a) Neu, B., Voigt, A., Mitlöhner, R., Leporatti, S., Gao, C.Y., Donath, E., Kiesewetter, H., Möhwald, H., Meiselman, H.J., and Bäumler, H. (2001) *J. Microencapsulation*, **18**, 385; (b) Leporatti, S., Voigt, A., Mitlöhner, R., Sukhorukov, G.B., Donath, E., and Möhwald, H. (2000) *Langmuir*, **16**, 4059.

118. Antipov, A.A., Sukhorukov, G.B., Donath, E., and Möhwald, H. (2001) *J. Phys. Chem. B*, **105**, 2281.

119. (a) Caruso, F., Lichtenfeld, H., Möhwald, H., and Griesig, M. (1998) *J. Am. Chem. Soc.*, **120**, 8523; (b) Caruso, F. and Möhwald, H. (1999) *Langmuir*, **15**, 8276.

120. (a) Caruso, F., Donath, E., and Möhwald, H. (1998) *J. Phys. Chem. B*, **102**, 2011; (b) Klitzing, R.V. and Möhwald, H. (1996) *Macromolecules*, **29**, 6901; (c) Klitzing, R.V. and Möhwald, H. (1995) *Langmuir*, **11**, 3554.

121. Möhwald, H. (2000) *Colloids Surf., A*, **171**, 25.

122. Voigt, A., Lichtenfeld, H., Sukhorukov, G.B., Zastrow, H., Donath, E.,

Baümler, H., and Möhwald, H. (1999) *Ind. Eng. Chem. Res.*, **38**, 4037.

123. (a) Lichtenfeld, H., Knapschinsky, L., Sonntag, H., and Shilov, V. (1995) *Colloids Surf., A*, **104**, 313; (b) Lichtenfeld, H., Knapschinsky, L., Dürr, C., and Zastrow, H. (1997) *Prog. Colloid Polym. Sci.*, **104**, 148; (c) Sukhorukov, G.B., Donath, E., Lichtenfeld, H., Knippel, E., Kinppel, M., Budde, A., and Möhwald, H. (1998) *Colloids Surf., A*, **137**, 253.

124. (a) Susha, A.S., Caruso, F., Rogach, A.L., Sukhorukov, G.B., Kornowski, A., Möhwald, H., Giersig, M., Eychmuller, A., and Weller, H. (2000) *Colloids Surf., A*, **63**, 39; (b) Rogach, A.L., Kotov, N.A., Koktysh, D.S., Susha, A., and Caruso, F. (2002) *Colloids Surf., A*, **202**, 135.

125. Estrela-Lopis, I., Leporatti, S., Moya, S., Brandt, A., Donath, E., and Möhwald, H. (2002) *Langmuir*, **18**, 7861.

126. Caruso, F., Schuler, C., and Kurth, D.G. (1999) *Chem. Mater.*, **11**, 3394.

127. Gittins, D.I. and Caruso, F. (2000) *Adv. Mater.*, **12**, 1947.

128. Qiu, X., Leporatti, S., Donath, E., and Möhwald, H. (2001) *Langmuir*, **17**, 5375.

129. Diaspro, A., Silvano, D., Krol, S., Cavalleri, O., and Gliozzi, A. (2002) *Langmuir*, **18**, 5047.

130. Georieva, R., Moya, S., Hin, M., Mitlohner, R., Donath, E., Kiesewetter, H., Möhwald, H., and Baumler, H. (2002) *Biomacromolecules*, **3**, 517.

131. Pastoriza-Santos, I., Scholer, B., and Caruso, F. (2001) *Adv. Funct. Mater.*, **11**, 122.

132. Antipov, A.A., Sukhorukov, G.B., Leporatii, S., Radtchenko, I.L., Donath, E., and Möhwald, H. (2002) *Colloids Surf., A*, **198**, 535.

133. Wang, D., Caruso, R.A., and Caruso, F. (2001) *Chem. Mater.*, **13**, 364.

134. Gao, C., Donath, E., Moya, S., Dudnik, V., and Möhwald, H. (1997) *Eur. Phys. J. E.*, **5**, 21.

135. Leporatti, S., Gao, C., Voigt, A., Donath, A., and Möhwald, H. (2001) *Eur. Phys. J. E.*, **5**, 13.

136. Gao, C., Leporatti, S., Moya, S., Donath, E., and Möhwald, H. (2001) *Langmuir*, **17**, 3491.

137. Gao, C., Leporatii, S., Moya, S., Donath, E., and Möhwald, H. (2003) *Chem. Eur. J.*, **9**, 915.

138. Antipov, A.A., Sukhorukov, G.B., Donath, E., and Möhwald, H. (2001) *J. Phys. Chem. B*, **105**, 2281.

139. Sukhorukov, G.B., Brumem, M., Donath, E., and Möhwald, H. (1999) *J. Phys. Chem. B*, **103**, 6434.

140. Gao, M., Lesser, C., Kirstein, S., Möhwald, H., Rogach, A.L., and Weller, H. (2000) *J. Appl. Phys.*, **85**, 2297.

141. Woo, W.-K., Shimizu, K.T., Jarosz, M.V., Neuhauser, R.G., Leatherdale, C.A., Rubner, M.A., and Bawendi, M.G. (2002) *Adv. Mater.*, **14**, 1068.

142. Locklin, J., Shinbo, K., Onishi, K., Kaneko, F., Bao, Z., and Advincula, R.C. (2003) *Chem. Mater.*, **15**, 1404.

143. He, J.-A., Mosurkal, R., Samuelson, L.A., Li, L., and Kumar, J. (2003) *Langmuir*, **19**, 2169.

144. Natansohn, A., Rochon, P., Gosselin, J., and Xie, S. (1992) *Macromolecules*, **25**, 2268.

145. (a) Viswanathan, N.K., Balasubramanian, S., Li, L., Kumar, J., and Tripathy, S.K. (1998) *J. Phys. Chem. B*, **102**, 6064; (b) He, J.-A., Bian, S.P., Li, L., Kumar, J., Tripathy, S.K., and Samuelson, L.A. (2000) *Appl. Phys. Lett.*, **76**, 3233.

146. Breit, M., Gao, M., von Plessen, G., Lemmer, U., Feldmann, J., and Cundiff, S.T. (2002) *J. Chem. Phys.*, **117**, 3956.

147. (a) Galeska, I., Hickey, T., Moussy, F., Kreutzer, D., and Papadimitrakopoulos, F. (2001) *Biomacromolecules*, **2**, 1249; (b) Sukhorukov, G.B., Montrel, M.M., Petrov, A.I., Shabarchina, L.I., and Sukhorukov, B.I. (1996) *Biosens. Bioelectron.*, **11**, 913.

148. (a) Onda, M., Lvov, Y., Ariga, K., and Kunitake, T. (1996) *J. Ferment. Bioeng.*, **82**, 502; (b) Caruso, F., Rodda, E., Furlong, D.F., Niikura, K., and Okahata, Y. (1997) *Anal. Chem.*, **69**, 2043.

149. Joly, S., Kane, R., Radzilowski, L., Wang, T., Wu, A., Cohen, R.E., Thomas, E.L., and Rubner, M.F. (2000) *Langmuir*, **16**, 1354.

150. Wang, T.C., Cohen, R.E., and Rubner, M.F. (2002) *Adv. Mater.*, **14**, 1534.

151. (a) Kotov, N.A., Magonov, S., and Tropsha, E. (1998) *Chem. Mater.*, **10**, 886; (b) Meier-Haack, J., Lenk, W., Lehmann, D., and Lunkwitz, K. (2001) *J. Membr. Sci.*, **184**, 233; (c) Stanton, B.W., Harris, J.J., Miller, M.D., and Bruening, M.L. (2003) *Langmuir*, **19**, 7038.

152. Kapnissi, C.P., Akbay, C., Schlenoff, J.B., and Warner, I.M. (2002) *Anal. Chem.*, **74**, 2328.

153. Cassagneau, T. and Fendler, J.H. (1998) *Adv. Mater.*, **10**, 877.

154. Hattori, H. (2001) *Adv. Mater.*, **13**, 51.

155. (a) Caruso, F., Yang, W., Trau, D., and Renneberg, R. (2000) *Langmuir*, **16**, 9595; (b) Ibraz, G., Dahne, L., Donath, E., and Möhwald, H. (2001) *Adv. Mater.*, **13**, 1324; (c) Dai, Z. and Möhwald, H. (2002) *Chem. Eur. J.*, **8**, 4751.

156. (a) Advincula, R., Baba, A., and Kaneko, F. (2000) *Colloids Surf., A*, **173**, 39–49; (b) Byrd, H., Holloway, C., Pogue, J., Kircus, S., Advincula, R., and Knoll, W. (2000) *Langmuir*, **26**, 10322–10328; (c) Advincula, R., Fells, E., and Park, M.-K. (2001) *Chem. Mater.*, **13**, 2870–2878; (d) Park, M.-K., Xia, C., Schütz, P., Caruso, F., and Advincula, R. (2001) *Langmuir*, **17**, 7670–7674; (e) Advincula, R., Locklin, J., Youk, J., Xia, C., Park, M.-K., and Fan, X. (2002) *Langmuir*, **18**, 877–883; (f) Katowa, T., Baba, A., Kaneko, F., Advincula, R., Shinbo, K., and Kato, K. (2002) *Colloids Surf., A*, **198–200**, 805–810; (g) Ishikawa, J., Baba, A., Kaneko, F., Advincula, R., Shinbo, K., and Kato, K. (2002) *Colloids Surf., A*, **198–200**, 917–922; (h) Fan, X., Locklin, J., Youk, J.H., Blanton, W., Xia, C., and Advincula, R. (2002) *Chem. Mater.*, **14**, 2184–2191; (i) Park, M.-K. and Advincula, R. (2002) *Langmuir*, **18**, 4532–4535; (j) Baba, A., Park, M.-K., Advincula, R., and Knoll, W. (2002) *Langmuir*, **18**, 4648–4652; (k) Park, M.-K. and Advincula, R. (2002) *Langmuir*, **18**, 4532–4535; (l) Kaneko, F., Kato, T., Baba, A., Shinbo, K., Kato, K., and Advincula, R. (2002) *Colloids Surf., A*, **198–200**, 805–810; (m) Ishikawa, J., Kaneko, F., Baba, A., Shinbo, K., Kato, K., and Advincula, R. (2002) *Colloids Surf., A*, **198–200**, 917–922; (n) Shinbo, K., Baba, A., Kaneko, F., Kato, T., Kato, K., Advincula, R., and Knoll, W. (2002) *Mater. Sci. Eng. C*, **22**, 319–325; (o) Shinbo, K., Ishikawa, J., Baba, A., Kato, K., Kaneko, F., and Advincula, R. (2002) *Jpn. J. Appl. Phys. Part I*, **41**, 2753–2758; (p) Advincula, R., Fan, X., and Park, M.-K. (2002) *J. Mater. Res.*, **17**, 1622–1633; (q) Kato, K., Kawashima, J., Baba, A., Shinbo, K., Kaneko, F., and Advincula, R. (2003) *Thin Solid Films*, **438–439**, 101–107; (r) Shinbo, K., Onishi, K., Miyabayashi, S., Takahashi, K., Katagiri, S., Kato, K., Kaneko, F., and Advincula, R. (2003) *Thin Solid Films*, **438–439**, 177–181; (s) Locklin, J., Shinbo, K., Onishi, K., Kaneko, F., Bao, Z., and Advincula, R. (2003) *Chem. Mater.*, **15**, 1404–1412; (t) Fan, X., Xia, C., Fulghum, T., Park, M.-K., Locklin, J. and Advincula, R. (2003) *Langmuir*, **19**, 916–923; (u) Tian, S.J., Baba, A., Liu, J., Wang, Z., Knoll, W., Park, M.-K., and Advincula, R. (2003) *Adv. Funct. Mat.*, **13**, 473–479; (v) Advincula, R., Park, M.-K., Baba, A., and Kaneko, F. (2003) *Langmuir*, **19**, 654–665; (w) Park, M.-K., Onishi, K., Locklin, J., Caruso, F., and Advincula, R. (2003) *Langmuir*, **19**, 8550–8554; (x) Kato, K., Watanabe, N., Katagiri, S., Shinbo, K., Kaneko, F., Locklin, J., Baba, A., and Advincula, R.C. (2004) *Jpn. J. App. Phys. Part I*, **43**, 2311–2314; (y) Patton, D., Locklin, J., Meredith, M., Xin, Y., and Advincula, R. (2004) *Chem. Mater.*, **16**, 5063–5070; (z) Park, M.-K., Deng, S., and Advincula, R. (2004) *J. Am. Chem. Soc.*, **126**, 13723–13731; (aa) Park, M.-K., Deng, S., and Advincula, R. (2005) *Langmuir*, **21**, 5272–5277; (ab) Advincula, M., Rahemtulla, F., Advincula, R., Ada, E., Lemons, J., and Bellis, S. (2006) *Biomaterials*, **27**,

2201–2212; (ac) Baba, A., Locklin, J., Xu, R., and Advincula, R. (2006) *J. Phys. Chem. B.*, **110**, 42–45; (ad) Baba, A., Jiang, G., Park, K., Park, J., Shin, H., and Advincula, R. (2006) *J. Phys. Chem.*, **110**, 17309–17314; (ae) Fulghum, T., Patton, A., and Advincula, R. (2006) *Langmuir*, **22**, 8397–8402; (af) Advincula, M., Rahemtulla, F., Advincula, R., Peterson, D., and Lemons, J. (2007) *J. Biomed. Mater. Res. B*, **80B**, 107–120; (ag) Jiang, J., Baba, A., and Advincula, R. (2007) *Langmuir*, **23**, 817–825; (ah) Waenkaew, P., Taranekar, P., Phanichphant, P., and Advincula, R. (2007) *Macromol. Rapid Commun.*, **28**, 1522–1527; (ai) Ravindranath, R., Ajikumar, P., Baba, A., Bahuleyan, S., Hanafiah, N., Advincula, R., Knoll, W., and Valiyaveettil, S. (2007) *J. Phys. Chem. B.*, **111**, 6336–6343; (aj) Jiang, G., Baba, A., Ikarashi, H., Xu, R., Locklin, J., Khan, K.R., Shinbo, K., Kato, K., Kaneko, F., and Advincula, R. (2007) *J. Phys. Chem. C.*, **111**, 18687–18694; (ak) Fulghum, T.M., Estillore, N.C., Vo, C.-D., Armes, S.P., and Advincula, R.C. (2008) *Macromolecules*, **41**, 429–435; (al) Huang, C., Jiang, G., and Advincula, R. (2008) *Macromolecules*, **41**, 4661–4670; (am) Sriwichai, S., Baba, A., Deng, S., Huang, C., Phanichphant, S., and Advincula, R. (2008) *Langmuir*, **24**, 9017–9023; (an) Fujie, T., Park, J., Murata, A., Estillore, N., Tria, M.C., Takeoaka, S., and Advincula, R. (2009) *ACS Appl. Mater. Interfaces*, **1**(7), 1404–1413.

4

Langmuir–Blodgett–Kuhn Multilayer Assemblies: Past, Present, and Future of the LB Technology

Débora T. Balogh, Marystela Ferreira, and Osvaldo N. Oliveira

4.1
Introduction

The possibility of molecular control offered by the Langmuir–Blodgett (LB) method has been exploited in a variety of fundamental studies in several areas, and in the identification of applications where predesigned molecular architectures may be advantageous. The main characteristics of the LB films are associated with their layer-by-layer (LbL) architecture, with intimate contact between adjacent layers, and with the mechanisms of adsorption as the molecules are assembled via intermolecular interactions including hydrophobic and dispersions forces [1]. The films are produced by transferring an ultrathin film, usually a compressed monolayer, from the air/water interface onto a solid substrate. Multilayers of a given material or more than one material may be deposited on the same substrate, which allows not only for control of molecular architectures but also to seek synergy between properties of distinct materials [2].

Many reviews and books have been dedicated to LB films [1, 3–5], and to avoid unnecessary repetition, we shall concentrate on specific issues, particularly in areas where there is promising outlook for further research and development. We take advantage of the availability of databases of publications and patents to identify the impact of the LB methodology on several areas, in addition to major trends for the near future. Details of experimental procedures to produce and characterize the films, and some historical notes, are included only as necessary to yield a self-contained chapter, with no aim of covering the literature in a comprehensive way. It is also worth stressing that the characterization of Langmuir monolayers at the air/water interface is considered an integral part of research associated with the LB methodology. However, since the specific topic of Langmuir monolayers may be covered in another chapter of this book, information on such monolayers will only be included in the context of the related LB films.

The chapter is organized as follows. Section 4.2 gives brief details of the history of the LB method, which is followed by a description of the basic procedures, characterization techniques, and most used types of molecules in Sections 4.3 and

Functional Polymer Films, First Edition. Edited by Wolfgang Knoll and Rigoberto C. Advincula.
© 2011 Wiley-VCH Verlag GmbH & Co. KGaA. Published 2011 by Wiley-VCH Verlag GmbH & Co. KGaA.

4.4. Some trends in research activity on LB films are discussed in Section 4.5, while Section 4.6 presents highlights of the research in LB films.

4.2
Historical Background

The formation of thin oil films on the water surface has been known since ancient times. Aristotle first reported the calming effect of oil on a water surface [6], and the first practical application of film formation on the air/water interface was the Japanese printing technique *sumi-nagashi* [7]. In the first scientific study in 1774, Benjamin Franklin described the spreading of an oil layer on the surface of a pond. In 1890s, Agnes Pockels prepared monolayers at the air/water interface and developed a prototype of a trough consisting of a recipient with movable barriers [8–10]. In 1899, Lord Rayleigh and coworkers [1] proposed that a drop of oil would spread on the water surface until the thickness of the film reached that of one molecule.

It was Irving Langmuir who developed the theoretical and experimental concepts for the fabrication and study of monolayers at the air/water interface, which is why these films are referred to as *Langmuir films (or Langmuir monolayers)*. His first publication on monolayers appeared in 1917. In 1919, when Irving Langmuir read before the Faraday Society a paper describing the transfer of monomolecular oleic acid films from an air/water interface onto several substrates, he acknowledged Miss Katharine Blodgett *"for carrying out most of the experimental work"* [11].

Katharine Burr Blodgett was born on 10 January 1898 in Schenectady, New York, birthplace of the General Electric (GE) company, daughter of the head of the patent department of GE. After receiving a master's degree from the University of Chicago in 1918, she was hired by Dr. Irving Langmuir, being the first woman to work as a research scientist at GE. In 1924, Blodgett took a leave from GE to be the first woman to earn a PhD in physics at Sir Ernest Rutherford's Cavendish Laboratory in Cambridge University [12, 13]. Back at GE, she worked in filaments from 1927 to 1932. Then, her research turned back to the transfer of films from the air/water interface onto solid substrates, with the aim of minimizing the friction in meters. In 1934, Katharine Blodgett published a short paper on the buildup of multilayer films of fatty acids [14]. The detailed process to build multilayers from the air/water interface she published in 1935 [15], and the basic procedures remain essentially the same until today, apart from the development of computer-controlled troughs. Rightly so, these deposited films became known as *Langmuir–Blodgett films*. Dr. Blodgett was an industrial researcher, and as such, specialized in solving problems. Therefore, after 1938 her attention was diverted from deposition of thin films to other fields, including development of smoke screens (in the World War II) and semiconducting coatings for X-ray tubes and gauges made of birefringent plastics.

The fact that amphiphilic molecules, such as fatty acids, have poor thermal and mechanical properties and do not have properties capable of generating more than a few applications (e.g., optical absorption in the visible region) led to a stagnant period of research after Blodgett's work. In 1970, the group of Prof. H. Kuhn in Göttingen, Germany, showed that LB films of an amphiphilic dye could be used in energy-transfer mechanisms [16]. This was indeed a turning point, as the field progressed enormously in the following years. In honor of the many contributions from Prof. Kuhn, actually responsible for the resurgence of interest in the method, the deposited films have sometimes been termed as *Langmuir–Blodgett–Kuhn films*. In 1972, Cemel *et al.* [17] produced LB films of molecules that could be further polymerized. In the 1980s studies on the fabrication of LB films of amphiphilic preformed polymers started to be carried out [18, 19], which were followed by the use of nonamphiphilic conducting polymers, bringing new possibilities for applications of LB films in opto-electronic devices [20]. Nowadays, LB films of many different materials can be built, such as phthalocyanines, porphyrins, macromolecules, and polymers, both synthetic and natural, without having *per se* an amphiphilic character. The nonspreading features of the parent molecules are normally circumvented by cospreading with amphiphilic substances or via interaction with molecules in the subphase.

The strong development of the LB technique generated high expectations of possible industrial applications. Indeed, in the 1980s and 1990s there was great expectation for applications, especially in devices and coatings, which generated approximately 1700 patents from 1980 to 2000 (according to the Derwent database). Though the number of patents involving LB films is still considerable, with over 600 patents filed in the 2001–2007 period, the expectation of large-scale industrial applications was not fulfilled due to various reasons. The first was the intrinsic difficulties in producing films at a large scale. There was then competition with other methods that also allow for fabrication of organized films with molecular control [4], as is the case of self-assembly (SA) methods, namely the electrostatically assembled LbL films, and the chemically adsorbed self-assembled monolayers (SAMs), in addition to other methods based on patterning substrates [21], and casting with friction [22]. It is now realized that the LB methodology may not be the most suitable film-forming technique for industrial applications, but is unique for some fundamental studies (e.g., mimicking membranes). In fact, LB films may be advantageous in building small biosensors with minute quantities of active biomolecules (e.g., enzymes, DNA) in a friendly environment provided by other biomolecules such as biopolymers and polysaccharides. Also, some properties are enhanced in LB multilayers in comparison to other types of film, including spin-coated or cast films. This is the case of thermochromic properties and optical storage, leading to a huge increase in the number of published papers in the 1990s and a continuous production in the field. This will be better illustrated in the following discussion.

An important development stemming from research in LB films was the creation of novel methods to produce nanostructured films with controlled architectures. Examples are the LbL [23–25] and SA methods [26, 27]. In fact, one of the most cited papers with LB as a keyword is associated with nanoassemblies using LbL films [28]. The so-called SAMs are produced with chemisorption of molecules onto solid substrates, yielding highly organized multilayer systems [26]. The LbL method is also based on material adsorption onto solids substrates, but performed with alternating layers of two or more substances that are physically adsorbed. The technique introduced by Decher and coworkers [23] was based on the alternated adsorption of nanometric layers of oppositely charged materials. In this method, a solid substrate is immersed in, for example, a diluted polycationic solution and after some time removed and washed to remove nonadsorbed material. Subsequently, the substrate is immersed in a polyanionic solution, promoting the adsorption of the polyanion over the polycation layer, forming a bilayer. The process is repeated as many times as desired to form a multilayer film. The adsorption process can be performed by electrostatic interactions or secondary interactions, such as hydrogen bonds or hydrophobic interactions, or even by specific interactions, such as in lock–key systems. Though the LbL method was originally created for polymers, now a variety of other materials are used, including solid particles [29].

4.3
Basic Procedures and Characterization Methods

The first step to fabricate an LB film is to obtain a Langmuir film by spreading a dilute solution of a substance on a subphase (usually ultrapure water) and compressing the film to a specific surface pressure. It should be mentioned that a special apparatus to prepare LB films is required because noncompressed films are not transferred onto solid substrates, for the establishment of surface pressure gradients at the meniscus is a prerequisite for film transfer. The Langmuir films are compressed by moving barriers to the desired surface pressure (measured by a Wilhelmy plate balance). After maintaining the surface pressure constant, a solid substrate is slowly raised out or dipped into the water. Multilayers are obtained by dipping/lifting the substrate repeatedly through the water interface, a procedure that may be performed as many times as needed. The LB films are nowadays fabricated in commercial systems, referred to as *Langmuir troughs*, whose modern versions are fully computer controlled. Figure 4.1 shows a schematic illustration of a Langmuir trough with a dipping well.

LB films of fatty acids and other amphiphilic molecules can form different arrangements depending on the deposition process. When the films are to be transferred onto a hydrophilic substrate, the latter is immersed in the water subphase prior to the spreading and compression of the material. After compression, the substrate is raised out of the water with the polar part (headgroup) being attached to the substrate and the nonpolar tail oriented perpendicular to the substrate. When the substrate is immersed again in the water containing the film, the nonpolar tail will attach to the previously transferred tail forming a so-called Y-type film (Figure 4.2). Under special cases, which depend on the material, substrate, and subphase conditions, transfer only occurs upon either withdrawing or immersing the substrate, leading respectively to the Z-type and X-type LB films. The latter types have a head–tail or tail–head configuration, as illustrated in Figure 4.2. Because nonamphiphilic molecules do not have defined polar head and nonpolar tails, the

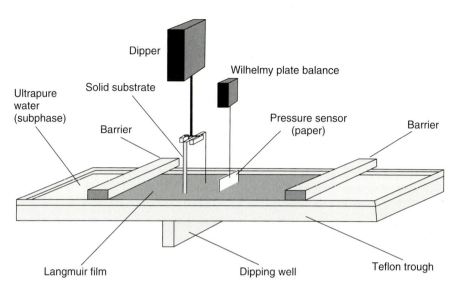

Figure 4.1 Illustration of a Langmuir trough with a dipping well, in which the dipper system, the movable barriers and the pressure sensor are shown.

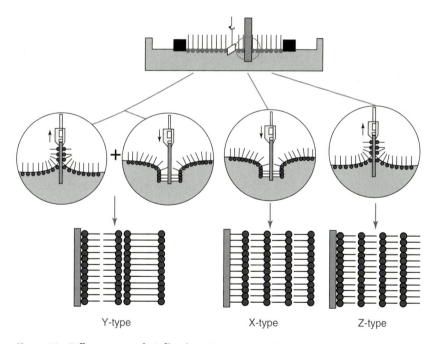

Figure 4.2 Different types of LB film deposition.

molecular arrangements in the film are not straightforward. In such cases, the type of deposition (rather than the film type) describes the films that are only transferred in the immersion (X-type), only in withdrawal (Z-type) or in both immersion and withdrawal (Y-type).

The main parameters controlled in the deposition of LB films are the pressure at deposition and the contents of the subphase, especially pH and incorporation of salts. For one of the most-studied materials, namely stearic acid, it is commonplace to add cadmium chloride to form cadmium stearate films, which are more stable than neat stearic acid films. The fundamental properties of LB films from amphiphilic materials are a high degree of order and accuracy in controlling film thickness. Because these amphiphilic materials have intrinsic poor mechanical properties, stability is not high.

The quality of film transfer for amphiphilic molecules can be obtained from the values of the transfer ratio (TR), which is the decrease in area occupied by the Langmuir film during deposition of one layer divided by the substrate area expected to be coated. A value of TR = 1 for both upstrokes and downstrokes represents an ideal Y-type film deposition. Experimentally, sometimes TR is less than 1 but constant for upstrokes and downstrokes, or higher for upstrokes than for downstrokes or *vice versa*. Thus, the films can be mixed Y- and X-types. When nonamphiphilic materials form metastable Langmuir films, transfer into LB films can still be done under optimized conditions. The measurement of TR in these cases is not straightforward, since the decrease in area can be due not only to the deposited film but also to the molecular rearrangements on the compressed Langmuir film.

The experimental challenge in the production of LB films is still the choice of the best conditions to transfer the Langmuir film onto the solid substrate. Various parameters may affect deposition, including the dipping speed, substrate surface, surface pressure at which the film is compressed and maintained during deposition, the period of time the Langmuir film is kept at the target pressure prior to deposition, the barrier speed during deposition and even the time the film is left to dry outside the water. Subphase properties such as pH and ionic force also have an impact on the deposition, especially in the case of charged materials. The structure of LB films can be different in the first layer compared to the subsequent layers, since the first one interacts with the substrate, which can become less important as the number of layers increases. For example, the arrangements of fatty acid molecules can turn from hexagonal to orthorhombic with increased number of layers in the LB film.

In addition to the TR, spectroscopic techniques are valuable to determine if the material was transferred homogeneously onto the solid substrate, for materials absorbing light at the ultraviolet, visible, or infrared wavelength regions. For homogeneous deposition, absorption of light increases linearly with the number of layers. If quantitative mass data is needed for each deposited layer, a quartz crystal microbalance (QCM) [30] can be used. The characterization of LB films, in terms of thickness, texture, surface properties, and order are performed as in thin films in general. There are several books and reviews on the matter [1, 3–5, 31–33], and therefore the techniques will be only mentioned briefly here. The thickness of the

LB films may be determined by ellipsometry, surface plasmon resonance (SPR), X-ray, neutron and electron diffractions, profilometry, electrical measurements (capacitance), and atomic force microscopy (AFM). The orientational order of molecules in the LB film can be investigated with some of the techniques above and also by Raman scattering, and electron spin resonance (EPR). Polarization modulation infrared reflection-absorption spectroscopy (PM-IRRAS), scanning electron microscopy (SEM), scanning tunneling microscopy (STM), transmission electron microscopy (TEM), AFM, and optical microscopy are also useful in film characterization. Table 4.1 lists some of the characterization methods employed in the evaluation of LB film properties.

4.4
Types of Materials Used in Producing LB Films

Almost any water-insoluble material, or partially soluble in the case of macro-molecules, can be used to build LB films, but a convenient method of spreading is required. Because the number of different types of material that can be used is immense, we obtained from the ISI Web of Science database some statistics shown in Figure 4.3, based on papers published from 1900 to 2007. Obviously, the distribution of the types of material does not consider possible overlaps, when more than one material is used in the same publication. Note that traditional amphiphilic materials, including lipids and fatty acids, account for only circa 11% of the literature, in contrast to polymers that are now the most studied. Also worth mentioning are the large number of papers associated with azobenzenes and phthalocyanines and porphyrins. We deliberately singled out nanotubes and quantum dots, for though the percentage of papers is still low, there may be a trend toward combining the LB method with other nanotechnology approaches.

The types of polymer encompass a broad range of materials, and the statistics includes mixtures with other materials, as illustrated in Figure 4.4. Perhaps the main reason why there is so much research on polymeric LB films lies in the advantages in terms of wide variability of properties and mechanical resistance. Furthermore, considerable research efforts were necessary to overcome the difficulties in producing polymer LB films, especially when preformed polymers are used. In contrast to the traditional amphiphilic molecules, obtaining stable Langmuir films from macromolecules is a challenge for several reasons. Indeed, the first difficulty appears already in preparing the spreading solution, as many macromolecules of interest – for example, conjugated polymers – are not readily soluble in volatile, organic solvents commonly used in Langmuir-film fabrication. This problem is circumvented by dissolving the polymer in a polar, water-soluble solvent, such as N,N-dimethylformamide or dimethylsulfoxide and then mixing with a water-insoluble solvent. The proportion of solvents is chosen based on the maximum amount of nonwater soluble solvent that can used in the mixture without imparting cloudiness to the solution.

In the order at which films are produced, the second difficulty is in obtaining spreading over the water surface without dissolution into the subphase. This can

Table 4.1 Techniques for characterization of multilayer LB films.

Name	Used for ...	Reference
Ellipsometry	Evaluation of film thickness	[1] and references therein.
Interference technique	Evaluation of film thickness	[1, 4] and references therein.
X-ray diffraction	Evaluation of film thickness and structural information	[1, 4, 32] and references therein.
Neutron diffraction and reflection	Evaluation of film thickness	[1, 4] and references therein.
Electrical measurements (capacitance)	Evaluation of film thickness	[1, 34]
Surface plasmon resonance	Evaluation of film thickness	[1, 4] and references therein.
Electron diffraction	Structural information	[1, 4] and references therein.
Optical, fluorescence, and electron microscopy	Structural information	[1, 4] and references therein.
Infrared and visible spectroscopy	Structural and analytical surface composition information	[1, 4] and references therein.
Raman spectroscopy	Structural information	[1, 4, 34] and references therein.
Surface analytical techniques (AFM, STM, SFG)	Structural information, surface morphology	[1] and references therein
Surface potential	Electrical properties	[1, 4] and references therein.
Conductance and impedance	Electrical properties	[1] and references therein.
Refractive index	Optical properties	[1] and references therein.
Contact angle and surface tension	Surface analysis	[4] and references therein.
Electrochemical techniques	Surface analysis, redox properties	[35]
Surface analytical techniques: (XPS or ESCA)	Quantitative surface composition	[36]

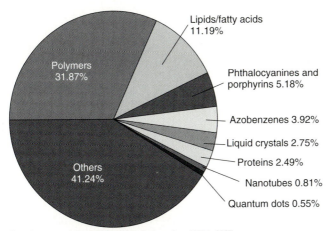

*Based on papers indexed in ISI – Web of science from 1900 to 2007

Figure 4.3 Published papers on LB films according to the type of material studied. Source ISI – Web of Science.

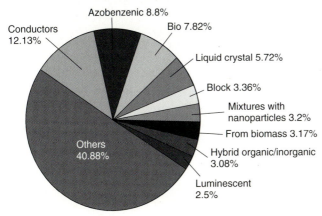

*Based on papers indexed in ISI – Web of science from 1900 to 2007

Figure 4.4 Published papers on polymeric LB films according to the type of polymer, and mixtures with other materials. Source ISI – Web of Science.

only be achieved if the right balance is reached between the hydrophilic and hydrophobic portions of the macromolecules. On one hand, polar groups are needed to allow efficient spreading, but the molecules should not be excessively polar, lest they dissolve. On the other hand, if the hydrophobic part is too dominant, spreading may not be effective, and lenses would be formed over the water surface. As for the transfer onto a solid substrate in the form of an LB film, the Langmuir film must be stable to withstand relatively high pressures at which deposition is carried out. However, if the film is too rigid, transfer may not occur, as has been shown to be the case for some polymers, because the necessary surface pressure gradient at the meniscus is not there. In such situations the touching Schaefer's method [37] of horizontal transfer could be used.

Thus, it is not surprising that a number of strategies were suggested in the literature for the fabrication of LB films from macromolecules. For conjugated polymers, for instance, Oliveira *et al.* [38] surveyed the main strategies used, which are basically classified into three approaches. The first, and most used one, is the LB preparation of preformed polymers, which could be synthesized by chemical or electrochemical polymerization of monomers. The second is the deposition of the LB film of a monomer or a precursor polymer that is subsequently converted into the polymer of interest, using UV irradiation or heat treatment. The third approach consists in spreading the monomer or a precursor polymer at the air/water interface where the polymerization reaction will take place, and then transfer the polymer formed onto a solid substrate.

Another important issue in LB films made with polymers is the difficulty in interpreting the data. For instance, the estimation of the area per molecule – so important in Langmuir films of amphiphilic molecules – is not straightforward for a polymer film. For monodisperse fractions, it is possible to estimate the characteristic surface concentration Γ^* that defines the boundary between the dilute and semidilute regimes, in analogy to c^* – the overlapping concentration for polymeric solutions. From this c^* the radius of gyration in two dimensions can be calculated. The majority of polymers, however, comprise chains with different lengths, with a distribution of molecular weights. Also, the polymer–solvent interactions lead to different polymer conformations in the spreading solution, thus affecting the molecular arrangements at the air/water interface upon compressing. Furthermore, polymers can adopt a random conformation at spreading and tend to coil upon compressing, as illustrated in Figure 4.5a. Depending on the chemical nature of the repeating units, the chain segments can adopt a loop-like structure

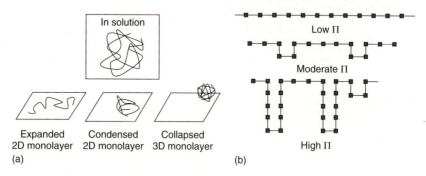

(a)   (b)

Figure 4.5 (a) The schematic drawings illustrate that polymer molecules may be coiled in solution, and then adopt distinct conformations at the air/water interface. These include a more extended conformation, most likely to occur for an expanded monolayer (bottom left), or more coiled conformations in the other two schemes, including a collapsed monolayer, with a 3D conformation for the polymer chains. (b) Pictorial representation of a polymer chain, which could be extended at low surface pressures, as depicted in the top drawing, or present kinks or ramifications at higher pressures. Reproduced from Ref. [38] with permission from Academic Press@2008.

Table 4.2 Examples of substances used to build LB multilayers and a few properties or possible applications investigated on these films.

Material	Investigated property/application
Lipids/fatty acids	Molecular recognition [42], phase separation [43], membrane modeling [44]
Phthalocyanines and porphyrins	Nonlinear optics [45], sensing [46], electrochromism [47]
Azobenzenes	Organization [48], light-induced morphology changes [49]
Liquid crystals	Anisotropy [50], phase change [51]
Proteins	Membrane modeling [52], adsorption processes [52], biosensors [53]
Nanotubes	Patterning [54]; alignment [55]
Quantum dots	Shape control [56], luminescence [57]
Polymers: conducting	Transistors [58], electroluminescence [59], molecular electronics [60]
Azobenzenic	Optical storage [61], nonlinear optics [62], photoresponsive membranes [63]
Liquid crystalline	Anisotropy [64], alignment [65]
Block	Surface morphology [66], patterning [67]
Mixed with nanoparticles	Shape [68], magnetic properties [69]
From biomass	Surface roughness [70], composites [71]

(Figure 4.5b). It is therefore difficult to estimate the mean molecular area and analyze the isotherms, while determining the best surface pressure for deposition is also tricky.

The LB method was primarily conceived to produce films with organic materials. Over the years, however, several inorganic compounds have been incorporated in LB films, such as semiconductors (PbS, CdS), minerals (hydroxyapatite, clays) or salts (calcium carbonate), usually in conjunction with organic materials, thus forming hybrids or composites [39]. The motivation for such studies was mainly the search for synergy between materials with distinct properties, as in composites of silver colloidal particles and fatty acid amines and hybrids of magnetic polyoxometalates and behenic acid [40, 41], or the identification of thin films suitable for coating to increase adhesion or biocompatibility.

Table 4.2 shows examples of substances used to build LB multilayers and a few properties or possible applications investigated on these films.

4.5
Trends in Research Activity on LB Films

The number of published papers associated with fabrication or use of LB films, indexed by the Chemical Abstracts, showed a remarkable increase in the 1980s

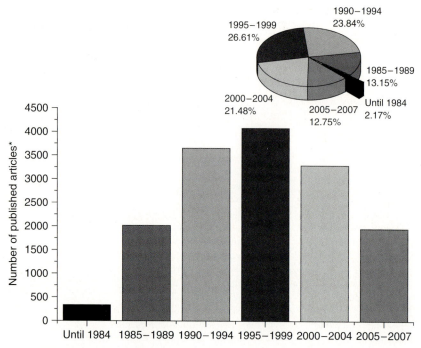

*Based on papers indexed in the chemical abstract

Figure 4.6 Number of published papers concerning LB films distributed by decades. Source Chemical Abstracts.

and 1990s, as indicated in Figure 4.6. This coincided with exploration of LB films from nontraditional amphiphilic molecules. In a subsidiary survey, we found that a similar distribution would be obtained if the data were collected from databases such as the ISI-Web of Science. It is worth noting that the number of published articles per year in the 2005–2007 period is equivalent to that of 2000–2004 (circa 2000), indicating a relatively constant production in the field in the present decade.

Research into LB films was first concentrated in Europe, mostly in England. In the period between 1985 and 1995, the Japanese and American contributions predominated, while over the last few years a strong Chinese participation has appeared, as shown in Figure 4.7.

Figure 4.8 shows that, according to data collected from ISI-Web of Science, research on LB films was initially mainly concentrated on physics, including applied physics and condensed matter. Later, LB films were studied in materials science and chemistry, with physical chemistry becoming the main area for such studies. In recent years, there has been a noticeable interest from other areas such as polymer science, optics, biophysics, and biochemistry.

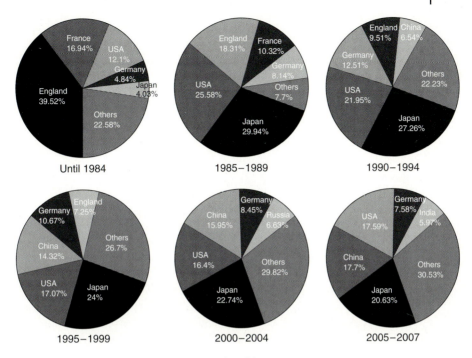

Figure 4.7 Published papers on LB films distributed by country/decade. Source ISI – Web of Science.

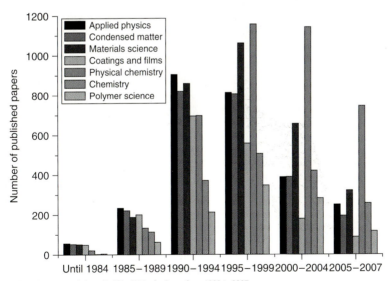

*Based on papers indexed in ISI – Web of science from 1900 to 2007

Figure 4.8 Published papers concerning LB films distributed according to the subject. Source ISI – Web of Science.

4.6
Highlights of Research in LB Films

In this section, we shall mention work in LB films in several areas, which were chosen either because considerable research has been performed in a particular topic or because we believe that there is a trend toward increased use of the LB method for the given topic. Among the areas highlighted are those involving polymer LB films and the application in biology-related problems.

4.6.1
Molecular-Recognition Processes

One of the most important properties of some biomolecules is their ability to recognize specific molecules, in a process that is termed *molecular recognition*. This ability is widely exploited in biosensing, as will be mentioned later, and serves to investigate a number of physiological processes at the molecular level. The LB method is ideally suited for studying molecular-recognition processes, both for Langmuir monolayers at the air/water interface and for LB films. In Langmuir monolayers, molecular recognition through "lock and key" interactions has been reported [72, 73], which includes compounds relevant to environmental impacts [74]. Molecular recognition is identified via changes in the monolayer properties, such as surface pressure, surface potential, or vibrational spectra [72–77], which occur only when a specific guest molecule is coupled to the monolayer. Therefore, a challenge in investigating molecular-recognition mechanisms is to distinguish changes caused by specific interactions from those of nonspecific interactions. This is particularly relevant for the cases where monolayer properties depend strongly on electrostatic interactions, which are affected by incorporation of charges and dipoles in the system, with no need to have specific binding.

By way of illustration, Figure 4.9 depicts the mechanism involving the steroid cyclophane in a monolayer, having as a guest molecule the fluorescent naphthalene, 6-(p-toluidino)naphthalene-2-sulfonate (TNS). At low surface pressures, an open conformation is observed but the reduction of the cross-sectional area for cyclophane upon compression leads to a transition into a cavity conformation at high surface pressures. The binding is reversible as demonstrated with several compression–decompression cycles [76].

With regard to LB films, molecular recognition has mainly being exploited in biosensing, as will be discussed in Section 4.6.2. Another example is the use of poly(acryroyloxymethyluracil-co-hexylacrylamide)s (poly (AU-co-HAAm)s) LB films that interact specifically with adenosine molecules, as demonstrated in results obtained with a QCM and AFM [78]. Another molecular-recognition process has been reported for the amphiphile AzoAde, comprising an azobenzene in the hydrophobic part and adenine in the hydrophilic portion of the molecule, while interacting with oligonucleotides with a homogeneous base (dA30, dT30, dG30, and dC30). The interaction was examined using surface-pressure isotherms and spectroscopy techniques at the air/water interface. There was little interaction between AzoAde

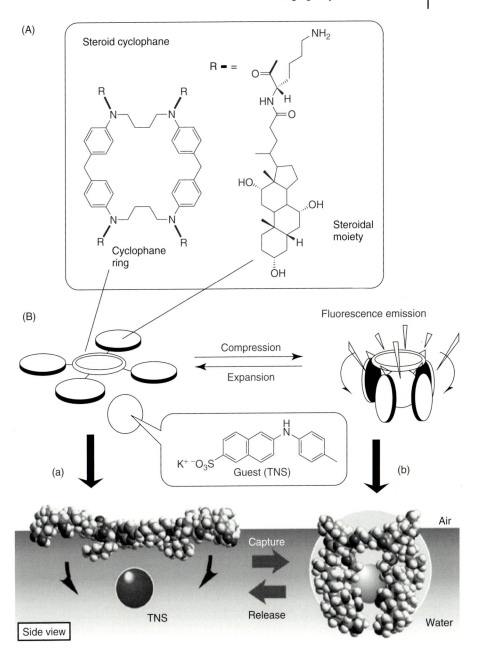

Figure 4.9 Chemical formula of the steroid cyclophane (A). Suggested models to explain conformational changes of cyclophane upon monolayer compression and reversible capture of the guest (6-(p-toluidino)naphthalene-2-sulfonate (TNS) (B). The models depict: (a) open conformation at low pressures and (b) cavity formation at high pressures. Reproduced from Ref. [76] with permission from The Royal Society of Chemistry.

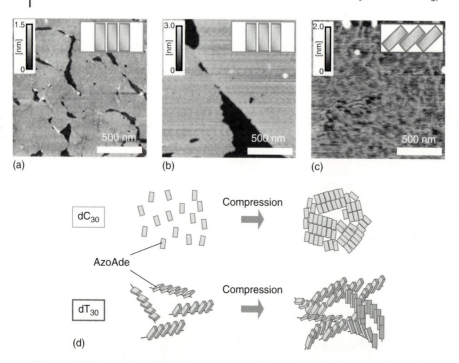

Figure 4.10 AFM images of AzoAde monolayers transferred onto freshly cleaved mica substrates from (a) TE buffer, (b) dC30, and (c) dT30 subphases, at a surface pressure of 15 mN/m. (d) Processes of morphology changes of AzoAde monolayers on dC30 and dT30 owing to compression. Reproduced from Ref. [79] with permission from ACS@2008.

and mismatched oligonucleotides (dA30, dG30, and dC30). However, AzoAde prepared on a dT30 subphase stoichiometrically assembled and interacted with dT30, subsequently forming a J-form assembly at the air/water interface. The AFM imaging of deposited LB films revealed the nanostructure of J-formed AzoAde monolayers on the dT30 subphase as well as the domain structures of H-formed monolayers on the other oligonucleotide subphases. Figure 4.10 shows the AFM images of one-layer LB films made with AzoAde, including changes after molecular recognition [79].

An excellent review on enzyme immobilization in LB films was presented by Girard-Egrot *et al.* in 2005 [80], where they drew attention to the need of an intimate association of LB structures containing proteolipids, or – glycans with the performance of the transducers, which allowed recognition and signal-transduction events in a single device.

4.6.1.1 Modeling Cell Membranes

The LB method has been useful for studying the mechanisms of action of several biologically relevant molecules in cell membranes, with the Langmuir monolayer or a deposited LB film serving as simplified membrane models. Several advantages exist in this approach, the most important being the possibility of determining the precise location of the guest molecules in the model membrane and the ability to probe interactions at distinct packing arrangements for the membrane. Again, molecular recognition is an important issue for some of the guest–host systems investigated.

Most of the studies with membrane models have been carried out only with Langmuir monolayers, especially those made with phospholipids that are known to be responsible for the structural framework of cell membranes. In order to mimic the constitution of cell membranes, zwitterionic phospholipids, namely dipalmitoyl phosphatidyl choline (DPPC) and dipalmitoyl phosphatidyl ethalonamine (DPPE), and the anionic phospholipids dipalmitoyl phosphatidic acid (DPPA) and dipalmitoyl phosphatidyl glycerol (DPPG) have been largely employed. A much smaller number of reports [81] have been made with deposited LB films, probably owing to the difficulties in depositing multilayers of phospholipids [82]. In Langmuir monolayers, the most-used characterization methods are the surface pressure and surface potential measurements, whose isotherms are usually affected by incorporation of biologically relevant molecules. Since the 1990s, several groups have studied interactions between lipid and proteins using surface-pressure and surface-potential isotherms [83–89], particularly the way the isotherms are affected by penetration of the proteins. A few years ago [90] it was found that phospholipid monolayers may respond cooperatively to the presence of pharmaceutical drugs, with measurable effects appearing for very low concentrations of the drugs. This cooperativity has later been observed for various other systems, including monolayers of sodium bis-2-ethylhexyl sulfosuccinate (AOT) [91] and other guest molecules such as proteins and peptides [92–99]. Molecular dynamics simulations also pointed to the possible cooperative response for chlorpromazine – an antipsychotic drug – interacting with DPPC monolayers [100]. Combined studies using Langmuir monolayers as cell membrane model and molecular simulation methods have proven that molecular packing of the membrane drives different modes of association for peptides that act in dimeric or nondimeric forms. With this type of study, it was possible to correlate differences in the primary structures with the different actions of peptides [98].

The insertion of the antibiotic griseofulvin (GF), an oral agent for treating dermatophytoses and a fungicide, into phospholipid membranes was studied by Corvis *et al.* [101]. The impact in film properties, assessed with physicochemical and enzymatic methods, was attributed to the GF retention in the monolayer induced by nonpolar interactions. The effects from GF on a DPPC monolayer are seen in the Brewster-angle microscopy (BAM) images in Figure 4.11. Significantly, the retention of GF in the membrane may be important to design GF formulations with increased bioavailability.

The investigation of interactions between guest molecules and model membranes consisting of Langmuir monolayers may be complemented with studies in the deposited LB films. Perhaps the first type of information inferred is whether the coupling between the guest molecule and the film-forming molecules is sufficiently

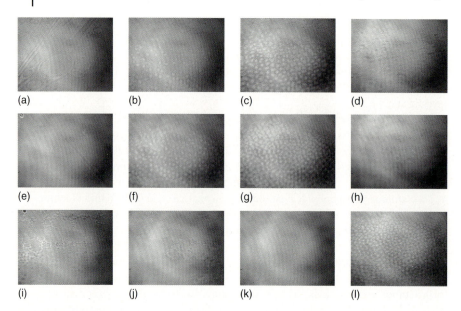

Figure 4.11 BAM images for neat DPPC monolayers are shown in (a–d), where the subphase was pure water. When the subphase contains 20 μM GF solution the images obtained are given in (e–h). In the images (i–l) the micrographs are for a DPPC/GF cospread monolayer. As for the area per molecule or pressure at which the pictures were taken: at 162, 169, and 178 Å^2 for (a, e, i), respectively, at 3.0 mN m^{-1} for (b, f, j), 4.0 for (c, g, k), and 10.0 mN m^{-1} for (d, h, l). Scale: the width of the snapshots corresponds to 400 μm. Reproduced from Ref. [101] with permission from ACS@2008.

strong for the transfer of both types of molecule onto the deposited LB film. Indeed, this was the case of chitosan – a polysaccharide obtained from the deacetylation of chitin extracted from the skin of arthropods [102] – interacting with a phospholipid monolayer. In a work using sum-frequency generation (SFG) and a QCM, it has been shown that chitosan can be transferred onto LB films, even though it only forms a subsurface below the monolayer at the air/water interface [103].

The motivation for a systematic study of the interaction between chitosan and membrane models arose from the many applications [103, 104] – some already in the market – of chitosan, whose activity as a bactericide and adequacy as drug-delivery carriers have been demonstrated [105, 106]. Yet, little is known about the molecular-level mechanisms for the action of chitosan. An example of the capabilities of the LB approach was provided by Caseli *et al.* [107], who showed that chitosan is able to sequester the protein β-lactoglobulin (BLG) originally adsorbed onto a negatively charged phospholipid monolayer at the air/water interface, thus preventing it from being deposited in the LB films. Figure 4.12 summarizes these findings, by indicating that chitosan diffuses from the aqueous subphase onto an already-formed phospholipid monolayer incorporating BLG, and removes BLG by complexation. The scheme shown in Figure 4.12 was supported by polarization-modulated infrared reflection-absorption spectroscopy (PM-IRRAS)

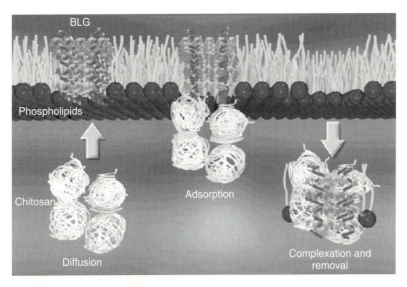

Figure 4.12 Model for chitosan action removing BLG from phospholipid monolayers. Reproduced from Ref. [107] with permission from ACS@2008.

data, which confirmed the absence of BLG after a given waiting time sufficient for the complexation with chitosan. Furthermore, results from QCM and infrared spectroscopy indicated that the deposited LB films contained BLG if transfer was performed immediately after spreading, that is, before chitosan could diffuse and remove the protein, but no BLG was transferred if deposition was carried out after a critical waiting time.

4.6.2
Sensing and Biosensing

The sensitivity of several types of sensors increases significantly if ultrathin films are used in the sensing units, for interfacial phenomena may be involved in the sensing. This is especially important when the method of detection is an electrical or an electrochemical measurement [108, 109]. Therefore, LB films have been widely employed in producing sensing units, with the added advantage of the possible control of molecular architecture, including synergistic combination of distinct materials in the same unit.

Of the many examples of LB films for sensing, we single out here two types: those using impedance spectroscopy as the principle of detection, leading for instance to taste sensors (electronic tongues), and the biosensors based on LB films containing biomolecules. The use of LB films in nanobioscience has been discussed in several publications [110–115].

Within the electronic tongue concept, the cross-sensitivity arising from the distinct responses of film-forming molecules to different analytes is exploited to obtain a fingerprint of a given liquid [116]. Highly sensitive electronic tongues have been achieved by combining impedance spectroscopy measurements and the use of LB or LbL films deposited onto interdigitated gold electrodes [108, 117–120], which were then used not only to identify the basic tastes – salty, sour, sweet, and bitter – at concentrations well below the human threshold [108, 117–120], but also to distinguish between liquids with very similar properties [121]. Indeed, with electronic tongues made with LB and LbL films, it has been possible to distinguish different types of wine [122, 123], coconut water [108], coffee samples [121, 124], and even correlate the electrical response to the human taste [125]. Furthermore, such sensors may be used in detecting trace amounts of pollutants [125, 126] and heavy metals [127, 128], being therefore useful for quality control in various types of industry. The importance of using impedance spectroscopy as the method of detection is associated with the strong dependence of the film electrical properties on small changes in the liquid. Using nanostructured LB or LbL films, on the other hand, is advantageous because it increases sensitivity as interfacial phenomena govern the electrical properties. In addition, with either the LB or LbL method, one may choose from a variety of materials that may be optimized for specific applications [108–129].

Electrochemical sensors for detecting dopamine have also been produced with LB films [109]. Gas and vapor sensors have been reported with LB films made with phthalocyanines or porphyrins, as the latter materials have their optical and electrical properties strongly affected by some vapors [125, 130–140]. This allowed the sensors to exhibit high sensitivity, in addition to good stability, and rapid response and recovery. In some cases, selectivity toward a specific type of vapor was also achieved.

With regard to biosensors, one of the main challenges is to preserve the activity of the immobilized molecules. In this context, the LB method has been shown to be suitable because the films normally contain entrained water, even after drying, which apparently is essential for keeping bioactivity. Furthermore, phospholipid LB films have been shown as adequate scaffolding for proteins, including enzymes, that are protected and may have their active sites exposed to the analytes. Indeed, there are reports of enhanced catalytic activity of enzymes in an LB film compared to that in a homogeneous solution [141].

Other examples of LB films applied as biosensors include antibody immunosensors using IgG monolayers transferred onto solid supports that preserve their antigen-recognition functionality [142, 143]. Biosensors made with convenient immobilization of enzymes are also numerous, including sensors for glucose [144–146] and choline [147]. LB films with enzymes immobilized with appropriate scaffolds were obtained with horseradish peroxidase and lipids for detecting H_2O_2 [148], and for cholesterol oxidase (Cox) or galactose oxidase incorporated in polyaniline and poly(3-hexyl thiophene) [149, 150]. Therefore, the majority of studies focused on biosensors that are promising for clinical diagnosis. Furthermore, biosensors have also been suggested for monitoring pollutants, as was the case of paraoxon (an organophosphorus compound), for which LB films were produced incorporating organophosphorus hydrolase (OPH) [151].

4.6.3
Signal Processing and Data Storage

LB films can be built with controlled thicknesses and layer structure, and in some cases, molecular order, with envisaged applications in signal processing and data storage [152–158]. Some examples are waveguiding, applications based on non-linear optical properties of the material (electro-optic, second-, and third-harmonic generation effects), optical storage, and surface-relief gratings (SRGs), in addition to photoluminescent, electroluminescent, photovoltaic devices, and transistors. Reviews on waveguiding and nonlinear optics applications of LB films can be found in the books of Petty [3, 62] Roberts [1], Ulman [4, 161] and in the reviews of Petty [31] and Kowel *et al.* [162], among others.

4.6.3.1 Optical Data Storage
The principal type of material explored in LB films for optical storage contains azoaromatic moieties, with nitrogen atoms linked by a double bond between two aromatic rings, generally phenyl, naphtyl, or heteroaromatic groups [61]. The main property exploited is the photoisomerization between the two isomers, namely the *cis* (or Z) and *trans* (or E) forms illustrated in Figure 4.13 for the azobenzene molecule.

The *trans* form is more thermodynamically stable, being thus predominant under room conditions. If the molecule is irradiated with light of a suitable wavelength, isomerization into *cis* takes place, but because the energy barrier is low the *cis* isomers are thermally converted back to the *trans* form following generally a first-order kinetics [163]. The *cis* isomer occupies a significantly higher volume than the *trans* isomer, also exhibiting different dipole moment and wavelength of maximum absorption. Following a *trans*–*cis*–*trans* isomerization cycle, the chromophores are reoriented, and may adopt any orientation. If excitation is performed with a linearly polarized light, the *trans* isomers possessing the dipole moment perpendicular to the light polarization do not absorb to undergo further photoisomerization. After some cycles, a larger number of *trans* isomers will have dipole moments oriented perpendicular to the light polarization, thus imparting anisotropy to the sample and hence dichroism and birefringence. After the source light is switched off, some relaxation occurs but a number of molecules remain oriented giving rise to a remnant birefringence. This photoinduced birefringence

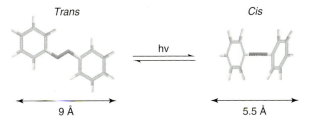

Figure 4.13 *Trans* and *cis* isomers of the azobenzene molecule.

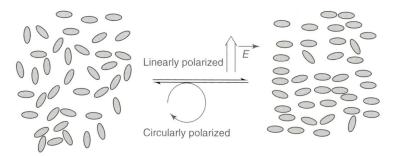

Figure 4.14 Illustration of a photoinduced molecular reorientation process.

is the basic property behind optical storage and SRG applications. Figure 4.14 illustrates the process of molecular reorientation induced by polarized light. The orientation can be erased using a circularly polarized light or by heating the sample.

A typical optical storage curve is illustrated in Figure 4.15. The transmitted signal (I) in Figure 4.15 is proportional to the birefringence (Δn) according to $\Delta n = \frac{\lambda}{\pi d} \sin^{-1} \sqrt{\frac{I}{I_0}}$ with λ being the wavelength of the incident radiation, d is the film thickness, and I_0 is the incident beam intensity. Before illumination of the film, the transmitted signal is zero (point A) since the chromophores in the film are randomly distributed (isotropic sample) and the sample is placed between two crossed polarizers. When the film is illuminated by a laser beam the transmission increases (writing process) and saturates at point B. After the laser is switched off, the transmission decreases to a nearly constant value, point C, that is, a considerable number of chromophores remained oriented. The isotropic distribution can be retrieved (erasing process) by illuminating with circularly polarized light at point D, or by heating the sample close to the glass-transition temperature for polymers. Generally, cycles of writing and erasing processes can be performed several times without significant influence on the photoinduced birefringence or degradation of the azo material, which is a desirable feature for optical-storage applications. Other desirable characteristics are short writing and

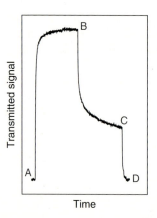

Figure 4.15 Typical optical-storage curve showing the initial point A where the chromophores are randomly distributed, and the writing process that saturates at point B, where the laser was switched off. Point C represents the signal from chromophores that remained oriented after relaxation, while in point D we depict the erasing process achieved by illuminating with circularly polarized light or by heating the sample.

reading times, high birefringence level, and stability of the residual birefringence [164].

Optical-storage experiments have been made in films of azoaromatic materials obtained with different methods, including casting, spin coating, vapor phase deposition (in the case of low molecular weight materials), dip coating, and sol-gel. Also used were film-forming methods that yield nanostructured films, namely the LbL and the LB methods. With the latter, one may study the photoinduced birefringence mechanism as close as possible to the molecular level [164], in addition to permitting fabrication of films with controlled thickness and architecture. One interesting feature of LbL films is the much longer time for writing information, in comparison with LB or spin-coated films, which is attributed to the strong electrostatic interactions between adjacent layers [165]. The electrostatic interactions hinder isomerization and reorientation of the chromophores.

As for the LB films, the photoinduced birefringence is normally higher than in cast films of the same azopolymers [166], though the amplitude of birefringence tends to decrease with increasing number of layers deposited, which is caused by a decrease in ordering in thick LB films. On the other hand, the number of layers has a negligible effect on the writing times and on the remnant birefringence in LB films [166]. The LB method is also important to design molecular architectures with controlled free volume for the chromophores, which is crucial for optical storage. For example, LB films from amphiphilic azochromophores, where an azobenzene moiety was attached to a long alkyl chain, are not amenable to optical storage [167], simply because the molecular packing in the LB film is such that no free volume is left for isomerization.

The attachment of azoaromatic molecules to polymer backbones (azopolymers) imparts several improvements in the optical storage properties of the material, combining the mechanical and thermal properties of the polymers with the optical characteristics of azoaromatic molecules. The optical properties are strongly affected by the structure of the polymeric backbone. For instance, polymers with high glass-transition temperatures, T_g, generally exhibit higher remnant birefringence and longer writing times owing to the polymer rigidity. Interactions between the polymer backbone and the chromophore and polymer chain packing, and consequently the free volume around the chromophore, also affect the writing and relaxation times of photoinduced birefringence, which range from a few seconds to several minutes [61]. These times are much longer than those of currently employed optical memories, therefore potential applications of azopolymers in optical storage should occur in devices such as the write-once memories.

In LB films of an azocrown ether [168] photoisomerization involved only the distortion of the crown, thus resulting in a small birefringence but with a very fast dynamics for the writing process. Indeed, the maximum birefringence could be achieved within only 1 s, a much shorter time than the minutes normally required for azobenzene-containing films, and this opens up a number of new possibilities for application. According to [168], the fast dynamics and small birefringence are due to the nanostructured nature of the LB films, in addition to the molecular architecture of the azocrown.

4.6.3.2 **Surface-Relief Gratings (SRGs)**

The *trans–cis–trans* cycles of the chromophores in the azopolymers can result in a mass transport due to the interaction of molecular dipoles with the electric field gradient of the laser light when an interference pattern of laser beams is used. The large scale molecular migration leads to regularly spaced SRGs on the film surface. The photoinscription of SRGs was described simultaneously by two research groups in the USA [169] and in Canada [170], where spin-coated films of azopolymers were used. A detailed description of the mechanisms and theoretical models proposed to explain the SRG formation can be found in Ref. [171].

The first photoinscription in LB films was performed by Mendonça *et al.* [166] on a 100-layer mixed LB film of HPDR13 and cadmium stearate (50 : 50 w/w). The LB film was 2.6 μm thick and the SRG photoinscribed, shown in the AFM image in Figure 4.16, had a peak-to-valley height in the range of 50–60 nm. The grating spacing can be adjusted by varying the angle between the laser beams according to the Bragg law: $\Lambda = \lambda/2 \sin(\theta/2)$ where Λ is the grating spacing, λ is the laser wavelength, and θ is the angle between the laser beams of the interference pattern. The maximum height that can be obtained depends on the film thickness (or number of deposited layers) since the depth of an SRG increases with the third power of the thickness for thin films [172]. As a large number of layers are required to achieve a significant peak-to-valley height, only a few reports on SRGs in LB films can be found.

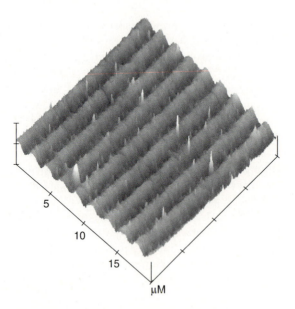

Figure 4.16 Surface-relief grating inscribed on a LB film. Reproduced from Ref. [166] with permission from ACS@2008.

4.6.3.3 **Photo- and Electroluminescent Devices**

The nanostructure, uniformity, and thickness control provided by the LB technique can be advantageous in applications such as light-emitting and photovoltaic devices. The possibility of inducing molecular orientation, as shown for phthalocyanines, porphyrins [173, 174], and some conjugated polymers, allows one to obtain devices with enhanced properties [175, 176].

Here, we highlight the electroluminescent devices, for the production of organic light-emitting diodes (OLEDs) or polymer light-emitting diodes (PLEDs) may bring a revolution to the display industry with large area, flexible displays [177] being used in various applications, including illumination [178]. Recent advances brought the efficiency in emitting close to that of fluorescent lamps, while the long-lasting problem of limited lifetime owing to photodegradation appears to have been solved with encapsulation [179]. Some applications are already in the market, such as in displays of small areas. A typical LED device comprises a multilayer structure, in which an emitting layer is accompanied by a hole transport and an electron transport layers, sandwiched between two electrodes, one of which needs to be transparent [180] as illustrated in Figure 4.17.

The main polymers investigated for PLEDs [181] include polyfluorenes and poly(p-phenylene-vinylene)s, which have also been used to produce LB films [182–185]. The thrust of the work on LB films – and indeed on LbL films – has been to understand the mechanisms of charge injection and transport through the multilayer structure, in addition to the hole–electron recombination that yields light emission [186]. There are also special cases where control of light emission is essential. For example, polymeric chains can be arranged in a way to produce anisotropic properties, such as polarized light emission. This was performed with LbL and LB techniques [175, 176]. The nanostructure produced by the LB

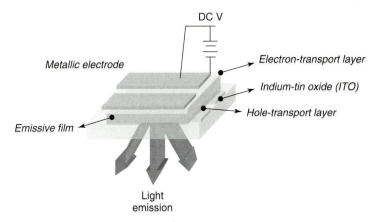

Figure 4.17 Typical structure of a multilayer LED device, containing a conductive solid support (glass or PET coated with ITO), an electron-transport layer (ETL), an emissive layer, a hole-transport layer (HTL) and a metallic contact.

technique led to more efficient electroluminescent devices, whose turn on voltage was lower (around 1–3 V) than for devices made with cast films.

In the case of photovoltaic devices, the LB technique may be exploited to build organized layers of polythiophenes, porphyrins, and phthalocyanines [187–189]. Recent research has shown that film properties are strongly influenced by their highly anisotropic nature [176, 190], with an essentially two-dimensional organization of their constituent molecules. Novel effects in charge transport at the molecular scale (approximately 1 nm) appear to be connected with very strong electron–phonon interactions, whose origins are still not understood [191].

4.7
Concluding Remarks

We have tried to provide the reader with a picture of latest developments and outlook for LB films. In terms of types of materials used, the relative importance of polymers is noteworthy, which is probably driven by the possible applications of polymer LB films. Though the order in such films is considerably less than in traditional amphiphilic materials, there has been ample evidence for the distinctive properties of polymer LB films owing to their LbL feature [108].

An analysis of the prospects for LB films in the years to come must necessarily include these films in the context of organic devices, in the now called *field of organic electronics*. Many of the ideas now being developed in organic electronics were inspired in earlier works on LB films, and indeed on other methods to form organized films. It suffices to say that LB films were considered essential building blocks for molecular electronics [192], as with rectifying molecular monolayers self-assembled on gold. For instance, switching was proven with LB films of bacteriorhodopsin [193] and ferroelectric polymers [194]. But perhaps the major market for the LB technology in the next few years is on biomolecules, especially if connected to investigations into physiological effects and medical applications. As shown in our survey of the research areas where the LB method is most used, relatively little research is done in Departments of Biology or Medicine. Yet, the LB method is now the most suitable means for a routine assessment of effects in membrane interactions.

Acknowledgments

We are grateful to Dr. M. Elisabete Zaniquelli for her helpful comments. This work was supported by FAPESP and CNPq (Brazil).

References

1. Roberts, G. (1990) *Langmuir–Blodgett Films*, Plenum Press, New York.
2. Oliveira, O.N. Jr., He, J.-A., Zucolotto, V., Balasubramanian, S., Li, L.,

Nalwa, H.S., Kumar, J., and Tripathy, S.K. (2002) Layer-by-Layer polyelectrolyte-based thin films for electronic and photonic applications, in *Handbook of Polyelectrolytes*, Chapter 1 vol. 1 (eds J. Kumar, S.K. Tripathy, and H.S. Nalwa), American Scientific Publishers, Los Angeles, pp. 1–37.

3. Petty, M.C. (1996) *Langmuir–Blodgett Films*, Cambridge University Press, Cambridge.

4. Ulman, A. (1991) *An Introduction to Organic Ultrathin Films from Langmuir–Blodgett to Self-Assembly*, Chapter 2, Academic Press, Boston.

5. Tredgold, R.H. (1994) *Order in Thin Organic Films*, Chapter 1, Cambridge University Press, Cambridge.

6. Fulford, G.D. (1968) Pouring holy oil on troubled water. *Isis*, **59**, 198–199.

7. Taishō, W.-E. (1990) *A Dictionary of Japanese Art Terms*, Bijutsu Co. Ltd, Tokyo, Japan.

8. Pockels, A. (1891) Surface tension. *Nature*, **43**, 437–439.

9. Pockels, A. (1892) On the relative contamination of water-surface by equal quantities of different substances. *Nature*, **46**, 418–419.

10. Pockels, A. (1893) Relations between the surface tension and relative contamination of water-surface. *Nature*, **48**, 152–154.

11. Langmuir, I. (1920) The mechanism of the surface phenomena of flotation. *Trans. Faraday Soc.*, **15**, 62–74.

12. Gaines, G.L. Jr. (1983) On the history of Langmuir–Blodgett films. *Thin Solid Films*, **99**, ix–xiii.

13. Davis, K.A. (1984) Katharine Blodgett and thin films. *J. Chem. Educ.*, **61**, 437–439.

14. Blodgett, K.B. (1934) Monomolecular films of fatty acids on glass. *J. Am. Chem. Soc.*, **56**, 495–495.

15. Blodgett, K.B. (1935) Films built by depositing successive monomolecular layers on a solid surface. *J. Am. Chem. Soc.*, **57**, 1007–1022.

16. Kuhn, H., Mobius, D., and Bucher, H. (1972) Spectroscopy of monolayer assemblies, in *Techniques of Chemistry*, vol. 1 (eds A. Weissberger and B. Rossiter), John Wiley & Sons, Inc., New York, pp. 577–702.

17. Cemel, A., Fort, T., and Lando, J.B. (1972) Polymerization of vinyl stearate multilayers. *J. Polym. Sci., Part A1*, **10**, 2061–2083.

18. Breton, M. (1981) Formation and possible applications of polymeric Langmuir–Blodgett films. A review. *Polym. Rev.*, **21**, 61–87.

19. Tredgold, R.H. and Winter, C.S. (1982) Langmuir–Blodgett monolayers of preformed polymers. *J. Phys. D: Appl. Phys.*, **15**, L55–L58.

20. Iyoda, T., Ando, M., Kaneko, T., Ohtani, A., Shimidzu, T., and Honda, K.C. (1987) Electron-microscopic evidence for the layered structure of a conducting polypyrrole Langmuir–Blodgett-film. *Langmuir*, **6**, 1169–1170.

21. Elshabini-Riad, A. and Barlow, F.D. (1998) *Thin Film Technology Handbook*, Chapter 2, McGraw Hill.

22. Tanigaki, N., Yase, K., Kaito, A., and Ueno, K. (1995) Highly oriented films of poly(dimethylsilylene) by friction deposition. *Polymer*, **36**, 2477–2480.

23. Decher, G. and Hong, J.D. (1991) Buildup of ultrathin multilayer films by a self-assembly process. 1. Consecutive adsorption of anionic and cationic bipolar amphiphiles on charged surfaces. *Macromol. Chem. Macromol. Symp.*, **46**, 321–327.

24. Decher, G. and Hong, J.D. (1991) Buildup of ultrathin multilayer films by a self-assembly process. 2. Consecutive adsorption of anionic and cationic bipolar amphiphiles and polyelectrolytes on charged surfaces. *Ber. Phys. Chem. Chem. Phys.*, **95**, 1430–1434.

25. Decher, G., Hong, J.D., and Schmitt, J. (1992) Buildup of ultrathin multilayer films by a self-assembly process. 3. Consecutively alternating adsorption of anionic and cationic polyelectrolytes on charged surfaces. *Thin Solid Films*, **210/211**, 831–835.

26. Whitesides, G.M., Mathias, J.P., and Seto, C.T. (1991) Molecular self-assembly and nanochemistry: a chemical strategy for the synthesis of nanostructures. *Science*, **254**, 1312–1319.

27. Hoeppener, S., Maoz, R., Cohen, S.R., Chi, L.F., Fuchs, H., and Sagiv, J. (2002) Metal nanoparticles, nanowires,

and contact electrodes self-assembled on patterned monolayer templates: a bottom-up chemical approach. *Adv. Mater.*, **14**, 1036–1041.

28. Decher, G. (1997) Fuzzy nanoassemblies: toward layered polymeric multi-composites. *Science*, **277**, 1232–1237.

29. Mamedov, A.A. (1994) Free standing layer-by-layer assemblies of magnetite nanoparticles. *Langmuir*, **13**, 5530–5533.

30. Buttry, D.A. and Ward, M.D. (1992) Measurement of interfacial processes at electrode surfaces with the electrochemical quartz crystal microbalance. *Chem. Rev.*, **92**, 1355–1379.

31. Petty, M.C. (1992) Possible applications for Langmuir–Blodgett films. *Thin Solid Films*, **210/211**, 417–426.

32. Schwartz, D.K. (1997) Langmuir–Blodgett film structure. *Surf. Sci. Rep.*, **27**, 241–334.

33. Lifshin, E. (1992) *Characterization of Materials–Materials Science and Technology: A Comprehensive Treatment*, vol **2a**, VCH publishers.

34. Caldwell, W.B., Campbell, D.J., Chen, K.M., Herr, B.R., Mirkin, C.A., Malik, A., Durbin, M.K., Dutta, P., and Huang, K.G. (1995) A highly ordered self-assembled monolayer film of an azobenzenealkanethiol on Au(111): electrochemical properties and structural characterization by synchrotron in-plane X-ray-diffraction, atomic-force microscopy, and surface-enhanced Raman-spectroscopy. *J. Am. Chem. Soc.*, **117**, 6071–6082.

35. Anderson, J.L., Coury, L.A. Jr., and Leddy, J. (1998) Dynamic electrochemistry: methodology and application. *Anal. Chem.*, **70**, 519R–589R.

36. Islam, A.K.M.M. and Mukherjee, M. (2008) Characterization of Langmuir–Blodgett film using differential charging in X-ray photoelectron spectroscopy. *J. Phys. Chem. B*, **112**, 8523–8529.

37. Ruiz, V., Nicholson, P.G., Jollands, S., Thomas, P.A., Macpherson, J.V., and Unwin, P.R. (2005) Molecular ordering and 2D conductivity in ultrathin poly(3-hexylthiophene)/gold nanoparticle composite films. *J. Phys. Chem. B*, **109**, 19335–19344.

38. Oliveira, O.N. Jr., Raposo, M., and Dhanabalan, A. (2001) *Handbook of Surfaces and Interfaces of Materials*, Chapter 1, vol. 4 (ed. H.S. Nalwa), Academic Press.

39. Mitzi, D.B. (2001) Thin-film deposition of organic-inorganic hybrid materials. *Chem. Mater.*, **13**, 3283–3298.

40. Sastry, M., Mayya, K.S., Patil, V., Paranjape, D.V., and Hegde, S.G. (1997) Langmuir–Blodgett films of carboxylic acid derivatized silver colloidal particles: role of subphase pH on degree of cluster incorporation. *J. Phys. Chem. B*, **101**, 4954–4958.

41. Coronado, E. and Mingotaud, C. (1999) Hybrid organic/inorganic Langmuir–Blodgett films. A supramolecular approach to ultrathin magnetic films. *Adv. Mater.*, **11**, 869–872.

42. Charych, D., Cheng, Q., Reichert, A., Kuziemko, G., Stroh, M., Nagy, J.O., Spevak, W., and Stevens, R.C. (1996) A 'litmus test' for molecular recognition using artificial membranes. *Chem. Biol.*, **3**, 113–120.

43. Kurnaz, M.L. and Schwartz, D.K. (1996) Morphology of microphase separation in arachidic acid/cadmium arachidate Langmuir–Blodgett multilayers. *J. Phys. Chem.*, **100**, 11113–11119.

44. Dori, Y., Bianco-Peled, H., Satija, S.K., Fields, G.B., McCarthy, J.B., and Tirrell, M. (2000) Ligand accessibility as means to control cell response to bioactive bilayer membranes. *J. Biomed. Mater. Res.*, **50**, 75–81.

45. Torre, G., Vazquez, P., Agullo-Lopez, F., and Torres, T. (1998) Phthalocyanines and related compounds: organic targets for nonlinear optical applications. *J. Mater. Chem.*, **8**, 1671–1683.

46. Mortimer, R.J. (1997) Electrochromic materials. *Chem. Soc. Rev.*, **26**, 147–156.

47. Guillaud, G., Simon, J., and Germain, J.P. (1998) Metallophthalocyanines gas sensors, resistors and field effect transistors. *Coord. Chem. Rev.*, **178**, 1433–1484.

48. Pedrosa, J.M., Romero, M.T.M., and Camacho, L.J. (2002) Organization of an amphiphilic azobenzene derivative in monolayers at the air/water interface. *Phys. Chem. B*, **106**, 2583–2591.

49. Matsumoto, M., Miyazaki, D., Tanaka, M., Azumi, R., Manda, E., Kondo, Y., Yoshino, N., and Tachibana, H. (1998) Reversible light-induced morphological change in Langmuir–Blodgett films. *J. Am. Chem. Soc.*, **120**, 1479–1484.

50. Overney, R.M., Meyer, E., Frommer, J., Guntherodt, H.J., Decher, G., Reibel, J., and Sohling, U. (1993) A comparative atomic force microscopic study of liquid-crystal films: transferred freely-suspended vs Langmuir–Blodgett. Morphology, lattice, and manipulation. *Langmuir*, **9**, 341–346.

51. Maliszewskyj, N.C., Mindyuk, O.Y., Heiney, P.A., Josefowicz, J.Y., Schuhmacher, P., and Ringsdorf, H. (1999) Structural phase transition in ultrathin films of disk-shaped molecules. *Liq. Cryst.*, **26**, 31–36.

52. Malmsten, M. (1999) Studies of serum protein adsorption at phospholipid surfaces in relation to intravenous drug delivery. *Colloids Surf. A Physicochem. Eng. Asp.*, **159**, 77–87.

53. Sukhorukov, G.B., Montrel, M.M., Petrov, A.I., Shabarchina, L.I., and Sukhorukov, B.I. (1996) Multilayer films containing immobilized nucleic acids. Their structure and possibilities in biosensor applications. *Biosens. Bioelectron.*, **11**, 913–922.

54. Hernandez-Lopez, J.L., Alvizo-Paez, E.R., Moya, S.E., and Ruiz-Garcia, J. (2006) Trapping, pattern formation, and ordering of polyelectrolyte/single-wall carbon nanotube complexes at the air/water and air/solid interfaces. *J. Phys. Chem. B*, **110**, 23179–23191.

55. Kim, Y., Minami, N., Zhu, W.H., Kazaoui, S., Azumi, R., and Matsumoto, M. (2003) Langmuir–Blodgett films of single-wall carbon nanotubes: layer-by-layer deposition and in-plane orientation of tubes. *Jpn. J. App. Phys.*, **42**, 7629–7634.

56. Panda, A.B., Acharya, S., Efrima, S., and Golan, Y. (2007) Synthesis, assembly, and optical properties of shape- and phase-controlled ZnSe nanostructures. *Langmuir*, **23**, 765–770.

57. Xu, J.M., J., X.J., Gattas-Asfura, K.M., Kerim, M., Wang, C.S., and Leblanc, R.M. (2006) Langmuir and Langmuir–Blodgett films of quantum dots. *Colloids Surf. A Physicochem. Eng. Asp.*, **284**, 35–42.

58. Sandberg, H.G.O., Frey, G.L., Shkunov, M.N., Sirringhaus, H., Friend, R.H., Nielsen, M.M., and Kumpf, C. (2002) Ultrathin regioregular poly(3-hexyl thiophene) field-effect transistors. *Langmuir*, **18**, 10176–10182.

59. Kraft, A., Grimsdale, A.C., and Holmes, A.B. (1998) Electroluminescent conjugated polymers: seeing polymers in a new light. *Angew. Chem. Int. Ed.*, **37**, 402–428.

60. Saxena, V. and Malhotra, B.D. (2003) Prospects of conducting polymers in molecular electronics. *Curr. App. Phys.*, **3**, 293–305.

61. Oliveira, O.N. Jr., dos Santos, D.S. Jr., Balogh, D.T., Zucolotto, V., and Mendonça, C.R. (2005) Optical storage and surface-relief gratings in azobenzene-containing nanostructured films. *Adv. Colloid Interface Sci.*, **116**, 179–192.

62. Cresswell, J.P., Petty, M.C., Ferguson, I., Hutchings, M., Allen, S., Ryan, T.G., Wang, C.H., and Wherrett, B.S. (1996) Langmuir–Blodgett deposition and second-order non-linear optics of several azobenzene dye polymers. *Adv. Mater. Opt. Electron.*, **6**, 33–38.

63. Takatoshi, K. (1998) Photoresponsive membrane systems. *J. Photochem. Photobiol. B: Biol.*, **42**, 12–19.

64. Penner, T.L., Schildkraut, J.S., Ringsdorf, H., and Schuster, A. (1991) Oriented films from polymeric amphiphiles with mesogenic groups: Langmuir–Blodgett liquid-crystals. *Macromolecules*, **24**, 1041–1049.

65. Vandevyver, M., Albouy, P.A., Mingotaud, C., Perez, J., and Barraud, A. (1993) Inplane anisotropy and phase-change in Langmuir–Blodgett films of disk-like molecules. *Langmuir*, **9**, 1561–1567.

66. Devereaux, C.A. and Baker, S.M. (2002) Surface features in Langmuir–Blodgett monolayers of predominantly hydrophobic poly(styrene)-poly(ethylene oxide) diblock copolymer. *Macromolecules*, **35**, 1921–1927.

67. Baker, S.M., Leach, K.A., Devereaux, C.E., and Gragson, D.E. (2000) Controlled patterning of diblock copolymers by monolayer Langmuir–Blodgett deposition. *Macromolecules*, **33**, 5432–5436.

68. Song, H., Kim, F., Connor, S., Somorjai, G.A., and Yang, P.D. (2005) Pt nanocrystals: shape control and Langmuir–Blodgett monolayer formation. *J. Phys. Chem. B*, **109**, 188–193.

69. Peng, X.G., Lu, R., Zhao, Y.Y., Qu, L.H., Chen, H.Y., and Li, T.J. (1994) Control of distance and size of inorganic nanoparticles by organic matrices in ordered LB monolayers. *J. Phys. Chem*, **98**, 7052–7055.

70. Pasquini, D., Balogh, D.T., Antunes, P.A., Constantino, C.J.L., Curvelo, A.A.S., Aroca, R.F., and Oliveira, O.N. Jr. (2002) Surface morphology and molecular organization of lignins in Langmuir–Blodgett films. *Langmuir*, **18**, 6593–6596.

71. Wegner, G. (1996) Nanoarchitectures and molecular composites as advanced materials. *Macromol. Symp.*, **104**, 29–30.

72. Leblanc, R.M. (2006) Molecular recognition at Langmuir monolayers. *Curr. Opin. Chem. Biol.*, **10**, 529–536.

73. Ariga, K., Hill, J.P., and Endo, H. (2007) Developments in molecular recognition and sensing at interfaces. *Int. J. Mol. Sci.*, **8**, 864–883.

74. Wang, C., Zheng, J., Oliveira, O.N. Jr., and Leblanc, R.M. (2007) Nature of the interaction between a peptidolipid Langmuir monolayer and paraoxon in the subphase. *J. Phys. Chem. C*, **111**, 7826–7833.

75. Wang, Y.C., Du, X.Z., Miao, W.G., and Liang, Y.Q. (2006) Molecular recognition of cytosine- and guanine-functionalized nucleolipids in the mixed monolayers at the air/water interface and Langmuir–Blodgett films. *J. Phys. Chem. B*, **110**, 4914–4923.

76. Volhart, D. (2008) Interfacial molecular recognition of non-surface-active species at Langmuir monolayers. *Curr. Opin. Colloid Interface Sci.*, **13**, 31–39.

77. Ariga, K., Hill, J.P., and Wakayam, Y. (2008) Supramolecular chemistry in two dimensions: self-assembly and dynamic function. *Phys. Status Solidi A: Appl. Mater.*, **205**, 1249–1257.

78. Sugiyama, N., Hirakawa, M., Zhu, H., Takeoka, Y., and Rikukawa, M. (2008) Molecular recognition of Langmuir–Blodgett polymer films containing uracil groups. *Colloids Surf. A Physicochem. Eng. Asp.*, **321**, 60–64.

79. Haruta, O., Matsuo, Y., Hashimoto, Y., Niikura, K., and Ijiro, K. (2008) Sequence-specific control of azobenzene assemblies by molecular recognition of DNA. *Langmuir*, **24**, 2618–2624.

80. Girard-Egrot, A.P., Godoy, S., and Blum, L.J. (2005) Enzyme association with lipidic Langmuir–Blodgett films: interests and applications in nanobioscience. *Adv. Colloid Interface Sci.*, **116**, 205–225.

81. Caseli, L., Furriel, R.P.M., Andrade, J.F., Leone, F.A., and Zaniquelli, M.E.D. (2004) Surface density as a significant parameter for the enzymatic activity of two forms of alkaline phosphatase immobilized on phospholipid Langmuir–Blodgett films. *J. Colloid Interface. Sci.*, **275**, 123–130.

82. Moraille, P. and Badia, A. (2003) Nanoscale stripe patterns in phospholipid bilayers formed by the Langmuir–Blodgett technique. *Langmuir*, **19**, 8041–8049.

83. Dhathathreyan, A. (1993) Interaction of atriopeptin III with lipids: a monolayer study. *Colloids Surf. A Physicochem. Eng. Asp*, **81**, 269–272.

84. Viitala, T., Albers, W.M., Vikholm, I., and Peltonen, J. (1998) Synthesis and Langmuir film formation of N-(epsilon-maleimidocaproyl) (dilinoleoylphosphatidyl) ethanolamine. *Langmuir*, **14**, 1272–1277.

85. Zaitsev, S.Y., Kalabina, N.A., Herrmann, B., Schaefer, C., and

Zubov, V.P. (1999) A comparative study of the photosystem II membrane proteins with natural lipids in monolayers. *Mater. Sci. Eng. C Biomim. Sens. Syst.*, **8-9**, 519–522.

86. Vaknin, D., Kjaer, K., Alsnielsen, J., and Losche, M. (1991) Structural-properties of phosphatidylcholine in a monolayer at the air-water interface–neutron reflection study and reexamination of X-ray reflection measurements. *Biophys. J.*, **59**, 1325–1332.

87. Piepenstock, M. and Losche, M. (1992) Steric and electrostatic aspects of antibody-binding to hapten functionalized lipid monolayers. *Thin Solid Films*, **210/211**, 793–795.

88. Wang, J.Y., Vaknin, D., Uphaus, R.A., Kjaer, K., and Losche, M. (1994) Fullerene films and fullerene dodecylamine adduct monolayers at air/water interfaces studied by neutron and X-ray reflection. *Thin Solid Films*, **242**, 40–44.

89. Diederich, A., Sponer, C., Pum, D., Sleytr, U.B., and Losche, M. (1996) Reciprocal influence between the protein and lipid components of a lipid-protein membrane model. *Colloid Surf. B Biointerfaces*, **6**, 335–346.

90. Haas, H., Caetano, W., Borissevitch, G.P., Tabak, M., Sachez, M.I.M., Oliveira, O.N. Jr., Scalas, E., and Goldmann, M. (2001) Interaction of dipyridamole with phospholipid monolayers at the air/water interface: surface pressure and grazing incidence X-ray diffraction studies. *Chem. Phys. Lett.*, **335**, 510–516.

91. Caetano, W., Ferreira, M., Oliveira, O.N. Jr., and Itri, R. (2004) Enhanced stabilization of aerosol-OT surfactant monolayer upon interaction with small amounts of bovine serum albumin at the air/water interface. *Colloid Surf. B Biointerfaces*, **38**, 21–27.

92. Losche, M. (1997) Protein monolayers at interfaces. *Curr. Opin. Solid State Mater. Sci.*, **2**, 546–556.

93. Ferreira, M., Dynarowicz-Latka, P., Minones, J., Caetano, W., Kita, K., Schalke, M., Losche, M., and Oliveira, O.N. Jr. (2002) On the origin of the plateau in surface-pressure isotherms of aromatic carboxylic acids. *J. Phys. Chem. B*, **106**, 10395–10400.

94. Hidalgo, A.A., Caetano, W., Tabak, M., and Oliveira, O.N. Jr. (2004) Interaction of two phenothiazine derivatives with phospholipid monolayers. *Biophys. Chem.*, **109**, 85–104.

95. de Souza, N.C., Caetano, W., Itri, R., Rodrigues, C.A., Oliveira, O.N. Jr., Giacometti, J.A., and Ferreira, M. (2006) Interaction of small amounts of bovine serum albumin with phospholipid monolayers investigated by surface pressure and atomic force microscopy. *J. Colloid Interface Sci.*, **297**, 546–553.

96. Maget, D. (1999) The monolayer technique: a potent tool for studying the interfacial properties of antimicrobial and membrane-lytic peptides and their interactions with lipid membranes. *Biochim. Biophys. Acta*, **1462**, 109–140.

97. McQuaw, C.M., Zheng, L., Ewing, A.G., and Winograd, N. (2007) Localization of sphingomyelin in cholesterol domains by imaging mass spectrometry. *Langmuir*, **23**, 5645–5650.

98. Kamilya, T., Pal, P., and Talapatra, G.B. (2007) Interaction of ovalbumin with phospholipids Langmuir–Blodgett film. *J. Phys. Chem. B*, **111**, 1199–1205.

99. Lourenzoni, M.R., Namba, A.M., Caseli, L., Degreve, L., and Zaniquelli, M.E.D. (2007) Study of the interactions of human defesins with cell membrane models: relationships with their structure and biological activity. *J. Phys. Chem. B*, **111**, 11318–11329.

100. Pickholz, M., Oliveira, O.N., Jr., and Skaf, M.S. (2007) Interactions of chlorpromazine with phospholipid monolayers: effects of the ionization state of the drug. *Biophys. Chem.*, **125**, 425–434.

101. Corvis, Y., Barzyk, W., Brezesinski, G., Mrabet, N., Badis, M., Hecht, S., and Rogalska, E. (2006) Interactions of a fungistatic antibiotic, griseofulvin, with phospholipid monolayers used as models of biological membranes. *Langmuir*, **22**, 7701–7711.

102. Kumar, M.N.V.R. (2000) A review of chitin and chitosan applications. *React. Func. Polym.*, **46**, 1–27.

103. Pavinatto, F.J., Caseli, L., Pavinatto, A., dos Santos, D.S. Jr., Nobre, T.M., Zaniquelli, M.E.D., Silva, H.S., Miranda, P.B., and Oliveira, O.N. Jr. (2007) Probing chitosan and phospholipid interactions using Langmuir and Langmuir–Blodgett films as cell membrane models. *Langmuir*, **23**, 7666–7671.

104. Pavinatto, F.J., Pavinatto, A., Caseli, L., dos Santos, D.S. Jr., Nobre, T.M., Zaniquelli, M.E.D., and Oliveira, O.N. Jr. (2007) Interaction of chitosan with cell membrane models at the air/water interface. *Biomacromolecules*, **8**, 1633–1640.

105. Dai, Y.N., Li, P., Zhang, J.P., Wang, A.O., and Wei, Q. (2008) Swelling characteristics and drug delivery properties of nifedipine-loaded pH sensitive alginate-chitosan hydrogel beads. *J. Biomed. Mater. Res. B*, **86B**, 493–500.

106. Chung, Y.C. and Chen, C.Y. (2008) Antibacterial characteristics and activity of acid-soluble chitosan. *Bioresour. Technol.*, **99**, 2806–2814.

107. Caseli, L., Pavinatto, F.J., Nobre, T.M., Zaniquelli, M.E.D., Viitala, T., and Oliveira, O.N. Jr. (2008) Chitosan as a removing agent of β-lactoglobulin from membrane models. *Langmuir*, **24**, 4150–4156.

108. Ferreira, M., Riul, A. Jr., Wohnrath, K., Fonseca, F.J., Oliveira, O.N. Jr., and Mattoso, L.H.C. (2003) High-performance taste sensor made from Langmuir–Blodgett films of conducting polymers and a ruthenium complex. *Anal. Chem.*, **75**, 953–955.

109. Ferreira, M., Dinelli, L.R., Wohnrath, K., Batista, A.A., and Oliveira, O.N. Jr. (2004) Langmuir–Blodgett films from polyaniline/ruthenium complexes as modified electrodes for detection of dopamine. *Thin Solid Films*, **446**, 301–306.

110. Girard-Egrot, A.P., Godoy, S., and Blum, L.J. (2005) Enzyme association with lipidic Langmuir–Blodgett films:

Interests and applications in nanobioscience. *Adv. Colloid Interface Sci.*, **116**, 205–225, and references therein.

111. Godoy, S., Chauvet, J.P., Boullanger, P., Blum, L.J., and Girard-Egrot, A.P. (2003) New functional proteo-glycolipidic molecular assembly for biocatalysis analysis of an immobilized enzyme in a biomimetic nanostructure. *Langmuir*, **19**, 5448–5456.

112. Caseli, L., Crespilho, F.N., Nobre, T.M., Zaniquelli, M.E.D., Zucolotto, V., and Oliveira, O.N. Jr. (2008) Using phospholipid Langmuir and Langmuir–Blodgett films as matrix for urease immobilization. *J. Colloid Interface Sci.*, **319**, 100–108.

113. Caseli, L., Moraes, M.L., Zucolotto, V., Ferreira, M., Nobre, T.M., Zaniquelli, M.E.D., Rodrigues-Filho, U.P., and Oliveira, O.N. Jr. (2006) Fabrication of phytic acid sensor based on mixed phytase-lipid Langmuir–Blodgett films. *Langmuir*, **22**, 8501–8508.

114. Cabaj, J., Idzik, K., Sołoducho, J., Chyla, A., Bryjak, J., and Doskocz, J. (2008) Well-ordered thin films as practical components of biosensors. *Thin Solid Films*, **516**, 1171–1174.

115. Katsuhiko, A., Takashi, N., and Tsuyoshi Michinobu, M. (2006) Immobilization of biomaterials to nano-assembled films (self-assembled monolayers, Langmuir–Blodgett films, and layer-by-layer assemblies) and their related functions. *J. Nanosci. Nanotechnol.*, **6**, 2278–2301.

116. Vlasov, Y.G., Legin, A.V., Rudnitskaya, A.M., D'Amico, A., and Di Natale, C. (2000) Electronic tongue: new analytical tool for liquid analysis on the basis of non-specific sensors and methods of pattern recognition. *Sens. Actuators B Chem.*, **65**, 235–236.

117. Constantino, C.J.L., Antunes, P.A., Venancio, E.C., Consolin, N., Fonseca, F.J., Mattoso, L.H.C., Aroca, R.F., Oliveira, O.N., and Riul, A. Jr. (2004) Nanostructured films of perylene derivatives: high performance materials for taste sensor applications. *Sensor Lett.*, **2**, 95–101.

118. Riul, A., Malmegrim, R.R., Fonseca, F.J., and Mattoso, L.H.C. (2003) An artificial taste sensor based on conducting polymers. *Biosens. Bioelectron.*, **18**, 1365–1369.

119. Ferreira, M., Constantino, C.J.L., Riul, A., Wohnrath, K., Aroca, R.F., Giacometti, J.A., Oliveira, O.N. Jr., and Mattoso, L.H.C. (2003) Preparation, characterization and taste sensing properties of Langmuir–Blodgett films from mixtures of polyaniline and a ruthenium complex. *Polymer*, **44**, 4205–4211.

120. Riul, A. Jr., dos Santos, D.S. Jr., Wohnrath, K., Di Tommazo, R., Carvalho, A.C.P.L.F., Fonseca, F.J., Oliveira, O.N. Jr., Taylor, D.M., and Mattoso, L.H.C. (2002) Artificial taste sensor: efficient combination of sensors made from Langmuir–Blodgett films of conducting polymers and a ruthenium complex and self-assembled films of an azobenzene-containing polymer. *Langmuir*, **18**, 239–245.

121. Riul, A., Soto, A.M.G., Mello, S.V., Bone, S., Taylor, D.M., and Mattoso, L.H.C. (2003) An electronic tongue using polypyrrole and polyaniline. *Synth. Met.*, **132**, 109–116.

122. Borato, C.E., Riul, A., Ferreira, M., Oliveira, Jr. O.N., and Mattoso, L.H.C. (2004) Exploiting the versatility of taste sensors based on impedance spectroscopy. *Instrum. Sci. Technol.*, **32**, 21–30.

123. Riul, A. Jr., de Sousa, H.C., Malmegrim, R.R., dos Santos, D.S. Jr., Carvalho, A.C.P.L.F., Fonseca, F.J., Oliveira, O.N. Jr., and Mattoso, L.H.C. (2004) Wine classification by taste sensors made from ultra-thin films and using neural networks. *Sensors Act. B: Chem.*, **98**, 77–82.

124. Ferreira, E.J., Pereira, R.C.T., Delbem, A.C.B., Oliveira, O.N. Jr., and Mattoso, L.H.C. (2007) Random subspace method for analysing coffee with electronic tongue. *Electron. Lett.*, **43**, 1138–1140.

125. Souto, J., Rodríguez-Méndez, M.L., de Saja- González, J., and de Saja, J.A. (1996) AC conductivity of gas-sensitive Langmuir–Blodgett films of ytterbium bisphthalocyanine. *Thin Solid Films*, **284**, 888–890.

126. Abdelmalek, F., Shadaram, M., and Boushriha, H. (2001) Ellipsometry measurements and impedance spectroscopy on Langmuir–Blodgett membranes on Si/SiO$_2$ for ion sensitive sensor. *Sens. Actuators B Chem.*, **72**, 208–213.

127. Martins, G.F., Pereira, A.A., Straccalano, B.A., Antunes, P.A., Pasquini, D., Curvelo, A.A.S., Ferreira, M., Riul, A., and Constantino, C.J.L. (2008) Ultrathin films of lignins as a potential transducer in sensing applications involving heavy metal ions. *Sens. Actuators B Chem.*, **129**, 525–530.

128. Pereira, A.A., Martins, G.F., Antunes, P.A., Conrrado, R., Pasquini, D., Job, A.E., Curvelo, A.A.S., Ferreira, M., Riul, A., and Constantino, C.J.L. (2007) Lignin from sugar cane bagasse: Extraction, fabrication of nanostructured films, and application. *Langmuir*, **23**, 6652–6659.

129. Aoki, P.H.B., Caetano, W., Volpati, D., Riul, A., and Constantino, C.J.L. (2008) Sensor array made with nanostructured films to detect a phenothiazine compound. *J. Nanosci. Nanotechnol.*, **8**, 1–8.

130. Zhao, J., Huo, L.-H., Gao, S., Zhao, H., Zhao, J.-G., and Li, N. (2007) Molecular orientation and gas-sensing properties of Langmuir–Blodgett films of copper phthalocyanine derivatives. *Sens. Actuators B Chem.*, **126**, 588–594.

131. Gu, C., Sun, L., Zhang, T., Li, T., and Zhang, X. (1998) High-sensitivity phthalocyanine LB film gas sensor based on field effect transistors. *Thin Solid Films*, **327**, 383–386.

132. Salleh, M.M., Akrajas, M., and Yahaya, M. (2002) Optical sensing of capsicum aroma using four porphyrins derivatives thin films. *Thin Solid Films*, **417**, 162–165.

133. Sauer, T., Caseri, W., Wegner, G., Vogel, A., and Hoffmann, B. (1990) Development of novel chemical sensor devices based on LB films from phthalocyaninato-polysiloxane polymers. *J. Phys. D: Appl. Phys.*, **23**, 79–84.

134. Fernandes, A.N. and Richardson, T.H. (2008) Conductometric gas sensing studies of tert-butyl silicon-[bis ethyloxy]-phthalocyanine LB films. *J. Mater. Sci.*, **43**, 1305–1310.

135. Miguel, G., Martín-Romero, M.T., Pedrosa, J.M., Muñoz, E., Pérez-Morales, M., Richardson, T.H., and Camacho, L. (2007) Improvement of optical gas sensing using LB films containing a water insoluble porphyrin organized in a calixarene matrix. *J. Mater. Chem.*, **17**, 2914–2920.

136. Açkbaş, Y., Evyapan, M., Ceyhan, T., Çapan, R., and Bekarodlu, Ö. (2007) Characterisation of Langmuir–Blodgett films of new multinuclear copper and zinc phthalocyanines and their sensing properties to volatile organic vapours. *Sens. Actuators B Chem.*, **123**, 1017–1024.

137. Çapan, R., Açkbaş, Y., and Evyapan, M. (2007) A study of Langmuir–Blodgett thin film for organic vapor detection. *Mater. Lett.*, **61**, 417–420.

138. Penza, M., Tagliente, M.A., Aversa, P., Re, M., and Cassano, G. (2007) The effect of purification of single-walled carbon nanotube bundles on the alcohol sensitivity of nanocomposite Langmuir–Blodgett films for SAW sensing applications. *Nanotechnology*, **18**, 185502.

139. Daly, S.M., Grassi, M., Shenoy, D.K., Ugozzoli, F., and Dalcanale, E. (2007) Supramolecular surface plasmon resonance (SPR) sensors for organophosphorus vapor detection. *J. Mater. Chem.*, **17**, 1809–1818.

140. Pedrosa, J.M., Dooling, C.M., Richardson, T.H., Hyde, R.K., Hunter, C.A., Martin, M.T., and Camacho, L. (2002) The optical gas-sensing properties of an asymmetrically substituted porphyrin. *J. Mater. Chem.*, **12**, 2659–2664.

141. Schmidt, T.F., Caseli, L., Nobre, T.M., Zaniquelli, M.E.D., and Oliveira, O.N. Jr. (2008) Interaction of horseradish peroxidase with Langmuir monolayers of phospholipids. *Colloids Surf. A Physicochem. Eng. Asp.*, **321**, 206–210.

142. Troitsky, V.I., Berzina, T.S., Pastorino, L., Bernasconi, E., and Nicolini, C. (2003) A new approach to the deposition of nanostructured biocatalytic films. *Nanotechnology*, **14**, 597–602.

143. Pastorino, L., Berzina, T.S., Troitsky, V.I., Fontana, M.P., Bernasconi, E., and Nicolini, C. (2002) Biocatalytic Langmuir–Blodgett assemblies based on penicillin G acylase. *Colloid Surf. B Biointerface*, **23**, 357–363.

144. Lee, Y.-L., Lin, J.Y., and Lee, S. (2007) Adsorption behavior of glucose oxidase on a dipalmitoylphosphatic acid monolayer and the characteristics of the mixed monolayer at air/liquid interfaces. *Langmuir*, **23**, 2042–2051.

145. Ohnuki, H., Saiki, T., Kusakari, A., Endo, H., Ichihara, M., and Izumi, M. (2007) Incorporation of glucose oxidase into Langmuir–Blodgett films based on Prussian blue applied to amperometric glucose biosensor. *Langmuir*, **23**, 4675–4681.

146. Kusakari, A., Izumi, M., and Ohnuki, H. (2008) Preparation of an enzymatic glucose sensor based on hybrid organic–inorganic Langmuir–Blodgett films: adsorption of glucose oxidase into positively charged molecular layers. *Colloids Surf. A Physicochem. Eng. Asp.*, **321**, 47–51.

147. Girard-Egrot, A.P., Morélis, R.M., and Coulet, P.R. (1998) Direct bioelectrochemical monitoring of choline oxidase kinetic behaviour in Langmuir–Blodgett nanostructures. *Bioelectrochem. Bioenerg.*, **46**, 39–44.

148. Kafi, A.K.M., Lee, D.-Y., Choi, W.-S., and Kwon, Y.-S. (2008) Fabrication and electrochemical characterization of HRP-lipid Langmuir–Blodgett film and its application as an H_2O_2 biosensor. *Thin Solid Films*, **516**, 3641–3645.

149. Matharu, Z., Sumana, G., Sunil, K.A., Singh, S.P., Vinay, G., and Malhotra, B.D. (2007) Polyaniline Langmuir–Blodgett-film-based cholesterol biosensor. *Langmuir*, **23**, 13188–13192.

150. Sharma, S.K., Singhal, R., Malhotra, B.D., Sehgal, N., and Kumara, A. (2004) Langmuir–Blodgett film based biosensor for estimation of galactose in milk. *Electrochim. Acta*, **49**, 2479–2485.

151. Cao, X., Mello, S.V., Leblanc, R.M., Rastogi, V.K., Cheng, T.-C., and DeFrank, J.J. (2004) Detection of paraoxon by immobilized organophosphorus hydrolase in a Langmuir–Blodgett film. *Colloids Surf. A Physicochem. Eng. Asp.*, **250**, 349–356.

152. Angelov, T., Radev, D., Ivanov, G., Antonov, D., and Petrov, A.G. (2007) Hydrophobic magnetic nanoparticles: synthesis and LB film preparation. *J. Optoelectron. Adv. Mater.*, **9**, 424–426.

153. Acharya, S., Kamilya, T., Sarkar, J., Parichha, T.K., Pal, P., and Talapatra, G.B. (2007) Photophysical properties of 4-methyl 3-phenyl coumarin organized in Langmuir–Blodgett films: formation of aggregates. *Mater. Chem. Phys.*, **104**, 88–92.

154. Çapan, R. (2007) Pyroelectric and dielectric characterisation of alternate layer Langmuir–Blodgett films incorporating ions. *Mater. Lett.*, **61**, 1231–1234.

155. Çapan, R., Ray, A.K., and Hassan, A.K. (2007) Electrical characterisation of stearic acid/eicosylamine alternate layer Langmuir–Blodgett films incorporating CdS nanoparticles. *Thin Solid Films*, **515**, 3956–3961.

156. Cavaliere-Jaricot, S., Etcheberry, A., Herlem, M., Noël, V., and Perez, H. (2007) Electrochemistry at capped platinum nanoparticle Langmuir–Blodgett films: a study of the influence of platinum amount and of number of LB layers. *Electrochim. Acta*, **52**, 2285–2293.

157. Chen, W., Kim, J., Xu, L.-P., Sun, S., and Chen, S. (2007) Langmuir–Blodgett thin films of $Fe_{20}Pt_{80}$ nanoparticles for the electrocatalytic oxidation of formic acid. *J. Phys. Chem. C*, **111**, 13452–13459.

158. Marczak, R., Sgobba, V., Kutner, W., Gadde, S., D'Souza, F., and Guldi, D.M. (2007) Langmuir–Blodgett films of a cationic zinc porphyrin-imidazole-functionalized fullerene dyad: formation and photoelectrochemical studies. *Langmuir*, **23**, 1917–1923.

159. Milner, S.T. (1991) Polymer brushes. *Science*, **251**, 905–914.

160. Dutta, A.K. and Belfort, G. (2007) Adsorbed gels versus brushes: viscoelastic differences. *Langmuir*, **23**, 3088–3094.

161. Ulman, A. (1991) *An Introduction to Organic Ultrathin Films from Langmuir–Blodgett to Self-Assembly*, Part 5, Academic Press, Boston.

162. Kowel, S.T., Selfridge, R., Eldering, C., Matloff, N., Stroeve, P., Higgins, B.G., Srinivasan, M.P., and Coleman, B.L. (1987) Future applications of ordered polymeric thin films. *Thin Solid Films*, **152**, 377–403.

163. Kumar', G.S. and Neckers, D.C. (l989) Photochemistry of azobenzene-containing polymers. *Chem Rev.*, **89**, 1915–1925.

164. Mendonça, C.R., Balogh, D.T., De Boni, L., dos Santos, D.S. Jr., Zucolotto, V., and Oliveira, O.N. Jr. (in press) Optically-induced processes in azopolymers, in *Molecular Switches*, vol. 2 (ed. B.L. Feringa), Wiley-VCH Verlag GmbH, Weinheim.

165. Zucolotto, V., Strack, P.J., Santos, F.R., Balogh, D.T., Constantino, C.J.L., Mendonça, C.R., and Oliveira, O.N. Jr. (2004) Molecular engineering strategies to control photo-induced birefringence and surface-relief gratings on layer-by-layer films from an azopolymer. *Thin Solid Films*, **453-454**, 110–113.

166. Mendonça, C.R., Dhanabalan, A., Balogh, D.T., Misoguti, L., dos Santos, D.S. Jr., Pereira-da-Silva, M.A., Giacometti, J.A., Zilio, S.C., and Oliveira, O.N. Jr. (1999) Optically induced birefringence and surface relief gratings in composite Langmuir–Blodgett (LB) films of poly[4′-[[2-(methacryloyloxy)ethyl]ethylamino]-2-chloro-4-nitroazobenzene] (HPDR13) and cadmium stearate. *Macromolecules*, **32**, 1493–1499.

167. dos Santos, D.S. Jr., Mendonça, C.R., Balogh, D.T., Dhanabalan, A., Cavalli, A., Misoguti, L., Giacometti, J.A., Zilio, S.C., and Oliveira, O.N., Jr. (2000) Chromophore aggregation hampers photoisomerization in Langmuir–Blodgett films of stearoyl

ester of Disperse Red-13 (DR13St). *Chem. Phys. Lett.*, **317**, 1–5.

168. Shimizu, F.M., Ferreira, M., Constantino, C.J.L., Skwierawska, A.S., Biernat, J.F., and Giacometti, J.A. (2008) Fast dynamics in the optical storage with Langmuir–Blodgett films of a diazocrown ether molecule. *J. Nanosci. Nanotechnol.*, **8**, 1–9.

169. Kim, D.Y., Tripathy, S.K., Li, L., and Kumar, J. (1995) Laser-induced holographic surface relief gratings on nonlinear optical polymer film. *Appl. Phys. Lett.*, **66**, 1166–1168.

170. Rochon, P., Batalla, E., and Natansohn, A. (1995) Optically induced surface gratings on azoaromatic polymer films. *Appl. Phys. Lett.*, **66**, 136–138.

171. Oliveira, O.N. Jr., Li, L., Kumar, J., and Tripathy, S.K. (2002) *Surface-relief gratings on azobenzene-containing films in photoreactive organic thin films*, Chapter 6 (eds Z. Sekkat and W. Knoll), Academic Press, San Diego, pp. 429–486.

172. Barrett, C.J., Rochon, P.L., and Natansohn, A.L. (1998) Model of laser-driven mass transport in thin films of dye-functionalized polymers. *J. Chem. Phys.*, **109**, 1505–1516.

173. Kalina, D.W. and Crane, S.W. (1985) Langmuir–Blodgett films of soluble copper octa(dodecoxymethyl) phthalocyanine. *Thin Solid Films*, **134**, 109–119.

174. Cook, M.J., Dumm, A.J., Daniel, M.F., Hart, R.C.O., Richardson, R.M., and Roser, S.J. (1988) Fabrication of ordered Langmuir–Blodgett multilayers of octa-n-alkoxy phthalocyanines. *Thin Solid Films*, **159**, 395–404.

175. Marletta, A., Gonçalves, D., Oliveira, O.N. Jr., Faria, R.M., and Guimarães, F.E.G. (2000) Highly oriented Langmuir–Blodgett films of poly(p-phenylenevinylene) using a long chain sulfonic counterion. *Macromolecules*, **33**, 5886–5890.

176. Olivati, C.A., Ferreira, M., Cazati, T., Balogh, D.T., Guimarães, F.E.G., Faria, R.M., and Oliveira, O.N. Jr. (2003) Anisotropy in the optical properties of oriented Langmuir–Blodgett films of

OC$_1$OC$_6$-PPV. *Chem. Phys. Lett.*, **381**, 404–409.

177. http://www.oled-display.net/flexible-oled (accessed 20 082008).

178. Cok, R.S. (2003) OLED area illumination lighting apparatus. US Patent 6, 565, 231.

179. Tolt, Z.L. (2007) Barrier, such as a hermetic barrier layer for O/PLED and other electronic devices on plastic. US Patent 2, 007, 003, 1674.

180. Olivati, C.A., Peres, L.O., Wang, S.H., Giacometti, J.A., Oliveira, O.N. Jr., and Balogh, D.T. (2008) Light-emitting diodes containing Langmuir–Blodgett films of copolymer of a poly(p-phenylene-vinylene) derivative and poly(octaneoxide). *J. Nanosci. Nanotechnol.*, **8**, 2432–2435.

181. Akcelrud, L. (2003) Electroluminescent polymers. *Prog. Polym. Sci.*, **28**, 875–962.

182. Ferreira, M., Constantino, C.J.L., Olivati, C.A., Vega, M.L., Balogh, D.T., Aroca, R.F., Faria, R.M., and Oliveira, O.N. Jr. (2003) Langmuir and Langmuir–Blodgett films of poly[2-methoxy-5-(n-hexyloxy)-p-phenylenevinylene]. *Langmuir*, **19**, 8835–8842.

183. Ferreira, M., Constantino, C.J.L., Olivati, C.A., Balogh, D.T., Aroca, R.F., Faria, R.M., and Oliveira, O.N. Jr. (2005) Langmuir and Langmuir–Blodgett (LB) films of poly[(2-methoxy,5-n-octadecyl)-p-phenylenevinylene] (OC$_1$OC$_{18}$-PPV). *Polymer*, **46**, 5140–5148.

184. Sluch, M.I., Pearson, C., Petty, M.C., Halim, M., and Samuel, I.D.W. (1998) Photo- and electroluminescence of poly(2-methoxy,5-(2′-ethylhexyloxy)-p-phenylene vinylene) Langmuir–Blodgett films. *Synth. Met.*, **94**, 285–289.

185. Ostergard, T., Paloheimo, J., Pal, A.J., and Stubb, H. (1997) Langmuir–Blodgett light-emitting diodes of poly(3-hexylthiophene): electro-optical characteristics related to structure. *Synth. Met.*, **88**, 171–177.

186. Olivati, C.A., Ferreira, M., Carvalho, A.J.F., Balogh, D.T., Oliveira, O.N. Jr., Von Seggern, H., and Faria, R.M.

(2005) Polymer light emitting devices with Langmuir–Blodgett (LB) films: enhanced performance due to an electron-injecting layer of ionomers. *Chem. Phys. Lett.*, **408**, 31–36.

187. Yang, S., Fan, L., and Yang, S. (2004) Langmuir–Blodgett films of poly(3-hexylthiophene) doped with the endohedral metallo-fullerene Dy@C-82: preparation, characterization, and application in photoelectrochemical cells. *J. Phys. Chem. B*, **108**, 4394–4404.

188. Brynda, E., Koropecky, I., Kalvoda, L., and Nespurek, S. (1991) Electrical and photoelectrical properties of copper tetra[4-tert-butylphthalocyanine] Langmuir–Blodgett-films. *Thin Solid Films*, **199**, 375–384.

189. Desormeaux, A., Max, J.J., and Leblanc, R.M. (1993) Photo-voltaic and electrical-properties of Al/Langmuir–Blodgett films/Ag sandwich cells incorporating either chlorophyll-a, chlorophyll-b, or zinc porphyrin derivative. *J. Phys. Chem.*, **9**, 6670–6678.

190. Liang, L. and Fang, Y. (2008) Photoluminescence of a mixed Langmuir–Blodgett film of C_{60} and stearic acid at room temperature. *Spectrochim. Acta A Mol. Biomol. Spectrosc.*, **69**, 113–116.

191. Riobóo, R.J.J., Souto, J., de Saja, J.A., and Prieto, C. (1998) Elastic properties of Langmuir–Blodgett films. A new Brillouin spectroscopic strategy. *Langmuir*, **14**, 6625–6627.

192. Girlando, A., Sissa, C., Terenziani, F., Painelli, A., Chwialkowska, A., and Ashwell, G.J. (2007) In situ spectroscopic characterization of rectifying molecular monolayers self-assembled on gold. *Chemphyschem*, **8**, 2195–2201.

193. Choi, H.G., Jung, W.C., Min, J., Lee, W.H., and Choi, J.W. (2001) Color image detection by biomolecular photoreceptor using bacteriorhodopsin-based complex LB films. *Biosens. Bioelectron.*, **16**, 925–935.

194. Ducharme, S., Reece, T.J., Othon, C.M., and Rannow, R.K. (2005) Ferroelectric polymer Langmuir–Blodgett films for nonvolatile memory applications. *IEEE Trans. Device Mater. Reliab.*, **5**, 720–735.

5
Self-Assembled Monolayers: the Development of Functional Nanoscale Films

Andrew C. Jamison, Pawilai Chinwangso, and T. Randall Lee

5.1
Introduction

The development of facile methods for the preparation of nanoscale films – films formed from a monolayer of a surfactant (adsorbate) or combination of surfactants – has led to advances in a wide variety of fields: nanoscale coatings for corrosion resistance [1, 2], organic molecular junctions for electronic applications [3, 4], anchored boundary lubricants for microelectromechanical systems (MEMSs) [5, 6], biological studies that require a surface, that is, either nonadhesive or adhesive to cells or proteins [7, 8], sacrificial coatings for nanolithography for a variety of applications requiring temporary surface protection [9–11], applications needing enhanced durability to create patterned surfaces immune to washing procedures for microscale patterning [12, 13], surface modification of nanoparticles, and/or nanoshells to enhance their stability [14, 15], methods of controlling domain formation of mixed-adsorbate films [16, 17], tethered bilayer lipid membranes that mimic cellular membranes for a variety of research applications [18, 19], rigid thin-film systems developed for crystal-growth studies [20, 21], thin films with attachment sites for macromolecules for specific binding or detection of biologically active species or chemical agents [22, 23], and many other similarly pragmatic research goals. However, key to the pursuit of such "real-world" applications has been the tedious, time-consuming analysis of the strengths and weaknesses of a variety of monolayer systems. The most prominent method for creating such thin films, through self-assembly, forming a self-assembled monolayer (SAM) on a compatible surface, has provided a readily accessible format for general research. Among the many adsorbate structure and surface combinations examined, SAMs formed from thiols on gold have proven to be the favored format for the examination of fundamental monolayer phenomena. For this reason, unless otherwise specified, this review will describe alkanethiols adsorbed on gold, where the alkanethiol adsorbates are abbreviated according to the number of carbons in their alkyl chain (e.g., $CH_3(CH_2)_{17}SH$ will be identified as C18SH).

Understanding the basic structural parameters of a typical SAM, along with the strengths and weaknesses associated with these parameters, is essential to gaining

Functional Polymer Films, First Edition. Edited by Wolfgang Knoll and Rigoberto C. Advincula.
© 2011 Wiley-VCH Verlag GmbH & Co. KGaA. Published 2011 by Wiley-VCH Verlag GmbH & Co. KGaA.

a grasp of the focus of much of the fundamental research that has occurred in this field of surface science. Some perspective on these parameters can be obtained by reviewing key developments since the inception of this field of study, while other aspects can only be elucidated by combing through the tedious details of fundamental SAM research. A review of the unique phenomena associated with these assemblies, such as interchain interactions, wetting behavior, and the impact of structure on surface energetics and frictional responses can provide the context for determining how ongoing SAM research can be linked to practical applications. Moreover, a survey of the structural manipulations of SAM adsorbates that have been pursued helps complete the overall picture of how such monolayer systems are being modified to successfully incorporate SAMs into the potential applications listed in the introductory paragraph. Prior reviews addressing SAM films already provide a significant amount of background for this field of research, so we have cited such resources where appropriate to limit repetition of information that is readily accessible. The focus of this review is the investigation of the nexus of fundamental SAM structure, unique SAM phenomena, and the evolving SAM adsorbate, which together illuminate the tie between long-term fundamental research and the SAM systems now being investigated for use in industry.

5.2
Historical Background for Monolayer Films

The recognition that thin films can form at phase boundaries (e.g., air/liquid, liquid/solid, or vapor/solid interfaces) is not a recent development. Mariners have long noted that oil can calm ripples on the surface of water. However, development of an understanding of the chemistry and physics of monolayer systems, the science of these thin films, has occurred during only the last 100 years or so. Reviewing the historical links between Langmuir–Blodgett (LB) films and SAMs provides the context for the nature of the projects currently being pursued in the study of SAMs.

In the 1890s, Agnes Pockels conducted experiments at her home with oils and a fatty acid (stearic acid) in a water trough [24]. Her research included detailed observations of the changes in surface tension as it related to the amount of oil per specific surface area. From her studies, she collected data that provide a reasonable estimate of the thickness of the stearic acid film formed on the surface of water. Lord Rayleigh, in his analysis of surface films, concluded that they were formed from a single layer of molecules [25], and Devaux advanced this hypothesis with his argument that such monomolecular films could be formed from the close packing of molecules on the surface of water [26].

In 1917, Irving Langmuir reviewed the observations of fellow scientists regarding the spreading of fatty acids on water [26]. Like several of his contemporaries, he concluded that the films formed at the air/water interface were comprised of fatty acid molecules in a single layer, a monolayer. To support this argument concerning the nature of the film, he outlined detailed surface-tension studies and provided a description of the physical and chemical interactions that led to the formation of these films. He also provided the details of his design of a trough that could be used to determine when the monolayer was occupying a minimum surface area, as tracked by the surface pressure, which would correlate with these surfactants aligning in the most efficient arrangement for a densely packed film: an arrangement in which the extended alkyl chains were perpendicular to the surface of water. The data that he collected with this device provided him a means of supporting his assertions regarding the monolayer nature of these thin films by enabling calculations of the surface area in relation to the amount of fatty acid introduced into the trough.

Further investigations utilizing the Langmuir trough, with the aid of Katherine Blodgett, led to additional detailed observations regarding fatty acid monolayers formed on water and improved methods of transfer of the resulting films to solid substrates by the careful application of pressure on the surface of the film during deposition [27]. A report by Blodgett in 1934 included an initial description of the deposition of multiple layers of fatty acids on a glass slide, and a subsequent report in 1935 provided a broader set of multilayer experiments with a variety of fatty acids and substrates [28]. Blodgett noted that when a fatty acid was transferred to a surface, forming a film in which all the molecules are oriented with the CH_3 groups facing outward, the resulting deposited film was wetted neither by water nor by oil. The collaboration by Langmuir and Blodgett was the basis for the eventual naming of these deposited films as "LB" films. Informative reviews of LB films by Tredgold and Roberts provide additional background about this monolayer system and also provide insight into the basic physics behind monolayer formation [29, 30].

The *relative* ease of transfer of monolayer films to solid substrates in a laboratory environment, along with the surprising change in surface characteristics for such substrates once coated, generated interest in these thin films for a variety of applications. Zisman, at the U.S. Naval Research Laboratory, further studied these monolayer films on solid substrates, but with an emphasis on coating metal surfaces, reflecting a concern for applications such as lubrication [31, 32], rust inhibition [33], and adhesion [34, 35]. In 1946, he reported a substantially different approach to monolayer film formation, noting that certain amphiphilic compounds such as long-chain carboxylic acids, were capable of spontaneously developing well-organized films on a metal surface directly from solution [34]. Zisman emphasized in later studies that the resulting films were autophobic: the film, once formed, was not subject to wetting by the amphiphilic solution that formed the film [36]. The monolayer research by Zisman and coworkers continued through the 1970s, adding extensively to the knowledge gained from work with LB Films, including studies on surfaces formed from adsorbates containing unique terminal groups, such as amphipathic highly fluorinated surfactants [37].

The development of computer chip technology in the 1970s brought a renewed interest in the manipulation of monolayer films, but on a different set of substrates. An indicator of this interest is a report in 1978 by Haller at IBM that describes films comprised of alkylsilane adsorbates on semiconductor surfaces [38]. The resulting monolayer films were not only bound to the substrate, but also provided a surface exposing amino acid groups available for surface bonding. In 1980, Sagiv reported that alkylsilanes intermixed with fatty acids (along with other adsorbate pairings) produced stable alkylsilane-dominated films [39]. But the alkylsilane route to monolayer development encountered a number of problems at its inception: stability issues for the alkylsilane molecules before film formation, inconsistencies in the reproducibility of the targeted monolayer films, challenges in determining the degree to which the substrate should be hydrated to enhance film development, and disruptions in the alignment of the extended alkyl chains on the substrate related to the cross-links of these chains being so close to the surface on which they were bound.

In 1983, about the same time that the term *"self-assembled monolayer"* was first appearing in print [40], Nuzzo and Allara [41] reported the use of bifunctional disulfides to form well-organized monolayer films on gold. The success of this initial work was soon followed by a report that alkanethiols were just as effective at forming well-organized films as the disulfides [42]. Subsequent analysis of the two adsorbate formats by Troughton *et al.* [43] provided evidence that alkanethiol-based films were more ordered than films formed from disulfides. The system of alkanethiols on gold has since become the prevalent monolayer system for general research. Several qualities that relate to both the chemical stability of gold and the unique interaction between gold and sulfur contribute to this preference in the scientific community. Most importantly, a freshly deposited layer of gold, which is resistant to oxidation under ambient conditions, provides surfaces that can be handled in the average laboratory without the need for investment in "clean-room" technology [43]. When the gold surface is exposed to a dilute solution of alkanethiol, these sulfur-based adsorbates spontaneously produce a thin film under standard conditions with the sulfur attaching preferentially to those of other functional groups [41, 44]. This affinity of gold for sulfur leads to a strong bond, with the Au$-$S bond enthalpy being $\sim$40$-$45 kcal per mol [45, 46]. Once formed, the extended alkyl chains are stabilized by van der Waals forces, adopting a predominantly *trans*-extended conformation under ambient conditions [47, 48]. This propensity for the alkanethiol adsorbates to reproducibly form highly ordered arrays on the surface of gold has provided the scientific community with a versatile and accessible system for studying organic thin films.

5.3
Self-Assembled Monolayer (SAM) Fundamentals

To understand the fundamental structural features that are required to form "ideal" self-assembled interfaces, a review of the basic constituents of SAM films is warranted. Prior reviews of these key structural components have typically focused on the following three segments: (i) the headgroup, which binds the adsorbate to the substrate, (ii) the spacer, which provides the final film with interchain van der Waals force attractions that assist in creating a well-organized film, and (iii) the tailgroup, which forms the SAM interface and exposes selected functional groups that can be used to dictate the manner by which SAMs interact with contacting media [48].

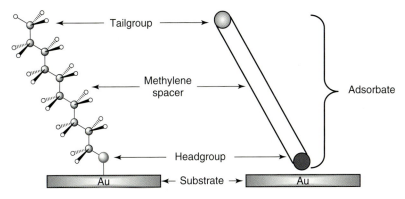

Figure 5.1 Simplified illustration of the four key components of a self-assembled monolayer. Adapted from Ref. [52].

In this review, we also focus on the substrate as a fourth critical component in the development of functional monolayer films, recognizing the breadth of literature indicating the substantial role that the substrate plays in film stability, packing geometry, and fundamental character [49–51]. Figure 5.1 provides an illustration of a typical alkanethiolate adsorbate as deposited in a film formed on a Au(111) surface [52]. Key to appreciating the following explanation of the relationship between the substrate and the resulting monolayer film is the recognition that the bonding arrangement of the headgroups, while providing a strong foundation for the film by binding to the thermodynamically favored bonding sites, does not preclude the extended alkyl chains from adopting structural alignments that maximize the interchain packing interactions of the overall system [53, 54].

5.3.1
Substrates

The prevalence of gold as the substrate of choice in the scientific community for fundamental studies of SAMs has not precluded investigations into a large variety of alternative substrates. Zisman, for example, investigated a broad array of surfaces (e.g., metals and metal oxides) in his research on organic thin films. Interestingly, the focus of Zisman's published work shifted to studies using gold as the substrate late in his career [55, 56]. It must be noted that the limitations of the available instrumental methods during Zisman's time, along with the methodology used to prepare the surfaces under study, hinder the direct comparison of Zisman's data with that collected on substrates prepared using current technologies. Nevertheless, the data as a whole provide a significant body of work, that is, a valuable resource for understanding fundamental interactions between various substrates and surfactants.

As indicated above, the renewed interest in thin-film research in the late 1970s was initiated with studies involving substrates such as silicon and gallium arsenide (GaAs) [38]. Research into semiconductor substrates has continued, but until recently, the number of studies involving these substrates was significantly fewer than that on gold. A variety of semiconductor surfaces have now been examined, including those of silicon [57–59], germanium [58, 60], indium phosphide [61, 62], and indium tin oxide [63]. However, GaAs is the substrate that has drawn the greatest interest over the past two years [64–69].

Early SAM research by Whitesides and coworkers, while primarily focusing on gold substrates, also included investigations of other "coinage" metals, such as silver and copper [70, 72]. Unfortunately, special handling procedures for these surfaces are generally necessary to minimize the formation of surface oxides. Despite this obstacle, silver [72, 74] and copper [72, 74–76] continue to be included in a large variety of studies, partially due to the prohibitive cost of incorporating gold in mass-produced products, and the relative conductivity of both copper and silver [77, 78], but also owing to the fact that in thiolates bonded to silver and copper, the sulfur is less susceptible to oxidation than in thiolates bonded to gold [79–81]. Additionally, copper has been the subject of renewed interest due to the ability of thiolate-based SAM films to protect the easily oxidized copper surface from degradation and corrosion [1, 75, 82], and the increasing interest in the utilization of copper connects in a variety of electronic devices [77]. The propensity of copper to oxidize readily upon exposure to air has also led to studies that revealed that oxidized copper surfaces provide a compatible platform for the development of fluorinated alkylsilane SAMs [76, 83, 84]. Other metal surfaces commonly utilized in SAM research include: aluminum [85, 86], mercury [87, 88], palladium [89, 90], and platinum [72, 89, 91]. For additional descriptions of the large number of substrates found in SAM research, the 1996 article by Ulman in *Chemical Reviews* provides a useful introduction [46], and the more recent review by Love *et al.* offers a longer-range perspective [92].

Copper and silver have also received attention as nanoscale plating metals owing to electrochemical methods that help reduce the propensity of these surfaces to oxidize under ambient conditions. Rogers *et al.* [93] first reported in 1949 on this approach for generating nanoscale metal films – the underpotential deposition (UPD) of a layer of a second metal on an electrode made of a more noble metal at potentials positive to that which would yield bulk electrodeposition. The utilization of this procedure in electrochemical studies involving SAMs led to the recognition that underpotentially deposited copper and silver on a gold substrate created a stronger bond for an alkanethiolate adsorbate than that formed directly on gold [94, 95]. Such thin layers of UPD copper or silver on gold appear to be more resistant to oxidation, which may be indicative of electron deficiency in this thin layer of surface atoms as compared to those of a substrate formed of pure copper or silver [96]. This method of enhancing the bonding of the adsorbate to the surface was first reported in 1996 by Jennings and Laibinis [95]. Since their initial report, an extensive amount of research has been published for UPD surfaces, including UPD copper, silver, bismuth, and mercury on gold [94, 96, 97].

Each of the individual metal substrates included in SAM research in the literature possesses their own unique set of parameters regarding the surface lattice, the corresponding adsorbate bonding sites, and the resulting surface structures. Articles

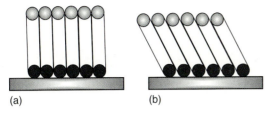

Figure 5.2 Illustration of the impact of the metal substrate on the packing arrangement of the overlying SAM: (a) thiols on silver, which exhibits a tighter surface lattice than gold, tilt less than (b) thiols on gold. Adapted from Ref. [47].

that provide insight into the structural parameters for a variety of SAMs produced on either copper and/or silver include the work of Whitesides and coworkers [71], Tao *et al.* [85, 98, 99], Rieley *et al.* [100], Floriano *et al.* [101], and Kondoh *et al.* [102, 103]. For alkanethiols on gold substrates, the film structure is characterized by an alkyl chain tilt from the surface normal of ∼27°. Silver substrates, with a similar structural array of smaller surface atoms, give rise to films with a reduced chain tilt, ∼12° according to one study [71]. But a second study, while agreeing with the tilt angle, noted that the silver substrate that formed upon vapor deposition was not predominantly (111), which was interpreted to indicate that the exact nature of the S–Ag bond likely influences the film structure [104]. A third report by Kondoh *et al.* in 2004 concluded that alkyl chain orientation was primarily determined by the density of the chains, which is substrate dependent, but for silver, the SAM surface structures were thought to be incommensurate with the substrate [103]. Assuming a tighter substrate lattice, the packing density of surface tailgroups would logically increase when the headgroup packing is commensurate with the surface structure. This phenomenon is illustrated in Figure 5.2.

The volume of literature related to the fundamental analysis of various surface structures for gold (e.g., (111), (110) or (100) lattices) and the organization of the film-forming alkanethiolate adsorbates on those surfaces is extensive [105–112]. This ample body of research not only appropriately reflects the importance of the substrate on the development of a well-organized film, but also the fact that the patterns of film development, including the size and nature of the domains that one would expect to be created, are intimately intertwined with the surface structure. For gold that is vapor-deposited on silicon wafers, the resulting surface possesses a predominantly (111) surface lattice [92]. But Au(100) and Au(110) also provide surfaces amenable to the formation of SAMs and have been studied in the form of carefully prepared single crystal surfaces. In a study by Camillone *et al.* [112] of these three surface lattice configurations on single-crystal gold (polished and annealed at ∼700 °C), the domain structures, periodicity, and packing density created from C22SH varied according to the nature of the lattice. For Au(111), the sulfur headgroups adopted a commensurate $(\sqrt{3} \times \sqrt{3})$R30° overlayer structure and appeared to adsorb on the threefold hollow sites, as shown in Figure 5.3. For

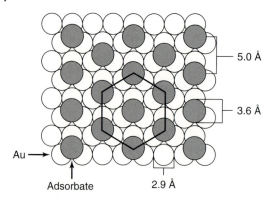

Figure 5.3 Surface structure of alkanethiolate SAMs on Au(111). The sulfur atoms (dark spheres) bind to the threefold hollow sites of the gold lattice (underlying white spheres), adopting a hexagonal arrangement with spacing between nearest-neighboring sulfurs being ~5.0 Å. This view from above shows only the sulfur headgroup of the adsorbate [51, 113, 114]. Adapted from Ref. [52].

Au(110), the sulfur headgroups formed a commensurate $c(2 \times 2)$ lattice, suggesting that the sulfur atoms removed the "missing row" reconstruction of the gold surface. In contrast, for Au(100), the sulfur headgroups adopted an incommensurate surface structure, with four kinds of domains present owing to the fourfold symmetry of the gold surface. Interestingly, the authors concluded that the SAMs formed on the Au(100) surface exhibited the highest chain density of the three surface structures studied.

In an article regarding the relationship between alkanethiolate oxidation and surface morphology for SAMs on gold, Lee *et al.* determined that the rate at which gold vapor is deposited on a silicon wafer can influence the quality of the resulting gold film [115]. This conclusion was reached after the authors noticed the kinetics of oxidation for the SAMs on gold varied depending on the morphology of the surface: an increase in surface oxidation corresponded to a decrease in the grain size of the gold. This interpretation was congruent with the already recognized importance that the evaporation rate could play on determining the nature of the resulting film: an accelerated rate of deposition led to an increase in the number of defect sites on the developing gold surface. According to the *Encyclopedia of Materials: Science and Technology*, the rate of deposition of the metal on the primed wafer should be maintained between 0.1 and 1.0 nm/s [48]. Following this deposition procedure, a polycrystalline gold surface exhibiting a surface lattice predominantly (111) in nature will develop with a grain size that averages approximately 50 nm in diameter.

Substrate grain size and surface defects not only impact the stability of the adsorbed film, but also the ability of an investigator to inspect the resulting monolayer via a number of instrumental techniques. Scanning tunneling microscopy (STM) and atomic force microscopy (AFM), along with other scanning probe microscopy methods, require that the surface areas under study be atomically flat over sufficiently large regions to enable scanning at a molecular level over an area approximately 100×100 nm [48]. One approach that researchers developed to address this need has been the creation of procedures for reproducibly preparing atomically flat metal surfaces. The fact that such substrates can be readily produced on sheets of mica was rapidly recognized in the wake of the introduction of scanning probe methods to the study of SAMs [116]. The initial work revolved around the vapor deposition of gold on a freshly exposed, carefully prepared mica surface, which was heated slowly to 300 °C during vapor deposition. With this work, Holland-Moritz *et al.* [116] found that atomically flat terraces that were about 1000 Å wide were readily formed. One example of the application of these surfaces was described by Leopold *et al.* [117] in a 2002 article that examined the impact of surface roughness on films formed from a carboxylic acid-terminated alkanethiolate SAM. The conclusion of the authors was that an atomically smooth surface produced a film that was less permeable than that formed on a gold surface prepared via standard vapor deposition procedures owing to a lower defect density for the resulting SAMs. A large number of articles that provide methods of preparation of atomically flat surfaces (or ultrasmooth surfaces) can be found in the literature [118–121].

5.3.2
Headgroups

The rationale for the eventual dominance of gold in current laboratory research is delineated in a 1989 article by Bain *et al.* [44] that examines the strengths of sulfur-containing adsorbates in the formation of well-organized films on gold. Key among these is the ability of the "soft" nucleophile (i.e. hard-soft acid-base chemistry) of the thiolate-forming surfactants to displace surface contaminates and preferentially create stable surface bonds when in competition with "hard" nucleophiles, allowing significant flexibility for the incorporation of other functional groups into the adsorbate structure. Additionally, the anchoring of the headgroup to the substrate in the fully formed film provides SAMs a level of stability unachievable under the transfer process used for the creation of LB Films on a solid surface when working with films that were initially formed at an air/water interface [122].

The recognition of the utility of the enhanced bond strength of the covalent interaction between sulfur and gold has led to research projects involving a variety of sulfur-based adsorbates: thiols, dialkyl or cyclic disulfides [19, 123, 124], dialkyl sulfides [43, 124, 125], *n*-alkyl xanthic acids (AXAs) [52], alkyl thioacetates [126], aliphatic dithiocarboxylic acids (ADTCAs) [127], dithiocarbamates or dithioesters [128–130], thioureas [131], thiophenols [132, 133], multidentate thiols [134–137], thiocyanates (forming thiolates) [138–140], and a number of other sulfur-based structures [129, 141]. A corresponding bonding capacity has been associated with compounds possessing selenium instead of sulfur in the headgroup [142–145], but no clear advantage has developed to motivate a shift toward its use [146]. Alkylsilane derivatives in the form of halides or hydroxides form strong covalent bonds with certain hydroxylated substrates, but have been the subject of much research related to problems associated with film formation for this monolayer format, as mentioned in Section 5.2. Recent articles have provided insight into alternative procedures for the deposition of well-organized alkylsilane monolayers, including one by Wang *et al.* [147, 148] where the self-assembly process occurs under strict anhydrous conditions, a procedure that significantly reduces the cross-linking of the chains, creating a densely packed film. Examples of headgroup/substrate combinations that have been the subject of recent research can be found in a review by Whitesides and coworkers [92]. A separate review by Schreiber [105] regarding the structures of SAM films provides additional detail concerning the relationship between headgroup bonding and surface organization.

A SAM film, once fully formed, often possesses characteristics that reveal the predominant orientation of the adsorbate's tailgroups in relation to the plane of the surface (see Figure 5.4). The tailgroup orientation influences the interfacial energy of the film, which is manifested in the contact-angle values for a variety of contacting liquids, as described in detail in Section 5.4.2. Examples that exhibit a possible structural influence from the headgroup–surface bonds upon the interfacial properties of the film can be found in the work of Colorado *et al.* [127] and the research of Moore [52]. The adsorbates under study in both instances possess multidentate dithiocarboxyl headgroups bound to the metal surface: the first study on ADTCAs and the second on AXAs. These unique headgroup structures give rise to SAMs that exhibit extreme variance in contact-angle values between sequential members of the homologous series (i.e. odd vs. even chain lengths) when hexadecane is used as the contacting liquid. The large variance in contact-angle

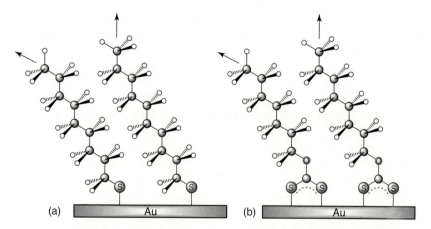

Figure 5.4 Illustration of the headgroup binding associated with the two sulfur atoms for an alkyl xanthate (b) as compared to a normal alkanethiolate with a single sulfur atom (a). This drawing also reveals the differences between the orientation of the terminal methyl groups for chains possessing an even or odd number of carbon atoms. Adapted from Ref. [52].

values for ADTCAs was partially attributed to the reduced bonding distance between the two sulfur atoms in the carboxylate-like headgroup structure – a calculated ~2.87 Å separation between the two sulfur atoms versus the ~5.0 Å distance between the bonding of alkanethiolates to the threefold hollow sites found on the surface of Au(111) [127]. The density of the headgroup population at the gold interface for the AXA SAMs as compared to the corresponding normal alkanethiolate SAMs appeared to indicate that a reduced alkyl chain density for these unique structures led to a larger tilt from the surface normal, creating the extreme systematic variance in the contact-angle values. However, the ellipsometric thickness measurements for both SAM systems (i.e. *n*-alkanethiolate and AXA) showed comparable film thicknesses. Additional studies of these unique adsorbate systems are currently underway to define more precisely the root causes of the contact-angle variations.

Another aspect related to the interaction of the adsorbate headgroup with the surface that has been the subject of extensive research is the propensity of the sulfur moiety, once bound to gold, to oxidize, leading to reduced film integrity. The formation of sulfonate species was noted in two reports in 1992: one by Tarlov and Newman [149], and the other by Hemminger and coworkers [150]. In the Tarlov/Newman study, the authors examined SAMs derived from a series of alkanethiols, ranging from C8SH to C18SH. SAMs with prolonged exposure to ambient conditions, but subject to at least a day's delay in analysis, were found to form sulfonates as determined by examining data gathered by static secondary ion mass spectroscopy (SIMS). SAMs tested immediately after formation in an ethanolic solution or after extended storage in solution showed no similar signs of oxidation for the sulfur headgroup. Additionally, the authors discovered that the sulfonate species were readily displaced by alkanethiol in solution. The Hemminger study explored a series of SAMs derived from C4SH to C18SH and found that the SAMs with shorter chains were subject to a greater level of headgroup oxidation. A subsequent report by Scott *et al.* [151] noted that both alkanesulfinates (RSO_2-) and alkanesulfonates (RSO_3-) were formed during the oxidation process. Additionally, two studies from 1998 indicated that ozone was the key active oxidizing agent [152, 153]. Such research results have provided impetus for the continuing efforts to develop an optimized SAM system: one where the headgroup/substrate combination not only exhibits a strong covalent bond, but also is either protected from or is resistant to degradation.

5.3.3
Spacers

In a study involving a series of alkanethiolate SAMs, Porter and coworkers demonstrated the vital importance of the spacer to the development of well-organized SAM films [47]. The goal of their work was to determine the influence of alkyl chain length on SAM formation and stability, utilizing a series of alkanethiol adsorbates extending from C2SH to C22SH. The conclusion reached by these investigators was that the thiols with extended alkyl chains produced densely packed monolayers with a high degree of conformational order or a "crystalline-like" phase. However, SAMs formed from short chains tended to be more disordered or "liquid-like." For the three techniques used to determine conformational order, the sharp transition between the well-ordered and disordered SAMs occurred anywhere from C6SH to C12SH, varying with the specific method used. The authors also noted a discernible chain tilt of 20–30° from the surface normal that was attributed to the mismatch between the spacing of the headgroup sulfurs and the alkyl chains, which are driven to pack densely by van der Waals forces. Additionally, the authors conducted electrochemical experiments that again supported the conclusion that the SAMs with short chains were relatively disordered, while the SAMs with long chains were well packed and relatively free of pinholes, creating films that provided a good barrier to electron transfer and ion penetration.

A comparative study by Laibinis *et al.* [71] examining the structural order of SAM films of alkanethiolates produced on gold, silver, and copper also revealed that a well-ordered monolayer of these adsorbates on gold possess "*trans* zig-zag extended" alkyl chains that not only exhibit a cant angle of ~27°, but also twist about the plane of the carbon backbone by ~53°. When compared to prior studies of the packing parameters of alkane chains in a solid state [46], the similarities make it apparent that these nanoscale coatings might exhibit a limited capacity to maintain order at elevated temperatures, as with the melting of paraffin waxes. Nuzzo *et al.* [53] noted in 1990 that the temperature dependence for the C22SH IR band intensities was similar to that of bulk hydrocarbons. The authors compared the IR bands of SAMs to those gathered by Snyder and coworkers for *n*-alkanes and polymethylene chains [154, 155]. The similarities in both intensity and peak position for the C–H stretching vibrations for these alkyl chain-based structures have also provided SAM researchers methods for determining the orientation and conformational order of these thin films through analysis by infrared spectroscopy.

Another observation regarding the influence of the alkyl chains in the spacer was made by Ringsdorf and coworkers in 1994 [156], who determined via STM that an annealing procedure could improve the molecular packing of a SAM, causing the formation of larger surface domains through a process of eliminating chain-tilt mismatches. A concurrent study by Delamarche *et al.* [157], while supporting the results of the Ringsdorf annealing study, also found that such annealing procedures increased oxidation of the sulfur headgroups and increased the vulnerability of the surface to desorption processes. These observations regarding the impact on SAM surfaces from moderate exposure to heat have significant implications not only for the stability of the film, but also for changes in frictional responses for SAM coatings in applications involving surface contact.

5.3.4
Tailgroups

In a report published in 1990, Nuzzo *et al.* [158] described a series of alkanethiol-based SAMs of the form $X(CH_2)_{15}SH$, where X represents an assortment of tailgroups: CH_3, CH_2OH, CO_2H, CO_2CH_3, and $CONH_2$. The authors concluded from this research that these adsorbates form well-organized SAMs with surfaces exposing the X functional groups at the interface, and that the X functional groups appear to have little impact on the underlying structure. There are, however, circumstances where the nature of the tailgroup can also have a notable influence on the structure of the SAM. In an article published in 2002, Pflaum *et al.* [159] compared the structural parameters and barrier properties of a series of SAMs produced from alkanethiols (C10SH-C15SH) and their CF_3-terminated analogs. While normal alkanethiol-based SAMs form long-range ordered domains exhibiting a $(\sqrt{3} \times \sqrt{3})R30°$ structure (along with a $2(\sqrt{3} \times \sqrt{3})R30°$ structure in the standing-up phase in the surfaces reviewed in this study), CF_3-terminated SAMs prepared for this study failed to exhibit any long-range order at room temperature, yet conformed to the underlying hexagonal symmetry. This variance was explained by the authors as being attributable to structural adjustments at the SAM surface related to the steric strain imposed on the SAMs by the larger terminal CF_3 groups (radius = ~ 2.7 Å) as compared to the smaller CH_3 groups (radius = ~ 2.0 Å) of the normal alkanethiol-based SAMs.

The role of an adsorbate's tailgroup on the integrity of the resulting SAM was also examined by Chidsey and Loiacono [160] in an investigation of several long-chain thiol-based adsorbates of similar length: a fluorinated alkanethiol $(CF_3(CF_2)_7(CH_2)_2SH)$, an unsubstituted alkanethiol (C10SH), a hydroxy-terminated alkanethiol $(HOCH_2(CH_2)_{10}SH)$, a nitrile-terminated alkanethiol $(NC(CH_2)_{10}SH)$, and a carboxylic acid-terminated alkanethiol $(HO_2C(CH_2)_{10}SH)$. The electrochemical tests conducted on these surfaces provided insight not only into the electron barrier properties of the monolayer films, but also an understanding of the influence of tailgroup functionality, revealing those that were most disruptive to the underlying structural order of the SAM film. The authors concluded that the fluorinated film, owing partially to the larger diameter of the fluorinated segment, was the best film for barrier properties and exhibited the lowest level of film defects. The remaining SAMs were rated according to decreasing film integrity in the following order: $CH_3 > CH_2OH > CN > CO_2H$.

An insightful example of how the tailgroup of the adsorbate can influence the fundamental character of the film in subtle ways can be found in studies of alkanethiol adsorbates with carboxylic acid tailgroups. In 1989, Bain and Whitesides [161] reported on "reactive spreading" – a phenomena that was associated with changes in the contact-angle values of a series of mixed-SAMs (carboxylic acid-terminated alkanethiols with methyl-terminated alkanethiols) for contacting liquids that were buffered. The reactive spreading concept came from the recognition that an increase in the pH of the aqueous contacting liquid was associated with a decrease in the contact angle on the surface of the mixed SAM. The authors concluded that the acidity of these carboxylic acid moieties when surface-bound and incorporated into the monolayer film was reduced when compared to carboxylic acids in solution. Subsequent studies by Lee *et al.* [162] noted further that carboxylic acid-terminated films exhibited reduced wetting when the contacting liquids had pH values from 3 to 7 as compared to liquids with higher or lower pH values. The authors attributed the unexpected increase in contact angle to changes in the conformational order of the acid moieties at the termini of the adsorbate chains, which was induced by the partial ionization of these chains. Cooper and Leggett [163] added further to this explanation in research involving the carboxylic acid-terminated monolayers for lithographic applications, noting that these films had added stability over normal alkanethiolate SAMs owing to hydrogen bonding between neighboring carboxylic acid groups.

There are also specific tailgroup phenomena that have been reported that relate to a reduction in the integrity of the film and/or have had an impact on the measurement of contact angles over time that are not as easily rationalized. Evans *et al.* [164] reported in a 1991 study involving hydroxy-terminated thiol adsorbates in a mixed-SAM format that such systems exhibited changes in contact-angle measurements over time. In these studies, X-ray photoelectron spectroscopy (XPS) revealed that the polar tailgroup was slowly buried into the surface of the SAM, and that this process was dependent on both temperature and chain length. In contrast, other polar tailgroups have been reported to exhibit behavior that is influenced by confinement within a densely packed and highly ordered monolayer film. For example, Moore reported that a series of cyano-terminated SAMs ($CN(CH_2)_nSH$, where $n = 14-17$) failed to register a peak for the CN stretching vibration in analysis by polarization-modulation infrared reflection adsorption spectroscopy (PM-IRRAS), indicating that the CN bond resided predominantly in the plane of the interface [52]. Since the peak was absent in the spectra of all four SAMs of the cyano-terminated series, the author concluded that strong electrostatic interactions between the terminal groups of neighboring chains were overriding the typical odd–even variance in terminal group alignment.

The exposed array of adsorbate tailgroups has found a broad variety of research purposes that range from the nonadhesive, chemically inert surfaces of terminally fluorinated films [5] to a variety of chemically reactive surfaces as detailed in a review by Chechik *et al.* [165]. In applications involving coating surfaces for contact with biological samples, such as cells or proteins, polyethylene glycol (PEC)-terminated films have provided resistance to adhesion. The desire to incorporate such nonadhesive characteristics into well-ordered ideal surface arrays has led researchers to develop alternative paths to a new generation of nonfluorinated, nonadhesive films [166, 167]. An example of the modification of this technology to enable its merger into an effective self-assembled film can be found in the research of Béthencourt *et al.* [7]. Instead of generating uneven surfaces of linear PEG-terminated adsorbates, the research team focused on developing a terminal ring for the adsorbate that incorporated the fundamental PEG repeating unit, allowing for the formation of an ideal surface array. Such innovations involving the manipulation of tailgroup design for specific environmental conditions has ensured that the adsorbate tailgroup will play a key role in the adoption of SAM systems for a variety of practical applications.

5.4
SAM Phenomena

5.4.1
Oriented Arrays

Recognition of the "ideal" nature of the surfaces formed from the oriented molecular array in a monolayer film is implicit in the description by Blodgett of the surface character of the deposited film, with all the CH_3 groups facing outward [28] and in the work of Zisman when he described the "autophobic" properties of his films [36]. Their observations, however, are accompanied by data from only a few analytical techniques. In contrast, the proliferation of semiconductor (computer chip) research in the 1970s led to the development of a substantial expansion of the number of analytical tools available for investigating surfaces. A series of biennial review articles detailing these developments in the field of analytical chemistry was initiated by Kane and Larrabee in 1977 [168]. The vast array of new surface instrumental techniques, in combination with improvements in the metal substrate (e.g., silicon wafer supports for vapor-deposited metals and atomically smooth surfaces), have enabled the development of deposition procedures for SAMs possessing reliably reproducible features, such as densely packed adsorbate arrays exhibiting unique interfacial characteristics that vary for individual adsorbates in a series of SAMs with monotonically increasing chain length (e.g., odd–even or parity effects).

5.4.2
Wettability

The wetting of a solid substrate is an important and integral component of several interfacial phenomena in surface science. Yet the value of the data gathered for contact-angle measurements is often misunderstood and underestimated. Contact-angle data trends can reveal important information about the nature of the SAM surface and the limitations for particular monolayer films in certain applications.

The contact angle is defined by a line tangent to the curvature of the surface of a liquid drop at the point where the base of the drop contacts the surface, forming an angle with the plane of the surface, such an angle being measured from the liquid side of the tangent line. For contact angles on SAMs, the measurement that is used to give a reliable, reproducible number is the angle formed upon expanding the drop – the advancing contact angle (θ_a), as illustrated in Figure 5.5. For more complete characterization, the contact-angle measured for a receding drop (θ_r) allows calculation of the hysteresis ($\Delta\theta = \theta_a - \theta_r$). Hysteresis data provide insight into the heterogeneity, roughness, and/or general nature of the surface [169].

The surface energy has been associated with the type and arrangement of the functional groups arrayed at the solid/vapor phase boundary and the roughness of the interface [45, 170–172]. Other factors also influence the surface energy as

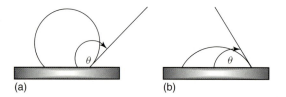

(a)   (b)

Figure 5.5 Illustration of the contact angles formed from probe liquids that exhibit a contact angle less than 90° when $(\gamma_{SV}/\gamma_{LV}) > (\gamma_{SL}/\gamma_{LV})$ (b) and a contact angle larger than 90° when $(\gamma_{SV}/\gamma_{LV}) < (\gamma_{SL}/\gamma_{LV})$ (a).

measured by contact angle: the relative crystallinity of the film along with the packing density or surface density of the adsorbates [71], the presence of other molecules intercalated in the SAM film [173], chemical interactions between the contacting liquid and the surface [43, 161, 174], structural parameters of the contacting liquid [175], the influence of the underlying substrate [173], and the presence of a dipole at or near the surface of the film [176]. Young's equation (Equation 5.1a) has been the primary tool of researchers in this field for developing an understanding of the interfacial forces associated with a SAM film [177].

$$\sigma_{LV} \cos \theta = \sigma_{SV} - \sigma_{SL} \tag{5.1a}$$

σ_{LV} is the surface tension at the liquid/vapor phase boundary, σ_{SV} for the solid/vapor phase boundary, and σ_{SL} for the solid/liquid phase boundary. The symbol θ represents the contact angle. Equation 5.1a provides the fundamental relationships between the surface forces for each interface and the corresponding contact angle. While this equation provides valid and useful insight into the forces involved at the SAM interface, the interpretation of wettability data and even the definition of the forces involved have been a matter of debate [169, 178]. Nevertheless, in accordance with most treatments [179, 180], our approach is to assume that σ (surface tension) can be used interchangeably with γ (surface free energy, for systems in thermal and mechanical equilibrium), to give a modified Young's equation (Equation 5.1b).

$$\gamma_{LV} \cos \theta = \gamma_{SV} - \gamma_{SL} \tag{5.1b}$$

The relationship between the "work of adhesion" (W_{SL}) and the surface free energies associated with the three phase boundaries were defined by Dupré as [181]:

$$W_{SL} = \gamma_{SV} + \gamma_{LV} - \gamma_{SL} \tag{5.2}$$

This equation defines the work required to separate a unit of solid surface area from a liquid in contact with that area in terms of the surface free energies associated with each of the interfaces in the system. Combining Equations 5.1b and 5.2 for a liquid drop with a contact angle on a solid surface produces the Young–Dupré equation:

$$W_{SL} = \gamma_{LV}(1 + \cos \theta) \tag{5.3}$$

This relationship indicates that the work of adhesion for a solid surface can be determined from the surface free energy of the contacting liquid and the associated contact-angle measurement. Calculations of the work of adhesion generally rely upon the use of surface tension data for the contacting liquid, which are readily available in the literature for most common probe liquids [182]. The forces that contribute to W_{SL} have been attributed to a number of sources: dispersive forces associated with the surface (γ_d) and acid/base interactions between the contacting liquid and tailgroups of the individual adsorbates (γ_{AB}) [183], along with a variety of other components of the work of adhesion [178].

Examples of how contact-angle measurements have been used to characterize systematic changes in adsorbate structure can be found in the literature. In particular, an analysis by Colorado and Lee [176] focused on three series of systematically varying SAMs in which the adsorbates were either terminally fluorinated to some degree or nonfluorinated. For the fluorinated adsorbates, Series 4.1 consisted of $F(CF_2)_n(CH_2)_mSH$ with $n = 1$ and $m = 12–15$, and Series 4.2 worked from the same generic formula except that $n = 1–10$ and $m = 15–6$, respectively. The normal alkanethiolate reference series consisted of C13SH through C16SH. This study was designed to examine the influence that an oriented dipole might have on the contact angles of polar probe liquids and on the associated adhesive forces.

For Series 4.1, the oriented dipoles of the terminal CF_3 moieties for the fluorocarbon/hydrocarbon ($R_F - R_H$) transition dipole gave rise to a surface that exhibited an odd-even trend for the contact-angle values of polar aprotic liquids that was inverse to that observed for the same liquids on n-alkanethiolate SAMs, as shown in Figure 5.6. These results are consistent with the effects of a local oriented dipole. For Series 4.2, where the $R_F - R_H$ dipole was systematically buried in the surface of a series of SAMs with a set number of carbon atoms in the chain, an analysis of the surface energetics revealed the unambiguous influence of the systematically buried $R_F - R_H$ dipole, as shown in Figure 5.7. The authors, working with contact-angle

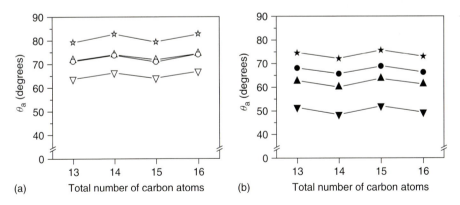

Figure 5.6 Contact-angle trends for (a) normal alkanethiolate SAMs and (b) CF_3-terminated alkanethiolate SAMs with a variety of probe liquids: dimethyl formamide ($\triangle$, $\blacktriangle$), acetronitrile ($\triangledown$, $\blacktriangledown$), nitrobenzene ($\circ$, $\bullet$), and dimethyl sulfoxide ($\star$, $\bigstar$). Reproduced by permission of John Wiley & Sons, Ltd. from Ref. [176].

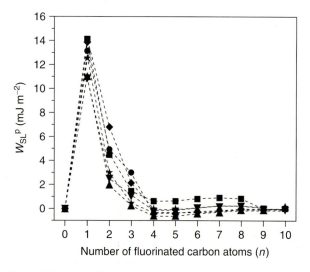

Figure 5.7 Graph of the calculated polar works of adhesion for SAMs possessing varying degrees of terminal fluorination. Symbols for contacting liquids are water (■), glycerol (♦), dimethyl formamide (DMF) (▲), acetronitrile (▼), nitrobenzene (●), and dimethyl sulfoxide (DMSO) (★). Reproduced by permission of John Wiley & Sons, Ltd. from Ref. [176].

data for an array of contacting liquids on fluorinated and nonfluorinated SAMs, calculated both the dispersive work of adhesion and the polar work of adhesion for the series, particularly for combinations where acid/base interactions were precluded. The calculated values clearly show the influence of the $R_F - R_H$ dipole on the series. The authors concluded that the work of adhesion should include a parameter for oriented dipoles (W_{SL}^{OD}) that arise from systems in which the conformation of the tailgroups is highly ordered (see Equation 5.4).

$$W_{SL} = W_{SL}^{d} + W_{SL}^{AB} + W_{SL}^{OD} \tag{5.4}$$

As noted above, tailgroup orientation can play a key role in measurements of contact angle. However, odd–even effects related to orientation have been noted not only for SAM wettability [127, 135, 184–187], but also for frictional properties [135, 186, 188, 189] and even chemical activity [190–192]. Generally, with regards to wettability and friction, such variances in surface activity are rationalized by the greater interfacial exposure of the CH_2 group lying just below the terminal moiety, leading to a change in the surface energy of that SAM, but other adsorbate-specific interactions might also play a role. For a series of alkanethiolate SAMs of monotonically increasing chain length, the chains with an odd number of carbon atoms experience this orientational effect, while the surfaces of even-numbered chains are dominated by the terminal group moiety, as shown in Figure 5.4 [188]. An extensive *Chemical Reviews* article by Tao and Bernasek [193] on the subject of odd-even effects in SAMs was published in 2007.

5.4.3
Frictional Properties

A variety of forces are involved in the formation of a well-organized SAM: chemical or physical bonding of the headgroup to the substrate, van der Waals interactions between the chains, other attractive forces between the tailgroups, repulsive forces between the tailgroups, interactions related to the presence of solvent or water molecules in or on the chain structure, and van der Waals forces related to the underlying substrate. All of these interactions also play a role in the nature of the nanoscale or microscale tribological or "frictional" properties of these ideal surfaces [163, 194, 196]. Additionally, SAMs exhibit unique surface interactions related to the highly ordered arrangement of the individual component molecules and to the tethered nature of this arrangement. As nanoscale lubricants, surface-bound SAMs cannot flow or move along the surface in response to the application of lateral forces. When used as boundary lubricants in applications (e.g., in MEMS), individual structural characteristics of these ideal arrays (conformational order, packing density, surface phase character, and the nature of the tailgroup) contribute specific properties to the surface that might not be beneficial when compared to an unbound lubricant. As an example, normal alkanethiolates that are surface-bound in a well-organized SAM, possessing no tailgroup functionality, exhibit an increased surface order at low temperatures (i.e. a more ordered film with greater crystallinity), which leads to reduced lateral forces for a contacting AFM probe by minimizing the contact area with the probe [135]. However, the same unbound normal alkane (less the thiolate moiety) possessing strong interchain van der Waals force-driven attractions, create a system that exhibits increased resistance to lateral motion through the lubricant as the temperature is decreased [196].

With the development of instrumental methods for observing phenomena related to nanoscale structures, such as AFM, interfacial force microscopy (IFM), or other surface force microscopies [194], the scientific community gained tools to reveal the role that minute features play when assembled as an oriented array on an ideal surface. Working with an atomically flat surface and a carefully prepared SAM, force measurements can be obtained that include acquisition of a lateral signal (converted to a lateral force value) or a normal signal (converted to a force measurement of a "load"). Through a series of cycles with varying loads, the compilation of force measurements can be interpreted within the context of the molecular properties of a specific SAM [197].

Fluorinated materials such as Teflon[1] exhibit properties that have traditionally been associated with low friction: they are generally chemically inert, possess resistance to wear, and are known for their nonadhesive characteristics. However, partially fluorinated SAMs exhibit characteristics related to the enforced orientation of the arrayed adsorbates. In 1997, Kim *et al.* [118] used AFM to examine SAMs derived from an adsorbate possessing a single terminally fluorinated carbon,

---

1) ®Trademark of E. I. du Pont de Nemours
    and Company.

13,13,13-trifluorotridecanethiol ($CF_3(CH_2)_{12}SH$), and compared the results with those obtained from a SAM derived from C13SH. The authors found that the SAM exposing a surface with such limited fluorination exhibited a frictional response that was three times greater than that of the SAM derived from the normal alkanethiol [118]. The difference in frictional response was attributed to the larger size of the CF_3 group as compared to the terminal CH_3 group of the nonfluorinated SAM. The enhanced frictional response associated with monolayers formed from adsorbates with terminal fluorocarbon(s) has been observed in several other studies under various conditions utilizing a variety of surfactant formats [195, 198–200].

In a 1999 article by Kim *et al.* [195], the authors concluded that the perturbation in film structure that would be attributable to the inclusion of a terminal fluorinated methyl group would be insufficient to account for the variance in frictional response as compared to the SAMs composed of normal alkanethiols. However, their analysis provided a more concise explanation of how the steric bulk of the larger CF_3 units could contribute to an enhanced frictional response: namely, more intimate molecular contact among the terminal groups can lead to long-range correlated lateral motion and thus additional modes of energy dissipation [201, 202]; alternatively, there might be an increase in the barrier to rotation for the bulky terminal groups [118, 185, 197, 203]. In another study by Kim *et al.* [195] with mixed SAMs containing both CF_3 and CH_3 terminated thiols, it was found that incorporation of even a relatively low level of the CF_3 component (~15%) led to a large increase in the frictional response of the resulting film, which remained approximately the same for the entire mixed-SAM series with systematically increasing CF_3 content. Within the same report, Kim and colleagues conducted tests with SAMs that were assembled from thiols possessing a terminal isopropyl moiety of the form $(CH_3)_2CH(CH_2)_nSH$, providing a means of further investigating the role of steric bulk on frictional responses measured by AFM. The collected data indicated that the terminal isopropyl structure created frictional responses for the associated SAM that fell between that for the terminal CF_3 and CH_3 groups [195, 204].

In a subsequent study, Perry *et al.* [186, 205] observed that phenyl-terminated SAMs exhibited an increase in frictional response when compared to normal alkanethiol SAMs, but not as much as with CF_3 tailgroups. The origin of the lower frictional response for SAMs with the bulkier phenyl group remains to be explained. At the very extreme of terminal group bulkiness, SAMs terminated with Buckminster fullerene (C_{60}) were investigated [186, 206]. Such studies were pursued owing to the anticipation that C_{60} would make an effective solid lubricant (similar to graphite), but the results indicated that C_{60} as a surface-bound molecule produces higher frictional responses than that of CH_3-terminated and phenyl-terminated SAMs.

A totally different approach to modifying the AFM frictional response engendered by SAMs is through the systematic modification of packing density. Lee *et al.* [207] synthesized a set of bidentate thiols in which one adsorbate possessed two extended alkyl chains from the branchpoint near the metal surface and another analogous adsorbate possessed only one extended alkyl chain. For this study, which included a normal alkanethiol of equivalent chain length for comparison, the single-chained alkanedithiol exhibited the highest frictional response. This observation was rationalized by the increased area of contact for the AFM probe tip in the more "liquid-like" SAM, owing to its lateral packing arrangement (i.e. predominantly lying down rather than standing up). These results were verified in a separate study involving a different unsymmetrical double-chained alkanedithiol that possessed two chains of varying length [135]. Based on frictional data collected from these unique alkanedithiol SAMs [135, 208, 209], the ultimate conclusion was that the enhanced AFM frictional response for loosely packed films was derived from various mechanisms of energy dissipation in such films, including the processes that are associated with increasing the effective contact area of the film, the local elastic modulus, and the nature of the conformational defects in the SAM at the sliding interface.

The degree to which variances in frictional response can be tied to the organization of the film was further demonstrated by the work of Li *et al.* [210]. In this report, mixed-SAMs formed from normal alkanethiols (C15SH and C16SH) of varying lengths, along with a CF_3-terminated hexadecanethiol ($CF_3(CH_2)_{15}SH$), were prepared to determine how frictional responses varied with minor perturbations in lateral surface packing. Combinations involving two alkanethiol chains of different lengths showed an increase in AFM frictional response as compared to SAMs composed solely of the individual thiols, with the nonfluorinated monothiol-based SAMs having small and indistinguishable frictional responses. The mixed-SAM combination of C16SH and the CF_3-terminated C16SH exhibited the greatest frictional response. Consistent with previous interpretations [118, 195, 197], the enhancement was attributed to an enhanced lateral packing density for this mixed SAM. Support for this hypothesis was provided by the mixed-SAM combination of C15SH and CF_3-terminated C16SH, which exhibited a significantly smaller frictional response when compared to the latter mixed film.

5.5
Manipulation of Adsorbate Structure

5.5.1
Cross-Linking of Adsorbates

The advantages associated with the polymerization (or other forms of cross-linking) of a SAM film have encouraged numerous investigations into this route of stabilizing these thin-film coatings for applications that require a robust protective layer. As with our analysis of the strengths and weaknesses of the individual SAM adsorbate, the routes to developing a cross-linked film can be divided into three segments: polymerization processes at or near the headgroup, cross-linking either through hydrogen bonds or covalent bonds between components in the spacer and polymerization between the tailgroups.

5.5.1.1 **Between Headgroups**

For one of the original self-assembling systems, alkylsilane monolayers, the surface-bonding mechanism is generally initiated in tandem with the cross-linking process. However, as already discussed, the positioning of these cross-links in close proximity to the surface links can add strain to the developing array, leading to disorder in the resulting film [46]. However, surface hydration can decouple these processes, leading to the generation of high quality films with a reduced level of surface bonding. When prepared by standard procedures and subjected to thermal stress under ambient conditions, the resulting film is more stable than that of a well-packed thiolate monolayer film on gold [211].

Examples of silane monolayers on surfaces other than mica, silicon dioxide, silicon, and a variety of semiconductor substrates, while uncommon, are known for a few select systems, such as that noted above for recent work on oxidized copper surfaces [76, 83, 84]. For standard alkylsilane film development, studies by Tripp and Hair [212, 213] determined that water is a necessary component for the initiation of surface bonding. As noted above, Wang *et al.* [147] found that the presence of water is not only unnecessary, but is often detrimental to the development of a well-organized film. With the latter procedure, the key to successful monolayer formation is the preparation of a fully hydroxylated surface, followed by strict anhydrous conditions when developing the film. However, while the adsorbates interact fully with the surface, they develop only limited cross-linking.

Examples of alkylsilane films developed on gold can also be found in the literature. Kurth *et al.* [214] worked with (3-glycidoxypropyl) trimethoxysilane films on gold substrates. This system, however, was then modified with an overlying set of monomers that when subsequently polymerized, created a complex thin-film system, where the contribution of the alkylsilane layer was largely indiscernible. A 1990 review by Ulman gives a glimpse of how alkylsilane films were viewed at the early stages of the development of self-assembly, while his 1996 *Chemical Reviews* article addressing SAM systems in general expands upon the earlier article [46, 215]. A review by Onclin *et al.* [40] provides a broader background for this approach to forming thin films along with insight into potential applications.

5.5.1.2 **Between Spacers**

Several methods have been developed that incorporate a cross-linking mechanism in the body of the adsorbate. Of these, two have received significant attention in the literature: hydrogen bonds between amide moieties in the backbone of the adsorbate and diacetylene moieties that can be used to cross-link neighboring adsorbates. A third process, the use of prepolymerized systems that form surface bonds upon exposure to the substrate, creates films that lack the surface order typically associated with SAM films.

In 1994, Lenk and coworkers introduced the concept of hydrogen bonding between amide linkages incorporated into the spacer of the adsorbate chains of SAM films [216]. For this particular system, the researchers were seeking to stabilize an extended fluorinated segment at the terminus of the chain. A similarly structured monolayer was the subject of a report in 1995 by Tam-Chang *et al.* [217] in which the amide moiety was positioned β to the thiolate sulfur. The authors reported that the resulting films varied little from the overall organizational structure of alkanethiolate SAMs. A variety of these amide-incorporating SAMs can be found in the literature, including amide-modified adsorbates for electrochemical studies that examined the influence of amide groups on barrier properties [218], on stability under electrochemical conditions [219], and on the properties of ferrocenylalkyl-modified films in which the amide is positioned at various points along the adsorbate chain [220].

In the first of a series of papers, Clegg *et al.* introduced an adsorbate with three amide groups designed to produce bonding in all directions within the 2D structure of a monolayer film [221]. The SAMs produced from these adsorbates exhibited improved stability against thermal desorption when compared to analogous alka-nethiolate SAMs. A subsequent study by Clegg and Hutchison [222] examined a series of alkanethiol adsorbates with a single amide moiety positioned near the headgroup ($CH_3(CH_2)_{n-1}(NHCO)(CH_2)_2SH$). For this project, the researchers systematically varied the alkyl chain above the amide group. The authors concluded that the inclusion of the amide moiety, while adding a region of well-organized hydrogen bonding, dominated the underlying surface order established by the sulfur–gold bonds, creating a need for an extended spacer (methylene units $\geqslant 15$) above the amide moiety to stabilize the SAM. Additional work by Clegg and colleagues [223] also included a study where the number of amide linkages was varied from one to three, leading to the observation that the SAMs with an even number of linkage sites in their adsorbates produced films that required longer overlying alkyl chains to create well-organized monolayers as compared to both SAMs with an odd number of linkages (one and three), where the amide moieties developed characteristic conformations independent from the nature of the hydrocarbon region.

One of the most prominent systems developed for cross-linking monolayer adsorbates is that of incorporating diacetylene units in the spacer and initiating cross-linking between neighboring chains via ultraviolet-radiation-induced radical polymerization. The application of this form of monolayer stabilization can be found in numerous articles involving LB Films [224, 225]. The first report of its use in SAMs was in 1994, when Batchelder *et al.* [226] described the successful utilization of the cross-linking procedure in a film composed of disulfide adsorbates of the form $[S(CH_2)_2COO(CH_2)_9C{\equiv}CC{\equiv}C(CH_2)_{13}CH_3]_2$. Crooks and colleagues reported in a series of papers [227–229] on their efforts to utilize carboxylic acid-terminated alkanethiol diactylenes as a route for preparing multilayer films and photoetch resists. In a 1997 report [230], the Crooks team conducted a stability study for a diacetylene monolayer film versus a SAM formed from the unpolymerized adsorbate and an analogous alkanethiol, concluding that the polymerized film withstood significantly harsher conditions in a series of tests for thermal, electrochemical, and chemical stability.

A variety of related studies have explored fundamental structural adjustments to evaluate their impact on the resulting film; these include: changing the tailgroup from a methyl group to a carboxylic acid, a hydroxyl group, and a terminal fluorinated segment [228–232], shifting the diacetylene moiety to an LB film supported by a C18SH monolayer [233], changing the length of the alkyl chain above and/or below the linking moiety [234, 235], and changing the chain length to observe the impact of odd/even effects on the number of carbons between the linking moiety and the headgroup [236]. For the study involving terminal fluorination, Cheadle *et al.* [232] found that the reduction in the methylene spacer located between the fluorocarbon tailgroup and the upper alkyne bond, along with an increase in the length of the fluorinated segment, led to disruptions in the polymerization. The authors suggested that possible contributions to this reduced cross-linking were the mismatch of the chain packing for the fluorinated segment, the methylene spacer, and the diacetylene units, along with the reduced mobility of the tailgroups due to the greater steric bulk of the fluorinated segments. Diacetylene systems are also subject to other problems that have limited their use outside of the laboratory environment: some diacetylene adsorbates are subject to degradation or polymerization prior to deposition [235], and substrates that lack atomic smoothness give rise to mismatch anomalies that can lead to inefficient cross-linking [234].

In a separate approach to cross-linking, Geyer *et al.* [237] exposed a SAM formed from 1,1′-biphenyl-4-thiol with a low-energy electron beam, which led to the formation of bonds between neighboring aromatic rings. The authors noted that the resulting film exhibited an increased etching resistance. A number of studies have been conducted that verify the electron-beam-induced cross-linking with a variety of SAM systems [12, 238, 239]. Recently, Turchanin *et al.* [13] noted a substantial improvement in the thermal stability of such cross-linked aromatic thiolate films over their noncross-linked counterparts. In tests conducted on SAMs composed of biphenyl adsorbates, the authors found that these more rigid adsorbates formed films that endured exposure to temperatures up to ~400 K – a temperature range slightly higher than the reported initial decomposition temperature for most normal alkanethiolate SAMs (~380 K), [53, 157]. However, a totally different regime was accessed for films that had been cross-linked by treatment with an electron beam, which produced films that withstood temperatures up to ~1000 K.

5.5.1.3 Between Tailgroups

A variety of electrochemical studies have focused on polymerizing the terminal groups of SAMs. In 1994, Willicut and McCarley [240] conducted experiments where ω-(*N*-pyrrolyl)-terminated alkanethiol adsorbates (with the tailgroup bonded to the chain through nitrogen) were electrochemically polymerized on a gold electrode. The authors noted the film exhibited enhanced stability, but a subsequent study by the same authors revealed that the pyrrole tailgroups suffer from irreversible oxidation and failed to polymerize, yet their participation in the surface polymerization did occur when a monomer source was present in the solution [241]. A concurrent study by Sayre and Collard [242] on 3-ethyl pyrrole-terminated thiol-based SAMs (with the tailgroup bonded to the chain through the carbon β to the nitrogen) revealed that this conformation of the pyrrole tailgroup also failed to produce surface polymerization however, the film efficiently polymerized when the monomer was present in the contacting solution.

The fact that hydrogen bonding can occur between tailgroups, as in the case with carboxylic acids, was noted in Section 5.3.4 [163]. Carey *et al.* [243] concluded in 1994 with an examination of 11-mercaptoundecanyl-1-boronic acid-based SAMs that these adsorbates also formed cross-links under conditions that caused dehydration of the surface, forming a surface structure similar to that of borate glass. Due, however, to the reversibility of the surface reaction, the applicability of these surfaces to real-world conditions is limited. In a recent study of a series of aromatic thiols, Barriet *et al.* [133] made a similar observation regarding the surface structures of *p*-mercaptophenylboronic acid-based self-assembled monolayers (MPBA SAMs) on gold. Specifically, the MPBA SAMs underwent cross-linking upon exposure to vacuum during analysis by XPS. In both reports, the resulting cross-links were subject to hydrolysis. What is apparent from these studies is that exposure of reactive groups at an interface greatly enhances the propensity for interchain bonding, providing a broader range of possible polymerization procedures as compared to potential reactive sites buried within the surface. One should also note that the conformational flexibility near the tailgroups (i.e. at the air/SAM interface) is markedly greater than that near the headgroups (i.e. at the SAM/solid interface).

The potential associated with tailgroup cross-linking is exemplified in the work of Ford *et al.* [244], where a 4-(mercaptomethyl)styrene-based SAM was transformed into a surface polymer. Both an azo initiator (Wako VA044) and laser irradiation were successfully used to initiate the surface polymerization. A laser desorption study found that the polymerized surface was resistant to laser removal, while the unpolymerized surface readily desorbed. The authors pointed to the potential of this system for producing stable negative resists for photolithography. Another example of a similar polymerization procedure was described by Duan and Garrett in 2001 [245]. In this report, the authors attached the styrene moiety to the end of a hexanethiol chain in the position para to the vinyl group, which proved to be an unfortunate choice of chain length because it was too short to stabilize the film via van der Waals attractions, affording disordered tailgroups that failed to polymerize efficiently. Exposure of the films to UV light from a low-pressure Hg lamp led to a conversion of only ~70% of the monomer units to polymer. Jordan *et al.* in 1999 [246] prepared styrene-terminated adsorbate structures *in situ* in an effort to conduct surface polymerization with these surface-bound monomers. The resulting surface organization restricted cross-linking of the chains leading to the formation of linear surface-bound polymers – a polymer brush. Having seen this report, Bartz *et al.* [247] applied this approach to a modified styrene adsorbate, creating temporary masks for specific sites on a surface that could be subsequently removed by cleavage at an ether linkage. The adsorbate structure is shown in Figure 5.8.

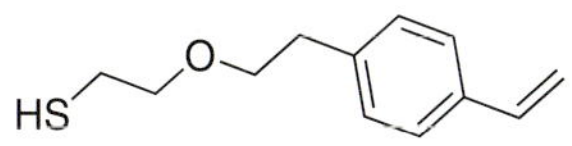

Figure 5.8 Structure of *p*-(2-mercapto-ethyl)-styrylethyl-ether [247].

5.5.2
Multidentate Adsorbates

5.5.2.1 Advantages of Multidentate Adsorbates

Another category of innovations in adsorbate structure intended to broaden the range of applications available for monolayer films is that of creating multiple bonds to the surface for each individual adsorbate. This approach brings a number of advantages to the self-assembled array. The first and most important benefit is that multiple surface bonds improve the overall stability of the monolayer via an entropy-driven process known as the "*chelate effect.*" The utilization of ligands that form multiple bonds to a multivalent metal ion is familiar to inorganic chemists, and the stability that such bonding brings to a metal–ligand complex is well recognized [248, 249]. However, the fact that the same dynamics that add to the stability of a metal–ligand complex can be readily employed to surfactant systems on solid substrates has largely been overlooked. For an alkanedithiol in which the two thiol moieties are appropriately incorporated into the adsorbate structure to provide simultaneous, unhindered access to the metal surface during film formation from solution, the proximity of the second thiol moiety to the surface upon binding of the first thiol moiety provides an enhanced probability that both will form a bond with the surface. Furthermore, when compared to two separate adjacent thiol chains, the odds that the dithiol will resist exchange and remain attached to the surface are significantly higher. A more in-depth explanation by Schlenoff *et al.* [250] of the advantages associated with adsorbates with multiple bonds to the surface can be found in the literature.

A second advantage associated with multidentate adsorbates is the ability to create films with precisely controlled surface density by manipulation of both the number of binding sites per adsorbate and the resulting density of surface chains associated with the SAM. One example of research involving surface density control can be found in the work of Shon *et al.* [251], where a set of SAMs prepared from specifically designed 2-monoalkylpropane-1,3-dithiol (m-CX) derivatives with the generic formula of $CH_3(CH_2)_nCH[CH_2SH]_2$ (where $n = 11$, 13, 14 and X = 14, 16, 17) were compared to SAMs formed from analogous single-chained alkanethiols (n-CX) and double-chained dithiols of equivalent length (d-CX), as shown in Figure 5.9. The authors noted that the SAMs formed from m-C16 exhibited a noticeably reduced film thickness, a reduced density of the alkyl chains, and reduced conformational order when compared to n-C16 and d-C16

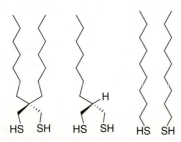

Figure 5.9 Examples of bidentate spiroalkanedithiols along with a pair of normal alkanethiols for comparison to the double-chained adsorbate [251].

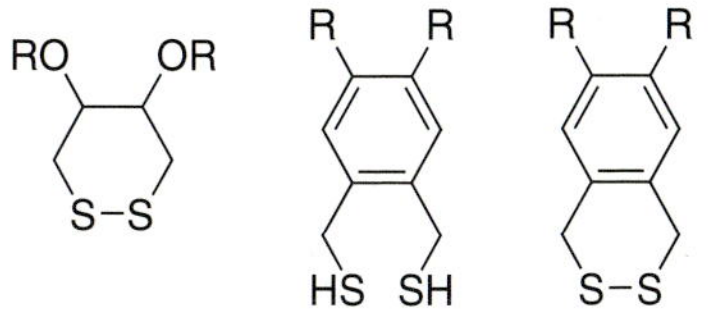

Figure 5.10 Examples of bidentate adsorbates [41, 134].

SAMs. In general, assemblies of well-ordered, well-packed hydrocarbon tailgroups yielded interfaces that exhibited low wettabilities and low frictional responses when compared to assemblies of disordered, loosely packed hydrocarbon tailgroups.

A third advantage associated with these multidentate adsorbates has been the ability to prepare a homogeneously mixed two-dimensional array of surfactant species that typically undergo phase separation when mixed as separate entities. Shon *et al.* reported such "mixed" SAMs of homogeneous chain distribution in 2000, where the two chains are incorporated into one alkanedithiol structure [209]. The authors compared the SAMs formed from these single mixed-chain adsorbates to SAMs generated by the coadsorption of mixtures of normal alkanethiols having analogous chain lengths. The mixed SAMs formed from two separate adsorbates were subject to local domain formation (or "islanding") of the tailgroups, a problem that was absent from the spiroalkanedithiol SAMs.

5.5.2.2 Multidentate Adsorbate Structures

A review of the multidentate thiol systems in the literature reveals that there are several structural formats that have been used to prepare bidentate and tridentate thiols. The initial sulfur-based adsorbate structures described by Nuzzo and Allara in 1983 [41], upon their discovery that such sulfur compounds would spontaneously form thin films from solution on specially prepared gold substrates, formed two bonds to the metal surface (see Figure 5.10). A subsequent report that covered an extended series of disulfides provided additional analysis of these film-forming adsorbates [252]. However, the stability advantage that a multidentate system might bring to these monolayers was not the predominant focus for these incipient disulfide adsorbates, and at least two shortcomings limited their usage: (i) a reduced packing density for the chains arose from the proximity of the two branchpoints on the six-sided disulfide ring, incorporating intrinsic mismatches for chain alignment for neighboring chains near the gold surface, which was compounded by the relative positioning of the bonds at the sp^3 carbons at the two branchpoints and (ii) the placement of oxygen atoms at the branchpoints, below the extended alkyl chains that form the bulk of the thin film likely created barriers to forming an ordered film.

A subsequent class of bidentate dithiol was reported by Garg and Lee in 1998 [253]. This article described studies of a series of 1,2-bis (mercaptomethyl)-4,5-dialkylbenzene derivatives used to form SAMs on gold. A quick comparison of the structure of these bidentate compounds to the original Nuzzo/Allara structure in Figure 5.10 reveals that the newer adsorbates incorporated an aromatic ring at the branchpoints and eliminated the intervening oxygen atoms between the chains and the base. The authors indicated in this initial report that the films formed from this series of symmetrical dithiols were both densely packed and highly ordered. Their preliminary thermal desorption study for this new form of chelation-stabilized SAM also provided evidence that such thin films had improved stability characteristics compared to those of normal alkanethiolate SAMs. In 2000, Garg and coworkers [134] extended their investigation of this SAM system to include a disulfide adsorbate analog to their original set of 1,2-bis(mercaptomethyl)-4,5-dialkylbenzene derivatives, as shown in Figure 5.10. Additionally, they included in their new SAM series linear disulfides that could form films analogous to the normal alkanethiols. Despite the fact that previously studied disulfide adsorbates produced films comparable to their thiol analogs [254], the aromatic disulfide analog in this study failed to form highly ordered films. The authors rationalized this discrepancy by pointing to the conformational constraints placed on the aromatic disulfide structure that hindered the ability of both sulfur atoms to make simultaneous contact with the gold surface during the initial lying down or physisorbed phase of the monolayer development process. With their report, the authors also pointed to fundamental film-forming problems associated with the rigidity of the central aromatic ring: the spacing imposed by the ring on the two alkyl chain substituents might introduce interchain repulsive contacts owing to the enforced orientation of the aromatic ring, leading to alternative packing arrangements that were not optimal for a well-organized film. In a third report on this system by Garg *et al.* in 2002 [255], the authors confirmed that these bidentate thiols possessed improved thermal stability as compared to a variety of monothiol adsorbates, including a number of adsorbates that incorporated aromatic rings in their structures.

A third approach to assembling bidentate thiols involves the synthesis of a single branchpoint adsorbate, as shown in Figure 5.9. In 1999, Shon and Lee [256] described an initial series of SAMs formed from these adsorbates, referring to them as *"spiroalkanedithiols."* The adsorbates for each SAM in this series possessed two alkyl chains of equal length, creating an overall SAM series with tailgroups ranging systematically from short alkyl chains to long alkyl chains having as many as 16 total carbon atoms. While the initial series provided evidence that the resulting SAMs were well organized, the data also indicated that SAMs formed from analogous normal alkanethiols exhibited a slightly higher degree of conformational order. The authors also noted that the time required to develop a fully formed film in solution was unusually long, with optimum results being recorded only after allowing 48 h in the dilute adsorbate solution. Nevertheless, in contrast to the dithiothreitol-based bidentate structures of Nuzzo and Allara, the spiroalkanedithiol structure created a surface-bound species that was resistant to desorption via intramolecular disulfide ring formation, which translated into improved thermal stability for this particular multidentate architecture.

Shon and Lee [257], concluded from displacement studies that spiroalkanedithiols with two extended alkyl tailgroups possessed a reduced conformational flexibility compared to their single-tailgroup bidentate analogs, offering a lower-energy pathway toward film formation for the latter dithiol and enhanced binding to the surface. The authors also noted that despite the fact that the SAMs derived from the double-chained spiroalkanedithiol exhibited a reduced thermodynamic stability compared to those derived from the single-chained bidentate dithiol analog, the double-chained system possessed a higher degree of conformational order, and both types of dithiol-based SAMs exhibited improved thermal stability compared to SAMs formed from analogous normal alkanethiols.

In 2004, Park et al. [136] reported the results from a series of adsorbates designed to provide further insight into the previous work reported by Shon and Lee. In addition to the single-chained bidentate dithiol series and comparable normal alkanethiols, the authors examined a series of 2-alkyl-2-methylpropane-1,3-dithiol derivatives. These new bidentate adsorbates possessed one extended alkyl tailgroup and a methyl group at the spiro branchpoint. The data collectively indicated that the SAMs formed from these new adsorbates exhibited only a slight reduction in conformational order when compared to SAMs formed from the bidentate dithiols with only one long-chain alkyl tailgroup and no methyl group. A subsequent report by Park et al. [137] in 2005 indicated that the thermal stability of SAMs formed from these two bidentate dithiols structures were approximately equivalent; however, the new dithiol structure exhibited enhanced stability when used to coat gold nanoparticles (see Section 5.5.2.3). Additional research involving this adsorbate architecture can be found in the literature [135, 207, 209, 258].

A fourth strategy for engineering self-assembling adsorbates with multiple bonds to the surface can also be found in the literature: trithiol adsorbates that are structurally similar to the spiroalkanedithiols. This form of tridentate structure was first reported in the 2005 article by Park et al. [137] as part of the series of adsorbates used to form SAMs with controlled packing density (see Figure 5.11). A comparison of the SAMs derived from the three types of multidentate adsorbates versus films formed from analogous normal alkanethiols revealed that the trithiol SAMs exhibited the lowest packing density for the series, forming monolayers possessing the least conformational order. However, owing to the presence of an additional thiol group on this molecule as compared to the dithiol adsorbates, the trithiols were the most thermally stable adsorbates examined in this preliminary trial.

Another class of tridentate adsorbate can be found in the literature [259], but this particular approach focused more on the manipulation of surface-bonding sites in order to adequately space attached helical peptides. In this work, Whitesell and Chang described oriented peptide films that were prepared on gold and

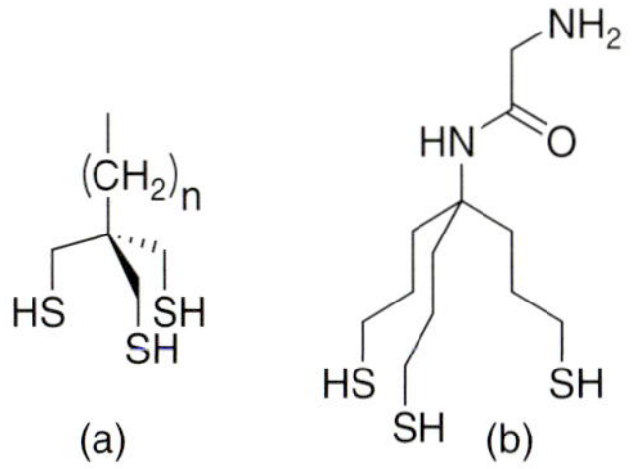

Figure 5.11 Examples of a tridentate adsorbate used to stabilize nanoparticles (a) and an aminotrithiol structure used to create oriented peptide films (b) [15, 137, 259].

indium-tin-oxide glass from an adsorbate with a trithiol headgroup (see Figure 5.11). This aminotrithiol adsorbate not only provided three thiol groups to afford a robust surface attachment, but also a functional tailgroup intended to create a route to surface polymerization or a means of controlling the density of surface-attachment sites via manipulation of the spacing between these adsorbates. This fundamental adsorbate structure was also the centerpiece of a second study, this time an electrochemical investigation by Fox *et al.* [260], which examined films derived from the aminotrithiol with various species attached to the termini of the tailgroups.

A fifth structure frequently associated with SAM research is that of a cyclic disulfide. The advantages of utilizing a multidentate linkage to bind a vulnerable moiety to the substrate have been incorporated into SAM projects with little fanfare by a number of researchers seeking ways to stabilize tethered macromolecules to the surface. In many but not all instances, these specialized adsorbates are dispersed in a monolayer film that includes simpler and shorter alkanethiol chains. An example of this approach can be found in the work of Schiller *et al.* [18], where the tethering of a complex adsorbate molecule to the surface of gold required a strong bond to the surface. A pentacyclic disulfide ring provided such a bond to an ultrasmooth gold film, giving the overlying lipid bilayer structures optimal assembly conditions.

5.5.2.3 Multidentate Adsorbates on Nanoparticles

Recent reports regarding multidentate thiols have moved beyond the fundamental investigations of these adsorbate systems and involve the analysis of pragmatic applications. In a 2008 study by Zhang *et al.* [14], the authors utilized a series of multidentate thiols along with an analogous normal alkanethiol, C16SH, to determine which adsorbate system would provide the best stability for large gold nanoparticles (diameters of 20–50 nm) in organic solution under ambient conditions. The nanoparticles were initially suspended in aqueous solution, with surfaces covered by citrate ions. Prior work established that thiols can displace citrate ions from the surface of gold nanoparticles during the process of SAM formation [261]. However, for larger nanoparticles, the newly developing thiol-modified gold nanoparticles can experience irreversible aggregation driven by the van der Waals attractions of the underlying metal [14, 262, 263]. Zhang and coworkers concluded from their research that a tridentate thiol provided the best stabilization for relatively large nanoparticles in solution and that the chelate effect was the primary contributor to this enhanced stability. Subsequently, Srisombat *et al.* [15] published the results of a chemical stability study involving smaller gold nanoparticles (~2 nm in diameter), where the particles were modified with a similar series of multidentate thiols along with an analogous normal alkanethiol, as illustrated in Figure 5.12. The results from this study, which tested the resistance to attack by cyanide of thiolate-modified nanoparticles, indicated that the tridentate thiol failed to provide the best resistance to chemical attack. The four adsorbates studied, *n*-octadecanethiol (**n-C18**), 2-hexadecylpropane-1,3-dithiol (**C18C2**), 2-hexadecyl-2-methylpropane-1,3-dithiol

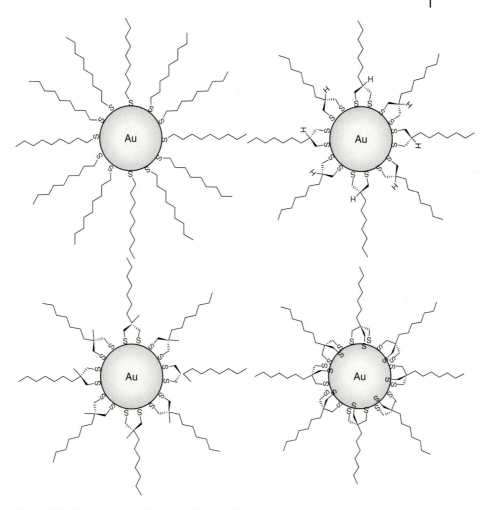

Figure 5.12 Nanoparticles with various thiolate adsorbates provided as a visual aid to illustrate the role that structure plays in protecting the nanoparticle core from chemical attack [15].

(**C18C3**), and 1,1,1-tris(mercaptomethyl) heptadecane (*t*-**C18**), exhibited the following trend in stability: **C18C3** ≫ **C18C2** > *t*-**C18** ≫ *n*-**C18**. The authors rationalized this trend on the basis of two separate contributing structural factors: the chelate effect being an important component of the added stability, but steric bulk provided a vital role by impeding the diffusion of cyanide through the low-dielectric hydrocarbon layer to the metal core of the nanoparticle. Detailed studies such as these are laying the groundwork for a variety of anticipated nanoparticle applications, as outlined in a review article by Daniel and Astruc [264].

5.5.3
Fluorinated Adsorbates

5.5.3.1 Impact of Fluorinated Segments on SAM Characteristics
The discovery of the advantages of fully fluorinated polymer systems (e.g., poly tetrafluoroethylene – PTFE or Teflon) encouraged the research community to investigate further the impact of fluorination on a variety of substrates. For fully fluorinated surfaces, advantages include inertness to chemical and biological agents, enhanced thermal stability as compared to hydrocarbon analogs, a reduced frictional response, and resistance to oxidation and corrosion [265–267]. For such fluorinated surfaces, by reducing the free energy of the interface and the associated critical surface tension, one also decreases the adhesive forces and wettability [32]. Zisman *et al.* [32], in their investigations of surface lubrication, tested potential applications of partially fluorinated surface active agents in the modification of polymer films, attempting to make traditional polymers behave like Teflon. This approach to solving issues of wear for polymer-based parts was considered after extensive research on a variety of interfaces involving partially fluorinated surfactant systems. Zisman and coworkers [37, 268–270] found that the adsorption of a partially fluorinated surfactant to the polymer surface by the incorporation of the additive during processing created low energy polymer surfaces – a result that concurred with his work on the partial fluorination of monolayer films. Consequently, the concept of imparting the characteristics of a Teflon surface by coating that surface with a thin layer of a fluorinated surfactant has an extensive research background.

Following the shift in monolayer research to the predominance of alkanethiolate/Au SAM systems, Chidsey and Loiacono [160] discovered that alkanethiol adsorbates with a terminal perfluorinated segment of eight carbons gave rise to films that possessed improved barrier characteristics when compared to those afforded by normal alkanethiolate SAMs, as highlighted in Section 5.3.4. Such fluorinated adsorbates are also known to provide additional advantages, such as greater rigidity for the individual adsorbate (enhancing the resistance of the film to localized disorder), improved thermal stability for the SAM film, and both hydrophobicity and oleophobicity [176, 271, 272].

A pair of articles by Schönherr *et al.* in 1996 [216, 273, 274] added an ester-stabilized SAM to the amide-containing, hydrogen-bond-stabilized system previously examined by the Ringsdorf team. The focus of the first article was the synthesis and characterization of the disulfide adsorbates, along with analysis of the resulting SAMs, while the second article included an examination of the thermal stability of the SAMs formed from these fluorinated adsorbates. However, a number of studies now lend support to the conclusion that SAMs formed from adsorbates possessing terminal fluorination exhibit enhanced stability when compared to nonfluorinated analogs, without the need for the incorporation of cross-links in the methylene spacers [271, 275]. An example of the enhanced thermal stability of fluorinated SAMs is provided by the work of Fukushima *et al.* [271], where terminally fluorinated SAMs survive exposure to 150 °C for 1 h in air with only a minor decrease in contact-angle values. While the enhanced stability of fluorinated SAMs has been attributed to phase transitions analogous to melting at temperatures less than $200\,^\circ$C [271], Biswas and coworkers [275] found evidence that the rigid helices of the perfluorinated segments uncoil at markedly higher temperatures. The coiled structure of perfluorocarbons is associated with the greater steric congestion arising from the substitution of fluorine for hydrogen and the consequent intrachain geometric constraints, as described in the literature [267, 276].

To further the understanding of the influence of structure on the fundamental character of SAMs formed from partially fluorinated adsorbates, a series of investigations was conducted by Lee and coworkers. By systematically adjusting the relative degree of fluorination of the spacer/tailgroup segment for a series of alkanethiol-based adsorbates, the research team reported a number of important findings regarding the impact of terminal fluorination on both structure and interfacial characteristics, as revealed by film-stability studies and a number of analytical techniques. Key to the development of their results was the diverse set of adsorbates prepared for analysis, including a set of alkanethiol-based SAMs where the degree of terminal fluorination was varied while the length of the spacer was held constant as shown in Figure 5.13a ($F(CF_2)_x(CH_2)_{11}SH$, where $x = 1$–10 or a subset thereof – Series 5.1) [277–279], another where the spacer was varied while the terminal fluorination was held constant as shown in Figure 5.13b ($CF_3(CF_2)_9(CH_2)_ySH$, where $y = 2$–$6, 11, 17, 33$ or a subset thereof – Series 5.2) [271, 277, 280–284], and a third where both the degree of terminal fluorination and the spacer were concurrently varied, while holding the total chain length constant as shown in Figure 5.13c ($F(CF_2)_x(CH_2)_ySH$, where $x = 1$–10 and $y = 16 - x$ or a subset thereof – Series 5.3) [3, 176, 184, 277, 279, 285, 286]. Key experimental data for all three series (e.g., XPS data, PM-IRRAS data, and ellipsometric measurements) can be found in the publication ACS Symposium Series 781 from 2001 [277]. Obviously, such extensive adsorbate research generated a number of efficient synthetic procedures, some of which can be found in the article by Graupe *et al.* from 1999 [287].

For the first SAM series, Series 5.1, which possessed systematically lengthened, terminally fluorinated segments on a methylene spacer held constant, the dispersive surface energies for the SAMs decreased as the fluorinated segment was increased for the adsorbates in the series extending from $x = 0$–6, and remained constant for the SAMs formed from adsorbates with longer fluorinated segments [279]. The researchers, Colorado and coworkers, found through a series of contact-angle measurements using *cis*-perfluorodecalin as the probe liquid that the decrease in dispersive surface energy correlated with a reduction in the surface density of the terminal CF_3 groups – an interfacial change that corresponded with an increase in the length of the fluorinated segment. They also noted that the interfacial wettabilities for the series exhibited a notable reduction in sensitivity to the underlying film structure as compared to what had been observed for an analogous series of nonfluorinated alkanethiolate SAMs. Additionally, contact-angle measurements for the series plateaued for SAMs formed from adsorbates with $n \geq 6$. The

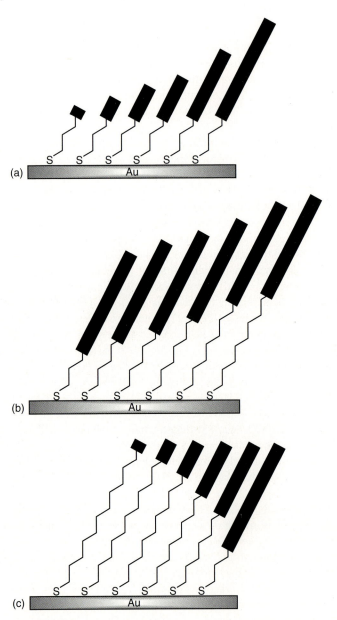

Figure 5.13 Illustration of a set of alkanethiol-based adsorbates where (a) the degree of terminal fluorination was varied while the length of the spacer was held constant ($F(CF_2)_x(CH_2)_{11}SH$; Series 5.1), (b) the spacer was varied while the terminal fluorination was held constant ($CF_3(CF_2)_9(CH_2)_ySH$; Series 5.2), and (c) both the degree of terminal fluorination and the spacer were concurrently varied while holding the total chain length constant ($F(CF_2)_x(CH_2)_ySH$; Series 5.3). See references in the text.

implication from this study is that the "low-energy" surface characteristics that many researchers have sought to incorporate into fluorinated thin films to mimic PTFE surfaces are diminished in monolayers with a terminally perfluorinated segment less than six carbons in length. A report from 2001 by Colorado *et al.* [277] provides an analysis of surface IR spectra for the C–H stretching frequencies of the underlying methylene spacer (held constant at 11 carbons) with a conclusion that the relative crystallinity of the alkyl chains is unperturbed by the changing length of the terminal fluorinated segment for the series. Finally, the last study that utilized this series explored the influence of the substitution of CF_2 units for CH_2 units on the attenuation length of photoelectrons in SAMs, which led to the unexpected conclusion that the attenuation lengths were indistinguishable for well-packed fluorocarbon and hydrocarbon films [278].

Similar terminally fluorinated adsorbates were used by Tamada *et al.* [282] to create a series of SAMs with 10 fluorinated carbons at the end of the chain; these SAMs also possessed methylene spacers that varied in length from 2 to 33 methylene units, and are identified above as Series 5.2. This study found that the relative conformational order at the chain termini diminished (as interpreted by AFM) with increasing spacer length. However, a separate study on the same series by Fukushima *et al.* [271] provided wettability data that indicated that the SAMs with the shortest methylene spacers afforded the lowest contact angles. These data were interpreted to indicate a reduced conformational order for the tailgroups of the SAMs with short methylene spacers. As the spacer was lengthened, the contact angles increased, reaching a maximum with $n = 11$, and then exhibiting roughly constant values for the longer methylene spacers, save for the film with 33 methylene units, for which the adsorbate exhibited signs of an increase surface disorder. Nevertheless, the results from a thermal stability study for the SAM series showed a discernible and steady improvement in resistance to thermal degradation as the methylene spacers of the adsorbates were lengthened.

A third study by Frey *et al.* [280], determined the orientation of the chains for the Series 5.2 SAMs on both gold and silver substrates by utilizing XPS, infrared reflection adsorption spectroscopy (IRRAS), and near-edge X-ray absorption fine structure (NEXAFS) spectroscopy. What these researchers discovered was that the upper fluorinated segments exhibited almost no tilt from the surface normal on both Au and Ag surfaces for the SAMs derived from adsorbates with only two methylene spacers. However, measurable tilts for the fluorinated segments developed as the methylene spacer was lengthened. The authors also found that the underlying spacer possessed chain tilt and twist characteristics that were comparable to those found in SAMs formed from nonfluorinated normal alkanethiol adsorbates.

Series 5.3 provided unique insight into the surface energy and interfacial interactions occurring on this set of fluorinated SAMs by holding the distance between the chain terminii and the underlying substrate effectively constant. For the research conducted by Graupe *et al.* [184] into the nature of the interfacial forces contributing to the contact-angle measurements of a series of probe liquids, the key difference from one member of the SAM series to another was the distance from the surface (air/SAM interface) of the R_F–R_H dipole along with its fundamental orientation to the substrate based upon the methylene count (odd versus even) below the fluorocarbon–hydrocarbon junction. This work was of fundamental importance because it offered the first unambiguous demonstration of the influence of surface dipoles on interfacial wettability. Stated differently, these studies showed that the work of adhesion due to dipole–oriented-dipole interactions was a key contributor to the surface forces experienced at SAM interfaces in contact with probe liquids, as subsequently rationalized in subsequent articles by Colorado and Lee [176, 279]. Additional studies involving this SAM series can also be found in the literature [285, 286] along with a series of partially fluorinated alkanethiols in which the chain length was held constant at 13 carbons instead of the 16 carbons used for all of the other projects highlighted in this paragraph [288].

The structural characteristics of fluorinated SAMs are slightly different from those described in Section 5.3 for alkanethiol-based SAMs. For alkanethiolate SAMs possessing extended segments of terminal fluorination, there is a variance in packing arrangements between the fluorinated and nonfluorinated segments when an extended underlying alkyl chain is present, or a mismatch between the overlying adsorbate packing and the surface lattice when the underlying alkyl chain is minimal. This type of variance in structures was broached by Ulman and Scaringe [289] in an article discussing the broader concern of the "commensurability of intra-assembly planes" for layered systems. This article focused on the problems that arise during the formation of stable two-dimensional (2D) assemblies from complex chains. As the mismatch between assembly layers increases, the cohesive energy decreases, leading to decreased monolayer stability and increased defect density. While the authors were examining the insertion of aromatic moieties into the structures that form 2D arrays, the fundamental concern applies also to extended segments of rigid fluorocarbons incorporated into the adsorbate structure of a SAM. For SAMs with a long spacer, this transition within the film is exemplified in the 2000 article by Frey *et al.* [280]. In this revealing study, it was found that an underlying spacer with 17 methylene units had a tilt angle of ~38°, while the overlying segment with 10 fluorinated carbons had a tilt angle of ~24°. Such a discontinuity in tilt between the layers in these adsorbates is a consequence of the larger cross-sectional area of the fluorinated chain (~5.6 Å) versus that of the alkyl chains (~4.2 Å) [46, 290].

This type of mismatch can also be found in studies involving SAMs with short spacers (e.g., two methylene units), such as the AFM study by Alves and Porter [290], where it was determined that the nearest-neighbor packing distance for the fluorinated adsorbates under study ($CF_3(CF_2)_7(CH_2)_2SH$ or F8H2SH) was 5.8 ± 0.2 Å, forming a (2×2) adlayer on the Au(111) surface with an average tilt angle of 20°. Such adsorbate spacing contrasts with the ~5.0 Å distance between the threefold hollow sites in which the sulfur headgroups are said to bind on the surface of Au(111) [112]. For these fluorinated adsorbates with only two methylene spacers between the headgroup and the fluorinated segment, the mismatch between surface lattice-binding sites and the nearest-neighbor spacing for the fluorinated chains might present a problem in nanoscale applications requiring surface registry. It is important to note, however, that this AFM study found that the domains formed by the partially fluorinated adsorbates are similar to those found formed by SAMs derived from analogous normal alkanethiols [291, 292].

The 2001 study by Tamada and coworkers [282] provides further support for the conclusion that the fluorocarbon SAMs on gold adopt different tilt angles and packing densities from analogous hydrocarbon SAMs. These authors determined that films formed from F10H2SH exhibited a two-dimensional hexagonal lattice with a lattice constant of $a = 5.9 \pm 0.1$ Å with an average tilt for the fluorinated segments of 16.4°. A separate study of the same type of adsorbate system by Liu *et al.* [293] utilized F6H2SH, F8H2SH, and F12H2SH to make determinations about the predominant packing arrangement of the corresponding SAMs. These investigators confirmed the Alves/Porter value of 5.8 Å for the spacing between nearest neighbor chains and calculated a tilt angle for the fluorinated segments of $12 \pm 2°$. However, rather than a commensurate $p(2 \times 2)$ lattice on the Au(111) surface, the authors proposed an incommensurate $c(7 \times 7)$ packing structure.

5.5.3.2 Developing Applications

One of the most commonly noted examples of the application of fluorinated SAMs is with the boundary lubrication of MEMSs or nanoelectromechanical systems (NEMSs). The impact of surface forces on microscale contacts is significant. For two contacting surfaces on a microscale mechanical device that rub against each other, the remedies for frictional wear are fundamentally different from those associated with macroscale systems. The lubricating systems that might provide a boundary film at the macroscale, and thereby reduce frictional wear, typically create adhesive forces for microscale systems that would lead to device failure, even for a lubricant with very low viscosity [5, 294]. The problems associated with microscale/nanoscale surface contact include systems that generate friction and systems in which no friction is generated, as is the case with microscale actuators. For systems where the contacting surfaces are extremely smooth, the interfacial contact can lead to adhesive forces that are significantly greater per surface area of contact than that experienced at a larger scale, owing to the larger surface-area-to-volume ratio of extremely small devices [295]. Interfacial interactions that become a concern for nonfrictional contact include electrostatic charges, van der Waals forces, and capillary action, which are sometimes denoted as "stiction" at the micro- or nanoscale (the static attractive forces that must be overcome to separate micromechanical surfaces). These forces can only be successfully subjugated by controlling them through nanoscale remedies.

Modification of small-scale contact surfaces with fluorinated SAMs possessing terminal fluorocarbon segments longer than six carbons is one possible route to reducing MEMS/NEMS adhesive and frictional problems. Such a reduction utilizing these adsorbates has proven to be beneficial and was found to provide superior performance over comparable nonfluorinated films in combating contact adhesion [295]. The recognition of the potential that terminally fluorinated SAMs hold for providing boundary lubrication solutions or for developing nonadhesive surfaces is apparent from the breadth of the studies conducted with films formed by fluorinated alkylsilanes on a variety of substrates, including aluminum [275], copper [76, 83], quartz [296, 297], glass [298, 299], germanium oxide [300], and SiO$_2$/silicon [301, 302].

Besides effectively modifying interfacial friction, alkylsilane films are more thermally stable than alkanethiolate SAMs, enduring temperatures up to 450 °C in an inert atmosphere [157, 303]. In thermal-stability tests comparing hydrocarbon and fluorocarbon alkylsilane films, the fluorinated system provided significantly better performance upon heating in air [304]. SAMs can also impart the hydrophobicity to the underlying substrate, which reduces problems associated with atmospheric moisture and the associated sticking due to capillary action [294]. This feature is particularly important for surfaces that are fully wet by water or are subject to oxidative processes that lead to enhanced hydrophilic character. However, a study by de Boer and coworkers that included contact adhesion measurements for microscale cantilever beams subject to controlled humidity conditions indicated that fluorinated SAMs do not perform as well as nonfluorinated SAMs at high levels of relative humidity, >90% [305]. Moore and coworkers [5] published a brief review in 2005 of the advantages that SAM systems offer to nanoscale boundary lubrication in MEMS devices. A recent review article by Maboudian can also be found in the literature [294].

Another potential application of fluorinated SAMs relates to methods of heat dissipation. In 2000, Das *et al.* [306] described the use of SAMs formed from C16SH to coat heat-exchange pipes made of gold-coated aluminum, copper, and copper-nickel tubing to improve the transfer of heat from the condensation of steam via "dropwise condensation". First described in an article by Schmidt *et al.* in 1930 [308], dropwise condensation is a process that depends on the contact surface having a low surface energy, thus water fails to wet the surface fully, forming drops instead. Most metal surfaces are high-energy surfaces that are fully wet by water, leading to "film condensation." Schmidt described an improvement in heat transfer for the dropwise-condensation process of five to seven times that of film condensation. For SAM-modified copper tubing prepared by Das and coworkers, the improvement in the heat-transfer coefficient was found to be a factor of 14. In 2006, Vemuri *et al.* [308] described the application of C18SH as a surface coating on copper tubing in a heat exchanger. The durability of the SAM system was tested and found to survive 2600 h of exposure to steam. These authors noted that one of the key advantages of such nanoscale coatings was their ability to modify the surface without creating a thermal barrier that would be detrimental to heat transfer, as was the case with most surface coatings, owing to their greater thickness. Surprisingly, the application of fluorinated alkanethiol adsorbates in such heat exchangers as promoters of dropwise condensation predates not only all of the published literature for the use of SAMs for this application, but also all papers describing alkanethiol-based SAMs. In 1974, a patent filed at the U.S. patent office described the utilization of adsorbates of the form $C_nF_{2n+1}(CH_2)_mX$, where n and m are numbers between 2 and 20, and X is a functional group that provides a bonding mechanism to the surface [309]. For this patent, X included both SH and S in the form of a disulfide. The stated purpose was to promote dropwise condensation in heat exchangers.

There is obviously a broad array of research projects currently involving or potentially involving fluorocarbon-containing adsorbates. A useful review by Barriet and Lee [310] focuses on the impact of fluorination on SAM films.

5.5.4
Mixed Adsorbates

5.5.4.1   Background for Mixed-Adsorbate Surfaces

The possible advantages of combining two adsorbates into one "mixed" SAM film were recognized early in the exploration of SAMs. In 1988, Troughton *et al.* [43] reported on their efforts to create "mixed" monolayer systems through the use of unsymmetrical dialkyl sulfides. The resulting films were poorly ordered and exhibited liquid-like characteristics. However, the film composed of both a carboxylic acid-terminated chain and a methyl-terminated chain (with the latter possessing an additional five methylene units) revealed that the longer alkyl chain prevented a contacting liquid (water) from detecting the presence of the buried polar termini. Concurrently, Bain and Whitesides [311] were proceeding with studies of mixed SAMs derived from the coadsorption of two distinct alkanethiols. In these pioneering studies, the authors also chose a lengthy alkanethiol (C22SH) combined with a shorter thiol exposing an alcohol terminus ($HO(CH_2)_{11}SH$). The parameters of the experimental procedure provided that the solution containing the mixture of adsorbates was maintained at a 1 mM total concentration of components, and that a series of solutions were prepared for a broad range of thiol mixtures. Recognizing that increased van der Waals attractions favors close packing of the longer alkyl adsorbate, it is not surprising that the data indicated a strongly preferred adsorption of C22SH, and that the transition between the predominance of one adsorbate over the other was sharp and unequivocal.

Subsequent studies by Bain and Whitesides [312] focused on a set of mixed SAMs composed of adsorbates of equal length but with tailgroups that were fundamentally different: one was polar (either –Br, –CO_2H, or –OH) and the other nonpolar (C11SH). The authors noted that these mixed systems formed SAMs that were well packed, which was not necessarily the case with the mixed SAMs having chains with substantial differences in length. The data also showed that the mixed SAMs having a constant chain length exhibited adsorbate compositions that were much closer to the solution composition than the former mixed SAMs. In fact, one particular combination created films with a surface composition exactly matching the solution composition (i.e. the combination that included the bromine-terminated thiol). To rationalize this result, the authors pointed to the work of Cassie, noting his claim that "if the two components of the surface act independently, then $\cos\theta_a$ is a linear function of the composition of the surface, in the absence of hysteresis," referring to the cosine of the contact angle for a probe liquid [170]. Accordingly, the data for the mixed-SAMs involving the bromine-terminated thiol exhibited a linear relationship for $\cos\theta_a$ when plotted as a function of the surface composition, accurately reflecting the nature of the composition, which was explicitly determined by XPS.

In a 1989 article, Bain and coworkers [313] continued their analysis of the parameters that influence the formation of mixed SAMs. The authors concluded that solubility effects could have a large impact upon preferential adsorption. For example, a hydroxy-terminated adsorbate was observed to adsorb almost execlusively over a methyl-terminated adsorbate when dissolved in a nonpolar hydrocarbon solvent; however, for the same components dissolved in ethanol, adsorption of the methyl-terminated component was preferred. A further study by Bain and Whitesides involving alkanethiols having chains of varying lengths revealed that the longer adsorbates were favored over the shorter ones [314] – a fact that was already apparent from their previous work [311].

One conclusion that might be drawn from these studies is that there exists a hierarchy for adsorbates that defines what leads one adsorbate to adsorb preferentially to another on a surface. Such was the conclusion of a 1994 study by Folkers *et al.* [315], where a similar set of adsorbates (a long-chain alkanethiol versus a shorter alcohol-terminated thiol) were allowed to form mixed SAMs from solution. The resulting SAMs exhibited phase domains that, over time, led to films dominated by one phase through displacement processes. The authors argued that all mixed-thiol systems would ultimately reflect this dominant adsorbate preference if allowed sufficient time to fully equilibrate, or, alternatively, establish a single phase formed from a homogeneous mixture of the adsorbates.

Other combinations of adsorbates possessing varying headgroups have also been competitively tested in efforts to better understand the phenomenon of preferential adsorption. Such an experiment was conducted by Bain and coworkers [44, 316], pitting an alkanethiol against a dialkyl disulfide, each having alkyl chains of approximately the same length. Studies of competitive adsorption found a 75 : 1 thiol:disulfide ratio in the resulting film. These investigations were followed by research conducted by Biebuyck *et al.* [254] that further examined the underlying causes for the large discrepancy in the composition of the final films formed from mixtures of thiols and disulfides. The two adsorbates exhibited similar film-forming rates, however, the authors noted that the process of film formation appeared to be ongoing, with adsorbate replacement occurring while the film was still exposed to the adsorbate solution. They also noted that this displacement/replacement process favored the less cumbersome adsorbate (i.e. the thiol), leading to its disproportionately larger presence. Additional reports on binary SAMs are numerous; however, most of the more recent work is application oriented [317–320].

5.5.4.2 New Concepts for Mixed-Adsorbate Systems

An exploration of methods for controlling the nature of phase boundaries in mixed-adsorbate SAM films is a logical next step for the development of applications for these functional nanoscale materials. Interest in the manipulation of the surface distribution of the adsorbates for the fabrication of organized two-dimensional (2D) patterns is driven by a number of research objectives, including the possible development of chemical routes to the creation of patterned circuits for nanoelectronics [321], the development of specialized biosensors or arrays for nanoscale biodiagnostics [322], and the preparation of SAM systems that create specific patterns on nanoparticles, enabling unique interparticle interactions [323].

The growing interest in nanoscale surface patterns is tied, at least in part, to the unrelenting drive for miniaturization of electronic circuitry, which has steadily pushed chip manufacturers to investigate new technologies or alternative methods for the preparation of computer components. This thrust has existed since the creation of the first integrated circuit in 1958 [324]. Recognition of the ensuing rapid pace of technological innovation was described in an article by Gordon Moore, one of the cofounders of Intel, who noted that an exponential trend in the pace of miniaturization (or affordably "cramming more components onto integrated circuits") was associated with the growth of computer technology – a relationship that became known as *"Moore's Law"* [325]. The inevitable limits of existing beam-lithography techniques used to create semiconductor devices are leading researchers to investigate a variety of "bottom-up" methods of circuitry preparation via chemical synthesis [326].

For mixed 2D systems on a liquid surface, the influence of the tension along the phase boundaries, the "line tension," has led to a variety of studies involving manmade lipid membranes in which the phases have been manipulated to determine the factors that contribute to the ultimate shapes that form [327]. Because the "substrate" for the membranes is fluid, manipulating the boundary tension of phases has generally involved fine tuning surfactant mixtures along with controlled application of surface pressure and the manipulation of the temperature [327, 328]. Such work has provided insight into the factors that impact the line tension along a domain boundary. Recent work, however, has revealed a growing interest in manipulating phase boundaries in 2D systems using unique phase-active boundary surfactants, which are amphipathic molecules possessing more than one type of phase-preferred moiety in their structure, as shown in Figure 5.14 [16, 17]. The specialized surfactants that provide this line-active surfactant intervention in a 2D environment are called *"linactants."*

For 2D systems on solid substrates, the concept of a phase-active adsorbate to stabilize the boundaries on a surface is complicated by the presence of strong interactions, either physical or chemical, between the nonfluid substrate and the monolayer-forming adsorbates. Consequently, the nature of the tension experienced at these boundaries by a linactant relate not only to the interactions between it and the two surfaces phases for which the linactant provides intervention, as restricted by adsorbate bonds to the substrate on either side of the boundary, but also the tension along the line defining the boundary as dictated by substrate–linactant interactions. Such systems, while subject to dynamic processes during film formation, should achieve relatively stable phase structures upon attaining equilibrium with the solution phase, providing a possible means for chemically developing nanoelectronic device features. Once perfected, phase boundaries subject to linactant intervention are expected to adopt 2D surface arrays that are analogous to 3D structures, such as bilayers, micelles, or microemulsions. The architecture of the initial thiol-based linactant molecules found in the literature possess two

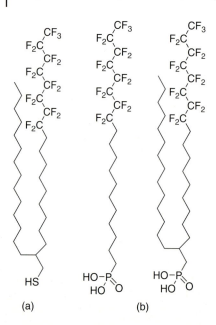

Figure 5.14 Illustration of proposed linactant molecules with (a) a thiol headgroup and (b) phosphonic acid headgroups [16, 17].

(a)   (b)

extended surface chains emanating from a branchpoint near the headgroup, thus allowing each chain to possess independent phase activity, such as one chain being oleophilic in nature and the other predominantly oleophobic. For 3D systems, if the compact polar end of the surfactant chain fails to create a substantial variance in solubility, the anticipated superstructures will fail to develop. However, owing to the adsorption process involved in SAM formation and the surface-bound orientation of adsorbate molecules, there appears to be a higher sensitivity to chain compatibility for surfactants assembling on a surface. Examples of the more subtle phase interactions found in nonlinactant-mediated binary systems include combinations of terminal group for alkyl chains of equal length [312, 313, 329] and long vs. short alkanethiolate adsorbates [330, 331]. Nevertheless, the resulting phase-segregated domains for such binary systems tend to have random size distributions.

A report by Zhang *et al.* in 2008 described a linactant structure with one chain exhibiting an extended fluorinated segment, while the other branch was a simple linear alkyl chain, both tied to a single thiol headgroup as shown in Figure 5.14a [16]. A subsequent study focused on the nature of the phase activity of binary SAMs involving this initial thiol-based linactant [332]. The collected data for a series of binary SAMs formed from adsorbate solutions of varying linactant content were consistent with the development of a homogeneous dispersion of a small number of linactant surface structures surrounded by a larger population of single-chained adsorbates (alkanethiol chains chosen because they were analogous to the linactant absent the fluorinated chain). The analysis of this particular linactant system is still incomplete, but the potential for such methods of surface manipulation is clear.

A better understanding of the factors contributing to the development of this technology in SAM research requires a brief review of Langmuir films, where phase-boundary manipulation in biocompatible systems is similar to that in SAM films [17, 333]. The study of phase manipulation derives in part from the potential for developing artificial cell membranes in which the specific cell-boundary characteristics might be more completely understood through investigations that manipulate cell membrane components and the resulting phase morphology, as highlighted by the work of Panda *et al.* in 2007 [334]. This research project examined bovine lung surfactants and their interaction with cholesterol and a cholesterol-palmitate derivative. The authors noted that cholesterol acts as a linactant, asserting that "it lowers the line tension between the fluid- and solid-phase regions of the monolayer, allowing for larger ratios of line length to surface areas, leading to noncircular solid domains."

In a more fundamental study focused on the effects of line tension (with no linactant intervention) on the organization of lipid membranes, Garcia-Saéz *et al.* [327] examined various parameters influencing the line tension of "raft-like" domains in model lipid membranes. The "raft" concept is based upon the belief that certain phase domains (rafts) within the cell membrane play functional roles in various cellular processes, including protein sorting and cell signaling. This paper utilized the theoretical models of Kuzmin *et al.* and Akimov *et al.* that linked the line tension for such systems "to physical properties of the membrane, like phase height mismatch, lateral tension, and spontaneous curvature" [335, 336]. Regarding their own experimental work, the authors observed that a "higher hydrophobic mismatch" between the ordered liquid phase of the raft-like domains and the disordered phase surrounding these generally circular regions led to an increase in line tension and to larger domains. They also noted that a difference in phase heights was a key contributor to line tension, as predicted by the theoretical models. However, while the phase manipulation in this study led to the creation of domains of varying sizes, the need to develop methods of phase boundary control that preserve the nature of the separate phases remains. To this end, the utilization of manmade linactants with phosphonic acid headgroups, as shown in Figure 5.14b, has already provided insight into how such structures might be useful in SAM research [17].

A recent article by Majumdar *et al.* [337] that focused on a mixed monolayer for studying the electrical behavior of a thiol-based nitro oligo(phenylene ethynylene) (nitroOPE) adsorbate, with the second adsorbate being dodecanethiol, illustrates the potential benefits of phase manipulation for ongoing projects in SAM research. The authors of this study noted that the nitroOPE adsorbates in the mixed system failed to exhibit the unique electronic characteristics that were present in the pure nitroOPE SAMs, and they concluded that such "behavior may be dependent on contacting a large area of pure nitro molecules." The results suggest that the key characteristics present in the single-component SAM, but absent with uncontrolled mixing of the two adsorbates, might be preserved with the development of effective methods of phase manipulation that lead to the creation of phase-segregated nitroOPE nanowells that preserve the critical adsorbate behavior found in the pure SAM.

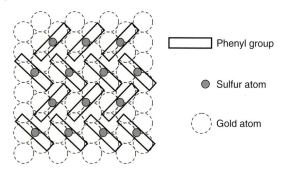

Phenyl group

Sulfur atom

Gold atom

Figure 5.15 Illustration of the herringbone structure associated with phenyl-terminated thiols (and other adsorbates incorporating an aromatic ring) on gold. Adapted from Ref. [186].

5.5.5
Aromatic Adsorbates

5.5.5.1 Impact of Aromatic Moieties on SAM Structure

The addition of aromatic rings into the structure of an adsorbate creates unique interactions between neighboring chains, initiating interchain alignment patterns not seen with normal alkanethiol SAMs. For adsorbates incorporating a single phenyl ring in their structure, the pattern that emerges has often been dubbed a "herringbone" structure, as illustrated in Figure 5.15 [186, 338–340]. In a study published in 2001, Lee and coworkers [186] examined a series of SAM films possessing a terminal unsubstituted phenyl ring (i.e. at a location that allows a more direct examination of the impact of the aromatic ring). Using AFM, they determined the lateral spacing for the individual phenyl-terminated adsorbates on the surface was 4.9 ± 0.2 Å. Such a packing arrangement would indicate little or no change from the packing density established by the sulfur–gold bonds on the Au(111) interface. Chang *et al.* [340], in their 1994 study of a series of SAMs incorporating aromatic rings into the spacer, analyzed in detail the configurations adopted by these adsorbates, concluding that the alkyl chains extending above the intervening aromatic rings also conformed to the $(\sqrt{3} \times \sqrt{3})R30°$ arrangement dictated by the underlying surface bonds.

In a 1993 study by Sabatani *et al.* [341], a series of aromatic thiol-based SAMs was prepared and subjected to a number of tests, including displacement with an alkanethiol, C18SH. For the aromatic thiols, the shortest adsorbate was thiophenol (TP) which was accompanied by thiols with additional aromatic rings *para* to the thiol moiety: *p*-biphenyl mercaptan (BPM) and *p*-terphenyl mercaptan (TPM), as shown in Figure 5.16. The conclusion drawn from the data was that the extended aromatic rings created SAMs with enhanced crystallinity compared to the thiophenolate SAMs. Additionally, displacement and wettability studies indicated that the BPM and TPM SAMs were more stable than the TP SAMs. These conclusions are also supported by observations made in 1997 by Tao *et al.* [338],

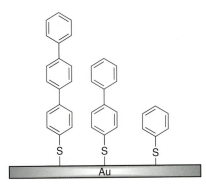

Figure 5.16 A series of rigid aromatic adsorbates: thiophenol (right), *p*-biphenyl mercaptan (BPM) (center), and *p*-terphenyl mercaptan (TPM) (left). Adapted from Ref. [341].

where the TP and BPM adsorbates were synthesized with extended alkoxy chains in the *para* position. The collected data indicated that the modified TP and BPM SAMs exhibited an enhancement in structural order when compared to the unmodified TP and BPM SAMs.

Ulman and coworkers [21, 342–346] have reported several studies of SAMs derived from aromatic thiols, with most focused on derivatives of BPM (aka 4-mercaptobiphenyl) having a diverse array of substituents in the terminal position: F, Cl, Br, I, OH, SH, CH_3, CF_3, NO_2, $N(CH_3)_2$, SCH_3, $CHOHCH_3$, $C(O)CH_3$, CO_2Et, $O(CH_2)_2OCH_3$, and H. One of the fundamental motivations driving this research is the fact that such rigid conjugated adsorbates create SAM surfaces that are not subject to the type of conformational disordering associated with the chain termini of alkanethiolate SAMs, as noted in an article from 2001 [21]. In this report, the authors described wettability studies for these BPM SAMs showing that their wetting characteristics are stable even after storage for an extended period of time. This added conformational reliability provides an opportunity to create stable SAM surfaces for conducting a number of experimental procedures that require an interface exposing an ordered array of a specific functional group (e.g., tailgroups-capable of hydrogen bonding for conducting crystal nucleation and growth). A review article by Ulman and colleagues in 2000 outlines the arguments of those who advocate this type of SAM architecture [20].

5.5.5.2 Toward Nanoscale Electronics

One concept that continues to generate interest in aromatic SAMs is their potential use in nanoelectronics. A recent study by Chen *et al.* [347] provides a glimpse of the research that is ongoing. The authors, utilizing *para*-substituted thiophenol adsorbates, tested SAMs with CF_3 versus CH_3 substituents to establish a better understanding of the impact that such structural changes might have on the rate of charge transfer into a SAM boundary layer. The reasoning behind the specific parameters chosen for this study can be found in a previous report by Zehner *et al.* [348]. The SAMs assembled by Chen on Au(111)/mica substrates provided evidence that the strong electron-withdrawing CF_3 group located in the terminal position for these aromatic adsorbates, enabled "significant electron transfer" from an overlying active layer (copper(II) phthalocyanine – CuPc) to the SAM network, as determined

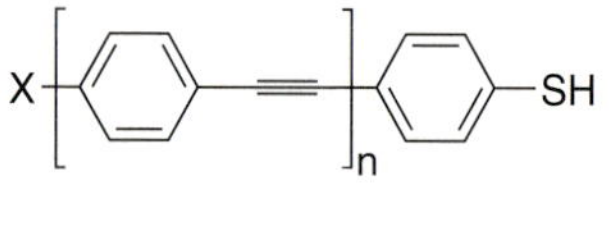

Figure 5.17 Illustration of the fundamental structure of oligo(phenylene ethynylene) thiol adsorbates, also known as *molecular wires* [349].

by synchrotron photoemission spectroscopy. Such an electron-transfer mechanism creates an accumulation of charge in the CF_3-modified aromatic SAM and a corresponding depletion in the overlying CuPc layer, providing an organic system with p-type doping analogous to that found in doped semiconductor surfaces. In contrast, the CH_3-terminated system exhibited only limited electron-transfer behavior.

The collection of adsorbate structures dubbed "molecular wires" that take advantage of the delocalization of electron density in organic systems possessing extended conjugation have been examined as a means of producing surfaces that manipulate the flow of electrons toward specific sites or that regulate the flow of electrons through a gate (see Figure 5.17). A 1995 article by Tour and coworkers [349] helped outline many of the initial structures being investigated. A subsequent 2001 paper by Tour and colleagues provides synthetic procedures for the assembly of a number of these specialized, conjugated adsorbates along with methods of fine tuning such structures to manipulate electron flow [146]. The article gives a lucid explanation of the generic concepts for the development of specific molecular electronic components, including formats for SAM-based random access memory (RAM) and a description of the "nanopore" structures created to test the devices.

The OPE structures that are the focus of this research deviate from the structure of the absorbates analyzed in Section 5.5.5.1 not only by the insertion of an alkyne moiety between the aromatic rings, but also by the utilization of substituent groups to modulate the flow of electrons through the adsorbate with a bias that favors a flow toward one end of the molecule (a negative differential resistance – NDR). Figure 5.18 provides an illustration of the author's concept of how such a RAM device would perform [350]. However, this research faces many challenges, including the sensitivity of the devices to temperature (the device in Figure 5.18 lost NDR characteristics above 260 K), the propensity of the OPE moieties to rotate away from a fully planar conformation (reducing orbital overlap and conduction efficiency), and electronic integration issues related to an impedance mismatch for typical SAM headgroups and substrates [146]. The impedance mismatch issue might be resolved with studies involving alternative headgroups and substrates, such as the combination of an isonitrile headgroup with a palladium substrate [351].

Efforts to integrate molecular electronics into a SAM matrix have also included research that recognized the possibility of horizontal electron flow (in the SAM plane) in contrast to the vertical flow within the structures described in the previous paragraph. The tendency of phenyl moieties in a SAM adsorbate to adopt a herringbone packing arrangement leads to configurations that do not favor electron flow between adsorbates. In 2003, Zareie *et al.* [352] described studies of two forms

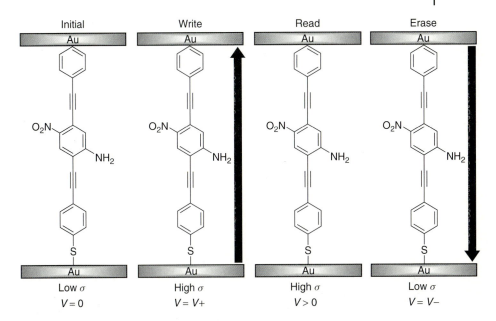

Initial	Write	Read	Erase
Low σ	High σ	High σ	Low σ
V = 0	V = V+	V > 0	V = V−

Figure 5.18 Illustration of the operational stages of a molecular random access memory device. The ability to store a high or low conductivity state (high or low σ), as controlled by strategically placed substituents on the central aromatic ring provides this molecular structure its "memory" characteristics. The flow of current is indicated by the arrows in the "write" and "erase" stages of the operation of this nanoscale device. The retained charge in the high-energy state for a single device would be a bit of memory for molecular electronics [350].

of the OPE structures described above: (i) one that contained two aromatic rings with an intervening alkyne and terminal thiol and (ii) one that substituted an anthracene moiety for the aromatic ring at the opposite end of the adsorbate from the thiol group (4-mercaptophenyl-9-anthryl acetylene – MPAA). The substitution of the anthracene led to surface alignments in which the anthracene moieties developed "overlap in a manner very similar to the interplanar stacking of graphite." The authors speculated that this overlap contributed to an improved conductivity for these adsorbates and a lowered energy gap. One of their rationalizations was that $\pi-\pi$ stacking interactions altered the energetics of the system, allowing electrons to tunnel between neighboring adsorbates. A subsequent STM study by Dou *et al.* [353] included MPAA along with two derivatives of this adsorbate, where either NO_2 or CO_2H were added to the anthracene in the position *para* to the alkyne moiety. The STM images collected for the SAMs formed from these adsorbates revealed that the stacking of the aromatic groups is the dominant organizational force on the nature of the array; however, subtle changes in structure are evident that might reflect the influence of the presence of a substituent on the anthracene moiety. A review article by Chen *et al.* [354] from 2007 provides additional insight into the challenges associated with molecular electronics.

5.6
Conclusions

Significant progress has been made in the development of functional SAM films. Methods for addressing various shortcomings have been pursued, including improving adsorbate–substrate binding through the modification of both the substrate (UPD metals) and the adsorbate (multidentate headgroups), exploring new applications for existing adsorbates (silane films on oxidized copper), developing alternative headgroup configurations to enhance nanoparticle stability (multidentate adsorbates), anchoring new tethered adsorbates more effectively (improved bidentate architectures), understanding friction at the molecular level (systematic adsorbate synthesis and analysis by AFM), improving interchain interactions to resist desorption (cross-linking between spacers), controlling barrier properties to modulate the flow of electrons across a junction (calibrated fluorination of the adsorbate chain), controlling the nature of the surface features of manmade lipid membranes and developing nanoscale electronics (studies of mixed SAMs and the development of linactants), synthesizing complex aromatic adsorbates for nanoscale electron flow (modified aromatic adsorbates), and creating durable films that can withstand exposure to high temperatures (electron-beam cross-linked films). These impressive advances in the study of SAM films are slowly being transferred from the laboratory to the factory, creating practical applications for a new generation of functional nanoscale films.

Acknowledgment

We thank the National Science Foundation (DMR-0447588) and the Robert A. Welch Foundation (Grant No. E-1320) for generously supporting this research.

References

1. Laibinis, P.E. and Whitesides, G.M. (1992) Self-assembled monolayers of n-alkanethiolates on copper are barrier films that protect the metal against oxidation by air. *J. Am. Chem. Soc.*, **114**, 9022–9028.
2. Telegdi, J., Rigo, T., and Kalman, E. (2004) Nanolayer barriers for inhibition of copper corrosion. *Corros. Eng., Sci. Technol.*, **39**, 65–70.
3. Beebe, J.M., Moore, H.J., Lee, T.R. and Kushmerick, J.G. (2007) Vibronic coupling in semifluorinated alkanethiol junctions: implications for selection rules in inelastic electron tunneling spectroscopy. *Nano Lett.*, **7**, 1364–1368.
4. Weiss, E.A., Kriebel, J.K., Rampi, M.-A., and Whitesides, G.M. (2007) The study of charge transport through organic thin films: mechanism, tools and applications. *Philos. Trans. R Soc. London A*, **365**, 1509–1537.
5. Lee, T.R., Barriet, D., and Moore, H.J. (2005) Structure and properties of fluorinated organic thin films: implications for MEMS devices. *NDSI Digest*, **2**, 93.
6. Chandross, M., Lorenz, C.D., Grest, G.S., Stevens, M.J., and Webb, E.B. III (2005) Nanotribology of anti-friction coatings in MEMS. *JOM*, **57**, 55–61.

7. Bethencourt, M.I., Barriet, D., Frangi, N.M., and Lee, T.R. (2005) Model glycol-terminated surfaces for adhesion resistance. *J. Adhes.*, **81**, 1031–1048.

8. Miller, J.S., Bethencourt, M.I., Hahn, M., Lee, T.R., and West, J.L. (2006) Laser-scanning lithography (LSL) for the soft lithographic patterning of cell-adhesive self-assembled monolayers. *Biotechnol. Bioeng.*, **93**, 1060–1068.

9. Dameron, A.A., Mullen, T.J., Hengstebeck, R.W., Saavedra, H.M., and Weiss, P.S. (2007) Origins of displacement in 1-adamantanethiolate self-assembled monolayers. *J. Phys. Chem. B*, **111**, 6747–6752.

10. Lee, T.-C., Hounihan, D.J., Colorado, R. Jr., Park, J.-S., and Lee, T.R. (2004) Stability of aliphatic dithiocarboxylic acid self-assembled monolayers on gold. *J. Phys. Chem. B*, **108**, 2648–2653.

11. Dameron, A.A., Charles, L.F., and Weiss, P.S. (2005) Structures and displacement of 1-adamantanethiol self-assembled monolayers on Au(111). *J. Am. Chem. Soc.*, **127**, 8697–8704.

12. Kuller, A., Eck, W., Stadler, V., Geyer, W., and Golzhauser, A. (2003) Nanostructuring of silicon by electron-beam lithography of self-assembled hydroxybiphenyl monolayers. *Appl. Phys. Lett.*, **82**, 3776–3778.

13. Turchanin, A., El-Desawy, M., and Golzhauser, A. (2007) High thermal stability of cross-linked aromatic self-assembled monolayers: nanopatterning via selective thermal desorption. *Appl. Phys. Lett.*, **90**, 053102/053101–053102/053103.

14. Zhang, S., Leem, G., Srisombat, L., and Lee, T.R. (2008) Rationally designed ligands that inhibit the aggregation of large gold nanoparticles in solution. *J. Am. Chem. Soc.*, **130**, 113–120.

15. Srisombat, L., Park, J.-S., Zhang, S., and Lee, T.R. (2008) Preparation, characterization, and chemical stability of gold nanoparticles coated with mono-, bis-, and tris-chelating alkanethiols. *Langmuir*, **24**, 7750–7754.

16. Zhang, S., Jamison, A.C., Schwartz, D.K., and Lee, T.R. (2008) Self-assembled monolayers derived from a double-chained monothiol having chemically dissimilar chains. *Langmuir*, **24**, 10204–10208.

17. Trabelsi, S., Zhang, S., Lee, T.R., and Schwartz, D.K. (2008) Linactants: surfactant analogues in two dimensions. *Phys. Rev. Lett.*, **100**, 037802/037801–037802/037804.

18. Schiller, S.M., Naumann, R., Lovejoy, K., Kunz, H., and Knoll, W. (2003) Archaea analogue thiolipids for tethered bilayer lipid membranes on ultrasmooth gold surfaces. *Angew. Chem., Int. Ed.*, **42**, 208–211.

19. Naumann, R., Schiller, S.M., Giess, F., Grohe, B., Hartman, K.B., Kaercher, I., Koeper, I., Luebben, J., Vasilev, K., and Knoll, W. (2003) Tethered lipid bilayers on ultraflat gold surfaces. *Langmuir*, **19**, 5435–5443.

20. Ulman, A., Kang, J.F., Shnidman, Y., Liao, S., Jordan, R., Choi, G.-Y., Zaccaro, J., Myerson, A.S., Rafailovich, M., Sokolov, J., and Fleischer, C. (2000) Self-assembled monolayers of rigid thiols. *Rev. Molec. Biotech.*, **74**, 175–188.

21. Ulman, A. (2001) Self-assembled monolayers of 4-mercaptobiphenyls. *Acc. Chem. Res.*, **34**, 855–863.

22. Ge, B. and Lisdat, F. (2002) Superoxide sensor based on cytochrome c immobilized on mixed-thiol SAM with a new calibration method. *Anal. Chim. Acta*, **454**, 53–64.

23. Nakamura, F., Mitsui, K., Hara, M., Kraemer, S., Mittler, S., and Knoll, W. (2003) Preparation of self-assembled monolayers containing anthryl groups toward hybridization of nucleotides. *Langmuir*, **19**, 5823–5829.

24. Pockels, A. (1891) Insoluble monolayers at the liquid-gas interface. *Nature*, **43**, 437.

25. Rayleigh, L. (1899) Investigations into capillarity. *Philos. Mag.*, **48**, 321.

26. Langmuir, I. (1917) Constitution and fundamental properties of solids and liquids. II. liquids. *J. Am. Chem. Soc.*, **39**, 1848–1906.

27. Blodgett, K.B. (1934) Monomolecular films of fatty acids on glass. *J. Am. Chem. Soc.*, **56**, 495.

28. Blodgett, K.B. (1935) Films built by depositing successive unimolecular layers on a solid surface. *J. Am. Chem. Soc.*, **57**, 1007–1022.

29. Roberts, G.G. (1985) An applied science perspective of Langmuir–Blodgett films. *Adv. Phys.*, **34**, 475–512.

30. Tredgold, R.H. (1987) The physics of Langmuir–Blodgett films. *Rep. Prog. Phys.*, **50**, 1609–1656.

31. Schulman, F. and Zisman, W.A. (1952) Surface chemical properties of solids coated with a monolayer of perfluorodecanoic acid. *J. Am. Chem. Soc.*, **74**, 2123–2124.

32. Bowers, R.C., Jarvis, N.L., and Zisman, W.A. (1965) Reduction of polymeric friction by minor concentrations of partially fluorinated compounds. *Ind. Eng. Chem. Prod. Res. Dev.*, **4**, 86–92.

33. Murphy, C.M. and Zisman, W.A. (1950) Structural guides for synthetic lubricant development. *Ind. Eng. Chem.*, **42**, 2415–2420.

34. Bigelow, W.C., Pickett, D.L., and Zisman, W.A. (1946) Oleophobic monolayers. I. films adsorbed from solution in nonpolar liquids. *J. Colloid Sci.*, **1**, 513–538.

35. Zisman, W.A. (1963) Influence of constitution on adhesion. *Ind. Eng. Chem.*, **55**, 18–38.

36. Hare, E.F. and Zisman, W.A. (1955) Autophobic liquids and the properties of their adsorbed films. *J. Phys. Chem.*, **59**, 335–340.

37. Jarvis, N.L. and Zisman, W.A. (1959) Surface activity of fluorinated organic compounds at organic liquid–air interfaces. I. Surface tension, parachor, and spreadability. *J. Phys. Chem.*, **63**, 727–734.

38. Haller, I. (1978) Covalently attached organic monolayers on semiconductor surfaces. *J. Am. Chem. Soc.*, **100**, 8050–8055.

39. Sagiv, J. (1980) Organized monolayers by adsorption. 1. formation and structure of oleophobic mixed monolayers on solid surfaces. *J. Am. Chem. Soc.*, **102**, 92–98.

40. Onclin, S., Ravoo, B.J., and Reinhoudt, D.N. (2005) Engineering silicon oxide surfaces using self-assembled monolayers. *Angew. Chem., Int. Ed.*, **44**, 6282–6304.

41. Nuzzo, R.G. and Allara, D.L. (1983) Adsorption of bifunctional organic disulfides on gold surfaces. *J. Am. Chem. Soc.*, **105**, 4481–4483.

42. Nuzzo, R.G., Zegarski, B.R., and Dubois, L.H. (1987) Fundamental studies of the chemisorption of organosulfur compounds on gold(111). Implications for molecular self-assembly on gold surfaces. *J. Am. Ceram. Soc.*, **109**, 733–740.

43. Troughton, E.B., Bain, C.D., Whitesides, G.M., Nuzzo, R.G., Allara, D.L., and Porter, M.D. (1988) Monolayer films prepared by the spontaneous self-assembly of symmetrical and unsymmetrical dialkyl sulfides from solution onto gold substrates: structure, properties, and reactivity of constituent functional groups. *Langmuir*, **4**, 365–385.

44. Bain, C.D., Troughton, E.B., Tao, Y.T., Evall, J., Whitesides, G.M., and Nuzzo, R.G. (1989) Formation of monolayer films by the spontaneous assembly of organic thiols from solution onto gold. *J. Am. Chem. Soc.*, **111**, 321–335.

45. Whitesides, G.M. and Laibinis, P.E. (1992) Wet chemical approaches to the characterization of organic surfaces: Self-assembled monolayers, wetting, and the physical-organic chemistry of the solid-liquid interface., *Langmuir*, **6**, 87–96.

46. Ulman, A. (1996) Formation and structure of self-assembled monolayers. *Chem. Rev.*, **96**, 1533–1554.

47. Porter, M.D., Bright, T.B., Allara, D.L., and Chidsey, C.E.D. (1987) Spontaneously organized molecular assemblies. 4. Structural characterization of n-alkyl thiol monolayers on gold by optical ellipsometry, infrared spectroscopy, and electrochemistry. *J. Am. Chem. Soc.*, **109**, 3559–3568.

48. Colorado, R. Jr. and Lee, T.R. (2001) *Encyclopedia of Materials*, Elsevier, pp. 9332–9344.

49. Fenter, P., Schreiber, F., Berman, L., Scoles, G., Eisenberger, P., and Bedzyk, M.J. (1998) On the structure

and evolution of the buried S/Au interface in self-assembled monolayers: X-ray standing wave results. *Surf. Sci.*, **412/413**, 213–235.

50. Chidsey, C.E.D., Liu, G.Y., Rowntree, P., and Scoles, G. (1989) Molecular order at the surface of an organic monolayer studied by low energy helium diffraction. *J. Chem. Phys.*, **91**, 4421–4423.

51. Strong, L. and Whitesides, G.M. (1988) Structures of self-assembled monolayer films of organosulfur compounds adsorbed on gold single crystals: electron diffraction studies. *Langmuir*, **4**, 546–558.

52. Moore, H.J. (2007) *Self-Assembled Monolayers on Gold Generated from Precisely Designed Organosulfur Adsorbates for the Preparation of Surfaces with Specific Interfacial Properties*, Doctor of Philosophy, University of Houston, Houston.

53. Nuzzo, R.G., Korenic, E.M., and Dubois, L.H. (1990) Studies of the temperature-dependent phase behavior of long chain *n*-alkyl thiol monolayers on gold. *J. Chem. Phys.*, **93**, 767–773.

54. Camillone, N. III, Chidsey, C.E.D., Liu, G.Y., and Scoles, G. (1993) Superlattice structure at the surface of a monolayer of octadecanethiol self-assembled on gold(111). *J. Chem. Phys.*, **98**, 3503–3511.

55. Bewig, K.W. and Zisman, W.A. (1965) The wetting of gold and platinum by water. *J. Phys. Chem.*, **69**, 4238–4242.

56. Bernett, M.K. and Zisman, W.A. (1970) Confirmation of spontaneous spreading by water on pure gold. *J. Phys. Chem.*, **74**, 2309–2312.

57. Katash, I., Luo, X., and Sukenik, C.N. (2008) *In situ* sulfonation of alkyl benzene self-assembled monolayers: product distribution and kinetic analysis. *Langmuir*, **24**, 10910–10919.

58. Chen, R. and Bent, S.F. (2006) Highly stable monolayer resists for atomic layer deposition on germanium and silicon. *Chem. Mater.*, **18**, 3733–3741.

59. Lenfant, S., Guerin, D., Van, F.T., Chevrot, C., Palacin, S., Bourgoin, J.P., Bouloussa, O., Rondelez, F.,

and Vuillaume, D. (2006) Electron transport through rectifying self-assembled monolayer diodes on silicon: Fermi-level pinning at the molecule–metal interface. *J. Phys. Chem. B*, **110**, 13947–13958.

60. Kosuri, M.R., Cone, R., Li, Q., Han, S.M., Bunker, B.C., and Mayer, T.M. (2004) Adsorption kinetics of 1-alkanethiols on hydrogenated Ge(111). *Langmuir*, **20**, 835–840.

61. Gu, Y. and Waldeck, D.H. (1998) Electron tunneling at the semiconductor–insulator–electrolyte interface. photocurrent studies of the *n*-InP-alkanethiol-ferrocyanide system. *J. Phys. Chem. B*, **102**, 9015–9028.

62. Schvartzman, M., Sidorov, V., Ritter, D., and Paz, Y. (2003) Passivation of InP surfaces of electronic devices by organothiolated self-assembled monolayers. *J. Vac. Sci. Technol., B*, **21**, 148–155.

63. Hillebrandt, H. and Tanaka, M. (2001) Electrochemical characterization of self-assembled alkylsiloxane monolayers on indium-tin oxide (ITO) semiconductor electrodes. *J. Phys. Chem. B*, **105**, 4270–4276.

64. McGuiness, C.L., Blasini, D., Masejewski, J.P., Uppili, S., Cabarcos, O.M., Smilgies, D., and Allara, D.L. (2007) Molecular self-assembly at bare semiconductor surfaces: characterization of a homologous series of n-alkanethiolate monolayers on GaAs(001). *ACS Nano*, **1**, 30–49.

65. Rodriguez, L.M., Gayone, J.E., Sanchez, E.A., Grizzi, O., Blum, B., Salvarezza, R.C., Xi, L., and Lau, W.M. (2007) Gas phase formation of dense alkanethiol layers on GaAs(110). *J. Am. Chem. Soc.*, **129**, 7807–7813.

66. McGuiness, C.L., Shaporenko, A., Zharnikov, M., Walker, A.V., and Allara, D.L. (2007) Molecular self-assembly at bare semiconductor surfaces: investigation of the chemical and electronic properties of the alkanethiolate – GaAs(001) interface. *J. Phys. Chem. B*, **111**, 4226–4234.

67. Zhou, C. and Walker, A.V. (2007) UV photooxidation and photopatterning of

alkanethiolate self-assembled monolayers (SAMs) on GaAs(001). *Langmuir*, **23**, 8876–8881.

68. Gassull, D., Ulman, A., Grunze, M., and Tanaka, M. (2008) Electrochemical sensing of membrane potential and enzyme function using gallium arsenide electrodes functionalized with supported membranes. *J. Phys. Chem. B*, **112**, 5736–5741.

69. Zhou, C. and Walker, A.V. (2008) UV photooxidation of a homologous series of *n*-alkanethiolate monolayers on GaAs(001): a static SIMS investigation. *J. Phys. Chem. B*, **112**, 797–805.

70. Laibinis, P.E. and Whitesides, G.M. (1992) ω-terminated alkanethiolate monolayers on surfaces of copper, silver, and gold have similar wettabilities. *J. Am. Chem. Soc.*, **114**, 1990–1995.

71. Laibinis, P.E., Whitesides, G.M., Allara, D.L., Tao, Y.T., Parikh, A.N., and Nuzzo, R.G. (1991) Comparison of the structures and wetting properties of self-assembled monolayers of *n*-alkanethiols on the coinage metal surfaces, copper, silver, and gold. *J. Am. Chem. Soc.*, **113**, 7152–7167.

72. Laiho, T. and Leiro, J.A. (2008) ToF-SIMS study of 1-dodecanethiol adsorption on Au, Ag, Cu and Pt surfaces. *Surf. Interface Anal.*, **40**, 51–59.

73. Sai, T.P. and Raychaudhuri, A.K. (2007) Adhesion behavior of self-assembled alkanethiol monolayers on silver at different stages of growth. *J. Phys. D: Appl. Phys.*, **40**, 3182–3189.

74. Telegdi, J., Otmacic-Curkovic, H., Marusic, K., Al-Taher, F., Stupnisec-Lisac, E., and Kalman, E. (2007) Inhibition of copper corrosion by self-assembled amphiphiles. *Chem. Biochem. Eng. Q.*, **21**, 77–82.

75. Sinapi, F., Lejeune, I., Delhalle, J., and Mekhalif, Z. (2007) Comparative protective abilities of organothiols SAM coatings applied to copper dissolution in aqueous environments. *Electrochim. Acta*, **52**, 5182–5190.

76. Hoque, E., DeRose, J.A., Hoffmann, P., Bhushan, B., and Mathieu, H.J. (2007) Chemical stability of nonwetting, low adhesion self-assembled monolayer films formed by perfluoroalkylsilanization of copper. *J. Chem. Phys.*, **126**, 114706/114701–114706/114708.

77. Whelan, C.M., Kinsella, M., Carbonell, L., Ho, H.M., and Maex, K. (2003) Corrosion inhibition by self-assembled monolayers for enhanced wire bonding on Cu surfaces. *Microelectron. Eng.*, **70**, 551–557.

78. Shackelford, J.F. (2000) *Introduction to Materials Science for Engineers*. Prentice Hall, Upper Saddle River.

79. Azzaroni, O., Vela, M.E., Fonticelli, M., Benitez, G., Carro, P., Blum, B., and Salvarezza, R.C. (2003) Electrodesorption potentials of self-assembled alkanethiolate monolayers on copper electrodes. An experimental and theoretical study. *J. Phys. Chem. B*, **107**, 13446–13454.

80. Liao, J.-D., Wang, M.-C., Weng, C.-C., Klauser, R., Frey, S., Zharnikov, M., and Grunze, M. (2002) Modification of alkanethiolate self-assembled monolayers by free radical-dominant plasma. *J. Phys. Chem. B*, **106**, 77–84.

81. Wang, M.C., Liao, J.D., Weng, C.C., Klauser, R., Shaporenko, A., Grunze, M., and Zharnikov, M. (2003) Modification of aliphatic monomolecular films by free radical dominant plasma: the effect of the alkyl chain length and the substrate. *Langmuir*, **19**, 9774–9780.

82. Jennings, G.K., Munro, J.C., Yong, T.-H., and Laibinis, P.E. (1998) Effect of chain length on the protection of copper by n-alkanethiols. *Langmuir*, **14**, 6130–6139.

83. Hoque, E., DeRose, J.A., Hoffmann, P., and Mathieu, H.J. (2006) Robust perfluorosilanized copper surfaces. *Surf. Interface Anal.*, **38**, 62–68.

84. Hoque, E., DeRose, J.A., Houriet, R., Hoffmann, P., and Mathieu, H.J. (2007) Stable perfluorosilane self-assembled monolayers on copper oxide surfaces: evidence of siloxy-copper bond formation. *Chem. Mater.*, **19**, 798–804.

85. Tao, Y.T. (1993) Structural comparison of self-assembled monolayers of *n*-alkanoic acids on the surfaces of silver, copper, and aluminum. *J. Am. Chem. Soc.*, **115**, 4350–4358.

86. Khatri, O.P. and Biswas, S.K. (2004) Thermal stability of octadecyl-trichlorosilane self-assembled on a polycrystalline aluminium surface. *Surf. Sci.*, **572**, 228–238.

87. Verleger, S., Rosenberg, N., Lieberman, I., and Richter, S. (2007) Strong mechanical stabilization and electrical passivation of metal–semiconductor contacts by self-assembled monolayer. *J. Phys. Chem. B*, **111**, 4481–4483.

88. Kiani, A., Alpuche-Aviles, M.A., Eggers, P.K., Jones, M., Gooding, J.J., Paddon-Row, M.N., and Bard, A.J. (2008) Scanning electrochemical microscopy. 59. effect of defects and structure on electron transfer through self-assembled monolayers. *Langmuir*, **24**, 2841–2849.

89. Williams, J.A. and Gorman, C.B. (2007) Alkanethiol reductive desorption from self-assembled monolayers on gold, platinum, and palladium substrates. *J. Phys. Chem. B*, **111**, 12804–12810.

90. Soreta, T.R., Strutwolf, J., and O'Sullivan, C.K. (2007) Electrochemically deposited palladium as a substrate for self-assembled monolayers. *Langmuir*, **23**, 10823–10830.

91. Rosario-Castro, B.I., Fachini, E.R., Hernandez, J., Perez-Davis, M.E., and Cabrera, C.R. (2006) Electrochemical and surface characterization of 4-aminothiophenol adsorption at polycrystalline platinum electrodes. *Langmuir*, **22**, 6102–6108.

92. Love, J.C., Estroff, L.A., Kriebel, J.K., Nuzzo, R.G., and Whitesides, G.M. (2005) Self-assembled monolayers of thiolates on metals as a form of nanotechnology. *Chem. Rev.*, **105**, 1103–1169.

93. Rogers, L.B., Krause, D.P., Griess, J.C. Jr., and Ehrlinger, D.B. (1949) Electrodeposition behavior of traces of silver. *J. Electrochem. Soc.*, **95**, 33–46.

94. Zamborini, F.P., Campbell, J.K., and Crooks, R.M. (1998) Spectroscopic, voltammetric, and electrochemical scanning tunneling microscopic study of underpotentially deposited Cu corrosion and passivation with self-assembled organomercaptan monolayers. *Langmuir*, **14**, 640–647.

95. Jennings, G.K. and Laibinis, P.E. (1996) Underpotentially deposited metal layers of silver provide enhanced stability to self-assembled alkanethiol monolayers on gold. *Langmuir*, **12**, 6173–6175.

96. Chen, I.W.P., Chen, C.-C., Lin, S.-Y., and Chen, C.-H. (2004) Effect of underpotentially deposited adlayers on sulfur bonding schemes of organothiols self-assembled on polycrystalline gold: sp or sp^3 hybridization. *J. Phys. Chem. B*, **108**, 17497–17504.

97. Lin, S.-Y., Tsai, T.-K., Lin, C.-M., Chen, C.-h., Chan, Y.-C., and Chen, H.-W. (2002) Structures of self-assembled monolayers of *n*-alkanoic acids on gold surfaces modified by underpotential deposition of silver and copper: odd–even effect. *Langmuir*, **18**, 5473–5478.

98. Tao, Y.-T., Lee, M.T., and Chang, S.C. (1993) Effect of biphenyl and naphthyl groups on the structure of self-assembled monolayers: packing, orientation, and wetting properties. *J. Am. Chem. Soc.*, **115**, 9547–9555.

99. Tao, Y.-T., Hietpas, G.D., and Allara, D.L. (1996) HCl vapor-induced structural rearrangements of *n*-alkanoate self-assembled monolayers on ambient silver, copper, and aluminum surfaces. *J. Am. Chem. Soc.*, **118**, 6724–6735.

100. Rieley, H., Kendall, G.K., Jones, R.G., and Woodruff, D.P. (1999) X-ray studies of self-assembled monolayers on coinage metals. 2. surface adsorption structures in 1-octanethiol on Cu(111) and Ag(111) and their determination by the normal incidence X-ray standing wave technique. *Langmuir*, **15**, 8856–8866.

101. Floriano, P.N., Schlieben, O., Doomes, E.E., Klein, I., Janssen, J., Hormes, J., Poliakoff, E.D., and McCarley, R.L. (2000) A grazing incidence surface X-ray absorption fine structure (GIXAFS) study of alkanethiols adsorbed on Au, Ag, and Cu. *Chem. Phys. Lett.*, **321**, 175–181.

102. Kondoh, H., Saito, N., Matsui, F., Yokoyama, T., Ohta, T., and Kuroda, H. (2001) Structure of alkanethiolate monolayers on Cu(100): self-assembly on the four-fold-symmetry surface. *J. Phys. Chem. B*, **105**, 12870–12878.

103. Kondoh, H., Nambu, A., Ehara, Y., Matsui, F., Yokoyama, T., and Ohta, T. (2004) Substrate dependence of self-assembly of alkanethiol: X-ray absorption fine structure study. *J. Phys. Chem. B*, **108**, 12946–12954.

104. Walczak, M.M., Chung, C., Stole, S.M., Widrig, C.A., and Porter, M.D. (1991) Structure and interfacial properties of spontaneously adsorbed *n*-alkanethiolate monolayers on evaporated silver surfaces. *J. Am. Chem. Soc.*, **113**, 2370–2378.

105. Schreiber, F. (2000) Structure and growth of self-assembling monolayers. *Prog. Surf. Sci.*, **65**, 151–256.

106. Yang, G. and Liu, G.-Y. (2003) New insights for self-assembled monolayers of organothiols on Au(111) revealed by scanning tunneling microscopy. *J. Phys. Chem. B*, **107**, 8746–8759.

107. Noh, J., Kato, H.S., Kawai, M., and Hara, M. (2006) Surface structure and interface dynamics of alkanethiol self-assembled monolayers on Au(111). *J. Phys. Chem. B*, **110**, 2793–2797.

108. Munuera, C., Barrena, E., and Ocal, C. (2007) Deciphering structural domains of alkanethiol self-assembled configurations by friction force microscopy. *J. Phys. Chem. A*, **111**, 12721–12726.

109. Lee, S., Bae, S.-S., Medeiros-Ribeiro, G., Blackstock, J.J., Kim, S., Stewart, D.R., and Ragan, R. (2008) Scanning tunneling microscopy of template-stripped Au surfaces and highly ordered self-assembled monolayers. *Langmuir*, **24**, 5984–5987.

110. Torrelles, X., Vericat, C., Vela, M.E., Fonticelli, M.H., Daza Millone, M.A., Felici, R., Lee, T.-L., Zegenhagen, J., Munoz, G., Martin-Gago, J.A., and Salvarezza, R.C. (2006) Two-site adsorption model for the $(\sqrt{3} \times \sqrt{3}) - R30°$ dodecanethiolate lattice on Au(111) surfaces. *J. Phys. Chem. B*, **110**, 5586–5594.

111. O'Dwyer, C., Gay, G., de Lesegno, B.V., and Weiner, J. (2004) The nature of alkanethiol self-assembled monolayer adsorption on sputtered gold substrates. *Langmuir*, **20**, 8172–8182.

112. Camillone, N. III, Chidsey, C.E.D., Liu, G.Y., and Scoles, G. (1993) Substrate dependence on the surface structure and chain packing of docosyl mercaptan self-assembled on the (111), (110), and (100) faces of single-crystal gold. *J. Chem. Phys.*, **98**, 4234–4245.

113. Dubois, L.H. and Nuzzo, R.G. (1992) Synthesis, structure, and properties of model organic surfaces. *Annu. Rev. Phys. Chem.*, **43**, 437–463.

114. Bondi, A. (1964) van der waals volumes and radii. *J. Phys. Chem.*, **68**, 441–451.

115. Lee, M.-T., Hsueh, C.-C., Freund, M.S., and Ferguson, G.S. (1998) Air oxidation of self-assembled monolayers on polycrystalline gold. The role of the gold substrate. *Langmuir*, **14**, 6419–6423.

116. Holland-Moritz, E., Gordon, J. II, Borges, G., and Sonnenfeld, R. (1991) Motion of atomic steps of gold(111) films on mica. *Langmuir*, **7**, 301–306.

117. Leopold, M.C., Black, J.A., and Bowden, E.F. (2002) Influence of gold topography on carboxylic acid terminated self-assembled monolayers. *Langmuir*, **18**, 978–980.

118. Kim, H.I., Koini, T., Lee, T.R., and Perry, S.S. (1997) Systematic studies of the frictional properties of fluorinated monolayers with atomic force microscopy: comparison of CF_3- and CH_3-terminated films. *Langmuir*, **13**, 7192–7196.

119. Priest, C.I., Jacobs, K., and Ralston, J. (2002) Novel approach to the formation of smooth gold surfaces. *Langmuir*, **18**, 2438–2440.

120. Gupta, P., Loos, K., Korniakov, A., Spagnoli, C., Cowman, M., and Ulman, A. (2004) Facile route to ultraflat SAM-protected gold surfaces by "Amphiphile splitting". *Angew. Chem., Int. Ed.*, **43**, 520–523.

121. Weiss, E.A., Kaufman, G.K., Kriebel, J.K., Li, Z., Schalek, R., and Whitesides, G.M. (2007) Si/SiO_2-templated formation of ultraflat metal surfaces on glass, polymer, and solder supports: their use as substrates for self-assembled monolayers. *Langmuir*, **23**, 9686–9694.

122. Swalen, J.D., Allara, D.L., Andrade, J.D., Chandross, E.A., Garoff, S., Israelachvili, J., McCarthy, T.J.,

Murray, R., Pease, R.F., Rabolt, J.F., Wynne, K.J., and Yu, H. (1987) Molecular monolayers and films. A panel report for the materials sciences division of the department of energy. *Langmuir*, **3**, 932–950.

123. Kunze, J., Leitch, J., Schwan, A.L., Faragher, R.J., Naumann, R., Schiller, S., Knoll, W., Dutcher, J.R., and Lipkowski, J. (2006) New method to measure packing densities of self-assembled thiolipid monolayers. *Langmuir*, **22**, 5509–5519.

124. Lavrich, D.J., Wetterer, S.M., Bernasek, S.L., and Scoles, G. (1998) Physisorption and chemisorption of alkanethiols and alkyl sulfides on Au(111). *J. Phys. Chem. B*, **102**, 3456–3465.

125. Zhang, M. and Anderson, M.R. (1994) Investigation of the charge transfer properties of electrodes modified by the spontaneous adsorption of unsymmetrical dialkyl sulfides. *Langmuir*, **10**, 2807–2813.

126. Tour, J.M. (2000) Molecular electronics. synthesis and testing of components. *Acc. Chem. Res.*, **33**, 791–804.

127. Colorado, R. Jr., Villazana, R.J., and Lee, T.R. (1998) Self-assembled monolayers on gold generated from aliphatic dithiocarboxylic acids. *Langmuir*, **14**, 6337–6340.

128. Morf, P., Raimondi, F., Nothofer, H.-G., Schnyder, B., Yasuda, A., Wessels, J.M., and Jung, T.A. (2006) Dithiocarbamates: functional and versatile linkers for the formation of self-assembled monolayers. *Langmuir*, **22**, 658–663.

129. Duwez, A.-S., Guillet, P., Colard, C., Gohy, J.-F., and Fustin, C.-A. (2006) Dithioesters and trithiocarbonates as anchoring groups for the "Grafting-To" approach. *Macromolecules*, **39**, 2729–2731.

130. Weinstein, R.D., Richards, J., Thai, S.D., Omiatek, D.M., Bessel, C.A., Faulkner, C.J., Othman, S., and Jennings, G.K. (2007) Characterization of self-assembled monolayers from lithium dialkyldithiocarbamate salts. *Langmuir*, **23**, 2887–2891.

131. Edwards, T.R.G., Cunnane, V.J., Parsons, R., and Gani, D. (1989) Construction of a stable flavin-gold electrode displaying very fast electron transfer kinetics. *J. Chem. Soc., Chem. Commun.*, 1041–1043.

132. Whelan, C.M., Barnes, C.J., Walker, C.G.H., and Brown, N.M.D. (1999) Benzenethiol adsorption on Au(111) studied by synchrotron ARUPS, HREELS and XPS. *Surf. Sci.*, **425**, 195–211.

133. Barriet, D., Yam, C.M., Shmakova, O.E., Jamison, A.C., and Lee, T.R. (2007) 4-mercaptophenylboronic acid SAMs on gold: comparison with SAMs derived from thiophenol, 4-mercaptophenol, and 4-mercaptobenzoic acid. *Langmuir*, **23**, 8866–8875.

134. Garg, N., Friedman, J.M., and Lee, T.R. (2000) Adsorption profiles of chelating aromatic dithiols and disulfides: comparison to those of normal alkanethiols and disulfides. *Langmuir*, **16**, 4266–4271.

135. Shon, Y.-S., Lee, S., Colorado, R. Jr., Perry, S.S., and Lee, T.R. (2000) Spiroalkanedithiol-based SAMs reveal unique insight into the wettabilities and frictional properties of organic thin films. *J. Am. Chem. Soc.*, **122**, 7556–7563.

136. Park, J.-S., Smith, A.C., and Lee, T.R. (2004) Loosely packed self-assembled monolayers on gold generated from 2-alkyl-2-methylpropane-1,3-dithiols. *Langmuir*, **20**, 5829–5836.

137. Park, J.-S., Vo, A.N., Barriet, D., Shon, Y.-S., and Lee, T.R. (2005) Systematic control of the packing density of self-assembled monolayers using bidentate and tridentate chelating alkanethiols. *Langmuir*, **21**, 2902–2911.

138. Dreesen, L., Volcke, C., Sartenaer, Y., Peremans, A., Thiry, P.A., Humbert, C., Grugier, J., and Marchand-Brynaert, J. (2006) Comparative study of decyl thiocyanate and decanethiol self-assembled monolayers on gold substrates. *Surf. Sci.*, **600**, 4052–4057.

139. Choi, Y., Jeong, Y., Chung, H., Ito, E., Hara, M., and Noh, J. (2008) Formation of ordered self-assembled

monolayers by adsorption of octylth-iocyanates on Au(111). *Langmuir*, **24**, 91–96.

140. Shen, C., Buck, M., Wilton-Ely, J.D.E.T., Weidner, T., and Zharnikov, M. (2008) On the importance of purity for the formation of self-assembled monolayers from thiocyanates. *Langmuir*, **24**, 6609–6615.

141. von Wrochem, F., Scholz, F., Schreiber, A., Nothofer, H.-G., Ford, W.E., Morf, P., Jung, T., Yasuda, A., and Wessels, J.M. (2008) Structure and conductance of aromatic and aliphatic dithioacetamide monolayers on Au(111). *Langmuir*, **24**, 6910–6917.

142. Monnell, J.D., Stapleton, J.J., Jackiw, J.J., Dunbar, T., Reinerth, W.A., Dirk, S.M., Tour, J.M., Allara, D.L., and Weiss, P.S. (2004) Ordered local do-main structures of decaneselenolate and dodecaneselenolate monolayers on Au(111). *J. Phys. Chem. B*, **108**, 9834–9841.

143. Shaporenko, A., Ulman, A., Terfort, A., and Zharnikov, M. (2005) Self-assembled monolayers of alka-neselenolates on (111) gold and silver. *J. Phys. Chem. B*, **109**, 3898–3906.

144. Shaporenko, A., Cyganik, P., Buck, M., Terfort, A., and Zharnikov, M. (2005) Self-assembled monolayers of aromatic selenolates on noble metal substrates. *J. Phys. Chem. B*, **109**, 13630–13638.

145. Monnell, J.D., Stapleton, J.J., Dirk, S.M., Reinerth, W.A., Tour, J.M., Allara, D.L., and Weiss, P.S. (2005) Relative conductances of alkaneseleno-late and alkanethiolate monolayers on Au(111). *J. Phys. Chem. B*, **109**, 20343–20349.

146. Tour, J.M., Rawlett, A.M., Kozaki, M., Yao, Y., Jagessar, R.C., Dirk, S.M., Price, D.W., Reed, M.A., Zhou, C.-W., Chen, J., Wang, W., and Campbell, I. (2001) Synthesis and preliminary test-ing of molecular wires and devices. *Chem. Eur. J.*, **7**, 5118–5134.

147. Wang, M., Liechti, K.M., Wang, Q., and White, J.M. (2005) Self-assembled silane monolayers: fabrication with nanoscale uniformity. *Langmuir*, **21**, 1848–1857.

148. Kim, S., Sohn, H., Boo, J.-H., and Lee, J. (2008) Significantly improved stability of n-octadecyltrichlorosilane self-assembled monolayer by plasma pretreatment on mica. *Thin Solid Films*, **516**, 940–947.

149. Tarlov, M.J. and Newman, J.G. (1992) Static secondary ion mass spectrometry of self-assembled alkanethiol monolay-ers on gold. *Langmuir*, **8**, 1398–1405.

150. Li, Y., Huang, J., McIver, R.T. Jr., and Hemminger, J.C. (1992) Character-ization of thiol self-assembled films by laser desorption Fourier transform mass spectrometry. *J. Am. Chem. Soc.*, **114**, 2428–2432.

151. Scott, J.R., Baker, L.S., Everett, W.R., Wilkins, C.L., and Fritsch, I. (1997) Laser desorption Fourier transform mass spectrometry exchange studies of air oxidized alkanethiol self-assembled monolayers on gold. *Anal. Chem.*, **69**, 2636–2639.

152. Zhang, Y., Terrill, R.H., Tanzer, T.A., and Bohn, P.W. (1998) Ozonolysis is the primary cause of UV photooxida-tion of alkanethiolate monolayers at low irradiance. *J. Am. Chem. Soc.*, **120**, 2654–2655.

153. Schoenfisch, M.H. and Pemberton, J.E. (1998) Air stability of alkanethiol self-assembled monolayers on silver and gold surfaces. *J. Am. Chem. Soc.*, **120**, 4502–4513.

154. Snyder, R.G., Maroncelli, M., Strauss, H.L., and Hallmark, V.M. (1986) Tem-perature and phase behavior of infrared intensities: the poly(methylene) chain. *J. Phys. Chem.*, **90**, 5623–5630.

155. Snyder, R.G., Strauss, H.L., and Elliger, C.A. (1982) Carbon-hydrogen stretching modes and the structure of n-alkyl chains. 1. long, disordered chains. *J. Phys. Chem.*, **86**, 5145–5150.

156. Delamarche, E., Michel, B., Gerber, C., Anselmetti, D., Guentherodt, H.J., Wolf, H., and Ringsdorf, H. (1994) Real-space observation of nanoscale molecular domains in self-assembled monolayers. *Langmuir*, **10**, 2869–2871.

157. Delamarche, E., Michel, B., Kang, H., and Gerber, C. (1994) Thermal sta-bility of self-assembled monolayers. *Langmuir*, **10**, 4103–4108.

158. Nuzzo, R.G., Dubois, L.H., and Allara, D.L. (1990) Fundamental studies of microscopic wetting on organic surfaces. 1. Formation and structural characterization of a self-consistent series of polyfunctional organic monolayers. *J. Am. Chem. Soc.*, **112**, 558–569.

159. Pflaum, J., Bracco, G., Schreiber, F., Colorado, R., Shmakova, O.E., Lee, T.R., Scoles, G., and Kahn, A. (2002) Structure and electronic properties of CH$_3$- and CF$_3$-terminated alkanethiol monolayers on Au(111): a scanning tunneling microscopy, surface X-ray and helium scattering study. *Surf. Sci.*, **498**, 89–104.

160. Chidsey, C.E.D. and Loiacono, D.N. (1990) Chemical functionality in self-assembled monolayers: structural and electrochemical properties. *Langmuir*, **6**, 682–691.

161. Bain, C.D. and Whitesides, G.M. (1989) A study by contact angle of the acid-base behavior of monolayers containing ω-mercaptocarboxylic acids adsorbed on gold: an example of reactive spreading. *Langmuir*, **5**, 1370–1378.

162. Lee, T.R., Carey, R.I., Biebuyck, H.A., and Whitesides, G.M. (1994) The wetting of monolayer films exposing ionizable acids and bases. *Langmuir*, **10**, 741–749.

163. Cooper, E. and Leggett, G.J. (1999) Influence of tail-group hydrogen bonding on the stabilities of self-assembled monolayers of alkylthiols on gold. *Langmuir*, **15**, 1024–1032.

164. Evans, S.D., Sharma, R., and Ulman, A. (1991) Contact angle stability: reorganization of monolayer surfaces? *Langmuir*, **7**, 156–161.

165. Chechik, V., Crooks, R.M., and Stirling, C.J.M. (2000) Reactions and reactivity in self-assembled monolayers. *Adv. Mat.*, **12**, 1161–1171.

166. Roberts, C., Chen, C.S., Mrksich, M., Martichonok, V., Ingber, D.E., and Whitesides, G.M. (1998) Using mixed self-assembled monolayers presenting RGD and (EG)$_3$OH groups to characterize long-term attachment of bovine capillary endothelial cells to surfaces. *J. Am. Chem. Soc.*, **120**, 6548–6555.

167. Mrksich, M. (1998) Tailored substrates for studies of attached cell culture. *Cell. Mol. Life Sci.*, **54**, 653–662.

168. Kane, P.F. and Larrabee, G.B. (1977) Surface characterization. *Anal. Chem.*, **49**, 221R–230R.

169. Johnson, R.E. Jr. and Dettre, R.H. (1969) Wettability and contact angles in *Surface Colloid Science*, vol. 2 (ed. E. Matijevic), Wiley-Interscience, New York, pp. 85–153.

170. Cassie, A.B.D. (1948) Contact angles. *Discuss. Faraday Soc.*, **3**, 11–16.

171. Wenzel, R.N. (1936) Resistance of solid surfaces to wetting by water. *J. Ind. Eng. Chem.*, **28**, 988–994.

172. Wenzel, R.N. (1949) Surface roughness and contact angle. *J. Phys. Colloid Chem.*, **53**, 1466–1467.

173. Miller, W.J. and Abbott, N.L. (1997) Influence of van der waals forces from metallic substrates on fluids supported on self-assembled monolayers formed from alkanethiols. *Langmuir*, **13**, 7106–7114.

174. Creager, S.E., and Clarke, J. (1994) Contact-angle titrations of mixed ω-mercaptoalkanoic acid/alkanethiol monolayers on gold. Reactive vs. nonreactive spreading, and chain length effects on surface pKa values. *Langmuir*, **10**, 3675–3683.

175. Lam, C.N.C., Kim, N., Hui, D., Kwok, D.Y., Hair, M.L., and Neumann, A.W. (2001) The effect of liquid properties to contact angle hysteresis. *Colloids Surf., A*, **189**, 265–278.

176. Colorado, R. Jr. and Lee, T.R. (2000) Physical organic probes of interfacial wettability reveal the importance of surface dipole effects. *J. Phys. Org. Chem.*, **13**, 796–807.

177. Young, T. (1805) An essay on the cohesion of fluids. *Philos. Trans. R Soc. London*, **95**, 65–87.

178. Kloubek, J. (1992) Development of methods for surface free energy determination using contact angles of liquids on solids. *Adv. Colloid Interface Sci.*, **38**, 99–142.

179. Morra, M., Occhiello, E., and Garbassi, F. (1990) Knowledge about polymer

surfaces from contact-angle measurements. *Adv. Colloid Interface Sci.*, **32**, 79–116.

180. Zisman, W.A. (1964) Relation of the equilibrium contact angle to liquid and solid constitution. *Adv. Chem. Ser.*, **43**, 1–51.

181. Dupré, A. (1869) *Théorie Méchanique de la Chaleur*, Gauthier-Villars, Paris.

182. Yaws, C.L., Yang, H.C., and Pan, X. (1991) 633 organic chemicals: surface tension data. *Chem. Eng. (N.Y.)*, **98**, 140–142, 144, 146, 148, 150.

183. Fowkes, F.M., Riddle, F.L. Jr., Pastore, W.E., and Weber, A.A. (1990) Interfacial interactions between self-associated polar liquids and squalane used to test equations for solid-liquid interfacial interactions. *Colloids Surf.*, **43**, 367–387.

184. Graupe, M., Takenaga, M., Koini, T., Colorado, R. Jr., and Lee, T.R. (1999) Oriented surface dipoles strongly influence interfacial wettabilities. *J. Am. Chem. Soc.*, **121**, 3222–3223.

185. Graupe, M., Koini, T., Kim, H.I., Garg, N., Miura, Y.F., Takenaga, M., Perry, S.S., and Lee, T.R. (1999) Self-assembled monolayers of CF_3-terminated alkanethiols on gold. *Colloid Surf. A: Phys. Eng. Asp.*, **154**, 239–244.

186. Lee, S., Puck, A., Graupe, M., Colorado, R. Jr., Shon, Y.-S., Lee, T.R., and Perry, S.S. (2001) Structure, wettability, and frictional properties of phenyl-terminated self-assembled monolayers on gold. *Langmuir*, **17**, 7364–7370.

187. Wenzl, I., Yam, C.M., Barriet, D., and Lee, T.R. (2003) Structure and wettability of methoxy-terminated self-assembled monolayers on gold. *Langmuir*, **19**, 10217–10224.

188. Wong, S.-S., Takano, H., and Porter, M.D. (1998) Mapping orientation differences of terminal functional groups by friction force microscopy. *Anal. Chem.*, **70**, 5209–5212.

189. Mikulski, P.T., Herman, L.A., and Harrison, J.A. (2005) Odd and even model self-assembled monolayers: links between friction and structure. *Langmuir*, **21**, 12197–12206.

190. Angelico, V.J., Mitchell, S.A., and Wysocki, V.H. (2000) Low-energy ion-surface reactions of pyrazine with two classes of self-assembled monolayers: influence of alkyl chain orientation. *Anal. Chem.*, **72**, 2603–2608.

191. Wolf, K.V., Cole, D.A., and Bernasek, S.L. (2001) Low-energy collisions of pyrazine and d_6-benzene molecular ions with self-assembled monolayer surfaces: the odd-even chain length effect. *Langmuir*, **17**, 8254–8259.

192. Wolf, K.V., Cole, D.A., and Bernasek, S.L. (2002) Chain length effects for cluster ion formation during high energy ion/surface collisions with self-assembled monolayer surfaces. *J. Phys. Chem. B*, **106**, 10382–10387.

193. Tao, F. and Bernasek, S.L. (2007) Understanding odd-even effects in organic self-assembled monolayers. *Chem. Rev.*, **107**, 1408–1453.

194. Mate, C.M. (2002) On the road to an atomic- and molecular-level understanding of friction. *MRS Bull.*, **27**, 967–971.

195. Kim, H.I., Graupe, M., Oloba, O., Koini, T., Imaduddin, S., Lee, T.R., and Perry, S.S. (1999) Molecularly specific studies of the frictional properties of monolayer films: a systematic comparison of CF_3-, $(CH_3)_2CH$-, and CH_3-terminated films. *Langmuir*, **15**, 3179–3185.

196. He, M., Blum, A.S., Overney, G., and Overney, R.M. (2002) Effect of interfacial liquid structuring on the coherence length in nanolubrication. *Phys. Rev. Lett.*, **88**, 154302-1–154302-4.

197. Kim, H.I., Koini, T., Lee, T.R., and Perry, S.S. (1998) Molecular contributions to the frictional properties of fluorinated self-assembled monolayers. *Tribol. Lett.*, **4**, 137–140.

198. DePalma, V. and Tillman, N. (1989) Friction and wear of self-assembled trichlorosilane monolayer films on silicon. *Langmuir*, **5**, 868–872.

199. Briscoe, B.J. and Evans, D.C.B. (1982) The shear properties of Langmuir-Blodgett layers. *Proc. R. Soc. London, A*, **380**, 389–407.

200. Chaudhury, M.K. and Owen, M.J. (1993) Adhesion hysteresis and friction. *Langmuir*, **9**, 29–31.

201. Lio, A., Charych, D.H., and Salmeron, M. (1997) Comparative atomic force microscopy study of the chain length dependence of frictional properties of alkanethiols on gold and alkylsilanes on mica. *J. Phys. Chem. B*, **101**, 3800–3805.

202. Xiao, X., Hu, J., Charych, D.H., and Salmeron, M. (1996) Chain length dependence of the frictional properties of alkylsilane molecules self-assembled on mica studied by atomic force microscopy. *Langmuir*, **12**, 235–237.

203. Graupe, M., Koini, T., Kim, H.I., Garg, N., Miura, Y.F., Takenaga, M., Perry, S.S., and Lee, T.R. (1999) Wettability and friction of CF_3-terminated monolayer films on gold. *Mater. Res. Bull.*, **34**, 447–453.

204. Colorado, R. Jr., Graupe, M., Kim, H.I., Takenaga, M., Oloba, O., Lee, S., Perry, S.S., and Lee, T.R. (2001) Interfacial properties of specifically fluorinated self-assembled monolayer films. *ACS Symp. Ser.*, **781**, 58–75.

205. Perry, S.S., Lee, S., Lee, T.R., Graupe, M., Puck, A., Colorado, R. Jr., and Wenzl, I. (2000) Molecular level interpretations of frictional force data collected with atomic force microscopy: chain length effects in self assembled organic monolayers. *Polym. Prepr.*, **41**, 1456–1457.

206. Lee, S., Shon, Y.S., Randall Lee, T., and Perry, S.S. (2000) Structural characterization and frictional properties of C_{60}-terminated self-assembled monolayers on Au(111). *Thin Solid Films*, **358**, 152–158.

207. Lee, S., Shon, Y.-S., Colorado, R. Jr., Guenard, R.L., Lee, T.R., and Perry, S.S. (2000) The influence of packing densities and surface order on the frictional properties of alkanethiol self-assembled monolayers (SAMs) on gold: a comparison of SAMs derived from normal and spiroalkanedithiols. *Langmuir*, **16**, 2220–2224.

208. Perry, S.S., Lee, S., Shon, Y.-S., Colorado, R. Jr., and Lee, T.R. (2001) The relationships between interfacial friction and the conformational order of organic thin films. *Tribol. Lett.*, **10**, 81–87.

209. Shon, Y.-S., Lee, S., Perry, S.S., and Lee, T.R. (2000) The adsorption of unsymmetrical spiroalkanedithiols onto gold affords multi-component interfaces that are homogeneously mixed at the molecular level. *J. Am. Chem. Soc.*, **122**, 1278–1281.

210. Li, S., Cao, P., Colorado, R. Jr., Yan, X., Wenzl, I., Shmakova, O.E., Graupe, M., Lee, T.R., and Perry, S.S. (2005) Local packing environment strongly influences the frictional properties of mixed CH_3- and CF_3-terminated alkanethiol SAMs on Au(111). *Langmuir*, **21**, 933–936.

211. Cohen, S.R., Naaman, R., and Sagiv, J. (1986) Thermally induced disorder in organized organic monolayers on solid substrates. *J. Phys. Chem.*, **90**, 3054–3056.

212. Tripp, C.P. and Hair, M.L. (1992) An infrared study of the reaction of octadecyltrichlorosilane with silica. *Langmuir*, **8**, 1120–1126.

213. Tripp, C.P. and Hair, M.L. (1995) Direct observation of the surface bonds between self-assembled monolayers of octadecyltrichlorosilane and silica surfaces: a low-frequency IR study at the solid/liquid interface. *Langmuir*, **11**, 1215–1219.

214. Kurth, D.G., Broeker, G.K., Kubiak, C.P., and Bein, T. (1994) Surface attachment and stability of cross-linked poly(ethylenimine)-epoxy networks on gold. *Chem. Mater.*, **6**, 2143–2150.

215. Ulman, A. (1990) Self-assembled monolayers of alkyltrichlorosilanes: building blocks for future organic materials. *Adv. Mater.*, **2**, 573–582.

216. Lenk, T.J., Hallmark, V.M., Hoffmann, C.L., Rabolt, J.F., Castner, D.G., Erdelen, C., and Ringsdorf, H. (1994) Structural investigation of molecular organization in self-assembled monolayers of a semifluorinated amidethiol. *Langmuir*, **10**, 4610–4617.

217. Tam-Chang, S.-W., Biebuyck, H.A., Whitesides, G.M., Jeon, N., and Nuzzo, R.G. (1995) Self-assembled monolayers on gold generated from alkanethiols

with the structure RNHCOCH$_2$SH. *Langmuir*, **11**, 4371–4382.

218. Clegg, R.S. and Hutchison, J.E. (1996) Hydrogen-bonding, self-assembled monolayers: ordered molecular films for study of through-peptide electron transfer. *Langmuir*, **12**, 5239–5243.

219. Zhang, H., Xia, H., Li, H., and Liu, Z. (1997) Effect of hydrogen bonding on the stability of thiol self-assembled monolayers (SAMs) on gold. *Chem. Lett.*, 721–722.

220. Sabapathy, R.C., Bhattacharyya, S., Leavy, M.C., Cleland, W.E. Jr., and Hussey, C.L. (1998) Electrochemical and spectroscopic characterization of self-assembled monolayers of ferrocenylalkyl compounds with amide linkages. *Langmuir*, **14**, 124–136.

221. Clegg, R.S., Reed, S.M., and Hutchison, J.E. (1998) Self-assembled monolayers stabilized by three-dimensional networks of hydrogen bonds. *J. Am. Chem. Soc.*, **120**, 2486–2487.

222. Clegg, R.S. and Hutchison, J.E. (1999) Control of monolayer assembly structure by hydrogen bonding rather than by adsorbate-substrate templating. *J. Am. Chem. Soc.*, **121**, 5319–5327.

223. Clegg, R.S., Reed, S.M., Smith, R.K., Barron, B.L., Rear, J.A., and Hutchison, J.E. (1999) The interplay of lateral and tiered interactions in stratified self-organized molecular assemblies. *Langmuir*, **15**, 8876–8883.

224. Lieser, G., Tieke, B., and Wegner, G. (1980) Structure, phase transitions and polymerizability of multilayers of some diacetylene monocarboxylic acids. *Thin Solid Films*, **68**, 77–90.

225. Tieke, B. and Lieser, G. (1982) Influences of the structure of long-chain diynoic acids on their polymerization properties in Langmuir-Blodgett multilayers. *J. Colloid Interface Sci.*, **88**, 471–486.

226. Batchelder, D.N., Evans, S.D., Freeman, T.L., Haeussling, L., Ringsdorf, H., and Wolf, H. (1994) Self-assembled monolayers containing polydiacetylenes. *J. Am. Chem. Soc.*, **116**, 1050–1053.

227. Taisun, K. and Crooks, R.M. (1994) Polymeric self-assembling monolayers. 1. synthesis and characterization of ω-functionalized n-alkanethiols containing a conjugated diacetylene group. *Tetrahedron Lett.*, **35**, 9501–9504.

228. Kim, T., Crooks, R.M., Tsen, M., and Sun, L. (1995) Polymeric self-assembled monolayers. 2. Synthesis and characterization of self-assembled polydiacetylene mono- and multilayers. *J. Am. Chem. Soc.*, **117**, 3963–3967.

229. Chan, K.C., Kim, T., Schoer, J.K., and Crooks, R.M. (1995) Polymeric self-assembled monolayers. 3. Pattern transfer by use of photolithography, electrochemical methods, and an ultrathin, self-assembled diacetylenic resist. *J. Am. Chem. Soc.*, **117**, 5875–5876.

230. Kim, T., Chan, K.C., and Crooks, R.M. (1997) Polymeric self-assembled monolayers. 4. Chemical, electrochemical, and thermal stability of ω- functionalized, self-assembled diacetylenic and polydiacetylenic monolayers. *J. Am. Chem. Soc.*, **119**, 189–193.

231. Kim, T., Ye, Q., Sun, L., Chan, K.C., and Crooks, R.M. (1996) Polymeric self-assembled monolayers. 5. Synthesis and characterization of ω- functionalized, self-assembled diacetylenic and polydiacetylenic monolayers. *Langmuir*, **12**, 6065–6073.

232. Cheadle, E.M., Batchelder, D.N., Evans, S.D., Zhang, H.L., Fukushima, H., Miyashita, S., Graupe, M., Puck, A., Shmakova, O.E., Colorado, R. Jr., and Lee, T.R. (2001) Polymerization of semi-fluorinated alkane thiol self-assembled monolayers containing diacetylene units. *Langmuir*, **17**, 6616–6621.

233. Li, Y., Ma, B., Fan, Y., Kong, X., and Li, J. (2002) Electrochemical and Raman studies of the biointeraction between escherichia coli and mannose in polydiacetylene derivative supported on the self-assembled monolayers of octadecanethiol on a gold electrode. *Anal. Chem.*, **74**, 6349–6354.

234. Mowery, M.D. and Evans, C.E. (1997) Steric and substrate mediation of

polymers formed within single molecular layers. *J. Phys. Chem. B*, **101**, 8513–8519.

235. Menzel, H., Mowery, M.D., Cai, M., and Evans, C.E. (1998) Vertical positioning of internal molecular scaffolding within a single molecular layer. *J. Phys. Chem. B*, **102**, 9550–9556.

236. Menzel, H., Horstmann, S., Mowery, M.D., Cai, M., and Evans, C.E. (2000) Diacetylene polymerization in self-assembled monolayers: influence of the odd/even nature of the methylene spacer. *Polymer*, **41**, 8113–8119.

237. Geyer, W., Stadler, V., Eck, W., Zharnikov, M., Golzhauser, A., and Grunze, M. (1999) Electron-induced cross-linking of aromatic self-assembled monolayers: negative resists for nanolithography. *Appl. Phys. Lett.*, **75**, 2401–2403.

238. Tai, Y., Shaporenko, A., Eck, W., Grunze, M., and Zharnikov, M. (2004) Depth distribution of irradiation-induced cross-linking in aromatic self-assembled monolayers. *Langmuir*, **20**, 7166–7170.

239. Eck, W., Kueller, A., Grunze, M., Voelkel, B., and Goelzhaeuser, A. (2005) Freestanding nanosheets from cross-linked biphenyl self-assembled monolayers. *Adv. Mat.*, **17**, 2583–2587.

240. Willicut, R.J. and McCarley, R.L. (1994) Electrochemical polymerization of pyrrole-containing self-assembled alkanethiol monolayers on Au. *J. Am. Chem. Soc.*, **116**, 10823–10824.

241. Willicut, R.J. and McCarley, R.L. (1995) Surface-confined monomers on electrode surfaces. 1. Electrochemical and microscopic characterization of ω-(N-Pyrrolyl)alkanethiol self-assembled monolayers on Au. *Langmuir*, **11**, 296–301.

242. Sayre, C.N., and Collard, D.M. (1995) Self-assembled monolayers of pyrrole-containing alkanethiols on gold. *Langmuir*, **11**, 302–306.

243. Carey, R.I., Folkers, J.P., and Whitesides, G.M. (1994) Self-assembled monolayers containing ω- mercaptoalkyl boronic acids adsorbed onto gold form a highly cross-linked, thermally stable borate glass surface. *Langmuir*, **10**, 2228–2234.

244. Ford, J.F., Vickers, T.J., Mann, C.K., and Schlenoff, J.B. (1996) Polymerization of a thiol-bound styrene monolayer. *Langmuir*, **12**, 1944–1946.

245. Duan, L. and Garrett, S.J. (2001) Self-assembled monolayers of 6-phenyl-n-hexanethiol and 6-(p-Vinylphenyl)-n-hexanethiol on Au(111): an investigation of structure, stability, and reactivity. *Langmuir*, **17**, 2986–2994.

246. Jordan, R., Ulman, A., Kang, J.F., Rafailovich, M.H., and Sokolov, J. (1999) Surface-initiated anionic polymerization of styrene by means of self-assembled monolayers. *J. Am. Chem. Soc.*, **121**, 1016–1022.

247. Bartz, M., Terfort, A., Knoll, W., and Tremel, W. (2000) Stamping of monomeric SAMs as a route to structured crystallization templates: patterned titania films. *Chem. -Eur. J.*, **6**, 4149–4153.

248. Cotton, F.A. and Wilkinson, G. (1988) *Advanced Inorganic Chemistry*, 5th edn, John Wiley & Sons, Inc., New York.

249. Bowser, J.R. (1983) *Inorganic Chemistry*, Brooks/Cole, Pacific Grove, CA.

250. Schlenoff, J.B., Li, M., and Ly, H. (1995) Stability and self-exchange in alkanethiol monolayers. *J. Am. Chem. Soc.*, **117**, 12528.

251. Shon, Y.-S., Colorado, R. Jr., Williams, C.T., Bain, C.D., and Lee, T.R. (2000) Low-density self-assembled monolayers on gold derived from chelating 2-monoalkylpropane-1,3-dithiols. *Langmuir*, **16**, 541–548.

252. Nuzzo, R.G., Fusco, F.A., and Allara, D.L. (1987) Spontaneously organized molecular assemblies. 3. preparation and properties of solution adsorbed monolayers of organic disulfides on gold surfaces. *J. Am. Chem. Soc.*, **109**, 2358–2368.

253. Garg, N. and Lee, T.R. (1998) Self-assembled monolayers based on chelating aromatic dithiols on gold. *Langmuir*, **14**, 3815–3819.

254. Biebuyck, H.A., Bain, C.D., and Whitesides, G.M. (1994) Comparison of

organic monolayers on polycrystalline gold spontaneously assembled from solutions containing dialkyl disulfides or alkanethiols. *Langmuir*, **10**, 1825–1831.

255. Garg, N., Carrasquillo-Molina, E., and Lee, T.R. (2002) Self-assembled monolayers composed of aromatic thiols on gold: structural characterization and thermal stability in solution. *Langmuir*, **18**, 2717–2726.

256. Shon, Y.-S. and Lee, T.R. (1999) Chelating self-assembled monolayers on gold generated from spiroalkanedithiols. *Langmuir*, **15**, 1136–1140.

257. Shon, Y.-S. and Lee, T.R. (2000) Desorption and exchange of self-assembled monolayers (SAMs) on gold generated from chelating alkanedithiols. *J. Phys. Chem. B*, **104**, 8192–8200.

258. Shon, Y.-S. and Lee, T.R. (2000) A steady-state kinetic model can be used to describe the growth of self-assembled monolayers (SAMs) on gold. *J. Phys. Chem. B*, **104**, 8182–8191.

259. Whitesell, J.K. and Chang, H.K. (1993) Directionally aligned helical peptides on surfaces. *Science*, **261**, 73–76.

260. Fox, M.A., Whitesell, J.K., and McKerrow, A.J. (1998) Fluorescence and redox activity of probes anchored through an aminotrithiol to polycrystalline gold. *Langmuir*, **14**, 816–820.

261. Giersig, M., and Mulvaney, P. (1993) Preparation of ordered colloid monolayers by electrophoretic deposition. *Langmuir*, **9**, 3408–3413.

262. Weisbecker, C.S., Merritt, M.V., and Whitesides, G.M. (1996) Molecular self-assembly of aliphatic thiols on gold colloids. *Langmuir*, **12**, 3763–3772.

263. Evans, D.F. and Wennerström, H. (1999) *The Colloidal Domain: Where Physics, Chemistry, Biology, and Technology Meet*, 2nd edn, John Wiley & Sons, Inc., New York.

264. Daniel, M.-C. and Astruc, D. (2004) Gold nanoparticles: assembly, supramolecular chemistry, quantum-size-related properties, and applications toward biology, catalysis, and nanotechnology. *Chem. Rev.*, **104**, 293–346.

265. Garbassi, F., Morra, M., and Occhiello, E. (eds) (1998) *Polymer Surfaces: From Physics to Technology Updated Edition*, John Wiley & Sons, Ltd, Chichester.

266. Krafft, M.P. (2001) Fluorocarbons and fluorinated amphiphiles in drug delivery and biomedical research. *Adv. Drug Delivery Rev.*, **47**, 209–228.

267. Grainger, D.W. and Stewart, C.W. (2001) Fluorinated coatings and films: motivation and significance. *ACS Symp. Ser.*, American Chemical Society, **787**, 1–14.

268. Hare, E.F., Shafrin, E.G., and Zisman, W.A. (1954) Properties of films of adsorbed fluorinated acids. *J. Phys. Chem.*, **58**, 236–239.

269. Shafrin, E.G and Zisman, W.A. (1960) Constitutive relations in the wetting of low-energy surfaces and the theory of the retraction method of preparing monolayers. *J. Phys. Chem.*, **64**, 519–524.

270. Shafrin, E.G. and Zisman, W.A. (1962) Effect of progressive fluorination of a fatty acid on the wettability of its adsorbed monolayer. *J. Phys. Chem.*, **66**, 740–748.

271. Fukushima, H., Seki, S., Nishikawa, T., Takiguchi, H., Tamada, K., Abe, K., Colorado, R., Graupe, M., Shmakova, O.E., and Lee, T.R. (2000) Microstructure, wettability, and thermal stability of semifluorinated self-assembled monolayers (SAMs) on gold. *J. Phys. Chem. B*, **104**, 7417–7423.

272. Colorado, R. Jr., (2002) Synthesis and characterization of self-assembled monolayers on gold generated from partially fluorinated alkanethiols and aliphatic dithiocarboxylic acids. Doctor of Philosophy, University of Houston, Houston.

273. Schoenherr, H. and Ringsdorf, H. (1996) Self-assembled monolayers of symmetrical and mixed alkyl fluoroalkyl disulfides on gold. 1. Synthesis of disulfides and investigation of monolayer properties. *Langmuir*, **12**, 3891–3897.

274. Schoenherr, H., Ringsdorf, H., Jaschke, M., Butt, H.J., Bamberg, E., Allinson, H., and Evans, S.D. (1996)

Self-assembled monolayers of symmetrical and mixed alkyl fluoroalkyl disulfides on gold. 2. Investigation of thermal stability and phase separation. *Langmuir*, **12**, 3898–3904.

275. Devaprakasam, D., Sampath, S., and Biswas, S.K. (2004) Thermal stability of perfluoroalkyl silane self-assembled on a polycrystalline aluminum surface. *Langmuir*, **20**, 1329–1334.

276. Bunn, C.W. and Howells, E.R. (1954) Structures of molecules and crystals of fluorocarbons. *Nature*, **174**, 549–551.

277. Colorado, R. Jr., Graupe, M., Shmakova, O.E., Villazana, R.J., and Lee, T.R. (2001) Structural properties of self-assembled monolayers on gold generated from terminally fluorinated alkanethiols. *ACS Symp. Ser.*, **781**, 276–292.

278. Colorado, R. Jr. and Lee, T.R. (2003) Attenuation lengths of photoelectrons in fluorocarbon films. *J. Phys. Chem. B*, **107**, 10216.

279. Colorado, R. Jr. and Lee, T.R. (2003) Wettabilities of self-assembled monolayers on gold generated from progressively fluorinated alkanethiols. *Langmuir*, **19**, 3288–3296.

280. Frey, S., Heister, K., Zharnikov, M., Grunze, M., Tamada, K., Colorado, R. Jr., Graupe, M., Shmakova, O.E., and Lee, T.R. (2000) Structure of self-assembled monolayers of semifluorinated alkanethiols on gold and silver substrates. *Isr. J. Chem.*, **40**, 81–97.

281. Ederth, T., Tamada, K., Claesson, P.M., Valiokas, R., Colorado, R. Jr., Graupe, M., Shmakova, O.E., and Lee, T.R. (2001) Force measurements between semifluorinated thiolate self-assembled monolayers: long-range hydrophobic interactions and surface charge. *J. Colloid Interface Sci.*, **235**, 391.

282. Tamada, K., Ishida, T., Knoll, W., Fukushima, H., Colorado, R. Jr., Graupe, M., Shmakova, O.E., and Lee, T.R. (2001) Molecular packing of semifluorinated alkanethiol self-assembled monolayers on gold: influence of alkyl spacer length. *Langmuir*, **17**, 1913–1921.

283. Alkhairalla, B., Boden, N., Cheadle, E., Evans, S.D., Henderson, J.R., Fukushima, H., Miyashita, S., Schonherr, H., Vancso, G.J., Colorado, R. Jr., Graupe, M., Shmakova, O.E., and Lee, T.R. (2002) Anchoring and orientational wetting of nematic liquid crystals on semi-fluorinated self-assembled monolayer surfaces. *Europhys. Lett.*, **59**, 410–416.

284. Frey, S., Heister, K., Zharnikov, M., and Grunze, M. (2000) Modification of semifluorinated alkanethiolate monolayers by low energy electron irradiation. *Phys. Chem. Chem. Phys.*, **2**, 1979–1987.

285. Weinstein, R.D., Moriarty, J., Cushnie, E., Colorado, R. Jr., Lee, T.R., Patel, M., Alesi, W.R., and Jennings, G.K. (2003) Structure, wettability, and electrochemical barrier properties of self-assembled monolayers prepared from partially fluorinated hexadecanethiols. *J. Phys. Chem. B*, **107**, 11626–11632.

286. Alloway, D.M., Hofmann, M., Smith, D.L., Gruhn, N.E., Graham, A.L., Colorado, R. Jr., Wysocki, V.H., Lee, T.R., Lee, P.A., and Armstrong, N.R. (2003) Interface dipoles arising from self-assembled monolayers on gold: UV-photoemission studies of alkanethiols and partially fluorinated alkanethiols. *J. Phys. Chem. B*, **107**, 11690.

287. Graupe, M., Koini, T., Wang, V.Y., Nassif, G.M., Colorado, R. Jr., Villazana, R.J., Dong, H., Miura, Y.F., Shmakova, O.E., and Lee, T.R. (1999) Terminally perfluorinated long-chain alkanethiols. *J. Fluorine Chem.*, **93**, 107–115.

288. Takenaga, M., Jo, S., Graupe, M., and Lee, T.R. (2008) Effective Van der Waals surface energy of self-assembled monolayer films having systematically varying degrees of molecular fluorination. *J. Colloid Interface Sci.*, **320**, 264–267.

289. Ulman, A., and Scaringe, R.P. (1992) On the formation of ordered two-dimensional molecular assemblies. *Langmuir*, **8**, 894–897.

290. Alves, C.A. and Porter, M.D. (1993) Atomic force microscopic characterization of a fluorinated alkanethiolate monolayer at gold and correlations to electrochemical and infrared reflection spectroscopic structural descriptions. *Langmuir*, **9**, 3507–3512.

291. Alves, C.A., Smith, E.L., and Porter, M.D. (1992) Atomic scale imaging of alkanethiolate monolayers at gold surfaces with atomic force microscopy. *J. Am. Chem. Soc.*, **114**, 1222–1227.

292. Alves, C.A., Smith, E.L., Widrig, C.A., and Porter, M.D. (1992) Scanning tunneling microscopy and atomic force microscopy of n-alkanethiolate monolayers spontaneously adsorbed at gold surfaces. *Proc. SPIE -Int. Soc. Opt. Eng.*, **1636**, 125–128.

293. Liu, G.-Y., Fenter, P., Chidsey, C.E.D., Ogletree, D.F., Eisenberger, P., and Salmeron, M. (1994) An unexpected packing of fluorinated n-alkanethiols on Au(111): a combined atomic force microscopy and X-ray diffraction study. *J. Chem. Phys.*, **101**, 4301.

294. Maboudian, R. (1998) Surface processes in MEMS technology. *Surf. Sci. Rep.*, **30**, 207–269.

295. Srinivasan, U., Houston, M.R., Howe, R.T., and Maboudian, R. (1998) Alkyltrichlorosilane-based self-assembled monolayer films for stiction reduction in silicon micromachines. *J. Microelectromech. Syst.*, **7**, 252–260.

296. Fadeev, A.Y., Soboleva, O.A., and Summ, B.D. (1997) Wettability of organosilicon and organofluorosilicon monolayers covalently grafted on quartz. *Colloid J.*, **59**, 222–225.

297. Pellerite, M.J., Wood, E.J., and Jones, V.W. (2002) Dynamic contact angle studies of self-assembled thin films from fluorinated alkyltrichlorosilanes. *J. Phys. Chem. B*, **106**, 4746–4754.

298. Banga, R., Yarwood, J., Morgan, A.M., Evans, B., and Kells, J. (1995) FTIR and AFM studies of the kinetics and self-assembly of alkyltrichlorosilanes and (Perfluoroalkyl)trichlorosilanes onto glass and silicon. *Langmuir*, **11**, 4393–4399.

299. Zybill, C.E., Ang, H.G., Lan, L., Choy, W.Y., and Meng, E.F.K. (1997) Monomolecular silane films on glass surfaces – contact-angle measurements. *J. Organomet. Chem.*, **547**, 167–172.

300. Hoffmann, P.W., Stelzle, M., and Rabolt, J.F. (1997) Vapor phase self-assembly of fluorinated monolayers on silicon and germanium oxide. *Langmuir*, **13**, 1877–1880.

301. Wu, K., Bailey, T.C., Willson, C.G., and Ekerdt, J.G. (2005) Surface hydration and its effect on fluorinated SAM formation on SiO$_2$ surfaces. *Langmuir*, **21**, 11795–11801.

302. Frechette, J., Maboudian, R., and Carraro, C. (2006) Thermal behavior of perfluoroalkylsiloxane monolayers on the oxidized Si(100) surface. *Langmuir*, **22**, 2726–2730.

303. Fontaine, P., Goguenheim, D., Deresmes, D., Vuillaume, D., Garet, M., and Rondelez, F. (1993) Octadecyltrichlorosilane monolayers as ultrathin gate insulating films in metal-insulator-semiconductor devices. *Appl. Phys. Lett.*, **62**, 2256–2258.

304. Srinivasan, U., Houston, M.R., Howe, R.T., and Maboudian, R. (1997) Proceedings of the 9th International Conference on Solid-State Sensors and Actuators – Transducers '97, p. 210.

305. de Boer, M.P., Knapp, J.A., Mayer, T.M., and Michalske, T.A. (1999) Role of interfacial properties on MEMS performance and reliability. *Proc. SPIE*, **3825**, 2–15.

306. Das, A.K., Kilty, H.P., Marto, P.J., Andeen, G.B., and Kumar, A. (2000) The use of an organic self-assembled monolayer coating to promote dropwise condensation of steam on horizontal tubes. *J. Heat Transfer*, **122**, 278–286.

307. Schmidt, E., Schurig, W., and Sellschopp, W. (1930) Versuche über die kondensation in film- und tropfenform. *Tech. Mech. Thermodynamik*, **1**, 53–63.

308. Vemuri, S., Kim, K.J., Wood, B.D., Govindaraju, S., and Bell, T.W. (2006) Long term testing for dropwise condensation using self-assembled monolayer coatings of *n*-octadecylmercaptan. *Appl. Therm. Eng.*, **26**, 421–429.

309. Deronzier, J.C., Foulletier, L., Huyghe, J., and Niezborala, J.M. (1975) Method for causing condensation in drops on heat exchanger tubes. US Patent 3878885.

310. Barriet, D. and Lee, T.R. (2003) Fluorinated self-assembled monolayers: composition, structure and interfacial properties. *Curr. Opin. Colloid Interface Sci.*, **8**, 236–242.

311. Bain, C.D. and Whitesides, G.M. (1988) Correlations between wettability and structure in monolayers of alkanethiols adsorbed on gold. *J. Am. Chem. Soc.*, **110**, 3665–3666.

312. Bain, C.D. and Whitesides, G.M. (1988) Formation of two-component surfaces by the spontaneous assembly of monolayers on gold from solutions containing mixtures of organic thiols. *J. Am. Chem. Soc.*, **110**, 6560–6561.

313. Bain, C.D., Evall, J., and Whitesides, G.M. (1989) Formation of monolayers by the coadsorption of thiols on gold: variation in the head group, tail group, and solvent. *J. Am. Chem. Soc.*, **111**, 7155–7164.

314. Bain, C.D. and Whitesides, G.M. (1989) Formation of monolayers by the coadsorption of thiols on gold: variation in the length of the alkyl chain. *J. Am. Chem. Soc.*, **111**, 7164–7175.

315. Folkers, J.P., Laibinis, P.E., Whitesides, G.M., and Deutch, J. (1994) Phase behavior of two-component self-assembled monolayers of alkanethiolates on gold. *J. Phys. Chem.*, **98**, 563–571.

316. Bain, C.D., Biebuyck, H.A., and Whitesides, G.M. (1989) Comparison of self-assembled monolayers on gold: coadsorption of thiols and disulfides. *Langmuir*, **5**, 723–727.

317. Capadona, J.R., Collard, D.M., and Garcia, A.J. (2003) Fibronectin adsorption and cell adhesion to mixed monolayers of Tri(ethylene glycol)- and methyl-terminated alkanethiols. *Langmuir*, **19**, 1847–1852.

318. Yang, W.R., Hibbert, D.B., Zhang, R., Willett, G.D., and Gooding, J.J. (2005) Stepwise synthesis of gly-glyhis on gold surfaces modified with mixed self-assembled monolayers. *Langmuir*, **21**, 260–265.

319. Nitahara, S., Terasaki, N., Akiyama, T., and Yamada, S. (2006) Molecular logic devices using mixed self-assembled monolayers. *Thin Solid Films*, **499**, 354–358.

320. Ekgasit, S., Yu, F., and Knoll, W. (2005) Displacement of molecules near a metal surface as seen by an SPR-SPFS biosensor. *Langmuir*, **21**, 4077–4082.

321. Ishida, T., Mizutani, W., Akiba, U., Umemura, K., Inoue, A., Choi, N., Fujihira, M., and Tokumoto, H. (1999) Lateral electrical conduction in organic monolayer. *J. Phys. Chem. B*, **103**, 1686–1690.

322. Mir, M., Alvarez, M., Azzaroni, O., and Knoll, W. (2008) Comparison of different supramolecular architectures for oligonucleotide biosensing. *Langmuir*, **24**, 13001–13006.

323. DeVries, G.A., Brunnbauer, M., Hu, Y., Jackson, A.M., Long, B., Neltner, B.T., Uzun, O., Wunsch, B.H., and Stellacci, F. (2007) Divalent metal nanoparticles. *Science*, **315**, 358–361.

324. Kilby, J. (1976) Invention of the integrated circuit. *IEEE Trans. Electron Devices*, **ED23**, 648–654.

325. Moore, G.E. (1965) Cramming more components into integrated circuits. *Electronics*, **38**, 4.

326. Zhang, S. (2007) *Stabilization of Monolayer-Protected Gold Nanoparticles, and the Growth of Gold Nanodendrites and Their Arrays*, Doctor of Philosophy, University of Houston, Houston.

327. Garcia-Saez, A.J., Chiantia, S., and Schwille, P. (2007) Effect of line tension on the lateral organization of lipid membranes. *J. Biol. Chem.*, **282**, 33537–33544.

328. Nag, K. and Keough, K.M.W. (1993) Epifluorescence microscopic studies of monolayers containing mixtures of dioleoyl- and dipalmitoylphosphatidylcholines. *Biophys. J.*, **65**, 1019–1026.

329. Stranick, S.J., Atre, S.V., Parikh, A.N., Wood, M.C., Allara, D.L., Winograd, N., and Weiss, P.S. (1996) Nanometer-scale phase separation in mixed composition self-assembled monolayers. *Nanotechnology*, **7**, 438–442.

330. Folkers, J.P, Laibinis, P.E., and Whitesides, G.M. (1992) Self-assembled monolayers of alkanethiols on gold:

comparisons of monolayers containing mixtures of short-and long-chain constituents with methyl and hydroxymethyl terminal groups. *Langmuir*, **8**, 1330–1341.

331. Atre, S.V., Liedberg, B., and Allara, D.L. (1995) Chain length dependence of the structure and wetting properties in binary composition monolayers of OH- and CH$_3$-terminated alkanethiolates on gold. *Langmuir*, **11**, 3882–3893.

332. Jamison, A.C. and Lee, T.R. (2009) *Lab Notes*, Department of Chemistry, University of Houston, Houston.

333. Trabelsi, S., Zhang, S., Lee, T.R., and Schwartz, D.K. (2007) Swelling of a cluster phase in Langmuir monolayers containing semi-fluorinated phosphonic acids. *Soft Matter*, **3**, 1518–1524.

334. Panda, A.K., Nag, K., Harbottle, R.R., Possmayer, F., and Petersen, N.O. (2007) Thermodynamic studies of bovine lung surfactant extract mixing with cholesterol and its palmitate derivative. *J. Colloid Interface Sci.*, **311**, 551–555.

335. Kuzmin, P.I., Akimov, S.A., Chizmadzhev, Y.A., Zimmerberg, J., and Cohen, F.S. (2005) Line tension and interaction energies of membrane rafts calculated from lipid splay and tilt. *Biophys. J.*, **88**, 1120–1133.

336. Akimov, S.A., Kuzmin, P.I., Zimmerberg, J., and Cohen, F.S. (2007) Lateral tension increases the line tension between two domains in a lipid bilayer membrane. *Phys. Rev. E: Stat., Nonlinear, Soft Matter Phys.*, **75**, 011919/011911–011919/011918.

337. Majumdar, N., Gergel-Hackett, N., Bean, J.C., Harriott, L.R., Pattanaik, G., Zangari, G., Yao, Y., and Tour, J.M. (2006) The electrical behavior of nitro oligo(phenylene ethynylene)'s in pure and mixed monolayers. *J. Electron. Mater.*, **35**, 140–146.

338. Tao, Y.-T., Wu, C.-C., Eu, J.-Y., Lin, W.-L., Wu, K.-C., and Chen, C.-h. (1997) Structure evolution of aromatic-derivatized thiol monolayers on evaporated gold. *Langmuir*, **13**, 4018–4023.

339. Jin, Q., Rodriguez, J.A., Li, C.Z., Darici, Y., and Tao, N.J. (1999) Self-assembly of aromatic thiols on Au(111). *Surf. Sci.*, **425**, 101–111.

340. Chang, S.-C., Chao, I., and Tao, Y.-T. (1994) Structure of self assembled monolayers of aromatic derivatized thiols on evaporated gold and silver surfaces: implication on packing mechanism. *J. Am. Chem. Soc.*, **116**, 6792–6805.

341. Sabatani, E., Cohen-Boulakia, J., Bruening, M., and Rubinstein, I. (1993) Thioaromatic monolayers on gold: a new family of self-assembling monolayers. *Langmuir*, **9**, 2974–2981.

342. Kang, J.F., Ulman, A., Liao, S., Jordan, R., Yang, G., and Liu, G.-y. (2001) Self-assembled rigid monolayers of 4'-substituted-4-mercaptobiphenyls on gold and silver surfaces. *Langmuir*, **17**, 95–106.

343. Kang, J.F., Liao, S., Jordan, R., and Ulman, A. (1998) Mixed self-assembled monolayers of rigid biphenyl thiols: impact of solvent and dipole moment. *J. Am. Chem. Soc.*, **120**, 9662–9667.

344. Liao, S., Shnidman, Y., and Ulman, A. (2000) Adsorption kinetics of rigid 4-mercaptobiphenyls on gold. *J. Am. Chem. Soc.*, **122**, 3688–3694.

345. Leung, T.Y.B., Schwartz, P., Scoles, G., Schreiber, F., and Ulman, A. (2000) Structure and growth of 4-methyl-4'-mercaptobiphenyl monolayers on Au(111): A surface diffraction study. *Surf. Sci.*, **458**, 34–52.

346. Shaporenko, A., Heister, K., Ulman, A., Grunze, M., and Zharnikov, M. (2005) The effect of halogen substitution in self-assembled monolayers of 4-mercaptobiphenyls on noble metal substrates. *J. Phys. Chem. B*, **109**, 4096–4103.

347. Chen, W., Gao, X.Y., Qi, D.C., Chen, S., Chen, Z.K., and Wee, A.T.S. (2007) Surface-transfer doping of organic semiconductors using functionalized self-assembled monolayers. *Adv. Funct. Mater.*, **17**, 1339–1344.

348. Zehner, R.W., Parsons, B.F., Hsung, R.P., and Sita, L.R. (1999) Tuning the work function of gold with self-assembled monolayers derived

from X–[C₆H₄–C≡C–]ₙC₆H₄–SH ($n = 0, 1, 2$; X = H, F, CH₃, CF₃, and OCH₃). *Langmuir*, **15**, 1121–1127.

349. Tour, J.M., Jones, L. II, Pearson, D.L., Lamba, J.J.S., Burgin, T.P., Whitesides, G.M., Allara, D.L., Parikh, A.N., and Atre, S. (1995) Self-assembled monolayers and multilayers of conjugated thiols, α, ω-dithiols, and thioacetyl-containing adsorbates. Understanding attachments between potential molecular wires and gold surfaces. *J. Am. Chem. Soc.*, **117**, 9529–9534.

350. Reed, M.A., Chen, J., Rawlett, A.M., Price, D.W., and Tour, J.M. (2001) Molecular random access memory cell. *Appl. Phys. Lett.*, **78**, 3735–3737.

351. Chen, J., Calvet, L.C., Reed, M.A., Carr, D.W., Grubisha, D.S., and Bennett, D.W. (1999) Electronic transport through metal – 1,4-phenylene di-isocyanide – metal junctions. *Chem. Phys. Lett.*, **313**, 741–748.

352. Hadi Zareie, M., Ma, H., Reed, B.W., Jen, A.K.Y., and Sarikaya, M. (2003) Controlled assembly of conducting monomers for molecular electronics. *Nano Lett.*, **3**, 139–142.

353. Dou, R.-F., Ma, X.-C., Xi, L., Yip, H.L., Wong, K.Y., Lau, W.M., Jia, J.-F., Xue, Q.-K., Yang, W.-S., Ma, H., and Jen, A.K.Y. (2006) Self-assembled monolayers of aromatic thiols stabilized by parallel-displaced π-π stacking interactions. *Langmuir*, **22**, 3049–3056.

354. Chen, F., Hihath, J., Huang, Z., Li, X., and Tao, N.J. (2007) Measurement of single-molecule conductance. *Annu. Rev. Phys. Chem.*, **58**, 535–564.

6
Polyelectrolyte Brushes: Twenty Years After

Patrick Guenoun

6.1
Introduction – Scope of the Review

Twenty years after their conception, polyelectrolyte brushes have gained full
recognition as an autonomous concept, whose zeroth order properties make a
clear consensus and made consequently the subject (at least partially) of several
reviews [1–5]. As mature objects, polyelectrolyte brushes are also elemental bricks
which are used to form more complex objects. The consequence of this matu-
rity is the need for theoreticians to develop sophisticated analyses in order to
cope with more precise measurements and to push calculations schemes at the
limits. This first order understanding reveals essential to control the making
and properties of the complex objects cited above. This review intends to show
how the concepts have been refined throughout these last years, how the experiments have gained in precision and how new objects have been made, based
on polyelectrolyte brushes. For this purpose, a first part gives some definitions
in order to precise key situations which are commonly studied in literature. A
second part depicts how the previously defined objects can be made, trying to
emphasize on good and bad aspects inherent to each method of fabrication.
Then, in a third part, the zeroth-order framework of observations and theories
that has been defined in the first 15 years of life of polyelectrolyte brushes will
be explained. The fourth part is devoted to more sophisticated achievements
which have recently deepen our understanding of how polyelectrolyte brushes
behave. At last, I will depict how application and fundamental science utilize
polyelectrolyte brushes as unique tools for bringing functions and properties to
objects.

6.2
Definitions

Polyelectrolyte brushes refer to charged polymer chains that are end-tethered to a
surface by one end. The surface density of anchoring points is high enough such

Functional Polymer Films, First Edition. Edited by Wolfgang Knoll and Rigoberto C. Advincula.
© 2011 Wiley-VCH Verlag GmbH & Co. KGaA. Published 2011 by Wiley-VCH Verlag GmbH & Co. KGaA.

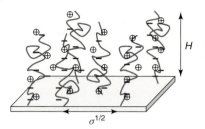

Figure 6.1 Scheme of a planar polyelectrolyte brush of height H made of end-tethered negative chains with positive counterions, each chain occupying a typical area σ on the plane.

as the chain conformation is perturbed compared to a single end-tethered chain conformation. This long statement calls immediately for a more precise explanation and additional definitions. The surface can be infinite, following the two dimensions (a plane), following only one dimension (the cylinder case will be the only one to be mentioned) or finite (usually a sphere). The first case is referred to as the *planar case* though the plane can be actually rough or fluctuating. Despite a couple of studies on cylinders [2, 6], the planar and spherical cases will be the cases that will be reviewed. This definition is of course essential when one refers to electrostatics where the space dimensionality induces qualitative and quantitative changes. Theory often assumes that no monomer adsorb onto the surface, a feature that is not always verified with actual systems. Referring to Figure 6.1, the average minimal distance (curvilinear distance along the surface) between anchoring points will be called $\sigma^{1/2}$, where σ is the area per chain. Next, one needs to define what is a charged chain since ionizable groups (in proportion, f, of the total number of monomers) can appear at fixed monomer positions along the chain (the quenched case) or form a fluctuating distribution along the chain (the annealed case), whose shape is usually controlled by the environment. The latter is usually (but not always) an aqueous solution in which the important parameters are the pH or the electrolyte solute nature and concentration. When ionized, a chain "releases" part of its counterions, meaning that the polyelectrolyte brush problem amounts to determining both the monomer and the counterion distribution (for theory and fundamental analysis). The latter distribution determines the chain configuration. This means that deciding whether the assembly is indeed a brush or not does not amount to comparing $\sigma^{1/2}$ with an unperturbed bulk chain dimension. Actually electrostatic interactions between chains may easily have a range exceeding $\sigma^{1/2}$, leading to conformation changes at large grafting distances. To conclude this part, it must be recalled that a bulk polyelectrolyte flexible chain of Kuhn monomer length, a, would be a rod-like molecule if all counterions were able to distribute randomly over all the available volume. However, in pure water, assumed to be a continuous fluid of dielectric constant ε, it is believed that only a proportion a/l_B of monovalent counterions are free to do so where $l_B = e^2/4\pi\varepsilon k_B T$ is the Bjerrum length, e, the electron charge, k_B, the Boltzmann constant and, T, the temperature. The Bjerrum length sets the amplitude of electrostatic interactions, and counterions that are too close along a rod (distance less than l_B) reduce the free energy by staying at close vicinity of the chain. This is qualitatively the condensation mechanism that is only demonstrated for a fixed rod geometry of the polyelectrolyte chain. This tells us that a highly inhomogeneous counterion distribution is expected along a path perpendicular to the chain.

6.3
How to Make Polyelectrolyte Brushes?

The two main routes for making polyelectrolyte brushes are physisorption and chemisorption. The first route relies on the spontaneous adsorption or self-assembly of heterogeneous polymers due to noncovalent bonds. This can be done by the adsorption of diblock copolymers on surfaces where one block (usually hydrophobic) adsorbs onto the surface, while the second charged dangling block constitutes the charged brush. This has been successfully done on both planar or spherical surfaces [7–9]. To reach large adsorbed amounts, adsorption usually takes place in the presence of electrolytes in order to screen the Coulomb interaction between adsorbing chains onto the surface. Self-assembly of charged-neutral diblock copolymers in the bulk has been also commonly used for making polyelectrolyte brushes. For instance, micelle or vesicle formation can occur and lead to self-assembled structures made of a hydrophobic core (the hydrophobic moiety of the diblock) and of charged arms [10–13]. It must be noted that these self-assembled structures are barely at equilibrium and usually consist of frozen structures, the core being unable to rearrange on experimental timescales [14], whatever is the stimulus used (though exceptions have been noted [15]). Another nonequilibrium method is the formation of Langmuir films of similar copolymers [16], where the brush is now formed at the air/solvent interface. Gibbs films, where an adsorbed layer may also form a brush in the presence of a copolymer solution, is *a priori* a more equilibrated situation but several complicated phenomena have been reported [17, 18] like the observation of buckled monolayers.

Chemisorption, that is, grafting of a charged chain by a covalent bond onto a substrate has also been developed as a powerful method for designing brushes. Chains can be grafted *to* the surface, and usually ionized by a subsequent process [19] though risks of cutting out chains exist and alternative methods have been employed [20]. Chains can also be grown *from* the surface (grafting-from process) starting from an initiator localized on the substrate. In this case, chains can be made directly by polymerizing charged monomers or ionized afterwards [21] and the chain densities obtained were claimed to be much higher than with grafting-to techniques, which is not completely true [20]. However, grafting-from techniques could be more practical to grow large molecular weight assemblies or heterogeneous chain compositions. Chemisorption has been mostly designed for planar surfaces but few examples exist where a spherical geometry has been reached [22].

6.4
Simple Observations and Theoretical Concepts

6.4.1
The Osmotic Regime

Back in 1991, Borisov and coworkers [23] and Pincus [24] predicted that a so-called osmotic regime should exist where the polyelectrolyte brush retains most of its counterions. A mean-field scaling assumption amounts to writing that the confinement entropy of the counterions and the elastic entropy of the chains are dominant over any other term in the free energy. When these electrostatic interactions are assumed to be screened out by the counterion cloud the brush height H scales as $Nf^{1/2}a$ where N is the number of monomers per chain, f, the degree of chain ionization (assumed to be here a quenched degree of freedom), and a, the monomer size. In this regime, the screening length ξ, defined, for monovalent counterions and monomers, as $\xi^{-2} = 8\pi l_B(fN/\sigma H)$, is such as $\xi < H$ if σ or $(fN)^{-1}$ are small enough (in practice not a very stringent condition), which justifies neglecting electrostatics within the brush at zeroth order. This comes from an Alexander–de Gennes like approximation where all chains end at the same distance H of the substrate and one recovers a strong chain deformation similar to a neutral brush (the N^1 scaling). A peculiar feature is the independence of H vs. the area per chain σ since, in this approximation both entropy terms of the free energy per chain are independent of the area per chain. Another important consequence of counterion confinement is that addition of external electrolyte at a concentration c_s (that we assume monovalent except in precise cases we will mention) does not perturb the structure up to concentrations on the order of the inner counterion concentration. The external osmotic pressure must overcome the inner counterion pressure for the external electrolyte to penetrate. In other words, there is an onset below which no effect of added salt is noticeable. Assuming perfectly flexible chains, it is then possible to calculate, by scaling arguments, the evolution of the height of the brush in the salt-added dominant regime, above the onset in and one finds $H \sim Na(a\sigma c_s)^{-1/3}$. In practice, this means that charged brushes should not be very sensitive to electrolyte addition, a very important point when a polyelectrolyte brush is used as a stabilizing layer for colloids. This is all the more true since the exponent 1/3 for a planar geometry becomes a much weaker exponent in spherical geometry.

6.4.2
Experiments on Planar Osmotic Brushes

The simple basic predictions depicted in the previous paragraph were partly verified by experiments on different configurations. Soap films made of opposing adsorbed layers of charged diblock copolymers (with sodium polystyrene(sulfonate) (PSSNa) as a polyelectrolyte) provided a first system to study the effect of salt addition. The latter was indeed found to be effective only above a certain threshold, whereas the 1/3 scaling was also verified [25]. The threshold value of the salt concentration enables one to estimate ξ which was shown to be much smaller than H, a confirmation of the osmotic regime. In the absence of salt, the layers were found to be quite extended, in agreement with the relation $H = \alpha Nf^{1/2}a$, where α is a numerical factor of order one. Using diblock copolymers with PSS and various counterions, Ahrens et al. [16] were able to build Langmuir films of variable surface densities. Their results corroborate the independence of the brush height with this density and the stoichiometric incorporation of counterions within the brush in the osmotic regime. On adding salt, a threshold in concentration was also found, above which the scaling $H \sim (\sigma c_s)^{-1/3}$ is confirmed [26]. Polyelectrolyte brushes grafted to a solid silicon surface were made by Tran *et al.* [27], using sulfonated chains but with a milder sulfonation method. Neutron reflectivity showed that indeed the brush height, without salt, was again independent of the surface density and linear in N and $f^{1/2}$. Of course, the latter variation is harder to check given the limited range authorized by the sulfonation and the theoretical limit imposed by the Manning condensation (see below).

6.4.3
Beside the Osmotic Regime

Though most experimental situations claim (more than they demonstrate) to be in the osmotic regime, it is important to depict other regimes. The osmotic regime breaks down when the polymer assembly is either too sparse and/or too little ionized. The first case of sparse chains is mostly specific for charged chains in interaction (in order to distinguish from "simple" isolated mushrooms) and has been depicted in detail by Borisov and coworkers [28]: from isotropic sticks when intramolecular electrostatic interactions dominate, one switches to oriented sticks for large intermolecular interactions for sparse, highly charged chains. The second case where few charges exist on each chain means that excluded volume interaction play a major role. The central question is then to what extent this assembly is distinct from a neutral brush because of the residual charges. In practice, the charge fraction borne by a chain lies in a very limited range. Too few charged monomers (the limit can be of 30% for PSSNa) make the polyelectrolyte (PE) chains insoluble in water for many chemical species since the polymer backbone is quite often hydrophobic. On the upper side, Manning's condensation limits the fraction of ionized monomers to a/l_B (more exactly it is a limit on the effective charge borne by the chain). Incorporating virial terms in the free energy, like an attractive second-order one representing a bad solvent situation of the un-ionized monomers and a third-order repulsive one, led Zhulina and coworkers [23, 29] and Ross and Pincus [30] to predict a collapsed state of the charged brush for low enough values of f. In the same range of f, for a repulsive virial term, one can predict a neutral brush behavior (in a good or theta solvent according to the virial value).

6.4.4
The Spherical Geometry

Scaling theory can also be applied to a spherical geometry where charged chains stem out from an inert nonadsorbing core. An osmotic regime is predicted for a large number of branches ($p > p^*$) where, in analogy with the planar case, the star radius R goes like $Nf^{1/2}a$ (see Figure 6.2), independent of p, the star retaining the majority of its counterions [31]. The limiting value p^* is given by: $p^* \sim f^{-1/2}(a/l_B)$, in practice an easy number to reach by self-assembly or chemical grafting. Of course, even when one sticks to a an Alexander–de Gennes approximation where

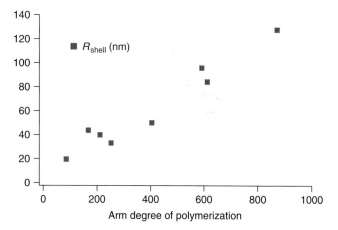

Figure 6.2 A plot of shell thicknesses R_{shell} of various fully ionized (quenched or annealed) spherical brushes taken from Refs. [37, 38] and from unpublished results in F. Muller's thesis (Université de La Rochelle, France) as a function of the polyelectrolyte degree of polymerization N. This plot demonstrates the linear scaling $R_{shell} \propto N^1$.

all chain ends are located at the same radius, the spherical geometry imposes an inhomogeneous radial profile of monomers that is predicted to follow a $c(r) \sim r^{-2}$ behavior, linked to a distribution of noncompact blobs [32]. Again, for an added salt concentration higher than the inner (average) counterion concentration, the star begins shrinking (see Figure 6.3) and its conformation is predicted by Borisov and Zhulina [32] to follow $R \sim N^{3/5} p^{1/5} f^{2/5} (C_S)^{-1/5}$. Charged stars, covalently linked at their center or self-assembled micelles of charged-neutral diblock copolymers are relevant experimental systems that provided the first examples of the above concepts. A concern, however, is that the aggregation number of the self-assembled objects (a number that fixes the p value) stays under control, for instance when salt is added. Fortunately, quite often, charged copolymer micelles behave as frozen objects on the experimental timescale [33], though some exceptions have been noted [15]. The large extension of the PE corona was noted in early experiments by Zhang *et al.* [11] and a very weak sensitivity to added salt, in possible agreement with the 1/5 exponent, was reported [34]. Neutron scattering within the charged corona was then successfully used as evidence to the transition between two regimes when salt was added [35], the onset in salt concentration being consistent with estimations of the inner counterion concentration. It also provided conclusive evidence that the r^{-2} profile is consistent with the data in absence of added salt [36].

6.4.5
Annealed Chains

So far, only charged chains whose distribution of charges is fixed (quenched chains) have been described. However, many practical situations are encountered

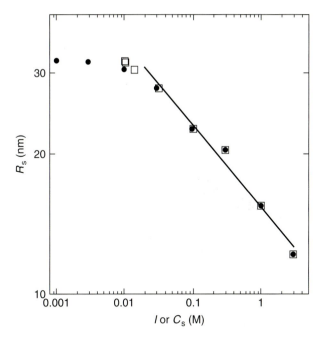

Figure 6.3 Spherical shell thickness of a polystyrene sulfonate brush as a function of added salt (ionic strength I or added salt concentration C_s), showing the onset in salt above which the brush contracts. (Reprinted from Ref. [39] with permission.)

where charged chains are polyacids or polybases, providing so-called annealed distributions of charges where the fraction of ionized charges along the chain is ruled by the pH close to the chain. The external pH then appears as a supplementary control parameter for the extension of a charged brush. The most salient difference of annealed brushes compared to quenched ones was the prediction by Israëls and coworkers [40] as well as Zhulina and coworkers [41] that a polyacid brush thickness, in the presence of added salt, would be non monotonous as a function of both ionic strength and area per chain. This occurs because an increase of both previous parameters amount to screening the monomer interactions, then favoring ionization, which goes through a maximum, whatever the external pH is. This can be generalized to spherical geometries where the number of branches play the role of grafting density. This non monotonous effect has been observed in spherical geometry for spherical lattices grafted by charged chains of a polybase [42] or micelles of poly(methacrylic acid) [43, 44] (Figure 6.4). In planar geometry, conclusive evidences came from anchored poly(methacrylic acid) chains [45] and for Langmuir films exhibiting polyacid dangling chains [46] but with a much weaker exponent in the latter case. On the contrary, other experiments on planar brushes did not provide evidence to this effect [14].

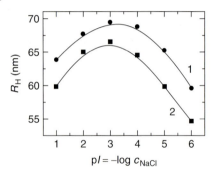

Figure 6.4 Hydrodynamic radius of spherical annealed brush as a function of the added salt concentration. A maximum is evidenced, showing that the ionization is first promoted by screening before screening induces chains contraction. (Reproduced with permission from Ref. [44].)

6.5
Beyond Simplicity

Of course, a full understanding and a complete appreciation of polyelectrolyte brushes potential needs to determine how monomers and counterions are actually distributed. In particular, in curved geometry like the spherical one, the decrease of the monomer density, $\rho(r)$, with the sphere of radius, r, has to be determined. Another way of assessing the monomer distribution is by scattering experiments on spherical brushes in order to get an indirect access to $\rho(r)$ through measurements in reciprocal space. As mentioned previously, by neutron scattering on charged diblock copolymer micelles, it was shown [36, 37] that $\rho(r)$ decreases as r^{-2}, supporting the idea of noncompact blobs. The experiment also demonstrates a strong extension of the chains which can be viewed as rod-like in first approximation [35].

6.5.1
Monomer Profile

One also needs to go beyond the simple scaling picture and to design experiments in that purpose. This was intended by the very early results on PE brushes by Miklavic and Marcelja [47] and Misra *et al.* [48] where the influence of added salt on the monomer profiles was demonstrated. As shown by Zhulina and coworkers [29], the analytical form of the monomer profile $\rho(z)$ in planar geometry is close to a Gaussian function in the absence of added salt, whereas the profile switches back to a less-extended parabola, characteristic of a neutral brush, for a large amount of added salt (above the onset we discussed before in Section 6.4). Note that this analytical self-consistent field approach is based on a weak charging approximation (no counterion condensation is assumed to occur) and that a somewhat more precise numerical self-consistent field (SCF) approach using a Poisson–Boltzmann assumption for the electrostatics confirms the findings [49]. This qualitative change in profile by salting in was suspected by neutron-scattering studies on chemically grafted brushes [19]. Neutron reflectivity was more informative by showing that, though the Gaussian shape is consistent with data, monomer adsorption on the substrate is a usual complication in order to confront with the models [27]. When this adsorption is minimal, or very localized at the anchoring surface, the Gaussian shape and the transition toward the parabola in added salt was evidenced for different surface densities of a diblock copolymer Langmuir film [50].

6.5.2
Counterion Distribution

The counterion distribution within the brush is also a big matter of debate. Of course in the osmotic regime, counterions are predicted to be roughly localized within the brush [51]. However, theoretical or experimental techniques able to directly evaluate or measure such a counterion distribution are scarce. Numerical results from a lattice SCF approach indeed confirm the equality of the counterion profile with the monomer profile in the osmotic regime [49]. An analytical SCF theory was built without any *a priori* assumption on the counterion distribution and confirmed the identity of profiles [52]. The same theory is able to describe the case of weakly charged brushes where counterions depart from the brush. On the experimental side, X-ray reflectivity is able to test the counterion profile and demonstrate counterion confinement in a Langmuir film of a charged diblock copolymer [26]. Neutron reflectivity with contrast conditions able to reveal counterions like tetramethyl ammonium (TMA), has also been used successfully to prove the osmotic assumptions [53]. The same TMA ions used in a bulk situation are able to test a spherical brush of charged diblock copolymer micelles by neutron scattering [54]. In this geometry also, a nice match is obtained between the counterion distribution and the monomer distribution. However, significant electrostatic ordering between charged copolymer micelles or charged stars was also measured [36, 55]. This means that a neat charge is always borne by the spherical brushes [33] since the ordering was measured for concentrations below the packing limit of the spheres: some counterions are free to leave the brush. Interestingly enough, this is not observed in planar geometry where surface force apparatus (SFA) measurements do not show any long-range electrostatic force between charged brushes [56]. Indeed, the condensation onto a plane is stronger than onto a sphere.

6.5.3
Counterion Condensation

As already mentioned, counterion condensation [57–60] is very likely to occur within charged brushes, limiting the effective ionization degree of the chains and localizing a proportion $1-1/\xi$ of counterions "very close" to the chain, where $\xi = l_b/a$ for monovalent species. The closeness of the counterions can be specified within the Poisson–Boltzmann theory around a cylinder of finite radius and of infinite length. This theory reproduces the condensation phenomenon [61, 62], at least qualitatively [63]. For more flexible chains, the concept of condensation can be maintained with quantitative changes in the condensation fraction [64].

Using X-ray scattering on spherical charged brushes, it was shown that, varying the contrast by exchanging counterions through dialysis, the Poisson–Boltzmann approximation around a rod was a valid model of the data [65] since chains in a brush are rather extended [35]. This naturally implies that most counterions are localized around the chains. A very powerful and pictorial way of assessing such a result is the use of numerical simulations like molecular dynamics [66] or Monte Carlo [67] methods since modern computers enable one to tackle a number of monomers and counterions like those embedded in a realistic charged micelle. For instance, when tuning the parameter ξ to 2.8, a value close to the experimental value for PSS chains in water, a strong condensation situation is reached as shown in Figure 6.5 that shows the equilibrium configuration obtained by Monte Carlo simulations on a micelle of 27 chains of length 130 monomers. The analysis of such a charged star whose polymer arms are coated by condensed counterions can also be made very quantitative. It has been shown that in the highly condensed regime, the counterion entropy, electrostatic interactions, and chain stretching are of the same order of magnitude [67]. This should call for more exact theoretical approaches that allow for counterion condensation in the treatment of electrostatic interactions.

6.5.4
Brush Extension

An example of new approaches is given by the prediction and simultaneous observation that osmotic brushes indeed increase in height when the surface density is increased, contrary to scaling predictions. This was observed by a careful check of reflectivity experiments carried out on Langmuir films of charged diblock copolymers under a neutron beam [50] or an X-ray beam [68]. Such a dependence of the osmotic thickness upon the grafting density came out from molecular

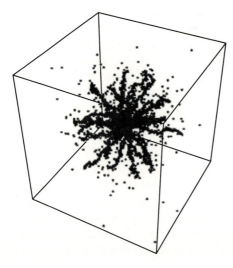

Figure 6.5 Monte-Carlo simulation of a spherical quenched brush in a regime above the Manning threshold where the counterions (in blue) are strongly localized along the chains (see Ref. [67] for details).

dynamics simulations [69, 70], with a possible *caveat* of dealing with relatively short chains. An analytical model taking into account the lateral segregation of counterions due to condensation and the covolumes of the species also described such an effect [71]. Recently, a density functional theory was designed and provided some confirmation of this brush thickness increase and predicted that no simple scaling would exist then [72]. Self-consistent field theory was also shown to predict such an increase even in the uncondensed regime, pointing to an essential role of the chains covolume at high grafting density [1]. More work is then clearly needed to fully understand the origin of the effect.

6.5.5
Salt Contraction

Though it is now well documented that charged brushes contract above some onset in added salt, some unresolved issues exist. A first one, related to the above condensation ideas, is the exact correspondence of this onset with the inner concentration of free, uncondensed, and then osmotically active counterions. In spherical geometry, when simple Manning condensation (calculated for an infinite rod) is taken into account, the agreement appears rather good [15, 35, 39, 73] within a factor of 2. On the contrary one can notice discrepancies for planar brushes where a good agreement with a Manning–Oosawa-like theory [14] has been reported while either an overcondensation [26, 56] or an undercondensation [27] with respect to Manning's scheme has been observed. The latter case was reported for a denser brush with more non ionizable (hydrophobic) monomers. Clearly, more experiments are needed to assess the validity of simple condensation theories in such a semidilute medium as a charged brush. A second partially unresolved issue is the status of a charged brush in high salt: is it fully identical to a neutral brush? Neutron-scattering experiments on charged spherical coronae of micelles do not confirm such a simple picture. For quenched chains, cases where the charged corona splits in two "phase-separated" regions have been described [15, 74]. Even when this phase separation is not observed, it can be shown that for high salt concentrations, the chains remain quite extended at short scale, maintaining a large persistence length [39]. This means that chains within a charged brush, even with a large amount of added salt, cannot be viewed as flexible "quasineutral" chains. In particular, the law of contraction of a spherical brush depends on how the persistence length vary with the amount of added electrolyte.

6.5.6
Electrostatic Collapse

The intensity of the electrostatic interaction can be easily varied by using multivalent counterions of valency Z instead of monovalent (the relevant coupling parameter is now $Z^2 e^2 / 4\pi \varepsilon k_B T$). Correlations beyond the mean-field level of the Poisson–Boltzmann approximation can introduce effective attractive interactions betweens chains. This has been predicted for bulk polyelectrolyte solutions [75] like DNA solutions and confirmed in several experiments [76, 77]. For polyelectrolyte brushes, such an attraction can lead to a collapse of the brush that has been predicted to occur for valencies larger than 2 [78–80] (for a valency of 1, even increasing $e^2/4\pi\varepsilon k_B T$ by means of Monte-Carlo simulations does not lead to collapse [67]) and is observed for spherical brushes where monovalent counterions were progressively replaced by trivalent ones [81].

6.6
Toward the Application of Polyelectrolyte Brushes

In this part, results that are more directly related to future applications of polyelectrolyte brushes as functional devices will be reviewed. However, these results contain also a great deal of very fundamental physics. The very first aspect is about the role of polyelectrolyte brushes as protective layers for preventing colloids from aggregating upon Van der Waals attraction [24, 82]. This was clearly demonstrated by adsorption of charged diblock copolymers onto polystyrene latices [9] which were then stabilized against large amounts of added salt. In order to assess the exact range and magnitude of forces exerted by the protective layers, (SFA) and atomic force microscope (AFM) were extensively used, as well as optical tweezers to hold grafted colloids [83]. Polyelectrolyte brushes are adsorbed onto opposing, planar (actually cylindrical but planar at the brush scale) mica surfaces [84, 85] for the SFA. This adsorption [7] can be a source of difficulty since it can vary during compression or when chemical conditions are changed [86]. In the absence of salt some results show the existence of a long range repulsion if enough counterions are able to leave the brush, usually correctly described by a double-layer model [86, 87] and a short-range repulsion due to the overlap or contraction of opposing segments of the two brushes. The latter repulsion is not very well described by simple counterion overlap [24]. This is probably due to an additional sign that the brushes were not fully in the osmotic regime. If brushes are dense enough, negligible numbers of counterions are outside the brush and no long-range repulsion is observed until the brushes are in contact [56]. When this happens scaling laws were found to apply very well, supporting the osmotic regime. With the addition of salt, SFA clearly confirmed the expected decrease of the interaction range of the charged brushes with a contraction law very sensitive to the chain surface densities [88]. On the theoretical side, a mean-field calculation of the force profiles in the added salt regime was shown to describe well the short distance part, where brushes are in contact [89]. In spherical geometries, the question of describing the force between charged stars was tackled through a comparison with molecular simulations that provides the picture that charged brushes in contact, mostly contract and do not overlap [2, 90]. A hint of such a behavior came from the existence of a very strong liquid order between concentrated charged micelles [36, 91], an order that persists at concentrations much higher than the close packing concentrations of the micelles, while for interpenetrating neutral stars, the order disappears in such a region [92]. Recent neutron-scattering experiments tend to confirm this contraction before interpenetration at higher volume fractions [93] which can be described by an adequate scaling theory [94]. This idea that charged brushes avoid overlap is of course very important when lubrication is concerned. This could be the reason for the very weak friction recently measured between such surfaces [95]. Molecular dynamics supports planar brushes shrinking upon approach in a model where counterions are explicit [96, 97]. Altogether the interpenetration of charged brushes in the osmotic regime (no added salt) is less than for neutral brushes [98].

A second very popular aspect is the use of charged brushes to complex oppositely charged species. While monovalent ions can be readily exchanged by other monovalent ions of the same size [65, 99], larger polyions or charged surfactants are irreversibly complexed at the outer fringe of a planar quenched PSS brush [99]. Polycarboxylic brushes also form stable complexes with small metal ions [100]. In spherical geometry, metallic ions can be immobilized within the charged corona and then reduced to make metallic nanoparticles whose catalytic activity was shown to be higher than for particles embedded in a microgel [101]. Spherically charged brushes are not only efficient carriers for neutral proteins like insulin [102] but also for charged proteins – even of a similar charge than the corona – that can bind to the charged corona. This is due to the entropic release of counterions [103] and proteins were shown to keep a large part of their activity. Mechanical devices can also be designed based on polyelectrolyte brushes since the switching-in thickness of the brush upon variation of an external parameter is the essential ingredient for a valve. An example of such a device is an AFM cantilever coated on one side by a cationic brush [104]. This asymmetric coating is able to generate an asymmetric deflection response to an alternate voltage bias. Similarly, an electrode coated by a polybase brush was shown to provide tunable redox activities due to the change in conformation of the brush that controls the transport of charged species to the electrode [105]. Closer to a valve is a planar surface coated with a mixed polyelectrolyte brush of polyacid and polybase polymers [106, 107]. At extreme pH values, the brush stretches away from the surface, whereas at intermediate values, a neutral polymer complex forms a reversible collapsed structure. Irreversible effects of the same kind can be obtained by transforming trivalent ions (used to make a spherical brush collapse) into divalent ions, then expanding back the corona to a more stretched conformation [108].

6.7
Conclusion

Two factors obviously concur to a widespread use of charged brushes as building blocks of sophisticated and/or stimulable polymer layers. On the one hand, as hopefully shown in this review, fundamental aspects are now relatively well understood leading to very predictive properties for a precise design. Secondly, building blocks or methods able to provide charged brushes of given characteristics are more currently available at a cheaper cost. As an example, one can cite the industrial scale synthesis of charged diblock copolymer illustrated by the MADIX® process [14, 38]. This possibility indicates one of the future avenues of research, which is the understanding of charged brushes for more polydisperse, and broadly speaking, more complex chain assemblies than so far studied. This will surely appear as an important future issue for both theory and experiments. Several recent instances clearly show that polydispersity is not necessarily an issue preventing the display of properties typical of polyelectrolyte brushes like a weak interpenetration [109]. Polyelectrolyte brushes remain an excellent benchmark for our actual fundamental understanding of polyelectrolyte physics since there is a constant need to describe complex situations that are anisotropic, including multivalent ions [110] or taking into account finite stretching of the chains [111]. Deepening this knowledge is necessary toward the conception and optimization of devices where polyelectrolyte brushes can contribute in a unique manner.

References

1. Guenoun, P., Argillier, J.F., and Tirrell, M. (2000) End-tethered charged chains at surfaces. *C. R. Acad. Sci., Sér. IV Phys., Astrophys.*, **1** (9), 1163–1169.
2. Rühe, J., Ballauff, M., Biesalski, M., Dziezok, P. *et al.* (2004) Polyelectrolyte brushes. *Adv. Polym. Sci.*, **165**, 79–150.
3. Naji, A., Seidel, C., and Netz, R.R. (2006) Theoretical approaches to neutral and charged polymer brushes. *Adv. Polym. Sci.*, **198**, 149–183.
4. Ballauff, M. and Borisov, O. (2006) Polyelectrolyte brushes. *Curr. Opin. Colloid Interface Sci.*, **11**, 316–323.
5. Toomey, R. and Tirrell, M. (2008) Functional polymer brushes in aqueous media from self-assembled and surface-initiated polymers. *Ann. Rev. Phys. Chem.*, **59**, 493–517.
6. Bendejacq, D., Joanicot, M., and Ponsinet, V. (2005) Pearling instabilities in water-dispersed copolymer cylinders with charged brushes. *Eur. Phys. J. E*, **17**, 83–92.
7. Amiel, C., Sikka, M., Schneider, J.W., Tsao, Y.H., Tirrell, M., and Mays, J.W. (1995) Adsorption of hydrophilic-hydrophobic block copolymers on silica from aqueous solutions. *Macromolecules*, **28**, 3125–3134.
8. Styrkas, D.A., Bütün, V., Lu, J.R., Keddie, J.L., and Armes, S.P. (2000) pH controlled adsorption of polyelectrolyte diblock copolymers at the solid/liquid interface. *Langmuir*, **16**, 5980–5986.
9. Hariharan, R., Biver, C., Mays, J.W., and Russel, W.B. (1998) Ionic strength and curvature effects in flat and highly curved polyelectrolyte brushes. *Macromolecules*, **31**, 7506–7513.
10. Khougaz, K., Astafieva, I., and Eisenberg, A. (1995) Micellization in block polyelectrolyte solutions. 3. Static light characterization. *Macromolecules*, **28**, 7135–7147.
11. Zhang, L., Barlow, R.J., and Eisenberg, A. (1995) Scaling relations and coronal dimensions in aqueous block polyelectrolyte micelles. *Macromolecules*, **28**, 6055–6066.
12. Guenoun, P., Davis, H.T., Mays, J.W., and Tirrell, M. (1996) Aqueous micellar solutions of hydrophobically modified polyelectrolytes. *Macromolecules*, **29**, 3965–3969.
13. Förster, S., Hermsdorf, N., Leube, W., Schnalblegger, H., Lindner, P., and Böttcher, C. (1999) Fusion of charged block copolymer micelles into toroid networks. *J. Phys. Chem. B*, **103**, 6657–6668.
14. Bendejacq, D., Ponsinet, V., and Joanicot, M. (2004) Water-dispersed

lamellar phases of symmetric poly(styrene)-block-poly(acrylic acid) diblock copolymers: model systems for flat dense polyelectrolyte brushes. *Eur. Phys. J. E*, **13**, 3–13.

15. Förster, S., Hermsdorf, N., Böttcher, C., and Lindner, P. (2002) Structure of polyelectrolyte block copolymer micelles. *Macromolecules*, **35**, 4096–4105.

16. Ahrens, H., Förster, S., and Helm, C.A. (1997) Polyelectrolyte brushes grafted at the air/water interface. *Macromolecules*, **30**, 8447–8452.

17. Fontaine, P., Daillant, J., Guenoun, P., Alba, M., Braslau, A., Mays, J.W., Petit, J.M., and Rieutord, F. (1997) Spontaneous buckling induced by the adsorption of charged copolymers at the air-water interface, *J. Phys. II France*, **7**, 401–407.

18. Dubreuil, F., Fontaine, P., Alba, M., Daillant, J., Mays, J.W., Zalczer, G., and Guenoun, P. (2005) Buckling of charged diblock copolymer monolayers at the air-water interface. *Europhys. Lett.*, **70**, 176–182.

19. Mir, Y., Auroy, P., and Auvray, L. (1995) Density profiles of polyelectrolyte brushes. *Phys. Rev. Lett.*, **75**, 2863–2866.

20. Tran, Y. and Auroy, P. (2001) Synthesis of poly(styrene sulfonate) brushes. *J. Am. Chem. Soc.*, **123**, 3644–3654.

21. Biesalski, M. and Rühe, J. (1999) Preparation and characterization of a polyelectrolyte monolayer covalently attached to a planar surface. *Macromolecules*, **32**, 2309–2316.

22. Guo, X., Weiss, A., and Ballauff, M. (1999) Synthesis of spherical polyelectrolyte brushes by photo-emulsion polymerization. *Macromolecules*, **32**, 6043–6046.

23. Borisov, O.V., Birshtein, T.M. and Zhulina, E.B. (1991) Collapse of grafted polyelectrolyte layer. *J. Phys. II (France)*, **1**, 521–526.

24. Pincus, P. (1991) Colloid stabilization with grafted polyelectrolytes. *Macromolecules*, **24**, 2912–2919.

25. Guenoun, P., Schalchli, A., Sentenac, D., Mays, J.W., and Benattar, J.J. (1995) Free-standing black films of polymers: a model of charged brushes in interaction. *Phys. Rev. Lett.*, **74**, 3628–3631.

26. Ahrens, H., Förster, S., and Helm, C.A. (1998) Charged polymer brushes: counterion incorporation and scaling relations. *Phys. Rev. Lett.*, **81**, 4172–4175.

27. Tran, Y., Auroy, P., and Lee, L.T. (1999) Determination of the structure of polyelectrolyte brushes. *Macromolecules*, **32**, 8952–8964.

28. Borisov, O.V., Zhulina, E.B., and Birshtein, T.M. (1994) Diagram of the states of a grafted polyelectrolyte layer. *Macromolecules*, **27**, 4795–4803.

29. Zhulina, E.B., Borisov, O.V., and Birshtein, T.M. (1992) Structure of grafted polyelectrolyte layer, *J. Phys. II (France)*, **2**, 63–74.

30. Ross, R. and Pincus, P. (1992) The polyelectrolyte brush: poor solvent. *Macromolecules*, **25**, 2177–2183.

31. Borisov, O.V. (1996) Conformations of star-branched polyelectrolytes. *J. Phys. II (France)*, **6**, 1–19.

32. Zhulina, E.B. and Borisov, O.V. (1996) Polyelectrolytes grafted to curved surfaces. *Macromolecules*, **29**, 2618–2626.

33. Cottet, H., Gareil, P., Guenoun, P., Muller, F., Delsanti, M., Lixon, P., Mays, J.W., and Yang, J. (2001) Capillary electrophoresis of associative diblock copolymers. *J. Chromatogr. A*, **939**, 109–121.

34. Mays, J.W. (1990) Synthesis of model branched polyelectrolytes. *Polym. Commun.*, **31**, 170–172.

35. Guenoun, P., Muller, F., Delsanti, M., Auvray, L., Chen, Y.J., Mays, J.W. and Tirrell, M. (1998) Rodlike behavior of polyelectrolyte brushes. *Phys. Rev. Lett.*, **81**, 3872–3874.

36. Muller, F., Delsanti, M., Auvray, L., Yang, J., Chen, Y.J., Mays, J.W., Demé, B., Tirrell, M., and Guenoun, P. (2000) Ordering of urchin-like charged copolymer micelles: electrostatic packing and polyelectrolyte correlations. *Eur. Phys. J. E*, **3**, 45–53.

37. Groenewegen, W., Egelhaaf, S.U., Lapp, A., and van der Maarel, J.R.C. (2000) Neutron scattering estimates of the effect of charge on the micelle structure in aqueous polyelectrolyte

diblock copolymer solutions. *Macromolecules*, **33**, 3283–3293.

38. Jacquin, M., Muller, P., Talingting-Pabalan, R., Cottet, H., Berret, J.F., Futterer, T., and Théodoly, O. (2007) Chemical analysis and aqueous solution properties of charged amphiphilic block copolymers PBA-*b*-PAA synthesized by MADIX. *J. Colloid Interface Sci.*, **316**, 897–911.

39. Muller, F., Guenoun, P., Delsanti, M., Demé, B., Auvray, L., Yang, J., and Mays, J.W. (2004) Spherical polyelectrolyte block copolymer micelles: Structural change in presence of monovalent salt. *Eur. Phys. J. E*, **15**, 465–472.

40. Israëls, R., Leermarkers, F.A.M., and Fleer, G.J. (1994) On the theory of grafted weak polyacids. *Macromolecules*, **27**, 3087–3093.

41. Zhulina, E.B., Birshtein, T.M., and Borisov, O.V. (1995) Theory of ionizable polymer brushes. *Macromolecules*, **28**, 1491–1499.

42. Wesley, R.D., Cosgrove, T., Thompson, L., Armes, S.P., Billingham, N.C., and Baines, F.L. (2000) Hydrodynamic layer of a polybase brush in the presence of salt. *Langmuir*, **16**, 4467–4469.

43. Guo, X. and Ballauff, M. (2001) Spherical polyelectrolyte brushes: a comparison between annealed and quenched brushes. *Phys. Rev. E*, **64**, 051406-1–051406-9.

44. Matejicek, P., Podhajecka, K., Humpolickova, J., Uhlik, F., Jelinek, K., Limpouchova, Z., Prochazka, K., and Spirkova, M. (2004) Polyelectrolyte behavior of polystyrene-block-poly(methacrylic acid) micelles in aqueous solutions at low ionic strength. *Macromolecules*, **37**, 10141–10154.

45. Kurihara, K., Kunitake, T., Higashi, N., and Niwa, M. (1992) Surface forces between monolayers of anchored poly(methacrylic acid). *Langmuir*, **8**, 2087–2089.

46. Currie, E.P.K., Sieval, A.B., Fleer, G.J., and Cohen Stuart, M.A. (2000) Polyacrylic acid brushes: surface pressure and salt-induced swelling. *Langmuir*, **16**, 8324–8333.

47. Miklavic, S.J. and Marcelja, S. (1988) Interactions of surfaces carrying grafted polyelectrolytes. *J. Phys. Chem.*, **92**, 6718–6722.

48. Misra, S., Varanasi, S., and Varanasi, P.P. (1989) A polyelectrolyte brush theory. *Macromolecules*, **22**, 4173–4179.

49. Israëls, R., Leermakers, F.A.M., Fleer, G.J., and Zhulina, E.B. (1994) Charged polymeric brushes: structure and scaling relations. *Macromolecules*, **27**, 3249–3261.

50. Romet-Lemonne, G., Daillant, J., Guenoun, P., Yang, J., and Mays, J.W. (2004) Thickness and density profiles of polyelectrolyte brushes: dependence on grafting density and salt concentration. *Phys. Rev. Lett.*, **93** (14), 148301-1–148301-4.

51. Wittmer, J. and Joanny, J.F. (1993) Charged diblock copolymers at interfaces. *Macromolecules*, **26**, 2691–2697.

52. Zhulina, E.B. and Borisov, O.V. (1997) Structure and interactions of weakly charged polyelectrolyte brushes: self-consistent field theory. *J. Chem. Phys.*, **107**, 5952–5967.

53. Tran, Y., Auroy, P., Lee, L.T., and Stamm, M. (1999) Polyelectrolyte brushes: counterion distribution and complexation properties. *Phys. Rev. E*, **60**, 6984–6990.

54. Groenewegen, W., Lapp, A., Egelhaaf, S.U., and van der Maarel, J.R.C. (2000) Counterion distribution in the coronal layer of polyelectrolyte diblock copolymer micelles. *Macromolecules*, **33**, 4080–4086.

55. Heinrich, M., Rawiso, M., Zilliox, J.G., Lesieur, P., and Simon, J.P. (2001) Small-angle X-ray scattering from salt-free solutions of star-branched polyelectrolytes. *Eur. Phys. J. E*, **4**, 131–142.

56. Balastre, M., Li, F., Schorr, P., Yang, J., Mays, J.W., and Tirrell, M.V. (2002) A study of polyelectrolyte brushes formed from adsorption of amphiphilic diblock copolymers using the surface force apparatus. *Macromolecules*, **35**, 9480–9486.

57. Alfrey, T., Berg, P.W., and Morawetz, H. (1951) The counterion distribution

in solutions of rod-shaped polyelectrolytes. *J. Polym. Sci.*, **7**, 543–547.

58. Fuoss, R.M., Katchalsky, A., and Lifson, S. (1951) The potential of an infinite rod-like molecule and the distribution of the counterions. *Proc. Natl. Acad. Sci. U.S.A.*, **37**, 579–589.

59. Oosawa, F. (1971) *Polyelectrolytes*, Marcel Dekker, New York.

60. Manning, G.S. (1969) Limiting laws and counterion condensation in polyelectrolyte solutions I. Colligative properties. *J. Chem. Phys.*, **51** (3), 924–933.

61. Le Bret, M. and Zimm, B. (1984) Distribution of counterions around a cylindrical polyelectrolyte and Manning condensation theory. *Biopolymers*, **23**, 287–312.

62. Belloni, L., Drifford, M., and Turq, P. (1984) Counterion diffusion in polyelectrolyte solutions. *Chem. Phys.*, **83**, 147–154.

63. Holm, C., Joanny, J.F., Kremer, K., Netz, R.R., Reineker, P., Seidel, C., Vilgis, T.A., and Winkler, R.G. (2004) Polyelectrolyte theory. *Adv. Polym. Sci.*, **166**, 67–111.

64. Muthukumar, M. (2004) Theory of counterion condensation on flexible polyelectrolytes: adsorption mechanism. *J. Chem. Phys.*, **120**, 9343–9350.

65. Muller, F., Fontaine, P., Delsanti, M., Belloni, L., Yang, J., Chen, Y.J., Mays, J.W., Lesieur, P., and Tirrell, M. (2001) Counterion distribution in a spherical charged sparse brush. *Eur. Phys. J. E*, **6**, 109–115.

66. Jusufi, A., Likos, C.N., and Löwen, H. (2002) Counterion-induced entropic interactions in solutions of strongly stretched osmotic polyelectrolyte stars. *J. Chem. Phys.*, **116**, 11011–11027.

67. Roger, M., Guenoun, P., Muller, F., Belloni, L., and Delsanti, M. (2002) Monte Carlo simulations of star-branched polyelectrolyte micelles. *Eur. Phys. J. E*, **9** (4), 313–326.

68. Ahrens, H., Förster, S., Helm, C.A., Kumar, N.A., Naji, A., Netz, R.R., and Seidel, C. (2004) Nonlinear osmotic brush regime: experiments, simulation, and scaling theory. *J. Phys. Chem. B*, **108**, 16870–16876.

69. Csajka, F.S. and Seidel, C. (2000) Strongly charged polyelectrolyte brushes: a molecular dynamics study. *Macromolecules*, **33**, 2728–2739.

70. Seidel, C. (2003) Strongly stretched polyelectrolyte brushes. *Macromolecules*, **36**, 2536–2543.

71. Naji, A., Netz, R.R., and Seidel, C. (2003) Non-linear osmotic brushe regime: simulations and mean-field theory. *Eur. Phys. J. E*, **12**, 223–237.

72. Jiang, T., Li, Z., and Wu, J. (2007) Structure and swelling of grafted polyelectrolytes: predictions from a nonlocal density functional theory. *Macromolecules*, **40**, 334–343.

73. Hariharan, R., Biver, C., and Russel, W.B. (1998) Ionic strength effects in polyelectrolyte brushes: the counterion correction. *Macromolecules*, **31**, 7514–7518.

74. van der Maarel, J.R.C., Groenewegen, W., Egelhaaf, S.U., and Lapp, A. (2000) Salt-induced contraction of polyelectrolyte diblock copolymer micelles. *Langmuir*, **16**, 7510–7519.

75. Rouzina, I.F. and Bloomfield, V.A. (1996) Macroion attraction by correlated counterion fluctuations: application to condensation of DNA. *Biophys. J.*, **70**, MP418–MP418.

76. Tang, J.X. and Janmey, P.A. (1996) The polyelectrolyte nature of F-actin and the mechanism of actin bundle formation. *J. Biol. Chem.*, **271**, 8556–8563.

77. Angelini, T.E., Liang, H., Wriggers, W., and Wong, G.C.L. (2003) Like-charge attraction between polyelectrolytes induced by counterion charge density waves. *Proc. Natl. Acad. Sci. U.S.A*, **100**, 8634–8637.

78. Csajka, F.S., Netz, R.R., Seidel, C., and Joanny, J.F. (2001) Collapse of polyelectrolyte brushes: scaling theory and simulations. *Eur. Phys. J. E*, **4**, 505–513.

79. Santangelo, C.D. and Lau, A.W.C. (2004) Effects of counterion fluctuations in a polyelectrolyte brush. *Eur. Phys. J. E*, **13**, 335–344.

80. Jiang, T. and Wu, J. (2008) Ionic effects in collapse of polyelectrolyte brushes. *J. Phys. Chem. B*, **112**, 7713–7720.

81. Mei, Y., Lauterbach, K., Hoffmann, M., Borisov, O.V., Ballauff, M., and

Jusufi, A. (2006) Collapse of spherical polyelectrolyte brushes in the presence of multivalent counterions. *Phys. Rev. Lett.*, **97**, 158301.

82. Rabin, Y., Fredrickson, G.H., and Pincus, P. (1991) Compression of grafted polyelectrolyte layers. *Langmuir*, **7**, 2428–2430.

83. Kegler, K., Salomo, M., and Kremer, F. (2007) Forces of interaction between DNA-grafted colloids: an optical tweezer measurement. *Phys. Rev. Lett.*, **98**, 058304.

84. Watanabe, H., Patel, S.S., Argillier, J.F., Parsonage, E.E., Mays, J., Dan-Brandon, N., and Tirrell, M. (1992) *Mater. Res. Soc. Proc.*, **249**, 255.

85. Kelley, T.W., Schorr, P.A., Johnson, K.D., Tirrell, M., and Frisbie, C.D. (1998) Direct force measurements at polymer brush surfaces by atomic force microscopy. *Macromolecules*, **31**, 4297–4300.

86. Abraham, T., Giasson, S., Gohy, J.F., and Jérôme, R. (2000) Direct measurements of interactions between hydrophobically anchored strongly charged polyelectrolyte brushes. *Langmuir*, **16**, 4286–4292.

87. Abe, T., Higashi, N., Niwa, M., and Kurihara, K. (1999) Density dependent jump in compressibility of polyelectrolyte brush layers revealed by surface forces measurement. *Langmuir*, **15**, 7725–7731.

88. Li, F., Balastre, M., Schorr, P., Argillier, J.F., Yang, J., Mays, J.W., and Tirrell, M. (2006) Differences between tethered polyelectrolyte chains on bare mica and hydrophobically modified mica. *Langmuir*, **22**, 4084–4091.

89. Tamashiro, M.N., Hernandez-Zapata, E., Schorr, P.A., Balastre, M., Tirrell, M., and Pincus, P. (2001) Salt dependence of compression normal forces of quenched polyelectrolyte brushes. *J. Chem. Phys.*, **115**, 1960–1969.

90. Jusufi, A., Likos, C.N., and Löwen, H. (2002) Conformations and interactions of star branched polyelectrolytes. *Phys. Rev. Lett.*, **88**, 018301.

91. Muller, F., Romet-Lemonne, G., Delsanti, M., Mays, J.W., Daillant, J., and Guenoun, P. (2005) Salt-induced contraction of polyelectrolyte brushes. *J. Phys. Condens. Matter*, **17**, S3355–S3361.

92. Grest, G.S., Fetters, L.J., Huang, J.S., and Richter, D. (1996) *Advances in Chemical Physics*, Vol. XCIV (eds I. Prigogine and S.A. Rice), John Wiley & Sons, Inc.

93. Korobko, A.V., Jesse, W., Egelhaaf, S.U., Lapp, A., and van der Maarel, J.R.C. (2004) Do spherical polyelectrolyte brushes interdigitate? *Phys. Rev. Lett.*, **93**, 177801.

94. Shusharina, N.P. and Rubinstein, M. (2008) Concentration regimes in solutions of polyelectrolyte stars. *Macromolecules*, **41**, 203–217.

95. Raviv, U., Giasson, S., Kampf, N., Gohy, J.F., Jérôme, R., and Klein, J. (2003) Lubrication by charged polymers. *Nature*, **425**, 163–165.

96. Hehmeyer, O.J. and Stevens, M.J. (2005) Molecular dynamics of grafted polyelectrolytes on two apposing walls. *J. Chem. Phys.*, **122**, 134909.

97. Arun Kumar, N. and Seidel, C. (2007) Interaction between two polyelectrolyte brushes. *Phys. Rev. E*, **76**, 020801.

98. Sirchabesan, M. and Giasson, S. (2007) Mesoscale simulations of the behavior of charged polymer brushes under normal compression and lateral shear forces. *Langmuir*, **23**, 9713–9721.

99. Tran, Y. and Auroy, P. (2001) Complexation and distribution of counterions in a grafted polyelectrolyte layer. *Eur. Phys. J. E*, **5**, 65–79.

100. Konradi, R. and Rühe, J. (2004) Interaction of poly(methacrylic acid) brushes with metal ions: an infrared investigation. *Macromolecules*, **37**, 6954–6961.

101. Mei, Y., Lu, Y., Polzer, F., Ballauff, M., and Drechsler, M. (2007) Catalytic activity of palladium nanoparticles encapsulated in spherical polyelectrolyte brushes and core-shell microgels. *Chem. Mater.*, **19**, 1062–1069.

102. Constancis, A., Meyrueix, R., Bryson, N., Huille, S., Grosselin, J.M., Gulik-Krzywicki, T., and Soula, G. (1999) Macromolecular colloids of diblock poly(amino acids) that bind

insulin. *J. Colloid Interface Sci.*, **217**, 357–368.

103. Wittemann, A. and Ballauff, M. (2006) Interaction of proteins with linear polyelectrolytes and spherical polyelectrolyte brushes in aqueous solution. *Phys. Chem. Chem. Phys.*, **8**, 5269–5275.

104. Zhou, F., Biesheuvel, P.M., Choi, E.-Y., Shu, W., Poetes, R., Steiner, U., and Huck, W.T.S. (2008) Polyelectrolyte brush amplified electroactuation of microcantilevers. *Nano Lett.*, **8**, 725–730.

105. Tam, T.K., Ornatska, M., Pita, M., Minko, S., and Katz, E. (2008) Polymer brush-modified electrode with switchable and tunable redox activity for bioelectronic applications. *J. Phys. Chem. C*, **112**, 8438–8445.

106. Shusharina, N.P. and Linse, P. (2001) Oppositely charged polyelectrolytes grafted onto planar surface: mean-field lattice theory. *Eur. Phys. J. E*, **6**, 147–155.

107. Houbenov, N., Mink, S., and Stamm, M. (2003) Mixed polyelectrolyte brush from oppositely charged polymers for switching of surface charge and composition in aqueous environment. *Macromolecules*, **36**, 5897–5901.

108. Plamper, F.A., Walther A., Müller, A.H.E., and Ballauff, M. (2006) Nanoblossoms: light-induced conformational changes of cationic polyelectrolyte stars in the presence of multivalent counterions. *Nano Lett.*, **7**, 167–171.

109. Raviv, U., Giasson, S., Kampf, N., Gohy, J.F., Jerome, R., and Klein, J. (2008) Normal and frictional forces between surfaces bearing polyelectrolyte brushes. *Langmuir*, **24**, 8678–8687.

110. Jiang, T., and Wu, J. (2008) Self-organization of multivalent counterions in polyelectrolyte brushes. *J. Chem. Phys.*, **129**, 084903.

111. Biesheuvel, P.M., de Vos, W.M., and Amoskov V.M. (2008) Semianalytical continuum model for nondilute neutral and charged brushes including finite stretching. *Macromolecules*, **41**, 6254–6259.

7

Preparation of Polymer Brushes Using "Grafting-From" Techniques

Zhiyi Bao, Ying Zheng, Gregory L. Baker, and Merlin L. Bruening

7.1
Introduction to Polymer Brushes and Their Formation

Polymer brushes are assemblies of polymer chains that are tethered by one end to a surface or an interface [1]. Typically, such assemblies present unique polymer architectures because the distance between neighboring anchored chains is less than the sum of their radii of gyration in solution, and the chains are forced to extend from the surface [2, 3]. Early applications of such brushes involved prevention of flocculation of colloidal particles [4–7]. More recently, polymer brushes have been employed to enhance the dispersion of nanomaterials in organic–inorganic hybrid nanocomposites [8–10], as well as in applications in gas and liquid separations [11, 12], protein purification [13–15], protein immobilization [16, 17], microelectronics [18], controlled cell growth [19, 20], biomimetic material fabrication [21], and drug delivery [22]. Brushes that respond to changes in pH [23], solvent [24–26], ionic strength [27, 28], or temperature [20, 28–30] can also be synthesized [31], and such materials could eventually prove useful in drug-delivery vehicles and temperature-controlled gates or switches [32–34].

The early processes for forming polymer brushes frequently employed block-copolymers where one block is adsorbed strongly to an interface to immobilize the polymer [35]. To achieve more stable interfaces, subsequent research investigated covalent linking of brushes to surfaces. In general, covalent attachment of brushes to surfaces can be performed by either "grafting-to" or "grafting-from" methods (Figure 7.1) [36]. The "grafting-to" approach refers to tethering preformed polymer chains onto surfaces, which facilitates formation of polymer brushes with well-defined molecular masses. This technique is described in more detail in Chapter 6. In contrast, the "grafting-from" method involves polymerization from initiators that are anchored to the surface. Since in the "grafting-from" technique the only limit to the polymer chain propagation is the diffusion of monomers to the reactive sites in the growing film, this method usually generates thicker coatings with higher grafting densities than the "grafting-to" approach [37]. (Film growth in the "grafting-to" method is often

Functional Polymer Films, First Edition. Edited by Wolfgang Knoll and Rigoberto C. Advincula.

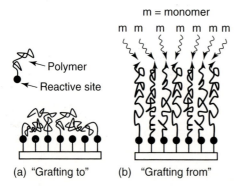

Figure 7.1 Schematic diagram showing the formation of polymer brushes using "grafting-to" (a) and "grafting-from" (b) techniques. (Figure courtesy of Dr. Jamie Dunn.)

limited by the ability of polymer chains to diffuse through previously bound chains to reactive sites.)

Methods such as UV-assisted grafting or plasma grafting where radicals are created on a surface might also be considered as "grafting-from" strategies, but these techniques are less controlled than polymerization from anchored initiators and will not be considered here [38, 39]. Because a variety of initiators can be attached to surfaces, a number of types of polymerizations can be performed including free radical, cationic, anionic, reversible addition–fragmentation chain transfer (RAFT), atom-transfer radical, ring-opening metathesis, and nitroxide-mediated polymerization (NMP). Controlled radical polymerization techniques [40, 41] initiated solely from the surface are particularly attractive because they allow (i) good control over the film thickness by variation of the polymerization time, (ii) simplified separation and purification since the active radicals are confined primarily to the surface and little polymer is formed in solution [42], (iii) tuning of grafting densities by controlling the surface coverage of the initiator [43, 44], and (iv) the ability to create a variety of brush architectures including linear tethered polymers, comb-like copolymers, cross-linked films, and block-copolymer coatings.

Below, we discuss several different types of surface-initiated polymerization and the resulting films. We begin with free-radical polymerization because this was the initial technique employed, and discussion of more controlled polymerization follows. Because atom-transfer radical polymerization (ATRP) is perhaps the most widely used method for forming brushes, this technique is discussed in more detail. In the section on ATRP, we also describe tailoring of brush density by varying the density of immobilized initiators on a surface and formation of brushes on a wide range of substrates such as porous membranes and nanoparticles. Growth of brushes in porous substrates is particularly intriguing for creating membranes that rapidly bind large amounts of proteins or other biomacromolecules.

7.2
Methods for Synthesis of Polymer Brushes

7.2.1
Surface-Initiated Free-Radical Polymerization

The first step in surface-initiated radical polymerization is the immobilization of the initiator. In an early example of free-radical polymerization from surfaces, Boven *et al.* [45] initiated polymerization of methyl methacrylate (MMA) from immobilized azo initiators. Sugawara and Matsuda [46] used a similar strategy to graft polystyrene (PS) from poly(vinyl alcohol) films and poly(acrylamide) from poly(ethylene terephthalate). The reactive azo initiators were attached to the surface by coating the substrate with poly(allylamine) that had been derivatized with azo functionalities.

Minko *et al.* [47–52] studied surface-initiated radical polymerization using both theoretical and experimental approaches. In their experimental work, azo or peroxide initiators were attached to solid substrates such as silicon wafers by either physisorption or chemical immobilization. Chemical immobilization consisted of priming surfaces with (3-glycidyloxypropyl)trimethoxysilane followed by reaction of the epoxide with 4,4′-azo-*bis*(4-cyanovaleric acid). Surface-initiated radical polymerization from these initiators was followed by *in situ* ellipsometric measurements of the amount of grafted chains. The resulting kinetics showed a linear dependence of the polymerization rate on the surface concentration of the initiator and an inverse square root dependence on the initiator concentration in solution, which is consistent with free-radical polymerization.

To simplify initiator attachment and achieve a high density of initiation sites, Rühe and coworkers [53–56] developed the one-step initiator-anchoring strategy shown in the first reaction of Scheme 7.1. Their initiator contains three important components: (i) an azo group that generates free radicals upon heating or UV irradiation, (ii) a chlorosilane that allows the initiator to be anchored to the surface through reaction with the silanol groups of silica, and (iii) an ester that can be hydrolyzed to detach the polymer brushes from the surface. The initiator was covalently linked to the surface, and following free-radical polymerization of styrene or other monomers, the ester bonds that connected the polymers to the surface were cleaved (Scheme 7.1). The molecular weights of the polymers were determined to allow comparison between free-radical polymerization in solution and at a surface, and the density of chains on the surface was calculated from the molecular weight and mass of the grafted polymer and compared to the density of initiators on the surface. These studies showed that the average distance between tethered PS chains was 2–3 nm, smaller than the radii of gyration of the corresponding polymer molecules [56].

Wittmer *et al.* [57] predicted significant differences between polymer brushes grown from surfaces and polymers generated in solution. They suggested that immobilized long chains are more efficient at adding monomers than short chains because they are more accessible to monomers, and thus polymer brushes formed

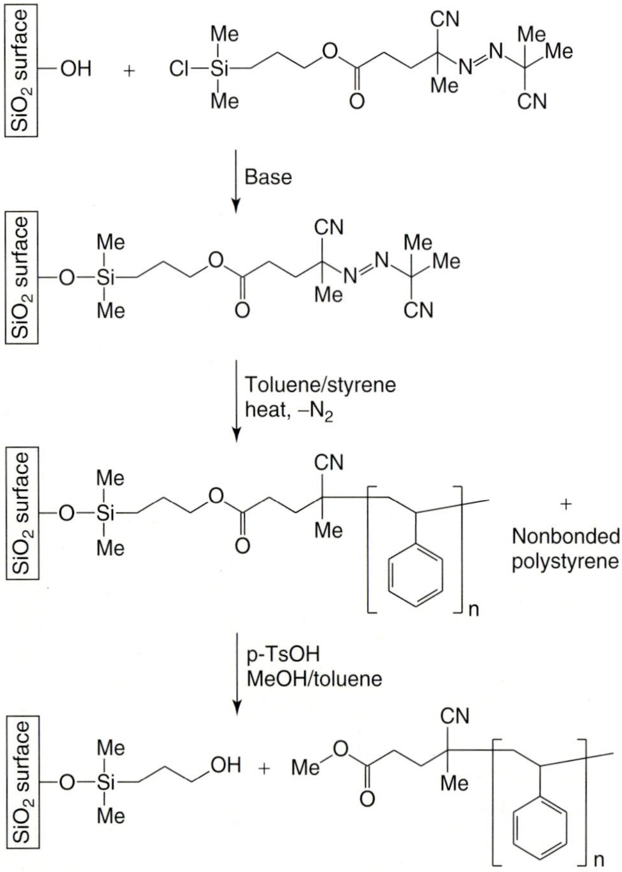

Scheme 7.1 Rühe's method for initiator attachment to silica, growth of polystyrene brushes, and cleavage of the polystyrene chains from the surface [55].

at the surface should have a higher polydispersity compared to the same reaction occurring in solution. The polydispersity indices (PDIs) of the detached polymer brushes prepared by Prucker and Rühe ranged from 1.5 to 2, close to the PDI expected for free-radical polymerization in solution. Consequently, they concluded that surface immobilization does not cause excessive broadening of the molecular weight distribution.

Rühe and coworkers [58] extended their surface free-radical polymerization strategy to the preparation of block-copolymer brushes where one block was synthesized by ring-opening polymerization (ROP). A poly(ε-caprolactone) macroinitiator containing azo groups was physisorbed on a silicon oxide surface and used to initiate the radical polymerization of a second monomer. This simple physisorbed macroinitiator system allows the creation of hydrophobic layers on hydrophilic surfaces.

7.2.2
Surface-Initiated Cationic Polymerizations

In the early 1980s, Vidal *et al.* [59, 60] used surface-initiated cationic polymerization to graft polyisobutylene on a silica surface. 2-(Chloromethylphenyl)ethyldimethyl chlorosilane was anchored to the silica surface, and the reaction of diethylaluminum chloride with the immobilized initiator produced carbocationic species that initiated the polymerization of isobutylene. High grafting rates were obtained with conversions as high as 30% in 60 min.

Jordan and coworkers [61, 62] reported the surface-initiated cationic polymerization of N-propionylethylenimine (PEI) from gold surfaces. Scheme 7.2 shows their strategy for coating gold nanoparticles, which is similar to their procedure for coating flat gold surfaces. They first formed a hydroxy-terminated self-assembled monolayer (SAM) and then exposed the monolayer to trifluoromethanesulfonic anhydride vapor to convert the hydroxy groups to triflate functional groups. On flat gold surfaces, subsequent polymerization of PEI for seven days under reflux yielded a 10 nm-thick polymer brush. Reaction of these films with *N,N*-di-*n*-octadecylamine

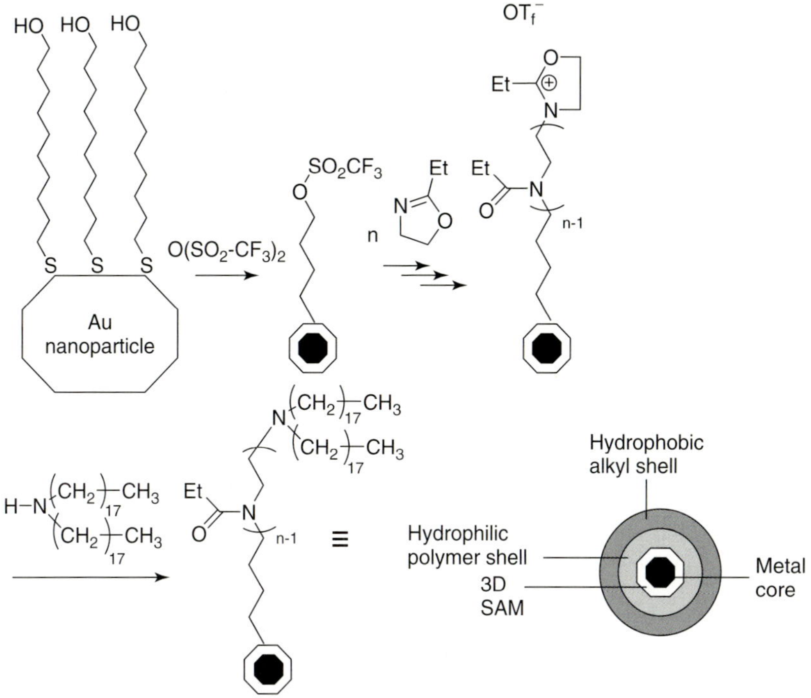

Scheme 7.2 Modification of gold nanoparticles with an amphiphilic coating using surface-initiated polymerization of 2-ethyl-2-oxazoline and termination with *N,N*-di-*n*-octadecylamine [62].

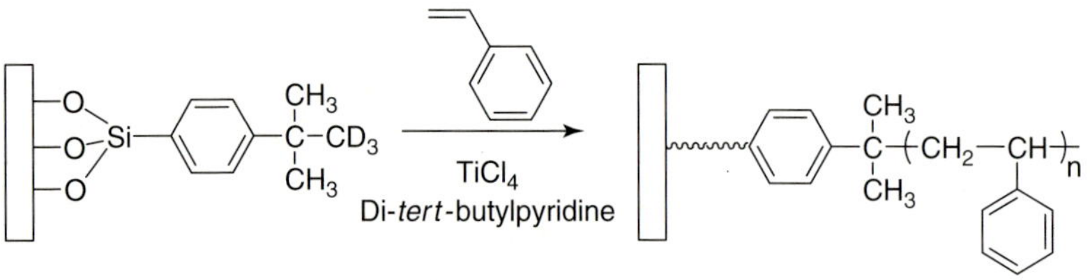

Scheme 7.3 Growth of polystyrene brushes by living cationic polymerization as described by Zhao and Brittain [63].

yielded a hydrophobic surface. However, the very slow polymerization rate in these studies may make this method difficult in practice.

Zhao and Brittain [63] reported the synthesis of 30–40 nm-thick PS brushes in one hour using surface-initiated cationic polymerization. 2-(4-Trichlorosilylphenyl)-2-methoxy-d$_3$-propane was immobilized on a silica substrate, and polymerization of styrene was initiated by addition of TiCl$_4$ (Scheme 7.3). The reaction was carried out at $-78\,^{\circ}$C to suppress the chain-transfer reactions characteristic of cationic polymerizations. The initiator efficiency estimated by attenuated total reflectance FTIR spectroscopy was $\sim$7%, but low initiator efficiencies, which may result from steric constraints, are typical of polymerizations from surfaces [42, 56].

More recently, Dronavajjala *et al.* [64] developed a simple strategy for cationic polymerization from a cyano-terminated SAM. Upon immersion of the SAM in a solution containing [Pd(C$_2$H$_5$CN)$_4$](BF$_4$)$_2$, the cyano group of the monolayer displaced a propanenitrile ligand to anchor a Pd(II) complex that would initiate polymerization of 4-methoxystyrene. A 24-h polymerization of saturated monomer vapor gave a film thickness of 24 nm.

Several studies also recently demonstrated cationic polymerization from clays [65, 66]. Overall, however, the stringent conditions associated with cationic polymerization and an inability to rapidly grow thick films will likely limit the practicality of this technique.

7.2.3
Surface-Initiated Anionic Polymerizations

Due to its lack of chain transfer and termination reactions, anionic polymerization provides the highest control over polymer architectures. Surface-initiated anionic polymerization has been used to grow polymer brushes from various small particles such as silica gel, carbon black, and silicon wafers [67]. Jordan *et al.* [68] used lithiated biphenyl SAMs to initiate the polymerization of styrene from gold substrates (Scheme 7.4a) and obtained uniform 18 nm thick films after a three-day reaction. Foster *et al.* [69] used diphenylethylene (DPE)-containing SAMs to generate polyisoprene-*b*-poly(ethylene oxide) (PI-*b*-PEO) copolymer brushes from silicon surfaces, creating a hydrophilic surface by introducing the PEO block (Scheme 7.4b). Again, only thin films (24 nm) were obtained. Later, Advincula *et al.*

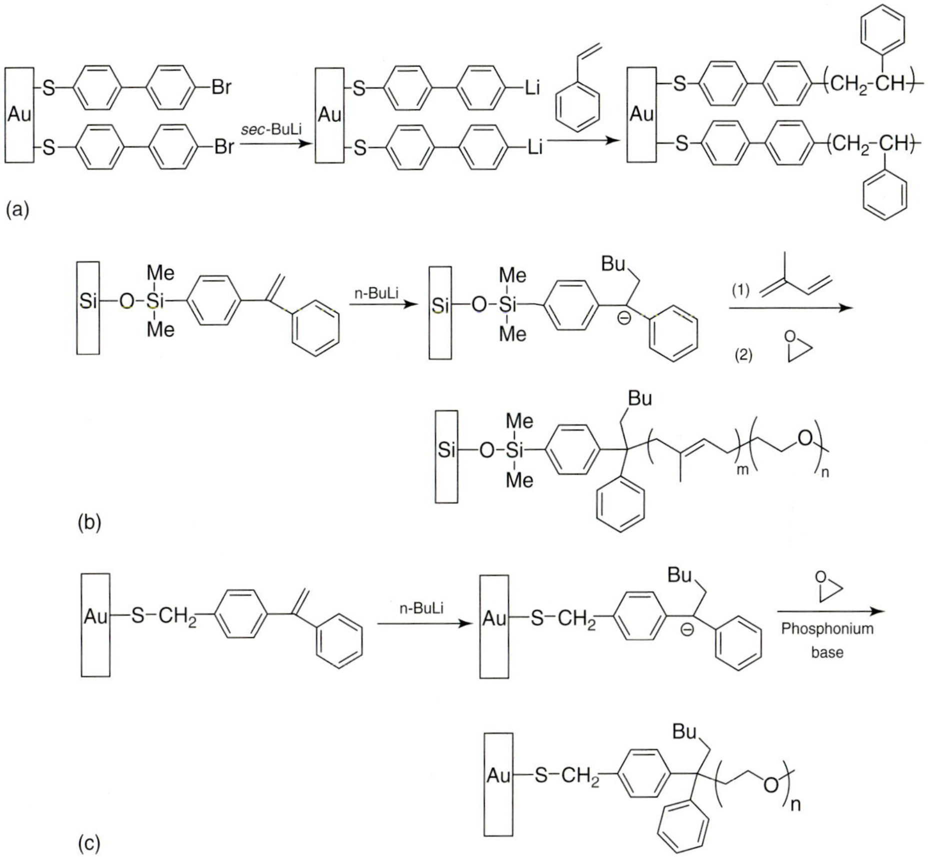

(a)

(b)

(c)

Scheme 7.4 Several strategies for growth of polymer brushes using anionic polymerization [68, 69, 73].

[70] activated DPE SAMs with *n*-butyllithium to initiate anionic polymerization of styrene. A five-day reaction gave a PS film that was only 23 nm thick, however, the formation of PS-*b*-PI copolymer brushes demonstrated the living nature of the polymerization. Polymer brushes were also grown from DPE derivatives anchored through quaternary ammonium tethers to clay surfaces [71]. Zheng *et al.* [72] grafted polybutadiene brushes to the surface of PS nanoparticles, and reported that the grafting density and degree of polymerization greatly affect the dispersion of nanoparticles in matrices and the mechanical properties of nanocomposites. Advincula and coworkers [73] also grew 14-nm thick PEO homopolymers (Scheme 7.4c) and polyisoprene-*b*-poly(methyl methacrylate) copolymer brushes on gold surfaces.

Zhou *et al.* [74] polymerized styrene by anionic polymerization from initiators that were immobilized using cation exchange. A linear relationship between the monomer concentration and the M_n of the cleaved polymers was consistent with a living anionic polymerization mechanism. Ingall *et al.* [75] polymerized acrylonitrile using a similar strategy. A SAM formed from 3-bromopropyltrichlorosilane was lithiated with lithium di-*tert*-butylbiphenyl and subsequent addition of monomer to the system initiated the anionic polymerization. Polymerization for eight days yielded a tethered poly(acrylonitrile) film with thicknesses up to 245 nm.

From these examples, we see that the traditional monomers used for anionic polymerizations can be applied to surface-initiated polymerization, however, the need to use long reaction times restricts some applications of anionic polymerization for the synthesis of polymer brushes. Stringent polymerization conditions are also required.

7.2.4
Surface-Initiated Ring-Opening Polymerization

Surface-initiated ROP is an attractive route for coating surfaces with thin layers of polycaprolactone (PCL), polylactide, and other polymers. Husseman and coworkers [40] prepared a SAM terminated with di(ethylene glycol) moieties, and using the pendent OH groups for initiation, they carried out the aluminum alkoxide catalyzed ROP of ε-caprolactone. This procedure yielded 70-nm thick PCL brushes after polymerization for a few hours at room temperature. Terminating the SAM with di(ethylene glycol) gave more reproducible polymer brush growth and better long-term stability than initiation from simple long-chain alcohol SAMs.

In related work, Choi and Langer [76] formed an oligo(ethylene glycol) terminated SAM on gold, and used tin(II) (2-ethylhexanoate)$_2$ to catalyze the ROP of L-lactide from gold and silicon substrates (Scheme 7.5). Poly(lactic acid) (PLA) is an important biodegradable polymer used in medical applications, and PLA brushes present a possible rout to well-defined surfaces with controlled release properties. Polymerization for three days at 40 °C provided PLA brushes up to 12 nm thick, and 70-nm thick PLA brushes were obtained from silicon surfaces after polymerization for three days at 80 °C. The PLA brushes were reported to be chiral and crystalline on the surface. ROP has also been utilized to modify Co nanoparticles [77]. Although ROP is attractive for modifying surfaces, the range of monomers that can be utilized with this technique is limited.

Ring-opening metathesis polymerization (ROMP) extends the scope of ROP and allows rapid formation of polymer brushes. Whitesides and coworkers [78] used a ruthenium catalyst immobilized on silicon wafers to grow brushes from norbornene-derived monomers (Scheme 7.6). The surface-bound catalytic sites were produced by forming a trichlorosilane SAM containing norbornene groups and then exposing the SAM to a solution of a Grubbs-type ROMP catalyst. Addition of the monomer initiated a rapid, but controlled polymerization that produced

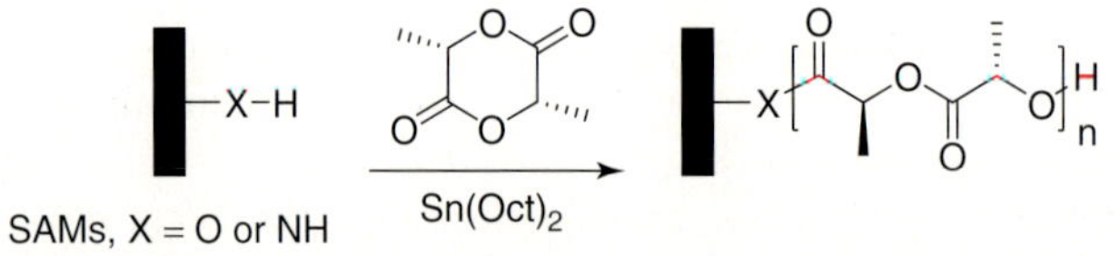

SAMs, X = O or NH   Sn(Oct)$_2$

Scheme 7.5   Surface-initiated ring opening polymerization of lactide [76].

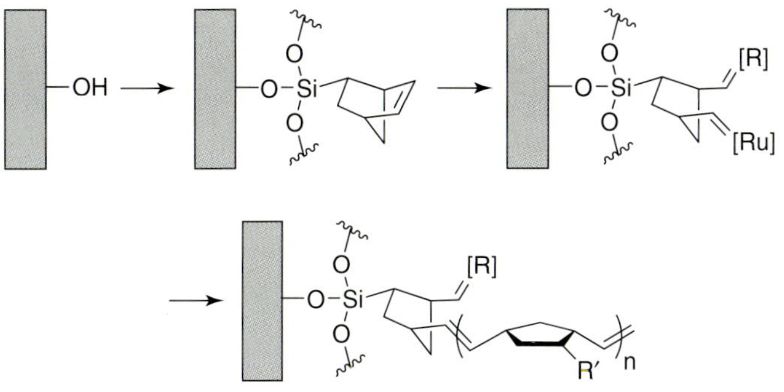

Scheme 7.6 Surface-initiated ring opening metathesis polymerization of functionalized norbornenes [78].

90-nm thick brushes in 30 min. The formation of block-copolymer brushes and the use of microcontact printing to produce patterned surfaces was also possible [78, 79]. In some cases, polymerization can proceed with vapor-phase monomers [80]. Grubbs and coworkers [81] also grew poly(norbornene) from silicon wafer surfaces using surface-initiated ROMP and an alternative initiator attachment scheme. A direct Si–C bond to the surface was used to anchor the initiator instead of the Si–O bond formed via condensation of chlorosilanes. The polymer brushes grown from these initiators had thicknesses up to 5.5 μm. More recently, Moon and Swager [82] used ROMP to grow brushes with poly(p-phenylene ethynylene) "molecular wire" side chains for applications in chemical sensing. Jennings and coworkers [83] grew poly(alkylnorbornene) films from surfaces and examined their barrier properties. Unfortunately, however, the rate of polymerization decreased as the length of the alkyl side chain increased.

7.2.5
Nitroxide-Mediated Polymerization (NMP)

The living character of NMP is based on the reversible capping of the active chain-end radical with a nitroxide leaving group. Husseman and coworkers [84] described the first example of NMP applied to the synthesis of polymer brushes (Scheme 7.7a). They first attached alkoxyamine initiators onto a Si surface and then heated the system to 120 °C to initiate radical polymerization. Under these conditions, over 100 nm of PS brush was produced in 16 h. The stable nitroxide radical, 2,2,6,6-tetramethylpiperidinyloxy (TEMPO), cleaves during the initiating process and reversibly caps the chain-end radicals to control radical propagation. The addition of free alkoxyamine initiator to the polymerization solution provides better control over the molecular weight, but induces polymerization in solution. Later, Hawker and coworkers [85] formed cross-linked, hollow nanoparticles by using NMP (Scheme 7.7b) to prepare random copolymer brushes of styrene and 4-vinylbenzocyclobutene on silicon nanoparticles. The benzocyclobutenes

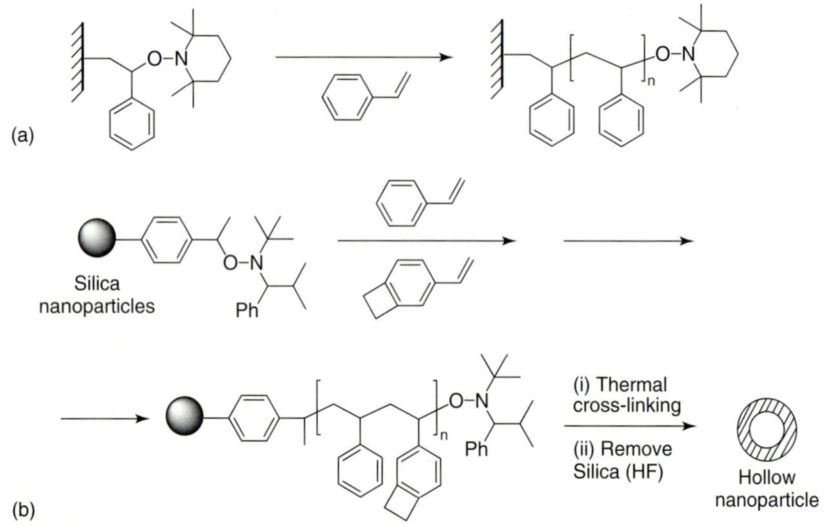

Scheme 7.7 (a) Growth of polystyrene brushes by nitroxide-mediated polymerization (NMP). (b) NMP-mediated polymerization of a random copolymer on Si and subsequent formation of cross-linked hollow nanoparticles [85].

can cross-link upon heating to 220 °C, and dissolution of the silica core using hydrofluoric acid, gives hollow cross-linked polymer spheres, which may be useful for drug delivery. Hawker and coworkers [86] also combined photolithography with NMP to yield patterned polymer brushes with well-defined hydrophobic and hydrophilic domains. To functionalize films, end-groups of polymer chains can be reacted with molecules such as biotin [87]. NMP can also be performed from initiators on clay to allow dispersion of exfoliated clays in a polymer matrix [88]. NMP is certainly a versatile technique for brush formation, but it does typically require temperatures >100 °C, which is incompatible with many polymeric substrates.

7.2.6
Reversible Addition-Fragmentation Chain-Transfer (RAFT) Polymerization

RAFT is another important technique for controlled radical polymerization. Chain growth is initiated using a conventional radical initiator such as 2,2′-azobisisobutyronitrile (AIBN), but propagation is mediated by a dithioester chain transfer agent that reversibly adds to chain ends to provide the polymerization its living character. Baum and Brittain [89] synthesized poly(methyl methacrylate) (PMMA), poly(N,N-dimethylacrylamide), and PS brushes from silica surfaces using surface-initiated RAFT polymerization as shown in Scheme 7.8. The initiator was anchored via formation of a monolayer containing an azo initiator or a dithiobenzoate group. Although RAFT polymerization is relatively slow compared to techniques such as ATRP and NMP, it is highly living, as demonstrated by easy

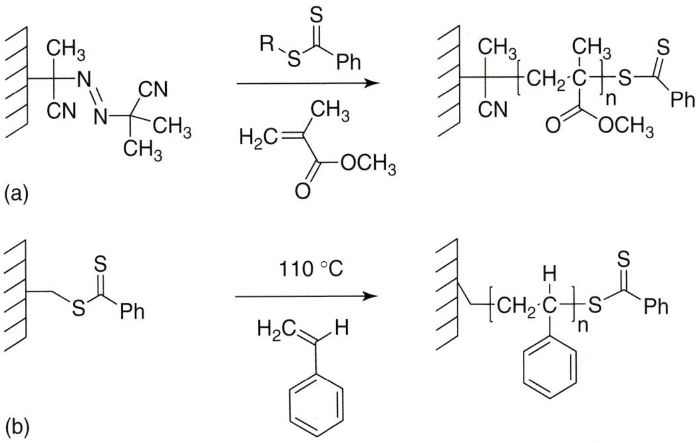

Scheme 7.8 Growth of polymer brushes by RAFT polymerization. (a) PMMA, (b) PS [89].

re-initiation of the polymer chains. Yuan and coworkers [90] performed RAFT from silicon wafers and were able to grow 20-nm thick films in 24 h, and RAFT polymerization has also been demonstrated from nanotubes, clays, and polymeric substrates [91–93]. Compared to ATRP, however, there are relatively few studies describing the use of RAFT polymerization for forming polymer brushes.

7.3
Surface-Initiated ATRP

7.3.1
Synthesis of Homopolymers

Because of its wide compatibility with various functionalized monomers, and its controlled nature, surface-initiated ATRP has become one of the most common "grafting-from" approaches for surface modification. Of the "grafting-from" methods described earlier, surface-initiated anionic and cationic polymerizations require rigorously dry conditions, and other controlled radical polymerization methods such as NMP and RAFT either require more complex initiator attachment steps or relatively high temperatures. The attachment of ATRP initiators to Au and silicon substrates is straightforward, as shown in Scheme 7.9.

In its most typical form, ATRP from a surface occurs when a halide atom is transferred from an initiator to a Cu complex with simultaneous oxidation of Cu(I) to Cu(II), as shown in Scheme 7.10 [94]. The radical is immobilized on the surface and if minimal chain transfer occurs, it is possible to avoid extensive polymerization in solution and grow polymer primarily at the surface. Under appropriate conditions, the equilibrium in Scheme 7.10 lies to the left and the chain is dormant most of the time. This results in a low concentration of radicals, so termination is minimized and the polymerization rate is nearly constant with

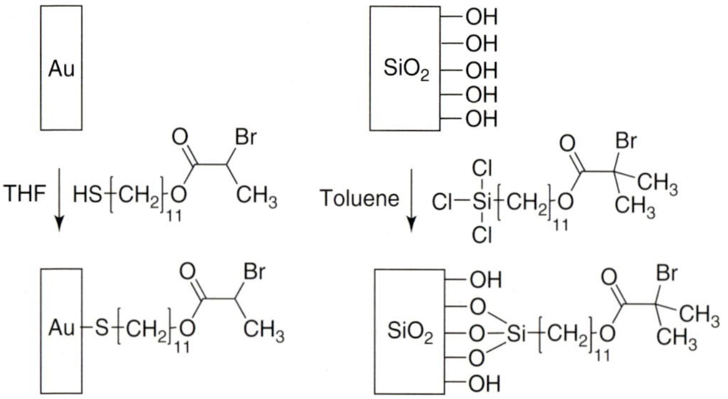

Scheme 7.9 Attachment of ATRP initiators to Au and SiO$_2$ surfaces.

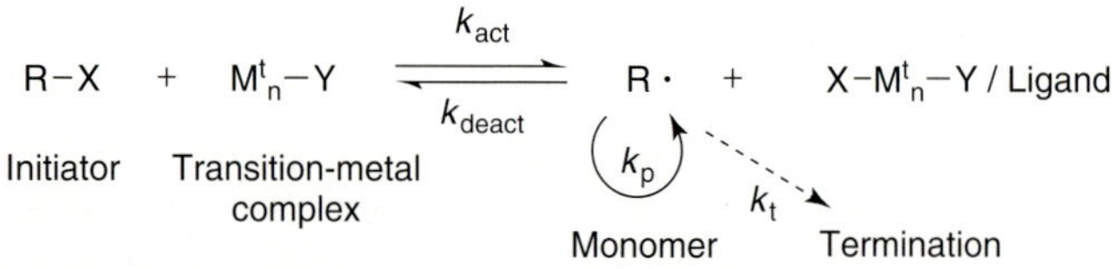

Scheme 7.10 Proposed mechanism of ATRP [94].

time. In polymerization from a surface, controlled polymerization is manifest by a constant rate of film growth. However, in many cases growth rates obtained when using ATRP from a surface are not constant, suggesting that termination does occur [95].

Fukuda and coworkers [96] published one of the first examples of surface-initiated ATRP. Using the Langmuir–Blodgett (LB) technique, they formed a well-ordered monolayer with aryl sulfonyl chloride head groups on a surface. Immersing the substrate in monomer and adding CuCl initiated the ATRP of MMA from the sulfonyl chlorides. The polymerization was not well controlled, but the addition of free initiator to the polymerization solution increased the Cu(II) concentration, the deactivation rate, and control of the polymerization. Concurrently, PMMA formed in solution and was characterized by conventional methods. However, the formation of polymer in solution necessitates extensive washing to remove physically adsorbed free polymer. By adding a Cu(II) complex instead of free initiator as a deactivator, Matyjaszewski *et al.* [97] achieved controlled growth of PS, PMMA, and poly(methacrylate) (PMA) brushes. A linear relationship between the polymer brush thickness and polymerization time confirmed controlled polymerization.

One of the drawbacks to ATRP is that growth rates are often slow because of the low concentration of radicals. There are, however, ways to greatly increase polymerization rates. Wang and Armes [98] demonstrated that the use of water as a polymerization solvent greatly accelerates ATRP, and polymerization of 2-hydroxyethyl methacrylate (HEMA) from surfaces yields 100-nm thick films in a few hours [99, 100]. Highly active catalysts can also greatly increase polymerization rates [101]. Bao and coworkers [102] showed that the use of a

Cu(I)1,4,8,11-tetramethyl-1,4,8,11-tetraazacyclotetradecane (Cu(I)-Me$_4$Cyclam) catalyst allows synthesis of 100 nm thick films of poly(*tert*-butyl acrylate) (PtBA) or
poly(2-hydroxyethylmethacrylate) (PHEMA) in 10 min or less.

7.3.2
Synthesis of Block-Copolymers

Zhao and Brittain [103] prepared diblock-copolymer brushes, PS-*b*-PMMA, using
sequential carbocationic polymerization and ATRP. Like Fukuda's example, the
addition of free initiator during ATRP was necessary to ensure a sufficient concentration of deactivating Cu(II) species, otherwise the polymerization was not
controlled. Matyjaszewski *et al.* [97] synthesized several block-copolymer brushes
such as PS-*b*-PMA, PS-*b*-PtBA, and PS-*b*-poly(acrylic acid) by sequential ATRP. For
example, a 10-nm PS film was exposed to a methyl acrylate solution containing
catalyst to give a PS-*b*-PMA block-polymer film. However, a significant fraction of
active chain-ends were either buried in the polymer brush or lost via termination
during growth of the PS block, since the initiator efficiency for the growth of the
second block was reduced. Growth of the 90-nm PMA took much longer (20 h)
than expected.

Kim *et al.* [104] used a simple but effective quenching and reinitiation (QR)
approach to grow PMA-*b*-PMMA-*b*-PHEMA triblock-copolymer brushes on Au
(Scheme 7.11). Polymerization was effectively stopped by quenching a growing
polymer brush with a concentrated CuBr$_2$/ligand solution, preserving the Br atoms
at the chain ends for subsequent re-initiation of the next polymer block. The
efficiency of the QR scheme was better than a simple solvent washing procedure,
which resulted in a higher loss of active chains.

Surface-initiated ATRP can also be applied to form patterned polymer brushes by
microprinting as has been demonstrated with other "grafting-from" methods. Shah
and coworkers [105] reported the use of surface-initiated ATRP to amplify patterned
initiator layers on gold films. PMMA, PHEMA, PtBA, and poly(dimethylaminoethyl
methacrylate) were grown from spatially patterned initiators, and then the pattern
was transferred into the substrates by using the brushes as barriers to wet chemical
etching of gold.

7.3.3
Variation of Brush Density

The properties of brushes in applications ranging from reduction of flocculation
to immobilization of biomolecules will likely depend on the brush density, which
can be controlled by varying the areal density of initiators on the substrate. Perhaps
the simplest way to control initiator density is to dilute the initiators on a surface.
Such a strategy is applicable to ATRP as well as most other polymerization
methods, including ROMP [106]. Dilution of initiators can be easily accomplished
by forming silane or thiol monolayers using both an initiator molecule and a
diluent [44], although possible phase segregation of initiator and diluent should

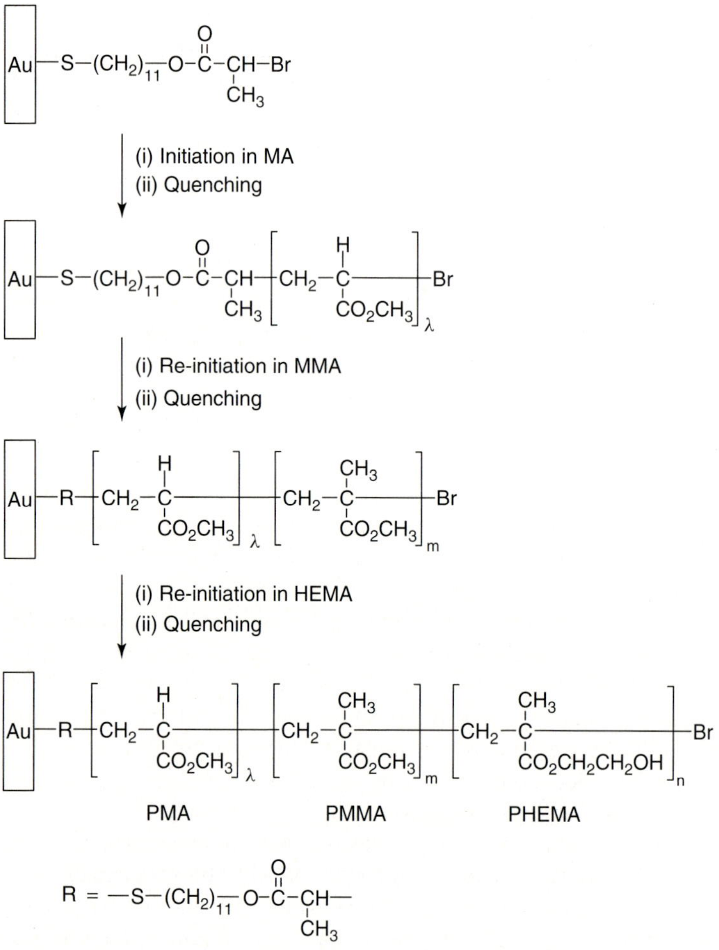

Scheme 7.11 Synthesis of PMA-*b*-PMMA-*b*-PHEMA triblock-copolymer brushes [104].

be kept in mind [107]. By using a high diluent to initiator ratio in the growth of PHEMA brushes, Bao and coworkers [43] were able to increase the swelling of PHEMA in water by an order of magnitude.

Genzer and coworkers [108] used mechanically assisted polymer assembly to produce polyacrylamide brushes of varying density on a cross-linked polydimethylsiloxane (PDMS) surface. After stretching the PDMS substrate and generating silanol (Si–OH) groups on the surface by exposure to UV/O_3, they attached a trichlorosilane ATRP initiator onto the surface from the vapor phase. They kept the substrate stretched until the poly(acrylamide) brushes were formed by ATRP at 130 °C and then released the strain, allowing the PDMS substrate to return to its former size. One advantage of this approach is that the brush grafting density can be controlled by altering the stretching extent of the PDMS substrate.

Several strategies were also developed to create a gradient of brush density [37, 109]. Such a gradient can be used to investigate how properties such as cell adhesion vary with brush density [110]. Bohn and coworkers [111] formed monolayers of hexadecanethiol spacers on Au prior to adsorption of a disulfide initiator. To create a gradient in the density of spacer molecules, they employed reductive desorption of the thiols using a potential that varied along the length of the sample. Subsequent adsorption of initiator in defects in the hexadecanethiol monolayer and ATRP yielded a polymer brush with a gradient in chain density. Variations of this method allowed formation of two-component brushes with gradients in composition [112]. Another strategy employed initiator attachment from the vapor phase to form a gradient [108, 113]. Initiator density decreased with the distance from the source of the vapor-phase initiator, and polymerization from these initiators yielded a gradient of brush densities. Luzinov and coworkers [114] used a temperature gradient to attach poly(glycidyl methacrylate) to a surface with a gradient of thickness. Subsequent attachment of initiators to the base polymer followed by ATRP yielded a gradient in brush density. Finally, a very simple strategy for forming gradients in brush thickness involves very slowly immersing the initiator-coated substrate into the polymerization solution [115]. Regions of the substrate that are exposed to monomer and catalyst for longer periods of time contain thicker brushes. A somewhat similar strategy using solution gradients in microchips has been employed to prepare gradients of statistical copolymers [116].

7.3.4
Surface-Initiated ATRP from Nonplanar Supports, Including Membranes

Surface initiated ATRP has also been applied to the synthesis of polymer brushes from nonplanar substrates such as nanoparticles, carbon nanotubes, and polymer supports. To list just a few examples, Huang and Wirth [117] synthesized poly-acrylamide brushes from porous silica gel by surface-initiated ATRP and used them for the separation of proteins by size exclusion. Armes and coworkers [118] synthesized poly(2-(N-morpholino)ethyl methacrylate) (PMEMA) from a silane ini-tiator on silica particles. These PMEMA-silica particles began aggregating at the lower critical solution temperature of PMEMA and redispersed upon cooling. Bon-tempo *et al.* [119] synthesized a variety of polymer brushes from PS microspheres using surface-initiated ATRP in aqueous media. Guerrini and coworkers [120] grew poly(2-hydroxyethyl acrylate) and other polymer brushes from cross-linked poly(styrene-*co*-2-(2-bromopropionyloxy) ethyl methacrylate) latex particles to form particles with a hydrophobic core and a hydrophilic shell.

Yan and coworkers [121] initiated ATRP from multiwalled carbon nanotubes (MWNTs) as shown in Scheme 7.12. To attach the ATRP initiator, the MWNT was treated sequentially with HNO_3, and $SOCl_2$, and finally ethylene glycol to produce a hydroxyl-covered surface. The ATRP initiators were readily anchored to the surface by reaction with α-bromoisobutyl bromide. ATRP of MMA provided a PMMA-covered MWNT, and sequential polymerization of MMA and HEMA yielded carbon nanotubes coated with amphiphilic PMMA-*b*-PHEMA polymer

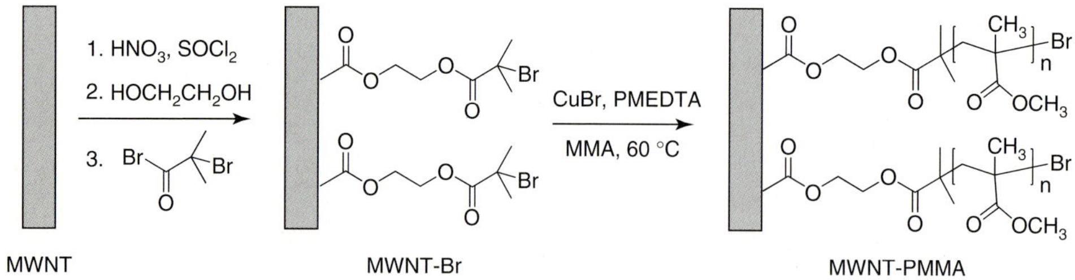

Scheme 7.12 ATRP of MMA from multiwalled carbon nanotubes (MWNTs) [121].

brushes. Other recent studies also demonstrated modification of nanotubes with polymer brushes [122, 123].

We are particularly interested in using surface-initiated ATRP to modify membranes [124]. ATRP from surfaces is attractive for such applications because it minimizes polymerization in solution to avoid plugging of membrane pores with physisorbed polymer. Additionally, film thickness can be readily controlled by varying polymerization time. When polymer brushes are grown only from the surface of a porous substrate (Figure 7.2a), they can be utilized as ultrathin membrane skins for applications such as gas and liquid separations. In this case the polymer film should be as thin as possible because permeation rates are typically inversely proportional to film thickness. Balachandra and coworkers [11] formed cross-linked poly(ethylene glycol dimethacrylate) films by ATRP from the surface of porous alumina, and showed that these films exhibit a CO_2/CH_4 selectivity of ~20. Most importantly, because these skins are only ~50 nm thick, they allow very high gas fluxes. Similarly, Sun and coworkers [12] grew PHEMA brushes on alumina and then derivatized these brushes with hydrophobic side chain to create pervaporation membranes that selectively pass organics over water. (In

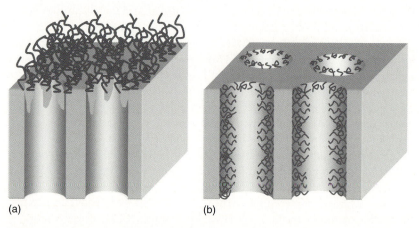

(a)   (b)

Figure 7.2 Cartoon of polymer brushes grown (a) on the surface and (b) in the pores of a porous substrate [124].

pervaporation, a liquid feed permeates through a membrane to a vapor phase.) Remarkably, trichloroethylene/water selectivities were about 500, and fluxes were typically an order of magnitude higher than those through commercial PDMS membranes.

Polymer brushes can also be grown in the interior of membranes [13, 125–127], and in this case the combination of the high surface area of the porous membrane and the ability of brushes to absorb multilayers of protein can yield materials with a high absorption capacity [128]. Moreover, because convective transport through membrane pores rapidly brings substrates to binding sites, membrane absorbers may allow much faster processing of solutions than packed columns [129, 130]. Sun *et al.* [13] showed that porous alumina membranes modified with polymer brushes containing metal-ion complexes can absorb up to 150 mg of protein per cm^3 of membrane. Moreover, when using Ni^{2+} complexes, these membranes can selectively adsorb proteins containing a polyhistidine tag [14]. Husson and coworkers [15] modified regenerated cellulose membranes with poly(acrylic acid) brushes and reported a lysozyme binding capacity as high as 100 mg/cm^3. More research in this area is still needed to develop high-capacity membranes in substrates that are highly permeable and selective.

7.4
Summary

Polymer brushes provide unique and promising coatings for a wide range of applications. Although there are a number of techniques for growth of these brushes, ATRP is currently the most versatile and frequently studied method because it occurs under gentle conditions (room temperature in some cases [42]), minimizes solution polymerization, and is compatible with a number of monomers. Under certain conditions ATRP can also be rapid, yielding 100-nm thick films in 10 min or less. One asset of polymerization from a surface in general, is that it can occur on substrates ranging from conventional polymers and Si to nanoparticles including carbon nanotubes. Growth of brushes on or in porous supports can provide highly selective gas-separation membranes or membrane absorbers capable of binding large amounts of macromolecules. However, it remains to be seen whether industrial application of polymer brushes will be viable. Growth of polymer brushes is still a relatively complicated process that typically includes initiator attachment, polymerization from degassed solutions, and rinsing of the substrate.

Acknowledgment

We thank the US National Science Foundation for supporting this work.

References

1. Milner, S.T. (1991) Polymer brushes. *Science*, **251**, 905–914.
2. Brittain, W.J. and Minko, S. (2007) A structural definition of polymer brushes. *J. Polym. Sci., Part A: Polym. Chem.*, **45**, 3505–3512.
3. Dutta, A.K. and Belfort, G. (2007) Adsorbed gels versus brushes: viscoelastic differences. *Langmuir*, **23**, 3088–3094.
4. Clayfiel, E.J. and Lumb, E.C. (1966) A theoretical approach, for polymeric dispersant action. 1. Calculation of entropic repulsion exerted by random polymer chains terminally adsorbed on plane surfaces and spherical particles. *J. Colloid Interface Sci.*, **22**, 269–284.
5. Clayfiel, E.J. and Lumb, E.C. (1966) A theoretical approach, for polymeric dispersant action. 2. Calculation of dimensions of terminally adsorbed macromolecules. *J. Colloid Interface Sci.*, **22**, 285–293.
6. Mackor, E. (1951) A theoretical approach, of the colloid-chemical stability of dispersions in hydrocarbons. *J. Colloid Sci.*, **6**, 492–495.
7. van der Waarden, M. (1950) Stabilization of carbon-black dispersions in hydrocarbons. *J. Colloid Sci.*, **5**, 317–325.
8. Velten, U., Shelden, R.A., Caseri, W.R., Suter, U.W., and Li, Y.Z. (1999) Polymerization of styrene with peroxide initiator ionically bound to high surface area mica. *Macromolecules*, **32**, 3590–3597.
9. Velten, U., Tossati, S., Shelden, R.A., Caseri, W.R., Suter, U.W., Hermann, R., and Muller, M. (1999) Graft polymerization of styrene on mica: formation and behavior of molecular droplets and thin films. *Langmuir*, **15**, 6940–6945.
10. Schadler, L.S., Kumar, S.K., Benicewicz, B.C., Lewis, S.L., and Harton, S.E. (2007) Designed interfaces in polymer nanocomposites: a fundamental viewpoint. *MRS Bull.*, **32**, 335–340.
11. Balachandra, A.M., Baker, G.L., and Bruening, M.L. (2003) Preparation of composite membranes by atom transfer radical polymerization initiated from a porous support. *J. Membr. Sci.*, **227**, 1–14.
12. Sun, L., Baker, G.L., and Bruening, M.L. (2005) Polymer brush membranes for pervaporation of organic solvents from water. *Macromolecules*, **38**, 2307–2314.
13. Sun, L., Dai, J., Baker, G.L., and Bruening, M.L. (2006) High-capacity, protein-binding membranes based on polymer brushes grown in porous substrates. *Chem. Mater.*, **18**, 4033–4039.
14. Jain, P., Sun, L., Dai, J., Baker, G.L., and Bruening, M.L. (2007) High-capacity purification of his-tagged proteins by affinity membranes containing functionalized polymer brushes. *Biomacromolecules*, **8**, 3102–3107.
15. Singh, N., Wang, J., Ulbricht, M., Wickramasinghe, S.R., and Husson, S.M. (2008) Surface-initiated atom transfer radical polymerization: a new method for preparation of polymeric membrane adsorbers. *J. Membr. Sci.*, **309**, 64–72.
16. Cullen, S.P., Liu, X., Mandel, I.C., Himpsel, F.J., and Gopalan, P. (2008) Polymeric brushes as functional templates for immobilizing ribonuclease a: study of binding kinetics and activity. *Langmuir*, **24**, 913–920.
17. Dai, J., Bao, Z., Sun, L., Hong, S.U., Baker, G.L., and Bruening, M.L. (2006) High-capacity binding of proteins by poly(acrylic acid) brushes and their derivatives. *Langmuir*, **22**, 4274–4281.
18. Niu, Q.J. and Frechet, J.M.J. (1998) Polymers for 193-nm microlithography: regioregular 2-alkoxycarbonylnortricyclene polymers by controlled cyclopolymerization of bulky ester derivatives of norbornadiene. *Angew. Chem. Int. Ed.*, **37**, 667–670.
19. Chen, C.S., Mrksich, M., Huang, S., Whitesides, G.M., and Ingber, D.E. (1997) Geometric control of cell life and death. *Science*, **276**, 1425–1428.
20. Mizutani, A., Kikuchi, A., Yamato, M., Kanazawa, H., and Okano, T.

(2008) Preparation of thermoresponsive polymer brush surfaces and their
interaction with cells. *Biomaterials*, **29**,
2073–2081.

21. Aksay, I.A., Trau, M., Manne, S.,
Honma, I., Yao, N., Zhou, L., Fenter,
P., Eisenberger, P.M., and Gruner,
S.M. (1996) Biomimetic pathways
for assembling inorganic thin films.
Science, **273**, 892–898.

22. Balazs, A.C., Singh, C., Zhulina, E.,
Gersappe, D., and Pickett, G. (1997)
Patterned polymer films. *MRS Bull.*,
22, 16–21.

23. Motornov, M., Sheparovych, R., Katz,
E., and Minko, S. (2008) Chemical
gating with nanostructured responsive
polymer brushes: mixed brush versus
homopolymer brush. *ACS Nano*, **2**,
41–52.

24. Granville, A.M., Boyes, S.G., Akgun,
B., Foster, M.D., and Brittain, W.J.
(2004) Synthesis and characterization
of stimuli-responsive semifluorinated
polymer brushes prepared by atom
transfer radical polymerization. *Macromolecules*, **37**, 2790–2796.

25. Kaholek, M., Lee, W.K., LaMattina, B.,
Caster, K.C., and Zauscher, S. (2004)
Fabrication of stimulus-responsive
nanopatterned polymer brushes by
scanning-probe lithography. *Nano Lett.*,
4, 373–376.

26. Boyes, S.G., Brittain, W.J., Weng, X.,
and Cheng, S.Z.D. (2002) Synthesis,
characterization, and properties of ABA
type triblock copolymer brushes of
styrene and methyl acrylate prepared
by atom transfer radical polymerization. *Macromolecules*, **35**, 4960–4967.

27. Moya, S.E., Azzaroni, O., Kelby, T.,
Donath, E., and Huck, W.T.S. (2007)
Explanation for the apparent absence
of collapse of polyelectrolyte brushes
in the presence of bulky ions. *J. Phys.
Chem. B*, **111**, 7034–7040.

28. Kizhakkedathu, J.N., Norris-Jones,
R., and Brooks, D.E. (2004) Synthesis
of well-defined environmentally responsive polymer brushes by aqueous
ATRP. *Macromolecules*, **37**, 734–743.

29. Alem, H., Duwez, A.S., Lussis,
P., Lipnik, P., Jonas, A.M., and
Demoustier-Champagne, S. (2008)
Microstructure and thermo-
responsive behavior of poly
(N-isopropylacrylamide) brushes grafted
in nanopores of track-etched mem-
branes. *J. Membr. Sci.*, **308**, 75–86.

30. Fu, Q., Rao, G.V.R., Ista, L.K., Wu,
Y., Andrzejewski, B.P., Sklar, L.A.,
Ward, T.L., and Lopez, G.P. (2003)
Control of molecular transport through
stimuli-responsive ordered mesoporous
materials. *Adv. Mater.*, **15**, 1262–1266.

31. Minko, S. (2006) Responsive polymer
brushes. *Polym. Rev.*, **46**, 397–420.

32. Ista, L.K., Mendez, S., Perez-Luna,
V.H., and Lopez, G.P. (2001) Synthesis of poly(N-isopropylacrylamide)
on initiator-modified self-assembled
monolayers. *Langmuir*, **17**, 2552–2555.

33. Jones, D.M., Smith, J.R., Huck,
W.T.S., and Alexander, C. (2002)
Variable adhesion of micropatterned thermoresponsive polymer
brushes: AFM investigations of poly
(N-isopropylacrylamide) brushes
prepared by surface-initiated polymerizations. *Adv. Mater.*, **14**, 1130–1134.

34. Xia, F., Ge, H., Hou, Y., Sun, T.L.,
Chen, L., Zhang, G.Z., and Jiang, L.
(2007) Multiresponsive surfaces change
between superhydrophilicity and superhydrophobicity. *Adv. Mater.*, **19**,
2520–2524.

35. Fleer, G.J., Cohen-Stuart, M.A.,
Scheutjens, J.M.H.M., Cosgrove, T.,
and Vincent, B. (1993) *Polymers at
Interfaces*, Chapman & Hall, London.

36. Edmondson, S., Osborne, V.L., and
Huck, W.T.S. (2004) Polymer brushes
via surface-initiated polymerizations.
Chem. Soc. Rev., **33**, 14–22.

37. Luzinov, I., Minko, S., and Tsukruk,
V.V. (2008) Responsive brush layers:
from tailored gradients to reversibly
assembled nanoparticles. *Soft Matter*, **4**,
714–725.

38. Pieracci, J.P., Crivello, J.V., and Belfort,
G. (2002) UV-assisted graft polymerization of N-vinyl-2-pyrrolidinone onto
poly(ether sulfone) ultrafiltration membranes using selective UV wavelengths.
Chem. Mater., **14**, 256–265.

39. Ulbricht, M. and Belfort, G. (1996)
Surface modification of ultrafiltration membranes by low temperature

plasma. II: graft polymerization onto polyacrylonitrile and polysulfone. *J. Membr. Sci.*, **111**, 193–215.

40. Husemann, M., Mecerreyes, D., Hawker, C.J., Hedrick, J.L., Shah, R., and Abbott, N.L. (1999) Surface-initiated polymerization for amplification of self-assembled monolayers patterned by microcontact printing. *Angew. Chem. Int. Ed.*, **38**, 647–649.

41. Huang, X. and Wirth, M.J. (1999) Surface initiation of living radical polymerization for growth of tethered chains of low polydispersity. *Macromolecules*, **32**, 1694–1696.

42. Kim, J.B., Bruening, M.L., and Baker, G.L. (2000) Surface-initiated atom transfer radical polymerization on gold at ambient temperature. *J. Am. Chem. Soc.*, **122**, 7616–7617.

43. Bao, Z.Y., Bruening, M.L., and Baker, G.L. (2006) Control of the density of polymer brushes prepared by surface-initiated atom transfer radical polymerization. *Macromolecules*, **39**, 5251–5258.

44. Jones, D.M., Brown, A.A., and Huck, W.T.S. (2002) Surface-initiated polymerizations in aqueous media: effect of initiator density. *Langmuir*, **18**, 1265–1269.

45. Boven, G., Folkersma, R., Challa, G., and Schouten, A.J. (1991) Radical grafting of poly(methyl methacrylate) onto silicon-wafers, glass slides and glass-beads. *Polym. Commun.*, **32**, 50–53.

46. Sugawara, T. and Matsuda, T. (1994) Novel surface graft-copolymerization method with micron-order regional precision. *Macromolecules*, **27**, 7809–7814.

47. Minko, S., Gafijchuk, G., Sidorenko, A., and Voronov, S. (1999) Radical polymerization initiated from a solid substrate. 1. Theoretical background. *Macromolecules*, **32**, 4525–4531.

48. Minko, S., Sidorenko, A., Stamm, M., Gafijchuk, G., Senkovsky, V., and Voronov, S. (1999) Radical polymerization initiated from a solid substrate. 2. Study of the grafting layer growth on the silica surface by

in situ ellipsometry. *Macromolecules*, **32**, 4532–4538.

49. Sidorenko, A., Minko, S., Gafijchuk, G., and Voronov, S. (1999) Radical polymerization initiated from a solid substrate. 3. Grafting from the surface of an ultrafine powder. *Macromolecules*, **32**, 4539–4543.

50. Luzinov, I., Minko, S., Senkovsky, V., Voronov, A., Hild, S., Marti, O., and Wilke, W. (1998) Synthesis and behavior of the polymer covering on a solid surface. 3. Morphology and mechanism of formation of grafted polystyrene layers on the glass surface. *Macromolecules*, **31**, 3945–3952.

51. Minko, S.S., Luzinov, I.A., Evchuk, I.Y., and Voronov, S.A. (1996) Synthesis and behaviour of the polymer covering on a solid surface.1. Attachment of the polymer initiator to the solid surface. *Polymer*, **37**, 177–181.

52. Luzinov, I., Evchuk, I., Minko, S., and Voronov, S. (1998) Synthesis and behavior of the polymer covering on a solid surface. 2. Effect of the adsorbent particle size on adsorption. *J. Appl. Polym. Sci.*, **67**, 299–305.

53. Prucker, O., Naumann, C.A., Rühe, J., Knoll, W., and Frank, C.W. (1999) Photochemical attachment of polymer films to solid surfaces via monolayers of benzophenone derivatives. *J. Am. Chem. Soc.*, **121**, 8766–8770.

54. Prucker, O. and Rühe, J. (1998) Polymer layers through self-assembled monolayers of initiators. *Langmuir*, **14**, 6893–6898.

55. Prucker, O. and Rühe, J. (1998) Synthesis of poly(styrene) monolayers attached to high surface area silica gels through self-assembled monolayers of azo initiators. *Macromolecules*, **31**, 592–601.

56. Prucker, O. and Rühe, J. (1998) Mechanism of radical chain polymerizations initiated by azo compounds covalently bound to the surface of spherical particles. *Macromolecules*, **31**, 602–613.

57. Wittmer, J.P., Cates, M.E., Johner, A., and Turner, M.S. (1996) Diffusive growth of a polymer layer by in situ polymerization. *Europhys. Lett.*, **33**, 397–402.

58. Stöhr, T. and Rühe, J. (2000) Monolayers of amphiphilic block copolymers via physisorbed macroinitiators. *Macromolecules*, **33**, 4501–4511.

59. Vidal, A., Guyot, A., and Kennedy, J.P. (1982) Silica grafted polyisobutylene and butyl rubber. 2. Synthesis and characterization of silica grafted butyl rubber. *Polym. Bull.*, **6**, 401–407.

60. Vidal, A., Guyot, A., and Kennedy, J.P. (1980) Silica-grafted polyisobutylene and butyl rubber.1. Synthesis and characterization of silica-grafted polyisobutylene. *Polym. Bull.*, **2**, 315–320.

61. Jordan, R. and Ulman, A. (1998) Surface initiated living cationic polymerization of 2-oxazolines. *J. Am. Chem. Soc.*, **120**, 243–247.

62. Jordan, R., West, N., Ulman, A., Chou, Y.M., and Nuyken, O. (2001) Nanocomposites by surface-initiated living cationic polymerization of 2-oxazolines on functionalized gold nanoparticles. *Macromolecules*, **34**, 1606–1611.

63. Zhao, B. and Brittain, W.J. (2000) Synthesis of polystyrene brushes on silicate substrates via carbocationic polymerization from self-assembled monolayers. *Macromolecules*, **33**, 342–348.

64. Dronavajjala, K.D., Rajagopalan, R., Uppili, S., Sen, A., Allara, D.L., and Foley, H.C. (2006) A simple technique to grow polymer brushes using in situ surface ligation of an organometallic initiator. *J. Am. Chem. Soc.*, **128**, 13040–13041.

65. Hasegawa, M., Kimata, M., and Takahashi, I. (2007) Mechanochemical polymerization of styrene initiated by the grinding of layered clay minerals. *Adv. Powder Technol.*, **18**, 541–554.

66. Haouas, M., Harrane, A., Belbachir, M., and Taulelle, F. (2007) NMR solid state characterization of formation of poly(epsilon-caprolactone)/maghnite nanocomposites by in situ polymerization. *J. Polym. Sci., Part B: Polym. Phys.*, **45**, 3060–3068.

67. Advincula, R. (2006) Polymer brushes by anionic and cationic surface-initiated polymerization in *Surface-Initiated Polymerization*, Advances in Polymer Science, Vol. 197 (ed. R. Jordan), Springer, Berlin, pp. 107–136.

68. Jordan, R., Ulman, A., Kang, J.F., Rafailovich, M.H., and Sokolov, J. (1999) Surface-initiated anionic polymerization of styrene by means of self-assembled monolayers. *J. Am. Chem. Soc.*, **121**, 1016–1022.

69. Quirk, R.P., Mathers, R.T., Cregger, T., and Foster, M.D. (2002) Anionic synthesis of block copolymer brushes grafted from a 1,1-diphenylethylene monolayer. *Macromolecules*, **35**, 9964–9974.

70. Advincula, R., Zhou, Q.G., Park, M., Wang, S.G., Mays, J., Sakellariou, G., Pispas, S., and Hadjichristidis, N. (2002) Polymer brushes by living anionic surface initiated polymerization on flat silicon (SiO$_x$) and gold surfaces: homopolymers and block copolymers. *Langmuir*, **18**, 8672–8684.

71. Fan, X.W., Zhou, Q.Y., Xia, C.J., Cristofoli, W., Mays, J., and Advincula, R. (2002) Living anionic surface-initiated polymerization (LASIP) of styrene from clay nanoparticles using surface bound 1,1-diphenylethylene (DPE) initiators. *Langmuir*, **18**, 4511–4518.

72. Zheng, L., Xie, A.F., and Lean, J.T. (2004) Polystyrene nanoparticles with anionically polymerized polybutadiene brushes. *Macromolecules*, **37**, 9954–9962.

73. Sakellariou, G., Park, M., Advincula, R., Mays, J.W., and Hadjichristidis, N. (2006) Homopolymer and block copolymer brushes on gold by living anionic surface-initiated polymerization in a polar solvent. *J. Polym. Sci., Part A: Polym. Chem.*, **44**, 769–782.

74. Zhou, Q.Y., Fan, X.W., Xia, C.J., Mays, J., and Advincula, R. (2001) Living anionic surface initiated polymerization (SIP) of styrene from clay surfaces. *Chem. Mater.*, **13**, 2465–2467.

75. Ingall, M.D.K., Honeyman, C.H., Mercure, J.V., Bianconi, P.A., and Kunz, R.R. (1999) Surface functionalization and imaging using monolayers and surface-grafted polymer layers. *J. Am. Chem. Soc.*, **121**, 3607–3613.

76. Choi, I.S. and Langer, R. (2001) surface-initiated polymerization of

L-lactide: coating of solid substrates with a biodegradable polymer. *Macromolecules*, **34**, 5361–5363.

77. Gurler, C., Feyen, M., Behrens, S., Matoussevitch, N., and Schmidt, A.M. (2008) One-step synthesis of functional Co nanoparticles for surface-initiated polymerization. *Polymer*, **49**, 2211–2216.

78. Kim, N.Y., Jeon, N.L., Choi, I.S., Takami, S., Harada, Y., Finnie, K.R., Girolami, G.S., Nuzzo, R.G., Whitesides, G.M., and Laibinis, P.E. (2000) Surface-initiated ring-opening metathesis polymerization on Si/SiO$_2$. *Macromolecules*, **33**, 2793–2795.

79. Kong, B., Lee, J.K., and Choi, I.S. (2007) Surface-initiated, ring-opening metathesis polymerization: formation of diblock copolymer brushes and solvent-dependent morphological changes. *Langmuir*, **23**, 6761–6765.

80. Feng, J.X., Stoddart, S.S., Weerakoon, K.A., and Chen, W. (2007) An efficient approach to surface-initiated ring-opening metathesis polymerization of cyclooctadiene. *Langmuir*, **23**, 1004–1006.

81. Juang, A., Scherman, O.A., Grubbs, R.H., and Lewis, N.S. (2001) Formation of covalently attached polymer overlayers on Si(111) surfaces using ring-opening metathesis polymerization methods. *Langmuir*, **17**, 1321–1323.

82. Moon, J.H. and Swager, T.M. (2002) Poly(p-phenylene ethynylene) brushes. *Macromolecules*, **35**, 6086–6089.

83. Berron, B.J., Graybill, E.P., and Jennings, G.K. (2007) Growth and structure of surface-initiated poly(N-alkylnorbornene) films. *Langmuir*, **23**, 11651–11655.

84. Husseman, M., Malmstrom, E.E., McNamara, M., Mate, M., Mecerreyes, D., Benoit, D.G., Hedrick, J.L., Mansky, P., Huang, E., Russell, T.P., and Hawker, C.J. (1999) Controlled synthesis of polymer brushes by "living" free radical polymerization techniques. *Macromolecules*, **32**, 1424–1431.

85. Blomberg, S., Ostberg, S., Harth, E., Bosman, A.W., Van Horn, B., and Hawker, C.J. (2002) Production of crosslinked, hollow nanoparticles by surface-initiated living free-radical polymerization. *J. Polym. Sci., Part A: Polym. Chem.*, **40**, 1309–1320.

86. Husemann, M., Morrison, M., Benoit, D., Frommer, K.J., Mate, C.M., Hinsberg, W.D., Hedrick, J.L., and Hawker, C.J. (2000) Manipulation of surface properties by patterning of covalently bound polymer brushes. *J. Am. Chem. Soc.*, **122**, 1844–1845.

87. Jhaveri, S.B., Beinhoff, M., Hawker, C.J., Carter, K.R., and Sogah, D.Y. (2008) Chain-end functionalized nanopatterned polymer brushes grown via in situ nitroxide free radical exchange. *ACS Nano*, **2**, 719–727.

88. Konn, C., Morel, F., Beyou, E., Chaumont, P., and Bourgeat-Lami, E. (2007) Nitroxide-mediated polymerization of styrene initiated from the surface of laponite clay platelets. *Macromolecules*, **40**, 7464–7472.

89. Baum, M. and Brittain, W.J. (2002) Synthesis of polymer brushes on silicate substrates via reversible addition fragmentation chain transfer technique. *Macromolecules*, **35**, 610–615.

90. Yuan, K., Li, Z.F., Lu, L.L., and Shi, X.N. (2007) Synthesis and characterization of well-defined polymer brushes grafted from silicon surface via surface reversible addition-fragmentation chain transfer (RAFT) polymerization. *Mater. Lett.*, **61**, 2033–2036.

91. Pei, X.W., Liu, W.M., and Hao, J.C. (2008) Functionalization of multiwalled carbon nanotube via surface reversible addition fragmentation chain transfer polymerization and as lubricant additives. *J. Polym. Sci., Part A: Polym. Chem.*, **46**, 3014–3023.

92. Yoshikawa, C., Goto, A., Tsujii, Y., Fukuda, T., Yamamoto, K., and Kishida, A. (2005) Fabrication of high-density polymer brush on polymer substrate by surface-initiated living radical polymerization. *Macromolecules*, **38**, 4604–4610.

93. Zhang, B.Q., Pan, C.Y., Hong, C.Y., Luan, B., and Shi, P.J. (2006) Reversible addition-fragmentation transfer polymerization in the presence of MMT immobilized amphoteric RAFT

agent. *Macromol. Rapid Commun.*, **27**, 97–102.

94. Matyjaszewski, K. and Xia, J.H. (2001) Atom transfer radical polymerization. *Chem. Rev.*, **101**, 2921–2990.

95. Kim, J.B., Huang, W.X., Miller, M.D., Baker, G.L., and Bruening, M.L. (2003) Kinetics of surface-initiated atom transfer radical polymerization. *J. Polym. Sci., Part A: Polym. Chem.*, **41**, 386–394.

96. Ejaz, M., Yamamoto, S., Ohno, K., Tsujii, Y., and Fukuda, T. (1998) Controlled graft polymerization of methyl methacrylate on silicon substrate by the combined use of the Langmuir-Blodgett and atom transfer radical polymerization techniques. *Macromolecules*, **31**, 5934–5936.

97. Matyjaszewski, K., Miller, P.J., Shukla, N., Immaraporn, B., Gelman, A., Luokala, B.B., Siclovan, T.M., Kickelbick, G., Vallant, T., Hoffmann, H., and Pakula, T. (1999) Polymers at interfaces: using atom transfer radical polymerization in the controlled growth of homopolymers and block copolymers from silicon surfaces in the absence of untethered sacrificial initiator. *Macromolecules*, **32**, 8716–8724.

98. Wang, X.S. and Armes, S.P. (2000) Facile atom transfer radical polymerization of methoxy-capped oligo(ethylene glycol) methacrylate in aqueous media at ambient temperature. *Macromolecules*, **33**, 6640–6647.

99. Huang, W.X., Kim, J.B., Bruening, M.L., and Baker, G.L. (2002) Functionalization of surfaces by water-accelerated atom-transfer radical polymerization of hydroxyethyl methacrylate and subsequent derivatization. *Macromolecules*, **35**, 1175–1179.

100. Jones, D.M. and Huck, W.T.S. (2001) Controlled surface-initiated polymerizations in aqueous media. *Adv. Mater.*, **13**, 1256–1259.

101. Matyjaszewski, K. (2002) From atom transfer radical addition to atom transfer radical polymerization. *Curr. Org. Chem.*, **6**, 67–82.

102. Bao, Z.Y., Bruening, M.L., and Baker, G.L. (2006) Rapid growth of polymer brushes from immobilized initiators. *J. Am. Chem. Soc.*, **128**, 9056–9060.

103. Zhao, B. and Brittain, W.J. (1999) Synthesis of tethered polystyrene-block-poly(methyl methacrylate) monolayer on a silicate substrate by sequential carbocationic polymerization and atom transfer radical polymerization. *J. Am. Chem. Soc.*, **121**, 3557–3558.

104. Kim, J.B., Huang, W.X., Bruening, M.L., and Baker, G.L. (2002) Synthesis of triblock copolymer brushes by surface-initiated atom transfer radical polymerization. *Macromolecules*, **35**, 5410–5416.

105. Shah, R.R., Merreceyes, D., Husemann, M., Rees, I., Abbott, N.L., Hawker, C.J., and Hedrick, J.L. (2000) Using atom transfer radical polymerization to amplify monolayers of initiators patterned by microcontact printing into polymer brushes for pattern transfer. *Macromolecules*, **33**, 597–605.

106. Jordi, M.A. and Seery, T.A.P. (2005) Quantitative determination of the chemical composition of silica-poly(norbornene) nanocomposites. *J. Am. Chem. Soc.*, **127**, 4416–4422.

107. Tamada, K., Hara, M., Sasabe, H., and Knoll, W. (1997) surface phase behavior of n-alkanethiol self-assembled monolayers adsorbed on Au(111): an atomic force microscope study. *Langmuir*, **13**, 1558–1566.

108. Wu, T., Efimenko, K., Vlcek, P., Subr, V., and Genzer, J. (2003) Formation and properties of anchored polymers with a gradual variation of grafting densities on flat substrates. *Macromolecules*, **36**, 2448–2453.

109. Genzer, J. and Bhat, R.R. (2008) Surface-bound soft matter gradients. *Langmuir*, **24**, 2294–2317.

110. Mougin, K., Ham, A.S., Lawrence, M.B., Fernandez, E.J., and Hillier, A.C. (2005) Construction of a tethered poly(ethylene glycol) surface gradient for studies of cell adhesion kinetics. *Langmuir*, **21**, 4809–4812.

111. Wang, X.J., Tu, H.L., Braun, P.V., and Bohn, P.W. (2006) Length scale

heterogeneity in lateral gradients of poly(N-isopropylacrylamide) polymer brushes prepared by surface-initiated atom transfer radical polymerization coupled with in-plane electrochemical potential gradients. *Langmuir*, **22**, 817–823.

112. Wang, X.J. and Bohn, P.W. (2007) Spatiotemporally controlled formation of two-component counterpropagating lateral graft density gradients of mixed polymer brushes on planar an surfaces. *Adv. Mater.*, **19**, 515–520.

113. Wu, T., Efimenko, K., and Genzer, J. (2002) Combinatorial study of the mushroom-to-brush crossover in surface anchored polyacrylamide. *J. Am. Chem. Soc.*, **124**, 9394–9395.

114. Liu, Y., Klep, V., Zdyrko, B., and Luzinov, I. (2005) Synthesis of high-density grafted polymer layers with thickness and grafting density gradients. *Langmuir*, **21**, 11806–11813.

115. Tomlinson, M.R. and Genzer, J. (2003) Formation of grafted macromolecular assemblies with a gradual variation of molecular weight on solid substrates. *Macromolecules*, **36**, 3449–3451.

116. Xu, C., Barnes, S.E., Wu, T., Fischer, D.A., DeLongchamp, D.M., Batteas, J.D., and Beers, K.L. (2006) Solution and surface composition gradients via microfluidic confinement: fabrication of a statistical-copolymer-brush composition gradient. *Adv. Mater.*, **18**, 1427–1430.

117. Huang, X.Y. and Wirth, M.J. (1997) Surface-initiated radical polymerization on porous silica. *Anal. Chem.*, **69**, 4577–4580.

118. Perruchot, C., Khan, M.A., Kamitsi, A., Armes, S.P., von Werne, T., and Patten, T.E. (2001) Synthesis of well-defined, polymer-grafted silica particles by aqueous ATRP. *Langmuir*, **17**, 4479–4481.

119. Bontempo, D., Tirelli, N., Feldman, K., Masci, G., Crescenzi, V., and Hubbell, J.A. (2002) Atom transfer radical polymerization as a tool for surface functionalization. *Adv. Mater.*, **14**, 1239–1241.

120. Guerrini, M.M., Charleux, B., and Vairon, J.P. (2000) Functionalized latexes as substrates for atom transfer radical polymerization. *Macromol. Rapid Commun.*, **21**, 669–674.

121. Kong, H., Gao, C., and Yan, D.Y. (2004) Controlled functionalization of multiwalled carbon nanotubes by in situ atom transfer radical polymerization. *J. Am. Chem. Soc.*, **126**, 412–413.

122. Kim, M.H., Hong, C.K., Choe, S., and Shim, S.E. (2007) Synthesis of polystyrene brush on multiwalled carbon nanotubes treated with KMnO$_4$ in the presence of a phase-transfer catalyst. *J. Polym. Sci., Part A: Polym. Chem.*, **45**, 4413–4420.

123. Wu, W., Tsarevsky, N.V., Hudson, J.L., Tour, J.M., Matyjaszewski, K., and Kowalewski, T. (2007) "Hairy" single-walled carbon nanotubes prepared by atom transfer radical polymerization. *Small*, **3**, 1803–1810.

124. Bruening, M.L., Dotzauer, D.M., Jain, P., Ouyang, L., and Baker, G.L. (2008) Creation of functional membranes using polyelectrolyte multilayers and polymer brushes. *Langmuir*, ASAP article. **24**, 7663–7673.

125. Singh, N., Chen, Z., Tomer, N., Wickramasinghe, S.R., Soice, N., and Husson, S.M. (2008) Modification of regenerated cellulose ultrafiltration membranes by surface-initiated atom transfer radical polymerization. *J. Membr. Sci.*, **311**, 225–234.

126. Singh, N., Husson, S.M., Zdyrko, B., and Luzinov, I. (2005) Surface modification of microporous PVDF membranes by ATRP. *J. Membr. Sci.*, **262**, 81–90.

127. Xu, F.J., Zhao, J.P., Kang, E.T., Neoh, K.G., and Li, J. (2007) Functionalization of nylon membranes via surface-initiated atom-transfer radical polymerization. *Langmuir*, **23**, 8585–8592.

128. Ulbricht, M. and Yang, H. (2005) Porous polypropylene membranes with different carboxyl polymer brush layers for reversible protein binding via surface-initiated graft copolymerization. *Chem. Mater.*, **17**, 2622–2631.

129. Yang, H. and Etzel, M.R. (2003) Evaluation of three kinetic equations in models of protein purification using ion-exchange membranes. *Ind. Eng. Chem. Res.*, **42**, 890–896.

130. Yang, H., Viera, C., Fischer, J., and Etzel, M.R. (2002) Purification of a large protein using ion-exchange membranes. *Ind. Eng. Chem. Res.*, **41**, 1597–1602.

8
Ultrathin Functional Polymer Films Using Plasma-Assisted Deposition

Renate Förch

8.1
Introduction

The gas-phase deposition of materials using plasma-assisted processes has been investigated and exploited for at least half a century. Nowadays, it has found a firm position in many areas of technology and device fabrication. In conjunction with other techniques, plasma technology is a key to meeting the challenges of sustainable growth, challenges of structural designs, and the increasing demand for smaller dimensions. Research in this area has opened up doors to many promising, ecologically efficient, and integrative solutions that will continue to provide satisfactory solutions for many technological challenges in terms of resources and environmentally friendly processes and products.

Plasma technology has already found applications in areas requiring high quality and productivity, environmental compatibility, and flexibility. It has revolutionalized growth in electronics, as well as automotive-, machine-, and tool-making industries. It has had a tremendous impact on growth areas such as energy technology, optics, and the textile industry as well as environmental and medical technology. With the continuing progress in the fundamental understanding of plasma physics, the potential of plasma technology is rapidly being realized and its importance in today's technological developments constantly growing. In contrast, the public awareness of this technology is still low, generally representing a typical "black-box" technology, that is, hidden behind plasma TV and plasma screens. Yet, all of us drive automobiles whose basic performance, such as drivability, safety, clean exhaust emission, and fuel economy that rely on thin films deposited in plasma-assisted processes. For example, thin-film techniques based on plasma technology play a key role in automobile electronics, sensors, and surface finishing. Today, the average automobile will contain many hundreds of components that have been treated by plasma-assisted thin-film processes.

Plasma chemistry is the term used to describe three main processes that take place in plasma-assisted processes. This includes the chemistry occurring in the plasma phase, the chemistry that leads to the deposition of a material, and the chemistry occurring during the surface modification of material. The reactions occurring are

Functional Polymer Films, First Edition. Edited by Wolfgang Knoll and Rigoberto C. Advincula.
© 2011 Wiley-VCH Verlag GmbH & Co. KGaA. Published 2011 by Wiley-VCH Verlag GmbH & Co. KGaA.

highly dependent on the temperature and the monomer feed. The nature of the layer formed is related to the type of excitation used, the power input, the nature of the precursor, and gas flow rates. However, it is often difficult to achieve a sufficiently high gas flux, such that processing at high speed under industrially viable conditions is still a challenge for many processes and precursor molecules. Plasmas are far from thermodynamic equilibrium and it is generally not easy to predict results. Fragmentation patterns of the precursor molecules can only be predicted, but cannot be conclusively modeled because of recirculation and the production of new "monomers" that partially replace the injected monomer. The new "monomers" formed during the reaction in turn influence the fragmentation processes and subsequent reaction taking place in the gas phase and at the surface. In addition, the components and geometry of the setup used have a significant impact on the distribution and density of the plasma species near the surface to be treated, which in turn influences the effectiveness and outcome of the plasma-assisted reaction. However, plasma chemistry and plasma processing remains a major component in an ever increasing number of areas of application. Recent advances in plasma physics, plasma modeling, and plasma chemistry help answer many previous questions such that processes and thin-film properties today are much more "predictable" than they were, for example, 20 years ago [1, 2].

One particular area that has recently seen a tremendous increase is the application of plasma-deposited thin films in the field of biomaterials, biosensing, and biomedical applications [3]. The diversity of plasma technology allows for multi-step processing to optimize sterility, adhesion, stability, and functionality within a single process. The literature describes newly developed films with unique characteristics that allow for controlled cell adhesion, differentiation, and proliferation. Plasma-deposited coatings using precursors that used to be considered too unstable for plasma polymerization are today being deposited with a high retention of the chemical structure and functionality such that properties and characteristics can be "predicted" within a reasonable margin of error.

The basic processes occurring in a plasma can be described by at least five steps: (i) ionization and fragmentation of the precursor [4], (ii) transport of reactive species to the surface, (iii) interaction and/or deposition at the surface, (iv) ablation and desorption [5], and finally (v) the generation of new monomer [6–8]. For fast deposition, a high flow of radicals is required and monolayer coverage can then be reached within a very short time. Deposition times of 1 or 2 s are often sufficient for the deposition of several tens of nanometer of material, a regime in which the deposition of material is extremely fast and economic [9]. The choice of precursors used for the deposition of materials has literally exploded in the past 15 years and today the literature shows a tremendous variety of films ranging from different inorganic materials [10–16] to materials of an inorganic–organic or metal–organic composites [17–23], all the way to hydrophobic coatings [24, 25] or hydrophilic and reactive polymer-like materials [26–30]. There is an enormous potential for the development of new materials in virtually all areas of technology.

8.2
Choice of Precursors

Precursors for the plasma polymerization of organic, functional thin films are generally chosen by a number of criteria.

1) They generally contain the functional group that is desired for a particular functionality. For example, this could be anhydride groups, esters, or acid chlorides that are, for example, chosen because of their high reactivity toward nucleophiles such as amines, typically present in proteins, antibodies, and integrins and are of particular interest for different biomedical and biomaterial applications.

2) Secondly, the precursor molecule must contain at least one other highly reactive group, such as a double bond or an epoxide, whose bond energy differs from that of the functional group to be retained and that can be used to initiate and carry the polymerization. It is thus the aim to run the plasma process in such a way that the excitation of the precursor molecules can be "preprogrammed" by matching the energy input and the choice of reactive groups. This approach is, however, still very objective and is generally dependent on trial and error.

3) Thirdly, the precursor must have a high enough vapor pressure to allow for sufficient flux of gaseous species.

4) Fourthly, the monomer must be stable under vacuum and at the temperatures achieved during plasma polymerization.

The choice of monomer is thus extremely broad and the material properties such as crystallinity, morphology [31–33], and chemical reactivity can be fine tuned, leaving almost no limits to the diversity of materials that can be deposited [3, 34–37]. In the case of more complex molecules, such as functional organometallics, the polymerization may be more difficult because of the multitude of possible reactions that can occur simultaneously.

8.3
Polymerization Mechanisms and Chemical Structure

Plasma deposition is generally marked by the large number of reactions taking place in the plasma phase as well as at the wall of the reactor. Research over the past few years has been able to elucidate some of the reactions taking place for very simple processes and gases [38–44], but exact mechanisms are still unknown. The reason for this is mostly because the reaction mechanisms taking place are dependent on the process conditions, the reactor geometry, and excitation energy used. These parameters influence the formation and the reactions of the radicals and ions formed, which have the largest impact on the chemical structure of the functional plasma polymer deposited.

The deposition of functional thin films that show a high retention of functional groups using plasma polymerization processes is believed to proceed mainly via

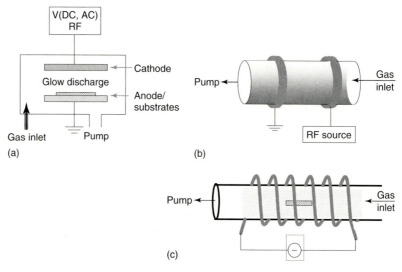

Figure 8.1 Schematic of some typical reactor configurations used for plasma polymerization, (a) parallel plate configuration, (b) external electrode configuration and (c) coil configuration.

radical polymerization processes, taking place at the gas phase near the surface. In order to "tailor" chemical reactions and thus film properties, the energy input into the system must be reduced in comparison to that used in other more conventional processes used for film deposition in microelectronics, optics, and barrier coatings. By careful control over the process conditions it is possible to minimize ion bombardment, UV reactions, and thus the well-described competitive ablation and polymerization (CAP) process [5], such that the film chemical structure, cross-link densities, and deposition rates can be "tuned."

The method most commonly adopted to reduce the energy input and to reach values allowing for the excitation of specific bonds is by a modulation of the applied field. The use of modulated plasma processes was first described by Yasuda in the 1980s [45], and was extensively investigated during the late 1990s up to the present day by groups worldwide. Pulsed plasma polymerization is believed to follow mechanisms in which the precursor molecules are activated during the "on" phase of the plasma pulse, see Figure 8.1, followed by a plasma "off" phase during which deposition occurs as a result of the polymerization of longer-lived radicals from the gas phase. Subsequent plasma on phases continuously regenerate these radicals. The modulation is typically defined by a duty cycle (DC) described as $DC = \frac{t_{on}}{t_{on}+t_{off}}$. The equivalent power over the pulse duration is given by: $P_{eq} = P_{peak} \times DC$. The thickness of the deposited layer is then determined by the length of time the modulation is continued and is typically in the range of 1–2 nm up to several hundred nanometers.

Thus, other typical plasma reactions such as ionization, UV emission, and electron and ion bombardment only occur during the plasma on phase. This

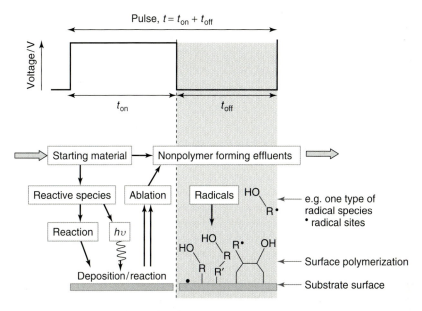

Figure 8.2 Schematic of the different mechanisms taking place during the t_{on} and t_{off} phase of pulsed plasma polymerization.

reduces temperature and ion-bombardment effects on the substrate and the growing film. For excitation of the plasma in the pulsed mode, alternating current (AC) plasmas run at frequencies <100 MHz are the most often described, with radio frequency of 13.56 MHz being the most widespread. Reactor geometries are either the RF-driven parallel plate reactors, Figure 8.2a, or RF-driven tubular reactors either with external electrodes, Figure 8.2b or an inductive coil, Figure 8.2c. Reactor design and preferences of each is a very diverse topic and a more detailed review is not in the scope of this work [46–48].

Experimental data has shown that films deposited contain a very high percentage of the desired functional group, suggesting a degree of functional-group retention up to an estimated 90% [49]. The remaining percentage can be associated with that fraction of the precursor molecules that dissociate in the plasma as a result of secondary reactions, wall effects, and ablation. The fraction of molecules that undergo dissociation are incorporated into the polymer network during the polymerization and lead to cross-linking within the structure, they also influence the density of the functional groups in the deposited layer and the stability of the layer in solution. The process conditions used during the deposition also influence the number of surface radicals available for post plasma reactions, which further influence the chemical structure of the deposits once these are removed from the reaction chamber [50].

The use of modulated plasma processes has become very popular and has led to the development of polymer-like materials with a very high density of functional groups. The deposited materials are most commonly referred to as "*plasma polymers.*" Conventionally, a *polymer* is defined as a large molecule (macromolecule) composed of repeating structural units connected by covalent chemical bonds. In the case of plasma polymers, the term *"polymer"* describes a macromolecular structure in which an infinite number of individual units are connected by covalent chemical bonds. In a typical plasma polymer the repeating structural unit is probably best described by the functional group, that is, randomly distributed throughout the structure and whose abundance has been proven experimentally by many researchers for specific types of precursors. These repeating functional groups are interconnected by covalent bonds through hydrocarbon chains of different length whose structure and composition depends on the process conditions used and the fragmentation processes of the precursor molecule. The polymer structure can be further characterized by different cross-link densities between chains, which is again dependent on the monomer fragmentation that occurs during the plasma polymerization process. A general rule of thumb is that the higher the fragmentation, the higher the cross-link density, and the lower the number of repeating functional groups. Thus, while it is possible to discuss the mechanisms of plasma-assisted polymerization in terms of radical, ionic, and radiation polymerization processes, the overall reaction mechanism is far more complex and much less well understood than conventional polymerization processes. It is in fact the complexity of the process that also allows for the diversity in materials that can potentially be formed by plasma polymerization.

8.4
Types of Functional Films

The types of films that have been studied by groups worldwide is extremely broad and the list of monomers investigated is long. A number of excellent reviews and articles have in the past described some of these in more detail [3, 51]. The monomers investigated can be divided into groups according to the types of functional group present within the molecule. Some representative Fourier transform infrared spectra and X-ray photoelectron spectra from the authors own work and from other groups are shown and discussed for some monomers below. References to work on similar and the same precursor molecules are given in all cases.

The literature describes the polymerization of cyclic monomers [52–56] and different aromatics [57–61] in which the ring structures are mostly retained as a function of the plasma conditions employed. Semiconducting layers have been reported using thiophenes and pyrrole and some of their derivatives [53, 62–65]. For example, a typical Fourier Transform Infra Red (FTIR) spectrum of thiophene plasma polymerized under conditions showing the largest retention of groups is shown in Figure 8.3 and is compared to the FTIR spectrum of unreacted thiophene. The spectra shown are from the authors own work, but are analogous to those published and discussed by Groenewoud and coworkers [64, 66–68]. The FTIR spectra show typical peaks at wave numbers of 705, 850, and $3100 \, \text{cm}^{-1}$ indicating that the thiophene structure has been partially retained in

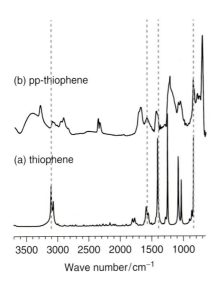

Figure 8.3 FTIR spectra of (a) pure thiophene and (b) plasma-polymerized thiophene (10/100 at 10 W input power).

(b) pp-thiophene

(a) thiophene

3500 3000 2500 2000 1500 1000

Wave number/cm^{-1}

the layers. Aliphatic structures as a result of fragmentation can also be observed around 2800–3000 cm^{-1} and by the peaks observed between 1400 and 1700 cm^{-1}. Reported conductivities are in the range of 10^{-7}–10^{-9} S/cm^2 and are influenced by iodine doping, vacuum, and long-term air exposure [67–70].

Plasma polymers containing oxygen functional groups can generally be deposited quite easily. Those that have been investigated in the past and that have been shown to retain their functional groups include epoxide [71, 72], ethers, and ethylene glycol related monomers [73–77], acids [78–82], aldehydes [83], alcohols [42, 84, 85], anhydrides [86, 87], and different esters [88–91].

Also of considerable interest are aminated plasma polymers, which are mostly deposited using allylamine [92–97], diaminohyclohexane [98, 99], ethylene diamine [100], and cysteamine [101]. The pulsed plasma polymerized layers for each of these monomers shows a high level of amine-group retention. The amine-rich surfaces have been shown to render otherwise inert materials compatible with biological molecules and fluids and the deposited layers show very promising properties for biomaterial sensing applications such as DNA [102–104] detection, protein adsorption [100, 105, 106], and cell adhesion [107].

Functional group retention is easily demonstrated using FTIR or by X-ray photoelectron spectroscopy (XPS) in the high-resolution mode. The two techniques complement one another and together reveal very important data on film chemical structure. Some examples from the literature of typical XPS C1s spectra for some common plasma polymers deposited using plasma conditions that allow for the highest functional group retention are shown in Figure 8.4. The acid functional groups in pp-acrylic acid appear at a binding energy of about 4–4.5 eV higher than the main hydrocarbon component (at ~289.3 eV in Figure 8.4a). The C 1s peak corresponding to carbon atoms of the anhydride group in plasma polymerized maleic anhydride films, generally cannot be distinguished from acid groups in the C 1s spectrum, but can be proven in the FTIR spectrum (see Figure 8.5

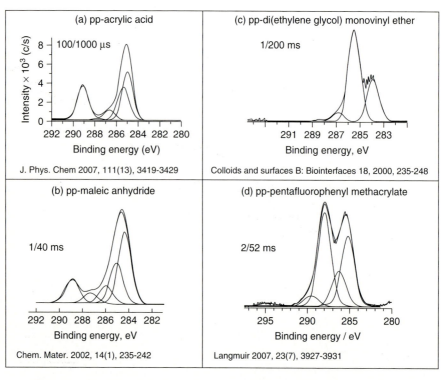

Figure 8.4 XPS C1s high-resolution spectra for a number of pulsed-plasma-deposited precursors as shown in the literature (see inserts). For each monomer there are a number of experimental conditions that may be used to deposit functional films with optimum functional group retention. The duty cycles shown in this figure are only representative (i.e. as quoted in the citation) and may vary for different experimental setups. (a) plasma polymerized acrylic acid, (b) plasma polymerized maleic anhydride, (c) plasma polymerized di(ethylene glycol) monovinyl ether and (d) plasma polymerized pentafluorophenyl methacrylate. (Copyright requested from Journals.)

below). Ether and alcohol groups are observed at a shift of about 1.5 eV from the main hydrocarbon peak and are clearly observed at high intensity in the C 1s spectrum of pp-diethylene glycol monovinyl ether in Figure 8.4c. The C 1s peak of pentafluorophenyl methacrylate (PFM) shows contributions from the fluorinated ring (C–F groups, shifted by about 2.5 eV) as well as the acrylate at a shift of about 4 eV. The aromatic ring is seen by the typical $\pi \rightarrow \pi^*$ shake-up satellite at a shift of about 9.5–10 eV from the main hydrocarbon peak. In all cases, there are always low-intensity contributions arising from monomer fragmentation. Sometimes, these are hidden behind contributions from other functional groups, often they lead to a broadening of the peak structure or additional peaks. For example, in Figure 8.4c the peak at 287 eV (3 eV higher than the main hydrocarbon peak, here at 284 eV) is typical for carbonyl groups. However, carbonyl groups are not present in the precursor, suggesting that this peak arises because of fragmentation during plasma polymerization.

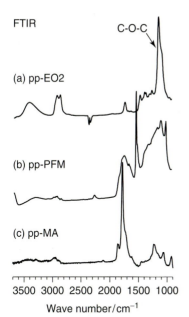

FTIR

C-O-C

(a) pp-EO2

(b) pp-PFM

(c) pp-MA

3500 3000 2500 2000 1500 1000

Wave number/cm^{-1}

Figure 8.5 Typical Fourier transform infrared spectra of (a) pp-di(ethylene glycol monovinyl ether) (DC = 10/300) (b) pp-pentafluorophenyl methacrylate (DC = 2/52), and (c) pp-maleic anhydride (DC = 1/41). The spectra show the level of functional group retention that can be achieved routinely using pulsed plasma polymerization methods.

Figure 8.5 shows typical FTIR spectra obtained for (a) plasma polymerized di(ethylene glycol) monovinyl ether, (b) the plasma polymerized active ester PFM, and (c) pulsed plasma polymerized maleic anhydride obtained in the authors own laboratory. The profiles shown in Figure 8.5 are representative and have been reproduced in the literature by different groups using different deposition systems. Thus, for films deposited from ethylene oxide and ethylene-glycol-related monomers, the FTIR and the XPS shows a high retention of the C–O–C and C–O–groups as can be seen by the FTIR band at 1100 cm^{-1} (C–O–C) and 3400 cm^{-1} (–OH), respectively. For this monomer it has been shown very clearly that the pulsed plasma-deposited films consist mostly of C–C and C–O bonds and film chemistry can be tuned such that relative peak intensities from the C1s peak profile are consistent with theoretical values. In the particular FTIR spectrum shown (Figure 8.5a), the FTIR band around 1750 cm^{-1} suggests the presence of some carbonyl and carboxylic acid groups within the film, which are probably the result of some fragmentation of the monomer and rearrangement of the structure during plasma polymerization. PFM is an active ester containing a double bond, an ester group, and a fluorinated benzene ring. Similar to acryloyl chlorides, described by Calderon and Timmons [108], only a small range of DCs allows for the successful plasma-assisted polymerization of these precursor [88, 89]. The FTIR spectra of the plasma polymer deposited distinctly show the presence of the ester group (1730 cm^{-1}) and the fluorinated aromatic ring (1525 cm^{-1}), Figure 8.5b. The much lower intensity bands at 2200, 3000, and 3400 cm^{-1} can be associated with some degree of fragmentation of the monomer and the presence of C=C, CH$_x$, and –OH probably as a result of post-plasma reaction in air.

Films deposited from maleic anhydride show the characteristic anhydride bands in the FTIR spectrum around 1780 and 1860 cm^{-1}. A low-intensity shoulder at lower wave number suggests the presence of some carboxylic acid groups that may arise both as a result of fragmentation processes during plasma deposition as well as a result of post-plasma reaction of the very reactive anhydride groups at ambient. The low-intensity bands around 3000 cm^{-1} can be associated with the hydrocarbon chains comprising the polymer backbone.

Beside the homopolymerization of only one precursor, it has been shown possible to carry out copolymerization processes in which two or more precursors are mixed to obtain films that partially exhibit the properties of each monomer [109–111]. Figure 8.6 shows FTIR spectra of a mixture of maleic anhydride and ethylene glycol dimethacrylate from the authors own work when deposited using different gas-flow ratios. From the FTIR bands it can be clearly seen that the product of the copolymerization shows the presence of bands representative of each monomer. The intensity of the bands (related to the relative abundance of the groups) of the respective precursor structure within the plasma polymer layer could be shown to increase with increasing gas flow ratios. Copolymerization has, for example, been described by the group of Short [79, 112, 113], Unger [109, 114], and Kikkawa [110, 115] amongst others. These groups report on improved cross-link densities and stability in aqueous media for some copolymers and the possibility to control dielectric properties and gas permeability of thin films [116].

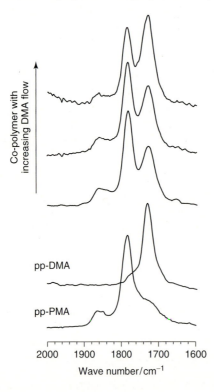

Figure 8.6 FTIR spectra for pure pp-MA, pure pp-DMA and mixtures of maleic anhydride (MA) and ethylene glycol dimethacrylate (DMA) with increasing amounts of DMA.

There have been many recent reports on nanocomposite materials comprising a metallic and an organic component. The group of Han *et al.* [117] investigated the effect on electrochemical properties as a function of functional-group control during the deposition of ferrocene and ferrocene-related monomers. They were able to show compositional changes in both per cent Fe content and the ratio of Fe-II/Fe-III with variations in the DCs. Other reports showed that the plasma polymerization of organometallics such as tetramethyltin led to nanoparticle inclusions within a polymer matrix. The size of the particles was shown to be a function of the DC employed during the deposition [22, 23]. More recently, there have been reports on gold [118], silver [18, 20], and copper [119, 120] inclusions in the form of nanoparticles in plasma polymer films that influence optical properties [121] and biomaterial and antimicrobial properties [122, 123]. These are obtained by the simultaneous polymerization and cosputtering techniques for both microwave and radio frequency (RF) excitation [18, 20, 118, 122].

8.5
Properties of Ultrathin Functional Plasma Polymers

Probably one of the most important properties of plasma-deposited films is the conformal coverage achieved over virtually any shape and size. This allows for the surface modification of virtually all shapes and sizes and represents a particular advantage over many wet-chemical processes of thin-film deposition where conformal coverage is not always achieved. Previous work on ultrathin SiO_2 films suggested that film growth on polymers occurs in a layer-by-layer (Frank–van der Merwe), not in a Volmer–Weber (island coalescence) mode [124]. Work by Jacobsen and coworkers [125, 126] using *in-situ* optical waveguide mode spectroscopy suggests that for some organic precursors (allylamine and maleic anhydride) the deposition proceeds via island formation. Thus, for certain monomers a "closed" film is only achieved after a few nanometers (>10 nm) of material have been deposited. It can thus be assumed that ultrathin functional plasma polymer films indeed have a pinhole structure. While this property may be undesirable for many application, it can, however, be used constructively to allow for the diffusion of reagents and reaction with lower-lying films, while at the same time allowing for the protection against another reagent or external influence [127]. This has recently been described by Chu *et al.* [128] who have used ultrathin antifouling films to protect an underlying reactive plasma polymer film, which was shown to undergo selective reaction with biological reagents, presumably via the pinholes in the outermost protective film.

Plasma polymer films have been shown to exhibit surface topographies ranging from ultraflat (rms < 1 nm) to a nanoscopic "cauliflower" or "grainy" structures [31, 129, 130] all the way to a roughness in the range of several tens of nanometers. The topography and morphology of the surface are influenced both by the substrate surface topography as well as the growth mode, which in turn is dependent on the type of discharge and precursor molecules [131].

While plasma polymers are shown to be very versatile, they can be deposited on almost any shape and material and offer many possibilities for the straightforward surface modification of otherwise inert materials, they still present a number of challenges that always need to be met and that often need to be tailored for a particular substrate–plasma polymer pair. One of these is the stability of the coatings in different environment [132]. Because of their high reactivity plasma polymers tend to change their chemical structure both in air and in solution. This can easily be monitored using XPS, FTIR, contact-angle measurements as well as different functionality test. The observed change in chemical structure proceeds both via chemical reactions in ambient and via macromolecular rearrangement of chains, to reduce surface energy. In air, the surfaces often show rapid hydrophobic recovery, which expresses itself at the apparent loss of functional groups from the surface as has been measured by XPS and FTIR accompanied by a loss of reactivity [3, 133–135].

Similar to many conventional polymers, plasma polymers have been observed to swell reversibly in humid environment [136] and in solution [30, 137]. This has been observed as changes in thickness (*d*) and refractive index (*n*) [138, 139] using optical and micromechanical techniques. Studies combining microcantilever sensors and surface plasmon resonance spectroscopy (SPR) suggested that the cross-link density of typical plasma polymers of allylamine had a profound effect on the films mechanical properties [136]. On the one hand, the change in thickness (as observed using SPR) of a high and a low cross-link density film was found to show a linear relationship to changes in humidity. However, using microcantilever sensors it could be seen that the deflection of the cantilever was a factor 2 higher for the high cross-linked film than that of the low cross-linked films. A linear relation between thickness evolution, as a result of swelling, and cantilever bending was not observed and was found to reach a maximum for a humidity of 40% in nitrogen. These observations were attributed to a decrease in the Young's modulus of the highly cross-linked films at higher humidity. Singamaneni *et al.* [130] used micromechanical sensor techniques to study the thermal properties of ultrathin plasma polymer films and observed a negative expansion coefficient for ultrathin plasma polymer films of continuous-wave-deposited acrylonitrile. This was found to be the opposite behavior to spin-cast polyacrylonitrile films.

The controlled swelling and collapse of polymer networks has generated considerable interest during the past few years and has led to the development of a number of different polymer hydrogels, which respond to external stimuli such as change in temperature, pH, ionic concentration, electric, or magnetic fields. The literature also shows considerable efforts in developing ultrathin functional films exhibiting such responsive properties using plasma-assisted polymerization processes. The advantage of plasma-assisted deposition to obtain thin hydrogel layers is the excellent control over film thickness, the conformal coverage of substrates as well as the optimal hydrogel–substrate adhesion that can be adapted depending on the materials in question by using simple plasma-assisted protocols. Thermally responsive plasma deposited films using N-isopropylacrylamide as the precursor have, for example, been reported by the group of Ratner and coworkers [140, 141],

Badyal and coworkers [142], Tamirisa and Hess [143], and Shard and coworkers [144]. Other precursors that have been investigated for their thermal responsiveness include related monomers such as N,N-diethylacrylamide [145]. Repetitive alternation between oleophobic and hydrophilic behavior has been reported for cationic fluorosurfactants complexed to maleic anhydride plasma polymer layers in response to their local liquid environment [146]. Bhattacharyya *et al.* [147] recently described the formation of a new class of thin-film hydrogels deposited from 1-amino-2-propanol and 2-(ethylamino)ethanol monomers in a pulsed plasma polymerization process, which was proven by a color change upon phase transition, contact angle, and capillary-rise measurements. Much of the work in this area has been stimulated by the need to develop new methodologies for biomaterials surface modification coupled with reliable and versatile biofunctionalization [141, 142, 148] and drug release [149].

8.6
Summary

This review aims to provide an overview of current topics of thin-film formation using plasma-assisted processes and has concentrated on aspects of thin-film formation using pulsed plasma deposition as well as describing the nature and some selected properties of reported films. The technique of plasma polymerization has been well documented since the 1960s, however, it has mainly been the tireless efforts of researchers over the past 15 years that has catapulted plasma polymerization into a new era of materials synthesis with novel applications and much more definable properties. The know-how acquired over the past years has impacted, and still is impacting, modern technological advancement in diverse applications such as automotive and electronics, textiles, environment, pharmaceuticals, packaging, and heath care. It is today routinely possible to deposit highly functional thin films on almost all types of substrate materials with a high degree of control. These films have been shown to be applicable both in ambient and in solution and have opened up a new area of expertise, which combines plasma-enhanced chemical vapor deposition processes, biomaterials, and microbiology. It is anticipated that research in this area will open up new opportunities for surface modification of implants, medical devices, and diagnostic tools for hospitals and the health-care system worldwide.

References

1. Favia, P., Creatore, M., Palumbo, F., Colaprico, V., and d'Agostino, R. (2001) Process control for plasma processing of polymers. *Surf. Coat. Technol.*, **142**, 1–6.

2. Friedrich, J.F., Retzko, I., Kuhn, G., Unger, W.E.S., and Lippitz, A. (2001) Plasma polymers with chemically defined structures in contact with metals. *Surf. Coat. Technol.*, **142**, 460–467.

3. Siow, K.S., Britcher, L., Kumar, S., and Griesser, H.J. (2006) Plasma methods for the generation of chemically reactive surfaces for biomolecule

immobilization and cell colonization – a review. *Plasma Process. Polym.*, **3** (6–7), 392–418.

4. Yasuda, H. and Yu, Q.S. (2004) Creation of polymerizable species in plasma polymerization. *Plasma Chem. Plasma Process.*, **24** (2), 325–351.

5. Yasuda, H. and Yasuda, T. (2000) The competitive ablation and polymerization (CAP) principle and the plasma sensitivity of elements in plasma polymerization and treatment. *J. Polym. Sci. Part A Polym. Chem.*, **38** (6), 943–953.

6. Hegemann, D., Schultz, U., and Fischer, A. (2005) Macroscopic plasma-chemical approach to plasma polymerization of HMDSO and CH₄. *Surf. Coat. Technol.*, **200** (1–4), 458–462.

7. Hegemann, D., Hossain, M.M., Korner, E., and Balazs, D.J. (2007) Macroscopic description of plasma polymerization. *Plasma Process. Polym.*, **4** (3), 229–238.

8. Stoffels, E., Stoffels, W.W., and Kroesen, G.M.W. (2001) Plasma chemistry and surface processes of negative ions. *Plasma Sources Sci. Technol.*, **10** (2), 311–317.

9. Zhang, J., van Ooij, W., France, P., Datta, S., Radomyselskiy, A., and Xie, H.Q. (2001) Investigation of deposition rate and structure of pulse DC plasma polymers. *Thin Solid Films*, **390** (1–2), 123–129.

10. Favia, P., Fracassi, F., and Dagostino, R. (1992) X-ray photoelectron-spectroscopy of plasma-polymerized films from tetramethylsilane-containing feeds. *J. Biomater. Sci. Polym. Ed.*, **4** (1), 61–73.

11. Fracassi, F., Dagostino, R., and Favia, P. (1993) Plasma-enhanced chemical vapor-deposition of organosilicon thin-films. *Abstr. Pap. Am. Chem. Soc.*, **205**, 191–POLY.

12. Fonseca, J.L.C., Tasker, S., Apperley, D.C., and Badyal, J.P.S. (1996) Plasma-enhanced chemical vapor deposition of organosilicon materials: a comparison of hexamethyldisilane and tetramethylsilane precursors. *Macromolecules*, **29**(5), 1705–1710.

13. Alexander, M.R., Short, R.D., Jones, F.R., Michaeli, W., and Blomfield, C.J. (1999) A study of HMDSO/O-2 plasma deposits using a high-sensitivity and -energy resolution XPS instrument: curve fitting of the Si 2p core level. *Appl. Surf. Sci.*, **137** (1–4), 179–183.

14. Hegemann, D., Vohrer, U., Oehr, C., and Riedel, R. (1999) Deposition of SiO$_x$ films from O-2/HMDSO plasmas. *Surf. Coat. Technol.*, **119**, 1033–1036.

15. Benitez, F., Martinez, E., Galan, M., Serrat, J., and Esteve, J. (2000) Mechanical properties of plasma deposited polymer coatings. *Surf. Coat. Technol.*, **125** (1–3), 383–387.

16. Johnson, E.M., Clarson, S.J., Jiang, H., Su, W., Grant, J.T., and Bunning, T.J. (2001) Plasma polymerized hexamethyldisiloxane (HMDS) barrier layers. *Polymer*, **42** (16), 7215–7219.

17. Lewis, H.G.P., Casserly, T.B., and Gleason, K.K. (2001) Hot-filament chemical vapor deposition of organosilicon thin films from hexamethylcyclotrisiloxane and octamethylcyclotetrasiloxane. *J. Electrochem. Soc.*, **148** (12), F212–F220.

18. Salz, D., Lamber, R., Wark, M., Baalmann, A., and Jaeger, N. (1999) Metal clusters in plasma polymer matrices – Part II. Silver clusters. *Phys. Chem. Chem. Phys.*, **1** (18), 4447–4451.

19. Choukourov, A., Pihosh, Y., Stelmashuk, V., Biederman, H., Slavinska, D., Kormunda, M., and Zajickova, L. (2002) RF sputtering of composite SiOx/plasma polymer films and their basic properties. *Surf. Coat. Technol.*, **151**, 214–217.

20. Despax, B. and Raynaud, P. (2007) Deposition of ''polysiloxane'' thin films containing silver particles by an RF asymmetrical discharge. *Plasma Process. Polym.*, **4** (2), 127–134.

21. Pihosh, Y., Biederman, H., Slavinska, D., Kousal, J., Choukourov, A., Trchova, M., Mackova, A., and Boldyryeva, A. (2006) Composite SiO$_x$/fluorocarbon plasma polymer films prepared by r.f. magnetron sputtering of SiO₂ and PTFE. *Vacuum*, **81** (1), 38–44.

22. Chen, X.L., Rajeshwar, K., Timmons, R.B., Chen, J.J., and Chyan, O.M.R. (1996) Pulsed plasma polymerization of tetramethyltin: nanoscale compositional control of film chemistry. *Chem. Mater.*, **8** (5), 1067–1077.

23. Hu, J.J., Zong, Y., Wang, Y.Z., Forch, R., and Knoll, W. (2005) Sn-containing composite thin films by plasma deposition of tetramethyltin. *Thin Solid Films*, **472** (1–2), 58–63.

24. Favia, P., Cicala, G., Milella, A., Palumbo, F., Rossini, R., and d'Agostino, R. (2003) Deposition of super-hydrophobic fluorocarbon coatings in modulated RF glow discharges. *Surf. Coat. Technol.*, **169**, 609–612.

25. Kolari, K. and Hokkanen, A. (2006) Tunable hydrophilicity on a hydrophobic fluorocarbon polymer coating on silicon. *J. Vac. Sci. Technol. A*, **24** (4), 1005–1011.

26. Friedrich, J., Kuhn, G., Mix, R., Fritz, A., and Schonhals, A. (2003) Polymer surface modification with monofunctional groups of variable types and densities. *J. Adhes. Sci. Technol.*, **17** (12), 1591–1617.

27. Razavi, A. (2006) Plasma surface modification of polymeric materials. *Surf. Eng. Manuf. Appl.*, **890**, 211–215.

28. Drews, J., Goutianos, S., Kingshott, P., Hvilsted, S., Rozlosnik, N., Almdal, K., and Sorensen, B.F. (2007) Plasma polymerized thin films of maleic anhydride and 1,2-methylenedioxybenzene for improving adhesion to carbon surfaces. *J. Vac. Sci. Technol. A*, **25** (4), 1108–1117.

29. Schiller, S., Hu, J., Jenkins, A.T.A., Timmons, R.B., Sanchez-Estrada, F.S., Knoll, W., and Förch, R. (2002) Chemical structure and properties of plasma-polymerized maleic anhydride films. *Chem. Mater.*, **14** (1), 235–242.

30. Forch, R., Zhang, Z.H., and Knoll, W. (2005) Soft plasma treated surfaces: tailoring of structure and properties for biomaterial applications. *Plasma Process. Polym.*, **2** (5), 351–372.

31. Carpenter, J. and Grundmeier, G. (2005) Chemical structure and morphology of thin bilayer and composite organosilicon and fluorocarbon microwave plasma polymer films. *Surf. Coat. Technol.*, **192** (2–3), 189–198.

32. Dutta, N.K., Tran, N.D., and Choudhury, N.R. (2005) Perfluoro(methylcyclohexane) plasma polymer thin film: growth, surface morphology, and properties investigated by scanning thermal microscopy. *J. Polym. Sci. Part B Polym. Phys.*, **43** (11), 1392–1400.

33. Grundmeier, G., Thiemann, P., Carpentier, J., Shirtcliffe, N., and Stratmann, M. (2004) Tailoring of the morphology and chemical composition of thin organosilane microwave plasma polymer layers on metal substrates. *Thin Solid Films*, **446** (1), 61–71.

34. Zhang, J., Feng, X.F., Xie, H.K., Shi, Y.C., Pu, T.S., and Guo, Y. (2003) The characterization of structure-tailored plasma films deposited from the pulsed RF discharge. *Thin Solid Films*, **435** (1–2), 108–115.

35. Cho, J., Denes, F.S., and Timmons, R.B. (2006) Plasma processing approach to molecular surface tailoring of nanoparticles: improved photocatalytic activity of TiO_2. *Chem. Mater.*, **18** (13), 2989–2996.

36. Panchalingam, V., Chen, X., Savage, C.R., Timmons, R.B., and Eberhart, R.C. (1994) Molecular tailoring of surfaces via pulsed RF plasma depositions. *Plasma Deposition Polym. Thin Films*, **54**, 123–141.

37. Oehr, C. (2003) Plasma surface modification of polymers for biomedical use. *Nucl. Instrum. Methods Phys. Res. B*, **208**, 40–47.

38. Candan, S., Beck, A.J., O'Toole, L., and Short, R.D. (1998) Effects of "processing parameters" in plasma deposition: acrylic acid revisited. *J. Vac. Sci. Technol. A*, **16** (3), 1702–1709.

39. Haddow, D.B., France, R.M., Short, R.D., Bradley, J.W., and Barton, D. (2000) A mass spectrometric and ion energy study of the continuous wave plasma polymerization of acrylic acid. *Langmuir*, **16** (13), 5654–5660.

40. Dhayal, M. and Bradley, J.W. (2005) Time-resolved electric probe measurements in the pulsed-plasma

polymerization of acrylic acid. *Surf. Coat. Technol.*, **194** (1), 167–174.

41. Swindells, I., Voronin, S.A., Fotea, C., Alexander, M.R., and Bradley, J.W. (2007) Detection of negative molecular ions in acrylic acid plasma: some implications for polymerization mechanisms. *J. Phys. Chem. B*, **111** (30), 8720–8722.

42. Otoole, L. and Short, R.D. (1997) An investigation of the mechanisms of plasma polymerization of allyl alcohol. *J. Chem. Soc. Faraday Trans.*, **93** (6), 1141–1145.

43. Steen, M.L., Butoi, C.I., and Fisher, E.R. (2001) Identification of gas-phase reactive species and chemical mechanisms occurring at plasma-polymer surface interfaces. *Langmuir*, **17** (26), 8156–8166.

44. Granier, A., Borvon, G., Bousquet, A., Goullet, A., Leteinturier, C., and van der Lee, A. (2006) Mechanisms involved in the conversion of ppHMDSO films into SiO$_2$-like by oxygen plasma treatment. *Plasma Process. Polym.*, **3** (4–5), 365–373.

45. Yasuda, H.K. (1985) *Plasma Polymerization*, Academic Press.

46. Moisan, M. and Wertheimer, M.R. (1993) Comparison of microwave and RF plasmas – fundamentals and applications. *Surf. Coat. Technol.*, **59** (1–3), 1–13.

47. Wertheimer, M.R. and Moisan, M. (1994) Processing of electronic materials by microwave plasma. *Pure Appl. Chem.*, **66** (6), 1343–1352.

48. Wertheimer, M.R., Martinu, L., and Moisan, M. (1997) Microwave and dual-frequency plasma processing. *Plasma Process. Polym.*, **346**, 101–127.

49. Friedrich, J.F., Mix, R., and Kuhn, G. (2003) Functional groups bearing plasma homo and copolymer layers as adhesion promoters in metal-polymer composites. *Surf. Coat. Technol.*, **174**, 811–815.

50. Swaraj, S., Oran, U., Lippitz, A., Schulze, R.D., Friedrich, J.F., and Unger, W.E.S. (2005) Surface analysis of plasma-deposited polymer films, 4(a) – In situ characterization of plasma-deposited ethylene films by XPS and NEXAFS. *Plasma Process. Polym.*, **2** (4), 310–318.

51. d'Agostino, R. (2005) Process control and plasma modification of polymers. *J. Photopolym. Sci. Technol.*, **18** (2), 245–249.

52. Tada, M., Yamamoto, H., Ito, F., Takeuchi, T., Furutake, N., and Hayashi, Y. (2007) Chemical structure effects of ring-type siloxane precursors on properties of plasma-polymerized porous SiOCH films. *J. Electrochem. Soc.*, **154** (7), D354–D361.

53. Iriyama, Y. and Hanawa, M. (2001) Plasma polymerization of pyrrole and structures and properties of the polymerized films. *Polym. J.*, **33** (5), 419–423.

54. Morales, J., Olayo, M.G., Cruz, G.J., and Olayo, R. (2002) Synthesis by plasma and characterization of bilayer aniline-pyrrole thin films doped with iodine. *J. Polym. Sci. Part B Polym. Phys.*, **40** (17), 1850–1856.

55. Mitu, B., Bauer-Gogonea, S., Leonhartsberger, H., Lindner, M., Bauer, S., and Dinescu, G. (2003) Plasma-deposited parylene-like thin films: process and material properties. *Surf. Coat. Technol.*, **174**, 124–130.

56. Park, Z.T., Choi, Y.S., Kim, J.G., and Boo, J.H. (2003) Electrochemical reliability of plasma-polymerized cyclohexane films deposited on copper in microelectronic devices. *J. Mater. Sci. Lett.*, **22** (13), 945–947.

57. Han, L.C.M., Timmons, R.B., Lee, W.W., Chen, Y.Y., and Hu, Z.B. (1998) Pulsed plasma polymerization of pentafluorostyrene: Synthesis of low dielectric constant films. *J. Appl. Phys.*, **84** (1), 439–444.

58. Han, L.M., Timmons, R.B., Bogdal, D., and Pielichowski, J. (1998) Ring retention via pulsed plasma polymerization of heterocyclic aromatic compounds. *Chem. Mater.*, **10** (5), 1422–1429.

59. Hynes, A. and Badyal, J.P.S. (1998) Selective incorporation of perfluorinated phenyl rings during pulsed plasma polymerization of perfluoroallylbenzene. *Chem. Mater.*, **10** (8), 2177–2182.

60. Mackie, N.M., Castner, D.G., and Fisher, E.R. (1998) Characterization of pulsed-plasma-polymerized aromatic films. *Langmuir*, **14** (5), 1227–1235.

61. Chowdhury, F.U.Z. and Bhuiyan, A.H. (2000) Dielectric properties of plasma-polymerized diphenyl thin films. *Thin Solid Films*, **370** (1–2), 78–84.

62. Ademovic, Z., Wei, J., Winther-Jensen, B., Hou, X.L., and Kingshott, P. (2005) Surface modification of PET films using pulsed AC plasma polymerization aimed at preventing protein adsorption. *Plasma Process. Polym.*, **2** (1), 53–63.

63. Groenewoud, L.M.H., Engbers, G.H.M., and Feijen, J. (2003) Plasma polymerization of thiophene derivatives. *Langmuir*, **19** (4), 1368–1374.

64. Ryan, M.E., Hynes, A.M., Wheale, S.H., Badyal, J.P.S., Hardacre, C., and Ormerod, R.M. (1996) Plasma polymerization of 2-iodothiophene. *Chem. Mater.*, **8** (4), 916–921.

65. Hosono, K., Matsubara, I., Murayama, N., Shin, W., and Izu, N. (2004) Effects of discharge power on the structure and electrical properties of plasma polymerized polypyrrole films. *Mater. Lett.*, **58** (7–8), 1371–1374.

66. Silverstein, M.S. and Visoly-Fisher, I. (2002) Plasma polymerized thiophene: molecular structure and electrical properties. *Polymer*, **43** (1), 11–20.

67. Groenewoud, L.M.H., Engbers, G.H.M., and Feijen, J. (2000) Plasma polymerization of thiophene derivatives. *Abstr. Pap. Am. Chem. Soc.*, **220**, 158–PMSE.

68. Groenewoud, L.M.H., Engbers, G.H.M., Terlingen, J.G.A., Wormeester, H., and Feijen, J. (2000) Pulsed plasma polymerization of thiophene. *Langmuir*, **16** (15), 6278–6286.

69. Kiesow, A. and Heilmann, A. (1999) Deposition and properties of plasma polymer films made from thiophenes. *Thin Solid Films*, **344**, 338–341.

70. Dams, R., Vangeneagden, D., and Vanderzande, D. (2006) Plasma deposition of thiophene derivatives under atmospheric pressure. *Chem. Vap. Deposit.*, **12** (12), 719–727.

71. Tarducci, C., Kinmond, E.J., Badyal, J.P.S., Brewer, S.A., and Willis, C. (2000) Epoxide-functionalized solid surfaces. *Chem. Mater.*, **12** (7), 1884–1889.

72. Chu, L., Knoll, W., and Föerch, R. (2008) Plasma polymerized epoxide functional surfaces for DNA probe immobilization. *Biosens. Bioelectron.*, **24** 118–122.

73. Zhang, B.M., Timmons, R.B., Knoll, W., and Förch R. (2003) Surface plasmon resonance studies of protein binding on plasma polymerized di(ethylene glycol) mono vinyl ether films. *Langmuir*, **19**, 4765–4770.

74. Beyer, D., Knoll, W., Ringsdorf, H., Wang, J.H., Timmons, R.B., and Sluka, P. (1997) Reduced protein adsorption on plastics via direct plasma deposition of triethylene glycol monoallyl ether. *J. Biomed. Mater. Res.*, **36** (2), 181–189.

75. Chu, L.Q., Knoll, W., and Forch, R. (2006) Pulsed plasma polymerized di(ethylene glycol) monovinyl ether coatings for nonfouling surfaces. *Chem. Mater.*, **18** (20), 4840–4844.

76. Shen, M.C., Martinson, L., Wagner, M.S., Castner, D.G., Ratner, B.D., and Horbett, T.A. (2002) PEO-like plasma polymerized tetraglyme surface interactions with leukocytes and proteins: in vitro and in vivo studies. *J. Biomater. Sci. Polym. Ed.*, **13** (4), 367–390.

77. Mar, M.N., Ratner, B.D., and Yee, S.S. (1999) An intrinsically protein-resistant surface plasmon resonance biosensor based upon a RF-plasma-deposited thin film. *Sens. Actuator. B*, **54** (1–2), 125–131.

78. Alexander, M.R., Payan, S., and Duc, T.M. (1998) Interfacial interactions of plasma-polymerized acrylic acid and an oxidized aluminium surface investigated using XPS, FTIR and poly(acrylic acid) as a model compound. *Surf. Interface Anal.*, **26** (13), 961–973.

79. Daw, R., Candan, S., Beck, A.J., Devlin, A.J., Brook, I.M., MacNeil, S., Dawson, R.A., and Short, R.D. (1998) Plasma copolymer surfaces of acrylic acid 1,7 octadiene: Surface characterisation and the attachment of ROS 17/2.8 osteoblast-like cells. *Biomaterials*, **19** (19), 1717–1725.

80. Alexander, M.R. and Duc, T.M. (1999) A study of the interaction of acrylic acid/1,7-octadiene plasma deposits with water and other solvents. *Polymer*, **40** (20), 5479–5488.

81. Chen-Yang, Y.W., Chen, C.W., Tseng, S.C., Wu, Y.Z., Yang, H.C., and Kau, J.Y. (2003) Acrylic acid plasma polymerization to modify the surface of expanded poly(tetrafluoroethylene). *Abstr. Pap. Am. Chem. Soc.*, **226**, U368–U368.

82. Voronin, S.A., Zelzer, M., Fotea, C., Alexander, M.R., and Bradley, J.W. (2007) Pulsed and continuous wave acrylic acid radio frequency plasma deposits: plasma and surface chemistry. *J. Phys. Chem. B*, **111** (13), 3419–3429.

83. Gong, X., Dai, L., Griesser, H.J., and Mau, A.W.H. (2000) Surface immobilization of poly(ethylene oxide): structure and properties. *J. Polym. Sci. Part B Polym. Phys.*, **38** (17), 2323–2332.

84. Rafik, M., Mas, A., Guimon, M.F., Guimon, C., Elharfi, A., and Schue, F. (2003) Plasma-modified poly(vinyl alcohol) membranes for the dehydration of ethanol. *Polym. Int.*, **52** (7), 1222–1229.

85. Otoole, L., Mayhew, C.A., and Short, R.D. (1997) On the plasma polymerization of allyl alcohol: An investigation of ion-molecule reactions using a selected ion flow tube. *J. Chem. Soc. Faraday Trans.*, **93** (10), 1961–1964.

86. Leich, M.A., Mackie, N.M., Williams, K.L., and Fisher, E.R. (1998) Pulsed plasma polymerization of benzaldehyde for retention of the aldehyde functional group. *Macromolecules*, **31** (22), 7618–7626.

87. Jenkins, A.T.A., Hu, J., Wang, Y.Z., Schiller, S., Föerch, R., and Knoll, W. (2000) Pulsed plasma deposited maleic anhydride thin films as supports for lipid bilayers. *Langmuir*, **16** (16), 6381–6384.

88. Francesch, L., Garreta, E., Balcells, M., Edelman, E.R., and Borros, S. (2005) Fabrication of bioactive surfaces by plasma polymerization techniques using a novel acrylate-derived monomer. *Plasma Process. Polym.*, **2** (8), 605–611.

89. Francesch, L., Borros, S., Knoll, W., and Forch, R. (2007) Surface reactivity of pulsed-plasma polymerized pentafluorophenyl methacrylate (PFM) toward amines and proteins in solution. *Langmuir*, **23** (7), 3927–3931.

90. Tarducci, C., Schofield, W.C.E., Badyal, J.P.S., Brewer, S.A., and Willis, C. (2001) Cyano-functionalized solid surfaces. *Chem. Mater.*, **13** (5), 1800–1803.

91. Teare, D.O.H., Schofield, W.C.E., Garrod, R.P., and Badyal, J.P.S. (2005) Poly(N-acryloylsarcosine methyl ester) protein-resistant surfaces. *J. Phys. Chem. B*, **109** (44), 20923–20928.

92. van Os, M.T., Menges, B., Förch, R., Timmons, R.B., Vancso, G.J., and Knoll, W. (1999) Pulsed plasma deposition of allylamine: aging and effects of solvent treatment. *Mater. Res. Soc. Symp. Proc.*, **544**, 45–50.

93. Muller, M. and Oehr, C. (1999) Plasma aminofunctionalisation of PVDF microfiltration membranes: comparison of the in plasma modifications with a grafting method using ESCA and an amino-selective fluorescent probe. *Surf. Coat. Technol.*, **119**, 802–807.

94. Harsch, A., Calderon, J., Timmons, R.B., and Gross, G.W. (2000) Pulsed plasma deposition of allylamine on polysiloxane: a stable surface for neuronal cell adhesion. *J. Neurosci. Methods*, **98** (2), 135–144.

95. Beck, A.J., Candan, S., Short, R.D., Goodyear, A., and Braithwaite, N.S.J. (2001) The role of ions in the plasma polymerization of allylamine. *J. Phys. Chem. B*, **105** (24), 5730–5736.

96. Lejeune, M., Bretagnol, F., Ceccone, G., Colpo, P., and Rossi, F. (2006) Microstructural evolution of allylamine polymerized plasma films. *Surf. Coat. Technol.*, **200** (20–21), 5902–5907.

97. Choukourov, A., Biederman, H., Slavinska, D., Hanley, L., Grinevich, A., Boldyryeva, H., and Mackova, A. (2005)

Mechanistic studies of plasma polymerization of allylamine. *J. Phys. Chem. B,* **109** (48), 23086–23095.

98. Choukourov, A., Biederman, H., Kholodkov, I., Slavinska, D., Trchova, M., and Hollander, A. (2004) Properties of amine-containing coatings prepared by plasma polymerization. *J. Appl. Polym. Sci.,* **92** (2), 979–990.

99. Choukourov, A., Biederman, H., Slavinska, D., Trchova, M., and Hollander, A. (2003) The influence of pulse parameters on film composition during pulsed plasma polymerization of diaminocyclohexane. *Surf. Coat. Technol.,* **174**, 863–866.

100. Kim, J., Park, H., Jung, D., and Kim, S. (2003) Protein immobilization on plasma-polymerized ethylenediamine-coated glass slides. *Anal. Biochem.,* **313** (1), 41–45.

101. Mutlu, S., Cokeliler, D., Shard, A., Goktas, H., Ozansoy, B., and Mutlu, M. (2008) Preparation and characterization of ethylenediamine and cysteamine plasma polymerized films on piezoelectric quartz crystal surfaces for a biosensor. *Thin Solid Films,* **516**, 1249–1255.

102. Zhang, Z.H., Chen, Q., Knoll, W., Föerch, R., Holcomb, R., and Roitman, D. (2003) Plasma polymer film structure and DNA probe immobilization. *Macromolecules,* **36** (20), 7689–7694.

103. Zhang, Z., Knoll, W., Föerch, R., Holcomb, R., and Roitman, D. (2005) DNA hybridization on plasma-polymerized allylamine. *Macromolecules,* **38** (4), 1271–1276.

104. Chu, L.Q., Forch, R., and Knoll, W. (2007) Surface-plasmon-enhanced fluorescence spectroscopy for DNA detection using fluorescently labeled PNA as "DNA indicator". *Ange. Chem. Int. Ed.,* **46** (26), 4944–4947.

105. Wagner, M.S., Shen, M., Horbett, T.A., and Castner, D.G. (2003) Quantitative analysis of binary adsorbed protein films by time of flight secondary ion mass spectrometry. *J. Biomed. Mater. Res. Part A,* **64A** (1), 1–11.

106. Forch, R., Chifen, A.N., Bousquet, A., Khor, H.L., Jungblut, M., Chu, L.Q.,

Zhang, Z., Osey-Mensah, I., Sinner, E.K., and Knoll, W. (2007) Recent and expected roles of plasma-polymerized films for biomedical applications. *Chem. Vap. Deposit.,* **13** (6–7), 280–294.

107. Hook, A.L., Thissen, H., Hayes, J.P., and Voelcker, N.H. (2006) Spatially controlled electro-stimulated DNA adsorption and desorption for biochip applications. *Biosens. Bioelectron.,* **21** (11), 2137–2145.

108. Calderon, J.G., and Timmons, R.B. (1998) Surface molecular tailoring via pulsed plasma-generated acryloyl chloride polymers: synthesis and reactivity. *Macromolecules,* **31** (10), 3216–3224.

109. Swaraj, S., Oran, U., Friedrich, J.F., Lippitz, A., and Unger, W.E.S. (2007) Surface chemical analysis of plasma-deposited copolymer films prepared from feed gas mixtures of ethylene or styrene with allyl alcohol. *Plasma Process. Polym.,* **4** (4), 376–389.

110. Kawahara, J., Nakano, A., Kunimi, N., Kinoshita, K., Hayashi, Y., Ishikawa, A., Seino, Y., Ogata, T., Sonoda, Y., Yoshino, T., Goto, T., Takada, S., Miyoshi, H., Matsuo, H., and Kikkawa, T. (2007) Plasma-enhanced co-polymerization of organo-siloxane and hydrocarbon for low-k/Cu interconnects. *Jpn. J. Appl. Phys. Part 1: Regul. Pap. Brief Commun. Rev. Pap.,* **46** (7A), 4064–4069.

111. Tarducci, C., Schofield, W.C.E., Badyal, J.P.S., Brewer, S.A., and Willis, C. (2002) Synthesis of cross-linked ethylene glycol dimethacrylate and cyclic methacrylic anhydride polymer structures by pulsed plasma deposition. *Macromolecules,* **35** (23), 8724–8727.

112. France, R.M., Short, R.D., Duval, E., Jones, F.R., Dawson, R.A., and MacNeil, S. (1998) Plasma copolymerization of allyl alcohol 1,7-octadiene: surface characterization and attachment of human keratinocytes. *Chem. Mater.,* **10** (4), 1176–1183.

113. Higham, M.C., Dawson, R., Szabo, M., Short, R., Haddow, D.B., and MacNeil, S. (2003) Development of a stable chemically defined surface for the culture of human keratinocytes under serum-free conditions for clinical use. *Tissue Eng.*, **9** (5), 919–930.

114. Oran, U., Swaraj, S., Friedrich, J.F., and Unger, W.E.S. (2006) Static ToF-SIMS analysis of plasma chemically deposited ethylene/allyl alcohol co-polymer films. *Appl. Surf. Sci.*, **252** (19), 6588–6590.

115. Kunimi, N., Kawahara, J., Kinoshita, K., Nakano, A., Komatsu, M., and Kikkawa, T. (2005) A novel organic low-k film deposited by plasma-enhanced co-polymerization. *Mater. Technol. Reliab. Adv. Interconnects*, **863**, 103–108.

116. Spanos, C.G., Badyal, J.P.S., Goodwin, A.J., and Merlin, P.J. (2005) Pulsed plasmachemical deposition of polymeric salt networks. *Polymer*, **46** (21), 8908–8912.

117. Han, L.C.M., Rajeshwar, K., and Timmons, R.B. (1997) Film chemistry control and electrochemical properties of pulsed plasma polymerized ferrocene and vinylferrocene. *Langmuir*, **13** (22), 5941–5950.

118. Dalacu, D., Brown, A.P., Klemberg-Sapieha, J.E., Martinu, L., Wertheimer, M.R., Najafi, S.I., and Andrews, M.A. (1999) Characterization of plasma/deposited Au/fluoropolymer nanocomposite films for nonlinear optical application. *Plasma Deposition Treat. Polym.*, **544**, 167–172.

119. Uglov, V.V., Kuleshov, A.K., Astashynskaya, M.V., Anishchik, V.M., Dub, S.N., Thiery, F., and Pauleau, Y. (2005) Mechanical properties of copper/carbon nanocomposite films formed by microwave plasma assisted deposition techniques from argon-methane and argon-acetylene gas mixtures. *Compos. Sci. Technol.*, **65** (5), 785–791.

120. Zhang, W., Zhang, Y.H., Ji, J.H., Zhao, J., Yan, Q., and Chu, P.K. (2006) Antimicrobial properties of copper plasma-modified polyethylene. *Polymer*, **47** (21), 7441–7445.

121. Martin, J., Kiesow, A., Heilmann, A., and Wannemacher, R. (2001) Laser microstructuring and scanning microscopy of plasma polymer-silver

composite layers. *Appl. Opt.*, **40** (31), 5726–5730.

122. Hegemann, D., Hossain, M.M., and Balazs, D.J. (2007) Nanostructured plasma coatings to obtain multifunctional textile surfaces. *Prog. Org. Coat.*, **58** (2–3), 237–240.

123. Balazs, D.J., Hollenstein, C., and Mathieu, H.J. (2005) Fluoropolymer coating of medical grade poly(vinyl chloride) by plasma-enhanced chemical vapor deposition techniques. *Plasma Process. Polym.*, **2** (2), 104–111.

124. Dennler, G., Houdayer, A., Latreche, M., Segui, Y., and Wertheimer, M.R. (2001) Studies of the earliest stages of plasma-enhanced chemical vapor deposition of SiO_2 on polymeric substrates. *Thin Solid Films*, **382** (1–2), 1–3.

125. Jacobsen, V., Menges, B., Scheller, A., Föerch, R., Mittler, S., and Knoll, W. (2001) In situ film diagnostics during plasma polymerization using waveguide mode spectroscopy. *Surf. Coat. Technol.*, **142**, 1105–1108.

126. Jacobsen, V., Menges, B., Forch, R., Mittler, S., and Knoll, W. (2002) *In-situ* thin film diagnostics using waveguide mode spectroscopy. *Thin Solid Films*, **409** (2), 185–193.

127. Janasek, D., Spohn, U., Kiesow, A., and Heilmann, A. (2001) Enzyme modified membranes protected by plasma polymer layers. *Sens. Actuators B Chem.*, **78** (1–3), 228–231.

128. Chu, L.Q., Knoll, W., and Forch, R. (2006) Biologically multifunctional surfaces using plasma polymerization methods. *Plasma Process. Polym.*, **3** (6–7), 498–505.

129. Michaeli, W., Fonteiner, I., and Goebel, S. (2000) Characterization of the layer growth of plasma-polymerized HMDSO coatings on polycarbonate. *Macromol. Mater. Eng.*, **284** (11–12), 30–34.

130. Singamaneni, S., Le Mieux, M.C., Jiang, H., Bunning, T.J., and Tsukruk, V.V. (2007) Negative thermal expansion in ultrathin plasma polymerized films. *Chem. Mater.*, **19** (2), 129–131.

131. Haidopoulos, M., Larrieu, J., Horgnies, M., Houssiau, L., and Pireaux, J.J. (2006) A comparative study between

inductively and capacitively coupled plasma deposited polystyrene films: chemical and morphological characterizations. *Surf. Interface Anal.*, **38** (9), 1266–1275.

132. Potter, W., Ward, A.J., and Short, R.D. (1994) The stability of plasma polymers. 1. A preliminary XPS study of the photostability of plasma-polymerized styrene and methyl- methacrylate. *Polym. Degrad. Stab.*, **43** (3), 385–391.

133. Chu, L.Q., Forch, R., and Knoll, W. (2006) Pulsed plasma polymerized maleic anhydride films in humid air and in aqueous solutions studied with optical waveguide spectroscopy. *Langmuir*, **22** (6), 2822–2826.

134. Drews, J., Launay, H., Hansen, C.M., West, K., Hvilsted, S., Kingshott, P., and Almdal, K. (2008) Hydrolysis and stability of thin pulsed plasma polymerised maleic anhydride coatings. *Appl. Surface Sci.*, **254** (15), 4720–4725.

135. Favia, P., Sardella, E., Gristina, R., Milella, A., and d'Agostino, R. (2002) Functionalization of biomedical polymers by means of plasma processes: plasma treated polymers with limited hydrophobic recovery and PE-CVD of -COOH functional coatings. *J. Photopolym. Sci. Technol.*, **15** (2), 341–350.

136. Igarashi, S., Itakura, A.N., Toda, M., Kitajima, M., Chu, L., Chifene, A.N., Forch, R., and Berger, R. (2006) Swelling signals of polymer films measured by a combination of micromechanical cantilever sensor and surface plasmon resonance spectroscopy. *Sens. Actuators B Chem.*, **117** (1), 43–49.

137. Zhang, Z., Chen, Q., Knoll, W., and Forch, R. (2003) Effect of aqueous solution on functional plasma polymerized films. *Surf. Coat. Technol.*, **174**, 588–590.

138. van Os, M.T., Menges, B., Forch, R., Knoll, W., Timmons, R.B., and Vancso, G.J. (1999) Thin film plasma deposition of allylamine; effects of solvent treatment. *Plasma Deposition Treat. Polym.*, **544**, 45–50.

139. van Os, M.T., Menges, B., Föerch, R., Vancso, G.J., and Knoll, W. (1999) Characterization of plasma-polymerized allylamine using waveguide mode spectroscopy. *Chem. Mater.*, **11** (11), 3252–3257.

140. Pan, Y.V., Wesley, R.A., Luginbuhl, R., Denton, D.D., and Ratner, B.D. (2001) Plasma polymerized N-isopropylacrylamide: synthesis and characterization of a smart thermally responsive coating. *Biomacromolecules*, **2** (1), 32–36.

141. Ratner, B.D., Cheng, X.H., Wang, Y.B., Hanein, Y., and Bohringer, K. (2003) Temperature-responsive polymeric surface modifications by plasma polymerization: cell and protein interactions. *Abstr. Pap. Am. Chem. Soc.*, **225**, U582–U582.

142. Teare, D.O.H., Barwick, D.C., Schofield, W.C.E., Garrod, R.P., Beeby, A., and Badyal, J.P.S. (2005) Functionalization of solid surfaces with thermoresponsive protein-resistant films. *J. Phys. Chem. B*, **109** (47), 22407–22412.

143. Tamirisa, P.A. and Hess, D.W. (2006) Water and moisture uptake by plasma polymerized thermoresponsive hydrogel films. *Macromolecules*, **39** (20), 7092–7097.

144. Bullett, N.A., Talib, R.A., Short, R.D., McArthur, S.L., and Shard, A.G. (2006) Chemical and thermo-responsive characterisation of surfaces formed by plasma polymerization of N-isopropyl acrylamide. *Surf. Interface Anal.*, **38** (7), 1109–1116.

145. Chu, L.Q., Zou, X.N., Knoll, W., and Forch, R. (2008) Thermosensitive surfaces fabricated by plasma polymerization of N,N-diethylacrylamide. *Surf. Coat. Technol.*, **202**, 2047–2051.

146. Lampitt, R.A., Crowther, J.M., and Badyal, J.P.S. (2000) Switching liquid repellent surfaces. *J. Phys. Chem. B*, **104** (44), 10329–10331.

147. Bhattacharyya, D., Pillai, K., Chyan, O.M.R., Tang, L.P., and Timmons, R.B. (2007) A new class of thin film hydrogels produced by plasma polymerization. *Chem. Mater.*, **19** (9), 2222–2228.

148. Cheng, X.H., Wang, Y.B., Hanein, Y., Bohringer, K.F., and Ratner, B.D. (2004) Novel cell patterning using microheater-controlled thermoresponsive plasma films. *J. Biomed. Mater. Res. Part A*, **70A** (2), 159–168.

149. Susut, C. and Timmons, R.B. (2005) Plasma enhanced chemical vapor depositions to encapsulate crystals in thin polymeric films: a new approach to controlling drug release rates. *Int. J. Pharm.*, **288** (2), 253–261.

9
Preparation of Polymer Thin Films by Physical Vapor Deposition

Hiroaki Usui

9.1
Introduction

Thin-film formation and interface control have played important roles in the recent development of organic devices. For example, the organic light-emitting diodes (OLEDs) were made possible by stacking nanometer-thick organic layers, enabling current flow through resistive organic materials, and confining excitons at the interface [1]. Most of the small molecules can be easily deposited by physical vapor deposition (PVD), which has been used as one of the standard techniques for fabricating solid-state organic devices such as OLED and thin-film transistors under high vacuum. The PVD has advantages in preparing highly pure thin films and multi-layer structures with high controllability. The technique is also compatible with the conventional semiconductor processes such as patterning through shadow masking and formation of electrical contacts by depositing metal electrodes.

Another class of organic polymer materials processing, that is, polymers, that has been deposited mainly by wet-coating processes from their solutions or dispersions, is desirable. Such processes, including a variety of coating and printing techniques, have advantages in productivity and cost effectiveness. However, it is not always possible to use wet-coating methods for constructing well-controlled microstructures that are required for devices. Moreover, organic solvents used in the wet-coating process can cause environmental problems through emission of volatile organic compounds (VOCs). In this respect, it is important to develop a technique for preparing polymeric thin films by vacuum-based dry-coating methods. Since polymers do not evaporate like metals or small organic molecules, the application of PVD for polymer deposition is also not straightforward. For one thing, the polymer volatility is limited. However, there are several methods that can still enable polymer deposition by PVD. This chapter introduces some examples of PVD of polymer thin films and their applications to organic devices. After giving a brief overview of polymer PVD in Section 1.2, recent examples of polymer PVD are introduced in the latter sections according to the type of chemical reaction involved in the film deposition.

Functional Polymer Films, First Edition. Edited by Wolfgang Knoll and Rigoberto C. Advincula.
© 2011 Wiley-VCH Verlag GmbH & Co. KGaA. Published 2011 by Wiley-VCH Verlag GmbH & Co. KGaA.

9.2
Overview of PVD of Polymers

9.2.1
PVD of Organic Molecules

PVD has a long history in preparing thin films of small organic molecules as well as metals and inorganic compounds. For organic molecules to be vapor deposited, they need to have sufficient thermal stability to endure the thermal evaporation process without decomposition. A typical example is vapor deposition of polyclic aromatic compounds such as phthalocyanines [2]. Epitaxial films of well-defined crystal structures have been explored to reveal the growth mechanism of organic thin films by PVD [3]. PVD can also be applied for deposition of linear molecules that consist of alkyl chains [4]. Preferential molecular orientation [5] and epitaxial growth [6] of fatty acids have been reported through crystallographic and microscopic investigations. One of the largest molecules that belong to this group is tetratetracontane (n-$C_{44}H_{90}$) [7].

Although tetratetracontane can be considered as a model compound of polyethylene (PE), there have been only a limited number of reports on PVD that use a polymer as the source material for evaporation [8]. All the materials mentioned above have well-defined molecular structures and can be vaporized without thermal decomposition, giving a definite value of vapor pressure under thermal equilibrium. On the other hand, most of the polymers do not evaporate due to the strong intermolecular force that comes from the large molecular size. Instead of evaporation, polymers undergo thermal degradation, such as depolymerization and random scission of chemical bonds when heated to such a temperature that causes weight loss of the material [9]. Therefore, a certain strategy is required to prepare polymer thin films by PVD, as schematically illustrated in Figure 9.1.

9.2.2
Vapor Deposition of Polymers

The most primitive PVD of polymer is to perform the conventional vapor deposition simply by using a polymer as the source material for evaporation, thereby directly evaporating the polymer as shown in Figure 9.1a. This method can be applied to those materials that have weak intermolecular interaction. PE, which has the simplest structure of all the polymers, is a typical example that can be directly evaporated in high vacuum [10]. Other polymers that have been deposited by the direct evaporation include poly(vinylidene fluoride) [11] and polytetrafluoroethylene (PTFE) or Teflon [12, 13]. Although the deposited films of these materials have a fragmented structure compared to their source materials, the PVD technique is reported to achieve high controllability in film formation. Examples of direct evaporation of PE and PTFE are described in Section 1.3.

Since the evaporation of polymer inevitably involves some extent of thermal degradation, the direct evaporation method can be applied only in limited cases.

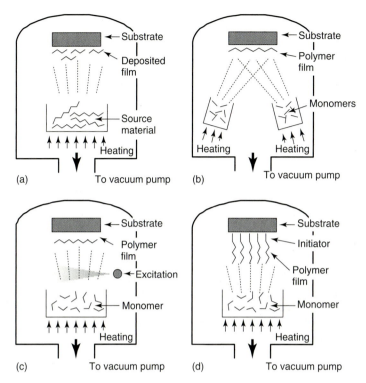

Figure 9.1 Four methods for physical vapor deposition of polymer thin films. (a) Direct evaporation, (b) stepwise reaction by coevaporation of two monomers, (c) chain reaction of single monomer assisted by electron or UV excitation, and (d) surface-initiated deposition polymerization.

On the other hand, the vapor-deposition polymerization is an *in-situ* technique to prepare polymer thin films by a bottom-up approach, namely by polymerizing small molecules that are vapor deposited onto the substrate surface. Vapor deposition of polymer thin films was reported for ring-opening polymerization of di-*p*-xylene [14] and 6-aminocaproic acid [15] in 1960s and 1970s, respectively. However, these depositions were achieved in vapor phase under low-vacuum conditions, and can be classified as chemical vapor deposition (CVD), rather than PVD. PVD of polymer was first achieved in 1980s to prepare polyimide by coevaporating diamine and carboxylic acid dianhydride monomers in high vacuum [16–18] in a manner illustrated in Figure 9.1b. Under the high-vacuum condition, where the mean free path of molecules is much longer than the dimension of the vacuum chamber, collisions of monomers do not occur in the volume of chamber but on the substrate surface after the monomers are adsorbed on its surface. The film growth is controlled by molecular kinetics rather than phase transition under thermal equilibrium. As a consequence, the PVD can provide unique controllability of microstructure such as molecular orientation. Using this technique, polyimide thin films can be produced by condensation polymerization by codeposition of two bifunctional monomers followed by annealing [19, 20]. The same method can be employed for depositing such polymers as polyamide [21, 22]. In a similar manner, polyurea (PU) can be deposited by coevaporation of diamine and diisocyanate monomers [23]. In this case, the reaction proceeds by polyaddition, which yields the product polymer in the as-deposited state without a post-deposition annealing process.

The coevaporation method of Figure 9.1b produces polymer thin films by stepwise reaction of bifunctional monomers (Type I or Type II). The nature of this type of reaction requires a precise balance and stoichiometry of monomer supply to achieve a high degree of polymerization. As far as the degree of polymerization is concerned, a chain reaction, such as radical polymerization, is favorable. This can be achieved by evaporating a vinyl or acrylic monomer, as illustrated in Figure 9.1c, where polymerization reaction is initiated by generating radicals in the course of deposition. This method is technically advantageous because higher molecular weight (MW) polymers can be easily obtained by using only one species of monomer. Vinyl or acryl monomers that react by radical polymerization are convenient for this type of deposition. Therefore, versatile design of polymer material is possible by developing monomers by attaching vinyl or acryl groups to arbitrary functional units. The radicals can be generated, for example, by installing a tungsten hot filament in the evaporation chamber [24–26], or by irradiating UV light to the substrate surface [27–29]. Some monomers can be thermally polymerized by post-deposition annealing [30]. A novel technique to initiate the polymerization by irradiating electron beam to the evaporated monomers [31, 32] is described in Section 1.5.

In most cases, the vapor-deposited films are physically adsorbed on the substrate surface without forming specific chemical bonds at the film/substrate interface. This means that PVD can form thin films on any kind of substrate material regardless of the chemical properties of the substrate surface. On the other hand, the film/substrate interface can cause such problems as poor adhesion, structural defects, electronic trapping, and charge-injection barrier. The surface-initiated deposition polymerization, which is illustrated in Figure 9.1d, was devised to solve this issue by forming covalent bonds at the film/substrate interface. This can be achieved by combining the vapor-deposition polymerization with a self-assembled monolayer (SAM) that works as a polymerization-initiating reagent [33]. Frank and coworkers [34, 35] applied this technique for depositing amino acids to form polypeptides, and revealed unique film-growth characteristics that have not been observed by the conventional solution process. They achieved film deposition in low-vacuum conditions, and the film-formation process can be characterized as in CVD, but the concept also works in high-vacuum conditions, as described in Section 1.6 [36].

9.2.3
Ionization-Assisted Vapor Deposition

In the PVD process, the film material is supplied to the substrate surface as a molecular flow in high vacuum without making molecular collisions in the volume of the vacuum chamber. As a consequence, the film-growth process is in a thermally nonequilibrium condition. This is in a great contrast to the CVD process, where the material is supplied as a viscous flow, and the film grows in contact with vapor under thermal equilibrium. Due to the thermally nonequilibrium nature of the process, the films by PVD frequently have imperfections such as microscopic voids, defects, dangling bonds, and nonstoichiometric products. One method to reduce such imperfections is to increase the substrate temperature during the film growth, which assists the surface migration of the molecules. As an alternative for supplying the migration energy by thermal activation, kinetic energy can be supplied to the impinging molecules by means of ion acceleration. The energy of molecules under thermal equilibrium is described by Boltzmann distribution, and is of the order of kT, where k is Boltzmann's constant and T is the temperature. On the other hand, charged particles such as ions can be accelerated to an arbitrary energy simply by applying a bias voltage. The effect of electric acceleration can be roughly estimated by simply putting $qV = kT$, where q is the ionic charge and V is the bias voltage. This equation shows that the electric acceleration of 1 eV gives a kinetic energy that corresponds to the thermal energy at a temperature as high as 10^4 K. According to this consideration, a variety of ion-based film-deposition techniques has been developed, including ion plating and plasma-assisted deposition. It is reported that the impingement of energetic ions influences the film-growth kinetics and leads to improvement of film characteristics [37].

The ionization-assisted vapor deposition (IAD), which is schematically illustrated in Figure 9.2, is one of the most effective ion-based film-deposition methods. IAD can be achieved by relatively simple modification to the standard PVD method. In IAD, a part of the evaporated material is ionized by electron impact using an electron source placed above the evaporator, and then accelerated toward the substrate by applying an ion acceleration voltage to the substrate. The fraction of ions is normally of the order of 1% of the evaporated molecules. However, the film characteristics, such as crystallinity, packing density, film morphology, and adhesion strength, are known to be largely influenced by the ion acceleration [38]. The IAD method does not resort to plasma discharge for generating the ions. Since

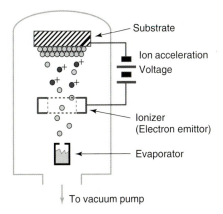

Substrate

Ion acceleration Voltage

Ionizer (Electron emittor)

Evaporator

To vacuum pump

Figure 9.2 Schematic diagram of IAD.

the film-growth proceeds in high vacuum without excessive molecular collisions, the IAD is also fitted for the deposition of organic compounds. Application of the IAD for direct evaporation of polymers is described in Section 1.3. IAD can also be applied for dipole orientation, as shown in Section 1.4.

9.2.4
Other Vacuum-Based Deposition Methods

Other well-known vacuum-based polymer deposition techniques include plasma polymerization and sputter deposition. These methods can be categorized rather as CVD, and are mentioned here only briefly. Plasma polymerization is a method that forms polymeric thin films by condensation from glow discharge of a monomer gas. As a requirement for sustaining the glow discharge, the plasma polymerization is operated in low-vacuum conditions, in the range of 0.1–10 Pa. Plasma polymerization can produce thin films of polymers that have a highly cross-linked or tight network structure. The films are characterized by high thermal and chemical stability, insolubility, strong adhesion to the substrate, low pinhole density, and so on. As a CVD-like process, uniform and conformal film formation is possible on a surface of complicated structure. On the other hand, the chemical structure of the film is not necessarily discernible. The film property is highly dependent on the glow-discharge conditions, yielding a variety of deposits that range from diamond-like carbon to oil-like liquid even from the same monomer material. A variety of monomer gases can be used for plasma polymerization. However, the molecular structures of the deposited films do not necessarily retain that of the monomer. The reaction of plasma polymerization is complicated, and involves both the polymer formation and ablation processes concurrently. Comprehensive descriptions of plasma polymerization have been compiled in books [39, 40], and the recent development has been reviewed in the literature [41].

Sputter deposition is a technique that atomizes a target by ion sputtering, that is, by bombardment of ions a in glow-discharge plasma. Since the atomization process in sputtering does not resort to thermal evaporation, sputter deposition has a film-formation capability of refractory materials. In addition, the particles sputtered from the target have kinetic energy much higher than the thermal energy, leading to improvement of film characteristics such as packing density, adhesion strength, and surface smoothness. There have been attempts to utilize the sputter deposition for film formation of polymers, especially fluoropolymers [42, 43]. Upon ion bombardment, a polymer target emits volatile fragments into the plasma instead of the polymer molecules. The sputtered small molecular units work as monomer gas, and produce thin films by a mechanism similar to that of the plasma polymerization. As a consequence, the polymer films deposited by sputter deposition have similar characteristics to those by plasma polymerization [44, 45]. Recent issues in sputter deposition of polymers have been reviewed in the literature [46].

The glow discharge involved in both the plasma polymerization and the sputter deposition causes complicated collisions between molecules and electrons, generating various ionic and radical species. This is why these methods are fitted

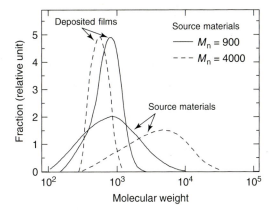

Figure 9.3 Molecular weight of PE source materials and deposited films for source molecular weight M_n of 900 and 4000.

for producing polymers of a robust network but are unable to control molecular structure precisely or the formation of high-MW polymers. In contrast to the plasma-assisted deposition methods, the PVD in high vacuum condition has an advantage in controllability of the molecular structure of the deposited films and the formation or retention of high-MW polymers relative to the other methods. The following sections describe recent examples of polymer PVD according to the type of reaction involved in the film-formation process.

9.3
Direct Evaporation of Polymers

9.3.1
Direct Evaporation of Polyethylene

Typical polymers that can be directly evaporated from the polymer source materials include PE [47, 48] and PTFE [49, 50]. Since these polymers have low solubility in organic solvents, PVD is a convenient option for film deposition that can be achieved without using a solvent. A linear PE having an average MW M_n of 900 can be evaporated at a temperature of 300 °C to yield a film growth rate of 10 nm/min, while a material having M_n of 4000 required an evaporation temperature of 340 °C to obtain a film growth rate of 8 nm/min. Figure 9.3 shows the MW distribution of PE source materials and deposited films for the source materials having M_n of 900 and 4000. The material having smaller M_n can be deposited without severe pyrolysis, while the higher polymer underwent considerable loss of MW upon evaporation. It should be noted that the deposited films had polydispersity values much smaller than those of the source materials, suggesting that the vapor-deposition process has some characteristics of distillation process or even zone refining.

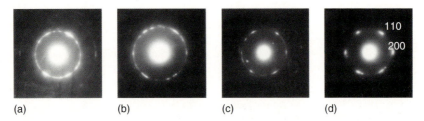

(a)   (b)   (c)   (d)

Figure 9.4 Electron-diffraction patterns of PE films deposited by IAD with ion-acceleration voltage of 0 (a), 200 (b), 500 (c), and 1000 V (d).

Reflecting on the small dispersity of MW, the vapor-deposited film has well-controlled properties in crystallographic orientation. Figure 9.4 shows electron-diffraction patterns of PE film deposited by IAD with ion acceleration voltage, that is, substrate bias voltage of 0, 200, 500, and 1000 V, respectively. With increasing ion energy, crystal growth, and preferential orientation was enhanced, which is one of the advantages reported for the ion-assisted film formation [38]. At the ion acceleration voltage of 1000 V, a single crystal-like diffraction was observed with the molecular chain aligning normal to the substrate surface.

9.3.2
Vapor Deposition of Polytetrafluoroethylene

Fluoropolymer thin films can be prepared by direct evaporation of PTFE [49]. Linear PTFE having M_n of 1100 can be evaporated at a temperature of 280 °C, while a source material having M_n of 8500 required an evaporation temperature of 470 °C, yielding rather slow film growth rate of 5 nm/min in either case. The latter source material had a melting point at 327 °C, while its deposited film had a melting point at 293 °C, suggesting a decrease of MW during the evaporation process. However, the vapor-deposited films showed the standard characteristics of PTFE. IR analyses of the deposited films showed typical spectra of PTFE without serious sign of decomposition. The deposited film showed preferential crystal orientation with the (100) plane parallel to the substrate surface. Figure 9.5 shows the dielectric loss of PTFE films deposited to different thicknesses by IAD under the ion acceleration (substrate bias) voltage V_a of 0 and 500 V. The film deposited without ion acceleration had a larger dielectric loss especially at low frequencies, suggesting poor electrical insulation. The value of tan δ was largely dependent on the film thickness because thinner films tend to have incomplete surface coverage, leaving pinholes in the film. However, the film deposited by IAD has smaller dielectric loss, and had a smaller dependence on film thickness, suggesting the formation of uniform thin films compared to the standard PVD. An electrochemical deposition test on the PTFE films revealed that the ion acceleration was effective in reducing the pinhole density [50].

Vapor deposition of PTFE, combined with a photolithographic technique, can be used to locally modify the surface energy of a substrate in arbitrary patterns,

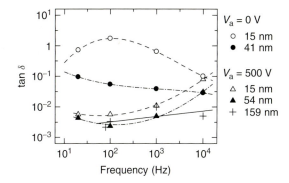

Figure 9.5 Dielectric loss of PTFE films of various thicknesses deposited by IAD under ion-acceleration voltage V_a of 0 and 500 V.

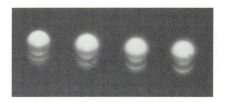

Figure 9.6 Optical micrograph and its schematic cross section of an array of spherical plastic microlenses prepared by using a PTFE deposited film as a template.

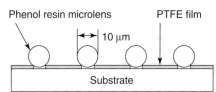

which can be applied for constructing a unique microstructure. Figure 9.6 shows an array of plastic microlenses prepared by utilizing the nonwetting characteristics of the PTFE thin film [51]. This structure was prepared first by depositing a PTFE thin film that has small openings at the positions where the microlenses are to be located. Such patterning can be easily achieved by the lift-off technique using a photoresist. In the next step, an appropriate amount of lens material (phenol resin in this case) was placed over the openings and then heated to form molten droplets on the surface of PTFE. The droplets align their positions to the openings on the PTFE film by themselves, and spontaneously form spheres due to the large contact angle on the PTFE surface. The droplets can then be thermosetted and cooled down to form solid microlenses on desired positions.

Vapor deposition of PTFE forms crystalline thin films reflecting the rigid molecular structure of the fluorinated alkyl chain. For obtaining amorphous fluoropolymer thin films, Teflon AF (DuPont) can be used as a material [52]. The vapor-deposited Teflon AF was found to have high durability against hot water vapor and UV irradiation compared to other conventional polymer coatings.

9.4
Stepwise Polymerization by Coevaporation of Bifunctional Monomers

9.4.1
Vapor-Deposition Polymerization of Polyimide

Although the direct evaporation scheme described in the previous section is a simple method, there are only a few polymers that can be evaporated by this method. Moreover, the direct evaporation has a limitation in MW of the film material. In general, polymer thin films need to be deposited by polymerization reaction on the substrate. The vapor-deposition polymerization illustrated in Figure 9.1b coevaporates two bifunctional monomers that react by condensation or polyaddition reaction. Unlike the CVD, PVD is operated in high-vacuum condition, where the mean free path of the evaporated monomers is much longer than the dimension of the vacuum chamber. As a consequence, the polymerization proceeds by collision of monomers that are migrating on the substrate surface, and not in the vapor phase.

A typical example of condensation polymerization is the formation of polyimide by coevaporation of diamine and cyclic dianhydrides [16, 17]. Polymide thin films having perylene units in the backbone can be prepared as shown in Scheme 9.1. Codeposition of 3,4,9,10-perylenetetracarboxylic dianhydride (PTCDA) and 1,5-diaminonaphthalene (DAN) yields a thin film of polyamic acid at room temperature, which can be annealed in air at 100 °C for 1 h to yield the polyimide. Although the reaction appears straightforward, there have been only limited numbers of reports on polyimide containing PTCDA units due to insolubility of the material [53]. In this respect, PVD can make the most of its advantage as a solvent-free process. Since the polymerization proceeds with a step reaction, it is important to balance the monomer supply on the substrate surface to obtain a high degree of polymerization, which can be achieved by adjusting the evaporation temperature of each monomer. A uniform amorphous polyimide thin film was obtained after annealing.

The polyimide thin film by coevaporation of PTCDA and DAN showed peculiar electrochemical and semiconducting characteristics of the perylene unit. Figure 9.7

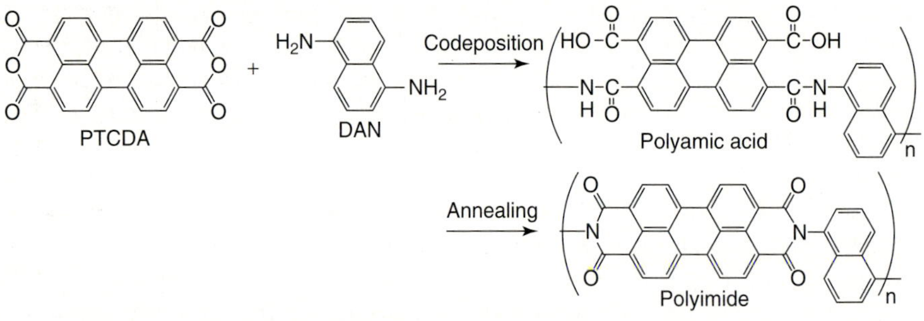

Scheme 9.1 Reaction scheme for vapor-deposition polymerization of polyimide having perylene unit in backbone.

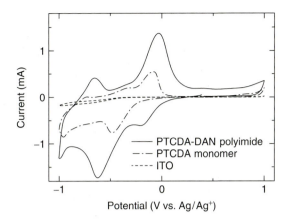

Figure 9.7 Cyclic voltammogram of PTCDA–DAN polyimide and PTCDA monomer films deposited on ITO surface.

shows a cyclic voltammogram of the PTCDA–DAN polyimide film deposited on an indium tin oxide (ITO) substrate. The measurement was made in 0.5 M Na_2SO_4 solution with an Ag/AgCl reference electrode. The film showed reversible two-step reduction at -0.6 and -1.0 V, changing its color from orange to purple and further to yellowish green, respectively. Similar electrochromism was also observed for a vapor-deposited film of PTCDA monomer. However, the monomer film started to disintegrate after repeating the redox cycle, while the polyimide had higher durability against the electrochemical process.

By evaporating 1,12-diaminododecane (DAD) in place of DAN, thin films of PTCDA–DAD polyimide were synthesized with a similar reaction as Scheme 9.1. This film showed a semiconducting nature and its characteristic was evaluated by displacement-current measurement. Figure 9.8 shows the schematic diagram and the result of the measurement. On an ITO substrate, a 30-nm thick Teflon AF insulating layer was vapor deposited, on which 100-nm thick polyimide was prepared by vapor-deposition polymerization of PTCDA and DAD, and then an aluminum electrode was deposited on its top. The current flow was measured under a ramp voltage of 0.6 V/s, as shown in Figure 9.8. The dashed rectangular curve shows the displacement current expected if the polyimide layer is totally insulating. A large increase in current at the voltage of -5 V, deviating from this line, indicates that negative charge can be injected from Al into the polyimide layer. This result indicates that the PTCDA–DAD polyimide works as an electron-transport material [54].

9.4.2
Vapor-Deposition Polymerization of Polyurea

The coevaporation type vapor-deposition polymerization can also be applied for polyaddition reaction, a typical example of which is deposition of PU by coevaporation of diamine and diisocyanates [22]. In this case, the product polymer is obtained in an as-deposited state without the post-deposition annealing that is

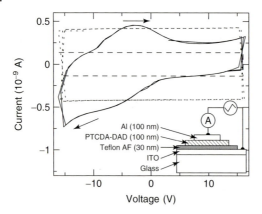

Figure 9.8 Displacement–current measurement of PTCDA–DAD polyimide. The inset shows the sample structure for the measurement. The outer dotted line represents the displacement current of a capacitor consisting of Teflon AF layer only, while the inner dashed line shows a calculated current expected when the polyimide is totally insulating.

required for condensation polymerization. This type of reaction has advantages not only in the simplicity of the process but also in elimination of by-products during the film formation.

PU is an optically transparent and thermally stable material. It is also characteristic in the dipole moment of the urea bond, which can be utilized to activate optical nonlinear, electro-optic, and piezoelectric functions. For this purpose, the urea bonds need to be aligned in a noncentrosymmetric orientation. The conventional method for aligning the dipole orientation, called *poling*, is to apply a high electric field to a preformed film under a temperature sufficiently high to cause molecular movement. The problem in the poling technique is that the dipole orientation slowly degrades even at room temperature due to the relaxation of molecules. On the other hand, a thermally stable polymer that has a rigid molecular structure is difficult to control by the poling technique. This problem can be solved if the dipole orientation can be built-in during the polymerization and film-formation process. The vapor-deposition polymerization by the IAD method has the possibility of producing a stable dipole orientation by achieving the polymerization and film growth simultaneously in a single process under an electric field.

Figure 9.9 shows the reaction scheme and the IAD deposition apparatus for preparing PU thin films [55, 56]. 1,3-di(4-piperidyl)propane (PIP) and 4,4′-diphenylmethane diisocyanate (MDI) were coevaporated from individual IAD sources to obtain PU on the substrate surface. The polymerization proceeds under the strong electric field formed by the ionic charges, which were generated by electron impact in the ionizer, and the bias voltage on the substrate. Efficient control of dipole orientation is expected, because the electric field exerts its force to the mobile end of the growing polymer chain. The dipole alignment was explored by using an attenuated total reflection (ATR) method, schematically illustrated in

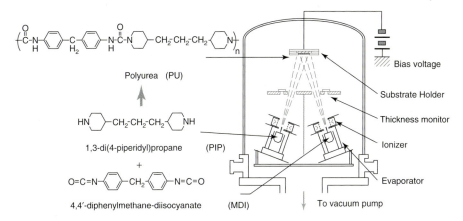

Figure 9.9 Reaction scheme and apparatus for deposition polymerization of polyurea. An ionizer and bias voltage on the substrate were employed to induce dipole orientation.

the inset of Figure 9.10 [57]. A film about 1 μm thick was deposited by IAD to construct a sandwiched structure between two semitransparent gold electrodes. He–Ne laser light was irradiated onto the PU film through a coupling prism from the back of the glass substrate at an angle that excites a waveguide mode in the film. When a voltage is applied between the electrodes, the optical constants and film thickness are modulated through the electro-optic and electrostriction effects, which can be detected as modulation of the reflectivity at the resonance angle.

Figure 9.10 shows the signal intensity of electrically modulated reflectivity for the films deposited with different substrate bias voltages. Larger modulation was observed for the films prepared under higher bias voltage, indicating that the dipole orientation was enhanced by the electric field during the film formation. The electro-optic coefficient r_{13}, r_{33}, and piezoelectric coefficient d_{33} calculated from the modulation signal are shown in Figure 9.11 for the films deposited under different bias voltage and ion current that was flowing into the substrate. Electrical potential on the substrate surface was also measured during the film growth by using an electrostatic monitor. In general, the surface potential, that is, the electric field on the substrate surface increased with increasing bias voltage and ion current, except when the ion current was exceedingly high to cause surface discharge. The result in Figure 9.11 indicates that the nonlinear characteristic of the film was enhanced by growing the polymer thin films in the presence of an electric field.

9.4.3
Other Polymers by Coevaporation Method

Other polymers that can be prepared by the coevaporation-type vapor-deposition polymerization include polyurethane and π-conjugated polymers. The polyurethane can be obtained with a single-step reaction by coevaporating diol and diisocyanate. The example shown in Scheme 9.2 prepares a polyurethane that has a

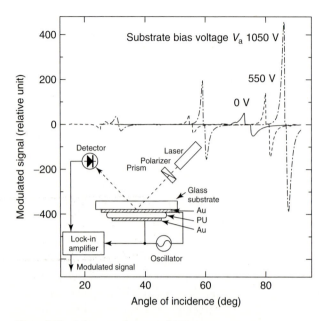

Figure 9.10 Schematic diagram of ATR measurement system (right) and the electrically modulated ATR spectra for the films deposited with different substrate bias voltage.

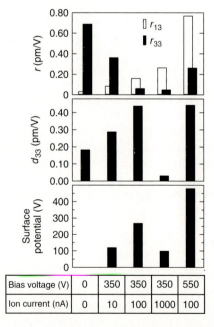

Bias voltage (V)	0	350	350	350	550
Ion current (nA)	0	10	100	1000	100

Figure 9.11 Electro-optic coefficient *r* and piezoelectric coefficient *d* calculated from the ATR measurement for films deposited with different substrate bias voltage and ion current. Surface electric potential of the substrate surface is also shown in the figure.

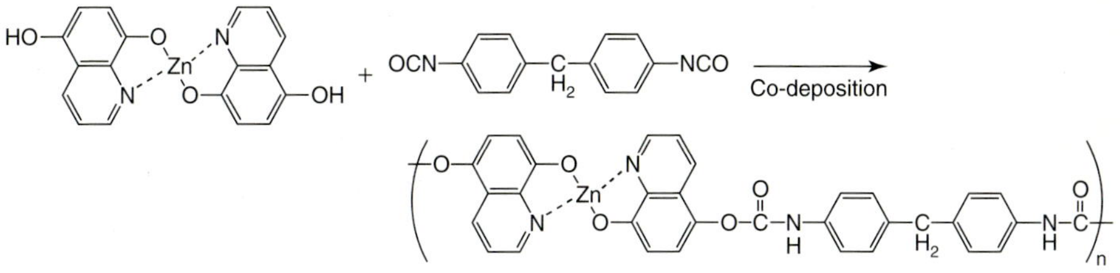

Scheme 9.2 Synthetic route for vapor-deposition polymerization of polyurethane of zinc complex.

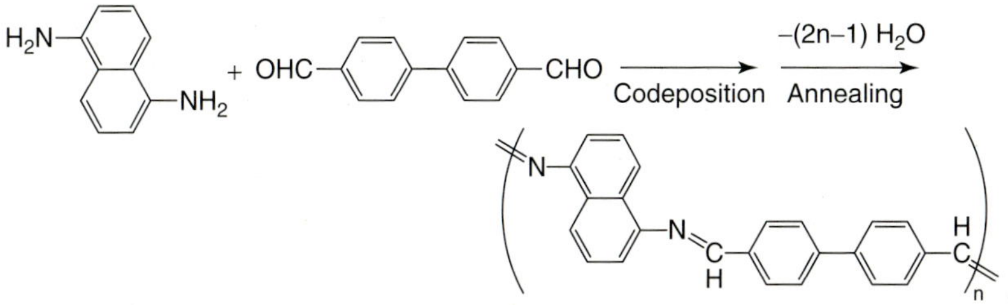

Scheme 9.3 Synthetic route for vapor-deposition polymerization of polyazomethine.

zinc complex in the backbone. This film can be applied as an electron-transporting light-emitting material [58]. As a π-conjugated polymer, polyazomethine can be prepared by coevaporation of 1,5-naphthalene diamine and 4,4′-biphenyl carbaldehyde, followed by annealing at 75 °C for 1 h [59] (Scheme 9.3). Deposition of polyoxadiazole is reported by coevaporation of aromatic dicarboxylicacid chloride and aromatic dicarbohydrazide derivatives [60]. These films can also be applied for the charge-transport layers of OLEDs. The advantage of vapor-deposition polymerization is that uniform thin films of the linear π-conjugated polymer can be prepared without adding side chains, which had been necessary to give film-formation capability by the conventional wet-coating processes.

9.5
Radical Polymerization of Vinyl Monomers

9.5.1
Radical Polymerization of Fluorinated Alkyl Acrylate Polymer

Vapor-deposition polymerization by radical reaction, illustrated in Figure 9.1c, has several advantages. First, it requires only a single evaporation source and does not require stoichiometric adjustment of two monomers. In addition, higher degrees of polymerization can be easily attained with radical reaction compared to the stepwise reaction. Moreover, the design of molecular structure can be achieved easily by developing vinyl or acryl derivative of functional monomers. On the other hand, polymerization does not proceed by simple vapor deposition without

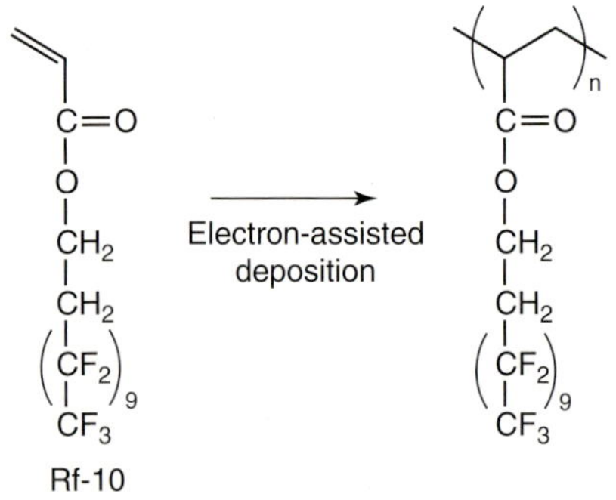

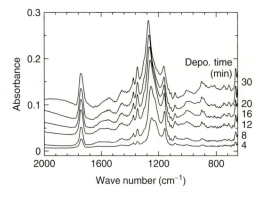

Scheme 9.4 Vapor-deposition polymerization of fluorinated alkyl acrylate polymer by radical reaction.

generating radicals. The radicals of the evaporated monomers can be generated by such means as thermal radiation [25], UV light [28], electron beam [31], or substrate heating [61]. The electron-assisted deposition can be readily achieved by installing an electron source, which consists of a hot-cathode filament and an anode, above the conventional evaporator. In this case, the deposition system is essentially the same as the IAD apparatus shown in Figure 9.2. The substrate bias voltage is not necessarily required to initiate the radical reaction.

One of the simple and useful examples of the radical-type deposition polymerization is the formation of fluoropolymer thin films by evaporating a fluorinated alkyl acrylate monomer, 1H, 1H, 2H, 2H-heneicosafluorododecylacrylate (Rf-10). (Scheme 9.4) Rf-10 can be evaporated at a temperature around 60 °C. However, since its saturated vapor pressure is high, the sticking coefficient of Rf-10 monomer is low unless the substrate temperature is cooled down below 0 °C. As a consequence, the film deposited by the conventional vapor deposition was unstable and re-evaporated readily at room temperature. On the other hand, stable thin films were obtained by using the deposition polymerization assisted by electron irradiation [62].

Figure 9.12 shows *in-situ* infrared absorption spectra measured in the reflection-absorption mode during the electron-assisted deposition of Rf-10 on a

Figure 9.12 *In-situ* IR spectra during film growth of Rf-10 by electron-assisted deposition polymerization. The ordinate of each spectrum has been shifted for clarity.

gold surface. The characteristic absorption bands of C–F stretching vibrations appeared at 1230 and around 1160 cm^{-1}. On the other hand, the bands associated to vinyl group (C=C at 1630 cm^{-1} and C–H around 990 cm^{-1}) were not observed as a consequence of vinyl polymerization. It is noteworthy that such evidence of polymerization was observed from the initial stage of film growth, indicating that the polymerization proceeds on the substrate surface instantly upon deposition of the monomers. Films were also deposited from monomers having shorter alkyl chains. In general, the film growth rate decreased with decreasing chain length since the smaller molecules are more volatile.

The product of Scheme 9.4 is a comb-shaped polymer having a flexible backbone. Films of a rigid molecular structure can be obtained by coevaporating a fluorinated alkyl acrylate monomer Rf-n with a bifunctional acryl compound such as zinc acrylate (ZnAc), which works as a cross-linking agent (Scheme 9.5). Polymerization proceeded even by irradiating electrons only to ZnAc monomers whose fraction was less than 10 wt%. The coevaporation with ZnAc was also effective in increasing the sticking probability of the fluoromonomers, and film growth was possible without cooling the substrate. Figure 9.13 shows the surface free energy of Rf-n homopolymer and Rf-n:ZnAc codeposited films for different alkyl chain length n of the fluoromonomer. In general, the surface energy increased with decreasing monomer chain length. A dynamic contact-angle measurement showed that the hysteresis of the contact angle, that is, the difference between advancing and receding contact angles, also increases with decreasing length of fluorinated alkyl side chain. These results suggest that a longer alkyl side chain gives higher stability in molecular packing [63]. The coevaporated films of Rf-n:ZnAc had slightly larger surface energy than the Rf-n homopolymer film due to the polar group introduced by the ZnAc unit. However, the codeposited film had smaller hysteresis of dynamic contact angle, suggesting that the cross-linking contributes to stabilize the molecular packing.

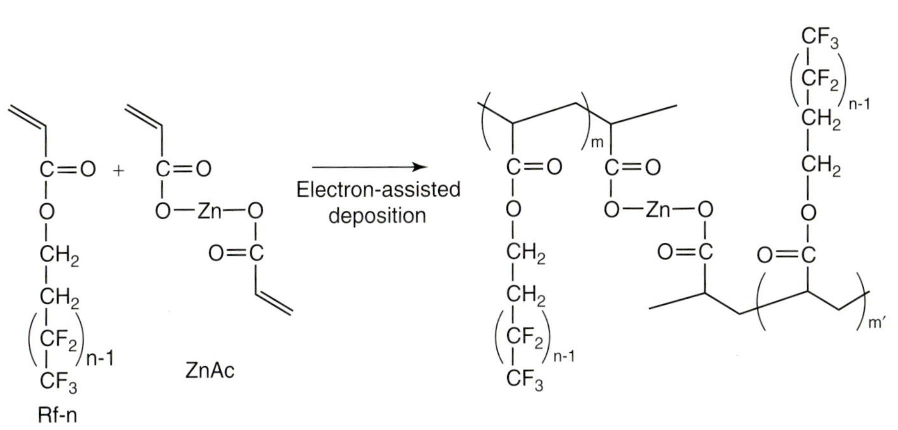

Scheme 9.5 Deposition of cross-linked fluoropolymer by co-evaporation of fluorinated alkyl acrylate with a cross-linking reagent.

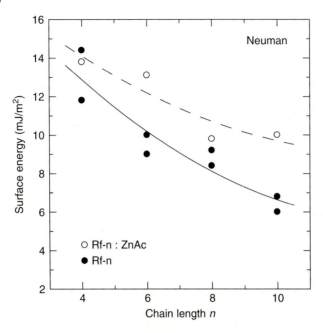

Figure 9.13 Surface free energy of Rf-n homopolymer films (closed circles) and Rf-n:ZnAc codeposited films (open circles) for different alkyl chain length *n*.

9.5.2
Radical Polymerization of Semiconducting Polymers and Their OLED Application

Deposition polymerization was effective especially in improving thermal stability of the deposited films. Figure 9.14 compares optical micrographs of 3-(*N*-carbazolyl)propyl acrylate (CPA) deposited by the conventional vapor deposition and by the electron-assisted deposition polymerization. The difference in these films was simply that the latter film was deposited by irradiating electrons to the evaporated monomers, while the former film was deposited in the same apparatus by turning off the electron source. The film by conventional vapor deposition (Figure 9.14a) had a considerable number of needle-like hillocks due to crystallization of CPA monomers. By annealing the film at 125 °C for 30 min, the film underwent severe coagulation due to crystallization (Figure 9.14b). On the other hand, the film prepared by deposition polymerization (Figure 9.14c) had a smooth surface, and was not damaged by the annealing. It was found that the thermal stability was improved by increasing the electron-irradiation current to the evaporated monomers, thereby enhancing the polymerization [64].

Radical polymerization of vinyl or acrylate monomers can be applied for constructing electronic devices such as OLED, taking advantage of the flexibility in molecular designing [61]. Figure 9.15 shows an example of an OLED structure and the monomers used for deposition. An ITO substrate was coated with

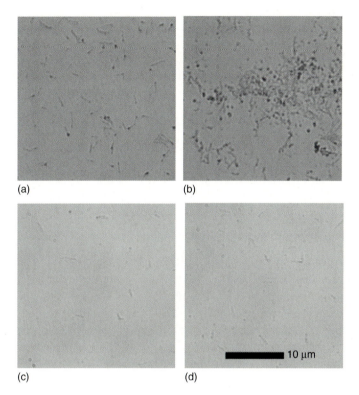

Figure 9.14 Differential interference optical micrographs of a CPA film deposited by conventional vapor deposition (a) in as-deposited state and (b) after annealing at 125 °C for 30 min. A CPA film by electron-assisted deposition polymerization is also shown (c) before and (d) after annealing, respectively.

a hole injection layer of poly(3,4-ethylene dioxythiophene)-poly(styrene sulfonate) (PEDOT:PSS), on which a hole-transport layer (HTL) was prepared by vapor deposition of N,N'-diphenyl-N,N'-bis(4-vinylphenyl) benzidine (DvTPD). DvTPD was also adopted as a host material of an emissive layer (EML) that was deposited successively on the HTL. The EML was doped with 5 wt% of either the commercial red phosphorescent dopant bis(1-phenylisoquinoline)acetylacetonate iridium(III) [Ir(piq)$_2$acac] or its styryl derivative bis(1-phenylisoquinolinate)-6-(4-vinylphenyl)acetylacetonate iridium(III) [Ir(piq)$_2$acac-vb]. It is expected that the latter dopant forms an EML consisting of a copolymer of the host and dopant molecules. After depositing these layers, the substrate was annealed in the vacuum chamber at 100 °C for 1 h to thermally polymerize these layers. It is expected that both HTL and EML can be continuously polymerized to form a seamless interface. The device was completed by depositing an electron-transport layer (ETL) of bathocuproin (BCP), an electron injection layer of LiF, and an aluminum cathode.

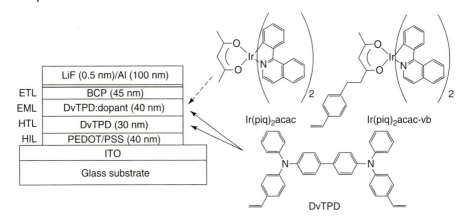

Figure 9.15 OLED structure and the materials used for vapor deposition.

Figure 9.16 shows the luminance–voltage characteristics prepared with and without annealing, that is, polymerizing the HTL and EML. When the conventional Ir(piq)$_2$acac was used as the dopant, the luminance slightly increased on annealing the device, namely by polymerizing the DvTPD layers. By using the polymerizable dopant of Ir(piq)$_2$acac-vb, the luminance stayed almost the same level as the nonannealed device using Ir(piq)$_2$acac dopant. Although the small difference in luminance of these devices reflects various factors such as the phosphorescent efficiency or purity of Ir(piq)$_2$acac-vb itself, the devices shown in Figure 9.16 can

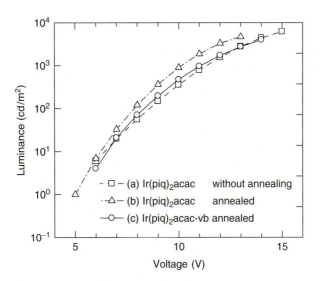

Figure 9.16 Luminance–voltage characteristics of OLEDs whose EML was prepared (a) without polymerization, (b) with polymerizing only the host material, and (c) with copolymerization of host and dopant.

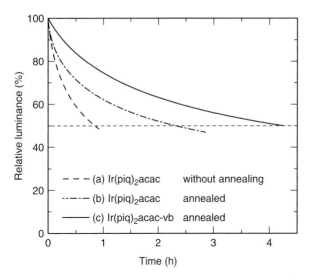

Figure 9.17 Luminance decay of OLEDs whose EML was prepared (a) without polymerization, (b) with polymerizing only the host material, and (c) with copolymerization of host and dopant. The device was operated at constant current starting from the initial luminance of 500 cd/m^2.

be regarded as having comparable luminescence characteristics. On the other hand, a substantial difference was observed in device lifetime by polymerizing the films. Figure 9.17 shows the luminance decay curves under a constant-current operation starting from an initial luminance of 500 cd/m^2 for the three OLEDs. The control device (Figure 9.17a) prepared without thermal polymerization had the shortest lifetime, while the lifetime increased by polymerizing the host material only (Figure 9.17b), which was prepared by annealing DvTPD:Ir(piq)$_2$acac EML. A further improvement in lifetime was achieved with the copolymerized EML (Figure 9.17c) that was prepared by annealing DvTPD doped with the polymerizable dopant Ir(piq)$_2$acac-vb. This result suggests that the polymerization increases thermal stability of the deposited film and leads to the stabilization of organic devices.

The effect of polymerization on electrical characteristics of the film has not yet been made clear. For molecular semiconductors, crystalline molecular packing is effective for $\pi-\pi$ stacking to facilitate the hopping conduction between molecules. On the other hand, polymer materials tend to have an amorphous structure, where molecules are oriented randomly, including free volume between the molecules. In this respect, polymerization is not necessarily convenient for improving the electrical characteristics of the materials. However, it was found that the deposition polymerization can lead to a better balance of holes and electrons. As a consequence, the carrier recombination zone was confined in the EML, resulting in higher quantum efficiency and well-controlled emission spectra [65].

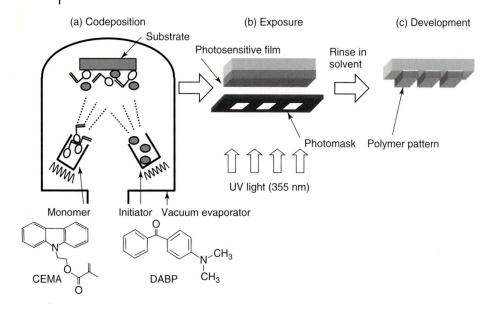

Figure 9.18 Pattern formation of polymer thin films by use of vapor-deposited photosensitive film.

9.5.3
Application to Patterning

Normally, the vapor-deposition polymerization polymerizes the evaporated monomers in the process of film deposition. However, post-deposition polymerization of a vapor-deposited monomer film is also possible, which can be applied for pattern formation of polymeric thin films. In general, patterning of vapor-deposited thin films is achieved by placing a shadow mask in front of the substrate. Although the shadow-mask technique is convenient, several problems, such as alignment accuracy and mask contamination, arise as the substrate size is enlarged. The vapor deposition of polymerizable monomers, such as vinyl and acrylate compounds, can be applied for the preparation of patterned polymer thin films in a procedure schematically shown in Figure 9.18. A photosensitive film was prepared by coevaporating 9H-carbazole-9-ethyl-methacrylate (CEMA) monomer with a photoinitiator, 4-dimethylamino benzophenone (DABP). The film was exposed to UV light of 355 nm for 1 min through a photomask in the air. It has been confirmed that the UV exposure induces polymerization of the codeposited film [66]. The film was then immersed in tetrahydrofuran (THF) to develop a negative pattern of the UV exposure by dissolving the nonpolymerized area.

Pattern formation was observed for the DABP concentration higher than 1% and the UV power higher than 12 mW/cm^2. Figure 9.19 shows the optical micrographs of dot and line/space patterns of 100-nm thick carbazole polymer thin films. Patterns of 10-μm size were clearly developed by this process. This method has an advantage of eliminating the shadow mask during the vapor-deposition process,

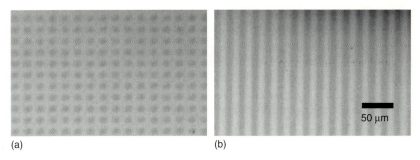

(a)   (b)

Figure 9.19 Photopatterned thin films of carbazole polymer having dot (a) and line/space (b) of 10 μm.

and forming the patterns in a similar process as the common photolithographic technique. It was also confirmed that the photopatterning process does not damage the electrical characteristics of the thin films [67].

9.6
Surface-Initiated Vapor-Deposition Polymerization

9.6.1
Surface-Initiated Vapor-Deposition Polymerization of Polypeptide

Most of the polymer thin films, whether it is formed by PVD or by wet coating, are physically adsorbed on the substrate surface without forming stable chemical bonds at the interface. As a consequence, the film/substrate interface involves such issues as insufficient adhesion, or electrical and optical imperfections. In principle, the surface-initiated deposition polymerization should be effective in solving this issue by grafting polymers on the substrate surface through chemically bound initiators. Figure 9.20 shows the procedure of surface-initiated vapor-deposition polymerization. It consists of two steps, modification of substrate surface with a polymerization initiator, followed by vapor deposition of monomers. The initiator can be attached to the substrate surface as a SAM if the substrate surface has appropriate chemical structures, such as pure metal or hydroxyl group. This step can be achieved simply by dipping the substrate into solutions of the SAM materials.

A typical example of the surface-initiated vapor-deposition polymerization is the formation of polypeptide thin films. A substrate that has hydroxyl groups can be modified with a SAM of trimethoxy(3-aminopropyl)silane (APS), on which amino acid N-carboxy anhydride (NCA) was vapor deposited. Starting from the amino end group of the SAM, a polypeptide thin film grows on the substrate by ring-opening addition polymerization [68, 69] (Scheme 9.6).

Figure 9.21 shows IR absorption spectra of (a) benzyl-L-serine NCA monomer, and the films deposited on (b) bare Al surface, and (c) on an Al surface that was modified with a SAM of APS. The Al surface has native oxide, which enables the modification by APS. Carbonyl peaks in the spectrum of the NCA monomer

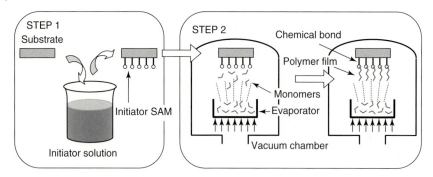

Figure 9.20 Concept of surface-initiated vapor-deposition polymerization, which consists of SAM formation on substrate surface in solution (STEP 1), and physical vapor deposition of monomer on the surface of SAM (STEP 2).

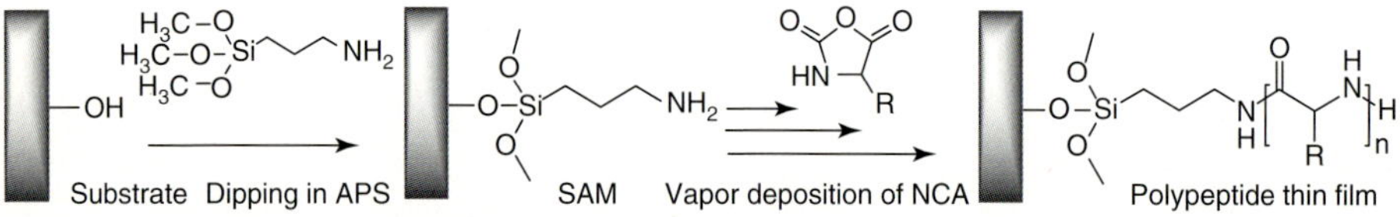

Scheme 9.6 Reaction scheme for surface-initiated vapor-deposition polymerization of polypeptide thin film.

appears at both 1860 and 1785 cm^{-1}. The lack of a change in location of the carbonyl peaks for the film deposited on the bare aluminum shows that no polymerization has occurred and only accumulation of the monomer has been carried out on the aluminum. On the other hand, the film deposited on the APS-modified surface showed the loss of the anhydride carbonyl peaks and development of new peaks at 1635 and 1530 cm^{-1} corresponding to amide I (backbone carbonyl stretch) and amide II (C–N stretch and N–H deformation) structures, respectively. Figure 9.22 shows optical micrographs of the films deposited on the bare and the APS-modified aluminum substrates after one week and one month of film formation. The film on bare Al crystallized gradually as the time passes by, whereas the films deposited on the SAM retained homogeneous morphologies. These results indicate that the SAM layer not only initiates the polymerization of NCA but also forms stable bonding between the film and the substrate.

It is also possible to prepare the polymerization initiator layer by PVD. In this case, the deposition condition of the initiator layer largely influenced the film growth morphology on its surface. Figure 9.23 shows scanning probe micrographs of 15-nm thick films of benzyl-L-glutamate *N*-carboxy anhydride (BLG-NCA) deposited on 2-aminoethanethiol (AET) initiator layers prepared on a gold surface [70]. The AET layer was prepared as (a) a SAM by the dipping process or by PVD to thickness of (b) 3.2, (c) 4.0, and (d) 4.5 nm, respectively. The thickness of AET films by PVD were measured with *in-situ* surface plasmon resonance spectroscopy. The BLG-NCA films deposited on SAM and thin PVD layer showed a highly uniform surface,

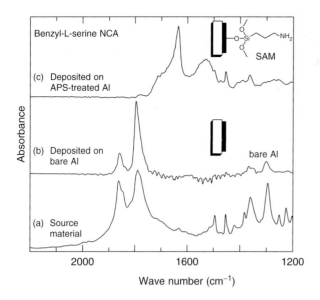

Figure 9.21 IR absorption spectra of (a) NCA monomer, and the films deposited (b) on bare Al surface, and (c) on APS-modified Al surface.

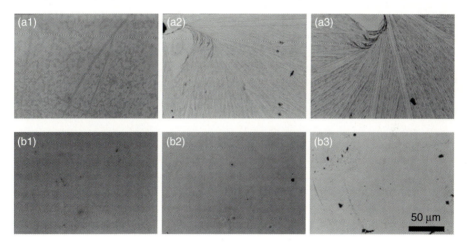

Figure 9.22 Optical micrographs of benzyl-L-serine films deposited on bare (a1–a3) and APS-modified (b1–b3) Al surface in as-deposited state (a1, b1), after one week (a2, b2), and after one month (a3, b3).

while the films deposited on thicker AET showed nonuniform island growth. It is considered that the AET molecules on the surface of thick PVD layers are not chemically bound to the gold surface, and the BLG-NCA polymers that grew on the free AET condense easily to form island structure. As the polymerization initiator layer, a monolayer tightly bound to the substrate surface is ideal.

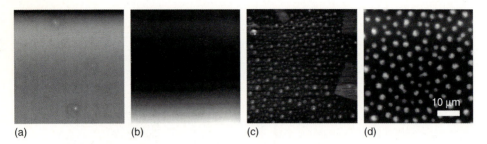

(a)   (b)   (c)   (d)

Figure 9.23 Scanning probe micrographs of the BLG-NCA films deposited on AET initiator layers by (a) SAM and by PVD with layer thickness of (b) 3.2, (c) 4.0, and (d) 4.5 nm.

9.6.2
Interface Control by Surface-Initiated Vapor-Deposition Polymerization

The surface-initiated vapor-deposition polymerization is effective not only to grow polymer thin films but also to control the interface between a polymer thin film and the substrate. Scheme 9.7 shows a concept to prepare a polymer bound to an oxide surface by combining a SAM layer of polymerization initiator with the vapor-deposition polymerization of a vinyl monomer. An oxide surface, such as ITO, was immersed in solutions containing APS, succinic anhydride, pentafluorophenol, and then Vazo 56 (DuPont), successively to form a SAM layer that has an azo end group as polymerization initiator [71]. The azo initiator can be activated by UV irradiation or by heating to generate a radical, which initiates radical polymerization of monomers deposited on its surface. Thin films of CPA showed excellent thermal stability only when the deposition was achieved on the SAM-modified surface activated by UV irradiation [72]. Surface-bound polymer thin films can be also prepared on metal surface by using a SAM of thiol instead of APS by a similar procedure, as shown in Scheme 9.7.

Formation of surface-bound polymer thin films is effective in preparing an organic device such as an OLED. A HTL was prepared on ITO electrode by the surface-initiated vapor-deposition polymerization of *N,N,N'*-triphenyl-*N'*-(4-vinylphenyl)-biphenyl-4,4'-diamine (vTPD) monomer, on which an EML of

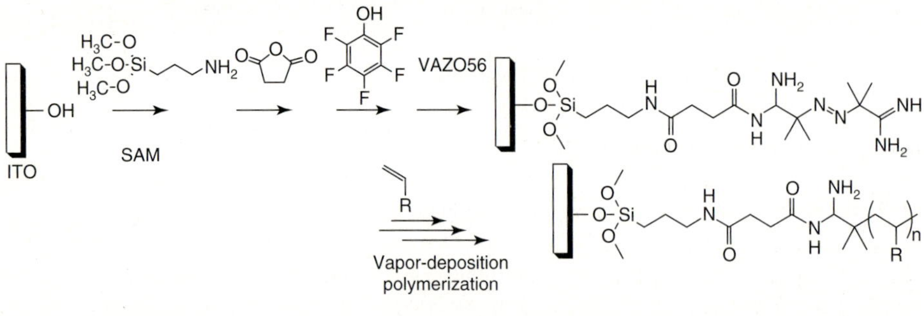

Scheme 9.7 Reaction scheme for preparing SAM of azo initiator on ITO surface.

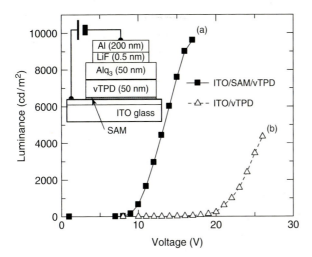

Figure 9.24 Voltage–luminance characteristics of OLEDs prepared (a) with and (b) without the initiator SAM on ITO surface.

tris(8-quinolinolato)-aluminum (Alq$_3$), an electron injection layer of LiF, and an aluminum cathode were vapor deposited successively to construct an OLED. Figure 9.24 compares the voltage–luminance characteristics of the OLEDs prepared with and without the SAM layer on the ITO surface. The SAM modification was effective in increasing the device current and reducing the onset voltage for luminescence [73]. It is considered that the surface-initiated deposition polymerization enables formation of a tight electrical contact at the interface.

It is reported that SAMs can be utilized as buffer layers on the surface of ITO electrode of OLEDs. The effect of SAM has been attributed to the reduction of the charge-injection barrier due to the interfacial dipole layer formed by the SAM [74, 75]. The SAM is also reported to play a role of improving wettability of organic materials on the ITO surface [76], enabling the formation of uniform organic layers [77]. In general, it is considered that the modification of surface free energy by SAMs influences the growth and nucleation of organic layers [78], leading to two-dimensional growth [79]. In fact, the surface energy of ITO substrate was modified by forming the SAM of azo initiator. However, it was confirmed that the growth morphology of an organic layer is primarily governed by the formation of chemical bonds through the SAM instead of mere difference in surface energy [80].

9.7
Conclusion

The PVD is a commonly accepted technique as a film-formation method for inorganic materials, and has played an important role in the development of semiconductor devices. On the other hand, polymer thin films is still the dominant material for large-area coatings of consumer products. Therefore, cost-effective and high-throughput wet-coating processes have been used with a good reason. However, PVD of polymers has assumed considerable importance in view of recent development of organic devices. Although PVD is not a well-known method for polymer coatings, a variety of polymers can be deposited very effectively with this method, as described in this chapter. The materials include fluoropolymers, polyimide, PU, π-conjugated polymer, variety of vinyl polymers, polypeptide, and so on. The PVD is especially useful for such applications that require nano-order thin films and their multilayered structures. For some polymers that have poor solubility, PVD can be an indispensable option for film deposition. The surface-initiated vapor-deposition polymerization is considered to be a new approach for interface control between inorganic substrates and the polymer layers. The solvent-free feature of PVD is also attractive from an environmental viewpoint to reduce emission of VOCs.

The biggest challenge with PVD is the cost issue and production. The vacuum evacuation, as well as the vapor deposition itself, requires considerable process time. In addition, vacuum systems are generally costly compared to wet-coating systems. On the other hand, vapor deposition of metals and oxides on roll-films is already a well-established technique in industry. Therefore, proper engineering might bring the PVD to an industrially significant technology for polymers as well. A key issue to improve production would be to balance the speed of monomer supply and the reaction rate. Although there still exist several issues that need further investigation and development, the PVD technique opens up unique and attractive aspects for the science and engineering of polymer thin films.

References

1. Tang, C.W. and VanSlyke, S.A. (1987) Organic electroluminescent diodes. *Appl. Phys. Lett.*, **51**, 913–915.
2. Ashida, M. (1966) Orientation overgrowth of metal-phthalocyanines on the surface of single crystals. I. Vacuum-condensed films on muscovite. *Bull. Chem. Soc. Jpn.*, **39**, 2625–2631.
3. Schreiber, F. (2004) Organic molecular beam deposition: growth studies beyond the first monolayer. *Phys. Status Solidi A*, **201**, 1037–1054.
4. Matsuzaki, F., Inaoka, K., Okada, M., and Sato, K. (1984) Molecular orientation in physical-vapor deposition of long-chain stearic acid. *J. Cryst. Growth*, **69**, 231–240.
5. Inoue, T., Yase, K., Inaoka, K., and Okada, M. (1987) Molecular orientation and growth mechanism of several fatty acids with different lengths. *J. Cryst. Growth*, **83**, 306–310.
6. Yase, K., Inoue, T., and Okada, M. (1990) Direct observation of cadmium arachidate thin films with lateral and normal molecular orientations by superconducting cryoelectron microscopy. *J. Electron Microsc.*, **39**, 454–458.
7. Valdri, G., Allessandrini, A., Muscatello, U., and Valdre, U. (1998) High-resolution imaging of n-alkane crystals by atomic force microscopy. *Philos. Mag. Lett.*, **78**, 255–261.
8. Kubono, A. and Okui, N. (1994) Polymer thin films prepared by vapor deposition. *Prog. Polym. Sci.*, **19**, 389–438.
9. Madorsky, S.L. (1952) Rates of thermal degradation of polystyrene and polyethylene in a vacuum. *J. Polym. Sci.*, **9**, 133–156.

10. Luff, P.P. and White, M. (1970) Structure and properties of evaporated polyethylene thin films. *Thin Solid Films*, **6**, 175–195.

11. Takeno, A., Okui, N., Kitoh, T., Muraoka, M., Umemoto, S., and Sakai, T. (1991) Preparation and piezoelectricity of β-form poly(vinylidene fluoride) thin film by vapor deposition. *Thin Solid Films*, **202**, 205–211.

12. Chow, R., Spragge, M.K., Loomis, G.E., Rainer, F., Ward, R.L., Thomas, I.M., and Kozlowski, M.R. (1994) Characterization of physically vapor deposited AF2400 thin films. *Mater. Res. Soc. Symp. Proc.*, **328**, 731–737.

13. Nason, T.C., Moore, J.A., and Lu, T.M. (1992) Deposition of amorphous fluoropolymer thin films by thermolysis of Teflon amorphous fluoropolymer. *Appl. Phys. Lett.*, **60**, 1866–1868.

14. Gorham, W.F. (1966) A new, general synthetic method for the preparation of linear poly-*p*-xylenes. *J. Polym. Sci. A*, **4**, 3027–3039.

15. Macchi, E.M. (1972) Crystallization and polymerization of 6-aminocaproic acid as simultaneous processes. *J. Polym. Sci. A*, **10**, 45–56.

16. Salem, J.R., Sequeda, F.O., Duran, J., Lee, W.Y., and Yang, R.M. (1986) Solventless polyimide films by vapor deposition. *J. Vac. Sci. Technol., A*, **4**, 369–374.

17. Takahashi, Y., Iijima, M., Inagawa, K., and Itoh, A. (1987) Synthesis of aromatic polyimide film by vacuum deposition polymerization. *J. Vac. Sci. Technol., A*, **5**, 2253–2256.

18. Grunze, M. and Lamb, R.N. (1987) Preparation and adhesion of ultrathin polyimide films on polycrystalline silver. *Chem. Phys. Lett.*, **133**, 283–287.

19. Iida, K., Nohara, T., Totani, K., Nakamura, S., and Sawa, G. (1989) Molecular orientation and photocurrent of alkyl-aromatic polyimide films prepared by vapor-deposition polymerization. *Jpn. J. Appl. Phys.*, **28**, 2552–2555.

20. Kubono, A., Higuchi, H., Umenoto, S., and Okui, N. (1993) Molecular orientation of alkyl-aromatic polyimide thin films prepared by vapour deposition polymerization. *Thin Solid Films*, **229**, 133–136.

21. Kubono, A., Okui, N., Tanaka, K., Umemoto, S., and Sakai, T. (1991) Highly oriented polyamide thin films prepared by vapor-deposition polymerization. *Thin Solid Films*, **199**, 385–393.

22. Kubono, A., Kanae, N., Umemoto, S., Sakai, T., and Okui, N. (1992) Molecular orientation of alkyl-aromatic polyamide thin films prepared by vapour deposition polymerization. *Thin Solid Films*, **215**, 94–97.

23. Takahashi, Y., Iijima, M., and Fukada, E. (1989) Pyroelectricity in poled thin films of aromatic polyurea prepared by vapor-deposition polymerization. *Jpn. J. Appl. Phys.*, **28**, L2245–L2247.

24. Tamada, M., Asano, M., Yoshida, M., and Kumakura, M. (1991) Formation of a thin film of poly(octadecyl methacrylate) using the physical vapor deposition technique. *Polymer*, **32**, 2064–2069.

25. Tamada, M., Omichi, H., and Okui, N. (1994) Vapor-deposition polymerization of N-methylolacrylamide. *Thin Solid Films*, **251**, 36–39.

26. Tamada, M., Omichi, H., and Okui, N. (1995) Preparation of polyvinylcarbazole thin film with vapor-deposition polymerization. *Thin Solid Films*, **268**, 18–21.

27. Tamada, M., Koshikawa, H., and Omichi, H. (1997) Real-time *in situ* observation of photo-induced vapor deposition polymerization of N-vinylcarbazole with Fourier transform IR reflection absorption spectroscopy. *Thin Solid Films*, **292**, 164–168.

28. Tamada, M., Koshikawa, H., Hosoi, F., Suwa, T., Usui, H., Kosaka, A., and Sato, H. (1999) UV polymerization of triphenylaminemethylacrylate thin film on ITO substrate. *Polymer*, **40**, 3061–3067.

29. Tamada, M., Koshikawa, H., Suwa, T., Yoshioka, T., Usui, H., and Sato, H. (2000) Thermal stability and EL (electroluminescent) efficiency of polymer thin film prepared from TPD-acrylate. *Polymer*, **41**, 5661–5667.

30. Tamada, M., Omichi, H., and Okui, N. (1996) Change of molecular orientation with post-polymerization of a thin film

of N-methylolacrylamide prepared with VDP. *Thin Solid Films*, **274**, 66–69.

31. Usui, H. (2000) Polymeric film deposition by ionization-assisted method for optical and optoelectronic applications. *Thin Solid Films*, **365**, 22–29.

32. Usui, H. (2000) Deposition of polymeric thin films by ionization-assisted method. *IEICE Trans. Electron.*, **E83-C**, 1128–1133.

33. Chang, Y.C. and Frank, C.W. (1998) Vapor deposition-polymerization of α-amino acid N-carboxy anhydride on the silicon(100) native oxide surface. *Langmuir*, **14**, 326–334.

34. Lee, N.H., Christensen, L.M., and Frank, C.W. (2003) Morphology of vapor-deposited poly(α-amino acid) films. *Langmuir*, **19**, 3525–3530.

35. Lee, N.H. and Frank, C.W. (2003) Surface-initiated vapor polymerization of various α-amino acids. *Langmuir*, **19**, 1295–1303.

36. Fulghum, T.M., Yamagami, H., Tanaka, K., Usui, H., Shigehara, K., and Advincula, R.C. (2002) Polypeptide ultrathin films by vacuum deposition: preparation and characterization. *Mater. Res. Soc. Symp. Proc.*, **711**, 251–256.

37. Takagi, T. (1982) Role of ions in ion-based film formation. *Thin Solid Films*, **92**, 1–17.

38. Takagi, T. (1987) Ionized cluster beam deposition and epitaxy in *Physics of Thin Films* (eds M.H. Francombe and J.L. Vossen), Academic Press, New York, pp. 1–58.

39. Yasuda, H. (1985) *Plasma Polymerization*, Academic Publishers, Orlando.

40. Biederman, H. and Osada, Y. (1992) *Plasma Polymerization Processes*, Elsevier, Amsterdam.

41. Yasuda, H.K. (2005) Some important aspects of plasma polymerization. *Plasma Process Polym.*, **2**, 293–304.

42. Harrop, R. and Harrop, P.J. (1969) Friction of sputtered PTFE [poly(tetrafluoroethylene)] films. *Thin Solid Films*, **3**, 109–117.

43. Morrison, D.T. and Robertson, T. (1973) Radio-frequency sputtering of plastics. *Thin Solid Films*, **15**, 87–101.

44. Tibbitt, J.M., Shen, M., and Bell, A.T. (1975) Comparison of rf (radio frequency) sputtered and plasma polymerized thin films of tetrafluoroethylene. *Thin Solid Films*, **29**, L43–L45.

45. Biederman, H., Ojha, S.M., and Holland, L. (1977) The properties of fluorocarbon films prepared by r.f. sputtering and plasma polymerization in inert and active gas. *Thin Solid Films*, **41**, 329–339.

46. Biederman, H. (2000) Organic films prepared by polymer sputtering. *J. Vac. Sci. Technol.*, **A18**, 1642–1648.

47. Usui, H., Yamada, I., and Takagi, T. (1986) Anthracene and polyethylene thin film deposited by ionized cluster beam. *J. Vac. Sci. Technol., A*, **4**, 52–60.

48. Usui, H., Numata, K., Dohmoto, H., Yamada, I., and Takagi, T. (1988) Characteristics of polyethylene thin films deposited by ionized cluster beam. *Mater. Res. Soc. Symp. Proc.*, **108**, 201–206.

49. Usui, H., Koshikawa, H., and Tanaka, K. (1995) Effect of substrate temperature on the deposition of polytetrafluoroethylene by an ionization-assisted evaporation method. *J. Vac. Sci. Technol.*, **A13**, 2318–2324.

50. Usui, H., Koshikawa, H., and Tanaka, K. (1998) Characteristics of polytetrafluoroethylene thin films prepared by ionization-assisted deposition. *IEICE Trans. Electron.*, **E-81-C**, 1083–1089.

51. Tamura, H., Kojima, R., and Usui, H. (2003) Preparation of plastic spherical microlenses by use of fluoropolymer stencil and oil-bath heating. *Appl. Opt.*, **42**, 4008–4015.

52. Usui, H., Kojima, R., Horie, S., Tanaka, K., and Ohishi, F. (2001) Insulation and passivation properties of vapor-deposited fluorinated polymer thin films. Proceedings of 2001 International Symposium on Electrical Insulating Material, Himeji, Japan, pp. 570–573.

53. Huang, W., Yan, D., Lu, Q., and Huang, Y. (2003) Synthesis and characterization of highly soluble fluorescent main chain copolyimides containing perylene units. *Euro. Polym. J.*, **39**, 1099–1104.

54. Usui, H., Watanabe, M., Arai, C., Hibi, K., and Tanaka, K. (2005)

Vapor-deposition polymerization of a polyimide containing perylene units characterized by displacement current measurement. *Jpn. J. Appl. Phys.*, **44**, 2810–2814.

55. Usui, H., Kikuchi, H., Tanaka, K., Miyata, S., and Watanabe, T. (1998) Deposition polymerization of polyurea thin films by ionization-assisted method. *J. Vac. Sci. Technol.*, **A16**, 108–113.

56. Usui, H., Kikuchi, F., Tanaka, K., Watanabe, T., and Miyata, S. (2002) Ionization-assisted deposition of azo-containing polyurea for NLO applications. *IEICE Trans. Electron.*, **E85-C**, 1270–1274.

57. Usui, H., Kikuchi, F., Tanaka, K., Watanabe, T., Miyata, S., Bock, H., and Knoll, W. (1999) *In-situ* orientation control of polyurea during film formation by ionization-assisted deposition. *Nonlinear Opt.*, **22**, 135–138.

58. Wang, X., Ogino, K., Tanaka, K., and Usui, H. (2004) Vapor deposition of polyurethane thin film having bis(hydroxyquinoline) zinc complex for organic LED. *IEICE Trans. Electron.*, **E87-C**, 2122–2124.

59. Wang, X., Ogino, K., Tanaka, K., and Usui, H. (2003) Novel polyimine as electroluminescent material prepared by vapor deposition polymerization. *Thin Solid Films*, **438/439**, 75–79.

60. Murata, H., Ukishima, S., Hirano, H., and Yamanaka, T. (1997) A novel fabrication technique and new conjugated polymers for multilayer polymer light-emitting diodes. *Polym. Adv. Technol.*, **8**, 459–464.

61. Kawakami, A., Otsuki, E., Fujieda, M., Kita, H., Taka, H., Sato, H., and Usui, H. (2008) Phosphorescent organic light emitting diode using vinyl derivatives of hole transport and dopant materials. *Jpn. J. Appl. Phys.*, **47**, 1279–1283.

62. Usui, H., Katayama, A., Honda, T., and Tanaka, K. (2003) Comb-like polymer thin films prepared by ionization-assisted deposition of acrylate compounds. *Mater. Res. Soc. Symp. Proc.*, **734**, 321–326.

63. Honda, K., Morita, M., Otsuka, H., and Takahara, A. (2005) Molecular aggregation structure and surface properties of poly(fluoroalkyl acrylate) thin films. *Macromolecules*, **38**, 5699–5705.

64. Katsuki, K., Kawakami, A., Ogino, K., Tanaka, K., and Usui, H. (2005) Preparation of carbazole polymer thin films by electron-assisted deposition of 3-(*N*-carbazolyl)propyl acrylate. *Jpn. J. Appl. Phys.*, **44**, 4182–4186.

65. Otsuki, E., Sato, H., Kawakami, A., Taka, H., Kita, H., and Usui, H. (2009) Vapor deposition polymerization of vinyl compounds and fabrication of OLED having double emissive layers. *Thin Solid Films*, **518**, 703–706.

66. Muroyama, M., Saito, I., Yokokura, S., Tanaka, K., and Usui, H. (2009) Selective patterning of organic light-emitting diodes by physical vapor deposition of photosensitive materials. *Jpn. J. Appl. Phys.*, **48**, 04C163/1–04C163/5.

67. Muroyama, M., Yokokura, S., Tanaka, K., and Usui, H. (2010) Polymeric film deposition by coevaporation of polymerizable monomer and initiator. *Jpn. J. Appl. Phys.*, **49**, 01AE03/1–01AE03/5.

68. Fulghum, T.M., Yamagami, H., Tanaka, K., Usui, H., Shigehara, K., and Advincula, R.C. (2002) Polypeptide ultrathin film preparation from amino acid *N*-carboxy anhydrides (NCA) using high vacuum physical vapor deposition techniques. *Polym. Mater. Sci. Eng.*, **86**, 196–197.

69. Ogura, K., Tanaka, K., Advincula, R.C., and Usui, H. (2006) *In-situ* surface plasmon observation of the growth process of polypeptide polymer brush by physical vapor deposition. *Polym. Prep.*, **47**, 72–73.

70. Duran, H., Ogura, K., Nakao, K., Vianna, S.D.B., Usui, H., Advincula, R.C., and Knoll, W. (2009) High-vacuum vapor deposition and *in situ* monitoring of *N*-carboxy anhydride benzyl glutamate polymerization. *Langmuir*, **25**, 10711–10718.

71. Hyun, J. and Chilkoti, A. (2001) Surface-initiated free radical polymerization of polystyrene micropatterns on a self-assembled monolayer on gold. *Macromolecules*, **34**, 5644–5652.

72. Katsuki, K., Bekku, H., Kawakami, A., Locklin, J., Patton, D., Tanaka, K., Advincula, R., and Usui, H. (2005)

Preparation of carbazole polymer thin films chemically bound to substrate surface by physical vapor deposition combined with self-assembled monolayer. *Jpn. J. Appl. Phys.*, **44**, 504–508.

73. Kawakami, A., Katsuki, K., Advincula, R.C., Tanaka, K., Ogino, K., and Usui, H. (2008) Interface control by surface-initiated deposition polymerization and its application to organic light-emitting devices. *Jpn. J. Appl. Phys.*, **47**, 3156–3161.

74. Ishii, H., Sugiyama, K., Ito, E., and Seki, K. (1999) Energy level alignment and interfacial electronic structures at organic/metal and organic/organic interfaces. *Adv. Mater.*, **11**, 605–625.

75. Appleyard, S.F.J., Day, S.R., Pickford, R.D., and Willis, M.R. (2000) Organic electroluminescent devices: enhanced carrier injection using SAM derivatized ITO electrodes. *J. Mater. Chem.*, **10**, 169–173.

76. Hsiao, C.-C., Chang, C.-H., Hung, M.-C., Yang, N.-J., and Chena, S.-A. (2005) Self-assembled monolayer modification of indium tin oxide anode surface for polymer light-emitting diodes with poly[2-methoxy-5-(2-ethylhexyloxy)-1,4-phenylene vinylene] for high

performance. *Appl. Phys. Lett.*, **86**, 223505-1–223505-3.

77. Malinsky, J.E., Jabbour, G.E., Shaheen, S.E., Anderson, J.D., Richter, A.G., Marks, T.J., Armstrong, N.R., Kippelen, B., Dutta, P., and Peyghambarian, N. (1999) Self-assembly processes for organic LED electrode passivation and charge injection balance. *Adv. Mater.*, **11**, 227–231.

78. Kim, D.H., Lee, H.S., Yang, H., Yang, L., and Cho, K. (2008) Tunable crystal nanostructures of pentacene on dielectrics possessing surface-order control. *Adv. Funct. Mater.*, **18**, 1363–1370.

79. Virkar, A., Mannsfeld, S., Oh, J.H., Toney, M.F., Tan, Y.H., Liu, G., Scott, J.C., Miller, R., and Bao, Z. (2009) The role of OTS density on pentacene and C_{60} nucleation, thin film growth, and transistor performance. *Adv. Funct. Mater.*, **19**, 1962–1970.

80. Umemoto, Y., Kim, S.-H., Advincula, R.C., Tanaka, K., and Usui, H. (2010) Effect of SAM modification on ITO surface for surface-initiated vapor-deposition polymerization of carbazole thin films. *Jpn. J. Appl. Phys.*, **49**, 04DK21-1–04DK21-5.

10
Electro-Optical Applications of Conjugated Polymer Thin Films

Nicholas Marshall, S. Kyle Sontag, and Jason Locklin

10.1
Introduction

Since the original reports of conducting polymers (CPs), which led to the 2000 Nobel Prize in Chemistry, the development and expectations for this unique class of modern materials have been high. It has been confidently predicted for many years that CPs would replace metals and inorganic semiconductors in a variety of applications and pave the way for the development of entirely new types of flexible electronic devices that take advantage of inherent mechanical flexibility in these conductive plastics. While the development of new applications for CPs has perhaps not proceeded at the optimistic rate predicted in the late 1970s and early 1980s, a number of promising industrial applications are beginning to emerge including a line of thin-film organic light-emitting diodes (OLEDs) from the likes of General Electric, Sony, and other display manufacturers, and commercially available printable solar-cell materials based on organic semiconductors. Other technologies, including organic thin-film transistors [1–3] (OTFTs) displays and "smart windows" based on organic electrochromic films [4–6] seem poised to break into the mainstream with an overwhelming number of literature reports and patents that report devices with excellent efficiency.

In most cases, inorganic materials significantly outperform organic electronic materials with respect to charge-carrier mobility and stability. The lower mobility found in organics is due to the lack of long-range crystallinity that can be practically achieved in an organic conductor or semiconductor, which results in a hopping-type charge-transfer mechanism dominating over band transport [7]. The relatively low stability of organic electronic materials compared to inorganics is due in part to the natural instability of most carbon-based compounds in the presence of oxygen and light [8, 9], but more importantly includes the extremely high reactivity of charge carriers in organic conductors and semiconductors. In general, the negative and positive polarons and related species that provide charge transport in an organic material will react readily with oxygen and water, often producing a degraded site that acts as a trap toward further charge propagation [10, 11]. While the stability and efficiency of the best organic-based photovoltaics, transistors, LEDs, and other

Functional Polymer Films, First Edition. Edited by Wolfgang Knoll and Rigoberto C. Advincula.
© 2011 Wiley-VCH Verlag GmbH & Co. KGaA. Published 2011 by Wiley-VCH Verlag GmbH & Co. KGaA.

devices continue to improve with research, it seems likely that they will always lag behind their inorganic counterparts in raw performance. One exception to this trend is in the developing field of electrochromics, where an applied potential causes a color or transparency change in the device. At the time of this writing, electrochromic conjugated polymers serve as the core of the most efficient reported electrochromic devices (ECDs) [12, 13].

Because of the limited charge-carrier mobility and stability of electroactive organic polymers, electronic applications of these materials have generally focused on a flexible-device paradigm that utilizes thin films of organic electroactive material, emphasizing the potential of electronic devices that can be incorporated into fabric or printed on paper or plastic sheets [14–17]. It is widely thought that the best chance for a general commercial application of electroactive polymers is in the production of printed or flexible electronics. In addition to their greater inherent mechanical flexibility, electroactive polymers can be processed into thin films more easily than inorganic materials due to their potential for solution processing [18]. The use of electroactive polymers in a thin-film configuration thus allows their useful properties to be leveraged and the impact of lower performance minimized. As a result, photovoltaics, transistors, and LEDs using modern organic electroactive materials tend to be designed around the use of multiple thin films in a laminar configuration.

Other applications of conjugated polymer thin films employ the polymer as a colorimetric or fluorescent indicator [19]. This technology is expected to be applied to the creation of on-site field test devices and simple assays, where the user can simply add analyte to the device and observe a colorimetric response or a change in fluorescence for a positive test. Several indicator systems have been created that can be chemically functionalized to enable the sensing of a tremendous range of biomolecules [20, 21], small molecules [22, 23], and ions [24].

Finally, in a less-popularized but highly practical application, electrodes modified with thin films of functional electroactive polymers show promise in chemical sensing via voltammetry, an application that directly takes advantage of the organic nature of the polymeric material. Since the precursor to the electroactive polymer is an organic molecule, the chemical sensor moiety can be built into the monomer through synthetic design, allowing the creation of a unique class of chemoselective sensor electrodes based on electrode-supported polymer films [25, 26].

It seems clear that the creation, characterization, and modification of electroactive polymer thin films is essential to the emerging industry of organic electronics. As new, powerful methods are introduced for the formation of such films and the subtleties of polymeric electronic structure are unraveled, research into materials with high-performance electronic and mechanical properties for practical devices will enjoy increasing prominence. This review aims at introducing many of the device architectures and techniques used to fabricate thin-film devices that involve conjugated polymers. It is not meant as an exhaustive review on the subject, but to provide a general overview of some of the tools and techniques that are used in synthesis and characterization of conjugated polymer materials. For more detailed descriptions of individual device architectures or classes of materials, the reader is referred to the many excellent textbooks and review articles that cover these in a more detailed manner. These can be found in the references for each subsection throughout the chapter.

10.2
Properties of Organic Semiconductors and Conjugated Polymers

10.2.1
Introduction to Charge Transport

In typical plastics such as polyethylene, which contains the repeat unit $(CH_2-CH_2)_n$, each carbon atom is surrounded by four σ bonds. As a result, the energy difference between the highest occupied molecular orbital (HOMO) and lowest unoccupied molecular orbital (LUMO) is large. Such plastics are electronically insulating, and depending on their crystallinity, transparent to visible light. However, when each carbon atom is sp^2 hybridized, each carbon atom is linked to its neighbors by three σ bonds (resulting from hybridization of 2s, 2p$_x$, and 2p$_y$ orbitals) and one π bond (from the remaining 2p$_z$ orbital). Multiple π bonds in close proximity to one another can overlap, forming extended molecular orbitals with a reduced HOMO energy relative to isolated π orbitals. This process is called *conjugation*, and materials with extended conjugation can absorb visible light and in some cases, have potential use as semiconductors. In conjugated polymers, a large number of individual π orbitals overlap to yield a final electronic structure that is somewhat similar to an inorganic semiconductor, with an occupied valence band and unoccupied conduction band.

Conjugated polymers in their native state can act as semiconductors. In inorganic semiconductors, charge transport is described by a band model, which serves as a useful first approximation for organic semiconductors as well. In the band model, the key parameter is the energy difference between the valence band and the conduction band. It is this energy difference, or bandgap, that determines the electrical properties of materials. For charge transport to occur in undoped conjugated polymers, an electron must be excited from the valence band to the conduction band. This process generates charge carriers that can move through the polymer. Most conjugated polymers have a large bandgap, which makes them insulating at room temperature. The band model of conjugated polymer electronic structure is qualitatively correct, showing that charge carriers must be generated in order for conjugated polymers to become conducting. However, the band model is unable to fully describe charge transport in organic semiconductors because it fails to account for the important phenomenon known as *polarization*.

The methods of generating charge carries in a conjugated polymer device depends on the ultimate device application. For example, charge carriers in solar cells are generated by the absorption of a photon. In an OTFT, an electrical bias is applied. In general, however, exposing the polymer chain to an oxidizing or reducing agent can also create these charge carriers. This redox reaction is often referred to as "*doping*," where the use of an oxidizing agent is p-doping and the use of a reducing agent is n-doping. For inorganic semiconductors, the term doping refers to the replacement of atoms in an inorganic lattice; however, in the case of conjugated polymers, doping is simply oxidation or reduction of the polymer backbone.

In contrast to inorganic semiconductors, where conductivity is associated with unpaired electrons in the conduction band, conjugated polymers become conducting due to *localized* charge carriers [27]. The geometry of a conjugated organic molecule (or localized region of a polymer chain) can be distorted, which costs energy (E_d). It is crucial to realize that upon distortion of an organic molecule, the HOMO level increases (by an amount ΔE) and the LUMO level decreases. Additionally, oxidation or reduction of a conjugated organic molecule has this same effect. Therefore, in the case of oxidation (p-doping), if the decrease in ionization energy (ΔE) is greater than the distortion energy (E_d), the charge localization process is adopted by a favorable relaxation to an equilibrium ionized state geometry (cost overall energy E_d). The oxidation process generates a pseudoparticle known as a *polaron*, which describes the lattice distortion around a charge. This distortion occurs over several monomer units in the polymer backbone. The charge is self-trapped by the deformation it induces in the chain. In conjugated molecules, the main polarization effect is on the π electron cloud. There is no metallic character in this system, since the LUMO is empty. In a polaron, the half-occupied level is *within* the bandgap. The lowest-energy ground state of conjugated polymer can be either degenerate or nondegenerate. For example, trans-polyacetylene (Scheme 10.1) has alternating C–C single and C=C double bonds. The two resonance structures drawn in the figure have the same total energy and are thus, degenerate. Other conjugated polymers, such as polythiophene (PT) and polyparaphenylene, have a nondegenerate ground state, which can be represented by the two nonidentical aromatic resonance structures, called the *benzenoid* and *quinoid states*. The quinoid structure represents the higher-energy state, where the ionization energy is lowered and the electron affinity is larger [28].

If a second electron is removed from the polaron, a bipolaron is formed (Scheme 10.1). Since the lattice relaxation is stronger around two charges (bipolaron) than around one (polaron), E_d is larger for a bipolaron. This causes the electronic states in a bipolaron to be further away from the edges of the bandgap than a polaron. In addition, it is more favorable to create a bipolaron than to create two polarons, due to the larger decrease in ionization energy (or energy gained by interaction with the lattice) relative to the energy of Coulombic repulsion between two positive charges. Theoretical calculations have shown that adjacent polarons are unstable and lead to the formation of bipolarons [28, 29]. Bipolarons are spinless, due to the levels within the bandgap being full or empty, depending on the doping mechanism.

In certain conjugated polymers, such as polyacetylene, degenerate structures can lead to the formation of solitons [30]. Two charges on a polyacetylene chain that would otherwise produce a bipolaron can readily separate, because the geometric structure between the two charges is the same energy as the structure outside of

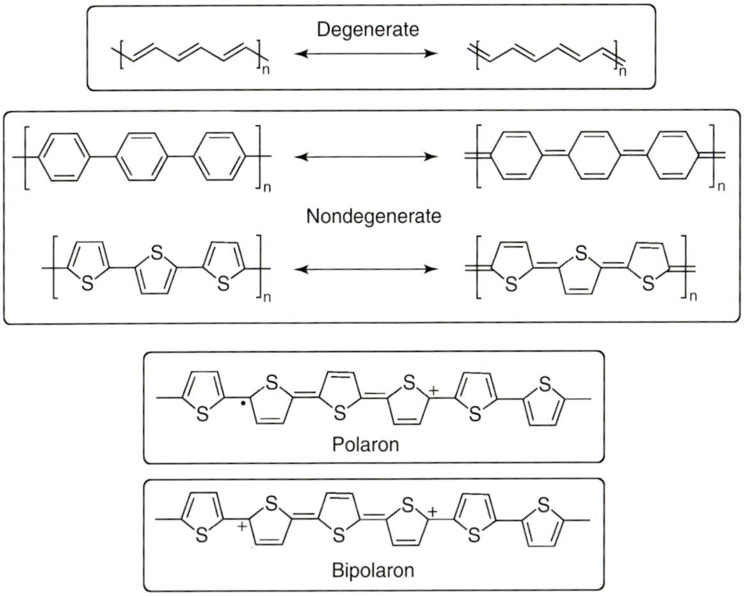

Scheme 10.1 Examples of degenerate and nondegenerate ground states. *trans*-Polyacetylene has two resonance structures of equal energy. Most conjugated polymers have a nondegenerate ground state comprised of a benzenoid and quinoid resonance structure. The resonance structures of each are shown for polyphenylene and polythiophene, where the benzenoid structure is on the left and the quinoid structure is on the right. Also, the chemical structure of a polaron and bipolaron is illustrated using a polythiophene backbone.

the two charges. Hence, these charges can propagate throughout the molecule as a solitary wave.

10.2.2
Parameters of Charge Transport

Charge transport is characterized by a quantity known as the *carrier mobility*, which is given by

$$\mu = \frac{V}{F} \tag{10.1}$$

where V is the velocity of a moving charge induced by an applied electric field and F is the amplitude of the applied field. This gives carrier mobility units of $cm^2\,V^{-1}\,s^{-1}$, which resembles the drifting of a charge. In the absence of an electric field, there is no drift, so the charge mobility is solely diffusive and is modeled by a diffusion equation

$$\langle x^2 \rangle = nDt \tag{10.2}$$

where $\langle x^2 \rangle$ is the mean square displacement of the charges, t is time, and n is an integer equal to 2, 4, or 6, that of which represents a one-, two-, or three-dimensional system, respectively. The charge mobility μ is related to the diffusion coefficient via the Einstein–Smoluchowski equation

$$\mu = \frac{eD}{k_{\mathrm{B}}T} \tag{10.3}$$

where D is the diffusion coefficient, e is the electron charge, k_{B} is the Boltzmann constant, and T is the temperature.

There is debate as to whether charge transport occurs by a localized or delocalized method. In inorganic semiconductors, charge transport occurs in delocalized states, where mobilities decrease with increasing temperature. However, for most organic semiconductors, the mobilities increase with increasing temperature. This observation supports the theory of localized states, in which charges hop from one localized state (trap) to another. The debate is centered on the fact that high-purity molecular crystals of rubrene [31–33], tetracene [33], and pentacene [33, 34] show higher mobilities with decreasing temperature, which suggests band-like transport in delocalized states. However, studies on the mean free path of charge carriers in these cases contradict delocalized transport [35, 36].

The carrier mobility of a material can be extracted from the electrical characteristics measured in a field effect transistor (FET) configuration. Other characterization methods include time of flight [37, 38] (TOF) and pulsed radiolysis time-resolved microwave conductivity measurements [39, 40], the Hall effect [32], as well as a diode configuration. In TOF charge transport is typically measured perpendicular to the substrate, so it is not ideal for measuring mobility in organic semiconductors that typically have the highest π-orbital overlap parallel to the substrate. FET devices allow for a more relevant measurement of mobility parallel to the substrate. For the commonly studied FET configuration, the current–voltage expressions derived for inorganic-based transistors are also applicable to organic field effect transistors (OFETs) in both the linear (Equation 10.4) and saturated (Equation 10.5) regime [41, 42].

$$I_{\mathrm{SD}} = \frac{W}{L}\mu C\,(V_{\mathrm{G}} - V_{\mathrm{T}})\,V_{\mathrm{SD}} \tag{10.4}$$

$$I_{\mathrm{SD}} = \frac{W}{2L}\mu C\,(V_{\mathrm{G}} - V_{\mathrm{T}})^2 \tag{10.5}$$

In these equations, I_{SD} and V_{SD} are the current and voltage bias between source and drain, respectively, V_{G} is the gate voltage, V_{T} is the threshold voltage, C is the capacitance of the gate dielectric, and W and L are the width and length of the conducting channel, respectively. Finally, μ is the field effect mobility, which is a property that depends on the molecular and crystal structure of the of the organic semiconductor material. The above equations assume that the field effect mobility is constant, however, that is, not always the case. With organic semiconductors, mobility has been reported to both increase and decrease with increasing gate voltage [42].

Other parameters beside carrier mobility used to characterize organic semiconductors in the FET configuration are the on/off ratio, threshold voltage, and subthreshold swing. The *on/off ratio* is defined as the drain–source current ratio between the on and off states. The threshold voltage is a parameter that evaluates the amount of traps (discussed below) in the organic semiconductor. Deep traps must be filled before any additionally induced charges can become mobile, hence the gate voltage must be higher than the threshold voltage. If organic semiconductors are to compete with amorphous silicon circuits in optoelectronic devices such as active matrix displays that require sharp turn-on and fast switching, charge mobilities of $>0.1\,\text{cm}^2\,\text{V}^{-1}\,\text{s}^{-1}$ and an on/off ratio of $>10^6$ are needed [43]. In summary, the use of an FET configuration is a common and useful way of characterizing organic semiconductors.

10.2.3
Improving Charge Transport

Over the last several years, there has been considerable interest in OFETs, however, charge transport in such devices is limited by localized states induced by defects and impurities within the organic layer at the interface. Evidence of this comes from the fact that the performance of OFETs is strongly dependent on the sample preparation [44, 45]. For example, impurities can cause a larger number of traps at the interface. It is generally understood that the FET mobility increases with chemical purity [46–48] and crystalline order [46, 49, 50], however, charge transport is also affected to varying degrees by temperature, pressure, charge-carrier density, and electric field [7].

Such defects act as charge carrier traps, and can be described by the commonly used multiple trapping and release model (MTR) [51]. The MTR model is the most widely used model to account for charge transport in amorphous silicon, but can be used for organic semiconductors, because it deals with localized states. In the MTR model, these localized electronic states are considered traps. For simplicity, two types of traps are considered. A deep trap has energy states further from the HOMO/LUMO level ($\gg k_B T$) than a shallow trap ($\sim k_B T$). Thus, in a deep trap, the charge cannot be released by thermal excitations. In contrast, a shallow trap can release a charge by thermal excitation (hence charge transport is highly temperature dependent), and is characterized by a finite trapping time τ_{tr}. The MTR model takes into account these shallow traps and their characteristic trapping time. All field-induced carriers contribute to the current flow at any moment of time, but their effective mobility, μ_{eff} is reduced relative to its trap-free value μ_0.

$$\mu_{eff} = \mu_0\,(T)\,\frac{\tau\,(T)}{\tau\,(T) + \tau_{tr}\,(T)} \tag{10.6}$$

Here, $\tau_{tr}(T)$ is the average trapping time in shallow traps and $\tau(T)$ is the average time a polaron spends diffusing between consecutive trapping events. When $\tau \gg \tau_{tr}$, the equation is dominated by diffusion processes, and when $\tau \ll \tau_{tr}$ the equation is dominated by trapping processes.

Polymers are highly attractive for OFETs, because thin films of these materials can be obtained through simple solution techniques such as drop casting, spin coating, and ink printing. Some examples are highlighted below, but for a more detailed account see the recent review by Arias [52]. Among all conjugated polymer semiconductors, charge-carrier transport has been most extensively investigated with poly(3-alkylthiophene)s (P3AT)s. The performance of poly(3-hexylthiophene) (P3HT) in OFETs was observed to approach that of amorphous silicon FETs, with field-effect mobilities of $0.05-0.1\,\text{cm}^2\,\text{V}^{-1}\,\text{s}^{-1}$ and on/off current ratios of $>10^6$ [53]. The high mobility is attributed to the formation of extended polaronic states as a result of local self-organization, in contrast to the variable-range hopping of self-localized polarons found in more disordered polymers. However, the mobilities are usually lower for conjugated polymers relative to oligomeric materials of the same molecular backbone, due to poor molecular ordering and semicrystallinity. It has been observed that the hole mobility is highly dependent on the regioregularity of the P3AT films [49, 50], where regioregular P3HT offers improved performance in FETs over their regioirregular counterparts. A clear correlation between the FET mobility of regioregular P3HT and its molecular weight (MW) was shown by Kline *et al.* [54], with mobility values ranging from 1.7×10^{-6} to 9.4×10^{-3} as MW is increased from 3.2 to 36.5 kDa.

The length of the alkyl side chain in P3ATs also plays a crucial role in thin-film charge transport. The field-effect mobility of holes in regioregular P3ATs was determined for a series of five alkyl chain lengths from 4 (*n*-butyl) to 12 (*n*-dodecyl) [55]. A nonmonotonic dependence of field-effect mobility on alkyl chain length was found. The average hole mobility varied from $1.2 \times 10^{-3}\,\text{cm}^2\,\text{V}^{-1}\,\text{s}^{-1}$ in poly(3-butylthiophene) and $1 \times 10^{-2}\,\text{cm}^2\,\text{V}^{-1}\,\text{s}^{-1}$ in P3HT to $2.4 \times 10^{-5}\,\text{cm}^2\,\text{V}^{-1}\,\text{s}^{-1}$ for poly(3-dodecylthiophene), demonstrating the optimal side chain length being six. These results are in similar agreement to earlier results demonstrated by Bao and Lovinger [56], where poor molecular ordering and low crystallinity was found for regioregular PTs with bulky or carboxylic acid-substituted side chains.

10.2.4
Optical Properties of Conjugated Polymers

The spectral features of conjugated polymers vary due to conformational changes of the polymer backbone and aggregation processes. For example, in a poor solvent chain planarization and interchain $\pi-\pi$ stacking interactions take place, which increases the conjugation length. This results in the appearance of fine vibronic structures on the low-energy side of the main $\pi-\pi^*$ transition band. Absorption spectra of neutral films of P3ATs [57] and poly(3-alkoxythiophenes) [58] show unresolved shoulders and side peaks along the higher wavelength side of the main $\pi-\pi^*$ absorption band. These features have been assigned to a distribution of different conjugation lengths or vibronic structures.

A general feature of CPs is that their characteristic absorption spectra change upon doping. Investigating the optical properties of CPs is thus important to develop an understanding of the basic electronic structure of these materials. This phenomenon, called *electrochromism*, is primarily due to the changes in the

CPs π-electronic character after doping. As discussed earlier, upon doping the band structure of the neutral polymer is modified, which generates lower-energy transitions than the typical $\pi-\pi^*$ transition and charge carriers. It is this effect of doping that facilitates electrochromic behavior. Derivatives of PT, poly(pyrrole), and poly(aniline) are the most widely studied [59].

A number of studies on the electrochromic potential of thiophene-based polymers have emerged. It is known that PT and its derivatives are air stable in both the undoped and doped states, which, combined with their outstanding electronic properties, make them the most commonly studied CP in the past two decades. Garnier and Tourillon [60, 61] compared the absorption spectra of doped and undoped PT and poly(3-methylthiophene) (P3MT). In P3MT, a disappearance of an absorption peak at 480 nm ($\pi-\pi^*$ interband transition) combined with the appearance of a peak from 650 nm to the near-IR was observed after doping, characteristic of free charge carriers. In addition, Chung et al. [62] studied in detail the absorption spectra of PT films at different voltages, which corresponded to different doping levels. As the doping level increased (with increasing voltage), the intensity of the $\pi-\pi^*$ transition decreased, and the absorption peak shifted to higher energy. The higher-energy band was attributed to the occurrence of bipolarons. At the highest doping level, the spectrum presents the characteristics of the free-carrier absorption of the metallic state [62, 63]. This switching is accompanied by a color change from red to blue-green (doped). Panero et al. investigated P3MT-coated indium tin oxide (ITO) glass electrodes and obtained an optical contrast of 30% [64], which demonstrates the optical switching potential of PT derivatives. The first studies on P3MT revealed that P3MT films on bulk Pt electrodes have a cycle life up to 1.2×10^5 charge–discharge cycles [60].

Further improvements followed, with the Reynolds group demonstrating the use of electrochemically prepared poly(3,4-ethylenedioxythiophene) (PEDOT) as a cathodically coloring material with optical contrasts around 60% [65–67]. It was also shown that varying the size of the alkylenedioxy ring has little effect on the extent of conjugation of the backbone, as revealed by opto-electrochemical measurements [68]. The spectral changes in the visible region associated with the doping and dedoping processes of PT derivatives have led to increasing interest in electro-optical systems such as display devices or electrochromic windows. For a more thorough description of electrochromics, the reader is referred to Beaujuge and Reynolds [69].

10.3
Synthetic Methods

10.3.1
Electrochemical Synthesis

The electrochemical synthesis of PTs has proven to be a robust method over the last few decades. After the first reported electrochemical synthesis of poly(pyrrole) [70],

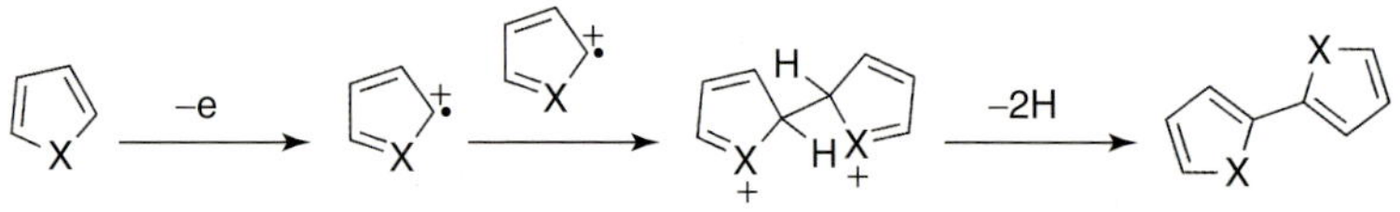

Scheme 10.2 Electrochemical polymerization mechanism of five-membered heteroaromatics.

the electrochemical method has been rapidly extended to other heteroaromatic and aromatic systems, including thiophene [71], benzene [72–74], and fluorene [75]. Electrochemical synthesis of CPs is typically done through an anodic [76] route, but can also be achieved cathodically [76, 77]. For the purposes of this section, we will be discussing the anodic route, which offers several advantages over the cathodic route, including the absence of catalyst, direct grafting of doped CP films onto the electrode surface, easy film thickness control, and *in situ* electrochemical characterization of the grafting process. Cyclic voltammetry is the most common electrochemical technique used in the synthesis of CP thin films.

In the proposed mechanism shown in Scheme 10.2 [78], the first electrochemical step is the oxidation of monomer to its radical cation, followed by the coupling of two radicals to produce a dihydrodimer dication. After loss of two protons and rearomatization, the dimer then undergoes oxidation, which reacts with another monomeric radical. This process is repeated until the oligomer becomes insoluble and precipitates onto the electrode surface. The electrochemical oxidation in the case of poly(pyrrole) requires 2.07–2.5 F/mol [79, 80], with excess charge being responsible for reversible doping of the polymer. In the case of the electropolymerization of benzene to generate polyphenylene, a Lewis acid is used to destabilize benzene to a lower oxidation potential. It is argued that the rate-limiting step in electropolymerization is not the diffusion of oligomer to the electrode surface, but a radical-coupling step, and that the initial stages of the film-forming reaction resembles the nucleation process also observed in the electrodeposition of metals [78, 81]. This is also observed in the electrochemical synthesis of P3MT at high scan rates [82]. The exact polymerization mechanism, with respect to the rate-limiting step and process of propagation, is still a matter of debate.

Electrochemical methods were further extended toward the polymerization of soluble CPs [83–85]. Within the class of soluble P3ATs, the highest conductivity of 750 S/cm was obtained for P3MT. P3ATs with alkyl carbons of four or more showed lower conductivities around 1–100 S/cm. The decrease in conductivity according to Hotta [85] is attributed to the influence of the hydrocarbon side chains on the electronic structure. There are very few reports demonstrating the immobilization of soluble CPs [86–88], which is accomplished through the use of an electroactive monolayer. For example, Zotti *et al.* [87] accomplished surface-initiated polymerization of pyrrole and thiophene-based monomers from oligothiophene monolayers on an ITO electrode. Because the resulting polymers were soluble, only surface-coupled polymers remained on the electrode surface.

Although the electrochemical technique is simple in execution, allows for control of film thickness by varying the time and oxidation potential, and can be used on a variety of monomers, it tends to produce films with highly irregular morphologies as well as regioirregular polymers. These drawbacks signify the need for a more versatile approach that affords more precise control of backbone structure and thus, electronic properties.

10.3.2
Cross-Coupling

Organometallic cross-coupling reactions [89] are the most powerful tools used in the synthesis of a wide array of CPs. The most common cross-coupling methods utilized in the synthesis of CPs include Kumada, Suzuki, Heck, Sonogashira, and Negishi reactions [89]. All of these cross-coupling reactions follow a general catalytic cycle, typically performed with a palladium or nickel catalyst. The catalytic cycle starts with oxidative addition, followed by transmetallation, and finally couples two species through reductive elimination. In the *oxidative addition* step, a M(0) catalyst inserts into an aryl-halogen bond, producing an Aryl-M(II)-X species, where X is a halogen and M is palladium or nickel. Upon insertion, the oxidation state of the metal increases to M(II). The *transmetallation* step of the catalytic cycle is unique depending on the cross-coupling reaction used. In the case of Kumada, an aryl-magnesium halide species that contains a source of nucleophilic carbon transmetallates with the aryl-M(II)-Br formed via the preceding oxidative addition step. The product of the transmetallation step is an aryl-M(II)-aryl. In *reductive elimination*, the M(II) catalyst couples the two aryl species, leaving an aryl–aryl species and eliminating M(0). This catalytic cycle is then repeated. Depending on the nature of the coordinating ligands, a *trans–cis* isomerization can precede the reductive elimination process. The use of bidentate ligands circumvents the *trans–cis* isomerization process, due to geometrical constraints about the metal center.

In the synthesis of CPs, this type of condensation polymerization can follow a step-growth or a chain-growth mechanism. In step growth, an A–A- and B–B-type monomer is used, where in chain-growth a bifunctional A–B-type monomer is used [90, 91]. Through the use of bifunctional A–B-type monomers, cross-coupling polycondensations can proceed by a unique catalyst-transfer mechanism. This mechanism gives a chain-growth polymerization, which has many advantages over a step-growth reaction. In the catalyst-transfer mechanism, the reductively eliminated M(0) species undergoes intramolecular oxidative addition in the nearest carbon–halogen bond, followed by transmetallation with monomer and intramolecular transfer of the catalyst to the elongated polymer endgroup. The propagating catalyst remains at the end of the polymer chain throughout the polymerization. A distinct advantage of this technique is that it introduces the ability to obtain near-monodisperse and regioregular polymers, to prepare end-functionalized polymers, and to create a diverse array of block copolymers. Such polymerizations are termed *catalyst-transfer polymerizations*. Since Kumada catalyst-transfer polymerization to form P3ATs is the most studied recently, it is practical to start with the synthesis of P3ATs using this method. Because P3AT synthesis can involve

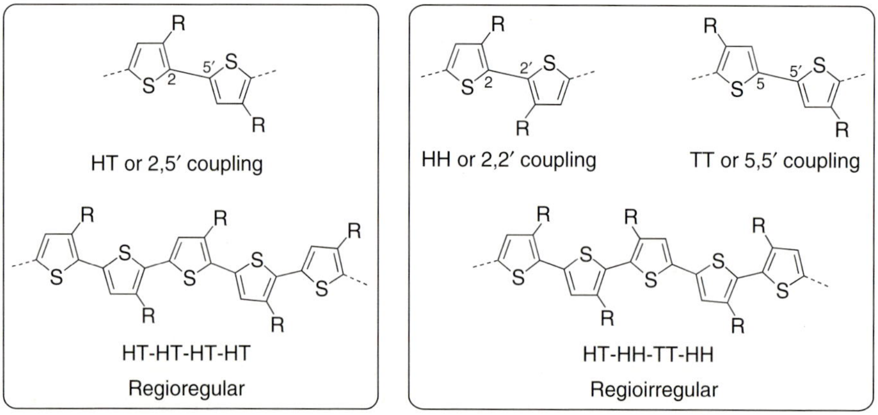

Scheme 10.3 Consistent head-to-tail coupling gives re-gioregular P3ATs, where a combination of head-to-head and tail-to-tail couplings give regioirregular P3ATs.

random couplings, it is necessary to review the concept of regioregularity before discussing the synthetic methods to achieve regioregular P3ATs.

10.3.3
Regioregularity

Among conjugated polymers, PTs are one of the most widely studied materials due to their use in a variety of applications. Electrochemically synthesized P3ATs create random couplings, which result in regioirregular polymers. The loss of regioregularity is due to a combination of head-to-head and tail-to-tail couplings (Scheme 10.3), which causes a twisting of the polymer backbone. This undesirable twisting of the backbone results in a loss of conjugation and subsequent degradation of electronic properties.

In 1992, McCullough *et al.* [92, 93] reported the first regioregular P3ATs by employing Kumada cross-coupling methods, in which structurally homogeneous poly(3-dodecylthiophene) upon doping with I_2 gave average conductivity values of 600 S/cm. In this method, regiospecific generation of 2-bromo-5-bromomagnesio-3-alkylthiophene is achieved through treatment of 2-bromo-3-alkylthiophene with lithium diisopropylamide (LDA) at $-78\,^\circ$C, followed by quenching with magnesium bromide ethyl etherate (MgBr$_2$Et$_2$O). Polymerization was initiated by addition of Ni(dppp)Cl$_2$ (dppp = diphenylphosphino propane) catalyst. It was found based on quenching studies that the resulting polymer was 98–100% regioregular. Typical number average MW (M_n) values of the polymers obtained through this method are between 20 and 40 kDa, with a polydispersity index (PDI) around 1.4.

This method was further improved using the Grignard metathesis (GRIM) method, in which cryogenic temperatures and highly reactive metals were not required [94, 95]. In the GRIM method, 2,5-dibromo-3-alkylthiophene is treated with 1 equiv. of a Grignard reagent (RMgX) to form a mixture of intermediates, in which the magnesium halogen exchange takes place at either the 2 or 5 position in a ratio of 85 : 15 to 75 : 25. Although the ratio of undesirable isomers is higher than the McCullough method, P3AT's with regioregularity of 99% or higher are still obtained. Typical M_n values of the polymers synthesized through the GRIM method are between 20 and 35 kDa, with a PDI around 1.2–1.4.

The Rieke method was reported [96] shortly after the McCullough method, in which 2,5-dibromo-3-alkylthiophene was treated with highly reactive zinc precursor (formed from *in situ* reduction of zinc chloride with lithium naphthalenide in THF). This highly reactive, "Rieke zinc" was found to react chemoselectively with 2,5-dibromo-3-hexylthiophene to yield quantitative mono-organozinc, with 90% regioselectivity. Use of both Ni(PPh$_3$)$_4$ (PPh$_3$ = triphenylphosphine) and Ni(dppe)Cl$_2$ (dppe = diphenylphosphino ethane) resulted in polymerization, however, Ni(dppe)Cl$_2$ afforded much better results, giving polymers with a regioregularity greater than 97% and Mn values around 24–34 kDa with a PDI of 1.4.

Other cross-coupling methods in the synthesis of regioregular P3ATs are the Stille [97] and Suzuki [98] palladium-catalyzed cross-coupling reactions. In the Stille reaction, the use of 3-hexyl-2-iodo-5-(tri-n-butylstannyl)thiophene with Pd(PPh$_3$)$_4$ resulted in polymers with greater than 96% regioregularity, with an M_n of 7.36 kDa and PDI of 1.36 when toluene was used as the solvent [99]. A major downside to the Stille reaction is the toxicity of the organotin reagents involved, combined with low MWs of the subsequent polymers. An alternative approach with the Suzuki reaction involves using a monomer with an iodide group and a boronic ester in the 2 and 5 positions, respectively [100]. In the presence of Pd(OAc)$_2$, poly(3-octylthiophene) was prepared with 96–97% regioregularity with a weight average MW value of 27 kDa. The Suzuki reaction encounters sluggish reaction rates because of the slow deboronation of thiophene boronic acids or esters during polymerization and the slow oxidative addition of the palladium catalyst to the electron-rich thiophene ring [100, 101]. An upside to polymerization with Suzuki and Stille reactions are that the monomers used in both is typically air stable.

Because of the large amount of success in the synthesis of regioregular P3ATs using Kumada catalyst-transfer polycondensation (KCTP), much attention has focused on the mechanism. It was originally proposed by Yokozawa *et al.* [102] that this condensation reaction proceeds via a chain growth mechanism (Scheme 10.4), followed by McCullough and coworkers [103, 104], who in addition claimed that KCTP is a quasiliving system. In the GRIM method, it was observed that P3ATs with relatively high MW are produced early in the reaction, which contradicts a step-growth mechanism. In addition, the M_n values of the polymers increased in proportion to monomer conversion, and can be controlled by the amount of nickel catalyst [102]. Chain-extension experiments were conducted by sequential monomer addition, and it was found that after the addition of a new portion of monomer at the end of the polymerization, an increase in MW of the resulting polymer was observed [104]. This strategy was used to create a triblock copolymer of poly(3-dodecylthiophene)-*b*-poly(thiophene)-*b*-poly(3-dodecylthiophene). In addition, it was also found that adding various Grignard reagents to the

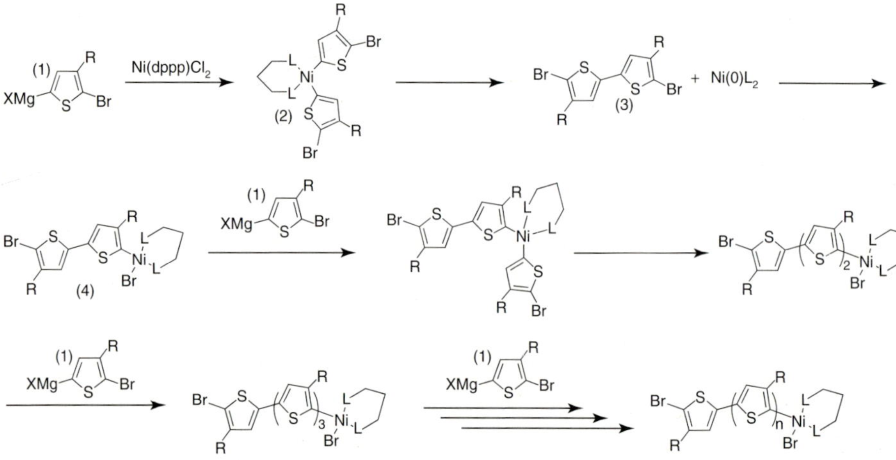

Scheme 10.4 Mechanism of Kumada catalyst-transfer polycondensation according to Yokozawa [91, 108].

nickel-terminated P3AT results in the formation of end-functionalized polymers via both the McCullough [105] and GRIM method [106, 107].

The mechanism in the KCTP synthesis of regioregular P3ATs is outlined in Scheme 10.4. The first step is the reaction of two equivalents of intermediate (1) with Ni(dppp)Cl$_2$ affording organonickel compound (2). Reductive elimination occurs immediately to generate Ni(0) and tail-to-tail dimer (3), of which Ni(0) proceeds to undergo intramolecular oxidative addition with (3) to generate (4). After (4) is generated, another monomer (1) transmetallates with complex (4). Growth of the polymer chain then proceeds via reductive elimination and oxidative addition to the elongated end-group. The cycle is repeated to yield regioregular P3ATs.

The KCTP reaction has also been extended toward the synthesis of poly(p-phenylene)s, poly(pyrrole), and poly(fluorene)s. Yokozawa investigated whether a monomer with no heteroatom in the aromatic ring would undergo KCTP [109]. Using p-hexyloxyphenylene as a monomer, it was found that LiCl was necessary to optimize the KCTP polymerization, because without LiCl, low MWs and broad polydispersities were observed. In contrast, Lanni and McNeil [110] have shown that under slightly different reaction conditions, the polymerization of this monomer is essentially unaffected by the presence of LiCl. It was suggested that the difference is due to the effect of LiCl on the reaction of the Grignard monomer with Ni(dppe)Cl$_2$. The synthesis of polypyrrole via KCTP also proceeded in a catalyst-transfer mechanism based on a linear relationship between Mn to monomer conversion and feed ratio of catalyst [111]. However, it was found that the use of dppe as a ligand gave a lower polydispersity than dppp as a ligand. Oligomeric byproducts were produced in the polymerization, which can be suppressed by lowering the polymerization temperature and adding additional dppe to the polymerization mixture to stabilize the catalyst. Yokozawa and coworkers [112] also produced block-copolymers of poly(p-phenylene)-*b*-poly(3-hexylthiophene). This is the first report of a block copolymer consisting of two different conjugated polymers. It was also reported that reversing the order of the polymerization resulted in a broad MW distribution of polymers. The π-electrons of the polymers are considered to assist the transfer of the nickel catalyst, and it would be difficult for the nickel catalyst to transfer to the terminal C–Br bond of the phenylene ring of the elongated unit when the thiophene block has a better pi-donating ability [112]. Poly(p-phenylene)-*b*-poly(pyrrole) as also synthesized in a similar manner [111]. Poly(fluorene)s have also been synthesized through KCTP by Geng and coworkers [113] via the GRIM method, in which a bromoiodofluorene monomer was used with i-PrMgCl/LiCl. In the presence of Ni(dppp)Cl$_2$ at 0 °C, poly(9,9-dioctylfluorene) was obtained in under 10 min with a M_n of 18.8–86 kDa and a PDI of 1.49–1.77. The above-mentioned methods are accomplished by creating an initiator *in situ*. When the initiator is created *in situ*, the initiator must first form before chain-growth polymerization can occur. By using an external initiator, where the initiator is synthesized prior to the polymerization, a more diverse array of possibilities is available, including the ability to achieve surface-initiated polymerizations. Chain-growth addition of monomer to the external initiator occurs immediately after exposing the initiator to a solution of monomer. The following section discusses such polymerizations.

10.3.4
External and Surface-Initiated Chain-Growth Polymerization of CPs

Dehalogenation polycondensation [114] of 2,5-dihalothiophene, 2,5-dihalo-3-alkylthiophene, and 1,4-dihalobenzene with Ni(COD)$_2$ (COD = 1,5-cyclooctadiene) and neutral PPh$_3$ ligand in solution was demonstrated by Yamamoto and has been widely adopted as a method of CP synthesis due to the ability of this method to obtain high MW [115]. All polycondensations employing PPh$_3$ as the ligand proceeded smoothly within the temperature range of 25–100 °C, with the best yields around 60–80 °C. The 2,2'-bipyridine (bpy) ligand was also employed, with similar results to that of PPh$_3$. This was utilized by Carter and coworkers [116] toward surfaces in a "grafting-through" approach, where thin patterned cross-linked polyacrylate network films on silicon wafers were prepared that incorporated 4-bromophenyl functionality at the surface, which acted as an attachment site for Yamamoto condensation polymerization of 2,7-dibromo-9,9-dihexylfluorene. Additionally, a surface-directed Yamamoto polycondensation reaction to graft poly(fluorene) from dibromofluorene-functionalized silica and quartz surfaces was accomplished [117]. Due to the uncontrolled nature of the polycondensation reaction, the resulting films were not smooth, however, poly(9,9-dihexylfluorene) was grafted over large areas with no delamination of the film. Poly(9,9-dihexylfluorene) was also formed in solution. Due to the desire for smooth films, as well as end-functionalized and block-copolymer films, a chain-growth polymerization is needed that will allow for the synthesis of these architectures.

Senkovskyy *et al.* [118] introduced external initiation of P3HT, where Ph-terminated P3HT was synthesized from (Ph)Ni(PPh$_3$)$_2$Br via reaction of phenyl bromide (PhBr) with Ni(PPh$_3$)$_4$. It was found that a relatively high regioregularity can be accomplished through the use of pure monomers initiated by (Ph)Ni(PPh$_3$)$_2$Br. It was also found that lowering the polymerization temperature helped suppress undesirable termination reactions inherent in the use of monodentate ligands. In further studies, a Ni(II) macroinitiator was immobilized through the use of photocross-linked poly(4-bromostyrene) films and reacted with Ni(PPh$_3$)$_4$. After thorough rinsing of the immobilized initiator to remove any unreacted catalyst, the initiator-layer was exposed to monomer, which resulted in selective grafting of P3HT from the immobilized initiator. The use of a PhBr functionalized monolayer on silicon, however, failed to give surface-grafted P3HT [119]. Further studies were done with monomers incorporating more than one thiophene unit from a (Ph)Ni(PPh$_3$)$_2$Br initiator to see how far intramolecular catalyst-transfer takes place [120]. It was found that an increase in the monomer molecular length somewhat decreased the fraction of Ph-terminated products. Additionally, the substitution pattern on the monomers was shown to be critical to the chain-growth performance. Monomers having an alkyl substituent ortho to the growing site gave better results, due to higher stability of the intermediate aryl-Ni complex. Boyd *et al.* [121] also observed this trend, where a reversed thiophene monomer (Grignard formed from 2-bromo-4-hexyl-5-iodothiophene) resulted in no polymerization from an external initiator. This observation was attributed to a combination of steric hindrance of the nucleophilic carbon and a lack of orthostabilization of the propagating nickel catalyst.

Externally initiated palladium catalyzed Suzuki polycondensation of 2-(7-bromo-9,9-dioctyl-9H-fluoren-2-yl) via (^{t}Bu$_3$P)Pd(Ph)Br was achieved by Yokozawa and coworkers [122], who demonstrated complete introduction of the initiator unit at the polymer end and the linear relationships of Mn to monomer conversion and feed ratio, which indicate that the polycondensation reaction proceeds via a chain-growth mechanism. The initiator (^{t}Bu$_3$P)Pd(Ph)Br is a Pd(II) complex derived from the oxidative addition of bromobenzene to Pd(0) ligated by two bulky tri-tert-butyl phosphines (^{t}Bu$_3$P) [123]. It was demonstrated previously that mono-ligated Pd(0) after reductive elimination undergoes preferential oxidative addition through intramolecular transfer across one fluorine unit [124, 125]. Recently, Beryozkina *et al.* [126] successfully polymerized poly[9,9-bis(2-ethylhexyl)fluorine] from a surface by immobilizing the initiator used by Yokozawa. The initiator was obtained by reacting a PhBr-functionalized monolayer with Pd(P^{t}Bu$_3$)$_2$. This allowed for selective grafting of poly(9,9-bis(2-ethylhexyl)fluorene) up to a thickness of 100 nm.

To further develop surface-initiated Kumada catalyst-transfer polycondensation (SI-KCTP), Sontag *et al.* [127] (Scheme 10.5) used a combination Ni(COD)$_2$ with four equivalents of PPh$_3$ ligand to produce a reactive Ni(COD)(PPh$_3$)$_2$ species, which readily undergoes oxidative addition with a thienyl bromide functionalized monolayer on gold. The surface-bound initiator was used to graft PT and

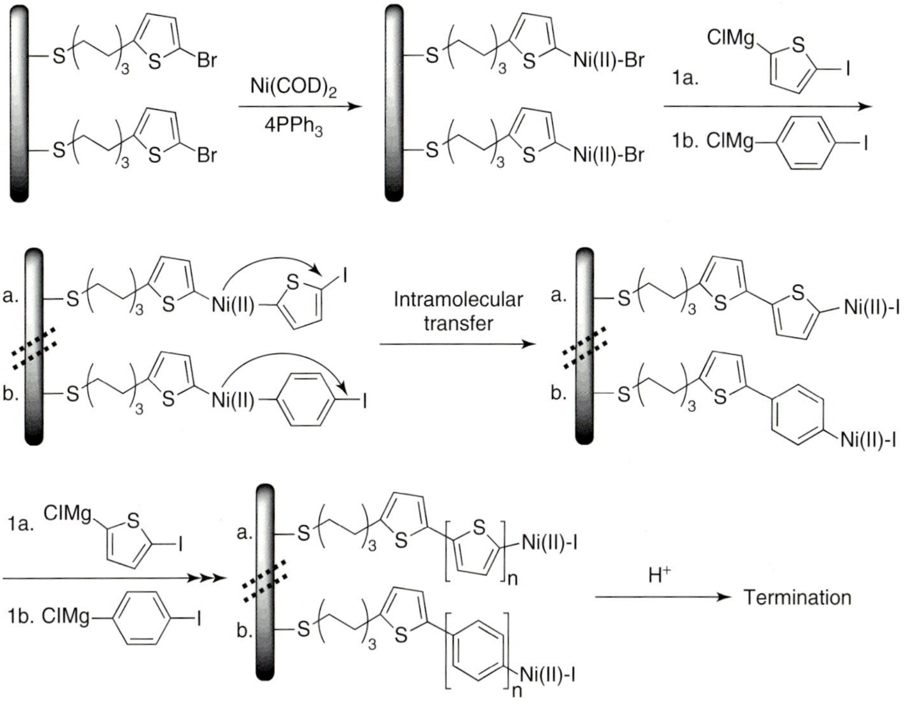

Scheme 10.5 SI-KCTP scheme for the synthesis of poly(thiophene) and poly(phenylene) according to Sontag *et al.* [127].

poly(p-phenylene) on the order of 10–40 nm thick via a SI-KCTP reaction. Electrochemical characterization verified the presence of a surface-bound Ni(II) initiator prior to polymerization and PT film after the polymerization. Further studies by Marshall *et al.* [128] on the polymerization of poly(p-alkoxyphenylenes) revealed that long alkoxy substituents on phenyl monomers significantly slow down the rate of reaction at the surface due to steric effects. The thickest film was obtained through the use of methoxy substituents on the ring, whereas hexyloxy substituents gave the thinnest films. In addition, it was found that the use of LiCl resulted in more uniform films, as revealed by atomic force microscopy (AFM).

To further develop KCTP from external initiators, Bronstein and Luscombe [129] found that a ligand exchange reaction between (Ph)Ni(PPh$_3$)$_2$Cl and dppp furnishes bidentate initiator (Ph)Ni(dppp)Cl. This bidentate initiator gave superior performance over the monodentate initiators used by Senkovskyy *et al.* [118]. A ligand-exchange reaction was also applied to initiators bound to silica particles, in which the PhBr-functionalized initiator was exposed to (Et)$_2$Ni(Bpy) to yield (Ph)Ni(bpy)Br, followed by ligand exchange with dppp to yield (Ph)Ni(dppp)Br [130]. This technique yielded a high-molecular-weight P3HT grafted from the silica particles.

10.3.5
Other Methods of Conjugated Polymer Synthesis

Chemical oxidative polymerization of electron-rich aryl monomers has been extensively used for the preparation of conjugated polymers, including PTs [131], polyaniline [132], and polyphenylenes [133]. Oxidative polymerization can also be used to prepare surface-grafted films, where Inaoka and Collard [86] accomplished chemical polymerization of 3-octylthiophene with $FeCl_3$ as an oxidizing agent. By modifying a glass slide with a thienyl-substituted silane and subjecting it to polymerization resulted in smooth, uniform, and strongly adhered poly(3-octylthiophene) films covalently immobilized to the substrate. The amount of polymer deposited on substrates modified with a thienyl-substituted silane varies inversely with the concentration of $FeCl_3$. The use of less oxidant resulted in more polymer deposition on the modified surface. This approach, however, is impractical due to low control over MW and PDI.

Jhaveri *et al.* [134] demonstrated the use of alkyllithium reagents for the synthesis of high-MW polymers, in which fluorene monomer was lithiated with t-BuLi *in situ*, and polymerized via Ni(dppp)Cl_2. Because the polydispersity of the resulting polymers was high, a step-growth mechanism was proposed. It was also demonstrated that this method works well for monomers that do not contain functionality sensitive to alkyllithiums [135].

Another important class of conjugated polymers are the poly(arylene-ethynylene)s and the poly(arylenevinylene)s. The enhanced stability and fluorescence of PPEs make them exceptional candidates in the field of organic electronics. The first synthesis of PPEs via coupling of diethynylbenzenes to dihaloarenes was first accomplished by Yamamoto and coworkers [136]. In this case, only low-MW oligomers with poor solubility were obtained. Attempts to improve on this method by incorporating soluble side chains was first attempted by Giesha and Schulz [137] via Sonogashira polycondensation of 1,4-dibromo-2,5-dialkoxybenzenes with diethynyl compounds. The polymers had a very low degree of polymerization (DP) of only 10–15 repeat units. Moroni *et al.* [138] then synthesized soluble PPEs by coupling a substituted diethynyl arene with a disubstituted halogenoaryl with various side chains in the presence of a catalytic amount of PdCl$_2$, Cu(OAc)$_2$, and PPh$_3$ in a triethylamine/THF mixture to obtain soluble PPEs with a DP of approximately 150, although there is some controversy regarding the actual DP being much smaller than reported [139]. Weder and coworkers [140] successfully produced high-MW PPE with ethylhexyoxy and octyloxy solubilizing groups, which are credited with the large DP of >200. Improvements in the synthetic procedure allowed Müllen and coworkers [141] to successfully synthesize soluble rod-like oligo and poly(1,4-phenyleneethynylene)s end-capped with various functional groups by varying the amount of end-capping agents. In summary, preparation of PPEs is best performed with aryl iodides, which react faster, in higher yields, and at lower reaction temperatures [139]. Cross-linking and structural defects plague this reaction, and the use of aryl iodides helps suppress side reactions mainly due to lower reaction temperatures. Additionally, electron-withdrawing substituents on the halide monomer improve the reaction.

Around the same time as the synthesis of soluble PPEs reported by Giesha and Schulz, Yamamoto and coworkers [142] independently reported the synthesis of soluble poly(aryleneethynylene)s (PAEs) by coupling 2,5-diiodo-3-hexylthiophene with a variety of diynes in the presence of Pd(PPh$_3$)$_4$ and CuI in triethylamine. Varying the temperature and reaction media further optimized the reaction, and the best results were with a 5 mol% of both Pd(PPh$_3$)$_4$ and CuI with a 1 : 1 mixture of triethylamine and toluene at 60 °C. Typical DP values were around 100 [143].

Bao *et al.* [144] applied the Heck reaction, where organic dihalides and divinyl-benzene compounds are coupled, to the synthesis of PPVs. The polymerization was carried out in DMF in the presence of Pd(OAc)$_2$ with the tertiary amine and triarylphosphine under nitrogen. GPC results demonstrated that incorporation of longer alkoxy chains on the monomer result in higher-MW polymer, where polymers with alkoxy chains of 16 carbons had an M_n of 18.7 kDa and a PDI of 2.88.

Acyclic diyne metathesis (ADIMET) is also a powerful tool used by Bunz and coworkers [145], in which 1,4-dipropynylbenzenes were metathesized to poly(p-phenyleneethynylene)s via an *in situ* catalyst system of molybdenum hex-acarbonyl and p-(trifluoromethyl)phenol. The advantages of this approach include a simple experimental setup with inexpensive and commercially available catalyst precursors that do not require a rigid exclusion of air and moisture. In addition, this method yielded defect-free polymer with a DP of nearly 100 units, which makes this method superior to the palladium-catalyzed method discussed above.

The McMurry reaction features the coupling of aromatic dialdehydes with low-valent titanium reagents to form olefinic units. This method was used to synthesize PPV via polymerization of 2,5-dihexylterephthalaldehyde in the presence of titanium tetrachloride and zinc dust with an average DP of 30 units [146]. Additionally, poly(3,4-dibutoxy-2,5-thienylenevinylene) (PTV) was also poly-merized via this method by coupling 3,4-dibutoxythiophene-2,5-dicarbaldehyde [147]. The Wittig–Horner reaction is another commonly used route to PPVs, for example, in a report by Hörhold and coworkers [148] where copolymers of alternating phenylene and arylene vinylene units were coupled from a dialde-hyde and a diethylphosphonate. This preparation yielded a stereoregular polymer with *trans*-only configuration, but when a diketone was used instead of the di-aldehyde, the phenyl-substituted double bonds consisted of both *cis* and *trans* configurations.

A unique approach by Grubbs and coworkers [149, 150] was taken with *living* ring-opening metathesis polymerization (ROMP) in the synthesis of polyacetylene, which offers precise control of the polymer MW, end groups, and polydispersity. ROMP has then been used in the synthesis of PPVs by Grubbs and coworkers [151], where bis(carboxylic ester) derivatives of bicyclo[2.2.2]octa-5,8-diene-*cis*-2,3-diol were polymerized, followed by pyrolytic acid elimination to form the PPV.

10.4
Characterization Methods

10.4.1
X-Ray Diffraction

X-rays are known to have total reflection on the surface of flat substances and are strongly attenuated at angles below a critical angle relative to the surface plane. Grazing-incidence X-ray diffraction (GIXD) was developed by Marra *et al.* [152] based on the total reflection phenomenon to study crystal surfaces. This method is commonly used to analyze in-plane crystal structures from a few nanometers to several hundred nanometers beneath the surface on solids in air. Organic thin films of monolayer or submonolayer are weak X-ray scatterers, so GIXD does not allow for a direct determination of atomic positions. However, information on the molecular packing and orientation based on fixed atomic coordinate models of organic molecules can be obtained.

In GIXD, the angle of incidence of the X-ray beam is generally kept below the critical angle, which limits the sample penetration depth to that of the evanescent wave, in the range of 50–100 Å. This is the general setup when sample substrates are disordered, since diffracted beams will scatter in all directions. When the incident beam is kept below the critical angle of the substrate, scattering of X-rays from the substrate is eliminated, which allows for sole measurement of the weak diffraction signal of a crystalline organic film. When the substrate is crystalline, it is not necessary to keep the incident X-ray beam below the critical angle as long as the substrate Bragg peaks are ignored in the analysis. The advantage of increasing the angle of incidence is the increased number of photons, which ensures that the whole thin film participates in the diffracted signal.

Bragg's law is used to analyze GIXD, where the GIXD patterns from two-dimensional crystals arise from a two-dimensional (2D) array of rods (Bragg rods) [153]. GIXD only requires that the X-ray beam impinging on the sample be at a grazing angle and does not place any restriction on the diffracted beam. By measuring all angles of the diffracted beam, more information from the sample can be obtained. The analysis of in-plane crystal structures only requires the use of a one-dimensional detector, however, by using a two-dimensional area detector out-of-plane structures can also be analyzed.

In three-dimensional crystals, diffraction from a set of *h,k,l* (Miller indices) crystal planes with an interplanar spacing d occurs only when the Bragg law is satisfied. Satisfaction of the Bragg law occurs under two conditions, when the scattering vector length q is equal to $2\pi(ha^* + kb^* + lc^*)$ and the normal to the planes bisect between the incident and outgoing beams. For a 2D crystal, there are no selection rules or restrictions on the scattering vector component q_z along the film normal. Thus, the Bragg scattering extends as continuous rods through the 2D reciprocal lattice points. Diffraction for a particular value of q_z of an (h, k) Bragg rod requires that the horizontal component q_{xy} of the vector q, is coincident with the reciprocal lattice vector $2\pi(ha^* + kb^*)$.

The intensity, $I(q_{xy}, q_z)$ in 2D GIXD is simultaneously acquired as a function of in-plane (q_{xy}) and out-of-plane (q_z) components of the scattering vector. Because the in-plane orientations of crystalline domains made by vacuum deposition are random, the in-plane component in 2D GIXD patterns is the sum of all the different orientations. Therefore, q_{xy} is represented as a combination of q_x and q_y without a distinction between the components, and q_z is measured directly. The angular 2θ positions of the Bragg peaks yield the repeat distances $d = 2\pi/q_{xy}$ for the 2D lattice structure. The Bragg peaks may be indexed by the two Miller indices h, k to yield the a, b unit cell. Their angular positions, $2\theta_{hor}$, yield the lattice plane spacing $d_{hk} = 2\pi/q_{hk}$. The intensity of a particular value of q_z in a Bragg rod is determined by the square of the molecular structure factor $|F_{h,k}(q_z)|^2$. The molecular structure factor is given by Equation 10.7, where f_j is the scattering factor of atom j. The vector $x_j a + y_j b$ specifies the position of atom j in the unit cell, and z_j is the atomic coordinate along the vertical direction. The measured intensity is then given by Equation 10.8.

$$F_{hk}(q_z) = \sum_j f_j e^{i[2\pi(hx_j + ky_j) + q_z z_j]} \tag{10.7}$$

$$I_{h,k}(q_z) = IV(q_z)\left|F_{h,k}(q_z)\right|^2 \tag{10.8}$$

For molecules of arbitrary shape, the overall crystalline structure can be established from an analogous 3D crystal structure if known. Precise information on the molecular chain orientation in a 2D crystal can be obtained from the positions of the maxima of the Bragg rods, assuming a uniform tilting of the chains.

GIXD has been used by various groups to study film orientation of P3HT films. GIXD studies of oriented P3HT solution-cast films show that these samples are characterized by well-organized lamellar structures, whereby stacks of planar thiophene main chains are uniformly spaced by the alkyl side chains [154]. Spin-coated and cast films were compared using GIXD, and it was found that the extent of conjugation of P3HT is homogeneous in spin-coated films, but inhomogeneous in cast films [155]. Sirringhaus et al. [50] through GIXD found that in samples with high regioregularity (>91%) and low MW the preferential orientation of ordered domains is with the (100)-axis (lamella layer structure) normal to the film and the (010)-axis ($\pi-\pi$ interchain stacking) in the plane of the film. In contrast, in samples with low regioregularity (81%) and high MW, the crystallites are preferentially oriented with the (100)-axis in the plane and the (010)-axis normal to the film. In films prepared by slow casting from a dilute solution the (100)-axis is normal to the film for all polymers. The explanation by the authors was that this might be a dynamic phenomenon during the rapid growth of spin-coated films affected by either regioregularity or MW. Kline et al. [54] also observed differences in GIXD data with different MWs of P3HT. The MW of regioregular P3HT has a substantial effect on the way that the chains pack on each other, and this, in turn, causes the mobility to vary by at least 4 orders of magnitude. Using GIXD, it was seen that a 3.2-kDa film has a more intense <100> diffraction peak than a 33.8-kDa film.

10.4.2
AFM

AFM is an important tool for the high-resolution profiling of surfaces. In an AFM, a sharp tip that interacts with a surface is positioned at the free end of a cantilever. A piezoelectric motor precisely moves either the tip or the sample to probe the surface. The topography of the sample can be evaluated with high precision and atomic-scale resolution in both the vertical and lateral dimensions. Two modes are generally used when probing surfaces, tapping mode and contact mode. Contact mode requires a constant contact between tip and sample, while tapping mode does not. In contact mode, the tip–sample interaction causes a quasistatic deflection. Tapping mode is the most commonly used technique for organic materials. In tapping mode, a change in the dynamic parameters of the cantilever tip occurs while it oscillates at or near its resonant frequency. An electric piezo stack vertically excites the cantilever, allowing the tip to bounce up and down. The cantilever motion is magnified by a deflection of a laser beam off the cantilever and into a photodiode detector, which generates a sinusoidal electric signal. The signal is then converted into a root mean square (RMS) value displayed as an AC voltage. When the tip encounters the surface, it is deflected, and the deflected laser beam reveals information about the surface, such as its height and morphology.

Kline *et al.* [54] demonstrated that AFM images of a 3.2 kDa film show a rod-like structure "nanorods," while a 31-kDa film showed an isotropic nodule structure. The images suggest that the lower-MW samples are more crystalline. The AFM data show that the rod-like structure has more grain boundaries than the isotropic nodule structure, which may explain why the more crystalline film has a lower mobility. It is likely that higher MWs of regioregular P3HT would have higher mobility because charge carriers can travel further on longer chains before they have to hop to a neighboring chain. The crystallinity of the samples was confirmed through GIXD. It was also observed that AFM images of the low-MW chloroform-cast, xylene-cast, and annealed chloroform-cast films show substantial differences in morphology [156]. The xylene-cast, the annealed chloroform-cast, and the drop-cast films are locally ordered and appear to be better connected with neighbors, whereas the small aggregates of nanorods in the chloroform-cast film appear loosely connected and randomly oriented. The nanorods in the xylene-cast, the annealed chloroform-cast, and the drop-cast films are also much longer than those of the chloroform-cast film. The AFM data shows that using processing conditions that give the chains more time to find an equilibrium position produces films with ordering of the nanorod structure and higher mobility.

AFM has also been used in bulk heterojunction (BHJ) solar cells to correlate morphology with performance, since the performance of a phase-separated mixture of donor and acceptor materials is known to be critically dependent on the morphology of the active layer [157]. Among a variety of techniques, AFM was used to resolve the morphology of spin-cast films of poly[2-methoxy-5-(3′,7′-dimethyloctyloxy)-1,4-phenylene vinylene] (MDMO-PPV, as donor) and 1-(3-methoxycarbonyl)propyl-1-phenyl-[6,6]-methanofullerene (PCBM, as acceptor) blends in three dimensions on a nanometer scale. All films were spun cast from chlorobenzene on ITO-covered glass substrates. The AFM height images revealed extremely smooth surfaces for both the pure films and the blends with a PCBM concentration ranging from 2 to 50 wt%. The surface becomes increasingly uneven for 67–90 wt%, accompanied with a reproducible phase contrast, indicating phase separation. It was observed that recombination of charges is reduced at the onset of phase separation at 67 wt% PCBM and 80 wt% PCBM is the optimum.

AFM was used to analyze blends of P3HT/PCBM after annealing at 150 °C for 30 min [158]. A comparison between samples annealed before and after aluminum deposition was made. After removal of the top aluminum layer, the underlying BHJ material was analyzed with AFM. A rougher surface indicates better contact between the BHJ material and the aluminum electrode, under the assumption that "clumps" of the former material were pulled off due to good contact. This rougher morphology was observed when annealing took place on the completed solar cell (after aluminum deposition), which allowed the power-conversion efficiency to approach 5%, much higher than previous reports.

10.4.3
KFM

In Kelvin probe force microscopy (KPFM), the sample and tip act as two conductors arranged as a parallel-plate capacitor with a small spacing. The contact potential difference is measured between the two materials via Equation 10.9, where ϕ_1 and ϕ_2 are the work functions of the conductors, including charges due to adsorbed layers. When two plates have different work functions, a periodic vibration gives an alternating current with the same frequency.

$$V_{CPD} = -\frac{(\phi_1 - \phi_2)}{e} \tag{10.9}$$

An AC voltage with adjustable DC offset is applied between a conducting AFM tip and the sample electrode and the resulting electrostatic force is detected by a lock-in amplifier, and a feedback circuit controls the DC tip potential until the CPD is compensated [159]. KPFM does not measure a current flow, but rather it records the electrostatic-force interaction between two objects (tip and the sample), which does not require a direct contact between the two objects. KPFM is well suited for the study of fragile and soft samples such as organic materials, and has a high voltage resolution around a few millivolts.

KPFM is ideal for measuring contact resistances between source and drain electrodes and the semiconducting material in OTFTs. Burgi *et al.* [160] produced a high-resolution map with 100 nm resolution of the electrostatic potential in the accumulation layer of P3HT OTFTs using KPFM, as well as investigated the time scale associated with charging or discharging the accumulation layer in FETs based on P3HT [161]. Nichols *et al.* [162] also used KPFM to simultaneously obtain high-resolution topography and potential images of pentacene OTFTs during device operation. Large potential drops at the source were observed. Puntambekar *et al.* [163] studied the potential drop between source and drain in top contact and bottom contact pentacene TFTs using KPFM. The potential drops were converted to resistances by dividing by the appropriate drain current values. The contact and the channel resistances decreased strongly with increasing gate bias, but did not depend strongly on the drain bias. These maps can be used to identify bottlenecks to charge transport and, in conjunction with simultaneous drain current measurements, allow resistance analysis of specific structures.

Chiesa *et al.* [164] established KPFM as a suitable technique for monitoring charge-separation processes in polymer photocells. The sub-100-nm lateral resolution of scanning KPFM allows characterization of the 3D structure of thin-film blends of poly-(9,9′-dioctylfluorene-*co*-benzothiadiazole) (F8BT) and poly-(9,9′-dioctylfluorene-*co*-bis-*N,N*′-(4-butylphenyl)-bis-*N,N*′-phenyl-1,4-phenylene-diamine) (PFB). It was observed at the boundary between the PFB-rich macrophase and the bulk region in the F8BT-rich macrophase that there is an interface region on the F8BT side of the boundary, which has a more negative potential than the bulk phases. This implies that part of the photogenerated charge does not recombine and remains trapped in the device under dark conditions. Under illumination, the authors observed a more negative surface potential in all regions as compared to the values measured in the dark. At a wavelength of 473 nm, where F8BT, but not PFB, is strongly absorbing, the surface potential is more negative in the F8BT than in the PFB macrophase, and it is the highest in the interface region between the two macrophases. KPFM allows a clear correlation between the topography and the potential distribution, and indicates that three regions, namely the PFB-rich, the "bulk," and the "interfacial" F8BT-rich phases, constitute the devices.

Hoppe *et al.* [165] conducted a comprehensive KPFM study on a classical organic solar-cell system consisting of MDMO-PPV/PCBM blends. Experiments were performed either in the dark or under continuous-wave (CW) laser illumination at 442 nm. Distinct differences in the energetics on the surface of films cast from chlorobenzene and toluene were identified, which combined with scanning electron microscopy suggest that surfaces of films deposited from toluene exhibit a morphologically controlled hindrance for electron propagation toward the cathode. The connectivity and percolation of both the electron- and hole-transporting phases with their respective electrodes appear to be a major factor for efficient organic solar cells.

Palermo *et al.* [166] made highly ordered nanoscopic crystals of 3″-methyl-4″-hexyl-2,2′:5′,2″:5″,2‴:5‴,2⁗-quinquethiophene-1″,1″-dioxide (T5OHM, electron acceptor) embedded in a regioregular P3HT (electron donor) matrix. Thin films of this blend were prepared in a one-step spin-coating deposition onto ITO/PEDOT substrates. A nanoscale phase-segregated structure characterized by T5OHM crystals in an amorphous P3HT grainy matrix was observed. Through KPFM, the T5OHM crystals exhibit an average potential, that is, more negative than the surrounding P3HT matrix, although the potential difference between the two phases is small and not uniform. When illuminating the sample with a white-light source, the surface potential changes dramatically, where the potential of all the T5OHM nanocrystals were much more negative with respect to the P3HT polymer. The contrast enhancement in the KPFM image demonstrates that photogenerated excitons are split at the interface between the two materials, leading to the accumulation of electrons in the T5OHM crystals and holes in the P3HT polymer matrix. The difference between the average values of the surface potential of the two phases increases from 25 mV to more than 75 mV, allowing a much easier identification of the T5OHM crystals from the polymer matrix.

Liscio *et al.* [167] through KPFM analysis of CHCl$_3$ cast *N,N′*-bis(1-ethylpropyl)-3,4:9,10-perylenebis(dicarboximide) (PDI) and regioregular P3HT blends proved that only the PDI clusters that are in physical contact with P3HT exhibit an appreciable charge transfer because of the existence of a complementary electron donor phase. Upon illumination with white light, the surface potential of the P3HT phase became more positive and displays a value close to that of the substrate (−10 mV), while the surface potential of PDI clusters in contact with P3HT became more negative at −90 mV.

10.5
Device Fabrication

10.5.1
Organic Light-Emitting Diodes

OLEDs and polymer light-emitting diodes (PLEDs) operate by the injection of holes (h$^+$) and electrons (e−$^-$) into an organic heterojunction consisting of one layer of an n-type semiconductor and one of a p-type semiconductor. This injection occurs in response to an external applied potential, and the charge carriers (negative and positive polarons) diffuse through the semiconducting layers to recombine into excitons at the interface. Typically, an emitting layer is placed at the interface that can accept both types of carriers. In this layer, the holes and electrons recombine into excitons, which can then decay by a variety of radiative and nonradiative processes. The energy of these excitons determines the wavelength of the emitted light, and the efficiency with which they undergo radiative decay strongly influences the efficiency of the resulting device.

A common system for OLEDs made from small-molecule vapor-deposited materials uses an ITO anode with a thin copper phthalocyanine (CuPc) layer and an aluminum cathode coated with LiF. Tris(8-hydroxyquinolato)aluminum serves as a combination electron transport layer and emissive layer, and *N,N′*-di(naphthalen-1-yl)-*N,N′*-diphenylbenzidine (α-NPD) is the hole-transport layer. CuPc and LiF are among a variety of materials used to coat injection electrodes and reduce injection barriers due to mismatches between the semiconductor band and the electrode work function. CuPc is also thought to aid charge injection by rendering the electrode surface smoother [168].

A typical example [169, 170] (Figure 10.1) of a PLED uses an ITO anode modified with the CP composite PEDOT-PSS and an Al cathode modified with an inorganic

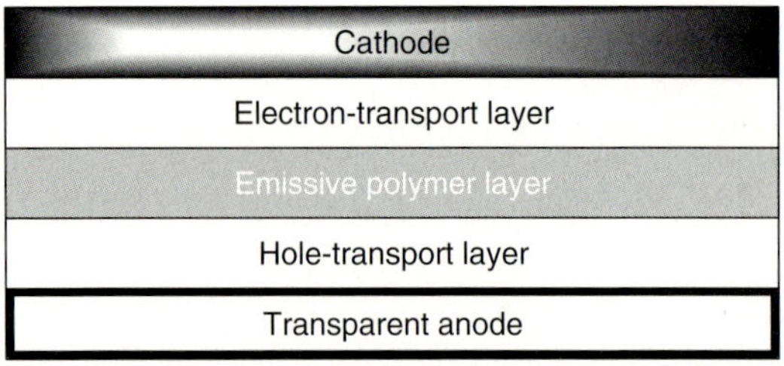

Cathode
Electron-transport layer
Emissive polymer layer
Hole-transport layer
Transparent anode

Figure 10.1 Diagram of a basic OLED with polymer emissive layer.

salt, electron-accepting polymer, or low work function metal. The emissive polymer layer, generally formed by spin casting on the anode, is sandwiched between the electrodes by vapor deposition of the cathode on top of the polymer emitting layer.

The efficiency of an OLED device can be analyzed in terms of the external quantum efficiency η_{EL}, which represents the rate of photon emission from the front of a device per electron injected at the cathode [171]. η_{EL} is given by the expression

$$\eta_{EL} = \xi \times \gamma \times \eta_{SE} \times \eta_{PL} \qquad (10.10)$$

The quantity ξ, called the *out-coupling efficiency*, is the fraction of emitted photons that exit the transparent side of the device, γ is the fraction of charge carriers that combine to form an exciton in the device, η_{SE} is the proportion of excitons which form as rapidly decaying singlets, and η_{PL} is the photoluminescent decay quantum yield from formed singlet excitons. To be precise, this expression applies only to devices in which the dominant mode of emission is photoluminescence due to the direct decay of singlet excitons. Emissive materials have been developed with a metal center in which spin-orbit coupling is significant, and in these materials, intersystem crossing (ISC) from a singlet to a lower-energy triplet exciton can be very rapid [172]. The phosphorescence resulting from radiative decay of this triplet state has been shown to occur with over 90% efficiency in several different materials [173–175]. Also, in an effort to develop white-appearing organic emitters, devices have been built that contain a mixture of phosphorescent and fluorescent emitter compounds, in which both decay pathways must be taken into account in estimating the device efficiency [176, 177].

The value of ξ is largely determined by the device geometry and the refractive index of the organic material [171]. Considerable losses occur due to total internal reflection of emitted light, but these losses can be considerably reduced by fabricating lens structures on the inside surface of the glass [178], or more simply by roughening the glass surface before device fabrication [179].

Most estimates of γ in state-of-the-art devices place it at nearly 1.0 in OLEDs with sufficiently thick organic layers composed of material with adequate carrier mobility [171]. The efficiency of injected charge combination into excitons in OLEDs is inherently high due to the low dielectric of the organic materials, which allows for strong attractions between opposite charge carriers as well as the inherently low electron mobility in typical hole-transport layers that prevents negative charge-carrier quenching at the anode. The deposition of a thin (circa 1.0 nm) hole-blocking layer between the emissive layer and the electron-transport layer generally is sufficient to prevent significant reduction in γ from reduction of holes at the cathode.

Spin statistics place a maximum value on η_{SE} of 0.25, since there exist three ways for polarons to combine into a triplet exciton but only one singlet configuration that can form. It is not conclusively demonstrated that simple statistics control the multiplicity of formed exciton states, but no device reported at this time gives an overall efficiency of greater than 0.25 from fluorescence although (as previously mentioned) that has been observed in some phosphorescence-based devices [179].

η_{PL} is largely determined by the rate of a variety of exciton-quenching processes that can occur in the emitter layer. Quenching by a variety of defects such as oxygen, water, or oxidative cross-linking sites is possible [180], as well as dissociation of excitons by the applied field [181] and energy transfer to the metal electrode [171]. All these factors can be minimized by the use of pure materials sealed away from water and oxidants, and the use of sufficiently thick emissive and transport layers. Additionally, singlet excitons can be quenched by polarons, a process that becomes increasingly important at higher currents [179, 182, 183].

Due to the high efficiencies reported in phosphorescence-based emitters, many devices have been created in which the emissive layer consists of a conjugated polymer host doped with a few per cent of a dye molecule that undergoes efficient exciton transfer from the host material. This strategy allows for the choice of a host layer based on its transport properties, stability, and other properties rather than its own radiative decay wavelength and efficiency. For the selection of a fluorescent dopant, the primary factor is the overlap of its absorption spectrum with the photoluminescence spectrum of the host to facilitate efficient Förster-type energy transfer from the host to the dopant [184]. However, triplet excitons do not undergo radiative decay in this scenario due to their inability to undergo FRET, and so considerable inherent inefficiency exists in OLEDs with fluorescent dye doping.

Phosphorescent dyes as dopants offer the potential of greater efficiency, since the large fraction of excitons that form as triplets from the recombination of charge carriers can in principle be converted into photons. However, the process of energy transfer to the dopant species is more complex in the case of phosphorescent dopants. First, triplet excitons are generally thought to move by a Dexter-type energy transfer process [184], which has a shorter range than Förster-type transfer necessitating a higher dopant concentration. Also, triplet exciton lifetimes are 3 orders of magnitude greater than singlet exciton lifetimes, increasing the probability of losses due to triplet–triplet annihilation, quenching at a charge-transport layer, or other processes [185]. Finally, the shorter range of energy transfer to phosphorescent dyes makes devices fabricated with them more susceptible to phase-separation processes that lead to aggregate formation of dopant molecules. This process removes the inner chromophores from contact with the host and renders them ineffective [186]. However, the greater potential of phosphorescent materials provides ample motivation for applied research to overcome these difficulties. One promising method for optimizing phosphorescence-based devices is to covalently attach dye molecules to the backbone of the host polymer [187]. Such attachment has been shown to yield efficient devices, and serves to prevent phase separation and keep host and dopant in close contact.

One of the most active current applications of OLEDs is in the area of lighting technology, which has necessitated the development of white organic light-emitting diodes (WOLEDs) [184]. These devices are generally produced by the combination of emitters of complementary colors, produced using either two or three different dopants. The dopants can be embedded in the same layer for a single-layer device [177, 188], or in several different layers in a lamellar architecture [189, 190]. The simplicity of the single-layer construction makes device construction easier but significantly complicates the photophysics of device operation, since energy transfer between all the dopants must be considered [184]. A triplet of red, blue, and green emitting dyes is the most common combination, where the red and green dyes are phosphorescent transition-metal complexes and the blue emitter is a fluorescent organic molecule.

The use of a fluorescent blue dopant rather than a phosphorescent one is due to the difficulty of finding a phosphorescent blue emitters that can readily accept triplet excitons from the conjugated polymer hosts [191, 192]. The triplet energy of the blue emitter must necessarily be the highest of the three, and in most common blue phosphorescent dyes is higher than the energy of a triplet exciton in any of the common host polymers. Devices utilizing blue phosphorescent dyes must rely upon thermally assisted, endothermic energy transfer from the host material, which is less efficient than the rate of generation of singlet excitons in the host material for transfer to a fluorescent emitter. To overcome the inherently lower rate of blue light emission in WOLEDs, the doping level of the blue emitter is generally many times higher than the concentration of the other dopants [184].

Currently, the most important issue facing developers of OLEDs is the longevity of the devices. The organic layers that comprise an OLED are far less durable than corresponding materials in inorganic LEDs, and a variety of degradation processes [11] occur over time causing dimming of emission intensity, the formation of dark spots, and unappealing alteration of color balance in WOLEDs. The most important degradation processes are caused by the presence of atmospheric oxygen and water, which directly react with excited states and charge-carrier sites in a device's transport and emission layers [9, 10]. These reactions destroy emissive chromophores and create trap sites that block further carrier transport. Considerable progress in limiting these undesirable reactions has been made by the development of encapsulating materials for OLED devices that remove [193] or form physical barriers to atmospheric oxidants [179, 194].

Also important in the degradation of OLEDs are a variety of physical processes that result in the organic material separating from the charge-injection electrodes and the formation of "dead zones" into which charge does not flow. These processes are largely caused by the heat of device operation [195], and can be minimized by the design of high-Tg host polymers that match each other in thermal expansion coefficient [196, 197]. The efficiency of device operation is also important in maximizing device lifetime, since a more efficient device can be operated at a lower current to produce the required light intensity. Lower operating currents result in cooler operating temperatures and reduce the rate of defect formation. With the advent of efficient doped-polymer emitters and research continuing into improvements in device durability, it is clear that OLED devices hold considerable promise for display and lighting applications in the near future.

10.5.2
Organic Photovoltaics

Organic polymer film-based solar cells convert incident light into an electrical current through the absorption of photons. The simplest possible organic solar cell is a thin film of organic material sandwiched between two electrodes (Figure 10.2a), one of which has a high work function and one a low work function. One electrode is transparent, generally the high work function side due to the availability of transparent conducting oxides. The difference in the electronic energy levels between these materials creates an electric field between them [198].

Upon the absorption of an incident photon, an electronic excited state is formed in the film. In this state, an electron is promoted to a higher-energy band in the organic material, leaving behind an unoccupied hole in lower-energy band. The hole acts as a quasiparticle and has a strong Coulombic attraction to the electron, the two forming a bound state that has a hydrogenic wavefunction [199]. The bound state is referred to as an *exciton*, and in common polymeric materials has a lifetime of several microseconds [200–202]. During this time, the exciton can diffuse randomly through the material and interact with other species. In the configuration shown above, under the influence of the field from the electrodes, the exciton can dissociate into its component hole and electron. In this case, the

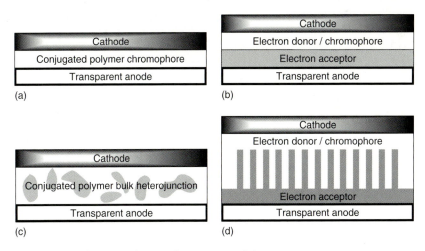

Figure 10.2 Architectures for OPV devices: (a) single-layer device; (b) bilayer heterojunction; (c) bulk heterojunction; and (d) ordered heterojunction.

hole will diffuse to the low work function metal, while the electron diffuses to the electrode with the higher work function. This process results in an electric current.

In reality, this process as described is extremely inefficient. The field produced by the difference in work function between two conductors is inherently low. Due to the low dielectric constant of organic materials, the interaction between the separated charges is extremely strong, although the formation of "hot" excitons from photons with greater energy than the bandgap helps to promote exciton reaction and dissociation [199]. Thus, the diffusion length of excitons in polymers is relatively small, being generally limited to the order of 10 nm [203–205]. Thus, the majority of excitons produced in an organic layer thicker than this diffusion length do not dissociate into their component charge carriers, but recombine to produce heat. The most efficient current implementations of organic photocells combine the organic chromophore layer with a second component comprised of a strong electron acceptor, forming a heterojunction. The second component acts to dissociate excitons by reacting with them directly, the electron transferring into the electron-acceptor phase while the hole remains in the chromophore phase. This charge-transfer process creates a *geminate pair* state in which the electron and hole are associated but confined to different phases. This state has a dipole moment and can be more easily dissociated by an electric field [199]. The use of an electron acceptor rather than a donor to dissociate the exciton is due to the requirements on the chromophore; the majority of known polymers with strong, broad absorbances through the solar spectrum are electron-rich materials that are good donors.

Since the diffusion length of the exciton is limited to a few nanometers, excitons must be generated very close to the donor/acceptor interface to ensure dissociation. Early implementations of polymer-based solar cells incorporating a heterojunction used a separate layer of electron-accepting material between the high work function anode and the electron-rich polymer [206] (Figure 10.2b), but this modification produces only moderate improvements in efficiency since most excitons are produced deep in the donor layer and far from the acceptor layer. It is not practical to use a single layer of acceptor material with a thickness less than the exciton diffusion length, since the absorbance of the layer is proportional to its thickness and a layer a few nanometers thick would transmit and waste most incident light. Various nanofabrication and self-assembly techniques have been used with inorganic semi-conductors, n-dopable organic materials, and carbon nanotubes as the electron acceptor to create a structured, thick region where the acceptor layer penetrates the donor layer with some form of nanoscale structure, such that all excitons generated are within one diffusion length of the donor/acceptor interface (Figure 10.2d). Efficient cells have been created using this "ordered heterojunction" strategy [207–210]. Technologies for structuring the interface in this way are relatively new, however, as they require nanoscale manipulation of the structure of the acceptor phase.

The BHJ strategy is widely used for fabrication of donor–acceptor PSCs in a configuration that overcomes the limitations of single heterojunctions without the need for complex nanoscale architecture [211]. In BHJ devices, donor and acceptor are mixed together, generally dissolved in some solvent. The organic layer in the photocell is then formed by spin casting the mixture onto one electrode. The components are selected such that they will phase separate from each other upon evaporation of the solvent. The phases remain in intimate contact with each other after evaporation, and careful control of the process leads to phase separation into domains only a few nanometers in diameter (Figure 10.2c). Annealing with solvent vapor or heat after fabrication can further increase the contact between the nanoscale domains and promote their formation from larger phase regions [212, 213]. The resulting nanoscale blend of materials efficiently dissociates excitons formed in the donor phase, since excitons form within regions only a few nanometers in diameter and surrounded by acceptors. Devices based on this technique have demonstrated quantum yield and power-conversion efficiency 10–60 times greater than the corresponding single heterojunction devices formed with two separate layers of donor and acceptor [206, 211]. The simplicity of the method and its ease of adaptation to large-scale manufacturing have sparked its widespread adoption in modern solar-cell research. The principal barrier to the wide use of this technique is the lack of rational techniques for achieving the desired nanoscale phase segregation. The structures of the donor and acceptor [214], the MW [54] and regioregularity [215] of polymeric components, the nature of the solvent [216, 217], the temperature at which film formation occurs [218], and many other external variables influence the structure of the final mixture, and therefore the efficiency of the resulting cell. Conditions for depositing the BHJ layer that achieve the desired nanoscale phase segregation must be determined empirically for each polymer blend. Furthermore, the layer once formed can be influenced by temperature and other factors, and is not necessarily stable once an ideal structure is produced [219].

Other efforts to improve the performance of organic photovoltaic devices (OPVs) have focused on increasing the efficiency of incident-light absorption. To produce an exciton in the polymer chromophore, a photon must have energy equal to or greater than the bandgap of the polymer. For the most common polymeric materials such as regioregular P3HT, this value is around $2.0\,eV$ [220], corresponding to a maximum absorbance at circa 500 nm. Photons below the absorption onset do not contribute to photocurrent. Therefore, by designing conjugated polymers with very small bandgaps, a broader section of the solar spectrum can be made accessible for energy generation. The most common strategy for producing low-bandgap conjugated polymers has been the synthesis of donor–acceptor-type materials, which are generally alternating copolymers of electron-rich and electron-poor conjugated moieties. Among the most common electron-rich systems used are oligothiophenes, either unsubstituted or substituted with alkyl or alkoxy groups [221–223]; N-alkylpyrroles [224]; 9,9-dialkylfluorenes [225]; and thieno[3,4-b]thiophenes [226, 227]. Electron-accepting repeat units which are frequently employed include 2,1,3-thiadiazoles [224] and cyanovinylenes [228, 229], with imide- [222] or trifluoromethyl-substituted phenylene units also having been used. Donor–acceptor polymers with absorbance spectra covering the entire visible light spectrum well into the infrared have been reported [230].

Many factors other than spectral width, however, affect the efficiency of photocurrent generation. In particular, charge-carrier mobility in the CP is essential to ensure efficient current production, and the morphology of the polymer powerfully influences mobility [231]. The best-performing organic semiconductors have been generally shown to form polycrystalline phases on deposition, with crystalline grains tens to hundreds of nanometers in size separated by amorphous grain boundaries. Charge-carrier mobility has been shown to be high within a single crystal of various organic semiconductors [232, 233], with the majority of interfering defects existing in the grain boundaries [54, 234, 235]. P3HT readily forms crystalline or semicrystalline films possessing high mobility from a variety of solvents upon spin casting, leading to its wide use in OPV devices despite its relatively high bandgap.

Photoexcitation of the acceptor material can also produce excitons and lead to photocurrent [236]. PCBM is by far the most popular organic acceptor at this time due to its superior stability and high electron mobility in its crystalline form. The most significant drawback to PCBM in this application is its poor absorbance in the solar spectrum. For this reason, the corresponding C_{70} and C_{84} methanofullerenes, which have a lower bandgap, have been fabricated into devices in the same configuration as PCBM. These devices based on the larger fullerenes show incremental improvements over the related C_{60} device [237]. Research into the use of other acceptor phases such as CdSe nanoparticles [238] and n-type semiconducting polymers is continuing in an attempt to balance mobility, absorbance, and stability. While it is possible to improve the absorbance of the acceptor phase by employing a material with a lower bandgap, it is still necessary to control morphology to maintain high electron mobility in the acceptor phase. Also, it is not desirable to reduce the bandgap energy in the acceptor by simply choosing a material with a lower-energy LUMO; an inherent energy loss exists in heterojunction devices as excited electrons transfer from the LUMO of the donor phase to the LUMO of the acceptor phase. The energy difference between the two unoccupied orbitals dissipates as heat. This energy loss is necessary to some extent, acting as a driving force for exciton dissociation, but should be minimized to produce the most efficient possible cell.

In principle, any material with (i) a sufficiently high electron affinity to dissociate the excitons produced in the donor material and (ii) appropriate solubility properties for producing the BHJ nanoscale structure could be used as an electron acceptor for the production of BHJ devices. However, research in this area has been almost exclusively dominated by fullerene derivatives due to their high electron affinity, good electron mobility, and extraordinary stability. Efforts to develop alternative acceptor materials with improved absorbance and electron transport properties is an area of active research, with both nanostructured inorganic n-type semiconductors [239] and novel organic polymers [240] and oligomers [241–243] being developed for applications in OPVs.

10.5.3
Organic Thin-Film Transistors

OTFTs are similar in concept to field effect transistorFETs based upon thin layers of inorganic semiconductors, generally referred to as metal oxide semiconductor

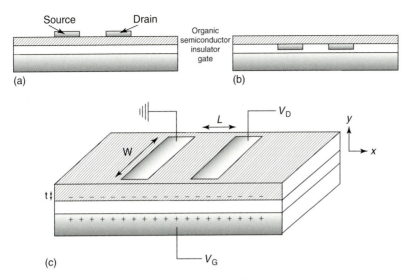

Figure 10.3 A top-contact (a) and bottom-contact (b) OTFT, with (c) important voltages and dimensions labeled. Reproduced by permission of the American Chemical Society from Ref. [245].

field effect transistors (MOSFETs). The construction and theory of TFTs has been extensively discussed elsewhere, and so we will present only an overview of their construction. OTFTs are being developed principally for use in flexible, low-cost electronics, an application complementary to the widespread use of inorganic semiconductor-based TFTs in high-speed, high-resolution flat-panel displays. This emphasis is due to the inherently lower performance of OTFTs, along with their potential for cheap manufacture and physical flexibility.

In an OTFT device, a flat electrode of metal or conducting oxide is coated with a dielectric substance. In the common *top-contact* configuration (Figure 10.3a), a layer of semiconducting organic material is deposited covering the dielectric, and two electrodes are printed or otherwise deposited on the semiconductor film. The *bottom-contact* geometry has the electrodes deposited directly on the dielectric (Figure 10.3b), resulting in the electrodes being sandwiched between the dielectric layer and the semiconductor. In these configurations, the bottom electrode is used as the gate, and the two electrodes in contact with the semiconductor film comprise the source and drain electrodes [244, 245].

For an undoped semiconductor with a moderate to high bandgap near room temperature, charge carriers do not flow between the source and drain electrodes in response to a voltage I_{SD}. However, upon application of a gate potential V_G, the band energies in the semiconductor are shifted as in Figure 10.4 [245].

In the case of a p-type semiconductor, the band energies of the semiconductor are raised sufficiently by applying a negative gate potential to allow electrons to move out of the HOMO into the source electrode, populating the HOMO band with holes and inducing charge transport. For n-type devices, a positive gate potential

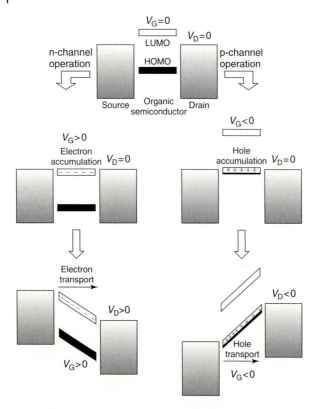

Figure 10.4 Band structure of the organic semiconductor and electrodes in an OTFT in the absence and presence of a gate voltage. Reproduced by permission of the American Chemical Society from Ref. [245].

is applied, lowering the band levels, until the LUMO enters into resonance with the Fermi level of the source electrode. Electrons transferred into the LUMO band from the source electrode serve as charge carriers in this case. A TFT operated with an applied potential such that the Fermi level of the source electrode resonates with one of the bands is said to be operating in *accumulation mode*. Application of the opposite potential to either device serves to increase the energy gap between the appropriate band and the source Fermi level, decreasing charge-carrier density in an operating configuration called *depletion mode*.

The ratio between the current measured when a gate voltage V_G is applied and the current when $V_G = 0$ is referred to as the *on/off ratio*, and is a fundamental figure of merit for an OTFT. For most targeted commercial applications, this ratio must be greater than 10^6. Also important is the response speed of the device, represented by the subthreshold swing S given by

$$S = \mathrm{d}(V_G)/\mathrm{d}(\log I_D) \tag{10.11}$$

and is a measurement of the initial increase of current upon application of a gate voltage; a lower value indicates a better-performing transistor. Frequently, S is normalized by the capacitance to give $C_d S$, which allows comparison between devices with different dielectric layers [245].

This band-based description gives a good qualitative picture of the electronic structure of the OTFT device, but several other factors affect the operation of such a device. First, the organic semiconductor layer inevitably contains defects that serve as traps for charge carriers. These sites must be filled for current to flow between the source and drain electrodes. Furthermore, the band structure of the system is distorted at the contacts between the electrodes and the semiconductor, due to the inevitable difference between the energy of the conducting band and the Fermi level of the resonant source electrode. This process results in charge transfer between the electrode and the semiconductor, producing an inherent bias at the electrode/source interface that affects the voltage at which charge begins to flow. Also, dopants can be introduced to the layer directly by reaction with atmospheric oxygen, water, or other contaminants. These incidental dopants will have the effect of introducing charge carriers at zero gate voltage; the device must then be operated in depletion mode to achieve an "off" state. Taking into account all these factors, thermal excitation of charge carriers, and other effects, the *gate voltage* V_T is defined as the value at which these effects are overcome. That is, at V_T charge begins flowing in response to a source–drain voltage as V_G moves from depletion mode to accumulation mode. At low source–drain voltages, current flows according to the Equation 10.4, which describes the source–drain current in the presence of abundant charge carriers in the semiconductor's conduction band.

At very low $(V_G - V_T) \gg V_D$ voltages, the drain current (Equation 10.4) is linear in V_D, and the mobility μ can be determined from the slope of the I–V curve. As V_D becomes larger, the quadratic term is no longer negligible and the I–V curve becomes concave downward. Physically, this curvature arises from the linear charge-carrier concentration gradient that occurs when V_D becomes comparable to $V_G - V_T$. When $V_D = V_G - V_T$, a condition called "*pinch-off*" is achieved, where the charge carrier concentration at the drain electrode drops to zero. At this point, the device is in saturation; further increases in V_D do not produce an increase in current. The current at this point is given by Equation 10.2. The saturated mobility μ_{sat} in Equation 10.5 can be obtained by fitting a plot of the square root of saturation current vs. V_G. μ_{sat} is not necessarily the same as the linear mobility in Equation 10.4, and is generally observed to be higher [245].

The nature of the source and drain electrodes is important to the properties of the device because of the importance of attaining a good match between the Fermi levels of the metal and the conduction band of the semiconductor. In the simplest possible picture, low-work-function metals such as calcium and aluminum should be best for electron injection into an n-type semiconductor. However, several reports have found that such devices actually require a more negative bias to function than the corresponding devices made with noble metals as the source electrode. This nonideal effect appears to be due to direct reaction of the semiconductor with the easily oxidized metal [246].

The majority of devices reported in the literature have been constructed with heavily p-doped silicon as the gate electrode, with a thick thermally induced SiO_2 layer (100–300 nm) as the dielectric to minimize leakage current. Many other types of material have been used as the dielectric, however, including a variety of inorganic oxides, polymer films, and even single monolayers of high-dielectric material. Also, modification of the surfaces of oxide dielectrics with self-assembled monolayers (SAMs) has been shown to substantially improve the properties of the resulting device over unmodified oxide dielectric layers [247].

The nature of the dielectric is important in determining the capacitance of the device and therefore the source–drain current according to Equation 10.1, but also plays a role in determining the structure of the semiconductor film by determining the preferred orientation of the semiconductor molecules. In turn, the orientation of molecules in the semiconductor film affects carrier mobility. Finally, silicon oxide and other dielectrics have been shown to contain a significant concentration of surface traps that can hinder charge transport. The impact of these traps is particularly large since it is known that the majority of charge carriers exist within 5 nm of the dielectric surface. A well-ordered SAM at the dielectric surface serves as a blocking layer, screening the charge carriers from surface traps.

An extremely wide variety of semiconducting oligomers and polymers have been used to fabricate OTFTs [248]. In devices based on small-molecule semiconductor, fullerenes and electron-poor phthalocyanines give among the best-performing n-type devices, and pentacene is dominant in p-type devices. In polymeric materials, alkyl-substituted PTs have been extensively explored for p-type function due to their high mobility, relative ease of synthesis, and commercial availability. The high charge-carrier mobility in PT films is generally attributed to the strong tendency of polymers in this family to form crystalline domains. Also investigated for p-type devices are polyfluorenes, both the homopolymer and alternating copolymer with various other aryl groups. The thieno[3,2-*b*]thiophene unit has shown excellent p-type transport properties when incorporated into conjugated polymer systems either alone or with a comonomer, with hole mobilities from 0.1 to 1 cm^2 V^{-1} s^{-1} and on/off ratios of 10^6 or better [249–251]. Many polymers containing electron-withdrawing groups have been reported that make efficient n-type semiconductors [240, 252–254], but relatively few have been fabricated into OTFT devices [255, 256], with small-molecule n-type semiconductors currently dominating the field [242, 257–259]. Generally speaking, the best-performing OTFTs have been created using vapor-deposition techniques to create the semiconductor layer [1, 257, 260]. Such techniques are not general, however, since they are completely inapplicable to polymeric semiconductors, and since many projected industrial applications for OTFT devices will be practical only if it is possible to pattern OTFTs rapidly through printing on a flexible substrate [261]. This type of fabrication requires good performance to be obtained from OTFTs deposited from solution by spraying or similar methods. Fortunately, several combinations of semiconductor material, flexible dielectric/gate electrode, and solution deposition conditions have been reported that yield adequate performance [262], although vacuum deposition still tends to produce superior results when it can be performed.

10.5.4
Sensors

In general, a sensor material must have a recognition method and a readout signal. Recognition with conjugated polymers has been performed with a wide variety of techniques, ranging from simple swelling of a polymer as a response to a good solvent, to covalent attachment of a functional group that selectively binds the desired analyte, to supramolecular recognition through molecular imprinting. Due to the electroactivity of many conjugated polymers, voltammetry is frequently used in conjugated polymer-based sensor electrodes to provide readout by monitoring the potential and intensity of redox processes in the polymer. Binding events are detected as a shift in voltage or current of a voltammetric peak. Some conjugated polymers are inherently solvatochromic, and thus have the inherent ability to sense solvents that produce a colorimetric response. Many conjugated polymers undergo bright fluorescence under certain wavelengths of light. In the presence of an analyte that binds strongly to the polymer, the chains aggregate and the fluorescence is quenched. In this way, fluorescence can be used as the readout for a sensor; more sophisticated architectures have also been developed that allow for a CP-based fluorescence sensor to be a "light-up" device, fluorescing in the presence of analyte rather than its absence.

A tremendous variety of electrochemical sensors have been simply and efficiently created by the electropolymerization of a functional CP monomer at an electrode surface. Functional PT, with a short PEG sidechain for cation sensing, was probably the first such device [263]. Polypyrroles have also been used as the electropolymerizable group in this type of application. A wide range of cation-sensing moieties have been employed, including crown ethers [25, 264], bpys [265, 266], and terpyridines [267]. Recognition groups for organic species have been incorporated as well, such as in a cyclodextrin-functionalized polypyrrole [268].

Electropolymerization has been used to prepare molecularly imprinted polymer (MIP) CP thin films for the sensing of neutral organic species [269]. Polyaniline is by far the most popular polymer for this application, due to its complex cross-linking, high redox activity, and variety of strong intermolecular forces that lead to strong imprinting of analyte. MIPs have been prepared that serve as efficient, selective electrochemical sensors for explosives [270], performance-enhancing drugs [269], and other significant chemical species [271, 272].

Chemical synthesis of the CP followed by spin casting of the sensor film on the electrode has opened the door for more sophisticated molecular architecture in the detection moiety and for a wider range of CP backbones [273]. Selective cation sensors have been produced by the copolymerization of a bithiophene-functionalized calixarene group with an alkoxy-substituted p-phenylene monomer [274]. The useful phenylenevinylene backbone has been incorporated into cation sensor films along with bpy groups in the polymer backbone, leading to enhanced signal. Sensor polymers with reporter groups have been synthesized as well. In this strategy, a group such as ferrocene with a strong and well-defined voltammetric response is bound into the polymer adjacent to the binding moiety [275]. The voltammetric peak corresponding to the reporter molecule is monitored rather than a peak from a process in the polymer backbone. The use of a reporter group decouples the CP's role as charge-carrier shuttle from its role as redox signaler, allowing the use of an optimized CP backbone for the given application regardless of the intensity of its redox peaks.

Another type of electrochemical sensor is produced when a CP is used as a charge-transport matrix with an embedded catalyst that facilitates an electrochemical reaction directly involving the analyte. A very active area of research at present is the use of CP films in the immobilization of enzymes that catalyze a redox reaction with the analyte as a reactant. The immobilization is usually performed by electropolymerization of the monomer in the presence of the enzyme. Sensors built in this way take advantage of the exquisite selectivity of natural enzymes for a given substrate, and the CP helps address issues with sensitivity caused in enzymatic biosensors due to the poor efficiency of charge transfer from the active site of an enzyme to the sensing electrode. Glucose sensors were among the first developed by this technique [276, 277], but at the present time a considerable variety of attached enzymes have been used to prepare sensors for their biological target including horseradish peroxidase [278], alcohol dehydrogenase [279], and penicillinase [280].

The dramatic color response of polydiacetylene (PDA) derivatives has led to the development of an entire class of optical sensors based on this unusual CP [19]. In general, the polymer formed by photopolymerization of monolayers of 1,3-diacetylenes is capable of existing in two phases, one of which is a bright red color and fluoresces, one of which is bright blue and does not fluoresce [281]. The "blue phase" is generally the native state of the PDA, in which it is formed upon polymerization. Several stimuli can induce a transition to the "red phase," including heat, mechanical stress, or the binding of a molecule or macromolecule to the polymerized SAM. Several different types of self-assembled structures of diacetylene monomer have been shown to give the distinctive blue-to-red transition, including Langmuir–Blodgett films [282], vesicles [283], and surface-bound monolayers [284]. When attached to a surface and functionalized with an appropriate recognition moiety such as biotin or an antibody [285], PDA layers have been shown to be effective sensing materials both through fluorescence detection and to the naked eye by colorimetric response.

A wide variety of fluorescent polymers have been synthesized and reported that are capable of sensitive and selective detection of analytes such as heavy-metal cations [24], DNA and other biomolecules [286], herbicides [287], and other species. However, most of these detection materials are studied in solution, which is relatively impractical for the construction of useful sensor devices. It is likely that some or all of these materials could be adapted to a thin-film configuration for use in a chip-based assay by grafting the polymers to a surface, but it is not guaranteed that the surface-bound polymer will have the same fluorescence properties or selectivity.

Conjugated polymers have an impressive array of applications in the preparation of functional materials for sensors, due to their inherent properties of electrical activity and fluorescence, which provide natural readout methods. Functional sensors, chemically designed to be selective for their analyte, are an emerging technology that should soon see rapidly increasing applications in industry, health care, and the life sciences.

10.5.5
Organic Electrochromics

The development of electroactive conjugated polymers, capable of reversible doping/dedoping processes upon chemical or electrochemical oxidation, has led to an application of these materials in the construction of ECDs. A basic ECD based on a polymer layer consists of a transparent ITO electrode in contact with a film of the electroactive polymer. Laminated on the polymer is a layer of electrolyte, which provides charge compensation for electrochemical doping of the polymer. Finally, a transparent inert layer generally seals the device and protects the interior of the cell. This structure is useful for applications such as "smart windows," where the amount of light transmitted by a window can be adjusted at the convenience of the user [288]. By patterning the surface into independently controlled pixels, displays can be manufactured [289–291]. Furthermore, by including multiple independently controlled layers of two or more types of electrochromic polymer in a "dual" configuration (Figure 10.5), the technology can be adapted to yield high-contrast or color displays [12, 69].

Since the active polymer layer in an ECD is in direct contact with the electrode, the most common polymer deposition technique is electropolymerization of the monomer on the bare ITO substrate [292, 293]. Spray coating or spin coating of soluble polymers has also been used to prepare devices [294].

Early electrochromics were based upon a thin layer of an inorganic oxide such as WO_3 [295] that undergoes a transition from its transparent native state to a blue color upon electrochemical reduction. Conjugated polymers and other organic materials hold an inherent advantage over inorganic oxides, however, for electrochromic applications. It is well known that the electronic structure and therefore the optical properties of conjugated polymers can be tuned by chemical substitution of the

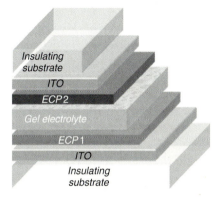

Figure 10.5 Diagram of a dual-type polymer electrochromic device designed for transmissive operation, with independent control of two complementary ECPs. Reproduced by permission of the American Chemical Society from Ref. [69].

conjugated backbone [67, 296] and control of the MW [297–299], regioregularity [298], and polydispersity of the polymer. This chemical tunability imparts a degree of flexibility to conjugated polymer-based ECDs that greatly exceeds that of similar devices based on inorganic materials. Additionally, conjugated polymers have been developed that possess order-of-magnitude better stability [300] and coloration efficiency [301] compared to state-of-the-art inorganic materials.

The majority of polymers used for electrochromic applications are colored in their native, neutral state. The application of an appropriate oxidizing potential electrochemically bleaches the polymer and renders it transparent, with a relatively flat absorbance spectrum. A slight yellow or brown color due to the uneven absorption spectrum of the oxidized polymer is common in some of the highest-contrast electrochromic polymers, and minimization of this residual color is an active area of research [12]. For widespread use in display applications, electrochromic polymers must be found that span the RGB color space to offer full-color displays [302]. Development is continuing of electroactive polymers that offer bright colors spanning the appropriate color space and offer good stability and contrast. Polymers with the opposite type of electrochromism have also been developed, where the neutral form is transmissive and the oxidized form colorless. High-contrast devices have been developed based on the combination of the two material types [303] (Figure 10.5).

The most important figure of merit for an ECD is its electrochromic contrast. This value is generally given as a per cent change in transmittance at the wavelength of maximum contrast between the neutral and colored state. Per cent change in total luminance has also been used as a measure of ECD performance, although it is more difficult to calculate. It is also important to quantify the efficiency of ECDs, generally done in terms of the coloration efficiency η, (Equation 10.12) which is a measure of the difference in transmittance at two different charge states normalized by the total amount of charge injected [304].

$$\eta(Q) = \log(T_n/T_c)/Q \tag{10.12}$$

Other important parameters for an ECD include its switching speed (for state-of-the-art organic ECDs [300], this value is less than a second) and its stability, generally characterized by the number of neutral/colored cycles that the device can undergo without degradation. The optical stability of the device is also significant; thin-film devices are generally stable in both colored and neutral states, or require only periodic applied voltages to maintain their state [304]. This color persistence is a desirable property for most applications.

The conjugated polymer PEDOT is one of the most thoroughly investigated electrochromic organic materials due to its intense blue native color in its reduced form and its near-transparency in the oxidized state [67, 68, 305]. PEDOT also is readily produced from the monomer by electropolymerization and has good stability and switching speed. The stability and switchability of PEDOT have been improved upon by using a (2-alkylpropylene) alkyl group instead of the ethylene group [68, 306]. This derivatization results in a soluble, swellable polymer that allows for more rapid ion diffusion during doping and dedoping. Red polymer electrochromics have been made based on the related poly(3,4)ethylenedioxypyrrole (PEDOP) [307, 308], and on P3ATs [309, 310]. Until recently, a stable green polymer with a clear oxidized state was not known; however, several substituted variants on poly(thieno[3,4-b]pyrazine) [302, 311] have been developed and appear to offer the desired properties.

Alternating copolymerization of various alternating electron-deficient and electron-rich monomer units can in principle give conjugated polymers with bandgap and resulting neutral-state color of nearly any desired value [312–314]. Polymers based on this strategy have been reported in blue, green, and red colors. The preparation of an alternating copolymer by electrosynthesis requires the preliminary synthesis of an alternating monomer. Alternatively, cross-coupling or other [315] synthetic methodologies may be used to prepare the polymer in solution from the individual monomers followed by film deposition through spin coating, spray coating, or other methods.

The technology exists for the adaptation of electrochromic polymers into smart windows and flexible displays. Industrial applications of these technologies will be dependent on the development of inexpensive transparent electrode materials, which is a very active area of materials and surface science at present. Additionally, improvements in transparency of the neutral state and in operational stability of ECDs would make industrial applications more desirable and speed their widespread adoption.

10.6
Concluding Remarks

The remarkable electronic properties of conjugated polymers are not directly competitive with inorganic semiconductors and metals, since the stability, conductivity, and mobility of inorganic materials are generally superior. However, conjugated polymers have a great deal of potential for complementary applications to these traditional electronic materials. This potential is due to the easy processability of conjugated polymers, the ability to tune their properties by functionalization using the techniques of organic synthesis, and particularly due to their ideal suitability for thin-film applications. The use of conjugated polymers in thin-film devices such as flexible displays, efficient color-tunable OLEDs, and printable OPVs efficiently takes advantage of the processability of the polymer, while potentially minimizing the impact of its lower charge-carrier mobility and stability.

Devices implementing these technologies are first becoming commercially available at the time of this writing, and the next decade should certainly produce several niche applications for organic thin-film electronics. However, the widespread industrial and commercial adoption of these devices will depend upon incremental advances in the stability and efficiency of organic electronic devices, particularly in the area of photovoltaics and OLEDs. Conjugated polymers can potentially be produced using renewable biomass feedstocks, further enhancing their desirability as a material if widespread industrial use is feasible. Recent advances in synthetic techniques and in the development of new conjugated polymer materials may eventually lead to great improvements in device performance as polymer morphology and electronic structure can be precisely controlled. As device engineering begins to take advantage of recent advances in thin-film technology for organic electronics, the field of devices based on thin-film architectures appears to have a bright future.

References

1. Schmidt, R., Oh, J.H., Sun, Y.-S., Deppisch, M., Krause, A.-M., Radacki, K., Braunschweig, H., Koenemann, M., Erk, P., Bao, Z., and Wuerthner, F. (2009) High-performance air-stable n-channel organic thin film transistors based on halogenated perylene bisimide semiconductors. *J. Am. Chem. Soc.*, **131**, 6215–6228.

2. Kim, J., Jeong, J., Cho, H.D., Lee, C., Kim, S.O., Kwon, S.-K., and Hong, Y. (2009) All-solution-processed bottom-gate organic thin-film transistor with improved subthreshold behaviour using functionalized pentacene active layer. *J. Phys. D Appl. Phys.*, **42**, 115107-1–115107-6.

3. Braga, D. and Horowitz, G. (2009) High-performance organic field effect transistors. *Adv. Mater.*, **21**, 1473–1486.

4. Zhao, L., Zhao, L., Xu, Y., Qiu, T., Zhi, L., and Shi, G. (2009) Polyaniline electrochromic devices with transparent graphene electrodes. *Electrochim. Acta*, **55**, 491–497.

5. Kang, J.-H., Oh, Y.-J., Paek, S.-M., Hwang, S.-J., and Choy, J.-H. (2009) Electrochromic device of PEDOT-PANI hybrid system for fast response and high optical contrast. *Sol. Energy Mater. Sol. Cells*, **93**, 2040–2044.

6. Beaupre, S., Breton, A.-C., Dumas, J., and Leclerc, M. (2009) Multicolored electrochromic cells based on poly(2,7-Carbazole) derivatives for adaptive camouflage. *Chem. Mater.*, **21**, 1504–1513.

7. Coropceanu, V., Cornil, J., da Silva, D.A., Olivier, Y., Silbey, R., and Bredas, J.L. (2007) Charge transport in organic semiconductors. *Chem. Rev.*, **107**, 926–952.

8. Cumpston, B.H. and Jensen, K.F. (1995) Photo-oxidation of polymers used in electroluminescent devices. *Synth. Met.*, **73**, 195–199.

9. Cumpston, B.H. and Jensen, K.F. (1996) Photo-oxidation of electroluminescent polymers. *Trends Polym. Sci.*, **4**, 151–157.

10. Cumpston, B.H., Parker, I.D., and Jensen, K.F. (1997) *In situ* characterization of the oxidative degradation of a polymeric light emitting device. *J. Appl. Phys.*, **81**, 3716–3720.

11. Sato, Y., Ichinosawa, S., and Kanai, H. (1998) Operation characteristics and degradation of organic electroluminescent devices. *IEEE J. Sel. Top. Quantum Electron.*, **4**, 40–48.

12. Sonmez, G. (2005) Polymeric electrochromics. *Chem. Commun.*, 5251–5259.

13. Somani, P.R. and Radhakrishnan, S. (2003) Electrochromic materials and devices: present and future. *Mater. Chem. Phys.*, **77**, 117–133.

14. Bao, Z., Rogers, J.A., and Katz, H.E. (1999) Printable organic and polymeric semiconducting materials and devices. *J. Mater. Chem.*, **9**, 1895–1904.

15. Bessiere, A., Marcel, C., Morcrette, M., Tarascon, J.M., Lucas, V., Viana, B., and Baffier, N. (2002) Flexible electrochromic reflectance device based on tungsten oxide for infrared emissivity control. *J. Appl. Phys.*, **91**, 1589–1594.

16. Sirringhaus, H., Kawase, T., Friend, R.H., Shimoda, T., Inbasekaran, M., Wu, W., and Woo, E.P. (2000) High-resolution ink-jet printing of all-polymer transistor circuits. *Science*, **290**, 2123–2126.

17. Brabec, C.J., Sariciftci, N.S., and Hummelen, J.C. (2001) Plastic solar cells. *Adv. Funct. Mater.*, **11**, 15–26.

18. Meijer, E.J., de Leeuw, D.M., Setayesh, S., van Veenendaal, E., Huisman, B.H., Blom, P.W.M., Hummelen, J.C., Scherf, U., Kadam, J., and Klapwijk, T.M. (2003) Solution-processed ambipolar organic field-effect transistors and inverters. *Nature Mater.*, **2**, 678–682.

19. Ahn, D.J. and Kim, J.-M. (2008) Fluorogenic polydiacetylene supramolecules: immobilization, micropatterning, and application to label-free chemosensors. *Acc. Chem. Res.*, **41**, 805–816.

20. Feng, F., He, F., An, L., Wang, S., Li, Y., and Zhu, D. (2008) Fluorescent conjugated polyelectrolytes for biomacromolecule detection. *Adv. Mater.*, **20**, 2959–2964.

21. Ambade, A.V., Sandanaraj, B.S., Klaikherd, A., and Thayumanavan, S. (2007) Fluorescent polyelectrolytes as protein sensors. *Polym. Int.*, **56**, 474–481.

22. Thomas, S.W., Joly, G.D., and Swager, T.M. III (2007) Chemical sensors based on amplifying fluorescent conjugated polymers. *Chem. Rev.*, **107**, 1339–1386.

23. Toal, S.J. and Trogler, W.C. (2006) Polymer sensors for nitroaromatic explosives detection. *J. Mater. Chem.*, **16**, 2871–2883.

24. Fan, L.-J., Zhang, Y., Murphy, C.B., Angell, S.E., Parker, M.F.L., Flynn, B.R., and Jones, W.E. (2009) Fluorescent conjugated polymer molecular wire chemosensors for transition metal ion recognition and signaling. *Coord. Chem. Rev.*, **253**, 410–422.

25. Fabre, B. and Simonet, J. (1998) Electroactive polymers containing crown ether or polyether ligands as cation-responsive materials. *Coord. Chem. Rev.*, **178–180**, 1211–1250.

26. Marsella, M.J., Carroll, P.J., and Swager, T.M. (1994) Conducting pseudopolyrotaxanes: a chemoresistive response via molecular recognition. *J. Am. Chem. Soc.*, **116**, 9347–9348.

27. Bredas, J.L. and Street, G.B. (1985) Polarons, bipolarons, and solitons in conducting polymers. *Acc. Chem. Res.*, **18**, 309–315.

28. Bredas, J.L., Themans, B., Andre, J.M., Chance, R.R., and Silbey, R. (1984) The role of mobile organic radicals and ions (solitons, polarons and bipolarons) in the transport properties of doped conjugated polymers. *Synth. Met.*, **9**, 265–274.

29. Bredas, J.L., Chance, R.R., and Silbey, R. (1982) Comparative theoretical study of the doping of conjugated polymers: polarons in polyacetylene and polyparaphenylene. *Phys. Rev. B: Condens. Matter*, **26**, 5843–5854.

30. Heeger, A.J. (2001) Semiconducting and metallic polymers: the fourth generation of polymeric materials. *J. Phys. Chem. B*, **105**, 8475–8491.

31. Podzorov, V., Menard, E., Borissov, A., Kiryukhin, V., Rogers, J.A., and Gershenson, M.E. (2004) Intrinsic charge transport on the surface of organic semiconductors. *Phys. Rev. Lett.*, **93**, 086602-1–086602-4.

32. Podzorov, V., Menard, E., Rogers, J.A., and Gershenson, M.E. (2005) Hall effect in the accumulation layers on the surface of organic semiconductors. *Phys. Rev. Lett.*, **95**, 226601-1–226601-4.

33. Ostroverkhova, O., Cooke, D.G., Hegmann, F.A., Anthony, J.E., Podzorov, V., Gershenson, M.E., Jurchescu, O.D., and Palstra, T.T.M. (2006) Ultrafast carrier dynamics in pentacene, functionalized pentacene, tetracene, and rubrene single crystals. *Appl. Phys. Lett.*, **88**, 162101-1–162101-3.

34. Ostroverkhova, O., Cooke, D.G., Shcherbyna, S., Egerton, R.F., Hegmann, F.A., Tykwinski, R.R., and Anthony, J.E. (2005) Bandlike transport in pentacene and functionalized pentacene thin films revealed by subpicosecond transient photoconductivity measurements. *Phys. Rev. B: Condens. Matter Mater. Phys.*, **71**, 035204-1–035204-6.

35. Cheng, Y.C., Silbey, R.J., da Silva Filho, D.A., Calbert, J.P., Cornil, J., and Bredas, J.L. (2003) Three-dimensional band structure and bandlike mobility in oligoacene single crystals: a theoretical investigation. *J. Chem. Phys.*, **118**, 3764–3774.

36. Kenkre, V.M. (2002) Finite-bandwidth calculations for charge carrier mobility

in organic crystals. *Phys. Lett. A*, **305**, 443–447.

37. Kepler, R.G. (1960) Charge carrier production and mobility in anthracene crystals. *Phys. Rev.*, **119**, 1226–1229.

38. LeBlanc, O.H. Jr. (1960) Hole and electron drift mobilities in anthracene. *J. Chem. Phys.*, **33**, 626.

39. Prins, P., Candeias, L.P., Van Breemen, A.J.J.M., Sweelssen, J., Herwig, P.T., Schoo, H.F.M., and Siebbeles, L.D.A. (2005) Electron and hole dynamics on isolated chains of a solution-processable poly(thienylenevinylene) derivative in dilute solution. *Adv. Mater.*, **17**, 718–723.

40. Hoofman, R.J.O.M., De Haas, M.P., Siebbeles, L.D.A., and Warman, J.M. (1998) Highly mobile electrons and holes on isolated chains of the semiconducting polymer poly(phenylene vinylene). *Nature*, **392**, 54–56.

41. Horowitz, G. (1998) Organic field effect transistors. *Adv. Mater.*, **10**, 365–377.

42. Horowitz, G., Hajlaoui, R., Bourguiga, R., and Hajlaoui, M. (1999) Theory of the organic field-effect transistor. *Synth. Met.*, **101**, 401–404.

43. Katz, H.E. and Bao, Z. (1999) The physical chemistry of organic field-effect transistors. *J. Phys. Chem. B*, **104**, 671–678.

44. Fichou, D. (2000) Structural order in conjugated oligothiophenes and its implications on opto-electronic devices. *J. Mater. Chem.*, **10**, 571–588.

45. Nelson, S.F., Lin, Y.Y., Gundlach, D.J., and Jackson, T.N. (1998) Temperature-independent transport in high-mobility pentacene transistors. *Appl. Phys. Lett.*, **72**, 1854–1856.

46. Garnier, F., Horowitz, G., Fichou, D., and Yassar, A. (1997) Role of mesoscopic molecular organization in organic-based thin film transistors. *Supramol. Sci.*, **4**, 155–162.

47. Jurchescu, O.D., Baas, J., and Palstra, T.T.M. (2004) Effect of impurities on the mobility of single crystal pentacene. *Appl. Phys. Lett.*, **84**, 3061–3063.

48. Katz, H.E., Torsi, L., and Dodabalapur, A. (1995) Synthesis, material properties, and transistor performance of highly pure thiophene oligomers. *Chem. Mater.*, **7**, 2235–2237.

49. Bao, Z., Dodabalapur, A., and Lovinger, A.J. (1996) Soluble and processable regioregular poly(3-hexylthiophene) for thin film field-effect transistor applications with high mobility. *Appl. Phys. Lett.*, **69**, 4108–4110.

50. Sirringhaus, H., Brown, P.J., Friend, R.H., Nielsen, M.M., Bechgaard, K., Langeveld-Voss, B.M.W., Spiering, A.J.H., Janssen, R.A.J., Meijer, E.W., Herwig, P., and De Leeuw, D.M. (1999) Two-dimensional charge transport in self-organized, high-mobility conjugated polymers. *Nature*, **401**, 685–688.

51. Le Comber, P.G. and Spear, W.E. (1970) Electronic transport in amorphous silicon films. *Phys. Rev. Lett.*, **25**, 509–511.

52. Arias, A.C., MacKenzie, J.D., McCulloch, I., Rivnay, J., and Salleo, A. (2010) Materials and applications for large area electronics: solution-based approaches. *Chem. Rev.*, **110**, 3–24.

53. Sirringhaus, H., Tessler, N., and Friend, R.H. (1998) Integrated optoelectronic devices based on conjugated polymers. *Science*, **280**, 1741–1744.

54. Kline, R.J., McGehee, M.D., Kadnikova, E.N., Liu, J., and Fréchet, J.M.J. (2003) Controlling the field-effect mobility of regioregular polythiophene by changing the molecular weight. *Adv. Mater.*, **15**, 1519–1522.

55. Babel, A. and Jenekhe, S.A. (2005) Alkyl chain length dependence of the field-effect carrier mobility in regioregular poly(3-alkylthiophene)s. *Synth. Met.*, **148**, 169–173.

56. Bao, Z. and Lovinger, A.J. (1999) Soluble regioregular polythiophene derivatives as semiconducting materials for field-effect transistors. *Chem. Mater.*, **11**, 2607–2612.

57. Salaneck, W.R., Inganaes, O., Themans, B., Nilsson, J.O., Sjoegren, B., Osterholm, J.E., Bredas, J.L., and Svensson, S. (1988) Thermochromism in poly(3-hexylthiophene) in the solid state: a spectroscopic study of temperature-dependent conformational defects. *J. Chem. Phys.*, **89**, 4613–4619.

58. Daoust, G. and Leclerc, M. (1991) Structure-property relationships in alkoxy-substituted polythiophenes. *Macromolecules*, **24**, 455–459.

59. Mortimer, R.J. (1999) Organic electrochromic materials. *Electrochim. Acta*, **44**, 2971–2981.

60. Garnier, F., Tourillon, G., Gazard, M., and Dubois, J.C. (1983) Organic conducting polymers derived from substituted thiophenes as electrochromic material. *J. Electroanal. Chem. Interfacial Electrochem.*, **148**, 299–303.

61. Tourillon, G. and Garnier, F. (1983) Effect of dopant on the physicochemical and electrical properties of organic conducting polymers. *J. Phys. Chem.*, **87**, 2289–2292.

62. Chung, T.C., Kaufman, J.H., Heeger, A.J., and Wudl, F. (1984) Charge storage in doped poly(thiophene): optical and electrochemical studies. *Phys. Rev. B: Condens. Matter*, **30**, 702–710.

63. Kaneto, K., Kohno, Y., and Yoshino, K. (1984) Absorption spectra induced by photoexcitation and electrochemical doping in polythiophene. *Solid State Commun.*, **51**, 267–269.

64. Panero, S., Passerini, S., and Scrosati, B. (1993) Conducting polymers: new electrochromic materials for advanced optical devices. *Mol. Cryst. Liq. Cryst. Sci. Technol. Sect. A*, **229**, 97–109.

65. Sapp, S.A., Sotzing, G.A., Reddinger, J.L., and Reynolds, J.R. (1996) Rapid switching solid-state electrochromic devices based on complementary conducting polymer films. *Adv. Mater.*, **8**, 808–811.

66. Kumar, A. and Reynolds, J.R. (1996) Soluble alkyl-substituted poly(ethylenedioxythiophenes) as electrochromic materials. *Macromolecules*, **29**, 7629–7630.

67. Groenendaal, L.B., Jonas, F., Freitag, D., Pielartzik, H., and Reynolds, J.R. (2000) Poly(3,4-ethylenedioxythiophene) and its derivatives: past, present, and future. *Adv. Mater.*, **12**, 481–494.

68. Kumar, A., Welsh, D.M., Morvant, M.C., Piroux, F., Abboud, K.A., and Reynolds, J.R. (1998) Conducting poly(3,4-alkylenedioxythiophene) derivatives as fast electrochromics with high-contrast ratios. *Chem. Mater.*, **10**, 896–902.

69. Beaujuge, P.M., and Reynolds, J.R. (2010) Color control in π-conjugated organic polymers for use in electrochromic devices. *Chem. Rev.*, **110**, 268–320.

70. Diaz, A.F., Kanazawa, K.K., and Gardini, G.P. (1979) Electrochemical polymerization of pyrrole. *J. Chem. Soc. Chem. Commun.*, 635–636.

71. Tourillon, G. and Garnier, F. (1982) New electrochemically generated organic conducting polymers. *J. Electroanal. Chem. Interfacial Electrochem.*, **135**, 173–178.

72. Satoh, M., Kaneto, K., and Yoshino, K. (1985) Electrochemical preparation of high-quality poly(p-phenylene) film. *J. Chem. Soc. Chem. Commun.*, 1629–1630.

73. Fauvarque, J.F., Petit, M.A., Pfluger, F., Jutand, A., Chevrot, C., and Troupel, M. (1983) Preparation of poly(1,4-phenylene) by nickel(0) complex-catalyzed electropolymerization. *Makromol. Chem. Rapid Commun.*, **4**, 455–457.

74. Aeiyach, S., Soubiran, P., Lacaze, P.C., Froyer, G., and Pelous, Y. (1989) A new method for obtaining polyparaphenylene films by electrochemical oxidation of (benzene-antimony pentafluoride) complexes in sulfur dioxide medium. *Synth. Met.*, **32**, 103–112.

75. Rault-Berthelot, J. and Simonet, J. (1985) The anodic oxidation of fluorene and some of its derivatives. Conditions for the formation of a new conducting polymer. *J. Electroanal. Chem. Interfacial Electrochem.*, **182**, 187–192.

76. Kathirgamanathan, P. and Shepherd, M.K. (1993) Anodic and cathodic electropolymerization of novel lightly colored poly(phenanthro[9,10-c]thiophen) and its electrical characterization. *J. Electroanal. Chem.*, **354**, 305–309.

77. Xu, Z., Horowitz, G., and Garnier, F. (1988) Cathodic electropolymerization of polythiophene on platinum and various semiconducting electrodes. *J. Electroanal. Chem. Interfacial Electrochem.*, **246**, 467–472.

78. Genies, E.M., Bidan, G., and Diaz, A.F. (1983) Spectroelectrochemical study of polypyrrole films. *J. Electroanal. Chem. Interfacial Electrochem.*, **149**, 101–113.

79. Waltman, R.J. and Bargon, J. (1984) Reactivity/structure correlations for the electropolymerization of pyrrole: an INDO/CNDO study of the reactive sites of oligomeric radical cations. *Tetrahedron*, **40**, 3963–3670.

80. Adams, R.N. (1969) Anodic oxidation pathways of aromatic hydrocarbons and amines. *Acc. Chem. Res.*, **2**, 175–180.

81. Asavapiriyanont, S., Chandler, G.K., Gunawardena, G.A., and Pletcher, D. (1984) The electrodeposition of polypyrrole films from aqueous solutions. *J. Electroanal. Chem. Interfacial Electrochem.*, **177**, 229–244.

82. Roncali, J. and Garnier, F. (1988) Electroactivity of transparent composite films from conducting poly(thiophenes). *J. Phys. Chem.*, **92**, 833–840.

83. Sato, M., Tanaka, S., and Kaeriyama, K. (1986) Soluble conducting polythiophenes. *J. Chem. Soc. Chem. Commun.*, 873–874.

84. Kaeriyama, K., Sato, M., and Tanaka, S. (1987) Electrochemical preparation of conducting polyalkythiophene films. *Synth. Met.*, **18**, 233–236.

85. Hotta, S. (1987) Electrochemical synthesis and spectroscopic study of poly(3-alkylthienylenes). *Synth. Met.*, **22**, 103–113.

86. Inaoka, S. and Collard, D.M. (1999) Chemical and electrochemical polymerization of 3-alkylthiophenes on self-assembled monolayers of oligothiophene-substituted alkylsilanes. *Langmuir*, **15**, 3752–3758.

87. Zotti, G., Zecchin, S., Vercelli, B., Berlin, A., Grimoldi, S., Groenendaal, L., Bertoncello, R., and Natali, M. (2005) Surface-initiated polymerization of thiophene and pyrrole monomers on poly(terthiophene) films and oligothiophene monolayers. *Chem. Mater.*, **17**, 3681–3694.

88. Rider, D.A., Harris, K.D., Wang, D., Bruce, J., Fleischauer, M.D., Tucker, R.T., Brett, M.J., and Buriak, J.M. (2009) Thienylsilane-modified indium tin oxide as an anodic interface in polymer/fullerene solar cells. *ACS Appl. Mater. Interfaces*, **1**, 279–288.

89. Diederich, F., Stang, P., and Wipf, P. (1998) *Metal-Catalyzed Cross-Coupling Reactions*, Wiley-VCH Verlag GmbH, Weinheim.

90. Yokozawa, T. and Yokoyama, A. (2009) Chain-growth condensation polymerization for the synthesis of well-defined condensation polymers and π-conjugated polymers. *Chem. Rev.*, **109**, 5595–5619.

91. Yokoyama, A. and Yokozawa, T. (2007) Converting step-growth to chain-growth condensation polymerization. *Macromolecules*, **40**, 4093–4101.

92. McCullough, R.D. and Lowe, R.D. (1992) Enhanced electrical conductivity in regioselectively synthesized poly(3-alkylthiophenes). *J. Chem. Soc. Chem. Commun.*, 70–72.

93. McCullough, R.D., Lowe, R.D., Jayaraman, M., and Anderson, D.L. (1993) Design, synthesis, and control of conducting polymer architectures: structurally homogeneous poly(3-alkylthiophenes). *J. Org. Chem.*, **58**, 904–912.

94. Loewe, R.S., Khersonsky, S.M., and McCullough, R.D. (1999) A simple method to prepare head-to-tail coupled, regioregular poly(3-alkylthiophenes) using Grignard metathesis. *Adv. Mater.*, **11**, 250–253.

95. Loewe, R.S., Ewbank, P.C., Liu, J., Zhai, L., and McCullough, R.D. (2001) Regioregular, head-to-tail coupled poly(3-alkylthiophenes) made easy by the grim method: investigation of the reaction and the origin of regioselectivity. *Macromolecules*, **34**, 4324–4333.

96. Chen, T.A. and Rieke, R.D. (1992) The first regioregular head-to-tail poly(3-hexylthiophene-2,5-diyl) and a regiorandom isopolymer: nickel versus palladium catalysis of 2(5)-bromo-5(2)-(bromozincio)-3-hexylthiophene polymerization. *J. Am. Chem. Soc.*, **114**, 10087–10088.

97. Espinet, P. and Echavarren, A.M. (2004) C-C coupling: the mechanisms of the Stille reaction. *Angew. Chem. Int. Ed.*, **43**, 4704–4734.

98. Miyaura, N. and Suzuki, A. (1995) Palladium-catalyzed cross-coupling reactions of organoboron compounds. *Chem. Rev.*, **95**, 2457–2483.

99. Iraqi, A. and Barker, G.W. (1998) Synthesis and characterization of telechelic regioregular head-to-tail poly(3-alkylthiophenes). *J. Mater. Chem.*, **8**, 25–29.

100. Guillerez, S. and Bidan, G. (1998) New convenient synthesis of highly regioregular poly(3-octylthiophene) based on the Suzuki coupling reaction. *Synth. Met.*, **93**, 123–126.

101. Jayakannan, M., van Dongen, J.L.J., and Janssen, R.A.J. (2001) Mechanistic aspects of the Suzuki polycondensation of thiophenebisboronic derivatives and diiodobenzenes analyzed by MALDI-TOF mass spectrometry. *Macromolecules*, **34**, 5386–5393.

102. Yokoyama, A., Miyakoshi, R., and Yokozawa, T. (2004) Chain-growth polymerization for poly(3-hexylthiophene) with a defined molecular weight and a low polydispersity. *Macromolecules*, **37**, 1169–1171.

103. Sheina, E.E., Liu, J., Iovu, M.C., Laird, D.W., and McCullough, R.D. (2004) Chain growth mechanism for regioregular nickel-initiated cross-coupling polymerizations. *Macromolecules*, **37**, 3526–3528.

104. Iovu, M.C., Sheina, E.E., Gil, R.R., and McCullough, R.D. (2005) Experimental evidence for the quasi-living nature of the Grignard metathesis method for the synthesis of regioregular poly(3-alkylthiophenes). *Macromolecules*, **38**, 8649–8656.

105. Langeveld-Voss, B.M.W., Janssen, R.A.J., Spiering, A.J.H., van Dongen, J.L.J., Vonk, E.C., and Claessens, H.A. (2000) End-group modification of regioregular poly(3-alkylthiophene)s. *Chem. Commun.*, 81–82.

106. Jeffries-El, M., Sauve, G., and McCullough, R.D. (2004) In-situ end-group functionalization of regioregular poly(3-alkylthiophene) using the Grignard metathesis polymerization method. *Adv. Mater.*, **16**, 1017–1019.

107. Jeffries-El, M., Sauve, G., and McCullough, R.D. (2005) Facile synthesis of end-functionalized regioregular poly(3-alkylthiophene)s via modified Grignard metathesis reaction. *Macromolecules*, **38**, 10346–10352.

108. Miyakoshi, R., Yokoyama, A., and Yokozawa, T. (2005) Catalyst-transfer polycondensation. Mechanism of Ni-catalyzed chain-growth polymerization leading to well-defined poly(3-hexylthiophene). *J. Am. Chem. Soc.*, **127**, 17542–17547.

109. Miyakoshi, R., Shimono, K., Yokoyama, A., and Yokozawa, T. (2006) Catalyst-transfer polycondensation for the synthesis of poly(p-phenylene) with controlled molecular weight and low polydispersity. *J. Am. Chem. Soc.*, **128**, 16012–16013.

110. Lanni, E.L. and McNeil, A.J. (2009) Mechanistic studies on Ni(dppe)Cl$_2$-catalyzed chain-growth polymerizations: evidence for rate-determining reductive elimination. *J. Am. Chem. Soc.*, **131**, 16573–16579.

111. Yokoyama, A., Kato, A., Miyakoshi, R., and Yokozawa, T. (2008) Precision synthesis of poly(N-hexylpyrrole) and its diblock copolymer with poly(p-phenylene) via catalyst-transfer polycondensation. *Macromolecules*, **41**, 7271–7273.

112. Miyakoshi, R., Yokoyama, A., and Yokozawa, T. (2008) Importance of the order of successive catalyst-transfer condensation polymerization in the synthesis of block copolymers of polythiophene and poly(p-phenylene). *Chem. Lett.*, **37**, 1022–1023.

113. Huang, L., Wu, S., Qu, Y., Geng, Y., and Wang, F. (2008) Grignard metathesis chain-growth polymerization for polyfluorenes. *Macromolecules*, **41**, 8944–8947.

114. Yamamoto, T. (2002) π-conjugated polymers with electronic and optical functionalities: preparation by organometallic polycondensation, properties, and applications. *Macromol. Rapid Commun.*, **23**, 583–606.

115. Yamamoto, T., Morita, A., Miyazaki, Y., Maruyama, T., Wakayama, H., Zhou, Z.H., Nakamura, Y.,

Kanbara, T., Sasaki, S., and Kubota, K. (1992) Preparation of π-conjugated poly(thiophene-2,5-diyl), poly(p-phenylene), and related polymers using zerovalent nickel complexes. Linear structure and properties of the π-conjugated polymers. *Macromolecules*, **25**, 1214–1223.

116. Beinhoff, M., Appapillai, A.T., Underwood, L.D., Frommer, J.E., and Carter, K.R. (2006) Patterned polyfluorene surfaces by functionalization of nanoimprinted polymeric features. *Langmuir*, **22**, 2411–2414.

117. Jhaveri, S.B., Peterson, J.J., and Carter, K.R. (2009) Poly(9,9-Dihexylfluorene) layers grown via surface-directed Ni(0) condensation polymerization. *Langmuir*, **25**, 9552–9556.

118. Senkovskyy, V., Khanduyeva, N., Komber, H., Oertel, U., Stamm, M., Kuckling, D., and Kiriy, A. (2007) Conductive polymer brushes of regioregular head-to-tail poly(3-alkylthiophenes) via catalyst-transfer surface-initiated polycondensation. *J. Am. Chem. Soc.*, **129**, 6626–6632.

119. Khanduyeva, N., Senkovskyy, V., Beryozkina, T., Horecha, M., Stamm, M., Uhrich, C., Riede, M., Leo, K., and Kiriy, A. (2009) Surface engineering using Kumada catalyst-transfer polycondensation (KCTP): preparation and structuring of poly(3-hexylthiophene)-based graft copolymer brushes. *J. Am. Chem. Soc.*, **131**, 153–161.

120. Beryozkina, T., Senkovskyy, V., Kaul, E., and Kiriy, A. (2008) Kumada catalyst-transfer polycondensation of thiophene-based oligomers: robustness of a chain-growth mechanism. *Macromolecules*, **41**, 7817–7823.

121. Boyd, S.D., Jen, A.K.Y., and Luscombe, C.K. (2009) Steric stabilization effects in nickel-catalyzed regioregular poly(3-hexylthiophene) synthesis. *Macromolecules*, **42**, 9387–9389.

122. Yokoyama, A., Suzuki, H., Kubota, Y., Ohuchi, K., Higashimura, H., and Yokozawa, T. (2007) Chain-growth polymerization for the synthesis of polyfluorene via Suzuki-Miyaura coupling reaction from an externally added initiator unit. *J. Am. Chem. Soc.*, **129**, 7236–7237.

123. Stambuli, J.P., Incarvito, C.D., Buehl, M., and Hartwig, J.F. (2004) Synthesis, structure, theoretical studies, and ligand exchange reactions of monomeric, T-shaped arylpalladium(II) halide complexes with an additional, weak agostic interaction. *J. Am. Chem. Soc.*, **126**, 1184–1194.

124. Weber, S.K., Galbrecht, F., and Scherf, U. (2006) Preferential oxidative addition in Suzuki cross-coupling reactions across one fluorene unit. *Org. Lett.*, **8**, 4039–4041.

125. Dong, C.-G. and Hu, Q.-S. (2005) Preferential oxidative addition in palladium(0)-catalyzed Suzuki cross-coupling reactions of dihaloarenes with arylboronic acids. *J. Am. Chem. Soc.*, **127**, 10006–10007.

126. Beryozkina, T., Boyko, K., Khanduyeva, N., Senkovskyy, V., Horecha, M., Oertel, U., Simon, F., Stamm, M., and Kiriy, A. (2009) Grafting of polyfluorene by surface-initiated Suzuki polycondensation. *Angew. Chem., Int. Ed.*, **48**, 2695–2698.

127. Sontag, S.K., Marshall, N., and Locklin, J. (2009) Formation of conjugated polymer brushes by surface-initiated catalyst-transfer polycondensation. *Chem. Commun.*, 3354–3356.

128. Marshall, N., Sontag, S.K., and Locklin, J. (2010) Substituted poly(p-phenylene) thin films via surface-initiated Kumada-type catalyst transfer polycondensation. *Macromolecules*, Article ASAP. doi: 10.1021/ma902710j.

129. Bronstein, H.A. and Luscombe, C.K. (2009) Externally initiated regioregular P3HT with controlled molecular weight and narrow polydispersity. *J. Am. Chem. Soc.*, **131**, 12894–12895.

130. Senkovskyy, V., Tkachov, R., Beryozkina, T., Komber, H., Oertel, U., Horecha, M., Bocharova, V., Stamm, M., Gevorgyan, S.A., Krebs, F.C., and Kiriy, A. (2009) "Hairy" poly(3-hexylthiophene) particles prepared via surface-initiated Kumada catalyst-transfer polycondensation. *J. Am. Chem. Soc.*, **131**, 16445–16453.

131. Xiao, R., Cho, S.I., Liu, R., and Lee, S.B. (2007) Controlled electrochemical synthesis of conductive polymer nanotube structures. *J. Am. Chem. Soc.*, **129**, 4483–4489.

132. Zujovic, Z.D., Laslau, C., Bowmaker, G.A., Kilmartin, P.A., Webber, A.L., Brown, S.P., and Travas-Sejdic, J. (2009) Role of aniline oligomeric nanosheets in the formation of polyaniline nanotubes. *Macromolecules*, **43**, 662–670.

133. Paul, G.S., Sarmah, P.J., Iyer, P.K., and Agarwal, P. (2008) Synthesis and study of CdS nanoparticle-doped poly(1,4-dihexyloxy benzene). *Macromol. Chem. Phys.*, **209**, 417–423.

134. Jhaveri, S.B., Peterson, J.J., and Carter, K.R. (2008) Organolithium-activated nickel (OLAN) catalysis: a new synthetic route for polyarylates. *Macromolecules*, **41**, 8977–8979.

135. Jhaveri, S.B. and Carter, K.R. (2008) Nickel-catalyzed coupling of aryl bromides in the presence of alkyllithium reagents. *Chem. Eur. J.*, **14**, 6845–6848.

136. Sanechika, K., Yamamoto, T., and Yamamoto, A. (1984) Palladium catalyzed carbon-carbon coupling for synthesis of π-conjugated polymers composed of arylene and ethynylene units. *Bull. Chem. Soc. Jpn.*, **57**, 752–755.

137. Giesa, R. and Schulz, R.C. (1990) Soluble poly(1,4-phenyleneethynylene)s. *Makromol. Chem.*, **191**, 857–867.

138. Moroni, M., Le Moigne, J., and Luzzati, S. (1994) Rigid rod conjugated polymers for nonlinear optics: 1. Characterization and linear optical properties of poly(aryleneethynylene) derivatives. *Macromolecules*, **27**, 562–571.

139. Bunz, U.H.F. (2000) Poly(aryleneethynylene)s: syntheses, properties, structures, and applications. *Chem. Rev.*, **100**, 1605–1644.

140. Steiger, D., Smith, P., and Weder, C. (1997) Liquid-crystalline, highly luminescent poly(2,5-dialkoxy-p-phenyleneethynylene). *Macromol. Rapid Commun.*, **18**, 643–649.

141. Francke, V., Mangel, T., and Muellen, K. (1998) Synthesis of α, ω-difunctionalized oligo- and poly(p-phenyleneethynylene)s. *Macromolecules*, **31**, 2447–2453.

142. Takagi, M., Kizu, K., Miyazaki, Y., Maruyama, T., Kubota, K., and Yamamoto, T. (1993) Synthesis of π-conjugated soluble poly(aryleneethynylene) type polymers and their properties. *Chem. Lett.*, 913–916.

143. Yamamoto, T., Honda, K., Ooba, N., and Tomaru, S. (1998) Poly(aryleneethynylene) type polymers composed of p-phenylene and 2,5-thienylene units. Analysis of polymerization conditions and terminal group in relation to the mechanism of the polymerization and chemical and optical properties of the polymer. *Macromolecules*, **31**, 7–14.

144. Bao, Z., Chen, Y., Cai, R., and Yu, L. (1993) Conjugated liquid-crystalline polymers – soluble and fusible poly(phenylenevinylene) by the Heck coupling reaction. *Macromolecules*, **26**, 5281–5286.

145. Kloppenburg, L., Song, D., and Bunz, U.H.F. (1998) Alkyne metathesis with simple catalyst systems: poly(p-phenyleneethynylenes). *J. Am. Chem. Soc.*, **120**, 7973–7974.

146. Rehahn, M. and Schlueter, A.D. (1990) Soluble poly(p-phenylenevinylenes) from 2,5-dihexylterephthalaldehyde using the improved McMurry reagent. *Makromol. Chem., Rapid Commun.*, **11**, 375–379.

147. Iwatsuki, S., Kubo, M., and Itoh, Y. (1993) Preparation of poly(3,4-dibutoxy-2,5-thienylene-vinylene) via titanium-induced dicarbonyl-coupling reaction of 3,4-dibutoxythiophene-2,5-dicarbaldehyde. *Chem. Lett.*, 1085–1088.

148. Rost, H., Teuschel, A., Pfeiffer, S., and Hörhold, H.H. (1997) Novel light emitting and photoconducting polyarylenevinylene derivatives containing phenylene arylamine and phenylene oxide units in the main chain. *Synth. Met.*, **84**, 269–270.

149. Gorman, C.B., Ginsburg, E.J., Sailor, M.J., Moore, J.S., Jozefiak, T.H., Lewis, N.S., Grubbs, R.H., Marder, S.R., and

Perry, J.W. (1991) Substituted poly-acetylenes through the ring-opening metathesis polymerization (ROMP) of substituted cyclooctatetraenes: a route into soluble polyacetylene. *Synth. Met.*, **41**, 1033–1038.

150. Jozefiak, T.H., Ginsburg, E.J., Gorman, C.B., Grubbs, R.H., and Lewis, N.S. (1993) Voltammetric characterization of soluble polyacetylene derivatives obtained from the ring-opening metathesis polymerization (ROMP) of substituted cyclooctatetraenes. *J. Am. Chem. Soc.*, **115**, 4705–4713.

151. Conticello, V.P., Gin, D.L., and Grubbs, R.H. (1992) Ring-opening metathesis polymerization of substituted bicy-clo[2.2.2]octadienes: a new precursor route to poly(1,4-phenylenevinylene). *J. Am. Chem. Soc.*, **114**, 9708–9710.

152. Marra, W.C., Eisenberger, P., and Cho, A.Y. (1979) X-ray total-external-reflection-Bragg diffraction: a structural study of the gallium arsenide-aluminum interface. *J. Appl. Phys.*, **50**, 6927–6933.

153. Vineyard, G.H. (1982) Grazing-incidence diffraction and the distorted-wave approximation for the study of surfaces. *Phys. Rev. B: Condens. Matter*, **26**, 4146–4159.

154. Prosa, T.J., Winokur, M.J., Moulton, J., Smith, P., and Heeger, A.J. (1992) X-ray structural studies of poly(3-alkylthiophenes): an example of an inverse comb. *Macromolecules*, **25**, 4364–4372.

155. Kobashi, M. and Takeuchi, H. (1998) Inhomogeneity of spin-coated and cast non-regioregular poly(3-hexylthiophene) films. Structures and electrical and photophysical properties. *Macro-molecules*, **31**, 7273–7278.

156. Kline, R.J., McGehee, M.D., Kadnikova, E.N., Liu, J., Frechet, J.M.J., and Toney, M.F. (2005) Dependence of re-gioregular poly(3-hexylthiophene) film morphology and field-effect mobility on molecular weight. *Macromolecules*, **38**, 3312–3319.

157. van Duren, J.K.J., Yang, X., Loos, J., Bulle-Lieuwma, C.W.T., Sieval, A.B., Hummelen, J.C., and Janssen, R.A.J. (2004) Relating the mor-phology of poly(p-phenylene viny-lene)/methanofullerene blends to solar-cell performance. *Adv. Funct. Mater.*, **14**, 425–434.

158. Ma, W., Yang, C., Gong, X., Lee, K., and Heeger, A.J. (2005) Thermally sta-ble, efficient polymer solar cells with nanoscale control of the interpenetrat-ing network morphology. *Adv. Funct. Mater.*, **15**, 1617–1622.

159. Jacobs, H.O., Knapp, H.F., and Stemmer, A. (1999) Practical aspects of Kelvin probe force microscopy. *Rev. Sci. Instrum.*, **70**, 1756–1760.

160. Burgi, L., Sirringhaus, H., and Friend, R.H. (2002) Noncontact potentiometry of polymer field effect transistors. *Appl. Phys. Lett.*, **80**, 2913–2915.

161. Burgi, L., Friend, R.H., and Sirringhaus, H. (2003) Formation of the accumulation layer in polymer field effect transistors. *Appl. Phys. Lett.*, **82**, 1482–1484.

162. Nichols, J.A., Gundlach, D.J., and Jackson, T.N. (2003) Potential imaging of pentacene organic thin-film transis-tors. *Appl. Phys. Lett.*, **83**, 2366–2368.

163. Puntambekar, K.P., Pesavento, P.V., and Frisbie, C.D. (2003) Surface po-tential profiling and contact resistance measurements on operating pentacene thin-film transistors by Kelvin probe force microscopy. *Appl. Phys. Lett.*, **83**, 5539–5541.

164. Chiesa, M., Burgi, L., Kim, J.-S., Shikler, R., Friend, R.H., and Sirringhaus, H. (2005) Correlation be-tween surface photovoltage and blend morphology in polyfluorene-based photodiodes. *Nano Lett.*, **5**, 559–563.

165. Hoppe, H., Glatzel, T., Niggemann, M., Hinsch, A., Lux-Steiner, M.C., and Sariciftci, N.S. (2005) Kelvin probe force microscopy study on conjugated polymer/fullerene bulk heterojunc-tion organic solar cells. *Nano Lett.*, **5**, 269–274.

166. Palermo, V., Ridolfi, G., Talarico, A.M., Favaretto, L., Barbarella, G., Camaioni, N., and Samorï, P. (2007) A Kelvin probe force microscopy study of the photogeneration of surface charges in

all-thiophene photovoltaic blends. *Adv. Funct. Mater.*, **17**, 472–478.

167. Liscio, A., De Luca, G., Nolde, F., Palermo, V., Mullen, K., and Samori, P. (2007) Photovoltaic charge generation visualized at the nanoscale: a proof of principle. *J. Am. Chem. Soc.*, **130**, 780–781.

168. Tutiš, E., Berner, D., and Zuppiroli, L. (2003) Internal electric field and charge distribution in multilayer organic light-emitting diodes. *J. Appl. Phys.*, **93**, 4594–4602.

169. Chen, Z., Niu, Q.L., Zhang, Y., Ying, L., Peng, J.B., and Cao, Y. (2009) Efficient green electrophosphorescence with Al cathode using an effective electron-injecting polymer as the host. *ACS Appl. Mater. Interfaces*, **1**, 2785–2788.

170. Saikia, G., Singh, R., Sarmah, P.J., Akhtar, M.W., Sinha, J., Katiyar, M., and Iyer, P.K. (2009) Synthesis and characterization of soluble poly(p-phenylene) derivatives for PLED applications. *Macromol. Chem. Phys.*, **210**, 2153–2159.

171. Kim, J.-S., Ho, P.K.H., Greenham, N.C., and Friend, R.H. (2000) Electroluminescence emission pattern of organic light-emitting diodes: Implications for device efficiency calculations. *J. Appl. Phys.*, **88**, 1073–1081.

172. Minaev, B., Minaeva, V., and Aagren, H. (2009) Theoretical study of the cyclometalated iridium(III) complexes used as chromophores for organic light-emitting diodes. *J. Phys. Chem. A*, **113**, 726–735.

173. Baldo, M.A., Lamansky, S., Burrows, P.E., Thompson, M.E., and Forrest, S.R. (1999) Very high-efficiency green organic light-emitting devices based on electro-phosphorescence. *Appl. Phys. Lett.*, **75**, 4–6.

174. Adachi, C., Baldo, M.A., Thompson, M.E., and Forrest, S.R. (2001) Nearly 100% internal phosphorescence efficiency in an organic light-emitting device. *J. Appl. Phys.*, **90**, 5048–5051.

175. Zhuang, W., Zhang, Y., Hou, Q., Wang, L., and Cao, Y. (2006) High-efficiency, electrophosphorescent polymers with porphyrin-platinum complexes in the conjugated backbone: synthesis and device performance. *J. Polym. Sci., Part A: Polym. Chem.*, **44**, 4174–4186.

176. Adamovich, V., Brooks, J., Tamayo, A., Alexander, A.M., Djurovich, P.I., D'Andrade, B.W., Adachi, C., Forrest, S.R., and Thompson, M.E. (2002) High-efficiency single dopant white electrophosphorescent light-emitting diodes. *New J. Chem.*, **26**, 1171–1178.

177. D'Andrade, B.W., Holmes, R.J., and Forrest, S.R. (2004) Efficient organic electrophosphorescent white-light-emitting device with a triple doped emissive layer. *Adv. Mater.*, **16**, 624–628.

178. Madigan, C.F., Lu, M.H., and Sturm, J.C. (2000) Improvement of output coupling efficiency of organic light-emitting diodes by backside substrate modification. *Appl. Phys. Lett.*, **76**, 1650–1652.

179. Shinar, J. and Shinar, R. (2008) Organic light-emitting devices (OLEDs) and OLED-based chemical and biological sensors: an overview. *J. Phys. D: Appl. Phys.*, **41**, 133001–133027.

180. Danielsen, P.L. (1987) Photoluminescence quenching in cis-polyacetylene (a lattice relaxation approach). *Synth. Met.*, **17**, 87–92.

181. Kersting, R., Lemmer, U., Deussen, M., Bakker, H.J., Mahrt, R.F., Kurz, H., Arkhipov, V.I., Baessler, H., and Goebel, E.O. (1994) Ultrafast field-induced dissociation of excitons in conjugated polymers. *Phys. Rev. Lett.*, **73**, 1436–1439.

182. Lee, M.K., Segal, M., Soos, Z.G., Shinar, J., and Baldo, M.A. (2005) Yield of singlet excitons in organic light-emitting devices: a double modulation photoluminescence-detected magnetic resonance study. *Phys. Rev. Lett.*, **94**, 137403-1–137403-4.

183. Segal, M., Baldo, M.A., Lee, M.K., Shinar, J., and Soos, Z.G. (2005) Frequency response and origin of the spin-1/2 photoluminescence-detected magnetic resonance in a π-conjugated polymer. *Phys. Rev. B: Condens. Matter Mater. Phys.*, **71**, 245201-1–245201-11.

184. Laquai, F., Park, Y.S., Kim, J.J., and Basche, T. (2009) Excitation energy transfer in organic materials: from fundamentals to optoelectronic devices. *Macromol. Rapid Commun.*, **30**, 1203–1231.

185. Partee, J., Frankevich, E.L., Uhlhorn, B., Shinar, J., Ding, Y., and Barton, T.J. (1999) Delayed fluorescence and triplet-triplet annihilation in π-conjugated polymers. *Phys. Rev. Lett.*, **82**, 3673–3676.

186. Noh, Y.-Y., Lee, C.-L., Kim, J.-J., and Yase, K. (2003) Energy transfer and device performance in phosphorescent dye doped polymer light emitting diodes. *J. Chem. Phys.*, **118**, 2853–2864.

187. Lee, C.-L., Kang, N.-G., Cho, Y.-S., Lee, J.-S., and Kim, J.-J. (2003) Polymer electrophosphorescent device: comparison of phosphorescent dye doped and coordinated systems. *Opt. Mater.*, **21**, 119–123.

188. Kido, J., Hongawa, K., Okuyama, K., and Nagai, K. (1994) White light-emitting organic electroluminescent devices using the poly(N-vinylcarbazole) emitter layer doped with three fluorescent dyes. *Appl. Phys. Lett.*, **64**, 815–817.

189. Sun, Y., Giebink, N.C., Kanno, H., Ma, B., Thompson, M.E., and Forrest, S.R. (2006) Management of singlet and triplet excitons for efficient white organic light-emitting devices. *Nature*, **440**, 908–912.

190. Tokito, S., Iijima, T., Tsuzuki, T., and Sato, F. (2003) High-efficiency white phosphorescent organic light-emitting devices with greenish-blue and red-emitting layers. *Appl. Phys. Lett.*, **83**, 2459–2461.

191. Adachi, C., Kwong, R.C., Djurovich, P., Adamovich, V., Baldo, M.A., Thompson, M.E., and Forrest, S.R. (2001) Endothermic energy transfer: a mechanism for generating very efficient high-energy phosphorescent emission in organic materials. *Appl. Phys. Lett.*, **79**, 2082–2084.

192. Holmes, R.J., Forrest, S.R., Tung, Y.J., Kwong, R.C., Brown, J.J., Garon, S., and Thompson, M.E. (2003) Blue organic electrophosphorescence using exothermic host-guest energy transfer. *Appl. Phys. Lett.*, **82**, 2422–2424.

193. Choulis, S.A., Choong, V.-E., Mathai, M.K., and So, F. (2005) The effect of interfacial layer on the performance of organic light-emitting diodes. *Appl. Phys. Lett.*, **87**, 113503-1–113503-3.

194. Savvate'ev, V.N., Yakimov, A.V., Davidov, D., Pogreb, R.M., Neumann, R., and Avny, Y. (1997) Degradation of nonencapsulated polymer-based light-emitting diodes: noise and morphology. *Appl. Phys. Lett.*, **71**, 3344–3346.

195. Do, L.-M., Kim, K., Zyung, T., Shim, H.-K., and Kim, J.-J. (1997) *In situ* investigation of degradation in polymeric electroluminescent devices using time-resolved confocal laser scanning microscopy. *Appl. Phys. Lett.*, **70**, 3470–3472.

196. O'Brien, D.F., Burrows, P.E., Forrest, S.R., Koene, B.E., Loy, D.E., and Thompson, M.E. (1998) Hole transporting materials with high glass transition temperatures for use in organic light-emitting devices. *Adv. Mater.*, **10**, 1108–1112.

197. Saragi Tobat, P.I., Spehr, T., Siebert, A., Fuhrmann-Lieker, T., and Salbeck, J. (2007) Spiro compounds for organic optoelectronics. *Chem. Rev.*, **107**, 1011–1065.

198. McGehee, M.D. and Topinka, M.A. (2006) Solar cells: pictures from the blended zone. *Nature Mater.*, **5**, 675–676.

199. Zhu, X.Y., Yang, Q., and Muntwiler, M. (2009) Charge-transfer excitons at organic semiconductor surfaces and interfaces. *Acc. Chem. Res.*, **42**, 1779–1787.

200. Guo, J., Ohkita, H., Benten, H., and Ito, S. (2009) Near-IR femtosecond transient absorption spectroscopy of ultrafast polaron and triplet exciton formation in polythiophene films with different regioregularities. *J. Am. Chem. Soc.*, **131**, 16869–16880.

201. Ohkita, H., Cook, S., Astuti, Y., Duffy, W., Tierney, S., Zhang, W., Heeney, M., McCulloch, I., Nelson, J.,

Bradley, D.D.C., and Durrant, J.R. (2008) Charge carrier formation in polythiophene/fullerene blend films studied by transient absorption spectroscopy. *J. Am. Chem. Soc.*, **130**, 3030–3042.

202. Ohkita, H., Cook, S., Astuti, Y., Duffy, W., Heeney, M., Tierney, S., McCulloch, I., Bradley, D.D.C., and Durrant, J.R. (2006) Radical ion pair mediated triplet formation in polymer-fullerene blend films. *Chem. Commun.*, 3939–3941.

203. Savenije, T.J., Warman, J.M., and Goossens, A. (1998) Visible light sensitization of titanium dioxide using a phenylene vinylene polymer. *Chem. Phys. Lett.*, **287**, 148–153.

204. Halls, J.J.M., Pichler, K., Friend, R.H., Moratti, S.C., and Holmes, A.B. (1996) Exciton diffusion and dissociation in a poly(p-phenylenevinylene)/C60 heterojunction photovoltaic cell. *Appl. Phys. Lett.*, **68**, 3120–3122.

205. Pettersson, L.A.A., Roman, L.S., and Inganas, O. (1999) Modeling photocurrent action spectra of photovoltaic devices based on organic thin films. *J. Appl. Phys.*, **86**, 487–496.

206. Granstrom, M., Petritsch, K., Arias, A.C., Lux, A., Andersson, M.R., and Friend, R.H. (1998) Laminated fabrication of polymeric photovoltaic diodes. *Nature*, **395**, 257–260.

207. Arango, A.C., Carter, S.A., and Brock, P.J. (1999) Charge transfer in photovoltaics consisting of interpenetrating networks of conjugated polymer and TiO_2 nanoparticles. *Appl. Phys. Lett.*, **74**, 1698–1700.

208. Bach, U., Lupo, D., Comte, P., Moser, J.E., Weissortel, F., Salbeck, J., Spreitzer, H., and Gratzel, M. (1998) Solid-state dye-sensitized mesoporous TiO_2 solar cells with high photon-to-electron conversion efficiencies. *Nature*, **395**, 583–585.

209. Coakley, K.M., Liu, Y., McGehee, M.D., Frindell, K.L., and Stucky, G.D. (2003) Infiltrating semiconducting polymers into self-assembled mesoporous titania films for photovoltaic applications. *Adv. Funct. Mater.*, **13**, 301–306.

210. Huisman, C.L., Goossens, A., and Schoonman, J. (2003) Preparation of a nanostructured composite of titanium dioxide and polythiophene: a new route towards 3D heterojunction solar cells. *Synth. Met.*, **138**, 237–241.

211. Yu, G., Gao, J., Hummelen, J.C., Wudl, F., and Heeger, A.J. (1995) Polymer photovoltaic cells: enhanced efficiencies via a network of internal donor-acceptor heterojunctions. *Science*, **270**, 1789–1791.

212. Deibel, C., Baumann, A., Wagenpfahl, A., and Dyakonov, V. (2009) Polaron recombination in pristine and annealed bulk heterojunction solar cells. *Synth. Met.*, **159**, 2345–2347.

213. Kong, H., Moon, J.-S., Cho, N.-S., Jung, I.-H., Park, M.-J., Park, J.-H., Cho, S., and Shim, H.-K. (2009) Thermal annealing induced bicontinuous networks in bulk heterojunction solar cells and bipolar field effect transistors. *Appl. Phys. Lett.*, **95**, 173301-1–173301-3.

214. Liu, C., Li, Y., Li, C., Li, W., Zhou, C., Liu, H., Bo, Z., and Li, Y. (2009) New methanofullerenes containing amide as electron acceptor for construction photovoltaic devices. *J. Phys. Chem. C*, **113**, 21970–21975.

215. Kim, Y., Cook, S., Tuladhar, S.M., Choulis, S.A., Nelson, J., Durrant, J.R., Bradley, D.D.C., Giles, M., McCulloch, I., Ha, C.-S., and Ree, M. (2006) A strong regioregularity effect in self-organizing conjugated polymer films and high-efficiency polythiophene:fullerene solar cells. *Nature Mater.*, **5**, 197–203.

216. Campbell, A.R., Hodgkiss, J.M., Westenhoff, S., Howard, I.A., Marsh, R.A., McNeill, C.R., Friend, R.H., and Greenham, N.C. (2008) Low-temperature control of nanoscale morphology for high performance polymer photovoltaics. *Nano Lett.*, **8**, 3942–3947.

217. Shaheen, S.E., Brabec, C.J., Sariciftci, N.S., Padinger, F., Fromherz, T., and Hummelen, J.C. (2001) 2.5% efficient organic plastic solar cells. *Appl. Phys. Lett.*, **78**, 841–843.

218. Ding, Y., Lu, P., and Chen, Q. (2008) Optimizing material properties of bulk-heterojunction polymer films for photovoltaic applications. *Proc. SPIE*, **7099**, 709919-1–709919-8.

219. Bertho, S., Janssen, G., Cleij, T.J., Conings, B., Moons, W., Gadisa, A., D'Haen, J., Goovaerts, E., Lutsen, L., Manca, J., and Vanderzande, D. (2008) Effect of temperature on the morphological and photovoltaic stability of bulk heterojunction polymer:fullerene solar cells. *Sol. Energy Mater. Sol. Cells*, **92**, 753–760.

220. Cunningham, P.D. and Hayden, L.M. (2008) Carrier dynamics resulting from above and below gap excitation of P3HT and P3HT/PCBM investigated by optical-pump terahertz-probe spectroscopy. *J. Phys. Chem. C*, **112**, 7928–7935.

221. Wienk, M.M., Turbiez, M.G.R., Struijk, M.P., Fonrodona, M., and Janssen, R.A.J. (2006) Low-bandgap poly(di-2-thienylthienopyrazine): fullerene solar cells. *Appl. Phys. Lett.*, **88**, 153511-1–153511-3.

222. Wei, Y., Zhang, Q., Jiang, Y., and Yu, J. (2009) Novel low bandgap EDOT-naphthalene bisimides conjugated polymers: synthesis, redox, and optical properties. *Macromol. Chem. Phys.*, **210**, 769–775.

223. Yue, W., Zhao, Y., Tian, H., Song, D., Xie, Z., Yan, D., Geng, Y., and Wang, F. (2009) Poly(oligothiophene-alt-benzothiadiazole)s: tuning the structures of oligothiophene units toward high-mobility "Black" conjugated polymers. *Macromolecules*, **42**, 6510–6518.

224. Bundgaard, E. and Krebs, F.C. (2006) Low-band gap conjugated polymers based on thiophene, benzothiadiazole, and benzobis(thiadiazole). *Macromolecules*, **39**, 2823–2831.

225. Perzon, E., Wang, X., Zhang, F., Mammo, W., Delgado, J.L., de la Cruz, P., Inganaes, O., Langa, F., and Andersson, M.R. (2005) Design, synthesis and properties of low band gap polyfluorenes for photovoltaic devices. *Synth. Met.*, **154**, 53–56.

226. Lee, B., Seshadri, V., and Sotzing, G.A. (2005) Water dispersible low band gap conductive polymer based on thieno[3,4-b]thiophene. *Synth. Met.*, **152**, 177–180.

227. Lee, K. and Sotzing, G.A. (2001) Poly(thieno[3,4-b]thiophene). A new stable low band gap conducting polymer. *Macromolecules*, **34**, 5746–5747.

228. Seshadri, V. and Sotzing, G.A. (2003) Low band gap poly(thieno[3,4-b]thiophene) consisting of cyanovinylene spacer units. *Polym. Prepr.*, **44**, 398–399.

229. Sotzing, G.A., Thomas, C.A., Reynolds, J.R., and Steel, P.J. (1998) Low band gap cyanovinylene polymers based on ethylenedioxythiophene. *Macromolecules*, **31**, 3750–3752.

230. Winder, C. and Sariciftci, N.S. (2004) Low bandgap polymers for photon harvesting in bulk heterojunction solar cells. *J. Mater. Chem.*, **14**, 1077–1086.

231. Coakley, K.M. and McGehee, M.D. (2004) Conjugated polymer photovoltaic cells. *Chem. Mater.*, **16**, 4533–4542.

232. Tripathi, A.K., Heinrich, M., Siegrist, T., and Pflaum, J. (2007) Growth and electronic transport in 9,10-diphenylanthracene single crystals-an organic semiconductor of high electron and hole mobility. *Adv. Mater.*, **19**, 2097–2101.

233. Xia, Y., Cho, J.H., Lee, J., Ruden, P.P., and Frisbie, C.D. (2009) Comparison of the mobility-carrier density relation in polymer and single-crystal organic transistors employing vacuum and liquid gate dielectrics. *Adv. Mater.*, **21**, 2174–2179.

234. Vissenberg, M. and Matters, M. (1998) Theory of the field-effect mobility in amorphous organic transistors. *Phys. Rev. B*, **57**, 12964–12967.

235. Street, R.A., Northrup, J.E., and Salleo, A. (2005) Transport in polycrystalline polymer thin-film transistors. *Phys. Rev. B*, **71**, 165202-1–165202-13.

236. Armstrong, N.R., Wang, W., Alloway, D.M., Placencia, D., Ratcliff, E., and Brumbach, M. (2009) Organic/organic' heterojunctions: organic light emitting

diodes and organic photovoltaic devices. *Macromol. Rapid Commun.*, **30**, 717–731.

237. Park, S.H., Roy, A., Beaupre, S., Cho, S., Coates, N., Moon, J.S., Moses, D., Leclerc, M., Lee, K., and Heeger, A.J. (2009) Bulk heterojunction solar cells with internal quantum efficiency approaching 100%. *Nature Photonics*, **3**, 297–303.

238. Ginger, D.S. and Greenham, N.C. (1999) Photoinduced electron transfer from conjugated polymers to CdSe nanocrystals. *Phys. Rev. B: Condens. Matter Mater. Phys.*, **59**, 10622–10629.

239. Huynh, W.U., Dittmer, J.J., and Alivisatos, A.P. (2002) Hybrid nanorod-polymer solar cells. *Science*, **295**, 2425–2427.

240. Meng, H. and Wudl, F. (2001) A robust low band gap processable n-type conducting polymer based on poly(isothianaphthene). *Macromolecules*, **34**, 1810–1816.

241. Jung, Y., Baeg, K.-J., Kim, D.-Y., Someya, T., and Park, S.Y. (2009) A thermally resistant and air-stable n-type organic semiconductor: naphthalene diimide of 3,5-bis-trifluoromethyl aniline. *Synth. Met.*, **159**, 2117–2121.

242. Song, D., Wang, H., Zhu, F., Yang, J., Tian, H., Geng, Y., and Yan, D. (2008) Phthalocyanato tin(IV) dichloride: an air-stable, high-performance, n-type organic semiconductor with a high field-effect electron mobility. *Adv. Mater.*, **20**, 2142–2144.

243. Sun, Y., Rohde, D., Liu, Y., Wan, L., Wang, Y., Wu, W., Di, C., Yu, G., and Zhu, D. (2006) A novel air-stable n-type organic semiconductor: 4,4′-bis[(6,6′-diphenyl)-2,2-difluoro-1,3,2-dioxaborine] and its application in organic ambipolar field-effect transistors. *J. Mater. Chem.*, **16**, 4499–4503.

244. Salleo, A. (2007) Charge transport in polymeric transistors. *Mater. Today*, **10**, 38–45.

245. Newman, C.R., Frisbie, C.D., da Silva Filho, D.A., Bredas, J.-L., Ewbank, P.C., and Mann, K.R. (2004) Introduction to organic thin film transistors and design of n-channel organic semiconductors. *Chem. Mater.*, **16**, 4436–4451.

246. Hill, I.G. and Kahn, A. (1998) Interface electronic properties of organic molecular semiconductors. *Proc. SPIE-Int. Soc. Opt. Eng.*, **3476**, 168–177.

247. DiBenedetto, S.A., Facchetti, A., Ratner, M.A., and Marks, T.J. (2009) Molecular self-assembled monolayers and multilayers for organic and unconventional inorganic thin-film transistor applications. *Adv. Mater.*, **21**, 1407–1433.

248. Bao, Z. and Locklin J. (2007) *Organic Field-Effect Transistors*, CRC Press, Boca Raton, FL.

249. Liu, Y., Wang, Y., Wu, W., Liu, Y., Xi, H., Wang, L., Qiu, W., Lu, K., Du, C., and Yu, G. (2009) Synthesis, characterization, and field-effect transistor performance of thieno[3,2-b]thieno[2′,3′:4,5]thieno[2,3-d]thiophene derivatives. *Adv. Funct. Mater.*, **19**, 772–778.

250. Yuen, J.D., Menon, R., Coates, N.E., Namdas, E.B., Cho, S., Hannahs, S.T., Moses, D., and Heeger, A.J. (2009) Nonlinear transport in semiconducting polymers at high carrier densities. *Nature Mater.*, **8**, 572–575.

251. McCulloch, I., Heeney, M., Bailey, C., Genevicius, K., MacDonald, I., Shkunov, M., Sparrowe, D., Tierney, S., Wagner, R., Zhang, W., Chabinyc, M.L., Kline, R.J., McGehee, M.D., and Toney, M.F. (2006) Liquid-crystalline semiconducting polymers with high charge-carrier mobility. *Nature Mater.*, **5**, 328–333.

252. Briseno, A.L., Mannsfeld, S.C.B., Shamberger, P.J., Ohuchi, F.S., Bao, Z., Jenekhe, S.A., and Xia, Y. (2008) Self-assembly, molecular packing, and electron transport in n-type polymer semiconductor nanobelts. *Chem. Mater.*, **20**, 4712–4719.

253. Zhu, Y., Alam, M.M., and Jenekhe, S.A. (2003) Regioregular head-to-tail poly(4-alkylquinoline)s: synthesis, characterization, self-organization, photophysics, and electroluminescence of new n-type conjugated polymers. *Macromolecules*, **36**, 8958–8968.

254. de Leeuw, D.M., Simenon, M.M.J., Brown, A.R., and Einerhand, R.E.F.

(1997) Stability of n-type doped conducting polymers and consequences for polymeric microelectronic devices. *Synth. Met.*, **87**, 53–59.

255. Babel, A. and Jenekhe, S.A. (2003) High electron mobility in ladder polymer field effect transistors. *J. Am. Chem. Soc.*, **125**, 13656–13657.

256. Babel, A. and Jenekhe, S.A. (2002) Electron transport in thin-film transistors from an n-type conjugated polymer. *Adv. Mater.*, **14**, 371–374.

257. Jones, B.A., Facchetti, A., Wasielewski, M.R., and Marks, T.J. (2008) Effects of arylene diimide thin film growth conditions on n-channel OFET performance. *Adv. Funct. Mater.*, **18**, 1329–1339.

258. Wang, Z., Kim, C., Facchetti, A., and Marks, T.J. (2007) Anthracenedicarboximides as air-stable N-channel semiconductors for thin-film transistors with remarkable current on-off ratios. *J. Am. Chem. Soc.*, **129**, 13362–13363.

259. Ando, S., Murakami, R., Nishida, J.-I., Tada, H., Inoue, Y., Tokito, S., and Yamashita, Y. (2005) n-type organic field-effect transistors with very high electron mobility based on thiazole oligomers with trifluoromethylphenyl groups. *J. Am. Chem. Soc.*, **127**, 14996–14997.

260. Gundlach, D.J., Royer, J.E., Park, S.K., Subramanian, S., Jurchescu, O.D., Hamadani, B.H., Moad, A.J., Kline, R.J., Teague, L.C., Kirillov, O., Richter, C.A., Kushmerick, J.G., Richter, L.J., Parkin, S.R., Jackson, T.N., and Anthony, J.E. (2008) Contact-induced crystallinity for high-performance soluble acene-based transistors and circuits. *Nature Mater.*, **7**, 216–221.

261. Forrest, S.R. (2004) The path to ubiquitous and low-cost organic electronic appliances on plastic. *Nature*, **428**, 911–918.

262. Yan, H., Chen, Z., Zheng, Y., Newman, C., Quinn, J.R., Dotz, F., Kastler, M., and Facchetti, A. (2009) A high-mobility electron-transporting polymer for printed transistors. *Nature*, **457**, 679–686.

263. Shi, L.H., Garnier, F., and Roncali, J. (1991) Electroactivity of poly(thiophenes) containing oxyalkyl substituents. *Synth. Met.*, **41**, 547–550.

264. Youssoufi, H.K., Hmyene, M., Garnier, F., and Delabouglise, D. (1993) Cation recognition properties of polypyrrole 3-substituted by azacrown ethers. *J. Chem. Soc. Chem. Commun.*, 1550–1552.

265. Cosnier, S., Deronzier, A., and Moutet, J.C. (1985) Oxidative electropolymerization of polypyridinyl complexes of ruthenium(II)-containing pyrrole groups. *J. Electroanal. Chem. Interfacial Electrochem.*, **193**, 193–204.

266. Lopez, C., Moutet, J.C., and Saint-Aman, E. (1996) Electrochemical recognition of chloride ions by a poly[tris-(2,2′-bipyridine)ruthenium(II)] modified electrode. *J. Chem. Soc. Faraday Trans.*, **92**, 1527–1532.

267. Kimura, M., Horai, T., Hanabusa, K., and Shirai, H. (1998) Fluorescence chemosensor for metal ions using conjugated polymers. *Adv. Mater.*, **10**, 459–462.

268. Lepretre, J.C., Saint-Aman, E., and Utille, J.P. (1993) Preparation of a poly(cyclodextrin-pyrrole) modified electrode. *J. Electroanal. Chem.*, **347**, 465–470.

269. Sun, X.X. and Aboul-Enien, H.Y. (1999) Internal solid contact electrode for the determination of clenbuterol in pharmaceutical formulations and human urine. *Anal. Lett.*, **32**, 1143–1156.

270. Riskin, M., Tel-Vered, R., Bourenko, T., Granot, E., and Willner, I. (2008) Imprinting of molecular recognition sites through electropolymerization of functionalized Au nanoparticles: development of an electrochemical TNT sensor based on π-donor/acceptor interactions. *J. Am. Chem. Soc.*, **130**, 9726–9733.

271. Sreenivasan, K. (2007) Synthesis and evaluation of molecularly imprinted polymers for nucleic acid bases using aniline as a monomer. *React. Funct. Polym.*, **67**, 859–864.

272. Deore, B. and Freund, S.M. (2003) Saccharide imprinting of poly(aniline boronic acid) in the presence of fluoride. *Analyst*, **128**, 803–806.

273. Reddinger, J.L. and Reynolds, J.R. (1998) A novel polymeric metallo-macrocycle sensor capable of dual-ion cocomplexation. *Chem. Mater.*, **10**, 3–5.

274. Crawford, K.B., Goldfinger, M.B., and Swager, T.M. (1998) Na+ specific emission changes in an ionophoric conjugated polymer. *J. Am. Chem. Soc.*, **120**, 5187–5192.

275. Ion, A., Ion, I., Popescu, A., Ungureanu, M., Moutet, J.C., and Saint-Aman, E. (1997) A ferrocene crown ether-functionalized polypyrrole film electrode for the electrochemical recognition of barium and calcium cations. *Adv. Mater.*, **9**, 711–713.

276. Aizawa, M. and Yabuki, S. (1985) Proceedings of the 51st Annual Meeting of the Japan Chemical Society, p. 6.

277. Umana, M. and Waller, J. (1986) Protein-modified electrodes. The glucose oxidase/polypyrrole system. *Anal. Chem.*, **58**, 2979–2783.

278. Tatsuma, T., Gondaira, M., and Watanabe, T. (1992) Peroxidase-incorporated polypyrrole membrane electrodes. *Anal. Chem.*, **64**, 1183–1187.

279. Pal, P., Nandi, D., and Misra, T.N. (1994) Immobilization of alcohol dehydrogenase enzyme in a Langmuir–Blodgett film of stearic acid: its application as an ethanol sensor. *Thin Solid Films*, **239**, 138–143.

280. Nishizawa, M., Matsue, T., and Uchida, I. (1992) Penicillin sensor based on a microarray electrode coated with pH-responsive polypyrrole. *Anal. Chem.*, **64**, 2642–2644.

281. Olmsted, J. and Strand, M. (1983) Fluorescence of polymerized diacetylene bilayer films. *J. Phys. Chem.*, **87**, 4790–4792.

282. Charych, D.H., Nagy, J.O., Spevak, W., and Bednarski, M.D. (1993) Direct colorimetric detection of a receptor-ligand interaction by a polymerized bilayer assembly. *Science*, **261**, 585–588.

283. Ma, G., Müller, A.M., Bardeen, C.J., and Cheng, Q. (2006) Self-assembly combined with photopolymerization for the fabrication of fluorescence turn-on vesicle sensors with reversible on-off

switching properties. *Adv. Mater.*, **18**, 55–60.

284. Kim, J.M., Ji, E.K., Woo, S.M., Lee, H., and Ahn, D.J. (2003) Immobilized polydiacetylene vesicles on solid substrates for use as chemosensors. *Adv. Mater.*, **15**, 1118–1121.

285. Kim, K.W., Choi, H., Lee, G.S., Ahn, D.J., Oh, M.K., and Kim, J.M. (2006) Micro-patterned polydiacetylene vesicle chips for detecting protein-protein interactions. *Macromol. Res.*, **14**, 483–485.

286. Pu, K.-Y., Cai, L., and Liu, B. (2009) Design and synthesis of charge-transfer-based conjugated polyelectrolytes as multicolor light-up probes. *Macromolecules*, **42**, 5933–5940.

287. Zhou, Q. and Swager, T.M. (1995) Method for enhancing the sensitivity of fluorescent chemosensors: energy migration in conjugated polymers. *J. Am. Chem. Soc.*, **117**, 7017–7018.

288. Furukawa, R., Xu, C., and Taya, M. (2006) Design of a passive-matrix smart window by the conjugated polymer electrochromics. *Mater. Res. Soc. Symp. Proc.*, **937E**, 0937-0M07–0937-0M12.

289. Andersson, P., Nilsson, D., Svensson, P.-O., Chen, M., Malmstrom, A., Remonen, T., Kugler, T., and Berggren, M. (2002) Active matrix displays based on all-organic electrochemical smart pixels printed on paper. *Adv. Mater.*, **14**, 1460–1464.

290. Berggren, M., Nilsson, D., and Robinson, N.D. (2007) Organic materials for printed electronics. *Nature Mater.*, **6**, 3–5.

291. Argun, A.A., Berard, M., Aubert, P.-H., and Reynolds, J.R. (2005) Back-side electrical contacts for patterned electrochromic devices on porous substrates. *Adv. Mater.*, **17**, 422–426.

292. Koyuncu, S., Gultekin, B., Zafer, C., Bilgili, H., Can, M., Demic, S., Kaya, I., and Icli, S. (2009) Electrochemical and optical properties of biphenyl bridged-dicarbazole oligomer films: electropolymerization and electrochromism. *Electrochim. Acta*, **54**, 5694–5702.

293. Yohannes, T., Carlberg, J.C., Inganaes, O., and Solomon, T. (1997) Electrochemical and spectroscopic characteristics of copolymers electrochemically synthesized from 3-methylthiophene and 3,4-ethylenedioxythiophene. *Synth. Met.*, **88**, 15–21.

294. Cirpan, A., Argun, A.A., Grenier, C.R.G., Reeves, B.D., and Reynolds, J.R. (2003) Electrochromic devices based on soluble and processable dioxythiophene polymers. *J. Mater. Chem.*, **13**, 2422–2428.

295. Niklasson, G.A. and Granqvist, C.G. (2007) Electrochromics for smart windows: thin films of tungsten oxide and nickel oxide, and devices based on these. *J. Mater. Chem.*, **17**, 127–156.

296. Sonmez, G. and Wudl, F. (2005) Completion of the three primary colours: the final step toward plastic displays. *J. Mater. Chem.*, **15**, 20–22.

297. Koyuncu, S., Zafer, C., Sefer, E., Koyuncu, F.B., Demic, S., Kaya, I., Ozdemir, E., and Icli, S. (2009) A new conducting polymer of 2,5-bis(2-thienyl)-1H-(pyrrole) (SNS) containing carbazole subunit: electrochemical, optical and electrochromic properties. *Synth. Met.*, **159**, 2013–2021.

298. Trznadel, M., Pron, A., Zagorska, M., Chrzaszcz, R., and Pielichowski, J. (1998) Effect of molecular weight on spectroscopic and spectroelectrochemical properties of regioregular poly(3-hexylthiophene). *Macromolecules*, **31**, 5051–5058.

299. Zen, A., Pflaum, J., Hirschmann, S., Zhuang, W., Jaiser, F., Asawapirom, U., Rabe, J.P., Scherf, U., and Neher, D. (2004) Effect of molecular weight and annealing of poly(3-hexylthiophene)s on the performance of organic field-effect transistors. *Adv. Funct. Mater.*, **14**, 757–764.

300. Aubert, P.-H., Argun, A.A., Cirpan, A., Tanner, D.B., and Reynolds, J.R. (2004) Microporous patterned electrodes for color-matched electrochromic polymer displays. *Chem. Mater.*, **16**, 2386–2393.

301. Sonmez, G., Meng, H., and Wudl, F. (2004) Organic polymeric electrochromic devices: polychromism with very high coloration efficiency. *Chem. Mater.*, **16**, 574–580.

302. Sonmez, G., Sonmez, H.B., Shen, C.K.F., and Wudl, F. (2004) Red, green, and blue colors in polymeric electrochromics. *Adv. Mater.*, **16**, 1905–1908.

303. Schwendeman, I., Hickman, R., Soenmez, G., Schottland, P., Zong, K., Welsh, D.M., and Reynolds, J.R. (2002) Enhanced contrast dual polymer electrochromic devices. *Chem. Mater.*, **14**, 3118–3122.

304. Argun, A.A., Aubert, P.-H., Thompson, B.C., Schwendeman, I., Gaupp, C.L., Hwang, J., Pinto, N.J., Tanner, D.B., MacDiarmid, A.G., and Reynolds, J.R. (2004) Multicolored electrochromism in polymers: structures and devices. *Chem. Mater.*, **16**, 4401–4412.

305. Cutler, C.A., Bouguettaya, M., and Reynolds, J.R. (2002) PEDOT polyelectrolyte based electrochromic films via electrostatic adsorption. *Adv. Mater.*, **14**, 684–688.

306. Reeves, B.D., Grenier, C.R.G., Argun, A.A., Cirpan, A., McCarley, T.D., and Reynolds, J.R. (2004) Spray coatable electrochromic dioxythiophene polymers with high coloration efficiencies. *Macromolecules*, **37**, 7559–7569.

307. Gaupp, C.L., Zong, K., Schottland, P., Thompson, B.C., Thomas, C.A., and Reynolds, J.R. (2000) Poly(3,4-ethylenedioxypyrrole): organic electrochemistry of a highly stable electrochromic polymer. *Macromolecules*, **33**, 1132–1133.

308. Walczak, R.M. and Reynolds, J.R. (2006) Poly(3,4-alkylenedioxypyrroles): the PXDOPs as versatile yet underutilized electroactive and conducting polymers. *Adv. Mater.*, **18**, 1121–1131.

309. Arbizzani, C., Mastragostino, M., Meneghello, L., Morselli, M., and Zanelli, A. (1996) Poly(3-methylthiophenes) for an all polymer electrochromic device. *J. Appl. Electrochem.*, **26**, 121–123.

310. Finden, J., Kunz, T.K., Branda, N.R., and Wolf, M.O. (2008) Reversible and

amplified fluorescence quenching of a photochromic polythiophene. *Adv. Mater.*, **20**, 1998–2002.

311. Sonmez, G., Sonmez, H.B., Shen, C.K.F., Jost, R.W., Rubin, Y., and Wudl, F. (2005) A processable green polymeric electrochromic. *Macromolecules*, **38**, 669–675.

312. Reynolds, J.R., Ruiz, J.P., Child, A.D., Nayak, K., and Marynick, D.S. (1991) Electrically conducting polymers containing alternating substituted phenylenes and bithiophene repeat units. *Macromolecules*, **24**, 678–687.

313. Irvin, J.A. and Reynolds, J.R. (1998) Low-oxidation-potential conducting polymers: alternating substituted para-phenylene and 3,4-ethylenedioxythiophene repeat units. *Polymer*, **39**, 2339–2347.

314. Sotzing, G.A., Reddinger, J.L., Katritzky, A.R., Soloducho, J., Musgrave, R., Reynolds, J.R., and Steel, P.J. (1997) Multiply colored electrochromic carbazole-based polymers. *Chem. Mater.*, **9**, 1578–1587.

315. Thompson, B.C., Kim, Y.-G., McCarley, T.D., and Reynolds, J.R. (2006) Soluble narrow bandgap and blue propylenedioxythiophene-cyanovinylene polymers as multifunctional materials for photovoltaic and electrochromic applications. *J. Am. Chem. Soc.*, **128**, 12714–12725.

11

Ultrathin Films of Conjugated Polymer Networks: A Precursor Polymer Approach Toward Electro-Optical Devices, Sensors, and Nanopatterning

Rigoberto C. Advincula

11.1
Introduction

Electrically conducting or π-conjugated polymers are organic semiconductor materials that continue to be the subject of intense research since the discovery of conductivity in polymer plastics, in 1961 by Hatano and coworkers [1]. Hatano *et al.* reported that a polyacetylene sample can have conductivity of the order of 10^{-5} S/cm. In 1977, a very important result was discovered by Shirakawa, MacDiarmid, Heeger (Nobel Prize in Chemistry 2000), and coworkers where they reported the conductivity increase of free-standing polyacetylene films of up to 12 orders of magnitude upon exposure to halogen vapors [2]. While behaving as insulators or semiconductors in the pristine form, conjugated polymers can reach metallic-like electrical conductivity when doped (in chemical terminology, when oxidized or reduced) [3]. Upon doping, the negatively charged ions are incorporated in order to compensate for the positive charges that are developed on the polymer chain as a result of polymer oxidation. The charges are referred to as *counteranions* or *dopants*. It was realized early that the applicability of polyacetylene is very limited due to its processing difficulties and the rapid decrease in conductivity upon exposure to air. Since then, other conducting polymers have been explored that are more environmentally stable and can be easily polymerized either chemically or electrochemically.

The research direction on conducting and π-conjugated polymers utilized for organic semiconductor and display materials tends to be an interdisciplinary approach [4]. Chemists, physicists, and engineers have investigated their synthesis, properties, and processing for fundamental reasons and their potential applications. Applications include display devices, sensors, touch screen panels, anticorrosion coatings, and so on. These polymers are often intractable and hard to process because of their semirigid to rigid backbone, but they have been made soluble by choosing proper substituents (usually alkyl groups) incorporated in their design and synthesis. The most well-studied polymers include: polyanilines, polypyrroles, polythiophenes, polyfluorenes, and so on (Scheme 11.1). These polymers can undergo physical or chemical changes when exposed to heat, light,

Functional Polymer Films, First Edition. Edited by Wolfgang Knoll and Rigoberto C. Advincula.

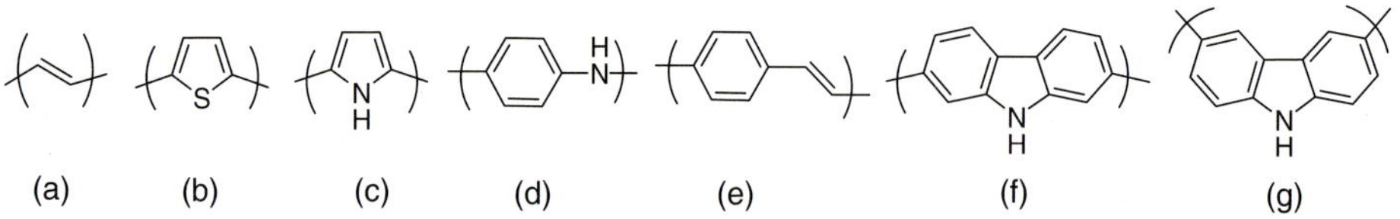

Scheme 11.1 Structures of conjugated polymers: (a) polyacetylene, (b) polythiophene, (c) polypyrrole, (d) polyaniline, (e) polyphenylene vinylene, (f) polycarbazole (2,7 link),
(g) polycarbazole (3,6 link).

electric fields, and various chemical moieties (dopants) giving rise to conductivity, fluorescence, reversible thermochromism, photochromism, electrochromism,
and even ionochromism. There has also been interest in the synthesis of conjugated dendrimers as reported by several groups including the author's group
[5]. These conjugated polymer dendrimers display greater solubility than their
linear derivatives. They can be prepared through a number of metal-mediated
cross-coupling reactions based on conjugated dendron structures that can be
linked to a core group (convergent approach). In addition, conducting polymers
can be blended with nanoparticle materials to show improved photophysical and
photochemical properties related to energy-transfer and electron-transport properties [6]. This author and coworkers have investigated the preparation of hybrid
thiophene dendron-nanoparticle materials [7]. These type of materials have found
applications as conducting polymers, polymer light-emitting diode (PLED) [8], field
effect transistor (FET) devices [9], and even lasers.

It is easy to see that a number of organic synthetic techniques leading to
novel molecular, macromolecular, and microstructural design has been employed.
However, these polymers can also be easily prepared either by chemical or
electrochemical methods. Chemical syntheses include either oxidative coupling of
heteroaromatic ring systems such as pyrrole, thiophene, indole, aniline, and so on,
utilizing chemical oxidants such as $FeCl_3$, I_2, $(NH_4)_2S_2O_8$, and so on [10]. However,
for regioregular polymers, metal-mediated coupling polymerizations of aromatic,
and halogenated aromatic systems are a must [11]. Although polymers with
side-chain alkyl derivatives are more processable, they often have inferior properties
because of the compromise in their doping or charge-transport properties, especially
for electrical-conducting applications.

Electrochemical oxidative polymerization has been utilized to precisely control
the oxidation potential for polymerization and investigate the electronic transport
properties of a conducting polymer. The conjugated polymers are formed *in situ*
as the monomers are polymerized (radical cation mechanism) and deposited on
a conductive working electrode. The mechanism for the anodic electropolymerization of polypyrrole, which can be done in a three-electrode cell, is shown in
Scheme 11.2. This usually results in brittle patchy films due to low degrees of
polymerization and poor mechanical and adhesion properties [12]. Very often, the
primary goal was to focus on electrical-conducting properties and therefore little
attention was given to the electro-optical properties and "fine patterning" of these

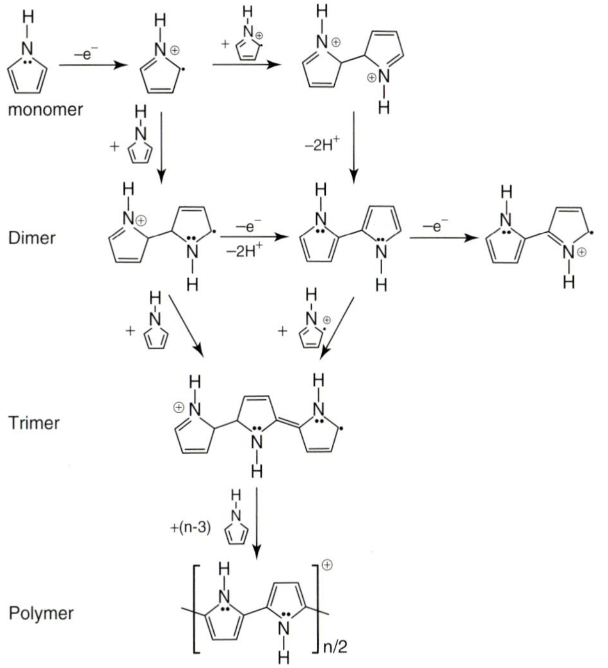

Scheme 11.2 The anodic polymerization (initiation and propagation) mechanism of pyrrole occurring at potentials above +0.6 V.

films on planar surfaces. The doping properties of these films were considered to be more important. Recently, the importance of optical film qualities has been emphasized for electrochromic applications [13]. For nondoped film applications such as light-emitting diodePLED and FET devices, the optical dielectric constants, charge-carrier properties, thickness, morphology (crystallinity or aggregation), and layer order of the films are also important. It will be necessary to emphasize on these parameters especially as it relates to the formation of "sandwich" devices [14].

Patterning poses a unique challenge for electropolymerization since electrodeposition does not afford very good deposition and solubility control. Numerous schemes on patterning have been reported based on photolithographic and soft-lithographic techniques [15]. This includes, photochemical lithography [16], micromolding [17], electropolymerization using modified electrodes [18], deposition using scanning electrochemical microscopy (SECM) [19], and printing techniques [20]. With the recent interest on microfabrication and nanolithography, it is important to translate the processing of these materials in an integrated microscale environment up to the nanometer level. Pixels and complex arrays for display and sensor applications have finite dimensional requirements. The important issues include feature size, resolution, stability, and adhesion, depending on the application. The goal then for most researchers is to apply these synthesis methods and improve their processability for practical applications.

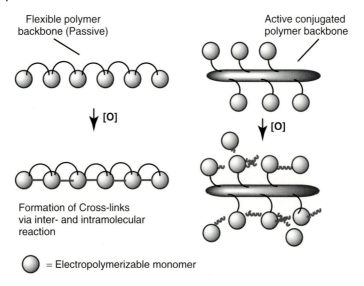

Flexible polymer
backbone (Passive)

Active conjugated
polymer backbone

↓ [O]

↓ [O]

Formation of Cross-links
via inter- and intramolecular
reaction

◯ = Electropolymerizable monomer

Figure 11.1 The precursor polymer approach to conjugated polymers with different polymeric backbones: passive and active.

For the last 10 years, this author and his coworkers have reported in a number of papers this novel method for depositing high optical quality ultrathin coatings of conjugated polymers on conducting substrates [21]. Using a "precursor polymer" route they have employed electrodeposition as a method to prepare ultrathin films mainly on conducting substrates such as Au, indium tin oxide (ITO) glass, and doped Si. They have demonstrated the viability of the "precursor polymer" approach based on single pendant electroactive monomer group using various polymer backbones (Figure 11.1). This involved the tethering of electroactive monomers to a polymer backbone that can be labeled as "passive" or "active" in terms of its contribution to the electro-optical properties of the film itself. Thus, by design the pendant electroactive monomer groups undergo electropolymerization or chemical oxidation resulting in a conjugated polymer network (CPN) film or coating. The network formation is based on both *inter-* and *intra*-molecular cross-linkages occurring between the pendant monomer units, that is, of the same chain or a different chain. The films formed are characterized by high optical quality (transparency), uniform coverage, good adhesion, smoothness in morphology, and controlled ion permeability. Moreover, by controlling the amount of conjugated species and doping, it should be possible to control electrical conductivity. Thus, the process is interesting for depositing *insoluble* cross-linked ultrathin coatings of conjugated polymers for practical electro-optical applications. A variety of combinations should be possible for the design of a precursor polymer backbone and the "electroactive monomer" side group to form the CPN film. It is also possible to *copolymerize* with small-molecule electroactive monomers with varying compositions in order to control the degree of cross-linking and linear polymer formation. In principle, a dendrimer or a hyperbranched macromolecule is also a precursor that enables a very high degree of cross-linking or the ability to "self-cross-link" at the surface, because of the very high end-group functionality [22]. The rest of this chapter reports on these various designs, combinations, methods of coating preparation, analysis, and patterning protocols.

11.2
Materials and Examples

The materials can be classified in terms of the polymeric backbone and/or the functional electropolymerizable group. Often the term refers to either one of the components above but it should be obvious which one is the precursor. There are many other papers and studies by others that are not fully cited in this review but each example will highlight a unique aspect of the approach. A number of these papers are cited in reference [45] of this chapter.

11.2.1
Polyfluorene Precursor Polymers

The precursor polymer approach to form a CPN was first demonstrated by the Advincula group on a series of poly(fluoren-9,9-diyl-*alt*-alkan-diyl) polymers, which was converted to conjugated poly(2,7-fluorene)s both by chemical and electrochemical oxidation into conjugated poly(2,7-fluorene) derivatives (Figure 11.2) [23]. The polymers were synthesized by Li-metal-activated nucleophilic substitution polymerization. These polyfluorene precursor polymers have interesting design properties in which the distance between adjacent fluorene groups can be varied by an alkylene spacer unit. This affected the thermal properties of the polymers where the T_g shifted to lower values with increasing spacer length. PLED devices were prepared by depositing electrochemically on selected areas of a substrate [24]. Efficient devices were prepared based on different hole-transport layers. The effectiveness of this experiment demonstrated the applicability of this technique to "site directing" and patterning. The group also reported a unique polyfluorene structure in which the "preformed" conjugated polyfluorene backbone has pendant carbazole units that were found to be capable of forming robust film structures by "electrografting" on ITO surfaces [25]. The ITO surface was first functionalized with a self-assembled monolayer (SAM) of silane-carbazole derivatives. The CPN films were then prepared either from solution or by first spin coating on the ITO substrates. These films showed improved fluorescence properties based on a decrease of exciplex formation due to chain-to-chain aggregation in polyfluorenes and have bandgap properties, E_g, suggesting good hole-transport properties.

Since then, the technique has been extended to new methods of synthesis, electrodeposition, characterization, patterning, and grafting of conjugated polymers on planar electrode surfaces, as will be described in this chapter. While a number of homopolymers and copolymers have been reported that contain one type of

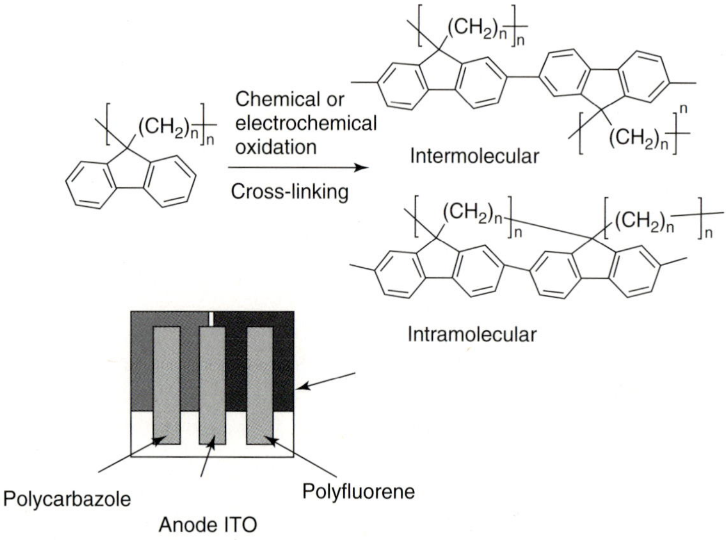

Figure 11.2 Cross-linking properties of precursor polyfluorenes: Synthesis of fluorene-containing polymers with n = 4, 6, 8, 10, 12. Chemical or electrochemical oxidation leads to the formation of cross-linked or a conjugated polymer network. In principle, site-directed patterning is possible by selectively addressable electrodes. (Reproduced from Ref. [24], with permission from The American Chemical Society.)

tethered monomer, the Advincula group have also reported the preparation of binary tethered monomer compositions pendant on a polymer backbone [26].

11.2.2
Polymethylsiloxane Precursor Polythiophenes and Polypyrroles

Using a low glass transition, T_g, flexible polymer backbone, the synthesis, electropolymerization, and ultrathin film formation of polymethylsiloxane modified polythiophene and polypyrrole precursor polymers have been reported [27]. The unique properties of the silicone polymers have been used toward depositing ultrathin films of conjugated polymers with *high optical quality* on surfaces. Like the previous work on polyfluorenes, the mechanism of electropolymerization has differences in parameters with solution polymerization involving small monomeric units. In terms of film formation, the emphasis was on determining the unique parameters affecting film formation, optical quality (refractive index and thickness, dielectric constants), morphology, and so on in order to optimize for specific electrical or luminescence properties. The polydimethylsiloxane backbone is a good polymer backbone because of its low T_g, inertness, low reactivity, and property contrast with other polyolefinic and methacrylate polymers. In essence, the high optical clarity of silicone polymers can be an advantage for coatings formation. A variety of electropolymerizable monomers can be grafted as shown in Figure 11.3. The synthesis of the polymers was done essentially by a Pt-catalyzed hydrosilylation

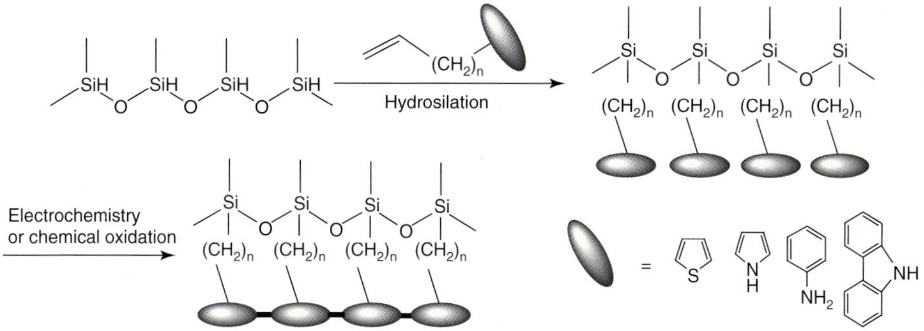

Figure 11.3 A general illustration of the polymethylsiloxane modified precursor polymer route with other electroactive monomers.

route with polyhydroxymethylsiloxane. These reactions can be easily followed by ^{1}H NMR and ^{13}C NMR and elemental analysis. The precursor siloxane polymer was then cross-linked by electropolymerization and monitored by cyclic voltammetry (CV) at various solvents, substrates, solution concentrations, and counter electrolytes. The electropolymerization was performed by sweeping the potential typically from -400 to $1500\,\text{mV}$ vs. Ag/Ag$^+$ reference electrode at various concentrations (Figure 11.4). It can be seen that there is a linear quality in the deposition process and that various concentrations affect the amount of material deposited. After several cycles, the substrate was taken out and rinsed with solvent, dried under dry nitrogen flow, and analyzed.

In this case, the polysiloxane precursor polymer can be cross-linked through the 2,5 linkages of thiophene by electrochemistry or oxidative chemical methods. The films are very uniform and are optically clear. This uniform cross-linked polymer film is essentially *insoluble* to solvents. The films also showed reversible color changes (electrochromism) during the redox process. For example, the precursor polythiophene turns to deep purple upon oxidation, and then changes back to orange during the corresponding reduction process. Different morphologies have been observed by atomic force microscopy (AFM) imaging on films at different cycling stages. Films were also prepared by copolymerizing with different compositions of the thiophene monomer.

For the pyrrole grafted to the polymethylhydrosiloxane to form poly(methyl-(10-pyrrol-3-yl-decyl)siloxane [28], interesting morphologies and optical behavior were also observed. This consisted of circular "nanodomains," 200–300 nm, which is believe to be phase-separated polysiloxane rich regions on the surface. By copolymerizing with pyrrole monomers at different composition ratios, these domains vary in size and distribution as a function of composition. The domain size decreased, while roughness was found to increase with increasing amount of pyrrole comonomer content. It will be interesting to compare the electrical conductivity of the resulting cross-linked polymers as a function of composition and correlation with morphology.

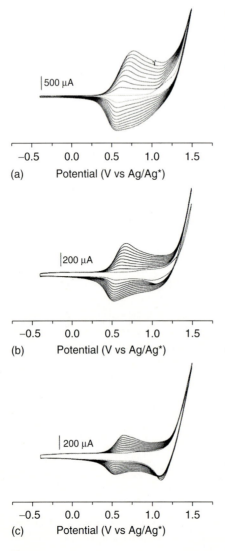

Figure 11.4 Cyclic voltammogram of the precursor polymer at different concentrations: (a) 0.2, (b) 0.1, and (c) 0.05 M with 0.1 M TBAH and methylene chloride. The polymerization was performed by sweeping the potential from −400 to 1500 mV vs the Ag/Ag$^+$ (0.01 M in ACN) reference electrode. (Reproduced from Ref. [25], with permission from The American Chemical Society.)

11.2.3
PMMA, PVK, and a Substituted Polyacetylene Polymer Backbone

The group has also prepared pyrrole grafted on a polymethylmethacrylate (PMMA) polymer backbone by first synthesizing the monomer (N-alkyl pyrrole methacrylate ester) followed by polymerization with AIBN to form poly 6-(N-pyrrolyl)hexyl methacrylate (PHMA)1 (Scheme 11.3) [29]. Likewise, by CV (100 mV/s), electropolymerization and cross-linking was done at different solution conditions on a flat ITO and Au electrodes for these polymers. The dramatic difference in morphology is highlighted by the roughness contrast between the pure precursor (PHMA)1 and that of polypyrrole from pyrrole monomer. The roughness is higher

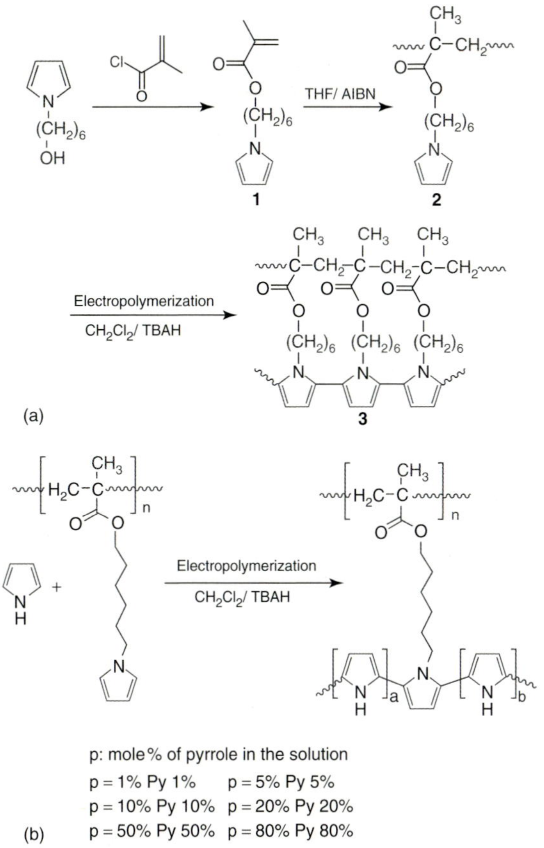

Scheme 11.3 (a) Selective polymerization of the functional groups in 6-(N-pyrrolyl) hexyl methacrylate (PHMA1) to form polymer (PPHMA2) and finally conjugated polymer 3. (b) Various copolymerization ratios of Pyrrole and precursor polymer PPHMA2. (Reproduced from Ref. [29], with permission from The American Chemical Society.)

by a factor of 20 for polypyrrole in the absence of the precursor (Figure 11.5). Copolymerization with other pyrrole monomers at different composition ratios was also done. It should be possible to study the composition dependence on the different morphologies for these polymers and the electrical and optical properties of these films. In principle, the more linear polymer sequences induced during electropolymerization, the higher the conjugation.

In another polymeric system, a comparative analysis of the copolymerization behavior between two tethered electroactive monomers: terthiophene and carbazole moiety with a PMMA backbone was investigated using electrochemical and hyphenated opto-electrochemical methods [30]. Five different precursor polymers were first synthesized and characterized using nuclear magnetic resonance (NMR),

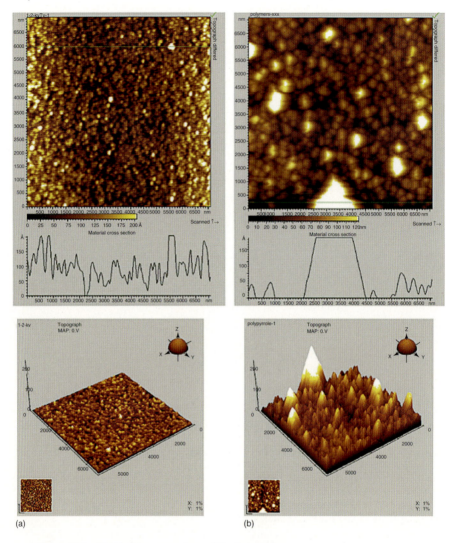

Figure 11.5 Morphology comparison of the two- and three-dimensional AFM images of (a) conjugated precursor (PHMA)1 and (b) the polypyrrole from pyrrole monomer. The roughness is higher by a factor of 20. (Reproduced from Ref. [29], with permission from The American Chemical Society.)

infra-red (IR) spectroscopy, and gel permeation chromatography (GPC). The polymers include homopolymers of individual electroactive groups (P3T, P-CBZ) and different compositions of 25, 50, and 75% (P3TC-25, P3TC50, and P3TC-75) with respect to the two electroactive groups. Since the oxidation potentials of terthiophene and carbazole lie very close to each other, highly cross-linked copolymer films of

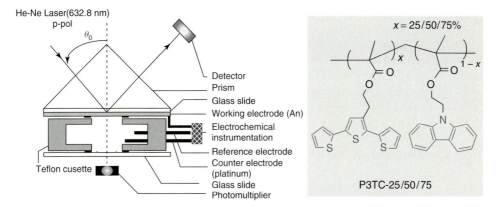

He-Ne Laser(632.8 nm)
p-pol
θ_0

Detector
Prism
Glass slide
Working electrode (An)
Electrochemical
instrumentation
Reference electrode
Counter electrode
(platinum)
Glass slide
Photomultiplier
Teflon cusette

$x = 25/50/75\%$

P3TC-25/50/75

Figure 11.6 The extent of copolymerization, electrochromic properties, and viscoelastic changes was quantitatively investigated using a number of hyphenated electrochemistry techniques for these precursor polymers including electrochemical surface plasmon resonance spectroscopy (EC-SPR). (Reproduced from Ref. [30], with permission from The American Chemical Society.)

varying extent were produced depending on the composition. The copolymerization extent was found to be dependent primarily on the amount of the terthiophene, which in this case provided for a more efficient carbazole polymerization and copolymerization than with just carbazole alone (homopolymer). The extent of copolymerization, electrochromic properties, and viscoelastic changes was quantitatively investigated using a number of hyphenated electrochemistry techniques: spectroelectrochemistry, electrochemical quartz crystal microbalance (EC-QCM) studies, and electrochemical surface plasmon resonance (EC-SPR) spectroscopy were reported (Figure 11.6). Each technique revealed a unique aspect of the electro-copolymerization behavior that was used to define structure–property relationships and the deposition/copolymerization mechanism.

11.2.3.1 Substituted Polyacetylenes with Cross-Linked Carbazole Units

Poly(N-vinylcarbazole) (PVK), which exhibits interesting electrical properties, has been studied extensively because of its possible applications in light-emitting diodes, polymer batteries, and for applications in various electrochromic devices [31, 32]. Polyacetylene and its substituted derivatives are well-known π-conjugated polymers with both high conductivity and interesting electro-optical properties [33]. This author and coworkers described the synthesis and electropolymerization of conjugated substituted polyacetylenes, poly(N-alkoxy-(p-ethynylphenyl)carbazole), or PPA-Cz-Cn with electropolymerizable carbazole side groups to form CPN films. The phenylacetylene monomer was functionalized with a carbazole group separated by an alkylene spacer. Polymerization of the monomer in solution is accomplished using a Rh metathesis catalyst to form a "precursor polymer." The electrochemical behavior and cross-linking of the carbazole side group was then investigated by CV and spectroelectrochemistry. A trend in the redox electrochemical behavior was observed with varying alkyl spacer length between the polyphenylacetylene backbone and carbazole side-group. The resulting film combined the electro-optical properties of the conjugated polyphenylacetylene polymer with polycarbazole units in a cross-linked electropolymerized film as evidenced by the CV and spectroelectrochemical behavior. Thus, this study emphasized the preparation of polymer materials with mixed π-conjugated species arising from the electrochemical cross-linking of a designed precursor polymer [34].

11.2.4
Electrodeposition of Polycarbazole Precursors on Electrode Surfaces and Devices

The electrochemical doping properties and morphology changes of electropolymerized PVK thin films were systematically investigated toward improved hole-injection/transport properties in light-emitting diodePLED devices (Figure 11.7) [35]. The conjugated network polycarbazole thin films (resulting from a poly(N-carbazole) (PCz) network) were prepared by electrodeposition of PVK and/or N-vinylcarbazole comonomer via precursor route and were investigated *in situ* by Electrochemical-Surface Plasmon Resonance Spectroscopy (EC-SPR). Distinct doping–dedoping properties and morphology transitions were observed with different compositions of PVK and Cz. By electrochemical doping of the cross-linked conjugated polycarbazole units, the electrochemical equilibrium potential (E_{eq}), which correlates to Fermi level (E_f) or the work function (ϕ_w) of the film, was adjusted in the vicinity of the anode electrode (Figure 11.8). The conjugated network PCz films were then used as a hole-injection/transport layer in a two-layer device. Remarkable enhancement of PLED properties was observed when optimal electrochemical doping was done with the films. The important insight gained was on understanding charge-transport phenomena between polymer materials and conducting oxide substrates.

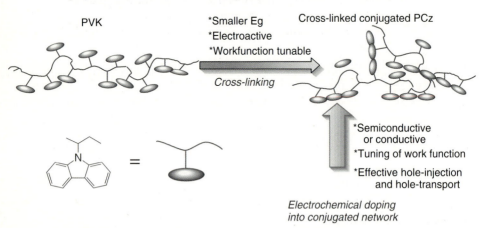

Figure 11.7 A strategy to use the precursor PVK route as both hole-injection layer and hole-transport layer. PVK is cross-linked to form oligocarbazole units.

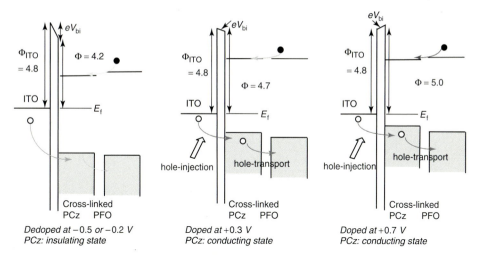

Figure 11.8 Schematic drawing of the proposed mechanism of the hole-injection/transport and the various relative energy levels. (Reproduced from Ref. [35], with permission from The American Chemical Society.)

The Advincula group has also reported the synthesis and electropolymerization of a precursor polymer with a binary molecular composition of thiophene and carbazole electroactive groups to form ultrathin films of CPN on flat ITO substrates [36]. While in the past the precursor polymer approach based on a single-pendant electroactive group has been demonstrated, in this work, interesting electrocopolymerization mechanism and properties of precursor polymers prepared with two different types of pendant electroactive groups (statistical copolymer) and compared behavior to their respective homopolymers. The presence of a smaller amount of carbazole induces the electropolymerization of the higher oxidation potential thiophene units via the reaction of a radical cation and a neutral molecule pathway. These electrochemically generated thin films gave unique optical, electrochemical, and morphological properties as a function of composition. Thus, the mechanism involved a "trigger" effect for the thiophenes based on the initial oxidation of the carbazole units in a radical cation to neutral monomer reaction pathway. Both copolymer and homopolymer electropolymerizations resulted in statistical intramolecular and intermolecular reaction between individual polymer chains resulting in templating and cross-linking. This was verified by CV, spectroelectrochemistry, and XPS, measurements. The morphology of the films correlated well with the deposition behavior.

11.2.5
Dendrimer Precursor Materials

In contrast to that of a linear polymer approach, dendrimers as precursor polymers have also been investigated for their very high end-group functionality. In this

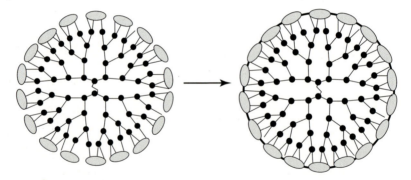

Figure 11.9 Dendrimeric precursor polymer approach with high end-group functionality.

approach, modified dendrimers are synthesized with electroactive functionality at the periphery as shown in Figure 11.9. This author and coworkers have reported on the use of Frechet polybenzyl ether dendrimers to form polycarbazole precursors through a convergent synthesis approach [22].

Conjugated-polymer-based electroactive dendrimers are of current interest for developing efficient electroluminescent display devices and other photonic applications. They have also been used to form nanoparticles objects, in this case, with an electropolymerizable surface [37]. While the formation, structure, and properties of these new electroactive materials pose many complex problems, their interesting structures should require further investigation. The electrogenerated dendrimers open an interesting perspective in the field of modified electrodes for electrocatalysis or electroanalysis, nanoparticles formation, and sensors. A particularly exciting approach concerns the entrapment of guest molecules during the electropolymerization process with possible electrochemical controlled release. This can be done by controlling the redox state of the electroactive moiety situated at the outermost layer or perhaps even for sensing application [38]. Furthermore, the three-dimensional scaffold of the dendrimers would exert some control over the structure and morphology of these materials and therefore would lead to interesting new properties based on controlled electron and ion transport. Conversely, this can lead to some new insight into dendrimer structure–property and electronic processes such as bandgap tunability, interfacial effects, and inter- and intramolecular cross-linking.

11.2.6
Micropatterning Strategies

Micro-contact printing is a soft-lithography approach that has been used for creating micrometer-sized patterns on flat surfaces [39]. By using an elastomeric stamp prepared from a photolithographically prepared master, various size resolutions and patterns have been prepared using amphiphilic inks capable of forming SAMs on a variety of surfaces. The pattern fidelity relies on differences in surface energies and specific amphiphile to substrate interactions (chemical bond).

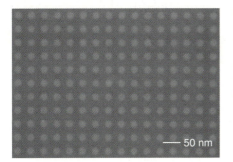

Figure 11.10 Fluorescence microscopy of the nonlithographically patterned film made under potentiostatic conditions showing high pattern fidelity. The structure of the polyfluorene copolymer is also shown. (Reproduced from Ref. [40], with permission from The American Chemical Society.)

— 50 nm

This author and coworkers have used selective-area electropolymerization to generate patterned conjugated and fluorescent polymer thin films [40]. Instead of using electroactive monomers, they have used a precursor polyfluorene polymer, which has fluorescent properties defined by the polymer main chain and electropolymerizability of the pendant carbazole units. The polymer preferably deposited on the alkyl SAM defined regions to form sharp patterns based on selective hydrophobicity and electron transport. The patterns have been characterized by AFM. The fluorescence micrograph confirmed that the polymer main chain was largely unaffected by the redox electrochemistry and it also showed highly regular patterning characteristics in a large area (Figure 11.10). This new combination of site-selective electropolymerization and the precursor approach may provide a new way to make patterned polymer light-emitting diodes, sensors, and printed electronics in the future.

11.2.7
Electronanopatterning of Precursor Polymers

The direct nanopatterning under ambient conditions to form conjugated polycarbazole patterns from ultrathin films of a "precursor polymer" and monomer has been reported by this author and his coworkers [41]. In contrast to previous reports on electrochemical dip-pen nanolithography using monomer ink or electrolyte-saturated films in electrostatic nanolithography [42], these features were directly patterned on spin-casted carbazole monomer and PVK polymer films under room temperature and humidity conditions. The nanopatterned electric field using a biased AFM tip induced polymerization and cross-linking between carbazole units in the films (Figure 11.11). Different parameters including writing speed and bias voltages were optimized to control line width and patterning geometry (Figure 11.12). The conducting property ($I–V$ curves) of these nanopatterns was also investigated using the conducting-atomic force microscopy (C-AFM) setup and the thermal stability of the patterns were evaluated by annealing the polymer/monomer film above the glass transition (T_g) temperature of the polymer. This was one of the first report in which *thermally stable* conducting nanopatterns were drawn directly on monomer or polymer film substrates using an electrochemical nanolithography technique under ambient conditions. Thus, this study

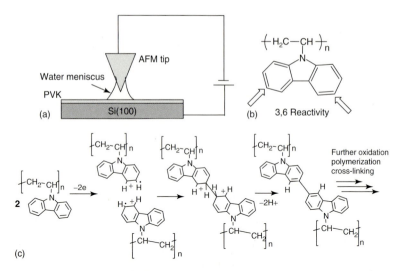

Figure 11.11 (a) Schematic diagram of electrochemical nanolithography, (b) chemical structure and possible polymerization sites of PVK, and (c) mechanism for electropolymerization (cationic) and cross-linking of PVK. (Reproduced from Ref. [41], with permission from The American Chemical Society.)

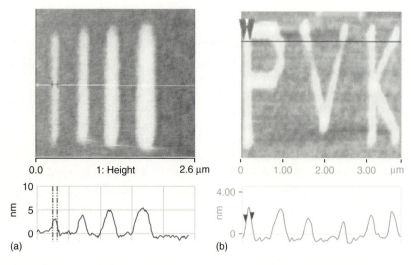

Figure 11.12 (a) Nanolines written on PVK film at constant tip speed of 1 μm/s with bias of −5, −7, −9 V and corresponds to the line width of 83, 128, 162, and 231 nm, respectively. (b) AFM image with height profile of the pattern "PVK" drawn at −7 V at a tip speed 1 μm/s. (Reproduced from Ref. [41], with permission from The American Chemical Society.)

should open up the investigation of other precursor polymer materials, electropolymerizable monomers, and ultrathin film assemblies with parameters of patterning relating, to T_g of the polymer, electropolymerizatibility, writing speed, and applied potential. A main advantage of this method is the use of ambient temperature and humidity conditions for these experiments.

11.2.8
Nanostructured Layers of Precursor Polymers

An important development is the control of ultrathin film thickness and nanostructuring to afford control of the electropolymerizability and electropatterning. In principle, this controlled "embedded monomer" approach can be based on the thickness, polymer microstructure, density of reactive groups, and monomer reactivity as prepared using a layer-by-layer approach. Recently, the Advincula group has reported on nanostructured ultrathin films of linear and dendrimeric cationic sexithiophenes, 6TNL and 6TND, respectively, alternated with anionic polycarbazole precursors, poly(2-(N-carbazolyl) ethyl methacrylate-co-methacrylic acid) or PCEMMA32 [43]. These films were successfully fabricated using the layer-by-layer self-assembly deposition technique. The two electro-optically active oligomers exhibited distinct optical properties and aggregation behavior in solution and films as studied by UV-vis and fluorescence spectroscopy. The stepwise increase of the 6TNL/PCEMMA32 and 6TND/PCEMMA32 layer combination

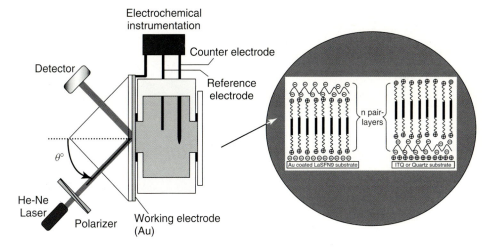

Figure 11.13 Nanostructured ultrathin films of cationic sexithiophenes, alternated with anionic polycarbazole precursor, poly(2-(N-carbazolyl) ethyl methacrylate-co-methacrylic acid) were successfully fabricated using the layer-by-layer self-assembly deposition technique. The stepwise increase of the layers was investigated using cyclic voltammetry (CV) and electrochemical surface plasmon resonance (EC-SPR) spectroscopy. (Reproduced from Ref. [43], with permission from The American Chemical Society.)

was confirmed by UV-vis spectroscopy and *in situ* SPR spectroscopy. The intralayer electrochemical polymerization and cross-linking behavior of the carbazole functionalized PCEMMA32 layers were then investigated using CV and EC-SPR spectroscopy (Figure 11.13). The increase in current with each cycle confirmed intralayer cross-linking followed by the doping–dedoping process within these films. The two types of films differed with respect to dielectric constant and thickness changes before and after electropolymerization, indicating the influence of the oligothiophene layers. This demonstrated for the first time the preparation of highly ordered organic semiconductors alternated with *in situ* electropolymerizable layers in ultrathin films.

To demonstrate electronanopatterning, nanostructured ultrathin films of pendant carbazole-modified polyelectrolytes have been fabricated using the layer-by-layer deposition technique in order to correlate the importance of highly ordered structures and well-defined layer composition in electrochemical cross-linking and nanopatterning [44]. Both UV-vis and SPR spectroscopy studies indicated differences in the film structure and bilayer formation based on the degree of electrostatic interaction and carbazole content. CV and SPR studies confirmed the relationship between film structure and the amount of carbazole units per layer affecting the degree of electrochemical cross-linking. CV, in particular, was used to determine the mechanism of electron transport and electrochemical cross-linking phenomena. Nanopatterning was demonstrated using current-sensing atomic force microscopy (CS-AFM) in which different applied voltages, writing speeds, and composition of polyelectrolytes were studied together with the formation of a complex nanopattern such as the "nanocar" in Figure 11.14.

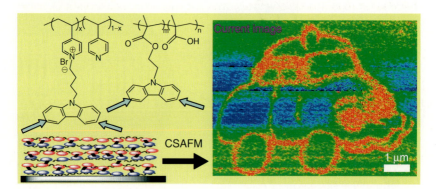

Figure 11.14 Nanostructured ultrathin films of pendant carbazole-modified polyelectrolytes were fabricated using the layer-by-layer deposition technique in order to correlate the importance of highly ordered structures and well-defined layer composition in electrochemical cross-linking and nanopatterning. Nanopatterning was demonstrated using current sensing AFM (CS-AFM) in which different applied voltages, writing speeds, and composition of polyelectrolytes were studied together with the formation of a complex nanopattern. (Reproduced from Ref. [44], with permission from The American Chemical Society.)

11.3
Conclusions

This author and coworkers have developed strategies for electrodepositing and patterning ultrathin films of conjugated polymers on flat electrode surfaces using the precursor polymer approach. This has resulted in the preparation of smooth, high optical quality films, which should be important for applications involving flat electrode surfaces in devices. Copolymerization with monomers, polymer backbone design, and grafting on modified surfaces are key points in this strategy. Novel methods of *in-situ* characterization and nanopatterning techniques have also been reported.

All the systems described in this chapter point to several advantages of the precursor approach: (i) as a methodology, it is superior to simple electrodeposition from electroactive monomers because of the high film quality and linearity of the deposition process, (ii) different precursor polymer chemical structures and compositions (copolymers) allow HOMO-LUMO control, (iii) the possibility of doing site-directed deposition and multilayer film formation allows for various composition combinations and pattern complexities, and (iv) the electrochemical method allows for control of the doping properties. Several disadvantages include decreased conductivity, synthesis of a precursor, and lack of crystallinity.

Nevertheless, the basic scientific and technological issues involved in this unique process needs to be understood: (i) understanding the first principles of precursor polymer design and synthesis will open up a new dimension in polymer molecular and microstructure engineering, (ii) control of competing cross-linking vs. higher degrees of polymerization is needed. This includes intra- vs. interreaction to form more ladder or network structures, (iii) control of aggregation or crystallinity as it affects function, (iv) the mechanism and kinetics of electropolymerization can have different parameters compared to solution polymerization involving diffusion of small monomer units, (v) there is a need to determine the unique parameters affecting thermal stability, optical quality (refractive index and thickness, dielectric constants), morphology, and so on, of films for optimization of specific semiconductor properties, and lastly (vi) electro-optical properties of films and patterns have to be evaluated and optimized for particular applications including patterning.

Finally, a number of work by others have reported similar approaches but not as comprehensively as the author has approached the topic [45].

Acknowledgment

The author acknowledges the contribution of the past and present members of the Advincula Research Group who have worked in this area: Seiji Inaoka, Chuanjun Xia, Prasad Taranekar, Suxian Deng, Jin-Young Park, Yushin Park, Cel Tria, Nicel Estillore, Roderick Pernites, Jane Felipe. The author would like to acknowledge funding from NSF DMR-1006776, CBET-0854979, CHE 10-41300, NHARP 003594-0018-2009 of Texas, and Robert A Welch Foundation, E-1551. Technical support from KSV Instruments (Biolin), Viscotek (Malvern Instruments), Metro

Ohm (Brinkmann Instruments), Agilent Technologies, and Optrel is also greatly acknowledged.

References

1. Hatano, M., Kambara, S., and Okamoto, S. (1961) *J. Polym. Sci.*, **51**, S26.
2. (a) Shirakawa, H., Louis, E.J., MacDiarmid, A.G., Chiang, C.K., and Heeger, A.J. (1977) *J. Chem. Soc. Chem. Commun.*, 578; (b) Chiang, C.K., Fincher, C.R., Park, Y.W., Heeger, A.J., Shirakawa, H., Louis, E.J., Gau, S.C., and MacDiarmid, A.G. (1977) *Phys. Rev. Lett.*, **39**, 1098; (c) Chiang, C.K., Fincher, C.R., Heeger, A.J., Louis, E.J., Gau, S.C., MacDiarmid, A.G., Park, Y.W., and Shirakawa, H. (1978) *J. Am. Chem. Soc.*, **100**, 1013.
3. Salaneck, W.R. and Brehdas, J.L. (1994) *Solid State Commun.*, **92**, 31 (Special Issue on Highlights in Condensed Matter Physics and Materials Science).
4. Salaneck, W., Lundstrom, I., and Ranby, B. (eds) (1993) *Conjugated Polymers and Related Materials*, Oxford University Press, Oxford.
5. (a) Xia, C., Fan, X., Locklin, J., Advincula, R.C., Gies, A., and Nonidez, W. (2004) *J. Am. Chem. Soc.*, **126**, 8735–8743; (b) Xia, C., Fan, X., Locklin, J., and Advincula, R.C. (2002) *Org. Lett.*, **4**, 2067–2070; (c) Moore, J.S. (1997) *Acc. Chem. Res.*, **30**, 402–413; (d) Gong, L., Hu, Q., and Pu, L. (2001) *J. Org. Chem.*, **66**, 2358–2367; (e) Deb, S.K., Maddux, T.M., and Yu, L. (1997) *J. Am. Chem. Soc.*, **119**, 9079–9080; (f) Meier, H. and Lehmann, M. (1998) *Angew. Chem., Int. Ed.*, **37**, 643–645; (g) Lupton, J.M., Samuel, I.D.W., Beavington, R., Burn, P.L., and Bassler, H. (2001) *Adv. Mater.*, **13**, 258–261; (h) Berresheim, A.J., Muller, M., and Mullen, K. (1999) *Chem. Rev.*, **99**, 1747–1786.
6. Mattoussi, H., Radzilowski, L.H., Dabbousi, B.O., Thomas, E.L., Bawendi, M.G., and Rubner, M.F. (1998) *J. Appl. Phys.*, **83**, 7965.
7. Locklin, J., Patton, D., Deng, S., Baba, A., Millan, M., and Advincula, R.C. (2004) *Chem. Mater.*, **16** (24), 5187–5193.
8. Segura, J. (1998) *Acta Polym.*, **49**, 319.
9. Bao, Z., Feng, Y., Dodabalapur, A., Raju, V.R., and Lovinger, A.J. (1997) *Chem. Mater.*, **9**, 1299.
10. (a) Allcock, H.R., Dodge, J.A., Van Dyke, L.S., and Martin, C.R. (1992) *Chem. Mater.*, **4**, 780–788; (b) Trchova, M., Stejskal, J., and Prokes, J. (1999) *Synth. Met.*, **101**, 840–841.
11. Loewe, R.S., Ewbank, P.C., Liu, J., Zhai, L., and McCullough, R.D. (2001) *Macromolecules*, **34**, 4324–4333.
12. (a) Tanaka, S., Sato, M., and Kaeriyama, K. (1988) *Synth. Met.*, **25**, 277–288; (b) Shirota, Y., Noma, N., Shimizu, Y., Kanega, H., Jeon, I.-R., Nawa, K., Kakuta, T., Yasui, H., and Namba, K. (1991) *Synth. Met.*, **41**, 3031–3036.
13. Sapp, S.A., Sotzing, G.A., and Reynolds, J.R. (1998) *Chem. Mater.*, **10**, 2101–2108.
14. Advincula, R., Inaoka, S., Roitman, D., Frank, C., Knoll, W., Baba, A., and Kaneko, F. (2000) Supramolecular assembly strategies using alternate adsorption of polyelectrolytes and dyes: Applications in PLED and LC display devices in *MRS Proceedings 558: Flat Panel Displays and Sensors – Principles, Materials, and Process*, (eds B. Chalamala, R. Friend, T. Jackson, and F. Libsch), Materials Research Society, Warrendale, PA, pp. 415–420.
15. (a) Schanze, K.S., Bergstedt, T.S., and Hauser, B.T. (1996) *Adv. Mater.*, **8**, 531–534; (b) Rozsnyai, L.F. and Wrighton, M.S. (1995) *Langmuir*, **11**, 3913–3920; (c) Zheng, X.-Y., Ding, Y., and Bottomley, L.A. (1995) *J. Electrochem. Soc.*, **142**, L226; (d) Gorman, C.B., Biebuyck, H.A., and Whitesides, G.M. (1995) *Chem. Mater.*, **7**, 526–529.
16. Persson, S.H.M., Dyreklev, P., and Inganaes, O. (1996) *Adv. Mater.*, **8**, 405–408.

17. Beh, W.S., Kim, I.T., Qin, D., Xia, Y., and Whitesides, G.M. (1999) *Adv. Mater.*, **11**, 1038–1041.

18. Huang, Z., Wang, P.-C., MacDiarmid, A.G., Xia, Y., and Whitesides, G. (1997) *Langmuir*, **13**, 6480–6484.

19. Kranz, C., Ludwig, M., Gaub, H.E., and Schuhmann, W. (1995) *Adv. Mater.*, **7**, 38–40.

20. Rogers, J.A., Bao, Z., Baldwin, K., Dodabalapur, A., Crone, B., Raju, V.R., Kuck, V., Katz, H., Amundon, K., Ewing, J., and Drzaic, P. (2001) *Proc. Natl. Acad. Sci. U.S.A.*, **98**, 4835–4840.

21. Inaoka, S., Roitman, D., and Advincula, R. (2001) Molecularly ordered π-conjugated polymer networks on conducting surfaces using the precursor polymer approach in *Forefront of Lithographic Materials Research*, (eds H. Ito, M. Khojasteh, and W. Li), Kluwer Academic Publishers, New York, NY, pp. 239–245.

22. Taranekar, P., Fulghum, T., Patton, D., Ponnapati, R., Clyde, G., and Advincula, R. (2007) *J. Am. Chem. Soc.*, **129**, 12537–12548.

23. Inaoka, S. and Advincula, R. (2002) *Macromolecules*, **35**, 2426–2428.

24. Inaoka, S., Roitman, D.B., and Advincula, R.C. (2005) *Chem. Mater*, **17**, 6781–6789.

25. Xia, C. and Advincula, R.C. (2001) *Chem. Mater.*, **13**, 1682–1691.

26. Taranekar, P., Baba, A., Fulghum, T.M., and Advincula, R. (2005) *Macromolecules*, **38**, 3679–3687.

27. Xia, C., Fan, X., Park, M.-K., and Advincula, R.C. (2001) *Langmuir*, **17**, 7893–7898.

28. Taranekar, P., Fan, X., and Advincula, R. (2002) *Langmuir*, **18**, 7943–7952.

29. Deng, S. and Advincula, R.C. (2002) *Chem. Mater.*, **14**, 4073–4080.

30. Taranekar, P., Fulghum, T., Baba, A., Patton, D., and Advincula, R. (2007) *Langmuir*, **23**, 908–917.

31. Burrows, P.E., Forrest, S.R., Sibley, S.P., and Thompson, M.E. (1996) *Appl. Phys. Lett.*, **69**, 2959.

32. Tamada, M., Omichi, H., and Okui, N. (1995) *Thin Solid Films*, **268**, 18.

33. Masuda, T. and Higashimura, T. (1984) *Acc. Chem. Res.*, **17**, 51–56.

34. Fulghum, T., Karim, S., Baba, A., Taranekar, P., Nakai, T., Masuda, T., and Advincula, R. (2006) *Macromolecules*, **39**, 1467–1473.

35. Baba, A., Onishi, K., Knoll, W., and Advincula, R.C. (2004) *J. Phys. Chem. B.*, **108**, 18949–18955.

36. Taranekar, P., Baba, A., Fulghum, T.M., and Advincula, R. (2005) *Macromolecules*, **38**, 3679–3687.

37. Taranekar, P., Park, J., Patton, D., Fulghum, T., Ramon, G., Bittner, E., and Advincula, R. (2006) *Adv. Mater.*, **18**, 2461–2465.

38. Taranekar, P., Baba, A., Park, J., Fulghum, T., and Advincula, R. (2006) *Adv. Funct. Mater.*, **16**, 2000–2007.

39. Kumar, A., Biebuyck, H.A., and Whitesides, G.M. (1994) *Langmuir*, **10**, 1498.

40. Xia, C., Advincula, R.C., Baba, A., and Knoll, W. (2004) *Chem. Mater.*, **16**, 2852–2856.

41. (a) Jegadesan, S., Advincula, R.C., and Valiyaveettil, S. (2005) *Adv. Mater.*, **17**, 1282–1285; (b) Jegadesan, S., Sindhu, S., Advincula, R.C., and Valiyaveettil, S. (2006) *Langmuir*, **22**, 780–786.

42. (a) Lim, J.H. and Mirkin, C.A. (2002) *Adv. Mater.*, **14**, 1474–1477; (b) Piner, R.D. and Mirkin, C.A. (1997) *Langmuir*, **13**, 6864–6868.

43. Sriwichai, S., Baba, A., Deng, S., Huang, C., Phanichphant, S., and Advincula, R. (2008) *Langmuir*, **24**, 9017–9023.

44. Huang, C., Jiang, G., and Advincula, R. (2008) *Macromolecules*, **41**, 4661–4670.

45. (a) Jang, S-Y., Sotzing, G., and Marquez, M. (2002) *Macromolecules*, **35**, 7293–7300; (b) Sebastian, R., Caminade, A., Marjoral, J., Levillain, E., Huchet, L., and Roncali, J. (2000) *Chem. Comm.*, 507. (c) Alvarez, J., Sun, L., and Crooks, R. (2002) *Chem. Mater.*, **14**, 3995–4001.

Part II
Patterning

Functional Polymer Films, First Edition. Edited by Wolfgang Knoll and Rigoberto C. Advincula.
© 2011 Wiley-VCH Verlag GmbH & Co. KGaA. Published 2011 by Wiley-VCH Verlag GmbH & Co. KGaA.

12
Nanopatterning and Functionality of Block-Copolymer Thin Films

Soojin Park and Thomas P. Russell

12.1
Introduction

A challenge in nanotechnology is controlling self-assembling systems to enable a specific functionality. Self-organizing block-copolymers (BCPs) offer a rich variety of periodic nanoscale patterns [1, 2]. BCPs consist of two or more chemically different polymer chains joined covalently at their ends. Due to the positive (nonfavorable) enthalpy and small entropy of mixing, dissimilar blocks tend to microphase separate into well-ordered arrays of domains, classically termed *microdomains*. The size scale of these microdomains, due to the connectivity of the blocks, is limited to molecular dimensions and, hence, are tens of nanometers in size or less. The product of the Flory–Huggins segment interaction parameter, χ, and the total number of segments in the BCP chain, N, dictate whether the BCP is phase mixed or microphase separated. At temperatures below an order-to-disorder transition (ODT) temperature, T_{ODT}, where segmental interactions are nonfavorable, since χ varies inversely with T, BCPs microphase separate into arrays of spherical, cylindrical, gyroid, or lamellar microdomains, depending on the volume fractions of the blocks, f and χN. Above the T_{ODT}, since χ is reduced, BCPs phase mix and are disordered [1, 2].

The self-assembly of BCPs into well-defined morphologies has opened numerous applications ranging from drug delivery to structural materials. In contrast to the bulk, the morphology of amorphous BCP thin films can be strongly influenced by surface and interfacial interactions and the commensurability between the film thickness, h, and the period of the microdomain morphology, L_0. With decreasing film thickness these parameters become increasingly important in defining the morphology. By controlling the orientation and lateral ordering of the BCP microdomains in thin films, unique opportunities in the use of BCPs in materials science (adhesive properties, lubrication, membranes, and coatings), lithography and microfabrication (addressable memory, magnetic storage, insulting foams), and device technologies (light-emitting diodes, photodiodes, and transistors) are beginning to emerge [3–8]. In this chapter, two aspects of BCP thin films will be addressed. First, the nanoscale patterning of BCP thin films will be addressed with

Functional Polymer Films, First Edition. Edited by Wolfgang Knoll and Rigoberto C. Advincula.
© 2011 Wiley-VCH Verlag GmbH & Co. KGaA. Published 2011 by Wiley-VCH Verlag GmbH & Co. KGaA.

discussion centering on fabrication of long-range ordered nanostructures of BCP thin films by applying various external fields. This is followed by a review of the functionality of nanosized patterns formed by BCPs, like scaffolds or templates for the fabrication of nanostructured materials and the generation of BCP arrays for potential addressable media.

12.2
Nanoscale Patterning with BCPs

BCPs have gained increasing attention as templates and scaffolds for the fabrication of high-density arrays of nanoscopic elements due to the size and tunability of the microdomains, the ease of processing on flat and patterned surfaces without introducing disruptive technologies, and the ability to manipulate their functionality [9–14]. For BCPs having cylindrical microdomains, it is necessary to control the orientation and lateral ordering of the microdomains to optimize the contrast in transfer applications and the density of elements. In addition, it is highly desirable that the process is independent of the underlying substrate. Approaches to control the orientation of the microdomains of BCPs include the use of solvent fields [15–18], electric fields [19, 20], chemically patterned substrates [21, 22], graphoepitaxy [23–25], epitaxial crystallization [26], controlled interfacial interactions [27, 28], thermal gradients [29], zone casting [30], and shear [31]. This section will address several methods used to control the orientation of the microdomains of BCP thin films.

12.2.1
Solvent Fields

The preparation of BCP thin films under various solvent evaporation conditions is an effective way to manipulate the orientation and lateral ordering of BCP microdomains in thin films. The solvent evaporation rate is one of the key factors that control these kinetically trapped microstructures. For instance, inverted phases consisting of spheres or cylinders of the majority fraction block in a polystyrene-*block*-polybutadiene-*block*-polystyrene (PS-*b*-PB-*b*-PS) copolymer, which are not predicted thermodynamically, were observed by the control of the solvent evaporation rate. Kim and Libera [15] demonstrated that cylindrical microdomains oriented normal to the surface could be obtained in a PS-*b*-PB-*b*-PS triBCP thin film for a thickness of ~100 nm when a certain evaporation rate was used. Various metastable morphologies can be achieved with different evaporation rates. With the appropriate preparation condition, the solvent concentration gradient became maximized along the direction perpendicular to the film surface, resulting in a cylindrical microdomain oriented normal to the film surface.

The same effect was also observed in a polystyrene-*block*-poly(ethylene oxide) (PS-*b*-PEO) BCP thin films and was attributed to a copolymer/solvent concentration gradient along the direction normal to the film surface giving rise to an

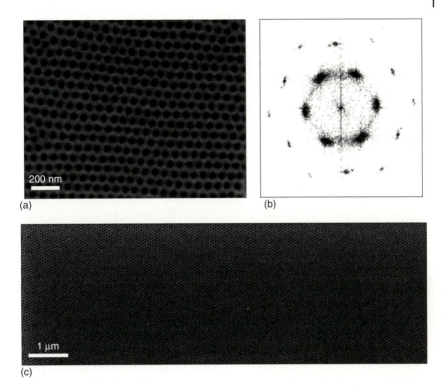

Figure 12.1 Highly ordered microstructure of a BCP. (a,c) Scanning electron microscopic (SEM) images of a 37-nm thick polystyrene-*block*-poly(2-vinylpyridine)-*block*-poly(t-butyl methacrylate) (PS-*b*-P2VP-*b*-PtBMA) film after solvent-vapor treatment and subsequent drying. (b) Fourier transform of an SEM image. (Reproduced with permission from *Nature Materials* [17].)

ordering front that propagated into the film during solvent evaporation. This orientation was independent of the substrate. However, the lateral ordering of the cylindrical microdomains was poor. Russell and coworkers [16, 18] have shown that evaporation-induced flow in solvent-cast BCP films can produce arrays of nanoscopic cylinders oriented normal to the surface with a high degree of ordering. Krausch and coworkers [17] showed that solvent annealing could markedly enhance the ordering of BCP microdomains in thin films, as shown in Figure 12.1.

Russell and coworkers [18] showed that, by controlling the rate of solvent evaporation and solvent annealing, nearly defect-free arrays of cylindrical microdomains normal to the surface in PS-*b*-PEO BCP thin films are produced that span the entire films thickness and have a high degree of long-range lateral order as shown in Figure 12.2. Moreover, the use of a cosolvent enables control over the characteristic length scales in the BCP structures even further. Recent results showed that cylindrical microdomains oriented normal to the film surface can be directly obtained by spin coating polystyrene-*block*-poly(4-vinylpyridine) (PS-*b*-P4VP) diBCPs from mixed solvents of toluene and tetrahydrofuran (THF) (Figure 12.3a) and a highly ordered cylindrical microdomain morphology over a large area (Figure 12.3c)

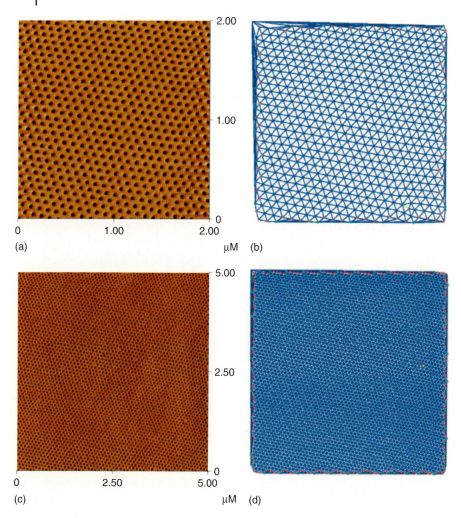

Figure 12.2 Scanning force microscopic (SFM) phase image of PS-*b*-PEO films (260 nm in thickness), spin cast from benzene and then annealed in a benzene/water atmosphere for 48 h. (a,b) is for a 2 μm × 2 μm area, whereas (c,d) is for a 5 μm × 5 μm area. The SFM images and the corresponding triangulation maps are shown. Only one defect is seen in these images. (Reproduced with permission from *Advanced Materials* [18].)

is formed after annealing the films in solvent mixture [32]. This process is independent of substrates, but strongly dependent on the quality of the solvent and solvent evaporation rate.

Solvent evaporation is a strong, highly directional field. Strong repulsion between BCP blocks combined with the directionality of solvent evaporation, where ordering is initiated at the surface of the film and propagates through the entire film, leading to a high degree of long-range lateral order with few defects. The power of this

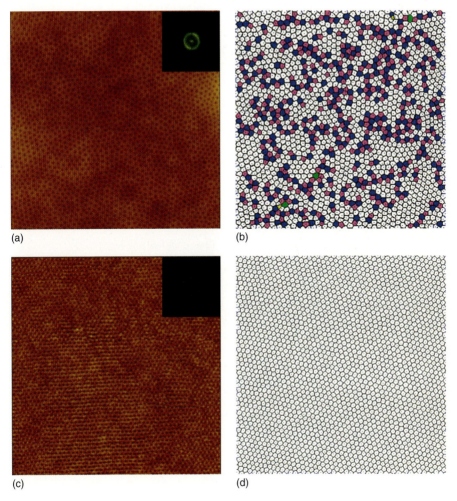

(a)

(b)

(c)

(d)

Figure 12.3 A PS-*b*-P4VP thin films obtained by spin coating and solvent annealing. (a) SFM height image of an as-spun film. (c) SFM height image of a highly ordered and oriented array of cylindrical microdomains after solvent annealing. Insets are the corresponding Fourier transform spectra. (b,d) The corresponding Voronoi diagrams. (Reproduced with permission from *ACS Nano* [32].)

technique can be better appreciated if a BCP film highly swollen with a good solvent for both blocks is considered. When the concentration of the solvent in the film is high enough, microphase separation is lost, and the solubilized copolymer segments mix throughout the film. As the solvent evaporates, a gradient in the solvent concentration into the film is obtained, with the concentration of the solvent at the air surface being lowest. As the solvent evaporates, the concentration of the solvent at the surface decreases to a point where the BCP microphase separates. If the blocks are highly immiscible, as is the case of PS-*b*-PEO and PS-*b*-P4VP, microphase separation occurs at a relatively high solvent concentration at the surface first and there is substantial mobility of the segments. The presence of the solvent mediates the surface energies of the blocks, and as such, both blocks are located at the surface; a lateral microphase separation occurs only at the surface, and the remainder of the film is disordered. Defects in the lateral ordering of the microdomains, which are energetically costly, are rapidly removed, since the chains are mobile. This produces an array of microdomains at the surface with long-range lateral order. Further solvent evaporation causes an ordering front to move into the film, the BCP microphase separates, templated by the existing microphase-separated morphology at the surface. Finally, the front propagates through the entire film, producing highly oriented and highly ordered nanostructures.

12.2.2
External Fields

12.2.2.1 Electric Field

Amundson *et al.* [33–35] showed that there is an electrostatic free energy penalty for interfaces not aligned with the field lines of an applied external electric field and, as a result, lamellae and cylinders tend to orient parallel to the field. The transition in orientation is first order [36] and occurs when the electric field can overcome the preferential orientation of the microdomains parallel to the film surface induced by surface interactions [20, 36]. Following bulk studies by Amundson and coworkers [35, 37], Morkved *et al.* [38] examined the effect of an applied electric field on BCP thin film morphology. Microfabricated electrodes were used to apply an electric field in the plane of the film during annealing. The resulting morphology exhibited inplane cylinders oriented normal to the electrode faces, that is, aligned along the electric field lines. Hence, as demonstrated by Thurn-Albrecht *et al.* [19, 20], application of the field perpendicular to the film plane induces an orientation of the cylindrical microdomains normal to the film surface [20]. They analyzed this situation theoretically and proposed that the origin of the alignment was the differential polarizability or dielectric constant between the microdomains. These arguments were supported by the experimental observation of the critical field strength required to overcome interfacial interactions. One advantage that the electric-field technique offers over microfabrication is that the microdomains in thick films (>30 μm) can be oriented [19]. Consequently, extremely large aspect ratio nanoscopic domains with a high degree of orientation can be obtained. Recently, Thurn-Albrecht *et al.* [6] have selectively etched films oriented by electric field and deposited ferromagnetic cobalt material into the resulting cylindrical holes-a step toward BCP-based fabrication of Terabit density memory storage media (Figure 12.4).

As another examples, Russell and coworkers [19, 39–44] studied this system in further detail and demonstrated that vertically ordered cylindrical poly(methyl methacrylate) (PMMA) microdomains could be achieved when the electric field was applied across the film (Figure 12.5). In addition, under a sufficiently high electric field, spherical microdomains will be deformed into ellipsoids. For a thin

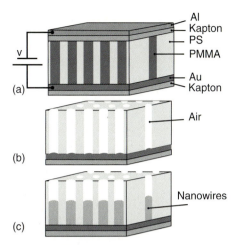

Figure 12.4 A schematic representation of high-density nanowire fabrication in a polymer matrix. (a) An asymmetric diBCP annealed at well above the glass-transition temperature of the copolymer between two electrodes under an applied electric field, forming a hexagonal array of cylinders oriented normal to the film surface. (b) After removal of the minor component, a nanoporous film is formed. (c) By electrodeposition, nanowires can be grown in the porous template, forming an array of nanowires in a polymer matrix. (Reproduced with permission from *Science* [6].)

BCP film with multiple layers of spheres, the ellipsoids can be sufficiently stretched that eventually they interconnect to form cylinders and penetrate through the film [42]. Conservation of volume requires that the diameters of the cylinders should be smaller than those of spheres. Such a sphere-to-cylinder transition may offer a simple route to generate cylinders with a high aspect ratio from the spherical microdomains [42].

12.2.2.2 Shear Field

BCPs in the bulk are routinely aligned by shear fields. Cylinder-forming BCPs that orient with their cylinder axes parallel to the flow direction have been organized into macroscopic single crystalline arrays [45]. Oscillatory shear can easily be applied between parallel plates with a well-defined separation distance, shear rate, and strain amplitude [46]. Roll casting, as discussed by Thomas and coworkers [47, 48], was one of the first techniques developed to translate shear induced orientation into thin films. Here, a BCP solution is placed between two counter-rotating rolls and the complex shear field induced by the rolls orients the BCP as the solvent evaporates, eliminating the need for high-temperature annealing. This technique has been used to align lamellae [48–50], cylinders [49, 50], spheres [51], and bicontinuous nanodomains [52], but is only applicable to relatively thick films. Flow of a BCP solution down an inclined substrate can also orient thick films containing about 10 layers of cylinders [16]. A stamp made from an elastomeric mold pressed onto a BCP solution on a solid substrate creates a flow field that has the potential for aligning the BCP [53]. Recently, Angelescu *et al.* [54] have

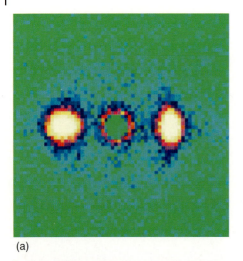

(a)

500 nm

(b)

Figure 12.5 (a) Small-angle neutron scattering (SANS) pattern obtained from a PS-*b*-PMMA ($f_{PMMA} = 30\%$) film (800 nm) annealed in an electric field of 25 V/mm. The two equatorial reflections indicate a strong orientation of the microphase structure normal to the substrate surface. (b) Cross-sectional transmission electron micrograph obtained from a PS-*b*-PMMA film (800 nm) annealed in an electric field of 25 V/mm. The BCP film is lying on top of a dark Au-film, which was used as the lower electrode. The upper electrode has been removed. Cylinders oriented normal to the substrate go all the way through the sample. (Reproduced with permission from *Advanced Materials* [19].)

demonstrated that molten BCP having a single layer of cylindrical microdomains could be aligned by shear. Here, a polydimethylsiloxane (PDMS) block is placed on a heated film. When the block is moved across the film, a lateral force is applied that shears the BCP film and orients the microdomains in the direction of shear.

12.2.2.3 Sequential, Orthogonal Fields

Three-dimensional control of ordering frequently requires that more than one direction of organization be controlled. Xu *et al.* [40] successfully achieved a three-dimensional control over the orientation of lamellar microdomains in PS-*b*-PMMA thin films by the combination of an orthogonal elongational flow field with an applied electric field. In their experiment, roll pressing was run at temperature below the ODT but above the glass-transition temperatures of both blocks where oriented nuclei of the microdomains formed. Subsequent annealing under an applied electric field across the film surface produced long-range ordered and highly oriented lamellar microdomains. As seen in Figure 12.6, there was

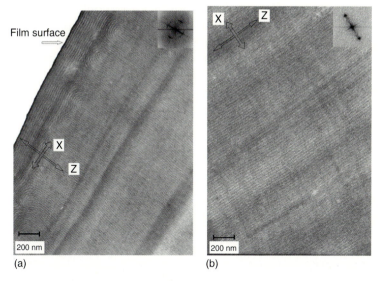

Figure 12.6 Cross-sectional transmission electron microscopic (TEM) images of (a) the copolymer/substrate interface and (b) the interior of the film. Insets are the Fourier transform spectra of the TEM micrograph. (Reprinted with permission from *Macromolecules* [40].)

near-perfect alignment parallel to the flow direction and normal to the surface. The lamellar microdomains near the substrate, however, still exhibited parallel orientations due to surface interactions. Alternatively, one could consider using a field-alignment technique, such as an electric field or a surface modification to force cylinders or lamellae to orient perpendicular to a substrate in combination with topographic ordering techniques, such as graphoepitaxy, to control order within the plane of the film. Russell and coworkers [18] combined solvent annealing and graphoepitaxy to allow the creation of long-range ordered cylindrical microdomains perpendicular to the substrate and in hexagonal arrays.

12.2.2.4 **Other Fields**
Magnetic fields are ideal for BCP structure control in that they may be applied without direct physical contact to the sample and even high magnetic field strengths are not damaging to polymers. The directors of small-molecule and polymeric liquid-crystalline nematics and cholesterics are routinely aligned in magnetic fields, but alignment of smectics generally requires prohibitively large fields [55]. Magnetic-field-induced alignment of classical coil–coil BCPs has not been demonstrated, presumably due to a lack of anisotropic magnetic susceptibility of the nanostructures. If a BCP contains one crystallizable block, a strong magnetic field can orient an entire crystal during crystallization resulting in a lamellar stack with the layers parallel to the magnetic field [56]. Segalman and coworkers [57] demonstrated a new route for creating 10-nm length-scale patterns in controllable directions by incorporating a diamagnetic BCP. Magnetic fields couple directly

to the liquid-crystalline director of the rod block of a rod-coil BCP allowing for control over the liquid crystalline length scale. This approach suggests a new route for the patterning that allows for arbitrary sample geometry and does not require special sample fabrication or surface damage [57]. As another approach, temperature gradients, successfully used by Hashimoto and coworkers [29] to make millimeter-thick single-crystal lamellar BCP domains, might also prove promising. It would be interesting to see the effect of inplane temperature gradients on the long-range order of laterally organized BCP thin-film morphologies such as spheres and hexagonal columns.

12.2.3
Graphoepitaxy

Graphoepitaxy is a process in which all topographic surface pattern is employed to control orientation of crystal growth in thin films [9, 58]. For BCPs, graphoepitaxy is achieved through patterning a hard substrate via standard photolithography, and then transferring the pattern by either chemical or physical etching. These templates can be used to control the ordering and orientation of the microdomains. Unlike atoms and spherical colloids that have fixed sizes, shapes, and, hence, spacings, the microdomains of BCPs can vary their sizes, spacing, and even their shapes or morphologies to accommodate the distance between template walls and other interface conditions imposed by the templates. Graphoepitaxy uses these topographic features on the surfaces, simple edges, or well-defined troughs to bias the lattice orientation of BCP microdomains, providing a way for controlling multilevel ordering where a *"bottom-up"* method such as self-assembly of BCPs is combined with a *"top-down"* lithographic method. The primary purpose of graphoepitaxy is to enhance the resolution of the conventional lithographic process by subdividing the patterned features and to improve the perfection of ordering of the dense periodic arrays of nanostructures naturally formed by BCPs. The ability to extend or compress polymer chains and to adjust the shape of the intermaterial dividing surface between the blocks gives distinct behavior in the guided self-assembly of BCPs in that they can be both more compliant and variable with respect to the template size than, for example, the assembly of "hard-sphere" colloids.

12.2.3.1 Creation of Templates
The choice of methods for creating templates depends on the dimensionality, length scale, and pattern requirements. Many direct-write methods are particularly suitable for fundamental guided block-copolymer research because these methods provide patterns in the deep submicrometer region with arbitrary geometries [59]. Imprint and soft lithography are very useful to mold polymers and to make chemical patterns.

The most widely used direct-write method for patterning structures with dimensions of 10–100 nm has been electron-beam lithography (EBL) [60]. PMMA is a widely used and reliable high-resolution positive resist [61]. Recently, hydrogen silsesquioxane (HSQ) has proven to be a good high-resolution negative electron-beam resist [62]. Following electron-beam exposure, HSQ forms silicon oxide that can be used as a topographical template for BCPs without further pattern transfer. EBL is very convenient for making templates of arbitrary geometry allowing researchers to precisely position individual microdomains and manipulate defects. A simple setup, such as a Lloyd's-mirror configuration with a HeCd laser ($k = 325$ nm), can create gratings with periodicities from 180 nm to 1.5 μm by simply varying the incident angle [63]. An extreme UV (EUV) source ($k = 13.5$ nm) produces gratings with a period close to 5 nm [64]. Besides grating patterns made by a single exposure, 2D grid patterns, and 3D structures can be achieved with multiple exposures or with multiple beams [65, 66]. These direct-write lithography methods create templates with high-precision patterns that is particularly important in the spatial registration and modulation of microdomains of BCPs.

Contact-molding techniques are able to make templates with sub-100 nm to micrometer features. Soft lithography and imprint lithography are the most prominent molding techniques. Soft lithography (also called *microcontact printing*) uses an elastomer mold, typically made from PDMS, to stamp materials or directly transfer a pattern [67, 68]. Imprint lithography can mold thermoplastic polymers (nanoimprint lithography) [69] or UV-curable monomers ("step-and-flash" imprint lithography) using rigid topographical masters [70]. The molds for soft lithography and imprint lithography are typically made from a silicon master written by EBL or interference lithography. Contact-molding techniques can be used not only to replicate the topographical pattern of a master made by direct-write lithography but can also be adapted to generate chemically patterned templates as well as to pattern BCPs directly.

12.2.3.2 Topographically Patterned Substrates

Kramer and coworkers [23, 71] were the first to demonstrate grapheoepitaxy with BCPs with the formation of long-range-ordered, in-plane spherical microdomain arrays on a grating substrate. A monolayer of P2VP spherical domains of a polystyrene-*block*-poly(2-vinylpyridine) (PS-*b*-P2VP) film was placed on the photolithographically patterned substrate, which was then annealed to generate ordered structures propagating several micrometers from the sidewalls of the grooves and the edges of the mesas (Figure 12.7). The effects of incommensurability are negligible in this case, since L_S (characteristic feature size of templates) $\gg L_0$ (bulk domain spacing), and the spacing of the spherical microdomains in the groove and on the mesa is indistinguishable from that on a smooth substrate. In addition to 1D confinement, well-ordered PS-*b*-P2VP spherical arrays were also created by Kramer and coworkers [72] in a 2D hexagonal-shaped topographical well whose diagonal width is in the range of a few micrometers. The spatial-confinement effect on the ordering behavior of 2D spherical BCP microdomains can be characterized by the translational correlation function and the orientational order correlation function [71, 73, 74]. When the microdomains in a thin film of a BCP have liquid-like packing, imposing a topographical constraint induces ordering from the confining

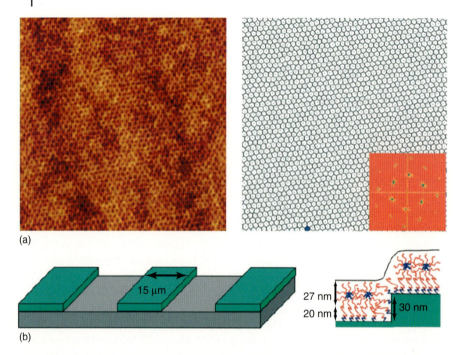

(a)

(b)

Figure 12.7 (a) SFM image (left) of a single grain of 2D periodic spheres in a PS-*b*-P2VP film on top of a mesa. Voronoi construction of the sphere array (right), demonstrating the absence of defects over the approximately 40 × 40 domain array. Inset: The sharp high-order diffraction peaks associated with hexagonal symmetry. (b) Perspective schematic of template and cross-sectional schematic of a P2VP (blue) brush on the SiO₂ substrate surface (green) and a monolayer of P2VP spheres (blue) encased in a PS matrix (red). (Reproduced with permission from *Macromolecules* [71].)

edges. The translational correlation length decreases algebraically with distance from an edge. The ordering starts from the template edge and propagates normal to the edge, similar to the surface-induced layer ordering of thin films on flat, unpatterned substrates. Both thermodynamics and kinetics play important roles in the in-plane ordering. Experiments by Kramer and coworkers [71] have demonstrated that the segregation energy associated with the minority block, χN_{min}, where χ is the Flory–Huggins parameter and N_{min} is the degree of polymerization of the minority block, can be used to describe the ordering of the system. A large value of χN_{min} gives a "polycrystalline" defect-laden structure because of very slow diffusion. Systems with an intermediate value of χN_{min} can form a near "single-crystalline" structure with few defects, whereas for a low χN_{min} value, fluctuations are more stable, so that dislocation pairs and disclination clusters are generated and the templated BCP forms hexatic and liquid phases. Therefore, given a particular molecular structure, there is a processing window to optimize both the diffusivity and microphase separation of the BCP thin film to reach the near single-crystalline state in a wide template.

In addition to the wide templates, results on the ordering of BCPs confined between narrow templates whose width is a few times L_0 have been reported. Lambooy *et al.* [75] and Koneripalli *et al.* [76] used neutron reflectivity to investigate how the domain periodicity varies when BCP with lamellar microdomains are confined by two rigid plates where one block of the BCP preferentially wets the interfaces. The separation distance between the plates was varied from several times to many times L_0. Lambooy *et al.* [75] found that the lamellar periodicity deviated from L_0 to satisfy the boundary conditions imposed by the hard confining surfaces. The periodicities derived from neutron-reflectivity experiments demonstrated that copolymer domains will contract or expand to make an integer number of lamellae fit into the confinement width L_S. Kellogg *et al.* [77] and Huang *et al.* [78] examined the orientation of symmetric PS-*b*-PMMA films sandwiched between two rigid substrates coated with random polystyrene-random-poly(methyl methacrylate) (PS-*r*-PMMA) copolymers to modify the interactions of the blocks with the walls of the confining surface. With no preferential wetting to the substrates, lamellae of PS-*b*-PMMA orient normal to the substrates when L_S is incommensurate with L_0 in order to remove the change in periodicity imposed by the confinement [77, 78]. Cheng *et al.* [25] demonstrated ordered microdomains of sphere-forming polystyrene-*b*-poly(ferrocenyl dimethylsilane) (PS-*b*-PFS) in templates with various widths (Figure 12.8). The number of rows n in the arrays is determined by the confinement width, L_S. Similar to lamellae confined between two rigid plates, defect-free arrays with n rows of domains form for a confinement width L_S when $(n - 0.5) L_0 < L_S < (n + 0.5) L_0$, as shown in Figure 12.8b. When $L_S \sim (n \pm 0.5) L_0$, both n-row and $(n + 1)$-row arrays are found at the same value of L_S, and this coexistence region increases with increasing L_S [25]. Sundrani *et al.* [79] and Xiao *et al.* [80] also observed similar behavior for cylinders of PS-*b*-PEP and PS-*b*-PMMA copolymers. The free energy of a confined spherical or cylindrical microdomain system can be treated analogously to that of lamellae confined between parallel surfaces [81, 82]. The ratio between the free energy per polymer chain F_c of the confined array of spheres in the template, and the free energy per polymer chain F_0 in the unconfined bulk BCP spherical microdomain morphology, is approximated as a function of normalized row spacing $\lambda (\lambda = L_S/nL_0)$ [83, 84]. A plot of free energy F_c versus confinement width L_S can be constructed for each value of n as shown in Figure 12.8c. F_c has its local minimum when L_S equals to the nL_0. At a given L_S, a confined BCP system will ideally select the integer value of n with the lowest free energy. A transition in the number of rows from n to $n + 1$ is expected to occur when $L_S = (n + 0.5) L_0$. If small energy fluctuations are available to the system, the coexistence of n rows and $n + 1$ rows for $nL_0 < L_S < (n + 1) L_0$ becomes more probable for large n, in agreement with data in Figure 12.8b. The increased overlap of energy between the n-row and an $(n \pm 1)$-row configurations at large confinement width indicates that precise registration is only possible if the confinement width is not very large ($L_S \sim 10 L_0$).

The ordering process for BCP microdomains in 1D templates using *ex situ* and *in situ* time-lapse SFM has been studied by Sibener and coworkers [85]. The process for microdomain rearrangement differs from the ordering process in an

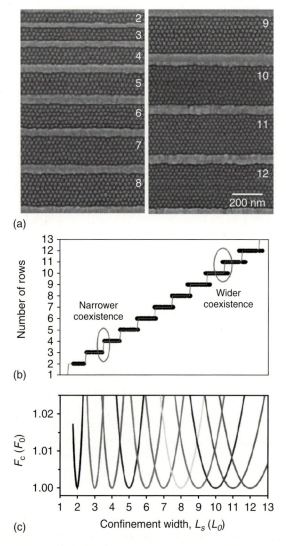

Figure 12.8 Spherical PFS domains within 1D templates of varying width. (a) SEM images of ordered arrays of PFS domains with number of rows in the groove $n = 2-12$. (b) Plot of n versus confinement width, L_S, showing the widths at which arrays with n rows are stable. The confinement width is expressed in terms of L_0, the equilibrium row spacing, which is 24.8 nm in this polymer. Regions of stability overlap at $L_S s \sim (n + 0.5)L_0$. (c) Free energy per polymer chain F_c of the confined array of spheres in the template and the free energy per polymer chain F_0 in the unconfined bulk BCP versus confinement width. The confined BCP system of given Ls will select the value of n with the lowest free energy. A transition in the number of rows from n to $n + 1$ occurs when $L_S \sim (n + 0.5)L_0$, in agreement with the experimental data of (a). The model also predicts wider coexistence regions at larger values of L_S. (Reproduced with permission from *Nature Materials* [25].)

unconfined film. The ordering of cylindrical microdomains begins with one or two cylinders aligned with the edge; then, at random locations along the groove, regions of parallel cylinders start to grow. The final stage of alignment is achieved by merging each misoriented region. Therefore, it takes a longer annealing time for cylinders in wider grooves to align parallel to the edge of the templates.

Even though the combination of BCP self-assembly and topographic patterned surfaces are effective method to control the orientation and the lateral ordering of the BCP thin films, they need to impose long-range order and eliminate defects. Ross, Thomas, Berggren, and coworkers [86] demonstrated that the self-assembly of a thin film of a sphere forming BCP is templated using an array of nanoscale topographical elements that act as surrogates for the minority domains of the BCP. The orientation and periodicity of the resulting array of spherical microdomains are governed by the commensurability between the BCP period and the template period. To template the BCP, they used a sparse 2D array of posts created by e-beam lithography of a 40-nm thick HSQ resist layer on a Si substrate. HSQ is a radiation-sensitive spin-on-glass that forms a silica-like material directly upon e-beam exposure. Development reveals the exposed posts, without requiring further etching or processing. A key requirement for this process is for the surface of the posts to exhibit preferential affinity toward one of the domains of the BCP, which was established by chemical functionalization of the template surface using hydroxyl-terminated homopolymer PS or PDMS brushes. Figure 12.9a illustrates the topographic and chemical design of the posts used in this study. The template was functionalized with a short PDMS homopolymer brush, and a post substitutes for a PDMS sphere in the close-packed array. The template can also be fabricated with an affinity toward the majority block by using a PS brush, which requires larger-diameter HSQ. The use of PDMS brush-coated posts, which required the ability to fabricate ~10-nm structures lithographically, was enabled by recent high-resolution development methods in e-beam lithography. Figures 12.9c,d, show that appropriately sized and functionalized posts could template the assembly of a BCP lattice. For comparison, Figure 12.9b shows the results of untemplated assembly for the same BCP. The template in Figure 12.9c consisted of ~12-nm diameter HSQ posts functionalized with PDMS (5 kg/mol) of thickness ~2 nm, resulting in a post diameter of ~16 nm. Figure 12.9d shows results form 20 nm diameter HSQ posts functionalized with PS (10 kg/mol) of ~5 nm thickness, resulting in a post diameter of 30 nm.

They next considered the more general problem of how a sparse template could be designed to ensure the formation of a single-grain BCP lattice of controlled period and orientation. For a close-packed template of period L_{post}, and a close-packed BCP microdomain array of period $L < L_{post}$, the commensurability between the BCP lattice and the template lattices depends on the ratio L_{post}/L. In the simplest case, where L_{post}/L is an integer, the lattice vectors of the template and the BCP sphere array are parallel, as seen in the SEM images of Figure 12.9, where $L_{post}/L = 3$, and θ, the angle between a post lattice basis vector and a BCP microdomain lattice basis vector, is zero. For noninteger values of L_{post}/L, however, a variety of commensurate BCP lattices with orientations $\theta \neq 0$ can occur. The problem of

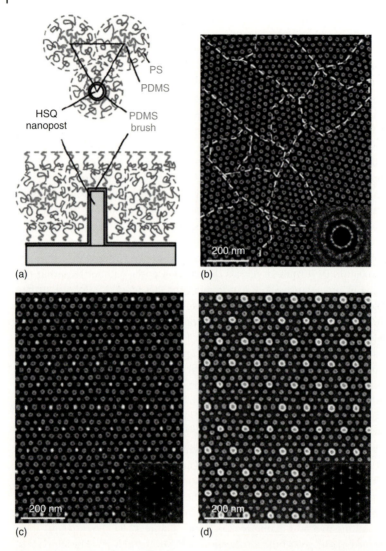

Figure 12.9 (a) Top-down and side-view schematics showing the arrangement of PS-*b*-PDMS BCP molecules in the region surrounding a single post made from cross-linked HSQ resist. The post and substrate surfaces have been chemically functionalized by a monolayer of short-chain PDMS brush. (b) SEM images of a poorly ordered monolayer of BCP spherical domains formed on a flat surface, that is, without templating. The boundaries between different grain orientations are indicated with dashed lines. The inset is a 2D Fourier transform of the domain positions that shows the absence of long-range order. (c and d) SEM images of ordered BCP spheres formed within a sparse 2D lattice of HSQ posts (brighter dots). The substrate and post surfaces were functionalized with a PDMS brush layer in (c), which corresponds to the schematic in (a), and with a PS brush layer in (d). The insets show the 2D Fourier transforms in which the low-frequency components originate from the post lattice. (Reprinted with permission from *Science* [86].)

this approach is that no preference for either orientation was observed for a sixfold symmetric post lattice. However, when the template was formed by removing posts from the original periodic post lattice and adding posts, that is, breaking symmetry, this problem can be solved. This method, which forms high spatial frequency arrays using a lower spatial frequency template, will be useful in nanolithography applications such as the formation of high-density microelectronic structures.

12.2.3.3 Chemically Patterned Substrate

Chemically patterned substrates with length scales comparable to BCP periodicity can be used to control the orientation and registry of the BCP microdomains. Theoretical [87–93] and experimental [21, 22, 94–100] results have indicated that under the appropriate surface grating and boundary conditions, lateral control over nanostructures can propagate micrometers away from the surface (deep into the film), thus providing true 3D control of the self-assembly process. Russell and coworkers [95, 97] and Krausch and coworkers [94, 96] introduced this concept by placing gold stripes on miscut silicon substrates to precisely guide the orientation of symmetric PS-b-PMMA BCPs. Both the commensurability and chemical affinity of the microdomains to the patterned substrate are important. When the commensurate condition is fulfilled, the PS and PMMA blocks are attracted to the gold and silicon oxide regions, respectively. Lamellar microdomains orient themselves perpendicular to the substrate plane and parallel to the striping. However, a mismatch in length scale of only ±10% is sufficient to cause the loss of domain orientation.

Nealey and coworkers [21, 22, 100–109] extended this concept by using soft X-rays to chemically patterned surfaces over large areas. When the attractive energy for each block to a chemically heterogeneous substrate is sufficient, and when the BCP period, L_0, matches the period of the surface pattern L_S, symmetric lamellar BCPs microphase-segregate into perpendicular lamellae that are ordered and directed with perfection over arbitrarily large areas (Figure 12.10). In this directed assembly, the additional interfacial interactions imparted via the surface pattern can stabilize the BCP morphologies and induce the stretching or compression of the individual polymer chains [103]. Figure 12.10b demonstrated that the final BCP structures were influenced by both L_S and the surface energy. A well-registered surface-directed BCP film can be formed if the interfacial energy gain from preferential wetting of each block is sufficient to compensate the strain energy that results from deviation between L_S and L_0. The perfect ordering on patterned PS brushes was achieved for 42.5 nm $< L_S <$ 52.5 nm or $L_S \sim L_0 \pm$ 10% L_0. In comparison, an imaging layer of P(S-r-MMA) copolymer with 50 vol% PS produces fairly low interfacial energy contrast between the chemically modified and unmodified stripes. The lower interfacial energy contrast provides less of a driving force for chain stretching or compression, so that defect-free assemblies can be achieved only over a narrow range of L_S (only $L_S = L_0$). For cylinder-forming PS-b-PMMA thin films [102], the cylindrical microdomains oriented in the plane of the film and formed defect-free periodic arrays over large areas in registration with the underlying chemical surface pattern only if three constraints were satisfied.

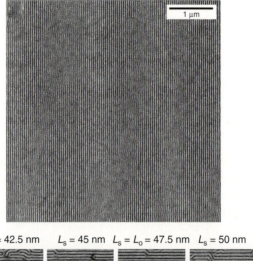

(a)

$L_S = 42.5$ nm $L_S = 45$ nm $L_S = L_0 = 47.5$ nm $L_S = 50$ nm $L_S = 52.5$ nm

PS-r-PMMA brush
(50 vol.% PS and 50
vol.% PMMA)

PS brush

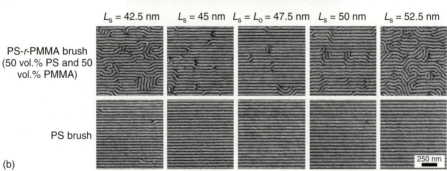

(b)

Figure 12.10 (a) Top-down SEM images of photoresist and PS-b-PMMA copolymer ($L_0 =$ 48 nm, film thickness of 60 nm) patterns. In the SEM images of the PS-b-PMMA copolymer the light and dark regions were PS and PMMA domains, respectively. The perfect epitaxial ordering of a block-copolymer pattern extended over a 5 μm × 5 μm area. In this case, the SAM surface was chemically patterned with a period of $L_S = 47.5$ nm that matched L_0. (Reproduced with permission from Nature [22].) (b) Directed assembly of lamellar PS-b-PMMA ($L_0 = 48$ nm) on surfaces chemically patterned with periodicities 42.5 nm $< L_S <$ 52.5 nm. The degree of interfacial energy contrast between the chemically modified and unmodified regions of the surface pattern and the corresponding BCP domains plays a significant role in directed self-assembly. The low-contrast P(S-r-MMA) 50 : 50 brush (top row) fails to provide perfect ordering for any L_S, but the high-contrast PS brush (bottom row) directs the assembly of well-ordered BCP domains for the entire range of L_S. (Reproduced with permission from Advanced Materials [103].)

These constraints were that the substrate pattern period (L_S) was commensurate with the intercylinder period (L_0) in BCP bulk; that the initial film thickness was quantized with respect to the thickness of a half-layer ($L/2$) or single layer of cylinders (L); and that the widths of adjacent stripes of the chemical surface pattern were nearly equal.

In working circuits, the final chip architecture requires a variety of designs from parallel lines to right angles, which cannot be fulfilled by BCPs with classic morphologies. Such sharp angles introduce severe curvature constraints on the

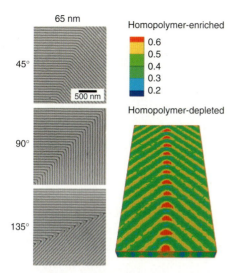

Figure 12.11 (a) Top-down SEM images of angled lamellae in a ternary PS-PMMA/PS/PMMA blend ($L_B = 70$ nm). The chemical surface patterns are fabricated with L_S values of 65 nm, and the lamellar domains of the BCP blend are self-assembled and registered around $45°$, $90°$, and $135°$ bends. The micrographs each depict a 2 μm × 2 μm area. (b) Redistribution of homopolymer facilitates assembly: concentration map of the homopolymers on the surface, where it is seen that the homopolymer is concentrated at the sharp edges to alleviate curvature constraints arising from the patterning. (Reprinted with permission from *Science* [21].)

copolymers. Nealey and coworkers [21] introduced a ternary blend to further extend their surface-directed method to this commercial process. Experimentally $45°$, $90°$, and $135°$ bends with a surface periodicity of 65 nm $< L_S <$ 80 nm were patterned (Figure 12.11). A BCP–homopolymer blend was found to assemble with perfection in both the linear and sharp corner sections, providing the first evidence that directed assembly could fabricate nonregular geometries. The simulations indicate that the bend corners have a higher homopolymer concentration, by 6–7 vol%, than the linear sections of the lamellae. The localized redistribution of homopolymer in the film swelled the domains in the corners in order to accommodate the dimensional differences between the linear period and the corner-to-corner period in this geometry [21, 110].

The prepatterning method, however, is disadvantageous in that it introduces lithographic step, nominally at the same feature density as that achieved by the BCP. Even though the BCP assembly can perform a substantial pattern quality rectification with respect to the prepattern, creating a template where each and every feature is exposed by e-beam is challenging because of the long writing times required over a large area. As an alternative, Ruiz *et al.* [111] demonstrated a directed assembly method for feature density multiplication and pattern quality rectification. With density multiplication, not only is the resolution increased with respect to the prepattern but also the exposure time is reduced with the decrease in the number of

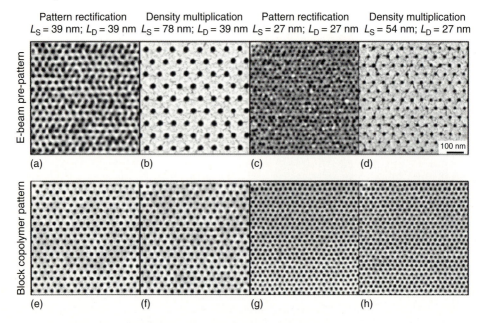

Figure 12.12 (a–d) SEM images of developed e-beam resist with $L_S = 39$, 78, 27, and 54 nm, respectively. (e–h) SEM images of the BCP film on top of the prepattern defined by the corresponding e-beam pattern above. The lattice pitch on the BCP samples is $L_p = 39$, 39, 27, and 27 nm, respectively. (Reprinted with permission from *Science* [111].)

written features. Figure 12.12 shows the improved quality of patterning afforded by directing the assembly of BCP films on chemically patterned surfaces in comparison to the lithographically defined patterns themselves. Figures 12.12a–d, shows SEM images of developed e-beam resist patterned at $L_S = 39, 78, 27$, and 54 nm, respectively. Figures 12.12e–h, shows SEM images of the thermally annealed PS-*b*-PMMA BCP films guided by the prepattern with the corresponding e-beam features above. The polymer pitch on the guided patterns (L_p) is 39, 39, 27, and 27 nm, respectively. Directed assembly may be implemented not only to improve the quality but also to substantially augment the capabilities of the lithographic process beyond current resolution limits.

By combining BCP in the strong segregation limit having spherical microdomains with chemically patterned surfaces Xiao, *et al.* [112] demonstrated that directed self-assembled addressable dot arrays with an areal density of 1.3–3.8 Tbit/in.^{2} can be easily achieved with a resolution scalability of at least a factor of 4. These arrays are at thermodynamic equilibrium, not kinetically trapped, and, as such, are stable. This approach has immediate application in fabricating patterned media with areal densities of 4 Tbit/in.^{2} and can easily be extended to much higher densities. They used a PS-*b*-PDMS BCP with spherical microdomains and a patterned substrate comprised of a semidense hole-tone pattern in an ultrathin preferential wetting layer. The substrate is pretreated with a 1–10-nm thick preferential wetting layer consisting of a polymer brush, with selective affinity to the major block of the copolymer. The surface energy of the minor component block is lower than that of the major component, forcing a preferential segregation of the minor component block to the free surface. The minor component also has a preferentially affinity to the shallow patterned hole area, which is formed by selectively removing the brush layer by EBL process. Under these conditions, the equilibrium, therefore thermodynamically stable, structure of a film having a thickness of 1.5 time the center-to-center distance of the spherical microdomains in the bulk ($1.5L_0$), will be a layer of the lower surface energy component on the surface with hemispherical caps of the minor component block covering the patterned features with an array of spherical microdomains imbedded within the thin film. The pinned hemispherical caps will guide arraying of spherical microdomains and, provided the separation distance of the lithographically placed feature is to within 10% of commensurability conditions, perfect ordering of the microdomains can be achieved with a resolution enhancement of at least a factor of 3–4, thereby, increasing the area density of the features to 4 Tbit/in^2.

12.2.3.4 Corrugated Substrate

Corrugated substrates are readily produced by ultrahigh vacuum annealing of suitably miscut silicon single crystals [95, 97, 113]. A surface faceting transition causes the surface to develop nanoscopic corrugations of well-defined mean spacing. The corrugations follow the respective crystal orientation and therefore are aligned over macroscopic distances.

Li *et al.* [114] examined symmetric BCP films on micrometer-scale corrugated substrates. Interestingly, they found that islands were formed preferentially over the corrugation troughs, creating a surface that was anticonformal with the substrate undulations. Moreover, this resulted in terrace-edge defects aligned perpendicular to the substrate corrugations. Fasolka *et al.* [115, 116] have also studied the effect of surface topography on BCP thin films and demonstrated a lateral patterning technique based on the thickness dependence of film morphology and domain orientation. This originally employed faceted silicon substrates exhibiting sawtooth profile corrugations of ~2 nm amplitude and periods in the 100 nm range [113]. Thin films of polystyrene-*block*-poly(butyl methacrylate) (PS-*b*-PBMA) deposited on these substrates exhibited a flat free surface, in accordance with predictions by Turner and Joanny [117]. This implies a periodic film thickness profile: thinner above corrugation peaks, thicker above troughs. Accordingly, if the average film thickness is chosen correctly, that is, such that these thickness modulations occur about a critical thickness at which a morphological transition takes place, a lateral domain pattern develops that mirrors the substrate topography, as demonstrated in the micrograph in Figure 12.13a. Here, the diblock film is, on average, about $L_0 = 3$ (~15 nm) thick. Near this film thickness, the hexagonally packed cylinder (HY) morphology gives way to perforated layer (PL). The corrugated substrate (~210 nm period) laterally modulates the film thickness such that HY (dots) are formed over the troughs whereas the thinner morphology, PL, is formed on the peaks. As a result,

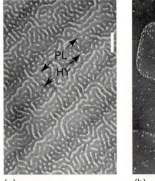
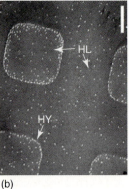

(a)   (b)

Figure 12.13 Patterning of BCP films via substrate topography. (a) TEM micrograph of PS-*b*-PBMA film peeled from corrugated substrate with a 210-nm period. PS domains appear dark. Thickness effects cause alternating HY and PL morphologies to form over corrugation troughs and peaks, respectively. Scale bar = 200 nm. (b) Dual-scale pattern of HY morphology. TEM micrograph of PS-*b*-PBMA film deposited on 30-nm high silicon posts with 1.5 mm diameter and 4 mm spacing. The HY morphology forms preferentially where the film is thinnest at the edges of the post. Areas without contrast are the HL morphology. Scale bar = 1 μm. (Reprinted with permission from *Physics Review Letters* [116].)

a complex lateral patterning with a unique motif develops, as shown. A dual-scale (nanometer, micrometer) pattern can develop if micrometer-sized features are used, as in Figure 12.13b. In this case the film surface is nearly conformal over the 30-nm high, 1.5-μm wide substrate columns. However, the film thins slightly as it bends over the column edges. Choosing the film thickness to be just below $L_0 = 2$, where the HL (laying down cylindrical structure) to HY transition occurs, results in the preferential formation of HY (dots) on the edges of the substrate features. Thus, through the use of substrate topography, diverse film patterns with a range of length scales can be created. This patterning technique has motivated a theoretical treatment by Podariu and Chakrabarti [118]. These researchers compare HY formation on flat and corrugated substrates, demonstrating that these features preferentially form over topographic peaks and edges where the film is thinner.

More recently, Park *et al.* [119] used faceted surfaces of commercially available sapphire wafers to guide the self-assembly of BCP microdomains into arrays with single-crystal textures over the entire wafer surface (cm² in area). Perfectly ordered arrays of BCP microdomains, with areal densities in excess of 10 Tbit/in.², have been produced. The sawtoothed substrate topography provides registered, directional guidance of the BCP self-assembly, that is, tolerant of surface defects, maintaining the lateral registry and ordering of the microdomains over the entire surface. The approach is highly parallel, applicable to different substrates and BCPs, provides unprecedented areal densities, and opens simple, yet versatile routes to ultrahigh density, addressable systems.

12.2.4
Epitaxy

Epitaxy is defined as the growth of a crystal of one phase on the surface of that of another phase in one or more strictly defined crystallographic orientations. The resulting mutual orientation is explained by a 1D or 2D structural matching in the plane of contact of the two species. The term epitaxy was introduced in an early theory of organized crystal growth based on structural matching [120, 121]. Pioneering work by Willems [122] and that by Fischer [123] in the 1950s, demonstrated epitaxy of homopolymers on alkali halide substrates. There are several detailed reviews covering these aspects in the literature [120, 124, 125].

Epitaxy was first introduced by De Rosa *et al.* [126] to control the spatial and orientational order of BCP microdomains. They employed epitaxy to control the molecular and microdomain orientation of a poly(ethylene-*block*-ethylenepropylene-*block*-ethylene) (PE-*b*-PEP-*b*-PE) semicrystalline BCP thin film [126]. A crystallizable organic solvent serves as a solvent for the semicrystalline BCP at temperatures above the solvent melting temperature. While the block is cooled below the melting point of solvent, the crystallizable solvent becomes a substrate. Epitaxy usually occurs between a semicrystalline BCP block and a crystalline substrate. On such a crystalline substrate, the semicrystalline block can be oriented. This provides an excellent way to form highly aligned edge-on crystalline lamellae in both lamellar and cylindrical microdomains formed from semicrystalline-amorphous BCPs. Epitaxial crystallization of the crystalline block on an organic crystalline substrate such as benzoic acid (BA) [26, 126] or anthracene (AN) [127], can result in precise control of the molecular orientation of the crystalline block and subsequent overall long-range order of the BCP microdomains. In the case of an amorphous BCP, the fast directional solidification of the solvent during microphase separation leads to an alignment of the BCP interface parallel to the fast growth direction of the solvent crystals [26, 128–130]. Both lamellar and cylindrical microdomains in a symmetric PS-*b*-PMMA and in an asymmetric polystyrene-*block*-polyisoprene (PS-*b*-PI) copolymer, respectively, are globally aligned by using either BA or AN as the crystallizable solvent [128, 129]. Typical examples of directionally solidified BCP microstructures are shown in Figure 12.14 [129].

12.2.5
Control of Interfacial Interactions

The interfacial interactions dictate the wetting layers at both the polymer/substrate and polymer/surface interfaces and the orientation of the microdomains in the film [14, 82, 95, 131–143]. If there is no preferential interactions of a block with the interfaces, the microdomains may align normal to the surface without any film thickness frustration. So, one way to achieve a perpendicular orientation of microdomains is to remove or balance all of the interfacial interactions, that is,

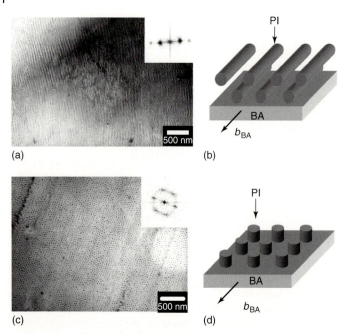

Figure 12.14 (a) TEM bright-field image of a thin film of PS-*b*-PI (45/12) BCP, directionally solidified with BA, and stained with OsO₄. The dark regions correspond to the stained PI microdomains. The cylindrical PI microdomains are well aligned along the fast growth direction of the BA crystals (crystallographic *b*-axis). Inset shows the fast Fourier transform (FFT) power spectrum of the TEM micrograph. Spot-like first reflection located on the meridian shows the nearly single crystal-like microstructure. (b) Schematic model of the microstructure of PS-*b*-PI processed with BA. Cylindrical PI microdomains are aligned along the *b*-axis of BA crystal. (c) Due to very thin film thickness, vertically undulated PI cylinders transform into hexagonally packed cylinders oriented perpendicular to the BA substrate. Inset shows the Fourier transform power spectrum of the TEM micrograph. Spot-like first reflections with sixfold symmetry show the nearly hexagonally packed microstructure. (d) Schematic model of the microstructure of PS-*b*-PI processed with BA. Cylindrical PI microdomains are oriented vertically to the substrate. (Reproduced with permission from *Polymer* [9] and *Macromolecules* [129].)

use nonpreferential or neutral surfaces. Under these conditions, the perpendicular orientation is favored over a parallel arrangement due to an entropic effect [144, 145]. By removing the oxide layer with buffered HF, the interactions of the substrate with PS and PMMA are equally nonfavorable, and the cylindrical microdomains will be oriented normal to the surface upon thermal annealing [137]. An alternative approach is to tune interfacial interactions through the use of a self-assembled monolayer (SAM) to modify the substrates [146–151]. Symmetric, neutral, and antisymmetric wetting of the BCP thin films on the SAM-modified substrates can be produced to direct the orientation of the microdomains in thin films of BCPs. In most cases, a mixed assembly of functional groups is used to control the interfacial interactions. The same end has been achieved by Kellogg *et al.* [136] by placing very high molecular weight random

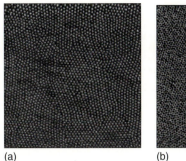

(a)   (b)

Figure 12.15 (a) SFM images of a spin-coated film of PS-*b*-PMMA diBCP having ~20-nm cylindrical microdomains of PMMA oriented normal to the film surface. The film, ~40 nm thick, was coated onto a surface where the interfacial interactions were balanced. (b) The same film after being exposed to UV radiation, followed by rinsing with alcohol. The nanoporous film of cross-linked PS has pores that are the same size as the original PMMA microdomains. The inset in (b) shows a field-emission electron micrograph of the film, demonstrating the uniformity in the size of the pores. The images are 2 μm × 2 μm. (Reproduced with permission from *MRS Bulletin* [152].)

copolymers, consisting of the same monomeric units as the diBCP, at the substrate. In order to avoid potential diffusion of the random copolymers into the diblock layer, Russell, Hawker, and coworkers [27, 78, 144, 152–155] anchored an end-functionalized random copolymer to the surface. After removal of the nonattached chains, the surface was a random copolymer brush, where the composition of the random copolymer could be varied in the synthesis and narrow molecular weight distributions of the brush length could be obtained by nitroxide-mediated synthetic routes. By simply varying the composition of the random copolymer, the interfacial interactions of the BCP with the modified substrate were balanced. Consequently, the microdomains can be oriented normal to the surface (Figure 12.15). Although this surface modification had proven to be exceptionally robust and easily applied to very large area surfaces, this surface-modification process was restricted to homogeneous oxide surfaces. Recently, several groups further developed a method in which the interfacial interactions of a surface could be easily manipulated in a rapid, robust manner through the use of a thin, cross-linked random copolymer film [28, 156, 157]. The interfacial interactions were dictated by the average composition of the cross-linked random copolymer film and the degree of cross-linking could be altered by changing the number of cross-linkable units incorporated into the copolymer. The cross-linked film was insoluble and compatible with further processing. Removing the requirement of chemical attachment to the underlying substrate made this process applicable to virtually any solid surface, and the substrate can be flexible or rigid, homogeneous or heterogeneous. The effectiveness in controlling the interfacial interactions was demonstrated with thin diBCP films on a wide range of substrates on which the orientation of the microdomains was perpendicular to the surface (Figure 12.16) [28].

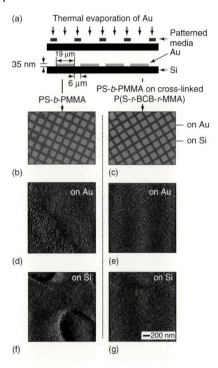

Figure 12.16 (a) Diagram of the evaporation process used to generate gold squares in a grid of silicon oxide, a chemically heterogeneous surface. In the images on the left (b,d,f), a diBCP was directly spin coated onto this heterogeneous surface. The reflection optical micrograph (b) reveals the grid underneath the copolymer; the scanning force micrograph of the copolymer on gold (d) shows that the cylindrical microdomains of the copolymer have a random orientation, while on silicon oxide (f), the microdomains orient parallel to the surface. The same surface is shown in the images on the right (c,e,g), where a cross-linked random copolymer film was used to balance interfacial interactions. The optical micrograph (c) shows the grid under the film, and (e), (g) scanning force micrographs show a BCP film on the gold (e) and silicon oxide (g) covered portions of the surface. In both cases, the cylindrical microdomains are oriented normal to the surface. (Reproduced with permission from *Science* [28].)

Controlling the interfacial interactions provides a simple, passive route to manipulating the orientation of copolymer microdomains in thin films. However, the thickness of the BCP films where such control can be achieved is limited. Usually, oriented microdomains are obtained in films with thickness approximately one period. For thicker films, a perpendicular orientation is lost and the microdomains adopt a random orientation. The addition of a small amount of homopolymer, that is, miscible with the minor component block and has a molecular weight higher than the minor component has been shown to produce a drastic increase in the persistence of the orientation of microdomains normal to the surface, enabling a directed self-assembly of the copolymers into arrays of highly oriented, high aspect ratio cylindrical microdomains over large areas [158–163]. The extended

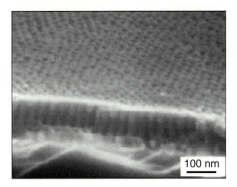

Figure 12.17 SEM image of the cross-section of a nanoporous film prepared from a mixture of PS-*b*-PMMA, having cylindrical microdomains, with a small amount of higher molecular weight PMMA homopolymer. The PMMA stretched along the axes of the cylindrical microdomains, has propagated the orientation of the microdomains over large distances from the surface. (Reproduced with permission from *Advanced Materials* [158].)

homopolymer along the axis of the cylinder, resulting from the confinement of the homopolymer to the centers of the cylindrical microdomains, acted as a molecular reinforcing agent and induced an orientation of the microdomains over a very long distance away from the substrate. Shown in Figure 12.17 is an example that PS-*b*-PMMA was mixed with a small amount of higher molecular weight PMMA, achieving persistence of the orientation over large distances [158]. Moreover, the addition of a homopolymer to a BCP is an easy means to control the size of cylindrical microdomains in thin films without perturbing their spatial order and orientation [159]. Removal of both of the homopolymer and corresponding block of the copolymer produce pores larger than that achieved from the pure copolymer. Thus, in one film, two distinct sizes of pores can be accessed [159].

12.2.6
Self-Directing Approaches

Combination of self-assembly at different length scales leads to structural hierarchies. It offers rich possibilities to construct nanostructured materials, nanoscale parts, and switching (responsive) properties based on the phase transitions of the self-assembled structures. In this section, we will review some results of cooperatively coupling inorganic particles, salts, and low molecular weight organic molecules with BCPs.

12.2.6.1 Nanoparticles
The organization of inorganic nanocomponents into self-assembled organic or biological materials has been interesting for decades to make functional hybrid materials [3, 164–167]. Theoretical arguments suggested that synergistic interactions between self-organized particles and a self-assembled matrix material can lead to hierarchically ordered structures [168–170]. Recently, Russell, Emrick, and

coworkers [171, 172] show that mixtures of PS-*b*-P2VP BCP and cadmium selenide (CdSe) nanoparticles exhibit cooperative, coupled self-assembly at the nanoscale. In thin films, the copolymers assemble into cylindrical microdomains, which dictate the spatial distribution of the nanoparticles; segregation of the particles to the interfaces mediates interfacial interactions and orients the copolymer domains normal to the surface, even when one of the blocks is strongly attracted to the substrate (Figure 12.18) [171].

Experiments on different substrates and film thicknesses showed that the perpendicular orientation is observed regardless of the nature of the substrate or the film thickness. This interplay between assembly processes was also applied to a wide variety of other systems, such as a blend of poly(ethylene glycol) (PEG)-tagged ferritin bionanoparticles, denoted ferritin-PEG, and a lamella-forming P2VP-*b*-PEO BCP. The ferritin-PEG bionanoparticles are incorporated into PEO microdomains, suppress crystallization, mediate interfacial interactions, and reorient the microdomains normal to the surface. This is an example of synthetic and biologically inspired systems, where a one-step hierarchical self-organization occurs via the interplay between distinct self-assembling processes, producing spatially ordered, organic–inorganic nanoparticle, and organic–bionanoparticle hybrid materials. This synergy represents a significant advance over other processes that rely on sequential fabrication steps to incorporate functionality into preorganized templates.

12.2.6.2 Inorganic Salts

Recently, Russell and coworkers [173–176] found that the orientation behavior was highly influenced by the amount of inorganic salts that was added into BCPs and the synergistic interactions between inorganic salts and BCPs were similar to those between BCPs and nanoparticles. Figure 12.19 exhibited the influence of inorganic salts on the orientation of lamellar microdomains in PS-*b*-PMMA thin films. For pure copolymers, after thermal annealing, several PS-*b*-PMMA lamellar layers adjacent to the interfaces remain orientated parallel to the interface, even under an external electric field (Figure 12.19b), due to the preferential interactions of PMMA with the substrate. In contrast to the films with lithium chloride (LiCl), after thermal annealing, lamellar microdomains do not completely cover the whole interfaces although some lamellae are still parallel to the interfaces, indicating the surface interaction in the system became weak (Figure 12.19c). Under an applied electric field, the complete alignment of lamellar microdomains normal to the surface can be achieved (Figure 12.19d). They proved the formation of lithium-PMMA complexes in PMMA microdomains after introducing LiCl into BCPs, resulting in the significantly increased dielectric contrast [174] and χ [175] between two blocks as well as the modified surface interactions [177].

Similar phenomena were also observed by Russell and coworkers in PS-*b*-PEO [176] and PS-*b*-P2VP [178] thin films. They found the orientation of the cylindrical microdomains strongly depended on the salt concentration and the ability of the ions complexed with PEO. The addition of salt gives rise to a change in the orientation of PEO cylinders, from parallel to perpendicular, in the solvent annealed

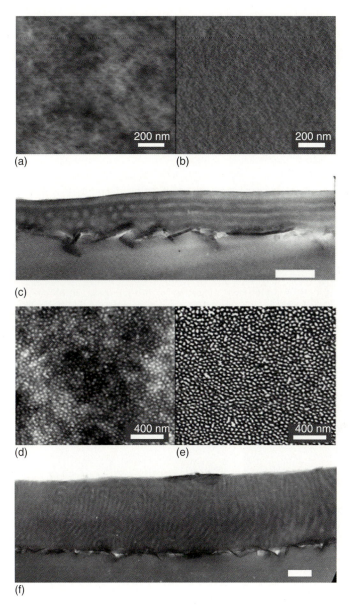

(a)

(b)

(c)

(d)

(e)

(f)

Figure 12.18 SFM topography (a) and phase images (b) of a PS-*b*-P2VP BCP film taken after spin coating and after thermal annealing at 170 °C for two days; (c) its corresponding cross-sectional TEM image. SFM topography (d) and phase images (e) of films prepared from a mixture of PS-*b*-P2VP BCP and CdSe nanoparticles after thermal annealing at 170 °C for two days; (f) its corresponding cross-sectional TEM image. (Reprinted with permission from *Nature* [171].)

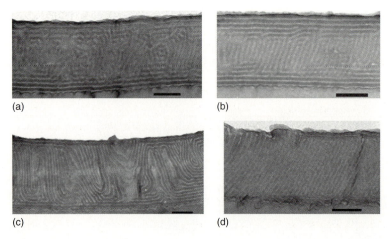

Figure 12.19 Cross-sectional TEM images of pure PS-*b*-PMMA thin films after thermal annealing at 175 ± 5 °C under N₂ for 24 h (a) and after applying a ∼40 V/μm dc electric field at 175 ± 5 °C under N₂ for 24 h (b); cross-sectional TEM images of PS-*b*-PMMA thin films with lithium-PMMA complexes after thermal annealing at 175 ± 5 °C under N₂ for 24 h (c) and after applying a ∼40 V/μm dc electric field at 175 ± 5 °C under N₂ for 24 h (d). Scale bar: 200 nm. (Reprinted with permission from *Physics Review Letters* [174] and *Macromolecules* [175].)

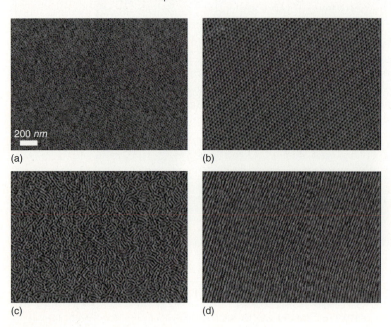

Figure 12.20 SFM images of PS-*b*-PEO thin films containing KI (a) as spun and (b) after solvent annealing and pure PS-*b*-PEO thin films (c) as-spun and (d) after solvent annealing. (Reprinted with permission from *Macromolecules* [176].)

films (Figure 12.20). The process shows large flexibility in the choice of salt used, including gold or cobalt salts, whereby well-organized patterns of nanoparticles can be generated inside the copolymer microdomains. Consequently, the added salt serves a dual role. The first is to orient and order the copolymer microdomains while the second is to serve as a precursor to the fabrication of metal nanoparticles.

12.2.6.3 Organic Molecules

Complexation of amphiphilic organic molecules to polymers by ionic inter-actions, coordination, or hydrogen bonding leads to polymeric comb-shaped supramolecules (complexes), which self-assemble at a length scale of a few nanome-ters. Simultaneously, BCPs provide self-assembly at an order of magnitude larger length scale. Thus, directed assembly of these polymeric supramolecules leads to the control of structures at several length scales and anisotropic properties [179–181]. The physical bonds within the supramolecules allow the control of the cleavage of selected constituents. They provide templates for mesoporous materials as well as nano-objects, and allow switching conductive or optical properties. Ikkala and coworkers [181, 182] used hydrogen bonding between 4-vinylpyridine monomer units and 3-pentadecyl phenol (PDP) to modify the morphology of PS-*b*-P4VP. The investigation of PS-*b*-P4VP and PDP in the bulk showed that the supramolecular assembling of P4VP and PDP changed the BCP morphology from spherical to cylindrical structures [182]. By dipping the complexes into a selective solvent, PDP can be easily removed, providing nanochannels in the matrix [183]. Figure 12.21 summarizes the recognition-driven supramolecular assemblies with comb-shaped architectures in polymers, the subsequent self-organization and preparation of functional materials as well as nano-objects [181, 182, 184–186]. Recently, perpen-dicularly oriented cylindrical microdomains were observed in a system where a PS-*b*-P4VP BCP supramolecular assembled with 2-(4′-hydroxybenzeneazo) benzoic acid (HABA) molecules [187]. Furthermore, it was found that the film can be reversibly switched from the perpendicular to parallel orientation and *vice versa* upon exposure to 1,4-dioxane and chloroform vapor, respectively. The alignment is insensitive to the composition of the confined surface. The modes of hydrogen bonding between the pyridine repeating units and HABA dictated the switch of the orientation.

More recently, Xu and coworkers [188] demonstrated a new paradigm to control the hierarchical assembly of nanoparticles through the synergistic coassembly of BCPs, small molecules, and readily available nanoparticles. Organizations of nanoparticles into one-, two-, and three-dimensional arrays with controlled inter-particle separation and ordering was achieved without any chemical modification of either the nanoparticles or BCPs. The ordering and distribution of small molecules between different BCP blocks were temperature dependent, leading to responsive materials where the spatial distribution of the nanoparticles could be varied, changing the local environment and the areal density of the nanoparti-cles. This approach is versatile; compatible with existing fabrication processes and enables a nondisruptive strategy for the generation of functional devices.

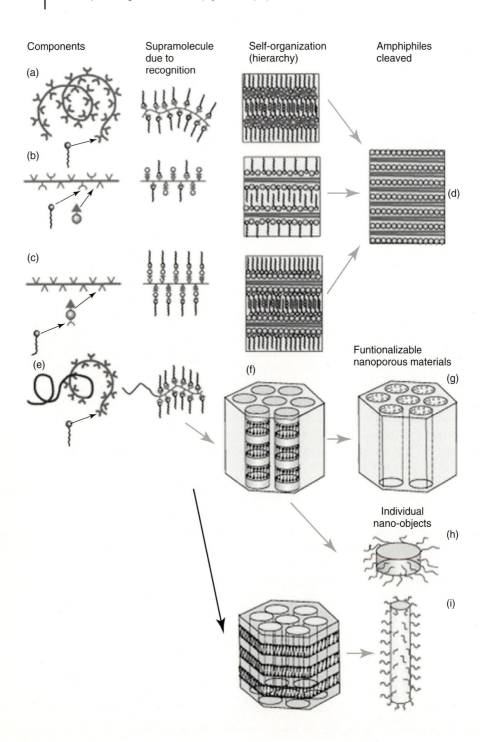

Components

Supramolecule due to recognition

Self-organization (hierarchy)

Amphiphiles cleaved

(a)

(b)

(c)

(d)

(e)

(f)

Funtionalizable nanoporous materials

(g)

Individual nano-objects

(h)

(i)

12.2.6.4 Homopolymer Addition

Since the molecular interactions for a given copolymer are dictated by their chemical structure, the segregation strength (χN) of BCPs can be tuned by changing the BCP molecular weight (N). The segment–segment interaction parameter (χ) on the other hand, can be increased by altering the chemical composition of one of the blocks or by the selective chemical modification of the segments. Additional structural control is to generate hierarchical morphologies through the supramolecular assembly of well-defined BCPs with homopolymers that can hydrogen bond with one of the blocks [189, 190]. Watkins and coworkers [189, 190] found that the domain segregation of low molar mass amphiphilic BCPs, including samples with relatively broad molar mass distributions, can be dramatically increased by simply blending with a homopolymer that strongly associates with the hydrophilic segment through hydrogen bonding. The selective association of homopolymer increases the effective segregation strength as a result of significantly increased effective interaction parameter in the blend. As an example, Figure 12.22 shows the small angle X-ray scattering (SAXS) profiles of Pluronic BCPs F108 and P105 as a function of added poly(acrylic acid) (PAA) concentration. The full width at half-maximum (FWHM) of the primary SAXS peak decreases in both systems as the PAA concentration is increased. The broad scattering peak from Pluronic BCP blends with less than 20% PAA by mass represents compositional fluctuations observed typically in disordered BCP melts. The Pluronic BCPs undergo a disorder-to-order transition (DOT) with the addition of PAA. In the Pluronic F108 BCP, which has the highest χN among the systems studied, addition of PAA above 25% by mass also leads to two higher-order scattering peaks at $\sqrt{2}q^*$ and $\sqrt{3}q^*$. These higher-order reflections represent long-range correlations of a body-centered cubic lattice with minimum lattice distortions, which is an indication of significantly improved domain segregation.

In addition to homopolymer, simple blending of different BCPs, such as A-B/B-C and A-B/C-D alloys, would seem to be attractive for combining physical properties and broadening the processing window [191, 192]. However, uniform long-range order has not been achieved in thin film blends of BCPs [193] because

---

Figure 12.21 Comb-shaped supramolecules and their hierarchical self-organization, showing primary and secondary structures. Similar schemes can, in principle, be used both for flexible and rod-like polymers. In the first case, simple hydrogen bonds can be sufficient, but in the latter case a synergistic combination of bondings (recognition) is generally required to oppose macrophase separation tendency. In (a–c), the self-organized structures allow enhanced processibility due to plasticization, and solid films can be obtained after the side chains are cleaved (d). Self-organization of supramolecules obtained by connecting amphiphiles to one of the blocks of a diBCP (e) results in hierarchically structured materials. Functionalizable nanoporous materials (g) are obtained by cleaving the side chains from a lamellae-within-cylinders structure (f). Disk-like objects (h) may be prepared from the same structure by cross-linking slices within the cylinders, whereas nanorods (i) result from cleaving the side chains from a cylinder-within-lamellae structure. Without loss of generality, (a) is shown as a flexible polymer, whereas (b) and (c) are shown as rod-like chains. (Reprinted with permission from *Science* [179].)

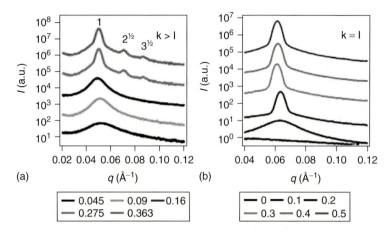

Figure 12.22 The change in SAXS profiles of bulk Pluronic F108 (a) and P105 (b) BCP melts obtained by blending with PAA at 80 °C. Scattering data in (a) and (b) are shifted in intensity for clarity. (Reprinted with permission from *Advanced Materials* [189].)

of the overwhelming tendency of such mixtures to exhibit macrophase separation [194, 195]. In an attempt to limit macrophase separation in these blends, Fredrickson, Kramer, Hawker, and coworkers [196] have exploited supramolecular (hydrogen-bonding) interactions in addition to the nonspecific dispersive interactions typically present in a BCP alloy. These attractive interactions between complementary hydrogen-bonding groups are designed to suppress macrophase separation in favor of microphase separation, thereby producing large-scale assembly of nanoscale features. By controlling the level of incorporation of hydrogen bonding units, the molecular weights, and compositions of the BCPs, and the relative amounts of the two BCPs in the alloy, a highly modular and tunable system can be developed that allows diverse families of ordered features to be achieved, including square arrays of cylinders.

In this study, they used the BCPs based on PEO-*b*-PS and PS-*b*-PMMA. Such a blend system combines the photodegradability of PMMA with the long-range ordering characteristics of the PEO based BCP under solvent annealing at controlled humidity [18]. The respective PS segments were modified with small fractions of randomly incorporated 4-hydroxystyrene and 4-vinylpyridine units [197]. 50-nm thick supramolecular BCP films were prepared by spin-coating polymer solutions in benzene onto silicon wafers, with the blends formulated by simply mixing the phenolic containing PEO A-B diblock with the corresponding PMMA B'-C diblock, containing various levels of 4-vinylpyridine substitution. In these samples, a 1:1 molar ratio of A-B chains to B'-C chains was maintained and the films were solvent-annealed under saturated toluene vapor in a controlled high-humidity environment [197].

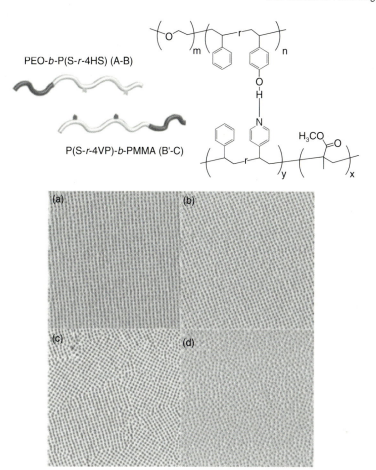

PEO-*b*-P(S-*r*-4HS) (A-B)

P(S-*r*-4VP)-*b*-PMMA (B'-C)

Figure 12.23 Hierarchical self-assembly and target morphology for a blend of supramolecular A-B and B'-C BCPs stabilized by hydrogen bonding and SFM phase images (2 μm × 2 μm) of solvent-annealed films from blends of supramolecular BCPs: ratio of phenolic to pyridyl group (a) 1 : 1, (b) 1 : 1.1, (c) 1 : 2, and (d) 1 : 3.5. (Reprinted with permission from *Science* [196].)

Figure 12.23 shows the SFM phase images of films from four representative blends, which indicate that the nanoscale morphology and grain size are affected by the composition of hydrogen-bonding components. For a blend having ratio of 1 : 3.5 (phenolic groups to pyridyl groups), vertical cylinders were obtained, but little order, either hexagonal or square packing was observed (Figure 12.23d). Upon decreasing the phenolic : pyridyl ratio to 1 : 2, distinct square arrays were observed, although the ordering was poor (Figure 12.23c). However, for blends with approximately equal numbers of phenolic and pyridyl units per chain, the corresponding thin films exhibited square arrays with a high degree of lateral order (Figure 12.23a). These square arrays will enable simplified addressability and circuit interconnection in integrated circuit (IC) manufacturing and nanotechnology.

12.2.7
Optical Alignment

Optical-alignment methods at a molecular level have been well established in liquid-crystal and the relevant systems [198, 199]. Directionally selective light excitation using linearly polarized light of photoisomerizable molecules on a liquid-crystalline polymer film produces patterned, oriented microdomains in the films. Recent investigations have revealed that such photoexcited collective molecular motions can lead to lateral mass transport over distances of micrometers [200, 201].

Seki and coworkers [202] proposed a new optical 3D (both out-of-plane and in-plane) alignment of nanocylinders of a BCP comprising liquid-crystalline photoresponsive block chains and PEO by applying the process of photoinduced mass migration. The key for the out-of-plane alignment (whether the cylinders are oriented normal or parallel to the substrate surface) is the control of the film thickness, while that for the in-plane alignment is the direction of the linear polarization during the illumination (Figure 12.24). Moreover, Ikeda and coworkers [203] addressed a noncontacted optical method by using a polarized laser beam to control a parallel patterning of PEO nanocylinders in an amphiphilic liquid-crystalline BCP film (Figure 12.25).

12.2.8
Convergence of Top-Down and Bottom-Up in BCP Design

The convergence of top-down and bottom-up fabrication in the same BCP architecture has been demonstrated using EBL by Bal *et al.* [204] and Spatz and coworkers [205, 206]. Through e-beam exposure of cylindrical PS-*b*-PMMA films, Bal *et al.* [204] have been able to create cylindrical nanochannels at defined locations from PS-*b*-PMMA on the substrate, and thereby demonstrated its potential to generate integrated magnetoelectronic devices. In this process, the PS matrix is cross-linked in the exposed area, thus making the BCP system behave as a negative tone e-beam resist in which the unexposed material is removed in a development step. In a similar manner, by using metal precursor loaded PS-*b*-P2VP BCP micellar monolayers as a negative e-beam resist, Spatz and coworkers [205, 206] created metallic nanodots in microscopically defined locations. In this system, the patternability is proposed to arise from cross-linking of the PS block as well as chemical modification of P2VP/metal salts due to e-beam exposure. Although EBL can give excellent spatial control of functional microdomains, this direct-write patterning process is not time efficient for large-area integration of functional devices. Techniques for rapid patterning of functional nanostructures are thus needed for real-time applications.

Ober and coworkers [207–209] have successfully developed a novel BCP system using poly(α-methylstyrene)-*block*-poly(4-hydroxystyrene) (PαMS-*b*-PHS) to achieve spatial control via high-resolution deep-UV lithographic processes. Through the incorporation of high-resolution PHS photoresist and PαMS in the block

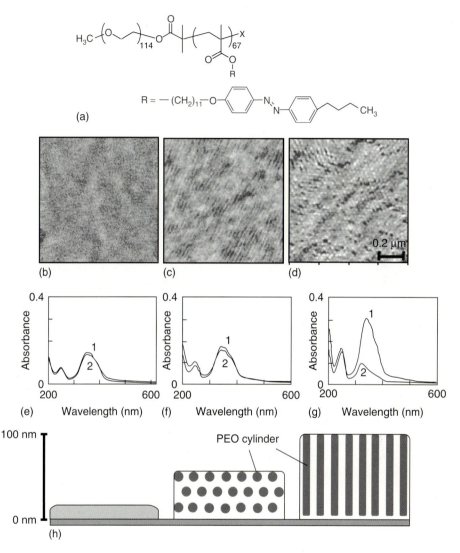

Figure 12.24 Thickness dependence of the azobenzene (Az) and cylinder alignment. (a) Chemical structure of the diBCP consists of PEO and poly(methacrylate) containing an Az unit, with the degree of polymerization of the PEO and the azopolymer being 114 and 67, respectively (denoted as p(EO₁₁₄-Az₆₇)). (b–d) Phase-mode SFM images (1 μm × 1 μm) of the p(EO₁₁₄-Az₆₇) film with different film thickness: (b) 20 nm, (c) 30 nm, and (d) 70 nm after annealing and exposure to hexane vapor. (e–g) UV-vis absorption spectra of the corresponding p(EO₁₁₄-Az₆₇) films are shown: (e) 20 nm, (f) 30 nm, and (g) 70 nm thickness for as-cast (1) films and after exposure to hexane vapor (2). (h) Schematic illustration of thickness dependence on cylinder alignment. (Reproduced with permission from *Advanced Materials* [202].)

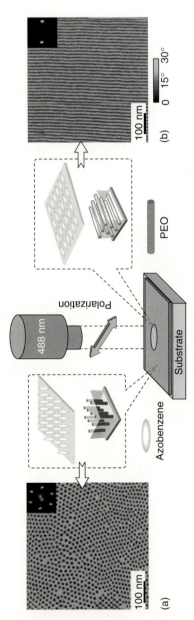

Figure 12.25 Scheme of liquid-crystalline (LC) alignment and microphase-separated structures in the irradiated and unirradiated area of the BCP films. Inset: SFM phase images of the annealed BCP films in unirradiated (a) and irradiated (b) area. (Reproduced with permission from *Journal of the American Chemical Society* [203].)

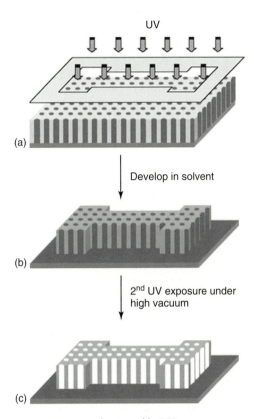

Figure 12.26 Novel patternable BCPs to achieve spatially controlled nanostructures. (a) An asymmetric PαMS-b-PHS copolymer/photoacid generator/cross-linker solution was spin coated on a silicon substrate and formed vertical PαMS cylinders due to rapid solvent evaporation. (b) 248-nm stepper exposure and subsequent development to form micropatterns with features as small as 400 nm. (c) Strong UV irradiation under high vacuum to remove PαMS, thus generating patterned nanochannels. (Reproduced with permission from *Advances in Polymer Science* [10].)

architecture, large-area uniform nanometer-sized cylinders in submicrometer-sized patterns were generated through simple fabrication processes (Figure 12.26). Additionally, this BCP was aligned with spin coating into a vertical orientation over a wide range of film thicknesses (40 nm to 1 μm), thereby avoiding tedious alignment procedures. Furthermore, this BCP system was designed such that the thermodegradable PαMS block could be removed to make a nanoporous material.

Russell and coworkers [210] demonstrated a convergence of top-down/bottom-up approach using PS-b-P4VP BCPs with a high degree of lateral order as a model system. Solvent annealing of PS-b-P4VP films in organic solvent exhibits the lateral ordering of BCP microdomains. Next, e-beam lithography was employed to prepare periodic patterns keeping the high degree of order in templates. Highly oriented PS-b-P4VP was cross-linked in e-beam-exposed regimes, while unexposed ones were transformed into nanoporous structures after immersion in the preferential

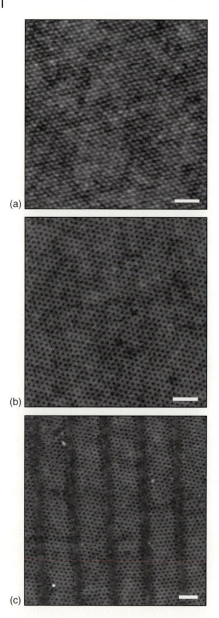

(a)

(b)

(c)

Figure 12.27 Height mode SFM images of solvent annealed (a), reconstructed films (b) and periodic microdomain arrays prepared from e-beam dose and reconstruction process (c). Scale bars are 200 nm (Reproduced with permission from *Small* [211].)

solvent of one of the blocks [211]. From this template, nickel (Ni) dotted arrays could be prepared via a thermal evaporation and lift-off process. This method does not need topographic or chemically patterned substrates to increase the order of BCP microdomains. When the as-spun PS-*b*-P4VP films were exposed to THF vapor at room temperature for 3 h, a significantly enhanced lateral order was obtained, as shown in the SFM image of Figure 12.27a. The solvent-annealed PS-*b*-P4VP films are characterized by a hexagonal array of circular domains with an average nearest-neighbor distance of 43.3 ±2.5 nm. When the well-developed PS-*b*-P4VP micelles were immersed in ethanol, a good solvent for P4VP and a nonsolvent for PS, for 20 min, surface reconstruction of the film occurred, producing a highly oriented array of nanoscopic pores (Figure 12.27b) without changing the separation distance of cylindrical microdomains. Solvent-induced surface reconstruction has been used to generate nanoporous templates without removing the minor component using post-treatment such as UV, ozone, and solvent [32]. By exposing the films to ethanol, P4VP was solvated, while the glassy PS matrix remained intact. Upon drying, a reconstruction of the film was observed where pores were opened in the positions of the original P4VP cores as the P4VP within the pores was transferred to the surface [212, 213]. The e-beam lithography was used to expose the spin-coated and/or well-developed film of PS-*b*-P4VP. Selected areas from e-beam can be cross-linked, in which BCPs might not move in ethanol, while P4VP blocks in the unexposed regimes should be reconstructed. Figure 12.27c shows SFM images of reconstructed films after e-beam exposure at four different doses. The single line exposures of 7 nC/cm gave line widths of 60 nm. It should be noted that the well-developed films exposed to e-beam can be immobilized in an appropriate solvent, and the precise control in location of nanoporous objects in periodic arrangements via surface reconstruction can be controlled.

Unlike other processes, the e-beam was used to cross-link the BCP film immediately after spin coating from a neutral solvent, *N*,*N*-dimethylformamide (DMF), for both blocks, when the BCP is very disordered [211]. When as-spun film was exposed to a dose of 8 nC/cm, such that the exposed regions were completely cross-linked. It should be noted that, over the course of 3 h, even cross-linked BCP thin films can migrate during solvent annealing in THF (vapors) a slightly selective solvent for PS and, therefore, it is difficult precisely to define a confinement area. When the films were exposed to solvent mixture, THF/cyclohexane (70/30, v/v) for 4 h, followed by immersion in ethanol for 20 min, the e-beam exposed regions unchanged, as shown in Figure 12.28. The lattice spacing between two micelles in the direction parallel to the e-beam pattern is equal to the original lattice spacing ($d = 37.6$ nm) observed on a flat substrate, as shown in Figure 12.27a. When the

Figure 12.28 SFM image of BCP films annealed in solvent vapor and followed by surface reconstruction process after cross-linking the BCP film immediately after spin coating. Scale bar is 100 nm (Reproduced with permission from *Small* [211].)

line width is incommensurate with the ideal period, *d*, the periodicity of the BCP micelle is compressed or expanded to fit within the patterns. This method opens a new route to prepare patterns having neutral walls, since disordered PS-b-P4VP films are cross-linked by e-beam over arbitrary patterns.

As another top-down/bottom-up approach, Kim *et al.* [214] demonstrated a practical method to fabricate a large area ordering of PS-*b*-PEO BCP microdomains by simply exposing solvent vapor to a prepatterned copolymer thin film prepared by spin casting on a microcontact printed SAMs surface. The dewetting of the BCP film during the solvent annealing for assisting the ordering of BCP domains was strictly confined into the micropatterned regions, leading to a regularly arrayed set of convex lens shaped spherical caps over 1 mm² area where each dewetted domain was composed of nearly single-crystal spherical PEO microdomains. The ordered hierarchical structure was conveniently formed by the combination of BCP self-assembly and controlled dewetting with scales from nanometer, micrometer to millimeter.

12.2.9
Zone Casting

Zone casting originally developed for the oriented growth of molecular crystals, has been used to achieve large-scale alignment of nanoscale domains in microphase-separated BCPs [215]. Kowalewski and coworkers [30] demonstrated a large-scale, long-range ordering of lamellae in thin films of poly(n-butyl acrylate)-*block*-poly(acrylonitrile) (PBA-*b*-PAN) BCPs by using a simple zone-casting technique. Zone casting was performed by depositing the copolymer solution from DMF onto a moving substrate with the aid of a syringe equipped with a flat nozzle (Figure 12.29a). To achieve the desirable solvent evaporation rate, the temperatures of the copolymer solution and of the substrate were controlled. It should also be noted that, by varying the casting condition, the orientation of the microdomains with respect to the casting direction could be controlled. Phase-contrast SFM images of films prepared by zone casting exhibited characteristic long-range ordered striped morphology with elongated rigid PAN domains appearing as brighter stripes alternating with more compliant and mechanically lossy PBA phase (Figure 12.29b).

12.3
Functionality of BCP Thin Films

BCP patterns show a high degree of order and symmetry over 10–100 nm length scales, however, achieving large-area defect-free patterns and specific orientation of anisotropic structures is a challenge [1, 2]. A uniform orientation and long-range lateral order of the microdomains is critical in device-related applications, for example, in the fabrication of displays or memory storage devices. As mentioned earlier, significant progress has been achieved in controlling BCP assembly via a number of strategies. Patterned BCP films can be used in their own right as flash

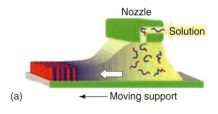

Figure 12.29 Long-range order in thin films of PBA-*b*-PAN BCPs prepared by zone-casting (a) and in nanostructured carbons prepared by subsequent pyrolysis. (b) SFM phase images (left, copolymer; right, carbon). (Reproduced with permission from *Journal of the American Chemical Society* [30].)

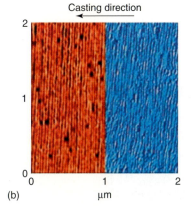

memory [216], photonic crystals [217], and optical waveguides [218–220], or as templates for transfer of pattern [6, 10, 19, 32, 80, 104, 152, 221], separation membranes [222, 223], and scaffolds for nanostructured inorganic materials [3, 4, 7, 171, 205, 206]. In this section, we will address the functionality of patterned BCP thin films.

12.3.1
Nanolithography

The fabrication of desired nanostructured patterns using BCPs has relied on achieving a mass thickness contrast, that is, patterning the polymer film thickness on the nanoscale, which is then transferred to the underlying substrate or replicated into another material [6, 10, 19, 32, 80, 104, 152, 221]. While one can rely on this aspect ratio to provide a natural etching contrast, other attempts have been made to enhance the contrast. A common feature of these approaches is to use a BCP where one block contains an inorganic element, like Si or Fe, or to load one of the phases with an etch-resistant inorganic component [224–228]. Harrison *et al.* [229] have published a comprehensive review of the transfer of patterns targeted to those familiar with traditional lithographic techniques. Park *et al.* [221] first demonstrated the use of BCP thin films as etch masks to transfer patterns to an underlying silicon nitride substrate. The film can be used as a positive mask by removing the minority block and then using a CF_4 reactive ion etching (RIE) to transfer the pattern of holes to an underlying silicon nitride substrate as shown in Figure 12.30. A negative mask (Figure 12.30c) can be made of the same BCP that will generate an array of posts instead of holes (Figure 12.30b). In this case, the polydiene is stained with

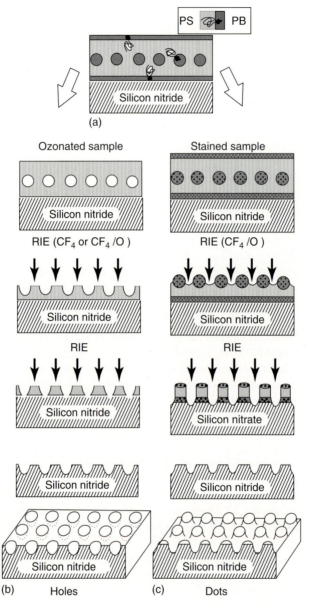

Figure 12.30 (a) Schematic illustration for BCP lithography. The sequential removal of one component and further etching of the sample replicate patterns on the order of tens of nanometers to a silicon nitride surface. Route (b) demonstrates a positive resist, while route (c) demonstrates a negative one. When an ozonated (b) and osmium-stained (c) BCP thin films were used, nano-sized holes and dotted structures were produced in silicon nitride, respectively. (Reproduced with permission from *Science* [221].)

osmium tetroxide so that the matrix will etch in CF_4, while the minority block is now etch resistant.

Similarly, PS-*b*-PMMA is a promising choice for nanolithographic applications since the nanodomains of PMMA may be degraded by radiation and then removed with acetic acid [6, 8, 12, 19, 152, 155, 212] to create a positive mask. While these etching techniques lead to a huge density of holes or pits on the surface, achieving a suitable aspect ratio has been a problem due to the spherical shape of the nanodomains that form the mask and the relatively fast etching rate of the polymeric mask. Alternatively, many of the etching techniques used above can be transferred to underlying metal layers. In these cases, however, the etching contrast between the developed and undeveloped regions must be quite large due to the slower etch rate of the metals. PFS have a significantly lower etch rate than most organic polymers, so selective etching of the PS domains in a PS-*b*-PFS BCP happens automatically and yields high aspect ratios [230–233]. Using a multilayer process, this pattern can be transferred first into a dielectric such as SiO_x and then into a magnetic material, such as cobalt, nickel, or iron, for use as magnetic data-storage media [24, 225, 230, 234]. For similar applications, Russell and coworkers [233] have used RIE to transfer the pattern from a nanoporous PS film into a FeF_2/Fe bilayer, while Asakawa and coworkers [8, 235] have used a spin-on-glass to create an intermediate mask from a similar nanoporous PS film and then used the glass dots as a mask to ion mill an underlying magnetic film. Ober and coworkers [208] have explored the use of small-molecule additives to enhance the phase-selective chemistry available in BCP systems, merging the BCP lithographic routes with more standard light-driven pattern development processes of the microelectronics industry. Spatz *et al.* [236] quarternized PS-*b*-P2VP with auric acid, and then deposited gold in the P2VP microdomain, to generate masks for nanolithography. Alternatively, Ti was grown on top of the PS matrix to yield sufficient contrast [237].

Without removing minor component, nanoporous BCP templates can be produced by surface reconstruction process [32, 212, 213]. Details for surface reconstruction will be addressed in Section 12.3.8. Park *et al.* reported that highly oriented cylindrical microdomains with a long-range lateral order in PS-*b*-P4VP BCP thin films are achieved by spin coating the copolymer from a mixed solvent (toluene/THF). Cylindrical microdomains normal to the surface is obtained directly after spin coating and this process is independent of underlying substrate. On annealing these films by solvent, arrays of cylindrical microdomains with a high degree of lateral order can be achieved with a little changing the interdomain distance. After dipping the films into a preferential solvent for P4VP, which is ethanol in this work, a nanoporous film can be produced by a fully reversible reconstruction process that leaves P4VP on the surface. Evaporation of Au (which interacts strongly with P4VP) on the surface of the film at a glancing angle leaves a nanoporous Au film that is ideal for RIE with high etch contrast for transfer of the pattern to the underlying substrate. The results described here are obtained by solvents and simple fabrication steps that are cost effective and fully compatible with current industrial processes.

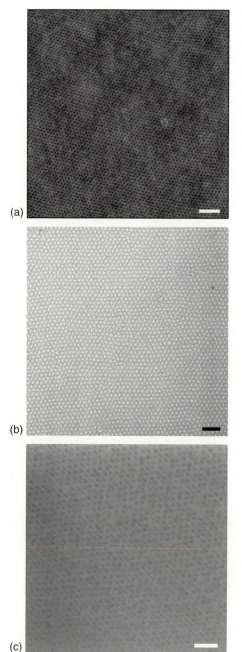

(a)

(b)

(c)

Figure 12.31 SFM, TEM, and SEM images of highly ordered PS-*b*-P4VP films with small molecular weight. (a) SFM image of surface reconstructed films. (b) TEM image of Au-evaporated films after surface reconstruction. (c) SEM images of highly ordered nanoporous templates fabricated on the silicon substrate via reactive ion etching (scale bar: 100 nm). (Reproduced with permission from *ACS Nano* [32].)

Figure 12.31 shows SFM and TEM images of PS-*b*-P4VP after spin coating and solvent annealing with a toluene/THF mixture, followed by swelling with alcohol and drying, to produce a nanoporous template [32]. After solvent annealing, a lateral order of cylindrical microdomains oriented normal to the film surface was significantly improved. When the films were immersed in a preferential solvent for the minor component, the nanopores were generated without changing the high degree of order (Figure 12.31a). Subsequently, a ~1-nm thick layer of gold was evaporated onto the surface of the films at a glancing angle, which yielded the TEM image in Figure 12.31b. The gold-coated films were exposed to a CF_4 RIE at 65 W for 40 s. Figure 12.31c shows SEM images of an array of holes that were etched into the underlying silicon oxide, which are identical to that seen in the original template. Top view (Figure 12.31c) indicates a highly ordered nanoporous structures with areal density of $\sim10^{11}/cm^2$ holes. It should be noted that the fidelity of the transfer could not be achieved without the use of the gold layer to enhance the etching contrast. In addition, alternative lift-off procedures by toluene/THF mixture could be used as effectively.

12.3.2
Shallow-Trench Array Capacitor

The high feature density provided by self-assembled BCP patterns present an avenue to IC performance benefits through controlled material nanostructuring. The following sections illustrate this idea using examples in which materials and devices incorporate designed nanostructuring in order to enhance properties such as surface area, optical refractive index, porosity, and feature density. The advantage provided by material nanostructuring is these applications stems from the intrinsic nanometer-scale uniformity and high feature density of self-assembled BCP patterns. The nanostructured materials can offer real performance, and the relatively loose requirements on the sublithographic patterning process in these applications mean easier incorporation of the process into future microelectronics technology generations [12].

Black *et al.* [238, 239] demonstrated the use of BCP self-assembly as a method for increasing capacitor surface area without introducing additional process complexity, creating high capacitance-density devices composed of arrays of nanometer-scale shallow trenches. This approach is a variation of a planar metal-oxide semiconductor (MOS) structure that combines the benefits of surface roughening and dynamic random access memory (DRAM) deep-trench devices. They have fabricated shallow-trench arrays by first transferring the self-assembled BCP pattern into a more rugged dielectric hard mask, which is used for further transfer into the device Si counterelectrode. The schematic process is shown in Figure 12.32a [240]. A silicon-gate plasma etch produces dense arrays of shallow trenches with aspect ratios of more than 5 to 1 (Figure 12.32b), and an average pore diameter of ~20 nm. A cross-sectional TEM image of part of a completed device (Figure 12.32d) shows three pores of the shallow-trench array lined with a 4.5-nm SiO_2 gate dielectric (light-colored) and filled with a tantalum nitride (TaN)

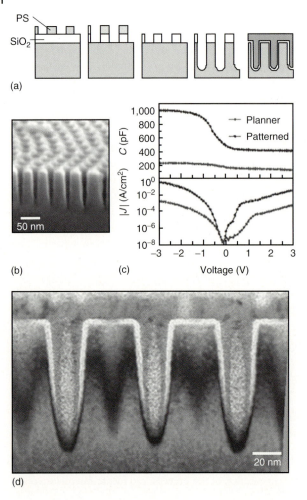

(a)

(b)

(c)

(d)

Figure 12.32 Shallow-trench-array capacitor. (a) Schematic process flow for shallow-trench-array decoupling capacitor fabrication. (b) 70°-angle SEM image of shallow-trench-array MOS capacitor bottom Si electrode. Etched pores have a depth of 100 nm. (c) Upper chart: Capacitance vs. voltage for planar (solid circles) and patterned (open circles) devices of the same lateral area. Lower chart: Leakage current per lateral device area for planar (sold circles) and patterned (open circles) devices. (d) Cross-sectional TEM image of completed shallow-trench-array MOS decoupling capacitor. (Reproduced with permission from *IBM J. Res. Dev.* [12].)

top electrode. Narrow pore diameters and relatively high trench aspect ratios make gate-electrode formation challenging, and they used TaN atomic-layer deposition (ALD) to conformally coat the gate oxide [241].

One deleterious effect of capacitor-electrode nanostructuring is a corresponding increase in device leakage current. The leakage current per lateral device area for the shallow-trench array capacitor is roughly 100 times higher than that for the planar device (Figure 12.32c) – a much greater increase than is explained by the increased device surface area alone (4.1 times). The excess current likely stems from higher charge tunneling rates in the high-curvature trench bottoms where electric fields are enhanced by a factor of 1.45 [242]. However, the shallow-trench array demonstrates a type of performance tradeoff that is often encountered in device design-in this case, the need to strike a balance between enhanced capacitance and appropriate leakage current levels. Dimensional tenability is a real advantage of self-assembly processes for nanostructuring surfaces. In addition to controlling the amount of surface area increase through the shallow-trench aspect ratio, they further controlled the surface area of the shallow-trench array by adjusting the pore diameter [240], a dimension that can be tuned either via the initial polymer template [243, 244] or by using a postetch widening process that controls trench diameter independently of pore separation [245].

Nanostructuring electrode surfaces using polymer self-assembly provides a realistic method for achieving increases of up to ~10 times in capacitance over that of structures with equivalent planar area by using relatively standard fabrication processes and without introducing novel dielectric materials. These shallow-trench-array devices are compatible with high-performance thin silicon-on-insulator (SOI) circuits, which require special consideration because of their smaller intrinsic circuit capacitance and lack of n-well capacitance. Because these large-area devices consist of many shallow trenches, their performance tolerates variations in the self-assembly process (pattern defects and other imperfections). Device electrical properties (in this case capacitance) therefore do not depend on statistical variations in the self-assembled pattern. Furthermore, our use of self-assembly in device fabrication does not require registration of the polymer pattern to other device lithographic levels.

12.3.3
Nanocrystal Flash Memory

Black *et al.* [12] described another example of a device performance benefit derived from controlling the nanostructure of a material using self-assembled BCP patterns. In this case the nanostructures material plays a critical role in an active electronic device, serving as the charge-storage node of a flash-memory transistor. This strategy is compatible with conventional semiconductor technologies and it has attractive applications in the fabrication of semiconductor devices [246]. Figure 12.33a illustrates the process of the fabrication of MOS capacitors with increased charge-storage capacity [246]. The device was fabricated by using a nanoporous PS-*b*-PMMA film as a mask for the etching of SiO_2 substrate and thereby creating a template (Figure 12.33b) for further patterning of the silicone electrode (Figure 12.33d). After removing the BCP and SiO_2 layers from the surface, a new thin SiO_2 layer was grown on the topographically patterned silicone substrate. This step was followed by the subsequent deposition of TaN onto the SiO_2 layer. The resulting capacitor shown in Figure 12.33d had more than 400% higher charge-storage capacity than an analogous capacitor formed on the smooth

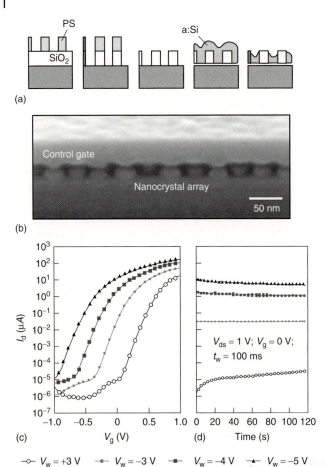

(a)

(b)

(c)

(d)

$-\!\circ\!-\ V_w = +3\ V$ $-\!\bullet\!-\ V_w = -3\ V$ $-\!\blacksquare\!-\ V_w = -4\ V$ $-\!\blacktriangle\!-\ V_w = -5\ V$

Figure 12.33 Nanocrystal flash memory. (a) Schematic process flow for nanocrystal floating-gate fabrication. (b) 70°-angle SEM of completed nanocrystal floating-gate structure. (c) Plot of I_d vs. V_g for a nanocrystal field-effect-transistor (FET) in erased state (open circles) and written states (filled symbols). Erase voltage was +3 V and write voltages were −3, −4, and −5 V. Write and erase times were 100 ms. (d) I_d vs. time for a nanocrystal FET in erased state ($V_w = +3$ V) (open circles) and written states ($V_w = -3, -4,$ and −5 V) (filled symbols). (Reproduced with permission from *IBM J. Res. Dev.* [12].)

silicone surface, although this increase was accompanied by the increased leakage current per lateral device area (Figure 12.33c).

12.3.4
Photonic Devices

It is well known that planar multilayered dielectric systems exhibit useful optical properties, as evidenced by the widespread use of various bandpass and notch optical filters. These "one-dimensional photonic crystals" selectively reflect a certain wavelength of light owing to their periodic structure and Fabry–Perot interference effects [247]. The reflected wavelength (position of the optical bandgap) is directly related to the layer periodicity and the dielectric contrast, that is, the difference in the index of refraction between the component materials in the layered system. Indeed, if designed correctly, as demonstrated recently by Fink and coworkers [248], the bandgaps of alternating multilayered systems shed their angular dependence. Such omnidirectional dielectric reflectors can serve as peerless waveguides capable of directing light though near hairpin turns without loss. The self-assembled morphologies of BCP systems make them intriguing candidates for photonic applications. This is especially true of BCP thin-film systems that spontaneously exhibit well-defined alternating layered structures. Fink *et al.* [249] have succinctly outlined the materials parameters involved in transforming BCP systems into effective photonic materials. Recently, Thomas and coworkers have utilized thin films of a high molecular weight PS-*b*-PI BCP as a narrow spectral-band selective feedback element for constructing a laser cavity [250]. With fluorescent organic laser dyes in a PMMA polymer matrix as a gain medium, optically pumped surface emitting lasing action has been demonstrated. Figure 12.34 shows a highly directional stimulated emission from the sample surface at a pump power greater than the lasing threshold, which is clear evidence for lasing. This BCP-based photonic structure opens up the possibility of creating all-organic, flexible, and self-assembled laser devices with fast and low-cost processing.

12.3.5
Planar Optical Waveguide

While polymer-based optical waveguide materials have been widely discussed [251], little attention has been paid to BCP systems [218] in regard to their

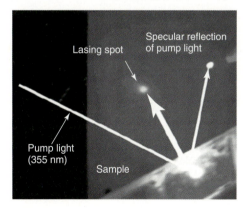

Figure 12.34 Photograph of the 410-nm lasing from the block-copolymer-based laser structure. A highly directional lasing output in the backward direction was observed on a white background. (Reproduced with permission from *Nano Letters* [250].)

applications in the field of optical elements. Kim *et al.* [218] demonstrated that thin films of BCPs with controlled microdomain orientation can be employed as planar optical waveguides and they investigated waveguiding properties by optical waveguide spectroscopy (OWS) [219, 220]. Thin films of mixtures of PS-*b*-PMMA and PMMA homopolymers with the cylindrical PMMA microdomains oriented normal to the film plane were used to couple waveguide optical modes, and typical nanofabrication processes occurring inside the layer were monitored by OWS. The confinement of the PMMA homopolymer to the microdomains markedly enhances the aspect ratio of the microdomain orientation by ∼10 times the bulk period [158], which makes the film appropriate as model BCP waveguide layers. Resonance coupling between surface plasmons (or plasmon surface polaritons) and incident photons can occur at a metal/dielectric interface, especially in the experimental setup known as the *Kretschmann configuration* (Figure 12.35) [219, 252]. If the thickness of the dielectric layer is further increased (in organic polymeric layers typically thicker than ∼200 nm), waveguide optical modes, in addition to surface plasmon resonance, can be observed [219, 220].

12.3.6
Incorporation of Nanocrystals and Nanoreactors

Quantum size effects and large surface to volume ratios contribute to the unusual properties of inorganic nanoparticles and studies of these effects have sparked great interests in novel fabrication processes of metal or metal oxide nanoparticles. The synthesis of nanoparticles from BCP microdomains is an effective route to control the size distributions, shapes, and spatial placements of nanoparticles. In the simplest rendition, one may use BCPs where one block has a higher affinity to the

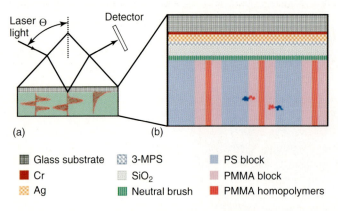

▦ Glass substrate	▨ 3-MPS	▧ PS block
■ Cr	░ SiO₂	░ PMMA block
▨ Ag	▥ Neutral brush	▥ PMMA homopolymers

Figure 12.35 (a) Schematic diagram of the OWS setup based on the Kretschmann configuration, and of the idealized field distributions of several guided modes in the waveguiding layer. (b) Schematic diagram of a thin film of PS-*b*-PMMA/PMMA homopolymer mixture with PMMA microdomains aligned normal to the substrate surface fabricated onto glass substrate. 3-MPS: 3-mercaptopropyl trimethoxysilane. (Reproduced with permission from *Advanced Materials* [218].)

inorganics. For instance, Lopes and Jaeger [7] demonstrated that metals deposited directly to the surface of a film would segregate into the underlying microdomains. The formation of ordered patterns is also possible by assembling BCP micelles upon casting, which has evoked a major interest in potential applications. Micellar cores offer a unique microenvironment, namely a nanoreactor, where inorganic precursors can be loaded and then processed by wet-chemical methods to produce nanoparticles with a narrow size distribution, in a similar way as is done with microemulsions [253]. The ordered deposition of gold and silver nanoclusters from micellar PS-*b*-PVP [254, 255] and PS-*b*-PAA [256] have been reported.

12.3.7
Nanoporous Membrane for Filtration of Viruses

Ultrafiltration membranes with small pore sizes have been used for the separation of viruses [257, 258]. However, they were not very effective, since the virus particles permeate through a small number of abnormally large-sized pores [222]. Track-etched polycarbonate (PC) and anodized aluminum oxide (AAO) membranes with uniform pore sizes have also been studied for the separation of viruses. While the pore-size distributions are narrow for these membranes, both types of membranes show a very low flux for virus separation [258]. Thus, a new type of membrane, providing both high selectivity and high flux, was needed for this purpose. Recently, Yang *et al.* [222] introduced a new membrane with an asymmetric film geometry, which shows both high selectivity and high flux. Figure 12.36 shows a schematic diagram of the fabrication of asymmetric nanoporous membranes. This membrane consists of a thin nanoporous layer, prepared from a PS-*b*-PMMA BCP template (~80-nm thick film with cylindrical pores of ~15 nm in diameter and a narrow pore-size distribution), and a supported membrane that provides mechanical strength. This asymmetric membrane showed ultrahigh selectivity while maintaining a high flux for the separation of human rhinovirus type 14 (HRV14), which has a diameter of ~30 nm [259]. This virus is a major pathogen for the common cold in humans. Since the pore diameter in the top layer can be tuned from 10 to 40 nm by changing the molecular weights of BCP or by adding a homopolymer miscible with the minor component of the BCP [160, 260], the cutoff size of the membrane filter could be precisely controlled.

With these pore sizes, these asymmetric membranes allow biomolecules like proteins, present in the unfiltered solution, to pass through the membrane, while only the viruses are screened. The unique characteristic of this new membrane filter eliminates the risk of contamination from viruses while processing biotherapeutic proteins such as vaccines and hormones. Therefore, this new membrane can be used to develop new types of blood-filtering systems, such as a hemodialysis membrane, that is, free of the risk of viral infection.

Peinemann *et al.* [223] reported similar results as an example of the combination of different thermodynamic effects by subjecting a concentrated microphase-separating BCP solution to a nonsolvent, which is miscible with the solvent and thus exchanges with it. The resulting integral asymmetric structure

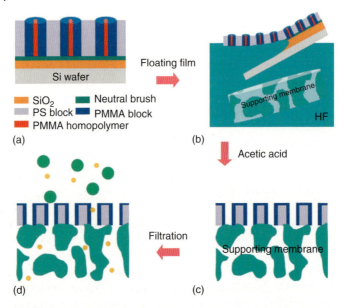

Figure 12.36 Schematic illustration of the procedure for the fabrication of asymmetric nanoporous membranes. PMMA cylindrical microdomains are oriented normal to the surface in films up to ~300 nm (a) and the film is transferred onto the polysulfone microfiltration membrane (b). When the films were immersed in a preferential solvent for PMMA homopolymer, porous thin films are prepared onto the supporting membrane (c) and can be used as a membrane for separating bio-molecules (d). (Reproduced with permission from *Advanced Materials* [222].)

of the BCP is a proof of principle to generate highly regular and narrow disperse membranes based on this concept. This work opens up a new and facile way to generate separating membranes with pore sizes on the mesoscopic length scale.

12.3.8
Nanoporous Templates

Porosity control of inorganic and organic nanoporous materials is increasingly critical for high-technology applications, such as filtration membranes, patterned templates, and photonic materials. Owing to their ability to form periodically ordered nanostructures, BCPs have received much attention as templates, scaffolds, and masks for nanoporous and mesoporous materials that have potential applications from optics to microelectronics. Several routes to generate nanoporous films via BCP self-assembly have been reported since Nakahama and coworkers [261] first demonstrated the formation of nanoporous polymer films from a siloxane-functionalized PS-*b*-PI system. The most common strategy to generate nanopores in polymer matrices is selective chemical or physical degradation and removal of minor microdomains. Numerous chemical means, such as ozonolysis [225, 230, 236, 262–264], thermal degradation [226, 265–267], UV degradation

[6, 19, 27, 268–271], "soft" chemical etch [272–275], as well as cleavable junction point [276–278], and physical means, such as volume contraction triggered by cross-linking [260], solvent-induced surface reconstruction [32, 212, 213], and extraction of homopolymer or organic molecules in a blend with BCP matrix [159, 187], have been employed to create nanoporous polymer films. Because of the uniform domain size and mild removal conditions, the precise control can be achieved to yield well-defined monolithic materials.

Smith and Meier [263] reported on the degradation of the polydiene component in PS-*b*-PB or PS-*b*-PI diBCPs by ozonolysis. This work demonstrated that the ozonolysis can effectively cleaves the double bonds and remove the polydiene component without negatively affecting the uncross-linked PS domains. Later, Park *et al.* [221] described tactics for preparing either pits or posts of silicon nitride using BCP masks and a combination of chemical modification and RIE. In the protocol for preparing holes in a silicon nitride substrate, a monolayer of spherical inclusions of polydiene was formed by annealing a PS-*b*-PI or PS-*b*-PB BCP on the substrate. Ozonolysis of the film degrades the PI and cross-links the PS matrix, leaving a thin, cross-linked film of PS containing nanoscopic spherical cavities. With RIE, the film is uniformly etched until the ion beam encounters the spherical cavities. At the location of the cavities, the etching front is advanced by an amount that is commensurate with the size of the spherical cavities and as the etching proceeds to the substrate, a pattern of holes that replicates the pattern in the film is transferred to the layer of silicon nitride with high fidelity.

Hedrick *et al.* [265–267] elegantly demonstrated an approach to generate porous materials, termed *nanofoams*, using polyimide-based materials. With the aim of preparing interlayer dielectrics with low dielectric constants to prevent "cross-talk" in microelectronic devices, polyimide-containing thermally labile blocks such as PMMA or poly(propylene oxide) (PPO) was used. Upon heating these materials above the decomposition temperatures of the minor blocks, the volatile by-products are removed, leaving behind voids in the polyimide with inherently high glass-transition temperature. Watkins and coworkers [275] used supercritical fluids (SCFs) to swell all of the microdomains in a triBCP of PEO-*b*-PPO-*b*-PEO. A BCP template containing hydrophobic PPO and hydrophilic PEO segments is prepared by spin coating from a solution containing p-toluene sulfonic acid as an acid catalyst. Tetraethyl orthosilicate (TEOS) was selectively infused into the hydrophilic SCF-swollen PEO matrix at 60 °C and, by calcination in air at 400 °C, TEOS was converted to silicon oxide and the PEO-*b*-PPO-*b*-PEO fully degraded. This leaves behind a mesoporous silica replica of the original microdomain morphology. These materials are being considered for low dielectric constant applications in advanced microelectronics.

In the work described above, many materials formed spherical microdomains in the thin film. A top layer of low surface energy block was always present and the spherical inclusions were buried in the film. Thus, the masks were generated in an anisotropic manner under etching conditions. An ideal mask, however, should have channels that were open on the surface of the film and extended through the entire film. Porous materials of this nature would allow for direct deposition (or growth) of materials in the holes, and following lift-off, leave the corresponding patterned surfaces. One way to achieve this porous mask using BCPs is to employ cylinder-forming materials that could be manipulated in such a way as to allow for the cylinder orientation perpendicular to the substrate. Subsequent removal of the component that formed the cylinders would leave the nanolithographic masks. The most effective, well-studied system is the PS-*b*-PMMA system by Russell and coworkers [19]. The random copolymer film was used to render the surface ambivalent toward adsorption of either the PS or PMMA block and thus resulted in a thermodynamic preference for the perpendicular orientation. Exposure to UV radiation cross-links PS block, degrades PMMA, and produces a PS film with an array of nanopores that penetrates through the film via acetic acid rinsing, which is a selective solvent for PMMA [19]. The conditions for optimization of order in the starting porous PS films have been detailed – the relevant variables being film thickness and annealing time and temperature [279]. This templating processes developed by Hawker and Russell has been used in a flash memory application by Black and coworkers [238, 240, 244, 280]. Pattern transfer into silicon dioxide by RIE creates a dielectric hard mask. The pattern in the oxide layer can then be transferred into silicon by chemical etching to produce silicon channels. This enables a higher storage capacity in the semiconductor capacitors without increase in the device area. Whereas, Ober and coworkers [208, 209] described an interesting combination of "phase-selective" cross-linking chemistry and polymer degradation to generate nanoporous thin films from patternable BCP systems. They have developed a novel copolymer system using PαMS-*b*-PHS to achieve the spatial control through a high-resolution deep-UV lithographic process. Cylindrical microdomains of PαMS oriented perpendicular to the substrate were observed. Because of the high T_g of PHS and the low ceiling temperature of PαMS, UV irradiation of these films to generate free radicals at 80 °C led to depolymerization of the PαMS. By combining the lithographic characteristics of PHS with the degradation of PαMS, the micrometer-sized patterns of PαMS-*b*-PHS thin films could be generated by negative-tone photoresist technology. This combination enables hierarchical structures formed in thin films from the macroscale down to the nanoscale. These materials are presently employed as a supported porous thin film to separate proteins in a selected molecular weight range.

Chemical etching is one of the methods for nanoporous BCP templates. Recently, Hawker and coworkers [280] prepared BCPs with a cross-linkable monomer intentionally incorporated into one of the blocks. Copolymers containing thermally cross-linkable groups, such as benzocyclobutene (BCB), can be annealed to induce ordered microdomains and then subsequently heated to cross-link the system, locking in the structure and eliminating the need for a multistep process. Thin films (~30 nm) of poly[(styrene-*random*-benzocyclobutene)-*b*-lactic acid] (PSBCB-*b*-PLA) annealed at 170 °C, followed by cross-linking at 200 °C, produced the perpendicular cylindrical microdomains with PLA as the minor component. By washing these films with a weak base, the PLA block was degraded and removed, producing a nanoporous cross-linked template. Nanoporous poly(4-fluorostyrene) templates on gold-coated silicon/silicon oxide substrates were prepared by the electric-field alignment of poly(4-fluorostyrene)-*block*-poly(D,L-lactide) (PFSt-*b*-PLA) thin films followed by mild degradation of the PLA phase via a dilute aqueous base [274]. Such "soft" chemical etch method is generally useful for the preparation of templates and nanostructures that are sensitive to more aggressive removal processes. Recently, Russell and coworkers [276, 281] prepared PS-*b*-PMMA copolymers containing $[4\pi + 4\pi]$ anthracene (AA) photodimer at the junction point between two blocks (PS-AA-PMMA), which can be cleaved to the parent homopolymer blocks upon heating above 130 °C or by UV irradiation at 280 nm. Microphase-separated cylindrical morphologies with the cylinders oriented normal to the surface in thin films were initially achieved by annealing spin-coated films in supercritical carbon dioxide at 80 °C. Heating and/or UV irradiation can cleave the copolymer junction point, effecting an *in-situ* conversion of the copolymer to its parent homopolymers. Subsequent washing with selective solvents removed one homopolymer, producing nanoporous templates. Venkataraman [277] and Russell and coworkers [278] also placed a cleavable juncture, a triphenylmethyl (trityl) ether linkage, between PS and PEO, to make a long-range ordered nanoporous thin film. Lee *et al.* [282] reported the preparation of nanoporous crystalline sheets of penta-p-phenylene (PPP) by using the rod-coil PPP and PPO BCPs with a cleavable juncture. The copolymers can self-assemble into layered microdomains that contain sheets of perforated crystalline PPP in which the perforations filled with PPO [283]. The PPO segment is covalently bonded to PPP through an ester linkage. Hydrolysis of this ester linkage with aqueous KOH and subsequent removal of the coil segments from the ordered structure resulted in a hexagonally ordered perforated layered crystalline structure. One of the easiest methods for fabrication of BCP nanoporous templates is to use a solvent-induced surface-reconstruction process. Russell and coworkers [212] reported solvent-induced surface reconstruction, a very simple technique for generating nanoporosity in thin films of PS-*b*-PMMA. By modifying the interfacial interactions, arrays of PMMA cylinders oriented normal to the surface in PS-b-PMMA thin films could be produced. By exposing the films to acetic acid, PMMA was solvated, while the glassy PS matrix remained intact. Upon drying, a film reconstruction was observed where pores were opened in the positions of the original PMMA cylinders as the PMMA within the pores was transferred to the surface. The pore generation is reversible (and can be cycled); heating the film regenerates the dense film. Extended this concept to PS-*b*-P4VP copolymers, Russell and coworkers [32] also made a highly ordered nanoporous template with areal density of $\sim 10^{11}/cm^2$ holes. They evaporated a thin layer of gold (~ 2 nm) on the surface of the reconstructed films to prevent thermal-induced reversible process and the porous structure could be stabilized up to 200 °C. In addition, this thin gold layer acted as an etching mask, so that an enhanced etching depth with an aspect ratio of $\sim 3 : 1$ during pattern transfer could be achieved. This combination of solvent annealing and surface reconstruction elegantly showed that by simple manipulation of BCP thin films, nanoporous materials can be generated with narrow pore-size distributions and a remarkable long-range order.

12.4
Summary and Outlook

The self-assembly of BCPs is a promising way to create a variety of different nanosized periodic patterns. This chapter summarizes efforts for nanopatterning from BCP thin films and their applications such as nanoflash memory, data storage, separation membrane, nanolithography, and nanoporous templates. However, a more complete understanding of the potential for interactions between different controlling methods is necessary for BCPs to fulfill their technical potential for nanofabrication. Although periodic patterns are useful in some applications, the ability to precisely control domain positions and the desired placement of defects will permit the encoding of additional spatial information into self-assembled systems. Combining solvent, external electric, optical, mechanical, and flow fields with the templates will bring more control factors into play and may allow better control over defect types and positions.

In closing, the self-assembly of BCP combined with various methods will be appropriate in a wide range of emerging applications with precise spatial control at the nanoscale. In principle, one can take any combination of controlling forces, for example, chemically and/or topographically patterned substrates, electric field, solvent field, and incorporation of inorganic materials to generate unprecedented microdomain control, leading us one step closer to realization of future applications.

References

1. Fredrickson, G.H. and Bates, F.S. (1996) Dynamics of BCPs: theory and experiment. *Annu. Rev. Mater. Sci.*, **26**, 501–550.
2. Bates, F.S. and Fredrickson, G.H. (1999) BCPs-designer soft materials. *Phys. Today*, **52**, 32–38.
3. Hamley, I.W. (2003) Nanotechnology with soft materials. *Angew. Chem. Int. Ed.*, **42**, 1692–1712.
4. Lazzari, M., Lopez-Quintela, M., and BCPs, A. (2003) as a Tool for Nano-material Fabrication. *Adv. Mater.*, **15**, 1583–1594.
5. Förster, S. and Antonietti, M. (1998) Amphiphilic BCPs in structure-controlled nanomaterial hybrids. *Adv. Mater.*, **10**, 195–217.
6. Thurn-Albrecht, T., Schotter, J., Kästale, A., Emley, N., Shibauchi, T., Krusin-Elbaum, L., Guarini, K., Black, C.T., Tuominen, M.T., and Russell, T.P. (2000) Ultrahigh-density nanowire arrays grown in self-assembled DiBCP templates. *Science*, **290**, 2126–2129.
7. Lopes, W.A. and Jaeger, H.M. (2001) Hierarchical self-assembly of metal nanostructures on DiBCP scaffolds. *Nature*, **414**, 735–738.
8. Naito, K., Hieda, H., Sakurai, M., Kamata, Y., and Asakawa, K. (2002) 2.5-inch disk patterned media prepared by an artificially assisted self-assembling method. *IEEE Trans. Magn.*, **38**, 1949–1951.
9. Park, C., Yoon, J., and Thomas, E.L. (2003) Enabling nanotechnology with self-assembled BCP patterns. *Polymer*, **44**, 6725–6760.
10. Li, M., Coenjarts, C.A., and Ober, C.K. (2005) Patternable BCPs. *Adv. Polym. Sci.*, **190**, 183–226.
11. Nie, Z. and Kumacheva, E. (2008) Patterning surfaces with functional polymers. *Nature Mater.*, **7**, 277–290.
12. Black, C.T., Ruiz, R., Breyta, G., Cheng, J.Y., Colburn, M.E., Guarini, K.W., Kim, H.-C., and Zhang, Y. (2007)

Polymer self-assembly in semiconductor microelectronics. *IBM J. Res. Dev.*, **51**, 605–633.

13. Fan, H.J., Werner, P., and Zacharias, M. (2006) Semiconductor nanowires: from self-organization to patterned growth. *Small*, **2**, 700–717.

14. Fasolka, M.J. and Mayes, A.M. (2001) BCP thin films: physics and applications. *Annu. Rev. Mater. Res.*, **31**, 323–355.

15. Kim, G. and Libera, M. (1998) Morphological development in solvent-cast polystyrene-polybutadiene-polystyrene (SBS) TriBCP thin films. *Macromolecules*, **31**, 2569–2577.

16. Kimura, M., Misner, M.J., Xu, T., Kim, S.H., and Russell, T.P. (2003) Long-range ordering of DiBCPs induced by droplet pinning. *Langmuir*, **19**, 9910–9913.

17. Ludwigs, S., Böker, A., Voronov, A., Rehse, N., Magerle, R., and Krausch, G. (2003) Self-assembly of functional nanostructures from ABC TriBCPs. *Nature Mater.*, **2**, 744–747.

18. Kim, S.H., Misner, M.J., Xu, T., Kimura, M., and Russell, T.P. (2004) Highly oriented and ordered arrays from BCPs via solvent evaporation. *Adv. Mater.*, **16**, 226–231.

19. Thurn-Albrecht, T., Steiner, R., DeRouchey, J., Stafford, C.M., Huang, E., Bal, M., Tuominen, M.T., Hawker, C.J., and Russell, T.P. (2000) Nanoscopic templates from oriented BCP films. *Adv. Mater.*, **12**, 787–791.

20. Thurn-Albrecht, T., DeRouchey, J., and Russell, T.P. (2000) Overcoming interfacial interactions with electric fields. *Macromolecules*, **33**, 3250–3253.

21. Stoykovich, M.P., Müller, M., Kim, S.O., Solak, H.H., Edwards, E.W., de Pablo, J.J., and Nealey, P.F. (2005) Directed assembly of BCP blends into nonregular device-oriented structures. *Science*, **308**, 1442–1446.

22. Kim, S.O., Solak, H.H., Stoykovich, M.P., Ferrier, N.J., De Pablo, J.J., and Nealey, P.F. (2003) Epitaxial self-assembly of BCPs on lithographically defined nanopatterned substrates. *Nature*, **424**, 411–414.

23. Segalman, R.A., Yokoyama, H., and Kramer, E.J. (2001) Graphoepitaxy of spherical domain BCP films. *Adv. Mater.*, **13**, 1152–1155.

24. Cheng, J.Y., Ross, C.A., Thomas, E.L., Smith, H.I., and Vancso, G.J. (2003) Templated self-assembly of BCPs: effect of substrate topography. *Adv. Mater.*, **15**, 1599–1602.

25. Cheng, J.Y., Mayes, A.M., and Ross, C.A. (2004) Nanostructure engineering by templated self-assembly of BCPs. *Nature Mater.*, **3**, 823–828.

26. De Rosa, C., Park, C., Thomas, E.L., and Lotz, B. (2000) Microdomain patterns from directional eutectic solidification and epitaxy. *Nature*, **405**, 433–437.

27. Mansky, P., Liu, Y., Huang, E., Russell, T.P., and Hawker, C.J. (1997) Controlling polymer-surface interactions with random copolymer brushes. *Science*, **275**, 1458–1460.

28. Ryu, D.Y., Shin, K., Drockenmuller, E., Hawker, C.J., and Russell, T.P. (2005) A generalized approach to the modification of solid surfaces. *Science*, **308**, 236–239.

29. Bodycomb, J., Funaki, Y., Kimishima, K., and Hashimoto, T. (1999) Single-grain lamellar microdomain from a DiBCP. *Macromolecules*, **32**, 2075–2077.

30. Tang, C., Tracz, A., Kruk, M., Zhang, R., Smilgies, D.-M., Matyjaszewski, K., and Kowalewski, T. (2005) Long-range ordered thin films of BCPs prepared by zone-casting and their thermal conversion into ordered nanostructured carbon. *J. Am. Chem. Soc.*, **127**, 6918–6919.

31. Villar, M.A., Rueda, D.R., Ania, F., and Thomas, E.L. (2002) Study of oriented BCPs films obtained by roll-casting. *Polymer*, **43**, 5139–5145.

32. Park, S., Wang, J.-Y., Kim, B., Xu, J., and Russell, T.P. (2008) A simple route to highly oriented and ordered nanoporous BCP templates. *ACS Nano*, **2**, 766–772.

33. Amundson, K. and Helfand, E. (1993) Quasi-static mechanical properties of lamellar BCP microstructure. *Macromolecules*, **26**, 1324–1332.

34. Amundson, K., Helfand, E., Quan, X., and Smith, S.D. (1993) Alignment of lamellar BCP microstructure in an electric field. 1. Alignment kinetics. *Macromolecules*, **26**, 2698–2703.

35. Amundson, K., Helfand, E., Quan, X.N., Hudson, S.D., and Smith, S.D. (1994) Alignment of lamellar block-copolymer microstructure in an electric-field. 2. Mechanisms of alignment. *Macromolecules*, **27**, 6559–6570.

36. Tsori, Y. and Andelman, D. (2002) Thin film DiBCPs in an electric field: transition from perpendicular to parallel lamellae. *Macromolecules*, **35**, 5161–5170.

37. Amundson, K., Helfand, E., Davis, D.D., Quan, X., Patel, S.S., and Smith, S.D. (1991) Effect of an electric field on BCP microstructure. *Macromolecules*, **24**, 6546–6548.

38. Morkved, T.L., Lu, M., Urbas, A.M., Ehrichs, E.E., Jaeger, H.M., Mansky, P., and Russell, T.P. (1996) Local control of microdomain orientation in DiBCP thin films with electric fields. *Science*, **273**, 931–933.

39. Xu, T., Hawker, C.J., and Russell, T.P. (2003) Interfacial energy effects on the electric field alignment of symmetric DiBCPs. *Macromolecules*, **36**, 6178–6182.

40. Xu, T., Goldbach, J.T., and Russell, T.P. (2003) Sequential orthogonal fields: a path to long-range, 3-D ordering in BCP thin films. *Macromolecules*, **36**, 7296–7300.

41. Xu, T., Zhu, Y., Gido, S.P., and Russell, T.P. (2004) Electric field alignment of symmetric DiBCP thin films. *Macromolecules*, **37**, 2625–2629.

42. Xu, T., Zvelindovsky, A.V., Sevink, G.J.A., Gang, O., Ocko, B., Zhu, Y., Gido, S.P., and Russell, T.P. (2004) Electric field induced sphere-to-cylinder transition in DiBCP thin films. *Macromolecules*, **37**, 6980–6984.

43. Xu, T., Hawker, C.J., and Russell, T.P. (2005) Interfacial interaction dependence of microdomain orientation in symmetric DiBCP thin films. *Macromolecules*, **38**, 2802–2805.

44. Xu, T., Zvelindovsky, A.V., Sevink, G.J.A., Lyakhova, K.S., Jinnai, H., and Russell, T.P. (2005) Electric-field alignment of asymmetric DiBCP thin films. *Macromolecules*, **38**, 10788–10798.

45. Keller, A., Pedemont, E., and Willmout, F.M. (1970) Macro-lattice from segregated amorphous phases of a three BCP. *Nature*, **225**, 538–539.

46. Mortensen, K., Almdal, K., Bates, F.S., Koppi, K., Tirrell, M., and Nordén, B. (1995) Shear devices for in situ structural studies of block-copolymer melts and solutions. *Physica B*, **213–214**, 682–684.

47. Albalak, R.J. and Thomas, E.L. (1994) Roll casting of BCPs and BCP-homopolymer blends. *J. Polym. Sci. B: Polym. Phys.*, **32**, 341–350.

48. Albalak, R.J. and Thomas, E.L. (1993) Microphase separation of BCP solutions in a flow field. *J. Polym. Sci. B: Polym. Phys.*, **31**, 37–46.

49. Albalak, R.J., Thomas, E.L., and Capel, M.S. (1997) Thermal annealing of roll-cast TriBCP films. *Polymer*, **38**, 3819–3825.

50. Cohen, Y., Albalak, R.J., Dair, B.J., Capel, M.S., and Thomas, E.L. (2000) Deformation of oriented lamellar block-copolymer films. *Macromolecules*, **33**, 6502–6516.

51. Prasman, E. and Thomas, E.L. (1998) High-strain tensile deformation of a sphere-forming TriBCP/mineral oil blend. *J. Polym. Sci. B: Polym. Phys.*, **36**, 1625–1636.

52. Dair, B.J., Avgeropoulos, A., Hadjichristidis, N., Capel, M., and Thomas, E.L. (2000) Oriented double gyroid films via roll casting. *Polymer*, **41**, 6231–6236.

53. Deng, T., Ha, Y.H., Cheng, J.Y., Ross, C.A., and Thomas, E.L. (2002) Micropatterning of BCP solutions. *Langmuir*, **18**, 6719–6722.

54. Angelescu, D.E., Waller, J.H., Adamson, D.H., Deshpande, P., Chou, S.Y., Register, R.A., and Chaikin, P.M. (2004) Orientation of BCP cylinders in single-layer films by shearing. *Adv. Mater.*, **16**, 1736–1740.

55. de Gennes, P.G. and Prost, J. (1993) *The Physics Of Liquid Crystals*, 2nd edn, Clarendon Press, Oxford University Press, Oxford, New York, p. 597.

56. Grigorova, T., Pispas, S., Hadjichristidis, N., and Thurn-Albrecht, T. (2005) Magnetic field induced orientation in diblock-copolymers with one crystallizable block. *Macromolecules*, **38**, 7430–7433.

57. Tao, Y., Zohar, H., Olsen, B.D., and Segalman, R.A. (2007) Hierarchical nanostructure control in rod-coil BCPs with magnetic fields. *Nano Lett.*, **7**, 2742–2746.

58. Smith, H.I. and Flanders, D.C. (1978) Oriented crystal growth on amorphous substrates using artificial surface-relief gratings. *Appl. Phys. Lett.*, **32**, 349–350.

59. Marrian, C.R.K. and Tennant, D.M. (2003) Nanofabrication. *J. Vac. Sci. Technol. A*, **21**, S207–S215.

60. Hatzakis, M. (1974) Electron beam processing systems. *Polym. Eng. Sci.*, **14**, 516–517.

61. Hu, W.C., Sarveswaran, K., Lieberman, M., and Bernstein, G.H. (2004) Sub-10 nm electron beam lithography using cold development of poly(methyl methacrylate). *J. Vac. Sci. Technol. B*, **22**, 1711–1716.

62. Word, M.J., Adesida, I., and Berger, P.R. (2003) Nanometer-period gratings in hydrogen silsesquioxane fabricated by electron beam lithography. *J. Vac. Sci. Technol., B*, **21**, L12–L15.

63. Cheng, J.Y., Ross, C.A., Smith, H.I., and Thomas, E.L. (2006) Templated self-assembly of BCPs: top-down helps bottom-up. *Adv. Mater.*, **18**, 2505–2521.

64. Solak, H.H., David, C., Govrecht, J., Golovkina, V., Cerrina, F., Kim, S.O., and Nealey, P.F. (2003) Sub-50 nm period patterns with EUV interference lithography. *Microelectron. Eng.*, **56**, 67–68.

65. Campbell, M., Sharp, D.N., Harrison, M.T., Denning, R.G., and Turberfield, A.J. (2000) Fabrication of photonic crystals for the visible spectrum by holographic lithography. *Nature*, **404**, 53–56.

66. Ullal, C.K., Maldovan, M., Wolgemuth, M., White, C.A., Yang, S., and Thomas, E.L. (2003) Triply periodic bicontinuous structures through interference lithography: a level-set approach. *J. Opt. Soc. Am.*, **A20**, 948–954.

67. Xia, Y.N. and Whitesides, G.M. (1998) Soft lithography. *Annu. Rev. Mater. Sci.*, **28**, 153–184.

68. Gates, B.D., Xu, Q.B., Stewar, M., Ryan, D., Wilson, C.G., and Whitesides, G.M. (2005) New approaches to nanofabrication: molding, printing, and other techniques. *Chem. Rev.*, **105**, 1171–1196.

69. Chou, S.Y., Krauss, P.R., and Renstrom, P.J. (1996) Sub-10 nm imprint lithography and applications. *J. Vac. Sci. Technol. B*, **14**, 4129–4133.

70. Resnick, D., Sreenivasan, S.V., and Wilson, C.G. (2005) Step and flash imprint lithography. *Mater. Today*, **8**, 34–42.

71. Segalman, R.A., Hexemer, A., Hayward, R.C., and Kramer, E.J. (2003) Ordering and melting of BCP spherical domains in 2 and 3 dimensions. *Macromolecules*, **36**, 3272–3288.

72. Hexemer, A., Stein, G.E., Kramer, E.J., and Magonov, S. (2005) BCP monolayer structure measured with scanning force microscopy Moire patterns. *Macromolecules*, **38**, 7083–7089.

73. Segalman, R.A., Hexemer, A., and Kramer, E.J. (2003) Effects of lateral confinement on order in spherical domain BCP thin films. *Macromolecules*, **36**, 6831–6839.

74. Segalman, R.A., Hexemer, A., and Kramer, E.J. (2003) Edge effects on the order and freezing of a 2D array of BCP spheres. *Phys. Rev. Lett.*, **91**, 196101.

75. Lambooy, P., Russell, T.P., Kellogg, G.J., Mayes, A.M., Gallagher, P.D., and Satija, S.K. (1994) Observed frustration in confined BCPs. *Phys. Rev. Lett.*, **72**, 2899–2902.

76. Koneripalli, N., Singh, N., Levicky, R., Bates, F.S., Gallagher, P.D., and Satija, S.K. (1995) Confined BCP thin films. *Macromolecules*, **28**, 2897–2904.

77. Kellogg, G.J., Walton, D.G., Mayes, A.M., Lambooy, P., Russell, T.P., Gallagher, P.D., and Satija, S.K. (1996) Observed surface energy effects in confined DiBCPs. *Phys. Rev. Lett.*, **76**, 2503–2506.

78. Huang, E., Russell, T.P., Harrison, C., Chaikin, P.M., Register, R.A., Hawker, C.J., and Mays, J. (1998) Using surface active random copolymers to control the domain orientation in DiBCP thin films. *Macromolecules*, **31**, 7641–7650.

79. Sundrani, D., Darling, S.B., and Sibener, S.J. (2004) Directing the self-assembly of nanoscale polymeric templates. *Langmuir*, **20**, 5091–5099.

80. Xiao, S., Yang, X., Edward, E.W., La, Y., and Nealey, P.F. (2005) Graphoepitaxy of cylinder-forming BCPs for use as templates to pattern magnetic metal dot arrays. *Nanotechnology*, **16**, S324–S329.

81. Turner, M.S. (1992) Equilibrium properties of a DiBCP lamellar phase confined between flat plates. *Phys. Rev. Lett.*, **69**, 1788–1791.

82. Walton, D.G., Kellogg, G.J., Mayes, A.M., Lambooy, P., and Russell, T.P. (1994) A free energy model for confined DiBCPs. *Macromolecules*, **27**, 6225–6228.

83. Abetz, V., Stadler, R., and Leibler, L. (1996) Order-disorder and order-order transitions in AB and ABC BCPs: description by a simple model. *Polym. Bull.*, **37**, 135–142.

84. Stadler, R., Auschra, C., Beckmann, J., Krappe, U., Voigt-Martin, I., and Leibler, L. (1995) Morphology and thermodynamics of symmetric poly(A-block-B-block-C) TriBCPs. *Macromolecules*, **28**, 3080–3091.

85. Zhang, Q., Lee, D.C., Yu, L., and Sibener, S.J. (2005) *Stability of Thin Films and Nanostructure*, Materials Research Society, Warrendale, PA.

86. Bita, I., Yang, J.K.W., Jung, Y.S., Ross, C.A., Thomas, E.L., and Berggren, K.K. (2008) Graphoepitaxy of self-assembled BCPs on two-dimensional periodic patterned templates. *Science*, **321**, 939–943.

87. Huang, K. and Balazs, A.C. (1991) Modeling copolymer adsorption on laterally heterogeneous surfaces. *Phys. Rev. Lett.*, **66**, 620–623.

88. Pereira, G.G. and Williams, D.R.M. (1998) DiBCP thin films on heterogeneous striped surfaces: commensurate, incommensurate and inverted lamellae. *Phys. Rev. Lett.*, **80**, 2849–2852.

89. Pereira, G.G. and Williams, D.R.M. (1998) Equilibrium properties of DiBCP thin films on a heterogeneous, striped surface. *Macromolecules*, **31**, 5904–5915.

90. Pereira, G.G. and Williams, D.R.M. (1999) DiBCP thin film melts on striped, heterogeneous surfaces: parallel, perpendicular and mixed lamellar morphologies. *Macromolecules*, **32**, 758–764.

91. Wang, Q., Nealey, P.F., and de Pablo, J.J. (2003) Simulations of the morphology of cylinder-forming asymmetric DiBCP thin films on nanopatterned substrates. *Macromolecules*, **36**, 1731–1740.

92. Balazs, A.C., Huang, K., Mcelwain, P., and Brady, J.E. (1991) Polymer adsorption on laterally heterogeneous surfaces: a Monte Carlo computer model. *Macromolecules*, **24**, 714–717.

93. Petera, D. and Muthukumar, M. (1998) Self-consistent field theory of DiBCP melts at patterned surfaces. *J. Chem. Phys.*, **109**, 5101–5107.

94. Heier, J., Kramer, E.J., Walheim, S., and Krausch, G. (1997) Thin DiBCP films on chemically heterogeneous surfaces. *Macromolecules*, **30**, 6610–6614.

95. Rockford, L., Liu, Y., Mansky, P., Russell, T.P., Yoon, M., and Mochrie, S.G.J. (1999) Polymers on nanoperiodic, heterogeneous surfaces. *Phys. Rev. Lett.*, **82**, 2602–2605.

96. Böltau, M., Walheim, S., Mlynek, J., Krausch, G., and Steiner, U. (1998) Surface induced structure formation of polymer blends on patterned substrates. *Nature*, **391**, 877–879.

97. Rockford, L., Mochrie, S.G.J., and Russell, T.P. (2001) Propagation of nanopatterned substrate templated ordering of BCPs in thick films. *Macromolecules*, **34**, 1487–1492.

98. Heier, J., Genzer, J., Kramer, E.J., Bates, F.S., Walheim, S., and Krausch, G. (1999) Transfer of a chemical substrate pattern into an island-forming DiBCP film. *J. Chem. Phys.*, **111**, 11101–11110.

99. Yang, X.M., Peters, R.D., Nealey, P.F., Solak, H.H., and Cerrina, F. (2000) Guided self-assembly of symmetric DiBCP films on chemically nanopatterned substrates. *Macromolecules*, **33**, 9575–9582.

100. Peters, R.D., Yang, X.M., Wang, Q., de Pablo, J.J., and Nealey, P.F. (2000) Combining advanced lithographic techniques and self-assembly of thin films of DiBCPs to produce templates for nanofabrication. *J. Vac. Sci. Technol. B*, **18**, 3530–3534.

101. Kim, S.O., Kim, B.H., Kim, K., Koo, C.M., Stoykovich, M.P., Nealey, P.F., and Solak, H.H. (2006) Defect structure in thin films of a lamellar BCP self-assembled on neutral homogeneous and chemically nanopatterned surfaces. *Macromolecules*, **39**, 5466–5470.

102. Edwards, E.W., Stoykovich, M.P., Solak, H.H., and Nealey, P.F. (2006) Long-range order and orientation of cylinder-forming BCPs on chemically nanopatterned striped surfaces. *Macromolecules*, **39**, 3598–3607.

103. Edwards, E.W., Montague, M.F., Solak, H.H., Hawker, C.J., and Nealey, P.F. (2004) Precise control over molecular dimensions of block-copolymer domains using the interfacial energy of chemically nanopatterned substrates. *Adv. Mater.*, **16**, 1315–1319.

104. Stoykovich, M.P. and Nealey, P.F. (2006) BCPs and conventional lithography. *Mater. Today*, **9**, 20–29.

105. Register, R.A. (2003) Materials science: on the straight and narrow. *Nature*, **424**, 378–379.

106. Edwards, E.W., Mueller, M., Stoykovich, M.P., Solak, H.H., de Pablo, J.J., and Nealey, P.F. (2007) Dimensions and shapes of BCP domains assembled on lithographically defined chemically patterned substrates. *Macromolecules*, **40**, 90–96.

107. Park, S.M., Stoykovich, M.P., Ruiz, R., Zhang, Y., Black, C.T., and Nealey, P.F. (2007) Directed assembly of lamellae-forming BCPs by using chemically and topographically patterned substrates. *Adv. Mater.*, **19**, 607–611.

108. Stoykovich, M.P., Edwards, E.W., Solak, H.H., and Nealey, P.F. (2006) Phase behavior of symmetric ternary BCP-homopolymer blends in thin films and on chemically patterned surfaces. *Phys. Rev. Lett.*, **97**, 147801–147804.

109. In, I., La, Y.-H., Park, S.-M., Nealey, P.F., and Gopalan, P. (2006) Side-chain-grafted random copolymer brushes as neutral surfaces for controlling the orientation of BCP microdomains in thin films. *Langmuir*, **22**, 7855–7860.

110. Daoulas, K.C., Mueller, M., de Pablo, J.J., Nealey, P.F., and Smith, G.D. (2006) Morphology of multi-component polymer systems: single chain in mean field simulation studies. *Soft Matter*, **2**, 573–583.

111. Ruiz, R., Kang, H., Detcheverry, F.A., Dobisz, E., Kercher, D.S., Albrecht, T.R., de Pablo, J.J., and Nealey, P.F. (2008) Density multiplication and improved lithography by directed BCP assembly. *Science*, **321**, 936–939.

112. Xiao, S., Yang, X., Park, S., Russell, T.P., and Weller, D. (2009) A novel approach to addressable 4 teradot/in² patterned media. *Adv. Mater.*, **21**, 2516–2519.

113. Mochrie, S.G.J., Song, S., Yoon, M.R., Abernathy, D.L., and Stephenson, G.B. (1996) Faceting of stepped silicon (113) surfaces: self assembly of nanoscale gratings. *Physica B*, **221**, 105–125.

114. Li, Z., Qu, S., Rafailovich, M.H., Sokolov, J., Tolan, M., and Turner, M.S. (1997) Confinement of BCPs on patterned surfaces. *Macromolecules*, **30**, 8410–8419.

115. Fasolka, M.J., Banerjee, P., Mayes, A.M., Pickett, G., and Balazs, A.C. (2000) Morphology of ultrathin supported DiBCP films: theory and experiment. *Macromolecules*, **33**, 5702–5712.

116. Fasolka, M.J., Harris, D.J., Mayes, A.M., Yoon, M., and Mochrie, S.G.J. (1997) Observed substrate topography-mediated lateral patterning of DiBCP films. *Phys. Rev. Lett.*, **79**, 3018–3021.

117. Turner, M.S. and Joanny, J.F. (1992) DiBCP lamellae at rough surfaces. *Macromolecules*, **25**, 6681–6689.

118. Podariu, I. and Chakrabarti, A. (2000) BCP thin films on corrugated substrates. *J. Chem. Phys.*, **113**, 6423–6428.

119. Park, S., Lee, D.H., Kim, B., Hong, S.W., Xu, J., Jeong, U., Xu, T., and Russell, T.P. (2009) Macroscopic 10-terabit per square-inch arrays from block copolymers with lateral order. *Science*, **323**, 1030–1033.

120. Frank, F.C. and van der Merwe, J.H. (1949) One-dimensional dislocations. II. Misfitting monolayers and oriented overgrowth. *Prog. R. Soc. Lond. A*, **198**, 216–225.

121. Swei, G.S., Lando, J.B., Rickert, S.E., and Mauritz, K.A. (1986) Epitaxial processes. *Encycl. Polym. Sci. Eng.*, **6**, 209–224.

122. Willems, J. (1955) Über orientierte Aufwachsungen von Paraffinen auf aromatischen Kohlenwasserstoffen und Alkalihalogeniden. *Naturwissenschaften*, **42**, 176–177.

123. Fischer, E.W. (1957) Stufen- und spiralförmiges Kristallwachstum bei Hochpolymeren. *Z. Naturforsch*, **12a**, 753–754.

124. Mauritz, K.A., Baer, E., and Hopfinger, A.J. (1978) The epitaxial crystallization of macromolecules. *J. Polym. Sci., Macromol. Rev.*, **13**, 1–61.

125. Wittmann, J.C. and Lotz, B. (1990) Epitaxial crystallization of polymers on organic and polymeric substrates. *Prog. Polym. Sci.*, **15**, 909–948.

126. De Rosa, C., Park, C., Lotz, B., Wittmann, J.-C., Fetters, L.J., and Thomas, E.L. (2000) Control of molecular and microdomain orientation in a semicrystalline BCP thin film by epitaxy. *Macromolecules*, **33**, 4871–4876.

127. Park, C., De Rosa, C., Fetters, L.J., Lotz, B., and Thomas, E.L. (2001) Alteration of classical microdomain patterns of BCPs by degenerate epitaxy. *Adv. Mater.*, **13**, 724–728.

128. Park, C., De Rosa, C., Lotz, B., Fetters, L.J., and Thomas, E.L. (2003) Molecular and microdomain orientation in semicrystalline BCP thin films. *Macromol. Chem. Phys.*, **204**, 1514–1523.

129. Park, C., De Rosa, C., and Thomas, E.L. (2001) Large area orientation of BCP microdomains in thin films via directional crystallization of a solvent. *Macromolecules*, **34**, 2602–2606.

130. Park, C., Cheng, J.Y., Fasolka, M.J., Mayes, A.M., Ross, C.A., Thomas, E.L., and Rosa, C.D. (2001) Double textured cylindrical BCP domains via directional solidification on a topographically patterned substrate. *Appl. Phys. Lett.*, **79**, 848–850.

131. Anastasiadis, S.H., Russell, T.P., Satija, S.K., and Majkrzak, C.F. (1989) Neutron reflectivity studies of the surface-induced ordering of DiBCP films. *Phys. Rev. Lett.*, **62**, 1852–1855.

132. Russell, T.P., Coulon, G., Deline, V.R., and Miller, D.C. (1989) Characteristics of the surface-induced orientation for symmetric diblock PS/PMMA copolymers. *Macromolecules*, **22**, 4600–4606.

133. Huinink, H.P., Brokken-Zijp, J.C.M., Van Dijk, M.A., and Mochrie, S.G.J. (2000) Asymmetric BCPs confined in a thin film. *J. Chem. Phys.*, **112**, 2452–2462.

134. Van Dijk, M.A. and Van Den Berg, R. (1995) Ordering phenomena in thin BCP films studied using atomic force microscopy. *Macromolecules*, **28**, 6773–6778.

135. Harrison, C., Adamson, D.H., Cheng, Z., Sebastian, J.M., Sethuraman, S., Huse, D.A., Register, R.A., and Chaikin, P.M. (2000) Mechanisms of ordering in striped patterns. *Science*, **290**, 1558–1560.

136. Kellogg, G.J., Walton, D.G., Mayes, A.M., Lambooy, P., Russell, T.P., Gallagher, P.D., and Satija, S.K. (1996) Observed surface energy effects in confined DiBCPs. *Phys. Rev. Lett.*, **76**, 2503–2506.

137. Kim, H.-C. and Russell, T.P. (2001) Ordering in thin films of asymmetric DiBCPs. *J. Polym. Sci. B: Polym. Phys.*, **39**, 663–668.

138. Menelle, A., Russell, T.P., Anastasiadis, S.H., Satija, S.K., and Majkrzak, C.F. (1992) The ordering of thin DiBCP films. *Phys. Rev. Lett.*, **68**, 67–70.

139. Bassereau, P., Brodbreck, D., Russell, T.P., Brown, H.R., and Shull, K.R. (1993) Topological coarsening of symmetrical DiBCP films: model 2D systems. *Phys. Rev. Lett.*, **71**, 1716–1719.

140. Factor, B.J., Russell, T.P., and Toney, M.F. (1991) Surface-induced ordering of an aromatic polyimide. *Phys. Rev. Lett.*, **66**, 1181–1184.

141. Russell, T.P., Menelle, A., Anastasiadis, S.H., Satija, S.K., and Majkrzak, C.F. (1993) Fluctuation effects in the ordering of thin DiBCP films. *Phys. Rev. Lett.*, **70**, 1352.

142. Green, P.F., Christensen, T.M., Russell, T.P., and Jerome, R. (1990) Finite chain length effects on the equilibrium surface composition of DiBCPs. *J. Chem. Phys.*, **92**, 1478–1482.

143. Mansky, P., Russell, T.P., Hawker, C.J., Mays, J., Cook, D.C., and Satija, S.K. (1997) Interfacial segregation in disordered BCPs: effect of tunable surface potentials. *Phys. Rev. Lett.*, **79**, 237–240.

144. Huang, E., Rockford, L., Russell, T.P., and Hawker, C.J. (1998) Nanodomain control in copolymer thin films. *Nature*, **395**, 757–758.

145. Pickett, G.T., Witten, T.A., and Nagel, S.R. (1993) Equilibrium surface orientation of lamellae. *Macromolecules*, **26**, 3194–3199.

146. Peters, R.D., Yang, X.M., Kim, T.K., Sohn, B.H., and Nealey, P.F. (2000) Using self-assembled monolayers exposed to X-rays to control the wetting behavior of thin films of DiBCPs. *Langmuir*, **16**, 4625–4631.

147. Laibinis, P.E., Whitesides, G.M., Allara, D.L., Tao, Y.T., Parikh, A.N., and Nuzzo, R.G. (1991) Comparison of the structures and wetting properties of self-assembled monolayers of n-alkanethiols on the coinage metal surfaces, Cu, Ag, Au[1]. *J. Am. Chem. Soc.*, **113**, 7152–7167.

148. Netzer, L. and Sagiv, J. (1983) A new approach to construction of artificial monolayer assemblies. *J. Am. Chem. Soc.*, **105**, 674–676.

149. Troughton, E.B., Bain, C.D., Whitesides, G.M., Nuzzo, R.G., Allara, D.L., and Porter, M.D. (1988) Monolayer films prepared by the spontaneous self-assembly of symmetrical and unsymmetrical dialkyl sulfides from solution onto gold substrates: structure, properties, and reactivity of constituent functional groups. *Langmuir*, **4**, 365–385.

150. Fadeev, A.Y. and McCarthy, T.J. (2000) Self-assembly is not the only reaction possible between alkyltrichlorosilanes and surfaces: monomolecular and oligomeric covalently attached layers of dichloro- and trichloroalkylsilanes on silicon. *Langmuir*, **16**, 7268–7274.

151. Genzer, J. and Kramer, E.J. (1997) Wetting of substrates with phase-separated binary polymer mixtures. *Phys. Rev. Lett.*, **78**, 4946–4949.

152. Hawker, C.J. and Russell, T.P. (2005) BCP lithography: merging "bottom-up" with "top-down" processes. *MRS Bull.*, **30**, 952–966.

153. Mansky, P., Russell, T.P., Hawker, C.J., Pitsikalis, M., and Mays, J. (1997) Ordered DiBCP films on random copolymer brushes. *Macromolecules*, **30**, 6810–6813.

154. Harrison, C., Chaikin, P.M., Huse, D.A., Register, R.A., Adamson, D.H., Daniel, A., Huang, E., Mansky, P., Russell, T.P., Hawker, C.J., Egolf, D.A., Melnikov, I.V., and Bodenschatz, E. (2000) Reducing substrate pinning of BCP microdomains with a buffer layer of polymer brushes. *Macromolecules*, **33**, 857–865.

155. Xu, T., Kim, H.-C., Derouchey, J., Seney, C., Levesque, C., Martin, P., Stafford, C.M., and Russell, T.P. (2001) The influence of molecular weight on nanoporous polymer films. *Polymer*, **42**, 9091–9095.

156. Han, E., In, I., Park, S.-M., La, Y.-H., Wang, Y., Nealey, P.F., and Gopalan, P. (2007) Photopatternable imaging layers for controlling BCP microdomain orientation. *Adv. Mater.*, **19**, 4448–4452.

157. Bang, J., Bae, J., Löwenhielm, P., Spiessberger, C., Given-Beck, S.A., Russell, T.P., and Hawker, C.J. (2007) Facile routes to patterned surface neutralization layers for BCP lithography. *Adv. Mater.*, **19**, 4552–4557.

158. Jeong, U., Ryu, D.Y., Kho, D.H., Kim, J.K., Goldbach, J.T., Kim, D.H., and Russell, T.P. (2004) Enhancement in the orientation of the microdomain in BCP thin films upon the addition of homopolymer. *Adv. Mater.*, **16**, 533–536.

159. Jeong, U., Kim, H.-C., Rodriguez, R.L., Tsai, I.Y., Stafford, C.M., Kim, J.K., Hawker, C.J., and Russell, T.P. (2002) Asymmetric BCPs with homopolymers: routes to multiple length scale nanostructures. *Adv. Mater.*, **14**, 274–276.

160. Jeong, U., Ryu, D.Y., Kho, D.H., Lee, D.H., Kim, J.K., and Russell, T.P. (2003) Phase behavior of mixtures of BCP and homopolymers in thin films and bulk. *Macromolecules*, **36**, 3626–3634.

161. Jeong, U., Ryu, D.Y., Kim, J.K., Kim, D.H., Wu, X., and Russell, T.P. (2003) Precise control of nanopore size in thin film using mixtures of asymmetric BCP and homopolymer. *Macromolecules*, **36**, 10126–10129.

162. Ahn, D.U. and Sancaktar, E. (2006) Perpendicularly aligned, size- and spacing-controlled nanocylinders by molecular-weight adjustment of a homopolymer blended in an asymmetric TriBCP. *Adv. Funct. Mater.*, **16**, 1950–1958.

163. Kim, D.H., Lau, K.H., Joo, W., Peng, J., Jeong, U., Hawker, C.J., Kim, J.K., Russell, T.P., and Knoll, W. (2006) An optical waveguide study on the nanopore formation in BCP/homopolymer thin films by selective solvent swelling. *J. Phys. Chem. B*, **110**, 15381–15388.

164. Haryono, A. and Binder, W.H. (2006) Controlled arrangement of nanoparticle arrays in block-copolymer domains. *Small*, **2**, 600–611.

165. Hamley, I.W. (2003) Nanostructure fabrication using BCPs. *Nanotechnology*, **14**, R39–R54.

166. Bockstaller, M.R., Mickiewicz, R.A., and Thomas, E.L. (2005) BCP nanocomposites: perspectives for tailored functional materials. *Adv. Mater.*, **17**, 1331–1349.

167. Balazs, A.C., Emrick, T., and Russell, T.P. (2006) Nanoparticle-polymer composites: where two small worlds meet. *Science*, **314**, 1107–1110.

168. Balazs, A.C. (2000) Interactions of nanoscopic particles with phase-separating polymeric mixtures. *Curr. Opin. Colloid Interf. Sci.*, **4**, 443–448.

169. Lee, J.Y., Shou, Z., and Balazs, A.C. (2003) Modeling the self-assembly of copolymer-nanoparticle mixtures confined between solid surfaces. *Phys. Rev. Lett.*, **91**, 136103.

170. Lee, J.Y., Shou, Z., and Balazs, A.C. (2003) Predicting the morphologies of confined copolymer/nanoparticle mixtures. *Macromolecules*, **36**, 7730–7739.

171. Lin, Y., Boeker, A., He, J., Sill, K., Xiang, H., Abetz, C., Li, X., Wang, J., Emrick, T., Long, S., Wang, Q., Balazs, A.C., and Russell, T.P. (2005) Self-directed self-assembly of nanoparticle/copolymer mixtures. *Nature*, **434**, 55–59.

172. He, J., Tangirala, R., Emrick, T., Russell, T.P., Boker, A., Li, X., and Wang, J. (2007) Self-assembly of nanoparticle-copolymer mixtures: a kinetic point of view. *Adv. Mater.*, **19**, 381–385.

173. Xu, T., Goldbach, J.T., Leiston-Belanger, J.M., and Russell, T.P. (2004) Effect of ionic impurities on the electric field alignment of DiBCP thin films. *Colloid Polym. Sci.*, **282**, 927–931.

174. Wang, J.-Y., Leiston-Belanger, J.M., Gupta, S., and Russell, T.P. (2006) Ion complexation: a route to enhanced BCP alignment with electric fields. *Phys. Rev. Lett.*, **96**, 128301–128304.

175. Wang, J.-Y., Leiston-Belanger, J.M., Sievert, J.D., and Russell, T.P. (2006) Grain rotation in ion-complexed symmetric DiBCP thin films under an electric field. *Macromolecules*, **39**, 8487–8491.

176. Kim, S.H., Misner, M.J., Yang, L., Gang, O., Ocko, B.M., and Russell, T.P. (2006) Salt complexation in BCP thin films. *Macromolecules*, **39**, 8473–8479.

177. Wang, J.-Y., Chen, W., Sievert, J.D., and Russell, T.P. (2008) Lamellae

orientation in BCP films with ionic complexes. *Langmuir*, **24**, 3545–3550.

178. He, J., Wang, J.-Y., Xu, J., Tangirala, R., Shin, D., Russell, T.P., Li, X., and Wang, J. (2007) On the influence of ion incorporation in thin films of BCPs. *Adv. Mater.*, **19**, 4370–4374.

179. Ikkala, O. and Ten Brinke, G. (2002) Functional materials based on self-assembly of polymeric supramolecules. *Science*, **295**, 2407–2409.

180. Ikkala, O. and Ten Brinke, G. (2004) Hierarchical self-assembly in polymeric complexes: towards functional materials. *Chem. Commun.*, 2131–2137.

181. Ten Brinke, G., Ruokolainen, J., and Ikkala, O. (2007) Supramolecular materials based on hydrogen-bonded polymers. *Adv. Polym. Sci.*, **207**, 113–177.

182. Ruokolainen, J., Makinen, R., Torkkeli, M., Makela, T., Serimaa, R., Ten Brinke, G., and Ikkala, O. (1998) Switching supramolecular polymeric materials with multiple length scales. *Science*, **280**, 557–560.

183. Maki-Ontto, R., De Moel, K., De Odorico, W., Ruokolainen, J., Stamm, M., Ten Brinke, G., and Ikkala, O. (2001) Hairy tubes: mesoporous materials containing self-organized cylinders with polymer brushes at the walls. *Adv. Mater.*, **13**, 117–121.

184. de Moel, K., Alberda Van Ekenstein, G.O.R., Nijland, H., Polushkin, E., Ten Brinke, G., Maeki-Ontto, R., and Ikkala, O. (2001) Polymeric nanofibers prepared from self-organized supramolecules. *Chem. Mater.*, **13**, 4580–4583.

185. Ruokolainen, J., Ten Brinke, G., and Ikkala, O. (1999) Supramolecular polymeric materials with hierarchical structure-within-structure morphologies. *Adv. Mater.*, **11**, 777–780.

186. Saito, R. (2001) Synthesis of discotic microgels by cross-linking of poly(styrene-block-4-vinylpyridine)/3-pentadecylphenol blend film. *Macromolecules*, **34**, 4299–4301.

187. Sidorenko, A., Tokarev, I., Minko, O., and Stamm, M. (2003) Ordered reactive nanomembranes/nanotemplates from thin films of BCP supramolecular assembly. *J. Am. Chem. Soc.*, **125**, 12211–12216.

188. Zhao, Y., Thorkelsson, K., Schilling, T., Mastroianni, A., Luthe, J.M., Wu, Y., Alivisators, A.P., and Xu, T. (2009) Small-molecule-directed nanoparticle assembly towards stimuli-responsive nanocomposites. *Nature Mater*, **8**, 979–985.

189. Tirumala, V.R., Romang, A., Agarwal, S., Lin, E.K., and Watkins, J.J. (2008) Well ordered polymer melts from blends of disordered triblock copolymer surfactants and functional homopolymer. *Adv. Mater.*, **20**, 1603–1608.

190. Tirumala, V.R., Daga, V., Bosse, A.W., Romang, A., Ilavsky, J., Lin, E.K., and Watkins, J.J. (2008) Well-ordered polymer melts with 5 nm lamellar domains from blends of a disordered BCP and a selectively associating homopolymer of low and high molar mass. *Macromolecules*, **41**, 7978–7985.

191. Abetz, V. and Goldacker, T. (2000) Formation of superlattices via blending of BCPs. *Macromol. Rapid Commun.*, **21**, 16–34.

192. Mao, H., Arrechea, P.L., Bailey, T.S., Johnson, B.J.S., and Hillmyer, M.A. (2005) Control of pore hydrophilicity in ordered nanoporous polystyrene using an AB/AC BCP blending strategy. *Faraday Discuss.*, **128**, 149–162.

193. Asari, T., Matsuo, S., Takano, A., and Matsushita, Y. (2005) Three-phase hierarchical structures from AB/CD DiBCP blends with complementary hydrogen bonding interaction. *Macromolecules*, **38**, 8811–8815.

194. Kimishima, K., Jinnai, H., and Hashimoto, T. (1999) Control of self-assembled structures in binary mixtures of A-B DiBCP and A-C DiBCP by changing the interaction between B and C block chains. *Macromolecules*, **32**, 2585–2596.

195. Jeon, H.G., Hudson, S.D., Ishida, H., and Smith, S.D. (1999) Microphase and macrophase transitions in binary

blends of DiBCPs. *Macromolecules*, **32**, 1803–1808.

196. Tang, C., Lennon, E.M., Fredrickson, G.H., Kramer, E.J., and Hawker, C.J. (2008) Evolution of BCP lithography to highly ordered square arrays. *Science*, **322**, 429–432.

197. Bang, J., Kim, S.H., Drockenmuller, E., Misner, M.J., Russell, T.P., and Hawker, C.J. (2006) Defect-free nanoporous thin films from ABC TriBCPs. *J. Am. Chem. Soc.*, **128**, 7622–7629.

198. Ichimura, K. (2000) Photoalignment of liquid crystal systems. *Chem. Rev.*, **100**, 1847–1873.

199. Ikeda, T. (2003) Photomodulation of liquid crystal orientations for photonic applications. *J. Mater. Chem.*, **13**, 2037–2057.

200. Natansohn, A. and Rochon, P. (2002) Photoinduced motions in azo-containing polymers. *Chem. Rev.*, **102**, 4139–4175.

201. Berg, R.H., Hvilsted, S., and Ramanujam, P.S. (1996) Peptide oligomers for holographic data storage. *Nature*, **383**, 505–508.

202. Morikawa, Y., Nagano, S., Watanabe, K., Kamata, K., Iyoda, T., and Seki, T. (2006) Optical alignment and patterning of nanoscale microdomains in a BCP thin film. *Adv. Mater.*, **18**, 883–886.

203. Yu, H., Iyoda, T., and Ikeda, T. (2006) Photoinduced alignment of nanocylinders by supramolecular cooperative motions. *J. Am. Chem. Soc.*, **128**, 11010–11011.

204. Bal, M., Ursache, A., Tuominen, M.T., Goldbach, J.T., and Russell, T.P. (2002) Nanofabrication of integrated magnetoelectronic devices using patterned self-assembled copolymer templates. *Appl. Phys. Lett.*, **81**, 3479–3481.

205. Spatz, J.P., Chan, V.Z.-H., Mößmer, S., Kamm, F.-M., Plettl, A., Ziemann, P., and Möller, M. (2002) A combined top-down/bottom-up approach to the microscopic localization of metallic nanodots. *Adv. Mater.*, **14**, 1827–1832.

206. Glass, R., Arnold, M., Blümmer, J., Küller, A., Möller, M., and Spatz, J.P. (2003) Micro-nanostructured interfaces fabricated by the use of inorganic BCP micellar monolayers as negative resist for electron-beam lithography. *Adv. Funct. Mater.*, **13**, 569–575.

207. Ober, C.K., Li, M., Douki, K., Goto, K., and Li, X. (2003) Lithographic patterning with BCPs. *J. Photopolym. Sci. Technol.*, **16**, 347–350.

208. Du, P., Li, M., Douki, K., Li, X., Garcia, C.B.W., Jain, A., Smilgies, D.M., Fetters, L.J., Gruner, S.M., Wiesner, U., and Ober, C.K. (2004) Phase selective chemistry in BCP thin films. *Adv. Mater.*, **16**, 953–957.

209. Li, M., Douki, K., Goto, K., Li, X., Coenjarts, C., Smilgies, D.M., and Ober, C.K. (2004) Spatially controlled fabrication of nanoporous BCPs. *Chem. Mater.*, **16**, 3800–3808.

210. Park, S., Kim, B., Yavuzcetin, O., Tuominen, M.T., and Russell, T.P. (2008) Ordering of PS-b-P4VP on patterned silicon surfaces. *ACS Nano*, **2**, 1363–1370.

211. Park, S., Yavuzcetin, O., Kim, B., Tuominen, M.T., and Russell, T.P. (2009) A simple top-down/bottom-up approach to sectored, ordered arrays of nanoscopic elements using block copolymers. *Small*, **5**, 1064–1069.

212. Xu, T., Stevens, J., Villa, J., Goldbach, J.T., Guarini, K.W., Black, C.T., and Hawker, C.J. (2003) BCP surface reconstruction: a reversible route to nanoporous films. *Adv. Funct. Mater.*, **13**, 698–702.

213. Park, S., Wang, J.-Y., Kim, B., and Russell, T.P. (2008) From nanorings to nanodots by patterning with BCPs. *Nano Lett.*, **8**, 1667–1672.

214. Kim, T.H., Hwang, J., Hwang, W.S., Huh, J., Kim, H.-C., Kim, S.H., Hong, J.M., Thomas, E.L., and Park, C. (2008) Hierarchical ordering of BCP nanostructures by solvent annealing combined with controlled dewetting. *Adv. Mater.*, **20**, 522–527.

215. Tracz, A., Jeszka, J.K., Watson, M.D., Pisula, W., Mullen, K., and Pakula, T. (2003) Uniaxial alignment of the columnar super-structure of a hexa(alkyl) hexa-peri-hexabenzocoronene on untreated glass by simple solution processing. *J. Am. Chem. Soc.*, **125**, 1682–1683.

216. De Blauwe, J. (2002) Nanocrystal non-volatile memory devices. *IEEE Trans. Nanotechnol.*, **1**, 72.

217. Fink, Y., Winn, J.N., Fan, S., Chen, C., Michel, J., Joannopoulos, J.D., and Thomas, E.L. (1998) A dielectric omnidirectional reflector. *Science*, **282**, 1679–1682.

218. Kim, D.H., Lau, K.H.A., Robertson, J.W.F., Lee, O.-J., Jeong, U., Lee, J.I., Hawker, C.J., Russell, T.P., Kim, J.K., and Knoll, W. (2005) Thin films of BCPs as planar optical waveguides. *Adv. Mater.*, **17**, 2442–2446.

219. Knoll, W. (1998) Interfaces and thin films as seen by bound electromagnetic waves. *Annu. Rev. Phys. Chem.*, **49**, 569–638.

220. Swalen, J.D. (1979) Optical wave spectroscopy of molecules at surfaces. *J. Phys. Chem.*, **83**, 1438–1445.

221. Park, M., Harrison, C., Chaikin, P.M., Register, R.A., and Adamson, D.H. (1997) BCP lithography: periodic arrays of $\sim 10^{11}$ holes in 1 square centimeter. *Science*, **276**, 1401–1404.

222. Yang, S.Y., Ryu, I., Kim, H.Y., Kim, J.K., Jang, S.K., and Russell, T.P. (2006) Nanoporous membranes with ultrahigh selectivity and flux for the filtration of viruses. *Adv. Mater.*, **18**, 709–712.

223. Peinemann, K.-V., Abetz, V., and Simon, P.F.W. (2007) Asymmetric superstructure formed in a BCP via phase separation. *Nature Mater.*, **6**, 992–996.

224. Hartney, M.A., Novembre, A.E., and Bates, F.S. (1985) BCPs as bilevel resists. *J. Vac. Sci. Technol. B.*, **3**, 1346–1351.

225. Cheng, J.Y., Ross, C.A., Thomas, E.L., Smith, H.I., and Vancso, G.J. (2002) Fabrication of nanostructures with long-range order using BCP lithography. *Appl. Phys. Lett.*, **81**, 3657–3659.

226. Temple, K., Kulbaba, K., Power-Billard, K.N., Manners, I., Leach, K.A., Xu, T., Russell, T.P., and Hawker, C.J. (2003) Spontaneous vertical ordering and pyrolytic formation of nanoscopic ceramic patterns from

poly(styrene-b-ferrocenylsilane). *Adv. Mater.*, **15**, 297–300.

227. Cao, L., Massey, J.A., Winnik, M.A., Manners, I., Riethmuller, S., Banhart, F., Spatz, J.P., and Moller, M. (2003) Reactive ion etching of cylindrical polyferrocenylsilane BCP micelles: fabrication of ceramic nanolines on semiconducting substrates. *Adv. Funct. Mater.*, **13**, 271–276.

228. Lammertink, R.G.H., Hempenius, M.A., Van Den Enk, J.E., Chan, V.Z.-H., Thomas, E.L., and Vancso, G.J. (2000) Nanostructured thin films of organic-organometallic BCPs: one-step lithography with poly(ferrocenylsilanes) by reactive ion etching. *Adv. Mater.*, **12**, 98–103.

229. Harrison, C., Dagata, J.A., and Adamson, D.H. (2004) *Lithography with Self-assembled BCP Microdomains*, John Wiley & Sons, London, pp. 295–323.

230. Cheng, J.Y., Ross, C.A., Chan, V.Z.H., Thomas, E.L., Lammertink, R.G.H., and Vancso, G.J. (2001) Formation of a cobalt magnetic dot array via BCP lithography. *Adv. Mater.*, **13**, 1174–1178.

231. Clendenning, S.B. and Manners, I. (2003) Lithographic applications of highly metallized polyferrocenylsilanes. *Macromol. Symp.*, **196**, 71–76.

232. Chan, W.Y., Cheng, A.Y., Clendenning, S.B., and Manners, I. (2004) Synthesis and lithographic applications of highly metallized cluster-based polyferrocenylsilanes. *Macromol. Symp.*, **209**, 163–176.

233. Liu, K., Baker, S.M., Tuominen, M., Russell, T.P., and Schuller, I.K. (2001) Tailoring exchange bias with magnetic nanostructures. *Phys. Rev. B*, **6305**, 060403.

234. Lammertink, R.G.H., Hempenius, M.A., Chan, V.Z.H., Thomas, E.L., and Vancso, G.J. (2001) Poly(ferrocenyldimethylsilanes) for reactive ion etch barrier applications. *Chem. Mater.*, **13**, 429–434.

235. Asakawa, K., Hiraoka, T., Hieda, H., Sakurai, M., and Kamata, Y. (2002) Nanopatterning for patterned media using BCP. *J. Photopolym. Sci. Technol.*, **15**, 465–470.

236. Spatz, J.P., Herzog, T., Möbmer, S., Ziemann, P., and Möller, M. (1999) Micellar inorganic-polymer hybrid systems-A tool for nanolithography. *Adv. Mater.*, **11**, 149–153.

237. Spatz, J.P., Eibeck, P., Möbmer, S., Möller, M., Herzog, T., and Ziemann, P. (1998) Ultrathin DiBCP/titanium laminates-A tool for nanolithography. *Adv. Mater.*, **10**, 849–852.

238. Black, C.T., Guarini, K.W., Milkove, K.R., Baker, S.M., Russell, T.P., and Tuominen, M.T. (2001) Integration of self-assembled DiBCPs for semiconductor capacitor fabrication. *Appl. Phys. Lett.*, **79**, 409–411.

239. Black, C.T., Guarini, K.W., Zhang, Y., Kim, H.J., Benedict, J., Sikorski, E., Babich, I.V., and Milkove, K.R. (2004) High-capacity, self-assembled metal-oxide-semiconductor decoupling capacitors. *IEEE Electron Device Lett.*, **25**, 622–624.

240. Guarini, K.W., Black, C.T., Zhang, Y., Kim, H., Sikorski, E.M., and Babich, I.V. (2002) Process integration of self-assembled polymer templates into silicon nanofabrication. *J. Vac. Sci. Technol. B*, **20**, 2788–2792.

241. Kim, H., Kellock, A.J., and Rossnagel, S.M. (2002) Growth of cubic TaN thin films by plasma enhanced-atomic layer deposition using inorganic metal precursor. *J. Appl. Phys.*, **92**, 7080–7085.

242. Ellis, R.K. (1982) Fowler–Nordheim emission from nonplanar surfaces. *IEEE Electron Device Lett.*, **3**, 330–332.

243. Nie, Z. and Kumacheva, E. (2008) Patterning surfaces with functional polymers. *Nature Mater.*, **7**, 277–290.

244. Guarini, K.W., Black, C.T., Milkove, K.R., and Sandstrom, R.L. (2001) Nanoscale patterning using self-assembled polymers for semiconductor applications. *J. Vac. Sci. Technol. B*, **19**, 2784–2788.

245. Black, C.T., Guarini, K.W., Breyta, G., Colburn, M.C., Ruiz, R., Sandstrom, R.L., Sikorski, E.M., and Zhang, Y. (2006) Highly porous silicon membrane fabrication using polymer self-assembly. *J. Vac. Sci. Technol. B*, **24**, 3188–3191.

246. Black, C.T., Guarini, K.W., Ruiz, R., Sikorsck, E.M., Babich, I.V., Sandstrom, R.L., and Zhang, Y. (2007) Polymer self-assembly in semiconductor microelectronics, Paper presented at the 2006 IEEE International Electron Devices Meeting Technical Digest, p. 16.3.

247. Jannopoulos, J.D., Mead, R., and Winn, J.N. (1995) *Photonic Crystals, Molding the Flow of Light*, Princeton University Press, Princeton, NJ.

248. Fink, Y., Winn, J.N., Fan, S., Chen, C., Michel, J., Joannopoulos, J.D., and Thomas, E.L. (1998) A dielectric omnidirectional reflector. *Science*, **282**, 1679–1682.

249. Fink, Y., Urbas, A.M., Bawendi, M.G., Joannopoulos, J.D., and Thomas, E.L. (1999) BCPs as photonic bandgap materials. *J. Lightwave Technol.*, **17**, 1963–1969.

250. Yoon, J., Lee, W., and Thomas, E.L. (2006) Optically pumped surface-emitting lasing using self-assembled block-copolymer-distributed Bragg reflectors. *Nano Lett.*, **6**, 2211–2214.

251. Ma, H., Jen, A.K.Y., and Dalton, L.R. (2002) Polymer-based optical waveguides: materials, processing, and devices. *Adv. Mater.*, **14**, 1339–1365.

252. Kretschmann, E., Ferrell, T.L., and Ashley, J.C. (1979) Splitting of the dispersion relation of surface plasmons on a rough surface. *Phys. Rev. Lett.*, **42**, 1312–1314.

253. Breulmann, M., Forster, S., and Antonietti, M. (2000) Mesoscopic surface patterns formed by BCP micelles. *Macromol. Chem. Phys.*, **201**, 204–211.

254. Spatz, J.P., Mössmer, S., Hartmann, C., Möller, M., Herzog, T., Krieger, M., Boyen, H.-G., Ziemann, P., and Kabius, B. (2000) Ordered deposition of inorganic clusters from micellar BCP films. *Langmuir*, **16**, 407–415.

255. Yun, S.-H., Yoo, S.I., Jung, J.C., Zin, W.-C., and Sohn, B.-H. (2006) Highly ordered arrays of nanoparticles in large areas from DiBCP micelles in hexagonal self-assembly. *Chem. Mater.*, **18**, 5646–5648.

256. Boontongkong, Y. and Cohen, R.E. (2002) Cavitated BCP micellar thin films: lateral arrays of open nanoreactors. *Macromolecules*, **35**, 3647–3652.

257. Urase, T., Yamamoto, K., and Ohgaki, S. (1994) Effect of pore-size distribution of ultrafiltration membranes on virus rejection in crossflow conditions. *Water Sci. Technol.*, **30**, 199–208.

258. Urase, T., Yamamoto, K., and Ohgaki, S. (1996) Effect of pore structure of membranes and module configuration on virus retention. *J. Membr. Sci.*, **115**, 21–29.

259. Rossmann, M.G. (1985) Structure of a human common cold virus and functional relationship to other picornaviruses. *Nature*, **317**, 145–153.

260. Jeong, U., Ryu, D.Y., Kim, J.K., Kim, D.H., Russell, T.P., and Hawker, C.J. (2003) Volume contractions induced by crosslinking: a novel route to nanoporous polymer films. *Adv. Mater.*, **15**, 1247–1250.

261. Lee, J.S., Hirao, A., and Nakahama, S. (1988) Polymerization of monomers containing functional silyl groups. 5. Synthesis of new porous membranes with functional groups. *Macromolecules*, **21**, 274–276.

262. Mansky, P., Harrison, C.K., Chaikin, P.M., Register, R.A., and Yao, N. (1996) Nanolithographic templates from DiBCP thin films. *Appl. Phys. Lett.*, **68**, 2586–2588.

263. Smith, D.R. and Meier, D.J. (1992) New techniques for determining domain morphologies in BCPs. *Polymer*, **33**, 3777–3782.

264. Haupt, M., Miller, S., Glass, R., Arnold, M., Sauer, R., Thonke, K., Möller, M., and Spatz, J.P. (2003) Nanoporous gold films created using templates formed from self-assembled structures of inorganic-BCP micelles. *Adv. Mater.*, **15**, 829–831.

265. Hedrick, J.L., Carter, K.R., Labadie, J.W., Miller, R.D., Volksen, W., Hawker, C.J., Yoon, D.Y., Russell, T.P., Mcgrath, J.E., and Briber, R.M. (1999) Nanoporous polyimides. *Adv. Polym. Sci.*, **141**, 1–43.

266. Hedrick, J.L., Labadie, J.W., Volksen, W., and Hilborn, J.G. (1999) Nanoscopically engineered polyimides. *Adv. Polym. Sci.*, **147**, 61–111.

267. Hedrick, J.L., Labadie, J., Russell, T.P., Hofer, D., and Wakharker, V. (1993) High temperature polymer foam. *Polymer*, **34**, 4717–4726.

268. Russell, T.P., Thurn-Albrecht, T., Tuominen, M.T., Huang, E., and Hawker, C.J. (2000) BCPs as nanoscopic templates. *Macromol. Symp.*, **159**, 77–88.

269. Kim, H.-C., Jia, X., Stafford, C.M., Kim, D.H., Mccarthy, T.J., Tuominen, M.T., Hawker, C.J., and Russell, T.P. (2001) A route to nanoscopic SiO$_2$ posts via BCP templates. *Adv. Mater.*, **13**, 795–797.

270. Jeoung, E., Galow, T.H., Schotter, J., Bal, M., Ursache, A., Tuominen, M.T., Stafford, C.M., Russell, T.P., and Rotello, V.M. (2001) Fabrication and characterization of nanoelectrode arrays formed via BCP self-assembly. *Langmuir*, **17**, 6396–6398.

271. Shin, K., Leach, K.A., Goldbach, J.T., Kim, D.H., Jho, J.Y., Tuominen, M.T., Hawker, C.J., and Russell, T.P. (2002) A simple route to metal nanodots and nanoporous metal films. *Nano Lett.*, **2**, 933–936.

272. Hillmyer, M.A. (2005) Nanoporous materials from BCP precursors. *Adv. Polym. Sci.*, **190**, 137–181.

273. Zalusky, A.S., Olayo-Valles, R., Wolf, J.H., and Hillmyer, M.A. (2002) Ordered nanoporous polymers from polystyrene-polylactide BCPs. *J. Am. Chem. Soc.*, **124**, 12761–12773.

274. Crossland, E.J.W., Ludwigs, S., Hillmyer, M.A., and Steiner, U. (2007) Freestanding nanowire arrays from soft-etch BCP templates. *Soft Matter*, **3**, 94–98.

275. Pai, R.A., Humayun, R., Schulberg, M.T., Sengupta, A., Sun, J.-N., and Watkins, J.J. (2004) Mesoporous silicates prepared using preorganized templates in supercritical fluids. *Science*, **303**, 507–510.

276. Goldbach, J.T., Russell, T.P., and Penelle, J. (2002) Synthesis

and thin film characterization of poly(styrene-block-methyl methacrylate) containing an anthracene dimer photocleavable junction point. *Macromolecules*, **35**, 4271–4276.

277. Yurt, S., Anyanwu, U.K., Scheintaub, J.R., Coughlin, E.B., and Venkataraman, D. (2006) Scission of DiBCPs into their constituent blocks. *Macromolecules*, **39**, 1670–1672.

278. Zhang, M., Yang, L., Yurt, S., Misner, M.J., Chen, J.-T., Coughlin, E.B., Venkataraman, D., and Russell, T.P. (2007) Highly ordered nanoporous thin films from cleavable polystyrene-block-poly(ethylene oxide). *Adv. Mater.*, **19**, 1571–1576.

279. Guarini, K.W., Black, C.T., and Yeung, S.H.I. (2002) Optimization of DiBCP thin film self-assembly. *Adv. Mater.*, **14**, 1290–1294.

280. Leiston-Belanger, J.M., Russell, T.P., Drockenmuller, E., and Hawker, C.J. (2005) A thermal and manufacturable approach to stabilized DiBCP templates. *Macromolecules*, **38**, 7676–7683.

281. Goldbach, J.T., Lavery, K.A., Penelle, J., and Russell, T.P. (2004) Nano- to macro-sized heterogeneities using cleavable DiBCPs. *Macromolecules*, **37**, 9639–9645.

282. Lee, M., Park, M.-H., Oh, N.-K., Zin, W.-C., Jung, H.-T., and Yoon, D.K. (2004) Supramolecular crystalline sheets with ordered nanopore arrays from self-assembly of rigid-rod building blocks. *Angew. Chem. Int. Ed.*, **43**, 6465–6468.

283. Lee, M., Cho, B.-K., and Zin, W.-C. (2001) Supramolecular structures from rod-coil BCPs. *Chem. Rev.*, **101**, 3869–3892.

13
Patterning by Photolithography

Anuja De Silva and Christopher K. Ober

13.1
Introduction

Patterning of thin films through photolithography has been the main driving force behind the rapid progress of the integrated circuit (IC) industry. The rapid improvement of the performance of semiconductor devices has been brought about by miniaturization through the reduction of the minimum feature size on the chip. In accordance with Moore's law [1], the continual scaling down of transistor dimensions has allowed more and more components on a chip, lowered the power consumption per transistor, and increased the speed of the circuitry.

In its original meaning, *"lithography"* is known as a method for printing using a plate or stone with a completely smooth surface. But in its modern terminology, the term lithography is generally applied to a number of methods for replicating a pre-determined master pattern on a substrate. Photolithography is the main technology used in the manufacture of semiconductor devices in the present IC industry [2–4]. With this technique, the master pattern on a mask can be transferred to an under-lying organic thin film by using UV light. This process is made possible by the pho-tosensitive organic film known as a *photoresist (or resist)*. As shown in Figure 13.1, exposure to radiation causes a photoinduced chemical change in the exposed area of the film that can be differentiated during the image-development step. The sol-ubility of the exposed area can either be enhanced to yield positive tone patterns or reduced to yield negative tone images. The resulting topographic pattern can then be transferred to the underlying silicon substrate by a reactive-ion etching process.

The resist material is first applied by spin coating to form $1-0.1\,\mu m$ thick films on silicon substrates. The photosensitive films are typically composed of polymers with specific structures and functionalities that impart good photoresist characteristics. To successfully enable small dimensions, the polymers used as photoresists must adequately address the multiple requirements of each step of the lithographic process. Suitable photoresist materials should have good solubility in a conventional spinning solvent and form defect-free uniform films upon spin coating. The resist thin film should show good adhesion to the underlying substrate. As for the silicon wafer, a typical substrate is hydrophilic, the polarity of the resist

Functional Polymer Films, First Edition. Edited by Wolfgang Knoll and Rigoberto C. Advincula.
© 2011 Wiley-VCH Verlag GmbH & Co. KGaA. Published 2011 by Wiley-VCH Verlag GmbH & Co. KGaA.

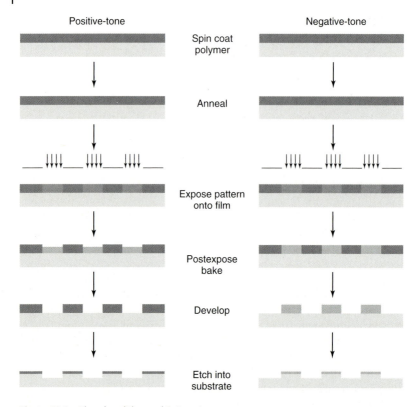

Figure 13.1 The photolithographic imaging process.

polymer needs to be modified to promote better adhesion [5]. As the contents of this chapter illustrate, transparency at the imaging wavelength is key to determining the photoresist platform [6]. If the resist polymer is too absorbing at a particular wavelength, it may be unable to conduct the patterning mechanism effectively. High glass-transition temperature (T_g) and high thermal stability of the resist materials are essential for the thermal processing steps that occur before and after exposure. Performing a postexposure bake treatment above the T_g of the material can have adverse effects such as pattern distortion and limiting resolution [7]. The chemistry of the resist materials needs to be optimized to obtain high development contrast and sensitivity depending on the efficiency of the solubility switching mechanism. A good photoresist should be capable of producing high-resolution images with minimum pattern fluctuations. The resulting topographic pattern can then be transferred to the underlying silicon substrate by a reactive-ion etching or related process. The ability of the photoresist to act as a barrier for the underlying substrate is determined by the structure of the photoresist. Two parameters, namely, the Ohnishi parameter and the ring parameter, have been found to describe the etch resistance of the system empirically. The Ohnishi parameter states that the number of carbon atoms must be maximized while the number of oxygen atoms is minimized in a resist composition [8]. The ring parameter predicts that more cyclic structures will impart a higher etch resistance to the photoresist [9]. This illustrates some important aspects of the photolithography process that need to be considered in the design of new photoresist materials.

As the minimum feature size continues to shrink, the requirements of photoresist materials have been adapted to new emerging technologies. As demonstrated by Raleigh's law, $R = k^*\lambda/NA$, the pitch R of the smallest features obtained is governed by the wavelength λ of light used during patterning. The constant k is a process-dependent parameter that can be decreased through improvements to the resist materials and processes as well as optical resolution enhancement techniques. In order to continually shrink feature sizes, industry has been moving to shorter wavelengths and is currently focused on applying 193-nm wavelength radiation to patterning. To extend this technology further, the numerical aperture (NA) of 193-nm projection imaging systems has been increased by employing immersion fluids [10]. The current state-of-the-art uses 193-nm water-immersion lithography to pattern 45-nm half-pitch structures.

In order to keep up with the demands of the evolving lithographic technologies, the photoresist materials have undergone various structural changes. The photoresist platform dependence on thin-film patterning performance has been of keen interest as it applies to current resist systems especially in the sub-50-nm regime. This chapter will focus on a brief a history of deep-UV lithography and describe the evolution of current resist platforms such as 193-nm immersion technology (193i). The 193-nm node, which was the first lithographic technology to pattern feature sizes that are substantially smaller than the wavelength of the imaging light, has put special demands on the photoresist material. The transition from 193-nm dry conditions to 193-nm water-immersion lithography has placed a number of additional performance requirements on the already high expectations of resist materials. Outstanding patterning materials based more typically on polymers, but also on small molecules that can meet the performance criteria of the semiconductor industry, is a subject of continuing focus. This chapter will also highlight the work on patterning materials for emerging lithography techniques such as extreme ultraviolet (EUV) ($\lambda = 13.4$ nm) lithography, high-index lithography, nanoimprinting, and double patterning of thin films to produce sub-30-nm feature sizes.

13.2
Deep-UV Lithography

13.2.1
Brief History

The resist system that allowed the patterning of submicrometer features through g-line and i-line technology for many years was the two component diazonaphthoquinone (DNQ)–novolac-based resist system [11]. This type of resist consists of

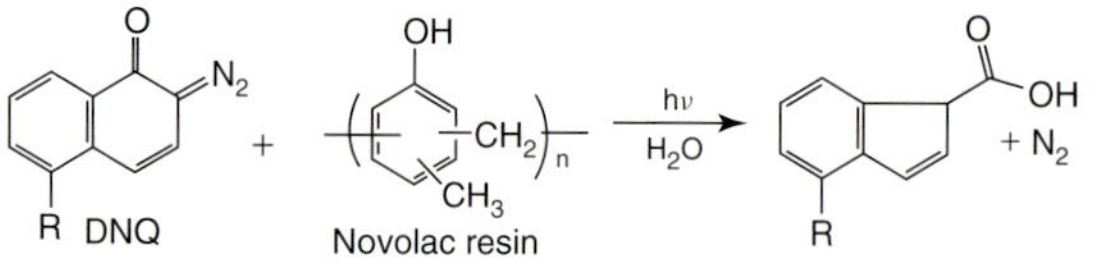

Scheme 13.1 Photoinduced dissolution inhibition mechanism of DNQ with novolac resin.

a photoactive DNQ dispersed in a novolac (a low molecular weight condensation product of formaldehyde and cresol isomers) matrix. The novolac resin is soluble in aqueous base due to the acidic character of the phenolic functionality. However, the solubility of novolac in the aqueous-base developer is determined by the lipophilic DNQ dispersed in the phenolic matrix, which acts as a dissolution inhibitor [12, 13]. Upon irradiation to UV light DNQ undergoes a series of reactions that lead to the formation of indene carboxylic acid (Scheme 13.1) [14, 15]. As a result, the exposed areas of the resist film dissolve much faster than the unexposed region in an aqueous-base developer resulting in positive-tone images. The structure and properties of the novolac resin in terms of the molecular weight [16], molecular weight distribution [17], ratio of various cresol isomers and the type of backbone linkages [16] have been studied in detail to control the dissolution behavior of the system. The optimization of the resist system along with improvements of the exposure tools have been used to manufacture features sizes as small as 350 nm [3].

13.2.2
Chemically Amplified Resist Technology

In order to meet the demands of higher resolution, the industry required a shift to deep-ultraviolet (DUV) patterning employing the 254-nm emission from Xe-Hg lamps or the 248-nm emission from krypton fluoride (KrF) excimer lasers [2]. The attempts to tailor the DNQ-novolac-based system to the DUV wavelength were futile due to its high absorbance in this spectral region that resulted in poor lithographic performance. Because of the weak energy sources at the 248 nm wavelength, a resist platform with photosensitivity that is orders of magnitude higher compared to the DNQ–novolac system was designed. The chemical amplification concept, initially proposed by Ito, Willson, and Fréchet, helped to overcome the sensitivity limitation by enabling one photon to initiate a cascading set of reactions [18]. This was realized with the aid of photoacid generators (PAGs) in the resist matrix [19]. 248-nm lithography employing chemical amplification positive-resist technology has been extended from the fabrication of minimum feature sizes of 250 nm down to 130 nm [6].

The very first chemically amplified resist system was designed on the basis of acid-catalyzed deprotection of poly(4-*tert*butoxycarbonyloxystyrene) (PBOCST). Scheme 13.2 illustrates the imaging chemistry of the lipophilic PBOCST polymer through the acid-catalyzed deprotection to form poly(4-hydroxystyrene) (PHOST) during the postexposure bake step around 100 °C. PHOST is soluble in aqueous-base developer due to its acidic character and gives rise to positive-tone

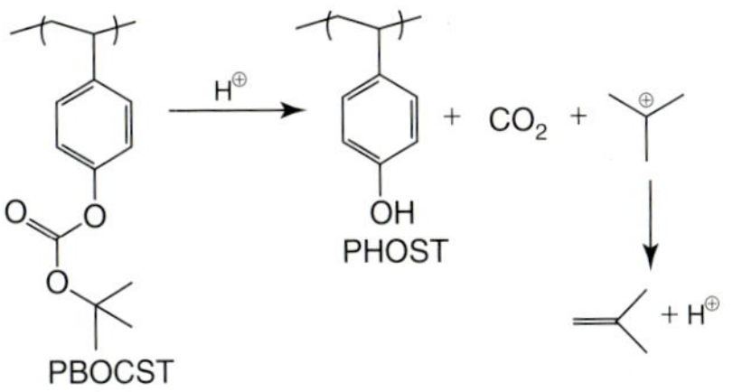

Scheme 13.2 Chemically amplified resist using t-BOC protected poly(4-hydroxystyrene).

features [19]. This reaction is truly catalytic as the photochemically generated acid is not consumed by one reaction but regenerated to carry out a number of deprotection reactions.

13.2.3
Photoresists for 248 nm Lithography

PHOST is the chosen resist for DUV lithography due to its structural features that impart good photoresist properties. Due to its low absorbance in the DUV spectral region and solubility-switching capability, both positive and negative 248-nm chemically amplified resists, are at least partially built on PHOST. The aromatic structure helps to increase the T_g of the polymer due to its rigid nature and also displays high thermal stability. The pKa of the phenolic component, which is ~10, ensures good solubility in the base developer without any swelling effects. The high hydrocarbon content and the ring character helps to increase plasma-etch resistance, which is crucial to successful pattern transfer to the silicon wafer.

Several DUV resist systems derived from PHOST typified by the APEX, environmentally stable chemical amplification photoresist (ESCAP), and KRS resists have been developed by the researchers at IBM (Scheme 13.3). In the APEX-type resist, the PHOST is only partially protected, resulting in a polymer that is more hydrophilic with better adhesion and dissolution behavior than its fully protected version [20]. APEX was the first CAR resist to be successfully integrated into the device manufacturing across the entire IC industry. The ESCAP resist system is based on a random copolymer of 4-hydroxystyrene with *tert*-butyl acrylate [21]. In comparison with the above-mentioned APEX-type systems, this copolymer has a higher thermal stability due to the higher activation energy of the *tert*-butyl ester deprotection reaction. The deprotected acrylate is more acidic than the deprotected

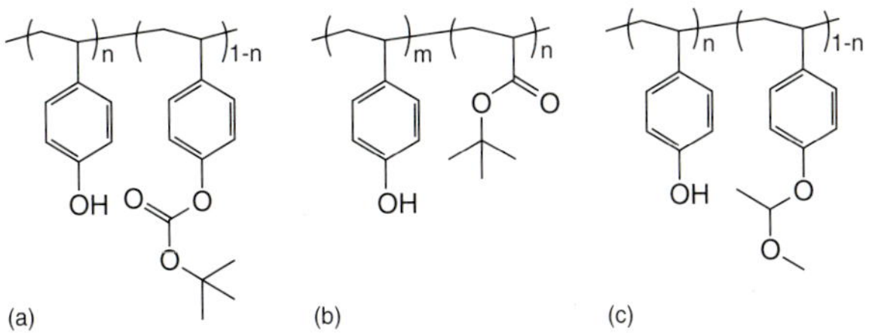

Scheme 13.3 Resist platforms developed for 248-nm lithography (a) APEX, (b) ESCAP, and (c) KRS.

PHOST, which results in faster dissolution and larger developer selectivity. The KRS-type resists are similar to the APEX resists, but rely on a highly reactive acetal based protecting group [22]. Due to the low activation of the acetal protecting group, the chemical amplification occurs at room temperature immediately following exposure. A KRS resist is insensitive to the temperature and time limits with respect to the postexposure bake step, thus providing processing advantages.

In addition to the resist systems discussed here, the acid-catalyzed thermolysis has been extended to several polymer backbones. Other protecting groups such as tetrahydropyranyl [23], trimethylsilyl [24], and *t*-butoxycarbonylmethyl [25] were also used in partial protection of PHOST. In order to address certain performance issues, each system is designed with a particular advantage such as high T_g, better adhesion, higher development contrast, improved sensitivity, or etch resistance.

Though most of the discussion has focused on positive-tone systems, acid catalyzed chemistry has been incorporated to negative-tone systems as well [26]. Acid-catalyzed cross-linking of PHOST can be attained with the aid of a cross-linking agent in the presence of PAG. The solubility of the exposed region in aqueous-base developer will be greatly reduced due to the increase in size and decrease of phenolic moieties, thereby producing a negative-tone image. Negative-tone images can also be obtained from the PBOCST system by changing the developer type. A nonpolar developer such as anisole has been used to selectively develop the unexposed region that consists of a t-BOC protected polymer [27]. This technology has been used in the production of 1-Mbit dynamic random access memory (DRAM) by IBM. A versatile resist platform has the ability to be modified between positive- and negative-tone fabrication. The selection of one specific tone over another may offer advantages depending on the actual manufacturing application. For example, positive-tone resists are compatible for printing contact holes and negative-tone resists have been well suited for certain device shapes encountered in the microprocessor fabrication [3].

13.3
193-nm Lithography

13.3.1
Polyacrylate-Based Resists

In keeping with the progression of Moore's law, photolithography moved to a shorter wavelength of 193 nm with the emission from an ArF excimer laser. In order to address the high optical absorbance at this wavelength by the aromatic groups of 248-nm resists, a different photoresist platform based on nonaromatic structures was developed [2, 4, 6]. The high absorption expected with almost all organic compounds at this wavelength provided a major challenge in resist-material design for 193-nm technology. The discovery of the transparency of acrylate based structures was a breakthrough in the initial photoresist design at this wavelength [28]. The *tert*-butyl methacrylate unit provided an acid-labile group for chemical

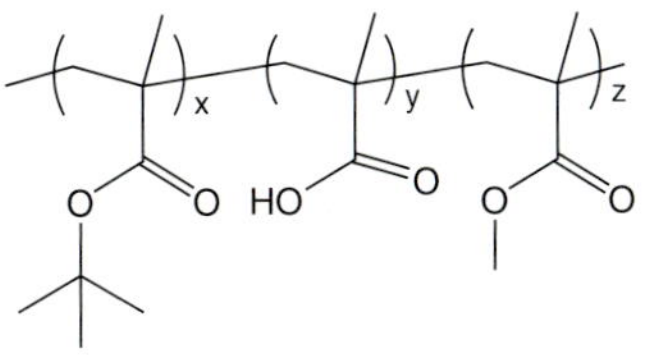

Scheme 13.4 IBM version 1 resist for 193-nm lithography.

amplification as carboxylic acid was employed as the solubility switching group in place of phenol. The first generation 193-nm positive resist developed by researchers at IBM as a tool-testing resist, employed a copolymer of t-butyl methacrylate, methacrylic acid, and methyl methacrylate (Scheme 13.4) [29]. Polymer systems built on ter- and tetrapolymers of monomers such as acrylic and methacrylic acid along with various acrylic esters were designed for 193-nm imaging. This major switch in resist platforms posed several design challenges to chemists in terms of photoresist performance. The carboxylic acid functionality, which is stronger than phenol in terms of acidity (pKa $\sim$ 6 for CO_2H *vs.* pKa $\sim$ 10 for phenolic OH), has encountered several issues during development caused by major swelling and very fast dissolution [30]. The poor dry-etch resistance of methacrylate polymers due to the lack of ring structures and oxygen-rich backbone was also a significant drawback of this system.

The issue of poor etch resistance in 193-nm resists was initially addressed by Fujitsu through the incorporation of alicyclic structure into polymethacrylates to improve etch resistance without sacrificing transparency [31, 32]. The etch resistance of these resist systems was greatly improved by incorporating ring structures such as adamantane, cyclohexanone, and nobornene units (Scheme 13.5). Improving the etch resistance of these resist systems came at the cost of increasing their hydrophobicity due to the bulky aliphatic pendant groups. This resulted in poor adhesion and development performance as well as increasing the T_g to undesirably high levels [33, 34]. In order to address these limitations, functional groups such as lactones have been ubiquitously used in alicyclic methacrylate polymers as a polarity-modulating group [35]. The design of successful polymers for 193-nm photoresist applications has been based on the delicate act of balancing the hydrophilic/hydrophobic nature of the polymer system.

Use of hexafluoroalcohol (HFA) pendant groups on norbornene as a replacement for carboxylic acid was proposed due its high transparency at the 193-nm wavelength (Scheme 13.5) [36]. Due to the similarities in pKa values to phenol, HFA was expected to have similar dissolution behavior to previous resist platforms such as ESCAP-based systems. A systematic study conducted by researchers at IBM supports the concept that a longer spacer between the methacrylate backbone and the HFA functionality results in a higher dissolution rate [37, 38]. A drawback of incorporating HFA groups to the resist system is the decrease in etch resistance due to incorporation of fluorine. But the impact of fluorine can be mitigated through the insertion of cyclic and hydrocarbon-rich aliphatic moieties through copolymerization [37, 38]. As anticipated, 193-nm resists can obtain ESCAP-resist-type swelling-free dissolution properties through the incorporation of HFA functionality and exhibit impressive patterning performance [39].

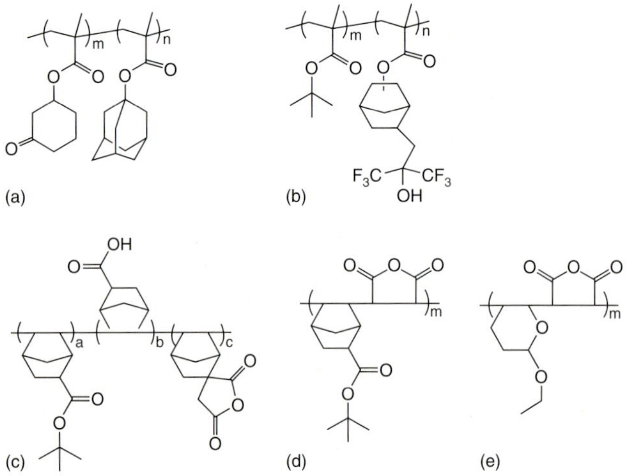

Scheme 13.5 (a) Acrylate polymer containing etch-resistant alicyclic groups. (b) Typical HFA containing ArF resist. (c) All norbornene-based ArF resist. (d) Cyclic olefin–maleic anhydride (COMA) polymer. (e) Vinyl ether–maleic anhydride (VEMA) polymer.

13.3.2
Poly(cycloolefin)-Based Resists

Photoresists based on poly(cycloolefin) derivatives comprise another class of ArF resists that has gained much attention (Scheme 13.5). The cycloaliphatic backbone of these polymers consisting of co- and terpolymers of various norbornenes contributes to dry-etch resistance, transparency, and thermal stability [40–42]. Carboxylic acid was incorporated as a solubility-switching group into the highly hydrophobic all-norbornene system. Though these resists have shown good imaging capabilities, the massive swelling caused during the dissolution of carboxylic acid was a major drawback of this system [43]. The nonlinear dissolution properties of this resist platform with respect to exposure dose caused significant performance issues despite its superior etch resistance.

13.3.3
Cyclic Olefin–Maleic Anhydride Polymers

Copolymers of alternating cycloolefin and maleic anhydride (COMA) consists of another widely studied resist platform for 193-nm lithography (Scheme 13.5) [44, 45]. This system is based on a generic structural motif that incorporates alicyclic structures directly in the polymer backbone and undergoes aqueous base dissolution without any swelling effects [46]. The ring opening (hydrolysis) of the anhydride by aqueous base during development has enabled high dissolution rates, even at low carboxylic acid composition. A large number of polymers were explored due to the large pool of cycloolefin monomers that were available for copolymerization with maleic anhydride but the poor yields of these materials remained a concern. The hydrolysis of the anhydride ring, which was beneficial during the polymer dissolution process posed a significant disadvantage during resist storage. The COMA resists suffered from premature ring rapture during polymer storage resulting in irreproducible lithographic performance and poor shelf life [47, 48]. As an alternative to the COMA system, vinyl ether monomers have been employed to form copolymers with maleic anhydride to yield an alternative resist system Vinyl ether–malefic anhydride (VEMA) [49, 50]. The replacement of cycloolefin with vinyl ethers improves copolymer yields and allows important resist properties such as T_g and adhesion to be fine tuned (Scheme 13.5).

Among the several polymer platforms described here, methacrylate-based polymers that incorporate cyclic etch-resistant moieties and HFA structures have emerged as the leading candidates to be developed as commercial 193-nm resists.

13.4
157-nm Lithography

Lithography with 157-nm fluorine lasers has been the rapidly emerging technology of choice for sub-100-nm feature sizes beyond the 193-nm wavelength. It was a fundamental extension of optical lithography to smaller feature sizes through the reduction in wavelength. The resist materials for this technology were faced with even more constraints due to the opacity of aromatic-based 248-nm resists or carbonyl-containing 193-nm resists at such short wavelengths [51]. Thus, 157-nm resists were built on fluoropolymers and silsesquioxane-based polymer backbones to provide high transparency at 157 nm [52, 53]. Functional groups that were explored for ArF resist applications such as HFA and fluorinated methcrylate and cyclic units were employed as common building blocks for this resist platform [54, 55]. But the high content of fluorine had an adverse effect on the etch resistance of the resist matrix. The etch resistance of these systems can be improved by introducing more alicyclic groups through polynorbornene backbone polymers and acid-labile moieties with adamantyl and tetrahydropyranyl groups [56, 57]. Despite significant efforts in terms of resist development, the semiconductor industry phased out 157-nm lithography due to many technical concerns. But the interesting photoresist platforms developed in this effort did not go to waste as they have been adapted to suit the current 193-nm immersion resist needs.

13.5
193-nm Immersion Lithography

193-nm immersion lithography (193i) has been accepted as the technology for patterning below 50 nm feature sizes. Immersion lithography is a lithography-enhancement technique that replaces the usual air gap between the final lens element and the photoresist surface with a liquid medium with a refractive index (RI) greater than one. Water, which has a RI of 1.44 at 193 nm, is the present immersion fluid of choice. Shifting from dry exposures to immersion medium enables higher resolution by increasing the acceptable lens NA to values greater than one [58]. Immersion lithography has developed with unprecedented speed and is currently being used for the patterning at the 45-nm node [59]. The development of this technology has been aided by the finding that no new photoresist platform is required. Most of the resists used in 193-nm dry lithography have been suitable as immersion resists after appropriate modifications.

13.5.1
Design of Topcoats for 193i Resists

In addition to meeting the resolution requirements, immersion resists need to address other material challenges that arise due to their contact with water [60, 61]. Controlling leaching, where small-molecule resist components, such as the PAG and the base quencher, cannot permeate into the water is a significant concern with this technique [62]. Leaching not only adversely affects the resist performance but also contaminates the water, and in turn, the optical equipment in contact with it. Immersion-related defects are another challenge to the 193i process. As the water between the lens and wafer forms a meniscus that moves with progressing exposure, various physical and chemical interactions between the water and resist surface can occur, leading to film pulling, water marks, and water absorption to the resist surface [63]. Therefore, controlling the water/resist interface is critical to minimizing defects and PAG extraction. Coating the resist film with protective topcoats has been the answer to address these resist constraints and enable immersion lithography [64, 65].

Several topcoat materials have been evaluated for their leaching behavior and surface properties. As leaching is expected to increase with exposure dose, an effective topcoat should provide a leaching barrier even at high exposures. The angle at which the water contacts the resist surface is another important parameter for evaluating topcoat performances. It has been shown that a high receding contact angle between water and the topcoat surface can limit immersion-related defects and minimize water absorption while allowing high scan rates (throughput) [66]. Developing topcoats that offer both low leaching behavior and hydrophobic surface properties have been a key focus for the development of 193i technology [64]. Two types of topcoat, one that is soluble in an organic solvent prior to development and the other type that is removed during normal development with aqueous base have been used successfully.

Topcoat materials have the advantage of decoupling the resist from surface-property considerations, as the topcoat design can be modified to adjust to immersion conditions without having to change the resist structure. Solubility in solvents that do not dissolve photoresists and transparency at the 193-nm wavelength are additional requirements for a good topcoat material. Due to its applicability to the resist-developing process, without the need for any additional processing

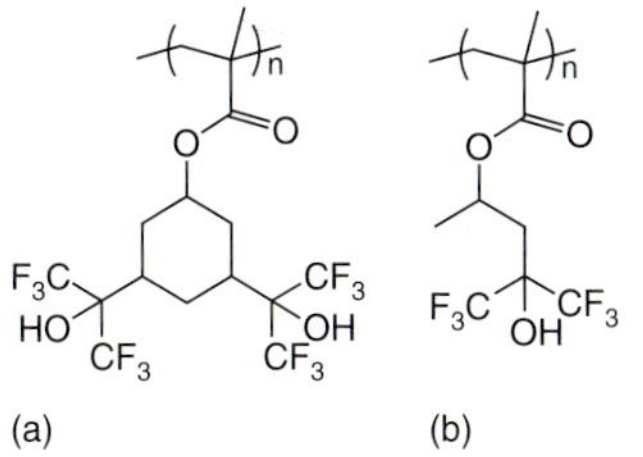

Scheme 13.6 (a) Bis hexa-fluoroalcohol cyclohexyl methacrylate topcoat material.
(b) Fluoromethacrylate-methacrylic acid polymer additive.

(a)

(b)

steps, base-developable topcoats have emerged as the mainstream approach for immersion lithography [67]. Achieving the careful balance between hydrophobicity required for high water contact angles and the base solubility required for top-coat removal is challenging [68, 69]. It has been observed that the hydrophobic topcoat tends to have a low dissolution rate in the base developer. However, a high dissolution rate in the developer is essential for the effective removal of the topcoats. The low surface energies required for high water contact angles in topcoat materials has been generally achieved through the incorporation of fluorinated groups into the polymer structure. The HFA group, which has been a compo-nent in193- and 157-nm resists is used to provide base solubility. Evaluation of several topcoat designs by researchers at IBM demonstrates that materials such as bis-hexafluoroalcohol cyclohexyl methacrylate (Scheme 13.6), which combine high T_g with good dissolution properties, give excellent results at the 45-nm node [64, 70]. Current research has applied the self-segregation concept to topcoats in an attempt to break the property tradeoffs that have limited the progress of base-soluble topcoats. A fluorinated polymer optimized for high water contact angles and an acidic polymer optimized for resist development are spin cast on the resist to segregate on the surface [71]. This creates a blended topcoat with each polymer optimized for its specific purpose but able to segregate in order to enhance performance.

13.5.2
Topcoat-Free Resists for 193i Technology

Topcoat-free resist materials are also being actively developed for immersion appli-cations [72]. Additives that control the interface between the resist material and the immersion liquid have enabled the topcoat-free technology through surface modi-fication of the resist films (Scheme 13.6). The hydrophobicity of the resist surface could be significantly improved through the addition of a small amount of additive [73]. Similar to topcoat materials, fluoropolymers are the main class of materials studied as additives. This has warranted an indepth study of structure–property relationship between various flourinated methacrylates and their contact-angle and dissolution properties [68]. As expected, the receding contact angles showed an increase with increasing carbon content. But similarities between materials with similar carbon content indicate the importance of chemical structure in the determination of their surface properties. A similar study mentioned earlier in this chapter, demonstrated that increasing the "spacer" length between the fluoroalcohol group and the methacrylate backbone accelerated the dissolution rate of the polymers in aqueous base [37]. Systematic studies have provided an insight into the contact angle/dissolution rate tradeoffs in these materials and enabled the design of appropriate additives.

13.6
Next-Generation Lithographic Technology

The IC industry is committed to extending Moore's law to the sub-30-nm regime in the near future. However, the ability to reach the resolution target while maintaining patterning performance is a serious challenge. The International Technology Roadmap for Semiconductors (ITRSs) has been continually revised to add new technologies that are being explored to achieve the expected resolution targets [74]. This section will elaborate on recent advances in techniques that are being considered as next-generation lithography candidates.

13.6.1
High-Index Immersion Lithography

Encouraged by the promising results of water immersion, immersion fluids' RI values greater than water are being considered to extend the capabilities of immersion lithography to sub-45-nm feature sizes [75, 76]. As the index of the fluid medium increases, the resolution of the system improves as lenses with higher NAs can be used despite having the wavelength of light fixed at 193 nm. Additional requirements for a high-RI immersion fluid include high transparency at the 193-nm wavelength and low viscosity to enable high scan rates.

An initial approach to raising the RI of water included adding organic or inorganic additives. Following this strategy, a RI close to 1.6 was attained through the solubility of inorganic salts. Toxicity due to high salt concentrations was an inherent limitation of this system [77]. Additives such as crown ethers were used to improve the solubility limit of benign inorganic salts such as $BaCl_2$ and $CaCl_2$. But increasing the additive concentration in water also tends to increase the absorbance to extremely high values [78]. Single-component organic fluids with RIs around 1.64, has been another promising approach for the second-generation immersion fluid [75, 79]. Degradation due to exposure and low surface energies of these fluids are issues that need to be addressed. The contact angle of such a high-RI fluid on a typical methacrylate resist is not high enough to prevent film pulling at predicted tool velocities [80]. Studies on designing topcoats with surface properties to suit high RI immersion fluids are currently underway [71]. Dispensing high-RI nanoparticles in a liquid host is the most viable approach to obtaining third generation high-index fluid with RI > 1.80. HfO_2 nanoparticles have been introduced as a potential candidate for high-RI applications through the consistent synthesis of nanoparticles less than 10 nm in size and the ability to be dispersed in solution [81]. More work remains in obtaining the required loadings of the nanoparticles in solution at the necessary purity level. However, there are concerns with high-RI fluids in terms of damage to the lens and other process control issues.

In addition to the high-RI fluids, high-RI lens materials, and high-RI resists, are being developed to enhance the imaging capability through further increase of NA [82]. Initial studies have demonstrated that LuAG with a RI of ~2.2, can be a strong candidate for making a high-RI lens. Developing a high-RI resist material is a significant materials challenge. Literature reports suggest that including atoms such as sulfur, bromine, or aromatic groups into a polymer structure increases its RI. But due to other resist-property constraints, phenyl groups that show relatively strong absorption at 193 nm and bromine that has issues with contamination preclude their use in any resist design. Therefore, incorporating sulfur into the resist has been actively pursued to develop high-RI resists [83, 84]. Though the refractive indices of sulfur-containing polymers were found to be greater than that of standard ArF resists, an increase in absorption was a significant limiting factor [84]. Despite having suffered delays in terms if development and significant materials challenges that need to be addressed, high-index immersion lithography remains a viable candidate for patterning thin films at sub-30-nm feature sizes [85].

13.6.2
EUV Lithography

The ultimate advancement in optical lithography is EUV lithography, which makes the enormous jump from 193 to 13.5 nm. This technique is expected to produce feature sizes in the sub-30-nm regime with a single-exposure technique due to the extremely short wavelength of the EUV radiation [86]. Due to the very short exposure wavelength, energy absorption is no longer dictated by atomic bonding but by the atomic composition [87]. Carbon and hydrogen are elements with high EUV transparency while oxygen and fluorine absorb strongly at this wavelength. Therefore, existing resist platforms have been modified as EUV resists without the need for developing a brand new resist system. Another concern for EUV resists is outgassing that could occur during exposure under vacuum [88]. Outgas species released from EUV resists before, during and after exposure may pose a source of contamination that could damage the optics of EUV exposure tools [89]. All EUV resists need to ensure that no decompositions that may lead to volatile components occur within the exposure chambers.

One of the main challenges faced by this technology is the low power of the EUV source. Hence, it is important for EUV resists to be highly sensitive in order to maximize their imaging capability. Most EUV resists have been based on previously established resist concepts such as PHOST, norbornene, and fluoropolymer structures [90, 91]. In addition, atoms such as silicon and boron, which have very high transparency at EUV have been incorporated to the resist system. For example, polysilane-based resists and boron-containing polymers have been specifically designed to enhance transparency [92, 93].

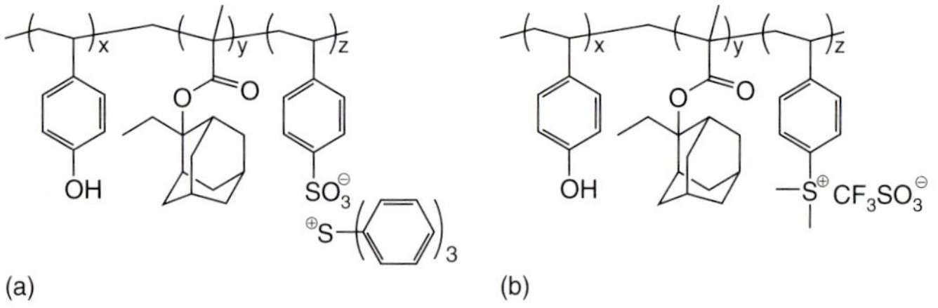

Scheme 13.7 PAG-attached resists for EUV lithography (a) anion bound and (b) cation bound.

In an attempt to improve sensitivity as well as control outgassing, PAG-attached photoresists were designed. Initially, this type of design consisted of a polymer based solely on a PAG-attached monomer unit [94]. A highly radiation sensitive group, based on a common PAG moiety such as a sulfonium salt, is introduced to the polymer structure. The unexposed polymer was polar due to its ionic character and soluble in polar solvents such as water. Upon exposure to radiation the polymer undergoes a polarity change as the sulfur–carbon bonds break to release sulfonium acid. Using slightly basic water as the developer, the more polar unexposed area can be dissolved to form negative-tone submicrometer feature sizes [95]. This system is an example of a very sensitive photoresist that does not depend on the chemically amplified technology.

Despite its high resolution potential this type of resist system could not be integrated to the existing fabrication process. Therefore, the PAG-attached monomers were copolymerized with conventional resist monomers such as PHOST and acrylate-based monomer units. This gives rise to a one-component resist system based on the CAR technology [96]. The PAG is attached to the monomer units in two different ways Scheme 13.7; through the cation or the anion of the PAG molecule [97, 98]. The cation bound PAG has the distinct advantage of showing no benzene outgassing from the polymer-bound PAG resist. As the benzene rings from the PAG cation comprise a significant percentage of the outgas, binding the cation to the polymer helps to significantly reduce resist outgassing. The anion is bound to the PAG to alleviate issues of PAG phase separation as well as to control PAG diffusion. All PAG-attached polymers show superior performance in terms of resolution, sensitivity, and outgassing with respect to conventional blend systems. Polymer-bound PAGs are currently one of the leading photoresist candidates capable of producing sub-30-nm feature sizes with performance advantages under EUV exposure.

13.6.3
Molecular-Glass Resists

As the drive to pattern at ever shrinking length scales continues, new imaging materials may be required to meet manufacturing demands. Typical photoresist materials consist of random copolymers with a molecular weight within the range of 5–15 kg/mol. As described through this chapter, photoresist materials have evolved with new functional groups and chemistry suited to each emerging wavelength. The influence of resist molecular weight as well as resist architecture becomes important in lithographic scales aiming at sub-30-nm resolution. The large size and polydispersity of polymeric resists have been identified as a potential reason for limiting resolution and adversely affecting the resist performance [99, 100]. This section introduces a relatively new concept in high-resolution patterning materials known as *molecular glass photoresists*, which are an alternative to conventional patterning materials.

Amorphous molecular materials termed "molecular glasses" (MGs) constitute a new class of functional materials being developed for electronic applications [101]. MGs combine the characteristic properties of small molecules such as high purity and well-defined structure with beneficial aspects of polymers such as high thermal stability and thin-film-forming properties. MGs possess structural features that inhibit crystallization and display high glass-transition temperatures despite their modest size. These molecules can be characterized by the presence of significant free volume and by disorder in both intermolecular distance and orientation [102, 103]. In contrast to polymers, MGs can be repeatedly synthesized with well-defined control of molecular weight, compositional, and stereochemical factors, so that a precise material is obtained after each synthetic step. The basic building block in a MG resist may be monodisperse and is as much as an order of magnitude smaller in size compared to conventional polymeric resists. The reduction in this "pixel" size is believed to be a fundamental improvement in the ability to obtain high-resolution patterns consistently. The development of MG photoresists has also been supported by theoretical work that shows that photoresist performance improves with the reduction of molecular weight of the resist [104]. The radius of gyration (R_G) of a photoresist polymer has also been directly correlated to Line edge roughness (LER) [105]. A smaller R_G can therefore consistently yield a smaller LER.

As most small molecules tend to crystallize, MGs need to be designed with specific design guidelines in mind. The stability of the amorphous phase in a small molecule can be explained by the crystal-growth velocity and the maximum crystal-growth temperature, which are kinetic parameters that relate to the transition from amorphous to the crystalline phase [106]. High maximum crystal-growth temperature and low minimum crystal-growth velocity are required for stable amorphous states of small organic materials. Therefore, MGs must be designed with nonplanar, irregular shapes that inherently resist crystallization. Molecular geometry also plays a crucial role in the glass-forming ability of molecular systems. Common glass-forming topologies include branched or star shapes, spiro links, ring, tetrahedral, and twin molecular structures [102]. When designing MG resist materials, structural features that increase T_g in addition to reducing crystal-growth rate must be incorporated [106]. The ability to demonstrate high T_gs (>100 °C) despite their modest size is an important requirement of MGs to be a viable alternative to polymeric resists. Structural features that decrease free volume and restrict rotation about any molecular axis are expected to raise T_g [107]. Inclusion of rigid and bulky groups such as tert-butyl, biphenyl, and fluorene moieties increases T_g by hindering translational, rotational, and vibrational motions of the molecule.

The presence of intermolecular interactions such as dipole–dipole interactions and hydrogen bonding has the ability to increase T_g by decreasing free volume.

The first MG resists based on branched phenylbenzene and triphenylene amine-based structures, were introduced by Shirota and coworkers [108] as non-chemically amplified resist systems. Despite their potential as high-resolution patterning materials, these MG systems were largely ignored due to their lack of sensitivity. Upon switching to the CAR technology MG resists have been able to demonstrate their high-resolution capabilities as well as improved resist properties [109–111]. The first MG resist to demonstrate 30-nm feature sizes under EUV conditions was based on a calix[4]resorcinarene derivative was developed by Ober and coworkers [112]. Since then, several dense, bulky, branched molecular systems, and variants of calix[4]resorcinarene ring molecules have been developed as MG resists capable of patterning in the sub-30-nm regime (Figure 13.2) [113, 114]. The high T_g of these photoresists has been identified as a key feature that enables patterning at these small feature sizes. Though most of these MG systems have been designed for EUV lithography, MG resists with saturated aliphatic functionalities have been designed for 193 nm lithography as well. Polyhedral oligomeric silesquioxane (POSS)-based MG resists, functionalized with chemically amplifiable bulky side groups have been successfully imaged under both dry and immersion conditions at the 193-nm wavelength [115]. PAG-attached MG resists are currently being developed as a single component resist system with promising results as a next-generation patterning material [116, 117].

13.6.4
Double Patterning

Double patterning is a patterning technology that can enhance feature density through multiple lithographic steps. As double patterning requires only the development of new patterning processes, it seems to be the most promising technique for future device fabrication. This is the primary lithography technique introduced for the sub-40-nm feature sizes employing existing ArF dry or water-immersion technology. Repetitive techniques such as dual trench (litho-etch-litho-etch) and dual-line (litho-litho-etch) processes are examples of double-patterning techniques [118, 119].

In the dual-trench process, the desired pattern is split in two parts, and the first set of features are printed and etched to the substrate with the exposed first resist, while the second set of images are printed and etched to the substrate with the recoated second resist. Two resists that can be either similar or different are used. Though there are a few technical issues to be overcome, the large number of process steps involved is a limiting factor. In comparison, litho-litho-etch is a more efficient process, which involves coating the second resist in the presence of the first, patterned resist without disturbing it. Several methods based on UV curing and chemically freezing the first resist have been put forth as possible solutions to this problem [120–122]. New techniques that protect the configuration of the first resist pattern without causing any damage during the second lithographic process

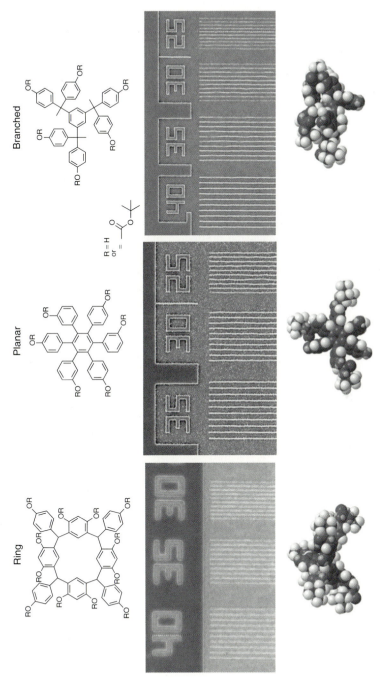

Figure 13.2 High-resolution EUV patterns with three different types of molecular-glass cores.

is an active area of research that is key to the progress of double patterning. Recent work reports that sub 40 nm feature sizes have been successfully demonstrated using a 193-nm wavelength for double-patterning technology [122].

13.6.5
Nanoimprint Lithography

Nanoimprint lithography (NIL) is an unconventional lithographic method, developed by Chou and coworkers [123, 124], with the ability to pattern nanometer-scale patterns. The main advantage of this technique is the ability to pattern sub-30-nm features over a large area with high-throughput and low-cost. As shown in Figure 13.3, this technique is based on two main steps. The imprint step transfers the nanoscopic patterns on the mold to the resist film cast on a substrate. During the imprint step, the resist is heated above its T_g to enable ready deformation into the shape of the mold. The second step is the pattern transfer, where a reactive-ion etch process transfers the thickness contrast obtained through the imprint step onto the underlying substrate. Unlike conventional lithographic techniques, NIL does not use light for imaging. Therefore, its resolving power is not limited by the wavelength constraints of the imaging radiation.

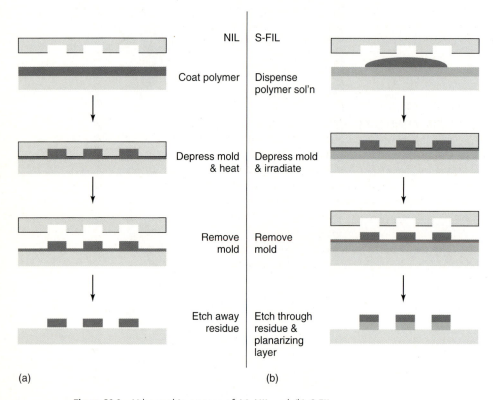

Figure 13.3 Lithographic process of (a) NIL and (b) S-FIL.

The mold in the NIL technique plays the same role as the photomask in photolithography. An important consideration for selecting mold materials include having sufficient hardness and durability for the nanoimprint application [125]. The thermal expansion coefficient is another important property as a thermal mismatch between the mold and the substrate could result in pattern distortion and stress buildup.

In this regard, a Si mold and Si substrate make good material candidates for the imprint process. PMMA as well as other typical photoresists have shown good performance as patterning materials for imprint lithography. The low thermal expansion coefficient and low pressure shrinkage coefficient makes PMMA an ideal imprint candidate [124]. The thermoplastic materials used in imprinting should be easily deformable under an applied pressure and temperature. The ability to maintain its mechanical integrity during mold–substrate separation as well as any subsequent pattern transfer steps is a requirement of a good NIL patterning material. NIL has emerged as a promising candidate in a variety of high-resolution patterning applications in the electronic as well as the biological field [125–127].

13.6.5.1 Step-and-Flash Imprint Lithography

Step-and-flash imprint lithography (S-FIL), a variation of NIL, was first reported by Willson and coworkers [128]. This process introduces a low-viscosity, photopolymerizable organosilicon fluid into the gap between the transparent template and the substrate. The organosilicon fluid spreads out under capillary action and is then cured upon exposure to UV radiation. This enables the formation of a solidified, silicon-rich replica of the template. Because this process requires a flat surface, a planarizing layer is used underneath the pattern to eliminate any substrate topography. The use of a low-viscosity UV-curing solution, in contrast to typical NIL, allows imprinting at room temperature with minimal applied pressure. Materials for S-FIL require properties not found in current patterning materials such as low viscosity and rapid photoinduced polymerization. Inorganic/organic hybrid materials derived from sol-gel chemistry and POSS-based structures have been used as promising UV-curable S-FIL patterning materials. Recent work shows how the imprinted film was further processed through thermally polymerizable, benzocyclobutane-type functional groups to enhance its mechanical properties [129]. Significant efforts from both academia and industry have been invested in S-FIL research and development since it was incorporated into the ITRS roadmap as a viable candidate for sub-30-nm lithography [130, 131].

13.7
Summary

The goals set by the IC industry for continually achieving high-resolution patterns have fueled the advances in lithographic patterning technologies. The current industry focus is on identifying the key technology for the 32 and 22 nm environments. This chapter gives a thorough review of the photolithography process, which has been the backbone of the semiconductor industry for the past few decades. The focus of this work has been on continuing evolution of the patterning materials through developments in chemistry. Due to the dynamic state of the industry requirements, some technologies fail while others advance more rapidly than expected. While there remains debate over the costs and the ability to manufacture in large volume, all next-generation lithography techniques described in this review have some real prospects for acceptance. Advances in the microelectronics industry have also enabled developments in other areas of thin-film patterning that are explored in other chapters of this volume. The authors hope that the reader finds this chapter useful in understanding the progress and the challenges of photolithography.

Acknowledgments

This review was made possible by research support from the Semiconductor Research Corporation (SRC) for funding our work in the area of lithography-patterning materials. The support from Intel Corporation and International Sematech is gratefully acknowledged. The authors would also like to thank the Cornell Nanoscale Facility and the Cornell Center for Materials Research for the use of equipment.

References

1. Bondyopadhyay, P.K. (1998) *Proc. IEEE*, **86**(1), 78–81.

2. Thompson, L.F., Willson, C.G., and Bowden, M.J. (1994) *Introduction to Microlithography*, 2nd edn, ACS, Washington, DC.

3. Wallraff, G.M. and Hindsberg, W.D. (1999) *Chem. Rev.*, **99**, 1801–1821.

4. Ueno, T. (1998) Advanced Chemically Amplified Photoresists, in *Microlithography: Science and Technology* (eds J.R. Sheats and B.W. Smith), Marcel Dekker Inc., pp. 429–514.

5. Kim, J.-B., Yun, H.-J., Kwon, Y.-G., and Lee, B.-W. (2000) *J. Photopolym. Sci. Technol.*, **13**(4), 629–634.

6. Ito, H. (2005) *Adv. Polym. Sci.*, **172**, 37–245.

7. Fryer, D.S., Nealey, P.F., and Pablo, J.J. (2000) *J. Vac. Sci. Technol., B*, **18**(6), 3376–3380.

8. Ohnishi, Y., Mizuko, M., Gokan, H., and Fujiwara, S. (1981) *J. Vac. Sci. Technol. B: Microelectron. Process. Phenom.*, **19**(4), 1141.

9. Kunz, R. *et al.* (1996) *Proc. SPIE: Int. Soc. Opt. Eng*, **2724**, 365.

10. Flagello, D.G. (2008) *Proc. SPIE: Int. Soc. Opt. Eng.*, **6827**, 68271N-1–68271N-11.

11. Dammel, R.R. (1993) *SPIE Tutorial Text TT11*, SPIE Optical Engineering Press, Bellingham.

12. Honda, K., Beauchemin, B.T., Hurditch, R.J., Blankeney, A.J., Kawabe, K., and Kokubo, T. (1990) *Proc. SPIE: Int. Soc. Opt. Eng*, **1262**, 493.

13. Kajita, T., Ota, T., Nemoto, H., Yumoto, Y., and Miura, T. (1991) *Proc. SPIE: Int. Soc. Opt. Eng.*, **1466**, 161.

14. Sus, O. (1994) *Justus Liebigs Ann. Chem.*, **556**, 65.

15. Packansky, J. and Lyerla, J.R. (1979) *IBM J. Res. Dev.*, **23**, 42.

16. Hanabata, M., Uetani, Y., and Furuta, A. (1989) *J. Vac. Sci. Technol.*, **B7**, 640.

17. Hanabata, M., Uetani, Y., and Furuta, A. (1988) *Proc. SPIE: Int. Soc. Opt. Eng.*, **920**, 349.

18. Ito, H. (2000) *IBM J. Res. Dev.*, **44**, 119–130.

19. Ito, H. and Willson, C.G. (1984) *Applications of Photoinitiators to the Design of Resists for Semiconductor Manufacturing*, ACS sym. Ser., (ed. T. Davidson), **212** (2), 11–23.

20. Woods, R.L., Lyons, C.F., Mueller, R., and Conway, J. (1988) Practical Half-Micron Lithography with a 10X KrF Excimer Laser Stepper. *Proceedings of the KTI Microelectronics Seminar*, pp. 341–359.

21. Ito, H. *et al.* (1994) *J. Photopolym. Sci. Technol.*, **7**, 433–437.

22. Huang, W.-S., Kwong, R., Katnani, A., and Khojasteh, M. (1994) *Proc. SPIE: Int. Soc. Opt. Eng.*, **2195**, 37–46.

23. Mertesdorf, C., Nathal, B., Munzel, N., Holzwarth, H., and Schacht, H. (1994) *Proc. SPIE - Int. Soc. Opt. Eng.*, **2195**, 246.

24. Murata, M., Takahashi, T., Koshiba, M., Kawamura, S., and Yamaoka, T. (1990) *Proc. SPIE: Int. Soc. Opt. Eng*, **1262**, 8.

25. Onishi, Y., Oyasato, N., Niki, H., Hayase, R., Kobayashi, H., Sato, Y., and Miyamura, M. (1992) *J. Polym. Sci. Technol.*, **5**, 47.

26. Reichmanis, E., Houlihan, F., Nalamasu, O., and Neehan, T. (1991) *Chem. Mater.*, **3**, 394.

27. Maltabes, J., Holmes, S.J., Morrow, J., Barr, R.L., HakeyRey, M., Reynolds, G., Brunsvold, W.R., Willson, C., Clecak, N.J., and MacDonald, S.A. (1990) *Proc. SPIE: Int. Soc. Opt. Eng*, **1262**, 2–7.

28. Kunz, R.R., Allen, R.D., Hinsberg, W.D., and Wallraff, G.M. (1993) *Proc. SPIE: Int. Soc. Opt. Eng.*, **1925**, 167–175.

29. Allen, R.D., Wallraff, G.M., Hinsberg, W.D., and Simpson, L.L. (1991) *J. Vac. Sci. Technol., B: Microelectron. Process. Phenom.*, **9**, 3357.

30. Ito, H. (2001) *IBM J. Res. Dev.*, **45**, 683–695.

31. Kaimoto, Y., Nozaki, K., Takechi, S., and Abe, N. (1992) *Proc. SPIE: Int. Soc. Opt. Eng.*, **1672**, 66–73.

32. Nozaki, K., Kaimoto, Y., Takahashi, M., Takechi, S., and Abe, N. (1994) *Chem. Mater.*, **6**, 1492.

33. Allen, R.D., Wallraff, G.M., DiPietro, R.A., and Kunz, R.R. (1994) *J. Photopolym. Sci. Technol.*, **7**, 507.

34. Allen, R.D., Wang, I.Y., Wallraff, G.M., DiPietro, R.A., Hofer, D.C., and Kunz, R.R. (1995) *J. Photopolym. Sci. Technol.*, **8**, 623.

35. Ito, H. (2008) *Proc. SPIE: Int. Soc. Opt. Eng.*, **6923**, 6923181-12.

36. Ito, H., Seehof, N., Sato, R., Nakayama, T., and Ueda, M. (1998) *Micro-and Nano-Patterning Polymers*, ed. H. Ito, E. Reichmanis, O. Nalamasu, and T. Ueno American Chemical Society, Washington, DC, p. 208.

37. Varanasi, P.R., Kwong, R.W., Khojasteh, M., Patel, K., Chen, K.-J., Li, W., Lawson, M.C., Allen, R.D., Sooriyakumaran, R., Brock, P., Sundberg, L.K., Slezak, M., Dabbagh, G., Liu, Z., Nishimura, Y., Chiba, T., and Shimokawa, T. (2005) *Proc. SPIE: Int. Soc. Opt. Eng.*, **5753**, 131–139.

38. Varanasi, P.R. *et al.* (2005) *J. Photopolym. Sci. Technol.*, **18**, 381–387.

39. Ito, H., Truong, H.D., Allen, R.D., Li, W., Varanasi, P.R., Chen, K.-J., Khojasteh, M., Huang, W.-S., Burns, S.D., and Pfeiffer, D. (2006) *Polym. Adv. Technol.*, **17**, 104–115.

40. Allen, R.D. *et al.* (1996) *J. Photopolym. Sci. Technol.*, **9**, 465–474.

41. Okoroanyanwu, U., Shimokawa, T., Byers, J., and Wilson, C.G. (1998) *Chem. Mater.*, **10**, 3319.

42. Klopp, J.M., Pasini, D., Byers, J.D., Willson, C.G., and Fre'chet, J.M.J. (2001) *Chem. Mater.*, **13**, 4147–4153.

43. Ito, H., Allen, R.D., Opitz, J., Wallow, T.I., Truong, H.D., Hofer, D.C., Varanasi, P.R., Jordhamo, G.M., Jayaraman, S., and Vicari, R. (2000) *Proc. SPIE: Int. Soc. Opt. Eng.*, **3999**, 2–12.

44. Houlihan, F.M., Wallow, T.I., Nalamasu, O., and Reichmanis, E. (1997) *Macromolecules*, **30**, 6517–6524.

45. Ito, H., Miller, D., and Sherwood, M. (2000), Fundamental Aspects of Norbornene-Maleic Anhydride Co- and Terpolymers for 193 nm Lithography:

Polymerization Chemistry and Polymer Properties. *J. Photopolym. Sci. Technol.*, **13**, 559.

46. Allen, R.D. *et al.* (1998) *J. Photopolym. Sci. Technol*, **11**, 475.

47. Rahman, M.D., Bar, J.-B., Cook, M., Durham, D.L., Kudo, T., Kim, W.-K., Padmanaban, M., and Dammel, R.R. (2000) *Proc. SPIE: Int. Soc. Opt. Eng.*, **3999**, 220.

48. Ito, H., Miller, D., and Sherwood, M. (2000) *J. Photopolym. Sci. Technol.*, **13**, 559.

49. Choi, S.-J., Kim, H.-W., Woo, S.-G., and Moon, J.-T. (2000) *Proc. SPIE: Int. Soc. Opt. Eng.*, **3999**, 54.

50. Choi, S.-J. *et al.* (2001) *J. Photopolym. Sci. Technol.*, **14**, 363.

51. Kunz, R.R., Bloomstein, T.M., Hardy, D.E., Goodman, R.B., Downs, D.K., and Curtin, J.E. (1999) *J. Vac. Sci. Technol., B: Microelectron. Process. Phenom.*, **17**, 3267.

52. Ito, H., Waliraff, G.M., Brock, P., Fender, N., Truong, H., Breyta, G., Miller, D.C., Sherwood, M.H., and Allen, R.D. (2001) *Proc. SPIE: Int. Soc. Opt. Eng.*, **4345**, 273–284.

53. Toriumi, M., Shida, N., Watanabe, H., Yamazaki, T., Ishikawa, S., and Itani, T. (2002) *Proc. SPIE: Int. Soc. Opt. Eng.*, **4690**, 191–199.

54. Ito, H., Truong, H.D., Okazaki, M., Miller, D.C., Fender, N., Brock, P.J., Wallraff, G.M., Larson, C.E., and Allen, R.D. (2002) *J. Photopolym. Sci. Technol.*, **15**, 591–602.

55. Takebe, Y., Eda, M., Okada, S., Yokokoji, O., Irie, S., Otoguro, A., Fujii, K., and Itani, T. (2004) *Proc. SPIE: Int. Soc. Opt. Eng.*, **5376**, 151.

56. Vohra, V.R., Douki, K., Kwark, Y.-J., Liu, X.-Q., Ober, C.K., Bae, Y.C., Conley, W., Miller, D., and Zimmerman, P. (2002) *Proc. SPIE: Int. Soc. Opt. Eng.*, **4690**, 84–93.

57. Takebe, Y., Eda, M., Okada, S., Yokokoji, O., Irie, S., Otoguro, A., Fujii, K., and Itani, T. (2004) *Proc. SPIE: Int. Soc. Opt. Eng.*, **5376**, 151–158.

58. McCallum, M., Kameyama, M., and Owa, S. (2006) *Microelectron. Eng.*, **83**, 640–642.

59. Petrillo, K., Patel, K., Chen, R., Li, W., Varanasi, P., Gil, D., Kimmel, K., Slezak, M., Dabbagh, G., Chiba, T., and Shimokawa, T. (2005) *Proc. SPIE: Int. Soc. Opt. Eng.*, **5753**, 52.

60. Hindsberg, W., Wallraff, G.M., Larson, C., Davis, B., Deline, V., Houle, F., Hoffnaggle, J., Sanchez, M., Medeiros, D., Dammel, R., and Conley, W. (2004) *Proc. SPIE: Int. Soc. Opt. Eng.*, **5376**, 21.

61. Conley, W., LeSuer, R.J., Fan, F.F., Bard, A.J., Taylor, C., Tsiartas, P., Willson, G., Romano, A., and Dammel, R. (2005) *Proc. SPIE: Int. Soc. Opt. Eng.*, **5753**, 64–76.

62. Dammel, R.R., Pawlowski, G., Romano, A., and Houlihan, F.M. (2005) *Proc. SPIE: Int. Soc. Opt. Eng.*, **5753**, 95–101.

63. Wallraff, G.M., Larson, C.E., Breyta, G., Sundberg, L., Miller, D., Gil, D., Petrillo, K., and Pierson, W. (2005) *Proc. SPIE: Int. Soc. Opt. Eng.*, **6153**, 61531M-1–61531M-10.

64. Allen, R.D., Brock, P.J., Sundberg, L., Larson, C.E., Wallraff, G.M., Hinsberg, W.D., Meute, J., Shimikawa, T., Chiba, T., and Slezak, M. (2005) *J. Photopolym. Sci. Technol.*, **18**, 615–619.

65. Padmanaban, M., Romano, A., Lin, G., Chiu, S., Timko, A., Houlihan, F., Rahman, D., Dammel, R., Turnquest, K., Rich, G., Schuetter, S., Shedd, T., and Nellis, G. (2006) *J. Photopolym. Sci. Technol.*, **19**, 555–563.

66. Schuetter, S., Shedd, T., Doxtator, K., Nellis, G., Peski, C.V., and Grenville, A. (2006) *J. Microlith. Microfab. Microsyst.*, **5**, 023002.

67. Wei, Y., Petrillo, K., Brandl, S., Goodwin, F., Benson, P., Housley, R., and Okoroanyanwu, U. (2006) *Proc. SPIE: Int. Soc. Opt. Eng.*, **6153**, 615306.

68. Sanders, D.P., Sundberg, L.K., Sooriyakumaran, R., Brock, P.J., DiPietro, R.A., Truong, H.D., Miller, D.C., Lawson, M.C., and Allen, R.D. (2007) *Proc. SPIE: Int. Soc. Opt. Eng.*, **6519**, 651904-1–651904-12.

69. Takebe, Y., Shirota, N., Sasaki, T., Murata, K., and Yokokoji, O. (2008)

Proc. SPIE: Int. Soc. Opt. Eng., **6923**, 69231U-1–69231U-9.

70. Allen, R.D., Breyta, G., Brock, P., DiPietro, R., Sanders, D., Sooriyakumaran, R., and Sundberg, L.K. (2006) *J. Photopolym. Sci. Technol.*, **19**, 615–619.

71. Sanders, D.P., Sundberg, L.K., Brock, P.J., Ito, H., Truong, H.D., and Allen, R.D. (2008) *Proc. SPIE: Int. Soc. Opt. Eng.*, **6923**, 692309-1–692309-12.

72. Wei, Y., Stepanenko, N., Laessig, A., Voelkel, L., and Sebald, M. (2006) *J. Microlith. Microfab. Microsyst.*, **5**, 033002.

73. Wang, D., Caporale, S., Andes, C., Cheon, K.-S., Xu, C.B., Trefonas, P., and Barclay, G. (2007) *J. Photopolym. Sci. Technol.*, **20**, 687–696.

74. *http://www.itrs.net/Links/2007ITRS/2007_Chapters/2007_Lithography.pdf.*

75. Miyamatsu, T., Wang, Y., Shima, M., Kusumoto, S., Chiba, T., Nakagawa, H., Hieda, K., and Shimokawa, T. (2005) *Proc. SPIE: Int. Soc. Opt. Eng.*, **5753**, 10–19.

76. French, R.H., Liberman, V., Tran, H.V., Feldman, J., Adelman, D.J., Wheland, R.C., Qiu, W., McLain, S.J., Nagao, O., Kaku, M., Mocella, M., Yang, M.K., Lemon, M.F., Brubake, L., Shoe, A.L., Fones, B., Fischel, B.E., Krohn, K., Hardy, D., and Chen, C.Y. (2007) High index immersion lithography with second generation immersion fluids to enable numerical apertures of 1.55 for cost effective 32 nm half pitches. *Optical Microlithography XX SPIE*, ML 6520–6559.

77. Lopez-Gejo, J., Kunjappu, J.T., Turro, N.J., and Conley, W. (2006) *Proc. SPIE: Int. Soc. Opt. Eng.*, **6153**, 61530C.

78. Costner, E., Taylor, J.C., Caporale, S., Wojtczak, W., Dewulf, D., Conley, W., and Willson, C.G. (2006) *Proc. SPIE: Int. Soc. Opt. Eng.*, **6153**, 61530B-1–61530B-11.

79. Wang, Y., Miyamatsu, T., Furukawa, T., Yamada, K., Tominaga, T., Makita, Y., Nakagawa, H., Nakamura, A., Shima, M., Kusumoto, S., Shimokawa, T., and Hieda, K. (2006) *Proc. SPIE: Int. Soc. Opt. Eng.*, **6153**, 61530A.

80. Harder, P. and Shedd, T. (2007) *Proc. SPIE: Int. Soc. Opt. Eng.*, **6533**, 653305.

81. Zimmerman, P.A., Byers, J., Rice, B., Ober, C.K., Giannelis, E.P., Rodriguez, R., Wang, D., O'Connor, N., Lei, X., Turro, N.J., Liberman, V., Palmacci, S., Rothschild, M., Lafferty, N., and Smith, B.W. (2008) *Proc. SPIE: Int. Soc. Opt. Eng.*, **6923**, 69230A-1–69230A-10.

82. Conley, W. and Socha, R. (2006) *Proc. SPIE: Int. Soc. Opt. Eng.*, **6153**, 61531L-1–615131L-9.

83. Blakey, I., Conley, W., George, G.A., Hill, D.J.T., Liu, H., Rasoul, F. and Whittaker, A.K. (2006) *Proc. SPIE: Int. Soc. Opt. Eng.*, **6153**, 61530H-1–61530H-10.

84. Blakey, I., Chen, L., Dargaville, B., Liu, H., Whittaker, A., Conley, W., Piscani, E., Rich, G., Williams, A., and Zimmerman, P. (2007) *Proc. SPIE: Int. Soc. Opt. Eng.*, **6519**, 651909.

85. McIntyre, G., Sanders, D., Sooriyakumaran, R., Truong, H., and Allen, R. (2008) *Proc. SPIE: Int. Soc. Opt. Eng.*, **6923**, 692304-1–692304-12.

86. Petrillo, K., Wei, Y., Brainard, R., Denbeaux, G., Goldfarb, D., Koay, C.-S., Mackey, J., Montgomery, W., Pierson, W., Wallow, T., and Wood, O. (2007) *J. Vac. Sci. Technol. B*, **25**, 2490–2495.

87. Kwark, Y.-J., Bravo-Vasquez, J.P., Chandhok, M., Cao, H., Deng, H., Gullikson, E., and Ober, C.K. (2006) *J. Vac. Sci. Technol. B: Microelectron. Process. Phenom.*, **24**(4), 1822–1836.

88. Yueh, W., Cao, H.B., Thirmala, V., and Choi, H. (2005) *Proc. SPIE: Int. Soc. Opt. Eng.*, **5753**, 765.

89. Naulleau, P., Goldberg, K., Anderson, E., Cain, J., Denham, P., Hoef, B., Jackson, K., Morlens, A., Rekawa, S., and Dean, K. (2005) *Proc. SPIE: Int. Soc. Opt. Eng*, **5751**, 56.

90. Brainard, R., Hassanein, E., Li, J., Pathak, P., Thiel, B., Cerrina, F., Moore, R., Rodriguez, M., Yakshinskiy, B., Loginova, E., Madey, T., Matyi, R., Malloy, M., Rudack, A., Naulleau, P., Wüest, A., and Dean, K. (2008)

Proc. SPIE: Int. Soc. Opt. Eng., **6923**, 692325-1–692325-14.

91. Wallow, T., Higgins, C., Brainard, R., Petrillo, K., Montgomery, W., Koay, C.-S., Denbeaux, G., Wood, O., and Wei, Y. (2008) *Proc. SPIE: Int. Soc. Opt. Eng.*, **6921**, 69211F-1–69211F-11.

92. Bravo-Vasquez, J.P., Kwark, Y.-J., and Ober, C.K. (2005) *Proc. SPIE: Int. Soc. Opt. Eng.*, **5753**, 732.

93. Dai, J., Ober, C.K., Kim, S.-O., Nealey, P.F., Golovkina, V., Shin, J., Wang, L., and Cerrina, F. (2003) *Proc. SPIE: Int. Soc. Opt. Eng.*, **5039**, 1164.

94. Wu, H. and Gonsalves, K.E. (2001) *Adv. Funct. Mater.*, **11**, 271–276.

95. Wu, H. and Gonsalves, K.E. (2001) *Adv. Mater*, **13**, 195–197.

96. Wang, M., Yueh, W., and Gonsalves, K.E. (2007) *Macromolecules*, **40**, 8220–8224.

97. Gonsalves, K.E., Thiyagarajan, M., and Dean, K. (2005) *Proc. SPIE: Int. Soc. Opt. Eng.*, **5753**, 771–777.

98. Wang, M., Gonsalves, K.E., Rabinovich, M., Yueh, W., and Roberts, J.M. (2007) *J. Mater. Chem.*, **17**, 1699–1706.

99. Yoshimura, T., Shiraishi, H., Yamamoto, J., and Okazaki, S. (1993) *Appl. Phys. Lett.*, **63**(6), 764–766.

100. Shiraishi, H., Yoshimura, T., Sakamizu, T., Ueno, T., and Okazaki, S. (1994) *J. Vac. Sci. Technol. B: Microelectron. Process. Phenom.*, **12**(6), 3895–3899.

101. Shirota, Y. (2005) *J. Mater. Chem.*, **15**, 75–93.

102. Strohriegel, P., and Grazulevicius, J.V. (2002) *Adv. Mater.*, **14**(20), 1439–1452.

103. Thorpe, M.F. and Tichy, L. (2001) *Properties and Applications of Amorphous Materials*, Kluwer Academic Publishers, Dordrecht, Boston.

104. Drygiannakis, D., Patsis, G.P., Raptis, I., Niakoula, D., Vidali, V., Couladouros, E., Argitis, P., and Gogolides, E. (2007) *Microelectron. Eng.*, **84**, 1062–1065.

105. Patsis, G.P. and Gogolides, E. (2006) *Microelectron. Eng.*, **83**, 1078–1081.

106. Naito, K. (1994) *Chem. Mater.*, **6**, 2343–2350.

107. Alig, I., Braun, D., Langendorf, R., Wirth, H.O., Voigt, M., and Wendorff, J.H. (1998) *J. Mater. Chem.*, **8**(4), 847–851.

108. Yoshiiwa, M., Kageyama, H., Shirota, Y., Wakaya, F., Gamo, K., and Takai, M. (1996) *Appl. Phys. Lett.*, **69**(17), 2605–2606.

109. Kadota, T., Kageyama, H., Wakaya, F., Gamo, K., and Shirota, Y. (2004) *Chem. Lett.*, **33**(6), 706–707.

110. Dai, J., Chang, S.W., Hamad, A., Yang, D., Felix, N., and Ober, C.K. (2006) *Chem. Mater.*, **18**, 3404–3411.

111. Yang, D., Chang, S.W., and Ober, C.K. (2006) *J. Mater. Chem.*, **16**, 1693–1696.

112. Chang, S.W., Ayothi, R., Bratto, D., Yang, D., Felix, N., Cao, H.B., Deng, H., and Ober, C.K. (2006) *J. Mater. Chem.*, **16**, 1470–1474.

113. Silva, A.D., Lee, J.K., Andre, X., Felix, N., and Ober, C.K. (2008) *Chem. Mater.*, **20**, 166.

114. Silva, A.D. and Ober, C.K. (2008) *J. Mater. Chem.*, **18**, 1903–1910.

115. Sooriyakumaran, R., Truong, H., Sundberg, L., Morris, M., Hinsberg, B., Ito, H., Allen, R., Huang, W.-S., Goldfarb, D., Burns, S., and Pfeiffer, D. (2005) *Proc. SPIE: Int. Soc. Opt. Eng.*, **5753**, 329.

116. Lawson, R.A., Lee, C.-T., Whetsell, R., Yueh, W., Roberts, J., Tolbert, L., and Henderson, C.L. (2007) *Proc. SPIE: Int. Soc. Opt. Eng.*, **6519**, 65191N-1–65191N-10.

117. Lawson, R.A., Lee, C.-T., Yueh, W., Tolbert, L., and Henderson, C.L. (2008) *Proc. SPIE: Int. Soc. Opt. Eng.*, **6923**, 69230K-1–69230K-10.

118. Maenhoudt, M., Versluijs, J., Struyf, H., Olmen, J.V., and Hove, M.V. (2005) *Proc. SPIE: Int. Soc. Opt. Eng.*, **5754**, 1508.

119. Liu, H.J., Hsieh, W.H., Yeh, C.H., Wu, J.S., Chan, H.W., Wu, W.B., Chen, F.Y., Huang, T.Y., Shin, C.L., and Lin, J.P. (2007) *Proc. SPIE: Int. Soc. Opt. Eng.*, **6520**, 65202J–65202J-1.

120. Owe-Yang, D.C., Yu, S.S., Chen, H., Chang, C.Y., Ho, B., Lin, J., and Burn, J. (2005) *Proc. SPIE: Int. Soc. Opt. Eng.*, **5753**, 171.

121. Hori, M., Nagai, T., Nakamura, A., Abe, T., Wakamatsu, G., Kakizawa, T., Anno, Y., Sugiura, M., Kusumoto, S., Yamaguchi, Y., and Shimokawa, T. (2008) *Proc. SPIE: Int. Soc. Opt. Eng.*, **6923**, 69230H-1.

122. Chen, K.-J.R., Huang, W.-S., Li, W.-K., and Varanasi, P.R. (2008) *Proc. SPIE: Int. Soc. Opt. Eng.*, **6923**, 69230G-1.

123. Chou, S.Y., Krauss, P.R., and Renstrom, P.J. (1995) *Appl. Phys. Lett.*, **67**, 3114.

124. Chou, S.Y., Krauss, P.R., and Renstrom, P.J. (1996) *J. Vac. Sci. Technol., B: Microelectron. Process. Phenom.*, **14**, 4129–4133.

125. Guo, L.J. (2004) *J. Phys. D: Appl. Phys.*, **37**, R123–R141.

126. Hoff, J.D., Cheng, L.J., Meyhofer, E., Guo, L.J., and Hunt, A.J. (2004) *Nano Lett.*, **4**, 853–857.

127. Cheng, X. and Guo, L.J. (2004) *Microelectron. Eng.*, **71**, 288.

128. Colburn, M., Johnson, S., Stewart, M., Damle, S., Bailey, T.C., Choi, B., Wedlake, M., Michaelson, T., Sreenivasan, S.V., Ekerdt, J., and Willson, C.G. (1999) *Proc. SPIE: Int. Soc. Opt. Eng.*, **3676**, 379–389.

129. Palmieri, F., Stewart, M.D., Wetzel, J., Hao, J., Nishimura, Y., Jen, K., Flannery, C., Li, B., Chao, H.-L., Young, S., Kim, W.C., Ho, P.S., and Willson, C.G. (2008) *Proc. SPIE: Int. Soc. Opt. Eng.*, **6151**, 61510J-1–61510J-8.

130. Bailey, T.C., Johnson, S.C., Sreenivasan, S.V., Ekerdt, J.G., Willson, C.G., and Resnick, D.J. (2002) *J. Photopolym. Sci. Technol.*, **15**, 481–486.

131. Resnick, D.J., Dauksher, W.J., Mancini, D.P., Nordquist, K.J., Bailey, T.C., Johnson, S.C., Stacey, N.A., Ekerdt, J.G., Willson, C., Sreenivasan, S.V., and Schumaker, N. (2003) *Proc. SPIE: Int. Soc. Opt. Eng.*, **5037**, 12–23.

14
Nanopatterning of Polymer Brush Thin Films by Electron-Beam Lithography and Scanning Probe Lithography

Tao Chen, Jianming Zhang, Andres Garcia, Robert Ducker, and Stefan Zauscher

14.1
Introduction

Novel methods for the synthesis and fabrication of nanopatterned polymer thin films and brushes, with controlled molecular architecture, chemical functionality, and size, have spawned opportunities for tailoring surface properties by imparting them with desirable energetic, mechanical, biological, optical, and electrical properties [1, 2]. Polymer brushes are ensembles of polymer chains that are tethered with one end to and densely packed on a surface or interface [1–3]. The ensuing lateral conformational restraint leads to steric repulsive interactions between the packed chains and causes the polymer molecules to extend away from the substrate surface [1, 2]. Such brushes are often confined to patterned, planar substrates that direct and localize the "top-down" or "bottom-up" synthesis of the brush. Patterning polymer brushes with submicrometer lateral resolution on surfaces is important for a wide range of applications [4] that range from biosensors [5], proteomic chips [6] to nanofluidic devices [7].

There are two approaches used to prepare polymer brushes. The grafting-from approach of surface-initiated polymerization (SIP) [8, 9] yields polymer brushes by direct synthesis from the surface. While the grafting-to approach adheres presynthesized polymers to a surface [10]. The grafting-from approach, however, offers greater control over grafting density and yields higher packing densities, as the polymer chains are grown from surface-bound initiators, eliminating many of the steric effects that are associated with attaching polymer chains to the surface. Self-assembled monolayers (SAMs), such as suitably functionalized alkylchlorosilane [11] and alkanethiol [12] SAMs, provide reliable initiators for SIP.

Although various synthesis methods, including anionic [13], cationic [14], plasma induced [15], condensation [16], photochemical [17], electrochemical [18], and ring-opening metathesis polymerization (ROMP) [19], have been use to prepare surface-grafted polymers, the synthesis of precisely patterned, surface-confined polymer brushes with controlled lengths, conformational geometries, and functionality still poses significant challenges and is an active area of research. More recently, controlled radical polymerizations have provided versatile approaches for

Functional Polymer Films, First Edition. Edited by Wolfgang Knoll and Rigoberto C. Advincula.
© 2011 Wiley-VCH Verlag GmbH & Co. KGaA. Published 2011 by Wiley-VCH Verlag GmbH & Co. KGaA.

the design of polymer brushes [20–28]. For example, atom-transfer radical polymerization (ATRP) affords advantages over other synthetic methods [20], because it leads to well-controlled brush growth under mild conditions, with defined molecular weight and relatively small polydispersity and, as a result of the "living" nature of the catalyst, readily allows synthesis of block-copolymers. ATRP is compatible with a variety of functionalized vinyl monomers such as styrene [29], acrylates [30], acrylamides [31], and acrylonitrile [32].

To date, fabrication of patterned polymer brushes has relied largely upon photolithography or soft lithography. These techniques have been extremely successful at the micrometer scale but have a number of inherent limitations that preclude their extension to the nanometer length scale. For nanopatterning of substrates, electron-beam lithography (EBL) has a long-standing tradition in the fabrication of solid-state devices [33]. More recently, scanning probe lithography (SPL) has become quite popular with researchers. This serial technique lends itself to prototyping a proof-of-concept demonstration and is widely accessible to researchers, as it uses the ubiquitous atomic force microscopes (AFMs) [34]. In this brief review we focus on the nanopatterning of polymer-brush thin films using EBL and SPL.

14.2
Electron-Beam Lithography

Electron beam lithography [35] was developed in the 1960s, building on existing scanning electron microscope (SEM) technology, and is today widely used in nanofabrication research. This technique uses a focused electron beam to modify the structure of the substrate surface and can routinely produce structures well below 100 nm, even sub-10 nm [36, 37]. Compared to photolithography and soft lithography, the lateral resolution achieved by EBL is substantially higher because the electron beam can be focused to a diameter of approximately 1 nm. Except for scanning probe microscopy-based lithography methods, the resolution of EBL has been unsurpassed by any other form of lithography [38]. The de Broglie wavelengths of electrons in EBL are substantially smaller than those of light in the UV and visible ranges, which enables the patterning of features that are significantly smaller than those accessible by photolithographic techniques. Although EBL has found widespread use in research, it has not yet become a standard technique for mass production, which results primarily from the slow patterning speed of the serial lithographic process. Despite several drawbacks, such as high cost of the instrumentation, the need of ultrahigh vacuum for operation, and the inherently serial patterning, EBL is one of the most powerful techniques to create well-defined features at the nanoscale.

14.2.1
Electron Beam Resist Lithography

Presently, EBL is almost exclusively used with resists, such as poly(methyl methacrylate) (PMMA), quite similarly to traditional photolithography. In this electron beam

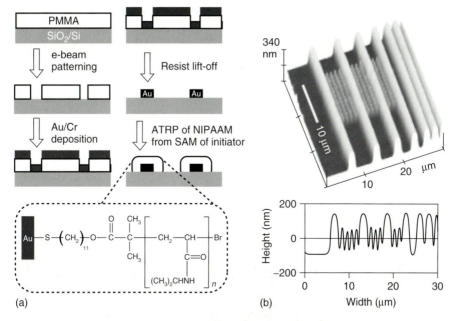

Figure 14.1 (a) Stepwise fabrication process of surface-confined pNIPAAM brush patterns by EBL and SI-ATRP. (b) Tapping-mode AFM height image showing micro-patterned lines and nanopatterned array of dots. (Adapted with permission from Ref. [41], Copyright 2004 Wiley-VCH.)

resist lithographic process, a focused electron beam is used to generate patterns on resist-coated surfaces. The area exposed by the electron beam is then chemically developed to reveal the underlying surface material. The patterned area can be etched or metallic thin films can be deposited and the remaining resist is finally lifted off. Present electron optics limit electron beam widths routinely to only a few nanometers, that is, the resolution limit is not so much determined by the beam size but rather by the size of the resist molecules and by secondary processes, like forward scattering, backscattering, secondary electrons traveling in the resist material, and proximity effects [39]. For example, sub-10-nm structures can be patterned on spin-coated PMMA exposed at electron beam energies of 100 keV [40].

The fabrication of finely patterned polymer brushes with a nanoscale resolution combining EBL and ATRP was first reported by Fukuda and coworkers [41]. Zauscher and coworkers [42] developed this technique to a "top-down/bottom-up" approach, where a silicon surface is first patterned with gold (Au) using lift-off EBL ("top-down") and the resulting pattern is then amplified by surface-initiated polymerization ("bottom-up") to obtain a polymer brush from immobilized thiol initiator using ATRP (Figure 14.1). A similar approach was used for the templated synthesis of polynucleotide brushes [43].

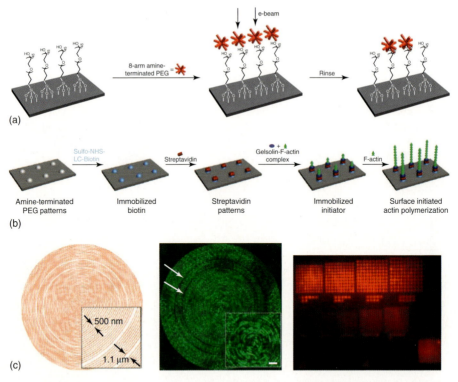

Figure 14.2 Fabrication of polymer nanopatterns using EBL. (a) Amino-functionalized polymer nanopatterns fabricated by EBL. (b) Biotin is attached using Sulfo-NHS-LC-Biotin. Streptavidin is then immobilized to the patterns to act as a linker between the biotin features and the biotinylated FX45-actin complex. Actin is subsequently polymerized from this surface-bound macroinitiator. (c) Fluorescent confocal images of actin grown from surface nanopatterns. (Adapted with permission from Ref. [44], Copyright 2007 RSC.)

This generic approach delivers patterns with well-defined feature dimension, shape, and interfeature spacing over large areas, and facilitates the fabrication of mixed polymer brushes with high lateral resolution by using silane- and thiol-containing initiators, immobilized onto silicon and gold surfaces, respectively.

Maynard and coworkers [44] used EBL for the localized cross-linking of a spin-coated polyethylene glycol (PEG) layer, yielding PEG islands with footprint sizes from 50 to 100 nm (Figure 14.2). Using biotin–streptavidin molecular recognition chemistry, they then immobilized biotinylated gelsolin-based actin initiators on the PEG islands for the subsequent SIP of actin brushes (Figure 14.2b,c).

14.2.2
Electron-Beam Chemical Lithography

Recently, an exciting EBL method, termed electron-beam chemical lithography (EBBCL), has been developed that allows direct control over surface chemistries.

Using EBCL, Grunze and coworkers [45, 46] fabricated nanopatterns by chemically modifying the endgroups of 4-nitro-1,10-biphenyl-4-thiol (NBT) SAMs on gold substrates. During EBCL the terminal nitro groups are converted to amine groups while the underlying aromatic layer is dehydrogenated and cross-linked, enhancing its stability. Lines, bearing amine functionality, and having widths ranging from 20 nm to 1 mm were patterned by EBCL. Subsequently, these amine-terminated organic nanostructures were used as templates for SIP using a surface-bound ATRP initiator to yield densely grafted poly N-isopropylacrylamide (pNIPAAM) brush nanopatterns. Using a similar synthesis approach, sub-50-nm polystyrene brush nanopatterns were fabricated, but in this case, the NBT SAM was first patterned by EBCL followed by diazotization and coupling of methylmalonodinitrile. This resulted in surfaces with well-defined areas of cross-linked initiator sites for polystyrene (PS) SIP (Figure 14.3) [47].

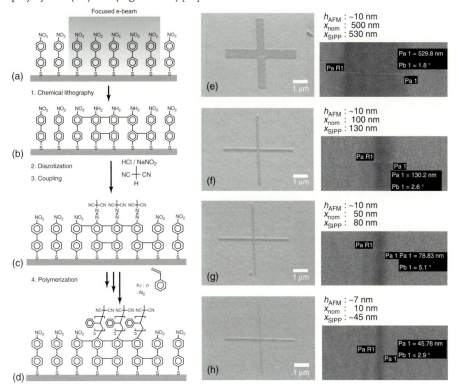

Figure 14.3 (a–d) Reaction scheme of preparing polymer-brush nanostructures. (a) Focused e-beam irradiation of NBT SAMs. (b) Selective local conversion to cAMBT (c) Diazotization and coupling of methylmalonodinitrile gives a SAM that bears an asymmetric azo-initiator (cAMBT) (d) Photochemical SIP of styrene yields region-selective formation of polymer brushes on the e-irradiated areas. (e–h) SEM images of nanostructured PS: individual crosses with initial line width (X_{nom}) of (e) 500, (f) 100, (g) 50, and (h) 10 nm and SEM analysis of the line width of the resulting structures created by SIP along with the height of the PS structures determined by AFM. (Adapted with permission from Ref. [46], Copyright 2007 Wiley-VCH.)

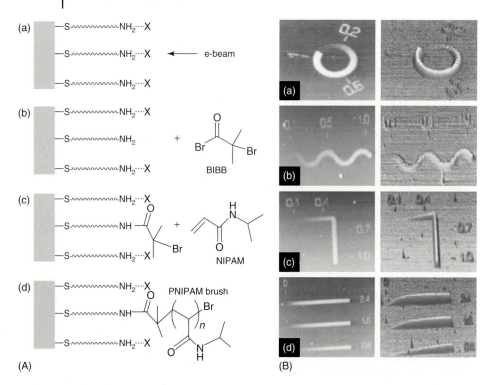

Figure 14.4 (A) EBCL/SI-ATRP with AUDT template: (a) electron-beam-induced activation of amino groups; (b) selective anchoring of the surface initiator (BIBB) to the activated sites; (c) polymerization; and (d) the resulting pNIPAAM brush. (B) AFM images of the gradient pNIPAAM patterns obtained by EBCL/SI-ATRP with AUDT SAMs as templates: (a) snail, (b) snake, (c) angle, and (d) stripes. (Adapted with permission from Ref. [50], Copyright 2008 Wiley-VCH.)

Steenackers *et al.* [48] found that polymer graft density, and thus brush height, can be finely controlled by adjusting the electron dose during pattern formation, where the height of the polymer brush asymptotically reaches a maximal height with increasing electron dose [49]. In an effort to overcome the need for custom-synthesized aromatic SAM resists, Ballav and coworkers [50] used aliphatic dodecanethiol (DDT). After exposure to the electron beam, the patterns were displaced with 11-aminoundecanethiol (AUDT). This resulted in amino-terminated AUDT templates with a size of ~50 nm in a background of methyl-terminated DDT. An ATRP bromoinitiator was then attached to the patterned AUDT templates for the SIP of pNIPAAM brushes (Figure 14.4).

EBCL has thus opened a new window of opportunities for the creation of polymer brush nanostructures with high resolution and fidelity. EBCL offers the capability to build at both the micro and nanoscale while imparting great control over the grafting density and height of the polymer brushes. Furthermore, the chemical modification of SAMs on surfaces provides templates for the fabrication of a wide variety of "smart" polymeric nanostructures.

14.3
Scanning Probe Lithography

SPL, using a scanning tunneling microscope (STM) or an AFM, provides access to patterning of nanometer-scale features [51, 52]. Among all the available nanofabrication techniques, SPL has unique advantages because of its simplicity and its dual capability to image and manipulate nanostructures on surfaces. SPL can be carried out under ambient conditions as well as in different solvent or buffer environments, with little or no sample preparation. SPL provides a broad arsenal of approaches for pattern generation, based on various chemical, physical, and electrical modifications of surfaces, including mechanical scratching [53–55], electrochemical anodization of silicon surfaces [56, 57], decomposition of SAMs [58–61], electric-field-induced chemical reactions [62]. In the past decade several nanopatterning methods, such as nanoshaving and nanografting based on mechanical scratching [53], anodization lithography based on an electrochemical oxidation [63], and "dip-pen" nanolithography (DPN) based in the diffusion of molecules from the tip onto surface [64], have been employed in the fabrication of nanopatterned polymer brushes as described below. As with photolithography and EBL, the majority of patterning efforts using SPL have been directed toward fabricating templates for subsequent modification with initiators and amplification to polymer brushes. A unique feature of many SPL techniques is the intimacy of contact between tip and substrate that provides the ability to pattern initiator directly or to trigger polymerization within a small reaction volume in the meniscus around the scanning probe tip [65].

14.3.1
Nanoshaving and Nanografting

While imaging with the AFM in contact mode, the force between the tip and the sample is a major concern as it may cause significant damage to the sample. Interestingly, this mechanical force can also be harnessed in a controlled way to specifically create nanostructures by nanoshaving [54]. Compared with other lithographic techniques, nanoshaving is a relatively simple and basic patterning method [66], where a soft resist SAM on a gold substrate is removed to create

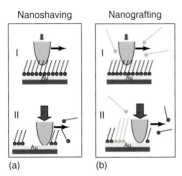

Figure 14.5 Schematic diagrams of basic mechanisms of (a) nanoshaving and (b) nanografting using AFM. (I) Below, threshold force and (II) above, threshold force.

patterns by the cantilever tip as it is dragged over the surface. The large contact pressure during patterning causes the displacement of molecules (Figure 14.5a). In nanoshaving, structures are formed either in air or in a solvent, and the sample is then immersed in a different thiol solution. Nanoshaving in a solvent generally produces structures with higher feature resolution as it minimizes the readsorption of the displaced species. While in nanografting [55], the same lithographic mechanism as in nanoshaving is applied, the cantilever and substrate SAM are now immersed in a thiol solution, from which new thiol self-assembles onto the scratched area (Figure 14.5b).

Various types of alkanethiols, differing in endgroup functionality and length of alkyl chains, have been employed. In pioneering these patterning approaches, Liu and coworkers [53, 55] investigated important parameters such as the scanning force and speed, the concentration of alkanethiol in solution, and the sharpness of

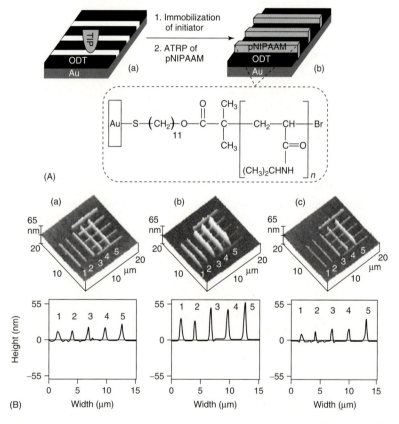

Figure 14.6 (A) Preparation of surface-confined, nanopatterned pNIPAAM brush by combining (a) nanoshaving and (b) SI-ATRP using a surface-tethered thiol initiator. (B) Contact mode AFM height images (20 μm × 20 μm) and corresponding typical height profiles of a pNIPAAM brush line nanopattern imaged at room temperature in (a) air, (b) MQ-grade water, and (c) a mixture of MeOH/water (1 : 1, v/v). (Adapted with permission from Ref. [68], Copyright 2004 ACS.)

the AFM tip. Their work suggests that the sharpness of the tip is a critical parameter for patterning, while the concentration of the thiol solution and the scanning speed are less important [53, 55, 67]. By using sharp tips, nanoshaving and nanografting can routinely generate sub-10-nm patterns.

The resulting patterns can now be used as templates for SIP. For example, stimulus-responsive pNIPAAM brush nanopatterns were prepared by nanoshaving and backfilling with an ATRP initiator SAM (Figure 14.6) [68, 69]. Kaholek *et al.* [68] showed that the brush height increases with increasing shaving time. This relationship can be understood by considering the extent of resist removal that will directly affect the initiator surface density that can be achieved by backfilling, and where high initiator densities lead to high brush heights [70]. The conformational and surface energetic state of the stimulus-responsive, nanopatterned pNIPAAM brush could be triggered reversibly by changing the solvent from water to a water–MeOH mixture.

14.3.2
Dip-Pen Nanolithography

DPN developed by Piner *et al.* [64] is a versatile technique to generate nanoscale patterns on surfaces (Figure 14.7). This method uses an AFM tip as a "pen," with molecules as an "ink," to write a pattern on a solid-state substrate as a "paper." The molecules on the AFM tip are transported to the substrate by a diffusional mechanism. DPN can generate thiol SAM patterns in a dry nitrogen environment [71], since it was demonstrated that a water meniscus is always present, even at 0% relative humidity [72]. During the DPN process, a water meniscus acts as a blocking layer for hydrophobic molecules such as 1-octadecanethiol (ODT). In the case of hydrophilic molecules, such as 16-mercaptohexadecanoic acid (MHA), it allows easy thiol transport to the gold surface [72].

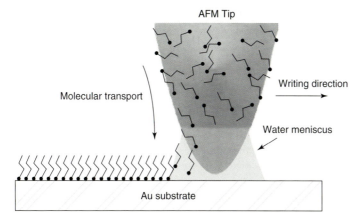

Figure 14.7 Schematic diagram of dip-pen nanolithography.
(Adapted with permission from Ref. [64], Copyright 1999 AAAS.)

Ahn *et al.* [73] used DPN to deposit nanoscale patterns of carboxylic-acid-terminated SAMs on gold substrates, which served, after conjugation with a free-radical initiator, as affinity templates for the SIP of pNIPAAM brushes. Liu *et al.* [69] combined DPN and ROMP to fabricate polymer-brush arrays on the nanometer length scale with great control over feature size, periodicity, and shape (Figure 14.8). In this approach, norbornenylthiol molecules were first patterned onto a gold substrate that was then passivated in a DDT solution and subsequently reacted with Grubbs' first generation catalyst and norbornyl monomers, yielding a polynorbornene brush array with 78-nm feature size.

In another approach, surface-bound gold nanostructures were fabricated by DPN and subsequently used as templates for the photoinitiated polymerization of methylmethacrylate, yielding a lateral resolution of only 20 nm [74]. Kaholek *et al.* [3] succeeded in fabricating polymer brush patterns with linewidths of 200 nm, by directly patterning the initiator thiol using DPN, and subsequent amplification of the pattern using ATRP of pNIPAAM. Maynor *et al.* [65] used a electrochemical dip-pen nanolithography (E-DPN) [75] to polymerize 3,4-ethylenedioxythiophene (EDOT). This approach yielded conductive poly-EDOT nanostructures with sub-100-nm dimensions on semiconducting and insulating surfaces.

14.3.3
Anodization Lithography

AFM anodization lithography, a form of field-induced scanning probe lithography (FISPL) [63, 76], is an electrochemical lithography process in which a voltage bias applied to an AFM tip, establishes a strong, localized electric field between the tip and substrate surface, and causes oxide growth on semiconducting silicon oxide substrates. The mechanism of AFM anodization lithography has been reported first by Gordon *et al.* [77] and Dagata *et al.* [78]. They suggested that the electric potential produces oxyanions, such as O^- and OH^- electrochemically at the air and oxide interface and that oxidation thus also promotes hydrolysis of Si–O bonds. The anodization lithography mechanism consists of two steps, (i) the electrolysis of the water meniscus and concomitant degradation of an organic resist and (ii) the subsequent formation of a silicon oxide layer through an anodization reaction [79]. Therefore, the factors affecting patterning are the applied electric potential between tip and surface, the relative humidity affecting the size of the water meniscus, the electronic state of tip and surface materials, and the patterning speed.

Figures 14.9a,b show the biased AFM tip coming into contact with the substrate with the subsequent generation of oxide features. An essential part of the approach is the presence of an inert resist on the substrate surface that is only removed where anodization pattering occurs. This thus insures that the patterned SiO_2 features can be functionalized with an initiator silane (Figures 14.9c,e). The lateral resolution of this approach is largely limited by process variables in the anodization lithography, such as the AFM tip radius, bias voltage, exposure time, and relative humidity. This field-induced technique is powerful, as it allows the step-and-repeat patterning of the substrate surface. For example, by combining anodization lithography with ROMP,

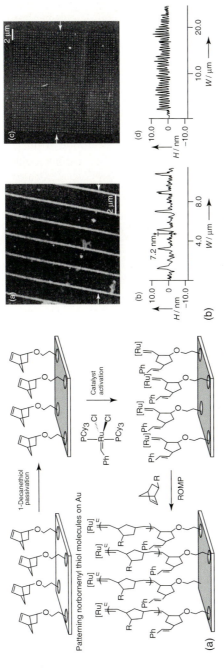

Figure 14.8 (a) Schematic of surface-initiated ROMP by DPN. (b) Topographic AFM image of polymer brush lines and dot arrays. (Adapted with permission from Ref. [69], Copyright 2003 Wiley-VCH.)

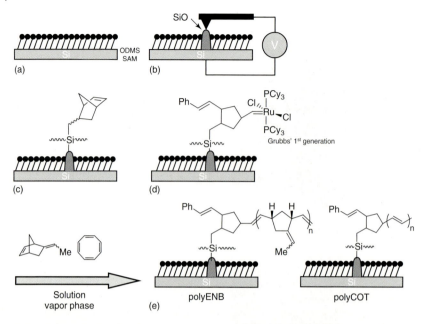

Figure 14.9 Stepwise fabrication of nanopatterned polyENB and polyCOT by using anodization lithography and ROMP. (a) Self-assembled monolayer of silane resist on cleaned silicon substrate, (b) anodization lithography, (c) immobilization of surface tether (norbornene-silane), (d) attachment of ROMP catalyst (only Grubbs first-generation catalyst is shown here), and (e) resulting polyENB and polyCOT brushes.

Lee *et al.* [80] fabricated spatially well-defined, poly-cyclooctatetraene (pCOT) and poly-5-ethylidene-2-norbornene (pENB) brushes with a feature size of about 200 nm (Figure 14.10). In this step and repeat process, anodic oxide patterns were generated next to already existing polycot nanopatterns and then amplified by ROMP of ENB, using Grubbs' first generation catalyst (Figure 14.10a). Although ROMP is a living polymerization, the relatively stable, chain-terminated ruthenium pCOT likely did not survive the extensive cleaning procedure that followed the first step; this allowed the height of the newly created polyENB pattern to be controlled to yield a similar height as that of the pCOT pattern (Figure 14.10b). Had the polycot survived the cleaning process, block-copolymer, and thicker brushes would have been produced on the initial pattern by the sequential addition of the monomers. This suggests the intriguing possibility that nanopatterned block-copolymer structures could be synthesized by more careful treatment of the ruthenium-terminated polymer brushes between subsequent ROMP.

14.3.4
Scanning Near-Field Optical Lithography

The main disadvantage of conventional photolithography is that the resolution of the feature size is ultimately diffraction limited to approximately $\lambda/2$ [81]. This

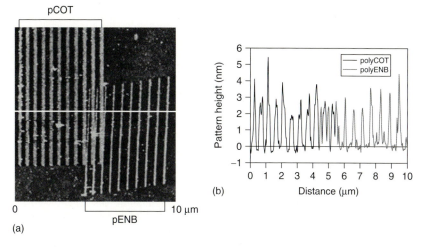

Figure 14.10 Dual polymer brushes fabricated on the same substrate by step and repeat processing according to the scheme shown in Figure 14.9. (a) Tapping-mode height image and (b) cross-sectional profile corresponding to the location of the white line in (a). (Adapted with permission from Ref. [80], Copyright 2006 Wiley-VCH.)

diffraction limit can be circumvented in scanning near-field optical lithography (SNOL) [82], using a scanning near-field optical microscope, where the sample is exposed to light through a small (∼50 nm) aperture in the near-field regime (10−5 nm) [83]. This exposes the sample to nondiffracting, evanescent light emerging from the aperture and affords photoexposure on the nanometer length scale. Ducker and Leggett [82] fabricated a trench with a resolution of l/30 (9 nm) in gold and also photochemically converted the surface chemistry of chloromethylphenylsiloxane to a carboxylic acid. A related technique is apertureless scanning near-field optical microscopy, an extension of the SNOM technique, where large, localized light intensities are achieved by local enhancement of far-field light through a nanometer-sized, pointed probe [84].

Wurtz et al. [85] reported the photopolymerization of a photoresist induced by near-field enhancement from a sharp probe. This method yielded 70-nm polymeric nanostructures and demonstrated that direct nanopolymerization is possible using near-field techniques. While these photochemical SNOL techniques have not yet been used for the nanoscale patterning and syntheses of polymer brushes on surfaces, it is clear that they provide powerful means to do so.

14.4
Conclusions

The ability to fabricate nanopatterned polymer-brush thin films relies to a large measure on the surface quality of the patterned substrate. While to date most patterning strategies have been adapted from established lithographic approaches, the potential of using lithographic methods to chemically modify and functionalize substrate surfaces has largely not been tapped. An exciting new development in this area is EBCL, which has evolved as one promising new strategy to exert direct control over surface chemistries and to fabricate topographic gradients in polymeric-brush nanostructures. Furthermore, scanning probe microscopes have undergone a remarkable development over the last two decades to become one of the leading technologies in manipulating matter and creating useful patterns at the nanoscale. More recently, significant improvements in the closed-loop substrate positioning and in cantilever microfabrication techniques have resulted in important gains in patterning speed, throughput, and resolution of scanning probe microscopes (SPMs). Also, attempts to shift from a serial to a parallel approach have been made using an array of probes, opening up the possibility for a much more automated and parallelized nanofabrication paradigm.

EBCL and SPL can effectively be combined with SIP to fabricate nanopatterned thin films with controlled shape, feature dimension, spacing, and chemical functionality. This fabrication approach allows a high level of lateral control in patterning surfaces with complex polymer-brush thin films. Yet many demonstrations of nanopatterned brush thin films are still at the level of proof-of-concept. To advance the field, the potential of a broader range of polymers needs to be tapped, and, in order to tune brush surface chemistry, improved strategies for their controlled, terminal chemical functionalization need to be developed. Access to a broader spectrum of stimulus-responsive polymers will be important for the hierarchical fabrication of functional brush nanostructures.

Acknowledgments

SZ gratefully acknowledges the National Science Foundation for support through a NSF DMR-0239769 CAREER AWARD, NSF DMR-0502953, and through NSF CMS-0609265 (NIRT).

References

1. Zhao, B. and Brittain, W.J. (2000) Polymer brushes: surface-immobilized macromolecules. *Prog. Polym. Sci.*, **25**, 677–710.

2. Milner, S.T., Witten, T.A., and Cates, M.E. (1988) Theory of the grafted polymer brush. *Macromolecules*, **21**, 2610–2619.

3. Kaholek, M., Lee, W.K., LaMattina, B., Caster, K.C., and Zauscher, S. (2005) *Polymerization, Nanopatterning and Characterization of Surface-Confined, Stimulus-Responsive Polymer Brushes*, in *Polymer Brushes*, (eds R.C. Advincula, W.J. Brittain, K.C. Caster, and J. Rühe),

Wiley-VCH Verlag GmbH & Co. KG, Weinheim pp. 381–402.

4. Love, J.C., Estroff, L.A., Kriebel, J.K., Nuzzo, R.G., and Whitesides, G.M. (2005) Self-assembled monolayers of thiolates on metals as a form of nanotechnology. *Chem. Rev.*, **105**, 1103–1169.

5. Blawas, A.S. and Reichert, W.M. (1998) Protein patterning. *Biomaterials*, **19**, 595–609.

6. Mrksich, M. and Whitesides, G.M. (1995) Patterning self-assembled monolayers using microcontact printing: a new technology for

biosensors. *Trends Biotechnol.*, **13**, 228–235.

7. Beebe, D.J., Moore, J.S., Yu, Q., Liu, R.H., Kraft, M.L., Jo, B.H., and Devadoss, C. (2000) Microfluidic tectonics: a comprehensive construction platform for microfluidic systems. *Proc. Natl. Acad. Sci. U.S.A.*, **97**, 13488–13493.

8. Edmondson, S., Osborne, V.L., and Huck, W.T.S. (2004) Polymer brushes via surface-initiated polymerizations. *Chem. Soc. Rev.*, **33**, 14–22.

9. Prucker, O. and Ruhe, J. (1998) Mechanism of radical chain polymerizations initiated by azo compounds covalently bound to the surface of spherical particles. *Macromolecules*, **31**, 602–613.

10. Hyun, J., Lee, W.K., Nath, N., Chilkoti, A., and Zauscher, S. (2004) Capture and release of proteins on the nanoscale by stimuli-responsive elastin-like polypeptide "switches". *J. Am. Chem. Soc.*, **126**, 7330–7335.

11. Becer, C.R., Haensch, C., Hoeppener, S., and Schubert, U.S. (2007) Patterned polymer brushes grafted from bromine-functionalized, chemically active surface templates. *Small*, **3**, 220–225.

12. Jordan, R. and Ulman, A. (1998) Surface initiated living cationic polymerization of 2-oxazolines. *J. Am. Chem. Soc.*, **120**, 243–247.

13. Jordan, R., Ulman, A., Kang, J.F., Rafailovich, M.H., and Sokolov, J. (1999) Surface-initiated anionic polymerization of styrene by means of self-assembled monolayers. *J. Am. Chem. Soc.*, **121**, 1016–1022.

14. Ingall, M.D.K., Honeyman, C.H., Mercure, J.V., Bianconi, P.A., and Kunz, R.R. (1999) Surface functionalization and imaging using monolayers and surface-grafted polymer layers. *J. Am. Chem. Soc.*, **121**, 3607–3613.

15. Chen, W., Fadeev, A.Y., Hsieh, M.C., Oner, D., Youngblood, J., and McCarthy, T.J. (1999) Ultrahydrophobic and ultralyophobic surfaces: some comments and examples. *Langmuir*, **15**, 3395–3399.

16. Husemann, M., Mecerreyes, D., Hawker, C.J., Hedrick, J.L., Shah, R., and Abbott, N.L. (1999) Surface-initiated

polymerization for amplification of self-assembled monolayers patterned by microcontact printing. *Angew. Chem. Int. Ed.*, **38**, 647–649.

17. Prucker, O., Naumann, C.A., Ruhe, J., Knoll, W., and Frank, C.W. (1999) Photochemical attachment of polymer films to solid surfaces via monolayers of benzophenone derivatives. *J. Am. Chem. Soc.*, **121**, 8766–8770.

18. Gurtner, C., Wun, A.W., and Sailor, M. (1999) Surface modification of porous silicon by electrochemical reduction of organo halides. *Angew. Chem. Int. Ed.*, **38**, 1966–1968.

19. Grubbs, R.H. and Chang, S. (1998) Recent advances in olefin metathesis and its application in organic synthesis. *Tetrahedron*, **54**, 4413–4450.

20. Matyjaszewski, K. and Xia, J.H. (2001) Atom transfer radical polymerization. *Chem. Rev.*, **101**, 2921–2990.

21. Patten, T.E. and Matyjaszewski, K. (1999) Copper(I)-catalyzed atom transfer radical polymerization. *Acc. Chem. Res.*, **32**, 895–903.

22. Cheng, N., Brown, A.A., Azzaroni, O., and Huck, W.T.S. (2008) Thickness-dependent properties of polyzwitterionic brushes. *Macromolecules*, **41**, 6317–6321.

23. Yoshikawa, C., Goto, A., Tsujii, Y., Fukuda, T., Yamamoto, K., and Kishida, A. (2005) Fabrication of high-density polymer brush on polymer substrate by surface-initiated living radical polymerization. *Macromolecules*, **38**, 4604–4610.

24. Bao, Z.Y., Bruening, M.L., and Baker, G.L. (2006) Rapid growth of polymer brushes from immobilized initiators. *J. Am. Chem. Soc.*, **128**, 9056–9060.

25. Wang, Y., Hu, S.W., and Brittain, W.J. (2006) Polymer brush grafted from an allylsilane-functionalized surface. *Macromolecules*, **39**, 5675–5678.

26. Matyjaszewski, K., Dong, H.C., Jakubowski, W., Pietrasik, J., and Kusumo, A. (2007) Grafting from surfaces for "Everyone": ARGET ATRP in the presence of air. *Langmuir*, **23**, 4528–4531.

27. Matyjaszewski, K., Jakubowski, W., Min, K., Tang, W., Huang, J.Y., Braunecker,

W.A., and Tsarevsky, N.V. (2006) Diminishing catalyst concentration in atom transfer radical polymerization with reducing agents. *Proc. Natl. Acad. Sci. U.S.A.*, **103**, 15309–15314.

28. Matyjaszewski, K., Tsarevsky, N.V., Braunecker, W.A., Dong, H., Huang, J., Jakubowski, W., Kwak, Y., Nicolay, R., Tang, W., and Yoon, J.A. (2007) Role of Cu-0 in controlled/"living" radical polymerization. *Macromolecules*, **40**, 7795–7806.

29. Zhou, F., Jiang, L., Liu, W.M., and Xue, Q.J. (2004) Fabrication of chemically tethered binary polymer-brush pattern through two-step surface-initiated atomic-transfer radical polymerization. *Macromol. Rapid Commun.*, **25**, 1979–1983.

30. Carrot, G., Diamanti, S., Manuszak, M., Charleux, B., and Vairon, I.P. (2001) Atom transfer radical polymerization of n-butyl acrylate from silica nanoparticles. *J. Polym. Sci. Part A: Polym. Chem.*, **39**, 4294–4301.

31. Xiao, D.Q., Zhang, H., and Wirth, M. (2002) Chemical modification of the surface of poly(dimethylsiloxane) by atom-transfer radical polymerization of acrylamide. *Langmuir*, **18**, 9971–9976.

32. Pietrasik, J., Dong, H.C., and Matyjaszewski, K. (2006) Synthesis of high molecular weight poly(styrene-co-acrylonitrile) copolymers with controlled architecture. *Macromolecules*, **39**, 6384–6390.

33. Menard, E., Meitl, M.A., Sun, Y.G., Park, J.U., Shir, D.J.L., Nam, Y.S., Jeon, S., and Rogers, J.A. (2007) Micro- and nanopatterning techniques for organic electronic and optoelectronic systems. *Chem. Rev.*, **107**, 1117–1160.

34. Ducker, R., Garcia, A., Zhang, J., Chen, T., and Zauscher, S. (2008) Polymeric and biomacromolecular brush nanostructures: progress in synthesis, patterning and characterization. *Soft Matter*, **4**, 1774–1786.

35. Campbell, S.A. (1996) *The Science and Engineering of Microelectronic Fabrication*, Oxford University Press, Oxford, pp. 203–226.

36. Lercel, M.J., Craighead, H.G., Parikh, A.N., Seshadri, K., and Allara, D.L.

(1996) Sub-10 nm lithography with self-assembled monolayers. *Appl. Phys. Lett.*, **68**, 1504–1506.

37. Seshadri, K., Froyd, K., Parikh, A.N., Allara, D.L., Lercel, M.J., and Craighead, H.G. (1996) Electron-beam-induced damage in self-assembled monolayers. *J. Phys. Chem.*, **100**, 15900–15909.

38. Djenizian, T. and Schmuki, P. (2006) Electron beam lithographic techniques and electrochemical reactions for the micro- and nanostructuring of surfaces under extreme conditions. *J. Electroceram.*, **16**, 9–14.

39. Broers, A.N., Hoole, A.C.F., and Ryan, J.M. (1996) Electron beam lithography: resolution limits. *Microelectron. Eng.*, **32**, 131–142.

40. Chen, W. and Ahmed, H. (1993) Fabrication of 5-7 nm wide etched lines in silicon using 100 keV electron-beam lithography and polymethylmethacrylate resist. *Appl. Phys. Lett.*, **62**, 1499–1501.

41. Tsujii, Y., Ejaz, M., Yamamoto, S., Fukuda, T., Shigeto, K., Mibu, K., and Shinjo, T. (2002) Fabrication of patterned high-density polymer graft surfaces. II. Amplification of EB-patterned initiator monolayer by surface-initiated atom transfer radical polymerization. *Polymer*, **43**, 3837–3841.

42. Ahn, S.J., Kaholek, M., Lee, W.K., LaMattina, B., LaBean, T.H., and Zauscher, S. (2004) Surface-initiated polymerization on nanopatterns fabricated by electron-beam lithography. *Adv. Mater.*, **16**, 2141–2145.

43. Chow, D.C., Lee, W.K., Zauscher, S., and Chilkoti, A. (2005) Enzymatic fabrication of DNA nanostructures: extension of a self-assembled oligonucleotide monolayer on gold arrays. *J. Am. Chem. Soc.*, **127**, 14122–14123.

44. Brough, B., Christman, K.L., Wong, T.S., Kolodziej, C.M., Forbes, J.G., Wang, K., Maynard, H.D., and Ho, C.M. (2007) Surface initiated actin polymerization from top-down manufactured nanopatterns. *Soft Matter*, **3**, 541–546.

45. Eck, W., Stadler, V., Geyer, W., Zharnikov, M., Golzhauser, A., and Grunze, M. (2000) Generation of surface amino groups on aromatic

self-assembled monolayers by low energy electron beams: a first step towards chemical lithography. *Adv. Mater.*, **12**, 805–808.

46. Golzhauser, A., Eck, W., Geyer, W., Stadler, V., Weimann, T., Hinze, P., and Grunze, M. (2001) Chemical nanolithography with electron beams. *Adv. Mater.*, **13**, 803–806.

47. Schmelmer, U., Paul, A., Kuller, A., Steenackers, M., Ulman, A., Grunze, M., Golzhauser, A., and Jordan, R. (2007) Nanostructured polymer brushes. *Small*, **3**, 459–465.

48. Steenackers, M., Kueller, A., Ballav, N., Zharnikov, M., Grunze, M., and Jordan, R. (2007) Morphology control of structured polymer brushes. *Small*, **3**, 1764–1773.

49. He, Q., Kueller, A., Schilp, S., Leisten, F., Kolb, H.A., Grunze, M., and Li, J.B. (2007) Fabrication of controlled thermosensitive polymer nanopatterns with one-pot polymerization through chemical lithography. *Small*, **3**, 1860–1865.

50. Schilp, S., Ballav, N., and Zharnikov, M. (2008) Fabrication of a full-coverage polymer nanobrush on an electron-beam-activated template. *Angew. Chem. Int. Ed.*, **47**, 6786–6789.

51. Binnig, G. and Rohrer, H. (1982) Scanning tunneling microscopy. *Helv. Phys. Acta*, **55**, 726–735.

52. Binnig, G., Quate, C.F., and Gerber, C. (1986) Atomic force microscope. *Phys. Rev. Lett.*, **56**, 930–933.

53. Liu, G.Y., Xu, S., and Qian, Y.L. (2000) Nanofabrication of self-assembled monolayers using scanning probe lithography. *Acc. Chem. Res.*, **33**, 457–466.

54. Xu, S. and Liu, G.Y. (1997) Nanometer-scale fabrication by simultaneous nanoshaving and molecular self-assembly. *Langmuir*, **13**, 127–129.

55. Xu, S., Miller, S., Laibinis, P.E., and Liu, G.Y. (1999) Fabrication of nanometer scale patterns within self-assembled monolayers by nanografting. *Langmuir*, **15**, 7244–7251.

56. Sugimura, H. and Nakagiri, N. (1996) Scanning probe anodization: nanolithography using thin films of anodically

oxidizable materials as resists. *J. Vac. Sci. Technol., A*, **14**, 1223–1227.

57. Legrand, B. and Stievenard, D. (1999) Nanooxidation of silicon with an atomic force microscope: a pulsed voltage technique. *Appl. Phys. Lett.*, **74**, 4049–4051.

58. Sugimura, H., Takai, O., and Nakagiri, N. (1999) Multilayer resist films applicable to nanopatterning of insulating substrates based on current-injecting scanning probe lithography. *J. Vac. Sci. Technol., B*, **17**, 1605–1608.

59. Sugimura, H., Okiguchi, K., and Nakagiri, N. (1996) Scanning probe lithography using a trimethylsilyl organosilane monolayer resist. *Jpn. J. Appl. Phys., Part 1*, **35**, 3749–3753.

60. Zamborini, F.P. and Crooks, R.M. (1998) Nanometer-scale patterning of metals by electrodeposition from an STM tip in air. *J. Am. Chem. Soc.*, **120**, 9700–9701.

61. Gorman, C.B., Carroll, R.L., He, Y.F., Tian, F., and Fuierer, R. (2000) Chemically well-defined lithography using self-assembled monolayers and scanning tunneling microscopy in nonpolar organothiol solutions. *Langmuir*, **16**, 6312–6316.

62. Maoz, R., Frydman, E., Cohen, S.R., and Sagiv, J. (2000) "Constructive nanolithography": inert monolayers as patternable templates for in-situ nanofabrication of metal-semiconductor-organic surface structures: a generic approach. *Adv. Mater.*, **12**, 725–713.

63. Snow, E.S. and Campbell, P.M. (1995) AFM fabrication of sub-10-nanometer metal-oxide devices with in-situ control of electrical-properties. *Science*, **270**, 1639–1641.

64. Piner, R.D., Zhu, J., Xu, F., Hong, S.H., and Mirkin, C.A. (1999) "Dip-pen" nanolithography. *Science*, **283**, 661–663.

65. Maynor, B.W., Filocamo, S.F., Grinstaff, M.W., and Liu, J. (2002) Direct-writing of polymer nanostructures: Poly(thiophene) nanowires on semiconducting and insulating surfaces. *J. Am. Chem. Soc.*, **124**, 522–523.

66. Kramer, S., Fuierer, R.R., and Gorman, C.B. (2003) Scanning probe lithography using self-assembled monolayers. *Chem. Rev.*, **103**, 4367–4418.

67. Schwartz, P.V. (2001) Meniscus force nanografting: Nanoscopic patterning of DNA. *Langmuir*, **17**, 5971–5977.

68. Kaholek, M., Lee, W.K., LaMattina, B., Caster, K.C., and Zauscher, S. (2004) Fabrication of stimulus-responsive nanopatterned polymer brushes by scanning-probe lithography. *Nano Lett.*, **4**, 373–376.

69. Liu, X.G., Guo, S.W., and Mirkin, C.A. (2003) Surface and site-specific ring-opening metathesis polymerization initiated by dip-pen nanolithography. *Angew. Chem. Int. Ed.*, **42**, 4785–4789.

70. Jones, D.M., Brown, A.A., and Huck, W.T.S. (2002) Surface-initiated polymerizations in aqueous media: effect of initiator density. *Langmuir*, **18**, 1265–1269.

71. Sheehan, P.E. and Whitman, L.J. (2002) Thiol diffusion and the role of humidity in "dip pen nanolithography". *Phys. Rev. Lett.*, **88**, 156104.

72. Rozhok, S., Piner, R., and Mirkin, C.A. (2003) Dip-pen nanolithography: what controls ink transport? *J. Phys. Chem. B*, **107**, 751–757.

73. Ahn, S.-J., Lee, W.-K., and Zauscher, S. (2003) in Fabrication of Stimulus-Responsive Polymeric Nanostructures by Proximal Probes, in *Bioinspired Nanoscale Hybrid Systems*, vol. 735 (eds U.S.G. Schmid, S.J. Stranick, S.M. Arrivo, and S. Hong), Materials Research Society, Warrendale, pp. C11.54.11–C11.54.16.

74. Zapotoczny, S., Benetti, E.M., and Vancso, G.J. (2007) Preparation and characterization of macromolecular "hedge" brushes grafted from Au nanowires. *J. Mater. Chem.*, **17**, 3293–3296.

75. Li, Y., Maynor, B.W., and Liu, J. (2001) Electrochemical AFM "dip-pen" nanolithography. *J. Am. Chem. Soc.*, **123**, 2105–2106.

76. Dagata, J.A., Schneir, J., Harary, H.H., Evans, C.J., Postek, M.T., and Bennett, J. (1990) Modification of hydrogen-passivated silicon by a scanning tunneling microscope operating in air. *Appl. Phys. Lett.*, **56**, 2001–2003.

77. Gordon, A.E., Fayfield, R.T., Litfin, D.D., and Higman, T.K. (1995) Mechanisms of surface anodization produced by scanning probe microscopes. *J. Vac. Sci. Technol., B*, **13**, 2805–2808.

78. Dagata, J.A., Inoue, T., Itoh, J., and Yokoyama, H. (1998) Understanding scanned probe oxidation of silicon. *Appl. Phys. Lett.*, **73**, 271–273.

79. Ahn, S.J., Jang, Y.K., Lee, H., and Lee, H. (2002) Mechanism of atomic force microscopy anodization lithography on a mixed Langmuir-Blodgett resist of palmitic acid and hexadecylamine on silicon. *Appl. Phys. Lett.*, **80**, 2592–2594.

80. Lee, W.K., Caster, K.C., Kim, J., and Zauscher, S. (2006) Nanopatterned polymer brushes by combining AFM anodization lithography with ring-opening metathesis polymerization in the liquid and vapor phase. *Small*, **2**, 848–853.

81. Dunn, R.C. (1999) Near-field scanning optical microscopy. *Chem. Rev.*, **99**, 2891–2928.

82. Ducker, R.E. and Leggett, G.J. (2006) A mild etch for the fabrication of three-dimensional nanostructures in gold. *J. Am. Chem. Soc.*, **128**, 392–393.

83. Betzig, E. and Trautman, J.K. (1992) Near-field optics: microscopy, spectroscopy, and surface modification beyond the diffraction limit. *Science*, **257**, 189–195.

84. Novotny, L. and Stranick, S.J. (2006) Near-field optical microscopy and spectroscopy with pointed probes. *Annu. Rev. Phys. Chem.*, **57**, 303–331.

85. Wurtz, G., Bachelot, R., H'Dhili, F., Royer, P., Triger, C., Ecoffet, C., and Lougnot, D.J. (2000) Photopolymerization induced by optical field enhancement in the vicinity of a conducting tip under laser illumination. *Jpn. J. Appl. Phys., Part 2*, **39**, L98–L100.

15
Direct Patterning for Active Polymers

Eunkyoung Kim, Jungmok You, Yuna Kim, and Jeonghun Kim

15.1
Introduction

Functional polymers are materials that can potentially be used to form active patterns in organic electronics. Polymers of this type include those with conductive, fluorescent, or photochromic properties (Figure 15.1) and their patterns are of great technological importance for charge or molecular transport and pattern recognition in electrical circuits, sensors, displays, bioengineering, and so on.

Several patterning methods have been attempted in organic electronics, including photolithographic methods such as selective photobleaching [1], photoinduced doping/dedoping [2], photochemical reactions [3], and chemical reactions [4]. Non-photolithographic methods, including microcontact printing [5], screening printing [6], selective electrochemical deposition, inkjet printing [7], and laser-induced thermal imaging [8] have also been investigated. These various methods have been continuously modified to meet many different requirements for specific applications.

The chemical and mechanical stability of functional polymers during the patterning process are particularly critical, as most of the functional polymers contain reactive units such as unsaturated bonds or heterocyclic structures. Therefore, establishing new methods that can both preserve polymer functionality and ensure a simple process is challenging. Recently, direct photopatterning approaches have attracted much attention for the formation of active patterns due to their simple procedures. To avoid decomposition of the functional group during the direct photopatterning step, the intensity of incident light has been reduced and thus functional polymers have been designed to be more photosensitive by the attachment of photoreactive groups such as acrylic moieties [9], cinnamoyl groups [10], bis-arylazide groups [11], and pendant tetrahydropyranyl (THP) groups [4].

Direct photopatterning for active pattern formation occurs through: (i) activity increase, (ii) activity decrease (detrimental), or (iii) changes in the activity to other activity on the light-exposed area. These direct photopatternings can be achieved by irreversible chemical reactions such as cross-linking or cleavage

Functional Polymer Films, First Edition. Edited by Wolfgang Knoll and Rigoberto C. Advincula.
© 2011 Wiley-VCH Verlag GmbH & Co. KGaA. Published 2011 by Wiley-VCH Verlag GmbH & Co. KGaA.

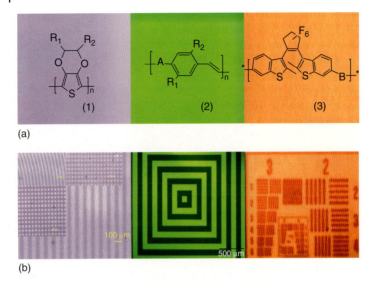

(a)

(b)

Figure 15.1 (a) Example of conductive (1), fluorescent (2), or photochromic polymers (3), where A and B are the aromatic bridging group, and (b) direct pattern from them.

(Equations 15.1–15.4 in Scheme 15.1) or by reversible reactions such as photoisomerization (5), as described in Scheme 15.1.

In the reaction described in (Equation 15.1), the polymerization at the acrylic or methacrylic unit results in a high molecular weight (M_w) insoluble network. The opposite case occurs in the oxidative cleavage reaction (Equation 15.2). In the dimerization approach (Equation 15.3), and the reaction of diacetylene-modified functional monomers afford high M_w network. The M_w change in the reactions described above results in a difference in solubility between the light-exposed and unexposed areas and allows lithographic patterning of the above system through wet- or dry-etching processes. Many functional polymers have been patterned through this type of photoexposure, followed by an etching or development step. Since strong acid or chemically reactive etchants are used in the etching step, etchless direct patterning is an alternative process that pertains to the activity of the polymer.

For the direct etchless photopatterning, the photoinduced property change other than M_w change is necessary. Thus, changes in alignment or structural distortion, oxidative quenching, or polarity change are all important properties for etchless direct patterning. For example, the cross-linking reaction (Equation 15.1) is often accompanied with a loss of conjugation via structural reorganization among the active groups, or distortion of the main active chain, to result in a large property change by photoirradiation, allowing for direct etchless photopatterning. Oxidative cleavage breaks the C=C double bonds, which are the conjugative linkers of a conductive or fluorescent polymer main chain. The activity of the light-exposed area of the polymer film is thus reduced by the bond cleavage, which allows etchless

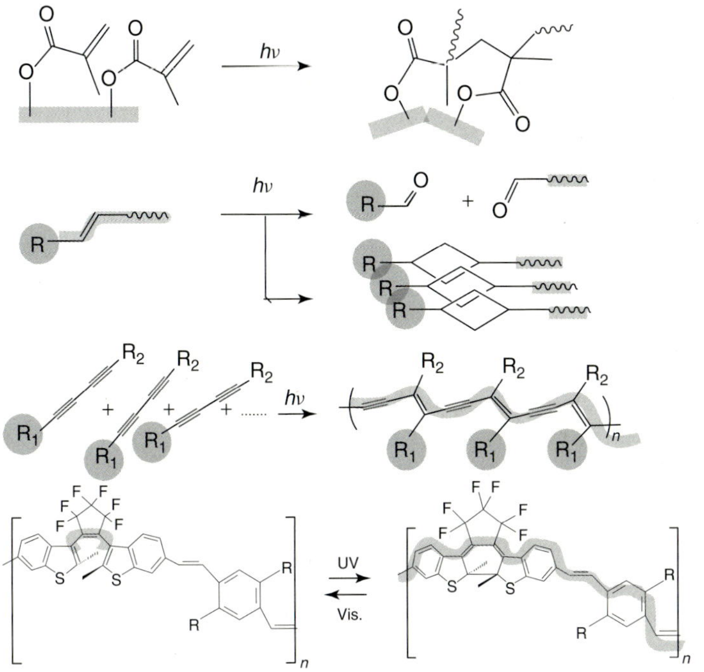

(1)

(2)

(3)

(4)

(5)

Scheme 15.1 Examples of chemical reactions for irreversible (1–4) and reversible patterning (5): R stands for functional group or functional polymer.

photopatterning. In the dimerization approach (Equation 15.3), the conjugative C=C linker is cyclized to reduce fluorescence and the conductivity of the polymer. Similarly, diacetylene-modified functional monomers afford not only conductive patterning through the main chain, but also aid in alignment of the functional side group (Equation 15.4), as schematically summarized in Scheme 15.1.

These photoreactions for generating patterns often proceed more effectively in the presence of photoinitiators, which play an important role as core materials in photoreactive polymers and composites for photopatterning. Photoinitiators can generate radicals and ions by exposure to the appropriate wavelength specific to that the initiator. Upon light absorption and reaction (Equation 15.6), the generated radicals and ions are able to polymerize unsaturated groups like vinyl and acrylate groups or epoxide groups.

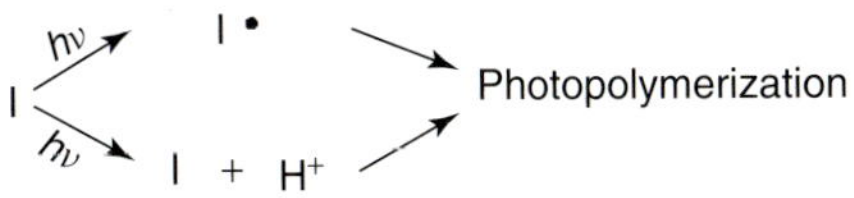

(15.6)

Furthermore, the direct patterning based on irreversible reactions also extends to the reaction of functional group with photoacid generators (PAGs) or photobase generators (PBGs), which are used widely in photolithography for passive patterning. In addition to photoinitiators, a photosensitizer or mixture of sensitizers can also be added to the photopatterning film to increase photoreactivity and to tune the wavelength of the incident light. The composition for photopatterning can be varied to tailor the properties of the patterned film, as is widely done in conventional photolithography or printing technologies in passive pattern generation.

In the case of reversible patterning, a photoreversible reaction can be utilized. For example, diarylethenes undergo reversible isomerization (Equation 15.5), that is, accompanied by fluorescence or conductivity changes [12]. The M_w change of the polymer from the photoreversible reaction would be negligible in general, before and after light exposure. Thus, the photoinduced property change from the reversible reaction is critical for the patterning. The patterning effect and reversibility from these reactions are directly related to the quantum yield of the photoconversion.

Although direct photopatterning, without going through an etching step, is beneficial for the patterning of functional polymers by virtue of its simplicity, there are not many examples known of this approach. Thus, in this chapter, we introduce some example of photopatterning of functional polymers that are followed by an etching step, but focus more on direct photopatterning methods that bypass the etching step.

15.2
Direct Photopatterning by Photocross-Linking

15.2.1
Patterning by Cross-Linking of Unsaturated Bond

Photocross-linking from unsaturated bonds such as ethylenenic, acrylic, methacrylic, acetylenic, and azide units are widely used to afford direct patterns. The $2\pi + 2\pi$ cross-linking in a vinylene or ethylene group results in a high-M_w polymer. Deng and Wong [13] used the cross-linking of ethylene groups in poly(phenylene-vinylene) (PPV) to create the red, green, and blue (RGB) light-emitting diodes (LED) patterns (Figure 15.2). As the fluorescence of the cross-linked PPVs is reduced, the emission should be different between the light-exposed and unexposed areas, to directly create emissive patterns. In addition, the difference in solubility between the light-exposed and unexposed areas allows lithographic patterning of the above system through a rinsing process. An LED device could be fabricated by cross-linking without use of a photoinitiator, as described in Section 15.2.3.

Diacetylene can be polymerized via a 1,4 addition reaction to form alternating ene−yne polymer chains upon UV irradiation, as shown in Equation 15.4 [14]. The cross-linking in diacetylene groups results in a high-M_w polymer and thus patterned fluorescence images are readily constructed by employing a photolithographic technique. Figure 15.3 shows a procedure for the generation of patterned fluorescence images with immobilized diacetylene supramolecules (**3**) on a solid substrate. First, the diacetylene-immobilized glass substrate is irradiated through a

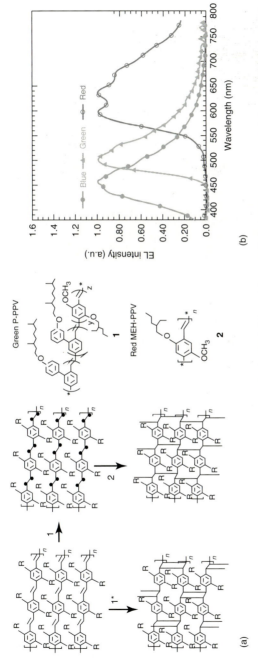

(a)

(b)

Figure 15.2 (a) Two possible cross-linking reactions (1 →2; diradical followed by cross-linking, 1*; 2π + 2π cross-linking) of the conjugated polymer and the molecular structures of the two PPV-related polymers studied; Green P-PPV (**1**) and Red MEH-PPV (**2**). (b) Electroluminescent spectra of RGB emissions from the three-color patterned PLED [13].

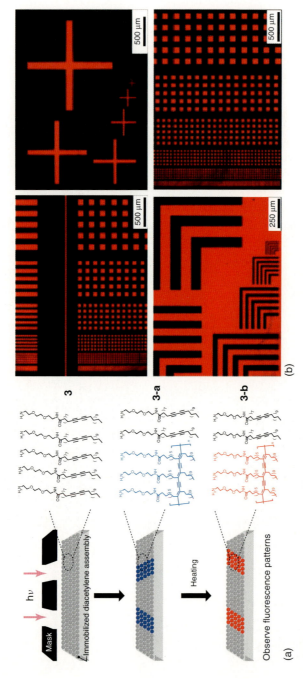

Figure 15.3 (a) Procedure for the generation of patterned fluorescence images with immobilized diacetylene supramolecules on the solid substrate. (b) Patterned fluorescent images obtained from photomasked irradiation of immobilized diacetylene vesicles on the aldehyde-modified glass substrate [14]. The **3-b** patterns were applicable to chip-based sensor systems.

photomask with 254 nm UV light (1 mW/cm²), which leads to photopolymerization of the immobilized diacetylene vesicles (3-a) but only in the exposed areas. The substrate is then heated at 100 °C for 10 s to induce a blue-to-red phase transition in the polymer. Since "red-phase" polydiacetylenes (3-b) are fluorescent and polymerization does not occur thermally, patterned fluorescence images are generated after the thermal treatment. Figure 15.3b shows patterned fluorescence images observed using fluorescence microscopy (red, bright areas correspond to areas exposed to UV light). The resolution of the patterned fluorescence images of 3-b approach the limit of fluorescence microscopy (<5 μm).

For a conductive polymer pattern, photocross-linkable acrylic and methacrylic units are substituted to thiophene or aniline, which can be polymerized to give a conductive film. There are several methods known for direct photopatterning followed by a wet-photopatterning process. For example, Figure 15.4 shows preparation of a cross-linkable conductive polymer that was prepared by FeCl₃ initiated polymerization of a mixture of the monomers of methyldioxythiophene-acrylate (MDOT-Acr) and methyldioxythiophenedecane (MDOTDec). In this example, the MDOT-Acr is a photoactive unit while the MDOTDec is introduced to increase the solubility of the resultant copolymer. After the film of cross-linkable conductive polymer,

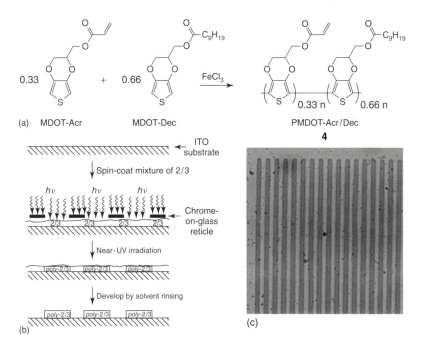

Figure 15.4 (a) Synthesis and structure of photocross-linkable conductive polymer. (b) Schematic illustration of photolithographic techniques using direct photocross-linking. (c) Optical microscope images of a film of 4 on an ITO electrode (pattern size is 5 μm) [15].

poly-methyldioxythiophene-acrylate (PMDOT-Acr) and poly-methyldioxythiophe-nedecane (PMDOTDec) (**4**), is exposed to near-UV light, the latent image is then developed by immersing the exposed film into 1,2-dichlorobenzene, to give a conductive pattern (Figure 15.4c).

On the other hand, an etchless direct photopatterning process utilizes a physical-property change accompanied by cross-linking. As cross-linking of the photoactive groups requires distortion of the main chain, which should be a core part of the functionality such as conductivity, absorption, or emission, the physical properties in the light-exposed area should be changed to generate patterns without going through an etching step.

Etchless direct photopatterning has been reported for a poly(3,4-ethylenedioxythiophene) (PEDOT) derivative having a methacrylate side group [16]. Figure 15.5a shows the electrochemical polymerization of 2-((2,3-dihydrothieno [3,4-b][1,4]dioxin-2-yl)methoxy)ethyl methacrylate (EDOT-EMA) and sidechain cross-linking of the resultant polymer upon UV exposure.

The PEDOT-EMA (**5a**) film was directly photopatterned by UV light through a photomask (Figure 15.5b), to give various conductive patterns (Figures 15.5d–f) with different shapes and linewidths. The UV-exposed area was bleached as the sidechain underwent photocross-linking, indicating that the conductivity of the UV exposed area was decreased. Indeed, the conductivity of the irradiated area was decreased from 5.6×10^{-3} to 7.2×10^{-4} S/cm, as determined from the slope of the $I-V$ curve in Figure 15.6a. Figure 15.6b shows the evolution of the film conductivity change (σ_t/σ_o) against the irradiation dose, where σ_t is the conductivity at the corresponding light dose and σ_o is the conductivity of the film before UV exposure. The conductivity change of the film of **5a** was correlated with the UV dose (D) according to the (Equation 15.7):

$$\sigma_t/\sigma_o = 0.98 \times \exp(-D/1.15) + 0.025 \tag{15.7}$$

The conductivity change by the sidechain cross-linking could be mainly ascribed to structural reorganization along the PEDOT main chain, accompanied by the cross-linking reaction. Figure 15.7 shows a transmission electron microscope (TEM) image of the film of **5a**. After UV exposure, the rigid rod-type polymer strips were bent and distorted. This type of morphology change could be ascribed to the distortion of the PEDOT main chain induced by the photoinduced cross-linking of the sidechain. The photodecomposition of the PEDOT chain, in the presence of a radical initiator in air, is unlikely to occur in this example, as the film of unsubstituted PEDOT was not patterned under the same condition.

The above method afforded direct patterns without the need for a wet etching or development process. However, the substrate for the polymer film deposition has to be conductive when the polymer was electrochemically deposited. In this context, an ethylenedioxythiophene (EDOT) derivative having a short methacrylate group is intriguing, as it could be deposited using vapor-phase polymerization (VPP). The VPP method generally affords PEDOT films with high film uniformity on any substrate [17]. Recently, the synthesis and photopatternability of a PEDOT bearing a short methacrylate sidechain was examined [18].

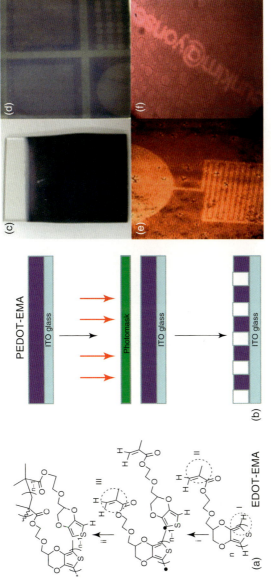

Figure 15.5 (a) Electrochemical polymerization of EDOT-EMA for film deposition and photocross-linking of PEDOT-EMA (**5a**) by UV light. (b) Scheme for direct photopatterning process using an electrochemically deposited film of **5a** by UV light. (c) Photograph of electrochemically polymerized film of **5a** on ITO glass and (d) Patterned film of **5a**. Optical microscope images of patterned film of **5a** with a pattern size of 50 μm (e) and 20 μm (f) [16].

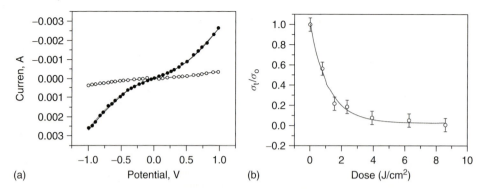

Figure 15.6 (a) *I*–*V* plot for a PEDOT-EMA film (thickness = 0.76 μm) before (●) and after (○) UV exposure between −1.0 and 1.0 V, with a potential step increase by 0.1 V/s. (b) Conductivity change against dose of UV irradiation [16].

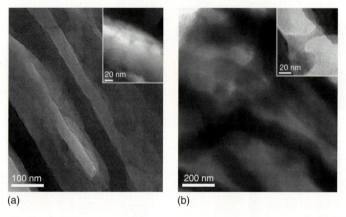

Figure 15.7 TEM image of the PEDOT-EMA (**5a**) film (a) before and (b) after UV exposure.

An EDOT substituted with a methacrylate group (EDOT-MA) was polymerized using vapor-phase polymVPP to form a patternable conductive polymer film on a flexible polyethylene terephthalate (PET) substrate with thickness of 40–200 nm. The PEDOT-MA (**5b**) film on PET showed a room-temperature conductivity (σ_{RT}) of 30–120 S/cm, depending on the oxidant layer thickness and was increased ∼30% when the film of **5b** was doped with aqueous solution of p-toluenesulfonic acid. The σ_{RT} of the film of **5b** was lower than that for VPP-PEDOT [17], possibly due to the presence of the methacrylate sidechain. Nevertheless, the conductivity of **5b** was much higher than that of **5a** ($\sigma_{RT} \sim 3.2$ mS/cm). Thin films of **5b** cross-linked successfully with only a few minutes of UV irradiation. Photoreaction of **5b** decreased the σ_{RT} to 1.7×10^{-3} S/cm, due to the photocross-linking of the sidechain. Figure 15.8 shows the effect of light exposure on the conductivity of the film. The conductivity decay was well correlated with the first exponential decay

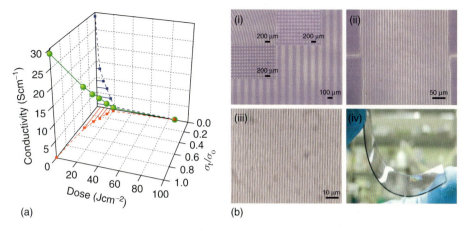

Figure 15.8 (a) The conductivity change of the PEDOT-MA (**5b**) film on PET according to UV dose. Green, blue, and red lines show total relation, the mean correlation of conductivity, and conductivity decrease according to dose, respectively. (b) Optical microscopic images of the **5b** film with pattern size of (i) 50 μm, insets show various patterns with different linewidths, (ii) 5 μm, (iii) 0.9 μm. (iv) Diffraction from the film of (iii).

curve, according to:

$$\sigma_t/\sigma_0 = 0.917 \times \exp(-D/5.29) + 0.066 \qquad (15.8)$$

The exponential slope in (Equation 15.8) was much larger than (Equation 15.7).

The conductivity change of the film of **5b** from photoirradiation is much larger than that of the photocross-linkable PEDOT with a longer sidechain (**5a**) [16] or PEDOT terminated with acrylates (**4**) [19]. This indicates that the conductivity change from the cross-linking is directly dependent on the distortion of the PEDOT main chain. When the cross-linking units are short, the twisting of the main chain should be severe, so that the conductive channels are distorted, to lower the conductivity of the cross-linked area. The main chain of the PEDOT with a longer sidechain (**5a**) or of the PEDOT terminated with acrylates **4** should be much less influenced by sidechain cross-linking and thus their conductivity change is small or negligible [16, 19].

The PEDOT-MA film prepared by VPP was patterned by UV exposure through a photomask to afford line patterns of 50–0.9 μm spacing, in which the light-exposed areas appeared as bleached (Figure 15.8b). When the line spacing of the patterns was below 5 μm, the patterned film became diffractive. In particular, the conductive line pattern with a spacing of 0.9 μm showed high diffraction to allow detection with the naked eye (Figure 15.8b). The σ_{RT} of the diffractive film with 0.9 μm line spacing were 8.91, 1.34, and 0.93 S/cm when patterned at light doses of 7.83, 15.6, and 90 J/cm², respectively. Thus, the line patterned film with 0.9 μm spacing provided the first example of a diffractive conductive polymer film that exhibited both flexibility and transparency.

An azide group is also cross-linkable and thus can be used for photopatterning of functional polymer films through a wet process. As shown in (Equation 15.9), the reaction mechanism of an azide group on a polymer involves photolysis at the $\pi - \pi^*$ band of the phenyl azide to generate a singlet nitrene, which rapidly either ring-expands to ketenimine [20], or intersystem crosses to the triplet nitrene, depending on the azide, substrate, and conditions [21].

$$R-\text{[aryl]}-N_3 \xrightarrow{h\nu} R-\text{[aryl]}-N: \xrightarrow{H-\overset{|}{C}-R} R-\text{[aryl]}-\overset{H}{N}-\overset{|}{C}-R$$

$$(15.9)$$

The ketenimine intermediate can participate in nucleophilic addition reactions with appropriate functional groups on the polymer chain to give cross-linking between polymer chains. The triplet intermediate, on the other hand, produces cross-linking only when it generates polymer free radicals that subsequently recombine. Using these types of cross-linking reactions for azide, a poly(3,4-ethylenedioxythiophene)-poly(styrenesulfonic acid) complex (PEDT:PSSH) was patterned by UV light, as shown in Figure 15.9, in the presence of bisfluorinated(phenyl azide) (bisFPA, **6**).

Successful photoinduced cross-linking can be achieved for film thicknesses from a few tens of nanometers to a few hundred nanometers, over the centimeter scale (Figure 15.9b, left photo) to sub-10-μm dimensions (right photo) [22]. This methodology could find widespread applications, including DNA immobilization, and in the fabrication of ion- and biochemical sensing films and membranes for organic electronics, as well as for polyelectrolyte-based devices.

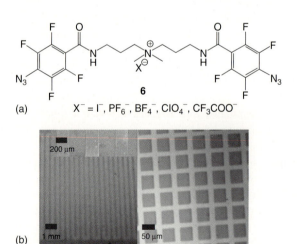

(a) $X^- = I^-, PF_6^-, BF_4^-, ClO_4^-, CF_3COO^-$

(b)

Figure 15.9 (a) Chemical structure of cross-linker (bisFPA, **6**) and (b) optical micrographs of shadow-mask photopatterned 60-nm thick PEDT:PSSH conducting polymer thin films with 5 w/w% of **6**.

15.2.2
Photopatterning by Dimerization of Anthracene

The photoreaction involving dimerization of functional groups requires rather high energy, as it requires covalent bonding of the separated functional units. In solution, a reverse reaction often occurs that dissociates the dimer to the corresponding monomeric units. Thus, the conversion to a dimer via direct photopatterning is rather low. Nevertheless, anthracene is known to undergo photodimerization under UV irradiation via $4\pi + 4\pi$ cycloaddition. The photodimerization of solid anthracene by ultraviolet (UV) irradiation has been examined by X-ray diffraction studies on single crystals of anthracene. When radiation of wavelengths less than about 3000 Å is excluded, dimerization has been reported to take place [23].

Thus, a solvent-free process is allowed for pattern and image formation for a highly fluorescent anthracene polymer (PMAn, **7**), which is linked by a methylene unit [24]. The emission of PMAn thin film was extinguished upon UV exposure, due to the photodimerization of the anthracene unit. Figure 15.10a shows the patterned fluorescence image prepared by UV irradiation of a PMAn film through a photomask with a gap-electrode pattern. The UV-exposed area is dark while the UV-shaded region is bright and fluorescent. As shown in Figure 15.10b, the gap between the dark and bright regions corresponds to the line gap of the mask. Figure 15.10c shows the intensity of fluorescence at the wavelength of 500 nm, probed at a patterned region in the white dotted bar of Figure 15.10b. The intensity of the 488-nm fluorescence in the UV-exposed area decreased about ~50% compared to the UV-shaded area. As a result of photodimerization of anthracene, the larger van der Waals intermolecular spacing (distance > 3.4 Å) between the anthracene units is replaced by the shorter intramolecular covalent bond (C–C ~ 1.54 Å), leading to a thickness change along the pattern (Figure 15.11).

The dimerization of anthracene to dianthracene induces a change in the conductivity of the polymer. To fabricate a conductive pattern via photodimerization of anthracene, selective UV irradiation through the photomask as well as the doping process are required [25]. For example, UV irradiation (λ > 350 nm) of poly(octamethylene-9,10-anthrylene), **8** leads to destruction of the aromaticity in

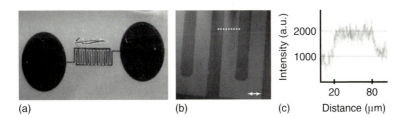

(a) (b) (c) Distance (μm)

Figure 15.10 (a) Photograph of a gap electrode pattern on a PMAn (**7**) film by UV exposure for 30 min (diameter of the circle = 0.3 cm). (b) A confocal macroscopic image of the gap electrode pattern of (a). The scale bar in (b) corresponds to 50 μm. (c) The intensity of fluorescence at the wavelength of 500 nm probed at a patterned region in the white dotted bar of (b).

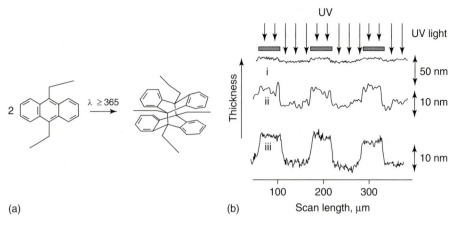

(a) (b)

Figure 15.11 (a) Photodimerization of anthracene polymers. (b) (i) Depth of the pattern formed from the film of **7** exposed to UV lamp under the photomask for 30 min (thickness of the film = 151 nm). (ii) Expanded scale in y-axis of (i). (iii) Depth of the pattern formed from the **7** film exposed to UV for 1 h (thickness of the film = 101 nm).

the anthracene ring as photodimerization proceeds on the UV exposed area. The remaining anthracene unit in the UV unexposed area could be doped with iodine vapor to enhance conductivity. Thus, UV patterning followed by an iodine-doping process would afford conductive patterns.

The resolution of this dry-patterning process should be easily optimized to give resolution to as high a level as can be achieved from photolithographic patterning of anthracene polymers, such as poly(N-neopentyl methacrylamide-co-9-anthrylmethyl methacrylate) (p(nPMA-AMMA)), (**9**) [26]. This type of optical, thickness, and selective electrical change induced by the photodimerization of anthracene in a polymer thin film could be applied to the preparation of various transistor and imaging patterns. Table 15.1 summarizes photopatterning by cross-linking and dimerization.

15.2.3
Applications of Photopatterning by Cross-Linking to Organic Electronics

Photopatterning by the cross-linking method can be applied to a number of different types of organic electronics [29]. These applications include electrochromic (EC) displays [16, 30] light-emitting displays (organic light-emitting diodes OLEDs) [27], biosensors [31], and organic electrodes [32].

EC cells were prepared by the photopatterning of a conductive oligomer (Figure 15.12a) [30] using thiophene/EDOT/phenylene (TPEEPT) diacrylate, followed by a wet-etching process, or by direct photopatterning of PEDOT-EMA without a wet-etching process (Figure 15.12b) [16].

Table 15.1 Photopatterning by cross-linking of unsaturated bond.

Photoreactive polymer	Abs$_{max}$, nm (eV)	Light (J/cm^2)/process^a	Pattern size (μm)	Application	Ref
4	520 (2.4)	U (UR^b)/W	5	Organic electrode and display	[15b]
5a	601 (2.1)	U (2.5)/D	5	Electrochromic display	[16]
5b	550 (2.2)	U (3.9)/D	0.9	Electrochromic display, patterned electrode	[18]
6	UR^b	LH (2.0)/W	10	Organic electrode and polyelectrolyte	[22]
7	415 (3.0)	U (0.63)/D	50	Bio-optoelectronics	[24]
8	265 (4.7)	HM (108)/D	$\sim 10^3$	Image formation	[25]
9	258 (4.8)	U (UR^b)/D	0.75	Image formation	[26]
10	600 (2.1)	U (>28.8)/W	$\sim$cm	Electrochromic display	[19]
11	392–399 (3.1–3.2)	U (2.4)/W	125	OLED	[27]
12	430 (2.9)	U (9 $\sim$ 54)/D	$\sim$cm	PLED	[13]
13	UR^b	U (1.3)/W	60	Bioengineering	[28]

^{a}Under UV using UV lamp (U), low-pressure Hg lamp (LH), or high-pressure mercury lamp (HM). Dry (D) or wet-etching process (W).
^{b}Unreported.

In the EC cell shown in Figure 15.12b, the patterned layer of **5a** is in contact with a solid polymer electrolyte layer, to produce an all-solid-state EC display. Upon application of +2 V, the pattern was erased because the conductive polymer was oxidized. The deep blue pattern reappeared upon application of −2 V, returning the color of the original pattern as the conductive polymer was reduced. The

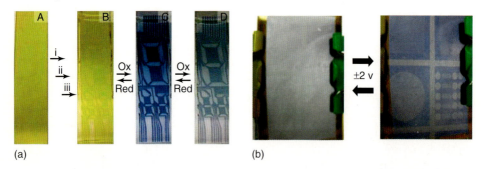

(a)

(b)

Figure 15.12 (a) Photopatterning process for a spray-cast film of TPEEPT diacrylate **(10)** (A). After (i) applying a mask, (ii) irradiating with UV light, and (iii) washing with tetrahydrofuran, the cross-linked and patterned film appears as illustrated in (B). Subsequent electrochemical oxidation affords the blue radical cation (C) and the pale blue/transparent dication (D) [30]. (b) Photographs of the patterned electrochromic (EC) cell after application of potential from +2 to −2 V for 10 s. (100 μm pattern) [16].

EC pattern of the all-solid-state EC cell was reversible within a step potential of ±2 V. During the alternative voltage application, EC switching in the UV-exposed area was not detectable. Thus, the photopatterning of the cross-linkable PEDOT film through sidechain cross-linking was shown to be an effective method for EC patterning.

The patterning by cross-linking reactions of the vinylene in PPV **(12)** was applied to a LED device as shown in Figure 15.13. The LED device could be fabricated by cross-linking and the blue light emission shifted without use of a photoinitiator.

An example of dry direct photopatterning of photocross-linked azide has not yet been reported; however, a photocross-linking followed by a wet-etching process was applied to create micropatterns for cell patterning (Figure 15.14). In this example, poly(acrylic acid) (PAA) conjugated with 4-azidoaniline, **13** was interwoven into PAA/polyacrylamide (PAM) multilayer films (Figures 15.14b,c) [28]. This patterning method can be used for biomaterials, making it a good application for bioengineering.

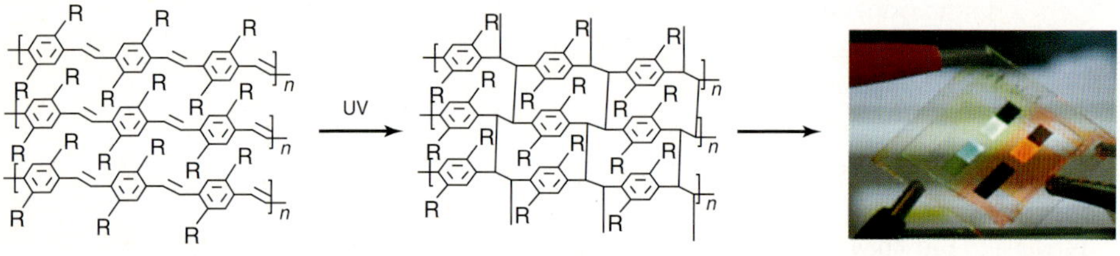

12

Figure 15.13 The patterning by cross-linking reactions of the vinylene in **12** and RGB color emission from a device [13].

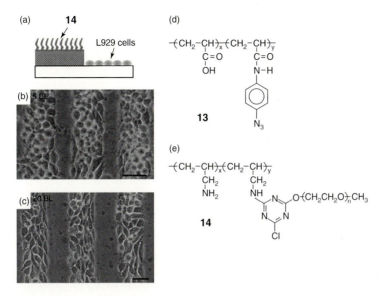

Figure 15.14 (a) Schematic illustration of L929 cells cultured on the micropatterned PAA/PAM multilayer film on which 13 was conjugated. (b) The microscopic image of L929 cells cultured on the 5-BL PEG conjugated PEM (14) pattern for six days (60/30 μm-wide stripes). (c) The microscopic image of L929 cells cultured on the 20-BL 14 pattern for six days (60/60-μm wide stripes). Scale bar = 50 μm. Structures of (d) 13 and (e) 14 [28].

15.3
Direct Photopatterning by Photo-Oxidation

15.3.1
Photo-Oxidation of C=C Double Bonds and Heteroatoms

The electro-optical properties and long-term stability of polymer-based devices are considerably affected by oxidative decomposition in air. This type of oxidation can be accelerated under light and utilized for photo-oxidative patterning.

UV exposure of conjugated polymers having C=C bonds or heterocycles leads to photo-oxidation or photobleaching in air [33]. The photo-oxidation of conjugated polymers thus affords fluorescence and conductive patterns by photo-oxidation. For example, UV exposure of conjugated polymer films in ambient air conditions resulted in photo-oxidation of a poly(1,4-phenylenevinylene) film [34]. The fluorescence intensity significantly decreased concomitantly with UV irradiation time. An fourier transform infrared spectroscopy (FT-IR) spectrum of the poly(1,4-phenylenevinylene) film showed that the carbonyl peak was increased, whereas the vinylene peak linked adjacent to the phenylene groups was decreased after UV exposure. However, UV irradiation of a poly(1,4-phenylenevinylene) film under nitrogen conditions did not lead to fluorescence quenching. Scurlock et al. [35]. reported that singlet oxygen was a reactive

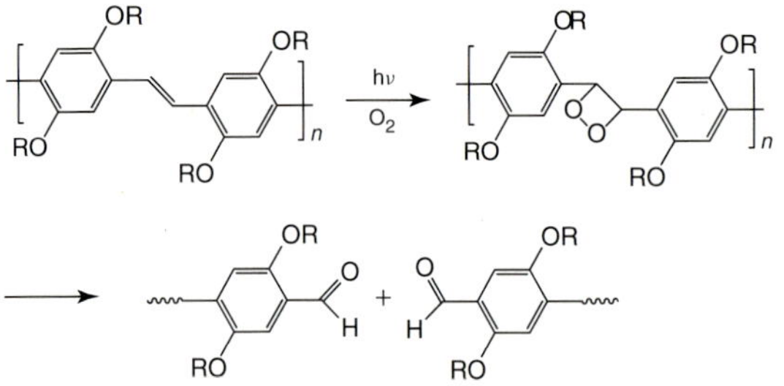

Scheme 15.2 The mechanism of photo-oxidation via singlet oxygen.

intermediate in the photoinduced oxidative degradation of a luminescent polymer, poly(2,5-bis(5,6-dihydrocholestanoxy)-1,4-phenylenevinylene) (BCHA-PPV), both in solution and in solid films. Upon UV exposure, the energy transfer from the BCHA-PPV triplet state to ground-state oxygen induced the generation of singlet oxygen. This singlet oxygen reacted with BCHA-PPV and then resulted in extensive chain scission of BCHA-PPV. This photodegradation is proposed to proceed in a $(2\pi + 2\pi)$ cycloaddition at the double bond that connects phenylene groups in the macromolecule through addition of singlet oxygen, as shown in Scheme 15.2.

This type of photo-oxidation creates a conjugated polymer that shortens the main chain and enhances solubility of the polymer. Yoshino *et al.* [36] reported optical patterning utilizing poly(*p*-phenylenevinylene) (PPV) and its derivatives by using this improved solubility. On the other hand, this photo-oxidation leads to a shift in the absorbance bands to the UV range, and ultimately leads to fluorescence quenching. Figure 15.15a shows that the emission of diphenylamino-s-triazine bridged *p*-phenylenevinylene polymer (DTOPV) (**15**) [37] was extinguished upon exposure to a UV source under ambient conditions [38]. After UV irradiation for 10 min $(7.8 \, \text{J/cm}^2)$, the fluorescence for the film of **15** was almost quenched by about 90% under ambient conditions. However, the fluorescence of the same film was only quenched about 10% when exposed to UV under argon conditions, that is, when oxygen was excluded. This supports the idea that the decrease in emission intensity mainly resulted from oxygen, which was essential for photo-oxidation. The small fluorescence quenching of **15** film in an argon atmosphere may be ascribed to photoinduced cycloaddition of the C=C double bonds [39].

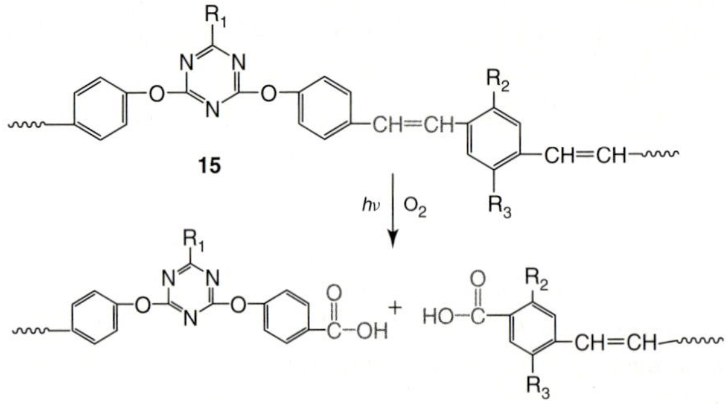

$$\tag{15.10}$$

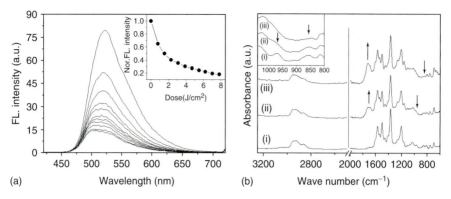

(a) Wavelength (nm)   (b) Wave number (cm⁻¹)

Figure 15.15 (a) Fluorescence spectral change for the film of **15** excited at 370 nm at different irradiation time, from top to bottom, 0, 1, 2, 3, 4, 5, 6, 7, 8, 9, 10 min under air, inset – normalized FL intensity change of the film at 522 nm at different dose. (b) FT-IR spectra for a film of **15** before irradiation (i) and after irradiation with a UV source for (ii) 15 min and (iii) 30 min (inset: magnification of the 1080–800 cm⁻¹ region).

With respect to the photopatterning process, this result indicates that photo-oxidation under air was much more efficient than cycloaddition of C=C bonds or *cis–trans* isomerization along the vinylene under oxygen exclusion conditions. Through photo-oxidation, the vinylene bond is broken to reduce the conjugation length of the PPV unit and is converted to a carbonyl group (Equation 15.10). The carbonyl group is also known to efficiently quench fluorescence [40]. Thus, a fluorescence pattern was clearly generated, without wet-chemical etching, from a thin film of **15** with micrometer (100 μm) to submicrometer (900 nm) wide lines that depended on the linewidths of the photomask (Figure 15.16a,b). The atomic force microscope (AFM) image shows that the surface of the fluorescence pattern

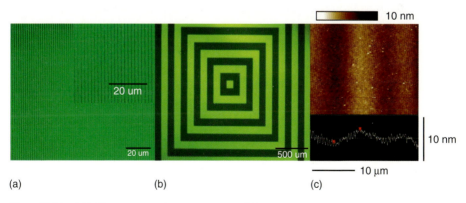

(a)   (b)   (c)

Figure 15.16 (a,b) Fluorescence microscope image of **15** film with 900 nm and 100-μm-wide line patterns, respectively, prepared by photo-oxidation of **15** under UV irradiation. (c) AFM image of the film of **15** with 5-μm line patterns.

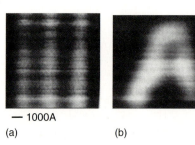

Figure 15.17 Transmission mode NSOM images of (a) gratings with a 0.14-pm gap and (b) a letter "A" written by NSOM on the surface of **16**.

— 1000A

(a)   (b)

was very smooth and uniform for both the UV-exposed and unexposed regions, with mean roughness (R_m) of 0.43 and 0.42 nm, respectively. The decrease of depth in the UV-exposed area was very small, typically 1.1 and 2.8 nm for high doses of 23.4 and 46.8 J/cm^2, respectively (Figure 15.16c).

Similarly, a pattern of conjugated polymer, poly(3,4-diphenyl-2,5-thienylene vinylene) (**16**), could be prepared using near-field scanning optical microscopy (NSOM), which leads to photo-oxidation at the nanometer scale [41]. A nanometer-scale NSOM could be used to write and read images on a conjugated polymer film with a submicrometer resolution, clearly breaking the diffraction limit (Figure 15.17).

UV irradiation on PEDOT resulted in the formation of sulfon groups, SO$_2$, through photo-oxidation (Equation 15.11). This reaction leads to a reduction in electrical conductivity because of structural distortion along the thiophene main chain, which eventually shortens the conjugation length [42].

$$\text{(15.11)}$$

The formation of SO$_2$ in the sulfur atom of PEDOT proceeds through singlet oxygen generated by energy transfer. This reactive oxygen attacks conjugated chains, resulting in photo-oxidation, as described above in Equation 15.10 In addition to sulfone formation, the photo-oxidation leads to chain scission, accompanied by the formation of carbonyl/carboxyl groups. However, the required energy for these processes is higher than that for the photo-oxidation of vinylene or ethylene bridges, mainly due to the aromatic stabilization in the heterocyclic aromatic compounds. Thus, although photo-oxidation at the heteroatom or heterocyclic structure could lead to a pattern, it is rarely used. Table 15.2 summarizes photopatterning by oxidation.

15.3.2
Applications of Photo-Oxidation

Photo-oxidation could be applied to surface modification to alter the chemical and physical properties of polymer surfaces. In general, special surface treatments with regard to chemical composition, wettability, roughness, conductivity, and

Table 15.2 Photopatterning by oxidation.

Photoreactive group	Abs$_{max}$, nm (eV)	Light (J/cm^2)/process^a	Pattern size (μm)	Application	Ref
15	522 (2.4)	U (7.8)/D	100	Cell patterning	[38]
16	540 (2.3)	HN (3.6 × 10^{-6})/W	0.1	Optical imaging	[41]
17	480 (2.6)	I (UR^b)/D or W	10^3–10^6	Optic device	[43a]
18	UR^b	I (UR^b)/W	10^3–10^6	Optic device	[43b]
19	470 (2.6)	U(3.7)/D	20	Patternable PLED	[44]

^{a}Under UV using UV lamp (U), He-Ne laser (HN), or incandescent lamp (I). Dry- (D) or wet-etching process (W).
^{b}Unreported.

16 **19**

cross-linking density are essential for the successful application of polymers for adhesion, biomaterials, microelectronic devices, and thin-film technology. In bioengineering, interactions between cells and their culture substrate are very important for controlling biomaterial function such as cell attachment, proliferation, differentiation, and protein secretion [45]. Recently, You *et al.* [38] examined the control of mesenchymal stem cell (MSC) properties, such as attachment and proliferation, utilizing a fluorescent polymer film. The water contact angles of the film were found to gradually decrease from 97.1 to 83.1° as the DTOPV(15) film was exposed to UV for 0 and 60 min, respectively (Figure 15.18b). Thus, the photo-oxidation of **15** significantly changed the surface properties and enhanced biocompatibility of the film, due to the photogenerated CO$_2$H group. This surface modification leads to alterations of cell attachment, patterning, alignment, and proliferation. The MSCs preferentially adhered to and spread out on the UV-exposed regions of the 100-μm wide line pattern (Figure 15.18d). An scanning electron microscope (SEM) image (Figure 15.18e) shows a well-aligned cell line on the patterned film arising from selective attachment of the cells to these UV-exposed lines.

Compared with the typical randomly attached spindle-shaped MSCs on the tissue culture polystyrene (TCPS) (Figure 15.19a). Figures 15.19b,d show the selective attachment and alignment of MSCs along the UV-exposed defined shape and pattern DTOPV film. Surprisingly, MSCs show selective attachment along circular line patterns and even in the bent letter pattern of "YONSEI", where the linewidth of the pattern was 100 μm wide. Through the investigation of several extracellular matrix proteins, this study revealed that hydrophilic proteins such as collagen and gelatin play an important role in the selective attachment of MSCs to the UV-exposed region. The proliferation of MSCs cultured on the UV-exposed region is also much

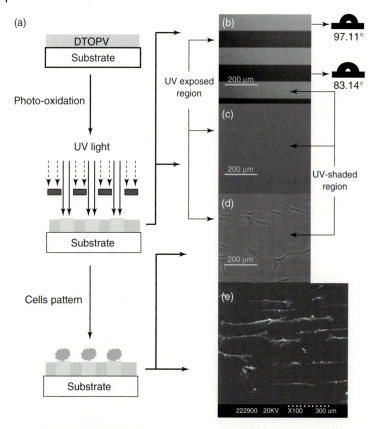

Figure 15.18 (a) Patterning procedure of a DTOPV (**15**) -coated substrate for cell pattern. (b) Fluorescent microscope image and water contact angle of the pattern of **15**. (c) and (d) Optical microscope images of the patterned surface of **15** (100-μm linewidth) before (c) and after (d) cells attachment. (e) Scanning electron microscope image of MSCs pattern attached to the patterned surface of **15** (100-μm linewidth).

greater than that on TCPS, indicating the potential application of photoreactive polymers as cell-culture substrates. Taking advantage of the emission from the DTOPV pattern, the cell location and pattern images were easily detected through a microscope, with or without an excitation probe beam. This cell patterning method could be applied to the enhancement of stem-cell differentiation, the development of drug screening, and cell imaging.

The photopatterning of polymers, including PPVs [46], polyphenylene ethylene (PPE) [1], didodecyl polydialkyl stilbenevinylene [47], and polyfluorenes [48], has been examined for its application to flexible electronic devices with simple processes. The direct photo-oxidation of poly(1-methoxy-4-dodecyloxy-p-phenylenevinylene) **17** or poly(3-dodecylthiophene) **18** was explored in order to manufacture optically patternable polymer light-emitting diodes (PLED) [43] (Figure 15.20).

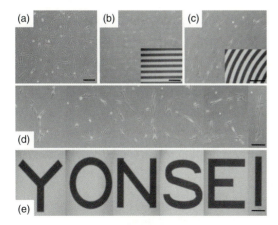

Figure 15.19 Optical microscope images of MSCs cultured on (a) tissue-culture polystyrene, (b) 50-µm-wide line pattern of **15** film, (c) 50-µm-wide curved line pattern of **15**, (d) 100-µm-wide "YONSEI" letter pattern with MSCs on **15**, and (e) fluorescent microscope image of the same letter pattern with cells. Fluorescent pattern images are inserted in parts (b) and (c). Scale bar = 200 µm.

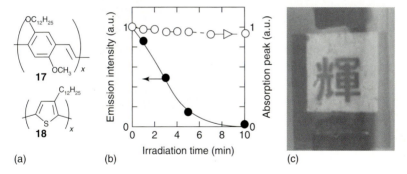

Figure 15.20 (a) Molecular structures of the polymers. (b) Dependence of emission and absorption intensity of the device from **17** on the irradiation time. (c) A photograph of a patterned emission from the device containing **17**.

Recently, a patterned electroluminescent device was fabricated with a multilayer structure of indium tin oxide (ITO) as an anode/PEDOT-PSS/poly[2-methoxy-5-(3,7-dimethyloctyloxy)-p-phenylene vinylene] (MDMO-PPV) **(19)**/hole, blocking 2,9-dimethyl-4,7-diphenyl-1,10-phenanthroline (BCP)/electron transporting LiF/aluminum as a cathode, in order to apply the direct pattern of emissive film to a PLED (Figure 15.21a) [44]. The layer of **19** (50 nm thick) coated on a PEDOT-PSS layer was directly photopatterned by UV irradiation through a photomask having 20-µm wide curved lines. The pattern was intact when a hole-blocking BCP and electron transporting LiF layer were deposited. Figure 15.21b shows a photograph of light emission from the device having a MDMO-PPV pattern layer, where this layer of **19** was inserted in the multilayered device of

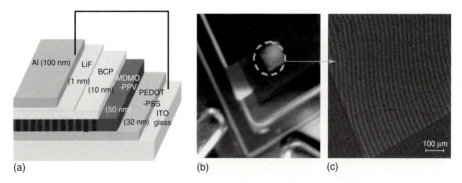

(a) (b) (c)

Figure 15.21 (a) Schematic representation of the EL patterned device. (b) A photograph of light emission from the device with a 20-μm line pattern at a driving voltage of 5 V. (c) EL micrograph for the device of (b) taken by an optical microscope under driving voltage of 5 V [44].

ITO/PEDOT-PSS/**19**/BCP/LiF/Al. Figure 15.21c shows an electroluminescence (EL) micrograph taken by an optical microscope at driving voltage of 5 V to show the 20-μm-wide curved line.

Selective photo-oxidation of two emissive polymer layers allows for three-color (RGB) polymeric light-emitting devices [49]. The fabrication procedure for producing the RGB emission areas is schematically illustrated in Figure 15.22. Two emissive layers are the red emitting polymer, poly(2-methoxy,5-(2-ethylhexyloxy)-1,4-phenylenevinylene) (MEH-PPV), and a blend of the blue-emitting polymer, poly (9,9′-dioctyl-fluorene) (PFO), and a green-emitting polymer, a phenyl-substituted poly (p-phenylene vinylene) derivative (P-PPV). The

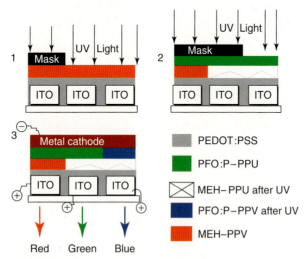

Figure 15.22 Schematic device fabrication procedures for producing the red-, green-, and blue-emitting areas [49].

first emissive layer is a red emitter, but could be converted to a nonemissive hole transporter by UV exposure. With the first layer converted to a hole transporter, the second layer could be determined either as a green emitter or a blue emitter depending on the selection of UV irradiation. The emission color is therefore defined by the presence or absence of UV exposure in each layer. This selective photo-oxidation method may provide a low-cost and high-throughput procedure to fabricate polymeric flat-panel display devices.

15.4
Direct Patterning by Photoinduced Acidification

Acidification by a PAG is a critical technology in the organic electronics industry, especially in terms of the chemical amplification (CA) concept [50], in which a single photochemical event induces a cascade of subsequent chemical transformations in a resist film, to manufacture a polymer pattern. For functional polymer patterning, photogenerated acids react with an acid-labile group of the functional polymer or dye dopants. The acid-labile groups are amine, polyaromatic, or alkoxy substituents. These photoinduced acidifications afford active patterns such as colored, fluorescent, or conductive pattern [51]. Thus, quinizarins [52], pyrene [53], pyridylbenzoxazoles [54], piperazine substituted bipyridazine fluorophores [55], and dimethyl-amino-substituted phenyl fluorophores [56] are introduced into photopolymer films for direct photopatterning.

Acid generators are categorized into two groups: (i) those with ionic groups, such as onium salts and (ii) those with nonionic groups, such as organohalides. The choice of PAG could depend on many factors, such as the nature of radiation, quantum efficiency of acid generation, solubility, and miscibility for a uniform film, thermal and hydrolytic stability, toxicity, strength, and size of generated acid, cost, and so on [57]. Figure 15.23 shows the structures of representative PAGs.

15.4.1
Fluorescence Patterns by Photoacidification

Using the acid-labile tert-Butoxycarbonyl (t-BOC) group, a quinizarin polymer (20) can be utilized for the generation of colored and fluorescent images by

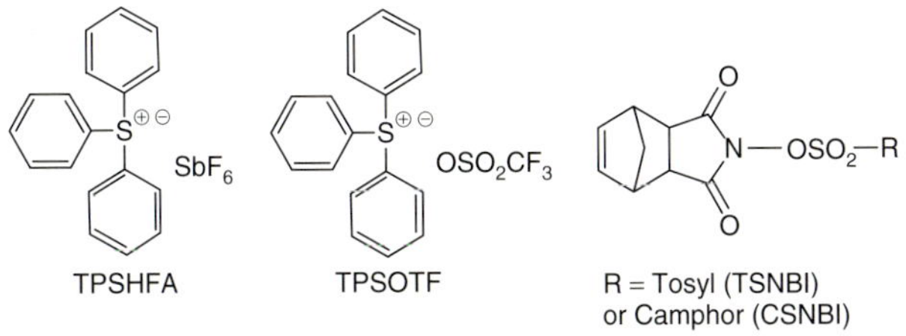

TPSHFA   TPSOTF   R = Tosyl (TSNBI)
                                            or Camphor (CSNBI)

Figure 15.23 Photoacid generators.

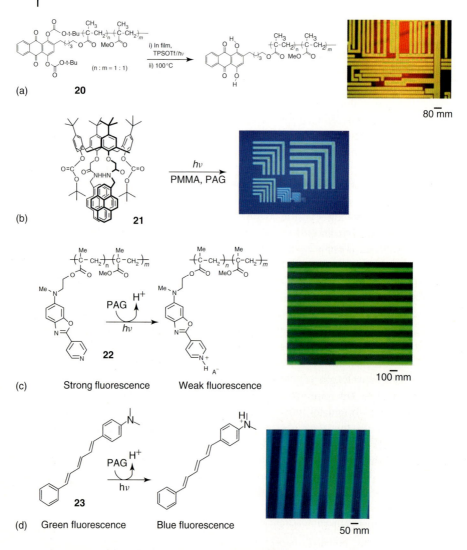

Figure 15.24 The schematic of photoacid-induced fluorescence change using (a) quinizarin, (b) t-BOC-protected pyrene-containing calixarene, (c) pendant pyridylbenzoxazole, and (d) DMA-DPH(**23**) and fluorescence pattern images after UV irradiation through a photomask.

photolithographic methods, which are induced by acid-catalyzed deprotection of the t-BOC group in the polymer film [52] (Figure 15.24a). The protecting t-BOC groups could be removed readily by the PAG, triphenylsulfonium triflate (TPSOTf), to regenerate quinizarin moieties, which recover color and fluorescence in the UV-exposed regions after the postexposure baking (PEB) step.

Similarly, using an acid-labile t-BOC group, an excimer emission approach was reported for fluorescent pattern imaging. Selective removal of t-Boc protecting groups in a polymer film imbedded, pyrene-containing calixarene derivative (21) results in the generation of a patterned fluorescence image, without employing wet developing processes, as shown in Figure 15.24b [53].

Generally, the fluorescence intensity is dependent on the structure of aromatic compounds. When aromatic compounds have extended π conjugation and electron donor substituents such as hydroxyl or amino groups at the ends of these compounds, fluorescence intensity is strong. However, when electron-withdrawing moieties such as a nitro group are substituted into aromatic compounds, fluorescence quenching occurs. In addition, fluorescence can be quenched when the fluorescent chromophore is protonated to give an electron-withdrawing group. As shown in Figure 15.24c, the benzoxazole chromophore (22) has an electron-donating group and an electron-rich pyridyl group at both end positions to give strong fluorescence. A fluorescence pattern image is obtained with the PAG, triphenylsulfonium hexafluoroantimonate (TPSHFA), which transforms the pyridyl group to an electron-withdrawing group [54].

In most cases, PAGs lead to the enhancement of fluorescence or the quenching of fluorescence. However, a new approach for the fluorescence patterning of thin films showing green and blue fluorescent regions was introduced using 1-(4′-dimethyl-aminophenyl)-6-phenyl-1,3,5-hexatriene (DMA-DPH) (23). The emission wavelength of 23 is significantly sensitive to the micropolarity of its immediate environment. Compared to the parent DPH, the spectral red shifts of 23 originate from charge delocalization in the extended con-jugated system of the molecule, induced by the electron-donating effect of the p-dimethylamino group. Figure 15.24d shows the photoacid-generator (TPSHFA) -induced protonation of the dimethylamino group in 23, which disrupted the electronic conjugation to shift the 23 emission maxima hyp-sochromically toward that of the parent DPH. Therefore, alternative green–blue fluorescence pattern images could be obtained with 23 after UV irradiation through a photomask [56]. This study of selective green and blue fluorescence pattern allows the potential opportunity of full-color display applications in OLEDs.

15.4.2
Conductive Patterns by Photoinduced Acidification

A soluble polyaniline (PANi) (24) with a photolabile, acid-labile, and thermolabile t-BOC group was synthesized to overcome its poor solubility in most common solvents, which has limited its industrial application. Figure 15.25a shows that this modified soluble form of PANI(t-BOC) was converted to an insoluble and electrically conductive polymer, upon photodoping with only a catalytic amount of PAGs such as N-(tosyloxy) norborneneimide TSNBI or (camphorsulfonyloxy)norborneneimide

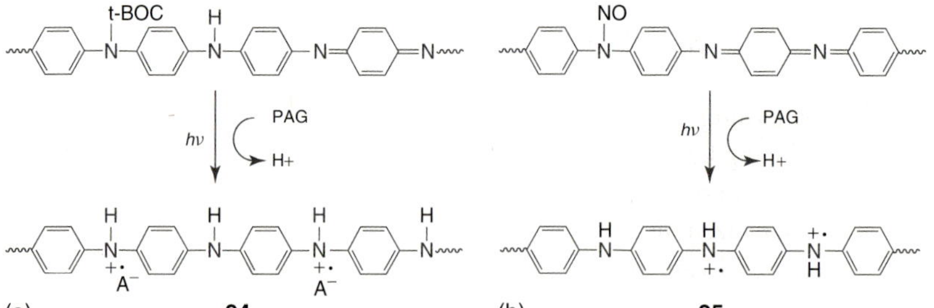

Figure 15.25 The schematic of the photoreaction of (a) PANI(t-BOC) and (b) PANI-NO with photoacid generator.

(CSNBI). This solubility difference also resulted in conductive patterns of high resolution through a conventional photolithography process. The photoacid-induced conversion recovered the original conductivity level of the doped PANI because the t-BOC protecting groups were easily removed in doping or acid-catalyzed reactions by CA [58].

The UV-exposed region showed a maximum conductivity of 1×10^{-3} S/cm. However, the subsequent external doping of UV exposed PANI by HCl vapor led to a significant increase in conductivity up to 3–5 S/cm. A conductive PANI pattern was also prepared with the change of nitrosated PANI(PANI-NO, **25**) as a PANI precursor to PANI with PAG, poly(vinyl chloride) (PVC), as shown Figure 15.25b. The **25** could be photochemically hydrolyzed with PVC as a PAG, and the conductivity of PANI thin films could be recovered without either degradation or cross-linking reactions [59]. Table 15.3 summarizes photopatterning by photoinduced acidification.

Table 15.3 Patterning by photoinduced acidification.

Photoreactive polymer	Abs$_{max}$, nm^a (eV)	Pattern size (μm)	Light (nm, J/cm^2)/process^b	Application	Ref
20	335 (3.7)	~10	U(254, 0.07)/D	Electronics device	[52a]
21	UR^c	~2	U/D	Display device	[53]
22	378 (3.3)	~100	U(360, 0.2)/D	Electronics device	[54]
23	400 (3.1)	50	HX(248, 0.01)/D	Display device	[56]
24	318 (3.9)	1–10	HX(250, UR^c)/W	Microelectronics	[58]
25	290^d (4.3)	UR^c	V(280, UR^c)/W	Electronic and optoelectronics	[59]

^{a}Dye + PAG film.
^{b}Under UV using UV lamp (U), Hg-Xe laser (HX), or incandescent lamp (I). Dry- (D) or wet-etching process (W).
^{c}Unreported.
^{d}In chloroform.

15.4.3
Application of Patterning by Photoinduced Acidification

The photopatterning in the presence of a PAG can be applied to organic electronic-inducing light-emitting displays (OLED). Meerholz and coworkers [27] examined the patterned multicolor OLED prepared from photopolymerizable oxetane-functionalized fluorene polymer (11) and a PAG. Oxetanes are commonly polymerized cationically via cross-linking in the presence of photoacid. Upon UV illumination, the initiator decomposes via a multiple-step mechanism and eventually generates protons H^+, which open the oxetane ring and initiate the polymerization. Rinsing the films with pure tetrahydrofuran (THF) led to the patterned RGB OLED device shown in Figure 15.26.

Recently, a novel approach to chemical processing of organic electronic fabrication based on fluorous solvents, in particular segregated hydrofluoroethers (HFTs), was developed for photolithographic patterning of organic electronic materials [60]. The combination of fluorous solvent as poor solvents for nonfluorinated organic materials, acid-sensitive semiperfluoroalkyl resorcinarene (26) processable in HTEs as an effective photoresist, and PAGs, could allow for multilevel patterning of organic electronic materials (Figure 15.27) [61].

This novel patterning method could be used to pattern polymer solar cells manufactured with a blend of poly(3-hexylthiophene) (P3HT) and [6,6]-phenyl-C61 butyric acid methyl ester (PCBM) in order to obtain high open-circuit voltages (V_{oc}) [62]. An array of 300 solar cells with a period of 50 μm, which indicated that the active region was 70% of the total cell area, showed a V_{oc} of 90 V and a power-conversion efficiency (PCE) of 0.3%.

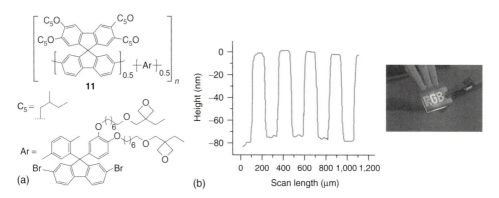

Figure 15.26 (a) Cross-section of a 73-nm-thick photostructured layer of the blue-emitting **11** illuminated through a grid with 250-μm spacing (125-μm openings). (b) Photograph of a RGB (red, green, blue) device. Dimensions of the glass substrate are 25 × 25 mm [27].

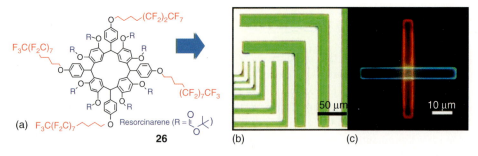

Figure 15.27 (a) Structure of **26** and (b) optical microscope image of patterned P3HT. (c) Fluorescent microscope image of overlaid patterns (feature width 5 μm) of poly(9,9-dioctylfluorene) (bottom) and $[Ru(bpy)_3]^{2+}(PF_6^-)_2$ (top).

15.5
Reversible Patterning

In most approaches for photochemical patterning, irradiation is responsible for irreversible modifications of the photosensitive polymers, resulting in irreversible patterning. Therefore, the generated patterns are neither correctable nor repatternable. Reversible patternings are highly desired as an effective basis for either environmentally friendly technologies or for smart surfaces. To create a reversible pattern, photochemically reversible modifications of properties is required. In addition, the patterned state should be stable and the pattern should be clearly erasable only when another stimuli such as light, heat, potential, or pressure, and so on was applied. In this context, photochemically reversible reactions such as ring closure–opening (Figure 15.28a), cross-linking and linking [63] (Figure 15.28b), and *cis–trans* isomerization reactions (Figure 15.28c) [64] are promising candidates for direct reversible photopatterning.

Reversible photochemistry is a fascinating area of investigation, because it has produced and continues to reveal completely unexpected phenomena, some of which are still unexplained as a result of a wide range of unexpected possible consequences. Furthermore, by coupling with a functional group, it could lead to multifunctionality with enhanced properties. Therefore, reversible patterning is rather new and still preliminary, but very promising. We describe here some examples of patterning, using the reactions described in Figure 15.28.

15.5.1
Reversible Patterning by Diarylethene Groups

Diarylethenes have attracted much attention as photoinduced switching materials as it is possible to switch their physical or chemical properties by stimulating them with light irradiation at an appropriate wavelength [12, 65]. A diarylethene

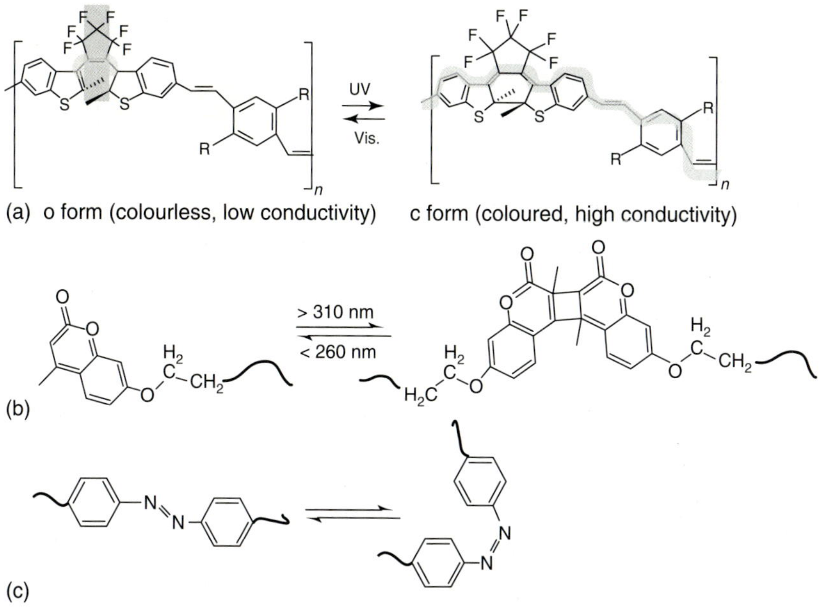

(a) o form (colourless, low conductivity)   c form (coloured, high conductivity)

(b)

(c)

Figure 15.28 Examples of photoreversible reactions for direct reversible photopatterning such as ring closure–opening (a), cross-linking and linking (b), and *cis–trans* isomerization reactions (c).

(BTF6) has been well known for its thermally irreversible photochromic properties (Equation 15.12). These reversible properties of diarylethene have been applied to introduce rewritable media or patterning methods by simple masking [66], two-photon laser patterning [67], and holography [65, 68]. Most diarylethene patterning work has been focused on utilizing its color or refractive-index-switching characteristics by selective photoirradiation, which is applicable to optical switches, data storage, and modulators [69–71].

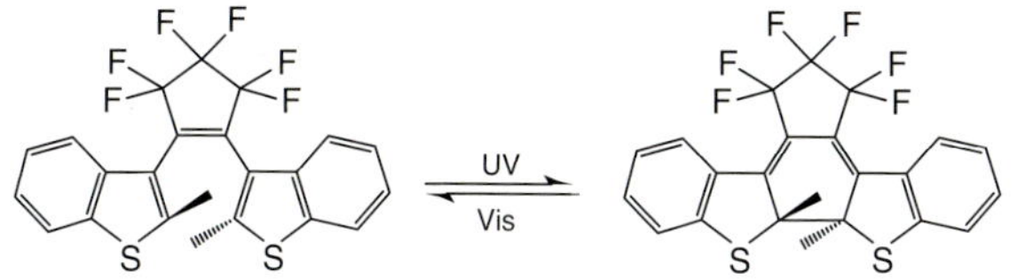

(15.12)

For reversible optical pattern generation, high optical contrast between patterned and unpatterned areas and switching stability are essential. The photopattern of a photochromic diarylethene containing film and the reversible and stable optical response of the film are shown in Figure 15.29 [72]. Photochromic films prepared by photocuring methods were applied in an erasable photon-mode recording with a 532-nm laser and erasing with a 365-nm laser. Cyclability of recording was determined by analyzing the optical response of the film, and reversibility was stable even after 1000 cycles.

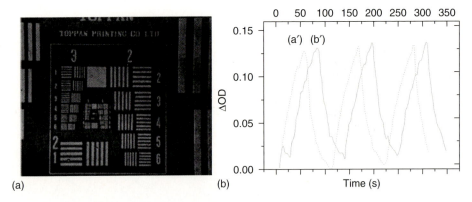

(a)

(b)

Figure 15.29 (a) Photograph of photochromic recording with a 532-nm laser on a colored film containing diarylethenes. (b) Reversible photochromic response of the film in (a) on the UV and visible light; (a') initial three cycles and (b') after 1000 cycles [72].

A photochromic conversion followed by a fluorescence change can be applied to various reversible pattern generations for which the reversible photochromic change provides a convenient light-activated method of converting a fluorescent state to a dark state. To obtain both photochromic- and fluorescent-patternable diarylethenes, two approaches have been tried. One is to design a diarylethene that has higher photocyclization yield to develop an efficient, nondestructive, fluorescence-based photochromic read-out system [65]. Figure 15.30a shows the photoconversion of 1,2-bis(2-methyl-1-benzothiophene-1,1-dioxide-3-yl)perfluoro-cyclopentene (BTFO4) (**27**), which is the sulfonyl derivative of 1,2-bis(2-methyl-1-benzothiophene-3-yl)perfluorocyclopentene (BTF6) [73]. It is a highly fluorescent photochromic diarylethene in the closed-ring form (**27-c**). For comparison, the fluorescence changes of the BTF6 system under the same conditions are shown in Figure 15.30b. The modulation of the fluorescence signal of **27** is carried out in ethyl acetate at room temperature, while alternating the illumination between UV 254 nm (unshaded regions) and visible 400 nm (shaded regions) wavelengths. The fluorescence signal was recorded as a function of time at 490 nm. As can be seen in Figure 15.30b, the fluorescence signal changes are much more pronounced in the BTFO4 system than they are in the BTF6 system. The fact that the fluorescence intensity of the **27** can be modulated in this way indicates that it can be applied to the reversible photochromic and fluorescence pattern generation with stable switching properties.

As an alternative approach to the creation of photochromic and fluorescent diarylethene molecules, combination of diarylethenes with fluorophores has been reported [65, 74, 75]. One example is the highly fluorescent diarylethene oligomer bridged by the fluorophore PPV (Figure 15.31) [74, 75].

As shown in Figure 15.31b, the red color indicates formation of the closed form of BTF-PPV (**28**). After irradiation with a visible light (>540 nm), the red color was

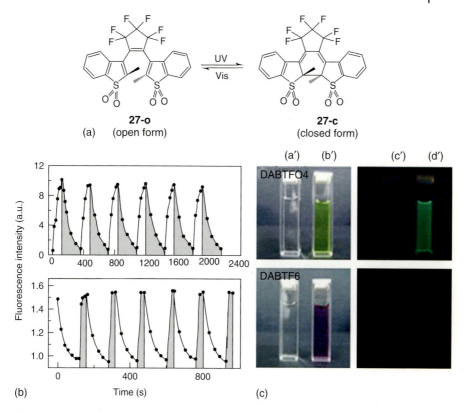

Figure 15.30 (a) Photochromic reaction of BTFO4 (**27**). (b) The modulation of the fluorescence signal of A: **27** and B: BTF6 in ethyl acetate (1.0×10^{-25} M) observed when alternating its illumination between UV (254 nm, unshaded regions) and visible (400 nm, shaded regions) wavelengths. (c) Photographs of solutions of **27** (1.0×10^{-4} M in ethyl acetate): (a') color of o-**27** (before UV irradiation under room light, (b') **27-c** (yellow in photostationary state after UV irradiation), (c') fluorescence of **27-o** (weak emission) after photoexcitation with a 365-nm UV lamp, and (d') fluorescence of **27-c** (green emission). Color and fluorescence of **27** are compared with those of **27-o** and **27-c** (bottom, weak emission), respectively [73].

bleached to colorless, indicating reversion to the open form with strong fluorescence because of the PPV unit. The solution after UV-light irradiation, to generate the closed form, showed very weak fluorescence intensity because the absorption energy of the closed form is used for ring opening and nonradiative processes [76]. This indicates that fluorescence may be controlled by photochromic conversion of diarylethene [65, 76]. This type of photochromism/fluorescence switching in **28** could be applied to the development of nondestructive optical-patterning and pattern-recognizing systems, by exploiting the difference in fluorescence between the photochromic and nonphotochromic sites. The principle of fluorescence readout modulated by the photochromic reaction has been demonstrated in many recent papers and appears to work efficiently in dissolved solutions or when doped in

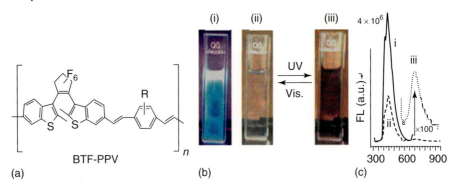

Figure 15.31 (a) Structure of BTF-PPV **(28)**. (b) Photographs for the solution of **28** (1×10^{-4} M in chloroform). (i) Under UV irradiation in a dark room, (ii) before UV irradiation under room light, and (iii) after UV irradiation under room light. (c) Fluorescence emission spectra before UV irradiation (excitation source for emission spectrum: (i) λ_{max} of the open form) and after UV irradiation (ii, iii) (excitation source for emission spectrum: λ_{max} of the closed form in the UV region (ii) and that in the visible region (iii)).

polymer matrices. However, in terms of patterning with fluorescent diarylethenes in high-density media, only a limited number of papers have dealt with neat amorphous films [77].

One of the difficulties in fluorescent photochromic patterning arises from the concentration quenching of the fluorescence emission. "Aggregation-induced enhanced emission (AIEE)" is one solution to this problem. Park and coworkers [78] have developed a polymeric fluorescent diarylethene that has AIEE properties (poly-(DCS-BTE, **29**) (Equation 15.13) and that provides reversible and bistable photochromism. The 1,4-bis(β-cyano-4′-methylstyryl)benzene (DCS) unit changes photoluminescence (PL) efficiency according to the fluorophore content (Figure 15.32a). When the solution or the film of **29** was irradiated with 290-nm light, BTE began to photoisomerize from open (**29-o**) to closed form (**29-c**), resulting in a new absorption band in the visible region (Figure 15.32b). Modulated by this photochromic ring closure, PL emission intensity from **29** significantly diminished, as shown in Figure 15.32c. In the PSS, additional intermolecular energy transfer occurs to further enhance fluorescence quenching, resulting in higher contrast in the film state compared to the solution state. The absorption and also the PL intensity were reversibly restored to open-form states, as shown in the inset graph in Figure 15.32c. Such a strongly enhanced fluorescence and photochromic switching in the film of **29** provides their possible application in efficient reversible photopatterning. By illumination of the **29** film through photomasks, a pattern was formed (Figure 15.33). The photochromic and fluorescent properties of the pattern were reversible and nondestructive, as confirmed from the repetitive patterning (290-nm light exposure) and erasing (550-nm light exposure) while monitoring excitation at 400 nm.

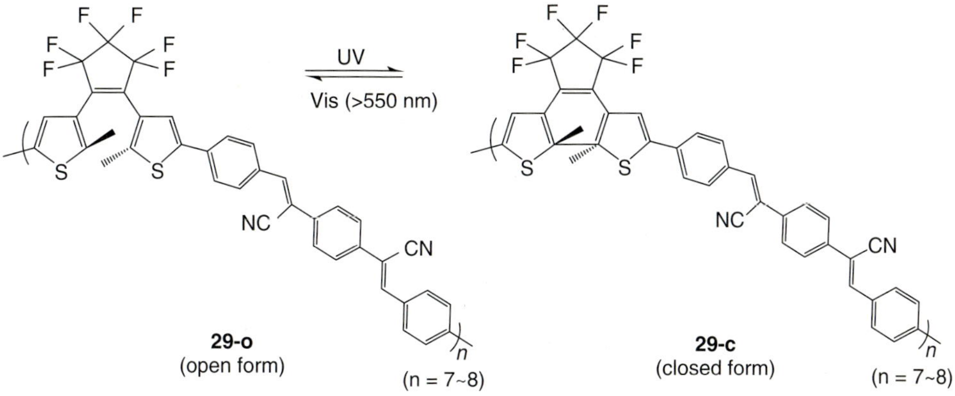

(15.13)

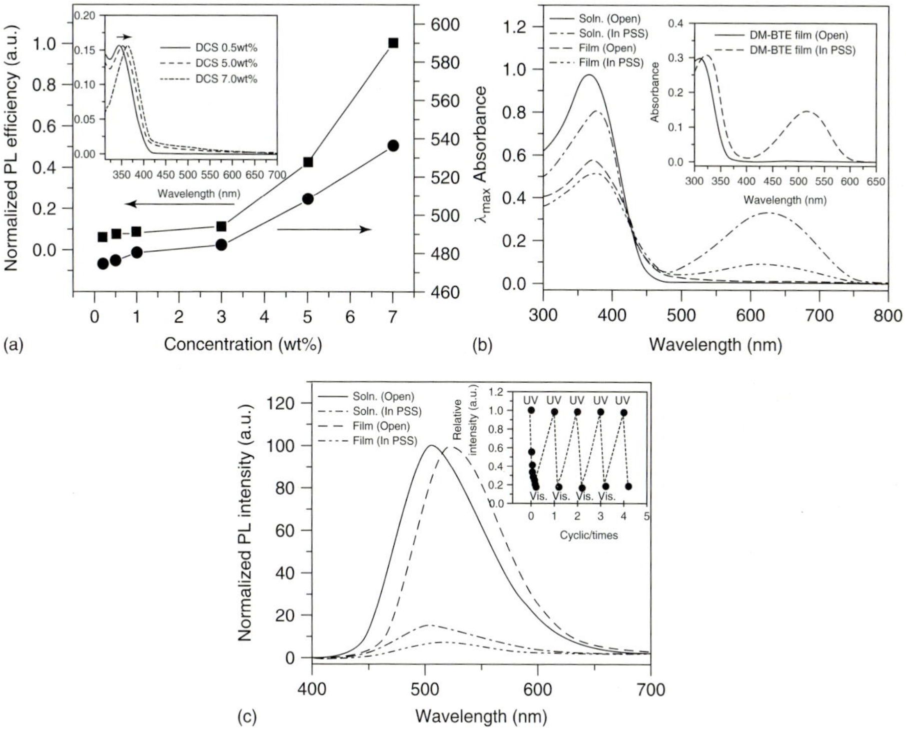

Figure 15.32 Normalized PL efficiency (■) and the wavelength of PL emission (●) in the spin-coated DCS/PMMA film (a) as a function of their concentration in PMMA matrix. Inset in (a) shows the bathochromic shift of UV-vis absorption in DCS/PMMA films with increasing concentration. UV-vis absorption (b) and PL spectra (c) of the **29-o** and at the 290 nm photostationary state (PSS). Inset in (b) shows the UV-vis absorption spectra of 6 wt% DM-BTE-loaded PMMA film, and inset in (c) shows the photochromic modulation of PL intensity in the **29** film.

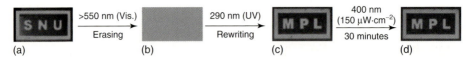

(a) (b) (c) (d)

Figure 15.33 Erasable fluorescence imaging on the spin-coated **29** film and its nondestructive readout capability. Writing (a), erasing (b), rewriting (c), and a continuous reading under irradiation with 400-nm light ($150 \mu W \, cm^{-2}$) for 30 min (d). The dark region represents the irradiated area with 290-nm UV light, and the real size of the photomasks is about 1 cm × 2 cm.

Therefore, to construct photochromic-fluorescent patterned film, self-quenching by coupling of fluorophores should be prevented. Introducing fluorophores having AIEE properties to a photochromic polymer can be a good strategy for realizing reversible photochromic and fluorescent photopatterning with high optical contrast.

The photoinduced electrochemical switching is one of the unique characteristics of diarylethenes that is not observed in other photochromic dyes such as spirobenzopyrans or fulgides. Reversible electrical properties of diarylethenes are induced by the open–closed isomerization by UV light. Since the π-electron delocalization is more extended in the closed form than in the open form, the diarylethene unit has been researched as a photochromic bridge for electrical switching. This π-electron delocalization can be further extended by introducing electrically active groups such as p-phenylenevinylenes [75, 79], thiophenes [80], and quinolines [81]. The electric conductivity change according to the open–close isomerization by light can be determined by measuring the *I–V* curve of the photocell, based on films containing photochromophores.

Kim and Lee [75] reported p-phenylenevinylene bridged diarylethene polymers, where BTF units are capable of reversibly manipulating the conjugation into the p-phenylenevinylene molecular wire through a polymer chain by irradiation at a selected wavelength. They attached a trimethyl silyl (TMS) group onto the p-phenylenevinylene ring (BTF-TPV) (**30**) for better solubility and film formation. The mechanism of reversible photoisomerization-induced electric conductivity change is described in Figure 15.34.

The polymer showed electrical conductivities of 2.5×10^{-8} and 3×10^{-9} S/cm for the colored cell (UV exposed, containing **30-c**) and the colorless cell (**30-o**), respectively (Figure 15.35a). This polymer has more efficient electrical connectivity than other photochromic polymers or photochromophore-dispersed polymer media [75]. Electrical modulation could be carried out optically using UV and visible-light sources with reversible patterning (Figure 15.35b).

15.5.2
Azo Compounds

Azo compounds are well-known photochromic materials that undergo *trans–cis* isomerization by UV light (Figure 15.28c). In contrast to diarylethene, azo compounds undergo reverse reactions thermally. Thus, the reverse reaction of *cis* to *trans* isomerization occurs by heat or visible light. The thermal reaction of azo

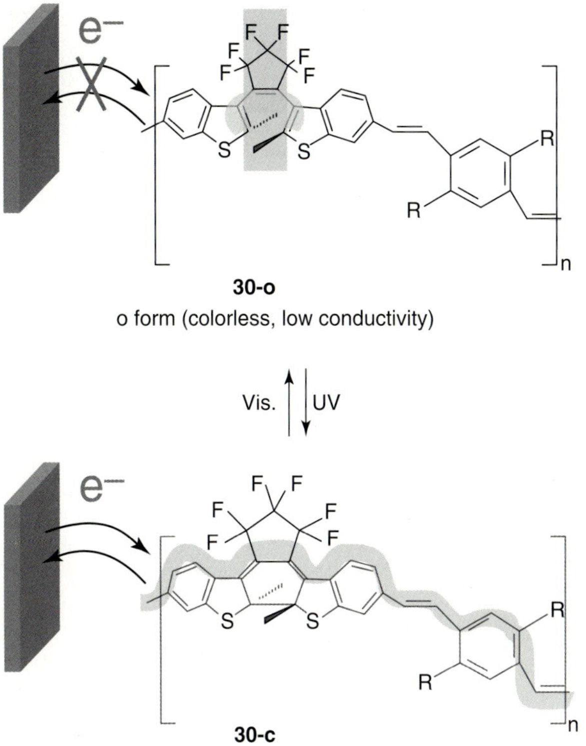

30-o

o form (colorless, low conductivity)

Vis. ↑↓ UV

30-c

c form (colored, high conductivity)

Figure 15.34 Reversible photoisomerization-induced conjugation length change of BTF-TPV (**30**).

compounds has been limited to the application of such isomerization reactions to organic devices that require long-term stability, such as in rewritable optical devices. Thus, thermal reversibility of azo polymers has been controlled by increasing the T_g of the polymer, to increase the stability of the pattern [81].

Fluorescent azobenzene dyes comprising fluorophores such as phenol, phenylphenol, and naphthol at sidechains have been introduced to show structural control for fine tuning of emission [82]. This study provides a new molecular design strategy for efficient fluorescent azobenzene molecules by suitably controlling the molecular geometry of the fluorophore attached to the azobenzene.

Fluorescence patterning was possible through photoinduced migration of a fluorescent squaraine group-functionalized azo derivative, **31** (Figure 15.36a), which has dynamic fluorescence and photoisomerization properties [83]. The photoresponse of polymethyl methacrylate (PMMA) thin films doped with **31** (1% w/w) created fluorescent gratings under a holographic recording setup equipped with a p-polarized Ar$^+$ laser beam (488 nm). A fluorescence confocal microscopic image showed fluorescent gratings with a 10% intensity contrast and a 1.3 μm periodic spacing (Figure 15.36b).

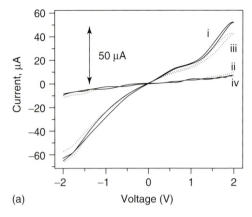

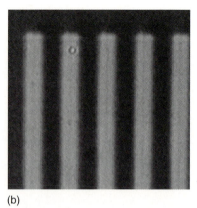

(a)   Voltage (V)   (b)

Figure 15.35 I–V plots of a cell containing a photochromic polymer film after UV- and visible-light exposure. (a) Neat **30** polymer: (i) as-prepared under UV irradiation; (ii) after irradiation with visible light for 30 min; (iii) second cycle, after UV irradiation for 30 min; and (iv) second cycle, after irradiation with visible light for 30 min, (b) A photograph of an electrode image written on a colored film of **30** through irradiation with visible light (532 nm laser), with the use of a mask. The dark area represents the masked (**30-c**) region, and the bright area the unmasked (**30-o**) region. The gap between the nearest two bright bars is 5 mm.

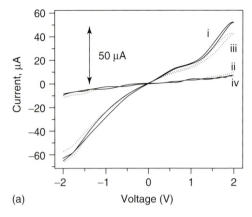

R= n-C$_4$H$_9$

31

(a)   (b)

Figure 15.36 (a) The structure of squaraine-functionalized azo benzene compound (**31**). (b) Confocal microscopic image of fluorescent gratings written in 1% w/w **31**-doped PMMA matrix ($\lambda_{exc} = 633$ nm, observed emission range, $\lambda_{obs} = 660–750$ nm).

Combination of conductive moieties such as thiophene [84], aniline [85], PPV [86], poly(p-phenyleneethynylene) [87] with azobenzene have been reported to control conductivity of the azopolymers by light. For example, copolymers of 3-hexylthiophene and the azobenzene moiety modified 3-hexylthiophene presented photocontrolled conductivity switching behavior, which originated from a generation of molecules photoexcited in the azobenzene moiety upon irradiation by UV light. However, most studies have dealt with limited strategies such as synthesis and electronic properties base on structural change. To our knowledge, reversible conductive patterning of polymer thin films based on conductive polymers containing

Table 15.4 Photoinduced reversible photopatterning.

Photoreactive group	Abs$_{max}$, nm (eV)	Light (nm, J/cm^2)^a	Pattern size (μm)	Application	Ref
29	360 (3.4)	U (290, UR^b) V(550, UR^b)	600	Erasable optical memory	[78]
30	410 (3.0)	U (325, UR^b) V (532,UR^b)	5	Organic electrode and electrochromic display	[75, 88]
31	491 (2.5)	Ar+ (488, 7.2)	~0.7	Optical data storage	[83]

^{a}All dry process, UV(U) and Vis(V) light for patterning and erasing, respectively.
^{b}Unreported.

an azobenzene moiety has been rarely introduced so far. Table 15.4 summarizes photoinduced reversible photopatterning.

15.5.3
Reversible Patterning by Photocross-Linking Followed by Reversible Reaction

Ionov *et al.* [89] reported on a chemical pattering based on stimulus responsive binary polymer brushes of two incompatible polymers showing different polarity, poly(2-vinylpyridine) (P2VP) and polyisoprene (PI). The thin film of the mixed polymer brush is exposed to a selective solvent and illuminated through a photomask, cross-linking the isoprene chains selectively in the illuminated areas. In the dark areas, the chains retain their capability of switching conformation and properties. The accordingly designed pattern can be developed by exposure to a solvent selective for one of the constitute polymer (PI or P2VP), which changes the chemical composition of the top layer in the dark areas as a result of phase segregation, but not that of the illuminated areas. Whenever the patterned mixed brush is exposed to a nonselective solvent, any contrast in the chemical composition of the top layer disappears and the image is erased. This process is reversible, and the pattern development and erasing can be repeated many times. Figure 15.37b shows the characteristic features of the patterned brush as result of interaction with different solvents. After photopatterning, the patterned brush is washed with ethanol, dried, and then exposed to water vapor. No pattern is developed on the surface (Figure 15.37bii). When treated again with acid (pH = 2) instead of ethanol, the image appeared on the brush (Figure 15.37bi) with the local condensation of water droplets. This patterning is original because water barely wets the surface of the illuminated areas (semispherical droplets), whereas water is more extensively spread over the surface of the dark areas. The image can be erased by washing with ethanol or neutral water (pH = 6.5).

Formation of regular arrays of microreactors and microchannels may be very important for analytical purposes. The lithography on mixed polymer brushes was used to trace quite complex arrays of microchannels, as shown in Figure 15.37c.

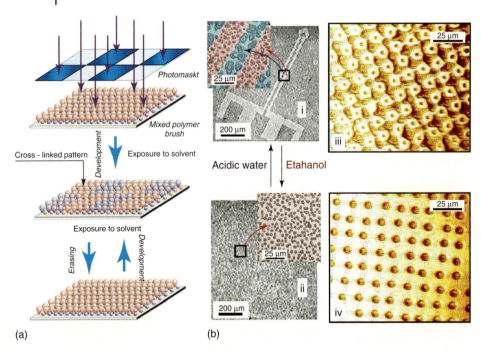

(a) (b)

Figure 15.37 (a) Scheme of photolithography on mixed polymer brushes. (b) Adsorption of water drops (optical microscopy) on the polymer brush with developed (i) and erased (ii) patterns. Insets demonstrate zoomed in details: hydrophilic and hydrophobic regions are artificially colored in blue and red, respectively. (c) Water-drop adsorption on the mixed brush patterned by UV irradiation through the grid: (iii) the grid was in direct contact with the brush; (iv) the grid was 100 µm away from the brush. Water drops are dark yellow. For convenience the grid is drawn with three parallel dashed lines indicating non-UV-irradiated areas.

A lattice of water droplets is formed in the case of irradiation through the mask, 100 µm away from the sample.

Recently, de Jong *et al.* [90] introduced light-driven dynamic pattern formation, which resulted in the spatial and temporal self-organization of the photochromic molecular material. Photochromic materials such as diarylethenes, azo molecules, and spiropyran molecules offer a convenient route to reversible change in not only molecular shape but also molecular properties [65].

15.5.4
Application of Reversible Patterning

The reversibility of the photoinduced electrical property allows the fabrication of patterned electrodes in a photon mode, without having to conduct an etching process or a wet process. By exposing the colorless or colored film to UV or visible light, respectively, an electrode image can be patterned on any substrate. An electrode image written on a colored film containing BTF-TPV(**30**) (10 wt% in polystyrene (PS)) was realized through irradiation with visible light (532 nm laser) with the use of a mask (Figure 15.35b) where the dark area represents the masked (**30-c**) highly conductive region, and the bright area the unmasked (**30-o**) resistive region. The electrode image was completely erased by exposing the electrode to UV or visible light, and was repatterned through a mask in the process described above.

This photon-mode electrode pattern generated from diarylethene photoreaction was utilized to control the growth of conducting polymers using **30** [88]. The pattern from **30** was simply obtained upon exposure to a UV light through a photomask and was photoerasable by visible light. As the **30** pattern has different electric conductivity according to the photoirradiation time or intensity, electrochemical polymerization of 1,4-bis(2-[3', 4'-ethylenedioxy]thienyl)-2-methoxy-5-2''-ethylhexyloxybenzene (BEDOT-MEHB) on the photopatterned film of **30** occurs selectively at the high conductive region (UV-exposed area), resulting an EC pattern.

30   **32**

Thus, the photopatterned film of **30** on an ITO glass was dipped into an electrolyte solution containing monomer BEDOT-MEHB, and the potential was scanned between −1.0 and 1.0 V *vs.* Ag/AgCl, resulting in a EC pattern. The deposited amount of conducting polymer (P(BEDOT-MEHB), **32**) after electrodeposition was different between the UV-exposed and unexposed spots, as shown in Figure 15.38.

Figure 15.38b shows tapping-mode AFM images for the film of **32** grown on different spots. Clearly, there is a qualitative difference in the morphology for the film of **32** grown in the area of **30** that was exposed to UV light (i) and that for the film of **32** grown on the bare ITO electrode (ii) or the area of **30** that was shaded (iii). The AFM image for the film of **32** grown in the area exposed to UV light (i) shows an irregular surface composed of small particles, and the AFM image for the film of **32** grown on the bare ITO is markedly rougher and shows larger particles that seem to have a broader size distribution. This could be due to the difference in current density during electropolymerization, as the photochromic layer is expected to have lower conductivity than that of the bare ITO.

To demonstrate a practical device with a prepatterned substrate, the authors fabricated a comblike electrode with a 50-μm gap using **30**. The photogenerated pattern was dipped into a solution of BEDOT-MEHB (same as above) and then exposed to three consecutive cycles in the potential range of −1 to 1 V. A digital photograph of the final image through an optical microscope is shown in

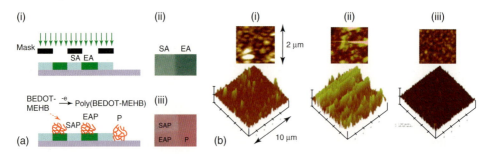

Figure 15.38 (a) Schematic illustration of the photo/electropatterning process of the conductive polymer layer (i). Digital photograph of the photopattern generated through UV exposure. SA and EA represent the UV-shaded and UV-exposed areas, respectively (ii). Digital photograph of the conductive polymer pattern generated through the electrochemical polymerization of 2 on the photochromic polymer film of (ii), with a dB of 59 nm (iii). SAP and EAP represent the electropolymerized area in the UV-shaded and UV-exposed areas, respectively. (b) Topography of the film of **32** in the UV-exposed area of **30** (EAP) (i), on the bare ITO (P) (ii), and in the UV-shaded area of **30** (SAP) (iii) as obtained through tapping-mode AFM.

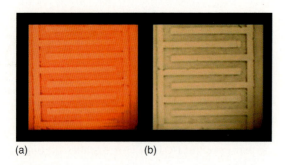

Figure 15.39 Digital photograph of an electrode pattern covered with **32** on top of the polymer **30** layer for (a) reduced and (b) oxidized state. The gap between the electrodes is 50 um (d_{32} = 40 nm, d_B = 59 nm).

Figure 15.39a,b for reduced and oxidized, respectively, in which the gap corresponds to 50 μm.

One of the applications of photochromic and fluorescent patterning is security applications. Kim and Kim [68] utilized a hologram media to pattern the photochromic and fluorescent diarylethene derivative, 1,2-bis[6-(3,4-ethylenedioxythienyl)-2-methyl-1-benzothiophen-3-yl]perfluorocyclopentene (BTFTT), **(33)**. Holograms record secure digitized data in two or three dimensions that cannot be easily copied. Holographic recording on a photochromic film affords an erasable 3D recording. **33** is colorless in visible light **(33-o)**, but it changes to dark blue upon exposure to a UV light source **(33-c)** as it undergoes photoisomerization (Equation 15.14) [71].

33-o
Colorless, fluorescent

33-c
Dark blue, non-fluorescent

(15.14)

The color of the photopolymer films containing photoreactive monomers and **33** was yellow–orange because of the color of the coumarin sensitizer. The photopolymer containing the chromophore gave a high diffraction efficiency and fast reactivity under a holographic recording system equipped with a 491-nm laser, ensuring that the holographic pattern could be efficiently recorded on the media. A hologram record is shown in Figure 15.40a. Interestingly, the fluorescence of the photopolymer film containing **33** is enhanced upon holographic recording. Thus, a fluorescent mark can be stored on the film and detected by a probe beam, as shown in Figure 15.40b. The fluorescent mark was erased upon exposure to a UV light, but recovered when exposed to visible light. Importantly, the film showed photochromism, as well as fluorescence switching, under a UV–vis light source. Thus, the film becomes dark under UV irradiation, hiding a patterned image, as shown in Figure 15.40c. The image is recovered when the film is exposed to a visible light. The photographs of the images and real objects (screw and letter "Y") are shown in Figure 15.40d. The holographic images were captured by a digital camera and could be detected under visible-light illumination. These results suggest that photopolymer film containing a photochromic fluorescent diarylethenes could be applied in multifunctional patterning process such as security media and anticounterfeiting systems.

During the photoisomerization of azobenzene, molecular motion occurs and thus topographic modification could be achieved from azo polymers. This topographic

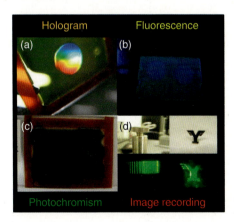

Figure 15.40 Detection of information recorded on the multifunctional photopolymer film containing **33** for security media.

modification of the surface was applied to dynamic holographic data storage through a direct and reversible topographic modification of the surface, resulting in the induction of surface relief gratings (SRGs) in conjunction with the light interference pattern [64]. This surface-relief grafting based on azo molecules provided potential opportunities for efficient polymer solar cells [91] and OLEDs [92].

Two different examples for practical application of the environment-responsive lithography are suggested. The first example demonstrates that this approach can be used for the fabrication of switchable microchannels to cause a valve to reversibly open or close upon an external stimulus. A switchable channel was prepared as shown in Figure 15.41. The mixed brush (1) was grafted between two hydrophilic channels (2) (Figure 15.41a) on a solid substrate. The switchable channel was fabricated (Figure 15.41b) by photocross-linking of the brush through the photomask (3) so that the irradiated areas (4) lose the switching properties and serve as walls of the channel. The bottom of the channel (6) is hydrophilic upon exposure to acidic solution and water can flow through the channel (Figures 15.41c,e). Upon heating above 80 °C or by changing pH (Figures 15.41d,f) the bottom surface of the channel switches into a hydrophobic state, closing the channel (5). This sample represents the general principle. Much more complicated systems can be fabricated using photolithography on mixed brushes.

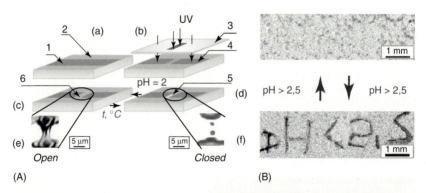

Figure 15.41 (A) Fabrication of a switchable channel employing environment-responsive lithography. (a) The mixed brush (1) was grafted between two hydrophilic channels (2) on a solid substrate. (b) The switchable channel was fabricated by photo-cross-linking of the brush through the photomask (3) so that the irradiated areas (4) lose the switching properties and serve as walls of the channel. (c) The bottom of the channel (6) is hydrophilic upon exposure to acidic solution and water can flow through the channel. (d) Upon heating above 80 °C or by changing of pH, the bottom surface of the channel (5) switches into hydrophobic state, closing the channel. The images (e) and (f) show open and closed states of the switchable channel, respectively, as they appear in optical microscopy. (B) Example of a smart sensor from the mixed brush grafted to Si wafer that displays the result of the analysis of acidic aqueous solution: the wafer was exposed to neutral water (top) and to water with pH 2.3 (bottom). The image appeared upon exposure to water vapor only if the sample had been treated with acidic water solution of pH < 2.5.

The second experiment presents fabrication of smart sensors that directly visualize the result of the test. For example, if the pattern is visualized at a particular pH, the smart surface can directly display the value of pH as an image (Figure 15.41B).

15.6
Conclusion

In this chapter, we focused on direct photopatterning either by irreversible or reversible photochemistry for functional polymer patterns. Direct photopatterning based on photocross-linking including photodimerization, photo-oxidation, and photoacidification remain the most promising methods for practical potential application. While the M_w change in the reactions allows lithographic patterning of the active polymer films through wet- or dry-etching processes, some examples of direct etchless photopatternings are developed based on the photoinduced property change other than the M_w change accompanied by the photoreactions. Thus, change in alignment or structural distortion, oxidative quenching, or polarity change are all important properties for etchless direct patterning process that pertains to the activity of the polymer. The irreversible photoreactions generate patterns that are neither correctable nor repatternable. Reversible patternings are highly desired as an effective basis for either environmentally friendly technologies or for smart surfaces. Thus, a novel challenge in the reversible functional polymer patterns has been proposed for a pioneering patterning method to manufacture a smart patterned surface.

Acknowledgment

We acknowledge the financial support of Seoul R&BD Program (10816), and the National Research Foundation (NRF) grant funded by the Korea government (MEST) through the Active Polymer Center for Pattern Integration (No. R11-2007-050-00000-0).

References

1. Kocher, C., Montali, A., Smith, P., and Weder, C. (2001) Patterning of oriented photofunctional polymer systems through selective photobleaching. *Adv. Funct. Mater.*, **11**, 31–35.

2. (a) Angelopoulos, M., Shaw, J.M., Kaplan, R., and Perreault, S.J. (1989) Conducting polyanilines: discharge layers for electron-beam lithography. *Vac. Sci. Technol.*, **B7**, 1519–1523;

(b) Angelopoulos, M. and Shaw, J.M. (1992) Conducting polymers as lithographic materials. *Polym. Eng. Sci.*, **32**, 153–158.

3. Holdcroft, S. (1997) *Handbook of Organic Conductive Molecules and Polymers*, vol. 4, John Wiley & Sons, Ltd, Chichester.

4. (a) Yu, J., Abley, M., Yang, C., and Holdcroft, S. (1998) Chemically amplified photolithography of a conjugated

polymer. *Chem. Commun.*, 1503–1504;
(b) Lowe, J. and Holdcroft, S. (1997)
Poly(3-(2-acetoxyethyl)thiophene): a
model polymer for acid-catalyzed lithography. *Synth. Met.*, **85**, 1427–1430; (c)
Yu, J. and Holdcroft, S. (2001) Chemically amplified soft lithography of a low
band gap polymer. *Chem. Commun.*,
1274–1275.

5. Kumar, A. and Whitesides, G. (1993)
Features of gold having micrometer to
centimeter dimensions can be formed
through a combination of stamping with
an elastomeric stamp and an alkanethiol
"ink" followed by chemical etching.
Appl. Phys. Lett., **63**, 2002–2004.

6. Garnier, F., Hadjlaoui, R., Yasser, A.,
and Srivastava, P. (1994) All-polymer
field-effect transistor realized by printing
techniques. *Science*, **265**, 1684–1686.

7. Bharathan, J. and Yang, Y. (1998)
Polymer electroluminescent devices
processed by inkjet printing: I. Polymer
light-emitting logo. *Appl. Phys. Lett.*, **72**,
2660–2662.

8. Wolk, M. and Baude, P. (2003) US
Patent 6 582, p. 876.

9. Lowe, J. and Holdcroft, S. (1995)
Synthesis and photolithography of
polymers and copolymers based
on poly(3-(2-(methacryloyloxy)ethyl)
thiophene). *Macromolecules*, **28**,
4608–4616.

10. Chittibabu, K.G., Li, L., Kamath, M.,
Kummar, J., and Tripathy, S.K. (1994)
Synthesis and properties of a novel poly-
thiophene derivative with a side-chain
NLO chromophore. *Chem. Mater.*, **6**,
475–480.

11. Cai, S.X., Keana, J.F.W., Nabity, J.C.,
and Wybourne, M.N. (1991) Conducting
polymers as deep-UV and electron beam
resists. Direct production of micrometer scale conducting structures from
poly(3-octylthiophene). *J. Mol. Electron.*,
7, 63–68.

12. Yun, C., You, J., Kim, J., Huh, J., and
Kim, E. (2009) Photochromic fluorescence switching from diarylethenes and
its applications. *J. Photochem. Photobiol. C: Photochem. Rev.*, **10**, 111–129.

13. Deng, X. and Wong, K.Y. (2009)
Cross-Linked conjugated polymers
for achieving patterned three-color and

blue polymer light-emitting diodes with
multi-layer structures. *Macromol. Rapid
Commun.*, **30**, 1570–1576.

14. Ahn, D.J. and Kim, J.M. (2008) Fluorogenic polydiacetylene supramolecules:
immobilization, micropatterning and
application to label-free chemosensors.
Acc. Chem. Res., **41**, 805–816.

15. (a) Schanze, K.S., Bergstedt, T.S.,
and Hauser, B.T. (1996) Photolithographic patterning of electroactive
polymer films and electrochemically modulated optical diffraction
gratings. *Adv. Mater.*, **8**, 531–534;
(b) Schanze, K.S., Bergstedt, T.S.,
Hauser, B.T., and Cavalaheiro, C.S.P.
(2000) Photolithographically-patterned
electroactive films and electrochemically modulated diffraction gratings.
Langmuir, **16**, 795–810.

16. Kim, J., Kim, Y., and Kim, E. (2009)
Electrochromic pattern formation by
photo cross-linking reaction of PE-
DOT side chains. *Macromol. Res.*, **17**,
791–796.

17. Winther-Jensen, B. and West, K.
(2004) Vapor-phase polymerization
of 3,4-ethylenedioxythiophene: a route
to highly conducting polymer surface
layers. *Macromolecules*, **37**, 4538–4543.

18. Kim, J., You, J., and Kim, E. (2010)
Flexible conductive polymer patterns
from vapor polymerizable and photo
cross-linkable EDOT. *Macromolecules*, **43**,
2322–2327.

19. Nielsen, C.B., Angerhofer, A., Abboud,
K.A., and Reynolds, J.R. (2008) Discrete
photopatternable π-conjugated oligomers
for electrochromic devices. *J. Am. Chem.
Soc.*, **130**, 9734–9746.

20. (a) Dunkin, I.R. and Thomson, P.C.P.
(1982) Pentafluorophenyl nitrene: a
matrix isolated aryl nitrene that does
not undergo ring expansion. *J. Chem.
Soc. Chem. Commun.*, 1192–1193;
(b) Meijer, E.W., Nijhuis, S., and
van Vroonhoven, F.C.B.M. (1988)
Poly-1,2-azepines by the photopolymerization of phenyl azides. Precursors
for conducting polymer films. *J. Am.
Chem. Soc.*, **110**, 7209–7210; (c) Poe, R.,
Schnapp, K.A., Young, M.J.T., Grayzar,
J., and Platz, M.S. (1992) Chemistry

and kinetics of singlet pentafluo-
rophenylnitrene. *J. Am. Chem. Soc.*, **114**,
5054–5067; (d) Schnapp, K.A., Poe, R.,
Leyva, E., Soundararajan, N., and Platz,
M.S. (1993) Exploratory photochemistry
of fluorinated aryl azides. Implications
for the design of photoaffinity labeling
reagents. *Bioconjug. Chem.*, **4**, 172–177.

21. (a) Scriven, E.F.V. (1984) *Azides and
    Nitrenes*, Academic, San Diego, CA;
    (b) Bräse, S., Gil, C., Knepper, K., and
    Zimmermann, V. (2005) Organic azides:
    an exploding diversity of a unique class
    of compounds. *Angew. Chem. Int. Ed.*,
    44, 5188–5240.

22. Khong, S.-H., Sivaramakrishnan,
    S., Png, R.-Q., Wong, L.-Y., Chia,
    P.-J., Chua, L.-L., and Ho, P.K.H.
    (2007) General photo-patterning of
    polyelectrolyte thin films via effi-
    cient ionic bis(Fluorinated Phenyl
    Azide) photo-crosslinkers and their
    post-deposition modification. *Adv. Funct.
    Mater.*, **17**, 2490–2499.

23. O'Donnell, M. (1968) Photo-dimerization
    of solid anthracene. *Nature*, **218**,
    460–461.

24. Rameshbabu, K., Kim, Y., Kwon, T.,
    Yoo, J., and Kim, E. (2007) Facile
    one-pot synthesis of a photo pattern-
    able anthracene polymer. *Tetrahedron
    Lett.*, **48**, 4755–4760.

25. Sinigersky, V., Müllen, K., Klapper,
    M., and Schopov, I. (2000) Photostruc-
    turing and consecutive doping of an
    anthracene-containing polymer: a new
    approach towards conductive patterns.
    Adv. Mater., **12**, 1058–1060.

26. Li, T., Chen, J., Mitsuishi, M., and
    Miyashita, T. (2003) Photolitho-
    graphic properties of ultrathin polymer
    Langmuir–Blodgett films containing
    anthracene moieties. *J. Mater. Chem.*,
    13, 1565–1569.

27. Müller, C.D., Falcou, A., Reckefuss, N.,
    Rojahn, M., Wiederhirn, V., Rudati, P.,
    Frohne, H., Nuyken, O., Becker, H.,
    and Meerholz, K. (2003) Multi-colour or-
    ganic light-emitting displays by solution
    processing. *Nature*, **421**, 829–833.

28. Chien, H.-W., Chang, T.-Y., and Tsai,
    W.-B. (2009) Spatial control of cellular

adhesion using photo-crosslinked micropatterned polyelectrolyte multilayer
films. *Biomaterials*, **30**, 2209–2218.

29. Cyr, P.W., Rider, D.A., Kulbaba, K., and
    Manners, I. (2004) Photopatternable
    metallopolymers: photocrosslinking and
    photolithography of polyferrocenylsi-
    lane methacrylates. *Macromolecules*, **37**,
    3959–3961.

30. Nielsen, C.B., Angerhofer, A., Abboud,
    K.A., and Reynolds, J.R. (2008) Discrete
    photopatternable π-conjugated oligomers
    for electrochromic devices. *J. Am. Chem.
    Soc.*, **130**, 9734–9746.

31. Kim, K.-W., Choi, H., Lee, G.S., Ahn,
    D.J., Oh, M.-K., and Kim, J.-M. (2006)
    Micro-patterned polydiacetylene vesicle
    chips for detecting protein-protein inter-
    actions. *Macromol. Res.*, **14**, 483–485.

32. Schanze, K.S., Bergstedt, T.S., and
    Hauser, B.T. (1996) Photolithographic
    patterning of electroactive polymer films
    and electrochemically modulated opti-
    cal diffraction gratings. *Adv. Mater.*, **8**,
    531–534.

33. Holdcroft, S. (1991) A photochemi-
    cal study of poly(3-hexylthiophene).
    Macromolecules, **24**, 4834–4838.

34. Yan, M., Rothberg, L.J.,
    Papadimitrakopoulos, F., Galvin,
    M.E., and Miller, T.M. (1994) Defect
    quenching of conjugated polymer lumi-
    nescence. *Phys. Rev. Lett.*, **73**, 744–747.

35. Scurlock, R.D., Wang, B., Ogilby, P.R.,
    Sheats, J.R., and Clough, R.L. (1995)
    Singlet oxygen as a reactive intermediate
    in the photodegradation of an electrolu-
    minescent polymer. *J. Am. Chem. Soc.*,
    117, 10194–10202.

36. Yoshino, K., Kuwabara, T., Iwasa, T.,
    Kawai, T., and Onoda, M. (1990) Optical
    recording utilizing conducting poly-
    mer, poly(p-phenylenevinylene) and its
    derivatives. *Jpn. J. Appl. Phys.*, **29** (Part
    2), L1514–L1516.

37. Yoo, J., Kwon, T., Sarwade, B.D., Kim,
    Y., and Kim, E. (2007) Multistate fluo-
    rescence switching of s-triazine-bridged
    p-phenylenevinylene polymers. *Appl.
    Phys. Lett.*, **91**, 241107–241109.

38. You, J., Heo, J.S., Lee, J., Kim, H.-S.,
    Kim, H.O., and Kim, E. (2009) A flu-
    orescent polymer for patterning of

mesenchymal stem cells. *Macromolecules,* **42**, 3326–3332.

39. Buchgraber, C., Spanring, J., Pogantsch, A., Turner, M., and Kern, W. (2007) Organosilanes as new reagents for the photopatterning of PPV type polymers. *Synth. Met.,* **147**, 91–95.

40. (a) DeAro, J.A., Gupta, R., Heeger, A.J., and Buratto, S.K. (1999) Nanoscale oxidative patterning and manipulation of conjugated polymer thin films. *Synth. Met.,* **102**, 865–868; (b) Rothbergay, L.J., Yan, M., Papadimitrakopoulos, F., Galvin, M.E., Kwock, E.W., and Miller, T.M. (1996) Photophysics of phenylenevinylene polymers. *Synth. Met.,* **80**, 41–58.

41. Wei, P.K., Hsu, J.H., Hsieh, B.R., and Fann, W.S. (1996) Surface modification and patterning of conjugated polymers with near-field optical microscopy. *Adv. Mater.,* **8**, 573–576.

42. Marciniak, S., Crispin, X., Uvdal, K., Trzcinski, M., Birgerson, J., Groenendaal, L., Louwet, F., and Salaneck, W.R. (2004) Light induced damage in poly(3,4-ethylenedioxythiophene) and its derivatives studied by photoelectron spectroscopy. *Synth. Met.,* **141**, 67–73.

43. (a) Tada, K. and Onoda, M. (1999) Photoinduced modification of photoluminescent and electroluminescent properties in poly(p-phenylene vinylene) derivative. *J. Appl. Phys.,* **86**, 3134–3139; (b) Tada, K. and Onoda, M. (2001) Photooxidation mechanism of polymer light emitting device and its application to optically patternable device. *Synth. Met.,* **121**, 1653–1654.

44. Kim, Y., Yun, C., Jadhav, P., You, J., and Kim, E. (2009) Emissive pattern formation by the photoreaction of poly(p-phenylene vinylene). *Curr. Appl. Phys.,* **9**, 1088–1092.

45. (a) Singhvi, R., Kumar, A., Lopez, G., Stephanopoulos, G., Wang, D., Whitesides, G., and Ingber, D. (1994) Engineering cell shape and function. *Science,* **264**, 696–698; (b) Mrksich, M., Chen, C., Xia, Y., Dike, L., Ingber, D., and Whitesides, G. (1996) Controlled cell attachment on contoured surfaces

with self-assembled monolayers of alkanethiolates on gold. *Proc. Natl. Acad. Sci. U.S.A.,* **93**, 10775–10778; (c) Ostuni, E., Chapman, R., Liang, M., Meluleni, G., Pier, G., Ingber, D., and Whitesides, G. (2001) Self-assembled monolayers that resist the adsorption of proteins and the adhesion of bacterial and mammalian cells. *Langmuir,* **17**, 6336–6343; (d) Yousaf, M., Houseman, B., and Mrksich, M. (2001) Using electroactive substrates to pattern the attachment of two different cell populations. *Proc. Natl. Acad. Sci. U.S.A.,* **98**, 5992–5996; (e) Guo, L., Kawazoe, N., Fan, Y., Ito, Y., Tanaka, J., Tateishi, T., Zhang, X., and Chen, G. (2008) Chondrogenic differentiation of human mesenchymal stem cells on photoreactive polymer-modified surfaces. *Biomaterials,* **29**, 23–32.

46. Pogantsch, A., Rentenberger, S., Langer, G., Keplinger, J., Kern, W., and Zojer, E. (2005) Tuning the electroluminescence color in polymer light-emitting devices using the thiol-ene photoreaction. *Adv. Funct. Mater.,* **15**, 403–409.

47. Krebs, F.C. and Spanggaard, H. (2005) Patterning of oriented photofunctional polymer systems through selective photobleaching. *Synth. Met.,* **148**, 53–59.

48. Cheng, X., Hong, Y.T., Kanicki, J., and Guo, L.J. (2002) High-resolution organic polymer light-emitting pixels fabricated by imprinting technique. *J. Vac. Sci. Technol. B,* **20**, 2877–2880.

49. Deng, X.Y., Wonga, K.Y., and Mo, Y.Q. (2007) Three-color polymeric light-emitting devices using selective photo oxidation of multilayered conjugated polymers. *Appl. Phys. Lett.,* **90**, 063505-1–063505-3.

50. Ito, H. (1997) Chemical amplification resists: history and development within IBM. *IBM J. Res. Dev.,* **41**, 69–80.

51. (a) Zhang, C., Vekselman, A.M., and Darling, G.D. (1995) Relief and functional photoimaging with chemically amplified resists based on di-tert-butyl butenedioate-co-styrene. *Chem. Mater.,* **7**, 850–855; (b) Schilling, M., Katz, H.E., Houlihan, F.M., Kometani, J.M., Stein, S.M., and Nalamasu, O. (1995) Photogenerated acid-catalyzed formation

of phosphonic/phosphoric acids by deprotection of esters in polymer films. *Macromolecules*, **28**, 110–115.

52. (a) Kim, J.-M., Chang, T.-E., Kang, J.-H., Han, D.K., and Ahn, K.-D. (1999) Synthesis of and fluorescent imaging with a polymer having t-BOC-protected quinizarin moieties. *Adv. Mater.*, **11**, 1499–1502; (b) Lee, C.-W., Yuan, Z., Ahn, K.-D., and Lee, S.-H. (2002) Color and fluorescent imaging of t-BOC-protected quinizarin methacrylate polymers. *Chem. Mater.*, **14**, 4572–4575.

53. Kim, J.-M., Min, S.J., Lee, S.W., Bok, J.H., and Kim, J.S. (2005) An excimer emission approach for patterned fluorescent imaging. *Chem. Commun.*, 3427–3429.

54. Kim, J.M., Chang, T.E., Kang, J.H., Park, K.H., Han, D.K., and Ahn, K.D. (2000) Photoacid-induced fluorescence quenching: a new strategy for fluorescent imaging in polymer film. *Angew. Chem. Int. Ed.*, **39**, 1780–1782.

55. Do, J., Kim, Y., Attias, A.-J., Kreher, D., and Kim, E. (2010) Patterning of pH sensitive fluorescent bipyridazine derivatives, *J. Nanosci. Nanotech.*, **10**, 6874–6878.

56. Pistolis, G., Boyatzis, S., Chatzichristidi, M., and Argitis, P. (2002) Highly efficient bicolor (green-blue) fluorescence imaging in polymeric films. *Chem. Mater.*, **14**, 790–796.

57. Ito, H. (2005) Chemical amplification resists for microlithography. *Adv. Polym. Sci.*, **172**, 37–245.

58. Lee, C.-W., Seo, Y.-H., and Lee, S.-H. (2004) A soluble polyaniline substituted with t-BOC: conducting patterns and doping. *Macromolecules*, **37**, 4070–4074.

59. Salavagione, H.J., Miras, M.C., and Barbero, C. (2006) Photolithographic patterning of a conductive polymer using a polymeric photoacid generator and a traceless removable group. *Macromol. Rapid Commun.*, **27**, 26–30.

60. Zakhidov, A.A., Lee, J.-K., Fong, H.H., DeFranco, J.A., Chatzichristidi, M., Taylor, P.G., Ober, C.K., and Malliaras, G.G. (2008) Hydrofluoroethers as orthogonal solvents for the chemical processing of organic electronic materials. *Adv. Mater.*, **20**, 3481–3484.

61. Lee, J.-K., Chatzichristidi, M., Zakhidov, A.A., Taylor, P.G., DeFranco, J.A., Hwang, H.S., Fong, H.H., Holmes, A.B., Malliaras, G.G., and Ober, C.K. (2008) Acid-sensitive semiperfluoroalkyl resorcinarene: an imaging material for organic electronics. *J. Am. Chem. Soc.*, **130**, 11564–11565.

62. Lim, Y.-F., Lee, J.-K., Zakhidov, A.A., DeFranco, J.A., Fong, H.H., Taylor, P.G., Ober, C.K., and Malliaras, G.G. (2009) High voltage polymer solar cell patterned with photolithography. *J. Mater. Chem.*, **19**, 5394–5397.

63. Jiang, J., Qi, B., Lepage, M., and Zhao, Y. (2007) Polymer micelles stabilization on demand through reversible photo-cross-linking. *Macromolecules*, **40**, 790–792.

64. Natansohn, A. and Rochon, P. (2002) Photoinduced motions in azo-containing polymers. *Chem. Rev.*, **102**, 4139–4176.

65. Matsuda, K. and Irie, M. (2004) Diarylethene as a photoswitching unit. *J. Photochem. Photobiol. C: Photochem. Rev.*, **5**, 169–182.

66. Cho, S.Y., Yoo, M., Shin, H.-W., Ahn, K.-H., Kim, Y.-R., and Kim, E. (2002) Preparation of diarylethene copolymers and their photoinduced refractive index change. *Opt. Mater.*, **21**, 279–284.

67. Pu, S., Tang, H., Chen, B., Xu, J., and Huang, W. (2006) Photochromic diarylethene for two-photon 3D optical storage. *Mater. Lett.*, **60**, 3553–3557.

68. Kim, J. and Kim, E. (2008) Holographic security media prepared from photochromic fluorescent films. *Proc. SPIE*, **7118**, 71180F1–71180F10.

69. (a) Kawata, S. and Kawata, Y. (2000) Three-dimensional optical data storage using photo chromic materials. *Chem. Rev.*, **100**, 1777–1788; (b) Corredor, C.C., Huang, Z.-L., and Belfield, K.D. (2006) Two-photon 3D optical data storage via fluorescence modulation of an efficient fluorene dye by a photochromic diarylethene. *Adv. Mater.*, **18**, 2910–2914; (c) Lee, H.W., Kim, Y.M., Jeon, D.J., Kim, E., Kim, J., and Park, K. (2002) Rewritable organic films for near-field recording. *Opt. Mater.*, **21**, 289–293; (d) Kim, J., Song, K.-B., Park, K.-H., Lee, H.W., and Kim, E.

(2002) Near-field optical recording of photochromic materials using bent cantilever fiber probes. _Jpn. J. Appl. Phys._, **41**, 5222–5225; (e) Kim, E., Choi, Y.-K., and Lee, M.-H. (1999) Photoinduced refractive index change of a photochromic diarylethene polymer. _Macromolecules_, **32**, 4855–4860.

70. Jeong, Y.-C., Yang, S.I., Kim, E., and Ahn, K.-H. (2006) A high-content diarylethene photochromic polymer for an efficient fluorescence modulation. _Macromol. Rapid Commun._, **27**, 1769–1773.

71. Lee, J., Kwon, T., and Kim, E. (2007) Electropolymerization of an EDOT-modified diarylethene. _Tetrahedron Lett._, **48**, 249–354.

72. Kim, E. and Cho, S.Y. (2002) Photochromic polymer films prepared by photocuring of fluoroalkyldiacrylate and diarylethene derivatives. _Mol. Cryst. Liq. Cryst._, **377**, 385–390.

73. (a) Jeong, Y., Park, D., Kim, E., Ahn, K., and Yang, S. (2006) Fatigue-resistant photochromic dithienylethenes by controlling the oxidation state. _Chem. Commun._, 1881–1883; (b) Jeong, Y., Yang, S., Kim, E., and Ahn, K. (2006) Fatigue-resistant photochromic dithienylethenes by controlling the oxidation state. _Tetrahedron_, **62**, 5855–5861.

74. Cho, H. and Kim, E. (2002) Highly fluorescent and photochromic diarylethene oligomer bridged by p-phenylenevinylene. _Macromolecules_, **35**, 8684–8687.

75. Kim, E. and Lee, H.W. (2006) Photo-induced electrical switching through a mainchain polymer. _J. Mater. Chem._, **16**, 1384–1389.

76. Youn, C. and Kim, E. (2008) Photochromic diarylethene materials. _News Inform. Chem. Eng._, **26**, 526–533.

77. Kim, M.-S., Kawai, T., and Irie, M. (2002) Fluorescence modulation in photochromic amorphous diarylethenes. _Opt. Mater._, **21**, 271–274.

78. Lim, S.-J., An, B.-K., and Park, S.Y. (2005) Bistable photoswitching in the film of fluorescent photochromic polymer: enhanced fluorescence emission and its high contrast switching. _Macromolecules_, **38**, 6236–6239.

79. Wigglesworth, T.J., Myles, A.J., and Branda, N.R. (2005) High-content photochromic polymers based on dithienylethenes. _Eur. J. Org. Chem._, **2005**, 1233–1238.

80. (a) Kim, E., Kim, M., and Kim, K. (2006) Diarylethenes with intramolecular donor–acceptor structures for photo-induced electrochemical change. _Tetrahedron_, **62**, 6814–6821; (b) Kwon, T. and Kim, E. (2008) Photoelectric organic nano-thin films prepared from diarylethene. _Curr. Appl. Phys._, **8**, 739–741.

81. Wang, Z.-P., Bo, S.-H., Zhao, W.-K., Liu, J.-L., Wang, J.-S., Qiu, L., Zhen, Z., and Liu, X.-H. (2007) Thermal stable retro-crosslinking polymers based on "Diels-Alder" reaction which containing a new AZO chromophore. _Ganguang Kexue Yu Guang Huaxue_, **25**, 327–332.

82. Smitha, P. and Asha, S.K. (2007) Structure control for fine tuning fluorescence emission from side-chain azobenzene polymers. _J. Phys. Chem. B_, **111**, 6364–6373.

83. Liu, L.-H., Nakatani, K., Pansu, R.T., Vachon, J.-J., Tauc, P., and Ishow, E. (2007) Fluorescence patterning through photoinduced migration of squaraine-functionalized azo derivatives. _Adv. Mater._, **19**, 433–436.

84. (a) Chen, S.-A. and Liao, C.-S. (1993) Photoelectric organic nano-thin films prepared from diarylethene. _Makromol. Chem. Rapid Commun._, **14**, 69–75; (b) Jousselme, B., Blanchard, P., Allain, M., Levillain, E., Dias, M., and Roncali, J. (2006) Structural control of the electronic properties of photodynamic azobenzene-derivatized π-conjugated oligothiophenes. _J. Phys. Chem. A_, **110**, 3488–3494.

85. Huang, K., Qiu, H., and Wan, M. (2002) Synthesis of highly conducting polyaniline with photochromic azobenzene side groups. _Macromolecules_, **35**, 8653–8655.

86. Matsui, T., Nagata, T., Ozaki, M., Fujii, A., Onoda, M., Teraguchi, M., Masuda, T., and Yoshino, K. (2001) Novel properties of conducting polymers containing azobenzene moieties in side chain. _Synth. Met._, **119**, 599–600.

87. Izumi, A., Nomura, R., and Masuda, T. (2001) Design and synthesis of stimuli-responsive conjugated polymers having azobenzene units in the main chain. *Macromolecules*, **34**, 4342–4347.

88. Kim, Y. and Kim, E. (2006) Conductive polymer patterning on a photoswitching polymer layer. *Macromol. Res.*, **14**, 584–587.

89. Ionov, L., Minko, S., Stamm, M., Gohy, J.-F., Jérôme, R., and Scholl, A. (2003) Reversible chemical patterning on stimuli-responsive polymer film: environment-responsive lithography. *J. Am. Chem. Soc.*, **125**, 8302–8306.

90. de Jong, J.J.D., Hania, P.R., Pugžlys, A., Lucas, L.N., de Loos, M., Kellogg, R.M., Feringa, B.L., Duppen, K., and van Esch, J.H. (2005) Light-driven dynamic pattern formation. *Angew. Chem. Int. Ed.*, **44**, 2373–2376.

91. Na, S.-I., Kim, S.-S., Jo, J., Oh, S.-H., Kim, J., and Kim, D.-Y. (2008) Efficient polymer solar cells with surface relief gratings fabricated by simple soft lithography. *Adv. Funct. Mater.*, **18**, 3956–3963.

92. Hubert, C., Fiorini-Debuisschert, C., Hassiaoui, I., Rocha, L., Raimond, P., and Nunzi, J.-M. (2005) Emission properties of an organic light-emitting diode patterned by a photoinduced autostructuration process. *Appl. Phys. Lett.*, **87**, 191105-1–19191105-3.

16
Nanopatterning of Photosensitive Polymer Films

Zouheir Sekkat, Hidekazu Ishitobi, Mamoru Tanabe, Tsunemi Hiramatsu, and Satoshi Kawata

16.1
Introduction

Photoinduced patterns of surface deformations in azobenzene-containing polymer films have attracted much attention because of possible applications in optical data storage and in micro/nanofabrication. It is well known that such patterns reflect the state of the incident-light polarization and the light intensity distribution [1–5]. *Trans* ↔ *cis* photoselective isomerization and molecular reorientation play important roles in the deformation process. Since photoisomerization was shown to enhance molecular mobility far below the glass-transition temperature (T_g) of azo-polymers in the beginning of the past decade [6–9], considerable exploration of sub-T_g photoinduced molecular movement was performed, especially targeting polymer structural effects, including T_g, the free volume and free-volume distribution, the mode of the attachment of the chromophore to a rigid or flexible chain, and the molecular weight [10]. Light-induced mass movement in azo-polymers, that is, surface-relief gratings (SRGs) [8, 9] triggered many studies to understand the mechanism of polymer migration, and most of the studies have focused on SRGs that are fabricated by the interference pattern of two coherent laser beams [11–14]. Yet, there are a few reports on surface deformations that are induced by a single focused laser beam [15–17].

To fabricate deformation structures with high spatial resolution, a small irradiation spot is required; a feature that can be achieved by focusing the laser beam by using a high numerical aperture (NA) objective lens. In this chapter, we highlight patterning of azo-polymer films with a resolution approaching 200 nm by irradiation with a single tightly focused laser beam with a high NA objective lens (NA = 1.4). We discuss the effect of the incident-light polarization and the position of the laser focus on the deformation pattern. In particular, we found that the deformation pattern is strongly dependent on the z-position of the focused laser spot. In addition to the well known *trans* ↔ *cis* surface deformation, it will be shown that a gradient force parenting to laser trapping pulls the polymer toward the laser focus. Then, we present a systematic study exploring the limits of the

Functional Polymer Films, First Edition. Edited by Wolfgang Knoll and Rigoberto C. Advincula.
© 2011 Wiley-VCH Verlag GmbH & Co. KGaA. Published 2011 by Wiley-VCH Verlag GmbH & Co. KGaA.

size of photoinduced deformation by changing the irradiation intensity and the exposure time. Two-photon patterning of the film samples will also be discussed.

To further reduce the size of the photoinduced surface deformation down to a few tens of nanometers, a near-field optical microscope with an aperture type probe, that is, tapered optical-fiber probe, can be used [18–20]. Near-field optics can confine photons to several tens of nanometers near the probe tip end and induce nanoscale deformation on the film surface. An apertureless type probe, that is, metal-coated tip that principally generates a smaller light spot than the aperture type probe has also been used [5, 21–23]. There are few reports on the mechanism of the near-field surface deformation because efforts were mainly devoted to making smaller fine structures on the surface of polymer films for nanofabrication. Recent reports showed surface deformations that are induced in an azo-polymer film by a combination of optical far- and near-field components of the irradiation light with a high irradiation intensity (typically, 30 kW/cm^2) [4]. In addition, irradiation with high laser intensities bleaches the azo dyes and produce deformation patterns that are different from those of photoinduced mass movement; a feature that adds complexity to the formation and understanding of the mechanism of the optical near-field-induced surface deformation.

In this chapter, we also report on patterning of azo-polymer films by tip-enhanced near-field, in which the incident laser power was chosen to induce the deformation by the near-field component only, wherein the far-field component had no contribution to the observed surface deformation. We compared the induced deformation patterns with and without the tip, and found that a nanoprotrusion was induced when the tip was inside the laser focus (vide infra). We also found that the deformation pattern was dependent on the incident-light polarization, a feature that implies the presence of not only the electric field parallel to the tip axis (E_z) but also the one nearer to incident-light polarization (E_x) under the tip end. We will go on to discuss near-field and two-photon nanofabrication.

16.2
Nanofabrication by a Tightly Focused Laser Beam

We prepared 100-nm thin films of poly(Disperse Red 1 methacrylate) (PMA-DR1, product No. 579009, Aldrich; $T_g = 82\,^\circ C$) by spin casting from a chloroform solution. The remaining solvent was removed by heating the films for an hour at 100 $^\circ$C. The chemical structure and the absorption spectrum of the film, that is, *trans*-DR1 are shown in Figure 16.1. Disperse Red 1 (DR1) is a nonlinear optical azo dye that is well known for its *trans* ↔ *cis* photoisomerization and for its ability to undergo efficient orientation and trigger important polymer movement when it is excited by polarized light [1]. The orientation effect is due to the highly anisometric nature of its polarizability tensor (rod-like molecule) [24]. The irradiation light source was a linearly polarized 460-nm light from a diode-pumped frequency-doubled laser (Sapphire 460 LP, Coherent Japan). The wavelength of the irradiation laser corresponds to the maximum absorption band of the film

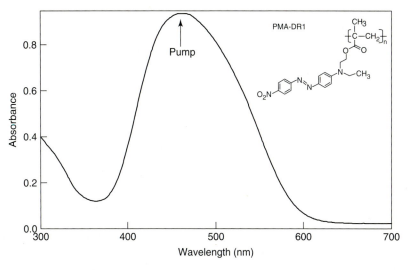

Figure 16.1 Chemical structure (inset) and absorption spectrum of the *trans*-DR1-PMA thin film. The wavelength of excitation is indicated.

sample. The laser beam was focused by an objective lens (NA = 1.4) (Plan Apo 60×, Nikon). The diffraction-limited spot diameters in the lateral (X or Y) and longitudinal (Z) axes are ~400 nm and 1.0 μm, respectively. Computer-controlled piezo stages (P-517 for X- and Y-axes and P-721 for Z-axis, Physik Instruments (PI)) were used to control the position of the focused laser spot in three dimensions. The induced surface deformation of the films was measured by an atomic force microscope (AFM) (SPA-400, SEIKO Instruments Inc.). The AFM was operated in the tapping mode using a Si cantilever to eliminate the mechanical deformation of the films by the cantilever itself.

Figure 16.2 shows AFM images of the surface deformation induced by (a,b) linear and (c) circular polarizations. The polarizations were controlled by half- and quarter-wave plates. The irradiation intensity and the exposure time were 12.5 mW/cm^2 and 30 s, respectively, and the laser beam was focused on the film surface. Irradiation with linearly polarized light induced the deformation pattern shown in Figures 16.2a,b. It is clearly shown in this figure that the polymer moved along the polarization direction from the center to the outside of the focused spot, thus producing two-sided lobes along the polarization direction and a pit at the center. Indeed, this polarization-dependent deformation was confirmed by an experiment in which the polarization direction of the irradiation light was rotated through an angle of 90° and the induced pattern followed the polarization of the light (see Figure 16.2a versus b). In contrast to irradiation with linear polarization, irradiation with circularly polarized light induced a deformation pattern in which the polymer moved from the center to the outside of the focused laser spot, thus forming a doughnut shape pattern (Figure 16.2c). For both linear and circular polarizations, the polymer migrates in the direction of the light gradient from high

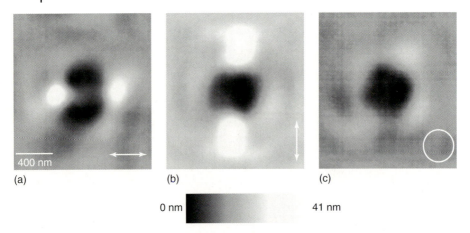

(a)   (b)   (c)

0 nm   41 nm

Figure 16.2 AFM images of the deformation induced by a tightly focused laser beam polarized (a) horizontally, (b) vertically, and (c) circularly, respectively.

to low light intensity regions, and the polarization dependence demonstrates that the light-induced polymer movement is anisotropically photofluidic [5].

The observed polarization dependence is consistent with the one obtained after irradiation with a low NA lens [15]. When a laser beam is tightly focused by a high NA objective lens, a non negligible electric field E_z component is created along the optical axis. The intensity distributions corresponding to E_x, E_y, and E_z at the focal position are different, and shown in Figure 16.3, and they should lead to different deformation patterns. However, in our experimental conditions, only E_x contributes appreciably to the deformation. With NA = 1.4 and wavelength = 460 nm, the maximum intensity corresponding to E_z and E_y are 7 and 200 times smaller than that of E_x, respectively.

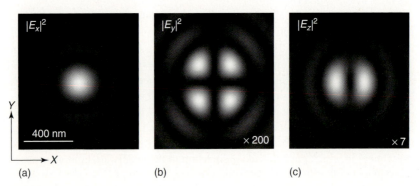

(a)   (b)   (c)

Figure 16.3 Calculated distributions of squared electric-field components created by a tightly focused linearly polarized laser beam. The components of the electric field of (a) E_x, (b) E_y, and (c) E_z are shown. The polarization direction is X, Z is perpendicular to the film, and Y is perpendicular to both X and Z. The distribution was calculated assuming a refractive index of the surrounding medium equal to 1.5.

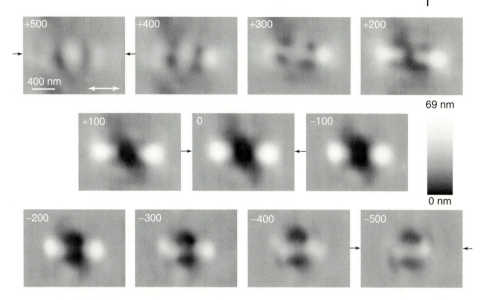

Figure 16.4 AFM images of the surface deformation induced by changing the Z-position of the focused laser spot. The Z-position was varied from −500 to +500 nm with an interval of 100 nm. The values inside each figure represent the Z-position of the focus (unit is nanometers). The polarization direction is indicated.

Figure 16.4 shows AFM images of the photoinduced deformation that have been obtained by changing the Z-position of the focused laser spot. The Z-position was controlled by the z-axis piezo stage that was attached to the objective lens. The irradiation started 500 nm under the film surface ($Z = -500$ nm), then the Z-position was moved to upper positions with an interval of 100 nm, and the next irradiation was done at a different lateral ($X - Y$) position. This procedure was repeated until the Z-position reached the 500 nm upper film surface ($Z = +500$ nm). For each irradiation, the irradiation intensity and the exposure time were 12.5 mW/cm^2 and 60 s, respectively. When the Z-position was just on the film surface ($Z = 0$ nm), the deformation pattern was the same as the one shown in Figures 16.2a,b. It is interesting to note that at distances larger than 200 nm above the film surface in air, the polymer formed a protrusion coming out toward the center of the laser focus and suggesting the existence of a gradient force [25] that pulls the polymer toward the region of maximum intensity (see Figure 16.5). This is optical trapping of a viscoelastic polymer showing nanoelasticity over 20 nm; that is, the maximum height of the protrusion obtained at $z = +500$ nm. For distances between 200 and 0 nm, the overlap of the laser intensity and the film are large enough to produce dips at the center as explained above. When the laser is focused into the glass substrate, there is no protrusion formed, because the polymer movement is blocked by the substrate.

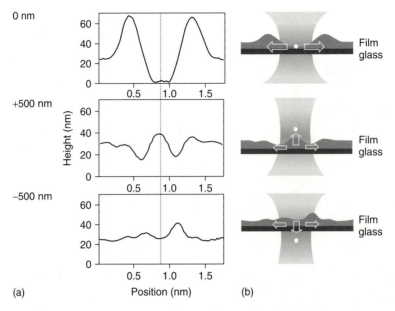

Figure 16.5 (a) Line plots of the surface deformation at $Z = +500$ nm, $Z = 0$ nm, and $Z = -500$ nm. The positions of each plot correspond to the directions that are between the arrows indicated in Figure 16.4. (b) Schematics describing the relationship between the Z-position of the focus and the film surface. The arrows oriented laterally and longitudinally in these schematics indicate the direction of the anisotropic photofluidity and the optical gradient force, respectively.

In a systematic set of experiments, we studied the dependence of the size of the photoinduced deformation on the intensity of the irradiation light and the exposure time. In those experiments, the irradiation light is linearly polarized and focused by a 1.4 NA objective lens. The deformation pattern was studied just at the laser focus, that is, $Z = 0$ nm, for three irradiation intensities (6.25, 12.5, 62.5 mW/cm²), and exposure times according to the series 1–500 s for 6.25 and 12.5 mW/cm², and 1–100 s for 62.5 mW/cm². The deformation patterns obtained at all intensities at all times of irradiation were the same, but the size of the deformation was different. Figure 16.6 shows the dependence of the height and the full width at half-maximum (FWHM) of the deformation pattern along the direction parallel to the light polarization on the irradiation intensity and the exposure time. The *height* is defined as the difference between the top of the side lobes and the bottom of the central pit as shown in Figure 16.6. As can be seen from this figure, the rate of the deformation of the height and FWHM decreased with the increasing irradiation dose, and the higher the irradiation intensity, the faster the increase of both the height and FWHM. The height increases more rapidly than FWHM that needs more time to reach saturation. The height increases rapidly at small irradiation doses, and saturates at larger irradiation doses near 90 nm, a value that corresponds to the film thickness. The minimum FWHM of the fabricated pattern is about 200 nm; a value that corresponds to the size of the diffraction limited laser spot.

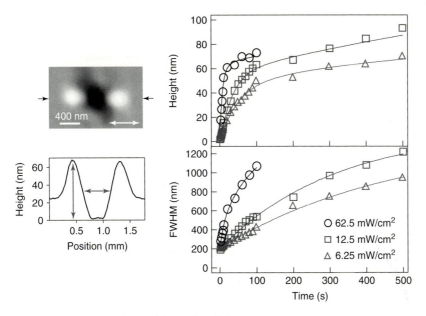

Figure 16.6 Size dependence of the surface deformation on the irradiation intensity and the exposure time. A typical deformation pattern and the corresponding line plot are shown in the left of the figure and the definitions of the height and FWHM are indicated. The points are experimental data, and solid lines are exponential empirical theoretical fits.

16.3
Near-Field Nanofabrication beyond the Diffraction Limit of Light

We prepared 50-nm thin films of PMA-DR1 as explained in Section 16.2, and the irradiation light source and the focused beam were as explained in Section 16.2 as well. An AFM tip (Silicon Cantilever CSG01, NT-MDT) was covered by silver with a vacuum evaporator. The diameter of the tip end was found to be 30 nm by using a scanning electron microscope (see Figure 16.7). The tip was brought close to the diffraction-limited focused spot on the film by an AFM (Bioprobe, Park Scientific Instruments) operating in a contact mode, and the enhanced near-field with a 30-nm diameter spot was generated in the vicinity of the tip via localized surface plasmons (see schematic in Figure 16.8). In this configuration, the tip was permanently in contact with the film surface. The irradiation light intensity was $3 \, mW/cm^2$ (versus $30 \, kW/cm^2$ in Ref. [4]). As explained in the previous section, a computer-controlled piezo stage was used to control the position of the laser spot, and the induced pattern was observed by an AFM that was operating in the tapping mode.

Figure 16.9 shows the AFM image of the surface deformation that was induced with and without the silver-coated tip. The irradiation intensity and the exposure time were $3 \, mW/cm^2$ and 10 s, respectively; an irradiation dose that permitted

Figure 16.7 SEM image of the silver-coated AFM tip.

100 nm

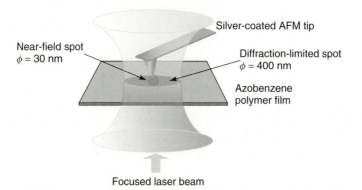

Near-field spot
$\phi = 30$ nm

Silver-coated AFM tip

Diffraction-limited spot
$\phi = 400$ nm

Azobenzene
polymer film

Focused laser beam

Figure 16.8 Schematic of the near-field surface-deformation configuration.

fabrication by only the near-field component of the light (*vide infra*). It can be clearly seen from Figure 16.9 that the protrusion was induced only with the tip present inside the focused spot, and no deformation was induced when the tip was away from it. The height of the protrusion was found to be 7 nm, and the lateral size was less than the diffraction-limited spot size, indicating that the surface deformation was induced by only the near-field component of the light.

It is well known that the component of the electric field parallel to the tip axis (E_z) is effectively enhanced near the tip end due to the local surface plasmons [26]. The polymer under the tip is pulled by the optical gradient force that is generated by E_z

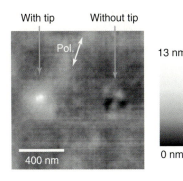

With tip   Without tip

Pol.

400 nm

13 nm

0 nm

Figure 16.9 AFM image of the surface deformation induced by irradiation with and without the silver-coated AFM tip. The polarization direction and the scale bar are indicated.

[25, 27] and formed the nanoprotrusion. This finding is in good agreement with the experimental results that were recently reported by us, in which we studied nanoscale polymer movement induced by a tightly focused laser beam, and we found that the deformation pattern was strongly dependent on the longitudinal focus position of the laser beam along the optical axis whereby the film is pulled by the optical gradient force, which is due to the focused beam, toward the focus when the beam is focused in air just on top of the sample surface [28]. In the present optical setup, the nanolight spot was placed just onto the film surface, and it was kept on it while the protrusion was being formed during irradiation. Here too, the film was pulled by the optical gradient force toward the nanosource of light; for example, the tip end, much like the case of a tightly focused beam on top of the film (*vide infra*). The field enhancement effect due to the metal tip allowed for the fabrication of near-field surface features with a light intensity that is as low as $3\,mW/cm^2$.

Figure 16.10 shows an expanded view of the surface deformation induced by the near-field irradiation (left side of Figure 16.9). It is found that the long axis of the deformation pattern was along a direction that is, not quite, but near the parallel to the incident-light polarization. This uniaxial anisotropic polymer movement suggests that the polymer moved along the polarization direction. The full width of half-maximum of the protrusion was found to be 65 and 47 nm in the direction nearly parallel (//) and perpendicular ($\perp$) to the incident-light polarization, respectively. Here too, the anisotropic photofluidity of the polymer tends to induce a polymer mass movement in the direction parallel to the incident-light polarization [5, 27]. This anisotropic polymer movement implies the presence under the tip end of a component of the electric field that is nearly parallel to the polarization direction [29]. The azo-polymer helps map the electric field just at the tip apex. In fact, the fabricated feature which is due to E_x is much smaller than the one due to E_z and it confirms the well-known phenomenon that only the longitudinal component E_z drives efficient oscillations of the surface plasmons polaritons at the metal tip and leads to a strong field enhancement, that is, enhancement of E_z, at the tip apex [26]. Two-photon patterning of azo-polymers is discussed next.

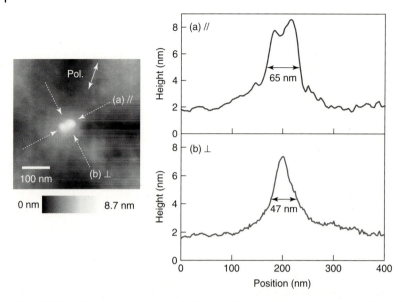

Figure 16.10 AFM image of the near-field surface deformation (expanded view of the left side of the Figure 16.9). The polarization direction and the scale bar are indicated. (a) and (b) show the line profiles in the directions near-parallel and near-perpendicular to the incident-light polarization, respectively. The FWHMs of the surface deformation in the parallel and perpendicular directions are also indicated.

16.4
Two-Photon Patterning

Recently, two-photon isomerization and reorientation of azo dyes in polymers have attracted much attention because of the scientific and technological extension into nanophotonics [30–46]. Basically, the photoreaction can be induced by tightly focused lasers into confined volumes – a resolution of 120 nm has been achieved for three-dimensional nanofabrication in photopolymerizable resins [47], and novel scientific information may be obtained, especially with regard to one versus two- or multiphoton reaction pathways – the one and two-photon transition dipole moments of a diarylethene derivative have been found to be perpendicular to each other in two-photon isomerization experiments [30]. In, this chapter, we show that azo-polymer films, are patterned by two-photon isomerization as well [48], and we discuss the effect of the incident-light polarization and the position of the laser focus (Z-position of the focus) on the deformation patterns induced by two-photon absorption. We discuss the mechanism of the induced surface deformation for two-photon versus one-photon absorption. The effect of photobleaching is also discussed by studying the effect of the irradiation wavelength, that is, 780- versus 920-nm irradiation, on the induced patterns.

We prepared 150-nm thin films of PMA-DR1 by spin coating as explained earlier. The films were irradiated by a linearly polarized near-infrared light from a Ti:sapphire laser (Spectral Physics, Mai Tai, pulse width = 130 fs; repetition

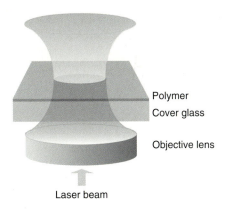

Figure 16.11 Schematic of the experimental setup.

Polymer

Cover glass

Objective lens

Laser beam

rate $= 80$ MHz) to induce patterns via a two-photon absorption process. The laser light was focused by an objective lens (NA $= 0.55$) from the bottom of a cover glass (see the schematic in Figure 16.11). The diffraction-limited spot diameter in the lateral and longitudinal axes are 2 and 12 μm, respectively. A computer-controlled piezo stage (Melles Griot, Nanoblock), on which the sample was placed, was used to control the position of the irradiation spot with respect to the sample surface. The surface topology of the deformed films was measured by an AFM (SEIKO Instruments Inc., SPA-400). The AFM was operated in the tapping mode.

Figure 16.12 shows an AFM image of the surface deformation induced by a focused, linearly polarized, 920-nm irradiation (irradiation intensity $= 61$ kW/cm²;

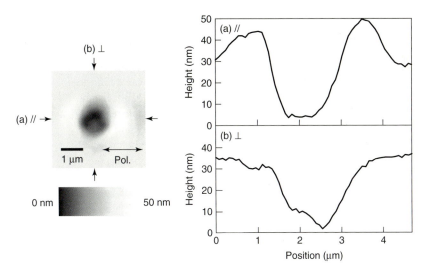

Figure 16.12 AFM images of the deformation induced by 920-nm light when the Z-position of the focused laser spot is +2 μm. The polarization direction is indicated by the arrow. The line plots (a) parallel and (b) perpendicular to the light polarization are shown. The positions of each plot correspond to the directions that are between the arrows indicated.

and exposure time $= 60$ s). This figure corresponds to a spot center 2 μm above the film surface in air. A similar behavior is observed when the center of the spot is 1 μm above and just at the film surface (*vide infra*). It can be clearly seen from Figure 16.12 that the polymer moves along the polarization direction from the center to the outside of the focused spot, thus producing two-sided lobes along the polarization direction and a pit at the center. This can also be seen at the AFM topographic line scans along and perpendicular to the direction of light polarization. This observed behavior, which is due to a photonic effect that is polarization dependent, and not to heat deposition, is much like that observed with single-photon isomerization (*vide infra*). At half the 920-nm fundamental wavelength, that is, 460 nm, the sample presents a strong absorption, suggesting that two-photon isomerization is the origin of the surface deformation.

The effect of dye photobleaching on the induced patterns was investigated by changing the irradiation wavelength of the laser light to 780 nm to confirm that the surface patterns induced by 920-nm irradiation were in fact due to two-photon isomerization and not to bleaching of the dye. Indeed, in a recent report, we showed that DR1 undergoes bleaching under irradiation at 780 nm, at the same range of light intensity, that is, GW/cm^2, via a multiphoton absorption process [33]. Furthermore, Figure 16.13 shows that the surface deformation induced by

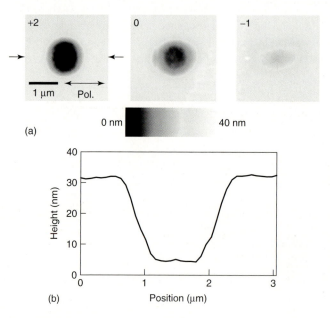

Figure 16.13 (a) AFM images of the surface deformation induced by 780-nm light by changing the Z-position of the focused laser spot. The Z-position was varied from -5 to $+5$ μm with an interval of 1 μm. Only images at $+2, 0, -1$ μm are shown. The values inside each figure represent the z-position of the focus (unit is μm). The polarization direction is indicated by the arrow. (b) Line plots of the surface deformations at $Z = +2$ μm. The position of the plot corresponds to the directions that are between the arrows indicated.

780-nm irradiation at the same light intensity as that of 920-nm irradiation formed dips at the center of focus, and the patterns are independent of the incident-light polarization. The same behavior was observed for all Z-positions of the focus. Only those at $+2\,\mu m$, and at the film surface, that is, $0\,\mu m$, and $-1\,\mu m$ are shown, and no deformation was observed when the Z-positions of the spot was smaller than $-2\,\mu m$. This finding confirms that the surface deformations shown on Figure 16.13 were induced by multiphoton bleaching of DR1. Note that while an ablated polymer surface by a high-intensity pulsed laser looks rough under AFM imaging, due to the deposition of fragments of the ablated polymer at the film surface [49], the AFM images of Figure 16.13 are smooth outside the irradiated area and reinforce photobleaching versus ablation as a mechanism of the dip formation.

Figure 16.14 shows the AFM images of the photoinduced patterns observed by changing the Z-position of the focused laser spot with irradiation at 920 nm. This irradiation, as well as that done at 780 nm that is discussed in the previous paragraph, started $5\,\mu m$ above the film surface ($Z = +5\,\mu m$) in air, then the Z-position was moved to lower positions, that is, toward the film, with an interval of $1\,\mu m$, and the next irradiation was done at a different lateral ($X - Y$) position. This procedure was repeated until the Z-position reached $-5\,\mu m$ ($Z = -5\,\mu m$), that is, in the other side of the film, inside the glass slide. For each irradiation, the irradiation intensity and the exposure time were 61 kW/cm^2 (peak intensity $= 5.9$ GW/cm^2) and 60 s, respectively.

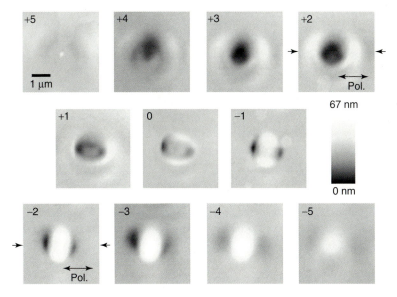

Figure 16.14 AFM images of the surface deformation induced at 920 nm light by changing the Z-position of the focused laser spot. The Z-position was varied from -5 to $+5\,\mu m$ with an interval of $1\,\mu m$. The values inside each figure represent the Z-position of the focus (unit is μm). The polarization direction is indicated by an arrow.

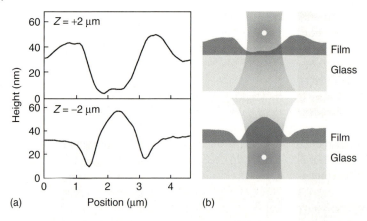

(a)　　　　　　Position (μm)　　　　(b)

Figure 16.15 (a) Line plots of the surface deformations at $Z = +2$ μm and $Z = -2$ μm. The positions of the each plot correspond to the directions that are between the arrows indicated in Figure 16.13. (b) Schematics describing the relationship between the Z-position of the focus and the film surface.

Figure 16.14 shows that when the Z-position of the focus is above the film surface in air ($Z > 0$ μm), the polymer moved along the polarization direction from the center to the outside of the focused spot, thus producing two-sided lobes along the polarization direction and a pit at the center (*vide supra*). A somewhat contrasting behavior is observed when the Z-position of the focus is below the film surface ($Z < 0$ μm), that is, the polymer moves from the outside to the center of the focused laser spot always along the direction of the polarization of the excitation light and formed a protrusion at the center and two dips, one at each side of the protrusion along the polarization direction (see also Figure 16.15). When the distance between the focus and sample surface is larger than 5 μm ($Z > 5$ μm, $Z < -5$ μm), no deformation is induced because the intensity of the light spot at the film is not large enough to induce a pattern.

At a low NA objective lens, that is, NA $= 0.55$ in the present experiments, the dominant component of electric field at the focus is the one along the polarization direction (E_x) when the incident light is linearly polarized along X. Indeed, electromagnetic calculations that we performed for the intensity distribution at 5 nm below the air ($n = 1.0$)/polymer ($n = 1.5$) interface when the beam focus is at $+2$ and -2 μm, show that the Y and Z components of the fourth power of the optical field are 4 and 5 orders of magnitude smaller that the fourth power of E_x, respectively (see Figure 16.16). So, during two-photon patterning, polymer movement proceeds in the direction of the light polarization (X) (*vide supra*).

Figure 16.16 also shows that the field distribution is nearly unchanged when the z-position of the focus is at $+2$ μm (in air) and -2 μm (in the cover glass). Since the field distribution in $X - Y$ plane is quasi-identical from both sides of the polymer film, that is, the laser is independently focused at the same distance from the film in air and glass, the presence of an intensity gradient along the

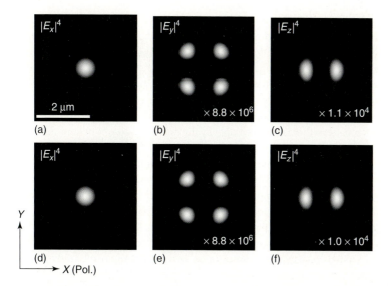

Figure 16.16 Calculated field distributions of the fourth power of the electric-field components created by a focused linearly polarized laser beam in the presence of the interface between air ($n = 1.0$) and polymer ($n = 1.5$). The distributions were calculated in the $X - Y$ inplane that is located at 5 nm below the interface. The upper and lower figures show the field distributions when the Z-position of the focus is $+2$ and -2 μm, respectively. Each components of electric field of (a,d) E_x, (b,e) E_y, and (c,f) E_z are shown. The polarization direction is X, and Z is perpendicular to the film surface, and Y is perpendicular to both X and Z.

Z-axis leads to the formation of the observed deformation. Note that when the same experiments were performed with one-photon absorption at 460 nm on the same polymer (see, e.g., Section 16.2), an optical gradient force, which is proportional to the Z-intensity gradient, was observed to be attractive, that is, a protrusion was formed when the focus was in the air and not in the glass. It is interesting to note that this behavior is opposite to that observed with two-photon isomerization. We do not yet have a clear explanation for this phenomenon, and more experiments are necessary in order to assess how the intensity gradient contributes to the formation of the patterns induced by one and two-photon isomerization. Systematic studies on irradiation wavelength and doses, and so on, are also needed in order to clarify the mechanism of the formation of the patterns, as well as for assessing the limit of the spatial resolution of two-photon surface nanofabrication on azo-polymers by a single tightly focused laser, especially at a high NA objective lens.

16.5
Conclusion

Both one- and two-photon absorption induced isomerization and patterns in azo-polymers. In both one- and two-photon cases, the polymer moves in the direction of the irradiation-light polarization. We also found that the induced surface pattern strongly depends on the Z-position of the laser focus suggesting a contribution, to the observed patterns, of a light intensity gradient along the Z-axis, that is, the direction of the propagation of the laser. We showed that nanomovement of photosensitive polymers occurs by polarization-sensitive photoisomerization at the nanoscale, and we studied polymer nanomovement induced by a tightly focused laser beam and a metal-tip-enhanced near-field irradiation. In particular, we observed a pattern that is induced only by the near-field component of the light with a resolution beyond the diffraction limit of light by controlling the irradiation light intensity. The optical gradient force that is generated by a strong E_z pulls the polymer toward the tip end, and anisotropic nanofluidity generated by the relatively small E_x moves the polymer along the polarization direction in the film plane. Future work with near-field nanofabrication on azo-polymers should focus on systematic studies at different irradiation doses, polarization states, and irradiation configurations with respect to the tip axis. Additional experimental and theoretical studies are needed to assess the contribution of the light intensity gradient and radiation forces and torques to the formation of patterns by one and two-photon isomerization [50–54].

References

1. Sekkat, Z. and Knoll, W. (2002) *Photoreactive Organic Thin Films*, Academic Press.
2. Koyayashi, K., Egami, C., and Kawata, Y. (2003) Optical storage media with dye-doped minute sphere on polymer films. *Opt. Rev.*, **10**, 262–266.
3. Hubert, C., Rumyantseva, A., Lerondel, G., Grand, J., Kostcheev, S., Billot, L., Vial, A., Bachelot, R., Royer, P., Chang, S.-H., Gray, S.K., Wiederrecht, G.P., and Schatz, G.C. (2005) Near-field photochemical imaging of noble metal nanostructures. *Nano Lett.*, **5**, 615–619.
4. Gilbert, Y., Bachelot, R., Vial, A., Lerondel, G., Royer, P., Bouhelier, A., and Wiederrecht, G.P. (2005) Photoresponsive polymers for topographic simulation of the optical near-field of a nanometer sized gold tip in a highly focused laser beam. *Opt. Exp.*, **13**, 3619–3624.
5. Karageorgiev, P., Neher, D., Schulz, B., Stiller, B., Pietsch, U., Giersig, M., and Brehmer, L. (2005) From anisotropic photo-fluidity towards nanomanipulation in the optical near-field. *Nature Mater.*, **4**, 699–703.
6. Sekkat, Z. and Dumont, M. (1992) Photoassisted poling of azo dye doped polymeric films at room temperature. *Appl. Phys. B*, **54**, 486–489.
7. Charra, F., Kajzar, F., Nunzi, J.M., Raimond, P., and Idiart, E. (1993) Light-induced second-harmonic generation in azo-dye polymers. *Opt. Lett.*, **18**, 941–943.
8. Rochon, P., Batalla, E., and Natansohn, A. (1995) Optically induced surface gratings on azoaromatic polymer films. *Appl. Phys. Lett.*, **66**, 136–138.
9. Kim, D.Y., Tripathy, S.K., Li, L., and Kumar, J. (1995) Laser-induced holographic surface relief gratings on nonlinear optical polymer films. *Appl. Phys. Lett.*, **66**, 1166–1168.
10. Sekkat, Z., Yasumatsu, D., and Kawata, S. (2002) Pure photoorientation of azo dye in polyurethanes and quantification of orientation of spectrally overlapping isomers. *J. Phys. Chem. B*, **106**, 12407–12417.
11. Lefin, P., Fiorini, C., and Nunzi, J.M. (1998) Anisotropy of the photoinduced translation diffusion of azo-dyes. *Opt. Mater.*, **9**, 323–328.

12. Labarthet, F.L., Bruneel, J.L., Buffeteau, T., and Sourisseau, C. (2004) Chromophore orientations upon irradiation in gratings inscribed on azo-dye polymer films: a combined AFM and confocal Raman microscopic study. *J. Phys. Chem. B*, **108**, 6949.

13. Natansohn, A. and Rochon, P. (2002) *Photoreactive Organic Thin Films*, Chapter 13, Academic Press, pp. 400–423, and references therein.

14. Oliveira, O.N., Li, J., Kumar, L., and Tripathy, S. (2002) *Photoreactive Organic Thin Films*, Chapter 14, Academic Press, pp. 430–480, and references therein.

15. Bian, S., Williams, J.M., Kim, D.Y., Lin, L., Balasubramanian, S., Kumar, J., and Tripathy, S. (1999) Photoinduced surface deformations on azobenzene polymer films. *J. Appl. Phys.*, **86**, 4498–4508.

16. Gilbert, Y., Bachelot, R., Royer, P., Bouhelier, A., Wiederrecht, G.P., and Novotny, L. (2006) Longitudinal anisotropy of the photoinduced molecular migration in azobenzene polymer films. *Opt. Lett.*, **31**, 613–615.

17. Grosjean, T. and Courjon, D. (2006) Photopolymers as vectorial sensors of the electric field. *Opt. Exp.*, **14**, 2203–2210.

18. Davy, S. and Spajer, M. (1996) Near filed optics: Snapshot of the field emitted by a nanosource using a photosensitive polymer. *Appl. Phys. Lett.*, **69**, 3306–3308.

19. Fukuda, T., Sumaru, K., Kimura, T., Matsuda, H., Narita, Y., Inoue, T., and Sato, F. (2001) Observation of optical near-field as photo-induced surface relief formation. *Jpn. J. Appl. Phys.*, **40**, L900–L902.

20. Patanè, S., Arena, A., Allegrini, M., Andreozzi, L., Faetti, M., and Giordano, M. (2002) Near-field optical writing on azo-polymethacrylate spin-coated films. *Opt. Commun.*, **210**, 37–41.

21. Iwata, F., Kobayashi, K., Sasaki, A., Kawata, Y., Egami, C., Sugihara, O., Tuchimori, M., and Watanabe, O. (2002) Nanometre-scale modification of a urethane-urea copolymer film using local field enhancement at an apex of a metal coated probe. *Nanotechnology*, **13**, 138–142.

22. Bachelot, R., H'Dhili, F., Barchiesi, D., Lerondel, G., Fikri, R., Royer, P., Landraud, N., Peretti, J., Chaput, F., Lampel, G., Boilot, J.-P., and Lahlil, K. (2003) Apertureless near-field optical microscopy: a study of the local tip field enhancement using photosensitive azobenzene-containing films. *J. Appl. Phys.*, **94**, 2060–2072.

23. Gillbert, Y., Fikri, R., Ruymantseva, A., Lerondel, G., Bachelot, R., Barchiesi, D., and Royer, P. (2004) High-resolution nanophotolithography in atomic force microscopy contact mode. *Macromolecules*, **37**, 3780–3791.

24. Sekkat, Z., Wood, J., Knoll, W., Volksen, W., and Miller, R.D. (1996) Light-induced orientation in a high glass transition temperature polyimide with polar azo dyes in the side chain. *J. Opt. Soc. Am. B*, **13**, 1713–1724.

25. Ashkin, A., Dziedzic, J.M., Bjorkholm, J.E., and Chu, S. (1986) Observation of a single-beam gradient force optical trap for dielectric particles. *Opt. Lett.*, **11**, 288–290.

26. Inouye, Y. and Kawata, S. (1994) Near-field scanning optical microscope with a metallic probe tip. *Opt. Lett.*, **19**, 159–161.

27. Novotny, L., Bian, R.X., and Xie, X.S. (1997) Theory of nanometric optical tweezers. *Phys. Rev. Lett.*, **79**, 645–648.

28. Ishitobi, H., Tanabe, M., Sekkat, Z., and Kawata, S. (2007) The anisotropic nanomovement of azo-polymers. *Opt. Exp.*, **15**, 652–659.

29. Ishitobi, H., Tanabe, M., Sekkat, Z., and Kawata, S. (2007) Nanomovement of azo polymers induced by metal tip enhanced near-field irradiation. *Appl. Phys. Lett.*, **91**, 091911.

30. Sekkat, Z., Ishitobi, H., and Kawata, S. (2003) Two-photon isomerization and orientation of photoisomers in thin films of polymer. *Opt. Commun.*, **222**, 269–276.

31. Sekkat, Z. (2004) Isomeric orientation by two-photon excitation: a theoretical study. *Opt. Commun.*, **229**, 291–303.

32. Ishitobi, H., Sekkat, Z., and Kawata, S. (2006) Ordering of azobenzenes by

two-photon isomerization. *J. Chem. Phys.*, **125**, 164718.

33. Ishitobi, H., Sekkat, Z., and Kawata, S. (2006) Photo-orientation by multiphoton photoselection. *J. Opt. Soc. Am. B*, **23**, 868–873.

34. Maeda, M., Ishitobi, H., Sekkat, Z., and Kawata, S. (2004) Polarization storage by nonlinear orientational hole burning in azo dye-containing polymer films. *Appl. Phys. Lett.*, **85**, 351–353.

35. Li, X., Chon, J.W.M., Evans, R.A., and Gu, M. (2008) Two-photon energy transfer enhanced three-dimensional optical memory in quantum-dot and azo-dye doped polymers. *Appl. Phys. Lett.*, **92**, 063309.

36. Gindre, D., Ka, I., Boeglin, A., Fort, A., and Dorkenoo, K.D. (2007) Image storage through gray-scale encoding of second harmonic signals in azo-dye copolymers. *Appl. Phys. Lett.*, **90**, 094103.

37. Gindre, D., Boeglin, A., Fort, A., Mager, L., and Dorkenoo, K.D. (2006) Rewritable optical data storage in azobenzene copolymers. *Opt. Exp.*, **14**, 9896–9901.

38. Gindre, D., Boeglin, A., Taupier, G., Crégut, O., Vola, J.-P., Barsella, A., Mager, L., Fort, A., and Dorkenoo, K.D. (2007) Toward submicrometer optical storage through controlled molecular disorder in azo-dye copolymer films. *J. Opt. Soc. Am. B*, **24**, 532–537.

39. Mendonça, C.R., Neves, U.M., De Boni, L., Andrade, A.A., dos Santos, D.S. Jr., Pavinatto, F.J., Zilio, S.C., Misoguti, L., and Oliveira, O.N. Jr. (2007) Two-photon induced anisotropy in PMMA film doped with disperse red 13. *Opt. Commun.*, **273**, 435–440.

40. Liu, S., Lin, K.S., Churikov, V.M., Su, Y.Z., Lin, J.T., Huang, T.-H., and Hsu, C.C. (2004) Two-photon absorption properties of star-shaped molecules containing peripheral diarylthienylamines. *Chem. Phys. Lett.*, **390**, 433–439.

41. Lai, N.D., Wang, W.L., Lin, J.H., and Hsu, C.C. (2005) Optical manipulation of third-harmonic generation via either one- or two-photon excitation in diarylethene-polymethylmethacrylate

polymer thin films. *Appl. Phys. B*, **80**, 569–572.

42. Magennis, S.W., Mackay, F.S., Jones, A.C., Tait, K.M., and Sadler, P.J. (2005) Two-photon-induced photoisomerization of an azo dye. *Chem. Mater. B*, **17**, 2059–2062.

43. Corredor, C.C., Belfield, K.D., Bondar, M.V., Przhonska, O.V., Hernandez, F.E., and Kachkovsky, O.D. (2006) One- and two-photon photochromism of 3,4-bis-(2,4,5-trimethyl-thiophen-3-yl) furan-2,5-dione. *J. Photochem. Photobiol. A*, **184**, 177–183.

44. Jung, Y., Kozenkov, V.M., Magnitskii, S.A., and Nagorskii, N.M. (2006) Optical orientation of azo dye molecules in a thin solid film upon nonlinear excitation by femtosecond laser pulses. *Quantum Electron.*, **36**, 1056–1057.

45. Belfield, K.D., Bonder, M.V., Corredor, C.C., Hernandez, F.E., Przhonska, O.V., and Yao, S. (2006) Two-photon photochromism of a diarylethene enhanced by Förster resonance energy transfer form two-photon absorbing fluorenes. *ChemPhysChem*, **7**, 2514–2519.

46. Dubrovkin, A.M., Jung, Y., Kozenkov, V.M., Magnitskii, S.A., and Nagorskiy, N.M. (2007) Nonlinear induce polarization dependent scattering in solid state azo-dye films. *Laser Phys. Lett.*, **4**, 275–278.

47. Kawata, S., Sun, H.-B., Tanaka, T., and Takada, K. (2001) Finer features for functional microdevices - micromachines can be created with higher resolution using two-photon absorption. *Nature*, **412**, 697–698.

48. Ishitobi, H., Shoji, S., Hiramatsu, T., Sun, H.-B., Sekkat, Z., and Kawata, S. (2008) Two-photon induced polymer nanomovement. *Opt. Exp.*, **16** (18), 14106–14114.

49. Hisakuni, H. and Tanaka, K. (1995) Optical microfabrication of chalcogenide glasses. *Science*, **270**, 974–975.

50. Juodkazis, S., Mukai, N., Wakaki, R., Yamaguchi, A., Matsuo, S., and Misawa, H. (2000) Reversible phase transitions in polymer gels induced by radiation forces. *Nature*, **408**, 178–181.

51. Friese, M.E.J., Nieminen, T.A., Heckenberg, N.R., and

Rubinsztein-Dunlop, H. (1998) Optical alignment and spinning of laser-trapped microscopic particles. *Nature*, **394**, 348–350.

52. La Porta, A. and Wang, M.D. (2004) Optical torque wrench: angular trapping, rotation, and torque detection of quartz microparticles. *Phys. Rev. Lett.*, **92**, 190801.

53. Liu, M., Ji, N., Lin, Z., and Chui, S.T. (2005) Radiation torque on a birefringent sphere caused by an electromagnetic wave. *Phys. Rev. E*, **72**, 056610.

54. Singer, W., Nieminen, T.A., Gibson, U.J., Heckenberg, N.R. and Rubinsztein-Dunlop, H. (2006) Orientation of optically trapped nonspherical birefringent particles. *Phys. Rev. E*, **73**, 021911.

Edited by
Wolfgang Knoll and
Rigoberto C. Advincula

Functional Polymer Films

Related Titles

Ariga, K. (ed.)

Organized Organic Ultrathin Films

Fundamentals and Applications

2012
ISBN: 978-3-527-32733-1

Barner-Kowollik, C., Gründling, T., Falkenhagen, J., Weidner, S. (eds.)

Mass Spectrometry in Polymer Chemistry

2011
ISBN: 978-3-527-32924-3

Friedbacher, G., Bubert, H. (eds.)

Surface and Thin Film Analysis

A Compendium of Principles, Instrumentation, and Applications

2011
ISBN: 978-3-527-32047-9

Chujo, Y. (ed.)

Conjugated Polymer Synthesis

Methods and Reactions

2011
ISBN: 978-3-527-32267-1

Mathers, Robert T./Michael A. R. Meier (eds.)

Green Polymerization Methods

Renewable Starting Materials, Catalysis and Waste Reduction

2011
ISBN: 978-3-527-32625-9

Leclerc, M., Morin, J.-F. (eds.)

Design and Synthesis of Conjugated Polymers

2010
ISBN: 978-3-527-32474-3

Kumar, C. S. S. R. (ed.)

Nanostructured Thin Films and Surfaces

2010
ISBN: 978-3-527-32155-1

Advinaia, R., et. al.

Polymer Brushes

ISBN: 978-3-527-31033-3

Matyjaszewski, K., Advincula, R. C., Saldivar-Guerra, E., Luna-Barcenas, G., Gonzalez-Nunez, R. (eds.)

New Trends in Polymer Sciences

2009
ISBN: 978-3-527-32735-5

Matyjaszewski, K., Müller, A. H. E. (eds.)

Controlled and Living Polymerizations

From Mechanisms to Applications

2009
ISBN: 978-3-527-32492-7

Elias, H.-G.

Macromolecules

2009
ISBN: 978-3-527-31171-2

Matyjaszewski, K., Davis, T. P.

Handbook of Radical Polymerization

E-Book
ISBN: 978-0-470-35609-8

Edited by Wolfgang Knoll and Rigoberto C. Advincula

Functional Polymer Films

Volume 2
Characterization and Applications

WILEY-VCH Verlag GmbH & Co. KGaA

The Editors

Prof. Dr. Wolfgang Knoll
AIT Austrian Institute
of Technology GmbH
Donau-City-Straße 1
1220 Vienna
Austria

Prof. Dr. Rigoberto Advincula
University of Houston
Department of Chemistry
136 Fleming Bldg.
Houston, TX 77204-5003
USA

Library of Congress Card No.: applied for

British Library Cataloguing-in-Publication Data
A catalogue record for this book is available from the British Library.

Bibliographic information published by the Deutsche Nationalbibliothek
The Deutsche Nationalbibliothek lists this publication in the Deutsche Nationalbibliografie; detailed bibliographic data are available on the Internet at <http://dnb.d-nb.de>.

Composition   Laserwords Private Ltd., Chennai
Printing and Binding   betz-druck GmbH, Darmstadt
Cover Design   Schulz Grafik-Design, Fußgönheim

Printed in the Federal Republic of Germany
Printed on acid-free paper

ISBN: 978-3-527-32190-2

Contents of Volume 1

Contents of Volume 2

Preface

Thick films, thin films, even ultrathin ones, or just industrial coatings? The appropriate classification is a question of what constitutes a barrier layer or a coating material and what the ultimate function is. For sure, there are still many unsolved technological problems and fundamental scientific interest in understanding the phenomena related to the behavior and the properties of a polymeric or organic film on a solid surface. In a truly industrial application, one thinks of a polymer film as a paint coating applied to a wall or a packaging product. On the other hand, we look at these films as a system to be probed with the latest surface analytical tools or to be imbued with unique properties and functions. Smart or stimuli-responsive coatings, nanostructured thin films or devices all evoke different responses to what an organic or polymeric film does. In an inevitably interdisciplinary approach, everyone has something to bring to the table: A physicist or a surface scientist is interested in the phenomena of adsorption, relaxation, mesophase separation; a materials chemist is interested in developing new synthetic methodologies or in studying polymerization in the confined dimensions of a thin film; a chemical engineer is interested in understanding the transport phenomena and barrier properties of new coatings; a biologist or biophysicist is interested in biomimetic systems that enable the replication of *in-vivo* conditions for quantitative measurements on surfaces; and finally a nanotechnology-oriented scientist or engineer thinks about the unique nanoscale dimension by which structure–property relationships can be derived. The relationship and findings with colloidal and core-shell particle structures is not remote and can be extrapolated. While this book addresses mostly some of the fundamentals and state-of-the-art of organic and polymer materials, there is room and interest for everyone to contribute and learn. A curious mind will always come out with a fresh new insight and an inspiration to learn, understand, and even invent. *This book is dedicated to such an individual.*

The collection is divided into two volumes: *Volume 1.* Preparation and Patterning; *Volume 2.* Characterization and Applications. The volumes are meant to address the main thrusts of this body of literature. The idea is to equip the reader with the basic knowledge and at the same time with the state-of-the-art for each theme, almost tracing the development of the field and the interdisciplinary nature of the endeavor. Taken together as a whole, the volumes should thoroughly provide the reader with a good understanding of what is collectively known as the *field of organic polymer thin films*. The editors and authors do not claim that this collection represents the most important works in the field, nor can it replace some of the revered classics in the fields. There are more up-to-date review articles that will always supplant the information in these chapters. The editors would also like to extend their apologies to any author or part of work that has not been invited to contribute or has not been cited. It is not intentional – rather there is always room for another project. Lastly, the editors heartily thank all the authors and researchers who have contributed to this book. We hope you will like what we have done here and you will be proud of your chapter.

Rigoberto Advincula and Wolfgang Knoll

List of Contributors

Rigoberto C. Advincula
University of Houston
Department of Chemistry
Department of Chemical and
Biomolecular Engineering
136 Fleming Bldg.
Houston, TX 77204-5003
USA

Omar Azzaroni
Universidad Nacional de La Plata
Instituto de Investigaciones
Fisicoquímicas Teóricas y
Aplicadas (INIFTA)
CONICET CC 16 Suc. 4
1900 La Plata
Argentina

Akira Baba
Niigata University
Center for Transdisciplinary
Research
8050 Ikarashi 2-nocho
Nishi-ku
Niigata 950-2181
Japan

Petra J. Cameron
University of Bath
Department of Chemistry
Bath BA2 7AY
UK

Kookheon Char
Seoul National University
The National Creative
Research Initiative Center
for Intelligent Hybrids
School of Chemical &
Biological Engineering
Seoul 151-744
Republic of Korea

Nam-Joon Cho
Stanford University
Division of Gastroenterology
School of Medicine
CCSR Bldg. Rm3110
269 Campus Dr.
Stanford, CA 94305-5025
USA

Géraldine Coullerez
ETH Zurich
Department of Materials
Laboratory for Surface
Science and Technology
Wolfgang-Pauli-Strasse 10
8093 Zurich
Switzerland

Hatice Duran
Max-Planck-Institute for
Polymer Research
Ackermannweg 10
55128 Mainz
Germany

and

TOBB University of
Economics and Technology
Turkey

Curtis W. Frank
Stanford University
Department of Chemical
Engineering
381 North South Mall #113
Stauffer III
Stanford, CA 94305-5025
USA

Antonio Francesco Frau
University of Houston
Department of Chemistry
Department of Chemical and
Biomolecular Engineering
136 Fleming Bldg.
Houston, TX 77204-5003
USA

Toshinori Fujie
Waseda University
Department of Life Science and
Medical Bioscience
Faculty of Science and
Engineering
2-2 Wakamatsu-cho
Shinjuku-ku
Tokyo 162-8480
Japan

George Fytas
Institute of Electronic Structure
and Laser-FO.R.T.H
P.O. Box 1527
71110 Heraklion
Greece

and

Max Planck Institute for
Polymer Research
Ackermannweg 10
55128 Mainz
Germany

and

University of Crete
Department of Materials
Science and Technology
Heraklion
Greece

Xingyu Gao
National University of Singapore
Department of Physics
2 Science Drive 3
Singapore 117542
Singapore

Claudio Gervasi
Universidad Nacional de La Plata
Instituto de Investigaciones
Fisicoquímicas Teóricas y
Aplicadas (INIFTA)
CONICET CC 16 Suc. 4
1900 La Plata
Argentina

Antonis Gitsas
Austrian Institute of
Technology (AIT)
Donau-City-Str. 1
1220 Vienna
Austria

Ganna Gorodyska
ETH Zurich
Department of Materials
Laboratory for Surface
Science and Technology
Wolfgang-Pauli-Strasse 10
8093 Zurich
Switzerland

H. Michelle Grandin
ETH Zurich
Department of Materials
Laboratory for Surface
Science and Technology
Wolfgang-Pauli-Strasse 10
8093 Zurich
Switzerland

Jennifer A. Irvin
Texas State University
Department of Chemistry
and Biochemistry
601 University Drive
San Marcos, TX 78666
USA

Yeongseon Jang
Seoul National University
The National Creative
Research Initiative Center
for Intelligent Hybrids
School of Chemical &
Biological Engineering
Seoul 151-744
Republic of Korea

Diethelm Johannsmann
Clausthal University of
Technology
Institute of Physical Chemistry
Arnold-Sommerfeld-Str. 4
38678 Clausthal-Zellerfeld
Germany

Futao Kaneko
Niigata University
Center for Transdisciplinary
Research
8050 Ikarashi 2-nocho
Nishi-ku
Niigata 950-2181
Japan

Wolfgang Knoll
Austrian Institute of
Technology (AIT)
Donau-City-Str. 1
1220 Vienna
Austria

K.H. Aaron Lau
Northwestern University
Department of Biomedical
Engineering
2145 Sheridan Road
Tech E210
Evanston, IL 60202
USA

Benoit Loppinet
Institute of Electronic Structure
and Laser-FO.R.T.H
P.O. Box 1527
71110 Heraklion
Greece

Renate L.C. Naumann
Austrian Institute of
Technology (AIT)
Donau-City-Str. 1
1220 Vienna
Austria

Yosuke Okamura
Waseda University
Department of Life Science and
Medical Bioscience
Faculty of Science and
Engineering
2-2 Wakamatsu-cho
Shinjuku-ku
Tokyo 162-8480
Japan

Ophir Ortiz
University of South Florida
Department of Chemical and
Biomedical Engineering
4202 E Fowler Ave
Tampa, FL 33620
USA

Erik Reimhult
ETH Zurich
Department of Materials
Laboratory for Surface
Science and Technology
Wolfgang-Pauli-Strasse 10
8093 Zurich
Switzerland

Marina Ruths
University of Massachusetts
Lowell
Department of Chemistry
1 University Avenue
Lowell, MA 01854
USA

Takahiro Seki
Nagoya University
Graduate School of Engineering
Department of Molecular Design
and Engineering
Furo-cho
Chikusa
Nagoya 464-8603
Japan

Martin Steinhart
Universität Osnabrück
Institut für Chemie
49069 Osnabrück
Germany

Gila E. Stein
University of Houston
Department of Chemical and
Biomolecular Engineering
4800 Calhoun Rd
Houston, TX 77204-4004
USA

Shinji Takeoka
Waseda University
Department of Life Science and
Medical Bioscience
Faculty of Science and
Engineering
2-2 Wakamatsu-cho
Shinjuku-ku
Tokyo 162-8480
Japan

Marcus Textor
ETH Zurich
Department of Materials
Laboratory for Surface
Science and Technology
Wolfgang-Pauli-Strasse 10
8093 Zurich
Switzerland

Ryan Toomey
University of South Florida
Department of Chemical and
Biomedical Engineering
4202 E Fowler Ave
Tampa, FL 33620
USA

Ajay Vidyasagar
University of South Florida
Department of Chemical and
Biomedical Engineering
4202 E Fowler Ave
Tampa, FL 33620
USA

Andrew T.S. Wee
National University of Singapore
Department of Physics
2 Science Drive 3
Singapore 117542
Singapore

Katie Winkel
Texas State University
Department of Chemistry
and Biochemistry
601 University Drive
San Marcos, TX 78666
USA

Bongjun Yeom
Seoul National University
The National Creative
Research Initiative Center
for Intelligent Hybrids
School of Chemical &
Biological Engineering
Seoul 151-744
Republic of Korea

Part III
Characterization

Functional Polymer Films, First Edition. Edited by Wolfgang Knoll and Rigoberto C. Advincula.
© 2011 Wiley-VCH Verlag GmbH & Co. KGaA. Published 2011 by Wiley-VCH Verlag GmbH & Co. KGaA.

17

Dynamics and Thermomechanics of Polymer Films

Benoit Loppinet and George Fytas

17.1
Introduction

In this chapter, we present experimental approaches used for the characterization of the thermally excited dynamics present in polymer-based films. The dynamics of bulk polymers in melts or in solutions has been intensively studied for many years creating an adequate understanding of the underlying mechanisms for macromolecular motion [1]. The macromolecular structure is at the origin of a number of properties specific to polymers, which includes: viscoelasticity inherent to polymer melts and concentrated solutions, rubber elasticity, and or cross-linked polymer networks. As they are particularly well adapted to the formation of thin layers and films, polymers are widely used as thin coatings on top of a surface, for example, to modify surfaces and interface properties. The general research effort in this area has mostly been directed toward the characterization of the structure and conformation of those layered systems. The dynamics of the thin polymer layers remain somewhat unexplored, despite its relevance for a number of specific properties like mechanical properties. Dynamic mechanical properties and the stability of nanostructured materials are of importance for a wide range of applications that comprise microelectronic, photonics, nanoelectromechanical systems, nanofluidics, and biomedical technologies. The ability to pattern substrates at the scale of tens of nanometers has largely evolved as a result of advances in the electronics and semiconductor industry [2]. The interest in particulate polymer films as a new-generation coatings has also just started [3]. It is our view that we have reached a point where we can nanofabricate these structures, but it is not known how the relevant materials behave in such structures. One broad objective of the research is to determine if, and at which scale, any continuum-level representations of a soft material become inadequate. The importance of developing an understanding of polymeric materials at the nanoscale cannot be overemphasized.

The main cause of the limited experimental work on the dynamic properties of thin films is probably the paucity of appropriate experimental tools. Indeed the small number of samples often imposes severe limits. In this chapter, we present a number of recently either developed or applied experimental techniques based on laser light scattering and fluorescence correlation that are well adapted to the study of thin polymer layers and discuss their applications to a few relevant topics. We describe first the concepts of the time and frequency domain dynamic light scattering (DLS) techniques and single-molecule fluorescence correlation spectroscopy (FCS). Selected applications on the dynamics and nanomechanics of different polymer-based coatings taken from the authors work are presented in Sections 17.3 and 17.4, respectively. We conclude this short review with perspectives and outlook.

17.2
Experimental Techniques

17.2.1
Dynamic Light Scattering

DLS designates a number of experimental techniques where the frequency shift between a continuous laser light beam incident on the sample and the light scattered at a finite angle is analyzed in order to provide information about the internal dynamics in the sample [4]. A large domain of frequency shift is experimentally accessible and can cover a wide range of time scales, from as fast as 10^{-12} s (Raman scattering) to as slow as 100 s. The fast dynamics is accessible through frequency-based detection (Raman, Brillouin) and the slow dynamics are accessed through real-time photon correlation spectroscopy (PCS). The basic techniques are by now very well established. DLS techniques bring information on the dynamics of the motion associated with the scattering process. In the case of polymeric systems, light scattering may arise as a result of the presence of density fluctuations, concentration fluctuations, or orientation fluctuations. These thermally excited fluctuations present in the sample (concentration, density, height) induce fluctuations of the scattered intensity or of its polarization or its frequency. The fluctuating signal is analyzed through PCS that provides the degree of correlation between dynamic properties over a period of time. The characteristic times obtained through the analysis of correlation functions relate to the transport coefficients of the specific motion, like diffusion coefficients or velocities.

An essential feature of DLS is the spatial and directional resolutions provided by the scattering wave vector q associated with the scattering process. DLS experiments provide the transport coefficients of the particular fluctuation with wavelength $2\pi/q$. The magnitude and direction of q are directly controlled through the scattering angle and the scattering geometry. The general scattering geometry is illustrated below for the standard bulk scattering and for the case of layered systems with the decomposition of the q-vector into an inlayer ($//$) and an out-of-layer ($\perp$) component. Directionalities affects on the dynamics are particularly relevant in the case of layered systems and can be explored through the control of the direction of the scattering wave vector q relative to the direction of the layer (Section 17.2.1.2). Many of the experimental results presented below rely on the measured dispersion

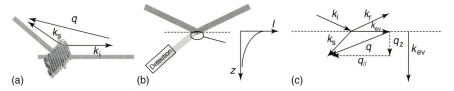

Figure 17.1 Scattering geometries: (a) standard geometry, (b) and (c) total internal reflection geometry of the evanescent wave DLS. k_i, k_s are the wave vectors of the incident, scattered light and q scattering wave vector define as $q = k_s - k_i$. In total internal reflection k_r is the wave vector of the reflected beam and k_{ev} is the wave vector of the evanescent field relation

$|k_{ev}| = (2\pi n_2 \lambda \sin\varphi/(n_2/n_1))$, $\kappa_{ev}|k_{ev}| = 2\pi\lambda n_2 (\sin^2\varphi - (n_2/n_1)^2)^{1/2}$ is the imaginary part of the evanescent wave vector describing the penetration into the medium of lower refractive index. The wave vector q is defined as $q = k_{ev} - k_s$ and can be decomposed in an inplane ($q_{//}$) and a perpendicular (q_z) components.

relation, which relates the frequency or rate of a classic wave to its wavelength. The simplest cases are represented by Brownian-like diffusion (Section 17.2.1.1) or propagating phonon-like motion (Section 17.2.1.2).

A specific geometry particularly well adapted to the study of interfacial dynamics is obtained at total internal reflection (TIR). Under TIR conditions, there is no light propagating in the medium of lower refractive index where the optical field is restricted to a region of the order of one wavelength (Figure 17.1). This evanescent field in the medium of lower refractive index provides a near-field probe that can be used as an incoming beam of a scattering experiment and provide characteristics of interfacial dynamics. It was first applied to colloidal diffusion [5] and later to polymer brushes [6] and application of EW-DLS to various soft-matter systems has recently been reviewed [7]. Evanescent near-fields are not restricted to interfaces between dielectrics but also occur at metal/dielectric interfaces where they are particularly amplified near-surface plasmon resonances. Plasmon-enhanced Raman scattering has been widely used but surprisingly Plasmon-enhanced Brillouin or Rayleigh DLS have remained mostly unexplored. The latter was recently demonstrated in Kretschmann geometry [8]. EW-DLS is not restricted to solid/liquid interfaces, it has recently been demonstrated at liquid/liquid or liquid-gas interfaces [9] and shown to present some advantages compared to the more standard surface light scattering.

An important aspect of the thin layered systems is the presence of surface and interfaces. If the scattering volume, defined by the intersection between the illuminated and the detection volume, contains surfaces and interfaces, then contributions from various surface excitations like surface waves are expected to be detected in DLS experiments. Surface waves can be visualized as height fluctuations also known as *ripples* or *ripplons*. The spectrum of thermally excited surface waves covers a broad range of wavelengths and frequencies. Mostly two types of surface waves are relevant to DLS experiments: capillary waves at the surface of liquids (typically in the kilohertz regime for a water/air interface) and surface acoustic waves at the surface of liquid or solids (mostly solid), typically in the gigahertz regime for solid/air interfaces (Section 17.4.1).

The case of acoustic waves in polymer film is treated in the next section. Capillary waves have experimentally been studied for a long time through surface quasielastic light scattering [10] and recently also through EW-DLS [9]. Regarding polymers, SQELS has mostly been used for the study of polymer adsorbed at the air/water interface and the determination of mechanical modulus of such thin polymeric films [11–13]. We will not treat this here, the recent reviews and books can be used for more information.

Slow capillary waves at the surface of polymer melts and glasses have recently been experimentally studied through X-ray PCS [14, 15]. The main motivation for this type of study is to isolate the dynamics of a film's top layer and compare it to the bulk dynamics, in order to identify possible confinement effects on the glass. The question of the glassy dynamic in thin layers has been debated for the last 15 years [16, 17]. This interesting part of polymer dynamics and more generally the confinement effect on the glass transition and the associated dynamics will be only partially treated from the point of view of elastic waves in Sections 17.4.1.1 and 17.4.2.2.

17.2.1.1 Microphoton Correlation Spectroscopy (μ-PCS) and EW- DLS

In the following section, we give some details regarding the microphoton correlation spectroscopy (μ-PCS) and EW-DLS techniques used for the study of thin polymer layers. μ-PCS is a simple variant of the standard PCS techniques and can be applied when no dynamic contribution to the intensity of the light scattered other than fluctuations in the film are present in the scattering volume, as in the case for films anchored on glass substrates immersed in pure solvent. A schematic presentation of the setup is depicted in Figures 17.2 [18]. EW-DLS requires the use of a prism to provide TIR conditions. The specific application to the dynamics of polymer brushes is schematically depicted in Figure 17.3 [19].

The same basic optical setup, data acquisition, and analysis were used for both μ-PCS and EW-DLS. The incident light source was a Nd: YAG laser at a wavelength

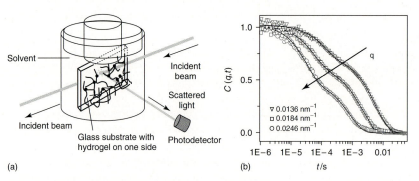

(a) (b)

Figure 17.2 Micro-PCS. (a) Scattering geometry and cell holder of the film (e.g., a supported hydrogel layer). The relaxation function $C(q,t)$ of the thermal concentration fluctuations for three different wave vectors in the hydrogel at ambient temperature is shown in (b) [18].

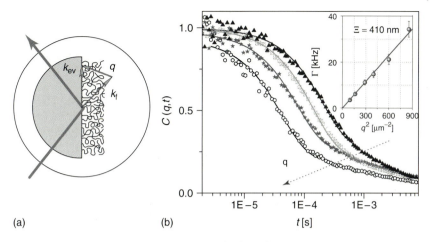

Figure 17.3 EW-DLS. (a) Scattering geometry sketching the semicylindrical lens with grafted polymer brush immersed in solvent. The relaxation function $C(q,t)$ of the thermal concentration fluctuations for different wave vectors in the hydrogel brush swollen in good solvent in (b) and the diffusive rate [19].

of $\lambda = 532$ nm with 150 mW power and a beam diameter of about 300 μm. The scattered light was collected for various wave vectors q by an optical fiber and detected by a photomultiplier tube connected to an ALV-5000 digital fast photon correlator. The value of q was controlled through the scattering angle θ. The output of the correlator is the time-averaged autocorrelation function, $G(q, t) = <I(q, t)I(q, 0)>/<I(q, 0)>^2$, of the scattered light intensity $I(q,t)$. In both μ-PCS and EW-DLS, the experimental $G(q,t)$ were recorded under (partial) heterodyne conditions due to the presence of strong static scattering contributions from the glass substrates and possibly from static inhomogeneities in the samples. The existence of a base line at long times allowed the determination of a normalized correlation function $C(q, t) = [G(q, t)-A]/bf^*$, where $f^* < 1$ is an instrumental factor and A is the base line of $G(q,t)$ at long times ($\sim$1 s), and b is the amplitude between the short time ($<$μs) and the long time values of $G(q, t)$. In heterodyne conditions, $C(q, t)$ directly relates to the desired normalized relaxation function for the concentration fluctuations in the sample and b relates to the contribution of the dynamic process I_d to the total intensity $I_c b \sim 2I_d/I_c$.

Figure 17.2 displays typical relaxation functions $C(q,t)$ of the concentration fluctuations in a thermoresponsive poly-N-isopropylacrylamide (PNIPAAm) swollen in ethanol at 25 °C. Two distinct decays are clearly seen and due to the well-separated time scales of the two processes, the normalized $C(q,t)$ was represented by the double-exponential decay function

$$C(q, t) = a \exp(-\Gamma_f t) + (1 - a) \exp(-\Gamma_s t) + a' \tag{17.1}$$

Using the two relaxation rates Γ_f, Γ_s for the fast and slow process and the amplitude a as adjustable parameters, respectively. To account for misnormalization and small base line fluctuations, we also allowed for a small nonzero value of the parameter a'. This simple expression represents well the experimental $C(q,t)$, as indicated by the solid lines in Figure 17.2. Both relaxation modes depend on the magnitude of the scattering wave vector q and the observed dependence of the relaxation rates is characteristic of a pure isotropic diffusive ($\Gamma \sim q^2$) behavior.

In the case of EW-DLS, glass prisms bearing grated brushes ($n_1 = 1.62$) were immersed in a solvent cell. It provided TIR conditions for incidence angle φ between grazing incidence (90°) and the critical angle $\varphi_c = 61.2°$. By varying φ from φ_c to 90° allowed a control of the penetration depth, with typical values for the field penetration depth ($1/\kappa_{ev}$) of the order of 500 nm. Normalized correlation functions $C(q,t)$ for a polystyrene (PS) brush swollen in a good solvent (dioxane) are shown in Figure 17.3 [19]. They were analyzed using a single-exponential decay and the rates obtained at different q are reported in the inset. The q^2 dependence confirmed the diffusive behavior of the measured concentration fluctuations.

17.2.1.2 Brillouin Light-Scattering Spectroscopy (BLS)

This technique records the spectrum of $I(q, \omega)$ of the light scattered inelastically by thermal phonons propagating in the sample. The scattering wave vector $q = k_i - k_s$ is defined by the wave vectors of the incoming (i) and scattered (s) photons with frequency shift ω. For homogeneous media over the length scale of about q^{-1}, the momentum conservation requires $q = k$, where k is the wave vector of the phonon involved in the scattering process. The frequency shift $\omega = \pm ck$, as a result of energy conservation, yields the phase velocity of the phonon with longitudinal, transverse polarization, or mixed polarizations. A high-resolution six-path tandem Fabry–Perot interferometer and a light-scattering setup that allows both q, temperature and strain variations has been developed [20] to record the spontaneous Brillouin light scattering (BLS) at hypersonic (1–500 GHz) frequencies. Choosing the polarization of the incident laser beam perpendicular (V) to the scattering plane (sagittal plane) and selecting the polarization of the scattered light either vertical (V) or parallel (H) to the scattering plane, both polarized (VV) and depolarized (VH) spectra can in principle be recorded. This chapter will focus on BLS from structured systems with spacing commensurate with the wavelength $(2\pi/k)$ of the phonon where a rich $I(q, \omega)$ is observed due to manifold of acoustic excitations. Hypersonic dispersion in homogeneous bulk polymers is reported elsewhere [21].

In the case of polymer films, the selection of the phonon propagation direction is a crucial advancement for the measurement of the thermomechanical properties along the main symmetry directions that is not possible otherwise. Further, for composite films, phononic gaps can be unidirectional. The two scattering geometries shown in Figure 17.4 enable the selective observation of phonon propagation with wave vector either parallel (transmission geometry in a) or normal (reflection geometry in b) to the film surface. In Figure 17.4b the magnitude of $q_{//} = 4\pi n/\lambda \sin(\alpha)$, where the incidence angle α is half of the scattering angle θ and n is the refractive index of the film. In the transmission geometry (Figure 17.4a) the magnitude of $q_\perp$ does not depend on n. Owing to the large refractive index

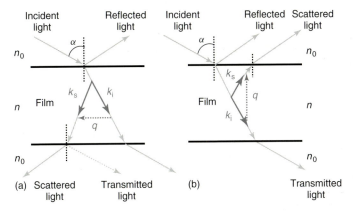

Figure 17.4 Selection of the propagation. (a) The Brillouin light-scattering configuration for probing elastic excitations either parallel (a) or normal (b) to the film (selection of the wave vector q). In (a) the magnitude of $q = 4\pi/\lambda \sin(\alpha)$ where the incidence angle α is half of the scattering angle θ.

difference between film and air, the variation of $q_\perp$ is restricted in a much smaller range (0.029–0.035 nm^{-1}) than $q_{//}$ (between 0.003 and 0.025 nm^{-1}) for $\lambda = 532$ nm and $n \sim 1.5$.

17.2.2
Fluorescence Correlation Spectroscopy (FCS)

FCS provides a complementary tool for the characterization of dynamics in thin layers. It measures fluctuations of the fluorescence intensity arising from a small volume with typical radial $x_0 \sim 300$ nm and axial $z_0 \sim 1\,\mu$m dimensions. The small volume is obtained through the focus of a microscope lens together with a confocal detection of the backscattered intensity. FCS detects number-density fluctuations of very few fluorescently labeled molecules in the small volume (Figure 17.5a). The small volume makes it particularly attractive for the study of thin samples. Moreover, a TIR setup using evanescent wave illumination has been implemented, reducing the z_0 dimension by a factor of the order of 5. Though a relatively old technique [22], FCS went through a renewed interest thanks to the advances of a new generation of confocal microscopes. However, this renewed interest has mostly benefited the biology field, and not so much polymer science.

In contrast to the varying probing length q^{-1} of PCS, x_0 and z_0 are essentially fixed and FCS is a single-molecule spectroscopic technique free of intermolecular interactions. The measurements are performed on a commercial FCS setup (Carl Zeiss, Jena, Germany) consisting of the module ConfoCor 2 and an inverted microscope model Axiovert 200. A 40× Plan Neofluar objective with a numerical aperture of 0.9 and oil as immersion liquid were used in this study. The fluorescence species are excited by lasers in the visible region. For detection, an avalanche photodiode capable of single counting is used. The temporal fluctuations of the fluorescence

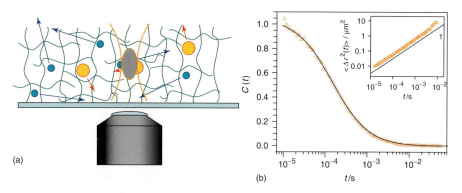

(a)

(b)

Figure 17.5 (a) Sketch of a fluorescence correlation spectroscopy setup defining the observation volume (yellow ellipsoid).The diffusion of fluorescently labeled molecules (orange and blue) in and out of the observation volume yields their translational mobility in the anchored hydrogel layer. (b) Normalized fluorescence intensity autocorrelation function (symbols) and the corresponding fits with Equation 17.2 (solid lines) for the tracer diffusion of Alexa 647 in a PNIPAAm hydrogel at 16.3 °C. The linear log-log plot (slope 1) of the tracer mean-squared displacement in the inset demonstrates the Fickian diffusion behavior in the swollen state.

intensity $I_f(t)$ as the labeled molecule diffuses into and out of the observation volume define the autocorrelation function $G(t) = < I_f(t)I_f(0)>/<I_f(0)>^2$, which in turn relates to the motional mechanism of the labeled molecule. For a Gaussian illumination profile and Gaussian particle displacement probability, $G_s(r,t) = (2\pi/3 < \Delta r^2(t) >)^{-3/2}\exp(-3r^2/<\Delta r^2(t)>)$, the fluorescence correlation function reads,

$$G(t) = 1 + N^{-1}\left[1 + (2 < \Delta r^2(t) > /3x_0^2]^{-1}\left[1 + (2 < \Delta r^2(t) > /3z_0^2]^{-1/2}\right.\right.$$

$$(17.2)$$

where N is the average number of labeled molecules in the observation volume. For free Fickian diffusion, the dye mean square displacement $<\Delta r^2(t)> = 6Dt$ with D being its tracer diffusion coefficient; the initial fast decay due to the triplet state is omitted in Equation 17.2. Figure 17.5b shows a normalized $G(t)$ for the Alexa 647 diffusion in a grafted PNIPAAm hydrogel at 16.3 °C. Under these good solvent conditions, free Fickian diffusion is observed (inset to Figure 17.5).

17.3
Dynamics

17.3.1
Hydrogel Layers Anchored to Solid Surfaces

Hydrogels have recently gained large exposure whose properties are already finding application in such diverse applications as drug-delivery systems, biosensors, and synthetic tissue [23, 24]. Despite such widespread usage, an intimate understanding of the mechanisms and detailed architectures underlying the known properties is still missing. Among hydrogels, PNIPAAm has proven of particular interest due to the high swelling ratio, high stability, and the presence of a well-defined, easily accessible lower critical solution temperature (LCST) exhibited at 32 °C. In particular, grafted PNIPAAm layers on solid substrates offer a platform for sensor applications. Of paramount importance for the latter is the penetration and mobility of analytes that in turn intimately depend on the hydrogels structure at different length scales. Parameters known to alter the behavior of gels are, among others, temperature, the solvent they are swollen in, pH and the specific preparation technique [25]. For a disordered soft system this structural elucidation requires a spatiotemporal study of the hydrogel/penetrant system, since the gel structure can impact differently the guest mobility due to specific host–guest interactions. The charged nature of hydrogels makes the system complex with rich behavior considering also its switching response to several external stimuli.

The relaxation of the concentration fluctuations in anchored PNIPAAm hydrogels was studied by μ-PCS (Section 17.2.1.1) [18], while the penetration and mobility of labeled tracers was probed by FCS (Section 17.2.2) [26]. The former yield valuable information on the local structure and osmotic forces and the latter allows for the elucidation of the diffusion mechanisms at different thermodynamic states of these responsible systems. The cooperative diffusion $D_{\mathrm{coop}} = \Gamma_f/q^2$ computed from the fast decay rate of the relaxation function $C(q,t)$ (Equation 17.1) increases with the PNIPAAm volume fraction ϕ as seen in Figure 17.6. The speed-up of the cooperative diffusivity is stronger in the anchored chemically cross-linked PNIPAAm than in the corresponding physical network

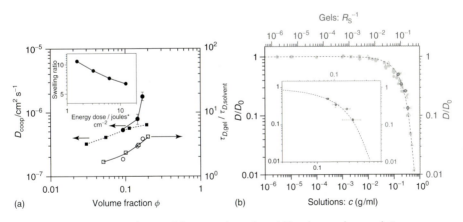

(a)   (b)

Figure 17.6 Cooperative and tracer diffusion. (a) D_{coop} and $\tau_{D,\mathrm{gel}}/\tau_{D,\mathrm{solvent}}$, the normalized (relatively to the neat solvent) molecular dye diffusion time, plotted as a function of the effective volume fraction ϕ for PNIPAAm gels obtained from the one-dimensional swelling ratio (inset). The dashed and solid lines are guides to the eye, while circles and squares denote data for gels, and semidilute homopolymer solutions, respectively. (b) Normalized tracer diffusion of various small molecules (size < 1 nm) in several polymer solutions in good solvents and Alexa in hydrogels (at 16 °C) with four different cross-linking densities represented by the swelling ratio R_s. The inset shows only the Alexa tracer diffusion in the hydrogels represented by a polynomial equation.

exemplified by PNIPAAm semidilute solutions (solid squares). The former undergo one dimensional swelling and $\phi \approx 1/R_s$ where R_s is the swelling ratio (inset of Figure 17.6a) that apparently leads to stronger concentration dependence of the gel osmotic pressure than the scaling prediction $(\sim \phi/\phi^*)^{3/4}$ for the physical PNIPAAm networks at concentrations above the overlap ϕ^*.

Conversely, the diffusion of a molecular tracer (Alexa 647 with hydrodynamic radius $R_h = 0.8$ nm) is insensitive to the type of PNIPAAm networking. This is not unexpected as the tracer diffusivity was found [27] to depend only on the concentration irrespective of the matrix molecular weight the mesh size $\xi > R_h$ as seen in Figure 17.6b. Such types of master curves are commonly represented either by a polynomial [28] or exponential [29]. The latter implies a dependence on the mesh size and hence a superposition only for a plot of D $vs. \phi/\phi^*$. The lack of polymer specificity favors, therefore, the polynomial representation $(D/D_0 = 1 - 5c + 8.9c^2 - 5.5c^3$ solid line in Figure 17.6) and renders the interpretation of the parameters the alternative exponential fitting $(D/D_0 = \exp(-9.9c^{-1.35}))$ less physically meaningful.

The slow diffusive process in the anchored gel layers (Figure 17.2) with diffusion coefficient $D_s = \Gamma_s(q)/q^2 (\sim 1.5 \times 10^{-8}$ cm^2/s) exhibits two pertinent differences compared to the slow dynamics in the reported polymer systems so far. It is "faster" and ergodic compared to the very slow, nonergodic process in a 3D gel, as seen in the well-resolved baseline of the functions of Figure 17.2. Within the scattering of the experimental data, D_s is found to be virtually insensitive to the variation of the cross-linking density and is at odds with the slow mode observed in the corresponding semidilute solutions [30]. In this case D_s was found to exhibit a strong decrease with increasing concentration, resembling the concurrent increase of the solution viscosity with concentration and was attributed to the self-diffusion of (polymer) clusters [30]. However, in the present case of the cross-linked PNIPAAm gel layers, with nominally infinite viscosity, an analog description is hardly conceivable. A common feature of the slow diffusive mode in both the semidilute solutions and the chemically cross-linked gels, however, is a large correlation length (in the submicrometer range). This can be phenomenologically assigned to long-wavelength concentration fluctuations in the grafted gels. The dynamics associated with these concentration inhomogeneities are frozen in the case of the 3D gels but could be the origin of the observed slow mode in the anchored gel layers, arising from a shorter correlation length. The relation between this ergodic slow mode in the anchored gel films and the nonergodic much slower process in the conventional 3D gels remains to be clarified. It would be helpful to examine the dynamics in different solvents and other types of gels, for example, formed by microgel particles [31].

17.3.2
Grafted Polymer Brushes

17.3.2.1 Concentration Dynamics

Adsorbed polymer chains at solid surfaces have long been used for steric stabilization of colloids, as the solvent swollen layer screens attractive interactions between solid particles. End-grafted chains represent an interesting specific case. If the density of grafting is high enough, the chains adopt extended configuration referred to as *polymer brush*. Research into polymer brushes has been very active over the last 30 years [32] due to the many opportunities offered for functionalized, smart modified advanced surfaces [33].

A number of preparation routes have been devised and structural characterization has focused on the static properties of the brush, like concentration profiles, and on novel surface properties, like modified wetting or pH responsive surfaces. Comparatively, the experimental studies of the dynamics are limited, while the theoretical consideration of the dynamics of adsorbed polymers was initiated more than two decades ago [34]. The first experimental report to unveil the dynamics of swollen brushes prepared by the grafting-to approach has utilized the EW-DLS technique [6]. The probed dynamics were later attributed to sliding of the physisorbed anchors [36]. To bypass this influence, PS brushes were covalently anchored on glass substrates through a grafting-from approach [19]. Combination of a large molecular weight and high grafting densities lead to thick brush possessing, however, significant polydispersity due to the used free-radical polymerization.

Experimental correlation functions for a thick dense brush swollen in good solvent (dioxane) using EW-DLS are shown in Figure 17.7 [19]. The fast diffusive dynamics was attributed to the cooperative mode of semidiluted polymer solutions. A hydrodynamic blob size can be then deduced from the diffusion coefficient using the Stokes–Einstein equation ($\xi = kT/6\pi\eta D$). The computed values were of the order of a few nanometers being very similar to the average distance between grafting points [19]. Different grafting densities were later considered [37] and the results supported the simple Alexander–De Gennes blob model, where the brush is envisaged as a linear array of blobs. In this simple model, the distance between grafting points imposes blob size and the average concentration within the brush.

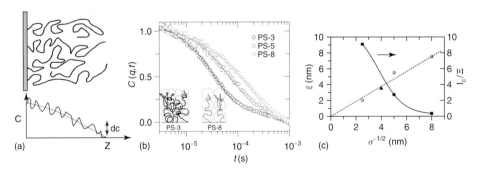

Figure 17.7 Concentration fluctuations within polystyrene brushes swollen in good solvent: (a) Schematic of concentration fluctuation within a grafted polymer brush, (b) normalized correlation functions for three different brushes of different grafting densities at a constant q showing the speed up of the relaxation with increasing grafting density. (c) Hydrodynamic blob size as a function of the average distance between grafted chains, confirming the blob picture for the dynamics of swollen brushes [35].

Interestingly, the q^2 dependence of the rate hints toward an isotropic diffusion, meaning no measurable difference between the inplane and out-of-plane diffusion. The lack of anisotropy is probably due to the fact that the blob size is much smaller than the brush height and at the length scale relevant for the diffusion, the system is almost isotropic.

One of the key interests of polymer brushes is their response to the external stimuli. As the macrophase separation characteristic of polymer solutions is prevented by the chemical graft, brushes react to the change of solvent quality by adjusting their height, and eventually collapse in the case of too low solvent quality. This behavior has long been considered and the average change, well documented. But little was known on the variation of the dynamics within the destabilized system. The consequence on the measured dynamics of the solvent-induced retraction was examined by EW-DLS experiments on the same grafted PS brushes. Cyclohexane was used as a theta solvent as it has a theta temperature for a linear chain of PS at 35 °C [19, 35]. The phase-separation temperature in solutions with conditions similar to the brushes (linear chains of molecular weight of 10^6 g/mol and concentration of 10%) is about 26 °C. The relaxation functions of the concentration fluctuations recorded at a single wave vector for different temperature are shown in Figure 17.8. The main signature of the lowered solvent quality on the relaxation functions relates to the occurrence of a second slower relaxation decay that is well resolved within the experimental window. For the different brushes of

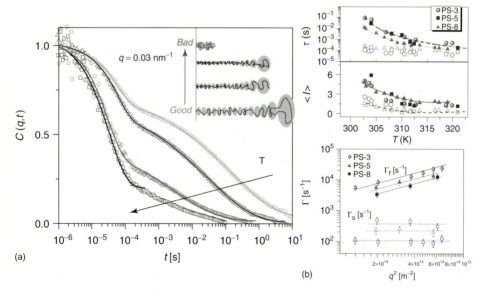

Figure 17.8 Brush dynamics in theta-solvent: As temperature is decreased the brush undergoes contraction. The dynamics reveal a second slower broad relaxation, the amplitude and time strongly depend on temperature (a). (b) The rate and amplitude obtained from a double KWW fit as a function of temperature for one q and as a function of q for three temperatures are displayed on the right, whereas the fast mode remains diffusive, the slow process appears q independent [19, 37].

Figure 17.8, this mode displays a nonexponential shape and exhibits strong slow down of the characteristic time with decreasing temperature. The temperature dependence of the slow relaxation time conforms to $\tau_s \sim (T - T_c)^\alpha$ with a critical temperature of $T_c = 26\,°C$. The amplitude of the slower mode increases with decreasing temperature, so that this relaxation channel becomes dominant at low temperatures toward T_c. As to the faster blob-like dynamics, the associated cooperative diffusion is not retarded, approaching T_c. Instead, this fast mode was found to speed up slightly with decreasing temperature. However, the association of the fast mode in terms of the mesh size (blob) is not straightforward in theta solvents.

The dynamics of the polymer brush (Figure 17.8) revealed an unknown unprecedented feature, as the existing results on the collapse transition indicate a smooth transition from a swollen brush to a collapsed film. The origin of this extra (slow) relaxation mode with diverging dynamics at about $26\,°C$ is not known. Possible motional mechanisms include; coupling between concentration and viscosity (viscoelastic modes) and cluster-like diffusion approaching the critical phase separation, in analogy to bulk polymer solutions near macrophase separation. Interestingly, the relative amplitude of the slow and to the fast relaxation decay in the thick brushes increases with evanescent-wave penetration depth. This uneven distribution of the two modes within the grafted layer implies that the slow mode is more pronounced in the outer part of the brush. The stronger concentration fluctuation at the outskirts of the brush would indicate an interfacial-like mechanism. Since then, computer simulations have been performed [38].

17.3.2.2 Probe Diffusivities Near to Soft Surfaces

Polymer brushes may be of interest as they provide interesting barrier, permeability and selectivity properties when in contact with different species with various sizes and chemistry [39]. In particular, the exclusion or inclusion of linear chains from brushes of the same polymer has been considered early on [32] and the theoretical predictions have been experimentally verified to some extend. More recently, brushes have been proposed as a means to trigger self-assembly of nanoparticles through a thermodynamically driven local microsegregation mechanism [41]. The dynamics of the process has only been very recently addressed through computer simulation [42].

Experimentally, EW-DLS was applied to study the penetrability and diffusivities of various colloidal probes in the vicinity of the PS brushes [40, 43]. Using the dependence of penetration depth of scattered intensities and the initial slope of normalized autocorrelation functions, both the positioning and diffusivities of the colloidal probes into the brushes could be deduced. Figures 17.9 and 17.10 illustrate the main results. The normalized intensity measured at different evanescent waves for different colloids are shown to display a different dependence on the penetration depth. The intensity related to the distribution of the colloids as $I_s(q, \kappa) \approx I_0(q) \int_z n(z) \exp(-2\kappa z)dz$. In a simple excluded volume model, it allows deduction of a depletion zone were the tracer can not penetrate. Different depletions were measured. Similarly, the early slope of the normalized correlation functions

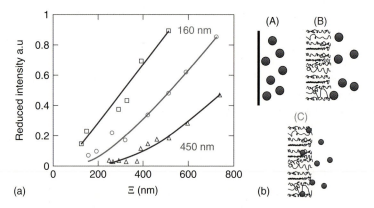

(a)

(b)

Figure 17.9 Different penetration of a large hard-sphere and a smaller softer particles within a PS brush deduced from the measured dynamic intensities of the tracer (a). (b) Brush with large colloids (120 nm) A: collapsed brush in decane, B: swollen brush in decalin, and C: brush with softer smaller colloids (40 nm) [40].

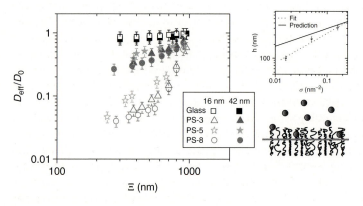

Figure 17.10 Apparent diffusivities of PS-nanogels in the vicinity of PS brushes of different grafting densities in toluene. The smaller nanogels appear to be much more slowed down with a similar type of penetrability in the brush. The diffusivities were obtained from the early slope analysis of the EW-DLS correlation functions [40].

could provide a measure of the average diffusivity of the tracer, weighted by the distribution of mobility over the penetration depth of the evanescent field as $\frac{\Gamma}{(q^2+\kappa^{-2})} = <D> = \int D(z)\exp(-2\kappa z)dz$, were Γ represents the measured early slope, q the wave vector and κ^1 the penetration depth of the evanescent field and $D(z)$ the z-dependent diffusion coefficient.

When different brushes were considered, the depletion zones were measured to depend on the brush as expected from the anticipated thickness (increasing grafting density lead to thicker brush). But the evolution of the diffusivities was very marked as the smaller particles were very strongly slowed down (Figure 17.10). The slow down being larger for the smaller species than for large ones was reminiscent of size-exclusion-type mechanisms. However, different local interaction as the three colloids had slightly different chemistry could also have contributed. Moreover, the contrast between the PS colloids and the PS brush was somewhat limited. Therefore, complementary measurements based on detection of single tracer, like FCS would certainly bring a useful confirmation of the EW-DLS observations.

17.4
Thermomechanical Properties

17.4.1
Thin Polymer Films

17.4.1.1 Inplane Elastic Properties
In contrast to thick films ($q_{//}h \gg 1$) where only a single doublet is present in the BLS spectrum (longitudinal phonon in VV and transverse phonon in VH) at each $q_{//}$, many acoustic excitations occur at a given $q_{//}$ when $q_{//}h = O(1)$. In this case the existence of boundaries has a substantial influence on the elastic-wave propagation. For isotropic supported thin polymer films, there are two kinds of dispersive modes that propagate between air/polymer film/glass substrate [44, 45]. These Love and Lamb modes are distinguished by their polarization, the displacements of the former being inplane and of the latter in sagittal plane. BLS is capable of measuring the elastic properties of thin hard (metal, semiconductors) films [46–49]. For free-standing thin polymer films, temperature-dependent BLS has revealed a thickness-dependent glass-transition temperature [50]. Despite the apparent virtue of BLS to obtain thermomechanical properties, this nondestructive technique has been restricted to free-standing polymer films [51–54] mainly due to drawbacks (heating, contribution of bulk scattering) from solid substrates. It is, however, the supported polymer thin films and coatings that are widely used technologically and other techniques [55–57] are already being employed to investigate several systems but not their mechanical properties. It was recently shown [58–60] that the utilization of the high-resolution q-dependent BLS can overcome the experimental difficulties and successfully infer the thermomechanical properties of thin polymer films on transparent substrates. A systematic study [58] of well-characterized PS and polymethylmethacrylate (PMMA) films on glass substrate over a large film thickness range (40–500 nm) led to the establishment of BLS as a powerful tool for studying thermomechanical properties at nanoscale, interface effects, mechanical, and phononic properties in complex composite structures [61]. We demonstrate first the model independent assessment of the elastic moduli and the estimation of T_g in the case of supported PMMA films.

Figure 17.11a shows exemplary BLS VV-spectra of PMMA film with thickness $h = 492$ nm at ambient temperature and $hq_{//} = 8.2$ selecting the direction of propagation $(q_{//})$ parallel to the film surface as indicated by the transmitted

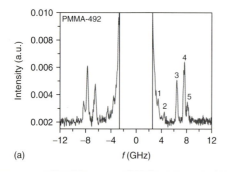

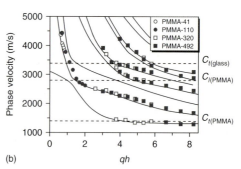

(a)

(b)

Figure 17.11 Inplane elastic excitations. (a) The BLS spectrum of the PMMA-492 film (thickness $h = 492$ nm) recorded at $q_{//} = 0.0167$ nm^{-1} and 25 °C shows five elastic excitations numbered from low to high frequencies. At the chosen frequency range, the longitudinal phonon peak of the glass substrate appears at about 15 GHz whereas, the longitudinal phonon of bulk PMMA at this temperature and $k_{//}$ would have the frequency 7.44 GHz ($c_l = 2800$ m/s) falling between modes 3 and 4. The frequency of the longitudinal phonon in the (thick) glass substrate ($\sim$15 GHz)

falls outside the chosen frequency range. (b) Dispersion diagram of the phase velocity $c = 2\pi f / k_{//}$ for the different elastic excitations in the supported PMMA thin films presented in the plot of c vs. the reduced thickness $k_{//}h$. The solid lines present the theoretical predictions for the Lamb modes based on the multilayer-scattering method using the density ($\rho_{PMMA} = 1180$ kg/m^3, $\rho_{glass} = 2500$ kg/m^3) and the sound velocities (dashed lines) of the bulk PMMA ($c_l = 2800$ m/s, $c_t = 1400$ m/s) and glass substrate ($c_l = 5665$ m/s, $c_t = 3320$ m/s).

geometry (cf. Figure 17.4a) on the top of the figure. The observed spectral features numbered from 1 to 5 with increasing frequency are due to the inelastic scattering of light by the film-guided phonons propagating parallel to the film/substrate interface. In spite of the existence of strong light scattering from the substrate, a manifold of modes are clearly observed. The identification of the observed modes requires both the measurement and the calculation of their dispersion relation (frequency *vs.* wave vector). The frequency values obtained by the representation of the BLS spectra by a sum of lorentzian line shapes are shown as effective sound velocities ($2\varpi f / q_{//}$) in the dispersion diagram of the Figure 17.11b encompassing four different film thicknesses and measurements at different values of $q_{//}$.

As expected, the film-guided modes are dispersive and hence the effective sound velocities depend on $hq_{//}$ and fall between the sound velocities (c_t) of the transverse phonon in the two materials as indicated by the dashed lines in Figure 17.11. The excellent superposition of the data from the PMMA films with different thicknesses suggests the very similar elastic constants of these thin films. For the theoretical calculation of the c vs. $hq_{//}$, a newly developed multilayer -scattering method [62] was employed. The simulation of the dispersion relation was performed using the densities and the sound velocities of the materials as fixed parameters. A very good agreement between experiment and theory is obtained only for the Lamb modes (solid lines in Figure 17.11) whereas, the computed dispersion curves (not shown) for the Love modes deviate systematically from the experimental data. Interestingly, at low $hq_{//}$ values there are experimental points, especially for the thinnest film above the glass transverse phase velocity, suggesting an appreciable elastic energy leak into the glass substrate. The good description of the experimental general Lamb modes for all films of the same material by using a single set of isotropic bulk elastic constants suggests no significant thickness dependence. A more accurate estimation of the elastic constants of the thinnest supported film seeking for the best agreement with the experiment led to the conclusion that within about 4% error both elastic constants assume the same values in the range 40–500 nm. This finding is consistent with a few earlier BLS experiments [63, 64] on thin supported films and nanostructures down to the size of about 100 nm. The represented approach illustrates the potential of this nondestructive optical technique to measure the elastic properties of thin films at a selected direction isothermally.

This successful detection of the general Lamb modes allows the study of the glass transition of thin supported films by BLS. Temperature-dependent measurements on the thickest and thinnest PMMA films led to the data of Figure 17.12. The variation of the sound velocity for the displayed modes is distinctly different at low and at high temperatures, conforming to a linear decrease with temperature. The intersection of the two (glass and rubbery) lines define the glass-transition temperature T_g (hatched areas in Figure 17.12). Both Lamb modes in thick PMMA-492 film display the same T_g value, which is expectedly T_g for bulk PMMA. For the thin PMMA-41 film the clearly observed 15 K decrease in T_g (from 387 to 372 K) is rather unexpected considering the favored interactions between PMMA segments and SiO$_2$ substrate. These are supposed to lead to higher energy barriers

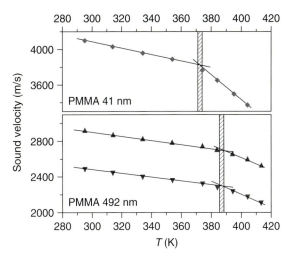

Figure 17.12 Glass transition in thin of polymer films. Phase velocities for a few general Lamb modes (at constant $k_{//}$) propagating in the PMMA films with two extreme thickness values as a function of temperature by heating the film from room temperature up to 413 K in successive steps. The solid lines denote the linear least-square fits for the glassy and rubbery states intersecting at the glass-transition temperature (shaded region).

for segmental dynamics compared to the thick PMMA film and therefore to an increase of T_g [58]. It should be mentioned that for PMMA films there is limited contradictory information and no reliable conclusion can be made [57, 64–66].

17.4.1.2 Out-of-Plane Elastic Properties

Mechanical properties and the stability of nanostructured materials are of paramount importance for a wide range of applications that comprise microelectronic, photonics, nanoelectromechanical systems, nanofluidics, and biomedical technologies. For supported polymer films, the characterization of the mechanical properties is of key importance in advancing the microlithography and other coating-based technologies. The demand for quantitative, noncontact and nondestructive probes drives the development of photoacoustic methods. Spontaneous BLS and impulsive-stimulated thermal scattering (ISTS), and similar pump-probe techniques are based on optical measurements of either spontaneous or laser-stimulated acoustic waves in the sample [20, 67, 68]. Directional selectivity of photoacoustic methods is of specific interest to the thin-film mechanics as many coatings exhibit substantial anisotropy of structure and properties. In Section 17.4.1.1 we demonstrated the utility of BLS to probe the inplane longitudinal and shear moduli in submicrometer-thick polymer films supported by a transparent substrate. ISTS has been used as an alternative method for thicker films [67] providing only indirect information on the out-of-plane elasticity, as studied acoustic excitations are characterized by finite inplane wave vector components $q_{//}$ [69]. Direct access to the out-of-plane elastic constants implies probing of acoustic waves with wave vectors $q_\perp$ normal to the film surface (Figure 17.4b). This measurement has been done by BLS for thick polymer films of a particular orientation relative to the incident laser beam [53]. A single-mode acoustic excitation, characteristic of the effective bulk medium, has been observed. Out-of-plane standing acoustic wave excitations in free-standing films have been known for decades [46]. These are shown schematically in Figure 17.13a. However, the same phenomenon in supported films has been reported only recently [70, 71] using a backscattering geometry typically utilized in all these experiments. In this geometry (the probing laser beam hits the coating side of the sample and the scattered light is monitored backwards), the corresponding wave vector has nonzero $q_{//}$, which complicates the spectra interpretation due to the inplane excitations. The scattering geometry of Figure 17.4b (refection mode) is the most appropriate for investigation of out-of-plane excitations because there is no inplane $q_{//}$ transfer.

For the proof of concepts, supported PS and PMMA films with thickness in the range 40 nm to 2.8 μm were fabricated by spin coating of Ø 25-mm microscope cover slides made by optical borosilicate glass. Figure 17.13 displays polarized BLS spectra of supported PS films with different thicknesses between 40 nm and 2.8 μm at $q_\perp = 0.037$ nm^{-1} and 20 °C. Since the light scattering from supported films on transparent substrates arises mainly from the elasto-optic effect [58], reduction in the scattering volume results to an increase of acquisition times. The thickest (2.8-μm thick) film spectrum is expectedly unimodal, corresponding to

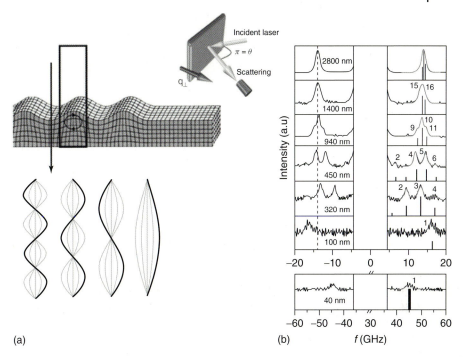

(a)   (b) f (GHz)

Figure 17.13 Standing elastic waves in supported films. (a) Schematic presentation of the Lamb modes with matter displacements normal to the substrate and four ($m = 1–4$) standing longitudinal excitations. (b) Experimental polarized BLS of supported polystyrene (PS) films with different thickness in the range 40 nm to 2.8 μm were recorded in the reflection geometry shown on the top. The integers and the red lines in the Stokes side of the spectra indicate the vibration order m (Equation 17.3) and the fit of up to four lorentzian line shapes, respectively. The vertical lines represent the computed amplitudes of the observed modes (Equation 17.4).

the longitudinal acoustic (LA) excitation propagating in the normal direction to the film plane. However, as film thickness decreases, the LA mode broadens and finally splits into the set of equidistant submodes leading to a remarkable multimodal pattern. Propagating through a thin film, a single LA phonon is localized [46] within the distance d from the film surface, where d is the film thickness. Due to the uncertainty principle, the phonon momentum forms $\Delta p \approx h/d$ – a wide distribution around the mean value $hq/2\pi$ with $\hbar$ being the Planck constant. The wave vector q is no longer a certain value, but has a distribution $\Delta q \approx 2\pi/d$, and the Brillouin peak has a width $\Gamma \approx c_l/d$.

Multiple reflections of an acoustic excitation from the film surface and interface can form *a standing wave* that can be observed when substantial contrast of the acoustic impedance ($Z = \rho c_l$) between the polymer film and the substrate exists [72]. Multiphase systems with small Z mismatch between phases do not support phonon localization and exhibit an effective medium behavior [73]. For supported PS, the contrast $\Delta Z = Z_{GLASS}/Z_{POLYMER} - 1$ is equal to 5.0, which is sufficient for standing-wave observation. The criterion for constructive interference of the excitations is $q_m d = \pi m + \pi/2$ where m is an integer and $\pi/2$ term is the phase shift due to the reflections on film boundaries. These submodes occupy the envelope of the broadened LA phonon mode, forming the fine structure with equidistant (Δf) frequencies $f_m = c_l q_m/2\pi$

$$f_m = (2m + 1)c_l/4d, \quad \Delta f = c_l/2d \tag{17.3}$$

More rigorously, the fine structure of the BLS spectrum at angle θ is described by:

$$P_s(\theta) \propto \sum_m \text{sinc}^2 \left[\frac{\pi d}{v} \left(q(\theta)\frac{v}{2\pi} - f_m \right) \right] \tag{17.4}$$

where the wave vector $q(\theta)$ is defined by the system geometry, and sinc is the cardinal sine function $\text{sinc}(x) = \sin(x)/x$. Equation 17.4 demonstrates that standing-wave modes f_m are localized in the $\text{sinc}^2(x)$ envelope around the LA mode with shift $q(\theta)\dfrac{v}{2\pi}$. Figure 17.13 clearly demonstrates very good agreement between the experimental and model spectra, concerning not only the eigenfrequencies but also the relative contribution of the individual modes to the BLS spectrum, notably with no adjustable parameter; the longitudinal sound velocity in PS $c_l = 2350$ m/s and the thickness d for the samples are experimentally known values.

17.4.2
Periodic Multilayer Polymer Films

Nanostructured multilayer composite films fabricated by layer-multiplying coextrusion [74–76] of incompatible polymers exhibit superior characteristics such as extraordinary optical effects, for example, broadband omnidirectional reflection and giant birefringence and improved barrier and mechanical behavior. These multilayers consist of hundreds or thousands of thin layers of two polymers in strictly alternating fashion. The distinctly different physical quantities of the constituent nanolayers and, in particular, the periodic modulation of density and elastic constants with numerous interfaces can modify the propagation of elastic waves in an unprecedented manner. Consequently, elastic excitations in these multinanolayers can be complex as the structure periodicity can influence the elastic properties. For thin bilayers, finite-size effects and confinement can modify polymer conformation that in turn impacts on different material properties such as heat conductivity. As phonons in dielectrics play a decisive role for heat transport, access to the phonon dispersion relations becomes an issue. On the experimental side, the microscopic dimensions of the stack of nanolayers strongly increase the signal-to-noise ratio while concurrently warranting information at nanoscale.

17.4.2.1 Effective Medium and Confinement Effects
For the inplane mechanical properties, BLS spectra were recorded in the transmission mode (Figures 17.14a and 17.11) [73]. For two symmetric PMMA/PC multilayer films with bilayer thickness $d = 25$ nm (1024 layers) and $d = 780$ nm (128 layers), polarized and depolarized spectra at $q_{//} = 0.0152$ nm^{-1} are shown

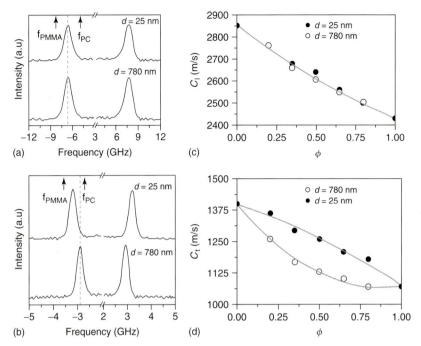

Figure 17.14 Confinement effect. (a) Polarized (VV) and (b) depolarized (VH) BLS at $q_{//} = 0.0152$ nm^{-1} and 25 °C for two periodic PC/PMMA multilayer films with symmetric composition but two extreme nanolayer film thicknesses (stated in the plot). The two arrows in the top of each figure denote the frequency of the longitudinal (a) and transverse (b) phonon (at 0.0152 nm^{-1} and 25 °C) in the pure PC and PMMA multilayer films (indicated by the subscripts). As indicated by the two dashed vertical lines only the transverse sound frequency depends on the individual nanolayer thickness. The variation of the longitudinal (c) and shear (d) sound velocity with PC composition in the PC/PMMA multilayer films with bilayer spacing $d = 25$ nm (solid symbols) and $d = 780$ nm (open symbols). The longitudinal sound velocity (a) is independent of nanolayer thickness and conforms to the effective medium Wood's law (solid line), whereas the shear sound velocity reveals confinement effects.

in the Figure 17.14a,b. All polarized spectra show a single doublet at the same frequency ($f = 6.35$ GHz) leading to the phase sound velocity ($c_l = 2\pi f/q_{//}$) of 2625 ± 30 m/s falling between c_{lPMMA} (2850 m/s) and c_{lPC} (2430 m/s) of the pure constituent films. This is indicative of an effective medium behavior irrespective of the bilayer thickness. The composition dependence of $c_l(\phi)$ depicted in Figure 17.14c,d for the two multilayer films with extreme bilayer thickness as indicated in the plot nicely conforms (solid line) to the effective medium Wood's equation

$$1/\rho c_l^2 = \phi/(\rho_{PC}\, c_{PC}^2) + (1 - \phi)/(\rho_{PMMA}\, c_{PMMA}^2) \tag{17.5}$$

with no adjustable parameter. Hence, confinement at nanoscale does not affect the longitudinal modulus $M = \rho c_l^2$ of these multilayer films. Based on Section 17.4.1.1

the presence of a single effective medium phonon (cf. Figure 17.11) even for the multilayer films with $dq_{//} = O(1)$ is an unexpected observation (see next section).

Figure 17.14 displays the VH spectra for the same two bilayer PC/PMMA films. Since the PMMA segments are almost optically isotropic, the intensity originates from the PC nanolayers but the transverse phonon propagates through the bilayers; the experimental c_t falls between the c_{tPMMA} and c_{tPC}. In clear contrast to the VV spectra, the peak position for the symmetric multilayers is no longer constant but shifts to lower frequencies with increasing bilayer thickness and consequently c_t decreases from 1270 m/s ($d = 25$ nm) to 1130 m/s ($d = 780$ nm). This trend for $\phi = 0.5$ applies to other compositions as seen in Figures 17.14c,d. Thus, confinement has an evident impact on the shear modulus despite its negligible influence on the longitudinal modulus possibly due to shearing forces during processing. This impacts the chain conformation more than the density. The shear modulus $G = \rho c_t^2$ increases by 60% in the bilayers with $d = 10$ nm compared to the bilayers with 780 nm ($G = 1.5$ GPa).

17.4.2.2 Interaction Elastic Waves and Structure

In addition to the dependence of the effective medium sound velocity on the phonon polarization, BLS spectra also carry information on the structure influence on phonon propagation. The interaction between wave and structure is anticipated for high $dq_{//}$ values. For the symmetric PC/PMMA multilayer films with bilayer thickness $d = 780$ nm and $d = 100$ nm, the VV spectra shown in a logarithmic intensity scale in Figure 17.15 are apparently very different. For the former with $q_{//}d/2 \sim 7$, the rich features are clearly visualized and up to five modes (2–6) are clearly resolved. For the second film with $q_{//}d/2 \sim 0.9$, the unimodal spectrum with a single longitudinal (5) (Figure 17.15) and transverse (1) (from the VH spectrum) expectedly displays a homogeneous-medium behavior. The absence of surface dispersive modes in the individual layers found in supported films (see Figure 17.11) is rationalized by the low elastic mismatch ΔZ of the bilayer along with their low c_t. For phonon propagation normal to the layers, Figure 17.15 at $q_{\perp} = 0.035$ nm^{-1} shows that the sound velocity ($2\pi f/q_{\perp}$) of the single acoustic longitudinal phonon is the same in both films as that of mode (5) for inplane propagation. This agreement indicates the mechanical isotropy of the present system, in contrast to polyimides (PIs) film (Section 17.4.3). Due to the small Z mismatch between PC and PMMA, the bilayers do not support phonon localization observed in Section 17.4.1.1 (Figure 17.11) for polymer films supported on glass substrates.

The identification of the spectral features of Figure 17.15 and the interpretation of the observed inplane propagation modes needs the comparison between experimental and theoretical dispersion diagrams. Finite element analysis (FEA) for perfectly bonded and ideally flat layers with uniform thickness predicts mainly quasilongitudinal (QL) and quasitransverse (QT) modes (with displacements fields primarily parallel or perpendicular to the wave vector). The details of the displacements field for $q_{//} = 0.025$ nm^{-1} in the symmetric film with $d = 780$ nm are shown in Figure 17.16a order from low to high frequency. There are three distinct

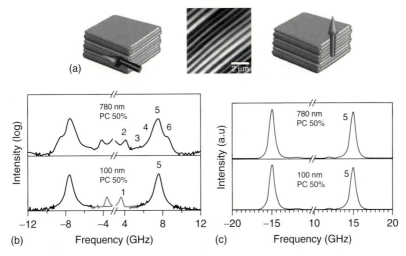

(a)

Intensity (log)

780 nm
PC 50%

5

2   4   6
    3

100 nm
PC 50%

5

1

−12   −8   −4   4   8   12

(b)   Frequency (GHz)

Intensity (a.u)

780 nm
PC 50%

5

100 nm
PC 50%

5

−20   −15   10   15   20

(c)   Frequency (GHz)

Figure 17.15 Direction dependence. Polarized BLS spectra at $q_{//} = 0.0181$ nm^{-1} and 25 °C for two periodic symmetric PC/PMMA multilayer films with bilayer thickness 780 mm and 100 nm at 25 °C and phonon propagation either (b) parallel ($q_{//} = 0.0181$ nm^{-1}) or (c) normal ($q_{\perp} = 0.035$ nm^{-1}) to the layers as indicated by the two cartoons in (a). Cross-sectional TEM image of the symmetric film with $d = 780$ nm is also shown in the top of the figure. The numbers indicate modes with increasing frequency and with modes (1) and (5) being respectively the transverse and longitudinal phonons in the VH and VV spectra in the left panel of Figure 17.14. The VV spectrum of the two films at $q_{\perp} = 0.035$ nm^{-1} perpendicular to the layers measures essentially the effective medium mode (5).

QT(1−3) and three QL(4−6) modes predicted for this sample, although only two QT modes (1,2) are resolved experimentally, probably due to the frequency proximity and/or structural imperfections. While the various QT and QL propagate with the same effective medium shear and longitudinal sound velocity, the displacements are localized primarily within the individual PC or PMMA layers. This strong intriguing prediction was proved by comparing the softening temperatures (cf. Figure 17.12) of the experimental phase velocities. Figure 17.16 shows this quantity for the symmetric $d = 780$ nm and $d = 25$ nm PC/PMMA films. For both effective medium modes 1 and 5, the latter sample displays a single T_g of 122 °C which is intermediate between the T_g of PMMA (105 °C) and PC (140 °C) layers. Thus, the propagation of these two phonons does not resolve the presence of individual layers displaying a homogeneous medium-like behavior. In contrast, there are three distinct T_g for the four modes in the film with $d = 780$ nm. Modes 1 and 4 have a T_g of about 135 °C, which is similar to T_g of PC, mode 6 has T_g of about 105 °C which is essentially the T_g of PMMA, while the T_g of mode 5 is about 122 °C. Therefore, modes 1 and 4 must propagate primarily in the PC layers, mode 6 primarily in the PMMA layers and mode 5 experiences both layers. In fact, the theoretical displacement fields captures reasonably well the thermomechanical of the $d = 780$ nm film. Figure 17.16 suggests that the lowest frequency QT and QL

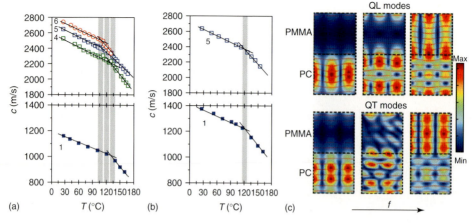

(a)   T (°C)   (b)   T (°C)   (c)

Figure 17.16 Heterogeneous temperature dependence. The softening transition temperatures (vertical shaded regions) for the observed modes in the experimental BLS spectra of the symmetric PC/PMMA multilayer film with (a) $d = 780$ nm (modes 1,4–6) and (b) $d = 25$ nm (modes the effective 1,5). Solid lines denote linear least-square fits at low and high temperatures intersecting at the softening transition temperatures. (c) Details of the displacement fields for the quasitransverse (QT) and quasi-longitudinal (QL) modes for the symmetric PC/PMMA with $d = 780$ nm at $q_{//} = 0.025$ nm^{-1}. These patterns suggest that the lowest frequency QT (1) and QL (4) should propagate in the PC layers, the QL (6) in the PMMA layers while QL(5) is not localized.

modes (1,4) should propagate in the PC layers, the highest frequency QL mode 6 in the PMMA layers, whereas the midfrequency QL mode (5) exhibits real localization.

17.4.3
Mechanical Anisotropy

The ability of BLS to measure elastic constants along different directions as presented in Section 17.4.1 can hardly be overestimated for anisotropic film samples frequently found in practical applications [77]. Mechanical anisotropic correlates with preferred orientation of structural units [78] and nondestructive probing of elastic properties has become a state-of-the-art approach in thin film [60, 79]. In comparison, the other parameter widely used for anisotropic characterization is optical birefringence which shows much less magnitude in some polymer films [80]. In a few studies of anisotropic supported films, the investigated films by ISTS are substantially thicker than 1 μm [69, 81]. For the important class of submicrometer coatings, exclusively used in microlithography, probing of elastic constants is performed on films with assumed isotropy and limited to one specific direction [49, 64]. BLS as introduced in Section 17.4.1 has very recently been applied [60] to the study of highly anisotropic supported films in the 0.1–20 μm. For thick films (>2 μm), the utilization of the transmission and reflection scattering modes (Figure 17.4) directly yields the acoustic longitudinal and transverse phonon along

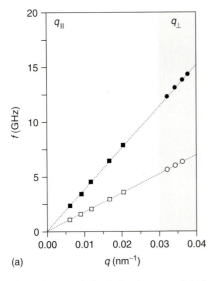

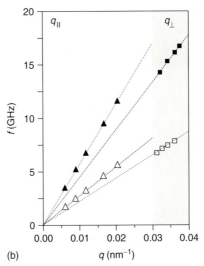

Figure 17.17 Mechanical anisotropy of thick films. Acoustic phonon propagating with longitudinal (solid symbols) and transverse (open symbols) polarization parallel ($q_{//}$) and normal ($q_\perp$ gray region) to the film surface of thick polystyrenes (PS) 2.8 μm) (a) and polyimide (BDPA-PDA, 7.2 μm) (b) films. The slopes of the six dashed lines yield the longitudinal (v_l) and the shear (v_t) parallel (‖) and normal ($\perp$) to the film.

the two symmetry directions as visualized in Figure 17.17 for PS and PI films. Clearly, both longitudinal and shear sound velocities are higher parallel than normal to the surface of PI films, while PS films are mechanically isotropic with the same sound velocities in both directions.

For films with submicrometer thickness, the simple unimodal BLS become multimodal and the resolved modes display significant dispersions (cf. Figures 17.11 and 17.13). In addition, the presence of anisotropy modifies the dispersion of the inplane excitations that involves the out-of plane sound velocities as well. Yet, the elastic properties can depend on thickness. Therefore, their estimation based only on the inplane propagation might be ambiguous. The fine structure of the out-of-plane LA mode detected in submicrometer coatings (Figure 17.13) offers an excellent opportunity to trace the size dependence of the corresponding elastic modulus. As shown in the BLS spectra of Figure 17.18, the representation of the standing Lamb modes (Section 17.4.1.2) by the sum of up to five lorentzians yields their eigenfrequencies and hence the frequency interval between adjacent resonance modes (Equation 17.3). The computed out-of-plane longitudinal sound velocity $v_l^\perp$ reveals no size dependence in the studied thickness interval (0.16–20 μm) and excellent agreement between $v_l^\perp$ values obtained by the two methods (Figures 17.17 and 17.18). The fine structure is found for the LA mode only. The low depolarized scattering signal may prevent observing the transverse acoustic phonon splitting and thereby access

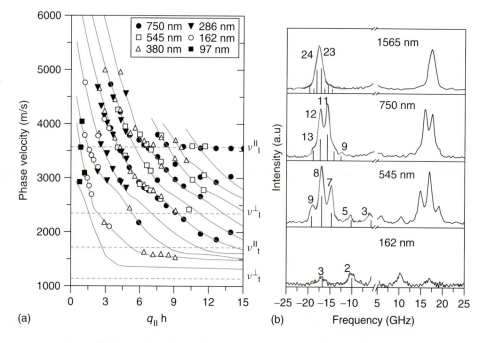

Figure 17.18 Standing longitudinal elastic excitations in supported polymer films. (a) Dispersive inplane elastic excitations in supported thin BPDA–PDA films. The plot contains data for six different thicknesses in the 0.1–0.75 μm range and for q in the 0.006–0.022 nm^{-1} range. Correspondence between the symbols and the thickness of coating is given in the inset legend. The red solid lines represent the theoretical model with all model parameters predetermined. The large disparity of the value of v_l and v_t in the two orthogonal directions is indicated by the dashed horizontal lines. (b) Thickness dependence (scattering angle $\theta = 150°$ ($q_\perp = 0.0443$ nm^{-1}) of the BLS spectra for the BPDA–PDA coatings. The spectra are well represented (red solid lines in the anti-Stokes side) by up to five lorentzian lines. The integers denote the vibration order m and the vertical lines represent the computed amplitudes of the observed modes.

to the corresponding sound velocity $v_t^\perp$ normal to the film for submicrometer films.

An alternative source of information is the dispersion diagram for inplane phonon propagation that possess mixed polarization, that is, displacement fields in the sagittal plane. As a result, their phase velocities are functions of both the inplane and out-of-plane ($v_l^{//} v_t^{//}$) and ($v_l^\perp, v_t^\perp$) intrinsic longitudinal and transverse sound velocity values. The dispersion of the inplane waveguide modes shown in Figure 17.18a is well represented (red solid lines) with all five ($v_l^{//} v_t^{//}, v_l^\perp, v_t^\perp, \rho$) values for PI films taken from the thick film BLS experiment (Figure 17.17); the values of the glass substrate are determined beforehand (cf. Figure 17.11). The tight correspondence of the experiment and the model for the whole thickness range (Figure 17.18) implies strong, size independent, mechanical anisotropy.

17.4.4
One-Dimensional Phononic Films

In the periodic multilayer films of Section 17.4.2.2, a single phonon was observed along the periodicity direction (with $q_\perp$). These 1D phononic structures should represent the mechanical analogs of photonic crystals that prohibit the propagation of light along high-symmetry directions due to interference effects [82]. Photonics in the visible that are mesoscopic structures with submicrometer lattice constants should control the propagation of elastic (acoustic) waves in the hypersonic (1–1000 GHz) range. The experimental realization of hypersonic phononics has been limited, being primarily focused on 2D and 3D systems [83–87]. In contrast, the few 1D systems investigated in the gigahertz range were polycrystalline [88], did not exhibit precisely controlled periodicity [73] or possess sufficiently high elastic contrast [53].

Nanocrystalline materials and amorphous polymers represent potential candidates for designing 1D phononic crystal due to their high elastic impedance contrast ΔZ and optical transparency, in contrast to silicon-based superlattices [89]. The hybrid phononic structure [90] depicted in Figure 17.12 is a multilayer stack of inorganic SiO_2 nanoparticles and PMMA nanolayers exemplifying high and low elastic modulus materials, respectively. Upon spinning and annealing the SEM image of a total of 20 repeat periods ensure reproducibility of the SiO_2 (65 nm) and PMMA (35 nm) layers. The BLS spectra along the two directions depicted schematically in Figure 17.19 are distinct. For phonon propagation parallel to the layers, the BLS spectrum resembles that of the PC/PMMA multilayers (Figure 17.14) for $q_{//}d/2 < 1$ and the single doublet yields an effective medium $v_1^{//}$ given by the slope of the blue circles in Figure 17.19a. Along the normal direction of the hypersonic periodic phononic structure, the momentum conservation is $\mathbf{q} = \mathbf{k} + \mathbf{G}$, where $\mathbf{G}$ is the reciprocal lattice vector ($G = \pi/d$) of the crystal As phonons with wave vectors $\mathbf{k}$ and $\mathbf{k} + \mathbf{G}$ belong to the same eigenmode within the phononic crystal, its phononic properties are revealed by recording the dispersion relation $\omega(q)$. Figure 17.19 shows polarized BLS of the 20-layer SiO_2–PMMA phononic crystal recorded at three different $q_\perp$ values near the Brillouin zone (BZ) boundary at $q_{BZ}(= \pi/d = 0.0314\ nm^{-1})$. For $q_\perp < q_{BZ}$, the BLS spectrum at $0.0301\ nm^{-1}$ displays a single doublet and then splits into a double doublet, when $q_\perp$ crosses q_{BZ} (Figure 17.19). Together, the spectra provide direct evidence for a stop band of the elastic wave due to the interference as a result of the discrete translational periodicity of the variable impedance structure along the film normal. This is clearly demonstrated in the dispersion relation of Figure 17.19.

This stop band with a central frequency at 12.6 GHz and a width of 4.5 GHz or a relative breadth of about 30% opens at the first BZ (vertical solid mine) and is hence called the *Bragg gap*. The propagation of hypersonic longitudinal phonons in the direction normal to the layers for frequencies within the marked blue region is forbidden. The inplane and out-of-plane (red circles) appears to fall on the same acoustic branch indicative of a single effective medium sound velocity along both directions expected for mechanically isotropic systems. The theoretical dispersion

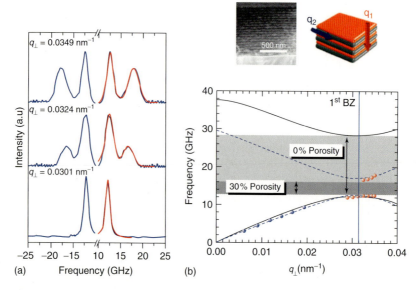

Figure 17.19 Hypersonic phononic crystal. Scanning electron microscopy image of a SiO$_2$–PMMA multilayer film cross-section and the schematic presentation of the scattering geometries for inplane ($q_{//}$) and out-of-plane ($q_\perp$) phonon propagation. (a) BLS spectra of the hybrid SiO$_2$/PMMA film at three different $q_\perp$ values near the first Brillouin zone (BZ) ($q_{BZ} = 0.0314$ nm^{-1}). The red solid lines indicate the fit of up to two lorentzian line shape. (b) Experimental dispersion for propagation parallel to the layers ($0 < q_{//} < 0.02$ nm^{-1}, blue circles) and along the direction of periodicity (0.029 nm$^{-1} < q_\perp < 0.033$ nm^{-1}, red points). The highlighted region indicates the observed hypersonic bandgap for 30% porosity SiO$_2$ particles. The double arrows show the much wider bandgap for SiO$_2$ particles with the elastic constants of the conventional glass (zero porosity).

diagram is computed by FEA calculations (Section 17.4.2.2) in an infinite ideal 1D periodic medium treating the porous SiO$_2$ layer as an effective medium. The experimental values obtained from individual SiO$_2$ and PMMA layers were used directly in the calculations thus performed with no adjustable parameter. The good agreement between the theoretical (dashed blue lines) and the experimental band dispersion relations is visualized in Figure 17.19 that also includes the calculations (solid black lines) for a PMMA/fused SiO$_2$ (zero porosity. For the latter both the gap midfrequency ($\sim$20 GHz) and the width ($\sim$80%) is increased due to the larger impedance contrast ΔZ. The porous SiO$_2$ retains a sizeable gap and furthermore lends itself as a possible structural scaffold for introducing secondary active media.

17.4.5
Particle-Shape Fluctuations (Vibration Modes)

Solid particles undergo thermal shape fluctuations or vibration motions in analogy to the standing waves in the supported polymer films (Sections 17.4.1.1 and 17.4.1.2). These localized modes [91] can be detected by Raman scattering [92, 93]

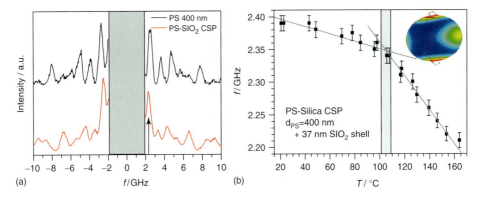

Figure 17.20 Particle music. (a) Eigenmode spectra of an uncoated polystyrene latex sphere with diameter $d = 400$ nm (black top) and the corresponding PS-SiO$_2$ core-shell particle with a 37-nm SiO$_2$ shell layer (red bottom). (b) The variation of the frequency of the strongest (1,2) mode (arrow in the (a)) with temperature indicates two linear regimes corresponding in the glassy and rubbery state of the polystyrene core. Particle-shape fluctuations due to the first-order second-harmonic (1,2) eigenmode for the pure PS sphere in air with the color (blue to red) showing increasing total displacement.

and BLS [94, 95] or can be excited by strong laser pulses [68, 87]. For turbid systems, for example, powders of mesoscopic ($d \sim \lambda$) particles, **q** is ill-defined due to strong multiple scattering. In this case, BLS can measure only localized in space (and hence is q independent) vibrational modes at frequencies $f(n.l)$ characterized by the angular momentum l on the nth order; the intensity of these modes does depend [91] on qd. The frequencies can be theoretically [95, 96] obtained from the density-of states spectra of a single sphere as a function of the c_t, c_l (or elastic moduli G and M) and the particle diameter. For PS spheres, the representation of up to 21 modes was obtained [95] using $M = 5.8$ GPa and $G = 1.54$ GPa. The lowest eigenfrequency $f(1,2) \sim 0.85\, c_t/d$ of a spherical particle depends only on c_t. Here, we present a recent application of the eigenmode to address the confinement effect in shape-persistent polymer colloids [96, 97].

Figure 17.20 shows high-resolution BLS eigenmode spectra for a core-shell particle consisting of a PS core with $d = 400 \pm 12$ nm and a thin SiO$_2$ nanoshell with thickness $L = 37 \pm 3$ nm and the parent plain PS spheres ($d = 400$ nm). Both spectra display rich music with the core-like $f(1,2)$ mode being the strongest (indicated by the arrow in Figure 17.20) line as the spectrum of the core-shell particle is very akin to the spectrum of the bare PS core. In the inverted case of the core-shell particle with SiO$_2$ core, mainly the soft shell-like modes were identified [96, 98]. The temperature dependence of $f(1,2)$ for the PS-SiO$_2$ particle is represented by two distinct straight lines with negative slope, as seen in Figure 17.20b. The characteristic kink in the f vs. T plot (cf. Figures 17.12 and 17.16) signifies the transformation of a glassy to rubbery state for the PS core occurring at about 107 °C. The relative drop of $f(1,2)$ between room temperature and T_g is less than 2% in the PS-SiO$_2$ particles and more than 6% in the bare PS cores. This is an unprecedented finding and might relate to the different vitrification conditions, that is, isochoric *vs.* isobaric. In the rubbery state, $f(1,2)$ at the highest examined temperature of 165 °C also assumes a much weaker decrease ($\sim$6%) compared to the corresponding reduction ($\sim$40%) in PS films (Figure 17.12 and Ref. [58]). This modified behavior of the confined mater might impact the physical aging (below T_g) and a size-dependent study can help elucidate the underlying confinement mechanisms.

17.5
Conclusion and Outlook

We have selected a few examples of polymer-based films and coatings to probe their dynamics at thermal equilibrium, measure their elastic constants and reveal novel phononic properties. This rich information is obtained from different light-scattering experimental techniques and FCS that provide spatiotemporal and single-molecule detection. The main message is that the behavior of soft materials in nanofabricated structures cannot be predicted from the materials properties in the bulk. And thin polymer layers and films are present in various states (dry or swollen) and forms in various advanced devices and structures enabled by significant progress in polymer and colloidal chemistries and processing methods.

The understanding of the novel properties depends on the development of both experimental techniques and theoretical concepts. Such research effort will accompany the movement toward a rational design of functional materials and optimal devices. Due to the thermal energy, the structure of soft materials is in perpetual motion and it is this dynamics that, in turn, reflects the complexity of structure, interactions, and environment. Two recent examples are mentioned to illustrate the point. The dynamics of solvent and temperature-swollen polymer films and the relation to the bulk replica remains a challenging endeavor. The connection between the internal dynamics and the tracer diffusion of probes in hydrogel layers that are the active components in many applications, for example, biosensoring.

Another consequence of thermally excited dynamics is the elastic or acoustic waves that cannot propagate in vacuum. Hence, composite structures and architectures such as polymer films, bilayers, or periodic hybrid materials dictate the shape of the wave-dispersion diagram. This experimental access by BLS has a dual impact. It allows the investigation of the material behavior in relatively small confined regions and yields the thermomechanics (Young's and shear moduli, T_g) at the nanoscale. It introduces amenability to tune novel phononic properties at gigahertz frequencies. Further, BLS is the only technique that allows the nondestructive and unperturbed measurement of dimension and direction-dependent mechanical properties in nanostructured materials.

The reviewed work is an exciting area of research with many potential applications and several new developments ahead. This will push the frontiers of soft-matter physics into new areas, for example, phononics and new material behavior. The importance of developing an understanding of polymeric materials at the nanoscale cannot be overemphasized. However, the fundamental understanding still needs foundation.

References

1. Colby, R. and Rubinstein, M. (2003) *Polymer Physics*, Oxford University Press.

2. Stoykovich, M.P., Móller, M., Kim, S.O., Solak, H.H., Edwards, E.W., De Pablo, J.J., and Nealey, P.F. (2005) Materials science: directed assembly of block copolymer blends into nonregular device-oriented structures. *Science*, **308** (5727), 1442–1446.

3. Akcora, P., Liu, H., Kumar, S.K., Moll, J., Li, Y., Benicewicz, B.C., Schadler, L.S., Acehan, D., Panagiotopoulos, A.Z., Pryamitsyn, V., Ganesan, V., Ilavsky, J., Thiyagarajan, P., Colby, R.H., and Douglas, J.F. (2009) Anisotropic self-assembly of spherical polymer-grafted nanoparticles. *Nature Mater.*, **8** (4), 354–359.

4. Berne, J. and Pecora, R. (2000) *Dynamic Light Scattering*, Dover.

5. Lan, K.H., Ostrowsky, N., and Sornette, D. (1986) Brownian dynamics close to a wall studied by photon correlation spectroscopy from an evanescent wave. *Phys. Rev. Lett.*, **57** (1), 17–20.

6. Fytas, G., Anastasiadis, S.H., Seghrouchni, R., Vlassopoulos, D., Li, J., Factor, B.J., Theobald, W., and Toprakcioglu, C. (1996) Probing collective motions of terminally anchored polymers. *Science*, **274** (5295), 2041–2044.

7. Sigel, R. (2009) Light scattering near and from interfaces using evanescent wave and ellipsometric light scattering. *Curr. Opin. Colloid Interface Sci.*, **14** (6), 426–437.

8. Plum, M.A., Steffen, W., Fytas, G., Knoll, W., and Menges, B. (2009) Probing dynamics at interfaces: resonance enhanced dynamic light scattering. *Opt. Exp.*, **17** (12), 10364–10371.

9. Stocco, A., Tauer, K., Pispas, S., and Sigel, R. (2009) Dynamics at the air-water interface revealed by evanescent wave light scattering. *Eur. Phys. J. E*, **29** (1), 95–105.

10. Langevin, D. (1992) Light Scattering by Liquid Surfaces and Complementary Techniques (ed. D. Langevin), Marcel Dekker: New York.

11. Earnshaw, J.C. (1996) Light scattering as a probe of liquid surfaces and interfaces. *Adv. Colloid Interface Sci.*, **68** (1–3), 1–29.

12. Cicuta, P. and Hopkinson, I. (2004) Recent developments of surface light scattering as a tool for optical-rheology of polymer monolayers. *Colloids Surf. A: Physicochem. Eng. Aspects*, **233** (1–3), 97–107.

13. Esker, A.R., Kim, C., and Yu, H. (2007) Polymer monolayer dynamics. *Adv. Polym. Sci.*, **209** (1), 59–110.

14. Koga, T., Li, C., Endoh, M.K., Koo, J., Rafailovich, M., Narayanan, S., Lee, D.R., Lurio, L.B., and Sinha, S.K. (2010) Reduced viscosity of the free surface in entangled polymer melt films. *Phys. Rev. Lett.*, **104** (6), 0066101.

15. Kim, A., Róhm, H., Lurio, L., Basu, J., Lal, J., Lumma, D., Mochrie, S., and Sinha, S. (2003) Surface dynamics of polymer films. *Phys. Rev. Lett.*, **90** (6), 068302.

16. Forrest, J. (2002) A decade of dynamics in thin films of polystyrene: where are we now? *Eur. Phys. J. E*, **8** (2), 261–266.

17. Alcoutlabi, M. and McKenna, G. (2005) Effects of confinement on material behaviour at the nanometre size scale. *J. Phys. Condens. Matter*, **17** (15), R461–R524.

18. Gianneli, M., Roskamp, R.F., Jonas, U., Loppinet, B., Fytas, G., and Knoll, W. (2008) Dynamics of swollen gel layers anchored to solid surfaces. *Soft Matter*, **4** (7), 1443–1447.

19. Yakubov, G.E., Loppinet, B., Zhang, H., Ruhe, J., Sigel, R., and Fytas, G. (2004)

Collective dynamics of an end-grafted polymer brush in solvents of varying quality. *Phys. Rev. Lett.*, **92** (11), 115501-1.

20. Still, T., Cheng, W., Retsch, M., Jonas, U., and Fytas, G. (2008) Colloidal systems: a promising material class for tailoring sound propagation at high frequencies. *J. Phys. Condens. Matter*, **20** (40), 404203.

21. Voudouris, P., Gomopoulos, N., Le Grand, A., Hadjichristidis, N., Floudas, G., Ediger, M.D., and Fytas, G. (2010) Does Brillouin light scattering probe the primary glass transition process at temperatures well above the glass transition? *J. Chem. Phys.*, **132** (7), 074906.

22. see Berne, J. and Pecora, R. (2000) *Dynamic Light Scattering*, Chapter 6, Dover.

23. Kim, S.Y. and Lee, S.C. (2009) Thermo-responsive injectable hydrogel system based on poly(n-isopropylacrylamide-co-vinylpnosphonic acid). i. biomineralization and protein delivery. *J. Appl. Polym. Sci.*, **113** (6), 3460–3469.

24. Qiu, Y. and Park, K. (2001) Environment-sensitive hydrogels for drug delivery. *Adv. Drug Delivery Rev.*, **53** (3), 321–339.

25. Tokarev, I. and Minko, S. (2009) Stimuli-responsive hydrogel thin films. *Soft Matter*, **5** (3), 511–524.

26. Giannelli, M., Beines, P.W., Roskamp, R.F., Koynov, K., Fytas, G., and Knoll, W. (2007) Local and global dynamics of transient polymer networks and swollen gels anchored on solid surfaces. *J. Phys. Chem. C*, **111** (35), 13205–13211.

27. Cherdhirankorn, T., Best, A., Koynov, K., Peneva, K., Muellen, K., and Fytas, G. (2009) Diffusion in polymer solutions studied by fluorescence correlation spectroscopy. *J. Phys. Chem. B*, **113** (11), 3355–3359.

28. Cheng, Y., Prud'homme, R.K., and Thomas, J.L. (2002) Diffusion of mesoscopic probes in aqueous polymer solutions measured by fluorescence recovery after photobleaching. *Macromolecules*, **35** (21), 8111–8121.

29. Gisser, D.J., Johnson, B.S., Ediger, M.D., and Von Meerwall, E.D. (1993) Comparison of various measurements of microscopic friction in polymer solutions. *Macromolecules*, **26** (3), 512–519.

30. Li, J., Li, W., Huo, H., Luo, S., and Wu, C. (2008) Re-examination of the slow mode in semidilute polymer solutions: the effect of solvent quality. *Macromolecules*, **41** (3), 901–911.

31. Cho, E.C., Kim, J., Fernandez-Nieves, A., and Weitz, D.A. (2008) Highly responsive hydrogel scaffolds formed by three-dimensional organization of microgel nanoparticles. *Nano Lett.*, **8** (1), 168–172.

32. De Gennes, P.G. (1980) Conformations of polymers attached to an interface. *Macromolecules*, **13** (5), 1069–1075.

33. Advicula, R.C., Brittain, W., Caster, K., and Ruehe, J. (eds) (2004) *Polymer Brushes: Synthesis, Characterization, Applications*, John Wiley & Sons, Inc.

34. de Gennes, P.G. (1987) Polymers at an interface; a simplified view. *Adv. Colloid Interface Sci.*, **27** (3–4), 189–209.

35. Michailidou, V.N., Loppinet, B., Vo, D.C., Prucker, O., Ruhe, J., and Fytas, G. (2006) Dynamics of end-grafted polystyrene brushes in theta solvents. *J. Polym. Sci., Part B: Polym. Phys.*, **44** (24), 3590–3597.

36. Semenov, A.N. and Anastasiadis, S.H. (2000) Collective dynamics of polymer brushes. *Macromolecules*, **33** (2), 613–623.

37. Michailidou, V.N., Loppinet, B., Prucker, O., Ruhe, J., and Fytas, G. (2005) Cooperative diffusion of end-grafted polymer brushes in good solvents. *Macromolecules*, **38** (21), 8960–8962.

38. Dimitrov, D.I., Milchev, A., and Binder, K. (2007) Polymer brushes in solvents of variable quality: molecular dynamics simulations using explicit solvent. *J. Chem. Phys.*, **127** (8), 084905.

39. Lee, H. and Penn, L.S. (2010) Polymer brushes make nanopore filter membranes size selective to dissolved polymers. *Macromolecules*, **43** (1), 565–567.

40. Filippidi, E., Michailidou, V., Loppinet, B., Ruhe, J., and Fytas, G. (2007) Brownian diffusion close to a polymer brush. *Langmuir*, **23** (9), 5139–5142.

41. Kim, J.U. and O'Shaughnessy, B. (2006) Nanoinclusions in dry polymer brushes. *Macromolecules*, **39** (1), 413–425.

42. Yaneva, J., Dimitrov, D.I., Milchev, A., and Binder, K. (2009) Nanoinclusions in polymer brushes with explicit solvent – a molecular dynamics investigation. *J. Colloid Interface Sci.*, **336** (1), 51–58.

43. Michailidou, V.N., Loppinet, B., Vo, C.D., Ruhe, J., Tauer, K., and Fytas, G. (2008) Observation of slow down of polystyrene nanogels diffusivities in contact with swollen polystyrene brushes. *Eur. Phys. J. E*, **26** (1–2), 35–41.

44. Landau, L. and Lifshitz, E. (1970) *Theory of Elasticity*, Series Course of Theoretical Physics, 2nd revised edn, vol. 7, Pergamon Press, Butterworth-Heinemann.

45. Farnell, G.W. and Adler, E.L. (1972) Elastic wave propagation in thin layers. *Physical Acoustic Principles and Methods*, Academic Press.

46. Sandercock, J.R. (1972) Structure in the Brillouin spectra of thin films. *Phys. Rev. Lett.*, **29** (26), 1735–1738.

47. Grimsditch, M., Bhadra, R., and Schuller, I.K. (1987) Lamb waves in unsupported thin films: a Brillouin-scattering study. *Phys. Rev. Lett.*, **58** (12), 1216–1219.

48. Hillebrands, B., Lee, S., Stegeman, G.I., Cheng, H., Potts, J.E., and Nizzoli, F. (1988) Evidence for the existence of guided longitudinal acoustic phonons in ZnSe films on GaAs. *Phys. Rev. Lett.*, **60** (9), 832–835.

49. Bandhu, R.S., Zhang, X., Sooryakumar, R., and Bussmann, K. (2004) Acoustic vibrations in free-standing double layer membranes. *Phys. Rev. B - Condens. Matter Mater. Phys.*, **70** (7), 075409-1–075409-4.

50. Forrest, J.A., Dalnoki-Veress, K., Stevens, J.R., and Dutcher, J.R. (1996) Effect of free surfaces on the glass transition temperature of thin polymer films. *Phys. Rev. Lett.*, **77** (10), 2002–2005.

51. Nizzoli, F., Hillebrands, B., Lee, S., Stegeman, G.I., Duda, G., Wegner, G., and Knoll, W. (1989) Determination of the whole set of elastic constants of a polymeric Langmuir-Blodgett film by Brillouin spectroscopy. *Phys. Rev. B*, **40** (5), 3323–3328.

52. Forrest, J.A., Dalnoki-Veress, K., and Dutcher, J.R. (1997) Interface and chain confinement effects on the glass transition temperature of thin polymer films. *Phys. Rev. E - Stat. Phys., Plasmas, Fluids Relat. Interdiscipl. Top.*, **56** (5, Suppl. B), 5705–5716.

53. Cheng, W., Gorishnyy, T., Krikorian, V., Fytai, G., and Thomas, E.L. (2006) In-plane elastic excitations in id polymeric photonic structures. *Macromolecules*, **39** (26), 9614–9620.

54. Forrest, J.A., Rowat, A.C., Dalnoki-Veress, K., Stevens, J.R., and Dutcher, J.R. (1996) Brillouin light scattering studies of the mechanical properties of polystyrene/polyisoprene multilayered thin films. *J. Polym. Sci. B: Polym. Phys.*, **34** (17), 3009–3016.

55. Keddie, J.L., Jones, R.A.L., and Cory, R.A. (1994) Size-dependent depression of the glass transition temperature in polymer films. *EPL (Eur. Lett.)*, **27** (1), 59.

56. Fryer, D.S., Nealey, P.F., and De Pablo, J.J. (2000) Thermal probe measurements of the glass transition temperature for ultrathin polymer films as a function of thickness. *Macromolecules*, **33** (17), 6439–6447.

57. Serghei, A., Huth, H., Schick, C., and Kremer, F. (2008) Glassy dynamics in thin polymer layers having a free upper interface. *Macromolecules*, **41** (10), 3636–3639.

58. Cheng, W., Sainidou, R., Burgardt, P., Stefanou, N., Kiyanova, A., Efremov, M., Fytas, G., and Nealey, P.F. (2007) Elastic properties and glass transition of supported polymer thin films. *Macromolecules*, **40** (20), 7283–7290.

59. Gomopoulos, N., Cheng, W., Efremov, M., Nealey, P.F., and Fytas, G. (2009) Out-of-plane longitudinal elastic modulus of supported polymer thin films. *Macromolecules*, **42** (18), 7164–7167.

60. Gomopoulos, N., Saini, G., Efremov, M., Nealey, P.F., Nelson, K., and Fytas, G. (2010) Non- destructive probing of mechanical anisotropy in polyimide films at nanoscale. *Macromolecules*, **43**, 1551–1555.

61. Sato, A., Knoll, W., Pennec, Y., Djafari-Rouhani, B., Fytas, G., and Steinhart, M. (2009) Anisotropic propagation and confinement of high frequency phonons in nanocomposites. *J. Chem. Phys.*, **130** (11), 111102.

62. Sainidou, R., Stefanou, N., Psarobas, I.E., and Modinos, A. (2005) A layer-multiple-scattering method for phononic crystals and heterostructures of such. *Comput. Phys. Commun.*, **166** (3), 197–240.

63. Sun, L., Dutcher, J.R., Giovannini, L., Nizzoli, F., Stevens, J.R., and Ord, J.L. (1994) Elastic and elasto-optic properties of thin films of poly(styrene) spin coated onto Si(001). *J. Appl. Phys.*, **75** (11), 7482–7488.

64. Hartschuh, R.D., Kisliuk, A., Novikov, V., Sokolov, A.P., Heyliger, P.R., Flannery, C.M., Johnson, W.L., Soles, C.L., and Wu, W.-L. (2005) Acoustic modes and elastic properties of polymeric nanostructures. *Appl. Phys. Lett.*, **87** (17), 173121.

65. Reiter, G. and Forrest, J. (2002) Special issue on properties of thin polymer films. *Eur. Phys. J. E*, **8** (2), 101.

66. Yamamoto, S., Tsujii, Y., and Fukuda, T. (2002) Glass transition temperatures of high-density poly(methyl methacrylate) brushes. *Macromolecules*, **35** (16), 6077–6079.

67. Rogers, J.A., Maznev, A.A., Banet, M.J., and Nelson, K.A. (2000) Optical generation and characterization of acoustic waves in thin films: fundamentals and applications. *Annu. Rev. Mater. Sci.*, **30**, 117–157.

68. Pelton, M., Sader, J.E., Burgin, J., Liu, M., Guyot-Sionnest, P., and Gosztola, D. (2009) Damping of acoustic vibrations in gold nanoparticles. *Nature Nanotechnol.*, **4** (8), 492–495.

69. Rogers, J.A., Dhar, L., and Nelson, K.A. (1994) Noncontact determination of transverse isotropic elastic moduli in polyimide thin films using a laser based ultrasonic method. *Appl. Phys. Lett.*, **65** (3), 312–314.

70. Zhang, X., Sooryakumar, R., Every, A.G., and Manghnani, M.H. (2001) Observation of organ-pipe acoustic excitations in supported thin films. *Phys. Rev. B*, **64** (8), 081402.

71. Ogi, H., Shagawa, T., Nakamura, N., Hirao, M., Odaka, H., and Kihara, N. (2008) Elastic constant and Brillouin oscillations in sputtered vitreous SiO_2 thin films. *Phys. Rev. B*, **78** (13), 134204.

72. Groenen, J., Poinsotte, F., Zwick, A., Sotomayor Torres, C.M., Prunnila, M., and Ahopelto, J. (2008) Inelastic light scattering by longitudinal acoustic phonons in thin silicon layers: from membranes to silicon-on-insulator structures. *Phys. Rev. B*, **77** (4), 045420.

73. Cheng, W., Gomopoulos, N., Fytas, G., Gorishnyy, T., Walish, J., Thomas, E.L., Hiltner, A., and Baer, E. (2008) Phonon dispersion and nanomechanical properties of periodic 1d multilayer polymer films. *Nano Lett.*, **8** (5), 1423–1428.

74. Baer, E., Hiltner, A., and Keith, H.D. (1987) Hierarchical structure in polymeric materials. *Science*, **235** (4792), 1015–1022.

75. Weber, M.F., Stover, C.A., Gilbert, L.R., Nevitt, T.J., and Ouderkirk, A.J. (2000) Giant birefringent optics in multilayer polymer mirrors. *Science*, **287** (5462), 2451–2456.

76. Wang, H., Keum, J.K., Hiltner, A., Baer, E., Freeman, B., Rozanski, A., and Galeski, A. (2009) Confined crystallization of polyethylene oxide in nanolayer assemblies. *Science*, **323** (5915), 757–760.

77. Shen, S., Henry, A., Tong, J., Zheng, R., and Chen, G. (2010) Polyethylene nanofibres with very high thermal conductivities. *Nature Nanotechnol.*, **5**, 251–255.

78. Kumar, S.R., Renusch, D.P., and Grimsditch, M. (2000) Effect of molecular orientation on the elastic constants of polypropylene. *Macromolecules*, **33** (5), 1819–1826.

79. Stafford, C.M., Harrison, C., Beers, K.L., Karim, A., Amis, E.J., Vanlandingham, M.R., Kim, H., Volksen, W., Miller, R.D., and Simonyi, E.E. (2004) A buckling-based metrology for measuring the elastic moduli of polymeric thin films. *Nature Mater.*, **3** (8), 545–550.

80. Krbecek, H., Kruger, J.K., and Pietralla, M. (1993) Poisson ratios and upper bounds of intrinsic birefringence from Brillouin scattering of oriented polymers. *J. Polym. Sci. B: Polym. Phys.*, **31** (11), 1477–1485.

81. Dhar, L., Rogers, J.A., Nelson, K.A., and Trusell, F. (1995) Moduli determination in polyimide film bilayer systems: prospects for depth profiling using impulsive stimulated thermal scattering. *J. Appl. Phys.*, **77** (9), 4431–4444.

82. Moldovan, M. and Thomas, E.L. (2008) *Periodic Materials and Interference Lithography for Photonics, Phononics and Mechanics*, Wiley-VCH Verlag GmbH.

83. Gorishnyy, T., Ullal, C.K., Maldovan, M., Fytas, G., and Thomas, E.L. (2005) Hypersonic phononic crystals. *Phys. Rev. Lett.*, **94** (11), 115501.

84. Cheng, W., Wang, J.J., Jonas, U., Fytas, G., and Stefanou, N. (2006) Observation and tuning of hypersonic bandgaps in colloidal crystals. *Nature Mater.*, **5**, 830–836.

85. Gorishnyy, T., Jang, J.H., Koh, C., and Thomas, E.L. (2007) Direct observation of a hypersonic bandgap in 2D single crystalline phononic structures. *Appl. Phys. Lett.*, **9**, 121915.

86. Still, T., Cheng, W., Retsch, M., Sainidou, R., Wang, J., Jonas, U., Stefanou, N., and Fytas, G. (2008) Simultaneous occurrence of structure-directed and particle-resonance- induced phononic gaps in colloidal films. *Phys. Rev. Lett.*, **100** (19), 194301.

87. Akimov, A.V., Tanaka, Y., Pevtsov, A.B., Kaplan, S.F., Golubev, V.G., Tamura, S., Yakovlev, D.R., and Bayer, M. (2008) Hypersonic modulation of light in three-dimensional photonic and phononic band-gap materials. *Phys. Rev. Lett.*, **101** (3), 033902.

88. Urbas, A.M., Thomas, E.L., Kriegs, H., Fytas, G., Penciu, R.S., and Economou, L.N. (2003) Acoustic excitations in a self-assembled block copolymer photonic crystal. *Phys. Rev. Lett.*, **90** (10), 108302/1–108302/4.

89. Parsons, L.C. and Andrews, G.T. (2009) Observation of hypersonic phononic crystal effects in porous silicon superlattices. *Appl. Phys. Lett.*, **95**, 241909.

90. Gomopoulos, N., Maschke, D., Koh, C.Y., Thomas, E.L., Tremel, W., Butt, H., and Fytas, G. (2010) One-dimensional hypersonic phononic crystals. *Nano Lett.*, **10** (3), 980–984.

91. Montagna, M. (2008) Brillouin and Raman scattering from the acoustic vibrations of spherical particles with a size comparable to the wavelength of the light. *Phys. Rev. B*, **77** (4), 045418.

92. Duval, E. (1992) Far-infrared and Raman vibrational transitions of a solid sphere: selection rules. *Phys. Rev. B*, **46** (9), 5795–5797.

93. Portales, H., Goubet, N., Saviot, L., Adichtchev, S., Murray, D.B., Mermet, A., Duval, E., and Pileni, M. (2008) Probing atomic ordering and multiple twinning in metal nanocrystals through their vibrations. *Proc. Natl. Acad. Sci. USA*, **105** (39), 14784–14789.

94. Kuok, M.H., Lim, H.S., Ng, S.C., Liu, N.N., and Wang, Z.K. (2003) Brillouin study of the quantization of acoustic modes in nanospheres. *Phys. Rev. Lett.*, **90** (25), 255502.

95. Cheng, W., Wang, J.J., Jonas, U., Steffen, W., Fytas, G., Penciu, R.S., and Economou, E.N. (2005) The spectrum of vibration modes in soft opals. *J. Chem. Phys.*, **123** (12), 1–4.

96. Still, T., Sainidou, R., Retsch, M., Jonas, U., Spahn, P., Hellmann, G.P., and Fytas, G. (2008) The "music" of core-shell spheres and hollow capsules: influence of the architecture on the mechanical properties at the nanoscale. *Nano Lett.*, **8** (10), 3194–3199.

97. Still, T., D'Acunzi, M., Vollmer, D., and Fytas, G. (2009) Mesospheres in nano-armor: probing the shape-persistence of molten polymer colloids. *J. Colloid Interface Sci.*, **340** (1), 42–45.

98. Still, T., Retsch, M., Jonas, U., Sainidou, R., Rembert, P., Mpoukouvalas, K., and Fytas, G. (2010) Vibrational eigenfrequencies and mechanical properties of mesoscopic copolymer latex spheres. *Macromolecules*, **43**, 3422–3428.

18
Investigations of Soft Organic Films with Ellipsometry

Diethelm Johannsmann

18.1
Introduction

Routine, reliable characterization is an issue of prime importance in the field of surface modification and organic overlayers. Ellipsometry certainly is one of the time-honored, well-proven techniques serving this purpose. The principles have been worked out around the turn of the last century [1]. Considerable advances – including major instrumental developments – have been made in the 1980s and 1990s. These developments were in part motivated by a widespread interest in ultrathin coatings, for instance created by molecular beam epitaxy (MBE) [2], the Langmuir–Blodgett (LB) technique [3], and self-assembly [4]. Today, ellipsometry has matured as far as instrumentation and techniques of analysis are concerned. There now is a solid understanding of the underlying physics, the instrumental options, and the limitations. Current developments mostly concern the combination of ellipsometry with other techniques such as imaging [5, 6], spectroscopy [7], electrochemistry [8, 9], nonlinear optics [10], and acoustics [11].

This chapter is neither intended as an introduction to ellipsometry, nor as a comprehensive review on recent applications of ellipsometry [12]. Rather, it is a (necessarily somewhat biased) view of the contribution that ellipsometry can bring to the field of organic thin films, where the latter comprise polymeric adsorbates, spin-cast films, brushes, or other assemblies with a thickness in the range from 1 Å to a few hundred nanometers. Special emphasis is placed on samples with a refractive index profile, $n(z)$, and on samples with weak optical contrast.

With regard to the basics of ellipsometry, handbooks supplied by the manufacturers of instruments often contain useful tutorials. A good introduction can also be found in Ref. [13]. The book by Azzam and Bashara is an authorative source of information [14]. The mathematics is presented as concisely and amenable as this can possibly be done without sacrificing correctness. Unfortunately, this book is not easily digested by scientists with a background in chemistry or chemical engineering. It is not a good tutorial for practitioners. Actually, explaining ellipsometry simply and, at the same time, correctly is somewhat of a challenge. "Ellipsometry" tends to be either a somewhat vague term behind a well-working

Functional Polymer Films, First Edition. Edited by Wolfgang Knoll and Rigoberto C. Advincula.
© 2011 Wiley-VCH Verlag GmbH & Co. KGaA. Published 2011 by Wiley-VCH Verlag GmbH & Co. KGaA.

instrument (the interiors of which are better left unexplored), or a conceptual framework of fascinating complexity, indeed. The latter view requires a physics background, affection with complex numbers, and fearlessness with regard to software and modeling. There is little safe ground in-between these two views and that is one of the drawbacks of ellipsometry. Both the acoustic instruments of surface analysis (such as the quartz crystal microbalance, QCM) and the competing optical technique of surface plasmon resonance (SPR) spectroscopy are simpler in this respect. They can be explained to the educated scientist in about an afternoon. Provided that the newcomer distrusts the outputs of automated instruments and performs the usual checks, he or she will quickly understand the rules of the game. Compared to QCM or SPR, ellipsometry is a mine field. There are sources of error that the novice would have never dreamt of. Artifacts are a particular challenge if the cell has windows, if the surface of interest has some roughness, if there is an ambient liquid, or if the underlying substrate is not characterized well.

Ellipsometry is still alive because it can do things that the competing instruments cannot do. One advantage is its applicability to a wide variety of substrates. A second one is that layer thickness and refractive index can be derived independently if the sample is thick enough. Further, ellipsometry can be combined with imaging [5] and spectroscopy [7]. SPR imaging is possible [15–17], but spectroscopy based on SPRs is confined to the (near-) infrared range due to the constraints imposed by the metal substrate [18]. The acoustic techniques have neither an imaging mode nor a spectroscopic mode (unless one calls an estimate of viscoelastic dispersion "spectroscopy").

This chapter is organized as follows: Section 18.2 gives a brief introduction into modeling. Sections 18.3–18.5 describe the fields of expertise of the author, which are multiple-angle total-internal-reflection (MA-TIR) ellipsometry, Fourier-transform ellipsometry, and the comparison of optical and acoustic reflectometry. Section 18.6 concludes the chapter.

18.2
Modeling

18.2.1
Importance of Angle Measurements

There is a golden rule in optics that says: "Never measure an intensity unless you absolutely have to." An impressive example is astronomy. Suppose one would be given the choice between measuring the position of a star or its brightness, one is always advised to go for position. The positions of the planets have been determined – and even predicted – with astounding preciseness at times when "the elements" (in Greece) were still fire, air, earth, and water. The brightness is a different matter. In order to determine brightness, start with calibrated lamps, filters, and detectors. You worry about fluctuations of power, alignment, transparency, and detector responsivity. Brightness is a clue to a star's distance and therefore is undoubtedly a second interesting bit of information. The astronomic distance scale is based on brightness measurements. Interestingly, there was an error margin of 50% on the astronomic distance scale as late as 1980. This lack of knowledge was only in part due to instrumental constraints but it highlights the ease by which angles are determined in comparison to intensities.

This same rule has wisely been put to work in SPR spectroscopy and – needless to say – in ellipsometry. Both techniques are essentially based on the measurements of angles, as opposed to intensities. Both techniques are variants of optical reflectometry and one may legitimately ask: how can you do reflectometry without measuring intensity? The easy answer is not the full answer. Of course reflectometry only requires the ratio of two intensities (those of the incident and of the reflected beam). However, that is not the essential trick. Both SPR spectroscopy and ellipsometry employ a scheme of measurement, where the reflectivity displays a sharp minimum at certain *angles*. In SPR spectroscopy, this is a certain angle of incidence, while in ellipsometry, the angles are certain orientations of the polarizers. Angles are determined with an accuracy of 10^{-4} (5×10^{-3} degree) with moderate effort. If a laser beams walks by 5×10^{-3} degree on some type of screen, this movement is well discernible by the eye. It is nothing of a challenge to achieve the same with a diode array. Going further requires careful thermal and mechanical engineering of the rails and rods carrying the optical components, thereby avoiding drifts and structural relaxations. The companies offering the current generation of instruments have gone through that exercise and the accuracy they provide easily surpasses 10^{-4} degree.

18.2.2
Null-Ellipsometry

Ellipsometry amounts to a measurement of angles because it is based on *interference*. At the detector, the amplitude of two beams are superimposed, which are the amplitudes of the beam polarized in the plane of incidence ("p" for "parallel") and the beam polarized perpendicular to this plane ("s" for German "senkrecht"). Nota bene: "p" and "s" refer to the plane of incidence, as opposed to the plane of the sample. The electric-field vector of s-light is parallel (!) to the plane of the sample.[1] Employing polarizers and a quarter-wave plate ("compensator"), one can always find relative orientations of these components that produce destructive interference at the detector. This scheme is termed "*null-ellipsometry*".[2] In order to determine these null-angles, the instrument typically varies the angles around the nulling position and determines the nulling position by fitting a parabola to the intensity–angle data. The nulling angles are the centers of the parabola. Since the measurement is background-free, it is intrinsically accurate.

---

1) In the chapter, the term "*polarization*" denotes the direction of the electric field vector (E). This usage differs from an older definition, where the polarization it denotes is the magnetic-field vector (B). The electric field interacts with matter much more strongly than the magnetic field, and therefore is of more practical interest than B.

2) There are other schemes of measurement, most notably the rotating analyzer setup. These differ in the details, but not in the principles of operation.

It is customary to convert the nulling angles to a second set of angles, termed Ψ and Δ. The definition of Ψ and Δ is historical. The equations for conversion depend on the configuration and are tabulated in Ref. [14]. While the nulling angles depend on the configuration, Ψ and Δ do not. Ψ and Δ are related to the reflectivity of the sample via

$$\rho = \frac{r_p}{r_s} = \tan \Psi \exp(i\Delta) = \frac{|r_p|}{|r_s|} \exp(i(\delta_p - \delta_s)) \tag{18.1}$$

Here, $r_p = |r_p|\exp(i\delta_p)$ and $r_s = |r_s|\exp(i\delta_s)$ are the complex amplitudes of reflection for p- and s-light. Importantly, ellipsometry determines a ratio of complex amplitudes. With regard to the phase, a ratio of amplitudes turns into a phase difference. The information on film thickness often is to a large extent contained in this phase lag (that is, in the parameter Δ). We elaborate on this fact in the following section.

18.2.3
Single-Layer Systems

In order to bring across that the film thickness is often encoded in the parameter Δ, consider the unrealistic (but still instructive) situation depicted in Figure 18.1. Let the substrate be almost perfectly reflective (which is realistic for metals), let the reflectivity of the film/air interface be almost zero for p-light (which is true if the angle of incidence is close to the Brewster angle), and let the reflectivity of the film/air interface for s-light be close to unity (which would require a refractive index much above 2, which is *unrealistic*). For such a situation, p-light is reflected at the bottom of the film, while s-light is reflected at the top. The two beams acquire a relative phase lag of

$$\Delta = \delta_p - \delta_s \approx \frac{4\pi}{\lambda} n_f d_f \cos \theta_i \tag{18.2}$$

Here, d_f is the film thickness, n_f is its refractive index, and θ_i is the angle of incidence evaluated inside the film. Here and in the following, the angle of incidence is quoted with respect to the surface normal. Given that both surfaces are partially reflective, the reality is not quite as simple. But the reader may trust the existing software with regard to proper treatment of partial reflectivities. Equation 18.2 brings across two essential points:

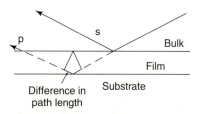

Figure 18.1 Beam propagation at a surface covered with a single layer.

- The measurement of a phase difference is essential in ellipsometry. The relative phase between the reflected s-light and p-light is closely related to film thickness.
- For thin films, it is *a priori* not possible to determine the thickness and the refractive index separately. We return to that topic in Section 18.4. Although suggested by Equation 18.2, measurement of Ψ and Δ does *not* amount to a measurement of the product $n_f\, d_f$.

Before referring the reader to the literature, we go one step further in the quantitative data analysis. We now do take partial reflectivities into account. The reflectivities at an interface between medium i and j are given by [19].

$$r_{s,ij} = \frac{q_i - q_j}{q_i + q_j}$$

$$r_{p,ij} = \frac{Q_j - Q_i}{Q_j + Q_i}$$

$$Q_j = \frac{q_i}{\varepsilon_i},$$

$$q_i^2 = \left(\varepsilon_i \left(\frac{\omega}{c}\right)^2 - q_x^2\right) \tag{18.3}$$

Here, q is the z-component of the wave vector, $\varepsilon = n^2$ is the dielectric permittivity (at optical frequencies), c is the speed of light, and q_x is the tangential component of the wave vector (which is the same for all layers). When the measurement occurs at the base of a prism, the angles of incidence around the critical angle, θ_c are of special interest (see Section 18.4). At the critical angle both r_s and r_p are equal to unity because the vertical component of q in the liquid phase, q_j, vanishes.

In the presence of a film, there are multiple reflections and the overall reflectivities are given by

$$r = \frac{r_{\text{bulk},f} + r_{f,\text{sub}} \exp\left(-2iq_f d_f\right)}{1 + r_{\text{bulk},f} r_{f,\text{sub}} \exp\left(-2iq_f d_f\right)} \tag{18.4}$$

The indices *bulk*, *f*, and *sub* denote the bulk, the film, and the substrate, respectively. Here, the beam reaches the sample through a bulk medium (liquid or air). The roles of the "substrate" and the "bulk" are interchanged if the measurement is carried out at the base of the prism because the beam then enters through the prism. d_f is the film thickness. Equation 18.4 applies to both s-light and p-light, but the reflectivities differ for the two polarizations.

Combining Equations 18.1 and 18.4 yields

$$\rho = \frac{r_p}{r_s} = \tan \Psi e^{i\Delta}$$

$$= \frac{\left(r_{p,\text{bulk},f} + r_{p,f,\text{sub}} \exp\left(-2iq_f d_f\right)\right) \left(1 + r_{s,\text{bulk},f} r_{s,f,\text{sub}} \exp\left(-2iq_f d_f\right)\right)}{\left(1 + r_{p,\text{bulk},f} r_{p,f,\text{sub}} \exp\left(-2iq_f d_f\right)\right) \left(r_{s,\text{bulk},f} + r_{s,f,\text{sub}} \exp\left(-2iq_f d_f\right)\right)} \tag{18.5}$$

Again, the critical angle in total internal reflection ellipsometry is of special importance. At the critical angle, q_f is close to zero and ρ approaches unity ($\tan \Psi \sim 1, \Delta \sim 0$).

We refer the reader to Ref. [14] for more details on modeling. Equation 18.5 is meant to convince the reader that there is no magic behind the software packages. The above relations are a condensed form of what is called "*Fresnel theory*." Fresnel applied conditions of continuity to dielectric interfaces and realized that he could calculate optical reflectivities that way. Again, the reflectivities are different for the different polarizations. Were the reflectivities not different for the two polarizations, there would be no point in making the waves interfere at the detector. The two waves traverse the sample in different ways and their interference at the detector therefore carries information about the sample.

Equation 18.5 is the final word on data analysis if

- all refractive indices are known;
- all surfaces are flat; and
- all media are isotropic.

We comment on problems connected to these conditions in the following paragraphs.

18.2.3.1 Refractive Indices

Calculation of the film thickness from the experimental data by means of Equation 18.5 requires knowledge of all refractive indices. We elaborate on the problem of independent determination of n_f and d_f in Section 18.2.4.

An unknown (or even variable) refractive index of the bulk, n_{bulk}, may be a severe problem when the measurement is carried out in buffer solution. In this case, the refractive index depends on salt concentration, analyte concentration, and temperature. A central advantage of evanescent ellipsometry (see Section 18.3) is the availability of the bulk refractive index from the critical angle.

The refractive index of the substrate usually is less problematic. It is either known or can be determined from a reference experiment on the bare substrate according to:

$$
\varepsilon_{sub} = n_{sub}^2 = \frac{q_x^2}{q_{bulk}^2} \left(\varepsilon_{bulk} - \frac{4q_x^2}{(\omega/c)^2} \frac{\rho}{(1+\rho)^2} \right)
$$

$$
= \tan^2 \theta_i \left(\varepsilon_{bulk} - \sin^2 \theta_i \frac{4\rho}{(1+\rho)^2} \right) \tag{18.6}
$$

Gold is an important substrate for studies on organic thin films for various reasons. Gold is reasonable inert. It can serve as an electrode. Importantly, the thiol-based coupling chemistry works very well on gold [20]. Self-assembled monolayers are easily produced. Unfortunately, the refractive index of gold surfaces is subject to aging. The underlying processes include contamination, oxidation, and structural annealing. It can therefore not necessarily be taken for granted that a metal substrate underneath a coating has the exact same optical properties as the same substrate before the coating process. A second caveat is the importance of nanoroughness for the reflectivity. Roughness leads to a coupling between the incident beam and surface plasmons, thereby reducing r_p [21]. The coupling efficiency may change when the (rough) surface is covered with a dielectric film.

Silicon wafers are a popular substrate for ellipsometry because they can be cleaned easily and reproducibly. Also, they are flat (close to atomically). With regard to analysis, there is a slight complication, because silicon wafers are covered with an oxide layer. One has to resort to modeling with (at least) two layers, where the first layer is the oxide and the second is the film of interest.

Glass surfaces can also be cleaned well and are very smooth. However, glass substrates are problematic for ellipsometry because the optical contrast vanishes when the refractive index of the film comes close to the refractive index of the glass. That may easily happen with conventional float glass. High-index glasses are available, but they are expensive and degrade when cleaned in aggressive solvents.

18.2.3.2 Roughness

Generally speaking, Fresnel theory assumes planar interfaces. Lateral homogeneity is often questionable for organic layers. That concerns both the substrate (in the case of gold) and the film itself. There are simple checks on heterogeneity. The first is the visibility of the laser spot on the sample surface. A clean, flat surface does not scatter the beam appreciably (which actually is a problem when aligning the sample, because the laser spot is invisible). A more elaborate scheme makes use of the residual intensity of the beam after nulling the ellipsometer. Ideally, the signal at the detector should vanish at the nulling angles (assuming a parallel beam and perfect optical components). The residual intensity is dominated by roughness effects and can be used as a quantitative measure of roughness. Finally, note that ellipsometric imaging is rather easy. One just replaces the detector by a camera and widens the beam. Figure 18.2 shows an example. In this particular case, the sample had slight scratches. These scratches were invisible to the naked eye. They only became apparent when the sample was placed in an ellipsometer and the polarizers were positioned close to the nulling angles.

Figure 18.2 Ellipsometric image taken from a sample with microroughness. (Reprinted with permission from Ref. [22].)

On gold surfaces there is a useful rule of thumb, which says that a rough film/air interface scatters s-light more strongly than p-light. This happens because p-light is predominantly scattered at the film/gold interface. At the film/air interface r_p is small because the angle of incidence is close to the Brewster angle. As a consequence, p-light is not scattered appreciably. Given that tan Ψ is equal to $|r_p|/|r_s|$, scattering typically decreases Ψ. For thin films on gold, the parameter Δ quantifies thickness, while Ψ quantifies roughness.

Given the importance of lateral structure formation in the field of organic coatings, one may wonder whether an ellipsometric setup might be advantageously expanded by combining it with static light scattering ("off-specular ellipsometry"). This is possible, but the technique is not widely used because it competes with grazing-incidence X-ray diffraction (GIXD) [23]. The latter technique requires a strong X-ray source (for instance a synchrotron), but it outperforms ellipsometry otherwise because the inverse scattering vector covers the nanoscopic scale. Static light scattering, on the other hand, can only access structures down to a scale of around 100 nm. The nanoscale is crucial for organic coatings because the lateral heterogeneities tend to be of the same size as the thickness (which is on the nanoscale, as well).

Roughness on the molecular scale is a separate issue. Such a type of heterogeneity typically occurs for submonolayer coverage. It goes unnoticed because such small heterogeneities do not noticeably scatter light. The problem connected to submonolayer samples is the choice of parameters for an "equivalent" continuous film. Evidently, there is no film with a thickness equal to (let's say) half the diameter of an adsorbate molecule. One may still postulate such a film in the frame of an effective medium theory [24]. While this approach is qualitatively viable, the details are uncertain because of local field effects. The polarizability (and hence the refractive index) of a medium depends on the interaction between neighboring molecules in a complicated way. Assuming an effective refractive index equal to the fractional coverage is a fair first approximation, but it is not quantitatively correct. In particular, the ellipsometric signal is not strictly proportional to the coverage.

18.2.3.3 Anisotropy and Birefringence

Equation 18.5 assumes all materials to be isotropic (or, more precisely, nonbirefringent; cubic crystals are nonbirefringent but still anisotropic). Birefringence is one of the Achilles heels of ellipsometry. If a material is birefringent, it changes the state of polarization of light traversing it. Most optical components are slightly birefringent due to mechanical stress. Even worse, the stress tends to relax over time, so that a careful calibration undertaken at some time does not apply some days later. It is therefore always advised not to have windows, lenses, or prisms between the first and the last polarizer. Even isotropic components change the state of polarization if they are mounted with a slight inclination with respect to the beam. Before the first and after the last polarizer one is free to place lenses and other components as needed.

Slightly birefringent films are not quite so much of a problem as birefringent optical components because they are thin. Two cases have to be distinguished.

Quite a few films of interest may have different refractive indices in the plane of the substrate and perpendicular to it. Typical such samples would be LB films [3]. These are composed of oriented molecules and it is certainly expected that orientation would induce uniaxial birefringence. For that case, the equations become a little more involved but there is no change in concept. Section 4.7.3.1 in Ref. [14] provides details. If, on the other hand, the sample displays in-plane birefringence, the number of equations doubles. For in-plane birefringence, a wave that has impinged onto the sample with s-polarization will leave it with some p-component and *vice versa*. There are four reflectivities (r_{pp}, r_{ss}, r_{sp}, and r_{ps}) rather than two. The set of equations dealing with in-plane anisotropy is called "4×4 *matrix formalism*" and is described in Section 4.7 in Ref. [14]. In-plane anisotropy is rare because there are not many mechanisms that produce it. One sometimes finds inplane anisotropy for polymeric LB layers, because the main chains may be oriented along the dipping direction [25]. Liquid-crystalline samples may display inplane anisotropy, as well.

Of course things may be even more complicated: For instance, the principal axes of the dielectric tensor may be inclined with respect to the plane of the sample. Or there may be optical rotation (for instance induced by a magnetic field, mediated by the optical Kerr effect [26]). These are special cases outside the scope of this chapter. Generally speaking, Fresnel theory covers it all. It is just a matter of complexity.

18.2.4
Multilayers and Refractive-Index Profiles

Many samples of practical interest are composed of more than one layer. Clearly, a two-layer system is described by four parameters (two thicknesses and two refractive indices) and one cannot recover all four from a measurement of Ψ and Δ, only. Predicting Ψ and Δ from the layer parameters is straightforward, but this is not the problem at hand.

If the layer system is deposited sequentially, one can always characterize the system at the intermediate steps. Another case, where theory has a clear (albeit unsatisfying) answer to the inversion problem is the thin-film limit. For layers much thinner than λ, one may expand Equation 18.5 to first order in film thickness, which yields [27]

$$d\rho \approx K \frac{(\varepsilon_f - \varepsilon_{bulk})(\varepsilon_f - \varepsilon_{sub})}{\varepsilon_f} d_f = K \frac{(n_f^2 - n_{bulk}^2)(n_f^2 - n_{sub}^2)}{n_f^2} d_f$$

$$K = \frac{2i(\varepsilon_{bulk} - \varepsilon_{sub})\varepsilon_{sub}q_x^2 q_{bulk}}{(q_{bulk} - q_{sub})^2 (\varepsilon_{sub}q_{bulk} + \varepsilon_{bulk}q_{sub})^2} \left(\frac{\omega}{c}\right)^2 \tag{18.7}$$

Here, $d\rho$ is the change of r_p/r_s induced by the deposition of the film. The coefficient K only depends on the substrate and the bulk medium. Note that there is a single coefficient K. There is no way to distinguish between the term containing the refractive indices and the film thickness, d_f, from separately analyzing Ψ and Δ.

Since Equation 18.7 is linear in d_f, it also holds in the integral sense:

$$d\rho \approx K \int_0^\infty \frac{\left(n_f^2(z) - n_{bulk}^2\right)\left(n_f^2(z) - n_{sub}^2\right)}{n_f^2(z)} dz \tag{18.8}$$

Here, $n_f(z)$ is the refractive index profile inside the film. The integral is the *ellipsometric moment* of the refractive-index distribution. In the thin-film limit, ellipsometry determines this ellipsometric moment, not the film thickness. Let it be emphasized, here, that most techniques for thickness determination share this or an analogous problem. SPR, for instance, also probes the ellipsometric moment [28]. The interpretation of data acquired with the QCM yields an "acoustic moment" [29]. This moment is the equivalent of the ellipsometric moment, where the role of the refractive index is taken by the acoustic impedance (see Section 18.5). Neutron reflectometry (NR) is the one exception [30]. NR does not need assumptions on material parameters, because the sample thickness exceeds the wavelength. Unfortunately, NR is not available on a routine basis to date.

If the refractive index of the substrate is much different from the refractive index of the film and the bulk (which is the case for metal surfaces) and if, further, the $n_f \sim n_{bulk}$, one may approximate Equation 18.8 by

$$d\rho \approx K\left(n_{bulk}^2 - n_{sub}^2\right) \int_0^\infty \frac{\left(n_f^2(z) - n_{bulk}^2\right)}{n_f^2(z)} dz \tag{18.9}$$

One may further Taylor-expand the integrand in Equation 18.8 to first order in the difference $(\Delta n(z) = n(z) - n_{bulk})$, leading to

$$d\rho \approx K \frac{2\left(n_{bulk}^2 - n_{sub}^2\right)}{n_{bulk}} \int_0^\infty \Delta n(z)dz \approx K \frac{2\left(n_{bulk}^2 - n_{sub}^2\right)}{n_{bulk}} \frac{dn}{dc} \int_0^\infty c(z)dz$$

$$= K \frac{2\left(n_{bulk}^2 - n_{sub}^2\right)}{n_{bulk}} \frac{dn}{dc} \Gamma \tag{18.10}$$

In the second step, it was assumed that the refractive-index increment, Δn, is proportional to the concentration of the adsorbate, c. The integral over the concentration yields the total adsorbed amount, Γ. These approximations lead to the conclusion that the output of ellipsometry is proportional to the *adsorbed amount*, irrespective of the state of swelling. The same statement holds for SPR, since SPR also senses the ellipsometric moment. This prediction can be tested by swelling and deswelling surface-attached polymer layers, such as pNIPAm (poly-N-isopropylacrylamide), the thickness of which can be tuned by temperature [31]. One finds that the apparent thickness *does* slightly depend on the degree of swelling. There are two approximations involved in the derivation of Equation 18.10, which is the thin-film limit and the proportionality of Δn to concentration, c. Both have limited applicability.

18.3
Multiple-Angle Ellipsometry and Total Internal Reflection Ellipsometry

This section is concerned with some general remarks on the use of multiangle ellipsometry (MAE) for the characterization of polymer brushes [22]. These samples are peculiar in a number of ways:

- There are synthetic constraints with regard to the substrate. Sample preparation on a gold surface (which would allow for the thickness measurement by SPR spectroscopy) often is of limited practical relevance. Typical substrates of interest are silica, titania, polystyrene, or PMMA. A technique is needed that is flexible with regard to the substrate. Ellipsometry fulfills this condition.
- The thickness can be anywhere between a few nanometers and a micrometer. The technique employed for characterization must span this range.
- The films are immersed in a liquid (often water or a buffer solution).
- The sample may contain a vertical gradient in composition or density.
- The sample may be laterally heterogeneous.

Ellipsometry is very well suited to the investigation of such materials. MAE on the base of a prism (Figure 18.3) has shown clear advantages. Evidently, measurements at more than one angle can provide a greater depth of information than single-angle measurements. The advantages of MAE come most strongly into play when the thickness of the sample is comparable to the wavelength of light, λ. The ellipsometric spectrum then contains fringes, which lead to unique and reliable fitting results, at least with regard to the sample thickness. Again, MAE cannot solve the fundamental problem of the ellipsometry in the thin-film limit, which is the dependence of the derived thickness on the assumed refractive index of the film. In the thin-film limit, the ellipsometric spectra are flat and featureless. In the absence of fringes, thickness and refractive index cannot both be separately extracted from the ellipsometric spectra. A second advantage of MAE on a prism base (MA-TIR ellipsometry, where TIR stands for "total internal reflection," Figure 18.3) is the availability of the bulk refractive index from the critical angle.

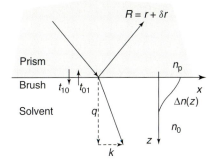

Figure 18.3 Geometry of multiangle ellipsometry on a prism base. (Reprinted with permission from Ref. [32].)

Importantly, the surface-attached layers investigated in Ref. [22] were *not* optically thin. For such films, one may, in principle, perform the fits with arbitrarily complex profiles. However, the solutions are not unique, and it is more meaningful to calculate certain effective parameters from the profiles. Typical parameters of interest would be the adsorbed amount (proportional to the ellipsometric moment), the film thickness (which now *is* available because the films are thick), and the sharpness of the film/solvent interface. We will elaborate on a model-free way to obtain the film thickness from MA-TIR ellipsometry data in Section 18.4. Here, we describe the derivation of the adsorbed amount, the thickness, and the interface sharpness based on fitting with a model profile, namely the error function. Evidently, the outcome of that procedure does entail an assumption on the profile. Fitting with different profiles would result in slightly different parameters.

Figure 18.4 shows a typical data set obtained on a polymer brush. Details are provided in Ref. [22]. The fit results shown as straight lines were obtained with error functions of the form

$$\varphi(z) = \varphi_0 \frac{1}{2}\left[1 - \operatorname{erf}\left(\frac{z - d}{w}\right)\right] \tag{18.11}$$

with $\varphi(z)$ the polymer volume fraction, d the point of inflection of the error function, w the width of the error function, and φ_0 a normalization constant. The adsorbed amount is the integral of Equation 18.11. The parameter w quantifies to what extent

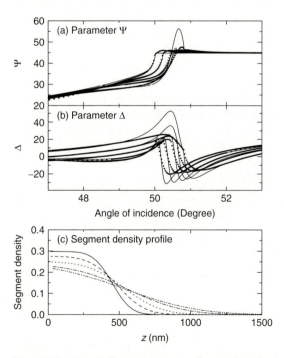

Figure 18.4 Typical ellipsometric spectra from multiangle ellipsometry. (Reprinted with permission from Ref. [22].)

the profile is box-like or not. A large parameter w corresponds to a smooth profile with an extended tail. The parameter d should not naively be identified with the film thickness. For instance, d may even become negative, if the refractive-index profile is close to an exponential. In order to rectify this situation, we define parameters $d*$ and $w*$, which coincide with d and w for box profiles (or profiles close to a box), but are reasonable indicators for thickness and interface width for all other profiles, as well. The definition is based on the normalized first and the normalized second moment of the refractive index profile, defined as:

$$Z_1 = \frac{A_1}{A_0} = \frac{\int_0^\infty z\varphi(z)dz}{\int_0^\infty \varphi(z)dz}, Z_2 = \frac{A_2}{A_0} = \frac{\int_0^\infty z^2\varphi(z)dz}{\int_0^\infty \varphi(z)dz} \tag{18.12}$$

Z_1 and Z_2 have the dimensions of a length and the square of a length. We define $d*$ and $w*$ as

$$d* = 2Z_1, \quad w* = 2Z_1\sqrt{\frac{3}{2}\frac{Z_2}{Z_1^2} - 2} \tag{18.13}$$

As shown in Ref. [22], this definition renders $d* \sim d$ and $w* \sim w$ in the limit of box profiles.

18.4
Fourier-Transform Ellipsometry

The analysis described in the previous section needs an assumption on the shape of the profile. It turns out that MA-TIR ellipsometry actually can be analyzed without resorting to such assumptions. The essential benefit is the availability of ellipsometric data over a wide range of incidence angles. The analysis builds on a methodology worked out in the context of scattering. Reflection is viewed as a special form of scattering, where the scattering vector is twice the z-component of the wave vector, $2q$. Note that the letter q here does not denote the scattering vector but half the scattering vector. In conventional scattering, intensities are converted to the autocorrelation function of the contrast distribution (which is $n(r)$ in optics). The conversion requires the Born approximation, that is, weak contrast and negligible multiple scattering. The contrast distribution $n(r)$ itself (that is, an image) cannot be recovered from the scattering intensity because of the phase problem. The phase problem also pertains to X-ray and neutron reflectometry. There is no way to explicitly convert the spectrum of the reflected intensity to the refractive index profile. Interestingly, such a direct conversion *is* possible in ellipsometry because the ellipsometric spectrum contains a ratio of complex *amplitudes* (including the phase difference). The algebra is somewhat involved, but the essential point is readily explained: the ellipsometric spectrum is (within known factors) equivalent to the Fourier transform of the refractive index profile $n(z)$. Back transformation yields $n(z)$.

Details of the mathematics are to be found in Ref. [32]. Briefly, the ellipsometric data are cast into a reduced complex coefficient

$$\delta\rho = \frac{R_p/R_s}{r_{p,10}/r_{s,10}} - 1 \tag{18.14}$$

where R_p and R_s are the reflection coefficients for p- and s-light of the brush-covered surface and $r_{p,10}$ and $r_{s,10}$ are the reflection coefficients of the bare prism surface. Here and in the following, the index "1" denotes the prism and the index "0" to denotes the bulk liquid. All quantities are functions of the incidence angle (and therefore of q, as well).[3]

The refractive index profile $\Delta n(z)$ can be Fourier transformed as

$$\Delta n(z) \approx \frac{1}{2\pi} \int_{-2q_{max}}^{2q_{max}} \tilde{n}\,(2q)\, e^{-2iqz} d(2q) \tag{18.15}$$

The Fourier transform $\tilde{n}\,(2q)$ is related to $\delta\rho(2q)$ via

$$\tilde{n}(2q) = \frac{\delta\rho(2q)}{iC(2q)} \tag{18.16}$$

where the function $C(2q)$ is given by

$$C\,(2q) = \frac{\omega}{c}\frac{n_0}{2q}\left[\left(r_{p,10}^{-1} - r_{s,10}^{-1}\right) - \left(r_{p,10} - r_{s,10}\right)\right] - \frac{4q}{n_0}\left[r_{p,10}^{-1} - r_{p,10}\right] \tag{18.17}$$

n_0 is the refractive index of the bulk liquid. Note that the reflectivities depend on q, as well. The use of Equations 18.15–18.17 to determine $n(z)$ is straightforward. Figure 18.5 shows an example. The sample was a polymer brush, the thickness of which could be tuned via the pH of the water phase. For details see Ref. [32]. There are two checks: firstly, $\Delta n(z)$ should be zero for $z < 0$ and secondly, $\Delta n(z)$ should be real-valued (unless the sample strongly adsorbs).

As always, when employing Fourier transforms, some limits of the technique are rooted in the limited range of available q-vectors. On the side of high q-vectors (incidence angles far away from the critical angle), the maximum q-vector limits the spatial resolution. The resolution along z is about 0.5 μm for the example shown in Figure 18.5. On the side of low q-vectors, there is a problem because all reflectivities become large close to the critical angle. Ultimately, there are multiple reflections very close to θ_c and the Born approximation is violated. For the same reason, scattering becomes large, close to θ_c. Scattering close to θ_c is a second limitation of Fourier-transform ellipsometry.

It turns out that Fourier-transform (FT) ellipsometry can provide a model-independent estimate of the film thickness [33]. In order do see that, remember that the film thickness is related to the first moment of the refractive index profile (see Equations 18.12 and 18.13). There is a simple relation between

3) We use a single q here, pertaining to the sample and the liquid alike. This is permissible because of the weak variability of the refractive index.

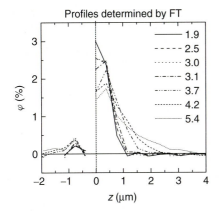

Figure 18.5 Segment density profiles of a polymer brush as calculated with FT ellipsometry in a model-free way. (Reprinted with permission from Ref. [32].)

the first moment of the refractive index profile, A_1, and the derivative of $\tilde{n}(2q)$ at $q = 0$:

$$\lim_{q \to 0} \frac{d\tilde{n}(2q)}{d(2q)} = \lim_{q \to 0} \left(\frac{d}{d(2q)} \int_0^\infty \Delta n(z) e^{2iqz} dz \right) = 2i \lim_{q \to 0} \left(\int_0^\infty z \Delta n(z) e^{2iqz} dz \right)$$

$$= 2i \int_0^\infty z \Delta n(z) dz = 2i A_1 \qquad (18.18)$$

At the critical angle, we have $\tan \Psi = 1$ and $\Delta = 0$. Close to the critical angle, we can therefore write

$$\tan \Psi \approx 1 + \alpha, \qquad \Delta \approx \beta \qquad (18.19)$$

with α and β small numbers. Equation 18.14 then reads

$$\begin{aligned}
\delta\tilde{n}(2q) &= \frac{1}{iC} \left(\frac{(1+\alpha)(1+i\beta)}{(1+\alpha_0)(1+i\beta_0)} - 1 \right) \\
&\approx \frac{1}{iC} \left((1+\alpha)(1+i\beta)(1-\alpha_0)(1-i\beta_0) - 1 \right) \\
&\approx \frac{(\beta - \beta_0)}{C} - \frac{i(\alpha - \alpha_0)}{C}
\end{aligned} \qquad (18.20)$$

Close to the critical angle, we can therefore write

$$A_1 = \frac{1}{2i} \lim_{q \to 0} \frac{d\tilde{n}(2q)}{d(2q)} = -\frac{1}{2} \frac{d}{d(2q)} \left(\frac{\alpha - \alpha_0}{C} \right) \qquad (18.21)$$

Equation 18.21 made use of the fact that C is real.

Using Equation 18.3, we convert from q to the angle of incidence.

$$q \approx \sqrt{n_{\text{bulk}}^2 \left(\frac{\omega}{c}\right)^2 - n_{\text{p}}^2 \left(\frac{\omega}{c}\right)^2 \sin^2 \theta_{\text{i}}}$$

$$\approx \frac{\omega}{c} \sqrt{n_{\text{bulk}}^2 - n_{\text{p}}^2 \sin^2 \left(\theta_{\text{c}} - \delta\theta\right)}$$

$$\approx \frac{\omega}{c} \sqrt{n_{\text{bulk}}^2 - n_{\text{p}}^2 \sin^2 \left(\arcsin^2\left(\frac{n_{\text{bulk}}}{n_{\text{p}}}\right) - \delta\theta\right)} \tag{18.22}$$

Nota bene: all angles are evaluated inside the prism ("inner angles"). $\theta_{\text{c}} = \arcsin(n_{\text{bulk}}/n_{\text{p}})$ is the critical angle. n_{p} is the refractive index of the prism. In line two, we have expressed the angle of incidence as a sum of θ_{c} and some small angle $\delta\theta$. Taylor expansion in $\delta\theta$ reveals that q is proportional to the square root of $\delta\theta$:

$$q \approx \frac{\omega}{c} \sqrt{2 n_{\text{bulk}} n_{\text{p}} \sqrt{1 - \frac{n_{\text{bulk}}^2}{n_{\text{p}}^2}} \sqrt{\delta\theta}} \tag{18.23}$$

One therefore plots $(\tan\Psi - \tan\Psi_0)/C$ versus $(\delta\theta)^{1/2}$ and estimates the layer thickness from the slope of this plot. Inserting Equation 18.23 into Equation 18.21 yields

$$A_1 = \frac{1}{4} \frac{\text{d}}{\text{d}q} \left(\frac{\alpha - \alpha_0}{C}\right)$$

$$= \frac{1}{4} \left(\frac{\omega}{c} \sqrt{2 n_{\text{bulk}} n_{\text{p}} \sqrt{1 - \frac{n_{\text{bulk}}^2}{n_{\text{p}}^2}}}\right)^{-1} \frac{\text{d}}{\text{d}\sqrt{\delta\theta}} \left(\frac{\tan\Psi - \tan\Psi_0}{C}\right) \tag{18.24}$$

Inserting this result into Equations 18.12 and 18.13 yields d^*. The zeroth moment (A_0, needed in Equation 18.12), is obtained via Equation 18.13 A final word of caution: this analysis requires smooth layers with low refractive-index contrast because scattering and multiple reflections close to θ_{c} will otherwise introduce artifacts.

18.5
Comparison of Optical and Acoustic Reflectometry

Given that both ellipsometry and the QCM are based on reflectometry (optical and acoustical, respectively) one suspects that the two techniques share certain common principles with respect to the data analysis. The detailed analysis confirms this. In the thin-film limit, the frequency shift of a quartz resonator induced by deposition of the sample is given by [29]

$$\frac{\Delta\tilde{f}}{f_{\text{F}}} \approx -\frac{-\omega}{\pi Z_{\text{q}}} \int_0^\infty \left[\frac{Z_{\text{f}}^2(z) - Z_{\text{bulk}}^2}{Z_{\text{f}}^2(z)}\right] \rho(z)\text{d}z \approx -\frac{\rho\omega}{\pi Z_{\text{q}}} \int_0^\infty \left[\frac{G_{\text{f}}(z) - G_{\text{bulk}}}{G_{\text{f}}(z)}\right] \text{d}z \tag{18.25}$$

Here, $\Delta \tilde{f} = \Delta f + i\Delta\Gamma$ is the complex frequency shift, Γ is the half-bandwidth at half-maximum, f_F is the resonance frequency on the fundamental, Z_q is the acoustic impedance of AT-cut quartz, $Z = (\rho G)^{1/2}$ is the acoustic impedance, ρ is the density, and G is the shear modulus. The acoustic impedance is the acoustic analog of the refractive index. In the second step, it was assumed that the density of the sample is about unity throughout. ρ can then be pulled out of the integral. Comparing Equations 18.25 and 18.10, one recognizes a similar structure of the integrand. This is no coincidence at all. The integrand is the contrast function. It is the same for all types of reflectometry, if the proper quantities are inserted for the respective type of wave.

Even though the structure of the equations is similar, the information contained in the "acoustic thickness" (or, more precisely, the acoustic moment of the impedance profile) and in the "optical thickness" (that is, the ellipsometric moment of the refractive index profile) is much different [34, 35]. The difference goes back to the fact that the contrast obtained with acoustic shear waves is usually much larger than the contrast in optics. While refractive indices typically vary in the range of a few per cent, shear moduli may easily vary by orders of magnitude even for rather dilute adsorbates. In optics the weight function (the integrand in Equation 18.10) is much smaller than unity and roughly proportional to the concentration. Therefore, the shift of the coupling angle is approximately proportional to the *adsorbed amount*. In acoustics, on the contrary, the weight function (integrand in

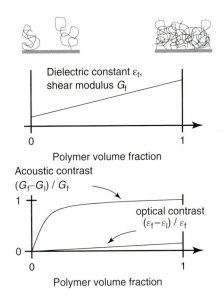

Figure 18.6 Comparison of the contrast function in ellipsometry and acoustic reflectometry. In acoustic reflectometry the contrast easily saturates, while the contrast function by an large is proportional to the concentration of the adsorbate in optics. (Reprinted with permission from Ref. [29].)

Equation 18.25) easily saturates to a value of unity even for dilute adsorbates. This happens because the shear modulus of the adsorbate is much higher than the shear modulus of the liquid (Figure 18.6). As a consequence, the Sauerbrey thickness (also "acoustic thickness") approaches the *geometric thickness* even for rather dilute polymer layers. If the adsorbate drags some solvent along in its shear movement, the trapped solvent appears as part of the film, as far as the acoustic properties are concerned. However, the trapped solvent does *not* increase the optical thickness. Swelling therefore effectively increases the acoustic thickness, while it affects the optical thickness to a much lesser extent [36]. The ratio of acoustic and optical thickness gives an estimate of the degree of swelling.

Evidently, the fact that the acoustic contrast is much higher than the optical contrast is beneficial for sensing [37]. The advantage that the QCM has over the competing optical techniques, is most strongly felt for dilute adsorbates. The QCM responds very sensitively to these because – pictorially speaking – a few polymer strands suffice to turn the film into a solid-like object.

18.6
Summary and Conclusions

Ellipsometry contains information unavailable with other techniques. This chapter has focused on samples with small refractive-index contrast. In the thin-film limit, ellipsometry determines the ellipsometric moment of the refractive-index distribution. It is by-and-large proportional to the adsorbed amount (as opposed to the geometric thickness). The geometric thickness is only available for thicker samples. Multiple-angle ellipsometry on the base of a high-index prism has a number of advantages. This includes the accessibility of the bulk refractive index and the occurrence for fringes, provided that the film is thick enough. In the presence for fringes, fitting usually yields unique results and robust estimates of the film thickness. Interestingly, there is a way to explicitly recover the refractive-index profile without resorting to any kind of model. The approach is termed *Fourier-transform ellipsometry*. Fourier-transform ellipsometry also entails a way of determining higher moments of the refractive-index profile (such as an effective thickness) from the derivatives of the ellipsometric spectrum close to the critical angle.

Ellipsometry is not exactly an easy technique. In operation, it is more difficult than SPR or the QCM. Its great strength is its flexibility and the multitude of ways by which it can be extended. This chapter as focused on MAE-TIR ellipsometry. Imaging, spectroscopy, and the various combinations with other techniques turn ellipsometry into a tool and a concept that will certainly stay with us.

Acknowledgments

Some of the results reported were obtained in the framework of collaboration with the Rühe group, now at Freiburg, Germany.

References

1. Drude, P. (1889) *Ann. Phys. Chem.*, **36**, 532.
2. Tsao, J.Y. (1992) *Materials Fundamentals of Molecular Beam Epitaxy*, Academic Press.
3. Roberts, G. (1990) *Langmuir–Blodgett Films*, Springer.
4. Ulman, A. (1998) *Thin Films: Self-Assembled Monolayers of Thiols: Self-assembled Monolayers of Thiols*, Academic Press.
5. Reiter, R., Motschmann, H., Orendi, H., Nemetz, A., and Knoll, W. (1992) *Langmuir*, **8**, 1784.
6. Erman, M. and Theeten, J.B. (1986) *J. Appl. Phys.*, **60**, 859.
7. Korte, E.H. and Roseler, A. (1998) *Analyst*, **123**, 647.
8. Widrig, C.A., Chung, C., and Porter, M.D. (1991) *J. Electroanal. Chem.*, **310**, 335.
9. Greef, R. (1993) *Thin Solid Films*, **233**, 32.
10. Bain, C.D. (1998) *Curr. Opin. Colloid Interface Sci.*, **3**, 287.
11. Domack, A., Prucker, O., Ruhe, J., and Johannsmann, D. (1997) *Phys. Rev. E*, **56**, 680.
12. Keddie, J.L. (2001) *Curr. Opin. Colloid Interface Sci.*, **6**, 102.
13. Tompkins, H.G. (2006) *A User's Guide to Ellipsometry*, Dover Publication Inc.
14. Azzam, R.M.A. and Bashara, N.M. (1987) *Ellipsometry and Polarized Light*, Elsevier, Amsterdam.
15. Brockman, J.M., Nelson, B.P., and Corn, R.M. (2000) *Ann. Rev. Phys. Chem.*, **51**, 41.
16. Rothenhausler, B. and Knoll, W. (1988) *Nature*, **332**, 615.
17. Yeatman, E. and Ash, E.A. (1987) *Electron. Lett.*, **23**, 1091.
18. Frutos, A.G., Weibel, S.C., and Corn, R.M. (1999) *Anal. Chem.*, **71**, 3935.
19. Lekner, J. (1987) *Theory of Reflection: of Electromagnetic and Particle Waves*, Springer.
20. Porter, M.D., Bright, T.B., Allara, D.L., and Chidsey, C.E.D. (1987) *J. Am. Chem. Soc.*, **109**, 3559.
21. Ogura, H. and Wang, Z.L. (1996) *Phys. Rev. B*, **53**, 10358.
22. Habicht, J., Schmidt, M., Ruhe, J., and Johannsmann, D. (1999) *Langmuir*, **15**, 2460.
23. Robinson, I.K. and Tweet, D.J. (1992) *Rep. Prog. Phys.*, **55**, 599.
24. Aspnes, D.E. and Theeten, J.B. (1979) *Phys. Rev. B*, **20**, 3292.
25. Schwiegk, S., Vahlenkamp, T., Xu, Y.Z., and Wegner, G. (1992) *Macromolecules*, **25**, 2513.
26. Ebert, H. (1996) *Rep. Prog. Phys.*, **59**, 1665.
27. Charmet, J.C. and Degennes, P.G. (1983) *J. Opt. Soc. Am.*, **73**, 1777.
28. Pockrand, I. (1978) *Surf. Sci.*, **72**, 577.
29. Johannsmann, D. (1999) *Macromol. Chem. Phys.*, **200**, 501.
30. Penfold, J. *et al.* (1997) *J. Chem. Soc. -Faraday Trans.*, **93**, 3899.
31. Wang, Z.H., Kuckling, D., and Johannsmann, D. (2003) *Soft Matter*, **1**, 353.
32. Biesalski, M., Ruhe, J., and Johannsmann, D. (1999) *J. Chem. Phys.*, **111**, 7029.
33. Biesalski, M., Johannsmann, D., and Ruhe, J. (2004) *J. Chem. Phys.*, **120**, 8807.
34. Plunkett, M.A., Wang, Z.H., Rutland, M.W. and Johannsmann, D. (2003) *Langmuir*, **19**, 6837.
35. Laschitsch, A., Menges, B., and Johannsmann, D. (2000) *Appl. Phys. Lett.*, **77**, 2252.
36. Voinova, M.V., Jonson, M., and Kasemo, B. (2002) *Biosens. Bioelectron.*, **17**, 835.
37. Wang, G., Rodahl, M., Edvardsson, M., Svedhem, S., Ohlsson, G., Höök, F., and KasemoRev, B. (2008) *Sci. Instrum.*, **79**, 075107.

19
Swelling Behavior of Thin Hydrogel Coatings

Ryan Toomey, Ajay Vidyasagar, and Ophir Ortiz

19.1
Introduction

A hydrogel is a cross-linked polymer network that swells with water, offering several routes toward soft materials with well-defined mechanical, physical, and biochemical properties [1–4]. First, hydrogels provide a soft three-dimensional scaffold capable of hosting a wide-array of functionalities, ranging from proteins to inorganic nanoparticles. Secondly, hydrogels can undergo substantial swelling and contraction in response to specific stimuli [5–8], making them excellent candidates for "smart" surfaces with sensing and actuating characteristics. While macroscopic gels have been hindered by slow response times, the collective diffusion of the network is the rate-limiting step; therefore, reducing their dimensions to the microscale significantly enhances performance, making them especially attractive in thin-film technologies [9].

A critical issue, nonetheless, in understanding and predicting the properties of surface-attached polymer networks is the impact of confinement on their swelling behavior [7, 10]. Chemical linkage to a substrate prevents swelling parallel to the substrate, effectively confining the volume change to one dimension, or normal to the surface. Such an effect will impact important properties, including structure, dynamics, and permeability of the network. The widely applied Flory–Rehner phenomenological model has met with some success in describing neutral, isotropic networks [11, 12]. Namely, the model predicts that the equilibrium swelling of a polymer network in a good solvent scales with the number of segments between cross-links to the 3/5 power. Two assumptions in the Flory–Rehner model are that the Flory interaction parameter χ is independent of polymer concentration and that swelling is uniform in all three directions.

Two key questions, therefore, include how surface attachment alters the scaling dependence of the network, and whether volume-phase transitions in stimuli-responsive networks can be modeled with a virial form of the Flory interaction parameter. For example, poly(N-isopropylacrylamide) (poly(NIPAAm)) undergoes a concentration-dependent hydrophilic/hydrophobic transition at roughly 32 °C [13–15]. This transition has been attributed to changes in the

Functional Polymer Films, First Edition. Edited by Wolfgang Knoll and Rigoberto C. Advincula.
© 2011 Wiley-VCH Verlag GmbH & Co. KGaA. Published 2011 by Wiley-VCH Verlag GmbH & Co. KGaA.

hydrogen-bonding tendency of water [16–20]. In poly(NIPAAm), it is thought that water molecules form ordered structures around both the hydrophilic amide moieties and the hydrophobic isopropyl groups to maximize favorable hydrogen-bonding associations. As temperature is increased, hydrogen-bonding interactions grow weaker until the demixing temperature is reached, wherein hydrophobic attractions between isopropyl groups dominate and collapse the polymer structure. Afroze *et al.* [21] determined that the critical point of poly(NIPAAm) occurs at a temperature of 29.5 °C and a volume fraction, ϕ, of 0.43. As the polymer concentration is reduced toward zero, the demixing temperature approaches 34 °C. To explain this phenomenon, the effective Flory interaction parameter χ can be expanded in powers of ϕ with a minimum of three terms, wherein each term is a function of temperature [22–24].

In this chapter, three issues will be covered. First, the effect of constraint on the scaling of swelling hydrogels will be discussed. Secondly, the appropriateness of a modified Flory–Rehner model with a virial expansion of χ will be discussed in order to capture the volume-phase transition of surface-attached poly(NIPAAm) networks. The third issue will be the implications of confined swelling on surface instabilities in thin-film configurations. With regard to the third point, as swelling is directed perpendicularly to the confining substrate, a biaxial stress is generated in the directions parallel to the surface. Upon swelling, the free surface can become unstable with respect to an out-of-plane undulation, analogous to a Raleigh wave.

19.2
Fabrication of Surface-Attached Networks and Characterization Techniques

To construct surface-attached networks, we photocross-linked thin films of statistical copolymers comprising methacryloyloxybenzophenone (MaBP) and dimethylacrylamide (DMAAm) or N-isopropylacrylamide (NIPAAm). An example structure is shown in Scheme 19.1. Irradiation of the films with UV light ($\lambda = 365$ nm) triggers the n, π^* transition in the MaBP segments, leading to a biradicaloid triplet state that is capable of abstracting a hydrogen from almost any kind of neighboring aliphatic C–H group, forming a stable C–C bond cross-link [25]. The approach can be used with essentially any technique of film formation that yields thin coatings of controllable thickness (e.g., spin or dip casting).

The swelling behavior of the surface-attached layers was characterized with multiple-angle ellipsometry or neutron reflection, which are two analytical techniques that yield interfacial segment profiles in thin swollen films. The basic principle behind both techniques involves directing a polarized light beam (ellipsometry [26]) or collimated neutron beam (neutron reflectometry [27]) toward a flat interface over a range of incident angles, θ, and measuring the change in polarization (ellipsometry) or the ratio of reflected intensity to incident intensity, R, (neutron reflectometry). We used two experimental configurations: (i) for measurements in the dry state and against vapors, the incident radiation entered from the air (or vapor) side and was transmitted through the underlying substrate and

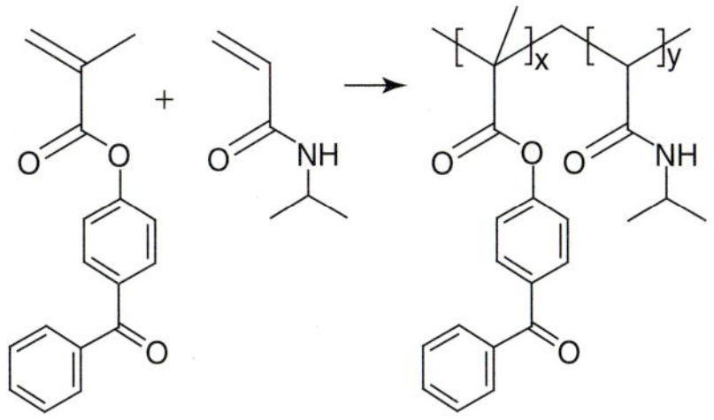

Scheme 19.1 Structure poly(NIPAAm-co-MaBP).

(ii) an "inverted" geometry with the incident beam entering the substrate and transmitted through the solvent. To model the ellipsometry data, model "delta" and "psi" profiles were generated used the formulism of Drude and compared to the measured data. To model the neutron-reflection data, a "model" reflectivity profile was generated using Parratt's recursion formalism and compared to the measured reflectivity profile. In both cases, the models were adjusted to obtain the best least-squares fit to the data using genetic optimization followed by the Levenberg–Marquardt nonlinear least-squares method.

19.3
Thermodynamics of Confined Hydrogels

When a dry, polymer network is exposed to solvent, the polymer segments swell until mechanical equilibrium is achieved. The combination of the polymer mesh and solvent molecules creates a nonlinear hyperelastic gel wherein the stress tensor (Π) is a function of both a Flory-like mean-field mixing energy and a Hookean elastic energy. The osmotic stress associated with mixing can be expressed in virial form as [28, 29]

$$P_{mix} \cong \frac{kT}{a^3} \left[\frac{\phi}{N} + \left(\frac{1}{2} - \chi \right) \phi^2 + \frac{1}{3} \phi^3 + \cdots \right] \tag{19.1}$$

where T is temperature, k is the Boltzmann constant, a^3 is the volume of a polymer segment, N is the number of segments between cross-links, ϕ is the polymer volume fraction, and $(1/2 - \chi)$ is the dimensionless second virial coefficient. The second virial coefficient, or the excluded volume integral, can be positive or negative and is interpreted as the interaction between two segments that are not immediately connected to one another. The first term in the virial expansion is negligible for large N, and the magnitude of the second virial coefficient dictates whether the network imbibes or expels solvent. For a value of χ less than $1/2$, the network imbibes solvent and equilibrium is determined when then the osmotic pressure of mixing equals the pressure of stretching.

To formulate an expression for swelling, it is assumed the subunits between cross-links can be modeled as Hookean springs [10, 12]. Upon swelling, a position x_j^o in the reference (nonswollen) state is transformed to x_i in the swollen state

$$x_i = \sum_j \alpha_{ij} x_j^o \tag{19.2}$$

where α_{iJ} is the deformation gradient tensor

$$\alpha_{iJ} = \frac{\partial x_i}{\partial x_J^\circ} \tag{19.3}$$

The corresponding stress associated with the distortion or stretching of the coils from the nonswollen state is

$$\sigma_{ij} = \left(\frac{B}{N}\frac{\phi}{\phi_o}\right)\delta_{ij} - \sum_i \frac{k_B^T}{a^3}\left(\frac{\phi}{N}\right)E_{iJ} \tag{19.4}$$

where ϕ_o is the polymer fraction at the time of cross-linking and B is the numerical prefactor associated with mixing entropy of the cross-links. The tensor **E** is the left Cauchy–Green tensor and is defined by

$$E_{iJ} = \sum_m \alpha_{im}\alpha_{Jm} \tag{19.5}$$

The relationship between the gel volume in the reference and swollen state is

$$\phi = \frac{\phi_o}{\det}\{\alpha_{iJ}\} \tag{19.6}$$

In the absence of external constraints, the surfaces of the equilibrium gel are stress-free and expansion of the network is complete when the mixing pressure exactly balances the elastic strain pressure and all components of the osmotic stress tensor $\Pi_{ij} = P_{mix}\delta_{ij} + \sigma_{ij}$ are zero everywhere throughout the gel. Under these conditions, **E** is diagonal and isotropic and the stretch ratio α is the same in all three dimensions

$$\mathbf{E} = \begin{bmatrix} \alpha & 0 & 0 \\ 0 & \alpha & 0 \\ 0 & 0 & \alpha \end{bmatrix} \tag{19.7}$$

Combining the dominant second-order term in the Flory–Huggins mean-field formulation (Equation 19.1) with the elastic stress (Equation 19.2) yields the homogeneous osmotic stress

$$\Pi = \frac{kT}{a^3}\left(\frac{1}{2} - \chi\right)\phi^2 - \frac{kT}{a^3}\left(\frac{1}{\alpha N}\right) \tag{19.8}$$

Setting $\Pi = 0$, the degree of linear expansion in a good solvent ($\chi = 0$) scales as $\alpha \cong (N_c)^{1/5}$. The overall volumetric swelling, S, is defined as $S = \alpha^3$ and consequently scales the number of segments between cross-links to the 3/5 power, or $S \cong N^{3/5}$.

If external constraints interfere with the swelling process, the constrained surfaces of the gel cannot expand to a stress-free state. If one side of the gel is clamped and cannot swell tangentially to the surface, the gel is restricted to one-dimensional swelling. In this case,

$$\mathbf{E} = \begin{bmatrix} \alpha & 0 & 0 \\ 0 & 1 & 0 \\ 0 & 0 & 1 \end{bmatrix} \tag{19.9}$$

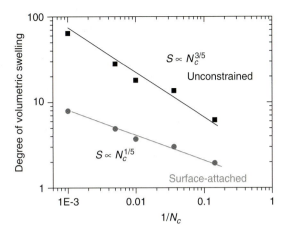

Figure 19.1 Comparison of the overall degree of swelling *S* between the surface-attached and unconstrained bulk poly(DMAAm-co-MaBP) networks. Adapted from Ref. [4].

And the osmotic pressure in the swelling direction (denoted by subscript *i*) is

$$\Pi_{ii} = \frac{kT}{a^3}\left(\frac{1}{2} - \chi\right)\phi^2 - \frac{kT}{a^3}\left(\frac{\alpha}{N}\right) \tag{19.10}$$

The gel expands until Π_{ii} is zero in the swelling direction, leading to $\alpha \cong N^{1/3}$, which is a stronger power law than unconstrained swelling. The stronger power law emerges as the osmotic stress due to mixing can no longer be relieved in the two directions parallel (denoted by the subscript *jj*) to the constraining surface. The overall volumetric swelling, *S*, is now defined as $S = \alpha$ and consequently scales the number of segments between cross-links to the 1/3 power, or $S \cong N^{1/3}$. The resultant stress in the constrained directions becomes:

$$\Pi_{jj} = \frac{kT}{a^3}\left(\frac{2}{N}\right)^{2/3} - \frac{kT}{2a^3}\left(\frac{2}{N}\right)^{4/3} \tag{19.11}$$

Figure 19.1 shows a comparison of the volumetric degree of swelling *S* as a function of cross-linking for both the surface-attached and unconstrained poly(DMAAm-MaBP) networks, as a function of cross-link density (1/*N*). It is readily seen that the unconstrained networks swell to a much higher degree than the surface-attached networks at all cross-link densities. As the unconstrained networks swell isotropically, this observation is consistent with the idea that higher degrees of freedom permit the network to swell to a greater extent than a gel, that is, mechanically or physically constrained. The fits are consistent with the 3/5 power-law prediction for unconstrained networks and the 1/3 power law for constrained networks.

Figure 19.2 shows the volume fraction profiles that best describe the 1, 0.2, and the 0.05% cross-linked samples in Scheme 19.1. For the 1% case, the data could be fit with the absence of an interface ($\sigma = 0$ nm). For the 0.2% MaBP and the 0.05% samples, the interfacial width took on values of $\sigma = 100$ nm and $\sigma = 200$ nm,

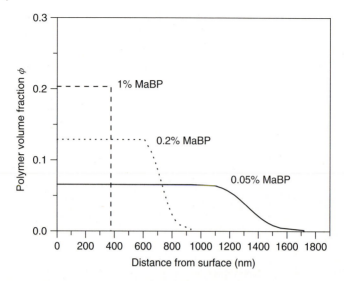

Figure 19.2 The polymer fraction profiles that represent the fits in Scheme 19.1 for the 1, 0.2, and 0.05 mol% poly(DMAAm-co-MaBP) samples. Reprinted from Ref. [4] with permission from the American Chemical Society.

respectively. While a finite interface width is expected under all conditions, features much less than the wavelength of light cannot be resolved with ellipsometry. The estimated precision in the interface width is ±50 nm. The physical explanation for the broad interfacial profiles most likely stems from a surface instability that emerges during the course of swelling. This instability was observed by Tanaka [30], and modeled by Onuki [10], who showed that the osmotic pressure in the directions parallel to the confining substrate can be relieved by periodic folding of the surface. This phenomenon will be further discussed later in the chapter.

19.4
Thermodynamics of Confined, Responsive Hydrogels

In order to analyze responsive polymers within the Flory–Huggins model, it is commonly found that the χ parameter depends on both temperature and concentration [24, 31–35], which can be traced to the role of the solvent. Water-soluble polymers induce a local solvation structure, where water molecules interact with the individual monomers and form a hydration sheath around the polymer [34]. The sheath, in turn, is strongly influenced by concentration. Although considerable effort has been devoted to clarifying the statistical mechanical underpinnings of hydration structures, the physical origin of these sheaths are still not well understood. Common microscopic models use a two-state description, wherein the monomers of the chain can exist in two distinct and interconverting hydrophilic and hydrophobic states [36–39]. These models, nonetheless, introduce unknown parameters that are difficult to verify through experiments. Experimentally, $\chi(T, \phi)$ can be obtained from colligative measurements. This approach does not provide physical understanding of behavior, but is does permit a connection between phase behavior of solutions and tethered assemblies.

To capture, at least phenomenologically, the behavior of responsive polymers, the effective Flory interaction parameter χ_{eff} must be expanded in powers of ϕ, up to order three, wherein each term is a function of temperature [40]:

$$\chi_{\text{eff}} = \chi_0(\phi, T) + \chi_1(\phi, T)\phi + \chi_2(\phi, T)\phi^2 \qquad (19.12)$$

The expansion in χ_{eff} leads to a modification of the second virial term in the mixing osmotic pressure

$$P_{\text{mix}} \cong \frac{kT}{a^3} \left\{ \frac{\phi}{N} + \left[\frac{1}{2} - \left(\chi_{\text{eff}}(T, \phi) + \frac{\partial \chi_{\text{eff}}}{\partial \phi} \right) \right] \phi^2 + \frac{1}{3}\phi^3 + \cdots \right\} \qquad (19.13)$$

which can be rewritten as

$$P_{\text{mix}} \cong \frac{kT}{a^3} \left\{ \frac{\phi}{N} + \left[\frac{1}{2} - \chi_0(T) \right] \phi^2 + \left[\frac{1}{3} - 2\chi_1(T) \right] \phi^3 \right.$$
$$\left. + \left[\frac{1}{4} - 3\chi_3(T) \right] \cdots \right\} \qquad (19.14)$$

where each term in the expansion can be positive or negative. Considering the appropriate values for χ_0, χ_1, and χ_2, to and setting $P_{\text{mix}} = 0$ can lead to three solutions, which is the necessary condition for a first-order phase transition.

The ϕ dependence of χ_{eff} gives rise to a notable difference from systems in which χ is invariant to concentration (or so-called classical behavior) [31]. The difference includes a shift in the position of the critical point and the shape of the coexistent curve enveloping the two-phase region. Whereas classical behavior always shows an upper critical solution temperature at vanishingly small concentrations, responsive polymers can have a critical point at any concentration, and they commonly display a lower critical solution temperature [33, 41]. A critical question, nonetheless, remains as to how the phase diagram of responsive polymers affects the behavior of the corresponding networks.

To aid in answering this question, we investigated cross-linked poly (NIPAAm-co-MaBP) coatings on solid quartz substrates by neutron reflection [42]. The reflectivity profiles at each temperature were initially modeled using a single slab of constant scattering length density, or a slab profile. The single-slab profiles at low temperatures, however, were inadequate, as shown in Figure 19.3 A functional form therefore was chosen that consists of three slabs, each of constant scattering length density, with smeared interfaces between the slabs by error functions. While this simple model was able to produce reasonable fits at all temperatures we do not exclude the possibility that more complicated models, which describe surface instabilities, will provide better fits [43, 44]. However, our attempts to use more sophisticated models, including cubic splines, showed that the fits, while quite sensitive to small changes in interfacial details, did not produce statistical differences in the overall thickness.

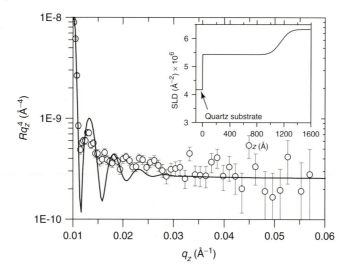

Figure 19.3 Neutron reflectivity data for surface-tethered poly(NIPAAm-co-MaBP(3%)) in D₂O environment at 15 °C. The solid line though the data corresponds to the best fit using a single slab scattering length density profile. The inset shows the scattering length density (SLD) distribution obtained from the best fit. Reprinted from Ref. [42] with permission from the American Chemical Society.

Figure 19.4 shows the volume fraction profiles of the poly(NIPAAm-co-MABP (3%)) layer exposed to bulk D₂O over a temperature range of 15–33 °C. The volume fraction profiles show a gradual contraction as temperature is raised from 15 to 29 °C. In this temperature regime, the average thickness decreased from 1100 to 750 Å. The average thickness $<z>$ is defined as $2 \int z\phi(z)\,dz / \int \phi(z)\,dz$. Moreover, in every case, a relatively diffuse interface was observed at the D₂O boundary, which was most likely the result of dangling ends that effectively behave as a brush. Between 29 and 31 °C, the average thickness abruptly changes from 714 (±35) to 339 (±34) Å. Above 31 °C, the layer thickness continues to contract, but is only very weakly dependent on temperature. No discernible difference in layer thickness was observed between 42 and 49 °C.

Figure 19.5 shows the corresponding change in the average thickness as a function of temperature. Above 15 °C, the thickness appears to decrease approximately linearly with temperature. At approximately 30 °C, the layer collapses to nearly its thickness as measured in the dry state. This state still contains 30–35% water, which corresponds to two to three D₂O molecules per segment. It is tempting to explain the nature of the collapse based on the experimentally determined phase diagram for linear poly(NIPAAm) solutions. As the demixing temperature of linear poly(NIPAAm) is a strong function of concentration with almost no effect on molecular weight, it is expected that cross-linking will have minimal effect on the phase behavior [45–47]. Afroze and coworkers [21] fitted experimental phase diagrams of linear poly(NIPAAm) with a quadratic form of a compositional dependent

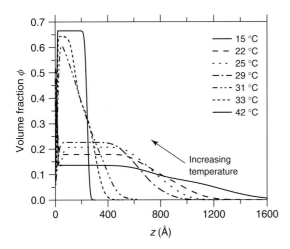

Figure 19.4 Volume fractions of the poly(NIPAAm-co-MaBP) in bulk D_2O as a function of the distance from the substrate and temperature. The data is the outcome of the best fit profiles obtained from fitting neutron reflectivities. Reprinted from Ref. [42] with permission from the American Chemical Society.

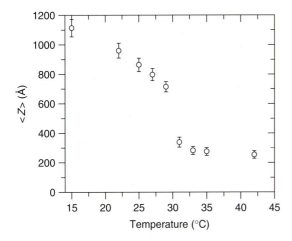

Figure 19.5 Variation of average thickness <z> surface-tethered poly(NIPAAm-co-MaBP (3%)) network as a function of temperature. Reprinted from Ref. [42] with permission from the American Chemical Society.

χ parameter, where the coefficients depend on temperature via

$$\chi_i(T) = A_i + B_i T \tag{19.15}$$

The A_i and B_i parameters are given in Table 19.1. Using these parameters, the phase diagram is drawn in Figure 19.6 showing the binodal boundaries.

Table 19.1 Parameters in Equation 19.14 for the χ parameter of linear poly(NIPAAm) solutions obtained by Afroze *et al.* [21].

i	A_i	B_i (K^{-1})
0	−12.947	0.04496
1	17.92	−0.0569
2	14.814	−0.0514

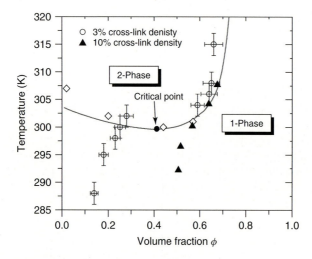

Figure 19.6 Experimental demixing temperature ($\diamond$) of uncross-linked poly(NIPAAm-co-MABP) and experimental swelling curve ($\bigcirc$: 3% MaBP) and ($\blacktriangle$: 10% MaBP) of the surface-tethered poly(NIPAAm-co-MABP) network. Also shown is the predicted binodal for linear poly(NIPAAm) at infinite molecular weight, from Ref. [21].

The binodal curve envelopes the two-phase region or the miscibility gap of poly(NIPAAm). Superimposed on the binodal are the experimentally determined demixing temperatures for linear poly(NIPAAm-co-MaBP(3%)), as determined by turbidity measurements, as well as the swelling isobars for two poly(NIPAAm-co-MaBP) networks. The networks represent a low degree of cross-link density (3% MaBP) and a high degree of cross-link density (10% MaBP). The polymer volume fractions for the networks were determined by neutron reflection (an example of which was shown in Figure 19.3 for the 3% MaBP network). At approximately 27 °C, the swelling of the 3% MaBP coating crosses into the miscibility gap of the polymer and the coating experiences a jump in concentration to approximately 65–70 vol% polymer. The 10% MaBP coating remains in the single-phase region and thus experiences no abrupt changes in concentration. Both coatings, independent of the cross-link density, collapse to approximately the same concentration (~65%) at temperatures above 30 °C and follow a similar trajectory with increasing temperature. Above 30 °C, the networks continue to expel water, but with a very weak dependence on temperature.

The extent of dilution in uncross-linked systems can be arbitrarily controlled; therefore, any part of the phase diagram can be accessed. Cross-linked systems, on the other hand, constrain the extent of dilution. The degree of dilution is determined by the competition between the osmotic pressure arising from contacts between segments and the entropic elasticity opposing dilution, and consequently a cross-linked system may or may not interfere with the two-phase region of the phase diagram. According to Figure 19.6, if the cross-link density is sufficiently high, chain elasticity prohibits the network from entering the miscibility gap, and the network therefore will move from a swollen to a less-swollen state in a more or less continuous manner. In the case of the lower 3% cross-link density, however, the deswelling of the network crosses the miscibility gap and experiences a swelling discontinuity. The significance of this finding is that the swelling discontinuity in surface-attached poly(NIPAAm) networks coincides with the two-phase region of uncross-linked poly(NIPAAm). Consequently, the two-phase region serves as a potential guide for anticipating volume-phase transitions in networks. For instance, a lower cross-link density is expected to result in a slightly higher demixing temperature and larger discontinuity in swelling (because the network would enter a wider region of the two-phase region). On the other hand, if the network is constrained to the region below the critical condition, the network is expected to show a smooth volume transition.

19.5
Swelling-Induced Surface Instabilities in Confined Hydrogels

In both Figures 19.2 and 19.4, we see the existence of extended profiles coatings of cross-linked poly(DMMAm) and poly(NIPAAm) exposed to water, which we now interpret as a near-the-surface instability. Periodic patterns, or surface buckling, can appear on the free surfaces of constrained hydrogel films [48–52]. Tanaka *et al.* reported that such patterns manifest upon swelling, where a planar surface loses stability and forms intersecting lines of cusps. Surface buckling, nonetheless, is remarkably general, the onset of which both Biot [53] and later Onuki [10] noted is mathematically analogous to Rayleigh waves.

Generally, in coatings of finite thickness, surface instabilities normally arises due to a mismatch or gradient in cross-linking or modulus [54]. This mismatch can be desirable; for instance, surface patterns may be generated in compressed (or swellable) elastic materials that have been modified with depthwise variation in properties (either chemically via UV, ion, or e-beam treatment or mechanically through the deposition of a disparate layer such as a metal) [55–59]. This mismatch can also be deleterious; for instance, multilayer structures may undergo near-the-surface buckling during drying, curing, or thermal cycling [60, 61].

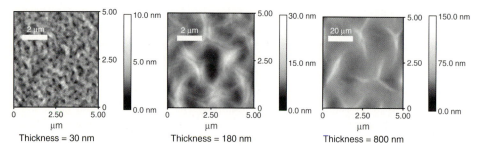

Figure 19.7 AFM scans of dry poly(NIPAAm-co-MaBP) coatings after exposure to water as a function of coating thickness.

In a constrained hydrogel coating, compression is brought about through swelling [62]. Normally, when exposed to a solvent, a hydrogel expands to a stress-free state. When constrained to a surface, however, inplane swelling is frustrated. As solvent penetrates the coating, expansion occurs perpendicularly to the constraining surface but not laterally, which generates inplane compressive stress, as described by Equation 19.10. Under sufficient stress, the free surface becomes unstable and deforms out of plane, which can lead to the formation of folds and cusps. Figure 19.7 shows AFM images of poly(NIPAAm) coatings after exposure to water, where surface features are quite apparent and the size depends strongly on the thickness of the coating. Figure 19.8 shows the variation of instability pattern height, width, and characteristic wavelength as a function of the dry thickness of the coating. As the film thickness was increased from 30 nm to 1.2 μm, there was a linear increase in height, width, and wavelength; moreover, the values were roughly independent of location. The linear scaling between wavelength and thickness is a well-established relationship, albeit with different methods of coating preparation [62, 63]. In this study, the wavelength is on the order of 25 times the dry thickness.

In order to determine the onset of the surface instability, we apply a first-order perturbation analysis to Flory–Rehner gels of finite thickness subject to initial, pre-critical homogeneous strain. The stress inside a gel can be described as $\Pi_{ij} = P_{mix}\delta_{ij} + \sigma_{ij}$, where P_{mix} and σ are described by Equations 19.1 and 19.8, respectively. The linear the linear deviation of the stress tensor for a small displacement $\mathbf{u} = (u, v)$ is

$$\delta\Pi_{ij} = \frac{kT}{a^3}\left\{ -K_{mix}(\nabla \cdot \mathbf{u})\delta_{ij} + \left(\frac{\phi}{\phi_0 N}\right) \right.$$
$$\left. \times \left[E_{ij}(\nabla \cdot \mathbf{u}) - \sum_m \left(E_{im}\frac{\partial u_j}{\partial x_m} + E_{jm}\frac{\partial u_i}{\partial x_m}\right)\right]\right\} \tag{19.16}$$

where

$$K_{mix} = \phi\left(\frac{\partial P_{mix}}{\partial \phi}\right) \tag{19.17}$$

For a two-dimensional gel infinitely long in the *y*-direction and clamped at $x = 0$, where x is the spatial position perpendicular to the confining substrate, the

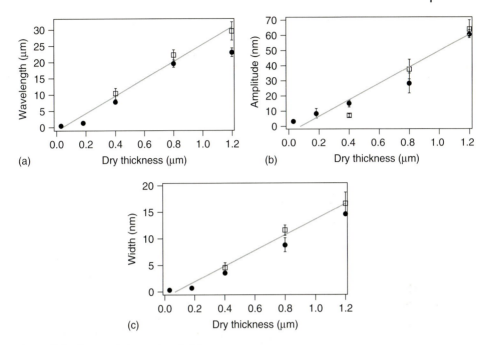

Figure 19.8 Characteristic wavelength (a), amplitude (b), and width (c) of the cusps formed in poly(NIPAAm-co-MaBP) coatings after exposure to water. The symbol (□) denotes a measurement in the center of the sample and (●) denotes a measurement at the edge.

precritical strain tensor is $E_{xx} = \alpha^2$, $E_{yy} = 1$, $E_{xy} = E_{yx} = 0$, where α satisfies the equilibrium condition $\Pi_{xx} = P_{ext}$ at the free edge. The planar surface is unstable to a small displacement $\mathbf{u} = (u, v)$ if mechanical equilibrium is satisfied [64],

$$\frac{\partial \Pi_{ij}}{\partial x_j^o} = 0 \tag{19.18}$$

The equilibrium relations for a small displacement can then be expressed as

$$\alpha \frac{(\delta \Pi_{xx})}{\partial x} + \frac{(\delta \Pi_{xy})}{\partial y} = \left(\alpha^2 + \beta^2\right)(\nabla \cdot \mathbf{u}) + \left(1 - \alpha^2\right)\frac{\partial^2 u}{\partial y^2} + 2\beta^2 \frac{\partial w}{\partial y} = 0 \tag{19.19}$$

$$\alpha \frac{(\delta \Pi_{xy})}{\partial x} + \frac{(\delta \Pi_{yy})}{\partial y} = \left(1 + \beta^2\right)(\nabla \cdot \mathbf{u}) - \left(1 - \alpha^2\right)\frac{\partial^2 u}{\partial x^2} - 2\beta^2 \frac{\partial w}{\partial x} = 0 \tag{19.20}$$

where $\beta^2 = \alpha \phi_o N K_{mix}$ and w is the rotation

$$w = \frac{1}{2}\left(\frac{\partial v}{\partial x} - \frac{\partial u}{\partial y}\right) \tag{19.21}$$

In general, the solution to Equations 19.19 and 19.20 for a sinusoidal deflection in height is

$$u(x, y) = \psi(x) \cos(qy) \tag{19.22}$$

$$v(x, y) = -\xi(x) \sin(qy) \tag{19.23}$$

where

$$\psi(x) = C_1 \sinh\left(\frac{qx}{\alpha}\right) + C_2 \cosh\left(\frac{qx}{\alpha}\right) + C_3 \sinh\left(\frac{qx}{\eta}\right) + C_4 \cosh\left(\frac{qx}{\eta}\right) \tag{19.24}$$

$$\xi(x) = \frac{C_1}{\alpha} \cosh\left(\frac{qx}{\alpha}\right) + \frac{C_2}{\alpha} \sinh\left(\frac{qx}{\alpha}\right) + C_3 \eta \cosh\left(\frac{qx}{\eta}\right) + C_4 \eta \sinh\left(\frac{qx}{\eta}\right) \tag{19.25}$$

And $\eta = \frac{\alpha^2 + \beta^2}{1 + \beta^2}$ and C_i are coefficients that must satisfy the boundary conditions. The boundary condition at the free edge is $\sum_j \Pi_{ij} n_j = n_i P_{ext}$, where n_i are the components of the normal vector at the free surface. As only small deflections are of interest, a small displacement **u** produces a normal vector with components $\mathbf{n} \cong (1, \partial u/\partial y)$. The linearized form of the boundary condition can then be written as

$$\delta \Pi_{xx}|_{x=\alpha l_o} = 0 \tag{19.26}$$

$$\left[\delta \Pi_{xy} + \left(\Pi_{yy}^s - P_{ext}\right) \frac{\partial u}{\partial y}\right]\Bigg|_{x=\alpha l_o} = 0 \tag{19.27}$$

and l_o is the unswollen thickness and Π_{yy}^s is the osmotic stress in the precritical ($\mathbf{u} = 0$) state. The two boundary conditions at $x = 0$ are displacement conditions and are

$$u(0, y) = 0 \tag{19.28}$$

$$v(0, y) = 0 \tag{19.29}$$

In order for the displacement u (Equations 19.17 and 19.18) to satisfy the boundary conditions as defined by Equations 19.26–19.29, the determinant of the coefficient matrix must be zero. For simplicity, it is assumed that $\chi = 0$, $B = 0$, and $N \gg 1$. In this limit, if the gel is allowed to swell homogeneously and quasistatically by lowering P_{ext} in infinitesimally small steps, the gel layer becomes unstable to a sinusoidal displacement in height at a critical degree of swelling of $\alpha^* \cong 3.3$ and the wavelength grows as

$$\lambda \cong \frac{1}{\log(\alpha - \alpha^*)} \tag{19.30}$$

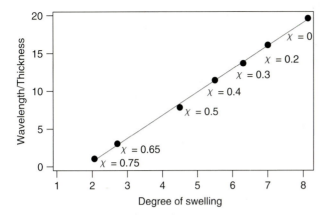

Figure 19.9 Wavelength/thickness ratio prediction based on a first-order perturbation analysis of a surface-confined Flory–Rehner gel. The model assumes that the top 1% of the layer is in equilibrium with the solvent and the remainder of the layer contains 1% solvent.

In this simple model, the divergence of the wave number at the critical degree of swelling is unrealistic. This can be corrected by taking the surface tension into account, which would lead to a finite wave number at the onset of the instability. Another consideration is that any model that assumes homogeneous distribution of solvent is inadequate [65]. At the start of solvent diffusion into the dry network, a very thin surface layer will quickly reach equilibrium with the external pressure, leading to an inhomogeneous distribution of solvent. The layer than grows as

$$\zeta \frac{\partial \alpha}{\partial t} = \frac{\partial_{xx}}{\partial x} \tag{19.31}$$

where ζ is the friction coefficient and t is time. This swollen layer is under mechanical constraint by virtue of the underlying unswollen layer, and the mismatch between the two layers can regularize the wavelength. If it is assumed, for instance, that the outer 1% of coating reaches equilibrium while the inner remains relatively unswollen ($\alpha = <1.1$), the inhomogeneity leads to a regularization of the wavelength, as shown in Figure 19.9. This agreement does necessarily diminish the importance of surface tension (which should be increasingly important as χ approaches 0.5), however it does lend strong support to the idea that inhomogeneous distribution of solvent can also explain near-the-surface instabilities.

19.6
Summary and Concluding Remarks

Hydrogel thin films permit a facile route toward smart surfaces with sensing and actuating characteristics. We have, in our studies, investigated the implications of constraints on the swelling of thin polymer hydrogels. Most notably, physically attaching a polymer network to the surface reduces the degrees of freedom in which the network can swell, affecting its swollen internal structure. Anisotropic stresses generated in the swelling of spatially constrained hydrogel lead to a stronger scaling dependence of linear expansion on the cross-link density, while the overall volume change is reduced with respect to the unconstrained hydrogel. The resultant residual stresses in the constrained directions give rise to a near-the-surface instability that has important implications for applications of hydrogel coatings. Finally, we showed how the phase diagram of uncross-linked responsive polymers can be used to understand the corresponding volume-phase transitions in surface-attached networks. It is worth noting that response can be engendered through a variety of environmental cues, including temperature, pH, and ionic interactions. The specific nature of these interactions and their relationship to mixing (as embodied in a virial expansion of the Flory interaction χ) and elasticity, however, will also challenge the interpretation of experiments and comparison to theory. Consequently, the binodal envelope of linear polymers may serve as a starting point by which to interpret the effects of confinement on volume-phase transitions in the corresponding networks.

Acknowledgments

R.T. acknowledges funding by the Camille and Henry Dreyfus Foundation, NSF DMR-0645574, and the USF College of Engineering.

References

1. Beebe, D.J., Moore, J.S., Bauer, J.M., Yu, Q., Liu, R.H., Devadoss, C., and Jo, B.H. (2000) Functional hydrogel structures for autonomous flow control inside microfluidic channels. *Nature*, **404**, 588.

2. Kuckling, D., Harmon, M.E., and Frank, C.W. (2002) Photo-cross-linkable PNIPAAm copolymers. 1. Synthesis and characterization of constrained temperature-responsive hydrogel layers. *Macromolecules*, **35**, 6377–6383.

3. Revzin, A., Russell, R.J., Yadavalli, V.K., Koh, W.G., Deister, C., Hile, D.D., Mellott, M.B., and Pishko, M.V. (2001) Fabrication of poly(ethylene glycol) hydrogel microstructures using photolithography. *Langmuir*, **17**, 5440–5447.

4. Toomey, R., Freidank, D., and Ruhe, J. (2004) Swelling behavior of thin, surface-attached polymer networks. *Macromolecules*, **37**, 882–887.

5. Harmon, M.E., Kuckling, D., and Frank, C.W. (2003) Photo-cross-linkable PNIPAAm copolymers. 2. Effects of constraint on temperature and pH-responsive hydrogel layers. *Macromolecules*, **36**, 162–172.

6. Peppas, N.A. (1991) Physiologically responsive hydrogels. *J. Bioact. Compat. Polym.*, **6**, 241–246.

7. Suzuki, A. and Kojima, S. (1994) Phase-transition in constrained polymer gels. *J. Chem. Phys.*, **101**, 10003–10007.

8. Tanaka, T. (1979) Phase-transitions in gels and a single polymer. *Polymer*, **20**, 1404–1412.

9. Zhao, B. and Moore, J.S. (2001) Fast pH- and ionic strength-responsive hydrogels in microchannels. *Langmuir*, **17**, 4758.

10. Onuki, A. (1993) Theory of phase-transition in polymer gels. *Adv. Polym. Sci.*, **109**, 63–121.

11. Flory, P. (1953) *Principles of Polymer Chemistry*, Cornell University Press.

12. Flory, P.J. and Rehner, J. (1943) Statistical mechanics of cross-linked polymer networks II swelling. *J. Chem. Phys.*, **11**, 521.

13. Heskins, M., Guillet, J.E., and James, E. (1968) Solution properties of poly(N-isopropylacrylamide). *J. Macromol. Sci. Chem.*, **2**, 1441–1455.

14. Tanaka, T. (1979) Phase-transitions in gels and a single polymer. *Polymer*, **20**, 1404.

15. Schild, H.G. (1992) Poly (N-Isopropylacrylamide) – experiment, theory and application. *Prog. Polym. Sci.*, **17**, 163–249.

16. Maeda, Y., Nakamura, T., and Ikeda, I. (2001) Changes in the hydration states of poly(N-alkylacrylamide)s during their phase transitions in water observed by FTIR spectroscopy. *Macromolecules*, **34**, 1391.

17. Meersman, F., Wang, J., Wu, Y.Q., and Heremans, K. (2005) Pressure effect on the hydration properties of poly(N-isopropylacrylamide) in aqueous solution studied by FTIR spectroscopy. *Macromolecules*, **38**, 8923.

18. Cheng, H., Shen, L., and Wu, C. (2006) LLS and FTIR studies on the hysteresis in association and dissociation of poly(N-isopropylacrylamide) chains in water. *Macromolecules*, **39**, 2325.

19. Katsumoto, Y., Tanaka, T., Sato, H., and Ozaki, Y. (2002) Conformational change of poly(N-isopropylacrylamide) during the coil-globule transition investigated by attenuated total reflection/infrared spectroscopy and density functional theory calculation. *J. Phys. Chem. A*, **106**, 3429–3435.

20. Okada, Y. and Tanaka, F. (2005) Cooperative hydration, chain collapse, and flat LCST behavior in aqueous poly(N-isopropylacrylamide) solutions. *Macromolecules*, **38**, 4465.

21. Afroze, F., Nies, E., and Berghmans, H. (2000) Phase transitions in the system poly(N-isopropylacrylamide)/water and swelling behaviour of the corresponding networks. *J. Mol. Struct.*, **554**, 55.

22. Moerkerke, R., Koningsveld, R., Berghmans, H., Dusek, K., and Solc, K. (1995) Phase-transitions in swollen networks. *Macromolecules*, **28**, 1103.

23. Solc, K., Dusek, K., Koningsveld, R., and Berghmans, H. (1995) "Zero" and "off-zero" critical concentrations in solutions of polydisperse polymers with very high molar masses. *Collect. Czech. Chem. Commun.*, **60**, 1661.

24. Erman, B. and Flory, P.J. (1986) Critical phenomena and transitions in swollen polymer networks and in linear macromolecules. *Macromolecules*, **19**, 2342–2353.

25. Prucker, O., Naumann, C.A., Ruhe, J., Knoll, W., and Frank, C.W. (1999) Photochemical attachment of polymer films to solid surfaces via monolayers of benzophenone derivatives. *J. Am. Chem. Soc.*, **121**, 8766.

26. Azzam, R.M. and Bashara, N.M. (1987) *Ellipsometry and Polarized Light*, Elsevier.

27. Russell, T.P. (1991) The characterization of polymer interfaces. *Annu. Rev. Mater. Sci.*, **21**, 249–268.

28. Flory, P.J. (1953) *Principles of Polymer Chemistry*, Cornell University Press.

29. Degennes, P.G. (1979) *Scaling Concepts in Polymer Physics*, Cornell University Press.

30. Tanaka, T., Sun, S.T., Hirokawa, Y., Katayama, S., Kucera, J., Hirose, Y., and Amiya, T. (1987) Mechanical instability of gels at the phase-transition. *Nature*, **325**, 796.

31. Baulin, V.A. and Halperin, A. (2003) Signatures of a concentration-dependent Flory chi parameter: swelling and collapse of coils and brushes. *Macromol. Theory Simul.*, **12**, 549–559.

32. Baulin, V.A. and Halperin, A. (2002) Concentration dependence of the Flory (x) parameter within two-state models. *Macromolecules*, **35**, 6432–6438.

33. SchaferSoenen, H., Moerkerke, R., Berghmans, H., Koningsveld, R., Dusek, K., and Solc, K. (1997) Zero and off-zero critical concentrations in systems containing polydisperse polymers with very high molar masses. 2. The system water-poly(vinyl methyl ether). *Macromolecules*, **30**, 410.

34. Bekiranov, S., Bruinsma, R., and Pincus, P. (1997) Solution behavior of polyethylene oxide in water as a function of temperature and pressure. *Phys. Rev. E,* **55**, 577–585.

35. Solc, K. and Koningsveld, R. (1992) Coalescence of upper and lower miscibility gaps in systems with concentration-dependent interactions. *J. Phys. Chem.,* **96**, 4056–4068.

36. Wagner, M., Brochardwyart, F., Hervet, H., and Degennes, P.G. (1993) Collapse of polymer brushes induced by N-clusters. *Colloid Polym. Sci.,* **271**, 621–628.

37. Karlstrom, G. (1985) A new model for upper and lower critical solution temperatures in poly(ethylene oxide) solutions. *J. Phys. Chem.,* **89**, 4962–4964.

38. Dormidontova, E.E. (2002) Role of competitive PEO-water and water-water hydrogen bonding in aqueous solution PEO behavior. *Macromolecules,* **35**, 987–1001.

39. Matsuyama, A. and Tanaka, F. (1990) Theory of solvation-induced reentrant phase-separation in polymer-solutions. *Phys. Rev. Lett.,* **65**, 341–344.

40. Baulin, V.A., Zhulina, E.B., and Halperin, A. (2003) Self-consistent field theory of brushes of neutral water-soluble polymers. *J. Chem. Phys.,* **119**, 10977–10988.

41. Solc, K., Dusek, K., Koningsveld, R., and Berghmans, H. (1995) "Zero" and "off-zero" critical concentrations in solutions of polydisperse polymers with very high molar masses. *Collect. Czech. Chem. Commun.,* **60**, 1661–1688.

42. Vidyasagar, A., Majewski, J., and Toomey, R. (2008) Temperature induced volume-phase transitions in surface-tethered Poly(N-isopropylacrylamide) networks. *Macromolecules,* **41**, 919–924.

43. Yim, H., Kent, M.S., Matheson, A., Stevens, M.J., Ivkov, R., Satija, S., Majewski, J., and Smith, G.S. (2002) Adsorption of sodium poly(styrenesulfonate) to the air surface of water by neutron and X-ray reflectivity and surface tension measurements: Polymer concentration dependence. *Macromolecules,* **35**, 9737–9747.

44. Yim, H., Kent, M., Matheson, A., Ivkov, R., Satija, S., Majewski, J., and Smith, G.S. (2000) Adsorption of poly(styrenesulfonate) to the air surface of water by neutron reflectivity. *Macromolecules,* **33**, 6126–6133.

45. Schild, H.G. and Tirrell, D.A. (1990) Microcalorimetric detection of lower critical solution temperatures in aqueous polymer-solutions. *J. Phys. Chem.,* **94**, 4352–4356.

46. Marchetti, M., Prager, S., and Cussler, E.L. (1990) Thermodynamic predictions of volume changes in temperature-sensitive gels. 1. Theory. *Macromolecules,* **23**, 1760–1765.

47. Fujishige, S., Kubota, K., and Ando, I. (1989) Phase-transition of aqueous-solutions of poly(N-Isopropylacrylamide) and poly(N-Isopropylmethacrylamide). *J. Phys. Chem.,* **93**, 3311–3313.

48. Tanaka, H., Tomita, H., Takasu, A., Hayashi, T., and Nishi, T. (1992) Morphological and kinetic evolution of surface patterns in gels during the swelling process – evidence of dynamic pattern ordering. *Phys. Rev. Lett.,* **68**, 2794–2797.

49. Matsuo, E.S. and Tanaka, T. (1992) Patterns in shrinking gels. *Nature,* **358**, 482–485.

50. Tanaka, T. (1978) Collapse of gels and the critical endpoint. *Phys. Rev. Lett.,* **40**, 820–823.

51. Hwa, T. and Kadar, M. (1988) Evolution of surface patterns on swelling gels. *Phys. Rev. Lett.,* **61**, 106–109.

52. Trujillo, V., Kim, J., and Hayward, R.C. (2008) Creasing instability of surface-attached hydrogels. *Soft Matter,* **4**, 564–569. doi: 10.1039/B713263h.

53. Biot, M.A. (1963) Surface instability in finite anisotropic elasticity under initial stress. *Proc. R. Soc. Lond. Ser. A,* **273**, 329–332–.

54. Genzer, J. and Groenewold, J. (2006) Soft matter with hard skin: from skin wrinkles to templating and material characterization. *Soft Matter,* **2**, 310–323.

55. Guvendiren, G., Burdick, J., and Yang, S. (2010) Kinetic Study of swelling-induced pattern formation and ordering in hydrogel films with depth-wise crosslinking gradient. *Soft Matter*. doi: 10.1039/b927374c.

56. Bowden, N., Brittain, S., Evans, A.G., Hutchinson, J.W., and Whitesides, G.M. (1998) Spontaneous formation of ordered structures in thin films of metals supported on an elastomeric polymer. *Nature*, **393**, 146–149.

57. Hayward, R.C., Chmelka, B.F., and Kramer, E.J. (2005) Template cross-linking effects on morphologies of swellable block copolymer and mesostructured silica thin films. *Macromolecules*, **38**, 7768–7783.

58. Huang, Z.Y., Hong, W., and Suo, Z. (2005) Nonlinear analyses of wrinkles in a film bonded to a compliant substrate. *J. Mech. Phys. Solids*, **53**, 2101–2118.

59. Huck, W.T.S. *et al.* (2000) Ordering of spontaneously formed buckles on planar surfaces. *Langmuir*, **16**, 3497–3501.

60. Hutchinson, J.W., Thouless, M.D., and Liniger, E.G. (1992) Growth and configurational stability of circular, buckling-driven film delaminations. *Acta Metallurg. Mater.*, **40**, 295–308.

61. Chopin, J., Vella, D., and Boudoud, A. (2008) The liquid blister test. *Proc. Natl. Acad. Sci. A*, **464**, 2887–2906.

62. Lin, C.-C., Yang, F., and Lee, S. (2008) Surface wrinkling of an elastic film: effect of residual surface stress. *Langmuir*, **24**, 13627–13631.

63. Kwon, O.H., Kikuchi, A., Yamato, M., Sakurai, Y., and Okano, T. (1999) Rapid cell sheet detachment from poly(N-isopropylacrylamide)-grafted porous cell culture membranes. *J. Biomed. Mater. Res. A*, **50**, 82–89.

64. Landau, L.D. and Lifshitz, E.M. (1973) *Theory of Elasticity*, Pergamon.

65. Ben Amar, M. and Pasquale, C. (2010) Swelling Instability of surface-attached gels as a model of soft tissue growth under geometric constraints. *J. Mech. Phys. Solids*, **58**, 935–954.

20
Scattering Techniques for Thin Polymer Films

Gila E. Stein

20.1
Introduction

Thin polymer films are widely investigated for applications in coatings, semiconductor patterning, and organic electronics. Thin polymer films are also interesting models for complex fluids under confinement, as film thickness is easily tuned to vary system size. Neutron and X-ray scattering can reveal details about the film structure that are essential for practical applications and invaluable for advances in fundamental polymer physics. This chapter provides an overview of synchrotron scattering techniques for thin polymer films, placing an emphasis on polymer systems that are relevant for microelectronics patterning and organic electronics. The techniques covered are largely reflection mode and include neutron reflectivity, soft X-ray reflectivity, grazing-incidence small-angle X-ray scattering (GISAXS), X-ray diffraction (XRD), and wide-angle X-ray scattering. Transmission scattering with hard and soft X-rays are also briefly discussed.

The first topics reviewed in this chapter are neutron and soft X-ray reflectivity from polymer/polymer interfaces. Experimental results from flexible Gaussian polymers and rigid conjugated polymers illustrate the challenges in calculating interfacial widths from reflectivity data. Neutron reflectivity requires isotopic labeling for contrast, which is usually achieved by partial deuteration of hydrocarbon polymers. Soft X-ray reflectivity, sometimes called *resonant reflectivity*, uses photon energies near the carbon absorption edge to exploit the intrinsic contrast due to resonances in the atomic scattering factors. The second topic reviewed in this chapter is GISAXS from block-copolymer thin films. GISAXS can reveal details of the lateral film structure that are missed by specular reflectivity, such as complex three-dimensional packing symmetries and depth-dependent domain orientations. Scattering experiments with hard X-rays are sensitive to spatial variations in electron density, and for different organic polymers the intrinsic contrast is very low. GISAXS is popular because the beam footprint is very long (on the order of 10 mm), which means a large volume of the film is sampled by the beam, and the increase in scattering centers will boost the signal to compensate for weak contrast. Section 20.4 discusses the structure of polymer semiconductors in thin

Functional Polymer Films, First Edition. Edited by Wolfgang Knoll and Rigoberto C. Advincula.
© 2011 Wiley-VCH Verlag GmbH & Co. KGaA. Published 2011 by Wiley-VCH Verlag GmbH & Co. KGaA.

films, emphasizing applications of GISAXS, grazing-incidence wide-angle X-ray scattering (GIWAXS), and reflection-mode XRD. X-ray scattering has revealed important structure–function correlations in transistors and solar cells based on polymer semiconductors. The chapter concludes with a brief discussion of transmission scattering with hard and soft X-ray sources. Transmission hard X-ray scattering has a small beam footprint and therefore samples a small volume, so the signal acquired is usually too small to detect. An exception is the scattering of X-rays by single crystals, because all the scattering power is concentrated into sharp diffraction peaks. Transmission scattering with soft X-rays, or resonant scattering, compensates for small sample volume with high contrast at energies near the carbon absorption edge.

20.2
Structure of Polymer/Polymer Interfaces Revealed by Reflectivity

The interface between different immiscible polymers is important for technological applications that include coatings, adhesives, lithography, and optoelectronic devices. Neutron or X-ray reflectivity can measure the interfacial width with sub-nanometer accuracy, and these techniques have been used to characterize interfaces in thin films of polymer bilayers and lamellar block-copolymers. Reflectivity results are often more meaningful when interpreted with the aid of theoretical models, so this section highlights the relevant polymer theory where appropriate.

Reflectivity measurements record the specular reflection from a thin-film "stack." This technique provides Angstrom scale depth sensitivity and is well suited to profile the perpendicular structure in thin polymer films, but does not offer detailed information regarding lateral structure. The physics of reflectivity are described in numerous articles and texts [1, 2], so here we include only a qualitative description of experimental geometry and data analysis. The reflectivity geometry is illustrated in Figure 20.1. Briefly, the incident angle (α) of the beam (highly collimated neutron or X-ray radiation) is varied from below the critical angle of the polymer ($\alpha_{c,p}$) up to approximately $20\alpha_{c,p}$, and the specular reflection is recorded at each point. The interface between two materials marks a transition in scattering length density, so part of the beam intensity is reflected at an interface, and part is transmitted across the interface at a refraction angle set by Snell's

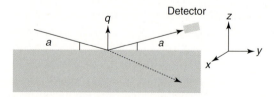

Figure 20.1 Schematic of a reflectivity experiment. Momentum transfer is perpendicular to the substrate, so reflectivity is well suited to characterize layered media.

Law. The partially reflected and transmitted beams within each layer interfere to produce periodic Kiessig fringes as a function of the reflection angle α. Data cannot be directly converted to a real-space measure of the film structure. Instead, the reflected intensity is modeled as a function of diffraction angle, and the model must incorporate the depth-dependent variations in scattering length density. These models are typically implemented by breaking the film into distinct layers characterized by unique scattering length densities, where the density transition between layers may be sharp (step function) or broad. Model parameters such as layer thickness and density are optimized through regression analysis to obtain agreement with experimental data. The reflected and transmitted wave amplitudes are described by dynamical scattering theory, and analytic solutions are available for the reflectivity geometry. These solutions are summarized in an article by Parratt [3].

Broad interfacial thickness will "damp" the signal at large scattering vectors such that the measured intensity is less than the ideal Fresnel reflectivity. The interface between two immiscible polymers nicely exemplifies this behavior: The unfavorable enthalpy of mixing is offset by a gain in configurational entropy, leading to a chemically diffuse interfacial boundary a few nanometers in width. Helfand and Tagami [4] derived an analytic solution for the mean-field composition profile at the interface between two immiscible homopolymers A and B. The profile for each component is predicted to take a hyperbolic tangent form:

$$\phi_A(z) = 0.5\{1 + \tan h(2z/\Delta_0)\}; \phi_B(z) = 1 - \phi_A(z) \tag{20.1}$$

Note that the interface plane is positioned at $z = 0$. The mean-field interfacial width Δ_0 is a function of the statistical segment length a_{st} and the Flory interaction parameter χ:

$$\Delta_0 = 2a_{st}/\sqrt{6\chi} \tag{20.2}$$

Neutron reflectivity has been used to measure the interface between polystyrene (PS) and poly(methyl methacrylate) (PMMA) homopolymers and block-copolymers [5, 6]. The thermodynamic properties of PS and PMMA are well characterized, so this system is a popular model for fundamental polymer physics. Each thin-film system was prepared with well-defined layers perpendicular to a silicon substrate, and scattering contrast was provided by selective deuterium labeling of either PS or PMMA. Interfacial widths were extracted by modeling the reflectivity data, and the results were compared with the predictions described by Equations 20.1 and 20.2. In all cases, the measured interfacial width was larger than predicted by approximately 2 nm, suggesting that mean-field theory does not capture all the essential physics in these systems. Similar results were reported for block-copolymers based on poly(ethylene-propylene) and poly(ethylene) constituents [7].

The discrepancy between mean-field theory and experiments suggests that finite molecular weight and/or thermal fluctuations are responsible for broadening the interface. The interfacial thickness between two immiscible polymers only weakly depends on molecular weight [8], while the copolymer interfacial width increases by approximately 1 nm [9]. As such, finite chain length alone cannot explain the broad interfaces measured by reflectivity. Semenov [9] and Shull et al. [10] proposed a simple fluctuation correction to the mean-field interfacial width based on a capillary wave model. Within this framework, the mean-square displacement of the interface position (relative to the average position at $z = 0$) is:

$$\langle \delta_z^2 \rangle = \frac{1}{2\pi\gamma_0} \ln\left(\frac{\lambda_{\max}}{\lambda_{\min}} \right) \tag{20.3}$$

The Helfand mean-field interfacial tension is $\gamma_0 = a_{\mathrm{st}}\chi^{0.5}/(v\sqrt{6})$, where v is the monomer volume [4]. The cut-offs $\lambda_{\max}$ and $\lambda_{\min}$ are the largest and smallest fluctuation wavelengths, respectively, and are determined by both physical constraints and experimental resolution. For immiscible homopolymers and block copolymers, the minimum wavelength $\lambda_{\min}$ is the intrinsic interfacial width Δ_0. For block copolymers, $\lambda_{\max}$ is limited by the block connectivity and is approximately equal to the domain periodicity d_0. However, $\lambda_{\max}$ in measurements of polymer bilayers is determined by the dispersive forces acting across the film (leads to thickness dependent capillary wave lengths) and the inplane coherence length of the radiation ($< 10\,\mu$m) [9–11]. Fluctuations are incorporated into the analysis of interfacial width by convolving the mean-field interface composition profile with a Gaussian distribution of interface positions based on Equation 20.3. The interfacial width Δ that would be measured in reflectivity is predicted as follows:

$$\Delta^2 = \Delta_0^2 + 2\pi \langle \delta_z^2 \rangle \tag{20.4}$$

The prediction for observed interfacial widths in Equation 20.4 accurately describes experimental measurements of block copolymer and homopolymer interfaces [9, 10], demonstrating that polymer interfaces are chemically diffuse at the molecular level with a longer wavelength roughness due to thermally excited fluctuations. The importance of thermal fluctuations is illustrated by reflectivity measurements from immiscible polymer bilayers of deuterated-polystyrene (dPS) and PMMA, where the thickness of the dPS layer was varied from a few nanometers up to a micrometer. The measured interfacial width increased with a logarithmic dependence on film thickness, consistent with capillary wavelengths that are determined by dispersive forces across the film [11]. This behavior is illustrated in Figure 20.2.

Reflectivity measurements from PS/PMMA thin films provided a fundamental understanding of the physics that determine the structure of polymer/polymer interfaces, most notably the importance of thermal fluctuations on the resulting interfacial area. These measurements also highlight the difficulty in relating measurements of interfacial thickness with molecular structure, as reflectivity cannot distinguish between different inplane length scales. With these factors in mind, reflectivity has emerged as a viable technique to characterize interfacial widths in thin films of semiconducting polymers. These studies are motivated by the need to understand complex structure–function relationships in organic electronic devices, such as the role of polymer interdiffusion on charge generation and transport. However, semiconducting polymers have extensive π-conjugation along the backbone that leads to rod-like conformations and intermolecular ordering, so

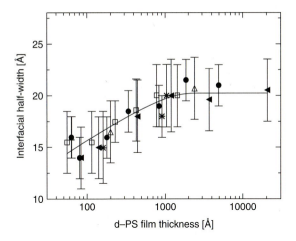

Figure 20.2 Dependence of dPS/PMMA interfacial half-width ($\Delta/2$) on dPS film thickness [11]. Interfacial widths were measured with neutron reflectivity. The plateau observed for the thickest films occurs when the dispersive capillary length exceeds the coherence length of the radiation source.

the mean-field models for Gaussian polymers presented in Equations 20.2–20.4 are not valid.

Neutron reflectivity was used by Higgins and coworkers to measure the interface between the semiconducting polymer poly(9,9-dioctylfluorene) (PFO) and a deuterated PMMA insulator [12]. Measurements were completed from "as-cast" and thermally annealed systems. The two annealing temperatures investigated were 136 and 165 °C, which produce crystalline and nematic PFO ordering, respectively. These annealing temperatures are above the glass-transition temperature of PMMA, which is approximately 105 °C. Examples of the reflectivity data are shown in Figure 20.3a. The density transition across the PMMA/PFO interface was modeled with an error function, and the calculated interfacial widths for different annealing temperatures and times are summarized in Figure 20.3b. Annealing at either temperature increases the interfacial width, although the broadening effect is more pronounced when PFO is in the nematic phase. The authors suggest that interfaces are broader in the nematic phase due to the absence of positional and orientational order. The measured interfacial widths reflect both chemical diffuseness and thermal fluctuations, and the authors estimate the intrinsic width and fluctuation contribution as 0.9 nm and 1.8 nm, respectively. These estimates are obtained by analyzing the data in Figure 20.3b using Equations 20.3 and 20.4.

The same reflectivity techniques were then applied to bilayers of deuterated PFO and poly(9,9-dioctylfluorene-co-benzothiadiazole) (F8BT), both semiconducting polymers with rigid π-conjugated backbones [13, 14]. The PFO/F8BT system is popular for optoelectronic devices such as light-emitting diodes. Thermal annealing was used to tune the width of the PFO/F8BT interface, and the root

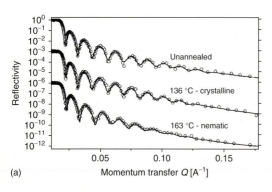

(a)

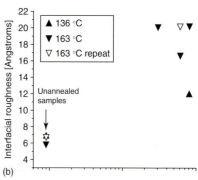

(b)

Figure 20.3 (a) Neutron-reflectivity data from bilayers of deuterated PMMA and PFO on silicon substrates (dots), and best fits to a bilayer model with roughness at the PMMA/PFO interface (solid line). The reflectivity is damped at large scattering vectors, which is characteristic of interfacial roughness. This damping effect increases with temperature. (b) The interfacial roughness is largest when PFO exhibits nematic ordering, but annealing at temperatures below the crystalline-to-nematic transition also increases the interfacial roughness relative to as-cast samples. Interfacial roughness is the half-width of the interface (or $\Delta/2$) for an error function composition profile [12].

mean square (RMS) roughness was calculated from modeling the reflectivity data with an error function interface composition profile. A variety of temperatures were explored that encompassed the crystalline to nematic transition for both PFO and F8BT. Roughnesses at the polymer/air and polymer/polymer interfaces are summarized in Figure 20.4. The PFO/F8BT RMS roughness in thick films (550 nm) can be tuned through annealing from 1 nm up to at least 25 nm, and in thin films (100 nm) from 1 nm up to at least 10 nm. The experiment cannot resolve interfacial widths that are greater than 10% of the total film thickness. For both thin and thick films, these roughnesses are significantly larger than reported for the PFO/PMMA interface annealed under similar conditions [12]. The authors qualitatively explain their results using Equations 20.1–20.4, although the theory is not strictly valid for rigid polymer chains, and both the PFO and F8BT were low molecular weight and highly polydisperse. Broad interfaces imply that the chains are very stiff (large a_{st}) and/or the polymers are somewhat miscible (small χ). At 280 °C, the 17 nm roughness is dominated by the intrinsic interfacial width. By constraining the segment length to 0.795 nm (the value for PFO), an estimate of $\chi \approx 0.01$–0.1 at 280 °C was obtained using Equation 20.2. It is difficult to calculate χ with greater accuracy when roughness due to thermal fluctuations cannot be quantified.

Ultimately, it is desirable to correlate the structure of conjugated polymer interfaces with optoelectronic function. A recent report from Higgins and coworkers compared interfacial roughness in deuterated PFO/F8BT bilayers with the Förster transfer from PFO to F8BT molecules [15]. They detected enhanced Förster transfer with increasing interfacial roughness in the range of 1–5 nm using photoluminescence measurements. Photoluminescence data were best explained

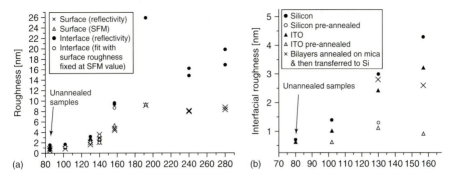

Figure 20.4 (a) RMS roughness of the PFO/F8BT interface measured with neutron reflectivity for different annealing conditions. Total film thickness was approximately 550 nm. Maximum RMS roughness that could be detected was 25 nm. (b) RMS roughness of the PFO/F8BT interface in thin bilayers of 100 nm total thickness. Maximum detectable RMS roughness was 10 nm.

by molecular mixing at the interface, rather than increased interfacial area due to thermal fluctuations or crystallization.

Neutron reflectivity requires partial deuteration of one of the constituents for contrast. However, many organic polymer pairs have intrinsic X-ray contrast at low photon energies due to $\pi - \pi^*$ and $\sigma - \sigma^*$ electronic transitions, eliminating the need for chemical modification. This technique has been termed *"resonant"* soft X-ray reflectivity, and uses photon energies near the carbon absorption edge (in the range of 270–300 eV) to provide sensitivity to the types and densities of different carbon bonds [16, 17]. In the case of PS and PMMA, strong contrast is observed at 285 eV and 289 eV from PS aromatic and PMMA carbonyl moieties, respectively. Reflectivity data for PS/PMMA bilayers are shown in Figure 20.5 as a function of photon energy. The interfacial widths for dPS and PMMA bilayers were measured with both resonant and neutron reflectivity, and the results agree within a few Angstroms [17]. This technique offers promise for characterizing conjugated polymer pairs, which have high densities of π-bonds that offer strong contrast at near-edge photon energies.

As an example, bilayers of the polymer semiconductor poly[2-methoxy-5-(2-ethylh exyloxy)p-phenylene vinylene] (MEH-PPV) and different conjugated polymer electrolytes (CPEs) were measured with resonant reflectivity [18]. The authors first measured the surface roughness of an 80 nm MEH-PPV films on silicon, then measured the interfacial roughness of bilayers prepared by sequential spin casting of an 80-nm thick MEH-PPV film from toluene and a 20-nm thick CPE film from methanol. The MEH-PPV surface roughness was 0.56 nm, the MEH-PPV/CPE interfacial roughnesses was 0.81 nm. To interpret this data, the authors assumed the as-cast films were sufficiently thin to prevent frozen-in thermal fluctuations. The chemical interdiffusion and initial surface roughness then add in quadrature per Equation 20.4, so the interdiffusion length was estimated as $\sqrt{(0.81^2 - 0.56^2)}$ nm^2 ≈ 0.6 nm. The sequential spin-casting process may alter the MEH-PPV surface,

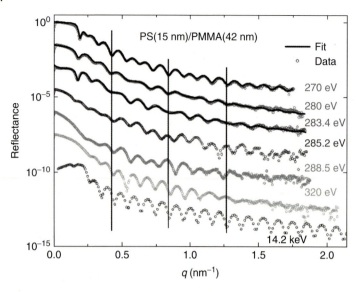

Figure 20.5 Resonant reflectivity from PS/PMMA bilayers. Varying the photon energy near the absorption edge provides contrast at the PS/PMMA interface [16, 17].

but these results still demonstrate that the polar solvent used for casting CPE films does not appreciably disturb the underlying MEH-PPV film.

20.3
Block Copolymer Thin Films Measured with Grazing-Incidence Small-Angle X-Ray Scattering

Thin films of block-copolymers provide a simple and inexpensive route to generate large-area nanoscale templates for applications in nanopatterning or nanoporous materials [19–22]. The physics of diblock copolymer melts in bulk are well understood through theory and experiment [23, 24]. The equilibrium phase behavior is determined by chain stretching and interfacial energy contributions to the total free energy. Below the order–disorder transition, a range of composition-dependent periodic morphologies are observed. Domain size and periodicity are controlled by molecular weight, and are on the order of 10 nm. Symmetric copolymers assemble into lamellar sheets, while asymmetric copolymers form morphologies with curved interfaces to reduce chain stretching in the matrix. Block-copolymers confined to a thin film may exhibit different phase behavior due to geometric confinement, surface energetics, and polymer–substrate interactions. The earliest studies of block-copolymer thin films used neutron reflectivity to study lamellar and cylindrical morphologies. These experiments have been extensively reviewed in the literature [22, 25, 26], but several important results are highlighted in the following paragraphs.

Neutron reflectivity from thin films of lamellar and cylindrical copolymers demonstrate that domain orientation is largely determined by the surface energies of the blocks at each boundary. Preferential attraction of one block to the interface induces layering of lamellar and cylindrical domains parallel to the boundaries [6, 7, 27–29]. Interfaces can induce ordering in thin films at temperatures above the bulk order–disorder transition, producing an exponentially damped sinusoidal concentration profile [7, 27, 29]. The structure in thin block copolymer films is often determined by commensurability between film thickness and the equilibrium domain size. For example, when lamellar films are confined between two hard walls, frustration can result if the wall separation is incommensurate with the equilibrium domain thickness. Neutron reflectivity demonstrates that frustration perturbs the lamellar periodicity despite the entropic penalty for stretching or compressing the chains [30, 31]. Substrate-supported block-copolymer thin films form "relief structures" at the free surface when film thickness and equilibrium domain size are incommensurate, so the film bifurcates into two distinct thicknesses that are commensurate with the block copolymer periodicity. Relief structure such as islands of excess material or depletion holes can be detected with neutron, soft X-ray, or hard X-ray reflectivity because they produce a surface layer with reduced density (air/polymer, rather than pure polymer).

As previously discussed, specular reflectivity is best suited to profile structure perpendicular to the substrate. Off-specular measurements are necessary to characterize lateral structure, which is why GISAXS measurements are common in the thin-film literature. GISAXS measurements are described in numerous articles and texts, so only a brief description of experimental geometry and data analysis is included in this chapter [32–34]. The GISAXS experimental geometry is illustrated in Figure 20.6. The sample is irradiated with hard X-rays at a fixed incident angle $\alpha_i \sim 0.1°$, that is below the critical angle for total external reflection from the substrate. The off-specular scattering $I(2\Theta, \alpha_f)$ is recorded with an area detector, and

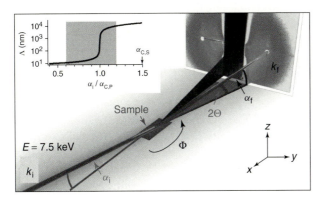

Figure 20.6 GISAXS experimental geometry. The beam impinges on the sample at a grazing angle of incidence α_i, which is on the order of 0.1°. The off-specular scattering is recorded with an area detector. Inset: X-ray penetration depth in a semi-infinite PS film as a function of incident angle α_i. The critical angle of a silicon substrate is marked as $\alpha_{c,s}$ [35].

the strong specular reflection is blocked by a lead beam stop. Varying the incident angle near the critical angle of the polymer allows for controlled X-ray penetration depths, so the near-surface structure can be probed independently from the bulk. X-ray penetration depth as a function of incident angle is illustrated in Figure 20.6 (inset) for a semi-infinite PS film. GISAXS provides a large scattering signal for polymer thin films due to the large beam footprint ($\sim 1 \, \mathrm{mm}^2$): The large sample volume compensates for the weak electron density contrast between different organic materials. In the absence of an orienting field, block copolymer domains assemble into micrometer-sized grains without a preferred inplane orientation, and the inplane scattering is characteristic of a powder. However, the out-of-plane orientation is often nearly perfect, so GISAXS patterns show the Bragg reflections that correspond with the preferred stacking sequence.

Analysis of GISAXS data can be highly complex because scattering from film structure occurs in combination with reflection from the substrate. The distorted-wave Born approximation (DWBA) is invoked for quantitative analysis of GISAXS data [36, 37]. In this framework, the electron density is calculated as the superposition of two terms: The first term describes a homogeneous thin film with no lateral density variations, and the second term describes the inplane disturbance. Scattering from the homogeneous sample is modeled with dynamical scattering theory (the Parratt recursions) while the disturbance is modeled kinematically. It is straightforward to calculate Bragg peak positions without modeling the scattering intensity, and this approach is usually sufficient to determine lattice symmetry and periodicity in thin copolymer films [35, 38, 39].

20.3.1
GISAXS Patterns of Cylindrical, Hexagonally Perforated Lamellar, and Gyroid Diblock Copolymer Phases

Lee *et al.* [38] used the DWBA to interpret the GISAXS patterns from thin films of strongly segregated poly(styrene-b-isoprene) (PS-PI) block-copolymers. PS-PI synthesized with 18 wt% PS self-assembled into cylindrical domains when cast on clean silicon substrates. The preferential wetting of PI at both interfaces drives a layering of cylindrical domains parallel to the interfaces. A GISAXS pattern acquired from the PS-PI cylinders is shown in Figure 20.7a. The ratio of peak positions along the inplane axis (2Θ) is 1 : 2, indicative of an inplane "fingerprint" pattern. The positions of the Bragg peaks are consistent with hexagonal (HEX) stacking of layers, which is the symmetry observed in bulk. The diffuse "ring" that connects Bragg peaks is characteristic of randomly oriented cylinders that coexist with the dominant parallel orientation. PS-PI synthesized with 37 wt% PS assembled into a structure identified as the hexagonally perforated lamellae (HPL) phase. The HPL symmetry is not stable in bulk diblock-copolymer films, but metastable (and possibly equilibrium) HPL structures have been reported for thin films. The thin-film HPL phase is comprised of parallel lamellar sheets that are perforated by hexagonally arranged cylindrical domains (the {003} plane parallel to the substrate). An example of the GISAXS pattern from a thin-film

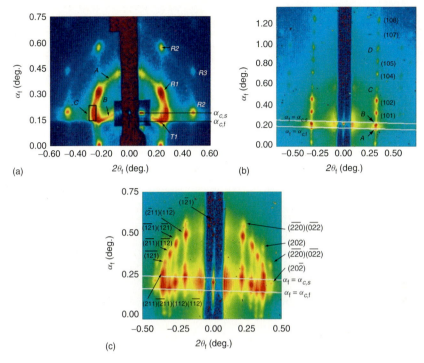

Figure 20.7 GISAXS patterns from (a) Parallel cylinders in an 800-nm thick film; (b) hexagonally perforated lamellae in a 450-nm thick film; and (c) gyroid phase in a 650-nm thick film [38].

HPL structure is shown in Figure 20.7b. The stacking sequence was identified as *ABC* and confirmed in a separate report [40]. PS-PI synthesized with 65 wt% PS produced a gyroid phase with the {121} plane oriented parallel to the substrate, and a representative scattering pattern is shown in Figure 20.7c.

20.3.2
Packing Symmetries of Spherical Domain Block Copolymers in Thin Films

Highly asymmetric block-copolymers form spherical microdomains. The body-centered cubic (BCC) symmetry is stable in bulk, while HEX symmetry is preferred by the monolayer. These different symmetries are the result of packing frustration, or a competition between stretching and interfacial energies: Stretching energy is minimized by uniform chain extension, while interfacial energy is minimized by uniform interfacial curvature. Both conditions cannot be satisfied simultaneously, so the equilibrium bulk and monolayer symmetries minimize spatial variations in domain thickness.

The BCC lattice does not have any planes with close-packed HEX symmetry, so it is not immediately obvious how the symmetry transitions from HEX

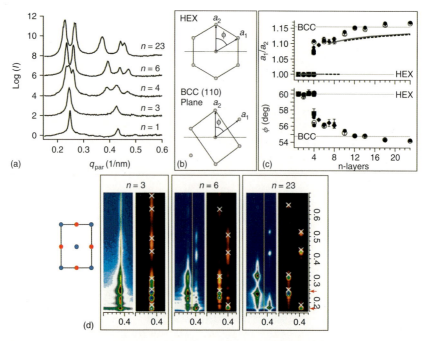

Figure 20.8 (a) Line profiles acquired by averaging GISAXS data in the z-direction (perpendicular to the substrate). Measurements were completed above the critical angle of the polymer. The number of layers is denoted by *n*. (b) Illustrating the difference between HEX symmetry and the BCC (110) plane. (c) Inplane symmetry as a function of the number of layers of spheres. Lines are SCFT predictions for $\chi N = 60$ (solid) and $\chi N = 50$ (dashed). (d) Experimental (blue) and simulated (black) GISAXS patterns for films 3, 6, and 23 layers thick. The incident angle is slightly above the critical angle for the polymer. As the number of layers is increased, the Bragg peaks (marked by white Xs) become sharper. This is analogous to "crystal truncation rods" in X-ray diffraction from finite-sized crystals [35, 41].

to BCC as film thickness is increased from a monolayer up to tens of layers (bulk-like). GISAXS measurements from thin films of poly(styrene-b-2-vinyl pyridine) (PS-P2VP) block-copolymer spheres identified a complex transition from HEX to BCC mediated by a face-centered orthorhombic (FCO) structure [35, 41]. The authors used a two-step process to interpret GISAXS data: First, the inplane peak positions were fit to a two-dimensional model to determine the inplane symmetry. Figure 20.8a–c illustrates the process. Second, two-dimensional GISAXS patterns were simulated with the DWBA by assuming various stacking sequences, and simulation results were compared with measured data. Examples are shown in Figure 20.8d. The monolayer HEX symmetry persisted through three layers of spheres with a close-packed ABA stacking sequence perpendicular to the substrate. At four layers, an equilibrium close-packed HEX phase coexisted with a metastable FCO phase. The FCO symmetry resembles a BCC lattice with the closest-packed {110} planes parallel to the substrate; close packing at the interfaces minimizes chain stretching. Further increases in film thickness produced a distortion of the lattice symmetry to approach asymptotically the BCC packing. Measurements below and above the critical angle demonstrate that structures are uniform through the film thickness. Experimental data were complemented by self-consistent-field theory (SCFT) calculations that demonstrate the origin of these thickness-dependent structures, namely, a competition between the packings preferred by the interior layers and interfaces [41]. More recently, packing frustration in thin films of PS-P2VP spheres (P2VP core, PS matrix) was tuned by blending with low molecular weight PS homopolymer [42]. The homopolymer accumulates at the interstitial regions to reduce the stretching energy of PS blocks, stabilizing close-packed structures and increasing the critical thickness for the HEX to FCO transition.

The GISAXS measurements from spheres illustrate an important consideration with interpreting the two-dimensional scattering patterns. The Bragg peaks are sharp along the inplane axis (q_{par}) due to the large grain size (a few micrometers), but peaks are highly elongated along the out-of-plane axis (q_z) due to the finite number of unit cells. The elongation produces significant overlap between adjacent peaks, so it is often difficult to fully characterize lattice symmetry without DWBA calculations. Figure 20.8d overlays the DWBA calculation with white Xs that mark Bragg peak centers.

20.3.3
Controlling Domain Orientations in Lamellar and Cylindrical Phases

Control over the domain orientation with respect to the substrate is critical for applications in semiconductor patterning. As previously mentioned, each block copolymer constituent has a different surface and/or substrate interfacial energy. This drives the formation of wetting layers at the interfaces, and induces parallel layering of cylindrical or lamellar domains with respect to the substrate. It is possible to manipulate energetics at the interfaces by coating substrates with a "neutral" film, adding surfactants, manipulating the copolymer architecture, or solvent annealing. The following paragraphs briefly describe the techniques used to control domain orientation, and how GISAXS may be used to evaluate the results.

Lamellar (LAM) and cylindrical (CYL) PS-PMMA block copolymers are popular for low-cost lithographic templates, because the PMMA block can be selectively removed in a chemical environment to produce a nanoporous mask for etching or deposition. PS and PMMA have similar surface energies, but PMMA preferentially wets the native oxide at the silicon substrate to promote a parallel domain orientation. However, a perpendicular orientation of PS-PMMA diblock-copolymer domains can be achieved if the substrate is chemically modified to be neutral to both segments [43–45]. The neutral coating is a random copolymer based on PS and PMMA constituents. GISAXS was used to characterize the orientation of cylindrical and lamellar domains on neutral PS-r-PMMA random copolymer brushes of varying PS compositions. The authors identified the brush compositions and film thicknesses that promote the desired perpendicular orientation [46]. Along similar lines,

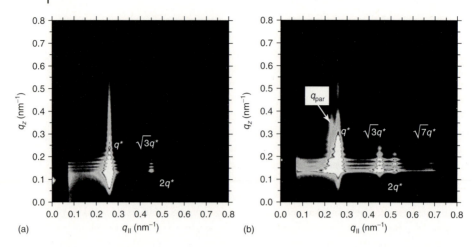

Figure 20.9 GISAXS measurement from PS-PMMA cylinders with oleic acid surfactant. (a) Incident angle below the critical angle for PS-PMMA, producing an X-ray penetration depth of approximately 10 nm. Perpendicular cylinders with HEX symmetry are detected. (b) Incident angle above the critical angle of PS-PMMA, so the X-rays penetrate through the full film thickness. Both parallel and perpendicular cylinders are detected [48].

neutral brushes based on poly(styrene-r-2-vinylpyridine-r-hydroxyethyl methacrylate) can control the domain orientation in thin films of PS-P2VP lamellar copolymers. The large surface energy difference between PS and P2VP favors the formation of a PS wetting layer at the air interface, and P2VP strongly adheres to the native oxide on silicon substrates. Depth-profiling GISAXS measurements demonstrate that a neutral coating can promote a perpendicular domain orientation near the substrate, while the parallel orientation persists at the air interface [47].

The addition of a surfactant can tune the surface energetics and promote a perpendicular domain orientation at the top of the film. For example, oleic acid produces neutral conditions at the surface of PS-PMMA thin films, so cylindrical or lamellar phases can be oriented normal to the air interface [48]. Depth profiling with GISAXS confirms the perpendicular orientation at the film surface, but demonstrates a mixed parallel and perpendicular morphology near the substrate. Representative measurements from below and above the PS-PMMA critical angle are shown in Figure 20.9a,b, respectively. Below the critical angle of the polymer, only the top 10 nm of the film is measured. The full film thickness is measured with incident angles above the polymer critical angle.

Perpendicular domain orientations have been observed in lamellar and cylindrical triblock-copolymers of poly(cyclohexylethylene-b-ethylene-b-cyclohexylethylene) (PCHE-PE-PCHE). The surface energy of the middle PE block is lower than the PCHE end segments. In PCHE-PE diblock copolymers, a PE wetting layers forms at the air interface and drives a parallel domain orientation. However, the PE midblock in a PCHE-PE-PCHE triblock must form loops to segregate to the air interface, and looping imposes an entropic penalty that must be offset by the reduction in surface energy. The difference in PCHE and PE surface energies is not enough to compensate for the conformational penalty, so the domains adopt a perpendicular orientation at the air interface [49]. Depth-profiling with GISAXS helped identify the conditions where the perpendicular orientation persists all the way through the film thickness.

Solvent annealing can also be used to control domain orientations. A dry polymer film is swollen with a good solvent for all the blocks, which typically disorders the structure, provides mobility, and can make the surface energies of different constituents comparable. As the solvent evaporates, an ordering front propagates from the free surface down into the film. The effectiveness of a solvent annealing protocol can be characterized with GISAXS. For example, the orientation of poly(isoprene-b-lactide) (PI-PLA) cylindrical block-copolymers was found to vary with the choice of solvent [50]. Films annealed in benzene adopted a parallel orientation, while chloroform successfully mediated the surface energies to provide a perpendicular orientation. The PI-PLA ordering could be improved by tuning film thickness and solvent volume fraction. Thin films of block-copolymers with PS and poly(ethylene oxide) (PEO) constituents exhibit a high degree of lateral order after annealing in humid benzene vapor [51, 52]. While PS has a lower surface energy than PEO, humidity seems to mediate the surface energies and promote the hydrophilic PEO to the film surface. In cylinder-forming PS-PEO diblock-copolymers, solvent annealing under humid conditions enables perpendicular orientation of PEO cylinders in a PS matrix, and GISAXS measurements demonstrate that the ordered morphology persists through the film thickness [51, 52]. The perpendicular orientation is facilitated by neutral substrate coatings [53]. Orientation of PS-PEO diblock-copolymers in solvent-annealed films can be further tuned by the addition of salts that associate selectively with the PEO phase [53, 54]. *In-situ* GISAXS was used to measure the structure of cylinder-forming PS-PEO films as a function of potassium iodide salt content. Films with low salt concentrations quickly disordered when swollen with benzene vapor, then reordered during solvent evaporation to form perpendicular cylinder arrays through the film thickness. As-cast films with high salt concentrations already exhibited perpendicular cylinder arrays, and did not disorder when swollen. The structure of PS-PMMA-PEO triblock-copolymers in thin films can also be controlled with solvent annealing under humid conditions [55, 56]. Humidity can produce half-spheres of PEO at the air interface in thin films of spherical-domain PS-PMMA-PEO triblock-copolymer prepared by solvent annealing [56].

The relative humidity during solvent annealing can dramatically influence the resulting block-copolymer architecture and orientation. For example, the role of humidity on solvent-annealed poly(ethylene oxide-b-methyl methacrylate-b-styrene) (PEO-PMMA-PS) triblock-copolymer thin films was investigated using GISAXS and complementary microscopies [55]. A PS-PMMA-PEO (13% PEO) triblock-copolymer was synthesized that assembled into an equilibrium lamellar structure as verified by bulk small-angle scattering. The same PS-PMMA-PEO triblock formed parallel lamellar sheets in thin films that were solvent annealed at low relative humidity (<70%), but assembled into an HPL phase with an ABC stacking sequence at high humidity (>90%). The LAM to HPL transition results from selective swelling in the PEO phase due to water. Along similar lines, humid air (98% relative humidity) has been used to control domain orientation in thin films of cylindrical poly(styrenesulfonate-b-methylbutylene) (PSS-PMB) block-copolymer electrolytes [57]. Depth profiling with GISAXS demonstrates a perpendicular cylinder orientation at the air interface and a parallel orientation near the substrate. The authors note that humidity is unlikely to mediate the surface energies and produce a neutral interface, so the mechanism resulting in perpendicular orientations is probably more complex.

20.3.4
Rod-Coil Block Copolymers in Thin Films

Lamellar rod-coil block-copolymers derived from poly(2,5-di(2'-ethylhexyloxy)-1,4-phenylenevinylene) (PPV) and PI constituents exhibit mixed-domain orientations in thin films [58]. PI forms a wetting layer at both the air and substrate interfaces, but the wetting layers do not induce parallel ordering: GISAXS measurements detect both parallel and perpendicular orientations throughout the film thickness, with no evidence of randomly oriented lamellae.

20.3.5
Templated Self-Assembly

The ordering of block-copolymer domains can be controlled with chemical or topographical templates [19, 20, 22]. Highly ordered crystals are necessary for certain applications, like manufacturing high-density bit-patterned media or integrated circuits. GISAXS is perfectly suited to characterize the quality of large-area arrays, and provides more reliable information with better statistics than scanning force microscopy or scanning electron microscopy. Implementation of the GISAXS measurements for single crystals is slightly different from films with randomly oriented grains: In addition to varying the incident angle for controlled penetration depths, the sample must be rotated along the inplane ϕ-axis to fully characterize the structure (see Figure 20.6). Rotation is necessary because scattering is only observed from planes that are parallel to the incident beam.

GISAXS has been used to characterize both positional and translational order in two-dimensional single crystals of PS-P2VP block-copolymer spheres [59]. Below the melting transition, a monolayer of optimal thickness assembles into a HEX lattice with 30-nm periodicity. In the absence of a template or external field, the monolayer has a grain size on the order of a micrometer. Large and oriented grains are produced when the monolayer is confined in 12-μm wide HEX wells, where the close-packed rows of the HEX lattice align with walls to minimize chain stretching at the interface. Block-copolymer arrays spanning the entire area of a silicon wafer were measured with GISAXS, and data for each rotation angle φ were compiled

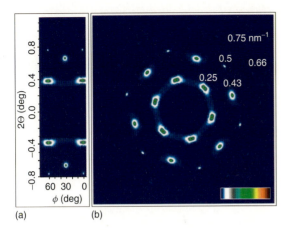

(a)   (b)

Figure 20.10 (a) $I(2\Theta, \phi)$ data map constructed from GISAXS measurements collected over an 80° range of in-plane rotation angles at a resolution of 0.25°. (b) A 60° segment from (a) shown in a radial plot so that it resembles a transmission powder diffraction pattern [59].

to build the data maps shown in Figure 20.10. The maps resemble a powder diffraction measurement from a single crystal with HEX symmetry, and it is clear that only one grain orientation is detected. The positional correlation function was calculated from line-shape analysis of the GISAXS data, and was determined to decay algebraically with a decay exponent of $\eta = 0.2$. This value is consistent with the Kosterlitz–Thouless–Halperin–Nelson–Young theory for 2D crystals, which predicts that long-wavelength phonons destroy positional order in two dimensions. When the monolayer thickness was less than optimal, the films did not form holes due to an absence of nucleating agents. Instead, the monolayer exhibited hexatic ordering with a positional correlation of approximately 0.5 μm [60]. Confining the film at a nonoptimal thickness was speculated to produce an entropic penalty that favored a more disordered lattice.

Single crystals of PS-PEO cylindrical block-copolymers have been generated with a sawtooth template and characterized with GISAXS [61]. The sawtooth profile is produced by heating miscut sapphire or silicon wafers, and the pitch can be varied in the range of 24–160 nm. Cylindrical morphologies of PS-PEO were cast on top of the sawtooth templates and solvent annealed by exposure to benzene under high humidity. For GISAXS measurements, the samples were rotated along the inplane ϕ-axis to collect scattering from different crystallographic planes. The authors detected a substantial increase in domain periodicity relative to the bulk equilibrium structure (from 36 to 63 nm). It is difficult to use GISAXS to characterize block copolymer ordering on these templates, however, because the GISAXS signal is dominated by diffraction from the sawtooth topography.

20.4
Thin Films of Organic Semiconductors Measured with X-Ray Scattering

Semiconducting polymers have delocalized π-bonds that are responsible for their optoelectronic properties, such as light absorption, charge generation, and charge transport. The extensive π-bonding also leads to intermolecular ordering with periodicity on the order of 1 nm, so thin films of polymer semiconductors are usually semicrystalline. Charge-carrier mobility is different along each crystallographic plane, so crystal structure and orientation is important for thin-film organic electronics [62]. XRD and GIWAXS can identify the structure and orientation of polymer crystals. XRD resembles a reflectivity measurement where the source and detector are located in specular plane, but a broader range of angles are scanned (tens of degrees). The resulting data are intensity as a function of diffraction angle, and peaks are observed that correspond with different Bragg planes oriented normal to the substrate. GIWAXS records the off-specular scattering associated with lateral structure using either a scanning or area detector. Like GISAXS, the penetration depth of X-rays is controlled with the incident angle, so thin films can be depth profiled to distinguish between near-surface and bulk structure.

20.4.1
Thin-Film Transistors

Field effect transistors (FETs) based on thin films of polymer semiconductors are widely investigated for low-cost, large-area plastic electronics. FETs conduct in the plane of the film, and current transport occurs within a few nanometers of the gate dielectric. The structure and orientation of polymer semiconductors within the film, and particularly near the gate dielectric, are therefore critical to device performance. Sirringhaus and coworkers [63, 64] used GIWAXS to measure the average crystal structure and orientation in poly(3-hexyl thiophene) (P3HT) thin-film transistors, and correlated the crystal orientation with charge-carrier mobility. P3HT self-organizes into a lamellar structure via $\pi - \pi$ interchain interactions, and thin films typically exhibit a mixture of lamellar crystallites and amorphous regions. Carrier mobility is highest along the (010) $\pi - \pi$ stacking axis, which is denoted by dimension "b" in Figures 20.11a,b. The authors determined that crystal orientation in spun-cast films depended on the regioregularity and molecular weight of the polymer. Low molecular weight and 96% regioregular P3HT assembled into crystals with the $\pi - \pi$ stacking axis in the plane of the film, while high molecular weight and 81% regioregular P3HT formed crystals with π-stacks oriented perpendicular to the substrate. The FET charge-carrier mobility in films with a high content of inplane π-stacking was roughly 2 orders of magnitude larger than films with perpendicular π-stacks. The authors suggested that slower solvent evaporation rates could produce the desired inplane (010) orientation in all the polymers studied. A series of publications from Kline and McGehee [65, 66] identified important structure–function relationships in thin films of regioregular P3HT. They considered the effects of P3HT crystallinity and crystal orientation

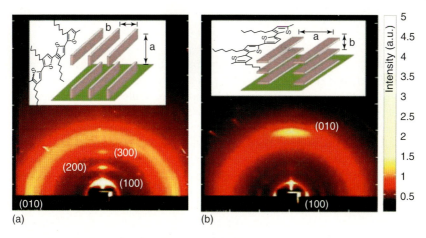

Figure 20.11 GIWAXS data from P3HT thin films. (a) Low molecular weight P3HT with 96% regioregularity. (b) High molecular weight P3HT with 81% regioregularity [63].

on FET mobility, using XRD and GIWAXS to measure the thin-film structure. Decreasing the molecular weight of regioregular P3HT increased the crystallinity, but carrier mobility dropped. The authors hypothesized that long chains bridge the ordered regions of high molecular weight P3HT, providing for more effective charge transport. Mobility improved when crystals were oriented with π-stacks in the plane of the film, and the extent of inplane π-stacking could be controlled by choice of solvent for film casting and the solvent evaporation rate, but was also affected by molecular weight [66]. GIWAXS measurements demonstrate that crystal orientation at the gate dielectric could be controlled by treating the dielectric with hexamethyldisilizane (HMDS) or octadecyltrichlorosilane (OTS) [67]. An OTS treatment produced a larger fraction of inplane π−π stacking than HMDS, and the FET mobility in low molecular weight devices increased by roughly 3 orders of magnitude.

20.4.2
Polymer Solar Cells

The active layer in a polymer solar cell is based on a thin film donor–acceptor heterojunction: The donor is a π-conjugated polymer donor, and the acceptor is usually a high electron affinity fullerene derivative. The thin film device is characterized by structure at multiple length scales. Conjugated polymers can assemble into crystallites via π−π interactions, and fullerenes can cluster into small domains. Donor–acceptor phases are usually distributed in an interpenetrating network with size and periodicity on the order of 10–100 nm, but bilayer devices with total film thickness of approximately 100 nm are also popular. The morphology of the active layer strongly influences optoelectronic properties, and a detailed understanding of these structure–function relationships is necessary to improve device efficiencies [68–71]. The structure of P3HT/fullerene bulk heterojunctions can be characterized with simultaneous GIWAXS and GISAXS [72]. The bulk heterojunction is prepared by arresting phase separation of a P3HT/fullerene blend, so the resulting nonequilibrium structure has a broad distribution of domain sizes with chemically diffuse interfaces. GIWAXS is sensitive to the P3HT intermolecular ordering, while GISAXS is sensitive to the average size of the fullerene clusters. The heterojunction structure in as-cast films was compared with devices annealed at temperatures above the P3HT glass transition. Thermal annealing increased the size of fullerene clusters and promoted P3HT intermolecular ordering. The highest power conversion efficiencies were acquired when the radii of fullerene clusters were approximately 20 nm and well-ordered P3HT crystallites were detected. The GISAXS data do not show a single diffraction peak characteristic of a single length scale, but are characterized by a diffuse intensity profile that decays with increasing diffraction angle. The diffuse GISAXS scattering reflects both lateral surface roughness and internal film structure, but is likely dominated by the former. Bulk heterojunctions based on P3HT and the acceptor poly((9,9-dioctylfluorene)-2,7-diyl-alt-[4,7-bis(3-hexylthien-5-yl)-2,1,3-benzothiadiaz ole]20,200-diyl) (F8TBT) have also been characterized with GISAXS [73]. Films were annealed at different temperatures prior to measurement. The vertical diffuse scattering indicated a modulation in electron density perpendicular to the substrate, consistent with enrichment of P3HT at both the film surface and substrate. The horizontal diffuse scattering associated with lateral structure became narrower with increasing annealing temperature, and the authors attribute this change with coarsening of inplane structure. XRD measurements from P3HT/F8TBT blends showed the evolution of intermolecular ordering ($\pi-\pi$ stacking of the conjugated polymers) with increasing annealing temperature [74]. X-ray measurements were complemented by absorbance and photoluminescence data, and the authors identified the optimal structure and annealing conditions for charge generation and separation.

20.5
Transmission X-Ray Scattering

Polymer thin films are rarely characterized with X-ray or neutron scattering in transmission mode, because the volume sampled by the beam at normal incidence is too small to produce a strong scattering signal. Techniques like neutron reflectivity, XRD, and grazing-incidence X-ray scattering produce a long beam footprint that samples a large volume of the film, thereby compensating for poor contrast and/or weak radiation fluxes. There are a few cases where transmission scattering from thin polymer films is feasible, however, and these are reviewed in the following paragraphs.

Resonant soft X-ray scattering using photon energies near the carbon absorption edge offers strong contrast between different organic constituents [75, 76]. This

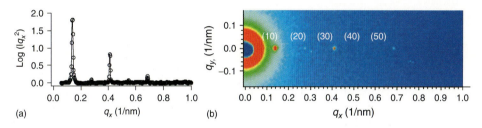

Figure 20.12 Transmission SAXS data from a lamellar block copolymer poly(styrene-b-methyl methacrylate) grating (G.E. Stein, Unpublished result). Pitch are film thickness are both 46 nm. (a) Line profile extracted from SAXS pattern in (b) shows five orders of diffraction peaks; the attenuation of the signal suggests a diffuse interface approximately 5 nm in width.

method has been used to measure the packing symmetry in thin films of PS-PI block-copolymers, where contrast results from the $\pi - \pi^*$ resonances in the PS aromatics or PI conjugated bonds [77]. Transmission measurements with soft X-rays require optically transparent substrates, so films must be free-standing or cast on thin silicon nitride membranes. Resonant soft X-ray scattering has also been used to measure single crystals of lamellar PS-PMMA diblock-copolymers that were assembled by epitaxy [78]. In that example, the photon energy was tuned to slightly below the π^* resonance for PS, and the diffraction data were modeled to calculate the PS-PMMA interfacial width in addition to domain sizes and shapes. The PS-PMMA interfacial widths extracted using Equations 20.1–20.4 matched predictions based on mean-field theory with corrections for finite molecular weight and capillary waves.

Transmission hard X-ray scattering is feasible from polymer films that exhibit single-crystal order, because the scattering signal from single crystals is concentrated into sharp peaks in reciprocal space. For example, transmission small-angle X-ray scattering has been used to characterize PS-PMMA lamellae assembled by epitaxy. A representative diffraction pattern is shown in Figure 20.12, where five diffraction peaks are detected with acquisition times of only 10 s. These highly ordered block-copolymer films cannot be characterized with laboratory X-ray sources, however, so synchrotron radiation is still required.

20.6
Summary

This chapter is a review of selected applications of synchrotron neutron and X-ray scattering for characterizing thin polymer films. An emphasis is placed on reflection mode techniques that include neutron reflectivity, resonant soft X-ray reflectivity, XRD, GISAXS, and GIWAXS. Reflectivity and XRD provide detailed information regarding structure perpendicular to the plane of the film, and are well suited to profile stratified media such as multilayered films or lamellar crystals.

Grazing-incidence scattering records the off-specular data and is sensitive to both inplane and out-of-plane ordering, so these techniques are widely employed for analysis of block-copolymer ordering, semicrystalline polymer films, and lateral phase separation in polymer blends.

References

1. Penfold, J., Richardson, R.M., Zarbakhsh, A., Webster, J.R.P., Bucknall, D.G., Rennie, A.R., Jones, R.A.L., Cosgrove, T., Thomas, R.K., Higgins, J.S., Fletcher, P.D.I., Dickinson, E., Roser, S.J., McLure, I.A., Hillman, A.R., Richards, R.W., Staples, E.J., Burgess, A.N., Simister, E.A., and White, J.W. (1997) Recent advances in the study of chemical surfaces and interfaces by specular neutron reflection. *J. Chem. Soc. -Faraday Trans.*, **93** (22), 3899–3917.

2. Tolan, M. (1999) *X-ray Scattering from Soft-Matter Thin Films*, Springer.

3. Parratt, L.G. (1954) Surface studies of solids by total reflection of X-rays. *Phys. Rev.*, **95** (2), 359–369.

4. Helfand, E. and Tagami, Y. (1971) Theory of interface between immiscible polymers. *J. Polym. Sci. B-Polym. Lett.*, **9** (10), 741–774.

5. Fernandez, M.L., Higgins, J.S., Penfold, J., Ward, R.C., Shackleton, C., and Walsh, D.J. (1988) Neutron reflection investigation of the interface between an immiscible polymer pair. *Polymer*, **29** (11), 1923–1928.

6. Anastasiadis, S.H., Russell, T.P., Satija, S.K., and Majkrzak, C.F. (1990) The morphology of symmetric diblock copolymers as revealed by neutron reflectivity. *J. Chem. Phys.*, **92** (9), 5677–5691.

7. Foster, M.D., Sikka, M., Singh, N., Bates, F.S., Satija, S.K., and Majkrzak, C.F. (1992) Structure of symmetrical polyolefin block copolymer thin-films. *J. Chem. Phys.*, **96** (11), 8605–8615.

8. Broseta, D., Fredrickson, G.H., Helfand, E., and Leibler, L. (1990) Molecular-weight and polydispersity effects at polymer/polymer interfaces. *Macromolecules*, **23** (1), 132–139.

9. Semenov, A.N. (1993) Theory of block-copolymer interfaces in the strong segregation limit. *Macromolecules*, **26** (24), 6617–6621.

10. Shull, K.R., Mayes, A.M., and Russell, T.P. (1993) Segment distributions in lamellar diblock copolymers. *Macromolecules*, **26** (15), 3929–3936.

11. Sferrazza, M., Xiao, C., Jones, R.A.L., Bucknall, D.G., Webster, J., and Penfold, J. (1997) Evidence for capillary waves at immiscible polymer/polymer interfaces. *Phys. Rev. Lett.*, **78** (19), 3693–3696.

12. Higgins, A.M., Jukes, P.C., Martin, S.J., Geoghegan, M., Jones, R.A.L., and Cubitt, R. (2002) A neutron-reflectometry study of the interface between poly(9,9-dioctylfluorene) and poly(methyl methacrylate). *Appl. Phys. Lett.*, **81** (26), 4949–4951.

13. Higgins, A.M., Martin, S.J., Jukes, P.C., Geoghegan, M., Jones, R.A.L., Langridge, S., Cubitt, R., Kirchmeyer, S., Wehrum, A., and Grizzi, I. (2003) Interfacial structure in semiconducting polymer devices. *J. Mater. Chem.*, **13** (11), 2814–2818.

14. Higgins, A.M., Martin, S.J., Geoghegan, M., Heriot, S.Y., Thompson, R.L., Cubitt, R., Dalgliesh, R.M., Grizzi, I., and Jones, R.A.L. (2006) Interfacial structure in conjugated polymers: characterization and control of the interface between poly(9,9-dioctylfluorene) and poly(9,9-dioctylfluorene-altbenzothiadiazole). *Macromolecules*, **39** (19), 6699–6707.

15. Higgins, A.M., Cadby, A., Lidzey, D.C., Dalgliesh, R.M., Geoghegan, M., Jones, R.A.L., Martin, S.J., and Heriot, S.Y. (2009) The impact of interfacial mixing on Forster transfer at conjugated polymer heterojunctions. *Adv. Funct. Mater.*, **19** (1), 157–163.

16. Wang, C., Araki, T., and Ade, H. (2005) Soft X-ray resonant reflectivity of low-Z material thin films. *Appl. Phys. Lett.*, **87** (21) 214109.

17. Wang, C., Araki, T., Watts, B., Harton, S., Koga, T., Basu, S., and Ade, H. (2007) Resonant soft X-ray reflectivity of organic thin films. *J. Vac. Sci. Technol. A*, **25** (3), 575–586.

18. Wang, C., Garcia, A., Yan, H., Sohn, K.E., Hexemer, A., Nguyen, T.-Q., Bazan, G.C., Kramer, E.J., and Ade, H. (2009) Interfacial widths of conjugated polymer bilayers. *J. Am. Chem. Soc.*, **131** (35), 12538–12539.

19. Segalman, R.A. (2005) Patterning with block copolymer thin films. *Mater. Sci. Eng. R-Rep.*, **48** (6), 191–226.

20. Darling, S.B. (2007) Directing the self-assembly of block copolymers. *Prog. Polym. Sci.*, **32** (10), 1152–1204.

21. Black, C.T., Ruiz, R., Breyta, G., Cheng, J.Y., Colburn, M.E., Guarini, K.W., Kim, H.-C., and Zhang, Y. (2007) Polymer self assembly in semiconductor microelectronics. *IBM J. Res. Dev.*, **51** (5), 605–633.

22. Hamley, I.W. (2009) Ordering in thin films of block copolymers: fundamentals to potential applications. *Prog. Polym. Sci.*, **34** (11), 1161–1210.

23. Bates, F.S. and Fredrickson, G.H. (1990) Block copolymer thermodynamics -theory and experiment. *Annu. Rev. Phys. Chem.*, **41**, 525–557.

24. Matsen, M.W. (2002) The standard Gaussian model for block copolymer melts. *J. Phys.-Condens. Matter*, **14** (2), R21–R47.

25. Krausch, G. (1995) Surface-induced self-assembly in thin polymer films. *Mater. Sci. Eng. R-Rep.*, **14** (1–2), 1–94.

26. Fasolka, M.J. and Mayes, A.M. (2001) Block copolymer thin films: physics and applications. *Annu. Rev. Mater. Res.*, **31**, 323–355.

27. Anastasiadis, S.H., Russell, T.P., Satija, S.K., and Majkrzak, C.F. (1989) Neutron reflectivity studies of the surface-induced ordering of diblock copolymer films. *Phys. Rev. Lett.*, **62** (16), 1852–1855.

28. Russell, T.P., Menelle, A., Anastasiadis, S.H., Satija, S.K., and Majkrzak, C.F. (1991) Unconventional morphologies of symmetrical, diblock copolymers due to film thickness constraints. *Macromolecules*, **24** (23), 6263–6269.

29. Menelle, A., Russell, T.P., Anastasiadis, S.H., Satija, S.K., and Majkrzak, C.F. (1992) Ordering of thin diblock copolymer films. *Phys. Rev. Lett.*, **68** (1), 67–70.

30. Lambooy, P., Russell, T.P., Kellogg, G.J., Mayes, A.M., Gallagher, P.D., and Satija, S.K. (1994) Observed frustration in confined block-copolymers. *Phys. Rev. Lett.*, **72** (18), 2899–2902.

31. Koneripalli, N., Levicky, R., Bates, F.S., Ankner, J., Kaiser, H., and Satija, S.K. (1996) Confinement-induced morphological changes in diblock copolymer films. *Langmuir*, **12** (26), 6681–6690.

32. Lazzari, R. (2002) Is GISAXS: a program for grazing-incidence small-angle X-ray scattering analysis of supported islands. *J. Appl. Crystallogr.*, **35** (Part 4), 406–421.

33. Muller-Buschbaum, P. (2003) Grazing incidence small-angle X-ray scattering: an advanced scattering technique for the investigation of nanostructured polymer films. *Anal. Bioanal. Chem.*, **376** (1), 3–10.

34. Renaud, G., Lazzari, R., and Leroy, F. (2009) Probing surface and interface morphology with grazing incidence small angle X-ray scattering. *Surf. Sci. Rep.*, **64** (8), 255–380.

35. Stein, G.E., Kramer, E.J., Li, X., and Wang, J. (2007) Layering transitions in thin films of spherical-domain block copolymers. *Macromolecules*, **40** (7), 2453–2460.

36. Vineyard, G.H. (1982) Grazing-incidence diffraction and the distorted-wave approximation for the study of surfaces. *Phys. Rev. B*, **26** (8), 4146–4159.

37. Holy, V., Kubena, J., Ohlidal, I., Lischka, K., and Plotz, W. (1993) X-ray reflection from rough layered systems. *Phys. Rev. B*, **47** (23), 15896–15903.

38. Lee, B., Park, I., Yoon, J., Park, S., Kim, J., Kim, K.W., Chang, T., and Ree, M. (2005) Structural analysis of block copolymer thin films with grazing incidence small-angle X-ray scattering. *Macromolecules*, **38** (10), 4311–4323.

39. Tate, M.P., Urade, V.N., Kowalski, J.D., Wei, T.C., Hamilton, B.D., Eggiman, B.W., and Hill-house, H.W. (2006) Simulation and interpretation of 2D diffraction patterns from self-assembled nanostructured films at arbitrary angles of incidence: from grazing incidence (above the critical angle) to transmission perpendicular to the substrate. *J. Phys. Chem. B*, **110** (20), 9882–9892.

40. Heo, K., Yoon, J., Jin, S., Kim, J., Kim, K.-W., Shin, T.J., Chung, B., Chang, T., and Ree, M. (2008) Polystyrene-b-polyisoprene thin films with hexagonally perforated layer structure: quantitative grazing-incidence X-ray scattering analysis. *J. Appl. Crystallogr.*, **41** (Part 2), 281–291.

41. Stein, G.E., Cochran, E.W., Katsov, K., Fredrickson, G.H., Kramer, E.J., Li, X., and Wang, J. (2007) Symmetry breaking of in-plane order in confined copolymer mesophases. *Phys. Rev. Lett.*, **98** (15) 158302.

42. Mishra, V., Hur, S.-M., Fredrickson, G.H., Kramer, E.J., Cochran, E.W., and Stein, G.E. (2010) Symmetry transitions in thin films of diblock copolymer/homopolymer blends. *Macromolecules*, **43**, 1942–1949.

43. Mansky, P., Liu, Y., Huang, E., Russell, T.P., and Hawker, C.J. (1997) Controlling polymer-surface interactions with random copolymer brushes. *Science*, **275** (5305), 1458–1460.

44. Bang, J., Bae, J., Lowenhielm, P., Spiessberger, C., Given-Beck, S.A., Russell, T.P., and Hawker, C.J. (2007) Facile routes to patterned surface neutralization layers for block copolymer lithography. *Adv. Mater.*, **19** (24), 4552–4557.

45. Han, E., In, I., Park, S.-M., La, Y.-H., Wang, Y., Nealey, P.F., and Gopalan, P. (2007) Photopatternable imaging layers for controlling block copolymer microdomain orientation. *Adv. Mater.*, **19** (24), 4448–4452.

46. Ham, S., Shin, C., Kim, E., Ryu, D.Y., Jeong, U., Russell, T.P., and H, C.J. (2008) Microdomain orientation of PS-b-PMMA by controlled interfacial interactions. *Macromolecules*, **41** (17), 6431–6437.

47. Ji, S., Liu, C.-C., Son, J.G., Gotrik, K., Craig, G.S.W., Gopalan, P., Himpsel, F.J., Char, K., and Nealey, P.F. (2008) Generalization of the use of random copolymers to control the wetting behavior of block copolymer films. *Macromolecules*, **41** (23), 9098–9103.

48. Son, J.G., Bulliard, X., Kang, H., Nealey, P.F., and Char, K. (2008) Surfactant-assisted orientation of thin diblock copolymer films. *Adv. Mater.*, **20** (19), 3643–3648.

49. Khanna, V., Cochran, E.W., Hexemer, A., Stein, G.E., Fredrickson, G.H., Kramer, E.J., Li, X., Wang, J., and Hahn, S.F. (2006) Effect of chain architecture and surface energies on the ordering behavior of lamellar and cylinder forming block copolymers. *Macromolecules*, **39** (26), 9346–9356.

50. Cavicchi, K.A., Berthiaume, K.J., and Russell, T.P. (2005) Solvent annealing thin films of poly(isoprene-b-lactide). *Polymer*, **46** (25), 11635–11639.

51. Kim, S.H., Misner, M.J., Xu, T., Kimura, M., and Russell, T.P. (2004) Highly oriented and ordered arrays from block copolymers via solvent evaporation. *Adv. Mater.*, **16** (3), 226–231.

52. Kim, S.H., Misner, M.J., and Russell, T.P. (2004) Solvent-induced ordering in thin film diblock copolymer/homopolymer mixtures. *Adv. Mater.*, **16** (23–24), 2119–2123.

53. Kim, S.H., Misner, M.J., and Russell, T.P. (2008) Controlling orientation and order in block copolymer thin films. *Adv. Mater.*, **20** (24), 4851–4856.

54. Kim, S.H., Misner, M.J., Yang, L., Gang, O., Ocko, B.M., and Russell, T.P. (2006) Salt complexation in block copolymer thin films. *Macromolecules*, **39** (24), 8473–8479.

55. Bang, J., Kim, B.J., Stein, G.E., Russell, T.P., Li, X., Wang, J., Kramer, E.J., and Hawker, C.J. (2007) Effect of humidity on the ordering of PEO-based copolymer thin films. *Macromolecules*, **40** (19), 7019–7025.

56. Tang, C., Bang, J., Stein, G.E., Fredrickson, G.H., Hawker, C.J., Kramer, E.J., Sprung, M., and Wang, J. (2008) Square packing and structural arrangement of ABC triblock copolymer spheres in thin films. *Macromolecules*, **41** (12), 4328–4339.

57. Park, M.J., Kim, S., Minor, A.M., Hexemer, A., and Bolsara, N.P. (2009) Control of domain orientation in block copolymer electrolyte membranes at the interface with humid air. *Adv. Mater.*, **21** (2), 203–208.

58. Olsen, B.D., Li, X., Wang, J., and Segalman, R.A. (2007) Thin film structure of symmetric rod-coil block copolymers. *Macromolecules*, **40** (9), 3287–3295.

59. Stein, G.E., Kramer, E.J., Li, X., and Wang, J. (2007) Single-crystal diffraction from two-dimensional block copolymer arrays. *Phys. Rev. Lett.*, **98** (8) 086101.

60. Stein, G.E., Lee, W.B., Fredrickson, G.H., Kramer, E.J., Li, X., and Wang, J. (2007) Thickness dependent ordering in laterally confined monolayers of spherical-domain block copolymers. *Macromolecules*, **40** (16), 5791–5800.

61. Park, S., Lee, D.H., Xu, J., Kim, B., Hong, S.W., Jeong, U., Xu, T., and Russell, T.P. (2009) Macroscopic 10-terabit-per-square-inch arrays from block copolymers with lateral order. *Science*, **323** (5917), 1030–1033.

62. Sirringhaus, H. (2005) Device physics of solution-processed organic field effect transistors. *Adv. Mater.*, **17** (20), 2411–2425.

63. Sirringhaus, H., Brown, P.J., Friend, R.H., Nielsen, M.M., Bechgaard, K., Langeveld-Voss, B.M.W., Spiering, A.J.H., Janssen, R.A.J., Meijer, E.W., Herwig, P., and de Leeuw, D.M. (1999) Two-dimensional charge transport in self-organized, high-mobility conjugated polymers. *Nature*, **401** (6754), 685–688.

64. Sirringhaus, H., Brown, P.J., Friend, R.H., Nielsen, M.M., Bechgaard, K., Langeveld-Voss, B.M.W., Spiering, A.J.H., Janssen, R.A.J., and Meijer, E.W. (2000) Microstructure-mobility correlation in self-organised, conjugated polymer field effect transistors. *Synth. Met.*, **111**, 129–132.

65. Kline, R.J., McGehee, M.D., Kadnikova, E.N., Liu, J.S., and Frechet, J.M.J. (2003) Controlling the field-effect mobility of regioregular polythiophene by changing the molecular weight. *Adv. Mater.*, **15** (18), 1519–1522.

66. Kline, R.J., McGehee, M.D., Kadnikova, E.N., Liu, J.S., Frechet, J.M.J., and Toney, M.F. (2005) Dependence of regioregular poly(3-hexylthiophene) film morphology and field-effect mobility on molecular weight. *Macromolecules*, **38** (8), 3312–3319.

67. Kline, R.J., Mcgehee, M.D., and Toney, M.F. (2006) Highly oriented crystals at the buried interface in polythiophene thin-film transistors. *Nature Mater.*, **5** (3), 222–228.

68. Guenes, S., Neugebauer, H., and Sariciftci, N.S. (2007) Conjugated polymer-based organic solar cells. *Chem. Rev.*, **107** (4), 1324–1338.

69. Blom, P.W.M., Mihailetchi, V.D., Koster, L.J.A., and Markov, D.E. (2007) Device physics of polymer: fullerene bulk heterojunction solar cells. *Adv. Mater.*, **19** (12), 1551–1566.

70. Mayer, A.C., Scully, S.R., Hardin, B.E., Rowell, M.W., and McGehee, M.D. (2007) Polymer-based solar cells. *Mater. Today*, **10** (11), 28–33.

71. Thompson, B.C. and Frechet, J.M.J. (2008) Organic photovoltaics -Polymer-fullerene composite solar cells. *Angew. Chem. Int. Ed.*, **47** (1), 58–77.

72. Chiu, M.-Y., Jeng, U.-S., Su, C.-H., Liang, K.S., and Wei, K.-H. (2008) Simultaneous use of small-and wide-angle X-ray techniques to analyze nanometerscale phase separation in polymer heterojunction solar cells. *Adv. Mater.*, **20** (13), 2573–2576.

73. McNeill, C.R., Abrusci, A., Hwang, I., Ruderer, M.A., Mueller-Buschbaum, P., and Green-ham, N.C. (2009) Photophysics and photocurrent generation in polythiophene/polyfluorene copolymer blends. *Adv. Funct. Mater.*, **19** (19), 3103–3111.

74. Flesch, H.-G., Resel, R., and McNeill, C.R. (2009) Charge transport properties and microstructure of polythiophene/polyfluorene blends. *Org. Electron.*, **10** (8), 1549–1555.

75. Mitchell, G.E., Landes, B.G., Lyons, J., Kern, B.J., Devon, M.J., Koprinarov, I. Gullikson, E.M., and Kortright, J.B.

(2006) Molecular bond selective X-ray scattering for nanoscale analysis of soft matter. *Appl. Phys. Lett.*, **89** (4) 044101.

76. Ade, H. and Hitchcock, A.P. (2008) NEXAFS microscopy and resonant scattering: composition and orientation probed in real and reciprocal space. *Polymer*, **49** (3), 643–675.

77. Virgili, J.M., Tao, Y., Kortright, J.B., Balsara, N.P., and Segalman, R.A. (2007) Analysis of order formation in block copolymer thin films using resonant soft X-ray scattering. *Macromolecules*, **40** (6), 2092–2099.

78. Stein, G.E., Liddle, J.A., Aquila, A.L., and Gullikson, E.M. (2009) Measuring the structure of epitaxially assembled block copolymer domains with soft X-ray diffraction. *Macromolecules*. doi: 10.1021/ma901914b.

21
Nanostructured Optical Waveguides for Thin-Film Characterization

Hatice Duran, K.H. Aaron Lau, Petra J. Cameron, Antonis Gitsas, Martin Steinhart, and Wolfgang Knoll

21.1
Introduction

The optical waveguide is the fundamental element that interconnects the various devices of an optical integrated circuit. Optical waves travel in waveguide in distinct optical modes. A mode is a spatial distribution of optical energy in one or more dimensions that remains constant in time and the number of modes can be supported depending on the thickness of the waveguiding layer and on the refractive index of waveguide components. Thus, optical waveguide spectroscopy (OWS) refers to the analysis of optical waveguide modes that are excited in a thin slab waveguide in order to measure with high sensitivity, the thickness and the (anisotropic) dielectric function of the thin film [1, 2]. It has been recognized as one of the most powerful analytical tools for observation of *interfacial* phenomena for the last three decades [3]. This method has at least three obvious advantages compared to others for thin-film analysis: (i) extreme sensitivity to interface modifications, because the intensity of the evanescent wave at the guiding-film surface is very strong [1, 4] and due to the sharp resonances of waveguide modes intrinsic to the physical phenomenon of optical waveguiding [1, 5], (ii) two polarizations, that is, transverse magnetic (TM- or p-polarized) and transverse electric (TE- or s-polarized) modes, can be excited by probing different components of the indicatrix of the thin-film material [1], which enables the characterization of lateral heterogeneities of thin films, (iii) the different electric-field distributions within the waveguiding film for different waveguide modes confer the ability to differentiate processes occurring in different regions of the film [1, 6]. Although optical waveguides were originally developed for applications in the telecommunications field [7], their mechanical stability, flexible geometry, noise immunity, and efficient light conductance over long distances make them well suited for implementation in thin film characterization applications [8]. OWS has applications in a variety of research areas including spectroscopic characterization of surfaces and interfaces [1, 9], chemical sensing [10], and biosensors [11, 12].

Functional Polymer Films, First Edition. Edited by Wolfgang Knoll and Rigoberto C. Advincula.
© 2011 Wiley-VCH Verlag GmbH & Co. KGaA. Published 2011 by Wiley-VCH Verlag GmbH & Co. KGaA.

Although optical waveguides can be created in numerous geometries such as gratings, channels and fiber-optics [1]; this chapter focuses on waveguides with planar geometry that are used to study thin films and interfaces. Besides, it is more compatible with organic and inorganic thin-film deposition technologies, most of which were designed for planar substrates [13]. In addition, they can be fabricated from a wider variety of materials, which makes them more compatible with established surface-modification and deposition techniques. The much higher density of total reflections (up to several thousand per centimeter of beam propagation) yields a concomitant increase in evanescent path length [10, 13]. Major developments in planar OWS began in the 1980s [14–16]. A typical planar OWS consists of a substrate, a thin light-coupling layer (usually metal) and a waveguiding layer with refractive index larger than that of the substrate while the covering material (clad or superstrate) is usually air [17]. The dielectric structure has to be effectively guided and the materials used must be compatible in order to adhere to one another and ensure the stability of the resulting structure [18]. The major methods used to launch light into the bound modes of OWS are prism, grating, endfire, and taper coupling [1]. Since the prism and grating coupling provide good control over mode selection and polarization, they are used more frequently for integrated optic waveguide-based spectroscopy [13]. For more comprehensive information, such as waveguide geometry, fabrication and analysis, other sources should be consulted [1, 9–11].

Here, we focus on prism-coupled OWS carried out in the Kretschmann configuration. In this geometry, incident light irradiates the metal layer through a prism. When a beam is directed onto the base of a glass prism at an angle greater than a critical angle (θ_c), given by Snell's law, the light is totally internally reflected (TIR) at the interface. The electric field couples with the charge density fluctuations of free electrons in the metal and excites a surface plasmon. At this instance, an evanescent wave is produced outside the film whose amplitude decays exponentially away from the metal surface due to the conservation of momentum and energy of the light. The evanescent wave propagates parallel to the interface. If an additional dielectric layer (with higher refractive index than its surroundings) is incorporated onto the metal film, the incident light excites leaky guided modes in this waveguide structure. Modes are excited if the tangential wave vector component matches the wave vector of a guided mode. The momentum at which the incident light beam is coupled into waveguide modes is dependent on the refractive index and the thickness of the substrate, superstrate, and waveguide material. Both surface plasmon resonance (SPR) (only in P-polarization) and waveguide modes (excited by both P- and S-polarized light) are observed as a sudden drop of reflectivity, as shown in Figure 21.1. The propagation in these waveguides can be analyzed on the basis of Maxwell's equations and of the boundary conditions, which will be explained in the following section in detail. The electric-field distributions and the reflectivity *vs.* incidence angle ($R\,vs.\,\theta$) trace shown in Figure 21.1 were calculated by solving the Maxwell equations for light incident on a one-dimensional layer system described by the thickness and refractive indices of the different layers.

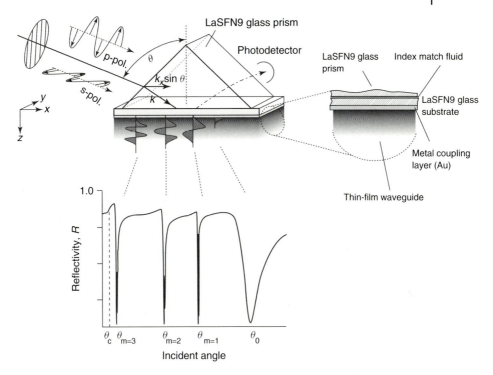

Figure 21.1 Schematic of prism coupling in the Kretschmann configuration and the waveguide-mode-detection scheme. θ_c is the critical angle for total internal reflection. Schematics of the waveguide-mode electric field distributions (transverse magnetic (TM) modes) are overlaid on the waveguiding film model. The R vs. θ trace on the lower left shows waveguide modes for a model thin film. The number of waveguide modes excited and their angle positions depend on the exact thickness and n_{eff}. The modes are indexed according to the number of modes of their field distribution in the waveguide structure. (Courtesy of Dr. Antun Peić.)

The propagation of a mode can be markedly affected by local modifications of the conditions of the guiding. Accordingly, the optical properties of the cladding in which the evanescent field lies can easily be modified. Perturbing the dielectric environment causes changes to the dispersion of the modes, a change in the propagation constant of the mode, a delay and even losses and eventually a shift in the angle necessary for their excitation. Their use for thin-film characterization is appealing as shifts in modes could be easily detected due to narrow line width. The physical parameters that can be detected with a given waveguide depend on the perturbation induced on the waveguides [19].

The perturbation can be achieved either (i) by modifying the geometrical parameters of the waveguides (thinning of the waveguide by fusion and pulling, chemical etching of the cladding), (ii) by modifying the refractive index of the cladding of the waveguide, and (iii) by curving the waveguide or by submitting it to microbending [2].

Enhancement of the sensitivity of the OWS can be obtained if the experimental conditions are such that the evanescent field reaches its maximal amplitude and confinement. Under these conditions, modification arising at the surface of the waveguide will then have a maximal perturbing effect [19]. The modification of the reflecting layer [13] with higher refractive index ($n > 4$ and $k \sim 0.5$) materials is one way [20]. Another possibility to achieve this is to use a nanoporous waveguiding matrix to provide a larger internal surface area for film modification [4]. In optical waveguides, minute changes in the waveguide's refractive index can result in large shifts in the waveguiding conditions. In the case of a nanoporous waveguide sensor, changes to effective refractive index (n_{eff}) due to analyte surface binding are amplified by the large internal surface area of the nanoporous structure, as even a relatively low density of analyte binding on the pore surfaces translates to a relatively large aggregate amount of bound material. Therefore, a nanoporous waveguide sensor can achieve extremely high sensitivities. Besides, the ability to differentiate processes that occur within the pore structure from those occurring on the top surface [21] such as the complicated processes occurring simultaneously inside and on top of a thin film of a block copolymer (BCP) mixture [22] (i.e. pore formation and reversible migration of poly(methyl methacrylate) (PMMA) block chains) by selective solvent swelling and a subsequent reannealing process or layer-by-layer deposition of dendrimers in nanoporous alumina [21], and the ability to functionalize the pore surface by established protocols [17, 23–26] that were not discerned by conventional spectroscopic analysis were successfully monitored by OWS. It has also been demonstrated that predictions of optical properties of hybrid systems (polymer–metal nanoparticles (NPs)) in terms of distribution of NPs is possible with OWS, and could guide the fabrication and testing of advanced polymer composite films [27]. Furthermore, measurements of the change in refractive index in the pores for a series of modes in different polarizations can help to built up a 3D picture of small-molecule diffusion (i.e. protein) in the film [28].

In the 1990s, thin film OWS based on highly porous glass materials prepared by the sol-gel process has been introduced [10, 29] for gaseous analytes. Waveguide sensing with nanoporous films prepared by lithographic approaches has also been reported [24, 30, 31]. The high surface area of the waveguiding material provided a means to concentrate a vast amount of analyte by adsorption from a dilute sample. The performance of this new design of sensor showed significant improvement over planar sensors for gaseous analytes in a way that the porous sol-gel waveguiding layer features both a large optical path length and gave faster response and recovery time. Following these studies, nanoporous and mesoporous films have received a great deal of attention for the development of analytical methodologies to investigate the films *in situ* for a range of applications including sensors and solar cells [4, 21, 24, 25, 28, 32, 33]. Special attention is given to the ability to quantify the filling of such films with guest molecules such as dyes [28], proteins [4, 21, 33], and NPs [21, 26].

The nanoporous systems introduced a range of nanoporous oxide structures (anodic aluminum oxide (AAO), TiO_2, SiO_2), BCPs poly(styrene)-b-poly(methyl methacrylate) (PS-b-PMMA), nanorod arrays (polymeric and Au). In the following sections, a detailed description of each waveguide system is provided according to pores geometrical organization (i.e. cylindrical, mesoporous, etc.). First, experimental techniques necessary for fabricating and applying nanoporous thin films as waveguide sensors are described. Then, the effective medium theory (EMT) used for OWS curve fitting is presented. We demonstrate the principles of waveguide sensing for both static and kinetic measurements, and the high sensitivity of the system, through a variety of examples. Finally, we discuss some advanced applications of the nanoporous waveguide sensing platform. We present the recent results on the application of nanorod arrays as a new way of waveguiding platform.

21.2
Experimental Techniques

21.2.1
Optical Waveguide Spectroscopy (OWS) Setup

We employed the Kretschmann configuration, shown in Figure 21.1, which uses a prism to couple light both in and out of the waveguide, and OWS to measure the waveguide response [4]. The nanoporous thin film acts as a 1D slab waveguide in which the light is confined only within the thickness of the film. Waveguide modes are indicated by minima appearing in reflectivity vs. incidence angle (R vs. θ) measurements as a laser ($\lambda = 633$ nm) is reflected off the metal (Au) layer. In an OWS measurement, different waveguide modes are excited in turn as the incidence angle (θ) is scanned in a $\theta-2\theta$ configuration. The different sets of waveguide modes were excited for light polarized in different orientations; transverse magnetic (TM) and transverse electric (TE) modes, that is, p- and s-polarized light. From the mode-coupling angles, the film thickness and refractive index of the waveguide were measured by Fresnel calculations [1, 5], which describe the slab waveguide sample prepared on glass substrates exactly.

The pore sizes were designed to be much smaller than the wavelength of the incident light (pore diameter <100 nm). Light propagation in such a material can then be described using conventional waveguide analysis by treating [4] the nanoporous material as an effectively homogeneous medium described by quasistatic EMT [34, 35].

21.2.2
Nanoporous Anodic Aluminum Oxide (Nanoporous AAO)

Thin films of nanoporous AAO suitable for OWS as planar slab waveguides were prepared based on a previously described method [4, 21, 25], whereby a ~1-μm Al thin film was deposited onto glass substrates and subsequently anodized to form nanoporous AAO. The porous AAO film ($d = \sim 80$ nm, $n_{633nm} = 1.64$) acts as the waveguiding core and the underlying thin Al ($n_{633nm} = 1.37 + i7.6$) and the surrounding air ($n = 1.0$) act as the cladding.

21.2.3
TiO$_2$ Particle Thin Films

A self-assembled monolayer of mercapto propionic acid (MPA) was formed on the Au-coated glass slide (LaSFN9) by leaving the slides overnight in a 1 mmol dm^{-3} MPA solution in ethanol. The slides were then rinsed in ethanol, dried in a stream of N$_2$, and attached to the dip-coater. The slide was first dipped into a colloidal TiO$_2$ solution (10 times diluted with respect to the 31.5 wt% stock solution in Milli-Q water); rinsed in Milli-Q water; dipped into phytic acid solution (40 mmol dm^{-3} acidified to pH 3 with perchloric acid); and again rinsed in Milli-Q water. The number of cycles dictated the film thickness.

21.2.4
Polymeric Nanorod Arrays

Self-curable monomers were loaded on top of a nanoporous alumina via spin coating (3000 rpm, 2 min). Then, it was kept under vacuum (5–10 mbar) at 120 °C for 12 h, to ensure complete pore wetting and degassing, before being subjected to the curing program. In the meantime, a 50-nm thick Au layer was evaporated on top of a high refractive index glass (LaSFN9, Hellma Optik, $\varepsilon = 3.406$). To ensure adhesion of the gold, a 2-nm Cr film was first deposited on the substrate. The gold surface was functionalized with 2-aminoethanethiol (Sigma-Aldrich) to achieve covalent binding of the polycyanurate nanorods on the gold surface after thermal curing. For that purpose, the gold-covered substrates were dipped in 5 mM 2-aminoethanethiol in absolute ethanol (Sigma-Aldrich) solution for 45 min. Afterwards, they were washed with copious amounts of ethanol and dried with N$_2$. Next, the monomer-filled AAO template was placed on top of the gold-deposited glass substrate with the open end of the pores directing toward the substrate, after carefully removing excess thickness monomer layer. The sample was compressed with a homemade press. Then, it was placed in a furnace tube purged with N$_2$ gas and subsequently cured at elevated temperature. After the thermal curing, the thick Al supporting layer was removed by using CuCl·2H$_2$O/HCl solution in an ice bath (0 °C). In a second step, the thin alumina layer was dissolved with a 10% H$_3$PO$_4$ aqueous solution, at room temperature overnight. The dimensions of the cured polymeric thermoset correspond to those of the template nanopores. The nanorods are oriented normal to the substrate surface.

21.3
Theoretical Descriptions

An equivalent mathematical definition of a waveguide mode is that it is an electromagnetic field that is a solution of Maxwell's wave equation. Here, we adopted the widely used Maxwell–Garnett (MG) family of EMTs [36] which is based on the Clausius–Mossotti relation [34, 35]. The EMT model based on the MG approximation is most appropriate for a composite system behaving like a single-component anisotropic material in its optical response assuming a composite mixture in terms of a spatially homogeneous electromagnetic response on a macroscopic scale. The theory holds under the hypothesis of a very low concentration of the dispersed component or that the scale of the dispersed component is much smaller than the electromagnetic wavelength in the host material. Therefore, the size of nanostructures (cylindrical domains and pores) are chosen such that they are much smaller than the incident light wavelength ($\lambda = 633$ nm) employed.

The theory predicts the effective value of any composite properties such as dielectric response (ε_{eff}^{MG}) using the expression [37],

$$\frac{\varepsilon_{eff}^{MG} - \varepsilon_h}{\varepsilon_h + P^h(\varepsilon_{eff}^{MG} - \varepsilon_h)} = f\frac{(\varepsilon_g - \varepsilon_h)}{\varepsilon_h + P^g(\varepsilon_g - \varepsilon_h)} \tag{21.1}$$

where ε_{eff}^{MG} is an effective dielectric constant; f is a volume fraction of the guest particles; superscripts g and h denote guest and host matrix. P is a geometrical factor related to the depolarization factor. Therefore, it takes into account the size, shape, and orientation of the inclusions [34, 38–40].

$$P_x + P_y + P_z = 1 \tag{21.2}$$

In the case of a cylindrical pores (AAO and PS-b-PMMA) which is structurally symmetric under rotation of the z-axis, the dielectric response can be defined

$$\varepsilon = \left\{\varepsilon_x = \varepsilon_y = \varepsilon^\perp, \varepsilon_z = \varepsilon''\right\} \tag{21.3}$$

$$\varepsilon_{eff(film)}^{MG} = \varepsilon_x = \varepsilon_y \text{ and } (P_x = P_y = 1/2); \ \varepsilon_z = \varepsilon_{eff(film)}(P_z = 0) \tag{21.4}$$

In the other case, where spherical inclusions embedded in host matrix are randomly oriented, the whole heterogeneous medium will be isotropic as in the case of mesoporous TiO$_2$ thin films.

$$\varepsilon_{eff(film)}^{MG} = \varepsilon_x \varepsilon_y = \varepsilon_z \text{ and } (P_x = P_y = P_z = 1/3) \tag{21.5}$$

From this dielectric function ($\varepsilon_{eff(film)}^{MG}$), the optical response to modification of pore walls (cylindrical or spherical) of nanoporous thin films with a layer of material can be calculated. OWS angular scans were fitted before and after the pore-wall modification using algorithms based on Fresnel equations in a multilayered system. Detailed information about the derivations specific to each pore geometry can be found in previous publications [4, 21, 22, 25, 27].

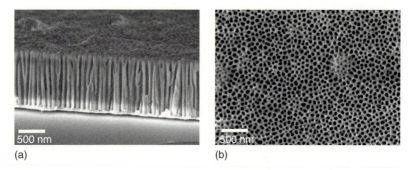

(a)   (b)

Figure 21.2 (a) Scanning electron microscope (SEM) cross-section of a nanoporous AAO sample and (b) the top view.

21.4
Cylindrical Nanostructures as Optical Waveguides

21.4.1
Nanoporous Anodic Aluminum Oxide (AAO)

Anodization of aluminum metal films is a widely established technique used to prepare AAO layers that are characterized by their straight, cylindrical pores with diameters in the range of 5–450 nm that self-organize and align with each other in a close-packed configuration (Figure 21.2) [41, 42]. Based on the simple AAO pore geometry and its availability, the development of nanoporous AAO waveguide sensing has been particularly mature [4, 21, 25, 43–45].

Label-free sensing in a nanoporous AAO waveguide is based on a change in the effective refractive index of the nanoporous layer (n_{eff}) as analytes bind to or otherwise adsorb on pore surfaces [4, 21]. Given a difference in the refractive indices between the analyte and the pore-filling medium (e.g., physiological buffer in biosensing), a concentration of surface-bound analyte displaces an equal volume of the pore-filling medium and changes the overall optical response within the pores and hence n_{eff} [4, 21]. In terms of the optical description of the nanoporous system, the sensing concept was anticipated by studies that characterized the optical properties of nanoporous templates infiltrated with pore-filling materials [38, 43].

In an early demonstration, a liquid flow cell was attached to the nanoporous AAO waveguide so as to measure, *in situ*, the adsorption of bovine serum albumin (BSA) on the pore surfaces of the AAO [4]. The waveguide mode angle minima were also tracked continuously to obtain real-time adsorption data (Figure 21.3). Shifts in the angle minima were converted to changes in n_{eff} by Fresnel calculations [1], which were then modeled by EMT to estimate the equivalent changes in analyte layer thickness. The authors reported a sensitivity of $\sim$0.01 nm change in the thickness of the analyte layer, which corresponds to a level of $\sim$0.1 ng/cm^2 in the surface mass density of proteins adsorbed. In comparison, a Mach–Zehnder interferometer demonstrated detection of a layer thickness change of 0.02 nm, which is already an order of magnitude better than obtainable with SPR [46].

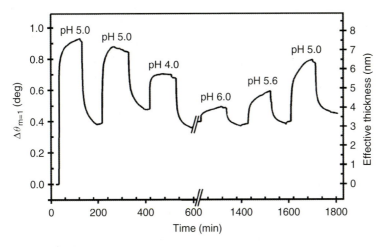

Figure 21.3 Kinetics of repeated BSA adsorption and desorption. pH values shown refer to those used during adsorption. Note that maximum adsorption is always at the isoelectric point (IEP) of BSA at pH 5.0. (From Lau *et al.*, Ref. [4].)

An important advantage in using AAO as a nanoporous layer to support optical waveguide modes is the availability of a large range of (oxy)silane molecules that can be used to functionalize the alumina surface with the desired chemical groups. For example, Lau *et al.* [25] functionalized the AAO pore surfaces with primary amines, which acted as the initiator chemical groups for surface-initiated polymerization of poly(γ-benzyl-L-glutamate) (PBLG), and characterized the polymerization process *in situ* over the course of 24 h. The oxysilane used was 3-aminopropyltriethoxy silane (APTES), and the functionalization was accomplished by exposing the AAO to APTES vapors at 100 °C under vacuum (2–3 mbar) for 3 h, followed by sonication in a solvent. The vapor-deposition process was essentially identical to that used for silanization of SiO_2 surfaces [47] and the associated chemical processes (hydrolysis, condensation, physisorption, and chemisorption) [47] on alumina are also thought to be similar to those on silica. In another study, the nanoporous AAO waveguide was also functionalized with amine groups to help define a suitable substrate for characterizing the layer-by-layer deposition of polyelectrolytes multilayers within a confined nanopore environment [21]. In addition to gas-phase silanization, *in-situ* APTES immobilization inside the AAO pores in methanol solution was monitored by OWS. The main difference between gas-phase and liquid-phase silanization is that the former gives a monolayer, while the latter gives a multilayered silane surface as shown in Figure 21.4. When the pore walls of the alumina membrane are covered with APTES molecules, the difference between an empty and silanized membrane is discerned at very high resolution giving a significant shift in the mode coupling angles ($\Delta\theta \sim 0.4°$). The increase in n_{eff} is calculated from the corresponding angle shifts and related to the thickness increase of the APTES layer using EMT. Around 2 nm of material was adsorbed, which corresponds to

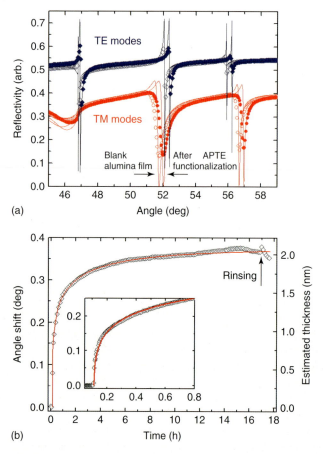

(a)

(b)

Figure 21.4 (a) TE and TM waveguide mode patterns of thin alumina film before (open symbols) and after functionalization with APTES (filled symbols) in 1 mM methanol. (b) *In-situ* kinetics of APTES coating followed using the rightmost guided mode in (a). Open diamonds are data, and the red solid line is fitting with two-step diffusion-limited model [33].

a double-layered silane film given the theoretical APTES monolayer thickness of ~1 nm [48]. Silanization kinetics were followed *in-situ* by tracking the temporal change of the angle (0.4°) of a waveguide mode in time (Figure 21.4b). Around 80% of the final adsorbed thickness is achieved within 2 h during the initial fast adsorption. In the subsequent slow adsorption regime, a slight, continuous increase in film thickness was observed, which could be correlated with physisorbed APTES multilayers. The significance of this kinetics curve is the very high signal-to-noise ratio for such a thin-film layer (~2 nm), which improves the sensitivity of OWS.

The morphology of the material deposited within the pores can be characterized in some detail by the anisotropy of n_{eff} obtained through measurements of different waveguide modes excited by incident light polarized in the directions normal ($\perp$) and parallel ($//$) to the plane of the AAO film. The dielectric constant, $\varepsilon_{eff}(= n_{eff}^2)$,

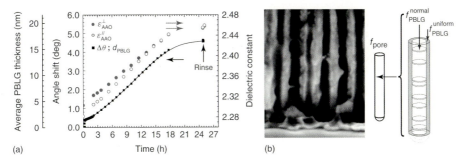

(a)

(b)

Figure 21.5 (a) Real-time waveguide mode angle shift and dielectric constant measurements. The average PBLG layer thickness estimated from the measured anisotropic dielectric constants is also indicated on the outer left axis. (b) SEM cross-sections of the PBLG structures protruding normal off the pore surfaces after 24–25 h polymerization (the native pore diameter is 60 nm), and the idealized three-dimensional PBLG structure, with both a uniform pore coating and nanostructures protruding normal off the pore surfaces. (Adapted from Lau *et al.*, Ref. [25].)

of the AAO (ε_{AAO}) can be described by EMT within the MG approach [25]:

$$\varepsilon^{\perp}_{AAO} = \varepsilon_{Al_2O_3} + f_{pore}(\varepsilon_{pore} - \varepsilon_{Al_2O_3}) \tag{21.6a}$$

$$\varepsilon''_{AAO} = \varepsilon_{Al_2O_3} \frac{\varepsilon_{Al_2O_3} + 1/2(1 + f_{pore})(\varepsilon_{pore} - \varepsilon_{Al_2O_3})}{\varepsilon_{Al_2O_3} + 1/2(1 - f_{pore})(\varepsilon_{pore} - \varepsilon_{Al_2O_3})} \tag{21.6b}$$

The anisotropy arises from the orientation of the AAO cylindrical pores that are aligned normal to the plane of the AAO layer. For electric fields polarized normal to the AAO top surface, hence parallel to the pore surfaces, there is no discontinuity as the traveling electric field of the guided light passes across pore surfaces. Therefore, $\varepsilon^{\perp}_{AAO}$ is simply the volume-averaged value of the alumina and the pore material. However, electric fields polarized parallel to the AAO layer and normal to the pore surfaces are heavily screened by the alumina-pore discontinuity and $\varepsilon''_{AAO} < \varepsilon^{\perp}_{AAO}$ [4, 21, 34]. This relationship is not changed for normal analyte surface binding, in which the analyte forms a conformal layer on the cylindrical pore surfaces. However, in the PBLG study mentioned above, as the degree of polymerization increased, the original anisotropy ($\varepsilon''_{AAO} < \varepsilon^{\perp}_{AAO}$) was reversed (Figure 21.5a) [25]. To negate the optical anisotropy of the AAO matrix, the PBLG material must have formed a microstructure that had an optical anisotropy that is the reverse of the AAO matrix. This was confirmed by *ex situ* scanning electron microscopy, which showed PBLG material spanning the width of the pores (Figure 21.5b).

Independent quantification of the layer thickness change on the internal pore surfaces and atop the AAO is also possible with a nanoporous waveguide sensor. At low analyte concentrations, the change to the atop thickness of the AAO layer due to analyte binding or deposition is relatively small and difficult to detect [4]. However, when the amount of material deposited becomes large, as in the characterization of layer-by-layer deposition of polyelectrolytes (Figure 21.6), the deposition atop the AAO may also be quantified by Fresnel multilayer calculations [21], as in conventional OWS [1].

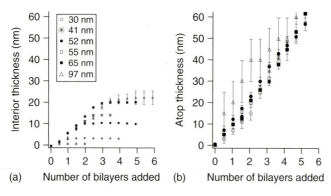

Figure 21.6 Deposition of polyelectrolyte dendrimers dissolved in 100 mM NaCl aqueous solution, within (a) and atop (b) nanoporous AAO waveguides with various pore diameters.

In general, the nanoporous waveguide sensor should be designed with a pore diameter much larger than the molecular size of the analyte to ensure efficient analyte transport within the pores. Figure 21.6a actually offers a counterexample, where, due to multilayer polyelectrolyte deposition, the pore diameter left open gradually decreased and polyelectrolyte deposition within the pores became inhibited after a few bilayers (while the deposition atop continued apace – Figure 21.6b). As expected, the onset of the inhibition was brought forward as the original pore diameter decreased. However, due to repulsive electrostatic effects [102], deposition was already inhibited when the remaining pore diameter was still significantly larger (e.g., ~25 nm for an original pore size of 65 nm) than the diameter of the polyelectrolyte dendrimer used (~7 nm).

The choice in pore size has also been discussed by Gopinath *et al.* [49] in using a nanoporous SiO_2 waveguide for the label-free detection of RNA aptamer–ligand interactions. Larger pore sizes not only aid in the transport of analytes, but also represent a larger total internal surface areas, and theoretically higher amplifications of the change in n_{eff} thus sensor sensitivities. However, it was shown that as pore sizes increased relative to the wavelength of the guided light, specificity in the waveguiding conditions also decreased [49], possibly due to both scattering effects and a broadened distribution in pore diameters.

In the studies described above, the optical absorptions (i.e. the imaginary component of n_{eff}) of both the AAO and the analyte are assumed to be negligible. However, if the analyte molecule can be highly excited at the chosen wavelength of guided light, then the waveguide response due to optical absorption may also be used for sensing purposes. For example, Yamaguchi *et al.* [50] characterized the adsorption of Fe[bathophenanthroline]$_3^{2+}$ via both the shifts in waveguide mode minima and the overall reflected intensity decrease as increasing amounts of Fe[bathophenanthroline]$_3^{2+}$ was incorporated into the AAO. In another study, changes in the waveguide response due to the hypothetical incorporation of Au nanopoarticles into a nanostructured BCP thin film was also investigated [27].

21.4.1.1 Other Types of Nanoporous AAO Optical Sensor Designs

High sensitivities can be achieved with the nanoporous AAO waveguide design but waveguide coupling requires a sophisticated optical setup. A potentially more "portable" nanoporous AAO sensor design may be realized by analyzing the thin-film interference pattern generated when light is simultaneously reflected, at normal incidence, off the AAO top film surface and the AAO/substrate interface. Alvarez *et al.* [44] used white-light illumination and measured the interference spectral pattern in order to measure changes in n_{eff} due to analyte surface binding (Figure 21.7). The wavelength (λ) of the peak maxima in the interference spectrum is governed by the Fabry–Perot relationship:

$$m\lambda = 2n_{eff}L \tag{21.7}$$

where *m* is the order of the interference peak and *L* is the thickness of the AAO film. Antibody binding immunoassays were performed to demonstrate the technique. Although the experimental setup is simple, the uncertainties in the bound protein analyte layer thicknesses were relatively high, on the order of 0.1–1 nm, as there is an inherent measurement tradeoff between time and spectral resolutions.

Nanoporous AAO layers have also been used in combination with surface-enhanced Raman spectroscopy (SERS) [51, 52]. SERS owes its high sensitivity to the intense Raman scattering induced by highly concentrated plasmonic fields generated over metal surfaces with extreme curvatures, such as at the cusps between adjoining metal NPs [53]. Ko and Tsukruk [51] reported that, by immobilizing ~20 nm Au NPs on the pore walls of an AAO membrane with the layer-by-layer polyelectrolyte deposition technique (Figure 21.8), the Raman signal intensity of their NPs architecture was enhanced by 5 orders of magnitude. The authors used a ~60-μm thick AAO layer with pore diameter ~240 nm and a 20-mW excitation laser with emission at $\lambda = 785$ nm. To observe the enhancement effect, the laser

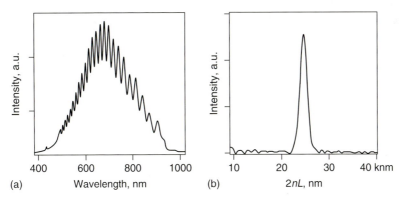

(a)   Wavelength, nm   (b)   2*nL*, nm

Figure 21.7 Optical characteristics of a porous alumina thin film. (a) White-light reflectance spectrum from a thin AAO film. (b) Fourier transform of the reflectance spectrum in (a) results in a single peak whose position along the *x*-axis corresponds to the effective optical thickness of the thin film, 2$n_{eff}L$. (From Alvarez *et al.*, Ref. [44].)

Figure 21.8 SEM images of porous alumina membranes decorated with Au nanoparticles. (Adapted from Ko *et al.*, Ref. [51].)

was directed parallel to the length of the pores, and the enhancement was attributed to the multiple reflections that can occur along the pore surfaces. The authors also performed a more extensive study that demonstrated the detection of the explosive organic compounds 2,4-dinitrotoluene (DNT) and trinitrotoluene (TNT), down to 0.1–0.05 ppt (Figure 21.9; approximately equivalent to femtomolar detection), beating previous reports by up to 4 orders of magnitude (1 ppb) [52]. Control of the polyelectrolyte and Au NPs deposition processes, as well as the choice of the NPs surfactant ligand used (cetyltrimethylammonium bromide (CTAB)), were also found to have significant effects on the sensitivity of the AAO-SERS detection scheme.

Nanoporous AAO films have also been used simply as a 3D matrix with a high internal surface area to enhance the sensitivities of other optical sensing techniques. For example, thin AAO layers have been prepared on top of SPR sensors to increase the total analyte binding surface area [54, 55]. Because the SPR used was only sensitive to optical changes within ~200 nm of the sensor

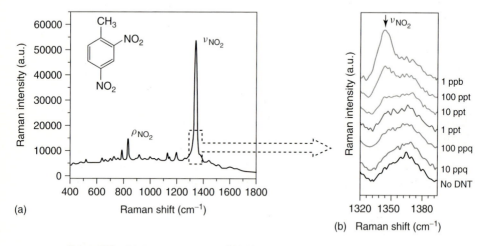

Figure 21.9 (a) Raman spectrum of 10 000 ppm 2,4-DNT on SERS substrate. The inset shows the chemical structure of 2,4-DNT. (b) Raman spectra of trace level 2,4-DNT on SERS substrate in the window of the strongest NO_2-stretching vibration. (From Ko *et al.*, Ref. [52].)

surface, the AAO films used have also been limited to ~200 nm in thickness. Sensitivity enhancements due to the 3D matrix of around an order of magnitude were reported. However, SPR involves the measurement of only a single plasmon resonance condition. Therefore, the AAO anisotropy in n_{eff} cannot be characterized by SPR, and independent measurements of the changes within the pores and atop the AAO can also not be made.

21.4.1.2 AAO Waveguide Fabrication Advances

Anodization of a thin aluminum film on a heterogeneous substrate has certain inherent challenges. Vacuum deposition of high-purity Al films suitable for AAO waveguiding (i.e. ~1 μm thick) with low surface roughness is technically demanding [56, 57], but is required for spatially uniform pore initiation during anodization. Furthermore, pore-ordering techniques developed for anodization of bulk Al, such as two-step anodization [58] or hard anodization [59], are impractical for thin Al films [45]. Therefore, techniques have been developed to transfer bulk anodized AAO thin films onto optical substrates for waveguide-sensing experiments. For example, the atomic-layer deposition of TiO_2 was used to fill the gap between a free-standing AAO thin film and a glass substrate on which the AAO was laid [60]. Further deposition of TiO_2 on the AAO bonded to the glass substrate can also be monitored with atomic-layer sensitivity by waveguide sensing [61]. A less resource-intense approach was taken by Lazzara *et al.* [45], where the free-standing AAO was bonded to the glass substrate with a spin-coated polymer adhesive. Two-step anodized, free-standing AAO films up to ~1 cm² were prepared and nanoporous waveguides with mode-minima sharpness higher than that obtained from direct anodization of a Al film were obtained.

21.4.2
PS-b-PMMA Block Copolymers

In recent years, the interest in making waveguides based on polymeric materials has grown rapidly due to their drastic cost reduction [2]. There are a number of different polymers that can be considered for use as waveguiding materials. The key properties are the index of refraction and the attenuation or optical loss. The other properties of interest are the thermal and mechanical stability of the waveguiding material. The concept of waveguide characterization of n_{eff} for sensing purposes has also been extended to the characterization of BCP thin-film nanostructure characterization. BCPs are composed of two or more immiscible polymer chains (blocks) that, driven by microphase separation, can self-assemble into ordered domains with periodicities in the range of 10–100 nm [62]. Kim *et al.* [63] first demonstrated the possibility of using a BCP thin film as an optical waveguide, and followed up with an investigation using the waveguiding technique to characterize the BCP thin-film nanostructure over several processing steps [22]. Similar to the work with nanoporous AAO waveguides, the nanostructure periodicity (46 nm) was chosen to be much smaller than the wavelength of the guided light (633 nm). The BCP system chosen should theoretically self-assemble

Figure 21.10 AFM height and phase images of the self-assembled PS-b-PMMA/PMMA blend. In the phase image, PS surfaces appear as darker areas. λ_{c-c} (46 nm) was measured from the fast Fourier transform (FFT) shown as the gray scale inset. (Taken from K. H. A. Lau' PhD dissertation 2008.)

into a cylindrical morphology, with one of the polymer blocks forming close-packed cylinder arrays within a matrix of the other polymer as shown in Figure 21.10. However, homogeneous self-assembly of BCP cylindrical domains uniformly oriented along the film normal is technically challenging for the relatively thick films used (thickness 400–500 nm), which were more than 10 times thicker than the nanostructure periodicity (46 nm).

Using waveguide measurements in the Kretschmann configuration, Kim *et al.* [63] were able to measure the anisotropy in n_{eff}. By comparing n_{eff} with the EMT theoretical values, the orientation of the cylindrical domains after different film annealing steps were characterized. The ability to characterize the selective dissolution of the cylindrical polymer domains (i.e. generation of a nanoporous polymer film) was also demonstrated. Although the results were only semiquantitative, the waveguiding experiments were non-destructive and able to characterize the internal morphology of the film (unlike, e.g., atomic force microscopy (AFM)), and could be performed relatively quickly with widely available laser sources (in contrast to, e.g., transmission electron microscopy (TEM) or grazing-incidence small-angle X-ray scattering (GISAXS)).

Independently, Garetz and coworkers [64, 65], developed the technique of guided-wave depolarized light scattering, which was also used to characterize the internal nanostructure of BCP thin films. Instead of measuring n_{eff}, these reports focused on the degree of cross-coupling between waveguide modes of different polarizations due to the presence of morphological domains oriented at different angles relative to the film normal [64]. The cross-coupling was also shown to be a function of the domain size [65].

21.4.3
Other Cylindrical Nanoporous Waveguide Materials

Silicon-based waveguide materials have been used in OWS quite extensively. Silicon oxynitride (SiON), silicon nitride (Si_2N_3), and silicon oxide (SiO_2) were the most common choice due to the tunability of their refractive-index contrast and their transparency over a wide wavelength range, including the visible. Propagation losses in the visible range as low as 0.1–0.2 dB/cm have been found in silicon nitride waveguides [66]. In an effort to develop porous waveguides for biosensing via OWS and electrochemical measurements, Reimhult *et al.* [67] used colloidal lithography for making very high aspect ratio structures in silicon nitride (Si_2N_3) dielectric materials. This newly designed porous silicon nitrate waveguide showed superior light-coupling efficiency and sensitivity. Awazu *et al.* [24] used the vapor-etching technique to create cylindrical holes through silica waveguides and they compared the efficiency of their waveguides with flat silica by using biotin–streptavidin (500 nM) binding. The same group also obtained porous silica waveguide by sequential swift-heavy ion (Au) irradiation and hydrofluoric acid (HF) vapor etching [23]. They have observed a 10-fold increase in sensitivity in terms of angular modulation compared with flat substrates.

21.5
Isotropic Mesoporous Waveguides

21.5.1
TiO$_2$ Foam Films

Sol-gel surfactant-templated mesoporous materials have a large internal surface area, that is, available for adsorption of molecules. Zhou and coworkers [32] used mesoporous TiO_2-P_2O_5 nanocomposite thin films placed on top of float glass substrates for chemical gas-sensing applications. They observed that for cubic-mesostructures mesoporous thin films with 55% porosity, the response time of the waveguide-based sensor is related with the film thickness: the thinner the film, the faster the response. Later, Zhou and coworkers produced nanoporous TiO_2 thin films on gold substrates waveguide fabricated by dip-coating gold-covered glass substrates from the colloidal TiO_2 solution. Gold substrates were dip coated into aqueous solution containing TiO_2 NPs and amphiphilic BCPs, and afterward they were calcined at elevated temperature ($\sim$400 °C) in order to improve the porosity.

21.5.2
OWS Combined with Electrochemical Measurements

Combining spectroscopy and electrochemistry to study thin films at electrode surfaces offers several advantages: spectroscopy can be used to differentiate non-faradaic and faradaic electrochemical processes is possible, something that is often problematic in conventional cyclic voltammetry (CV) techniques. In addition, the spectral response provides information that is complementary to that obtained from conventional current or voltage response, such as the structure of the molecular film [13]. To maximize the sensitivity of electrochemical-OWS, Cameron *et al.* [28] created a 1.6-μm thick nanoporous TiO_2 thin film deposited on a Au electrode, that is, in direct contact with the buffer-filled porous network in

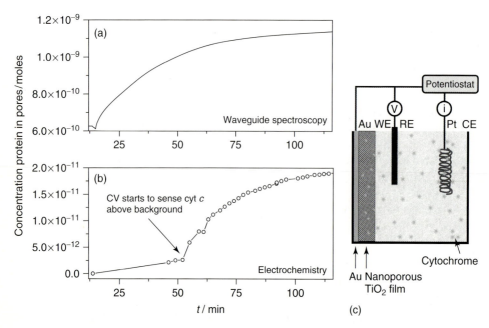

Figure 21.11 Adsorption from 100 μM cytochrome c in PBS in a nanoporous TiO₂ thin film, simultaneously measured by (a) waveguide-mode tracking and (b) cyclic voltammetry (CV). The schematic (c) shows the electrochemical setup and the diffusion of cytochrome c to the Au electrode at the bottom of the nanoporous TiO₂ film [21, 28].

order to monitor *in-situ* adsorption of the redox active cytochrome c protein. The nanoporous titania film supported on a gold substrate acts as the working electrode, the cell is completed by a platinum wire counterelectrode and a silver-wire pseudoreference electrode (Figure 21.11). Analogous to the previous examples on silanization, PBLG and layer-by-layer polyelectrolyte assemblies, cytochrome c adsorption within the nanoporous TiO₂ thin film can be measured by waveguide angle tracking. However, a redox current (due to the oxidation and reduction of the metal center of the cytochrome c) would only be measured when the protein has diffused through the porous network to the bottom of the film, where it can exchange electrons with the Au electrode. In addition, the kinetics of adsorption and the time lag between registering the waveguide response and the electrochemical signal would give valuable information about the diffusion characteristics within the nanoporous film. Results for adsorption from 0.1 mM cytochrome c in phosphate-buffered saline (PBS) are shown in Figure 21.11.

The presence of the electrochemical signal (Figure 21.11b) showed conclusively that cytochrome c was able to permeate the nanoporous network composed of TiO₂ particles only 6 nm in diameter. Moreover, this electrochemical signal lagged behind the waveguide signal (Figure 21.11a) by ~30 min. Analysis of the adsorption kinetics reveals that the diffusion coefficient of cytochrome c within the nanoporous network was $\sim 1 \times 10^{-16} \, m^2/s$, which is 5 orders of magnitude lower than in solution when there is no geometric confinement [28]. Moreover, comparison of Figure 21.11a,b also showed that waveguide sensing recorded 2 orders of magnitude higher amounts of cytochrome c than the electrochemical measurement. This indicated that electron transfer occurred only via the bottom Au electrode and not via the TiO_2 matrix. It also demonstrated the sensitivity of the nanoporous waveguide sensing technique imparted by the vast internal pore surface area over which surface processes can occur.

21.5.3
Dye-Uptake Measurements for Dye-Sensitized Solar Cells (DSSCs)

Dye-sensitized solar cells (DSSCs) are in the second generation of photovoltaic devices, with the most efficient devices obtaining efficiencies of >11% [68, 69]. DSSCs consist of a film of mesoporous nanocrystalline titanium dioxide (nc-TiO_2), that is, sensitized with a dye molecule that absorbs strongly over the visible spectrum. Many advances in the understanding of the processes occurring in DSSCs have been made over the last 10 years, but there still remain several issues that are poorly understood [70–72]. One such issue is the lack of knowledge of the dye loading and dye distribution throughout the nanocrystalline titania. Efficient cells require high dye coverage; they also require the dye to be homogeneously spread throughout the approximately 10-μm film. Typically, dyeing of the nc-TiO_2 film is done by immersing the film in $3 \times 10^{-4} \, mol \, dm^{-3}$ dye solution for approximately 12 h. The temperature of the nc-TiO_2 (and hence the dye bath) is one important parameter that controls both dye packing and solar-cell efficiency [73, 74].

Coadsorbates such as deoxycholic acid [75–77] and phosphinate amphiphiles [78] are thought to improve dye packing in a way that improves efficiency, but there is little information on the adsorption kinetics of dyes with coadsorbates. Dyeing times can be drastically reduced by constant agitation of the dye bath but further systematic studies relating dye coverage to dyeing conditions are required. Dye uptake is currently measured by one of two methods. The first method involves monitoring the falling concentration of dye in the solution above the nc-TiO_2 film by UV-Vis spectroscopy [79]. This method is most accurate when very low concentration dye baths are used and so does not give realistic information about dye uptake in normal highly concentrated dye baths. In addition, information about the diffusion of dye into the film can only be inferred by the absence of dye in the bulk solution, which gives no information about the diffusion front or the distribution of dye throughout the nc-TiO_2 film. A second method for measuring dye coverage is to completely desorb the dye from the nc-TiO_2 film by submerging it in an aqueous or ethanolic solution of base and then measuring the desorbed dye concentration by UV-Vis spectrometry [80]. This method gives the total concentration of dye in the film, but again no information about adsorption/desorption kinetics or the homogeneity of the dye distribution.

OWS, on the other hand, is an ideal technique to probe the dyeing of titania films. The films have a relatively high refractive index ($n > 2.1$) ideal to act as a waveguiding layer [28]. In addition, the pores and crystals are all nanosized and do not scatter the incident laser light to any great degree [81]. The technique has already been used to couple light into nc-TiO$_2$ films in complete DSSCs, increasing the optical field intensity at certain wavelengths and promoting high injection rates [82, 83]. Peić *et al.* [81] are using OWS as a diagnostic tool to help build a 3D picture of the dye distribution inside the nc-TiO$_2$. OWS allows the real-time monitoring of dye adsorption and allows the effect of systematic variations on dyeing temperature or coadsorbate concentrations to be investigated. It has been shown that the change in extinction coefficient, k, measured by OWS as dye diffuses into the nc-TiO$_2$ is directly related to the amount of active dye inside the film. A 3D understanding of dye distribution is being build up by modeling the field distribution and signal intensities of different modes and relating them to an advancing diffusion front in the films. Better understanding of the parameters that control dye uptake will lead to more efficient dyeing techniques. This is of particular interest to the fledgling differential scanning calorimetry (DSC) manufacturers who are designing continuous reel-to-reel production lines for DSC manufacture and wish to have rapid and reliable online dyeing.

21.6
Nanostructured Nanorod Arrays by Templating Strategies

Nanostructured arrays based on anodic alumina layers have been long known in electrochemistry [84], however, they gained an enormous popularity as template structures for the fabrication of different kinds of nanosized objects from various materials. Nanorods, nanotubes, or nanowires were reported from diverse materials, such as: metals, semiconductors, oxides, and polymers [85–90]. Further progress in nanofabrication techniques has opened up the possibility to fabricate metallic nanorods in macroscopic size arrays with their long axes aligned perpendicular to the substrate [91]. In previous sections, we have discussed functionalized AAO membranes and other inorganic nanoporous scaffolds employed as optical waveguide for visible light [4] and various sensing applications [24, 25, 28, 92]. On the other hand, the template-based fabrication of released nanorod arrays that are stable under real-time conditions is still challenging [93]. Very recently, the integration of optically transparent nanorod arrays into device architectures enabling OWS has been accomplished, as will be explained below with a few examples.

21.6.1
Plasmonic Metallic Nanoarrays

Zayats and coworkers [94, 95] demonstrated an improvement in biosensing technology using a plasmonic metamaterial, that is, capable of supporting a guided mode in porous nanorod layers. They used an array of parallel gold nanorods (embedded to alumina matrix) oriented normally to a glass substrate. They obtained plasmonic gold nanorod metamaterials by electrodeposition into thin AAO. In this way, they fabricated nanorod arrays with a length in the range of 50–400 nm and diameters in the range of 10–40 nm with interrod distances varying from about 100 nm to about 170 nm. MG calculations of the optical properties of metal nanorod arrays were reported that showed strongly anisotropic optical behavior that can be beneficial for spectroscopic, sensing, and imaging applications [95]. One of the most important properties of their metal nanoarrays is the possibility of wide-range tuning of the geometrical parameters of the structure so as to control the electromagnetic interaction between the rods (i.e. changing the interrod distance). They also reported that the optical response of the nanorods was governed by a collective plasmonic mode resulting from the strong electromagnetic coupling between the dipolar longitudinal plasmon resonances supported by individual nanorods. In their metamaterial based sensor, the figure of merit (FOM) reaches a value of 330, which is twofold higher than the conventional SPR-based sensors.

21.6.2
Polymeric Nanorod Arrays

Following the encouraging results of the waveguiding ability of the nanoporous AAO templates, in a recent research study, Gitsas *et al.* [96] used free-standing polymeric nanorod arrays (PCNs) on gold substrates by template replication that reproduced the inverse structure of the AAO as an optical waveguiding platform. The nanorods, with diameter of 60 nm and length of 650 nm were formed by thermal polymerization of cyanate ester monomers (CEMs) within porous alumina templates. CEM is chosen due to: (i) its self-curing nature, (ii) its low viscosity (liquid at room temperature) which gives ease of processing for nanomolding, (iii) its high chemical and mechanical stability upon cross-linking, which gives sufficiently long lifetime, and most importantly (iv) its residual $-OCN$ groups are still eligible as functional sites for further surface modification (i.e. attachment of biomolecules), and (v) its high dielectric constant ($\varepsilon = 3.41$–3.75 at 1 MHz) due to the aryl ether linkage that increase the polarizability of the film. The details of the template-assisted fabrication of PCNs together with their thermal, mechanic, and dielectric characterizations can be found in a previous publication [97]. They used the simple example of solvent exchange around the PCNs, which effectively changes the dielectric constant of the interfaces between nanorod and surrounding medium, resulting in detectable shifts of the waveguide modes. Three solvents were chosen as interrod space-filling medium: water ($n = 1.333$), ethanol ($n = 1.361$), and isopropanol ($n = 1.378$). Solvents were chosen in a way that PCNs can keep their chemically stability for extended time duration. They have similar refractive indices, which makes them more attractive for verifying the achievable sensitivity for a biosensor system. Guided optical modes were observed as sharp minima in reflectivity *vs.* incidence-angle scans (R *vs.* θ) (Figure 21.12). The refractive indices of ethanol and isopropanol are higher that water and thus the effective refractive index of the sample is altogether increased. This was detected as a shift of the waveguiding modes to higher angles; $\Delta\theta_{m1} = 0.79°$, $\Delta\theta_{m2} = 1.67°$ for TM modes and $\Delta\theta_{m1} = 0.75°$, $\Delta\theta_{m2} = 1.38°$ for TE modes for the exchange of pore-filling medium from water to ethanol. Changes in the external media

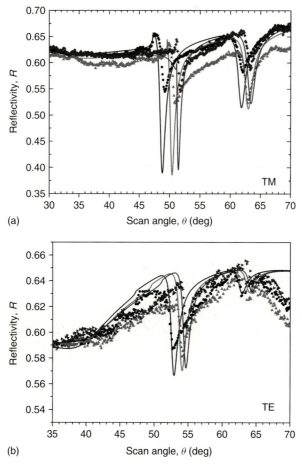

(a)

(b)

Figure 21.12 OWS R vs. θ scans of the nanorod array in different solvent environments in (a) TM-polarization and (b) TE-polarization. Milli-Q water (circles); ethanol (up triangles); isopropanol (rhombi). Fresnel calculations are plotted in the figure as solid lines.

produced larger waveguide minimum angle shifts for higher-order modes. The highest FOM $= (\Delta n/\Delta\theta)(1/\Gamma)$ (where Γ is the full-width of the waveguide mode dip at half-minimum and $\Delta\theta$ is the mode shift for Δn refractive index change) was 98 for the TE modes at $m = 2$ for the ethanol–isopropanol system. This value reached 5 for localized surface plasmon resonance spectroscopy (LSP) [98] and 26 for the sensors based on long-range SPR [99].

Recently, optical waveguides composed of patterned silicon oxide nanorod arrays prepared by means of BCP (poly(styrene)-b-poly(2-vinylpyridine) (PS-P2VP)) lithography were reported [100]. In this study, Ta_2O_5 was used as the waveguiding layer while SiO_2 rod arrays were used to increase the surface area available for analyte binding. The average diameter and height of the nanorods were reported to be 60 and 32 nm, respectively. On the other hand, in previous studies [95, 96], nanorod arrays acted as waveguiding layers with increased effective surface area for bioaffinity. Nevertheless, the group reported 114% higher response of the nanostructures waveguide in comparison with flat chips for BSA protein adsorption due to increase in the surface area.

21.7
Conclusions

We have demonstrated sensing and characterization of a variety of surface processes relevant to biosensing using nanoporous waveguiding thin films. This approach has opened up the area of label-free, high-sensitivity biosensing while advancing conventional OWS and SPR-based sensors on label-free thin-film detection. We have shown simple examples of pore-surface modification with very thin silane films and protein binding. However, at the same time, other interesting systems and sensitivity enhancements may be explored in combination with fluorescent- and NPs labeling. The optical properties of the nanoporous thin films as well as the quantitative description of analyte absorption/binding were analyzed by Fresnel equations and EMT. In designing sensing architectures with nanoporous waveguide platforms, attention should be paid to the size ratio between the pores $(D < \lambda/10)$ and the analyte. Some of the reviewed research studies state that porous waveguide-based sensors are superior to SPR sensors, taking into account the change in reflectivity as the comparison index [20, 23, 24, 94]. These techniques compare the sharpness and degree of angular shift of nanoporous waveguides with conventional SPR or flat surfaces. However, sharpness alone cannot be a measure for improved sensitivity [99, 101] since the signal-to-noise ratio, different sources of noise and instrumental imperfections should also be considered in FOM and limit of detection (LOD) calculations.

Acknowledgments

We wish to thank G. Glasser at the Max-Planck-Institute for Polymer Research, Mainz, for scanning electron microscopy work and A. Peić for technical support. H.D. gratefully acknowledges financial support from European Union for the Marie Curie Intra-European Fellowship (MEIF-CT-2005-024731).

References

1. (a) Knoll, W. (1997) Guided wave optics for the characterization of polymeric thin films and interfaces, in *Handbook of Optical Properties-Optics of Small Particles, Interfaces, and Surfaces,* Chapter 13 (eds R.E. Hummel and P. Wissmann), CRC-Press, Boca Raton, FL, pp. 373–400; (b) Knoll, W. (1998) *Ann. Rev. Phys. Chem.*, **49**, 569.

2. Hunsperger, R.G. (2009) *Integrated Optics: Theory and Technology*, 6th edn, Chapters 2 and 5, Springer, New York, pp. 17–31 and 85–104.

3. Matsuda, N. (2005) *Talanta*, **65**, 1065.

4. Lau, K.H.A., Tan, L.S., Tamada, K., Sander, M.S., and Knoll, W. (2004) *J. Phys. Chem. B*, **108**, 10812–10818.

5. Fowles, G.R. (1989) *Introduction to Modern Optics*, 2nd edn, Dover Publications, Inc., New York, p. 328.

6. Beines, P.W., Klosterkamp, I., Menges, B., Jonas, U., and Knoll, W. (2007) *Langmuir*, **23**, 2231–2238.

7. Barlow, H.M. (1985) *J. Phys. D: Appl. Phys.*, **18**, 1511–1520.

8. Lechuga, L.M. (2005) Optical biosensors, in *Biosensors and Modern Biospecific Analytical Techniques*, (Chapter 5), Comprehensive Analytical Chemistry, Vol. 44 (ed. L. Gorton), Elsevier B. V., pp. 209–247.

9. Mendes, S.B., Li, L., Burke, J.J., Lee, J.E., Dunphy, D.R., and Saavedra, S.S. (1996) *Langmuir*, **12**, 3374–3376.

10. (a) Yang, L., Saavedra, S.S., Armstrong, N.R., and Hayes, J. (1994) *Anal. Chem.*, **66**, 1254–1263; (b) Yang, L., Saavedra, S.S., and Armstrong, N.R. (1996) *Anal. Chem.*, **68**, 1834–1841.

11. Voros, J., Graf, R., Kenausis, R.G.L., Bruinink, A., Mayer, J., Textor, M., Wintermantel, E., and Spencer, N.D. (2000) *Biosens. Bioelectron.*, **15**, 423–429.

12. Aulasevich, A., Roskamp, R.F., Jonas, U., Menges, B., Dostálek, J., and Knoll, W. (2009) *Macromol. Rap. Commun.*, **30**, 872–877.

13. Bradshaw, J.T., Mendes, S.B., and Saavedra, S.S. (2005) *Anal. Chem.*, **77**, 29A–36A.

14. Lee, D.L. (1986) *Electromagnetic Principles of Integrated Optics*, John Wiley & Sons, Inc., New York.

15. Bohn, P.W. (1987) *Trends Anal. Chem.*, **6**, 223–233.

16. Rabolt, J.F. and Swalen, J.D. (1988) Advances in non-linear spectroscopy, in *Advances in Spectroscopy: Spectroscopy of Surfaces*, vol. 16, Chapter 1 (eds R.E. Hester and J.R.H. Clark), John Wiley & Sons, Ltd, Chichester, pp. 1–36.

17. Yimit, A., Rossberg, A.G., Amemiya, T., and Itoh, K. (2005) *Talanta*, **65**, 1102–1109.

18. Adams, M.J. (1981) *An Introduction to Optical Waveguides*, Chapter 3, John Wiley & Sons, Inc., New York, pp. 51–74.

19. de Fornel, F. (2001) *Evanescent Waves from Newtonian Optics to Atomic Optics*, Chapter 6, Springer-Verlag, Berlin, pp. 111–129.

20. Fujimaki, M., Rockstuhl, C., Wang, X.M., Awazu, K., Tominaga, J., Fukuda, N., Koganezawa, Y., and Ohki, Y. (2008) *Nanotechnology*, **19**, 095503–095509.

21. Lau, K.H.A., Cameron, P.J., Duran, H., Kandil, A.I.A., and Knoll, W. (2009) Nanoporous thin films as highly versatile and sensitive waveguide biosensors, in *Surface Design: Applications in Bioscience and Nanotechnology*, Chapter 4.5 (eds R. Förch, H. Schönherr, and A.T.A. Jenkins), Wiley-VCH Verlag GmbH, Weinheim, 383–399.

22. Kim, D.H., Lau, K.H.A., Joo, W., Peng, J., Jeong, U., Hawker, C.J., Russell, T.P., and Knoll, W. (2006) *J. Phys. Chem. B*, **110**, 15381–15388.

23. Fujimaki, M., Rockstuhl, C., Wang, X.M., Awazu, K., Tominaga, J., Koganezawa, Y., Ohki, Y., and Komatsubara, T. (2008) *Opt. Exp.*, **16**, 6408–6416.

24. Awazu, K., Rockstuhl, C., Fujimaki, M., Fukuda, N., Tominaga, J., Komatsubara, T., Ikeda, T., and Ohki, Y. (2007) *Opt. Exp.*, **15**, 2592–2597.

25. Lau, K.H.A., Duran, H., and Knoll, W. (2009) *J. Phys. Chem. B*, **113**, 3179–3189.

26. Zourob, M., Mohr, S., Brown, B.J.T., Fielden, P.R., McDonnel, M., and Goddart, N.J. (2003) *Sens. Actuators, B*, **90**, 296–307.

27. Lau, K.H.A., Knoll, W., and Kim, D.H. (2007) *Macromol. Res.*, **15**, 211–215.

28. Cameron, P.J., Jenkins, A.T.A., Knoll, W., Marken, F., Milsom, E.V., and Williams, T.L. (2008) *J. Mater. Chem.*, **18**, 4304–4310.

29. MacCraith, B.D., McDonagh, C.M., O'Keefe, G., Keyes, E.T., Vos, J.G.,

O'Kelley, B., and McGilp, J.F. (1993) *Analyst*, **118**, 385–388.

30. Fujimaki, M., Rockstuhl, C., Wang, X., Awazu, K., Tominaga, J., Ikeda, T., Ohki, Y., and Komatsubara, T. (2007) *Microelectron. Eng.*, **84**, 1685–1689.

31. Reimhult, E., Kumar, K., and Knoll, W. (2007) *Nanotechnology*, **27**, 275303.

32. Qi, Z.-M., Honma, I., and Zhou, H. (2006) *J. Phys. Chem. B.*, **110**, 10590–10594.

33. Duran, H., Lau, K.H.A., Luebbert, A., Jonas, U., Steinhart, M., and Knoll, W. (2008) Biopolymers for biosensors: Polypeptide nanotubes for optical biosensing, in *Polymers for Biomedical Applications*, ACS Symposium Book Series, vol. 977, Chapter 22 (eds A. Mahapatro and A. Kulshrestha), Oxford University Press, 371–390.

34. Aspnes, D.E. (1982) *Thin Solid Films*, **89**, 249–262.

35. Choy, T.C. (1999) *Effective Medium Theory: Principles and Applications*, Oxford University Press, New York.

36. Maxwell-Garnett, J.C. (1904) *Philos. Trans. Roy. Soc. London Ser. A*, **203**, 385–420.

37. Skryanbin, I.L., Radchik, A.V., Moses, P., and Smith, G.B. (1997) *Appl. Phys. Lett.*, **70**, 2221–2223.

38. Foss, C.A., Tierney, M.J., and Martin, C.R. (1992) *J. Phys. Chem.*, **96**, 9001–9007.

39. Hornyak, G.L., Patrissi, C.J., and Martin, C.R. (1997) *J. Phys. Chem. B*, **101**, 1548–1555.

40. Granqvist, C.G. and Hunderi, O. (1978) *Phys. Rev. B*, **18**, 2897–2902.

41. O'Sullivan, J.P. and Wood, G.C. (1970) *Proc. R. Soc. London. Ser. A, Math. Phys. Sci. (1934-1990)*, **317**, 511–543.

42. Jessensky, O., Muller, F., and Goesele, U. (1998) *Appl. Phys. Lett.*, **72**, 1173–1175.

43. Saito, M., Shibasaki, M., Nakamura, S., and Miyagi, M. (1994) *Opt. Lett.*, **19**, 710–712.

44. Alvarez, S.D., Li, C.P., Chiang, C.E., Schuller, I.K., and Sailor, M.J. (2009) *ACS Nano*, **3**, 3301–3307.

45. Lazzara, T.D., Lau, K.H.A., and Knoll, W. (2010) *J. Nanosci. Nanotechnol.*, **10**, 4293–4299.

46. Weisser, M., Tovar, G., Mittler-Neher, S., Knoll, W., Brosinger, F., Freimuth, H., Lacher, M., and Ehrfeld, W. (1999) *Biosens. Bioelectron.*, **14**, 405–411.

47. Jonas, U. and Krueger, C. (2002) *J. Supramol. Chem.*, **2**, 255–270.

48. Wieringa, R.H., Siesling, E.A., Werkman, P.J., Angerman, H.J., Vorenkamp, E.J., and Schouten, A.J. (2001) *Langmuir*, **17**, 6485–6490.

49. Gopinath, S.C.B., Awazu, K., Fujimaki, M., Sugimoto, K., Ohki, Y., Komatsubara, T., Tominaga, J., Gupta, K.C., and Kumar, P.K.R. (2008) *Anal. Chem.*, **80**, 6602–6609.

50. Yamaguchi, A., Hotta, K., and Teramae, N. (2009) *Anal. Chem.*, **81**, 105–111.

51. Ko, H. and Tsukruk, V.V. (2008) *Small*, **4**, 1980–1984.

52. Ko, H., Chang, S., and Tsukruk, V.V. (2009) *ACS Nano*, **3**, 181–188.

53. Stiles, P.L., Dieringer, J.A., Shah, N.C., and Van Duyne, R.P. (2008) *Annu. Rev. Anal. Chem.*, **1**, 601–626.

54. Cloutier, S.G., Lazareck, A.D., and Xu, J. (2006) *Appl. Phys. Lett.*, **88**, 013904–013903.

55. Koutsioubas, A.G., Spiliopoulos, N., Anastassopoulos, D., Vradis, A.A., and Priftis, G.D. (2008) *J. Appl. Phys.*, **103**, 094521–094526.

56. Biring, S., Tsai, K.-T., Sur, U.K., and Wang, Y.-L. (2008) *Nanotechnology*, **19**, 015304.

57. Lita, A.E. and Sanchez, J.J.E. (1999) *J. Appl. Phys.*, **85**, 876–882.

58. Masuda, H. and Fukuda, K. (1995) *Science*, **268**, 1466–1468.

59. Lei, Y. and Chim, W.-K. (2005) *Chem. Mater.*, **17**, 580–585.

60. Tan, L.K., Gao, H., Zong, Y., and Knoll, W. (2008) *J. Phys. Chem. C*, **112**, 17576–17580.

61. Gao, H., Zong, Y., Kheng, T.L., and Chong, M.A.S. (2008) Highly sensitive waveguide-based porous anodic alumina nanosensors for monitoring atomic layer deposition. 2008 2nd IEEE International Nanoelectronics Conference, Vols. 1-3, pp. 589–592.

62. Fredrickson, G.H. and Bates, F.S. (1996) *Ann. Rev. Mater. Sci.*, **26**, 501–550.

63. Kim, D.H., Lau, K.H.A., Robertson, J.W.F., Lee, O.J., Jeong, U., Lee, J.I., Hawker, C.J., Russell, T.P., Kim, J.K., and Knoll, W. (2005) *Adv. Mater.*, **17**, 2442–2446.

64. Garetz, B.A., Newstein, M.C., Wilbur, J.D., Patel, A.J., Durkee, D.A., Segalman, R.A., Liddle, J.A., and Balsara, N.P. (2005) *Macromolecules*, **38**, 4282–4288.

65. Fang, Z., Newstein, M.C., Garetz, B.A., Wilbur, J.D., and Balsara, N.P. (2007) *J. Opt. Soc. Am. B: Opt. Phys.*, **24**, 1291–1297.

66. Netti, N.C., Charlton, M.D.B., Parker, G.J., and Baumberg, J.J. (2000) *Appl. Phys. Lett.*, **76**, 991–993.

67. Reimhult, E., Kumar, K., and Knoll, W. (2007) *Nanotechnology*, **18**, 275303–275309.

68. Gratzel, M. (2005) *Inorg. Chem.*, **44**, 6841–6851.

69. Yum, J.H., Chen, P., Gratzel, M., and Nazeeruddin, M.K. (2008) *ChemSusChem*, **1**, 699–707.

70. Hamann, T.W., Jensen, R.A., Martinson, A.B.F., Ryswyk, H.V., and Hupp, J.T. (2008) *Energy Environ. Sci.*, **1**, 66–78.

71. Imahori, H., Hayashi, S., Hayashi, H., Oguro, A., Eu, S., Umeyama, T., and Matano, Y. (2009) *J. Phys. Chem. C*, **113**, 18406–18413.

72. Rawling, T., Austin, C., Bucholz, F., Colbran, S.B., and McDonagh, A.M. (2009) *Inorg. Chem.*, **48**, 3215–3227.

73. Agrios, A.G. and Hagfeldt, A. (2008) *J. Phys. Chem. C*, **112**, 10021–10026.

74. Baik, C., Kim, D., Kang, S., Kang, S.O., Ko, J., Nazeeruddin, M.K., and Gratzel, M. (2009) *J. Photochem. Photobiol. A-Chem.*, **201**, 168–174.

75. Hara, K., Dan-oh, Y., Kasada, C., Ohga, Y., Shinpo, A., Suga, S., Sayama, K., and Arakawa, H. (2004) *Langmuir*, **20**, 4205–4210.

76. Ikeda, M., Koide, N., Han, L., Pang, C.L., Sasahara, A., and Onishi, H. (2009) *J. Photochem. Photobiol. A-Chem.*, **202**, 185–190.

77. Wang, Z.S., Cui, Y., Dan-oh, Y., Kasada, C., Shinpo, A., and Hara, K. (2007) *J. Phys. Chem. C*, **111**, 7224–7230.

78. Wang, M.K., Li, X., Lin, H., Pechy, P., Zakeeruddin, S.M., and Gratzel, M. (2009) *Dalton Trans.*, **45**, 10015–10020.

79. Ishihara, T., Tokue, J., Sano, T., Shen, Q., Toyoda, T., and Kobayahi, N. (2005) *Jpn. J. Appl. Phys. Part 1*, **44**, 2780–2782.

80. Jang, S.R., Lee, C., Choi, H., Ko, J.J., Lee, J., Vittal, R., and Kim, K.-J. (2006) *Chem. Mater.*, **18**, 5604–5608.

81. Peić, A., Staff, D., Risbridger, T., Menges, B., Peter, L.M., Walker, A.B., and Cameron, P.J. (2011) Real time optical waveguide measurements of dye adsorption inside nanocrystalline TiO$_2$ films with relevance to dye sensitized solar cells, *J. Phys. Chem. C*, **115**, 613–619.

82. Durr, M., Menges, B., Knoll, W., Yasuda, A., and Nelles, G. (2007) *Appl. Phys. Lett.*, **91**, 021113–021116.

83. Ruhle, S., Greenwald, S., Koren, E., and Zaban, A. (2008) *Opt. Exp.*, **16**, 21801–21806.

84. Takahashi, H. and Nagayama, M. (1978) *Electrochim. Acta*, **23**, 279–286.

85. Metzger, R.M., Konovalov, V.V., Sun, M., Xu, T., Zangari, G., Xu, B., Benakli, M., and Doyle, W.D. (2000) *IEEE Trans. Magn.*, **36**, 30–35.

86. Ding, G.Q., Shen, W.Z., Zheng, M.J., and Fan, D.H. (2006) *Appl. Phys. Lett.*, **88**, 103106-1–103106-3.

87. Sanz, R., Johansson, A., Skupinski, M., Jensen, J., Possnert, G., Boman, M., Vazquez, M., and Hjort, K. (2006) *Nano Lett.*, **6**, 1065–1068.

88. Steinhart, M., Wehrspohn, R.B., Goesele, U., and Wendorff, H.W. (2004) *Angew. Chem. Int. Ed.*, **43**, 1334–1344.

89. Martín, J. and Mijangos, C. (2009) *Langmuir*, **25**, 1181–1187.

90. Kim, D.H., Karan, P., Göring, P., Leclaire, J., Caminade, A.-M., Majoral, J.-P., Goesele, U., Steinhart, M., and Knoll, W. (2005) *Small*, **1**, 99–102.

91. Evans, P., Hendren, W.R., Atkinson, R., Wurtz, G.A., Dickson, W., Zayats, A.V., and Pollard, R. (2006) *Nanotechnology*, **17**, 5746–5753.

92. Fujimaki, M., Rockstuhl, C., Wang, X., Awazu, K., Tominaga, J., Ikeda, T.,

Ohki, Y., and Komatsubara, T. (2007) *Microelectron. Eng.*, **84**, 1685–1689.

93. Grimm, S., Giesa, R., Sklarek, K., Langner, A., Goesele, U., Schmidt, W., and Steinhart, M. (2008) *Nano Lett.*, **8**, 1954–1959.

94. Kabashin, A.V., Evans, P., Pastkovsky, S., Hendren, W., Wurtz, G.A., Atkinson, R., Pollard, R., Podolsky, V.A., and Zayats, A.V. (2009) *Nature Mater.*, **8**, 867–871.

95. Wurtz, G.A., Dickson, W., O'Conner, D., Atkinson, R., Hendren, W.R., Evans, P., Pollard, R., and Zayats, A.V. (2008) *Opt. Exp.*, **16**, 7460–7470.

96. Gitsas, A., Yameen, B., Lazzara, T.D., Steinhart, M., Duran, H., and Knoll, W. (2010) Polycyanurate nanorod arrays for optical-waveguide-based biosensing, *Nano Lett.*, **10**, 2173–2177.

97. Yameen, B., Duran, H., Best, A., Jonas, U., Steinhart, M., and Knoll, W. (2008) *Macromol. Chem. Phys.*, **209**, 1673–1685.

98. Sherry, L.J., Chang, S.H., Schatz, G.C., and VanDuyne, R.P. (2005) *Nano Lett.*, **5**, 2034–2038.

99. Dostálek, J., Kasry, A., and Knoll, W. (2007) *Plasmonics*, **2**, 97–106.

100. Popa, A.M., Wegner, B., Scolan, E., Voirin, G., Heinzelmeann, H., and Pugin, R. (2009) *Appl. Surf. Sci.*, **256S**, S12–S17.

101. Unger, A. and Kreiter, M. (2009) *J. Phys. Chem. C*, **113**, 12243–12251.

102. Lazzara, T.D., Lau, K.H.A., Abou-Kandil, A.I., Caminade, A.-M., Majoral, J.-P., and Knoll, W. (2010) *ACS Nano*, **4**, 3909–3920.

22
Electrochemical Surface Plasmon Resonance Methods for Polymer Thin Films

Akira Baba, Futao Kaneko, Rigoberto Advincula, and Wolfgang Knoll

22.1
Introduction

The electrochemical and optical quantitative analysis of π-conjugated polymer ultrathin films upon oxidation/reduction and doping/dedoping is of great interest for a number of applications. These include electrochromic displays [1], battery electrodes [2], sensors, and so forth. Many of these applications make use of the fact that the conductivity or color of the film can be controlled by its redox state. Because of the many factors affecting the transition between doped and dedoped states of these polymer films, it is very difficult to obtain quantitative analysis of each electrochemical event. A number of experimental *in-situ* techniques on the doping/dedoping process of conjugated polymers have been introduced, such as quartz crystal microbalance (QCM) [3, 4], Fourier transform infrared (FTIR) [5], scanning probe microscopy [6], and so on.

Optical techniques such as surface plasmon resonance (SPR), waveguide spectroscopy, or ellipsometry are well-established methods for the characterization of surfaces, interfaces, and thin films [7]. SPR, in particular, uses evanescent waves to probe thin films. This technique allows for the determination of optical constants and the thickness of layered systems. The combination of SPR with electrochemical measurements has been demonstrated as a powerful analytical technique for the simultaneous characterization and manipulation of electrode/electrolyte interfaces [8, 9]. In electrochemical-surface plasmon resonance spectroscopy (EC-SPR) measurements, the gold substrate that carries the optical surface mode, is also used as the working electrode in the electrochemical experiments. One advantage of using the EC-SPR technique is that the electrochemical and optical properties are simultaneously obtained on surfaces and ultrathin films at the nanometer scale. Our group and other groups have been applying the EC-SPR technique for the characterization of conducting polymer thin films [10–15]. This involved the *in-situ* monitoring of the film swelling/contraction and of electrochromic properties during electropolymerization or during the anion doping/dedoping process of deposited conducting polymers. This technique has also been applied in biosensor development, for example, by integration of redox enzymes with conducting

Functional Polymer Films, First Edition. Edited by Wolfgang Knoll and Rigoberto C. Advincula.

polymers as a bioelectrocatalytic layer [16]. Other combination methods have also been reported for the characterization of conducting polymers, for example, using surface-plasmon-enhanced light scattering (SPLS) and the QCM technique.

In this chapter, recent developments and applications on EC-SPR methods for conducting polymer thin films are overviewed.

22.2
Electrochemical Surface Plasmon Spectroscopy

As schematically shown in Figure 22.1, the SPR setup combines the three-electrode electrochemical cell with a Kretschmann configuration for the excitation of surface plasmons (SPs). When a p-polarized laser beam is reflected at the incident angle θ from the base of a prism of refractive index n_p above θ_c, a strong nonradiative electromagnetic wave is excited at the resonant angle. This SP propagates at the metal/electrolyte interface, with the coupling angle being given by the energy- and momentum-matching conditions between photons and SPs. As the energy of the incident light is transferred toward excitation of the SP, a decrease of the reflectivity in the angular scan is observed. The resonance character of this excitation gives rise to an enhancement of the electric field at the interface by more than an order of magnitude, which is the origin of the remarkable sensitivity enhancements, for example, in scattered light [17], fluorescence [18, 19], or Raman spectroscopies [20, 21] when working with SP light. If the formation of a thin

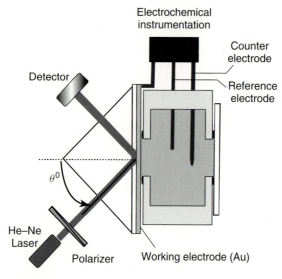

Figure 22.1 A schematic diagram of the combined electrochemical-surface plasmon resonance spectroscopy setup.

dielectric coating is observed *in situ*, for example, during the electropolymerization process of conducting polymers at the metal electrode/electrolyte interface, kinetic information can be obtained. This includes film deposition and diffusional transport of mass adjacent to the interface by cycling potentials. The experiment often involves monitoring the reflected intensity at a fixed angle of observation θ_{obs}. The Au/glass substrates used for the excitation of SPs also serve as a working electrode in the electrochemical measurements, with a counter and a reference electrode. For these experiments, the gold film thickness is chosen for optimum excitation of the SPs with both wavelengths (42–45 nm).

First, the EC-SPR study during the electropolymerization of conducting polymer is shown. Electrochemical-quartz crystal microbalance (EC-QCM) measurements, which enable the study of mass changes of depositing materials, are separately carried out for the comparison to EC-SPR measurements. Electropolymerization of aniline on the gold surface was achieved by applying potential cycling between −0.2 and 0.9 V at a scan rate of 20 mV/s. Shown in Figure 22.2a are the mass change as a function of time during the seven electropolymerization cycles and corresponding SPR minimum angle shifts measured after each potential cycling [22]. Each arrow shows the starting point of each potential cycle, that is, −0.2 V. In observing the mass change, the increase in each of the cycles mostly corresponds to the oxidation of aniline to form the polyaniline (PANI) film and the doping of anions into the deposited PANI film. The decrease of the curve is mostly due to the dedoping of the anions from the deposited PANI film. The trace of the SPR minimum angle shift was similar to the mass change at −0.2 V, that is, at the end of each potential cycle. This indicates that the optical thickness corresponds to the acoustic mass. The kinetic measurements for SPR were performed at a fixed angle, 58.0°, slightly below the angle of the reflectivity minimum of the blank gold substrate as shown in Figure 22.2b. In the kinetic measurement, the behavior of the reflectivity has a large difference with mass change. In the case of the SPR reflectivity, the curve is very sensitive to the thickness, the real (ε') and the imaginary (ε'') part of the dielectric constant. The increase in reflectivity below about +0.2 V was seen in Figure 22.2, whereas no increase was observed in the mass. This clearly indicates that the dielectric constant dramatically changes in this potential range (−0.2 to +0.2 V).

22.3
Evaluation of Polymer Thin Films by EC-SPR

EC-SPR has been applied for *in-situ* optical/electrochemical investigations in order to determine the complex dielectric constants of conjugated polymer thin films. Figure 22.3 shows the SPR curves of a poly(3,4-ethylenedioxythiophene) (PEDOT) film at several constant potentials. The PEDOT films were deposited by in-situ electropolymerization. From the amount of charge applied, the thickness of this film at an open-circuit potential (OCP) was estimated to be about 22.5 nm using the correlation with the QCM data. For both wavelengths, the resonance angle and

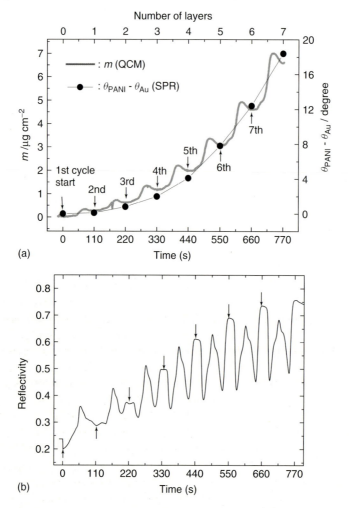

Figure 22.2 *In-situ* monitoring of mass change during electropolymerization of aniline by potential cycling between −0.2 and 0.9 V at a scan rate of 20 mV/s, and corresponding SPR minimum angle shifts measured after each potential cycling (a). The reflectivity as a function of time during the electropolymerization (b). A current efficiency $\Delta m/\Delta Q$ plot during electropolymerization. From Ref. [22] with permission.

the shape of the SPR curves changed significantly in going from the doped state to the dedoped state of the polymer film. For all wavelengths, the resonance peaks shifted to higher angles, however, at $\lambda = 594$ nm, the curve became broader and the reflectivity values at the minimum of the resonance increased upon dedoping. Only for $\lambda = 1152$ nm does the resonance dip became deeper and sharper upon dedoping.

Fitting of the experimental SPR reflectivity curves was done by Fresnel calculations using the thickness values obtained by the QCM 2.5 min after changing

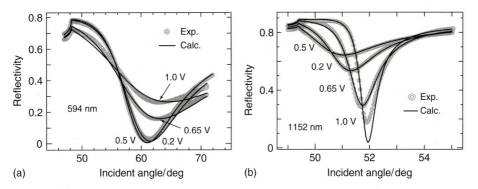

Figure 22.3 Experimental SPR reflectivity curves at different applied potential as indicated, measured at (a) $\lambda = 594$ and (b) $\lambda = 1152$ nm. The solid lines are the Fresnel simulations. From Ref. [23] with permission.

the potential. As demonstrated in Figure 22.3, all fitting curves show excellent agreement with the experimental results, with parameters summarized in Figure 22.4. Drastic changes, both in the real part and in the imaginary part of the dielectric constants were determined at each wavelength. The error bars in Figure 22.4 are based on the density fluctuation of the polymer upon doping/dedoping, that is, 1.45–1.55. Based on these calculations, the accuracy for the determination of the complex dielectric constant is estimated to be better than ± 0.04 for the real part and ± 0.08 for the imaginary part of the dielectric constant, that is, EC-SPR measurements are able to determine the complex dielectric constants of the conducting polymer films in the doped/dedoped state rather independently of the uncertainty of the film density. The tendency of the dielectric constant with wavelength is almost in good agreement with the result, which is

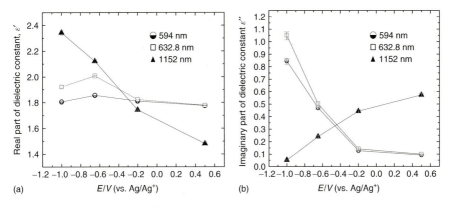

Figure 22.4 The (a) real part, ε', and (b) imaginary part, ε'', of the dielectric constants as a function of the applied potential, measured at $\lambda = 594$, 632.8, and 1152 nm, respectively. From Ref. [23] with permission.

obtained from the conventional optoelectrochemistry measurement by using the Kramers–Kronig (K–K) relation as a function of wavelength and of potential. Thus, electrochromic phenomena of PEDOT films can be characterized by this EC-SPR technique with high sensitivity, which may lead to new designs for SPR-based sensors/biosensors based on the electrochromism of thin conducting polymer films.

22.4
Electrochemical SPR-SPPL

The combination of SPR and surface-plasmon-enhanced photoluminescence spectroscopy (SPPL) with electrochemistry was demonstrated for the detection of weak photoluminescence materials such as PEDOT films. In this case, SPs were excited at the metal/dielectric interface, upon total internal reflection of polarized light from a He–Ne laser at a wavelength of 594 nm. SPPL from the deposited film was detected by a photomultiplier after passing through a low-wavelength cut-off filter ($\lambda = 610$ nm), as schematically shown in the inset of Figure 22.5.

The response of an ultrathin PEDOT film to a sequence of potential steps applied is shown in Figure 22.5. The potential, the current, the SPR reflectivity, and the SPPL intensity simultaneously at several potentials, that is, at 0.5 V: 5 min $\rightarrow$ −1.0 V: 30 min $\rightarrow$ open circuit: 30 min $\rightarrow$ 0.5 V: 5 min. Here, the electrochemical surface-plasmon-enhanced photoluminescence spectroscopy (EC-SPPL) intensity increased at an applied potential of −1.0 V. An interesting observation is that the SPPL decreased after the circuit was opened open circuit (OC). After 30 min of open circuit, the OCP reached about −0.5 V. In this case, the reflectivity did not decrease significantly within 30 min of open circuit as compared to the reflectivity at −0.5 V applied as shown in Figure 22.5. By taking all these simultaneous observations into consideration, one can come to the conclusion that this is attributed to the ionic charge carriers migrating to a new equilibrium state spontaneously because the dedoped state of the PEDOT film is not stable. However, the electrons were not emitted from the valence band because a potential was not applied, so that the dielectric constant did not significantly change. On the other hand, the PL intensity still continued to decrease even after 30 s of open circuit. This suggests that the migrating ionic charge carriers might quench the photoluminescence. When a potential of 0.5 V was applied to the PEDOT film after the 30 min at open circuit, a large current was observed and the SPPL and the reflectivity decreased significantly. This is further evidence that ionic charge carriers migrated into the PEDOT film under the open-circuit conditions. A mechanistic model obtained from these observations is schematically shown in Figure 22.6. These results show that the EC-SPPL method is a sensitive tool in detecting weak photoluminescence from a conjugated polymer film and the setup allows for the electrochemical control of three independent signals.

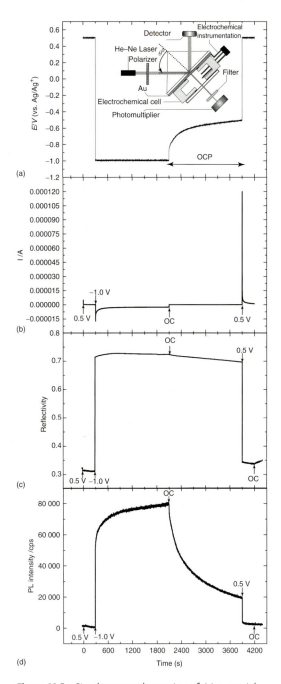

Figure 22.5 Simultaneous observation of (a) potential, (b) current, (c) SPR reflectivity, and (d) SPPL at several potentials (0.5 V: 5 min → −1.0 V: 30 min → open circuit: 30 min → 0.5 V: 5 min). From Ref. [24] with permission.

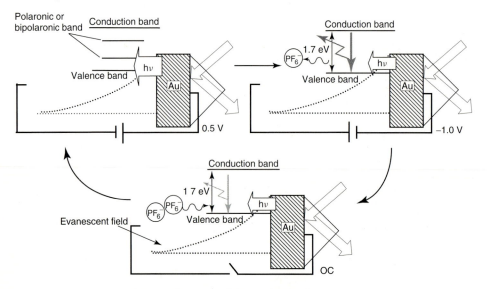

Figure 22.6 Schematic illustration of the mechanism for SPPL from an ultrathin PEDOT film under different potentials applied. From Ref. [24] with permission.

22.5
Electrochemical SPR Microscopy

The combination of surface plasmon resonance (SPR) microscopy [25] with an electrochemical setup has been shown to be a powerful tool for the investigation of spatiotemporal pattern growth [26]. The background and a detailed theory of SPR microscopy are well documented [25]. Basically, SPR microscopy (based on surface plasmon resonance) allows one to monitor laterally resolved changes of the thickness and of the dielectric constant of a polymer coating on top of a thin (noble) metal film. Thus, any patterned film grown by electropolymerization on the metal surface, can be *in situ* detected by EC-SPR microscopy.

A series of EC-SPR microscopy pictures were taken at 0 V during the electropolymerization of pyrrole onto μconducting polymer (μCP) octadecane thiol-self-assembled monolayer (ODT-SAM)/Au substrate for up to three potential cycles from 0 to 0.9 V at a scan rate of 20 mV s^{-1}, as shown in Figure 22.7. The SPR microscopy pictures were taken at an angle slightly lower than the minimum reflectivity angle of the surface plasmon resonance of the μCP ODT SAM/gold substrate. This means that the area where polypyrrole is deposited turns brighter as the film grows. As can be seen in Figure 22.7, the patterned growth was observed already clearly after one potential cycling. This result corresponds to the EC-SPR results, and, hence, we conclude that the polypyrrole predominantly grew on the bare gold areas. After the second potential cycle,

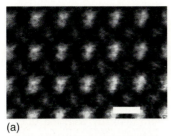

(a)

Figure 22.7 EC-SPR microscopy images taken after electropolymerization by scanning the potential from 0 to 0.9 V at a scan rate of 20 mV/s: after one cycle (a), two cycles (b), and three cycles (c) at 0 V. The scale bar corresponds to 40 μm.

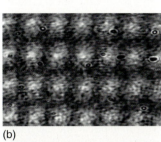

(b)

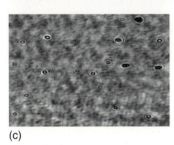

(c)

it was observed that the areas where polypyrrole was grown became larger in their lateral dimensions. It was reported before that a hydrophobic SAM surface enhances the lateral growth of polypyrrole between two electrodes, whereas a hydrophilic SAM surface reduces the lateral growth [27, 28]. From our SPR microscopy observations, we conclude that polypyrrole in this case could be grown not only in the vertical direction but also in the lateral direction. The whole area became brighter after the third potential cycle. This suggests that the polypyrrole had grown over the entire surface, or a precipitation of oligomers occurred over the whole surface. In the case of electropolymerization at a flat ODT-SAM/Au substrate, no polypyrrole was deposited until the third potential cycle as discussed above. However, in the case of electropolymerization onto a μCP-patterned ODT-SAM/Au substrate, polypyrrole was deposited over the entire surface, as can be seen in Figure 22.7. Since μCP ODT-SAMs seem to have more pinholes than flat ODT-SAM, it is also possible that, besides this effect, the polypyrrole grew from the pinholes in the μCP ODT-SAM. Once polypyrrole grows on a SAM/Au electrode, the rate of growth can exceed that of bare gold [29].

22.6
Simultaneous Electrochemical–Atomic Force Microscopy-SPR

A combined electrochemical-surface plasmon resonance spectroscopy–atomic force microscope (EC-SPR–AFM), which provides simultaneous and *in-situ* optical, electrochemical, and surface morphological information of electrochemically deposited conducting polymer ultrathin films, has been demonstrated [30]. With independent and separate measurements using electrochemistry, SPR and atomic force microscope (AFM), it is possible to obtain three types of data, but the results are collected under basically different conditions. Especially in surface-science studies at the nanometer scale, very careful measurements are required in order to interpret the data and to obtain quantitative results. The combined technique leads to "real-time" dynamic and simultaneous acquisition of the dielectric (optical), surface morphological, and current data of depositing conducting polymer thin film at the electrode/electrolyte interface.

A schematic diagram of the combined EC-SPR–AFM instrumentation is shown in Figure 22.8. An attenuated total reflection (ATR) setup (Optrel GbR, Germany) using a wide circle $\theta-\theta$ goniometer was utilized for the excitation of SPs in the Kretschmann configuration. In order to remove the vibration, the system was set on a Sigma optical table. An AFM head (Pico-SPM, Molecular Imaging, Phoenix, AZ now Agilent Technologies) was mounted on a *XYZ-θ* adjustable stage. An equilateral triangular LaSFN9 prism was used with its base optically coupled to a LaSFN9 glass substrate using index-matching oil. An evaporated-Au ($\sim$45 nm)/LaSFN9 glass substrate/LaSFN9 prism was clamped against the Teflon cell using a metal prism-holder with an O-ring providing a liquid-tight seal. The size of the LaSFN9 glass slide was 38 $\times$ 26 mm and the thickness $\sim$1.5 mm. The volume of the Teflon cell was 0.5 ml. The holder was then magnetically attached to the electrochemical-AFM system. SPs are excited at the Au/electrolyte interface, upon total internal reflection of a polarized laser light beam. A He–Ne laser with a wavelength of $\lambda = 632.8$ nm (2 mW power) irradiated from the bottom side was used in all measurements. The Au/glass substrates used for the excitation of SPs served as a working electrode in the electrochemical measurements, with a Pt counter and a Ag reference electrode. The exposed electrode area was 2.0 cm^2. The electrochemical experiments were controlled using a Picostat instrument (Molecular Imaging, Phoenix, AZ). In the AFM measurements, the sample surface was scanned with a sharp tip situated on the probe cantilever, while a potential was applied to the samples. Contact-mode AFM images with a top-down 6-μm size scanner can be recorded, while SPR angular or kinetic measurements are measured at various applied potentials simultaneously. The monomer 0.01 M 3,4-ethylenedioxythiophene (EDOT) (Aldrich) was used as received. An acetonitrile solution was prepared containing tetrabutylammonium hexafluorophosphate (0.1 M) as the supporting electrolyte.

Simultaneous observation during electropolymerization at constant potential was also tested. The difference in the initial stages of film growth was also investigated using the EC-SPR–AFM system by potentiostatic electropolymerization as shown

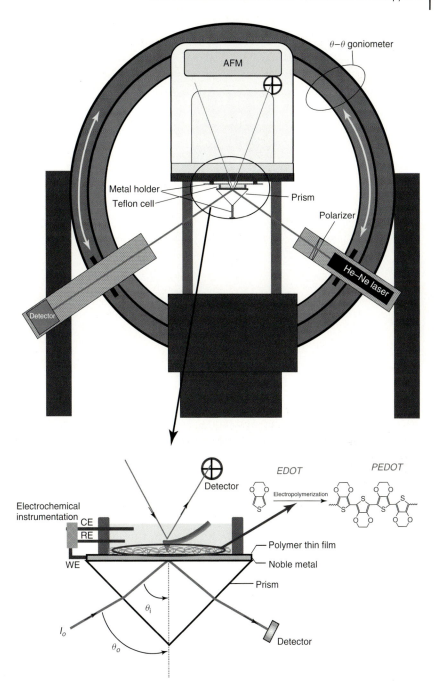

Figure 22.8 A schematic diagram of the combined electrochemical-SPR–AFM instrumentation. From Ref. [30] with permission.

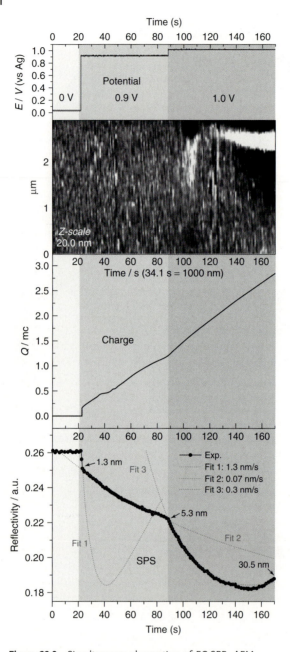

Figure 22.9 Simultaneous observation of EC-SPR–AFM during electropolymerization at constant potentials. (a) Potential, (b) AFM morphology, (c) amount of charge, and (d) the SPR reflectivity curve. Three dashed curves in SPR reflectivity show the theoretical fitting to experimental SPR curves. From Ref. [30] with permission.

in Figure 22.9. In this experiment, a potential of 0 V was applied for 22 s, then 0.9 V from 22 to 88 s, and 1.0 V from 88 to 170 s. Figure 22.9a-d shows the potential, the AFM morphology, the amount of charge, and the SPR reflectivity curve. At 0 V, all three signals showed no change. Then, the SPR curve decreased and the amount of charge increased once a potential of 0.9 V was applied. When 1.0 V was applied, the SPR signal decreased further and the amount of charge increased more steeply, indicating that the speed of electrodeposition and oxidation of EDOT was accelerated. In order to evaluate this observation quantitatively, the theoretical fitting to experimental SPR curves was done for the prism/Au/PEDOT/acetonitrile system using Fresnel calculations. As can be seen, the theoretical fitting showed good agreement with the experimental result. In this calculation, the dielectric constant of the PEDOT film was assumed to be constant $\hat{\varepsilon} = (1.88 + 0.1)$, which was estimated from previous studies [23, 24] and this experiment. The fitting was tried as a function of the deposition rate. The calculated curve was also shown in Figure 22.9 (three dashed curves in the SPR reflectivity). As shown in this fit, the deposition rate can be divided into three parts, that is, two parts in the initial deposition at 0.9 V and one part in the subsequent deposition at 1.0 V. This calculation indicated that the PEDOT film initially grows on the gold electrode until it reaches a thickness of $d = 1.3$ nm (for 1 s) at a deposition rate of 1.3 nm/s and then slows down to 0.07 nm/s. This is an interesting observation in that initially the film formation occurs rapidly but that the subsequent growth proceeds more gradually. If 1.0 V was applied continuously, the deposition rate increased to 0.3 nm/s. However, if the potential of 0.9 V was applied, the average roughness observed by the AFM increased very slightly, indicating that the deposited film was rather smooth. This result agrees well with the calculated thickness of $d = 5.3$ nm at the end of the period while the potential of 0.9 V was applied. Once the deposition rate was increased at 1.0 V, a certain increase of the surface roughness of the PEDOT film was also observed. The electrodepostition at constant potential shows that there is no obvious border for the surface morphological change in the AFM image, unlike the cyclic voltammetry experiment as shown in Figure 22.9. In this case, the potentiostatic method seemed to require an "induction period" which would have required the build-up of charge that eventually resulted in the nucleation of both polymerization and gradual deposition to the surface. In general, cyclic voltammetry electrodeposition of polymer films has been found to be more reproducible and homogeneous [31]. For example, previous AFM studies differentiating polypyrrole deposition have shown different morphologies especially at the early stages of deposition. The nucleation mechanism can be very different for the two processes, involving differentiation between two- *vs.* three-dimensional or slow *vs.* rapid nucleation and growth mechanisms [32]. For potentiostatic methods, typical chronoamperograms show a constant current when initially applied, which then abruptly decreases (sharply) followed by a gradual rise. In this case, this is related to a nucleation process involving two stages. The slow initial stage is related to the nucleation on the bare sections of the electrode surface involving a double-layer charging effect. The second stage involves rapid deposition of the polymer corresponding to autocatalytic growth. The magnitude of the current change and the slope reveals much about the polymer growth mechanism under various conditions [33, 34]. For cyclic-voltammetry methods, the formation of oxidized radical cation species (for anodic polymerization) results in *simultaneous polymerization and deposition* once the appropriate peak oxidation potential (E_p) is reached. Subsequent cycles result in lowering this oxidation potential due to the formation of more conjugated monomeric species and the possible appearance of new oxidation and reduction peaks [35]. The rapid nucleation and deposition results in a very abrupt change in morphology or appearance of a polymer, as observed (Figure 22.9). It should be noted that both processes could be dramatically influenced by solvent, concentration, temperature, monomer, and so on, conditions since these are essentially polymerization processes that result in precipitation and adsorption of polymers toward an interface.

The combined EC-SPR–AFM allowed a very important observation and differentiation of these two common methods for depositing electropolymerized films.

22.7
Application to Bio/Chemical Sensors

As introduced, many studies can be done using EC-SPR techniques. Based on these studies, several applications such as biosensors and chemical sensors have been reported.

Applications of conducting polymers to biosensors have been reported from many groups [36, 37]. Further application of conducting-polymer-based biosensors using an EC-SPR system has also been reported by several groups. Willner and coworkers [38] reported that electropolymerized PANI was used as mediator for the β-nicotinamide adenine dinucleotide (NADH) sensing. We have also introduced the electrochemical behavior of PANI and sulfonate polyaniline (SPANI) layer-by-layer ultrathin films and their efficiency to electrocatalyze the oxidation of NADH in neutral solutions [39]. Since the PANI/SPANI multilayer films prepared in the work can maintain very good electroactivity at pH 7.1, it should be possible to electrocatalyze the oxidation of NADH. Figure 22.10 shows the cyclic voltammograms of a PANI/SPANI multilayer film measured in 0.1 M phosphate buffered saline (PBS) buffer (pH 7.1) in the absence and presence of different amounts of NADH. As shown in Figure 22.10, upon the addition of NADH, the anodic peak current increases, while the cathodic peak current decreases significantly, indicating clearly the catalytic capability of PANI/SPANI multilayer films for the oxidation of NADH. As is also apparent, the anodic catalytic peak currents increase gradually with the increase of the concentration of NADH. The catalytic behavior of PANI/SPANI multilayer film can also be characterized by *in-situ* EC-SPR measurements. Figure 22.10 also shows the SPR curves of PANI/SPANI multilayer film measured in 0.1 M PBS buffer (pH 7.1) in the presence of different concentrations of NADH with the potential held at −0.2 and 0.35 V, respectively. If the potential is at −0.2 V, the film is in its reduced state.

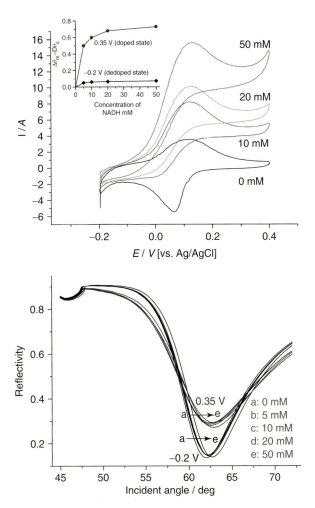

Figure 22.10 (a) Cyclic voltammograms of a PANI/SPANI multilayer film measured in 0.1 M PBS buffer (pH 7.1) in the absence and presence of different amounts of NADH. (b) SPR curves of PANI/SPANI multilayer film measured in 0.1 M PBS buffer (pH 7.1) in the presence of different concentrations of NADH with the potential held at −0.2 and 0.35 V.

Upon the addition of NADH, both the critical angle and the minimum angle shift toward higher values. The higher the concentration of NADH, the larger the shift. These changes are caused by the change of the refractive index of the solution as a result of adding NADH. But if a potential of 0.35 V is applied at which the film is in its oxidized state, a different behavior is found upon the addition of NADH: the critical angle shifts to higher values, but the minimum angle shifts to lower values. This is the net results of the refractive-index change of the solution and the refractive-index change of the multilayer film caused by the reduction of the multilayer film by its electrocatalyzed oxidation of NADH.

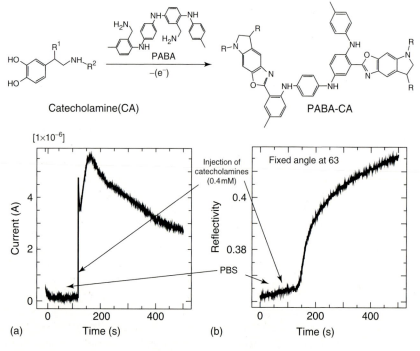

Figure 22.11 Simultaneous electrochemical (a) and SPR optical (b) detection of catecholamine (0.4 mM) at a constant potential, 0.65 V.

Another example of conducting-polymer-based biosensor applications using EC-SPR has been introduced for the *in-situ* detection of catecholamine [40]. A simultaneous detection of optical and electrochemical signals from the specific adsorption of catecholamines on poly(3-aminobenzylamine) (PABA) using EC-SPR. Since PABA is a PANI derivative, bearing benzylamines in the structure that can act as specific adsorption sites to catecholamines (Figure 22.11). PABA was electropolymerized from electroactive 3-aminobenzylamine monomers. Electroactivity of PABA was confirmed from cyclic voltammetry in PBS solution, indicating that the signals can be enhanced on the event of specific adsorption. Large signal enhancements both in optical and electrochemical signals were obtained in the case of PABA mediator, while PANI thin film showed little enhancement. Figure 22.11 shows simultaneous observation of the current change and SPR reflectivity change on the injection of 0.4 mM catecholamine. Both the current and reflectivity increased when 0.4 mM glucose was injected into buffer solution. In a controlled experiment using PANI, which does not have specific adsorption site to catecholamines, both the current and SPR reflectivity did not show an obvious increase as compared to the PABA mediator. Hence, the large increase in both the signals in Figure 22.11 indicates the possibility to develop a high-sensitivity EC-SPR catecholamine sensor using electropolymerized PABA film.

22.8
EC-SPR Method – Grating-Coupling Surface Plasmon Excitation

As introduced in the previous section, the most frequently used technique to excite SPs is the prism-coupling method based on an ATR configuration; in contrast, the grating-coupling technique has not been widely used for SPR sensors or plasmonic devices [41–46]. One major advantage of using the grating-coupling excitation method is the fact that a prism is not necessary; hence, inexpensive and disposable plastics can be used as the substrates, allowing for more flexible configurations [47, 48]. In the following section, recent developments of EC-SPR using the grating-coupling SP excitation are introduced.

For the excitation of SPR, the wave vector of the incident light must match the wave vector of the SPR, as schematically illustrated in Figure 22.12. In the grating-coupling SP excitation method, the wave vector of the incident light is increased by the grating vector as the wave vector of the light in air is smaller than that of SPR [49]. The excitation condition is defined as

$$k_{sp} = k_{px} + G = \frac{2\pi}{\lambda} \sqrt{\varepsilon_m(\omega)} \sin\theta + \frac{2\pi}{\Lambda} m \qquad (22.1)$$

Here, Λ is the diffraction grating pitch, λ is the wavelength, m is the diffraction order, and $\varepsilon_m(\omega)$ is the wavelength-dependent dielectric constant of metal given by the classical Drude's free-electron model. As can be obtained from this equation, when the grating pitch Λ increases, the distance between SP dispersion modes decreases, indicating the condensed state of the SP dispersion relation.

22.9
Combination of Electrochemical-Quartz Crystal Microbalance

A simultaneous measurement technique of EC-SPR and EC-QCM has been introduced to study the electropolymerization and characterization of conducting polymers on metallic grating surfaces [50, 51]. A schematic diagram of the combined EC-SPR-QCM instrumentation is shown in Figure 22.13.

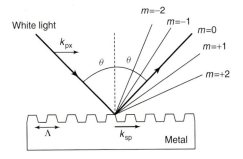

Figure 22.12 Schematic image of grating-coupling of optical wave to surface plasmon.

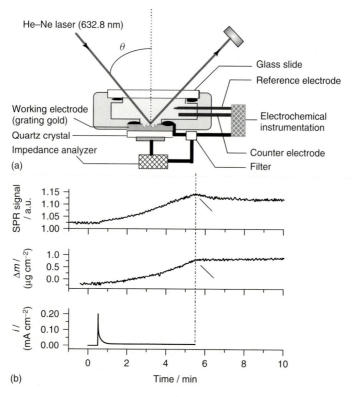

Figure 22.13 (a) Schematic diagram of the combined electrochemical-SPR-QCM instrumentation, and (b) current transient during electropolymerization of pyrrole from 0.1 M pyrrole, 0.1 M KCl, and accompanying changes of mass and reflectivity (10.5°) for a potential step from OCP to 0.5 V (SCE) and back to OCP (arrow). From Ref. [51] with permission.

The SPR measurements for the electropolymerization of pyrrole were carried out at a fixed incident angle, which was chosen prior to the experiment from the angular SPR spectra of the bare gold surface. The electropolymerization of pyrrole was initiated by stepping the potential from the OCP to +0.5 V standard calomel electrode (SCE) after 30 s. The current transient shows the typical large peak due to double-layer charging. Afterwards the current slowly increases after about 200 s. The cell was switched back to OCP after 332 s (Figure 22.13, arrow). Both the SPR and QCM kinetic curves show an increase during the electropolymerization, indicating the gradual deposition of polypyrrole layer on the Au surface. After the current potential was switched back to OCP, the QCM kinetic curve was constant, while the SPR kinetic curve decreased slightly. This result indicates that the dielectric constant of the polypyrrole film slightly changed at OCP because the doping level slightly decreases. On the other hand, the mass of the polypyrrole film was constant after the electropolymerization. Hence, the technique is especially important to study both the "optical thickness" and "acoustic thickness" simultaneously.

22.10
Electrochemical SPR under High Pressure

The EC-SPR technique at high pressure has been demonstrated for the study of conducting polymer thin films [52, 53]. In the setup, grating-coupling SPR was combined with a high-pressure electrochemical cell that enabled the investigation of the pressure effect and bias voltages on polymer thin films. The set-up is schematically shown in Figure 22.14. The body of the cell is made from stainless steel. Two windows and three pressure fittings are mounted into the cell. The optical axes of the sapphire windows are oriented so as to avoid birefringence effects. Gold wire rings are used as seals between the steel body and the sapphire windows. The temperature inside the chamber is controlled by a thermostat. For the SP excitation by the grating coupler, the sample, that is, the fused silica-grating/gold/polymer film, was mounted inside the pressure chamber and fixed by a screw ring from the back side of the cell. A small slit in the top of the cell allowed a $\lambda = 633$ nm light beam of a He–Ne laser to fall onto the detector after being reflected off the fused silica-grating/gold/polymer film system.

Examples for SPR curves of PEDOT films prepared at 0.1 MPa but then recorded at different potentials that were held constant during the angular reflectivity scan are shown in Figure 22.14. These films, grown at the same ambient pressure but investigated at different pressures of 0.1 and 20 MPa, respectively, show virtually identical SPR curves at the respective potentials and also show the same change of the reflectivity curve as one induces a redox reaction in the films upon changing the potential from $E = 0$ V to $E = 0.45$ and 0.9 V, respectively. On the other hand, films grown at 20 MPa but then investigated at different potentials and also at the two different pressures, that is, 0.1 and 20 MPa, respectively, also show an identical behavior for the two pressures, however, seem to exhibit virtually no optical response as one sweeps through the potential window of 0–0.9 V. The experiments indicate the importance of pressure as a processing parameter and thermodynamic variable important for a complete picture of the structural and functional features of thin polymer films.

22.11
Conclusions

In this review, we introduced the useful combination of electrochemical measurement and SPR, which was further combined with a variety of surface-sensitive techniques such as AFM, QCM, and so on, for the study of conducting polymer thin films. Their applications to biosensors and chemical sensors were also introduced. Besides the combined techniques with EC-SPR shown in this review, other novel combinations such as with scanning electrochemical microscopy [54, 55], microfluidic devices [56], and electrochemically tunable SP-enhanced diffraction gratings [57] have been reported for the study of polymer thin films or sensor applications. In these studies, two SP excitation methods, that is, the prism-coupling method

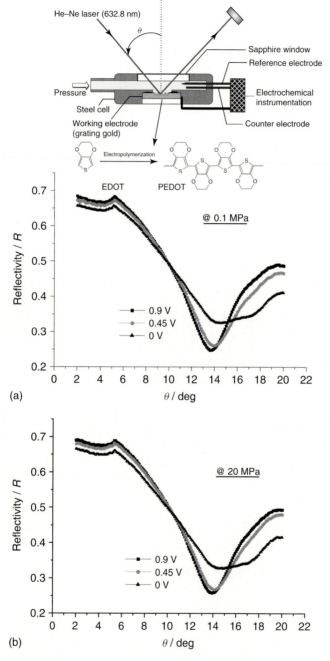

Figure 22.14 Surface plasmon angular scans of PEDOT films prepared at $p = 0.1$ MPa (a) and $p = 20$ MPa (b), respectively, at three different potentials applied, as indicated.

based on the Kretschmann configuration and the grating-coupling method, were demonstrated. Moreover, electrochemical localized surface plasmon resonance (LSPR) has recently been studied for possible applications such as biosensors [58], tunable electrochemical switches [59, 60], and so forth.

The introduced method should provide new opportunities both for basic studies and for sensor applications using conducting polymers.

Acknowledgments

We would like to thank all collaborators in this work at the Max-Planck-Institute for Polymer Research, the National University of Singapore, the University of Houston, and Niigata University. R.C.A would like to acknowledge funding from: NSFDMR-10-06776, CBET-0854979, and Robert A. Welch Foundation, E-1551.

References

1. Kobayashi, T., Yoneyama, H., and Tamura, H. (1984) *J. Electroanal. Chem.*, **161**, 419.

2. MacDiarmid, A.G., Mu, S.L., Somarisi, N.L.D., and Mu, W. (1985) *Mol. Cryst. Liq. Cryst.*, **121**, 187.

3. Orata, D. and Buttry, D.A. (1987) *J. Am. Chem. Soc.*, **109**, 3574.

4. Cui, S.-Y. and Park, S.-M. (1999) *Synth. Met.*, **105**, 91.

5. Zimmermann, A. and Dunsch, L. (1997) *J. Mol. Struct.*, **410**, 165.

6. Nyffenegger, R., Ammann, E., Siegenthaler, H., and Haas, O. (1995) *Electrochim. Acta*, **40**, 1411.

7. Knoll, W. (1998) *Annu. Rev. Phys. Chem.*, **49**, 569.

8. Tadjeddine, A., Kolb, D.M., and Kötz, R. (1980) *Surf. Sci.*, **101**, 277.

9. Iwasaki, Y., Horiuchi, T., Morita, M., and Niwa, O. (1997) *Electroanalysis*, **9**, 1239.

10. Baba, A., Advincula, R.C., and Knoll, W. (2001) Simultaneous observation of the electropolymerization process of conducting polymers by surface plasmon resonance spectroscopy, surface plasmon enhanced light scattering and cyclic voltammetry, in *Novel Methods to Study Interfacial Layers*, Studies in Interface Science, Vol. 11 (eds D. Möbius and R. Miller), Elsevier Science, p. 55.

11. Raitman, O., Katz, E., Willner, I., Chegel, V., and Popova, G. (2001) *Angew. Chem. Int. Ed.*, **40**, 3694.

12. Baba, A., Advincula, R.C., and Knoll, W. (2002) *J. Phys. Chem. B*, **106**, 1581.

13. Baba, A., Park, M.-K., Advincula, R.C., and Knoll, W. (2002) *Langmuir*, **18**, 4648.

14. Chegel, V., Raitman, O., Katz, E., Gabai, R., and Willner, I. (2001) *Chem. Commun.*, 883.

15. Kang, X., Jin, Y., Chen, G., and Dong, S. (2002) *Langmuir*, **18**, 1713.

16. Raitman, O.A., Katz, E., Bückmann, A.F., and Willner, I. (2002) *J. Am. Chem. Soc.*, **124**, 6487.

17. Kretschmann, E. (1972) *Opt. Commun.*, **5**, 331.

18. Atteidge, J.W., Daniels, P.B., Deacon, J.K., Robinson, G.A., and Davidson, G.P. (1991) *Biosens. Bioelectron.*, **6**, 201.

19. Liebermann, T. and Knoll, W. (2000) *Colloids. Surf. A*, **171**, 115.

20. Ushioda, S. and Sasaki, Y. (1983) *Phys. Rev. B*, **27**, 1401.

21. Knobloch, H., Duschl, C., and Knoll, W. (1989) *Chem. Phys.*, **91**, 3810.

22. Baba, A., Tian, S., Stefani, F., Xia, C., Wang, Z., Advincula, R.C., Johannsmann, D., and Knoll, W. (2004) *J. Electoanal. Chem.*, **562**, 95.

23. Baba, A., Lübben, J., Tamada, K., and Knoll, W. (2003) *Langmuir*, **19**, 9058.

24. Baba, A. and Knoll, W. (2003) *J. Phys. Chem. B*, **107**, 7733.

25. Rothenhäusler, B. and Knoll, W. (1988) *Nature*, **332**, 615.

26. Flätgen, G., Krischer, K., Pettinger, B., Doblhofer, K., Junkes, H., and Ertl, G. (1995) *Science*, **269**, 668.

27. Nishizawa, M., Shibuya, M., Sawaguchi, T., Matsue, T., and Uchida, I. (1991) *J. Phys. Chem.*, **95**, 9042.

28. Nishizawa, M., Miwa, Y., Matsue, T., and Uchida, I. (1993) *J. Electrochem. Soc.*, **140**, 1650.

29. Sayre, C.N., and Collard, D.M. (1997) *Langmuir*, **13**, 714.

30. Baba, A., Knoll, W., and Advincula, R. (2006) *Rev. Sci. Instrum.*, **77**, 064101.

31. Lyons, M.E.G. (1994) *Electroative Polymer Electrochemistry, Part 1: Fundamentals*, Plenum Press, New York, p. 164.

32. Hernandez-Perez, T., Morales, M., Batina, N., and Salmon, M. (2001) *J. Electrochem. Soc.*, **148**, C369.

33. Wurm, D.B. and Kim, Y.-T. (2000) *Langmuir*, **16**, 4533.

34. Norris, I.D., Kane-Maguire, L.A.P., and Wallace, G.G. (2000) *Macromolecules*, **33**, 3237.

35. Schottland, P., Zong, K., Gaupp, C.L., Thompson, B.C., Thomas, C.A., Giurgiu, I., Hickman, R., Abboud, K.A., and Reynolds, J.R. (2000) *Macromolecules*, **33**, 7051.

36. Gerard, M., Chaubey, A., and Malhotra, B.D. (2002) *Biosens. Bioelectron.*, **17**, 345.

37. Wallace, G.G. and Kane-Maguire, L.A.P. (2002) *Adv. Mater.*, **14**, 953.

38. Raitman, O.A., Katz, E., Buckmann, A.F., and Willner, I. (2002) *J. Am. Chem. Soc.*, **124**, 6487.

39. Tian, S., Baba, A., Liu, J., Wang, Z., Knoll, W., Park, M.-K., and Advincula, R. (2003) *Adv. Funct. Mater.*, **13**, 473.

40. Baba, A., Mannen, T., Ishigami, R., Ohdaira, Y., Shinbo, K., Kato, K., Kaneko, F., Fukuda, N., and Ushijima, H. (2009) JP Patent P2009-180704A.

41. Barnes, W.L., Dereux, A., and Ebbesen, T.W. (2003) *Nature*, **424**, 814.

42. Homola, J., Koudela, I., and Yee, S.S. (1999) *Sens. Actuators B*, **54**, 16.

43. Chien, F.C., Lin, C.Y., Yih, J.N., Lee, K.L., Change, C.W., Wei, P.K., Suna, C.C., and Chen, S.J. (2007) *Biosens. Bioelectron.*, **22**, 2737.

44. Chiu, N.F., Lin, C.W., Lee, J.H., Kuan, C.H., Wu, K.C., and Lee, C.K. (2007) *Appl. Phys. Lett.*, **91**, 083114.

45. Massenot, S., Chevallier, R., de Bougrenet de la Tocnaye, J.-L., and Parriaux, O. (2007) *Opt. Commun.*, **275**, 318.

46. Baba, A., Kanda, K., Ohno, T., Ohdaira, Y., Shinbo, K., Kato, K., and Kaneko, F. (2010) *Jpn. J. Appl. Phys.*, **49**, 01AE02.

47. Singh, B.K. and Hillier, A.C. (2006) *Anal. Chem.*, **78**, 2009.

48. Singh, B.K. and Hillier, A.C. (2008) *Anal. Chem.*, **80**, 3803.

49. Raether, H. (1988) *Surface Plasmons*, Springer, Berlin.

50. Bailey, L.E., Kambhampati, D.K., Kanazawa, K.K., Knoll, W., and Frank, C.W. (2002) *Langmuir*, **18**, 479.

51. Bund, A., Baba, A., Berg, S., Johannsmann, D., Lubben, J., Wang, Z., and Knoll, W. (2003) *J. Phys. Chem. B*, **107**, 6743.

52. Jokob, T. and Knoll, W. (2003) *J. Electroanal. Chem.*, **543**, 51.

53. Baba, A., Kleideiter, G., Jakob, T., and Knoll, W. (2004) *Macromol. Chem. Phys.*, **205**, 2267.

54. Szunerits, S., Knorr, N., Calemczuk, R., and Livache, T. (2004) *Langmuir*, **21**, 9236.

55. Fortin, E., Defontaine, Y., Mailley, P., Livache, T., and Szunerits, S. (2005) *Electroanalysis*, **17**, 495.

56. Choi, S. and Chae, J. (2009) *Biosens. Bioelectron.*, **25**, 527.

57. Tian, S., Armstrong, N.R., and Knoll, W. (2005) *Langmuir*, **21**, 4656.

58. Hiep, H.M., Endo, T., Saito, M., Chikae, M., Kim, D.K., Yamamura, S., Takamura, Y., and Tamiya, E. (2008) *Anal. Chem.*, **80**, 1859.

59. Leroux, Y.R., Lacroix, J.C., Chane-Ching, K.I., Fave, C., Félidj, N., Lévi, G., Aubard, J., Krenn, J.R., and Hohenau, A. (2005) *J. Am. Chem. Soc.*, **127**, 16022.

60. Leroux, Y., Lacroix, J.C., Fave, C., Stockhausen, V., Lidj, N.F., Grand, J., Hohenau, A., and Krenn, J.R. (2009) *Nano Lett.*, **9**, 2144.

23
Characterization of Molecularly Thin Polymer Layers with the Surface Forces Apparatus (SFA)

Marina Ruths

23.1
Introduction: Polymer Layers Adsorbed on and Confined between Solid Surfaces

In solution, polymers form coils whose size depends on a balance between entropy and the strength of interaction between the polymer segments and the solvent, the so-called solvent quality. The solvent quality also affects the structures formed by polymer molecules adsorbed on surfaces. In a "theta solvent" or at "theta conditions" (θ), the interactions (van der Waals attraction and hydrogen bonding) between segments and solvent molecules are of the same magnitude as the segment–segment attraction [1]. The size of the polymer coil is then characterized by its radius of gyration, R_g, the average root mean square distance of a segment from the center of mass, which is proportional to $(N/6)^{1/2}$ for an ideal, freely jointed chain (N is the number of segments in the chain). In a "good solvent" the segment–solvent interactions are stronger than the segment–segment attraction. This allows an expansion of the polymer coil in solution (to a size described by the Flory radius R_F, proportional to $N^{3/5}$), and an extension into solution of polymer chains adsorbed to a surface [1–5]. The opposite behavior is seen for a "poor solvent." As a result of this balance between segment–segment, segment–solvent, and segment–surface interactions, a larger amount of polymer may adsorb on a surface from a poor solvent to form a more compact layer close to the surface. In many systems, such effects can be observed simply by changing the temperature around the so-called theta or Flory temperature. A special example of a system at theta conditions is a polymer melt, where the polymer segments are interacting only with other segments of the same chemical composition.

The first polymer molecules to adsorb onto a surface from a solution typically have many segments in contact with the substrate, but at equilibrium adsorption, large sections of the molecules protrude into the solvent. It is commonly observed that only a limited adsorbed amount and layer thickness can be obtained by adsorption from solution, since after some chains have become attached, incoming chains have to diffuse against a concentration gradient to reach the surface. In addition, there is an entropy loss arising from the change in conformation of both the incoming and the adsorbed polymer to accommodate another chain. For these reasons, typically only a few milligrams of polymer per square meter adsorbs onto a solid surface from a good or θ solvent, regardless of whether the chains become chemically bound to the surface or only physisorbed [6].

As two surfaces with adsorbed polymer layers are brought toward one another, some amount of steric repulsion is commonly observed due to compression (chain elasticity) and osmotic effects. However, if the adsorbed amount of polymer is low, the interaction may have an attractive component due to "bridging," that is, adsorption of a polymer chain to two opposing surfaces (Section 23.3.1). In the case of nonadsorbing polymer, an attractive "depletion" force can arise at small separation distance between two surfaces as a result of the osmotic pressure of the solution (Section 23.3.2). In cases where a chain end-group or block has different chemical properties from the rest of the chain, it may become preferentially adsorbed, which gives rise to structures called polymer *"mushroom"* and *"brush"* layers (Section 23.4), with a distinctive, repulsive interaction. In systems where no solvent is present, for example, dried adsorbed layers, no long-range forces are observed. However, a stronger adhesion than that expected from van der Waals interactions alone may develop between contacting polymer layers above their glass-transition temperature, because of molecular rearrangement and entanglement, and viscoelastic effects upon separation. This enhanced adhesion typically increases with increasing contact time and separation rate (Section 23.5). All of these types of interactions can be measured with the Surface Forces Apparatus (SFA) (Section 23.2) and compared to theoretical models and computer simulations, as illustrated below. Modeling of the conformation of polymers at surfaces and the interactions of adsorbed homopolymer layers and brush layers has attracted significant interest, and only a few examples of the available models are discussed within the scope of this chapter.

23.2
Force Measurements with the Surface Forces Apparatus (SFA)

Interaction forces between two surfaces can be measured in a well-defined geometry with the SFA. In the most common type of SFA, two transparent, back-silvered surfaces face one another in a crossed-cylinder configuration (geometrically equivalent to a sphere near a flat surface). The radius of curvature, R, is typically 1–2 cm and is measured for undeformed surfaces (i.e., with the surfaces out of contact) at each contact position used in the experiment. The separation distance between the surfaces, D, is measured by multiple beam interferometry with an accuracy of 0.1 nm [7–9] as the base of a double-cantilever leaf spring supporting the lower surface is moved with respect to the base of the rigidly mounted upper surface by means of motor-driven mechanical stages. Depending on the strength of the interaction forces and the stiffness of the cantilever spring, forces normal to the surfaces can be measured with the motor controls in most SFA setups (Mark II [10], Mark IV [11], SFA3 [12], SFA2000 [13], and other models [14]), but very fine control of the separation distance (to 0.1 nm) and corresponding force measurements are done with a piezoelectric tube incorporated in the mount for the upper surface (Figure 23.1). This piezoelectric tube may be sectored to enable lateral movement of the upper surface [15]. A larger range of separation rates than that possible with the motor controls and piezoelectric tube can be accessed by attaching a piezoelectric bimorph actuator to the leaf spring holding the lower surface [16, 17]. Instruments have also been developed where the force is measured with a piezoelectric bimorph sensor (MASIF [18]), eliminating the need for transparent substrates. However, the optical technique has certain advantages such as the direct measurements of film thickness and the shape of the surfaces in and out of contact (deformations due to flattening, protrusions or particles, and measurements of the radius of curvature, R), and the capability for measurements of the optical properties of confined layers and films [7–9].

The interaction force normal to the surfaces, F, is calculated from the deflection of the leaf spring supporting the lower surface, that is, from the difference between measured separation distance D and the value expected from a calibration of

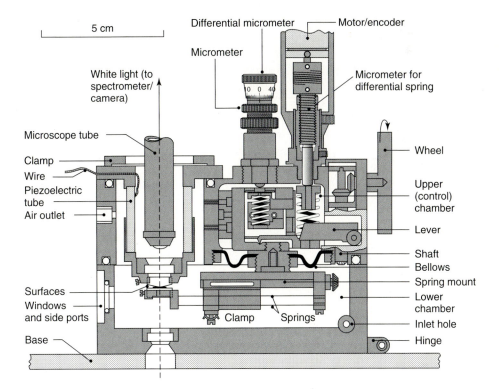

Figure 23.1 Schematic illustration of a Surface Forces Apparatus (the SFA3 [12]), showing the two half-cylindrical surfaces mounted in a crossed configuration. The lower surface can be mounted on a double-cantilever spring as shown above, or on a similar spring attached to a bimorph slider [19] for lateral movement. The upper surface can be mounted on a piezoelectric upper mount [10] as shown above, or on a friction-detecting device [20]. Redrawn with permission from Ref. [12].

distance *vs.* motor movement at large separation where no force acts between the surfaces. In most experiments, the sensitivity in the normal force is about 10^{-7} N, but the maximum sensitivity is 10^{-8} N [10–13, 21], obtained by using very soft cantilever springs. Normal forces are commonly presented as $F(D)/R$ to allow comparison between different experiments (done with curved surfaces of different R) and with the energy between flat plates, $W(D)$, which is related to the force between a sphere and a flat surface (or between two crossed cylinders) according to the Derjaguin approximation [22, 23],

$$\frac{F(D)}{R} = 2\pi W(D) \tag{23.1}$$

The most commonly used surface substrate is muscovite mica, a mineral that can be cleaved from larger blocks into atomically smooth, several square centimeter large sheets with a well-known crystal structure [24] and a thickness of a few micrometers. The disadvantage of mica substrates is that they are chemically inert (unreactive). To obtain surfaces with different structure and composition, mica substrates have been covered with various physisorbed organic or inorganic layers such as Langmuir–Blodgett or self-assembled monolayers [25–27]; adsorbed homopolymer [14, 28–35], end-functionalized polymer [36–40] or block-copolymer [36, 41–46]; or electron-beam- or chemical-vapor-deposited Al_2O_3 [47] or SiO_2 [48]. It is also possible to render the mica substrates more reactive through water-vapor plasma treatment, which creates hydroxyl- and other reactive groups on the surface [49, 50]. This enables chemical bonding of silane-based self-assembled monolayers to the mica, and is also beneficial for the deposition of thin, smooth gold layers [50–52], which opens up the possibility for chemical bonding of thiol-functionalized molecules. In addition, the mica substrates can be entirely replaced by back-silvered thin sapphire (Al_2O_3) crystals [53, 54]; thin, flexible sheets of quartz or borosilicate glass [55], see also [39]; or micrometer-thick sheets of glassy polymers [56, 57], allowing a wide variety of substrates to be studied.

23.3
Forces in Systems with Adsorbing or Nonadsorbing Homopolymer

Uncharged homopolymers (polymers containing only one type of segment) interact with surfaces through van der Waals and hydrogen-bonding interactions. The physisorption of individual segments of a homopolymer chain is reversible and highly dynamic. The adsorbed and free segments thus exchange rapidly, but the exchange with free chains in the solution is slow, since the polymer remains bound to the surface as long as one segment along the chain is adsorbed [58]. The adsorption energy per segment is typically on the order of kT, and a polymer chain is said to adsorb as trains (segments in contact with the surface), loops (sections extending from the surface, anchored at both ends by trains), and tails (freely dangling chain ends). If the polymer solution contains chains of different lengths, longer chains eventually replace shorter ones on the surface, since longer chains have more attachments points and are thus less likely to become desorbed. Classic scaling models [4, 58, 59] predict that at semidilute, good solvent conditions, the adsorbed layer thickness is proportional to the extension of the longest loops, that is, to the polymer coil size in solution, which is characterized by the Flory radius R_F (proportional to $N^{3/5}$). The segment density scales as $D^{-4/3}$ from the surface [4, 58, 59]. Mean-field theory [60] predicts an rms average layer thickness proportional to $N^{1/2}$ when the polymer is free to exchange with polymer in solution (see below). It has been shown [61, 62] that under dilute, good solvent conditions, the outer region of the segment density profile is dominated by segments in tails, which determine the characteristic layer thickness.

Interaction forces between adsorbed polymer layers arise from a balance between intermolecular forces and the entropy of mixing. Theoretically and experimentally, there are two distinctly different types of equilibrium interactions between adsorbed polymer layers: "true" equilibrium, where polymer is desorbing and migrating (diffusing) out from the confining gap between the surfaces into the surrounding bulk solution, where it can gain conformational entropy, and "restricted" ("constrained") equilibrium where the total amount of confined polymer is kept constant (unchanged from the equilibrium amount adsorbed on the surface before the confinement), but the distribution of adsorbed and nonadsorbed polymer segments within the gap is changing as the gap thickness is changed [4, 59, 60].

At "true" equilibrium, some theories for interactions between layers of adsorbed homopolymer predict a monotonic attraction due to bridging and depletion interactions (cf. Sections 23.3.1 and 23.3.2) as the surfaces are brought closer [59, 60]. Experimentally, it is difficult to reach a true equilibrium situation, since the time needed for total desorption of a high molecular weight polymer is very long, especially when confined between two surfaces. In general, some kind of restricted equilibrium is investigated. Theoretically, at restricted equilibrium [59, 60], the total amount of polymer between the surfaces equilibrates across the gap, so that the individual polymer chains are no longer associated only with the surface they were originally adsorbed on. Experimentally, even this condition is difficult to reach, and one often measures the interactions between two adsorbed, separate polymer layers, which is an even more restricted condition than that assumed in the theory for restricted equilibrium. In such cases, the range and magnitude of the interactions are often more strongly dependent on chain length and adsorption time than expected theoretically for restricted equilibrium interactions. The measured interactions may also be dependent on the approach and separation rate of the surfaces, and previous compressions and decompressions of the polymer layers (history effects).

At restricted equilibrium, scaling theories for interactions between adsorbed layers under good solvent conditions predict a monotonic (steric-entropic) repulsion [59], proportional to D^{-2} at large separations and to $D^{-5/4}$ at small D. In a poor solvent, the short-range repulsion is expected to be proportional to D^{-2} [59], but interactions in a poor solvent are expected to contain additional, attractive regimes at larger distances [63], and attraction is also predicted in a θ solvent at $0.5\,R_g < D < $ circa $2\,R_g$ [64]. In mean-field models, the configuration-dependent interaction potential in scaling theories is replaced with a mean potential resulting from the distribution of chain configurations [60, 62]. The predicted interaction varies from attractive (bridging) at low adsorption density (produced from low segment concentration in the solution) to repulsive at high adsorption density [60, 62–67], with an additional attractive component in a poor solvent. A final hard-wall repulsion from one remaining monolayer of polymer is expected [60]. Monotonically repulsive interactions, and also attraction followed by repulsion at smaller gap distances, are commonly measured between high molecular weight polymer layers [14, 28–30, 33], but it is difficult to establish whether they correspond to the type of restricted equilibrium situation investigated in the models.

23.3.1
Bridging Interactions

Bridging occurs when segments from a chain adsorbed on one surface are able to reach over to another surface and adsorb on it. This requires available adsorption sites on the opposing surface, either due to low adsorption density (a low adsorbed amount also facilitates the diffusion of chains through the opposing layer) or rearrangements on compression. Once attached to the opposing surface, the bridging polymer chain would gain conformational entropy if the two surfaces came closer together, and the result is an attractive force acting between the two surfaces. Depending on the extension of the polymer, this attraction may be long range, typically much longer than the van der Waals attraction or the depletion interaction (Section 23.3.2) in a system. In addition, in a poor solvent, there are attractive segment–segment interactions. These may affect the range of the attraction, since segment–segment attraction is possible as soon as there is some overlap between the adsorbed layers, whereas bridging requires chains to reach the opposing surface [64]. At small separation distances and restricted equilibrium, repulsion is observed due to the osmotic interaction of the adsorbed, highly confined polymer.

Different functional forms of bridging forces have been predicted theoretically [60, 64], and examples can be found of measured bridging attraction varying exponentially or linearly with separation distances, as well as cases that cannot be as easily categorized [14, 28, 30, 33]. In a simple model [23], it is assumed that one segment after the other of a bridging polymer chain adsorbs onto either of the surfaces (each changing the length of the "bridge" by $\Delta D = -l$ and the energy by $\Delta w = -\varepsilon$), which suggests that the energy varies as $w(D) = -\varepsilon(L_c - D)/l$, where $L_c = Nl$ is the contour length of the chain, and $(L_c - D)/l$ is the number of segments of the bridging chain that have adsorbed onto the surfaces. Taking into account the density of bridging chains from one surface, Γ, and assuming that both surfaces contribute chains that form bridges (hence a factor 2), the interaction energy between flat plates is $W(D) = -2\Gamma\varepsilon(L_c - D)/l$. The force between a sphere and a flat surface (or between two crossed cylinders) is obtained by using the Derjaguin approximation (Equation 23.1), and increases linearly (becomes more strongly attractive) with decreasing D [23]:

$$\frac{F(D)}{R} = -4\pi\varepsilon\Gamma\frac{(L_c - D)}{l} \tag{23.2}$$

Examples of SFA measurements of bridging forces between adsorbed, high molecular weight homopolymer layers are shown in Figure 23.2. Figure 23.2a shows examples of interactions between polystyrene layers adsorbed on mica from cyclohexane at different temperatures [28]. The polystyrene was a "monodisperse standard" with $M_w = 900\,000$ g/mol (where M_w is the weight-average molecular weight) and $R_g \approx 25$ nm. The theta temperature for polystyrene in cyclohexane is 34.5 °C. The forces measured at adsorption saturation (curves (i) and (ii), where the data points have been omitted for clarity) indicated a long-range attraction with an almost linear part. The adsorbed amount (determined from the compressed layer thickness) was larger in the layer adsorbed under poor solvent conditions (curve (i)), and the attraction in this system was stronger, as expected if there was a contribution from segment–segment attraction. Partially saturated adsorption, giving rise to force curve (iii), was obtained by keeping the mica substrates at a close separation during the adsorption to slow down the diffusion of polymer to the surfaces. The shorter interaction range and stronger attraction in contact in case (iii) was consistent with an initial adsorption situation with many segments per chain in contact

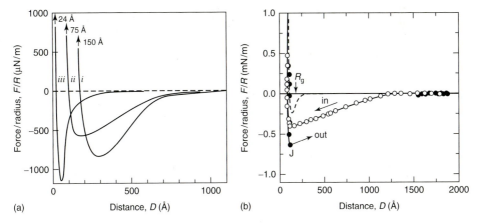

(a)   Distance, D (Å)   (b)   Distance, D (Å)

Figure 23.2 Bridging attraction between high molecular weight polystyrene layers adsorbed on mica. The measured force, F, is normalized by the radius of curvature, R, and shown as a function of separation distance, D, between the mica substrates. (a) $M_w = 900\,000$ g/mol polystyrene adsorbed from cyclohexane solution above and below the theta temperature of 34.5 °C. Curves (i) and (ii) show interaction forces at poor (26 °C) and good (37 °C) solvent conditions, respectively (data points omitted for clarity). Curve (iii) shows interactions at 26 °C and low coverage (1.1 mg/m², which was 20–30% of the saturation value), with a shorter-ranged, stronger attractive force. Adapted with permission from Ref. [28]. Copyright 1984 American Chemical Society. (b) Bridging attraction measured on the first approach (open symbols) and separation (filled symbols) of $M_w = 400\,000$ g/mol polystyrene adsorbed on mica at 23 °C from cyclopentane (theta temperature 19.5 °C). The dashed curve is the result from a self-consistent field calculation that accounted for the presence of loops but not tails [66]. "J" indicates a jump of the surfaces from one stable position to another (here, a jump out to a large separation distance). Solid curves are drawn as a guide only. Reprinted with permission from Ref. [33]. Copyright 1997 American Chemical Society.

with the substrate (so that the polymer's extension into solution was smaller), but a low overall adsorbed amount with adsorption sites available on the surface.

Figure 23.2b shows interaction forces on the first approach of polystyrene layers ($M_w = 400\,000$ g/mol, $R_g = 18$ nm, $M_w/M_n = 1.06$, where M_n is the number-average molecular weight and M_w/M_n is the polydispersity) adsorbed to an equilibrium thickness at $23\,°C$ from cyclopentane [33] (near-theta conditions, theta temperature $19.5\,°C$ [68]). On a slow, first approach (2.5 min between measured points, i.e., an approach rate of <1 Å/s), a long-range, linear attraction was observed. If the approach was done faster (not shown), a shorter interaction range was observed and the attractive force was not linear [33]. After the first approach, the adsorbed layers became irreversibly compressed (on the time scale of the experiment, a few days). Shorter adsorption times gave a shorter-ranged, linear interaction with a thinner "hard-wall" (compressed layer) thickness and a stronger attraction on separation, consistent with a different distribution of adsorbed trains, loops, and tails than that at equilibrium adsorption. A force curve calculated using self-consistent field theory [33, 66] is shown as a dashed line in Figure 23.2b. This model accurately described the repulsive regime (hard-wall thickness), and accounted for attractive interactions due to bridging of loops, but did not take into account the contribution from tails. The measured long-range interaction in Figure 23.2b suggested that the contribution of tails to bridging may be significant.

23.3.2
Depletion Interactions

In systems containing sufficiently high concentrations of nonadsorbing polymers or large aggregates of self-assembled molecules (e.g., micelles), an attractive force may arise at small surface separations (smaller than the coil or aggregate diameter) from the difference in osmotic pressure of the solution remaining in the gap between the surfaces and the surrounding (bulk) solution containing a high concentration of polymer [58, 69–75]. When the surfaces are brought closer together than the average diameter of the polymer coils ($2R_g$ or $2R_F$), the remaining coils between the surfaces are expelled from the gap, and the surfaces are then pulled closer together to compensate for the local change in number concentration of solute (polymer). This phenomenon is called *depletion attraction* and is proportional to the number concentration of the solute (polymer coils).

Between two flat surfaces at a separation below $2R_g$, the depletion force per area is the osmotic pressure of the solution, that is, $\pi_{osm} = -\rho kT$, where ρ is the number density of polymer coils. For flat surfaces, the depletion force is thus independent of the separation distance D, and acts only at separations below $2R_g$. Integrating the force between flat plates from $2R_g$ to D to get the corresponding interaction energy, and using the Derjaguin approximation (Equation 23.1), one obtains the depletion force between a sphere and a flat surface (or between two interacting crossed cylinders) [23]:

$$\frac{F(D)}{R} = -2\pi\rho kT(2R_g - D), \quad \text{for } D < 2R_g \tag{23.3}$$

which is seen to increase (become more strongly attractive) with decreasing D. The strongest attraction (at contact, i.e., $D = 0$) corresponds to the osmotic pressure multiplied by the contact area (the area over which the gap distance between a sphere and a flat surface is too low to accommodate polymer coils) [23, 27, 76]. Depletion attraction is maximized for a large number of small coils at the semidilute concentration limit, that is, right when the polymer coils start to overlap. Furthermore, the polymer should not interact strongly with the confining surfaces, although it is seen from Figures 23.3b–d that depletion can also occur between adsorbed polymer layers. Depletion attraction in such systems has also been suggested theoretically, provided that all adsorption sites are filled [72].

In highly concentrated systems, one may also observe "depletion stabilization" [72, 73] due to compression of polymer coils, that is, a weakly repulsive force at a separation distance of about twice the polymer coil diameter. Albeit found over a different distance regime, depletion interactions are similar to solvation (oscillatory) forces [23], where alternating repulsive barriers and attractive minima can be observed as the separation distance is decreased such that only discrete numbers of semiordered solvent layers can be accommodated between the surfaces.

Examples of measurements of depletion interactions with the SFA are shown in Figure 23.3. Figure 23.3a shows the interactions between lipid bilayer-covered mica surfaces across an aqueous solution (25 °C, good solvent conditions) of polyethylene glycol (polyethylene oxide, $M_w = 8000$ g/mol) [77]. The hydrodynamic radius (proportional to $N^{1/2}$) and the Flory radius ($R_F = 8$ nm) are indicated in Figure 23.3a. The depth of the attractive minimum is stronger than the van der Waals attraction shown as a dotted curve, and increases with increasing polymer concentration. On compression, the attraction is preceded by a weakly repulsive regime that may result from compression of the polymer coils, that is, depletion stabilization, or from very weak adsorption of the polymer [77]. The hard-wall repulsion (the strong repulsion at the smallest distances) appeared at a gap size similar to that measured in the pure bilayer system, without polymer. Depletion interactions in aqueous solution have also been measured for dextran [78] between bare mica surfaces.

Figures 23.3b–d show repulsive and attractive forces measured in a study [76] of hydrocarbon surfaces (adsorbed monolayers of dihexadecyldimethyl-ammonium acetate surfactant on mica) across a saturated hydrocarbon polymer, poly(ethylene-propylene) ($-[CH_2CH(CH_3)CH_2CH_2]_n-$, $M_w = 5800$ and $33\,000$ g/mol, $R_g \approx 3$ and 8 nm, respectively [79]), in a good solvent, n-tetradecane. Supporting experiments using light-scattering showed an attractive interaction between surfactant-coated particles in decane, linearly proportional to the concentration of poly(ethylene-propylene) in the solution and of a magnitude as expected for depletion forces [80]. Experiments were done at concentrations of about 1/10 of the overlap concentration (Figures 23.3b–d) and at the overlap concentration, where the attractive "outer minimum" was somewhat deeper (not shown). A small amount of polymer adsorbed although the surfaces was covered with hydrophobic monolayers, but despite this, depletion interactions

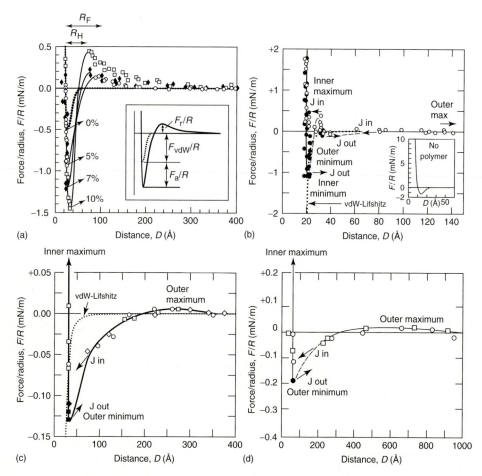

Figure 23.3 Depletion attraction and stabilization. Solid curves are drawn as guides only. (a) Interaction forces between lipid bilayer-covered mica surfaces across aqueous solutions of $M_w = 8000$ g/mol polyethylene glycol at 25 °C. The depletion attraction is increasing with increasing polymer concentration in wt%. The dotted curve shows the calculated van der Waals attraction. Reprinted with permission from Ref. [77]. Copyright 1996 American Chemical Society. (b–d) Interactions between surfactant monolayers on mica in poly(ethylene-propylene) ($M_w = 5800$ and 33 000 g/mol) solutions in *n*-tetradecane (good solvent). $D = 0$ is at monolayer–monolayer contact. (b) $M_w = 5800$ g/mol. A small amount of polymer

adsorbs, indicated by the repulsive maximum at $D = 32$–35 Å, and a hard-wall repulsion at $D \approx 19$ Å. (Inset) Interactions between the surfactant monolayers in pure *n*-tetradecane (no polymer). (c) Detail of (b), showing the outer minimum (depletion attraction), van der Waals interaction, and small outer maximum (depletion stabilization) at $D \approx 200$–300 Å. (d) $M_w = 33000$ g/mol, showing only the detail of the outer minimum and maximum, as in (c). Open symbols denote approach and filled symbols separation of the surfaces, and "J" indicates a jump of the surfaces from one stable position to another. Reprinted with permission from Ref. [76]. Copyright 1996 American Chemical Society.

were observed. As the surfaces were brought toward one another from a large distance, a weak repulsion ("outer maximum") was observed, followed by a weak "outer minimum." These two regions (Figures 23.3c,d) were ascribed to depletion interactions, whereas the force barrier on further compression and the inner minimum (the depth of which is well described by van der Waals attraction, cf. Figure 23.3b) were ascribed to rearrangement of the adsorbed polymer [76]. The position of the outer maximum and the onset of the attractive regime were shifted to larger distance with higher molecular weight (Figure 23.3d), but the magnitude of the forces was quite similar to that in the lower molecular weight system (Figure 23.3c).

23.4
Forces in Systems with End-Adsorbed Polymer

If a polymer chain contains an end-group or a block of segments that is different from the rest of the chain, this group may preferentially adsorb on a surface due to electrostatic interaction, chemical bonding, or different solubility in the solvent. End-adsorbed polymers are attached to the surface at only one point, and the extension of the chain, which arises from a balance between elastic energy and conformational entropy, is dependent on the grafting density, that is, on the grafting distance s between end-groups adsorbed on the surface. At low coverage, where there is no overlap between neighboring chains, the thickness of the adsorbed layer is proportional to $N^{1/2}$ in a theta solvent and to $N^{3/5}$ in a good solvent [3, 81]. At higher grafting densities, the chains avoid overlapping one another and will therefore extend further into solution with an adsorbed layer thickness proportional to N [3, 4, 81]. In order of increasing overlap (decreasing grafting distance, i.e., distance between attachment points), these model regions of different stretching are called *"pancake," "mushroom,"* and *"brush"* layers [4]. Typically, there is very little interdigitation between two interacting brush layers (especially for long chains), so that their interaction forces in a good solvent show very little hysteresis between the compression (approach) and separation, and the interaction forces when compressing two brush layers can be based on those obtained when compressing one brush layer against a wall. Based on self-consistent mean-field theory by Dolan and Edwards [2], the interaction forces between a sphere and a flat surface (or between two crossed cylinders), each with a low coverage (Γ) of polymer mushrooms, can be approximated by an exponentially decaying function [23, 40]

$$\frac{F(D)}{R} = 144\pi\Gamma kTe^{-D/Rg}, \quad 2R_g < D < 8R_g \tag{23.4}$$

where Γ is the surface coverage ($\Gamma = s^{-2}$, i.e., the number of molecules per square meter) and D the separation distance between the surfaces. Equation 23.4 is valid when the coverage is sufficiently low so that polymer coils attached to one surface interact only with bare areas on the opposing surface but not with opposing polymer coils. In a reconsideration of the original work by Dolan and Edwards, it has been suggested that the exponential term should be of the form $e^{-\sqrt{3}D/R_g}$, and that the resulting equation is valid in the range $R_g < D < 5R_g$ [40]. In cases where the coverage of polymer mushrooms is higher (but not high enough to induce chain stretching), so that the interaction occurs between polymer coils attached to opposing surfaces, the factor R_g in Equation 23.4 is replaced by $2R_g$ [40]. This model has been successfully applied to low molecular weight systems where the grafting density was very low and in good solvents where R_g was replaced by R_F [37, 38, 40]. Other descriptions of the interactions of layers of polymer mushroom layers have also been put forward based on scaling theory [4].

Several increasingly advanced models have been developed to describe the segment density of densely grafted (strongly stretched) chains in the direction normal to a substrate, and the resulting range and magnitude of the interactions between two brushes or one brush and a bare surface. Loops formed by chains where both ends adsorb to the same surface can also be considered a type of brush layer [82]. One of the earliest models for the interactions of monodisperse brush systems, the Alexander–de Gennes model [3, 4, 81], assumes that the segment density profile is uniform, i.e., a step function. In this approximation, scaling analysis gives the force between a sphere and flat surface (or between two crossed cylinders) covered with polymer brushes as [4]

$$\frac{F(D)}{R} = \frac{16\pi k_B T L_0}{35s^3}\left[7\left(\frac{2L_0}{D}\right)^{5/4} + 5\left(\frac{D}{2L_0}\right)^{7/4} - 12\right], \quad \text{for } D < 2L_0$$

(23.5)

where s is the grafting distance and L_0 is the height of one brush layer ($L_0 = R_F^{5/3}s^{-2/3}$). The first term arises from the osmotic pressure and the second from chain elasticity. Comparisons of this model to experimental data on monodisperse polymer brushes [36, 37, 41, 42] have confirmed its prediction of the brush height increasing linearly with N for a fixed grafting density. In many cases, it also describes the force between two confined brushes well, especially at larger compression (in the region dominated by osmotic interactions).

Mean-field theories developed later have shown that the density profile of strongly stretched monodisperse brushes is parabolic [83–87]. According to the model by Milner *et al.* [83, 84], the interaction force between monodisperse brush layers on crossed cylinders (or a sphere and a flat surface) is obtained from the interaction energy per brush layer for flat plates, $f(u)$, by using the Derjaguin approximation (Equation 23.1) and accounting for the compression of two polymer layers (hence the factor $2 \times 2\pi$):

$$\frac{F(D)}{R} = 4\pi[f(L) - f(L_0)] \quad \text{for } D < 2L_0$$

(23.6a)

using

$$f(u) = \frac{5}{9}f_0\left(\frac{1}{u} + u^2 - \frac{u^5}{5}\right)$$

(23.6b)

$$L_0 = \left(\frac{12}{\pi^2}\right)^{1/3} N \left(\frac{\sigma w}{v}\right)^{1/3} \qquad (23.6c)$$

$$f_0 = \frac{9}{10} \left(\frac{\pi^2}{12}\right)^{1/3} N(\sigma^5 w^2 v)^{1/3} = \frac{9}{10} \frac{(N\sigma)^2 w}{L_0} \qquad (23.6d)$$

where L_0 is the height of one uncompressed brush layer with $\sigma = s^{-2}$ (cf. Γ in Equation 23.4), N is the number of segments, and f_0 is the free energy per unit area of the uncompressed brush (f_0 and $f(u)$ are given in units of kT/area). $u = D/(2L_0)$ for two contacting brush layers, whereas $u = D/L_0$ if only one brush layer is being pressed against a hard wall. w is the excluded volume parameter and v is a parameter related to the statistical segment length, which can be obtained from experimental data on polymers in solution [1, 39, 84, 88]. It is seen from Equation 23.6b that $f(L_0)$ is a constant for a given system, $f(L_0) = (5/9)f_0 \times 1.8$ ($D = 2L_0$, i.e. $u = 1$). These more refined models have been compared to experimental force curves [36, 39, 85, 89], and a model for polydisperse brushes has also been developed [90]. Even a small polydispersity results in a significantly changed segment density profile, where the chain ends segregate along the brush according to chain length. The ends of the short chains are located closer to the bottom, while the longest chains reach to the top. The segment density profile and chemical potential for a polydisperse brush system can be determined numerically based on the molecular weight distribution [39, 84, 90].

Examples of SFA measurements of forces between brush-covered surfaces interacting across a good solvent are shown on semilogarithmic scales in Figure 23.4. Figures 23.4a,b show interactions between low molecular weight polymers in the mushroom and brush grafting density regimes, and Figures 23.4c,d show interactions between higher molecular weight brush layers. Figures 23.4a,b show interactions in water between surfaces covered with monodisperse, end-adsorbed polyethylene glycol (polyethylene oxide) chains with a lipid end-group, inserted into the outer layer in a lipid bilayer on each mica substrate. The grafting density of the chains was varied by selecting different concentrations and surface areas during Langmuir–Blodgett deposition, and was given as the mole% of lipid that carried a polymer chain. The electrostatic double-layer forces [23, 40, 77] in these systems were subtracted from the data to show only the steric forces. The 0.3% case in Figure 23.4a corresponded well to the assumption of very low coverage in Equation 23.4, whereas the 3.0% case represented polymer mushrooms on both surfaces interacting with one another, and required replacement of R_g with $2R_g$ in Equation 23.4. Experimentally determined decay lengths were in good agreement with predicted values of R_g. Figure 23.4b shows interactions in the same system but at a higher grafting density (4.5%), in the brush regime. These data were well described with the Alexander–de Gennes model (Equation 23.5), but could, as expected, not be described with the Dolan–Edwards model for a low density of polymer mushrooms (Equation 23.4). Figure 23.4c shows interaction forces in toluene of end-functionalized polystyrene with a low polydispersity of 1.02, adsorbed on

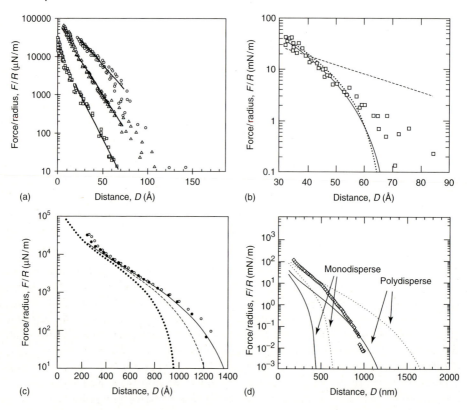

Figure 23.4 Interaction forces between layers of end-adsorbed polymer in good solvents. (a,b) Forces measured in water between low molecular weight ($N = 45$, $M_w \approx$ 2000 g/mol, $M_w/M_n = 1.1$) polyethylene glycol chains with a lipid end-group inserted in lipid bilayers on mica to achieve different grafting densities. (a) 0.3% (squares), 1.5% (triangles), and 3.0% (circles) of the lipid molecules in the outer lipid layer carried a polymer chain. The solid lines are fits of Equation 23.4. Adapted with permission from Ref. [40]. Copyright 2010 American Chemical Society. (b) A higher grafting density of the same polymer as in (a): 4.5% of the lipid molecules carried a polymer chain. The experimental data was compared to calculations with Equation 23.4 (dashed line), 23.5 (dotted curve), and 23.6 (solid curve). Adapted with permission from Ref. [37]. (c) Analysis [85] of data for 1.14×10^5 g/mol end-functionalized polystyrene physisorbed on mica, interacting across toluene (experimental data from Ref. [36]). The dotted and dashed curves show the calculated force according to Equations 23.5 and 23.6, respectively, and the solid curve shows the force according to the extended model [85, 90] that accounts for the polydispersity $M_w/M_n = 1.02$. Reprinted with permission from Ref. [85]. (d) Measured (symbols) and theoretically predicted (curves) interaction forces on compression of two polystyrene brush layers in toluene with $M_n = 7 \times 10^5$ g/mol, $M_w/M_n \sim 2$, and grafting density 1.4×10^{16} m^{-2}. Curves were calculated for monodisperse brushes according to Equation 23.6, and for polydisperse brushes according to Ref. [90], with an excluded volume of $w = (0.23$ nm$)^3$ (solid curves) and $w = (0.32$ nm$)^3$ (dotted curves). Reprinted with permission from Ref. [39]. Copyright 2000 American Chemical Society.

mica (experimental data from [36]). Filled and open symbols show data on compression and decompression, respectively. The dotted curve shows the calculated force assuming a step-function segment density profile (Equation 23.5), and the dashed curve shows the corresponding result for a monodisperse polymer with a parabolic density profile (Equation 23.6). The solid curve was calculated using the extended model [85, 90] that accounts for the polydispersity $M_w/M_n = 1.02$. Ref. [85] and Figure 23.4c show that even a small polydispersity may affect the brush height by as much as 10% compared to the ideal case of a perfectly monodisperse system (theoretically, the brush height is predicted to exceed that of the monodisperse by a factor $\sim(M_w/M_n - 1)^{1/2}$ [85]). Larger polydispersities are typical for most synthetic polymers and will have a very strong effect on the segment density, brush height, and interaction forces. Figure 23.4d shows interaction forces [39] in dry toluene (good solvent conditions) between high molecular weight polydisperse polystyrene brush layers formed through a "grafting-from" technique on quartz-glass substrates, involving radical chain polymerization from an initiator layer adsorbed on the substrate [91, 92]. In contrast to physisorbed systems, where the distance between grafted, long chains is expected to scale with molecular weight as $N^{3/5}$ for polymer mushrooms [3], different, high grafting densities for high molecular weights could be chosen by varying the polymerization conditions [91, 92]. The grafting densities in this type of system significantly exceed those obtainable by the physisorption and "grafting-to" of functionalized long chains from solution shown in Figure 23.4c. Dry layer thicknesses of more than 100 nm of polystyrene with a molecular weight around 1×10^6 g/mol and a grafting distance between chains as low as 3 to 4 nm could be formed on large planar surfaces [93]. As expected, calculations for monodisperse brushes using Equation 23.6 (giving the two curves on the left in Figure 23.4d) did not describe the interactions in these polydisperse systems. Instead, a Flory–Schulz molecular weight distribution was used as the starting point for calculations ([39] following Refs. [85, 90]) of the segment density profile and the interaction forces shown on the right side of Figure 23.4d. In these high molecular weight polydisperse systems ($M_n \approx 700\,000$ g/mol, $M_w/M_n \sim 2$), the segment density was almost exponential [39], and thus differed very strongly from the step function or parabolic density profile assumed for monodisperse brushes. At large compression, the prediction for the polydisperse system merged with that for the monodisperse, since in this limit, the interactions are dominated by the osmotic pressure. At large compression (small D), the measured forces exceeded the theoretical ones. In this region of highest pressure (largest force), the substrates (especially the glue layer holding the quartz-glass substrates to which the polystyrene brushes were anchored) may deform, which was not accounted for when dividing the measured force by the radius of curvature of the undeformed surfaces (F/R), and the region of highest pressure in Figure 23.4d ($F/R \geq 100$ mN/m) may thus be overestimated by a factor 2 to 3. However, on closer inspection, the slope of the force curve exceeded the slope of the calculated curves for polydisperse systems in a larger range of film thicknesses (distances), possibly due to three-body interactions due to the high concentration of polymer segments encountered in the compressed brushes [39]. A steeper increase of the interaction force with compression than the one predicted from self-consistent field theory has also been found in a molecular dynamics study on monodisperse brushes [94], where it was concluded that the deviations were caused by a monomer density above the semidilute regime.

The main differences in the interactions between polymer brushes and adsorbed homopolymer layers are the reduction or absence of hysteresis in the force–distance curve on approach and separation of brush layers, and the lack of interpenetration and entanglement due to the strong stretching of the chains. Bridging is also avoided with polymer brush layers, which makes them highly important for the modification of surfaces.

23.5
Interactions between Dry Polymer Layers: Adhesion

The adhesion of polymer layers in the absence of solvent is of importance for the function of adhesives and surface coatings. Some applications require a high adhesion, whereas in others it is important with a temporary adhesion and rapid release of two surfaces (two polymer layers, or polymer in contact with another material) [95]. Layers of homopolymer, end-functionalized polymer, and block-copolymer adsorbed from solution collapse when the solvent is removed, and form smooth, unstructured layers [43, 46, 96] or a large variety of microstructures [97, 98], depending on their composition. Typically, no long-range forces (other than the van der Waals attraction) are measured on approach of dry layers, but for contacting polymer layers above their glass-transition temperature, a number of interesting effects are seen on separation. The strength (and sometimes the distance range) of the adhesion measured on separation is often observed to exceed the value expected from van der Waals interactions [99–104], and may depend on the polymer chain length, mobility, and layer thickness, and on the time in contact and separation rate. The polymer layers may separate along their original contact interfaces (adhesive failure) or within one or both of the layers (cohesive failure). The latter is associated with a transfer of material from one surface to the other or with crazing, that is, strand formation and breaking. In both adhesive and cohesive failure, there can be significant additional energy dissipation (viscoelastic loss) within the films underneath the immediate contact interface, ahead of an opening crack, or even in the supporting materials [95, 99–105]. After the separation has occurred, the layers may relax back to their original conformation, or remain altered (over time scales of practical relevance) with chains pulled out, bumps, or material transferred toward the contact through flow along each surface. Many of these phenomena are also known for liquid or amorphous films of small molecules, but can be observed on different time and rate scales for polymers.

Some well-controlled experiments on polymer adhesion have been performed using half-spherical caps of elastomers in contact with a flat surface [106, 107]. The elastomer conformed to the substrate so that molecular contact was obtained and effects of surface roughness, which typically acts to decrease the adhesion [27, 108, 109], were avoided. Similarly, the smooth, elastically deformable substrates in the SFA make it possible to bring adsorbed layers in molecular contact, provided that the layers themselves are smooth and compliant. The adhesion and linearly elastic deformation of the bare substrates in the SFA is well known [110, 111], and deviations from this response can be ascribed to events within the adsorbed layers [25, 31, 34, 35, 46]. SFA experiments provide information on the adhesion hysteresis [31, 34] and pull-off force (the force at the point of separation between the surfaces) of dry layers [31, 46, 56]. Bulk contact mechanics models [112, 113] have been used to relate the observed changes in contact area and the measured forces to the surface energy, and values obtained for glassy polymers [46, 56] and surfactant monolayers [25, 26] have been found to agree well with values determined by other techniques. Details of these models and how they relate to SFA measurements are discussed in Refs. [23, 46], and only some basic equations are given here. According to the Derjaguin–Muller–Toporov theory (DMT) [112], the relationship between the pull-off force and the surface energy, γ, is given by (cf. Equation 23.1 and Refs. [23, 112]):

$$\frac{F(D)}{R} = -2\pi W(D) = -4\pi\gamma \tag{23.7}$$

whereas according to the Johnson–Kendall–Roberts theory (JKR) [23, 113],

$$\frac{F(D)}{R} = -\frac{3}{2}\pi W(D) = -3\pi\gamma \tag{23.8}$$

Recent extensions of these models address the situation of thin, compliant films confined between substrates of higher rigidity [114–118].

Examples of the adhesion between dry polymer layers are shown in Figure 23.5. Figure 23.5a shows the effective surface energy $\gamma_{\text{eff}} = -F/(3\pi R)$ (cf. Equation 23.8) between two circa 2-μm thick layers of poly(butylmethacrylate) ($M_{\text{w}} = 93\,200$ g/mol, glass-transition temperature $\sim$25 $^\circ$C) adsorbed on mica from methyl ethyl ketone solution [34]. At low temperatures (well below the glass-transition temperature), the polymer layers were unable to interdigitate, and the measured surface energy was similar to that determined by other methods, $\gamma = 32$ mJ/m^2. At the chosen separation rate, 2 mN/s, a maximum in the adhesion was found slightly above the glass-transition temperature. At the highest temperatures in Figure 23.5a, the polymer layers interdigitated readily, but the increased chain mobility also allowed them to separate more easily than in the intermediate temperature range. Figure 23.5b shows the normalized adhesion between dry poly-2-vinylpyridine–polybutadiene block-copolymer brush layers adsorbed on mica from toluene (a good solvent for polybutadiene, so that the polybutadiene block forms the outer region of the adsorbed and dried layer [96]). The M_{w} of the polyvinylpyridine and polybutadiene blocks were 23\,700 g/mol (polydispersity 1.05) and 38\,500 g/mol (polydispersity 1.07), respectively. Separation of the surfaces was done after different contact times (0.01, 1, 100, and 500 s) and at different separation rates (corresponding to the rate of change in the normal force) [46]. The large range of separation rates was obtained by attaching piezoelectric bimorph

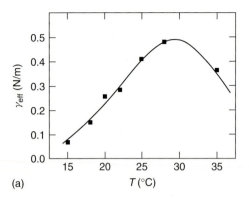

(a)

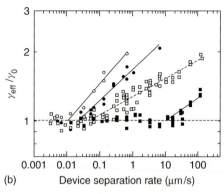

(b)

Figure 23.5 Adhesion between dry polymer layers. (a) Effective surface energy $\gamma_{eff} = -F/3\pi R$ between 2-μm thick poly(butyl methacrylate) layers on mica as a function of temperature at a separation rate (rate of change in normal force) of 2 mN/s. A maximum is seen in the adhesion slightly above the glass-transition temperature of circa 25 °C. Adapted with permission from Ref. [34]. Copyright 1998 American Chemical Society.

(b) Normalized adhesion at 23 °C (γ_{eff}/γ_0, $\gamma_0 = -F_0/3\pi R = 55$ mN/m) between dry poly-2-vinylpyridine–polybutadiene block-copolymer layers (thickness 2.8 nm) as a function of device separation rate (proportional to the rate of change in normal force). The contact times before separation were 0.01 s (filled squares), 1 s (open squares), 100 s (filled circles), and 500 s (open circles). Reprinted with permission from Ref. [46]. Copyright 1998 American Chemical Society.

elements (one actuator and one detector) to the leaf springs supporting the lower surface in the SFA [17, 44, 45]. The pull-off force was normalized by the pull-off force at slow separation so that Figure 23.5b shows an effectively increased surface energy $\gamma_{eff} = -F/(3\pi R)$ normalized by the constant value obtained at the slowest separations, $\gamma_0 = -F_0/(3\pi R) = 55$ mJ/m². This value of γ_0 was higher than that expected for polybutadiene (31–43 mJ/m²). A similar observation of a high γ_0 had been made previously for dry layers of poly-2-vinylpyridine–polyisoprene [42]. The interference fringe patterns (not shown) in the system in Figure 23.5b indicated that the surfaces separated from a small point, not from the finite contact area assumed in the JKR theory (Equation 23.8), and the value of γ_0 may thus be an overestimate. Such a situation where the separation occurs once the contact area has decreased to zero is better described by Equation 23.7, which would give $\gamma_0 \approx 41$ mJ/m². It was proposed that segmental mobility in the polybutadiene may be sufficient to cause this type of separation [46]. Similar experiments on dry poly-2-vinylpyridine–polystyrene layers (not shown), where both the anchoring block and the exposed polystyrene were glassy at room temperature, showed no dependence on contact time and rate within the range shown in Figure 23.5b. The separation occurred from a finite contact area, and γ_0 was 37 mJ/m², in good agreement with values measured by other techniques. This is an example of how the optical detection of separation distances and surface shapes in the SFA can be helpful when studying the interaction between surfaces.

23.6
Importance of Polymer Interactions for Applications

In many applications it is highly important that adsorbed molecules (surface coatings or lubricants) remain localized at a solid surface when exposed to different solvents or to external pressure from another surface. For example, to avoid aging effects in suspensions or to promote adhesion, it is often desirable to modify interfaces uniformly and permanently with selectively adsorbed or grafted (end-adsorbed) polymer. Polymer layers are often used to stabilize suspensions of colloidal particles against coagulation due to van der Waals forces. If bridging can be avoided, as is typically the case with dense polymer layers in a good solvent (especially polymer brushes), a steric repulsion arises due to compression and osmotic effects as the particles come close together (cf. Section 23.4). The dielectric properties of the polymer may also be chosen to be similar to those of the solvent ("refractive-index matching"), which strongly reduces the van der Waals attraction to a point where the particles stay dispersed due to thermal motion. The opposite effect, a precipitation of colloidal material in solution, can be achieved through bridging by high molecular weight polymers (cf. Section 23.3.1). In tribological (friction and lubrication) applications, polymer brush layers are of importance for the prevention of adhesion and wear, and experimental studies have shown that covalently bound lubricant layers are more effective in protecting sliding surfaces against wear than physisorbed molecules. It has recently been found that layers of charged polymers (polyelectrolytes) in aqueous solutions may give ultralow friction of potential importance for the lubrication of biological systems (e.g., joints) [119]. Since it has been suggested that there is a connection between friction and adhesion (or adhesion hysteresis) [23, 27], this implies further use of versatile polymer coatings to control interactions in other solvents and between other types of surfaces.

Acknowledgments

I would like to thank J. Israelachvili, S. Granick, D. Johannsmann, J. Rühe, and W. Knoll for collaborations and many helpful discussions. The data shown in Figure 23.4a were kindly provided by F. Li and F. Pincet. Financial support through NSF CAREER Award NSF-CMMI #0645065 is gratefully acknowledged.

References

1. Flory, P. (1953) *Principles of Polymer Chemistry*, Cornell University Press, Ithaca, NY.
2. Dolan, A.K. and Edwards, S.F. (1974) Theory of the stabilization of colloids by adsorbed polymer. *Proc. R. Soc. Lond. A*, **337**, 509–516.
3. Alexander, S. (1977) Adsorption of chain molecules with a polar head – a scaling description. *J. Phys. (Paris)*, **38**, 983–987.
4. de Gennes, P.G. (1987) Polymers at an interface: a simplified view. *Adv. Colloid Interface Sci.*, **27**, 189–209.

5. See, for example, Chakraborty, A.K. and Tirrell, M. (1996) Polymer adsorption. *MRS Bull.*, **21**, 28–32.

6. Dijt, J.C., Cohen Stuart, M.A., and Fleer, G.J. (1994) Kinetics of adsorption and desorption of polystyrene on silica from decalin. *Macromolecules*, **27**, 3207–3218.

7. Israelachvili, J.N. (1973) Thin film studies using multiple-beam interferometry. *J. Colloid Interface Sci.*, **44**, 259–272.

8. Horn, R.G. and Smith, D.T. (1991) Analytic solution for the three-layer multiple beam interferometer. *Appl. Opt.*, **30**, 59–65.

9. Heuberger, M., Luengo, G., and Israelachvili, J. (1997) Topographic information from multiple beam interferometry in the Surface Forces Apparatus. *Langmuir*, **13**, 3839–3848.

10. Israelachvili, J.N. and Adams, G.E. (1978) Measurement of forces between two mica surfaces in aqueous electrolyte solutions in the range 0–100 nm. *J. Chem. Soc., Faraday Trans. 1*, **74**, 975–1001.

11. Parker, J.L., Christenson, H.K., and Ninham, B.W. (1989) Device for measuring the force and separation between two surfaces down to molecular separations. *Rev. Sci. Instrum.*, **60**, 3135–3138.

12. Israelachvili, J.N. and McGuiggan, P.M. (1990) Adhesion and short-range forces between surfaces. Part I: new apparatus for surface force measurements. *J. Mater. Res.*, **5**, 2223–2231.

13. Israelachvili, J., Min, Y., Akbulut, M., Alig, A., Carver, G., Greene, W., Kristiansen, K., Meyer, E., Pesika, N., Rosenberg, K., and Zeng, H. (2010) Recent advances in the surface forces apparatus (SFA) technique. *Rep. Prog. Phys.*, **73**, 036601-1–036601-16.

14. Klein, J. (1983) Forces between mica surfaces bearing adsorbed macromolecules in liquid media. *J. Chem. Soc., Faraday Trans. 1*, **79**, 99–118.

15. Klein, J. and Kumacheva, E. (1998) Simple liquids confined to molecularly thin layers. I. Confinement-induced liquid-to-solid phase transitions. *J. Chem. Phys.*, **108**, 6996–7009.

16. Israelachvili, J.N., Kott, S.J., and Fetters, L.J. (1989) Measurements of dynamic interactions in thin films of polymer melts: the transition from simple to complex behavior. *J. Polym. Sci.: Polym. Phys.*, **27**, 489–502.

17. Dhinojwala, A. and Granick, S. (1996) New approaches to measure interfacial rheology of confined fluids. *J. Chem. Soc. Faraday Trans.*, **92**, 619–623.

18. (a) Parker, J.L. (1992) A novel method for measuring the force between two surfaces in a surface forces apparatus. *Langmuir*, **8**, 551–556; (b) Parker, J.L. (1994) Surface force measurements in surfactant systems. *Prog. Surf. Sci.*, **47**, 205–271.

19. Luengo, G., Schmitt, F.-J., Hill, R.S., and Israelachvili, J. (1997) Thin film rheology and tribology of confined polymer melts: Contrast with bulk properties. *Macromolecules*, **30**, 2482–2494.

20. Homola, A.M., Israelachvili, J.N., Gee, M.L., and McGuiggan, P.M. (1989) Measurements of a relation between the adhesion and friction of two surfaces separated by molecularly thin liquid films. *J. Tribol.*, **111**, 675–682.

21. Kumacheva, E. (1998) Interfacial friction measurements in Surface Forces Apparatus. *Prog. Surf. Sci.*, **58**, 75–120.

22. Deryagin, B.V. (1934) Untersuchungen über die Reibung und Adhäsion. IV. Teorie des Anhaftens kleiner Teilchen. *Kolloid-Z.*, **69**, 155–164.

23. Israelachvili, J.N. (2011) *Intermolecular and Surface Forces*, 3rd edn, Academic Press.

24. Radoslovich, E.W. (1960) The structure of muscovite, $KAl_2(Si_3Al)O_{10}(OH)_2$. *Acta Cryst.*, **13**, 919–932.

25. Chen, Y.-L., Helm, C.A., and Israelachvili, J.N. (1991) Molecular mechanisms associated with adhesion and contact angle hysteresis of monolayer surfaces. *J. Phys. Chem.*, **95**, 10736–10747.

26. Peanasky, J., Schneider, H.M., Granick, S., and Kessel, C.R. (1995) Self-assembled monolayers on mica for experiment utilizing the surface forces apparatus. *Langmuir*, **11**, 953–962.

27. See, for example, Ruths, M. and Israelachvili, J.N. (2010) Surface forces and nanorheology of molecularly thin films, in *Springer Handbook of Nanotechnology*, 3rd edn, Chapter 29 (ed. B. Bhushan), Springer-Verlag, Berlin and Heidelberg, pp. 857–922.

28. Israelachvili, J.N., Tirrell, M., Klein, J., and Almog, Y. (1984) Forces between two layers of adsorbed polystyrene immersed in cyclohexane below and above the θ temperature. *Macromolecules*, **17**, 204–209.

29. Patel, S.S. and Tirrell, M. (1989) Measurement of forces between surfaces in polymer fluids. *Annu. Rev. Phys. Chem.*, **40**, 597–635.

30. Hu, H.W., Van Alsten, J., and Granick, S. (1989) Influence of solvent quality on surface forces between polystyrene layers adsorbed on mica. *Langmuir*, **5**, 270–272.

31. Schmitt, F.-J., Yoshizawa, H., Schmidt, A., Duda, G., Knoll, W., Wegner, G., and Israelachvili, J. (1995) Adhesion energy hysteresis and friction between ultrathin polyglutamate films measured with the surface forces apparatus. *Macromolecules*, **28**, 3401–3410.

32. Klein, J. (1996) Shear, friction and lubrication forces between polymer-bearing surfaces. *Annu. Rev. Mater. Sci.*, **26**, 581–612.

33. Ruths, M., Israelachvili, J.N., and Ploehn, H.J. (1997) Effects of time and compression on the interactions of adsorbed polystyrene layers in a near-θ solvent. *Macromolecules*, **30**, 3329–3339.

34. Luengo, G., Pan, J., Heuberger, M., and Israelachvili, J.N. (1998) Temperature and time effects on the "adhesion dynamics" of poly(butyl methacrylate) (PBMA) surfaces. *Langmuir*, **14**, 3873–3881.

35. Maeda, N., Chen, N., Tirrell, M., and Israelachvili, J.N. (2002) Adhesion and friction mechanisms of polymer-on-polymer surfaces. *Science*, **297**, 379–382.

36. Taunton, H.J., Toprakcioglu, C., Fetters, L.J., and Klein, J. (1990) Interactions between surfaces bearing end-adsorbed chains in a good solvent. *Macromolecules*, **23**, 571–580.

37. Kuhl, T.L., Leckband, D.E., Lasic, D.D., and Israelachvili, J.N. (1994) Modulation of interaction forces between bilayers exposing short-chained ethylene oxide headgroups. *Biophys. J.*, **66**, 1479–1488.

38. Kuhl, T.L., Leckband, D.E., Lasic, D.D., and Israelachvili, J.N. (1995) Modulation and modeling of interaction forces between lipid bilayers exposing terminally grafted polymer chains, in *Stealth Liposomes*, Chapter 8 (eds D. Lasic and F. Martin), CRC Press, Boca Raton, FL, pp. 73–91.

39. Ruths, M., Johannsmann, D., Rühe, J., and Knoll, W. (2000) Repulsive forces and relaxation on compression of entangled, polydisperse polystyrene brushes. *Macromolecules*, **33**, 3860–3870.

40. Li, F. and Pincet, F. (2007) Confinement free energy of surfaces bearing end-grafted polymers in the mushroom regime and local measurement of the polymer density. *Langmuir*, **23**, 12541–12548.

41. Hadziioannou, G., Patel, S., Granick, S., and Tirrell, M. (1986) Forces between surfaces of block copolymers on mica. *J. Am. Chem. Soc.*, **108**, 2869–2876.

42. Watanabe, H. and Tirrell, M. (1993) Measurement of forces in symmetric and asymmetric interactions between diblock copolymer layers adsorbed on mica. *Macromolecules*, **26**, 6455–6466.

43. Watanabe, H., Matsuyama, S., Mizutani, Y., and Kotaka, T. (1995) Adhesion of thin, dry block copolymer layers adsorbed on mica. *Macromolecules*, **28**, 6454–6461.

44. Dhinojwala, A. and Granick, S. (1997) Surface forces in the tapping mode: solvent permeability and hydrodynamic thickness of adsorbed polymer brushes. *Macromolecules*, **30**, 1079–1085.

45. Cho, Y.-K., Dhinojwala, A., and Granick, S. (1997) Apparent hydrodynamic thickness of densely grafted polymer layers in a theta

solvent. *J. Polym. Sci.: Polym. Phys.*, **35**, 2961–2968.

46. Ruths, M. and Granick, S. (1998) Rate-dependent adhesion between polymer and surfactant monolayers on elastic substrates. *Langmuir*, **14**, 1804–1814.

47. Berman, A., Steinberg, S., Campbell, S., Ulman, A., and Israelachvili, J.N. (1998) Controlled microtribology of a metal oxide surface. *Tribol. Lett.*, **4**, 43–48.

48. Vigil, G., Xu, Z., Steinberg, S., and Israelachvili, J. (1994) Interactions of silica surfaces. *J. Colloid Interface Sci.*, **165**, 367–385.

49. Parker, J.L., Claesson, P.M., Cho, D.L., Ahlberg, A., Tidblad, J., and Blomberg, E. (1990) Plasma modification of mica. *J. Colloid Interface Sci.*, **134**, 449–458.

50. Ruths, M., Heuberger, M., Scheumann, V., Hu, J., and Knoll, W. (2001) Confinement-induced film thickness transitions in liquid crystals between two alkanethiol monolayers on gold. *Langmuir*, **17**, 6213–6219.

51. Levins, J.M. and Vanderlick, T.K. (1995) Impact of roughness on the deformation and adhesion of a rough metal and smooth mica in contact. *J. Phys. Chem.*, **99**, 5067–5076.

52. Chai, L. and Klein, J. (2007) Large-area, molecularly smooth (0.2 nm rms) gold films for surface forces and other studies. *Langmuir*, **23**, 7777–7783.

53. Horn, R.G., Clarke, D.R., and Clarkson, M.T. (1988) Direct measurement of surface forces between sapphire crystals in aqueous solutions. *J. Mater. Res.*, **3**, 413–416.

54. Ducker, W.A., Xu, Z., Clarke, D.R., and Israelachvili, J.N. (1994) Forces between alumina surfaces in salt solutions: non-DLVO forces and the implications for colloidal processing. *J. Am. Ceram. Soc.*, **77**, 437–443.

55. Horn, R.G., Smith, D.T., and Haller, W. (1989) Surface forces and viscosity of water measured between silica sheets. *Chem. Phys. Lett.*, **162**, 404–408.

56. Mangipudi, V., Tirrell, M., and Pocius, A.V. (1994) Direct measurement of molecular level adhesion between

poly(ethylene terephthalate) and polyethylene films: determination of surface and interfacial energies. *J. Adhes. Sci. Technol.*, **8**, 1251–1270.

57. Merrill, W.W., Pocius, A.V., Thakker, B.V., and Tirrell, M. (1991) Direct measurement of molecular level adhesion forces between biaxially oriented solid polymer films. *Langmuir*, **7**, 1975–1980.

58. de Gennes, P.G. (1981) Polymer solutions near an interface: 1. Adsorption and depletion layers. *Macromolecules*, **14**, 1637–1644.

59. de Gennes, P.G. (1982) Polymers at an interface. 2. Interaction between two plates carrying adsorbed polymer layers. *Macromolecules*, **15**, 492–500.

60. Scheutjens, J.M.H.M. and Fleer, G.J. (1985) Interaction between two adsorbed polymer layers. *Macromolecules*, **18**, 1882–1900.

61. Semenov, A.N. and Joanny, J.-F. (1995) Structure of adsorbed polymer layers: loops and tails. *Europhys. Lett.*, **29**, 279–284.

62. Ploehn, H.J. and Russell, W.B. (1989) Self-consistent field model of polymer adsorption: matched asymptotic expansion describing tails. *Macromolecules*, **22**, 266–276.

63. Klein, J. and Pincus, P. (1982) Interaction between surfaces with adsorbed polymers: poor solvents. *Macromolecules*, **15**, 1129–1135.

64. Ingersent, K., Klein, J., and Pincus, P. (1990) Forces between surfaces with adsorbed polymers. 3. θ solvent. Calculations and comparisons with experiments. *Macromolecules*, **23**, 548–560.

65. Evans, E.A. (1989) Force between surfaces that confine a polymer solution: derivation from self-consistent field theories. *Macromolecules*, **22**, 2277–2286.

66. Ploehn, H.J. (1994) Compression of polymer interphases. *Macromolecules*, **27**, 1627–1636.

67. Ennis, J. and Jönsson, B. (1999) Interactions between surfaces in the presence of ideal adsorbing block copolymers. *J. Phys. Chem. B.*, **103**, 2248–2255.

68. Saeki, S., Kuwahara, N., Konno, S., and Kaneko, M. (1973) Upper and lower critical solution temperatures in polystyrene solutions. II. *Macromolecules*, **6**, 589–593.

69. Asakura, S. and Oosawa, F. (1958) Interaction between particles suspended in solutions of macromolecules. *J. Polym. Sci.*, **33**, 183–192.

70. Joanny, J.F., Leibler, L., and de Gennes, P.G. (1979) Effects of polymer solutions on colloid stability. *J. Polym. Sci.: Polym. Phys.*, **17**, 1073–1084.

71. Vincent, B., Luckham, P.F., and Waite, F.A. (1980) The effect of free polymer on the stability of sterically stabilized dispersions. *J. Colloid Interface Sci.*, **73**, 508–521.

72. Feigin, R.I. and Napper, D.H. (1980) Stabilization of colloids by free polymer. *J. Colloid Interface Sci.*, **74**, 567–571.

73. Feigin, R.I. and Napper, D.H. (1980) Depletion stabilization and depletion flocculation. *J. Colloid Interface Sci.*, **75**, 525–541.

74. Evans, E. and Needham, D. (1988) Attraction between lipid bilayer membranes in concentrated solutions of nonadsorbing polymer: comparison of mean-field theory with measurements of adhesion energy. *Macromolecules*, **21**, 1822–1831.

75. Fleer, G.J. and Tuinier, R. (2005) Concentration and solvency effects on polymer depletion and the resulting pair interaction of colloidal particles in a solution of non-adsorbing polymer. *Polymer Prepr.*, **46**, 366.

76. Ruths, M., Yoshizawa, H., Fetters, L.J., and Israelachvili, J.N. (1996) Depletion attraction versus steric repulsion in a system of weakly adsorbing polymer –effects of concentration and adsorption conditions. *Macromolecules*, **29**, 7193–7203.

77. Kuhl, T., Guo, Y., Alderfer, J.L., Berman, A.D., Leckband, D., Israelachvili, J., and Hui, S.W. (1996) Direct measurement of polyethylene glycol induced depletion attraction between lipid bilayers. *Langmuir*, **12**, 3003–3014.

78. Perez, E. and Proust, J.E. (1985) Effects of a non-adsorbing polymer on colloid stability: force measurements between mica surfaces immersed in dextran solution. *J. Phys. Lett.*, **46**, L79–L84.

79. Mays, J.W. and Fetters, L.J. (1989) Temperature coefficients of unperturbed dimensions for atactic polypropylene and alternating poly(ethylene-propylene). *Macromolecules*, **22**, 921–926.

80. Tong, P., Witten, T.A., Huang, J.S., and Fetters, L.J. (1990) Interactions in mixtures of colloid and polymer. *J. Phys. (Paris)*, **51**, 2813–2827.

81. de Gennes, P.G. (1980) Conformations of polymers attached to an interface. *Macromolecules*, **13**, 1069–1075.

82. Milner, S.T. and Witten, T.A. (1992) Bridging attraction by telechelic polymers. *Macromolecules*, **25**, 5495–5503.

83. Milner, S.T., Witten, T.A., and Cates, M.E. (1988) Theory of the grafted polymer brush. *Macromolecules*, **21**, 2610–2619.

84. Milner, S.T., Witten, T.A., and Cates, M.E. (1988) A parabolic density profile for grafted polymers. *Europhys. Lett.*, **5**, 413–418.

85. Milner, S.T. (1988) Compressing polymer "brushes": a quantitative comparison of theory and experiment. *Europhys. Lett.*, **7**, 695–699.

86. Zhulina, E.B., Borisov, O.V., and Priamitsyn, V.A. (1990) Theory of steric stabilization of colloid dispersions by grafted polymers. *J. Colloid Interface Sci.*, **137**, 495–511.

87. Zhulina, E.B., Borisov, O.V., Pryamitsyn, V.A., and Birshtein, T.M. (1991) Coil-globule transition in polymers. 1. Collapse of layers of grafted polymer chains. *Macromolecules*, **24**, 140–149.

88. See, for example Brandrup, J., Immergut, E.H., Grulke, E.A., Abe, A., and Bloch, D.R. (eds) (2005) *Polymer Handbook*, 4th edn, John Wiley & Sons, Inc. (online edition).

89. Klein, J., Kumacheva, E., Mahalu, D., Perahia, D., and Fetters, L.J. (1994) Reduction of frictional forces between solid surfaces bearing polymer brushes. *Nature*, **370**, 634–636.

90. Milner, S.T., Witten, T.A., and Cates, M.E. (1989) Effects of polydispersity in the end-grafted polymer brush. *Macromolecules*, **22**, 853–861.

91. Prucker, O. and Rühe, J. (1998) Synthesis of poly(styrene) monolayers attached to high surface area silica gels through self-assembled monolayers of azo initiators. *Macromolecules*, **31**, 592–601.

92. Prucker, O. and Rühe, J. (1998) Mechanism of radical chain polymerizations initiated by azo compounds covalently bound to the surface of spherical particles. *Macromolecules*, **31**, 602–613.

93. Prucker, O. and Rühe, J. (1998) Polymer layers through self-assembled monolayers of initiators. *Langmuir*, **14**, 6893–6898.

94. Murat, M. and Grest, G.S. (1989) Interaction between grafted polymeric brushes: a molecular dynamics study. *Phys. Rev. Lett.*, **63**, 1074–1077.

95. See, for example, Brown, H.B. (1996) Adhesion of polymers. *MRS Bull.*, **21**, 24–27.

96. Watanabe, H., Shimura, T., Kotaka, T., and Tirrell, M. (1993) Synthesis, characterization, and surface structures of styrene–2-vinylpyridine–butadiene three-block polymers. *Macromolecules*, **26**, 6338–6345.

97. Kelley, T.W., Schorr, P.A., Johnson, K.D., Tirrell, M., and Frisbie, C.D. (1998) Direct force measurements at polymer brush surfaces by atomic force microscopy. *Macromolecules*, **31**, 4297–4300.

98. Fasolka, M.J. and Mayes, A.M. (2001) Block copolymer thin films: physics and applications. *Annu. Rev. Mater. Res.*, **31**, 323–355.

99. Johnson, K.L. (1996) Continuum mechanics modeling of adhesion and friction. *Langmuir*, **12**, 4510–4513.

100. Gent, A.N. and Kinloch, A.J. (1971) Adhesion of viscoelastic materials to rigid substrates. III. Energy criterion for failure. *J. Polym. Sci. A-2*, **9**, 659–668.

101. Gent, A.N. and Schultz, J. (1972) Effects of wetting liquids on the strength of adhesion of viscoelastic materials. *J. Adhes.*, **3**, 281–294.

102. Gent, A.N. (1996) Adhesion and strength of viscoelastic solids. Is there a relationship between adhesion and bulk properties? *Langmuir*, **12**, 4492–4496.

103. Brown, H.R. (1991) The adhesion between polymers. *Annu. Rev. Mater. Sci.*, **21**, 463–489.

104. Brown, H.R. and Yang, A.C.M. (1992) The use of peel tests to examine the self-adhesion of polyimide films. *J. Adhes. Sci. Technol.*, **6**, 333–346.

105. Baljon, A.R.C. and Robbins, M.O. (1996) Energy dissipation during rupture of adhesive bonds. *Science*, **271**, 482–484.

106. Chaudhury, M.K. and Whitesides, G.M. (1991) Direct measurement of interfacial interactions between semispherical lenses and flat sheets of poly(dimethyl siloxane) and their chemical derivatives. *Langmuir*, **7**, 1013–1025.

107. Mangipudi, V.S., Huang, E., Tirrell, M., and Pocius, A.V. (1996) Measurement of interfacial adhesion between glassy polymer using the JKR method. *Macromol. Symp.*, **102**, 131–143.

108. Maugis, D. and Gauthier-Manuel, B. (1994) JKR–DMT transition in the presence of a liquid meniscus. *J. Adhes. Sci. Technol.*, **8**, 1311–1322.

109. Creton, C. and Leibler, L. (1996) How does tack depend on time of contact and contact pressure? *J. Polym. Sci.: Polym. Phys.*, **34**, 545–554.

110. Horn, R.G., Israelachvili, J.N., and Pribac, F. (1987) Measurement of the deformation and adhesion of solids in contact. *J. Colloid Interface Sci.*, **115**, 480–492.

111. McGuiggan, P.M., Wallace, J.S., Smith, D.T., Sridhar, I., Zheng, Z.W., and Johnson, K.L. (2007) Contact mechanics of layered elastic materials: experiment and theory. *J. Phys. D: Appl. Phys.*, **40**, 5984–5994.

112. Derjaguin, B.V., Muller, V.M., and Toporov, Yu.P. (1975) Effect of contact deformations on the adhesion of particles. *J. Colloid Interface Sci.*, **53**, 314–326.

113. Johnson, K.L., Kendall, K., and Roberts, A.D. (1971) Surface energy and the contact of elastic solids. *Proc. R. Soc. Lond. A*, **324**, 301–313.

114. Sridhar, I., Johnson, K.L., and Fleck, N.A. (1997) Adhesion mechanics of the surface force apparatus. *J. Phys. D: Appl. Phys.*, **30**, 1710–1719.

115. Johnson, K.L. and Sridhar, I. (2001) Adhesion between a spherical indenter and an elastic solid with a compliant elastic coating. *J. Phys. D: Appl. Phys.*, **34**, 683–689.

116. Sridhar, I., Zheng, Z.W., and Johnson, K.L. (2004) A detailed analysis of adhesion mechanics between a compliant elastic coating and a spherical probe. *J. Phys. D: Appl. Phys.*, **37**, 2886–2895.

117. Reedy, E.D. Jr. (2006) Thin-coating contact mechanics with adhesion. *J. Mater. Res.*, **21**, 2660–2668.

118. Reedy, E.D. Jr. (2007) Contact mechanics for coated spheres that includes the transition from weak to strong adhesion. *J. Mater. Res.*, **22**, 2617–2622.

119. Raviv, U., Giasson, S., Kampf, N., Gohy, J.-F., Jérôme, R., and Klein, J. (2003) Lubrication by charged brushes. *Nature*, **425**, 163–165.

24

Biomimetic Thin Films as a QCM-D Sensor Platform to Detect Macromolecular Interactions

Nam-Joon Cho and Curtis W. Frank

24.1
Introduction

Biomimetic thin films on solid supports have a wide range of applications including biosensors, artificial photosynthesis platforms, and biocompatible, nonfouling substrates [1–6]. In this chapter, we discuss their utility as sensor platforms to investigate biomacromolecular interactions with real-time monitoring. From antibody binding [7] to enzymatic reactions [8], these platforms can be used together with a variety of surface-sensitive analytical techniques to detect interaction kinetics and associated changes in thin-film properties. One important consideration for detection of a specific macromolecular recognition event is the structural properties of the platform itself. Thus, the ability to control these properties is an important design parameter.

As one type of biomimetic thin film, solid-supported phospholipid structures can be formed in a controlled fashion in order to mimic key architectural features of the cell membrane [1–6]. The most popular method to form these cell-membrane-mimicking platforms is the spontaneous adsorption of phospholipid vesicles onto solid substrates. Depending on experimental parameters, including pH, phospholipid composition, ion type, temperature, and the substrate's surface chemistry, this self-assembly process can follow a number of pathways that result in different thin-film structures [9–12]. In cases where conditions favor the rupture of adsorbed vesicles, a complete planar bilayer can form [13]. The bilayer has a well-defined surface area [13] and high degree of lateral lipid mobility [14] – the latter a characteristic of biological membranes. For other systems, adsorbed vesicles do not rupture and instead form a layer of intact vesicles, thereby preserving some degree of membrane curvature.

Based on these two fundamental platforms – a planar bilayer and a vesicle layer – more complex platforms can also be developed [7, 15]. A number of phospholipid and other membrane component (e.g., cholesterol) functionalization possibilities including complementary DNA strand pairing [16, 17] and biotin–streptavidin chemistry [7, 15] permit the linkage of additional structures to these platforms. One possibility presented herein is the tethered vesicle array

Functional Polymer Films, First Edition. Edited by Wolfgang Knoll and Rigoberto C. Advincula.
© 2011 Wiley-VCH Verlag GmbH & Co. KGaA. Published 2011 by Wiley-VCH Verlag GmbH & Co. KGaA.

where vesicles are linked to a planar bilayer [7, 15]. An alternative method to these bottom-up model membrane-fabrication approaches is the use of protein-containing membranes derived from cellular extracts [18], which can also self-assemble on the solid support.

As sensor platforms, biomimetic thin films must be used in conjunction with a surface-sensitive measurement technique in order to characterize their self-assembly and resultant physical properties and for subsequent detection of macromolecular interactions [19]. A number of optical- and acoustic-based sensors, including quartz crystal microbalance with dissipation monitoring (QCM-D) [13], surface plasmon resonance (SPR) [1], ellipsometry [20], and reflectometry [21] have proven effective at measuring these types of interactions with high sensitivity and in real time. The focus of the rest of this chapter will be on the unique measurement capabilities of the QCM-D sensor for characterization of biomimetic thin films and for monitoring macromolecular recognition events.

24.2
Brief Overview of Quartz Crystals

A quartz crystal has the necessary combination of mechanical, electrical, chemical, and thermal properties to behave as a piezoelectric material [22]. The converse piezoelectric effect, which is utilized in the QCM, is caused by a polarizing electric field that produces a mechanical strain [23]. Due to the flexibility of the frequency excursion and range of temperature stability, AT-cut quartz is used most frequently in QCM devices [22]. As depicted in Figure 24.1, the mechanical response excites the thickness-shear mode of the crystal when a voltage is applied in a perpendicular direction. Since the shear deformation of the quartz is purely elastic, quartz resonators work well as sensors. Standing-wave properties are determined from the velocity of wave propagation and dimensional constraints [22]. The wavelength can easily be converted to a frequency. Quartz crystals exhibit resonance with the resonant frequency of the crystal typically on the order of megahertz, depending mainly on the thickness of the crystal itself. When a uniform mass is added to the active sensing area of the crystal, the resonant frequency decreases due to an increase of the standing wavelength as a result of an increase in the effective resonator thickness. Frequency-change measurements are sensitive to surface-mass changes (the sensitivity factor varies as the square of the frequency) [23–25]. As a result of these useful properties, quartz resonators can be found in common devices such as watches and computers to give an accurate time base, and as signal generators or reference systems in electronic devices [22]. In 1959, Sauerbrey [26] first demonstrated that there was a proportional frequency decrease when mass was added to the quartz crystal substrate in the gas phase. Throughout the 1960s and 1970s, the QCM technique gained importance with the introduction of a number of QCM-based devices that could monitor thin-film thickness in vacuum and air.

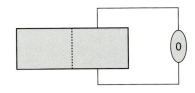

Figure 24.1 The quartz crystal is a resonator device that exhibits shear-wave behavior. When a foreign mass is added to the quartz crystal, the resonance frequency decreases with increasing standing wavelength due to an increase in effective thickness.

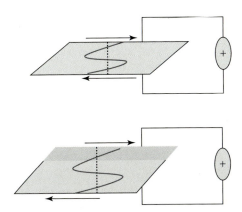

A notable advance in this field was made by Nomura and Minemura [27] when they demonstrated that the QCM can be reliably operated in contact with liquid, thus providing the starting point for the development of a new class of bioanalytical tools.

24.3
QCM Methodologies: Steady-State versus Transient Behavior

In addition to measuring resonant-frequency changes due to mass adsorption, researchers at Chalmers University of Technology [23, 28] have developed a method to characterize the viscoelastic properties of thin films by measuring a second parameter, energy dissipation. There are two common approaches, steady-state and transient behavior, to obtain QCM data, along with three different methods that are generally used to acquire QCM parameters. Steady-state techniques include compensated phase-locked oscillation (CPLO) and impedance analysis. In CPLO, the oscillator stays locked on the resonance frequency, permitting its continuous tracking. The CPLO resonator system readily facilitates direct, undistorted real-time measurement of both resonance frequency and accompanying resistance. Impedance analysis is also a steady-state method to acquire the full impedance curve, from which one can extract resonance frequency and equivalent circuit parameters such as capacitance, inductance, and resistance. In addition, the resonance peak bandwidth can be related to the quality factor, Q. However, many situations exist where a measurement of the absolute dissipation factor of the crystal oscillation is important as well [29, 30].

Rodahl *et al.* developed a transient method in which the frequency and transient decay time are measured simultaneously [23, 25, 28]. The decay time, equivalent circuit parameters, and quality factor, Q, are all related and provide comparable information such that [23, 25, 28]:

$$Q = \pi f \tau = \frac{2\pi f L}{R} \tag{24.1}$$

The basic principle is that when the driving power to a resonator is switched off at time $t = 0$, the voltage over the resonator, $U(t)$, decays in an exponentially damped sinusoidal fashion according to the following equation:

$$U(t) = A_0 \exp(-t/\tau) \sin(2\pi f t + \varphi) \tag{24.2}$$

where U is the voltage over the resonator, A_0 is the amplitude, τ is the decay time, and φ is the phase [25, 28].

Several theories have been used to describe the behavior of quartz crystals operating in the thickness-shear mode, unloaded and loaded with rigid, viscous, and viscoelastic layers, as described extensively in literature reviews [26–28, 31–40]. There have been two main theoretical approaches; the Sauerbrey model [26] was the first mechanical model, followed by the Lu and Lewis [41] model based on an analytical solution using one-dimensional acoustic waves. Voinova *et al.* [31] used a simplified version of an acoustic-mechanical model to derive changes in resonant frequency and dissipation with overlayer loading, that is, not dependent on the quartz crystal's physical properties. Another electrical approach is the transmission-line model [42] in which the equivalent circuit serves as an approximation. Kanazawa and coworkers [35, 43] described a combination of the mechanical wave model coupled with the electrical properties of the quartz resonator to yield the complete physical, or electromechanical model (EMM). The following sections summarize two representative fitting models, the Sauerbrey and Voigt–Voinova models.

24.3.1
Sauerbrey Model

Sauerbrey [26] first recognized the potential usefulness of QCM technology and demonstrated the extremely sensitive nature of these piezoelectric devices toward mass changes at the surface of QCM electrodes. The results of his work are embodied in the Sauerbrey equation, which relates the mass change per unit area at the QCM electrode surface to the observed change in oscillation frequency of the crystal with three key assumptions: (i) the added foreign mass is treated as an equivalent mass change of the quartz itself; (ii) foreign mass is uniformly distributed; and (iii) the adlayer is a rigid film. Under such conditions, the following equation has been derived:

$$\Delta f_n = -n \frac{2\rho_f f_n^2}{\rho_q v_q} t_f \tag{24.3}$$

where Δf_n is the change in the resonance frequency at the nth harmonic when the mass is added, f_n is the resonance frequency of the crystal at the nth harmonic with no film, ρ_f and ρ_q are the densities of the film and quartz, respectively, t_f is the added film thickness, and n is the harmonic number, which is the number of half-wavelengths of the thickness-shear waves in the crystal.

24.3.2
Voigt–Voinova Model

While the frequency is the main quantity measured in traditional QCM experiments in order to calculate mass change, many recent studies suggest that adsorbed soft films do not follow this linear relationship between frequency and mass change. Such deviations are mainly due to trapped solvent between and within adsorbed macromolecules, inducing a friction loss within the soft film that leads to damping of the crystal's oscillation.

Thus, the linear relationship between the adsorbed mass and the change in frequency is not necessarily valid for viscoelastic films that exhibit additional energy dissipations as well as frequency-overtone-dependent responses. As discussed below in our discussion of QCM-D applications, this type of information is especially important for understanding interactions involving structurally complex biomembranes [13, 44]. If the system under investigation exhibits frequency and dissipation overtone dependence in the measured interval (5–55 MHz), measurements at several harmonics will allow the set of data (at $n = 1, 3, 5, 7, 9, 11,$ and 13) to be compared with the theoretical representations (with several unknown parameters) that must be applied in such situations.

For the QCM measurement, the inverse of Q is obtained and is referred to as the *energy dissipation*, D. The Voigt–Voinova model has been employed to evaluate the experimentally measured Δf and ΔD for the bound mass in order to gain insight into the physical properties of the adsorbed film on the solid support. The adsorbed layer is represented by a frequency-dependent complex shear modulus, which is defined by

$$G = G' + iG'' = \mu_f + 2\pi \text{ if } \eta_f = \mu_f(1 + 2\pi \text{ if } \tau_f) \tag{24.4}$$

where G' describes energy storage, G'' describes energy dissipation, f is the oscillation frequency, μ_f is the elastic shear (storage) modulus, η_f is the shear viscosity, and τ_f is the characteristic relaxation time of the film ($\tau_f = \mu_f/\eta_f$). The governing viscoelastic functions for the Voigt element are defined using these variables. In this viscoelastic model, the adsorbed film is represented by a single Voigt element consisting of a spring that corresponds to the shear rigidity and a dashpot that represents viscosity, as shown in Figure 24.2.

The Voigt–Voinova model includes these viscoelastic effects and assumes uniform thickness, uniform film density, liquid with Newtonian characteristics, and no slip. Further, it is assumed that the film viscosity is independent of overtone frequencies. Then, Δf and ΔD can be related to the film density, viscoelastic

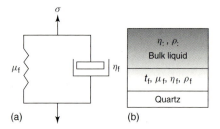

Figure 24.2 Voigt–Voinova model used to analyze viscoelastic properties of an adsorbed film. (a) The film is subject to an oscillating shear stress, σ, and is treated like a Voigt element with shear viscosity, η_f, and shear modulus, μ_f. (b) Schematic illustration of the geometry of a quartz crystal covered by a film with a thickness, t_f, in contact with a semi-infinite Newtonian liquid. Reprinted from Ref. [45].

properties, and thickness in terms of a Taylor expansion [31]:

$$\Delta f \cong \frac{1}{2\pi \rho_q t_q} t_f \rho_f \omega \left(1 + \frac{2t_f^2 \chi}{3\delta^2 (1+\chi^2)} \right) \tag{24.5}$$

$$\Delta D \cong \frac{2t^3 \rho_f f}{3\pi f_{ro} \rho_q t_q} \frac{1}{\delta^2 (1+\chi^2)} \tag{24.6}$$

where ρ_q and ρ_f are the density of the quartz and film, respectively; t_q and t_f are the thickness of the quartz and film, respectively; f is the measured frequency; χ, is related to the ratio of the storage modulus (μ_f) and the loss modulus (η_f), which is the inverse of tan δ. Using two different overtones and Equations 24.4 and 24.5, the changes in the frequency and dissipation can be fit to the film viscosity (η_f), shear modulus (μ_f), thickness (t_f), and density (ρ_f), thereby gaining insight into the film properties.

24.4
QCM-D Analysis of Biomimetic Thin Films: Assembly and Applications

The QCM-D sensor's ability to monitor in real time the mass and viscoelastic properties of biomimetic thin films enables characterization of these structures and the detection of macromolecular interactions that alter their physical properties. In this section, we discuss how QCM-D can be utilized to investigate the physical properties of four different biomimetic thin films that are composed of phospholipids and serve as representations of cellular membranes. Then, we present a cutting-edge biological application for each of these thin films when used as a QCM-D sensor platform.

Depending on the specific targeted macromolecular interaction of interest, the choice of platform can be critical to detection and analysis of subsequent interactions. Moreover, proper characterization of the thin film is necessary because the activity of many enzymes and other macromolecules is significantly affected by the structural properties of the model membrane.

24.4.1
Design and Characterization of Biomimetic Thin Films

A key advantage of the QCM-D technique is the wide range of substrate materials, including inorganic and thiol-modified surfaces that can be coated onto the quartz crystal sensor chip with evaporation, deposition, and self-assembly methods [46]. This versatility allows for advanced QCM-D detection systems that combine acoustic sensing with electrochemical [47] and optical sensing measurements [21]. Bottom-up design is one promising nanofabrication strategy to assemble biomimetic thin films on these substrates. The design principles are based on the development of biomimetic membrane structures composed of phospholipids, which are the fundamental building blocks of cellular membranes. The formation of these structures is guided by phospholipid molecule self-assembly. This self-assembly process is primarily driven by the interactions between phospholipid vesicles and solid substrates [13, 44].

The pioneering QCM-D studies of Kasemo and coworkers [13, 44] identified for the first time the surface-specific kinetics of these vesicle–substrate interactions. By characterizing the mass and viscoelastic properties of the lipid mass adsorbed on various substrates, different adsorption kinetics were revealed, which resulted in the assembly of one of three possible lipid structures: (i) planar monolayer; (ii) planar bilayer; or (iii) layer of adsorbed, unruptured vesicles.

While the formation of a lipid monolayer can occur on hydrophobic surfaces such as thiolipid-modified gold, we focus our attention here on QCM-D characterization of the self-assembly process for planar bilayers and vesicle layers. As shown in Figure 24.3, the self-assembly of each structure begins with the adsorption of unruptured vesicles onto the substrate. In the case of planar bilayer formation, which can occur on hydrophilic substrates including silica and mica, intact vesicles continue to adsorb on the substrate until reaching a critical surface coverage. At this point, which is indicated by maxima in the frequency and dissipation responses (Figure 24.3c, see arrow), the combination of vesicle–vesicle and vesicle–surface interactions overcomes the vesicles' membrane tension, inducing vesicle rupture [13, 44]. The rupturing process creates islands of planar bilayers with hydrophobic edges, which propagate continued vesicle rupture until a complete planar bilayer forms on the substrate and the edges are minimized. For modeling purposes, the SLB's thickness can be calculated from the QCM-D frequency response by the Sauerbrey relationship [26] since the adlayer is rigid, as indicated by the minor energy dissipation response. However, to characterize the properties of the adlayer throughout the structural transformation, it is necessary to use a viscoelastic model to account for intermediate stages where the frequency and dissipation responses are overtone dependent, as shown in Figure 24.3d [45]. This type of QCM-D analysis provides useful information about the quality of the planar bilayer platform, which can be influenced by viscoelastic structural elements such as unruptured vesicles still present on the substrate. In general, comparison of the calculated thickness obtained with the Sauerbrey and Voigt–Voinova models,

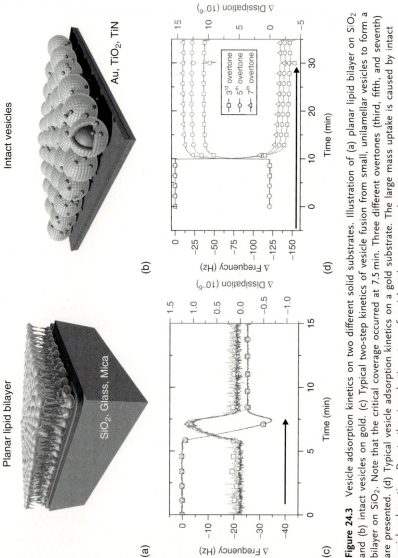

Figure 24.3 Vesicle adsorption kinetics on two different solid substrates. Illustration of (a) planar lipid bilayer on SiO_2 and (b) intact vesicles on gold. (c) Typical two-step kinetics of vesicle fusion from small, unilamellar vesicles to form a bilayer on SiO_2. Note that the critical coverage occurred at 7.5 min. Three different overtones (third, fifth, and seventh) are presented. (d) Typical vesicle adsorption kinetics on a gold substrate. The large mass uptake is caused by intact vesicle adsorption. Due to the viscoelastic nature of vesicles, the overtones do not merge.

which should be approximately equal for a rigid bilayer, can be used to assess the suitability of the planar bilayer as a sensor platform [48].

By contrast, the physical and chemical properties of some substrates, such as bare gold [11] and titanium oxide [49], do not create strong enough vesicle–substrate interactions to promote vesicle rupture. For these substrates, vesicle adsorption continues until an irreversibly adsorbed, single layer of vesicles fully covers the surface (Figure 24.3d). While the degree of surface coverage is greater in the case of a complete vesicle layer than at the point of critical vesicle surface coverage during planar bilayer formation, the QCM-D frequency and dissipation responses of both systems are analytically similar.

Despite the facile self-assembly of the vesicle layer, achieving reproducibility of its mass and viscoelastic properties requires significant experimental care since these adlayer properties depend on a number of parameters including flow rate, vesicle size, and vesicle concentration. To facilitate the development of a vesicle layer with more controllable features, researchers have utilized a number of different chemical linking methods [7, 15]. Since nonspecific protein adsorption does not occur on a planar bilayer, it can be functionalized to promote specific protein–ligand interactions, which serve as links between the planar bilayer and vesicles in order to form a tethered vesicle array [7]. Among other tethering possibilities, oligonucleotide hybridization [50], cholesterol-tagged complementary DNA [51], and streptavidin–biotin [15] have been successfully employed to form ordered arrays of vesicles on a planar bilayer. As illustrated in Figure 24.4, the self-assembly of biotin–streptavidin-tethered vesicle arrays can be controlled by the linker density of biotinylated lipids in the planar bilayer. As the percentage of biotinylated lipid increases, the QCM-D demonstrates an observed mass increase during the subsequent binding of streptavidin and vesicles, the latter of which also contain a fraction of biotinylated lipids. Thus, while the vesicle adsorption behavior is similar to the case on solid substrates that do not promote vesicle rupture, a greater control over the adsorption process is gained with tethering methods.

So far, we have presented three different bottom-up, design-inspired platforms, each having certain advantages for biological studies, as discussed below. Investigations with biomimetic, phospholipid-based platforms can be enhanced with an analog platform based on cellular-derived biological membranes. Such platforms are based on the top-down approach [6, 52] of utilizing the rich diversity of protein components in biological membranes to form a functional system for detection of protein–protein interactions. By combining these two approaches, specific membrane–protein interactions can be understood separately as occurring primarily via lipid–protein or protein–protein interactions. As with biomimetic platforms, these cellular membrane platforms can be formed by self-assembly methods. Figure 24.5 illustrates the adsorption of microsomes – vesicle-like structures extracted from human liver-derived cell membrane – onto a silicon oxide substrate. The adsorption kinetics demonstrates monotonic behavior that results in the assembly of a membrane protein-containing platform with stable mass and visoelastic properties that permit its use together with the QCM-D sensor for investigating protein–protein interactions. Together, a wide range of biomimetic thin films

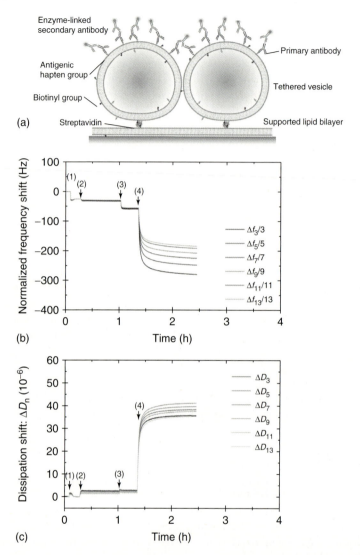

Figure 24.4 Planar bilayer-tethered vesicle array. (a) Illustration of a vesicle array tethered to a planar bilayer. Ligands such as a primary antibody can bind to the surface of the vesicles tethered by streptavidin-biotin to the underlying bilayer. In the case of the primary antibody used in Ref. [7], 3 mol% DNP-X-PE lipid was incorporated into the tethered vesicles in order to detect antibody binding. Self-assembly of the planar bilayer-tethered vesicle array causes normalized changes in (b) resonant frequency and (c) energy dissipation of the quartz crystal as a function of time. The assembly steps include the introduction of (1) osmotically shocked vesicles, (2) blocking buffer, (3) streptavidin, and (4) tethering vesicles, as noted in the figure. Reprinted from Ref. [7].

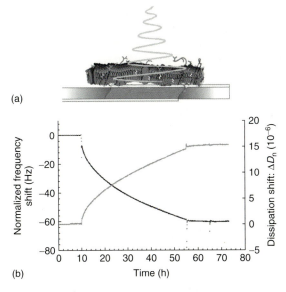

(a)

(b)

Time (h)

Figure 24.5 Human-liver-derived cell membrane on a SiO$_2$-coasted quartz crystal. (a) Schematic of human liver derived cell membrane on SiO$_2$-coated quartz crystal. (b) Changes in resonant frequency and dissipation are presented as a function of time for human liver membrane assembly on a SiO$_2$-coated quartz crystal.

can be utilized as platforms either by themselves or together with cellular-based platforms to provide insight into biological interactions with membranes.

24.4.2
Two-Dimensional, Planar Bilayer Platform

Planar bilayers are popular model systems for mimicking biological membranes because their physical properties can be adjusted in a controlled fashion by changing the lipid composition or incorporating other components such as sterols [6, 52]. In addition to the well-defined surface area afforded by its two-dimensional structure, fluid-phase planar bilayers also exhibit a high degree of lateral lipid diffusivity, a feature that is necessary for certain ligand binding events in cellular systems. These properties make the platform attractive for detection and quantification of a wide range of macromolecular interactions. Here, we report a study by Jackman *et al.* [8] that employed a planar bilayer on the QCM-D sensor to elucidate the binding mechanism of a member of the secretory phospholipase A$_2$ (sPLA$_2$) family, a class of enzymes that functions at aggregated phospholipid interfaces such as membranes.

The relationship between sPLA$_2$'s mechanistic actions during its membrane binding step and its physiological functions, which include active roles in biological processes such as signal transduction and lipid metabolism, is not well understood.

The main advantage of the planar bilayer for this study was the ability to change the platform's lipid composition without altering its mass or viscoelastic properties. As shown in Figure 24.6, the membrane binding interaction of sPLA$_2$ to a series of bilayer platforms with different surface charges was investigated. After an initial period of sPLA$_2$ adsorption, as indicated by the frequency decrease, an unexpected desorption behavior was observed for negatively charged membranes. However, only minor energy-dissipation responses were observed for all platforms, indicating that membrane binding of sPLA$_2$ did not significantly affect the lipid bilayer's viscoelastic properties.

While enzyme adsorption was similar for most lipid compositions, there was wide variance in the desorption behavior. Most interesting was the correlation identified by the QCM-D sensor between the mass of desorption and the lipid composition, with increased desorption for more negatively charged bilayers. Based on the adsorption and desorption behaviors as well as bilayer-disrupting activity, three distinct types of membrane-binding interaction were identified depending on lipid composition. The authors concluded that: (i) adsorption of sPLA$_2$ to model membranes is not primarily driven by electrostatic interactions; (ii) lipid desorption can proceed sPLA$_2$ adsorption, resulting in nonhydrolytic bilayer disruption; and (iii) this desorption is driven by electrostatic interactions. Compared to past biochemical studies on this enzyme, the QCM-D sensor platform allowed for determination of its membrane kinetics in real time and with a high measurement sensitivity that enabled detection of sPLA$_2$'s dynamic, two-step binding mechanism.

24.4.3
Intact Vesicle Platform

While planar bilayers offer numerous advantages for detecting and quantifying ligand binding interactions, they also have disadvantages for studying integral, membrane-spanning receptors and coreceptors due to the close proximity of the bilayer to the underlying substrate. Further, some macromolecular interactions are sensitive to membrane topology and some minimal level of membrane curvature can be required in these cases. As a soft film, that is, less rigidly attached to the substrate, an intact vesicle layer is thus a useful platform for studies involving integral membrane proteins and/or membrane curvature-dependent binding interactions. In addition, the corresponding curvature of the intact vesicles can be adjusted by preparing different-sized vesicle populations.

Our group investigated how vesicle size affects an amphipathic, alpha-helical (AH) peptide's vesicle-rupturing ability in order to better understand the mechanism of virus-particle lysis [53]. Previously, we described a novel method that employs the AH peptide to destabilize a layer of intact vesicles adsorbed on various substrates such as gold and titanium oxide. The AH peptide interaction causes vesicle rupture, transforming the adsorbed vesicles into a planar bilayer [5]. In order to investigate this membrane-curvature-dependent interaction, we first created an

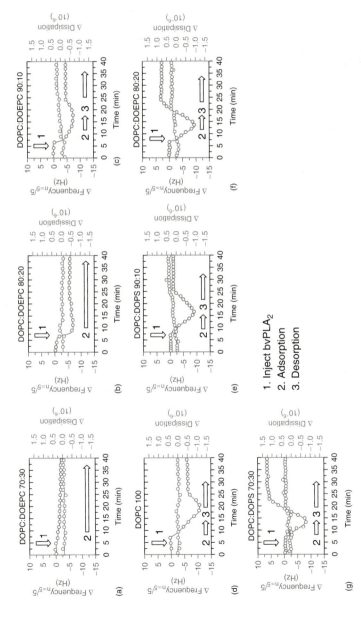

Figure 24.6 QCM-D monitoring of enzyme binding to charged planar bilayers. QCM-D adsorption kinetic versus time for the binding of secretory phospholipase A2 isoform. The enzyme-containing solution was injected (arrow 1) and, depending on the bilayer lipid composition, there was adsorption (arrow 2) and, in some cases, subsequent desorption (arrow 3) behavior. DOPC is a zwitterionic lipid, DOPS is negatively charged, and DOEPC is positively charged. (a) DOPC:DOEPC 70 : 30, (b) DOPC:DOEPC 80 : 20, (c) DOPC:DOEPC 90 : 10, (d) DOPC 100, (e) DOPC:DOPS 90 : 10, (f) DOPC:DOPS 80 : 20, and (g) DOPC:DOPS 70 : 30. Reprinted from Ref. [8].

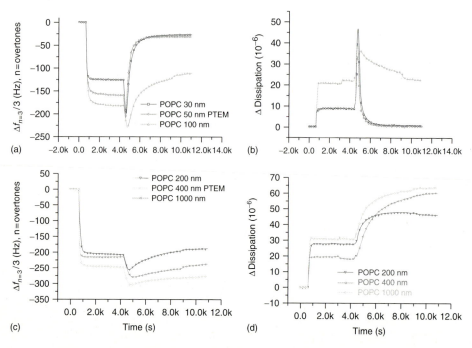

Figure 24.7 Screening of an AH peptide's antiviral activity with the intact vesicle platform on the QCM-D sensor. (a) QCM-D frequency response for various vesicle-size distributions up to 100 nm upon AH peptide addition (vesicles added at first arrow at 1.0 ks; AH peptide added at second arrow at 4.2 ks). (b) Corresponding energy dissipation changes. (c) QCM-D frequency response for vesicle size distributions greater than 100 nm upon AH peptide addition. (d) Corresponding energy dissipation changes. Note that the graphs below present the average vesicle effective diameters as measured by dynamic light scattering. Reprinted from Ref. [53].

intact vesicle platform on a gold-coated quartz crystal, employing different-sized vesicles to measure the AH peptide's lysis activity.

As shown in Figure 24.7, QCM-D measurements revealed that the AH peptide's lysis potency has three distinct regimes depending on the size distribution of the vesicles; (i) complete vesicle rupture and bilayer formation for vesicles with average diameters ranging from 59 to 67 nm (Figure 24.7a,b, black and red traces), (ii) incomplete vesicle rupture and bilayer formation for vesicles with an average diameter greater than 90 nm (Figure 24.7a,b, green trace), and (iii) no vesicle rupture or bilayer formation for significantly larger vesicles (>200 nm, Figure 24.7c,d).

For complete bilayer formation, as described in regime (i), the characteristic QCM-D frequency and dissipation values correspond to that of a planar bilayer. While complete bilayer formation can be achieved in the case of small vesicles, the AH peptide demonstrated less lysis potency against larger vesicles, as evidenced by the incomplete bilayer formation of regime (ii). In the case of much larger vesicles with average diameters greater than 200 nm, the AH peptide did not exhibit any vesicle lysis potency, as seen in regime (iii), indicating that AH peptide binding to the vesicle platform is the predominant interaction rather than vesicle rupture and bilayer formation (Figure 24.7c,d). The QCM-D analysis suggests that the degree of vesicle rupturing is related to vesicle size, with more complete rupturing occurring as the average vesicle effective diameter decreases. The size dependence of vesicle rupturing helps to explain the peptide's observed effect on the infectivity of a wide range of viruses. From these QCM-D experiments, we determined that a significant number of viruses have diameters of their lipid envelope within a size range such that the AH peptide demonstrates lysis potency.

Moreover, AH peptide-mediated vesicle rupture is followed by membrane fusion that results in planar bilayer formation. By analyzing the viscoelastic properties of these two distinct layers, a layer of "soft" vesicles and a "rigid" bilayer, we also created a model system to permit the study of film behavior in the region of nonlinear mass and frequency change (non-Sauerbrey) [45].

24.4.4
Bilayer-Tethered Vesicle Platform

In addition to the intact vesicle platform, there are several advantages to utilizing the bilayer-tethered vesicle assembly that include (i) controlling the density of tethered vesicles by specific chemical or biological links for lab-on-a-chip applications and (ii) the planar bilayer support acting as a barrier to nonspecific protein adsorption. Taking advantage of these useful properties, Patel *et al.* [7] investigated antibody binding to functionalized dinitrophenyl (DNP) hapten lipid headgroups in a bilayer-tethered vesicle membrane platform.

In order to check the stability of the platform, the authors first used QCM-D monitoring to follow the sequential processes of planar bilayer assembly and biotin–streptavidin tethering of vesicles with a fraction of lipids that possess DNP hapten headgroups. Subsequently, the interaction between a monoclonal rat anti-DNP IgG$_1$ antibody and the DNP hapten headgroups was monitored by QCM-D, as shown in Figure 24.8. The measured frequency and dissipation responses upon binding were compared to an independent measure of the amount of bound antibody obtained through the use of an *in situ* ELISA assay in order to validate the model system. At saturation, the surface mass density of bound antibody was approximately $900 \, \text{ng} \, \text{cm}^{-2}$.

Of note, converting the signal generated upon binding to the actual amount of bound antibody has been a challenge for viscoelastic systems such as the tethered vesicle assembly [7, 15]. In order to solve this problem, an empirical relationship between the amount of bound antibody and the corresponding QCM-D response was first established. Then, the results were examined using the Voigt–Voinova model in order to describe the QCM-D response under a variety of theoretical loading conditions. By comparing models that describe the response of the quartz when loaded by either a single homogeneous viscoelastic film or by a two-layered viscoelastic film, the authors found that a homogeneous, one-layer model accurately predicts the amount of antibody bound to the tethered vesicles near antibody-surface saturation. However, a two-layer model must be invoked to

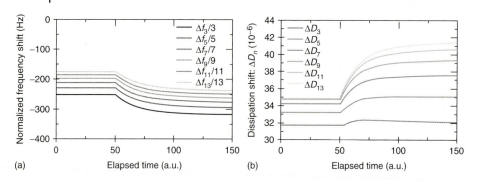

Figure 24.8 Antibody binding to a planar bilayer-tethered vesicle platform. A monoclonal rat anti-DNP IgG$_1$ antibody binds DNP-X-PE lipids within the tethered vesicles. Normalized changes in (a) resonant frequency and (b) dissipation factor for measured overtones are plotted as a function of time. Subsequent antibody binding to tethered vesicles composed of 3 mol% DNP-X-PE. The baseline corresponds to the bilayer-tethered vesicle platform on a bare SiO$_2$-coated quartz crystal. Reprinted from Ref. [7].

accurately describe the kinetic response of the dissipation factor, which suggests that the binding of the antibody results in a stiffening of the top layer of the film.

24.4.5
Biological Membrane-on-a-Chip Platform

Biomimetic thin films are typically created by bottom-up strategies and can include modifying the phospholipid composition and incorporating transmembrane proteins into the bilayer system in order to study specific macromolecular recognition events [5, 6]. New lipid-based platforms have enabled a wide array of new research on biological membranes by incorporating various membranous components into model membranes. However, replicating the complex transmembrane conformations and proper protein attachments of biologic membranes with model systems has proven difficult. Thus, there is increasing demand for alternative sensor platforms that can mimic conventional biological responses such as those triggered by properly functioning protein receptors.

In order to achieve this goal of a platform that could host functional protein receptors, our group has designed a novel "membrane-on-a-chip" system [18] that can capture in real-time with QCM-D monitoring the binding dynamics of a viral peptide to cell-derived membranes assembled by the "top-down" approach as shown in Figure 24.9. To mimic the in *vivo* actions of this viral peptide, we used membranes prepared from Huh7 cells, which come from a human liver tumor-derived cell line that is widely used for hepatitis C virus replication studies [54, 55]. The kinetics of biomembrane platform formation and viral peptide binding to the platform were monitored by QCM-D. By utilizing the "top-down" cell-derived membrane, we were able to determine that peptide binding is significantly enhanced by the presence of native protein receptors. The direct evidence for this peptide–protein interaction

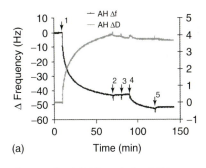
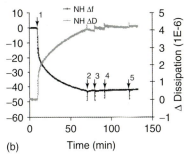

(a) Time (min)   (b) Time (min)

Cellular membrane
1. Inject Huh7
2. PBS buffer wash
3. PBS buffer wash
4. Inject AH/NH
5. PBS buffer wash

Figure 24.9 QCM-D analysis of AH peptide binding to a quartz crystal coated with a cell-derived membrane. (a) Binding of AH peptide to liver cell-derived membranes adsorbed on a SiO$_2$-coated quartz crystal. Compared to model, lipid-only bilayers, the saturated membrane mass is greater, presumably due to the presence of protein components. As expected, a higher viscoelastic energy dissipation is also observed with the protein-containing membranes. (b) By comparison, a control NH peptide does not bind to the membranes. Reprinted from Ref. [18].

was captured by eliminating protein receptors through prior treatment of the cellular membranes with trypsin [18]. A control peptide, whose interaction with membranes involves trypsin-insensitive glycosaminoglycans, was not altered by the protease treatment. Standard biochemical membrane flotation assays supported the validity of the QCM-D results, validating the platform's use as a novel bioassay. The experimental results demonstrate the potential of QCM-D technology for studying these types of protein–membrane interactions as well as a broad range of problems involving membrane proteins or lipids with biomimetic thin films. In order to further develop membrane-on-a-chip technology with the "top-down" approach, conventional bioseparation techniques may be a powerful tool to create functional platforms to study biomacromolecular interactions and dynamics. The main advantage of this membrane-on-a-chip biosensor is that it is simple and quick to use, and shows the real-time kinetics of the interactions happening on a membrane derived from living cells that is comparable to the results obtained by traditional biochemical analyses.

24.5
Conclusion

QCM-D biomimetic thin-film platforms have been utilized to study a number of bioanalytical problems due to the system's superior detection advantages. These advantages include label-free detection, real-time monitoring of binding kinetics, significant flexibility for designing the thin-film platforms, and high measurement sensitivity. One of the challenges with QCM-D measurements of biomimetic and biological membrane platforms is the difficulty of interpreting data for viscoelastic films. However, recent technological and theoretical developments associated with characterizing platform self-assembly and macromolecular recognition events have led to new applications for QCM-D biomimetic thin film systems as biosensors. In this chapter, we have discussed how four different biomimetic thin films can be formed on the quartz crystal, and then utilized these platforms to study macromolecular interaction events. In particular, the focus was on phospholipid-based biomimetic thin films that mimic cell membranes in order to demonstrate how this technology can be used to reveal insight into a range of macromolecular interaction events associated with important biological processes. The development of biomimetic thin films for sensing application is in its nascence, and there is much room for the field to grow, both in terms of developing more advanced platforms and employing different analytical tools. Thus, there is great potential to design novel biomimetic thin films that may have a wide range of nanobiotechnology applications from fundamental mechanistic studies to clinical diagnostics.

References

1. Knoll, W. (1998) Interfaces and thin films as seen by bound electromagnetic waves. *Annu. Rev. Phys. Chem.*, **49**, 569–638.
2. Tanaka, M. and Sackmann, E. (2005) Polymer-supported membranes as models of the cell surface. *Nature*, **437**, 656–663.
3. Tamm, L.K. and McConnell, H.M. (1985) Supported phospholipid bilayers. *Biophys. J.*, **47**, 105–113.
4. Salafsky, J., Groves, J.T., and Boxer, S.G. (1996) Architecture and function of membrane proteins in planar supported bilayers: a study with photosynthetic reaction centers. *Biochemistry*, **35**, 14773–14781.
5. Sackmann, E. (1996) Supported membranes: scientific and practical applications. *Science*, **271**, 43–48.
6. Sackmann, E. and Tanaka, M. (2000) Supported membranes on soft polymer cushions: fabrication, characterization and applications. *Trends Biotechnol.*, **18**, 58–64.
7. Patel, A.R., Kanazawa, K.K., and Frank, C.W. (2009) Antibody binding to a tethered vesicle assembly using QCM-D. *Anal. Chem.*, **81** (15), 6021–6029.
8. Jackman, J.A., Cho, N.J., Duran, R.S., and Frank, C.W. (2010) Interfacial binding dynamics of bee venom phospholipase A(2) investigated by dynamic light scattering and quartz crystal microbalance. *Langmuir*, **26**, 4103–4112.
9. Hook, F., Kasemo, B., Nylander, T., Fant, C., Sott, K., and Elwing, H. (2001) Variations in coupled water, viscoelastic properties, and film thickness of a Mefp-1 protein film during adsorption and cross-linking: a quartz crystal microbalance with dissipation monitoring, ellipsometry, and surface plasmon resonance study. *Anal. Chem.*, **73**, 5796–5804.
10. Hook, F., Rodahl, M., Brzezinski, P., and Kasemo, B. (1998) Measurements using the quartz crystal microbalance technique of ferritin monolayers on methyl-thiolated gold: dependence of energy dissipation and saturation coverage on salt concentration. *J. Colloid Interface Sci.*, **208**, 63–67.
11. Reimhult, E., Hook, F., and Kasemo, B. (2002) Temperature dependence of formation of a supported phospholipid bilayer from vesicles on SiO_2. *Phys. Rev. E Stat. Nonlin. Soft. Matter. Phys.*, **66**, 051905.
12. Reimhult, E., Hook, F., and Kasemo, B. (2002) Intact vesicle adsorption and supported biomembrane formation from vesicles in solution: influence of surface chemistry, vesicle size, temperature, and osmotic pressure. *Langmuir*, **19**, 1681–1691.
13. Keller, C.A. and Kasemo, B. (1998) Surface specific kinetics of lipid vesicle adsorption measured with a quartz

crystal microbalance. *Biophys. J.*, **75**, 1397–1402.

14. Cremer, P.S. and Boxer, S.G. (1999) Formation and spreading of lipid bilayers on planar glass supports. *J. Phys. Chem. B*, **103**, 2554–2559.

15. Patel, A.R. and Frank, C.W. (2006) Quantitative analysis of tethered vesicle assemblies by quartz crystal microbalance with dissipation monitoring: binding dynamics and bound water content. *Langmuir*, **22**, 7587–7599.

16. Benkoski, J.J. and Hook, F. (2005) Lateral mobility of tethered vesicle-DNA assemblies. *J. Phys. Chem. B*, **109**, 9773–9779.

17. Chung, M., Lowe, R.D., Chan, Y.H., Ganesan, P.V., and Boxer, S.G. (2009) DNA-tethered membranes formed by giant vesicle rupture. *J. Struct. Biol.*, **168**, 190–199.

18. Cho, N.J., Cheong, K.H., Lee, C., Frank, C.W., and Glenn, J.S. (2007) Binding dynamics of hepatitis C virus' NS5A amphipathic peptide to cell and model membranes. *J. Virol.*, **81**, 6682–6689.

19. Hook, F., Kasemo, B., Grunze, M., and Zauscher, S. (2008) Quantitative biological surface science: challenges and recent advances. *ACS Nano*, **2**, 2428–2436.

20. Richter, R.P. and Brisson, A.R. (2005) Following the formation of supported lipid bilayers on mica: a study combining AFM, QCM-D, and ellipsometry. *Biophys. J.*, **88**, 3422–3433.

21. Cho, N.J., Wang, G., Edvardsson, M., Glenn, J.S., Hook, F., and Frank, C.W. (2009) Alpha-helical peptide-induced vesicle rupture revealing new insight into the vesicle fusion process as monitored in situ by quartz crystal microbalance-dissipation and reflectometry. *Anal. Chem.*, **81**, 4752–4761.

22. Bottom, V.E. (1982) *Introduction to Quartz Crystal Unit Design*, Van Nostrand Reinhold Company, New York.

23. Rodahl, M., Hook, F., and Kasemo, B. (1996) QCM operation in liquids: an explanation of measured variations in frequency and Q factor with liquid conductivity. *Anal. Chem.*, **68**, 2219–2227.

24. Rodahl, M., Hook, F., Fredriksson, C., Keller, C.A., Krozer, A., Brzezinski, P.,

Voinova, M., and Kasemo, B. (1997) Simultaneous frequency and dissipation factor QCM measurements of biomolecular adsorption and cell adhesion. *Faraday Discuss.*, **107**, 229–246.

25. Rodahl, M., Hook, F., Krozer, A., Brzezinski, P., and Kasemo, B. (1995) Quartz-crystal microbalance setup for frequency and Q-factor measurements in gaseous and liquid environments. *Rev. Sci. Instrum.*, **66**, 3924–3930.

26. Sauerbrey, G. (1959) Verwendung von schwingquarzen zur waegun duenner schichten und zur mikrowaegung. *Z. Physik.*, **155**, 206.

27. Nomura, T. and Minemura, A. (1980) Behavior of a piezoelectric quartz crystal in an aqueous-solution and the application to the determination of minute amount of cyanide. *Nippon Kagaku Kaishi*, **1980**, 1621–1625.

28. Rodahl, M. and Kasemo, B. (1996) A simple setup to simultaneously measure the resonant frequency and the absolute dissipation factor of a quartz crystal microbalance. *Rev. Sci. Instrum.*, **67**, 3238–3241.

29. Rodahl, M. and Kasemo, B. (1996) On the measurement of thin liquid overlayers with the quartz-crystal microbalance. *Sens. Actuators A-Phys.*, **54**, 448–456.

30. Buttry, D.A. and Ward, M.D. (1992) Measurement of interfacial processes at electrode surfaces with the electrochemical quartz crystal microbalance. *Chem. Rev.*, **92**, 1355–1379.

31. Voinova, M.V., Rodahl, M., Jonson, M., and Kasemo, B. (1999) Viscoelastic acoustic response of layered polymer films at fluid-solid interfaces: continuum mechanics approach. *Phys. Scr.*, **59**, 391–396.

32. Kanazawa, K.K. and Reed, C.E. (1990) A new description for the viscoelastically loaded quartz resonator. *J. Appl. phys.*, **68** (5), 1993–2001.

33. Kanazawa, K.K. and Melroy, O.R. (1993) The quartz resonator: electrochemical applications. *IBM J. Res. Dev.*, **37**, 157–171.

34. Kanazawa, K.K. and Gordon, J.G. (1985) Frequency of a quartz microbalance in contact with liquid. *Anal. Chem.*, **57**, 1770–1771.

35. Kanazawa, K.K. (1997) Mechanical behaviour of films on the quartz microbalance. *Faraday Discuss.*, **107**, 77–90.

36. Kanazawa, K. and Gordon, J. (1985) The oscillation frequency of a quartz resonator in contact with a liquid. *Anal. Chim. Acta*, **175**, 99–106.

37. Lucklum, R. and Hauptmann, P. (2001) Thin film shear modulus determination with quartz crystal resonators: a review. IEEE International Frequency Control Symposium and PDA Exhibition, pp. 408–418.

38. Lucklum, R. and Hauptmann, P. (2000) The quartz crystal microbalance: mass sensitivity, viscoelasticity and acoustic amplification. *Sens. Actuators B: Chem.*, **1**, 1–3.

39. Lucklum, R., Behling, C., and Hauptmann, P. (1999) Thin film shear modulus determination with quartz crystal resonators. *Sens. Mater.*, **11**, 111–130.

40. Lucklum, R., Behling, C., Cernosek, R.W., and Martin, S.J. (1997) Determination of complex shear modulus with thickness shear mode resonators. *J. Phys. D: Appl. Phys.*, **30**, 346–356.

41. Lu, C.S. and Lewis, O. (1972) Investigation of film-thickness determination by oscillating quartz resonators with large mass load. *J. Appl. Phys.*, **43**, 4385–4390.

42. Krimholtz, R., Leedom, D.A., and Matthaei, G.L. (1970) New equivalent circuits for elementary piezoelectric transducers. *Electron. Lett.*, **6**, 398–399.

43. Reed, C.E., Kanazawa, K.K., and Kaufman, J.H. (1990) Physical description of a viscoelastically loaded at-cut quartz resonator. *J. Appl. Phys.*, **68**, 1993–2001.

44. Keller, C.A., Glasmastar, K., Zhdanov, V.P., and Kasemo, B. (2000) Formation of supported membranes from vesicles. *Phys. Rev. Lett.*, **84**, 5443–5446.

45. Cho, N.J., Kanazawa, K.K., Glenn, J.S., and Frank, C.W. (2007) Employing two different quartz crystal microbalance models to study changes in viscoelastic behavior upon transformation of lipid vesicles to a bilayer on a gold surface. *Anal. Chem.*, **79**, 7027–7035.

46. Höök, F. and Kasemo, B. (2007) The QCM-D technique for probing biomacromolecular recognition reactions. *Springer Ser. Chem. Sens. Biosens.*, **5**, 425–447.

47. Briand, E., Zach, M., Svedhem, S., Kasemo, B., and Petronis, S. (2010) Combined QCM-D and EIS study of supported lipid bilayer formation and interaction with pore-forming peptides. *Analyst*, **135**, 343–350.

48. Kanazawa, K. and Cho, N.J. (2009) Quartz crystal microbalance as a sensor to characterize macromolecular assembly dynamic. *J. Sens.*, **2009**, 824947.

49. Reimhult, E., Hook, F., and Kasemo, B. (2002) Vesicle adsorption on SiO_2 and TiO_2: dependence on vesicle size. *J. Chem. Phys.*, **117**, 7401–7404.

50. Yoshina-Ishii, C., Miller, G.P., Kraft, M.L., Kool, E.T., and Boxer, S.G. (2005) General method for modification of liposomes for encoded assembly on supported bilayers. *J. Am. Chem. Soc.*, **127**, 1356–1357.

51. Stadler, B., Falconnet, D., Pfeiffer, I., Hook, F., and Voros, J. (2004) Micropatterning of DNA-tagged vesicles. *Langmuir*, **20**, 11348–11354.

52. Sackmann, E. (1996) Supported membranes: scientific and practical applications. *Science*, **271**, 43–48.

53. Cho, N.J., Dvory-Sobol, H., Xiong, A., Cho, S.J., Frank, C.W., and Glenn, J.S. (2009) Mechanism of an amphipathic alpha-helical peptide's antiviral activity involves size-dependent virus particle lysis. *ACS Chem. Biol.*, **4**, 1061–1067.

54. Elazar, M., Liu, P., Rice, C.M., and Glenn, J.S. (2004) An N-terminal amphipathic helix in hepatitis C virus (HCV) NS4B mediates membrane association, correct localization of replication complex proteins, and HCV RNA replication. *J. Virol.*, **78**, 11393–11400.

55. Elazar, M., Cheong, K.H., Liu, P., Greenberg, H.B., Rice, C.M., and Glenn, J.S. (2003) Amphipathic helix-dependent localization of NS5A mediates hepatitis C virus RNA replication. *J. Virol.*, **77**, 6055–6061.

25
Electrochemical Impedance Spectroscopy (EIS)

Renate L.C. Naumann

25.1
Basic Principles

Electrochemical impedance spectroscopy (EIS) [1–3] is a well-known technique that has attracted considerable attention for use with polymer films. It is considered noninvasive as the film is only exposed to small alternating sinusoidal voltages used as excitation signals. Equally, small alternating currents as a response to the perturbation can give information about the dielectric properties of the film. These signals can be applied in a broad frequency range, which makes it possible to cover a wide range of time constants.

In the case of a polymeric film on a metal or semiconductor support, EIS is usually performed in an electrochemical cell consisting of the conducting support as the working electrode, a counter electrode, and a reference electrode (Figure 25.1). The AC voltage is applied between the working and the reference electrode. An electrolyte solution carries the electric current which is measured between the working and the counter electrode.

The sinusoidal voltage is given by

$$e = E \sin \omega t \tag{25.1}$$

where ω is the angular frequency $\omega = 2\pi f$, if f is the conventional frequency in hertz.

This voltage can be pictured as a rotating vector (phasor) quantity, the length of which is the amplitude E and the frequency of rotation is ω (Figure 25.2). The voltage as a function of time is the component E of the phasor $\hat{E}$. Given the current as a response to the sinusoidal voltage, it can be represented as a separate phasor $\hat{I}$, rotating at the same frequency with $\hat{E}$. The two phasors are usually phase shifted by a phase angle, ϕ (Figure 25.3). Consequently, the current is also phase-shifted according to

$$i = I \sin \omega t \tag{25.2}$$

These principles will now be applied to some simple circuits. Consider first a pure resistor and a sinusoidal voltage applied across it. According to Ohm's law the

Functional Polymer Films, First Edition. Edited by Wolfgang Knoll and Rigoberto C. Advincula.
© 2011 Wiley-VCH Verlag GmbH & Co. KGaA. Published 2011 by Wiley-VCH Verlag GmbH & Co. KGaA.

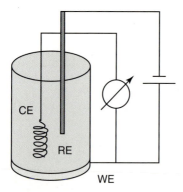

Figure 25.1 The electrochemical cell composed of the substrate covered by the polymer film used as working electrode (WE), the counter electrode (CE), and reference electrode (RE).

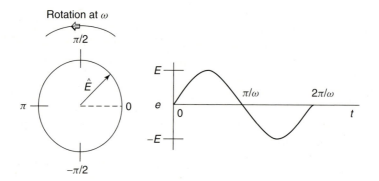

Figure 25.2 The sinusoidal alternating voltage.

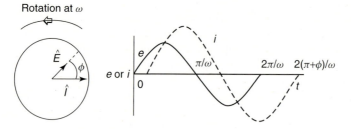

Figure 25.3 The sinusoidal alternating current phase shifted *vs.* the sinusoidal alternating voltage.

current will be

$$i = \left(\frac{e}{R}\right) \sin \omega t \qquad (25.3)$$

or in phasor notation

$$\hat{I} = \frac{\hat{E}}{R} \qquad (25.4)$$

The phase angle is zero and the current is in phase with the voltage.

Now consider a pure capacitance, C, the fundamental equations are

$$q = Ce \qquad (25.5)$$

$$i = \frac{dq}{dt} = C\frac{de}{dt} \qquad (25.6)$$

with

$$e = E \sin \omega t \qquad (25.7)$$

$$i = \omega C \cos \omega t \qquad (25.8)$$

with

$$i = I \sin(\omega t + \phi) \qquad (25.9)$$

and

$$\phi = \frac{\pi}{2} \qquad (25.10)$$

$$\Delta i = \omega CE \qquad (25.11)$$

$$\Delta i = \frac{E}{X_c} = \omega CE \qquad (25.12)$$

$$X_c = \frac{1}{\omega C} \qquad (25.13)$$

$$i = \frac{E}{X_c} \sin\left(\omega t + \frac{\pi}{2}\right) \qquad (25.14)$$

X_c is the capacitive reactance with the dimension of a resistance.

These relations can be more conveniently expressed in complex notation. First, we calculate the capacitive reactance of a capacitor

$$i = C\frac{de}{dt} \qquad (25.15)$$

$$e = Ee^{j\omega t} \qquad (25.16)$$

$$i = j\omega CEe^{j\omega t} \qquad (25.17)$$

$$i = Ie^{j\omega t} \qquad (25.18)$$

$$I = j\omega CE \qquad (25.19)$$

$$I = \frac{E}{Z_c} = j\omega CE \qquad (25.20)$$

$$Z_c = \frac{1}{j\omega C} \qquad (25.21)$$

Consider now E and I as complex quantities

$$\hat{E} = E' + jE'' \tag{25.22}$$

$$\hat{I} = I' + jI'' \tag{25.23}$$

According to Ohm's law the two quantities are connected by

$$\hat{Z} = \frac{\hat{E}}{\hat{I}} \tag{25.24}$$

where $\hat{Z}$ is the impedance with the real and imaginary part

$$\hat{Z} = Z' - jZ'' \tag{25.25}$$

and the phase angle

$$\tan\phi = \frac{Z''}{Z'} \tag{25.26}$$

with

$$Z' \geq 0 \tag{25.27}$$

The inverse of the impedance is the admittance, Y

$$\hat{Y} = \frac{1}{\hat{Z}} = \frac{\hat{I}}{\hat{E}} \tag{25.28}$$

which can be expressed in terms of real and imaginary parts of the impedance

$$\hat{Y} = \frac{1}{\hat{Z}} = \frac{Z'}{(Z')^2 + (Z'')^2} - j\frac{Z''}{(Z')^2 + (Z'')^2} \tag{25.29}$$

The conditions of two elements connected in parallel (for $e = \text{const}$) are

$$\frac{1}{R_{tot}} = \frac{1}{R_1} + \frac{1}{R_2} \tag{25.30}$$

$$C_{tot} = C_1 + C_2 \tag{25.31}$$

and the conditions for two elements in series are

$$R_{tot} = R_1 + R_2 \tag{25.32}$$

$$\frac{1}{C_{tot}} = \frac{1}{C_1} + \frac{1}{C_2} \tag{25.33}$$

With these relationships it is possible to set up the equations for different circuits. For example, a capacitance in series with a resistance, the total impedance will be

$$Z_{tot} = Z_c + Z_R = \frac{1}{j\omega C} + R \tag{25.34}$$

$$|Z| = \sqrt{((Z_c)^2 + (Z_R)^2)} = \sqrt{\left(-\frac{1}{\omega^2 C^2} + R^2\right)} \tag{25.35}$$

and the phase angle ϕ, is given by

$$\tan\phi = \frac{Z''}{Z'} = \frac{X_c}{R} = \frac{1}{\omega RC} \tag{25.36}$$

whereas the total impedance of a capacitance in parallel to resistance will be described by

$$\frac{1}{Z_{tot}} = \frac{1}{Z_c} + \frac{1}{Z_R} = j\omega C + \frac{1}{R}$$

$$Z_{tot} = \frac{R}{1+(\omega CR)^2} + j\frac{\omega CR^2}{1+(\omega CR)^2} = \frac{R}{1+j\omega CR} \qquad (25.37)$$

$$Z_{tot} = \frac{R}{1+j\omega CR}$$

$$\tan\phi = \frac{Z''}{Z'} = \omega CR \qquad (25.38)$$

Equations for other circuits can be set up accordingly.

25.1.1
Data Presentation

Impedance spectra can be represented in different formats. A simple equivalent circuit is represented in several of these formats in Figure 25.4. This circuit is

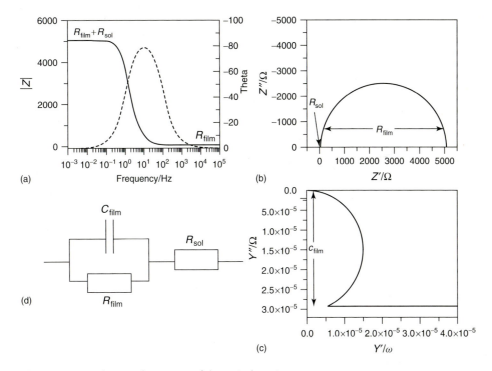

Figure 25.4 Simulations of EI spectra of the equivalent circuit (d) in different presentations, (a) Bode plot, (b) Nyquist plot, and (c) frequency-normalized admittance plot.

often used to represent a dielectric film immersed in a solution and consists of a resistor, R_{sol} connected in series with an RC element (R_{film}, C_{film}). The Bode plot is the magnitude of the impedance $|Z|$ and the phase angle *vs.* the frequency in hertz both in a logarithmic scale (Figure 25.4a). The magnitude has a limiting value of R_{sol} at high frequencies and a limiting value at the sum $R_{sol} + R_{film}$ at low frequencies. The transition between high-frequency and low-frequency asymptotes has a slope of -1 on the log-log scale. This indicates the increase of the capacitive resistance due to charging of the capacitance C_{film}. The complex plane, or Nyquist plot, is the real part of the impedance Z' vs. the imaginary part Z'' (Figure 25.4b). It yields a half-circle, the diameter of which represents the resistance R_m, while the intercept on the x-axis yields R_{sol}, the resistance of the solution. Finally, the admittance plot is the real part *vs.* the imaginary part of the admittance normalized to the circular frequency ω (Figure 25.4c). It also yields a half-circle, the diameter of which represents the capacitance of the dielectric film, C_{film}.

The impedance plots shown in Figure 25.4 are ideal plots, which are rarely seen in reality. Real systems are mostly characterized by suppressed arcs instead of ideal semicircles as seen in Figure 25.4. The reason is that a real system is composed of a mixture of elements whose time constants $\tau = RC$ are not sufficiently separated to allow for separate semicircles. One way out of this dilemma is the introduction of a constant-phase element (CPE) instead of a capacitor, which is characterized by a distribution of time constants indicated by the exponent α. For example, the expression of the real part of the impedance

$$Z' = \frac{R}{1 + (\omega\tau)^2} \tag{25.39}$$

where

$$\tau = RC \tag{25.40}$$

is replaced by

$$Z' = \frac{R\left[1 + (\omega\tau)^{1-\alpha} \cos\frac{\pi}{2}(1-\alpha)\right]}{\left[1 + (\omega\tau)^{1-\alpha} \cos\frac{\pi}{2}(1-\alpha)\right]^2 + \left[(\omega\tau)^{1-\alpha} \sin\frac{\pi}{2}(1-\alpha)\right]^2} \tag{25.41}$$

As α approaches unity the CPE approaches a simple capacitance.

25.1.2
Data Analysis

Although analytical solutions are available for many sets of equations, data analysis is usually performed by complex nonlinear least squares fit or CNLS-fit programs [4–6]. Such programs are commercially available, for example, LEVM by J. Ross Macdonald [6] or EquivCrt by B.A Boukamp [4]. Others are integrated into the electrochemical instrumentation, for example, ZPlot®, ZVIEW® available from Scribner Assoc. in association with frequency analyzers from Solartron Analytical, Farnborough, Hampshire, GU14 ONR, England, or FRA, the frequency-response analyzer from ECO Chemie B.V. the Netherlands. Most of these programs employ the Marquardt–Levenberg algorithm. For details see Boukamp [7], who pointed out, however, that in order to obtain a reliable CNLS-analysis, the system must be in equilibrium over the entire frequency range. The most effective way to test for possible nonequilibrium problems are the Kramers–Kronig transformations [8]. Careful preanalysis of an assumed equivalent circuit can lead to more appropriate equivalent circuits, see below [9].

25.2
Polymer Films

25.2.1
Corrosion-Protective Coatings

EIS is extensively employed to assess the electrical properties of polymer films used as corrosion protective coatings [10]. The dielectric film is usually considered in terms of a simple RC circuit with resistance R_{film} and capacitance C_{film} of the film connected in parallel (Figure 25.5c), however, often with the capacitance replaced by a CPE (Figure 25.5d). The two quantities are usually obtained from Bode plots. The resistance can be normalized over the measurement area and film thickness d and can be reported as the resistivity ρ (in Ω cm):

$$\rho = \frac{R_{film} \times A}{d} \tag{25.42}$$

The capacitance can be expressed as the dimensionless relative dielectric constant ε_r of the coating:

$$\varepsilon_r = \frac{C_{film} \times d}{\varepsilon_0 \times A} \tag{25.43}$$

where d is the thickness of the film, A the area, and ε_0 is the dielectric constant of free space (8.85×10^{-12} F m^{-1}).

The experimental setup consists of a plastic tube pressed on the film that contains the electrolyte and the reference electrode. If necessary the film can be glued to the metallic support.

For a coating providing good barrier properties impedance values in the order of $|Z| = 10^9–10^{14}$ were measured. Care has to be taken to use a setup designed to cover the appropriate frequency range. Three different types of polymer films were measured: Tedlar® (polyvinylfluoride) from Dupont®, OptiLiner™, and SupraLiner™ (polyethyleneterephthalate) from Saint Gobin Performance Plastics and 3M™ Polyurethane Protective Tape 8671. They were recommended as calibration standards that closely match the electrical properties of many coating materials [10].

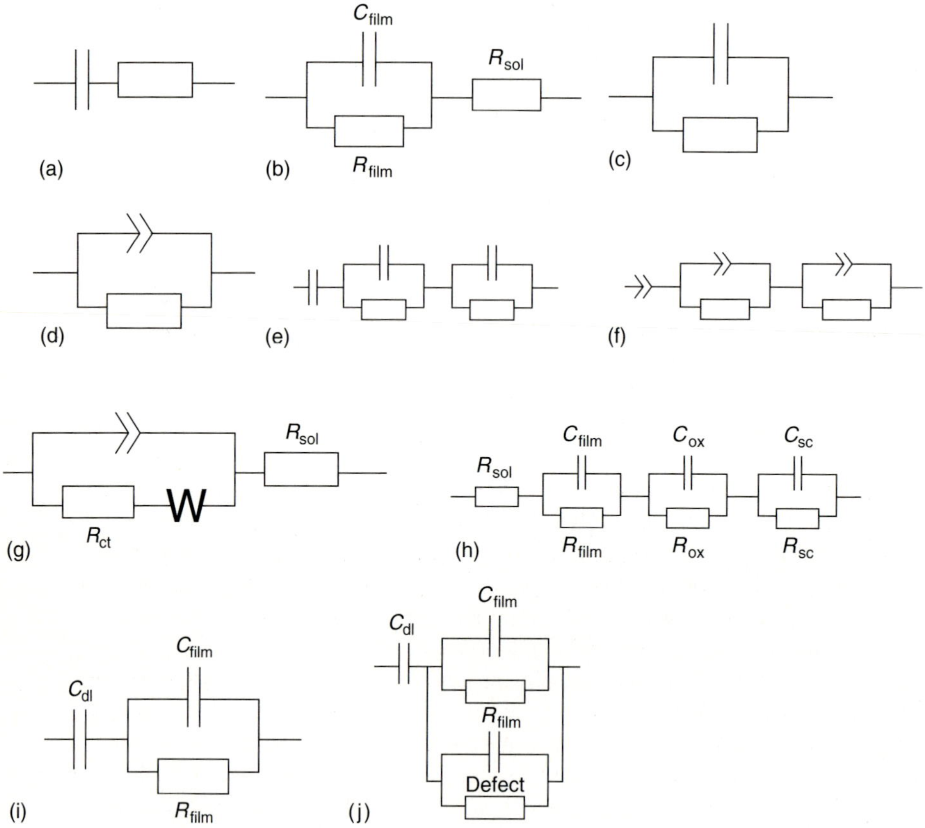

Figure 25.5 Various equivalent circuits, (a) is the equivalent circuit of a bare electrode immersed in an electrolyte solution with the capacitance of the double layer connected in series with the resistance of the electrolyte, (b) the simplest example of a polymer film immersed in solution, (c) and (d) a polymer film alone (e) and (f) two processes taking place at different frequencies (g) Randles equivalent circuit (h) equivalent circuit of a polymer film on a semiconductor surface (i) a lipid film on a lubricating water layer (j) a lipid film with defects. All other circuits are described in the text, $\gg$ stands for a constant phase element, W stands for a Warburg resistance, R_{ct} is a charge transfer resistance, C_{dl} a double layer capacitance, R_{sc} and C_{sc} are the resistance and capacitance of the space charge region, the other symbols are self-explanatory.

25.2.2
Ionic Conducting Films

Ionically conductive polymeric films (polymeric electrolytes) are used in such electrochemical devices as fuel cells, lithium batteries, and electrochromic windows [11]. EI spectra are recorded while the films are sandwiched between two stainless steel or Li electrodes. This allows measurements under blocking or nonblocking conditions. Nonblocking conditions mean that charge transfer with the electrode, for example, the Li electrode, is also allowed. The limiting value of ionic conductivity of the thin film used for the above-mentioned applications is equal to 10^{-4} S cm^{-1}. The values observed of long-time current densities of lithium batteries are about one order of magnitude lower than the values predicted on the basis of the Ohmic drop in the electrolyte layer. This phenomenon can be explained by the formation of a passivation layer at the interface between the polymeric electrolyte and the lithium anode [12]. Results of EIS measurements are usually given as conductivities σ_{dc} (in Ω^{-1} cm^{-1}). Valuable information can be obtained, if conductivities are measured as a function of temperature. Arrhenius plots of ionic conductivity σ_{dc} vs. temperature T are then analyzed with respect to the activation energy Q according to the Arrhenius equation

$$\sigma_{dc} T = \sigma_0 \exp\left(-\frac{Q}{k_B T}\right) \tag{25.44}$$

where σ_0 is the pre-exponential factor.

Examples of polymeric electrolytes are films made from polyethyleneoxide (PEO) and polymethylmethacrylate (PMMA) in the presence of electrolytes such as NaI or LiClO$_4$, sometimes mixed with inorganic (ceramic) grains. Such films, also called *blend-based polymer electrolytes* are complicated multiphase systems. EI spectra were analyzed [13] using the Almond–West formalism [14], which discriminates between the enthalpy of conduction due to charge carrier mobility and concentration. For details the reader should refer to the original literature.

Impedance spectra are usually presented as conductivity vs. frequency plots in the log-log presentation. They are mostly characterized by a high-frequency and low-frequency region. These two phenomena correspond to the dielectric relaxation in the bulk and to the response of the electrode/electrolyte interface. The dc response of the system is observed for the middle frequencies more clearly seen in the Nyquist plots. Here, two or three semiarcs can be assigned to different processes including ionic conductivity of the bulk and passivation of the interface [11, 15, 16]. They are modeled by equivalent circuits consisting of a number of parallel (RQ) subcircuits connected in series (Figure 25.5f).

Another important field is solid-state ionics with materials such as solid electrolytes, glasses of specific composition, or doped ceramics of oxidic origin [17, 18]. The complexity of these systems is like that of the polymeric systems described above and requires even more complicated equivalent circuits, which are hard to analyze. Progress made in the application of impedance spectroscopy to such materials was described by Boukamp [9]. It includes a careful validation of the data with respect to the equilibrium condition. In order to obtain a reliable CNLS-analysis of the EIS data the system must be in equilibrium over the entire frequency range. The most effective way to test for possible nonequilibrium problems are the Kramers–Kronig transformations [8]. Another important contribution to improve the reliability of EIS data is the preanalysis by deconvolution of the overall spectra. This method consists of a partial analysis of consecutive regions of the impedance. The partial CNLS fits are each subtracted from the overall CNLS fit. The frequency dispersion of the data is thus treated separately starting with the high-frequency region. The quality of the fit is optimized step-by-step. Complex equivalent circuits

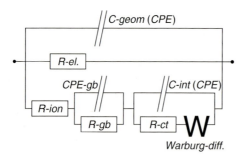

Figure 25.6 An example of an equivalent circuit that has been assembled during the subtraction procedure. Reprinted with permission from B. A. Boukamp [7].

can be assembled on the basis of the subtraction procedure providing access to different properties of the systems. An example of a complex equivalent circuit thus obtained is depicted in Figure 25.6, see Ref. [9].

Recent advantages in solid-state ionics, also described by Boukamp, include the Fourier transform (FT) analysis of time-domain relaxation data. Low-frequency impedances are thus accessible, which are required in order to characterize diffusion processes of ions in solid-state ionic materials [17].

An interesting new application of EIS is the temperature dependence of the protonic conductivity of solid acid compounds such as $CsHSO_4$ and CsH_2PO_4. These materials have possible applications as electrolytes in fuel cells operating at medium temperatures (100–300 °C) where they undergo a phase transition to a state of superprotonic conductivity. Examples of the application of EIS to determine this property were detected recently in solid acid phosphites [9].

25.2.3
Electron-Conducting Films

The best-known electron-conducting polymers are polyaniline (PAn), polypyrrole, polythiophene, and polyacetylene. They are usually prepared by anodic oxidation of the monomer, for example, aniline in acidic solutions [19]. Electropolymerization is considered to be an autocatalytic process. In the case of PAn, the polymer is nonconducting and colorless in the fully reduced state and conducting and intensely colored in the partially oxidized state. The reduced state is stable at potentials below 0.3 V and the oxidized state is stable between 0.4 and 0.9 V whereas at potentials >0.9 V PAn is "overoxidized," that is, irreversibly oxidized [20]. In a first approximation, impedance spectra of these polymers are often represented by a simple Randles equivalent circuit consisting of CPE, a charge transfer resistance R_{ct}, a Warburg element W, and the solution resistance R_{sol} (Figure 25.5g) [21].

A CPE is used to represent the nonhomogeneous nature of the electrode material and the distribution of relaxation times, with the impedance given by

$$Z_{CPE} = (1/\sigma) (j\omega)^{-\alpha} \tag{25.45}$$

where σ and α are positive constants. A CPE describes a capacitor when $\alpha = 1$, in which $\sigma = C$, and tends to a resistor as $\alpha \to 0$.

Electron exchange with the electrode gives rise to a faradaic current. The linearized function of this current is taken into account as the charge transfer resistance, R_{ct}.

In order to maintain electroneutrality, ions must penetrate the polymer. Diffusion is represented by a Warburg element, with frequency dispersion obtained from the solution of the one-dimensional Fick's second law for diffusion in a finite region, with blocking boundary for ions at the metal/polymer interface. The resulting frequency dispersion is

$$Z(\omega) = \frac{B^2}{C} \frac{\coth\left(B\left(j\omega\right)^{1/2}\right)}{B\left(j\omega\right)^{1/2}} \tag{25.46}$$

C is the intrinsic capacitance, $B = \left(\delta^2/D\right)^{1/2}$ where δ is the width of the film and D the ionic diffusion coefficient.

Impedance spectra of polypyrrole films were recorded at different potentials and in the presence of different cations and anions. Warburg elements could be seen as straight lines in the low-frequency range but only for thin films. They are due to the diffusion of ions within the polymer taken up for reasons of electroneutrality. Diffusion coefficients were calculated according to Equation 25.46. Oxidation of the polymer backbone generates charge carriers, which confer electronic conductivity to the polymer. The transition between insulator and conductor properties could be deduced from the charge-transfer resistance as a function of potential [21].

In this context it should be mentioned that conducting polymer films can adopt mixed ionic–electronic conduction properties [9, 22]. This was shown by EIS in an early example of the investigation of polypyrrole films. Their conductivity depends not only on the kind and concentration of electrolyte solutions, but also of the time of immersion. Data were analyzed in terms of the thickness of the polymer and diffusion coefficients of the ions in question.

Conducting polymers are a type of polymer containing conjugated double bonds in the backbone of their macromolecule. This allows free movement of electrons within the conjugation length. To be electrically conducting the polymers must be doped with counterions opposite in sign to charges present in the polymer. This was investigated by EIS in the case of various polythiophenes [23]. Spectra were recorded at different potentials in the presence of TBAPF$_6$ and LiClO$_4$. Only in the case of TBAPF$_6$ could both anions and cations be trapped inside the polymer. The uptake of ClO$_4$ ions was shown to be rate limiting for the doping process in the case of poly(3-methoxythiophene) [24].

Doping was also induced by mixing polyimide and polypyrrole to obtain a polypyrrole/polyimide composite [25]. This was concluded from an increase in the pseudocapacitance with increasing cathodic potentials applied to the polymer film. The increased pseudocapacitance makes such materials interesting as electrical storage devices (supercapacitors).

It usually takes a long time to measure an electrochemical impedance (EI) spectrum over the entire frequency range. Stationary systems, therefore, are accessible by EIS measurements. Processes that are faster than the measurement time, however, are obscured, as, for example, earlier phases of aniline polymerization. This problem was overcome by FTEIS, which made it possible to follow electrochemical processes of aniline polymerization as a function of potential in real time [26].

Another recent development is scanning probe impedance spectroscopy. Electrical properties of polymer films could be observed on the nanoscale [27].

25.2.4
Conductive Films on Semiconductor Supports

In order to obtain the electrical properties of dielectric films on semiconductor surfaces such as silicon or ITO (indium tin oxide), the capacitance of the space-charge region of the semiconductor has to be taken into account [28]. This is done by first estimating the flat-band potential by a Mott–Schottky plot (C^2 vs. potential). The space-charge capacitance between potentials of 0 and -1 V vs. Ag/AgCl was calculated at the range of 1.3–1.9 nF. The equivalent circuit of the adsorbed organic film shown in Figure 25.5h consists of an RC element (R_{sc}, C_{sc}) representing the space charge region, an RC element (R_{ox}, C_{ox}) representing the silicon oxide layer, a circuit representing the adsorbed film (R_{film}, C_{film}) and the resistance of the solution, R_{sol}. If the adsorbed film is defect-free, the capacitance of the layer is determined by the film, as long as $C_{film} \ll C_{sc}$. C_{film} can be estimated from the thickness and the dielectric constant of the film according Equation 25.43 rearranged to

$$C_{film} = \frac{\varepsilon_0 \varepsilon_r A}{d} \qquad (25.47)$$

25.3
Stratified Films

EIS has played a prominent role in the development of polymer-supported lipid films on solid surfaces [29–36]. These stratified layer structures can be well represented by equivalent circuits consisting of the same electric elements described above. The evaluation, following the same principles, has some additional characteristic features that will be described in some detail below.

25.3.1
Solid-Supported (s)BLMs

The simplest arrangement a lipid bilayer membrane (BLM) can adopt on a planar support is a supported BLM or sBLM, that is, the two leaflets of a BLM directly attached to the metallic or semiconductor support [35, 37]. For example, this is achieved by so-called hybrid lipid bilayers consisting of a self-assembled monolayer of long-chain alkanethiols onto which a second monolayer of phospholipids is assembled by vesicle spreading [38]. This structure is generally presented by the circuit shown in Figure 25.5c or d. Another possibility is the formation of a complete phospholipid bilayer by vesicle spreading on a hydrophilic surface. In this case the hydrophilic head-groups of the lower leaflet are separated from the support by a thin (1–2 nm) lubricating water layer. This structure is usually represented by the equivalent circuit shown in Figure 25.5i. It differs from the circuit given in Figure 25.5c by a capacitance representing the electrical double layer of the water film (C_{dl}) beneath the dielectric film. The capacitance of the film, C_{film}, is $0.5\ \mu F\ cm^{-2}$ according to Equation 25.47 with $\varepsilon_r = 2$. This is much smaller than C_{dl} (30 to ideally 70 $\mu F\ cm^{-2}$ on a gold electrode). According to Equation 25.33, with C_{film} and C_{dl} arranged in series, the overall capacitance is dominated by C_{film}. However, if the lipid layer has defects, another equivalent circuit has to be taken into account (Figure 25.5j), with two RC elements arranged in parallel. In this case, according to Equation 25.31, the overall capacitance is dominated by C_{dl}. This illustrates that the EIS spectrum is very sensitive to defects in the structure of a lipid film.

25.3.2
Polymer-Supported BLMs and tBLMs

Polymer-supported BLMs are a very important tool for the investigation of lipid bilayers [39]. Many physical techniques were successfully applied. However, in order to perform electrical measurements they have to meet the requirements of the Giga-seal known from electrophysiological studies. This is very difficult to achieve considering that the structure is mostly represented by the equivalent circuit Figure 25.5j rather than Figure 25.5i. In this case, C_{dl} represents the capacitance of the polymer connected in series with the RC circuit of the film. The same problem then arises as for the sBLM if the lipid film is not absolutely defect-free. EIS measurements on polymer-supported BLMs are therefore relatively sparse.

In order to overcome this problem, tethered bilayer lipid membranes (tBLMs) were introduced where hydrophilic spacer groups are covalently attached to the lower leaflet of the lipid bilayer [30, 33, 40, 41]. EIS measurements of these structures showed resistances and capacitances of the lipid bilayer of 3–12 MΩ cm and 0.5 $\mu F\ cm^{-2}$, well comparable to the properties of a freely suspended BLM. Ion channels and ion carriers were inserted in these tBLMs and ion transport were measured by EIS.

25.3.3
Ion Transport through Channels Incorporated into a tBLM

Impedance spectra obtained from channels and carriers incorporated into tBLMs or sBLMs are usually analyzed in terms of the equivalent circuit Figure 25.5i. Ion transport is indicated by a decrease of the resistance of the lipid film as a function of kind and concentration of ions, and as a function of the potential difference applied across the entire layer structure [33]. The resistance can be transferred into a conductance, the value of which is characteristic for the particular ion. This gives information about the specificity of the channel. Ion transport through channels in sBLMs or tBLMs can thus conveniently be measured in terms of electrical properties. However, information as to the mechanism of the transport process cannot be obtained from these data.

Attempts have been made to interpret the fitted parameter values of impedance spectra in terms of potential-dependent rate equations [42–44]. Guidelli and coworkers [44] interpreted the multiple arcs of a $-\omega Z''$ vs. $\omega Z'$ plot in terms of the layer structure depicted in Figure 25.7. They added a very important aspect to the discussion, that of considering the stratified metal/polymer/lipid interface in terms of an electrified interface. A specific potential drop is thus attributed to every layer. However, modeling ion transport by a resistor is an oversimplification from which relevant kinetic information cannot be obtained [45].

Gervasi and Vallejo [42] developed a mathematical treatment of ion transport through gramicidin channels in sBLMs. The impedance as a function of frequency is calculated from potential-dependent rate equations taking into account concentrations of ions transported through the channel. Fitting the data to the equations yields the rate constants. These authors claim to be able to distinguish between different models of ion transport.

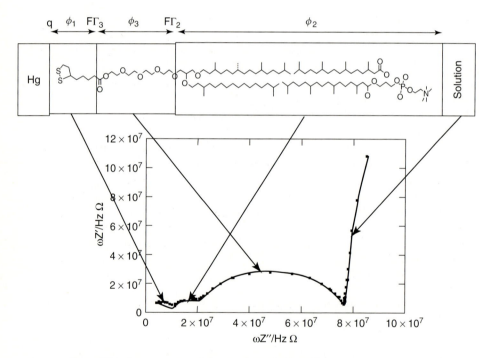

Figure 25.7 Representation of the correlation between the semicircles of the $\omega Z'$ versus $\omega Z''$ plot and the different substructural elements of the tBLM. The location of the charges and the potential differences across the different dielectric slabs are reported on top of the figure. Reprinted with permission from L. Becucci *et al.*, [44].

These model calculations, however, do not account for the potential drop across different layers of the film structure and the concentration gradients throughout the system. Accordingly, we have designed a model that combines a mathematical description of processes across membranes and passive electrical elements [46]. Parameters, such as concentration and volume in different parts of the layer structure, can then be taken directly into account, together with the electrical potential and other electrical parameters. We use a simulation program with integrated circuit emphasis (SPICE) to explore the behavior of this model and to simulate electrochemical impedance spectra. In the case of passive ion transport across tBLMs measured as a function of potential we were able to discriminate between the solubility-diffusion and the pore model of ion transport.

References

1. Macdonald, J.R. (1987) *Impedance Spectroscopy: Emphasizing Solid Materials and Systems*, John Wiley & Sons, Inc., New York.
2. Bard, A.J. and Faulkner, L.R. (2001) *Electrochemical Methods, Fundamentals and Applications*, 2nd edn, John Wiley & Sons, Inc.
3. Orazem, M.E. and Tribollet, B. (2008) *Electrochemical Impedance Spectroscopy*, John Wiley & Sons, Inc., Hoboken, NJ.
4. Boukamp, B.A. (1986) A nonlinear least-squares fit procedure for analysis of emittance data of electrochemical systems. *Solid State Ionics*, **20**, 31–44.
5. Boukamp, B.A. (1986) A package for impedance admittance data-analysis. *Solid State Ionics*, **18-9**, 136–140.
6. Macdonald, J.R. (1990) Impedance spectroscopy – old problems and new developments. *Electrochim. Acta*, **35**, 1483–1492.
7. Boukamp, B.A. (2004) Impedance spectroscopy, strength and limitations. *Tech. Mess.*, **71**, 454–459.
8. Boukamp, B.A. (1995) A linear Kronig–Kramers transform test for immittance data validation. *J. Electrochem. Soc.*, **142**, 1885–1894.
9. Boukamp, B.A. (2004) Electrochemical impedance spectroscopy in solid state ionics: recent advances. *Solid State Ionics*, **169**, 65–73.
10. Bonitz, V.S., Hinderliter, B.R., and Bierwagen, G.P. (2006) Commercial polymer films as calibration standards for EIS measurements. *Electrochim. Acta*, **51**, 3558–3565.
11. Borkowska, R., Siekierski, M., and Przyluski, J. (1996) An electrochemical impedance spectroscopy study of thin polymeric films. *Appl. Surf. Sci.*, **92**, 447–451.
12. Jamnik, J. (1998) Influence of boundaries on ionic conductance. *Interfacial Sci. Ceram. Joining*, **58**, 443–455, 480.
13. Siekierski, M. and Wieczorek, W. (1993) Description of AC response of Blend-based polymer electrolytes. *Conduct. Polym. – Trans. Phenom.*, **122**, 215–228, 269.
14. Almond, D.P. and West, A.R. (1986) Entropy effects in ionic-conductivity. *Solid State Ionics*, **18-19**, 1105–1109.
15. Borkowska, R., Laskowski, J., Plocharski, J., Przyluski, J., and Wieczorek, W. (1993) Performance of acrylate-poly(ethylene oxide) polymer electrolytes in lithium batteries. *J. Appl. Electrochem.*, **23**, 991–995.
16. Siekierski, M. and Wieczorek, W. (1993) Application of the universal power-law to the studies of AC conductivity of polymeric electrolytes. *Solid State Ionics*, **60**, 67–71.
17. Zhou, W., Bondarenko, A.S., Boukamp, B.A., and Bouwmeestter, H.J.M. (2008) Superprotonic conductivity in MH(PO3H). *Solid State Ion.*, **179**, 380–384.

18. Berkemeier, F., Abouzari, M.S., and Schmitz, G. (2007) Thickness dependent ion conductivity of lithium borate network glasses. *Appl. Phys. Lett.*, **90**, 113110–113113.

19. Lohrengel, M.M. and Genz, O. (1995) Mechanism of the redox process of conducting polymers. *Ionics*, **1**, 304–310.

20. Johnson, B.J. and Park, S.-M. (1996) Electrochemistry of conducting polymer XIX. Oxidation of aniline at bare and polyaniline-modified platinum electrodes. *J. Electrochem. Soc.*, **143**, 1269–1276.

21. Mostany, J. and Scharifker, B.R. (1997) Impedance spectroscopy of undoped, doped and overoxidized polypyrrole films. *Synth. Met.*, **87**, 179–185.

22. Jamnik, J., Maier, J., and Pejovnik, S. (1999) A powerful electrical network model for the impedance of mixed conductors. *Electrochim. Acta*, **44**, 4139–4145.

23. Fall, M., Diagne, A.A., and Dieng, M.M. (2005) Electrochemical impedance spectroscopy. *Synth. Met.*, **155**, 569–575.

24. Ding, H., Pan, Z., Pigani, L., Seeber, R., and Zanardi, C. (2001) P- and n-doping processes in polythiophene with reduced bandgap. An electrochemical impedance spectroscopy study. *Electrochim. Acta*, **46**, 2721–2732.

25. Iroh, J.O. and Levine, K. (2003) Capacitance of the polypyrrole/polyimide composite by electrochemical impedance spectroscopy. *J. Power Sources*, **117**, 267–272.

26. Layson, A., Gadad, S., and Teeters, D. (2003) Resistance measurements at the nanoscale: scanning probe ac impedance spectroscopy. *Electrochim. Acta*, **48**, 2207–2213.

27. Hong, S.Y. and Park, S.M. (2007) Electrochemistry of conductive polymers 40. Earlier phases of aniline polymerization studied by Fourier transform electrochemical impedance spectroscopy. *J. Phys. Chem. B*, **111**, 9779–9786.

28. Vlasak, R., Klueppel, I., and Grundmeier, G. (2007) Combined EIS and FTIR-ATR study of water uptake and diffusion in polymer films on semiconducting electrodes. *Electrochim. Acta*, **52**, 8075–8080.

29. Knoll, W., Koper, I., Naumann, R., Schiller, S., and Duran, R.S. (2004) Functional tethered bimolecular lipid membranes. *Abstr. Pap. Am. Chem. Soc.*, **227**, U853.

30. Lang, H., Duschl, C., and Vogel, H. (1994) A new class of thiolipids for the attachment of lipid bilayers on gold surfaces. *Langmuir*, **10**, 197–210.

31. Pace, R.J., Braach-Maksvytis, V.B.L., King, L.G., Osman, P.D.J., Raguse, B. *et al.* (1998) The gated ion channel biosensor – a functioning nano-machine. *Methods Ultrasensitive Detect.*, **3270**, 50–59, 228.

32. Plant, A.L. (1993) Self-assembled phospholipid alkanethiol biomimetic bilayers on gold. *Langmuir*, **9**, 2764–2767.

33. Raguse, B., Braach-Maksvytis, V., Cornell, B.A., King, L.G., Osman, P.D.J. *et al.* (1998) Tethered lipid bilayer membranes: formation and ionic reservoir characterization. *Langmuir*, **14**, 648–659.

34. Schiller, S.M., Naumann, R., Lovejoy, K., Kunz, H., and Knoll, W. (2003) Archaea analogue thiolipids for tethered bilayer lipid membranes on ultrasmooth gold surfaces. *Angew. Chem. Int. Ed.*, **42**, 208–211.

35. Steinem, C., Janshoff, A., Ulrich, W.P., Sieber, M., and Galla, H.J. (1996) Impedance analysis of supported lipid bilayer membranes: a scrutiny of different preparation techniques. *Biochim. Biophys. Acta Biomembr.*, **1279**, 169–180.

36. Stelzle, M., Weissmuller, G., and Sackmann, E. (1993) On the application of supported bilayers as receptive layers for biosensors with electrical detection. *J. Phys. Chem.*, **97**, 2974–2981.

37. Purrucker, O., Hillebrandt, H., Adlkofer, K., and Tanaka, M. (2001) Deposition of highly resistive lipid bilayer on silicon-silicon dioxide electrode and incorporation of gramicidin studied by ac impedance spectroscopy. *Electrochim. Acta*, **47**, 791–798.

38. Silin, V.I., Wieder, H., Woodward, J.T., Valincius, G., Offenhausser, A. *et al.* (2002) The role of surface free energy on the formation of hybrid bilayer

membranes. *J. Am. Chem. Soc.*, **124**, 14676–14683.

39. Tanaka, M. and Sackmann, E. (2005) Polymer-supported membranes as models of the cell surface. *Nature*, **437**, 656–663.

40. Naumann, R., Schiller, S.M., Giess, F., Grohe, B., Hartman, K.B. *et al.* (2003) Tethered lipid Bilayers on ultraflat gold surfaces. *Langmuir*, **19**, 5435–5443.

41. Schiller, S., Naumann, R., Lovejoy, K., and Knoll, W. (2003) Thiolipids for tethered bilayer membranes. *Abstr. Pap. Am. Chem. Soc.*, **225**, U539–U540.

42. Vallejo, A.E. and Gervasi, C.A. (2002) Impedance analysis of ion transport through gramicidin channels in supported lipid bilayers. *Bioelectrochemistry*, **57**, 1–7.

43. Steinem, C., Janshoff, A., Galla, H.J., and Sieber, M. (1997) Impedance analysis of ion transport through gramicidin channels incorporated in solid supported lipid bilayers. *Bioelectrochem. Bioenerg.*, **42**, 213–220.

44. Becucci, L., Moncelli, M.R., Naumann, R., and Guidelli, R. (2005) Potassium ion transport by valinomycin across a Hg-supported lipid bilayer. *J. Am. Chem. Soc.*, **127**, 13316–13323.

45. Naumann, R., Walz, D., Schiller, S.M., and Knoll, W. (2003) Kinetics of valinomycin-mediated K+ ion transport through tethered bilayer lipid membranes. *J. Electroanal. Chem.*, **550**, 241–252.

46. Robertson, J.W.F., Friedrich, M.G., Kibrom, A., Knoll, W., Naumann, R.L.C. *et al.* (2008) Modeling ion transport in tethered bilayer lipid membranes. 1. Passive ion permeation. *J. Phys. Chem. B*, **112**, 10475–10482.

26
Characterization of Responsive Polymer Brushes at Solid/Liquid Interfaces by Electrochemical Impedance Spectroscopy

Omar Azzaroni and Claudio Gervasi

26.1
Introduction

Characterization of thin and ultrathin polymer films has always attracted great and widespread interest, both for fundamental intellectual and technological challenges, to chemists, materials scientists, physicists, and engineers. This interest stems from their usefulness to a wide range of technological applications that include photolithography, liquid-crystal displays, sensors, or antireflection coatings, among others. Furthermore, as very thin films are incorporated into device applications, it is also necessary to have a better understanding of their physicochemical properties. A more detailed knowledge of the characteristics of polymers deposited on solid substrates is critical for understanding and improving the performance of polymers in numerous applications and is also necessary for further work directed toward the molecular design of soft surfaces. As is well known in the science and technology of surfaces, unraveling and quantifying the physicochemical details of processes occurring within a thin polymer layer is often a challenge. In this regard, impedance spectroscopy is increasingly recognized as a very powerful technique for studying and characterizing thin polymer films deposited on solid supports. Herein, we review some of the basic aspects of electrochemical impedance spectroscopy (EIS) as well as a critical examination of key examples from the literature illustrating its use to investigate polymer brushes. From these examples it is hoped that the reader will get an understanding not only of the information that can be obtained from EIS but also of the way in which this readily available electrochemical technique can open new opportunities to explore the physical properties of surface-confined macromolecular architectures.

26.2
Electrochemical Impedance Spectroscopy – Basic Principles

EIS is a versatile method for probing the features of surface-modified conducting supports [1]. A small-amplitude perturbing sinusoidal voltage signal $\Delta V(t)$ is

Functional Polymer Films, First Edition. Edited by Wolfgang Knoll and Rigoberto C. Advincula.
© 2011 Wiley-VCH Verlag GmbH & Co. KGaA. Published 2011 by Wiley-VCH Verlag GmbH & Co. KGaA.

applied to the electrode surface, and the resulting current response $\Delta I(t)$ is measured. A quite complex response in the time domain is obtained when the perturbation is not sinusoidal, but of some other function of time t. The response is found as the solution of (a set of) differential equations. A relatively simple method to find a general solution is the method of Laplace transforms. In formal notation, with $\mathcal{L}$ = the Laplace operator and s = the Laplace parameter:

$$\Delta V = \Delta V(t), \; \mathcal{L}(\Delta V) = \Delta V(s)$$
$$\Delta I = \Delta I(t), \; \mathcal{L}(\Delta I) = \Delta I(s)$$

For small perturbations, that is, a linear system, the impedance is the ratio of the Laplace transforms of potential and current oscillations with time around the steady-state values;

$$Z(s) = \frac{\Delta V(s)}{\Delta I(s)}$$

With $s = j\omega$ the complex impedance can be written as (Equation 26.1)

$$Z(j\omega) = \frac{\Delta V(j\omega)}{\Delta I(j\omega)} = Z_{re}(\omega) + jZ_{im}(\omega) \tag{26.1}$$

where $j = \sqrt{-1}$, $\omega = 2\pi f$ (in radians/second), and f is the excitation frequency (in hertz). The complex impedance can be represented as the sum of the real, $Z_{re}(\omega)$, and imaginary $Z_{im}(\omega)$ components that originates from the interfacial relaxation processes.

One approach to model the impedance of the processes taking place at the interface uses circuit analogs or equivalent circuits that describe the experimental impedance spectra. A most popular equivalent circuit is that initially introduced by Randles [2], commonly used to model interfacial phenomena, that includes the ohmic resistance of the electrolyte solution (R_s), the Warburg impedance (Z_W) resulting from the semi-infinite diffusion of electroactive species from the bulk electrolyte to the electrode interface, the double layer capacitance (C_{dl}) and the electron transfer resistance (R_{et}), that represents the charge transfer process of a redox probe at the surface (Figure 26.1).

The parallel connection between the elements C_{dl} and $Z_W + R_{et}$ indicates that the displacive current (associated with the charging process of the interfacial capacitance on the gold surface) and the faradaic current (due to diffusion and charge transfer of the electroactive species) are considered independent of one

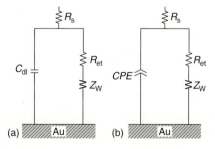

Figure 26.1 Typical Randles equivalent circuit at a gold electrode commonly used in impedance-spectroscopy studies including (a) the double-layer capacitance or (b) a constant phase element (CPE).

another and so the total current through the working interface is the sum of the two distinct contributions [3]. Provided that all of the current must pass through the uncompensated resistance of the electrolyte solution, R_s, is inserted as a series element in the circuit.

Constant-phase elements (CPEs) are used extensively in equivalent electrical circuits for fitting experimental impedance data exhibiting frequency dispersion. The CPE behavior is generally attributed to distributed surface reactivity, surface inhomogeneity, roughness or fractal geometry, electrode porosity, and to current and potential distributions associated with electrode geometry. The frequency dispersion is generally attributed to a "capacitance dispersion" expressed in terms of a CPE.

$$\text{CPE} = A^{-1}(j\omega)^{-n} \tag{26.2}$$

The CPE is represented by two parameters, A and n [3].

A complex-plane representation for impedance data according to the Randles circuit (Nyquist plot) is shown in Figure 26.2. A typical EIS response describes a semicircle region followed by a straight line. The semicircle region (obtained at high frequencies) corresponds to the electron-transfer-limited processes, whereas the straight line with $-45°$ slope (low frequencies) represents semi-infinite linear diffusion or Warburg diffusion. Hence, the electron-transfer kinetics and the diffusional characteristics can be extracted from the EIS data. The semicircle diameter represents R_{et} whereas the intercept of the semicircle with the Z_{re}-axis for $\omega \to \infty$ corresponds to the solution resistance, R_s.

Even though simple inspection of Nyquist plots for experimental data gives a broad hint of the type of processes taking place at the electrode surface, the computational fitting of the experimental data in accordance to the theoretical impedance of the equivalent circuit allows quantitative values to be obtained for the macroscopic circuit parameters that are related to the interfacial processes.

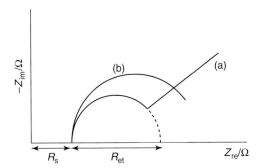

Figure 26.2 Simplified schematic representation of a Nyquist plot describing the impedance spectra of (a) a modified electrode in which the impedance response is governed by diffusion of the redox probe and the interfacial electron transfer and (b) a modified electrode in which the impedance response is mostly dominated by the interfacial electron-transfer process.

26.3
Electrochemistry as a Tool to Characterize Thin Polymer Films

Electrochemical measurements were employed in the past in highly specialized fields. More recently, this situation has changed quite rapidly, being applied toward interdisciplinary research. The two central aspects of this change are: (i) the use of nonelectrochemical techniques in electrochemical studies (e.g., *in situ* techniques) and (ii) a tendency to use electrochemical tools in numerous scientific fields, with particular emphasis on surface and materials science [4]. Among the different electrochemical techniques, EIS has become particularly valuable to many scientists in the soft-matter community as an alternative characterization method complementing the information achieved using other techniques. EIS has been recently employed to study a wide range of soft-matter-based systems [5], but this chapter will focus on surveying recent progress made in the field of polymer brushes since they represent key building blocks to construct highly functional thin films. Due to the increasing interest from the macromolecular and materials-science community, particular emphasis will be placed on the application of EIS to study their responsive properties under different environmental conditions. When a polyelectrolyte (PE) brush grafted to a conductive surface comes into contact with an electrolyte solution the resulting system acts as a "modified electrode." Here, EIS serves as a sensitive probe of structural changes in the macromolecular array and provides an initial assessment of brush permeability. Consequently, the technique may be used to characterize not only structural features of the brush but their impact on permeation and reaction of redox probe molecules at the substrate/brush interface, as well. The conformation of the polymer chains in the brush depends strongly on external conditions; especially the ionic strength of the surrounding medium. Thus, conformational changes of the brush may slow down permeability and electrochemical reaction rates. EIS is ideally suited to the study of these steps of the global process in a single experiment, while separating the individual contributions when they exhibit different relaxation time constants.

The typical reaction considered to measure the impedance response of the modified electrode is the reduction of ferricyanide at potentials sufficiently cathodic to allow the anodic reaction to be ignored, yet sufficiently anodic to avoid reduction of oxygen (as a side reaction). Under these conditions, the reaction

$$Fe(CN)_6^{3-} + e^- \; \underset{k_a}{\overset{k_c}{\rightleftarrows}} \; Fe(CN)_6^{4-} \tag{26.3}$$

provides an example of a first-order reaction involving mass-transfer-limited species undergoing a subsequent charge-transfer step.

The barrier effect of a grafted film or deposited multilayer is usually qualitatively analyzed in terms of the voltammetric response of the bare electrode relative to the response of the film-covered electrode (Figure 26.3). Thus, the voltammogram of the $Fe(CN)_6^{3-/4-}$ redox system presents a quasireversible behavior at the bare electrode, while the blocking properties of the modified electrode surface bring about a change in shape of the system's response characterized by an almost

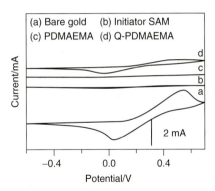

Figure 26.3 Cyclic voltammograms of $Fe(CN)_6^{3-/4-}$ on different electrodes: bare gold, initiator-modified gold, poly (dimethylamino)ethyl methacrylate (PDMAEMA) functionalized electrode, and a quaternized PDMAEMA brush. Reprinted with permission from Ref. [6] by American Chemical Society.

featureless voltammetric trace with peaks that may become no longer detectable (see Figure 26.3) [6].

In this situation EIS can still be used to obtain quantitative information on the system. If the electron-transfer reaction takes place at bare spots on the electrode surface, the fractional coverage may be estimated by comparing the charge-transfer resistance of the modified surface with the charge-transfer resistance of the substrate surface without modification. If the voltammetric response, although no longer quasireversible, still allows determinations of peak currents, then one may be tempted to calculate the fractional coverage in terms of the ratio of peak currents (I_p) for the bare and covered electrodes. However, it has been demonstrated that this approach is inappropriate for describing the fractional coverage because of the dominance of radial diffusion near each defect site and, as a result, peak currents that are not a simple function of the exposed area [7]. Consequently, values for fractional coverage obtained from impedance spectroscopy data are more reliable. Moreover, the high-frequency measurement of the electrode's impedance, allowing the double-layer capacitance (C_{dl}) to be obtained, is also a fruitful method to follow the percentage of covered surface [8].

Charge-transfer dynamics at the electrochemical interface are strongly influenced by the nature of the electrode surface and the structure of the electrical double layer. Recent studies of electrodes modified by grafted PE brushes have shown that electron transfer to a species in solution is retarded either by reducing the active area of the electrode or by preventing the redox species from approaching the electrode closely [6]. This phenomenon improves the measurement of fast charge-transfer kinetics resulting from a reaction of the type of Equation (26.3) which otherwise is difficult to observe on bare metal electrodes because it is diffusion limited over most of the polarizable range of the electrode. The impedance of an electrode covered with thin permeable polymer films undergoing heterogeneous electron transfer is usually described in terms of an equivalent circuit derived from the model by Randles [9, 10]. Effects of solution resistance, double-layer charging, and currents due to diffusion or to other processes occurring in the surface film can be observed more explicitly using impedance spectroscopy as compared to chronoamperometry and cyclic voltammetry. If the electron-transfer reaction takes place at bare spots on the electrode surface, the fractional coverage may be estimated by comparing the charge-transfer resistance of the modified surface with the charge-transfer resistance of the surface without modification.

26.4
Probing the Responsive Properties of Polymer Brushes through EIS Measurements

Polymer brushes are increasingly used in many physical systems such as colloid stabilization, polymeric surfactants, polymer compatibilizers, and copolymer microphases [11]. A polymer brush consists of end-tethered polymer chains stretched away from the substrate so that in the given solvent the brush height (h) is large compared to the end-to-end distance of the same nongrafted chains dissolved in the same solvent. In polymer brushes the distance between grafting points (d) is smaller than the chain end-to-end distance. Typically, these systems are characterized by a balance of elastic and excluded-volume interactions. There is also increasing interest in PE brushes, where electrostatic interactions between charges on the polymer and ions in solution play a dominant role. To some extent, PE brushes have been considered as a new class of material provided that strong segment–segment repulsions and the electrostatic interactions bring about completely new physical properties contrasted with noncharged polymer chains. The range of possible electrostatic interactions in PE brushes introduces new opportunities for manipulating the brush structure and its corresponding properties. For example, PEs in solution change from an extended conformation in low ionic strength solutions to a coiled conformation in solutions of high ionic strength. When PEs are covalently attached to a surface to form a PE brush, the polymer chains show a strongly extended (stretched) conformation in pure water, as a result of both repulsion between neighboring chains and repulsion between monomers. Conversely, when polymer brushes are placed in electrolyte solutions, the charges of the pendant groups in the polymer chains are screened, and minimization of the electrostatic repulsions leads to entropically more favorable collapsed conformations. This collapse is comparable to transitions from extended to coil-like conformations commonly observed for PEs in solution. As a result of the constraints on the degrees of freedom related to conformational changes, the PE brush collapse is accompanied by a decrease in film thickness and water content.

In this context, EIS has emerged as a very versatile tool to explore diverse responsive characteristics of PE brushes in different environments. For example, it is well known that poly (N-isopropylacrylamide) (PNIPAM)-based polymers precipitate in water in response to an increase of temperature. Hence, it is reasonable to suppose that the interfacial properties of Au electrodes modified with PNIPAM brushes will display thermocontrolled electrochemical properties [12]. Figure 26.4 describes the variation of the electron-transfer resistance (ETR), as obtained from modeling the impedance response, upon cycling the temperature of the electrolyte solution below and above the lower critical solubility temperature (LCST) of PNIPAM. The low interfacial ETR ($R_{et} \sim 100\,k\Omega$) at 25 °C reveals that the polymer is in an expanded coil state, whereas the high interfacial ETR

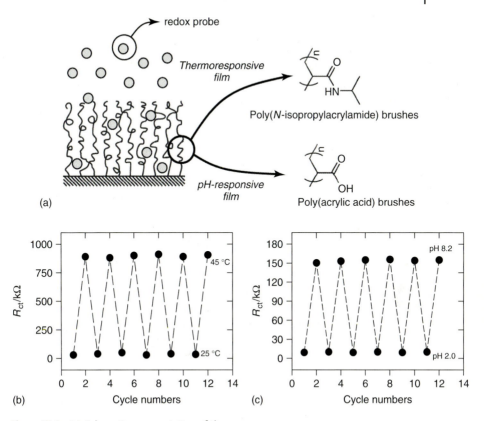

Figure 26.4 (a) Schematic representation of the thermo- and pH-responsive polymer brushes. Interfacial electron-transfer resistance variation upon switching (b) the temperature of PNIPAM-modified Au electrodes and (c) the solution pH of PAA-modified Au electrodes. Reprinted with permission from Ref. [12] by Royal Society of Chemistry.

($R_{et} \sim 890$ kΩ) at 45 °C strongly suggests that the insulating polymer brush is in a collapsed globule conformation. The interfacial ETR shows excellent reversibility, thus indicating that the polymer brushes are very stable. The resulting "ON" and "OFF" cycles of R_{et} can be interpreted by the competition between intermolecular and intramolecular hydrogen bonding below and above the LCST, which controlled the interfacial properties and the permeation of ions or molecules (e.g., Fe(CN)$_6^{3-/4-}$) through the film. This means that the responsive polymer interfaces are capable of converting environmental information effectively into an electrochemical signal, and this phenomenon may find applications in the construction of "smart" macromolecular platforms.

Grafting polyacrylic acid (PAA) brushes onto conducting substrates may also lead to the facile creation of electrode supports with pH-dependent electron-transport properties. At pH ~2, the R_{et} of Fe(CN)$_6^{3-/4-}$ is ~1.8 kΩ. Along with the increase of pH, the interfacial ETR increased to $\sim$160 kΩ (at pH $\sim$8). The increase of the interfacial ETR can be attributed to the ionization of carboxylate groups, which enhance the negative charge density of the surface, repelling Fe(CN)$_6^{3-/4-}$ from the interface. When the pH was repeatedly cycled between 2.0 and 8.2, the variation of R_{et} revealed two well-defined interfacial configurations in which at low pH carboxylate groups are protonated, whereas at high pH the carboxylate groups are fully protonated. In the latter case, the ionic repulsion among COO$^-$ groups in the polymer chain and the anionic redox probes is a key factor setting the high ETR sensed by EIS. Furthermore, the interfacial charge density of PAA brush-modified electrodes can be modulated via changes in the environmental pH, which may lead to novel configurations to manipulate the ion permeability in chemically modified electrodes.

Minko and coworkers [13] described a novel approach to create electrochemical gating systems using polymer brushes grafted to an electrode surface. These authors explored the switchable properties of mixed and homopolymer brushes arising from morphological transitions in the polymer brush layer. The derivatization of indium tin oxide (ITO) electrodes with poly(2-vinyl pyridine) (P2VP) and poly(dimethylsiloxane) (PDMS) brushes enabled the construction of electrode platforms displaying channels formed in the nanostructured thin film in which the precise tuning of their ionic permeability can be manipulated in response to pH changes. In comparison to a homopolymer brush system, the mixed brush showed much broader variation of ion transport through the thin film. Both polymers, P2VP and PDMS, were segregated forming nanosize phases that scale with the mean chain end-to-end distances. Changes in the surrounding environment, such as the quality of the solvent, or pH can promote switching between various phase-segregated morphologies and these structural changes can alter the electrical properties of the brush-modified electrodes.

The EIS spectra of P2VP and P2VP-PDMS-modified ITO electrodes were recorded in the presence of a soluble redox probe, [Fe(CN)$_6$]$^{3-/4-}$ in order to monitor the permeability of the polymer brush thin films in different states for the diffusional redox species. The experimentally obtained impedance spectra (Figure 26.5) were fitted by a theoretical equivalent circuit in order to derive the corresponding R_{et} values. This equivalent circuit included the following components: ohmic resistance of the bulk electrolyte solution, R_s, electron-transfer resistance, R_{et} (dependent on the characteristics of polymer brush anchored on the electrode surface), CPE (reflecting the interfacial capacitance distributed in the modifying thin film), and the diffusional Warburg impedance, Z_W (characterizing the diffusional charge transport through the bulk solution) (Figure 26.1b). Of particular importance is the value of R_{et} provided that this parameter obtained at different pHs reflects the various states of the polymer brushes.

The Faradaic impedance spectra of a ITO electrode modified with a P2VP homopolymer brush at different pH values reflects a decrease of R_{et} for the interfacial electron transfer upon acidification of the electrolyte solution.

This interfacial electron transfer originates from the pH-induced swelling of the polymer brush and results in the formation of ion channels. The R_{et} values derived

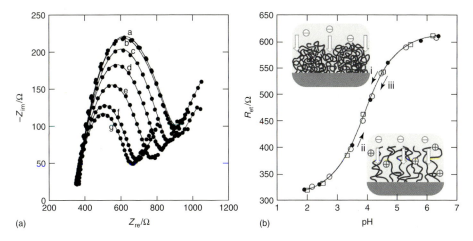

(a) Z_{re}/Ω (b)   pH

Figure 26.5 (a) Nyquist plot for the impedance measurement of P2VP brush-modified ITO electrodes. (b) Titration curve depicting the changes in R_{et} upon variation of the environmental pH. Reprinted with permission from Ref. [13] by American Chemical Society.

from the fitting with an equivalent circuit of the impedance spectra were plotted as a function of pH to yield a titration curve. This plot eloquently illustrates the structural rearrangements taking place in the polymer brush upon changes in the pH of the solution. The opposite titration (from low to high pH) performed by adding NaOH to the electrolyte solution in the electrochemical cell showed no hysteresis in the pH-induced conformational change of the P2VP brush, that is: the titration curve for the shrinking process of the polymer brush upon increasing the pH value coincides with the titration curve obtained upon acidification of the solution. In this case, EIS provides valuable information regarding the reversible character of the structural changes in the homopolymer brush, which swells and shrinks as the pH decreases and increases, respectively. The open and closed states of the "chemical gate" made of the homopolymer P2VP brush were characterized by variations in R_{et} values between 610 and 320 Ω, respectively.

Similar experiments performed in bicomponent-mixed polymer brush (PDMS/P2VP) also revealed a marked pH-dependent behavior of R_{et} (Figure 26.6). Acidification of the solution causes the electron-transfer resistance values to decrease, thus demonstrating the opening of the "chemical gate" on the electrode support. Thereafter, the electrolyte solution was titrated with NaOH to raise the acidic solution to a neutral pH. The titration curve shows hysteresis with the first titration curve from the first acidification experiment. However, the next acidification cycle repeated the titration curve obtained upon the second "alkaline" titration. All other titration experiments result in the same highly reproducible dependence of R_{et} versus pH. This interesting set of experiments highlights the versatility of EIS as a tool to study the emergence of hysteresis in interfacial processes involving structural changes of polymer brushes. The reorganization

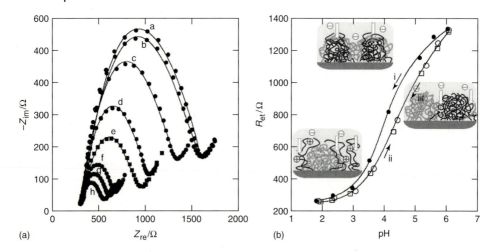

Figure 26.6 (a) Nyquist plot for the impedance measurement of P2VP-PDMS brush-modified ITO electrodes. (b) Titration curve depicting the hysteresis and changes in R_{et} upon variation of the solution pH. Reprinted with permission from Ref. [13] by American Chemical Society.

of the nanostructured polymer brushes upon changes in the pH value results in the reversible shrinking and swelling of the brush that, in turn, is electronically transduced in the closing and opening of the chemical gate on the interface.

These authors also demonstrated that the range R_{ct} changes measured by EIS is significantly larger for the bicomponent-mixed polymer brush compared to the monocomponent brush. Most of this difference corresponds to the closed state of the gate (the maximum value of R_{et} is ~1350 Ω for the mixed brush and ~610 Ω for the monocomponent brush). This was ascribed to the fact that the nonpolar component of the mixed brush contributes to the high electron-transfer resistance in the closed state of the chemical gate, resulting in twofold higher resistance for the interfacial electron transfer compared to that in the monocomponent brush.

In a similar vein, Zhou *et al.* [6] also employed EIS to probe the responsive properties of cationic brushes constituting quaternized poly[(dimethylamino)ethyl methacrylate] (Q-PDMAEMA) tethered onto a gold surface in different electrolytes containing a 1 mM $K_3[Fe(CN)_6]/K_4[Fe(CN)_6]$ (1 : 1) mixture as a redox probe. Conformational transitions in PE brushes were studied as a function of changes in the electrolyte concentration.

Experimental Nyquist spectra exhibit a single loop in the high-frequency region followed by a straight line with a 45° slope (Warburg-like region) at lower frequencies (Figure 26.7). However, it must be noticed that the lowest frequency of measurement was $\omega = 0.1$ Hz, that is, relatively high. The equivalent circuit shown in Figure 26.2a was used to account for the phenomena that influence the impedance response of the studied system. In turn, this circuit is a derivation from that developed by Jennings and coworkers [14] for pH-responsive copolymer

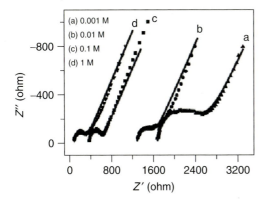

Figure 26.7 Nyquist plots of Q-PDMAEMA brushes at different NaNO$_3$ concentrations in the presence of 1 mM [Fe(CN)$_6$]$^{3-/4-}$ (1 : 1) mixture. In all cases the measured data points are shown as symbols with the calculated fit to the equivalent circuit as a solid line. Reprinted with permission from Ref. [6] by American Chemical Society.

films on gold and shown in Figure 26.8a. The essential difference between the two electrical analogs is the parallel connection between the Warburg impedance and the interfacial capacitance C_i. This modification indicates that, for this, as well as for a second comparable system [15], the current due to interfacial double-layer charging and the faradaic current owing to the electrochemical reaction of the redox probe, are considered to be independent of one another. Surprisingly, the film resistance R_f in series connection with the interfacial impedance is retained, although this resistance results from ionic permeation through a polymeric. Thus, it would have been perhaps more realistic to consider this resistance as an ohmic resistance related to an aqueous electrolytic path inside the brush. Moreover, both relaxation time constants in the model at high frequencies must be similar to account for a single capacitive loop (or two highly overlapped contributions) as shown in the experimental spectra.

On the other hand, it is asserted that in the absence of an electroactive redox species the circuit in Figure 26.8a transforms into the circuit in Figure 26.8b [16]. Thus, for the special case of no redox probes present R_i approaches infinity and thus,

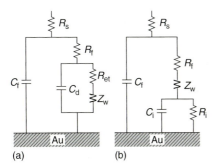

(a) (b)

Figure 26.8 Equivalent circuits used to model the impedance spectra of gold electrodes modified with polymer brushes in the presence (a) and in the absence (b) of redox probes in solution.

the interfacial conductance is purely capacitive. The model in Figure 26.8b is used when R_f is much larger than the combined impedance of R_i and C_i in parallel connection, so only the time constant due to the polymer is observed in the impedance spectrum. The Warburg impedance Z_W in Figures 26.8a,b is claimed to be related to ion transport through the film. However, it remains unclear how, for a system that can be modeled with the equivalent circuit Figures 26.8b, diffusional ion transport can be modulated with the applied sinusoidal potential (Warburg impedance element) if the diffusing species is not electroactive (redox species absent).

The difficulties in the analysis described above derive from the use of circuit analogs to model experimental impedance spectra [17, 18]. Consequently, a word of caution must be expressed concerning the use of equivalent-circuit modeling under certain experimental conditions. In order to represent certain processes believed to take place in the system additional elements are usually incorporated in the basic Randles circuit or the type of connection (series or parallel) is altered, expecting to obtain a better agreement between theory and experiment. However, using multicomponent models to fully describe impedance spectra poses a risk, in view of the fact that it becomes increasingly difficult to prove the utility and interpretation of each parameter [19].

Despite some limitations, as described above, a number of interesting observations were made by Zhou *et al.* [6] that were derived from their impedance study. Swelling is associated with fast transport of redox species to the substrate surface, while polymer brushes in collapsed states exhibit hindered transport. Some salts ($NaNO_3$) cause brush collapse due to charge screening, while others such as those with more hydrophobic anions (ClO_4^-, PF_6^-, and Tf_2N^-) induce brush collapse via solubility changes. The collapsed brushes exhibit intrinsically different resistance as probed with impedance. Charge-screened brushes retain good permeability to electroactive probes. Strongly coordinating hydrophobic anions lead to insoluble brushes, resulting in increased impedance.

26.5
Molecular Transport within Polymer Brushes Studied by EIS

Recently, an approach based on the derivation of the theoretical impedance transfer function to unambiguously describe the impedance response was adopted to characterize gold electrodes modified with poly(methacryloyloxy)-ethyl-trimethyl-ammonium chloride (PMETAC) brushes [20]. Experiments were performed in the presence of $K_3[Fe(CN)_6]/K_4[Fe(CN)_6]$ (1 : 1) mixture as a redox probe in ClO_4^- solutions with different anion concentrations. Figure 26.9 displays experimental and fitted impedance spectra for gold electrodes modified with a surface-tethered PMETAC brush in ClO_4^--containing solutions with concentrations of 1 mM $KClO_4$, 0.1 M $KClO_4$, and 1 M $NaClO_4$, respectively. All spectra exhibit a high-frequency capacitive contribution and a distinctive contribution at low frequencies that suggests Warburg-type behavior in the electrolyte with lower concentration, while this contribution becomes related to

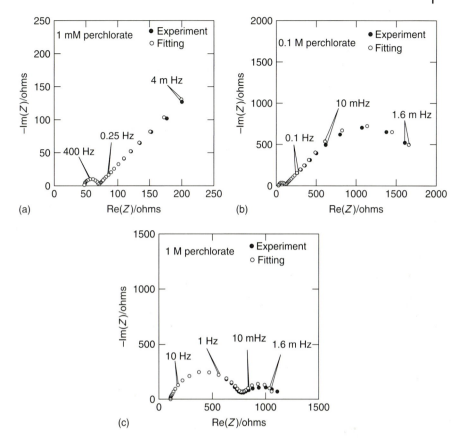

Figure 26.9 Nyquist plots of impedance data for PMETAC-modified electrodes in (a) 1 mM, (b) 0.1 M, and (c) 1 M ClO_4^- a 1 mM $K_3[Fe(CN)_6]/K_4[Fe(CN)_6]$ (1 : 1) mixture. Experimental data (●) and fittings (O) according to Equation 26.23 were depicted in the same plots for the sake of comparison. Reprinted with permission from Ref. [20] by American Chemical Society.

finite-length diffusion with an increase in the anion concentration (Figure 26.9). Thus, the high-frequency contribution can be associated with the relaxation of the charge-transfer process at the surface and the charging of the interfacial capacitor, while the second contribution at low frequencies can be related to ionic transport of the electroactive probe through the PE brush.

The increase in impedance values observed for increasing perchlorate concentrations can be ascribed to the ClO_4^--driven collapse of the PMETAC brush, that may lead to a scenario in which the partial blockage/coverage (θ) of electroactive area is sensitively increased and, as such, the passage of Faradaic current is strongly blocked. It is worth mentioning that the increased blockage is not due to an increase in surface coverage of grafted PE chains at the metal/film interface. The density of anchoring/grafting points of the polymer chains does not change when the brush collapses. The increased blockage results from an increase in film rigidity/density owing to the strong conformational changes of the polymer chains in the presence of strong ion-pairing interactions. Assuming, as a first approximation, that partial coverage does not affect significantly linear conditions for the mass-transport step at a planar electrode, the effect of the partial coverage is simply to reduce the active area. All impedances are inversely proportional to the active area; thus, if the capacitance of the covered surface can be neglected with respect to the double-layer capacitance of the active surface, all impedance values are inversely proportional to the fractional area of active surface $(1 - \theta)$. In particular, these circumstances enable the observed larger charge transfer resistances R_{ct} (Ω) and smaller double layer capacitances C_{dl} (F) for increasing θ, that is, for an increasing degree of collapse of the PE brush to be explained.

Regarding the second time constant at low frequencies, it results from a uniformly accessible electrode to mass transfer through a PE brush of finite thickness. The diffusion impedance response of the brush-containing electrode, when the resistance of the brush to diffusion is much larger than that of the bulk electrolyte, can be approximated by the diffusion impedance of the brush. Moreover, the thickness of the diffusion layer depends significantly on the timescale of a transient experiment, or equivalently on the frequency scale of impedance experiments. If there is a finite length associated with the diffusion layer beyond which the electrode process can have no effect on concentrations, then there is a frequency range for sufficiently low frequencies where impedance response departs from a pure Warburg behavior. Noteworthy, for the experiments shown in Figure 26.9, the lowest measurement frequency was set at a lower value than that used in comparable experiments from the literature [6]. Thus, it can be anticipated that the brush layer represents a diffusion-limiting barrier of finite thickness for electroactive molecular probes (Figure 26.10). Also, when additional negative charges are present in the form supporting electrolyte anions, the diffusion constant of the probe anion is expected to further decrease [21]. In fact, several arguments were advanced in the literature indicating that ion diffusion rather than electron hopping is the charge-transport mechanism in $[Fe(CN)_6]^{3-}$–coordinated PMETAC brushes [22].

Figure 26.11 displays experimental and fitted impedance spectra for gold electrodes modified with surface-tethered PMETAC brushes in Cl^-- and NO_3^--containing solutions with concentrations of 1 M, respectively. Again, when chloride or nitrate replace perchlorate, the impedance response, although exhibiting qualitatively similar dynamics, is quantitatively less affected by the anion concentration, in the 1 mM to 1 M concentration range, as compared to ClO_4^-.

These experimental facts suggest that increasing Cl^- and NO_3^- concentration determines a less-efficient collapse of the brush structure and in turn, the electrode active area becomes less blocked, as compared to perchlorate ion. As previously discussed, the distinctive action of different anionic species in solution originates from ion-pairing interactions between the QA^+ monomer units and the different counterions.

The electrochemical reaction can formally be written as

$$O + e^- \underset{k_c}{\overset{k_a}{\rightleftharpoons}} R \qquad\qquad (26.4)$$

Molecular transport inside the brush layer

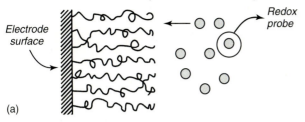

(a)

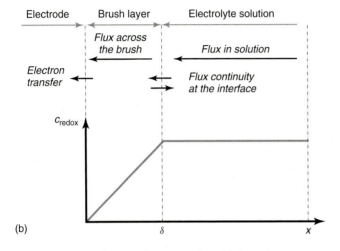

(b)

Figure 26.10 (a) Schematic illustration of a gold electrode modified with polymer brushes in the presence of redox probe molecules in solution. (b) Qualitative description of the steady-state concentration profile of redox species under finite-length diffusion conditions (only a concentration profile for the reactive species is considered).

where O is the $[Fe(CN)_6]^{3-}$ species and R represents the $[Fe(CN)_6]^{4-}$ species.

The potential (V) dependence of the rate constants can be expressed by an exponential law:

$$k_a = k_a^0 \exp\left[b_a(V - V_r)\right] \tag{26.5}$$

$$k_c = k_c^0 \exp\left[b_c(V - V_r)\right] \tag{26.6}$$

where $b_c = -\alpha_c F/RT$ and $b_a = \alpha_a F/RT$, k^0 is a constant independent of V, α is the transfer coefficient, and V_r the Nernst equilibrium potential.

Charge balance is given by:

$$I_f = -A_e F\left[k_c C_O(0) - k_a C_R(0)\right] \tag{26.7}$$

where $C_O(0)$ and $C_R(0)$ represent concentrations of the oxidized and reduced species at the surface and A_e is the apparent electrode area. The negative sign arises from the assumed convention in which the cathodic current is negative.

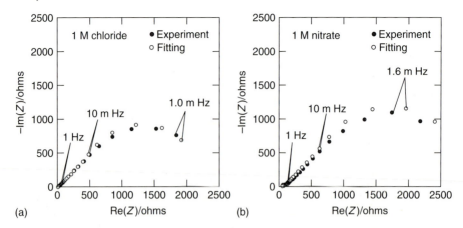

(a)

(b)

Figure 26.11 Nyquist plots of impedance data for PMETAC-modified electrodes in (a) 1 M KCl and (b) 1 M KNO$_3$ containing 1 mM K$_3$[Fe(CN)$_6$]/K$_4$[Fe(CN)$_6$] (1 : 1) mixture. Experimental data (•) and fittings (O) according Z_f (Equation 26.23), R_s, and C_{dl}, were depicted in the same plots for the sake of comparison. Reprinted with permission from Ref. [20] by American Chemical Society.

When a small ac perturbation signal $\Delta V = \tilde{V}\exp(j\omega t)$ is applied, the current and concentrations oscillate around steady-state values: $I_f = I^{dc} + \Delta I$, $C = C^{dc} + \Delta C$ (for both O and R), where the superscript dc indicates a parameter that changes only slowly with time (i.e. either a steady-state term or one that does not change with the frequency of the perturbation ω), and the symbol Δ indicates a parameter oscillating periodically with time t. The resulting oscillations with time may be written as: $\Delta I = \tilde{I}\exp(j\omega t)$ and $\Delta C = \tilde{C}\exp(j\omega t)$. $\tilde{V}$ and $\tilde{I}$ represent the phasors of potential and current. Taking $\tilde{V}$ as the reference signal and considering the phase shift ϕ between them results $\tilde{E} = E_o$ and $\tilde{I} = I_o\exp(j\phi)$, where E_o and I_o are the amplitudes of the applied potential and the resulting current waves, respectively.

For simplicity, we write a single generic expression for O and R mass balances according the second Fick's law, as:

$$\frac{dC}{dt} = -\nabla J = D\frac{\partial^2 C}{\partial x^2} \tag{26.8}$$

for the flux J along the x coordinate perpendicular to the electrode surface. D refers to the diffusion coefficient of the redox probe.

Or, equivalently,

$$\frac{d\Delta C}{dt} = D\frac{\partial^2 \Delta C}{\partial x^2} \tag{26.9}$$

which in the Fourier domain can be expressed as

$$j\omega\Delta C(x, \omega) = D\frac{d^2 \Delta C(x, \omega)}{dx^2} \tag{26.10}$$

with $j = \sqrt{-1}$

The general solution of the second-order differential (Equation 26.10) is

$$\Delta C(x,\omega) = P \exp\left(x\sqrt{\frac{j\omega}{D}}\right) + Q \exp\left(-x\sqrt{\frac{j\omega}{D}}\right) \tag{26.11}$$

subject, in this case, to the following boundary conditions corresponding to a finite-length diffusion in a system having a transmissive boundary,

$$x = 0 \quad A_e F \Delta J = \Delta I_f \tag{26.12}$$

which expresses the flux continuity condition at the interface, and

$$x = \delta \quad \Delta C = 0 \tag{26.13}$$

which indicates that transfer of electroactive species O and R is possible at $x = \delta$ (thickness of the diffusion layer), while for $x \geq \delta$, C remains unaltered and equal to C^{dc}.

Using Equations 26.11 and 26.13 we can write,

$$\Delta C(0,\omega) = -2Q \exp\left(-\delta\sqrt{\frac{j\omega}{D}}\right) \sinh\left(-\delta\sqrt{\frac{j\omega}{D}}\right) \tag{26.14}$$

and calculate ΔJ at the surface in the Fourier domain

$$\Delta J(0,\omega) = -D \left.\frac{d\Delta C(x,\omega)}{dx}\right|_{x=0} = 2QD\sqrt{\frac{j\omega}{D}} \exp\left(-\delta\sqrt{\frac{j\omega}{D}}\right)$$

$$\times \cosh\left(-\delta\sqrt{\frac{j\omega}{D}}\right) \tag{26.15}$$

Let us call $M(0,\omega)$ the mass-transfer function, defined as [25].

$$M(0,\omega) = \frac{\Delta C(0,\omega)}{\Delta J(0,\omega)}$$

Considering the boundary condition at the surface (Equation 26.12), results

$$M(0,\omega) = \frac{\Delta C(0,\omega)}{\Delta J(0,\omega)} = \frac{A_e F \Delta C(0,\omega)}{\Delta I_f} = \frac{1}{D\sqrt{\frac{j\omega}{D}}} \tanh\left(\delta\sqrt{\frac{j\omega}{D}}\right) \tag{26.16}$$

In order to calculate the reaction impedance Z_f, Equation 26.7 describing the rate of charge transfer should be linearized, according to Taylor-series expansion retaining only terms with first-order derivatives, giving:

$$\frac{1}{Z_f} = \frac{\Delta I_f}{\Delta V} = \left(\frac{\partial I_f}{\partial V}\right)_{dc} + \left(\frac{\partial I_f}{\partial C_O(0)}\right)_{dc} \frac{\Delta C_O(0)}{\Delta I_f} \frac{\Delta I_f}{\Delta V}$$

$$+ \left(\frac{\partial I_f}{\partial C_R(0)}\right)_{dc} \frac{\Delta C_R(0)}{\Delta I_f} \frac{\Delta I_f}{\Delta V} \tag{26.17}$$

Derivatives in Equation 26.17 correspond to stationary conditions and may be obtained from Equation 26.7.

$$\left(\frac{\partial I_f}{\partial V}\right)_{dc} = \frac{1}{R_{ct}} = A_e F\left[C_O^{dc}(0)\left(\frac{k_c \alpha_c F}{RT}\right) + C_R^{dc}(0)\left(\frac{k_a \alpha_a F}{RT}\right)\right] \tag{26.18}$$

$$\left(\frac{\partial I_f}{\partial C_O(0)}\right)_{dc} = -A_e F k_c \tag{26.19}$$

$$\left(\frac{\partial I_f}{\partial C_R(0)}\right)_{dc} = A_e F k_a \tag{26.20}$$

and Equation 26.17 becomes

$$\frac{1}{Z_f} = \frac{1}{R_{ct}} + \left(\frac{\partial I_f}{\partial C_O(0)}\right)_{dc} \frac{M_O(0)}{A_e F} \frac{1}{Z_f} + \left(\frac{\partial I_f}{\partial C_R(0)}\right)_{dc} \frac{M_R(0)}{A_e F} \frac{1}{Z_f} \tag{26.21}$$

For simplicity we assume $D_O = D_R = D$, and so $M_O(0) = M_R(0) = M(0)$ that can be calculated according to Equation 26.16.

Consequently, the reaction impedance can be identified as

$$Z_f = R_{ct} + R_{ct} \frac{(k_c - k_a)}{D\sqrt{\frac{j\omega}{D}}} \tanh\left(\delta\sqrt{\frac{j\omega}{D}}\right) \tag{26.22}$$

which, after rearranging reduces to

$$Z_f = R_{ct} + \frac{\sigma}{\sqrt{\omega}} \tanh\left(B\sqrt{j\omega}\right)(1-j) \tag{26.23}$$

where the so-called mass transfer coefficient σ contains the contributions of the forms O and R and $B = \delta/\sqrt{D}$.

Finally, the electrode impedance consists of the electrolyte resistance R_s connected in series with a parallel connection of the double-layer capacitance C_{dl} and the reaction impedance Z_f.

Fitting experimental impedance spectra in Figures 26.9 and 26.11 to the theoretical model allows the parameters described in Table 26.1 to be estimated.

Decreasing C_{dl} values (expressed in Farads) were observed for increasing blocking of the active surface area as explained above. A noteworthy fact is that this decrease was also reported in the literature, although it was interpreted in terms of partial exclusion of water from the brush and a decrease in charge density [13]. A most relevant parameter obtained from the fitting procedure is the diffusion coefficient, D, that provides insightful information about the molecular transport of probe species within the macromolecular environment provided by the polymer brush. As a consequence, EIS can be used as a valuable tool to study the effect of different

Table 26.1 Fitting parameters R_s, C_{dl} and those from the reaction impedance Z_f according to Equation 26.23 for the experimental spectra shown in Figures 26.9 and 26.11.

Electrolyte	R_s (ω)	C_{dl} (μF)	R_{ct} (ω)	B (s$^{1/2}$)	δ (cm)	D (cm^2 s^{-1})
KClO$_4$ 1 M	1070	8.7	6440	10.47	23 × 10^{-7}	4.8 × 10^{-14}
KCl 1 M	8.4	23.6	15.4	13.34	32 × 10^{-7}	5.75 × 10^{-14}
KNO$_3$ 1 M	53.8	18.0	24	14.4	30 × 10^{-7}	4.3 × 10^{-14}

experimental variables, For example: solvent, ionic strength or temperature, on the diffusion of probe species into the brush layer.

26.6
Time-Resolved EIS Measurements on Responsive Polymer Brushes

So far, we have described the use of EIS as a tool to characterize polymer-brush-modified surfaces in a wide frequency range. EIS capabilities can be extended to perform time-resolved experiments to explore, for example, barrier properties of polymer films, that is, considering only relaxation processes exhibiting small time constants. Recently, Jennings and coworkers [23] proposed the use of "single-frequency" measurements as an alternative means to monitor the pH-activated evolution of the barrier properties of copolymer brushes containing polymethylene (99%) with randomly distributed carboxylic groups (1%), PM-COOH. At pH 4, the side chains of PM-COOH were in an uncharged state (-COOH) and the film was hydrophobic due to the prevalence of methylene functionality (99%) along the chains. The film impedance dominates the entire frequency region, indicating that it is much greater than the combined impedance due to the polymer/metal interface (Figure 26.12a). At pH 11, deprotonation of the carboxylic acid side chains increases the hydrophilicity of the film. The increasing charge of the film with pH results in markedly reduced film resistance. While EIS can be used to evaluate the film performance and extract useful properties such as the film resistance and capacitance, each spectrum requires about 10 min to

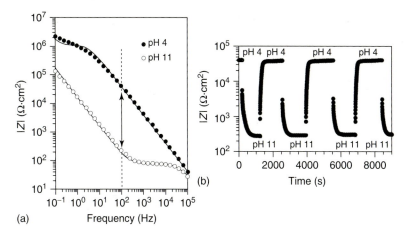

Figure 26.12 (a) Bode plot describing the impedance spectra of PM-COOH brushes with 1% acid content on Au electrodes at pH 4 and 11. (b) Impedance as a function of time when the electrolyte solution is cycled between pH 4 and 11. Reprinted with permission from Ref. [23] by American Chemical Society.

accumulate, precluding a detailed analysis of kinetics on the timescale of several seconds to a few minutes. Consequently, a more rapid sampling method is proposed that consists in evaluating the impedance response at a fixed frequency, after a pH change. Single-frequency measurements can probe the impedance change as a function of time by sampling every few seconds, which enables a more effective monitoring of the rate of film response as the film changes from one state to another. As indicated in Figure 26.12a by the switching of solution pH from 4 to 11, the actual value of impedance modulus is altered by more than 2 orders of magnitude, the largest difference within the studied frequency range. Figure 26.12b shows the measured impedance modulus upon exposure of PM-COOH brushes to pH 4 and 11 buffer solutions in a cyclic manner. The impedance at 100 Hz changes by a factor of 150 when the pH of the contacting solution is switched from 4 to 11. As expected, for each cycle the impedance increases upon exposure to the pH 4 buffer and decrease upon exposure to the pH 11 buffer solution. In this case, single-frequency EIS measurements provide real-time information regarding the response rate and reversibility of the chemical and structural changes taking place inside the brush upon protonation and deprotonation at different pH values.

26.7
Concluding Remarks

We have shown that EIS is a very useful technique to explore, monitor and, more important, quantify the responsive properties of polymer brushes. However, its successful application requires some caution in order to describe a realistic scenario. Worth highlighting is that a good fit of the impedance response is not sufficient measure to validate the proposed scenario. As discussed by Orazem and Tribollet [24], impedance spectroscopy is not a standalone technique. In many cases, additional observations and measurements are needed to validate the suggested physical model. This question is particularly critical when circuit analogs are used to provide quantitative information of the physical processes taking place at the solid/liquid interface in the presence of the polymer film. Consequently, the recommended approach for data analysis and interpretation involves developing a theoretical expression for the impedance in terms of microscopic variables and fitting the experimental results to this model. Within this framework, we consider that EIS may open up new opportunities to characterize the physical properties of polymer brushes as well as the dynamics of their responsive behavior.

Acknowledgments

C.A.G. gratefully acknowledges the Comisión de Investigaciones Científicas y Técnicas Buenos Aires (CICBA) for his position as a member of the Carrera del Investigador Científico. O.A. is a CONICET fellow and acknowledges financial support from the Alexander von Humbodlt Stiftung (Germany), the Max Planck

Society (Germany), Agencia Nacional de Promoción Científica y Tecnológica (ANPCyT projects: PRH 2007-74 – PIDRI No. 74, PICT-PRH 163) and the Centro Interdisciplinario de Nanociencia y Nanotecnología (CINN – Argentina). This work was partially funded by UNLP, CONICET, and CICPBA.

References

1. Barsoukov, E. and Mac Donald, J.R. (2005) *Impedance Spectroscopy: Theory, Experiment and Applications*, Wiley-Interscience, New York.
2. Randles, J.E.B. (1947) Kinetics of rapid electrode reactions. *Discuss. Faraday Soc.*, **1**, 11–19.
3. Katz, E. and Willner, I. (2003) Probing biomolecular interactions at conductive and semiconductive surfaces by impedance spectroscopy: routes to impedimetric immunosensors, DNA sensors, and enzyme biosensors. *Electroanalysis*, **15**, 913–947.
4. Rubinstein, I. (ed.) (1995) *Physical Electrochemistry Principles, Methods, and Applications*, Marcel Dekker Inc., New York.
5. Vallejo, A.E. and Gervasi, C.A. (2006) On the use of impedance spectroscopy for studying bilayer lipid membranes in *Advances in Planar Lipid Bilayers and Liposomes*, vol. 3, Chapter 10 (eds H.T. Tien and A. Ottova), Elsevier, Amsterdam, pp. 331–353.
6. Zhou, F., Hu, H., Yu, B., Osborne, V.L., Huck, W.T.S., and Liu, W. (2007) Probing the responsive behavior of polyelectrolyte brushes using electrochemical impedance spectroscopy. *Anal. Chem.*, **79**, 176–182.
7. Finklea, H.O., Snider, D.A., and Fedyk, J. (1993) Characterization of octadecanethiol-coated gold electrodes as microarray electrodes by cyclic voltammetry and ac impedance spectroscopy. *Langmuir*, **9**, 3660–3667.
8. Devos, O., Gabrielli, C., and Tribollet, B. (2006) Simultaneous EIS and in situ microscope observation on a partially blocked electrode application to scale electrodeposition. *Electrochem. Acta*, **51**, 1413–1422.
9. Doblhofer, K. (1980) Electrodes covered with thin, permeable polymer films. *Electrochem. Acta*, **25**, 871–878.
10. Freger, V. and Bason, S. (2007) Characterization of ion transport in thin films using electrochemical impedance spectroscopy. I. Principles and theory. *J. Membr. Sci.*, **302**, 1–9.
11. Advincula, R.C., Brittain, W.J., Caster, K.C., and Rühe, J. (2004) *Polymer Brushes: Synthesis, Characterization, Applications*, VCH-Wiley Verlag GmbH, Weinheim.
12. Zhou, J., Wang, G., Hu, J., Lu, X., and Li, J. (2006) Temperature, ionic strength and pH induced electrochemical switching of smart polymer interfaces. *Chem. Commun.*, 4820–4822.
13. Motornov, M., Sheparovych, R., Katz, E., and Minko, S. (2008) Chemical gating with nanostructured responsive polymer brushes: mixed brushes versus homopolymer brushes. *ACS Nano*, **2**, 41–52.
14. Bai, D., Habersberger, B.M., and Jennings, G.K. (2005) pH-responsive copolymer films by surface-catalyzed growth. *J. Am. Chem. Soc.*, **127**, 16486–16493.
15. Yu, B., Zhou, F., Hua, H., Wang, C., and Liu, W. (2007) Synthesis and properties of polymer brushes bearing ionic liquid moieties. *Electrochem. Acta*, **53**, 487–494.
16. Bai, D., Ibrahim, Z., and Jennings, G.K. (2007) pH-responsive random copolymer films with amine side chains. *J. Phys. Chem. C*, **111**, 461–466.
17. Silverman, D.C. (1991) On ambiguities in modeling electrochemical impedance spectra using circuit analogues. *Corrosion*, **47**, 87–89.
18. Harrington, D.A. and van den Driessche, P. (2004) Equivalent circuits for some surface electrochemical

mechanisms. *J. Electroanal. Chem.*, **567**, 153–166.

19. Raguse, B., Braach-Maksvytis, V., Cornell, B.A., King, L.G., Osman, P.D., Pace, R.J., and Wieczorek, L. (1998) Tethered lipid bilayer membranes: formation and ionic reservoir characterization. *Langmuir*, **14**, 648–659.

20. Rodríguez Presa, M.J., Gassa, L.M., Azzaroni, O., and Gervasi, C.A. (2009) Estimating diffusion coefficients of probe molecules into polyelectrolyte brushes by electrochemical impedance spectroscopy. *Anal. Chem.*, **81**, 7936–7943.

21. Reznik, C., Darugar, Q., Wheat, A., Fulghum, T., Advincula, R.C., and Landes, C.F. (2008) Single ion diffusive transport within a poly(styrene sulfonate) polymer brush matrix probed

by fluorescence correlation spectroscopy. *J. Phys. Chem. B*, **112**, 10890–10897.

22. Spruijt, E., Choi, E.-Y., and Huck, W.T.S. (2008) Reversible electrochemical switching of polyelectrolyte brush surface energy using electroactive counterions. *Langmuir*, **24**, 11253–11260.

23. Bai, D., Hardwick, C.L., Berron, B.J., and Jennings, G.K. (2007) Kinetics of pH response for copolymer films with dilute carboxylate functionality. *J. Phys. Chem. B*, **111**, 11400–11406.

24. Orazem, M.E. and Tribollet, B. (2008) *Electrochemical Impedance Spectroscopy*, Chapter 23, John Wiley & Sons, Inc., New Jersey.

25. Jacobson, T. and West, K. (1995) Diffusion impedance in planar, cylindrical and spherical symmetry. *Electrochim. Acta*, **40**, 255–262.

27

X-Ray Photoelectron Spectroscopy of Ultrathin Organic Films

Xingyu Gao and Andrew T. S. Wee

27.1
Introduction

X-ray photoelectron spectroscopy (XPS), also known as electron spectroscopy for chemical analysis (ESCA), has long been an established surface-analytical technique for materials characterization. It provides quantitative information regarding elemental composition, empirical formula, chemical state, electronic state, as well as spatial distribution and structure on an elemental/chemical specific basis. During the last two decades, organic molecular thin films have been widely studied for their applications in organic electronics, while XPS of organic materials has been an active field of study. This chapter will focus on the application of XPS to organic molecular films on various substrates. The interface between the organic film and an inorganic substrate plays a key role in determining the properties of the film. Depending on the organic molecules and the substrates, the interface can be quite different from the organic multilayer with regard to the structure, physical, and chemical properties. XPS, as a powerful surface-chemical analytical tool, is ideal for studying these differences.

Probably the most widely used reference for XPS is the "Handbook of X-ray Photoelectron Spectroscopy" by Physical Electronics [1]. Here we present an overview of XPS principles. In XPS measurements, a sample is irradiated by X-ray photons of energy $h\nu$ and the emitted photoelectrons from a certain core-level of an element with a specific binding energy (BE) E_i will be collected by an electron analyzer at kinetic energy E_k. According to energy conservation, $h\nu = E_k + E_i + \varphi$, where φ is the workfunction of the analyzer. Therefore, E_i, the characteristic binding energy of photoelectrons emitted from a particular energy level in an atom, can be determined with known $h\nu, \varphi$, and measured E_k. Since the number of photoelectrons from specific elements is proportional to the number of atoms present, quantitative information about the particular element can be obtained. The XPS experiment records electron numbers per energy interval as a function of electron energy. Since there are photoemission peaks from different energy levels of different elements, the quantitative data obtained can be mainly summarized into two categories: (i) the position (binding energy values) of the XPS peaks and

Functional Polymer Films, First Edition. Edited by Wolfgang Knoll and Rigoberto C. Advincula.
© 2011 Wiley-VCH Verlag GmbH & Co. KGaA. Published 2011 by Wiley-VCH Verlag GmbH & Co. KGaA.

(ii) the intensity of the XPS features. If the energy levels are truly atom-like, these two quantities are independent: the first reveals the chemical species of the target element and the latter contains information regarding the quantity of this element. However, photoemission peaks in the bulk often have non-negligible linewidths and peaks can overlap especially for features belonging to the same element but at different chemical states. Therefore XPS peaks overlapping with each other need to be resolved through a proper fitting process to determine their binding energies and intensities. In addition, the angular dependence or photon-energy dependence of XPS is useful for determining unique element-specific structure information.

As a nondestructive method, XPS depth profiling has been widely used to obtain spatial distribution of chemical composition either by varying the emission angle or excitation X-ray photon energy to achieve different probe depths. More detailed depth profiling can even reveal the packing of the molecular film on the surface, as will be demonstrated later. As depth profiling simply deals with the vertical chemical distribution of the film in the surface region, the structural information of the film both inplane and out-of-plane can be studied by photoelectron diffraction effects in angular resolved XPS.

In the following sections, we will first discuss how the binding-energy shifts of photoemission from both the thin organic molecular films and the substrate could provide information about both the chemical state as well as identify the band bending and dipoles at the interface. Band-bending phenomena are very common in semiconductor interfaces where XPS can be used in combination with ultraviolet photoemission spectroscopy (UPS) to obtain a complete picture of the electronic structure of the interface under investigation. In the final section, we will review research on angle-resolved XPS to study depth profiling as well as photoelectron diffraction for quantitative structural information.

27.2
Binding Energy

The XPS core level peaks occur at specific binding energies due to the quantization of atomic energy levels, and the energies are unique to each element. Core-level chemical shifts, band bending, sample charging, and final-state screening shifts are four factors that determine the exact value of binding energy measured in XPS experiments. Different binding energies for the same element in different chemical states are explained by the so-called chemical shifts, and band binding occurs due to Fermi-level equilibration across the interface. These two shifts are important quantities containing valuable information regarding the interface of organic thin films.

The final-state screening shifts are due to differences of polarization energies of the molecules at the interface and within the film. The sudden presence of the hole left behind by the photon-ionization process leads to a very fast (femtosecond timescale) polarization of all surrounding matter, thus "screening" the Coulomb attraction between the photoelectron and its hole. The *polarization energy* is defined as the difference between the gas phase and condensed phase ionization energies of a particular element or molecule. Charging occurs when positive charges accumulate on the surface during XPS measurements when there is not enough electron current to compensate the loss of photoelectrons from the sample surface. This is often observed on poorly conductive samples resulting in an apparent increase in binding energy. Charging and final-state screening effects are artifacts that need to be considered, and compensation for these effects during data evaluation is necessary. As the shifts due to charging depend on the intensity of photoemission, which is proportional to photon flux, charging effects can be easily distinguished. For example, the high binding energy cutoff (secondary edge) of both XPS and UPS can be used to identify charging shifts as the photon intensity of X-ray and UV light is very different [2]. As charging effects do not discriminate between photoelectrons from the film or substrate, identical shifts to higher binding energies for both photoemission peaks is an indication of charging.

Using photoemission techniques alone, it is difficult to distinguish between band bending and final-state screening, both of which exist at the interface and are film-thickness dependent. One way to investigate these phenomena is to measure several interfaces composed of the same film on substrates with different work functions. If band bending is the cause of the observed shifts, their magnitude will depend proportionally on the work function of the substrate (if interface dipoles are absent). If final-state screening is responsible, the shifts would depend on the screening capabilities of the substrate materials. In the following, we will focus our discussion on chemical shifts and band-bending shifts, although it should be remembered that charging and final-state screening shifts have to be carefully accounted for during the experiments.

To study the different chemical and electronic properties of the organic/substrate interface, a simple but effective general approach is to study the evolution of these properties as a function of film thickness. For monolayer films, the contribution from the interface is more visible than that from thicker films, providing the surface sensitivity of the applied technique and its ability to isolate the film signal from that of the substrate. With increasing thickness, the contribution from the multilayers will become dominant, whereas that from the interface will decrease. Such an evolution will reveal information about the interface that is often different from the bulk. XPS is therefore ideal for interface studies due to its high surface sensitivity of several monolayers: this means that the XPS features of the interface will dominate at the monolayer regime, and will be replaced stepwise by features from the multilayers as the film thickness increases. During this evolution, both XPS features related to film and substrate are valuable for interface studies. For example, new XPS features of substrate elements after film deposition reveals element-specific information about chemical reactions at the molecule/substrate interface. Monitoring the XPS bulk component of substrate during film growth on semiconductor substrates gives band-bending information at the substrate side. Comparison of the XPS features of the elements from the thin film at different thickness allows us to distinguish the chemically reactive components at the interface due to different binding energies from those of the multilayers. The binding-energy shifts of the multilayer XPS components also give information regarding band alignment at the organic-film side. Hence, combining XPS data of elements from both substrate and the film will help identify the chemical-reaction mechanisms between molecules and the substrate, as well as deliver information on interface band alignment. Often, different elements in either substrate or molecular film at the interface could well have different roles in the chemical reaction, and this can be revealed by the energy shifts of their corresponding XPS features.

27.2.1
Chemical Shifts

As the binding energies of XPS peaks are characteristic of each element, the exact binding energy value and even the shapes of the peaks are very sensitive to the chemical environment/states of the emitting atom. Different binding energies for the same element at different chemical states are so-called chemical shifts, which is the reason why XPS is also known as *ESCA*. In this section, we will focus on the binding-energy shifts to show how XPS reveals the chemical properties at the interface of organic molecular films on various substrates.

To illustrate this, we used the copper phthalocyanine (CuPc) molecule, a commonly used stable organic semiconductor, grown on the diamond (100) surface to illustrate how thickness-dependent XPS reveals information regarding the interface chemical and electronic properties. CuPc molecule has a planar structure comprising four aromatic rings around a porphyrin-like central ring with a Cu atom at its center (shown in Figure 27.1), while bare diamond has a 2×1 reconstruction in which surface atoms with unsaturated valences pair up to form dimers. Therefore, bare diamond is expected to react with CuPc molecules that also possess numerous unsaturated bonds. Details of the reaction at the interface are revealed from the following thickness-dependent XPS study.

First, we follow the evolution of C 1s XPS spectra of bare diamond with increasing CuPc thickness as shown in Figure 27.2. Even though carbon exists both in CuPc as well as the substrate, XPS is able to tell the difference. We start with the bare diamond that exhibits a pronounced surface state with binding energy 0.9 eV lower than the main peak due to the π-bonded dimers on the (100) surface. The main peak from bulk diamond is located 0.2 eV higher than that of hydrogenated diamond.

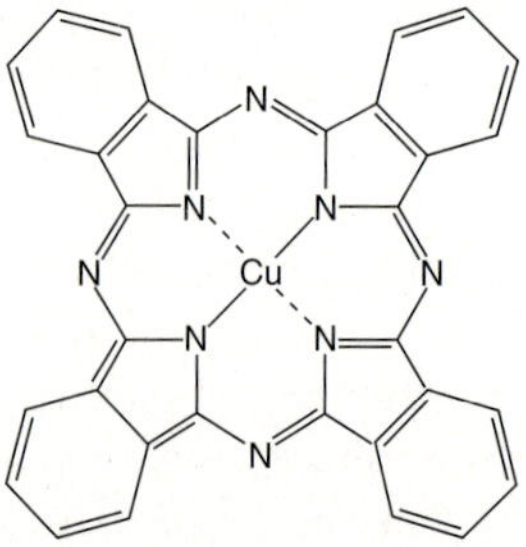

Figure 27.1 Chemical structure of CuPc molecule.

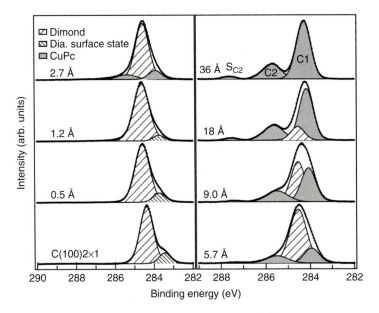

Figure 27.2 C 1s XPS spectra (photon energy, 350 eV) of CuPc on bare diamond with increasing thickness. C 1s spectra are all normalized to the same height for better viewing. Solid lines through the experimental points demonstrate the results of the least-squares fitting [4].

Upon deposition of CuPc molecules, the intensity of the dimer-related surface state decreases and completely disappears with the emergence of C1 and C2 peaks from CuPc at a thickness of 2.7 Å. The gradual decrease of the surface state peak in Figure 27.1 also indicates a chemical reaction involving the breaking of dimer π-bonds between CuPc and diamond dimers. However, no additional components related to newly formed bonds between diamond and CuPc are observed, possibly due to overlap with the dominant diamond component. XPS is usually able to distinguish new features related to modified chemical states of the substrate atoms at the interface. Indeed, for CuPc submonolayer grown on Si(111)-(7 × 7), two new Si 2p features emerge at the high binding energy side +1.33 and +2.04 eV relative to the bulk component, respectively [3]. On a diamond surface, subsequent deposition leads to further attenuation of the diamond peak, while the CuPc components become dominant. The well-formed CuPc components (C1, C2, and SC2), especially the $\pi-\pi^*$ shake-up satellite of C2 carbon (SC2), clearly indicate the integrity of CuPc molecules deposited after all diamond dimers are passivated by the first CuPc monolayer. The ratio of C1–C2 on bare diamond increases from 1.3 at low CuPc coverage (2.7 Å) to 3.0 at high coverage (36 Å). The variation of C1/C2 ratio is related to the change of CuPc molecular orientation due to the different attenuation of the pyrrolic (C2) and aromatic carbon (C1) signals, respectively. When CuPc molecules lie flat on the surface, the C 1s XPS signals from pyrrolic and aromatic carbon atoms are attenuated in the same way; for molecules standing up on the surface, the electrons from the pyrrolic C atoms (C2) are attenuated by aromatic rings (C1) closest to the vacuum interface. Therefore, the apparent increase in the C1/C2 ratio as a function of film thickness indicates a change in CuPc molecular orientation from lying down to standing up, which are also supported by independent angle-dependent NEXAFS (near edge x-ray absorption fine structure) measurements.

Unlike carbon, nitrogen is an element present only in the CuPc molecules. Hence, the XPS study of N gives unambiguous information about the adsorbed molecules. Figure 27.3 reports the N 1s XPS spectra of bare diamond (100) with increasing CuPc thickness. The spectral shape of submonolayer CuPc contrasts strongly with those of the multilayer films. The initial deposition of 0.15 Å CuPc leads to the formation of two distinct peaks at around 397.8 eV (peak N1) and 398.8 eV (peak N2). Subsequent depositions leads to a gradual decrease of N2 intensity until it disappears at a thickness of 2.7 Å. Beyond 2.7 Å, the spectrum is dominated by peak N1 with a small satellite peak at 1.7 eV higher binding energy (N–S), consistent with XPS data of bulk CuPc films on other substrates [5]. Therefore the N1 component is attributed to nitrogen atoms of unreacted CuPc molecules, while N2 is related

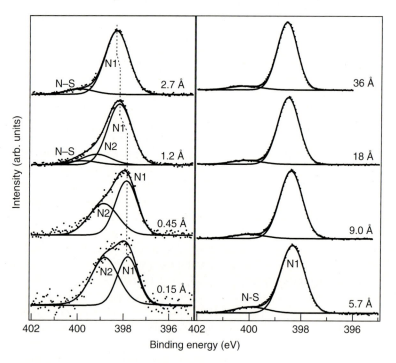

Figure 27.3 N 1s XPS spectra (photon energy, 500 eV) of CuPc on bare diamond with increasing thickness. N 1s spectra are all normalized to the same height for better viewing. Solid lines through the experimental points demonstrate the results of the least-squares fitting. The dashed line indicate the shifts of N1 component as film thickness increases [4].

to the interfacial states of nitrogen atoms covalently bonded to diamond surface dimers. It should be noted that the N1 and N2 components for the submonolayer may originate from two inequivalent types nitrogen atoms (aza-bridging and pyrrolic) within the same CuPc molecule, although a specific assignment of the two components is difficult. Figure 27.3 indicates that a chemical reaction occurs between diamond and CuPc molecules. The chemical reaction breaks the highly strained π bonds of surface dimers and directly couples the dimer atoms to C and N atoms of CuPc molecules through cycloaddition reactions. Although a detailed reaction scheme and configuration of CuPc on diamond are not clear using only XPS data, the realization of covalent bonding between nitrogen atoms and diamond dimers suggests a lying-down geometry of CuPc molecules upon adsorption, in agreement with the evolution of C1/C2 ratio in the C 1s XPS spectra.

27.2.2
Band Bending

When a semiconductor is in contact with another semiconductor or metal with different Fermi levels, there will be an exchange of charge carriers and this causes band realignment at the interface region, forming a common Fermi level. When band bending happens, the core levels will display the same amount of binding-energy shifts. In other words, the binding-energy shifts of the core levels can be used to monitor the band bending, which can be used in combination with UPS to obtain a complete picture of the interface electronic structure of organic films. This will give us valuable understanding relevant to the performance of organic electronic devices such as organic light-emitting diodes.

We again use the CuPc/diamond (001) system as an example to show how XPS can reveal information about band bending at the film interface. From the previous XPS data, it is clear that the diamond bulk peak and peak C1 in Figure 27.2 as well as nitrogen N-1 peak in Figure 27.3 display systematic shifts with increasing film thickness. These three components are not involved in the chemical reaction previously discussed, so their shifts are only related to the band realignment at the interface. The binding-energy shifts as a function of CuPc thickness of C1, N1, and diamond bulk peak extracted from Figures 27.2 and 27.3 are summarized in Figure 27.4. It is obvious from Figure 27.4 that the binding-energy shifts of core levels occur most significantly within the submonolayer regime (below 4 Å) and become moderate in multilayers, in agreement with the scenario of charge transfer induced by the chemical reaction at the interface. Moreover, the "band bending" direction inside the CuPc film indicates that electrons are transferred from CuPc to diamond most probably via the newly formed covalent bonds, leaving the CuPc layers at the interface positively charged (electron depletion). The charge-transfer direction is further corroborated by the downward band bending in the diamond surface region (electron accumulation), as indicated by the subtle increase in diamond C 1s binding energy after CuPc deposition.

Both the diamond side and thin-film side have significant band bending, thus the change in work function due to band bending can be calculated as the sum of the

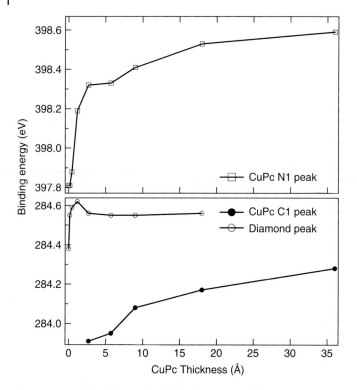

Figure 27.4 The binding-energy shifts of C1, N1, and diamond peak (Figures 27.2 and 27.3) as a function of CuPc thickness on bare diamond [4].

diamond C 1s peak binding-energy shift and the CuPc N1 peak binding-energy shift [4]. This identification of band-bending contribution to work-function change will enable us to isolate the interface dipole contribution as shown in Figure 27.5, where the thickness-dependent work function was derived from the UPS measurements [4]. Thus, a complete energy-level-alignment diagram can be depicted in Figure 27.6. More details about this diagram can be found in Ref. 4.

In another example, we turn to the surface-transfer doping of hydrogenated diamond by the adsorption of F4-TCNQ (tetrafluoro-tetracyanoquinodimethane) molecules. F4-TCNQ is widely used to control p-doping in hole-transporting organic layers, or as an optimizer to tune the hole-injection barrier at organic/metal interfaces. Unlike the bare diamond surface, hydrogenated diamond is fully passivated by hydrogen termination and is thus inert to chemical reaction. Figure 27.7a,b reports the evolution of XPS during the deposition of F4-TCNQ for N 1s and C 1s [6]. Initial deposition of up to 0.5 Å leads to the formation of a pronounced peak at 397.50 eV (N-1) and a broader component at higher binding energy. At 1.0 Å coverage, the higher binding-energy peak (N-2, 398.90 eV) becomes stronger than N-1. Further deposition results in a continuous increase in N-2 peak intensity and a decrease of N-1. The higher binding-energy peak N-2 is attributed

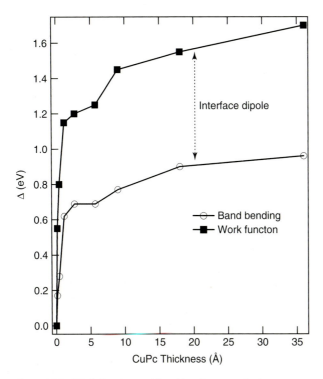

Figure 27.5 Work-function and band-bending magnitudes as a function of CuPc thickness on bare diamond surface. The band bending is calculated as the sum of diamond C 1s peak binding-energy shift and CuPc N1 peak BE shift [4].

to neutral multilayer F4-TCNQ, and the lower binding-energy peak N-1 to anion molecules in direct contact with diamond with their C≡N groups withdrawing electrons from diamond. Peak N-S at 2.60 eV higher binding energy than peak N-2 is attributed to shake-up processes. The loss of electrons (accumulation of holes) in the diamond-surface region is further corroborated by the C 1s spectra shown in Figure 27.7b. A substantial shift (0.65 eV) of the diamond peak to lower binding energy is immediately observed at the initial coverage of 0.2 Å. This shift continues to increase with subsequent deposition and saturates at −0.80 eV at a coverage of 1.0 Å. This indicates an upward band bending of 0.80 ± 0.05 eV at the diamond-surface region due to hole accumulation that acts to balance the negatively charged anion molecules. Further deposition leads to virtually no change of the diamond peak position, and three new components (C-1, C-2, and C–S) related to carbon atoms in F4-TCNQ can be clearly resolved after several molecular layers are formed. C–S, similar to N–S, is also related to shake-up processes. The appearance of the anion F4-TCNQ interlayer species, together with the upward band bending inside diamond, clearly reveals electron transfer from diamond to surface acceptors (F4-TCNQ molecules), or p-type surface transfer doping. Combined with the UPS data, a complete energy-level-alignment diagram is also obtained [6].

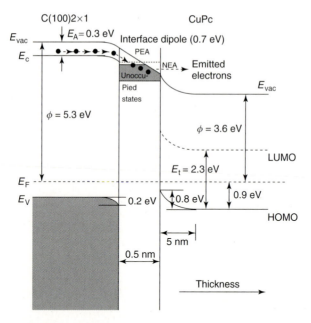

Figure 27.6 Schematic energy level diagram of CuPc on bare diamond surface [4].

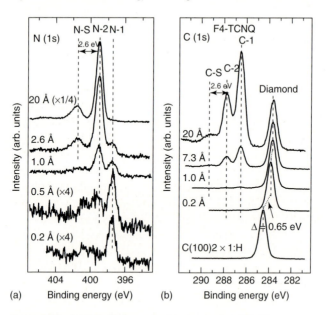

Figure 27.7 (a) N 1s XPS spectra (photon energy, 500 eV) and (b) C 1s XPS spectra (photon energy, 350 eV) of F4-TCNQ on diamond with increasing coverage. C 1s spectra are all normalized to the same diamond peak intensity for better viewing [6].

27.3
Angle-Resolved XPS

27.3.1
Depth Profiling

Compositional depth profiling in ultrahigh vacuum generally has two approaches: sputtering techniques or depth-reconstructive spectroscopy. Since the sputtering process (such as in secondary ion mass spectrometry (SIMS)) destroys submonolayer depth information, it is not favored especially for organic thin films. As maybe the most popular depth-reconstructive spectroscopy, angle-resolved XPS permits identification of elements and their chemical states and quantification of elemental compositions in the analyzed layer. The depth of analysis is determined by the attenuation function of the emitted electrons. The study of the distribution of chemical species near the surface as a function of depth (depth profiling) using angle-resolved XPS can be applied to the ordered organic films such as self-assembled monolayer (SAM) to determine the alignment of the organic molecules.

In homogeneous materials, the intensity contribution from a depth z is, to a first approximation,

$$I(z) = I(z = 0) \exp(-z/\lambda \cos\theta), \tag{27.1}$$

in which θ is the angle between the surface normal and the analyzer direction (photoemission angle, which is different from the so-called electron take-off angle (ETOA) defined as the angle between the surface and analyzer direction: $\theta = 90° - (\text{ETOA})$, and λ is the inelastic mean free path (IMFP) of the photoelectrons. As a function of kinetic energy, λ is of the order of several angstroms, which explains the surface sensitivity of XPS. Although additional contributions such as multiple scattering, elastic scattering and nonisotropic effects cause the effective attenuation length to be actually somewhat smaller than the IMFP and the decay function to deviate from the exponential law [7], it is clear that significant contributions to the measured intensities arise also from locations deeper than $z = \lambda$. By changing the emission angle, the relative depths of elements can be obtained, where larger emission angle means smaller analysis depth. For homogeneous films on flat substrates, Equation 27.1 is widely used to determine the film thickness. There have been many experimental and theoretical efforts to determine the IMFP in various organic systems. Perhaps the most accurate theoretical model was proposed by Tanuma *et al.* [8] to evaluate λ using empirical parameters:

$$\lambda = E_k / \left[E_p^2 \left\{ \beta \ln(0.191\rho^{-0.5} E_k) - \frac{(1634 - 0.91 E_p)}{829.4 E_k} + \frac{(4429 - 20.8 E_p)}{829.4 E_k^2} \right\} \right] \tag{27.2}$$

where E_k is the kinetic energy of the photoelectron, E_p is the free electron plasmon energy, and β is an empirical parameter. E_p and β are given by the following two equations:

$$E_p = 28.8 \left(\frac{N_v \rho}{M} \right)^{0.5}$$

(27.3)

$$\beta = -0.10 + 0.994(E_p^2 + E_g^2)^{-0.5} + 0.069\rho^{0.1}$$

(27.4)

where M and N_v are the molecular weight and the number of valence electrons of the overlayer molecule, respectively. ρ (g/cm^{-3}) is the density of the overlayer, and E_g is the band gap energy (all energies are expressed in electron volts). Equation 27.2 is widely used to calculate λ.

For nonhomogeneous samples, the signal intensity of an element i at an excitation energy $h\nu$ is proportional to the Laplace transformation of the depth-dependent concentration of i, $\rho_i(z)$,

$$I_i \propto F_i \delta_i(h\nu) \int_0^\infty \rho_i(z) \exp(-z/\lambda(E_k) \cos\theta) dz$$

where E_k is the kinetic energy of the photoelectron, with corresponding photo-ionization cross section $\sigma_i(h\nu)$ and IMFP $\lambda(E_k)$, and F_i denotes the detector efficiency at E_k. Any solution of the above equation for $\rho_i(z)$ is an inverse Laplace transform of I_i. However, inverse-transforming finite sets of data will produce a set of solutions from smooth to highly oscillatory.

Many algorithms have been developed for different hypothetical depth profile types to obtain $\rho_i(z)$ from the ARXPS (angle-resolved XPS) data. The simplest assumes a homogeneous flat substrate with a thin homogeneous overlayer. For multicomponent films, more complex depth-profile models are used, such as an exponential profile, rectangular profile, trapezoidal profile, and even profiles based on diffusion models. Other approaches employ multilayer models in which the maximum sampling depth of the sample is divided into a number of parallel layers and the photoelectron intensity is treated as the sum of intensities arising from each layer. The composition in each layer is assumed either to be uniform or to vary linearly. In some studies, the depth profiles are generated by the use of certain inversion algorithms, for example, a linear combination of exponential functions. A detailed review on computational methods in ARXPS to create a concentration depth profile has been lucidly presented by Cumpson [9].

In general, high-resolution depth profiles can reveal the arrangement of different elements within organic molecule films on substrate, and this has been successfully applied to many SAM and polymer samples to explain how these molecules are oriented. As an example, analysis of angle-dependent XPS data from cystamine, (NH$_2$CH$_2$S–)$_2$ on Au(111), at 17 angles by the simplex algorithm produced the elemental depth profiles in Figure 27.8 [10]. A schematic molecular model for cystamine/Au(111) is presented at the top of Figure 27.8 [10]. It is clear that the trends in the depth profiles agree well with atomic positions and interfaces in the model. Specifically, the cystamine/Au interface lies at 10 Å. Also, the peak for sulfur lies at the Au surface, and the carbon peak lies more distant from this interface, in agreement with chemisorption involving an S–Au covalent bond. The nitrogen

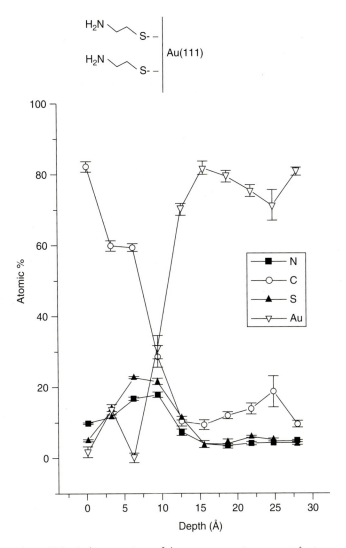

Figure 27.8 Scale comparison of the elemental depth profiles for cystamine, $(NH_2CH_2S-)_2$ on Au(111), with a schematic structure show how the molecules are attached to the substrate. Note the S/Au interface near 10 Å, the position of the C maximum at 0 Å near the organic surface, and the correct stoichiometry between the N and S profiles, in agreement with the molecular model. (From Ref. [10].)

peak is localized within the organic adsorbate, but could not be further resolved to a specific intramolecular conformational position. There is a possibility that nitrogen adopts a range of orientations, both toward and away from the gold surface. The concentrations of sulfur and nitrogen were approximately stoichiometric for this adsorbate at depths less than 10 Å. This example shows how high-resolution depth profiling using XPS can reveal the molecular orientation at the surface.

As the depth sensitivity of XPS can also be varied by the variation of the excitation energy to change IMFP of the photoelectrons with different kinetic energies, excitation energy XPS (ERXPS) as an alternative to ARXPS can also deliver depth information from surfaces with vertical concentration gradients in the XPS sampling region [11]. Recently, Merzlikin *et al.* [11] have developed a new data-treatment algorithm for ERXPS. With atomically flat surfaces, the accuracy of ERXPS is comparable to that of ARXPS. They also found that both ERXPS and ARXPS are affected by surface roughness, but ERXPS appears to be more robust. In ERXPS, the depth coordinate turns out to be extended, but the elemental concentration in the topmost surface layer was correctly analyzed even for an extremely rough substrate where ARXPS failed to detect any surface enrichment at all [11]. With ARXPS, interface roughness smoothes out the angular dependence of the intensities, which results in an underestimation of the layer thickness. ERXPS offers the advantage of being performed at normal emission where deviations of the photoemission angle (by sample misalignment or surface roughness) have only a minor influence on the depth coordinate. Compared with single experiments at low excitation energies, ERXPS has the advantage of utilizing a larger set of experimental information for the description of properties at the external surface and offers an attractive analysis method for the outmost surface layers of thin films irrespective of surface roughness. As synchrotron-radiation studies have become more popular, ERXPS should have a bright future for depth profiling.

27.3.2
Photoelectron Diffraction

Depth profiling by ARXPS focuses on out-of-plane information and thus ignores the inplane arrangement, however, ARXPS is also governed by photoelectron diffraction that could provide both inplane and out-plane information. Photoelectron diffraction is the consequence of coherent interference of the directly emitted component of the photoelectron waves emitted from a core level of an atom with other components elastically scattered by surrounding atoms. The resulting variations of photoemission intensity with emission angle and energy provide a means of determining the local surface structural environment on an element-specific basis. Photoelectron diffraction has now developed into an established method for the quantitative determination of the local structural environment of adsorbates on surfaces, primarily exploiting low-energy electron backscattering, but with some higher-energy forward-scattering work providing valuable and complementary information. Typically, for electron energies greater than about 500 eV, the forward-scattering phase shift is small and the scattering cross-section is strongly peaked in the forward direction, so one can determine interatomic directions in adsorbed molecules by searching for these forward-scattering, zero-order diffraction peaks in the photoelectron angular distribution. This is the basis of the traditional method of X-ray photoelectron diffraction (XPD) – in which the angular distribution of photoelectrons is measured following excitation by an X-ray source (commonly Mg or Al Kα with photon energies of 1253.6 and 1486.6 eV) such that most photoelectrons have a relatively high kinetic energy. Of course, a key requirement for exploiting this idea is that the emitter atom lies below the scatterer atoms relative to the detector. This means it can be used to determine certain interatomic directions in adsorbed molecules and in crystals and thin films. However, these forward-scattering peaks alone cannot provide information on the location of atoms behind the emitter relative to the detector. While the photoemission from the adsorbates on top of the substrate encounters no substrate atoms on the path to the detector as scatter for forward scattering, XPD cannot provide information on the location of adsorbates relative to the underlying solid. By contrast, photoelectrons with lower photoelectron energies (below about 500 eV) have the largest elastic backscattering cross-sections. In such a case, photoelectrons backscattered by substrate atoms can provide information on adsorbate–substrate registry and bond lengths. In this scattering geometry, all the scattering paths involve path-length differences relative to the directly emitted component of the photoelectron wave. As the wavelength (photon energy) of the electrons determines the phase differences of different wave components, the detected photoemission in any particular direction is modulated as the photoelectron energy is changed. For backscattering photoelectron diffraction, one thus has the option of extracting the structural information from either angle-scan or energy-scan measurements.

While intrinsically element specific, the use of core-level shifts in photoelectron binding energies provides a means of even greater specificity in the local structural information. This may be used to distinguish atoms of the same element in different environments, not only in chemically distinct species, but also associated with subtle changes in the chemisorption character at different binding sites. With the increasing availability of high-resolution synchrotron-radiation beamlines around the world, the potential for further expansion in the application of the method is excellent. However, extracting complete quantitative structural solutions from the experimental data requires significant computational effort and large data sets to ensure uniqueness in the solution. Several recent reviews on photoelectron diffraction have been published [12, 13].

Photoelectron diffraction has been widely applied in various systems; however, large molecules with more than a few atoms are generally difficult. In particular, modeling these systems is a rather tedious and time-consuming task since for these systems a large unit cell has to be used. Many scattering events at neighboring atoms within the molecule have to be included and thus long computation times are expected. Furthermore, the structure of a molecule may be distorted due to adsorption at the surface, leading to a different molecular configuration. Thus, structure determination of large nonsymmetric molecules adsorbed at surfaces is difficult. For highly symmetric molecules, the analysis may be simplified by symmetry arguments. In particular, the use of intramolecular forward scattering in adsorbed molecules is ideal for establishing the molecular orientation. A good example is the study of C_{60} molecules adsorbed on different Al and Cu surfaces by XPD, which show clear intramolecular zero-order forward-scattering peaks that depend on different crystal faces, providing clear evidence that C_{60} has well-defined

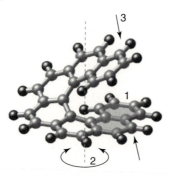

Figure 27.9 Structure of the M-enantiomer of heptahelicene with the three structural parameters indicated for the analysis of the XPD results: (1) the number of C-6 rings oriented parallel to the surface plane, (2) the azimuthal orientation, and (3) the molecular pitch of the M-heptahelicene molecule [15].

preferred orientations on specific crystal faces [14]. More details of the adsorption geometry can be revealed by single-scattering cluster (SSC) calculations [14].

The XPD study of the orientation of chiral heptahelicene on copper substrates is a promising example of large nonsymmetric organic molecules. Figure 27.9 illustrates the geometry of the molecule in the gas phase. It comprises seven C-6 rings, and the six-rings contain six common C–C axes leading to 30 carbon atoms in total. The remaining bonds of carbon atoms are saturated with hydrogen atoms. The molecular structure contains a chirality that is a consequence of the repulsive steric interaction between terminal aromatic rings. In Figure 27.9, the three different arrows indicate the three possible modifications to the molecular skeleton on Cu(111): the number of C-6 rings oriented parallel to the surface plane, the azimuthal orientation, and the molecular pitch of the M-heptahelicene molecule.

The measured C 1s XPD pattern from a 1/3 ML (monolayer) coverage film of M-heptahelicene on Cu(111) and Cu LMM Auger-electron diffraction pattern from the clean Cu(111) surface are shown in Figures 27.10a,b, respectively. The plot of the azimuthal diffraction anisotropy $A = (I_{max} - I_{min})/I_{max}$ for helicene C 1s and Cu(111) substrate in Figure 27.10c,d are also shown as a function of polar emission angle, where I_{max} and I_{min} are the maximum and minimum intensities observed at a particular polar emission angle. While the diffraction anisotropies in the substrate pattern are in the 40–50% range, the helicene C 1s pattern exhibits anisotropies less than 2%, which requires very careful data analysis such as background subtraction and accurate calibration of the relative efficiencies of the channeltrons of the electron-energy analyzer [15].

In Figure 27.10a, the C 1s diffraction pattern exhibits a six-fold rotational symmetry, with the strongest anisotropy appearing in a wheel-type feature close to the center of the plot. As the free heptahelicene molecule only possesses a two-fold rotational (C2) symmetry axis, this observation immediately tells us that the molecules exist in at least three distinct orientations on the surface. Considering that the forward scattering dominates along directions of near-neighbor emitter–scatterer

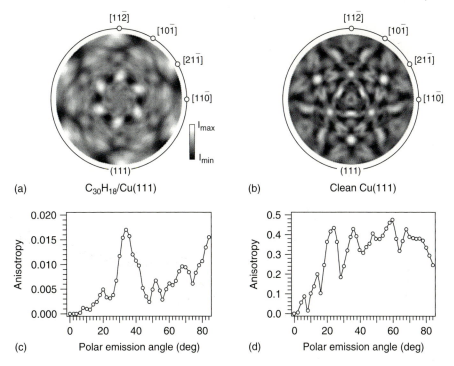

(a) C₃₀H₁₈/Cu(111) (b) Clean Cu(111)

(c) Polar emission angle (deg) (d) Polar emission angle (deg)

Figure 27.10 (a) Angle-scanned X-ray photoelectron diffraction pattern for C 1s emission from the M-enantiomer of heptahelicene on Cu(111). (b) Cu LMM Auger electron-diffraction pattern from the clean Cu(111) surface. (c) and (d) Diffraction anisotropy as a function of the polar emission angle determined from the patterns in (a) and (b), respectively [15].

pairs in an XPD experiment, it is also clear that the adsorbed molecules take distinct well-defined orientational configurations. To obtain quantitative information from the diffraction experiments, theoretical calculations are used to simulate the diffraction patterns. This requires a starting structural model as input for the diffraction calculations. However, the large number of atomic coordinates for such big molecules requires some assumptions about the molecular conformation to simplify the computation process. Most of the atomic coordinates of the molecules used in the calculations come from X-ray diffraction analysis from crystalline samples, with some modifications of the molecular skeleton as shown in Figure 27.9: the number of C-6 rings oriented parallel to the surface plane, the azimuthal orientation, and the molecular pitch of the M-heptahelicene molecule. The best-fit SSC calculation after the optimization of all parameters reproduces the main features of the experimental diffraction pattern quite well, as shown in Figure 27.11 [15]. From the best fit, it was determined that the molecules adsorb in a geometry with their first three C-6 rings oriented parallel to the <111> faces and successively spiral away from the surface from the fourth ring on (see Figure 27.11a). On Cu(111), the coexistence of six azimuthal molecular orientations are also identified. Following a

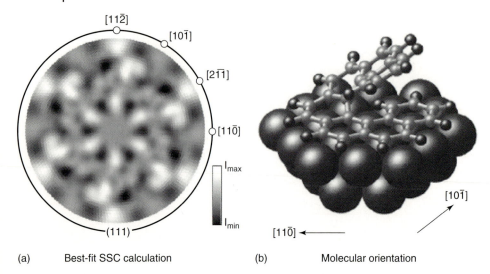

(a)   Best-fit SSC calculation   (b)   Molecular orientation

Figure 27.11 (a) Best-fit single-scattering cluster XPD calculation, obtained for a M-heptahelicene molecule adsorbed with one terminal phenanthrene group (three C-6 rings) parallel to the Cu(111) surface. (b) Schematic drawing of the molecular orientation of M-heptahelicene on the Cu(111) surface that gives the best fit pattern in shown in the (a). Due to the six-fold rotational symmetry of the first Cu(111) surface layer, the helicene molecules take six azimuthal orientations related by subsequent rotations of 60° [15].

similar methodology, useful structural information about the same molecules on Cu(332) has also been obtained by XPD [15].

Photoelectron diffraction can also be presented as photoelectron intensity as a function of kinetic energy by varying the photon energy. The principle of this so-called scanned-energy mode photoelectron diffraction is very similar to that of the extended X-ray absorption fine structure (EXAFS) where the variation of X-ray absorption is due to local diffraction effects, though there are some differences even when averaging over large photoemission angles [13]. The advantages, however, are obvious. EXAFS demands no other absorption edges in a range of several hundred electron volts after the studied absorption edge at the higher photon energy side in the X-ray absorption spectrum, otherwise the presence of any other absorption edges will practically make the data analysis impossible to obtain structural parameters from the oscillation of X-ray absorption in the extended range. For light elements like C, N, and O, EXAFS is thus generally difficult due to overlaps of many energy levels from different elements. Scanned-energy mode photoelectron diffraction can overcome this problem by monitoring the photoemission of the same energy level at different photon energies thereby excluding any interference from other energy levels or elements. Moreover, the resolved chemical-shifted components of the same elements can be studied in this mode to reveal different structural environments of the same elements with different chemical states. By varying the photoemission angle in scanned-energy mode photoelectron diffraction, abundant choices of data sets with different diffraction conditions allow the quantitative determination of structural information of the photoelectron emitter atoms. Unlike XPD, the information obtained from the scanned-energy mode is less transparent and generally requires theoretical simulation based on optimization of an input structural model. As the kinetic energy of electrons in this mode is usually several hundred electron volts, multiple scattering has to be taken into account.

In the past, Woodruff's group [16] has successfully studied molecules such as amino acids and thymine on Cu(110) by scanned-energy mode photoelectron diffraction. Their study of thymine on Cu(110) is an excellent example to demonstrate the ability of this technique for quantitative analysis of an adsorbed molecular structure. The structure of the thymine molecule is shown in Figure 27.12, where there are two different N atoms in different environments. XPS of N 1s shows two clear peaks with a photoelectron binding energy difference of about 1.7 eV, whereas O 1s shows a single peak. This implies that the local bonding environments of the two O atoms are similar, while the two N atoms are in quite different environments.

The photoelectron diffraction results are shown in Figure 27.13, as well as the theoretical simulations for the best-fit structure [16]. The diffraction results for the two N 1s components are obviously quite different. The theoretical calculations based on realistic multiple-scattering calculations of a series of model trial

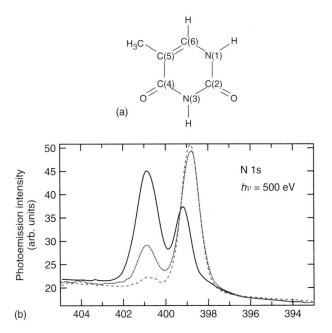

Figure 27.12 (a) Schematic molecular structure of thymine showing labeling convention for the different constituent atoms. (b) XPS of the N 1s from thymine deposited on Cu(110) and room temperature (solid line), at 530 K (long dashed line), and deposited at room temperature and subsequently heated to 520 K (short dashed line) [16].

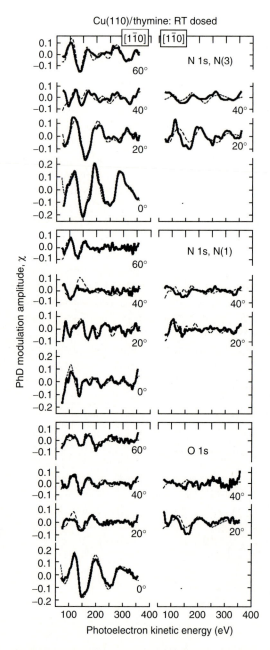

Figure 27.13 Comparison of experimental (full lines) N 1s and O 1s PhD modulation spectra from thymine deposited at room temperature onto Cu(110) with theoretical simulations (dashed lines) for the best-fit structure. Spectra were recorded at a range of angles of incidence from normal to grazing in each of the two principle azimuths ([001] and [1$\bar{1}$0]) [16].

structures are necessary to obtain quantitative structural information [16]. This allows the precise determination of the structural coordination parameters of the photoelectron emitter atoms (N, O) inside the adsorbed molecules [16], such as their distances to neighboring atoms or the orientation angles, similar to EXAFS. In this way, the structure of adsorbed thymine on Cu(110) can be quantitatively determined.

27.4
Photodegradation of Organic Films

XPS measurements are commonly regarded as a nondestructive surface-analysis technique. However, sample-surface degradation during XPS measurement has been reported, especially for organic films, including polymers and ultra thin organic films such as Langmuir–Blodgett (LB) layers and SAMs. This sample surface degradation could be caused by X-ray irradiation and more likely photoelectrons generated from the sample itself. The degradation of organic films is a real problem during the XPS experiments, especially for the experiments using high-intensity flux and long exposure time. The use of a low X-ray flux during analysis has been suggested to limit damage, and constant changing of the exposed sample area under X-ray irradiation could help, but some film decomposition is usually unavoidable. The extent of beam-induced degradation varies strongly from one sample to another, and depends on the properties of the organic sample and the XPS experiment (photon energy, photon flux, etc.). Even the same film on different substrates can display different extents of degradation. For example, it was found that the damage is stronger in the case of the SAM monolayer adsorbed on Au than on Si due to the fact that Au produces a higher electron yield than Si [17]. Therefore, one should be mindful of this degradation during XPS experiments on organic thin films. As both the photoemission intensity and binding energy can change due to the degradation, it is thus necessary to monitor both quantities carefully throughout the experiments. Otherwise, any damage process that does not result in significant changes to the film stoichiometry may remain unnoticed. Other independent analyzes techniques such as IR spectroscopy can also be used to monitor the sample degradation during the XPS experiments.

27.5
Conclusion

XPS is one of the most important and versatile analytical techniques in the study of organic thin films. Information on the chemical states of the surfaces of thin films with different thicknesses allows the understanding of interfacial phenomena. The determination of band bending and interface dipoles is essential for obtaining a detailed picture of the interface electronic structures. Depth profiling using XPS is a powerful nondestructive method for studying film thickness, uniformity, interfacial diffusion, and molecular orientation. For highly ordered organic molecules, which are generally much more complex than most inorganic systems, photoelectron diffraction can be applied to resolve quantitative structural information related to the photoelectron emitter atoms.

References

1. Moulder, J.F., Stickle, W.F., Sobol, P.E., and Bomben, K.D. (1995) *Handbook of X-ray Photoelectron Spectroscopy, A Reference Book of Standard Spectra for Identification and Interpretation*, Physical Electronics, Inc.
2. Schlaf, R., Merritt, C.D., Crisafulli, L.A., and Kafafi, Z.H. (1999) Organic semiconductor interfaces: discrimination between charging and band bending related shifts in frontier orbital line-up measurements with photoemission spectroscopy. *J. Appl. Phys.*, **86**, 5678–5686.
3. Qi, D.-C., Gao, X., Wang, L., Chen, S., Loh, K.P., and Wee, A.T.S. (2008) Tailoring the electron affinity and electron emission of bare diamond (001) 2 × 1 by surface functionalization using organic semiconductor. *Chem. Mater.*, **20**, 6871–6879.
4. Wang, L., Chen, S., Qi, D.-C., Gao, X., and Wee, A.T.S. (2007) Shielding copper atoms by distortion of phthalocyanine ring on Si(111). *Surf. Sci.*, **601**, 4212–4216.
5. Peisert, H., Knupfer, M., and Fink, J. (2002) Electronic structure of partially fluorinated copper phthalocyanine (CuPCF$_4$) and its interface to Au(100). *Surf. Sci.*, **515**, 491–498.
6. Qi, D.-C., Chen, W., Gao, X., Wang, L., Chen, S., Loh, K.P., and Wee, A.T.S. (2007) Surface transfer doping of diamond (100) by tetrafluoro-tetracyanoquinodimethane. *J. Am. Chem. Soc.*, **129**, 8084–8085.
7. Werner, W.S.M. (2003) Surface sensitivity–electron transport in solids in *Surface Analysis by Auger and X-ray Photoelectron Spectroscopy* (eds D. Briggs and J.T. Grant), IM Publications and Surface Spectra Ltd., Chichester, p. 235.
8. Tanuma, S., Powell, C.J., and Penn, D.R. (1993) Calculations of electron inelastic free paths. V. Data for 14 organic compounds over the 50–2000 eV range. *Surf. Interf. Anal.*, **21**, 165–176.
9. Cumpson, P.J. (1995) Angle-resolved XPS and AES – Depth-resolution limits and a general comparison of properties of depth profile and depth-profile reconstruction methods. *J. Electron Spectrosc. Relat. Phenom.*, **73**, 25–52.
10. Williams, J.M. and Beebe, T.P. (1997) High-resolution algorithm for quantitative elemental depth profiling by angle-resolved X-ray photoelectron spectroscopy. *J. Vac. Sci. Technol. A*, **15**, 2122–2133.
11. Merzlikin, S.V., Tolkachev, N.N., Strunskus, T., Witte, G., Glogowski, T., Wöll, C., and Grünert, W. (2008) Resolving the depth coordinate in photoelectron spectroscopy – comparison of excitation energy variation *vs.* angular-resolved XPS for the analysis of a self-assembled monolayer model system. *Surf. Sci.*, **602**, 755–767.
12. Westphal, C. (2003) The study of the local atomic structure by means of X-ray photoelectron diffraction. *Surf. Sci. Rep.*, **50**, 1–106.
13. Woodruff, D.P. (2007) Adsorbate structure determination using photoelectron diffraction: methods and applications. *Surf. Sci. Rep.*, **62**, 1–38.
14. Fasel, R., Aebi, P., Agostino, R.G., Naumovic, D., Osterwalder, J., Santaniello, A., and Schlapbach, L. (1996) Orientation of adsorbed C-60 molecules determined via X-ray photoelectron diffraction. *Phys. Rev. Lett.*, **76**, 4733–4736.
15. Fasel, R., Cossy, A., Ernst, K.-H., Baumberger, F., Greber, T., and Osterwalder, J. (2001) Orientation of chiral heptahelicene C$_{30}$H$_{18}$ on copper surfaces: an X-ray photoelectron diffraction study. *J. Chem. Phys.*, **115**, 1020–1027.

16. Allegretti, F., Polcik, M., and Woodruff, D.P. (2007) Quantitative determination of the local structure of thymine on Cu(110) using scanned-energy mode photoelectron diffraction. *Surf. Sci.*, **601**, 3611–3622.

17. Laibinis, P.E., Graham, R.L., Biebuyck, H.A., and Whitesides, G.M. (1991) X-ray-damage to CF_3CO_2-terminated organic monolayers on Si/Au-principle effect of electrons. *Science*, **254**, 981.

Part IV
Applications

28
Self-Assembled Multifunctional Polymers for Biointerfaces

Géraldine Coullerez, Ganna Gorodyska, Erik Reimhult, Marcus Textor, and H. Michelle Grandin

28.1
Introduction

When a synthetic material is brought into contact with a biological system *in vitro* or *in vivo*, its surface will quickly become covered by biomolecules including proteins, lipids, and saccharides, which will – depending on the specific case – mediate the attachment of cells and/or micro-organisms such as bacteria and fungi. Such interfacial processes are key aspects for the performance of biomaterials and devices ranging from implants and tissue-engineering constructs, to sensors and diagnostic devices, carriers for drug delivery, to pharmaceutical packaging directly contacting protein solutions, or whole blood. The nonspecific adsorption of proteins is often the first step of cascade events taking place when foreign materials are exposed to biological fluids or extracellular matrix and the kinetics and degree to which this occurs depend on several factors related to both material properties and the type of proteins involved. On the one hand, the physical (hydrophobicity/hydrophilicity, surface energy, charge, and topography) and chemical (chemical functional groups, specific receptor sites) surface properties [1–4] define the possible types of interactions with biomolecules. On the other hand, the primary, secondary, and tertiary structure of a particular protein is paramount in determining to what extent conformational changes, denaturation, and loss of functionality takes place once the protein has adsorbed to the substrate surface. In response, it has been demonstrated that application of an interfacial polymer film, for example, is a feasible means to greatly reduce nonspecific interactions with proteins and other biomolecules. Such biologically "noninteractive" (also called "*nonfouling*") surfaces are furthermore a prerequisite for the successful design of interfaces that present specific biochemical cues and elicit specific responses.

For many bio-oriented applications, resistance to nonspecific adsorption of proteins and other bioentities at the material interface is by itself an important requirement [2, 5]. Surfaces with such properties have great practical significance in areas including blood-contacting devices (reduction of fibrinogen adsorption and risk of blood clotting on cardiovascular implants such as stents), containers for

Functional Polymer Films, First Edition. Edited by Wolfgang Knoll and Rigoberto C. Advincula.

controlled drug release (increasing blood circulation time), nonfouling surfaces for food and drug handling and packaging, as well as marine applications to prevent biofouling on equipment [6, 7].

While certain applications require nonfouling surface properties, in others one aims at additionally introducing specific biointeractive ligands. Typical examples are *in vitro* bioaffinity sensors and biochips for the analysis of genes and proteins in biomedical diagnostics, which monitor specific interactions while keeping non-specific adsorption as low as possible [8, 9]. Such cases involve the immobilization of ligands on top of the "inert" background, either in one step by assembly of multifunctional polymers, or in a subsequent immobilization step. Density and con-servation of active conformation and optimum orientation of recognition units such as proteins/antibodies or enzymes are particular challenges in this field [10, 11].

For applications requiring passivation and/or presentation of specific bioactive elements, nanometer-thin polymer coatings, applied to various types of materials (from polymers and ceramics to metallic implant materials), can be exploited to control the interactions with biofluids [12, 13]. In fact, Zauscher and Chilkoti [14] argue, in the preface to a recent in-focus special issue of Biointerphases on biological applications of polymer brushes, that the most promising technical advances in using polymer brushes have been associated with their biointerfacial applications. The main attractiveness of the polymer-coating approach is the ability to control the physical and chemical properties of the surface and to tailor the architecture of the interface at the molecular level. The chemical nature of the chosen polymer coating, as well as the conformation of single chains, has a significant influence on the performance of the system. Design aspects typically cover three levels: sufficiently strong anchorage of the molecules/polymers to the underlying substrate, nonfouling surface chemistry as the background, and sufficiently stable immobilization and optimum presentation of the biointeractive entities.

In order to understand the interaction between biomaterials and the surrounding interfacial environment, it is important that the surface physicochemistry of the substrate and the adsorbed polymer coating be thoroughly understood. Thus, the chemical composition, molecular structure, orientation, morphology, stability, and spatial distribution of the biofunctional species present at the interface must be characterized. A wide range of surface analytical tools, including *ex situ* and *in situ* techniques, have been used to characterize biointerfaces at the necessary level of resolution and sensitivity [15–23]. These include X-ray photoelectron spectroscopy (XPS), static time-of-flight secondary ion mass spectrometry (ToF-SIMS), ellipsom-etry, atomic force microscopy (AFM), optical waveguides light mode spectroscopy (OWLS), quartz crystal microbalance with dissipation (QCM-D), and many others. As each technique has its strengths and weaknesses, one often requires the use of complementary techniques for a more complete characterization of the surface and subsequent interactions with biofluids. The reader is referred to the book chapter by Konradi *et al.* [24] for a more detailed discussion of characterization techniques used in the study of biomaterial surface coatings.

In deciding upon a strategy for modifying a particular surface for a specific application one must consider the substrate-surface properties, regarding the anchorage concept, the environmental conditions under which the coating must be stable (time frame, pH, storage conditions, etc.), and the degree of passivation or biointeractiveness required. Each step in this process must be carefully designed and characterized and the final choice will certainly differ from one application to another. Therefore, it is the aim of this chapter to provide an overview of selected chemistries, primarily those developed in our group, for tailoring the surface properties and biorecognition elements at the molecular level, with special emphasis on self-assembly approaches and multifunctional polymer thin films and their use for selected biomedical applications, for example, diagnostic assays, cell-selective surfaces, and bacterial adhesion studies. After a short overview of the surface-grafting approaches of multifunctional polymer thin films (Section 28.2) we present how these polymer brushes at biointerfaces are used for bio-oriented applications (Section 28.3) Three approaches of surface modifications will be addressed in the following section; (i) passivation of the surface to eliminate nonspecific protein adsorption and cell attachment, (ii) immobilization of biological cues to bind desired biomolecules and cells, and (iii) methods to create adhesive surface patterns in a nonfouling background for bio-oriented applications.

28.2
Immobilization and Conformation of Polymers at Biointerfaces

28.2.1
Surface Immobilization of Polymers via the "Grafting-To" Method

28.2.1.1 Physisorption of Block- and Graft-Copolymers
There are a number of methods to immobilize polymer chains at surfaces. The simplest technique is that of *physisorption*, which is based on a physical interaction between a particular part of the polymer with the surface. In this case, a specific sequence of the polymer prefers to interact directly with the surface rather than staying solvated in solution, thereby lowering the free energy of the system, while the fraction of the molecule that is responsible for the solubility remains in contact with the solvent and becomes accommodated at the polymer layer/liquid interface (Figure 28.1a). The mechanism of adsorption depends on the nature of the less-soluble fragment, the overall structure of the polymer as well as the nature of the substrate. For example, amphiphilic *block*-copolymers, such as poly(ethylene oxide)-poly(propylene oxide)-poly(ethylene oxide), Pluronics™, adsorb via hydrophobic interactions to hydrophobic surfaces [25]. In this case, in aqueous solution, the hydrophobic poly(propylene oxide) (PPO) block adsorbs on the surface while the poly(ethylene oxide) (PEO) chains are preferentially solvated and point toward the solution. This method of polymer adsorption is fast and simple, however, the physisorbed blocks can be displaced by other adsorbates or can become unstable under certain environmental conditions.

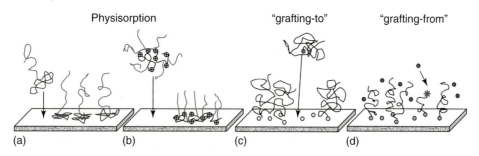

Figure 28.1 Surface modifications to produce polymeric interfaces via physisorption by (a) hydrophobic and (b) electrostatic interactions, and via chemisorption by (c) "grafting-to" and (d) "grafting-from" approaches.

Another example of this category of self-assembled surface systems is the family of poly(ethylene glycol)-poly(propylene sulfide) (PEG-PPS) di- or triblock copolymers for the functionalization of hydrophobic as well as gold surfaces; in the latter case, immobilization relies on the formation of multivalent thioether–gold bonds [26, 27].

Alternatively, *block*-copolymers with one component being a polyelectrolyte or *graft*-copolymers with a polyelectrolyte backbone (Figure 28.1b) spontaneously physisorb from solution onto oppositely charged surfaces through multiple electrostatic interactions [28, 29]. Once adsorbed, the presence of the multiple charges provides additional stability to the polymer coating due to the cooperative assistance supplied by the vicinity of other charges. Flat conformation and uniform coverage is possible provided there is a regular charge distribution along the adsorbing chain [30], and the surface presents a sufficiently high counter charge and a low roughness in comparison to the molecular dimension of the assembly system [31–34]. Clearly, the pH [35–38] and ionic strength [35, 36, 39–43] of the assembly solution will affect the molecular conformation of the polyelectrolyte in solution prior to its physisorption and thus affect the adsorption process and the architecture of the surface-assembled polymer [39, 44–47]. In addition, the isoelectric point of the surface and the surface charge density will determine the molecular architecture and properties of the adlayer [48]. Examples for such systems include poly(ethylene imine)-*graft*-poly(ethylene oxide) (PEI-PEO) [49, 50] or poly(L-lysine)-*graft*-poly(ethylene glycol) (PLL-*g*-PEG) [51], this latter example will be discussed in detail below. The stability of polyelectrolyte-based assembly systems is limited by the electrostatic nature of the interaction with the surface (the single-charge-interaction energy is of the order of kT) [44, 52, 53] and strongly depends on the properties of the solution in contact with the adlayer as well as on the (pH-dependent) surface charge density of the (oppositely charged) substrate. It has been demonstrated that pH [35, 37, 54] and ionic-strength variations [55–57] may result in modification or desorption of the adsorbed polymers. Despite this, physisorbed polymers are employed in a number of biomedical applications, as discussed throughout this paper.

28.2.1.2 Self-Assembled Monolayers (SAMs)

In contrast to the systems presented in Section 28.2.1.1, covalent binding of polymer chains at the interface enhance stability of the tethered layers. Polymeric films covalently attached to the surface from a solution of the corresponding polymers or monomers are synthesized by one of two techniques: "grafting-to" or "grafting-from." The *"grafting-to"* method (Figure 28.1c) involves the formation of a chemical bond, either by direct reaction of molecules on the target surfaces after an initiation step [58–60], or by the chemisorption of suitable functional groups that strongly interact with a particular type of surface. A particularly attractive system is that of *self-assembled monolayers (SAMs)*, where single-chain oligomers (~5–15 monomers) consisting of an anchoring group, a spacer, and a terminal functionality can self-assemble from solution in a highly ordered and close-packed monomolecular structure [61]. The most cited examples are alkanesilane SAMs for hydroxylated substrate surfaces [62], and alkanethiol SAMs on gold and silver surfaces [63–65]. Publications by Ulman [61, 66] provide detailed information on SAM formation and their properties.

SAMs present the important advantage of relatively high adhesion strength on substrates due to the "grafting-to" functionality of its molecules [67–69], thereby resulting in better adlayer stability against thermal [64, 70], chemical [71], and mechanical external stimuli [72]. Molecular order and stability of SAMs also arise from its inherent chemical structure [73], orientation, and dense packing of the molecules favored by hydrophobic interactions between alkane chains [74–77]. The end-functionalization of SAMs with biointeractive moieties is also possible either by introducing specific reactive functional groups and subsequent modification with biofunctional molecules [78–80], or by direct coassembly [81] of two (or more) differently functionalized molecules. As an example, alkanethiol SAMs have been end-functionalized with short oligo(ethylene glycol) (OEG) chains with the aim to produce protein-resistant surfaces [82, 83].

28.2.1.3 Chemisorption of End-Functionalized Polymers

The grafting of longer hydrophilic chains, such as poly(ethylene glycol) (PEG), to the surface through "grafting-to" self-assembly suffers from the problem that high chain surface densities are difficult to achieve due to repulsive forces increasing rapidly with chain surface density. However, this can to some extent be overcome by carrying out the reaction under so-called "cloud-point conditions" (higher temperature and/or higher ionic strength of the assembly solution), whereby for polymers that show an inverse solubility/temperature relationship, the polymer chains lose coordinated water and partly collapse, thus enabling a closer packing at the surface. Aldehyde-PEGs, for example, have been covalently bound to aminated plasma-functionalized surfaces under such conditions [84].

Regarding end-functionalized SAMs, adsorption or coadsorption of molecules does not always result in a homogeneously ordered and close-packed SAM due to steric-hindrance effects, particularly for bulky end groups [85–87]. Steric repulsion by the already grafted functional groups can inhibit further grafting to an adsorbed SAM resulting in disordered and less stable SAMs [88, 89].

Furthermore, using single chains with a single anchorage foot may not provide sufficient thermodynamic stability or kinetic inertness for certain applications [90]. The self-assembled surface systems PEG-PPS di- or triblock copolymers for the functionalization of gold surfaces is an example of an assembled monolayer where improved stability is achieved through the concomitant surface binding of multiple thioether groups per adsorbate molecule. Moreover, it was shown that thioether sulfur atoms are much less prone to spontaneous oxidation when compared to thiolate SAMs that oxidize over a longer time frame (months) with a loss of adhesion and stability [26, 27].

28.2.2
Surface Immobilization of Polymers via the "Grafting-From" Method: Surface-Initiated Polymerization of Polymer Brushes

For the "grafting-from" procedure (Figure 28.1d), functional groups and surface-bound initiators are required to initiate the polymerization process at the surface [91]. Monomers present in solution can penetrate through already grafted polymer layers easily and layer thicknesses in the range of hundreds of nanometers can be achieved. A frequently used polymer-deposition technique is via plasma activation of the surface for which no chemical initiator is required [92–95]. Polymeric substrates can be functionalized by plasma and glow discharge in the presence of O_2 or N_2. However the method is not adapted to the functionalization of metal-oxide surfaces of technological importance for biological applications. A more attractive approach combines instead the method of SAMs of an initiator with a number of different surface polymerization mechanisms including anionic and cationic polymerization [96], ring-opening polymerization (ROP) [97], conventional radical [98], and controlled radical polymerization such as atom-transfer radical polymerization (ATRP) [91, 99, 100]. ATRP is the most often used technique because of its versatility of monomer type and functional groups under mild reaction conditions, and as it affords good control of polymer graft density and thickness [101]. Polydispersity is often close to that found in solution, for example, ATRP: $M_n/M_w \sim 1.2$. A more detailed description on surface-initiated polymer brushes and their properties can be found in reviews by Zhao and Brittain [102], Stuart and coworkers [103], Tsukruk and coworkers [104], and Huck and coworkers [105] and in the book on polymer brushes edited by Advincula *et al.* [106].

The "grafting-from" approach has several attractive features as higher brush surface densities of up to 8.5 µg/cm^2 [98] can be obtained in comparison to physisorption approaches with typical densities of the order of <4 µg/cm^2 [103]. The polymer film thickness increases linearly with polymer molecular weight, and is easily tuned by the initiator surface density and polymerization conditions, for example, time and polymer type [98]. The brush density is controlled by using a mixed SAM consisting of an initiator and a nonfunctional coadsorbate. Different hydrophilic monomers have been employed by ATRP to prepare biologically relevant "grafting-from" brushes, for example, 2-hydroxyethyl methacrylate [107, 108] to produce hydrophilic brushes or acrylamide monomers to tune transitions in surface energy and cell adhesion and detachment [109, 110]. Chilkoti and coworkers [111] have prepared polymer brushes by ATRP that present short oligoethylene side chains, for example, poly(oligo(ethylene glycol) methyl methacrylate) (pOEGMA), that are highly effective in reducing protein adsorption and cell adhesion on gold and silicon oxide surfaces. In comparison with linear or branched PEGs, the surface density in ethylene glycol (EG) can be significantly greater. The brush film thickness must, however, be at least 10 nm for optimal nonfouling properties [99, 112]. Recently, surface-initiated polymerization has emerged as an alternative grafting method of biomolecules such as single-stranded DNA and proteins [113, 114]. Garcia and coworkers [115, 116] have further shown that such oligo(ethylene glycol) methyl methacrylate (OEGMA) polymer brushes provide the ability to tether peptide bioligands, thus promoting bone-cell differentiation *in vitro* and improved integration *in vivo* for titanium implants. Nevertheless, "grafting-from" methods may suffer from competitive bulk-phase polymerization, surface-recombination reactions, and nonhomogeneous lateral brush chain-length distributions. Furthermore, there is the disadvantage that this approach requires specialized polymer synthesis knowledge and is not easily transferable to an industrial and larger-scale production environment.

28.2.3
Conformations of Polymers at Interfaces

The conformation of the polymer chains tethered on a surface can be rather different, depending on the solvent quality, affinity of the chain monomers to the surface and crowding conditions on the surface. Figure 28.2 shows, schematically, the possible conformational regimes for polymers at interfaces. If the distance

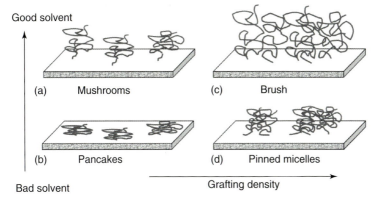

Figure 28.2 Different morphologies formed by polymers tethered at surfaces: (a) mushroom conformation of single chains under good solvent conditions, (b) pancake conformation of single tethered chains, occurring when there is strong attraction between the surface and the chain monomers, (c) brush conformation for high grafting densities, leading to extension of the chains away from the surface, and (d) pinned micelles in poor solvent.

between the chains is larger than the chain size, there are no interchain interactions between neighboring chains. In this case, the polymer chains will adopt a "mushroom" conformation (Figure 28.2a). If the chain monomers do not have an affinity to the surface, one chain end is attached to the surface and the rest roughly forms a random coil. If there is strong interaction between the surface and the chain monomers, this will result in the chain laying flat on the surface, a situation referred to as *"pancake"* conformation (Figure 28.2b). When the surface becomes more crowded, the chains will start to mutually interact, eventually resulting in a stretching of the individual chains away from the surface leading to a so-called "brush" formation in good solvent (Figure 28.2c) or formation of pinned micelles in poor solvent (Figure 28.2d).

The stretched-chain conformation in the brush regime is the result of a balance between excluded-volume interactions, free energy of the segments, and the loss of conformational entropy upon stretching (the latter being associated with the elastic free energy of the chain). The brush regime is critical in the design of antifouling surfaces, and will be discussed further below. An important point to be made is that there is no sharp transition, rather a smooth transition between the mushroom and the brush regime. The usual characteristic parameter of grafted polymer layers is the inverse value of the distance between grafting points (D). The grafting density (σ) is defined as $\sigma = (h\rho N_A)/M_n$, where h is the thickness of grafted polymer layer, ρ is the bulk density of the grafted layer, and N_A is Avogadro's number. Another way to determine grafting density is $\sigma = 1/D^2$ [117]. The parameter commonly used in the literature for quantitative characterization of the transition between the mushroom regime and the brush regime is the reduced tethered density $\Sigma = \sigma \pi R_g^2$, where R_g is the radius of gyration of grafted chains at a specific temperature and solvent [118]. Three regimes have been proposed to characterize tethered polymer chains in terms of grafting density: (i) a substrate dominated (mushroom) regime $\Sigma < 1$, (ii) a chain dominated (mushroom-to brush transition) regime with moderate repulsion effect $1 < \Sigma > 5$, and (iii) a high stretching ("true brush") regime with strong chain repulsion at $\Sigma > 5$ in good solvent and $\Sigma > 10$ for poorly miscible brushes [119].

Recently, a new variety of physisorbed polymer mushrooms and brushes, which are inspired by the ubiquitous polymer-coated membranes found in various cells in biology, are being investigated [120–124]. Instead of tethering the polymers directly to the substrate surface it is immobilized to a membrane with very weak van der Waals and charge interactions to the substrate, thereby allowing for substantial mobility of the scaffold itself (Figure 28.3). While the local interactions are too weak to afford much stability to hard mechanical perturbations, the membrane that anchors the polymers can be made to extend over truly macroscopic dimensions and thus yield a high stability for the overall system in face of changes in ionic strength, pH, and other chemical perturbations, even for a hydrated, cushioned liquid-crystalline layer. For example, PEG or lipopolysaccharides (LPSs) modified membranes can be formed using well-known protocols for liposome fusion on oxide surfaces such as glass [121, 122, 125]. Alternatively, a supported phospholipid bilayer (SPB) membrane can first be formed by the standard protocols and biopolymers,

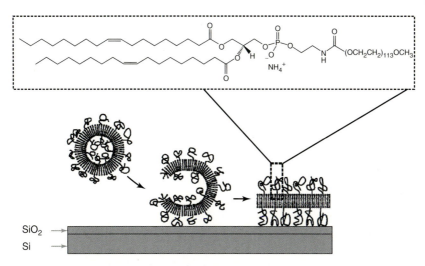

Figure 28.3 Schematic of the envisioned process for polymer-coated membranes. Liposomes incorporating a low percentage of lipids with PEG chains attached to their headgroup self-assemble on a SiO$_2$ surface and fuse into a PEG-supported lipid bilayer (SLB) self-assembled layer.

like hyaluronan, can be subsequently attached by high-affinity biomimetic specific binding like biotin–avidin coupling or CD44 [124]. These "grafting" methods have been shown to result in surface polymer modifications spanning the range from end-grafted mushrooms to side-chain-grafted mushrooms and brushes to very high end-grafted brushes.

In terms of mechanistic aspects of protein adsorption to polymeric mushroom- or brush-type surfaces, the following types of interaction/adsorption have been proposed: (i) primary adsorption at the substrate surface due to proteins diffusing through the interfacial brush to the underlying substrate (invasive mechanism), particularly relevant for small (protein) molecules that can penetrate through an insufficiently densely grafted chain surface [103, 126]; (ii) secondary adsorption at the outer edge of the brush due to a thin polymer thickness, providing only a weak screening and allowing protein–brush interactions via Van der Waals and/or electrical double-layer attraction, mostly relevant for larger proteins [127, 128]; and (iii) tertiary adsorption of proteins after compression of the polymer-brush film (compressive mechanism) due to extended interactions with the substrate, which is a variation of primary adsorption for very large proteins (compression instead of diffusion) [129]. Protein resistance is obtained when the polymer interface is able to exclude all three processes.

Protein adsorption assays, for example, based on the highly sensitive surface plasmon resonance (SPR), OWLS, or fluorescence monitoring techniques, are widely used for determining a surfaces' nonfouling capability, generally reported as a percent reduction in protein adsorption compared to a bare (control) substrate. In terms of types of proteins, a wide range of different solution compositions are reported and used in the literature ranging from single-protein solutions (e.g., albumin, fibrinogen) to blood plasma or serum, or whole blood. While single-protein adsorption studies have contributed to better understanding mechanistic aspects of protein–surface interactions, they lack the compositional complexity of body fluids; the results cannot be claimed, therefore, to represent general resistance to unspecific protein adsorption. For example, single proteins of a given charge can be excluded from surfaces by electrostatic repulsion, while the same surface is likely to adsorb proteins of opposite charge always found in more complex biological fluids (serum, bacterial broth) [130, 131]. Resistance to serum adsorption, evaluated with media containing proteins of different size/shape and charge, is therefore more relevant for judging the nonfouling character of surfaces.

28.3
Surface Strategies for Bio-Oriented Applications

This section aims to provide an overview of state-of-the-art chemistry, with special emphasis on self-assembly approaches and multifunctional polymer thin films, for tailoring the biomaterial surface properties and introducing bioactive moieties at the molecular level. Examples of selected biomedical applications, for example, diagnostic assays, cell-selective surfaces, and bacterial adhesion studies will also be presented. Three approaches to surface modifications will be addressed; (i) passivation of the surface to eliminate all nonspecific protein adsorption and cell binding, (ii) immobilization of biological cues to bind desired biomolecules and cells, and (iii) methods to create adhesive patterns.

28.3.1
Surface Passivation

28.3.1.1 Poly(Ethylene Glycol) Nonfouling Surfaces
As mentioned in the introduction, bioinert (or nonfouling) surface coatings can circumvent certain problems associated with applications in areas such as diagnostics and medical devices. Based on (mostly) empirically derived design criteria, bioinert surfaces should be hydrophilic and electrically neutral (including zwitterionic) and should have hydrogen-bond acceptor but not donor properties [132, 133]. Regarding the latter criterion, there are exceptions, however, notably polyacrylamide (PAAm). Accordingly, various surface chemistries have been reported to impart low fouling or even nonfouling protein resistance to biomaterials such as poly(hydroxyethyl)methacrylate (PHEMA) [134], PAAm [135], phosphorylcholine [136], N-vinylpyrrolidone (NVP) [137], poly(methyl-oxazoline) (PMOXA) [138, 139], or glycomimetic polymers [140, 141]. In comparison to other chemistries, PEG, and PEG-like materials in brush conformation, are by far the most frequently used and remain the "gold standard" in the context of eliminating nonspecific adsorption, exceeding in some cases the adsorbed mass-detection limits of currently available analytical methods [142, 143]. The nonfouling properties of PEG

have been explained in terms of theoretical considerations, with hydration of PEG chains being a necessary requirement. Grafted onto solid surfaces, PEG chains are extensively hydrated, which prevents adsorption by "interfacial energy matching" such that highly hydrated biological molecules perceive little energetic differences between the solution and the surface [144]. In addition to the high degree of hydration, steric excluded-volume effects, as a result of the high mobility of PEG chains, and entropic and osmotic barrier properties are also considered to be key factors in preventing spontaneous protein adsorption [144–146]. Optimization of the grafting density and PEG-chain length (molecular weights), affecting both the packing and the conformation of the polymer chains on the surfaces, are critical in the development of nonfouling coatings. There is a relationship between polymer layer structure and fouling that can be approached *a priori* with theoretical tools. For polypeptoid polymer brushes immobilized on titanium oxide surfaces, Statz *et al.* [146] showed that the optimal conditions for surface modification are such that the bound surface coverage is larger than the minimal coverage needed for *thermodynamic* prevention of protein adsorption. In general, sufficiently long PEG chains with brush or helical conformations provide the required steric repulsion and hydration [83]. In addition, PEG-layer thickness should exceed the Debye length of electrostatic forces arising from ionizable groups on the substrate underneath the PEG layer since electrostatic adsorption of proteins can occur if such charges "shine through" the PEG layer [131]. Many experimental and theoretical studies have shown that both PEG length or molecular weight and surface density are important; the product of the two, that is, the EG monomer surface density (EG density) correlates reasonably well with the normalized protein adsorbed mass data as shown in Figure 28.4. Excellent resistance is usually seen at EG monomer densities higher than 25 EG units/nm^2, corresponding to PEG surface densities

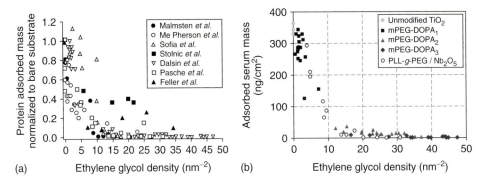

Figure 28.4 Protein adsorbed mass as a function of EG surface density measured on various PEG-grafted surfaces. (a) Protein adsorbed mass is normalized to the adsorbed mass on a bare substrate (from the same source) and is an average of the adsorbed masses of all serum proteins presented in the different studies [5, 147, 149–153]. (b) Adsorbed mass of human serum versus ethylene glycol surface density calculated from the mass per unit area of mPEG-DOPA polypeptoids on TiO$_2$ and PLL-g-PEG on Nb$_2$O$_5$ substrates [5].

of $\geq 1.0, 0.5$, and 0.2 chains/nm^2 for PEG molecular weights of 1000, 2000, and 5000 g/mol, respectively. These values compare favorably with the theoretical value of 0.5 chains/nm^2 for PEG chains with an average of 25 EG monomers [147, 148].

Many approaches have been tested to immobilize PEG-based polymers of different molecular structure on surfaces, that is, polymer brushes, comb-copolymers, or hydrogels, using the aforementioned physical or chemical adsorption methods. Nearly all types of substrates can be coated, from polymers to metal oxides, depending on the chemistry used [83]. Covalent attachment of reactive PEGs are preferentially done under cloud-point conditions (marginal solvation) to form dense elongated brushes [84]. Alternative approaches make use of molecular-assembly systems possessing two domains of different properties that can assemble at interfaces via electrostatic or hydrophobic interactions. One particularly useful nonfouling PEG-coating is a poly(L-lysine) (PLL) PEG cationic graft copolymer consisting of PEG chains grafted to amino-terminated side chains of a PLL backbone, known hereafter as *PLL-g-PEG* (Figure 28.5a). The PLL-g-PEG technology offers the possibility to tailor in a controlled manner the PEG surface coverage. The grafting ratio g (ratio of the number of lysine monomers to the number of PEG side chains) that defines the PEG content in the polymer, is easily varied by using appropriate ratios of the reagents during the copolymer synthesis. Synthesis of the PLL-g-PEG copolymer also enables variation of the molecular weights of both PLL and PEG. Therefore, a large variety of polymer architectures, from very sparse grafting densities and short side chains to high PEG densities with long side chains, can be prepared. Grafting ratios ranging from 2 to 20 and PEG chain lengths ranging from 1 to 5 kDa have been systematically varied to investigate the polymer-architecture-dependent antifouling properties [147]. The PLL-g-PEG with PEG chains of 2 kDa grafted to a PLL backbone of 20 kDa at a grafting ratio Lys/PEG of approximately 3.5 has been shown to be a preferred polymer structure for rendering metal oxide surfaces resistant to protein adsorption from human blood serum [154, 155]. Other architectures with grafting ratios above 4−5 were found to be less effective and a steady increase of protein-adsorbed mass has been reported and correlated with lower PEG-chain surface densities [147, 156]. A complete and uniform coverage of the surface by the copolymer is a further requirement to tune the protein-repellent behavior. From the point of view of polymer structure, a too dense grafting of long side chains to the PLL backbone was found to affect adversely the adsorption of the molecules due to increased rigidity of the ("overcrowded") PEG copolymer [157], eventually leading to an unstable polymer layer of only submonolayer coverage, which did not resist protein adsorption [147, 157].

Thanks to the polycationic character of the PLL backbone, the copolymers can be applied to a number of negatively charged surfaces such as many metal oxides (e.g., Nb_2O_5, TiO_2, SiO_2, Ta_2O_5) or negatively charged tissue-culture polystyrene (TCPS) by a simple dip-and-rinse procedure (Figure 28.5b). The stability of the adlayer is provided by multiple electrostatic interactions mediated by the free amino-terminated lysine units, which are positively charged under physiological conditions at pH below the pKa of 9 of the copolymer. The fact that many charges per molecule interact simultaneously with the surface leads to an increased

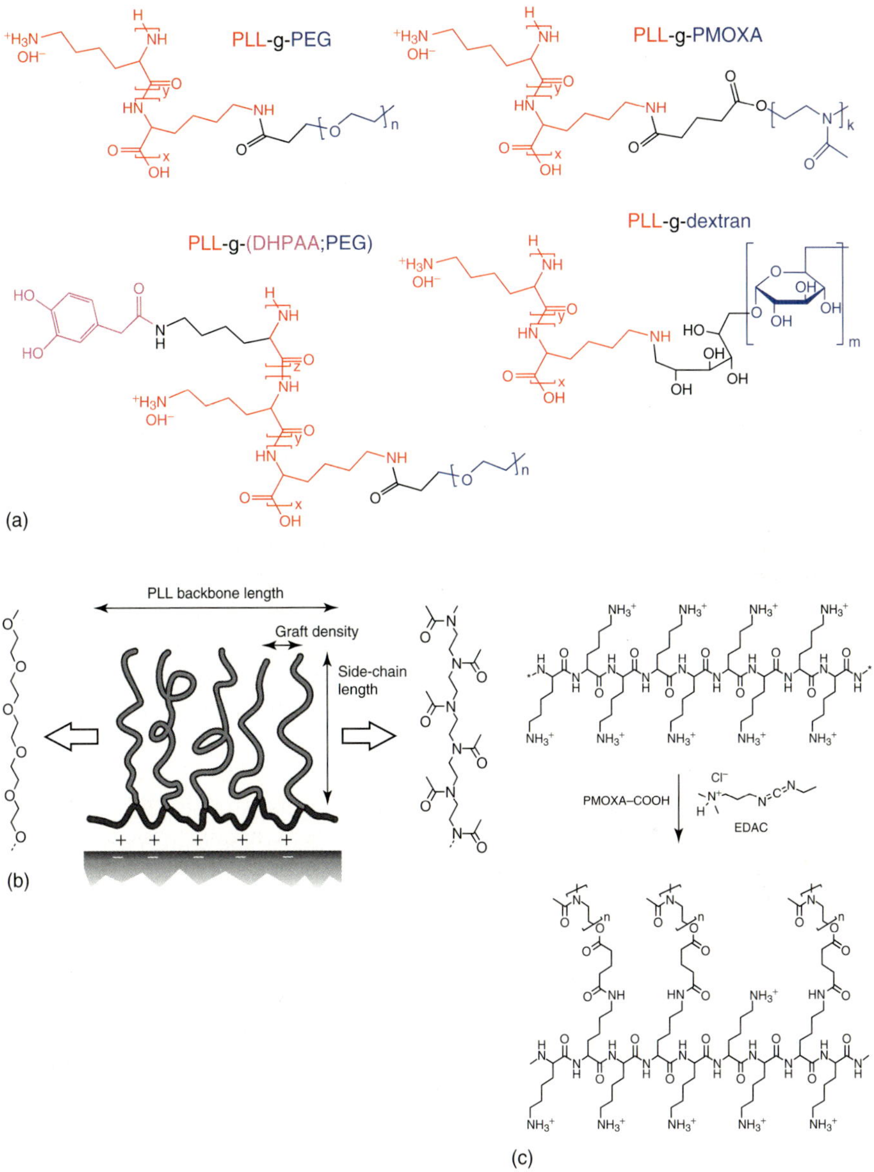

(a)

(b)

(c)

Figure 28.5 (a) Chemical structure of polycationic poly-L-lysine (PLL) copolymers grafted with different nonfouling polymers as side chains: poly(ethylene glycol) (PLL-g-PEG); poly(2-methyl-2-oxazoline) (PLL-g-PMOXA); dextran (PLL-g-dextran); PEG in combination with 3,4-dihydroxyphenylacetic acid (PLL-g-(DHPAA; PEG)) for surface immobilization exploiting both electrostatic and coordinative immobilization. (b) Cartoon of polyelectrolyte copolymers immobilized at surfaces through electrostatic interaction with PEG (left) and PMOXA (right) side chains as examples. (c) Synthesis of PLL-g-PMOXA via standard 1-Ethyl-3-(3-dimethylaminopropyl)carbodiimide (EDAC) conjugation chemistry.

thermodynamic stability and/or kinetic inertness of the copolymer assembled on the surface. The adhesive interface is believed to be dynamic and mediated by these reversible electrostatic bonds, thus allowing for some mobility of the polymeric molecules at the surface, probably an important property with respect to the formation of densely packed, confluent monolayers. Even if single-charged groups are desorbed from the surface, there are still enough groups to keep the copolymer in the adsorbed state. While protein resistance on PLL-*g*-PEG-coated metal-oxide surfaces in serum was demonstrated over a timeframe of 48 h, PLL-*g*-PEG-coated TCPS surfaces were shown to remain resistant to cell attachment for 6 days and an extended stability of 21 days was found for PLL-*g*-PEG coatings on polystyrene microspheres and on microcapsules in serum-supplemented cell-culture medium [154, 158, 159].

Upon adsorption, the PEG side chains are forced to orient toward the aqueous phase as shown by investigations of the interfacial architecture of the PLL-*g*-PEG adlayers with reflection-absorption infrared spectroscopy (RAIRS) and angle-dependent XPS [155]. Surface-force measurements, together with fits to steric barrier-layer models, have further shown that, depending on the PLL/PEG ratio, such adsorbed PLL-*g*-PEG layers can form stretched and relatively densely packed protein-repellent PEG brushes [130]. The entropic penalty for confining and stretching the PEG chains is then compensated for by the energy gain from electrostatic binding of the polyelectrolyte backbone to the surface. In addition, the excellent protein resistance of this type of surface is also the result of the high degree of hydration (approx. 80 wt% of the PEG brush surface). Protein adsorption would require an energetically and kinetically unfavorable disruption of this hydrated structure [160, 161].

PLL-*g*-PEG adlayers with optimized architectures have been shown to provide efficient protein-resistance against albumin, fibrinogen, and human blood serum, even exceeding the detection limit of OWLS ($<2\,\text{ng/cm}^2$) [51, 147]. Furthermore PLL-*g*-PEG adlayers were shown to increase the signal-to-noise ratio in diagnostics [51], to slow down opsonization and cellular recognition in drug-delivery applications [159, 162], and to reduce undesirable bacterial colonization of titanium surfaces, of relevance in view of infection being a significant problem in hospital settings [163].

28.3.1.2 Polyoxazoline Nonfouling Surfaces

Although displaying excellent nonfouling properties for *in vitro* applications, the PEG technology has limitations for long-term applications, particularly *in vivo*, the primary reason being that PEG is subject to oxidative degradation, which has been reported to cause a loss of function [94, 164, 165]. Therefore, substantial efforts have been made into developing alternative bioinert polymers with comparable protein resistance but higher stability in biofluids. Biologically inspired approaches include, for example, polysaccharides and nonionic peptidomimetic polymers as PEG substitutes [166]. By analogy to the PLL-*g*-PEG system, new graft-copolymers were designed in our group with alternative water-soluble grafts, that is, using poly(2-oxazoline) [138] and dextran [141].

The polyoxazoline (POX) polymers are interesting for biomedical applications as they are known to be highly water soluble, nontoxic, and biocompatible, and can be used for the fabrication of artificial membranes showing very favorable circulation times of liposomes and polymerosomes in blood [167–170]. In addition, the POX derivatives have a structure that is isomeric to that of polypeptides (peptidomimetic) with amide-bonded side groups, thereby making them less prone to biological degradation by proteases [171]. This polymer was recently reported as an excellent nonfouling polymeric coating against serum-protein and bacterial adhesion [138, 172]. The use of brush-like, comb-copolymers that consist of PLL and poly(2-methyl-2-oxazoline) (PLL-g-PMOXA) (Figures 28.5a,c) have been shown to provide protein resistance to metal oxide surfaces, reducing nonspecific adsorption of proteins from full human serum to less than $2 \, \text{ng/cm}^2$, which is as good as the best PLL-g-PEG-based coating [138]. Oxazoline-based materials show additional advantages as the polymer can be synthesized via living cationic polymerization (Figure 28.5c) providing a straightforward route for the incorporation of various functional groups along the polymer chain, either at the termini or within the main chain by an appropriate choice of the initiator/terminator or reactive functional monomer, as well as for the synthesis of a variety of polymer constructs to tune the solubility or biofunctionality in a more versatile manner than with the PEG chemistry.

28.3.1.3 Glycocalyx Mimetics: Polymers with Oligosaccharide Grafts

Mimicking the biological membrane surface is another alternative for conferring antifouling properties to surfaces. In nature, the highly hydrated glycocalyx, that surrounds certain cells, possesses antiadhesive properties and is thought to prevent undesirable nonspecific adsorption of proteins and cell adhesion. Carbohydrate-rich structures grafted to surfaces, for example, dextran, hyaluronic acid, or heparin, have been reported to be an alternative to PEG for preparing low-fouling brush-type surfaces [173]. Dextran, for example, is a natural polysaccharide, that is, nontoxic, water-soluble, and electrically neutral, showing favorable characteristics for applications in the biomedical field and for protein-resistant surfaces [140, 174, 175]. Zhu and Marchant [176] have synthesized a poly(vinylamine) backbone tagged with glycosylated dendrons to prepare dense surface coatings that suppress platelet adhesion. More recently, comb-like structures of PLL-g-dextran graft-copolymers (Figure 28.5a) have been synthesized and shown to spontaneously assemble as nonadhesive layers on negatively charged metal oxides, successfully suppressing adhesion of proteins from human serum [141].

As mentioned in Section 28.1 (Figure 28.3), the mimicking of cell membranes can be taken one step further by tethering synthetic or biological polymers to a self-assembled lipid membrane weakly physisorbed to the substrate. While not providing the long-term and mechanical robustness necessary for most industrial applications, such platforms are starting to generate special interest as model systems to study biological polymer systems and as modifications to biosensors. Interestingly, supported lipid membranes in themselves have been shown to produce highly protein-repellent surfaces with few defects, which are believed to be due to a combination of the special high hydration of the lipid headgroup and the fluid molecular interface of the lipid membrane in the liquid-crystalline state [177]. By adding the polymer brush to an already protein-repellent lipid membrane a highly biomimetic and nonfouling interface can be obtained. Since the same specific interactions can be added through functional groups or, for example, specific oligosaccharide binding sequences as can be done to covalently grafted polymer films, the same applications in biosensing and biointerface science can be addressed as for the latter.

The tethering of polymers to a biomimetic liquid crystal was shown to lead to novel mechanical behavior of such systems. At mushroom and low brush densities, local compression of the polymer membrane makes it possible to "squeeze out" the polymer, in analogy with so-called escape transitions, and leave only the lipid membrane intact at the interface [122]. Breaking of the membrane at any point is recovered through the self-healing and self-spreading properties of this coating. The significance of such mechanics for cell-surface recognition and colloidal interactions remains to be explored, but an interesting concept of using the polymer coating as a molecular size filter (removing large proteins) to only allow small proteins to reach binding sites at the membrane interface has been demonstrated, which most likely utilizes the ability of the polymer film to continuously restructure itself [125].

28.3.1.4 Bioinspired Anchorage Strategies for the Attachment of Polymers to Interfaces

Despite the simplicity of the molecular assembly process and the possibility to design surfaces that are protein repellent under physiological conditions, interfacial electrostatic binding of PLL-*g*-PEG and its analogs, may suffer from stability problems in some aqueous media such as at low pH (isoelectric point (<IEP) of surface), above pH 10, or in media of higher ionic strength, which compromise the nonfouling character of the surface. Interfacial electrostatic binding is compromised when the density of positive charges on the PLL backbone is reduced with increasing pH or when the negative surface charge of the metal oxide is decreased upon lowering the pH. In high ionic strength media, electrostatic forces are screened, resulting in desorption of the PLL-*g*-PEG layer. Furthermore, during long-term contact with bodily fluids, proteins may exchange with the polymers that are not irreversibly bound to the surface ("Vroman effect") [178].

Alternative strategies have therefore been addressed with the aim of providing greater polymer stability, for instance, by the use of anchoring strategies involving the formation of robust chemical bonds between the polymer and the surface, such as silane–metal oxide or thiol–gold interactions [128, 179]. Biologically inspired coatings based on mussel adhesive proteins (MAPs) as anchoring groups, for example, amino acid L-3,4-dihydroxyphenylalanine (DOPA), a catechol derivative, have provided a new route to prepare coatings with favorable adhesion strength under both dry and wet conditions and compatibility with many types of substrate materials of interest for biomedical and diagnostic applications including metals, metal oxides, and a range of polymers [180, 181]. For example, Messersmith and coworkers demonstrated the usefulness of a DOPA–PEG copolymer (preferentially with a multiple DOPA anchor) immobilized on TiO_2 in reducing the nonspecific adhesion of proteins and cells to the coated surface [5, 180–183]. DOPA has been shown to directly coordinate to Ti cations of the surface and to form a strong, but reversible bond of 22.2 kcal/mol, as demonstrated by single-molecule AFM force spectroscopy [184]. Interestingly, DOPA-Lys tandem sequences, a motif also found in MAPs, have been shown to adhere even more strongly to titanium oxide surfaces [166]. Another biomimetic, but also catechol-chemistry-related anchorage approach has been reported by Zürcher, *et al.* [185] exploiting the strong chelating properties of anachelin-based siderophores used by cyanobacteria to harvest iron cations (Fe[III]). PEG-anacat polymers with a single binding moiety derived from anachelin showed increased layer stability in buffer solution when assembled under cloud-point conditions on TiO_2 surfaces in comparison to DOPA or dopamine. Moreover, the former is much less prone to oxidation and crosslinking in aqueous solution than the latter two.

Following up on this discovery, Amstad, Reimhult and coworkers [186, 187] exploited a number of catechol-based PEG dispersants for the stabilization and functionalization of <10 nm diameter, superparamagnetic iron oxide (Fe_3O_4) nanoparticles using an easy "grafting-to" protocol (Figure 28.6). They demonstrated grafting densities of close to three PEG(5 kDa)/nm^2, that is, more than a factor of 3 higher than previously demonstrated for nanoparticles. Interestingly, the highly curved surfaces of nanoparticles were shown to yield grafting densities at least a factor of 2 higher than for flat surfaces, although a direct comparison is challenging. The ultrahigh PEG densities led to unsurpassed core-shell nanoparticle stability in aqueous media upon repeated heating to 90 °C as judged by dynamic light scattering, transmission electron microscopy (TEM), and thermogravimetric analysis (TGA). Moreover, they facilitate particle handling markedly due to the possibility to store stabilized, freeze-dried single nanoparticles in powder form to later be redispersed when needed. Hence, the essentially irreversible affinity awarded these dispersants by the developed anchor groups pave the way for a variety of new applications of such oxide nanoparticles, for example, in the field of magnetic resonance imaging (MRI) where excellent stability of <30 nm individual particle dispersions under harsh conditions and controlled shell interfacial chemistry are stringent requirements. Furthermore, using irreversibly binding dispersants grafted to a particle core carrying functional groups allow careful tailoring of targeting ligands, as was demonstrated *in vitro* using QCM-D and fluorescence-activated cell sorting (FACS) by targeting a VCAM-chimera presenting a surface of interest to future applications in early recognition of atherosclerotic cardiovascular diseases with vascular cell adhesion molecule 1 (VCAM) antibody functionalized superparamagnetic iron oxide nanoparticle (SPIONs) [187].

Finally, the advantages of the electrostatically driven, spontaneous self-assembly and molecular organization of the polycationic PEG-copolymer on negatively charged surfaces were combined with the high adhesion strength, kinetic inertness, and expected long-term stability of the multivalent catechol-surface anchorage concept by synthesizing PLL-graft-(3,4-dihydroxyphenylacetic acid, DHPAA; PEG)

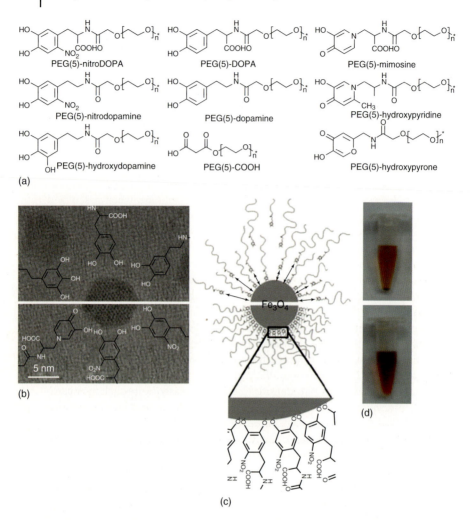

(a)

(b)

(c)

(d)

Figure 28.6 Stabilization of iron oxide nanoparticles. (a) Chemical structures of investigated dispersants with a degree of polymerization of n ≈ 114 for PEG(5). (b) High-resolution transmission electron microscopy of iron-oxide nanoparticles air-dried on a carbon supported Cu grid. The hexagonal pattern of the inverse spinel structure of Fe₃O₄ oriented in the (010) direction is clearly visible, proving the high crystallinity of these nanoparticles. Dispersants with different anchor groups have been grafted to as-synthesized iron-oxide nanoparticles. (c) Stabilization of iron oxide cores with PEG(5)-dopamine that adsorbed reversibly (top) and the irreversibly adsorbing PEG(5)-nitroDOPA (bottom). (d) While PEG(5)-dopamine stabilized nanoparticles agglomerated and thus visibly precipitated (top), PEG(5)-nitroDOPA-coated nanoparticles were stable even after being kept in 4-(2-hydroxyethyl)-1-piperazineethanesulfonic acid (HEPES) containing 150 mM NaCl for 20 h at 90 °C (bottom). The initial nanoparticle concentration was 1 mg of iron oxide/ml; thus, the color difference between PEG(5)-dopamine and PEG(5)-nitroDOPA-stabilized nanoparticle dispersions is a result of the precipitation of PEG(5)-dopamine stabilized nanoparticles [186].

copolymers (Figure 28.5a) with different fractions of DHPAA grafted to the PLL backbone. These polymers were applied to TiO_2 surfaces under different assembly conditions and showed straightforward assembly behavior (within a specific window of DHPAA grafting ratio), excellent layer stability and very low fibrinogen adsorption, particularly when the copolymers were adsorbed from aqueous solutions of higher temperature (50–70 °C) and low ionic strength. The outstanding nonfouling character of the polymeric thin films was further demonstrated for up to 100 days in culture tests with cyanobacteria Lyngbya [188].

28.3.2
Bioactive Surfaces

In certain bioapplications, for example, diagnostic assays, biomaterial devices, and drug delivery, the controlled immobilization of bioactive molecules, for example, enzymes, antibodies, sugars, peptides, and nucleic acids, onto surfaces is essential. Correspondingly, a further step in the development of PLL-g-PEG antifouling polymers was the end-functionalization of the PEG chains to prepare surfaces with the capacity to initiate specific molecular interactions with biomolecules or cells in the adjacent medium. This and other self-assembly systems have numerous advantages including ease of surface preparation and the possibility to control the concentration of the functional groups either at the precursor state in solution when grafting the PEG side chains to the PLL backbone or via controlled dilution with nonfunctionalized PLL-g-PEG and coassembly during the surface-preparation step. Functionalized and methoxy-terminated PEGs are reacted to produce polymers with the desired degree of functionality. Furthermore, the use of longer PEG chain tethers (i.e. 3.4 kDa) presenting the functional biomolecules and shorter methoxy-PEG (i.e. 2 kDa) has been reported to improve presentation and interfacial accessibility of the functional group [189, 190].

28.3.2.1 Biotin-Functionalized Surfaces for (Strept)avidin-Based Immobilization of Biotionylated Biomolecules

Because the streptavidin–biotin-based linkage (instead of streptavidin, the neutral protein NeutrAvidin is often used) is a common surface-bioengineering approach, owing to the high affinity of the interaction even under mild aqueous conditions, biotin was introduced as a biofunctional end group on the PEG chains [9, 191]. It is one of the strongest noncovalent affinity interactions in nature ($K_{aff} = 10^{13} - 10^{15}$ M^{-1} in solution) as a result of the shape specificity of the biotin-binding pocket. Varying the amount of PEG-biotin on the surface, by coadsorbing PLL-g-PEG and biotinylated PLL-g-PEG from mixed solutions in different molar ratios, or by varying the fraction of biotin-terminated-PEG chains to the total number of grafted PEG chains per copolymer, was shown to result in quantitative control of the surface density of bound streptavidin. Corresponding quantitative surface data were obtained by *in situ* OWLS measurements along with *ex situ* XPS and radiometry measurements [9, 191]. This system can, therefore, be used to quantitatively tailor the surface density of bioligands, which is an important aspect for biosensing applications as well as for studying fundamental aspects of cell–surface interactions [9]. Furthermore, this enables optimized detection in terms of sensitivity and specificity for systems requiring different surface concentrations of bioactive ligands for different target molecules. The PLL-*g*-PEG/PEG-biotin approach has been used to coat planar metal-oxide substrates for the specific binding of (strept)avidin and subsequent immobilization of biotinylated antirabbit immunoglobulin (αRIgG) antibodies for immunoassays as well as enzymes for biosensing applications [9, 192]. It was further used for the production of biotinylated poly(lactic-co-glycolic acid) (PLGA) microparticles and polyelectrolyte microcapsules for potential drug-delivery applications as well as immobilization on surfaces [193].

Different genetically engineered variants of the enzyme lactamase with one cystein introduced at different locations of the enzyme's surface were produced by site-directed mutagenesis. The cysteine thiol group was subsequently biotinylated with a dithiothreitol (DTT)-cleavable biotinylation reagent, which allowed to surface-immobilize the enzyme variants in different orientation on the PLL-*g*-PEG/PEG–biotin surface. All tested enzyme variants turned out to be fully active upon immobilization at the polymeric surface. Furthermore, no significant effect of different enzyme orientations could be detected, probably because the enzymes were attached to the surface through long, flexible PEG chain linkers. *In situ* OWLS and colorimetric analysis of enzymatic activity were used to distinguish between specific and unspecific enzyme adsorption, to sense quantitatively the amount of immobilized enzyme, and to determine Michaelis–Menten kinetics. It was found that the turnover rate was not limited by the catalytic activity of the enzyme (k(cat)) but by impaired substrate (Nitrocefin) diffusion and/or restricted accessibility of the active site [192].

28.3.2.2 NTA-Functionalized Polymeric Surfaces for Selective and Reversible Binding of Oligo-Histidine-Tagged Proteins

Bioconjugation through chelate-Ni^{2+}-histidine tag coupling is a standard technique for the separation and purification of proteins. Commonly used chelates include iminodiacetic acid [194] or nitrilotriacetic acid (NTA) [195]. Chelate–Ni^{2+}–histidine-based coupling offers several distinct advantages over biochemical recognition elements such as biotin/(strept)avidin systems: (i) His-tags bound to either the C- or N-terminus are commercially available for a large number of proteins; (ii) the formation of a histidine–Ni^{2+} complex is a fast reaction, and reversible when the surface is exposed to a competitor chelator such as ethylenediaminetetraacetic acid (EDTA) or imidazole; (iii) the spatially defined attachment of His-tags on proteins/antibodies offers the opportunity for their immobilization in controlled orientations. Large libraries of His-tagged receptor molecules, such as single-chain antibody variable-region fragments are commercially available and have been immobilized in microarray formats from expression supernatants [196].

Further examples of exploiting the NTA–Ni^{2+} conjugation technique to immobilize proteins on planar chips have been reported [197], including NTA-functionalized dextran [198, 199], NTA-terminated alkanethiol SAMs [200, 201], as well as solid-supported lipid bilayers containing a fraction of NTA-functionalized lipids [202–204]. A critical, largely unsolved issue in the context of NTA-based biosensor platforms remains the elimination of nonspecific protein adsorption.

To this end, successful approaches combining the advantage of using PEG-based, nonfouling surface chemistry, and chelation-based immobilization schemes have been reported [205, 206]. A major limitation of these approaches, however, is the relatively complex, multistage surface-chemistry preparation that makes it demanding to achieve quantitatively controlled surface quality and reproducibility from batch to batch.

A convenient one-step surface preparation scheme has been reported based on PLL-g-PEG with a fraction of the PEG chains carrying a terminal NTA group (Figure 28.7a) [207]. The NTA-functionalized PLL-g-PEG surfaces proved to be resistant to nonspecific adsorption in contact with human serum, while allowing the specific and reversible surface binding of the 6 × His-tagged green fluorescent protein (GFP)uv-6His in its native conformation (see Figure 28.7b, for example). The amount of GFPuv-6His immobilized on the polymeric surface increased with increasing NTA surface density. In terms of stability of the bound protein, a sufficiently high NTA surface density is crucial, since only multiple Ni^{2+}-binding per histidine tag provides sufficient binding strength. The latter can be further increased by the use of higher histidine oligomers, for example, 8–12× rather than the conventionally used 6 × His tag. Bound His-tagged proteins can be conveniently regenerated for reuse by exposing the chip surface to a solution containing either EDTA (that removes both Ni^{2+} and the His-tagged protein) or imidazole (that removes the His-tagged protein only) (Figure 28.7c shows an example of surface regeneration as followed by OWLS). Furthermore, micropatterns consisting of NTA-functionalized PLL-g-PEG in a background of PLL-g-PEG were produced using the molecular assembly patterning by lift-off (MAPL) technique (Figure 28.7d) [207]. Exposure to Ni^{2+} and GFPuv-6His resulted in a protein pattern of excellent contrast, as judged by fluorescence microscopy.

In a related, comparative study, the attachment, activity, stability, and regeneration capacity of β-lactamase immobilized to five different types of interfacial architectures were investigated [208]. The platforms were based on biotin–NeutrAvidin binding matrices of biotin-terminated alkanethiol SAMs and PLL-g-PEG/PEG-biotin polymeric adlayers (Figure 28.7b) and studied with both OWLS and SPR. The authors used a novel, real-time enzymatic activity assay that combines an affinity and a catalytic sensor in a connected flow system, and offers several advantages, in particular, the online activity detection for surface-attached enzymes, quantitative control of the surface density of the enzyme, and the option of using a miniaturized, portable fiber-optic absorbance spectrometer (FOAS) device as an add-on device with other *in situ* interfacial detection techniques, such as OWLS, the quartz crystal microbalance (QCM), or impedance spectroscopy (IS). With regard to the enzymatic activity of biotin–β-lactamase, this study highlighted the importance of steric crowding of surface-bound enzymes and mass-transport processes that strongly influenced the kinetic parameters of the catalytic reaction with Nitrocefin.

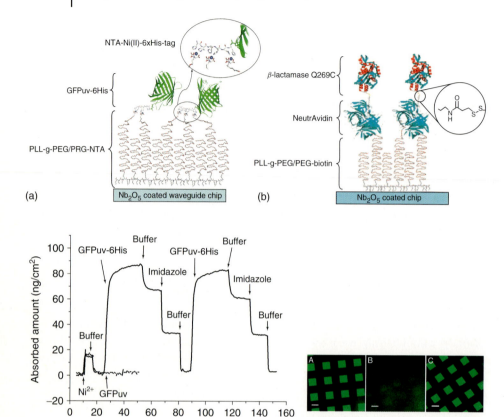

Figure 28.7 (a) Schematic view of the molecular assembly architecture of the polycationic polymer PLL-*g*-PEG/PEG-NTA assembled on the negatively charged niobium oxide surface. Green fluorescent protein (GFPuv-6His) is reversibly bound through the NTA-Ni-histidine docking site. (b) Schematic view of the molecular assembly architecture of the polycationic polymer PLL-*g*-PEG/PEG-biotin assembled on the negatively charged niobium oxide surface, NeutrAvidin adsorbed to the biotinylated polymer, and biotinylated β-lactamase specifically attached to the NeutrAvidin interfacial layer through a disulfide-containing, cleavable linker. (c) A typical OWLS adsorption curve of the PLL-*g*-PEG-NTA on Nb$_2$O$_5$-coated waveguide chip: activation of the surface with Ni^{2+}, adsorption of the GFPuv- 6His on the Ni^{2+}, and desorption by rinsing with imidazole solution. (d) Patterns of the GFPuv-6His on Nb$_2$O$_5$ analyzed by confocal laser scanning microscopy (CLSM). The scale bar represents 60 μm. A: 6His-GFP immobilized on MAPL patterned substrate after charging with Ni^{2+}. B: As A, rinsed with imidazole removing 6His-GFP. C: As B, reloaded with 6His-GFP [207].

28.3.2.3 **RGD-Peptide Surfaces**

Peptides present another bioactive moiety of interest for the surface modification of biomaterials. One of the most investigated peptide sequences is RGD (arginine-lysine-aspartic acid), a sequence found in fibronectin, an extracellular matrix protein, which is known to be involved in integrin-mediated cell adhesion

[209, 210]. Although a range of polymer systems have been explored for the immobilization of peptide sequences, suppression of nonspecific adsorption of proteins from surrounding media that interfere with the direct study of peptide–cell interactions, remains a challenge. The nonfouling PLL-g-PEG platform has been used to covalently bind small peptides to the polymer via a vinyl sulfone–cysteine coupling reaction to the PEG side chains [158]. We have used this system to investigate different cell behavior on biochemically modified dental implant materials [4]. In doing so, it was found that the number of attached osteoblastic (bone-forming) cells and their cellular footprint (spreading area) increased with increasing RGD-peptide surface density (see Figure 28.8). Such monomolecular thin films have the further advantage that rough surfaces can be (bio)chemically modified without alteration of surface topography. This is a substantial advantage in metal implants such as dental-root implants, since surface topography in the micro- and submicrometer range have been shown to increase bone-cell differentiation [211].

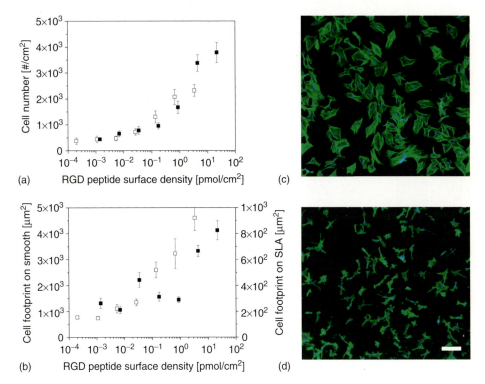

(a)

(b)

(c)

(d)

Figure 28.8 Number (a) and cell footprint (b) of rat calvarial osteoblasts attached to PLL-g- PEG/PEG-RGD monolayers on titanium (Ti) as a function of RGD peptide surface density. White squares refer to cells on the smooth surfaces and black squares to cells on rough, particle-blasted and acid-etched (so-called sandblasted and acid-etched (SLA)) titanium surfaces. (c) and (d) Fluorescence microscopy images show rat calvarial osteoblast morphology on smooth (c) and bare SLA Ti (d) surfaces, respectively (scale bar corresponds to 100 μm), green stain for actin cytoskeleton and blue stain for nucleus) [4, 212].

We have further found that these PLL-g-PEG/PEG-RGD surfaces, while preventing nonspecific protein adsorption and enabling osteoblast, fibroblast, and epithelial cell attachment, are also effective in significantly blocking bacterial adhesion [213]. One of the major impediments in the use of biomedical implants is failure due to implant-associated infections [214–217]. Bacteria, such as *Staphylococcus aureus*, *Staphylococcus epidermidis, Escherichia coli, Salmonella typhimurium, Pseudomonas*

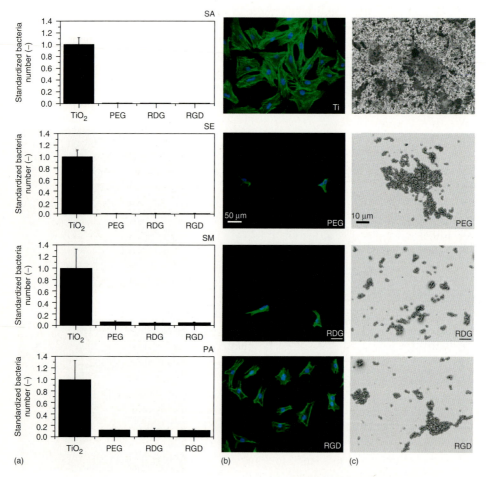

Figure 28.9 (a) Graphs showing the effect of PLL-g-PEG ("PEG") and RGD-peptide-modified PLL-g-PEG ("RGD") and control surface "RDG" (scrambled RDG peptide) coatings on smooth titanium surfaces on the density of adherent *S. aureus* (SA), *S. epidermidis* (SE), *Strep. mutans* (SM), and *P. aeruginosa* (PA) in comparison with the bare (uncoated) Ti surface ("Ti").

(b) Fluorescent images of rat calvarial osteoblasts stained for actin (green) and nuclei (blue). (c) Scanning electron micrographs of *S. aureus* (right column) on the same type of polymer-coated titanium surfaces and bare titanium control. Data refer to 24 h of culture for both osteoblast and bacteria study. Peptide density, in each case, was approximately 3.0 pmol/cm² [213, 220].

fluorescens, and *Pseudomonas aeruginosa*, compete with host cells in a race to be the first to adhere to and colonize the surface [218]. If the bacteria arrive first and are able to form a mature biofilm, then they become protected from the host immune system, as well as from the action of antibiotics, which often results in a failure of the implant and the need for revision surgery [219]. In the work of Harris and coworkers [163, 213], it was shown that PLL-*g*-PEG-RGD surfaces could reduce the amount of S. aureus, S. epidermidis, Strep. mutans, and P. aeruginosa, the most relevant implant-associated infection bacteria, by 88–98% depending on the type of bacteria (Figure 28.9).

28.3.2.4 Carbohydrate-Functionalized Surfaces

Carbohydrate ligands (frequently referred to as *sugars* or *saccharides*) are important tools for mechanistic studies in cell biology [221]. Located at outer membranes of mammalian cells, carbohydrates are primary signaling molecules, which make them relevant targets in the development of diagnostics and pharmaceutical applications, for example, vaccines and therapeutics [222]. For instance, microbial adhesion to cells is a carbohydrate-mediated cell-recognition process that causes the pathogens, that is, virus or bacteria to invade the host tissues and induce infection; protein-sugar-specific interactions generate the adhesive forces [223]. Therefore, functional surfaces that present multiple copies of carbohydrates provide a potential means for the specific detection and study of such binding events. Since PLL-*g*-PEG coatings have the ability to prevent nonspecific protein adsorption and bacterial attachment, we used this approach for surface glycoengineering wherein PEG side chains were end-functionalized with oligosaccharides (Figure 28.10a) [224, 225]. Copolymers were tagged with molecularly well-defined synthetic determinants of high-mannose N-glycans, which are of importance in the N-linked glycosylation pathway and common glycoproteins in mammalian organisms, as well as on the surface of infectious agents such as the human immunodeficiency virus (HIV) or ebola virus [226, 227]. PLL-*g*-PEG/PEG-mannose was assembled onto negatively charged surfaces and successively tested for its ability to sense, specifically, lectins and bacteria. Using OWLS, specific interactions with the plant lectin Concanavalin A (ConA) were monitored, thus demonstrating the functionality of the ligands for biointeractions (Figure 28.10b). The surfaces were further used to detect type 1 fimbriated *Escherichia coli* bacteria (*E. coli* K-12 strain), which express mannose-specific adhesin FimH at the end of the bacterial fimbria recognizing specifically α-mannose. A dependence of the bacterial surface accumulation to the carbohydrate molecular patterns was observed, that is, higher numbers of *E. coli* attached to the branched trimannose Man(α1–3)(Man(α1–6))Man compared to monomannose, while larger and complex oligomannoses exposing Man(α1–2) Man at their non reducing end showed lower binding capacity (Figures 28.10c,d). Furthermore, differences in bacterial attachment were observed under various flow conditions, depending on the type of mannoside ligands immobilized. These studies indicate that controlled surface chemistries with the sugar-tagged PLL-*g*-PEG have great potential to selectively trap micro-organisms on surfaces for medical diagnostic applications or for mechanistic studies of biofilm formation [224].

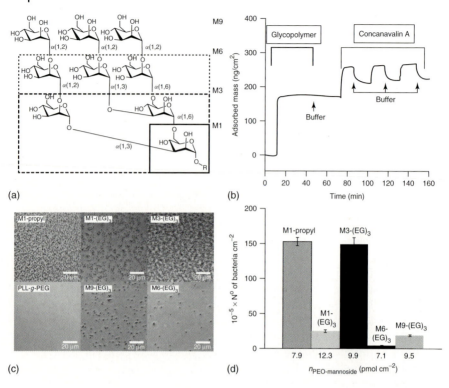

Figure 28.10 (a) Structures of the mannosides that were covalently coupled to the PEG-graft copolymer and assembled on niobia-coated glass substrates. The oligomannosides (M9, M6, M3) are all substructures of the glycan part of natural N-linked high-mannose glycoproteins. (b) Optical waveguide lightmode spectroscopy (OWLS) is used to monitor specific interactions with the plant lectin Concanavalin A (ConA), thus demonstrating the functionality of the mannosides for biointeractions. (c) Light-optical micrographs of bacteria adhering to the mannoside-functionalized glass slides and the nonfunctionalized PLL-g-PEG-coated control surface (field of view 93 × 93 μm²). (d) Number of adherent *E. coli* bacteria per square centimeter [224, 228].

Carbohydrates also play a crucial role in the body's defenses against disease-causing viruses and bacteria, since most of these microbes have unique carbohydrate markers on their surfaces recognized as foreign material by antigen-presenting cells (APCs) in the immune system. Surface-based approaches with functional microparticles (1–10 μm) decorated with pathogen-associated molecular patterns have therefore gained much attention to target receptors of APCs for both vaccine delivery and immunomodulation since they can be internalized in a manner analogous to pathogens. Wattendorf *et al.* challenged this hypothesis using stealth microparticles with mannose-enriched surface coatings (PLL-g-PEG/PEG-mannose) for a systematic investigation of receptor-specific targeting and preferred uptake by professional phagocytes, such as macrophages or dendritic cells carrying mannose receptors (MRs) (Figures 28.11a,b). Specific interactions with the plant lectin ConA demonstrated the functionality of the

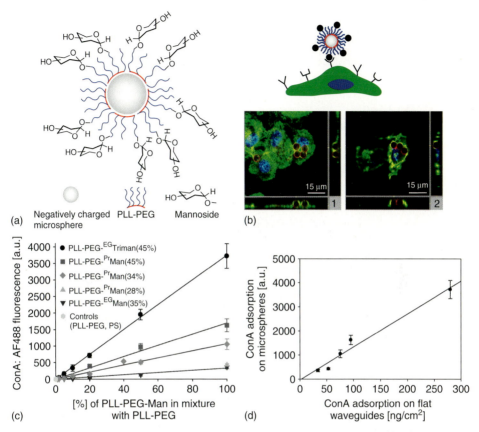

(a) Negatively charged microsphere   PLL-PEG   Mannoside

(b)

(c) ConA: AF488 fluorescence [a.u.]

- ● PLL-PEG-^{EG}Triman(45%)
- ■ PLL-PEG-^{Pr}Man(45%)
- ◆ PLL-PEG-^{Pr}Man(34%)
- ▲ PLL-PEG-^{Pr}Man(28%)
- ▼ PLL-PEG-^{EG}Man(35%)
- ● Controls (PLL-PEG, PS)

[%] of PLL-PEG-Man in mixture with PLL-PEG

(d) ConA adsorption on microspheres [a.u.]

ConA adsorption on flat waveguides [ng/cm²]

Figure 28.11 (a) Schematic representation of carbohydrate-modified microspheres for receptor-specific targeting of human-antigen-presenting cells (macrophages or dendritic cells); the surface of carboxylated polystyrene (PS) microspheres is modified through adsorption of a PLL-*g*-PEG mannoside copolymer. (b) Example of intracellular localization of PLL-*g*-PEG-mannose coated microspheres by confocal microscopy; (1) 45% Man-propyl, (2) 34% Man-propyl. Macrophages were incubated with surface-modified microspheres (red) for 4 h, fixed and stained for the actin cytoskeleton (green) and the nucleus (blue). (c) The microspheres are incubated with the mannose-binding lectin ConA to prove the availability of the mannose type carbohydrates; ConA adsorption is tuned with the surface density of PLL-*g*-PEG mannose and PLL-*g*-PEG trimannose in a mixture with ligand-free PLL-*g*-PEG. (d) The ConA adsorption on microspheres (a.u ± SD) correlates strongly with the amount of lectin quantified by OWLS (nanograms per square centimeter, single measurement) [229].

ligands on the particle surface and the results, obtained by measurements of ConA binding to mannose-presenting planar waveguides, showed a strong correlation between the ConA binding to flat surfaces and binding to particles (Figures 28.11c,d). A particular advantage of this self-assembly system is the feasibility to vary and quantitatively control the sugar surface density through either the percentage of saccharide-functionalized PEG side chains in the polymer or by coassembly from mixed solutions of nonfunctionalized and sugar-functionalized PLL-*g*-PEG present at different molar ratios. Mannose density was identified as a major factor for phagocytosis, though with limited efficiency, thus strengthening the hypothesis that MRs can only act as phagocytosis receptors in combination with yet unidentified partner receptors [159].

28.3.3
Micropatterning by Combining Lithographical and Self-Assembly Techniques

Micro- and nanoengineered surfaces have become an important tool for many biorelated research areas from DNA and protein arrays to fundamental studies of cell behavior in response to chemical and physical surface cues. The general objective is to fabricate heterogeneous surfaces exhibiting adhesive, biologically interactive patterns or microstructures in a background that resists protein adsorption or cell attachment. The influences of, for example, substrate stiffness, ligand density, and cell shape, on cell behavior are now being explored by adapting the micro- and nano-fabrication technology of the semiconductor industry to the world of biology [230–234]. Engineered surfaces of this nature may further prove to be pivotal for tissue engineering as well as being used to organize cells on transducers for cell-based sensing and/or drug screening. In this section, we will describe three straightforward methods for fabricating 2D patterns and 3D microstructures, which exploit self-organization of organic molecules or functional polymers at surfaces, and allow for the spatially defined presentation of specific (bio)chemical ligands and their application to cell-based assays.

Among the different micropatterning techniques, microcontact printing (μCP) is the most widely used, primarily thanks to the simplicity of the printing step and the ability to apply the technique without the need for cleanroom facilities ("soft" lithography). It was first adapted to generate patterns of SAMs [235–237], followed by proteins [238], as well as polymers [239] and has been used and further adapted for a number of bioapplications ranging from neuron guidance [240] to printing of functional DNA arrays [241]. The other examples for techniques based on photolithography for generating 2D patterns for biointerfacial studies, are selective molecular assembly patterning (SMAP) [242] and MAPL [243]. Although SMAP and MAPL are slightly more demanding (both in time and cost), they can reproducibly generate precise patterns at the micrometer and nanoscale over comparatively large areas, for example, full 4-in. wafers. Furthermore, when combined with some of the molecular-assembly techniques introduced above, the MAPL technique also enables quantitative control of ligand densities within the adhesive patches and independent of pattern size and geometry, a topic of interest for the study of protein–, cell–, and bacteria–surface interactions, where multivalency is a key aspect of affinity or avidity.

SMAP, developed by Michel *et al.* [242], involves in a first step the preparation of a micropattern with SiO_2-TiO_2 chemical contrast by sputtering thin films of TiO_2 on a SiO_2/wafer substrate, followed by patterning using photolithography. In a second step, alkane-phosphate SAMs adsorb selectively onto the TiO_2 regions, due to the affinity of the phosphate group to titania, while leaving the SiO_2 uncoated for the subsequent adsorption of PLL-g-PEG. Thus, the chemical contrast is transferred into a hydrophobic/hydrophilic contrast whereby further protein adsorption takes place only on the hydrophobic SAMs and not on the nonfouling PEG background. In this way, cells can be made to interact exclusively with the protein-coated patterns to study fundamental aspects of cell adhesion or the relationship of cell shape/degree of spreading and cell function (Figure 28.12) [244]. Figure 28.12 also demonstrates the ability of lithography to produce complex pattern shapes with combined large and small pattern dimensions, which is more difficult with simpler techniques such as μCP. Furthermore, the SMAP technique can also be

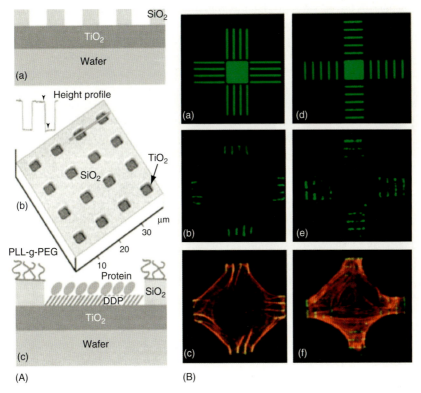

Figure 28.12 (A) Schematic illustration of the SMAP methodology. A sample exhibiting a material contrast is produced using common photolithographic techniques. Schematic view of the surface after surface-modification procedure. (B) Cell behavior on subcellular SMAP patterns. (a) and (d) show fluorescence microscopy images of fibronectin, a prominent cell-adhesion-promoting protein present in serum, on SMAP patterns. (b) and (e) show focal contacts, visualized by immunostaining for vinculin, a protein present on the cytoplasmatic side of focal contacts, are located exclusively on the adhesive fibronectin lines and in the peripheral regions of the cell. (c) and (f) show stress fibers visualized by immunostaining for f-actin. Their anchoring points colocalize with focal contact sites and are thus also subjected to the constraints of the patterns [244].

combined with nanopatterning using either nanoparticle-based (colloidal) [245] or electron-beam lithography [246]. Although nearly perfect and complex patterns can be produced across large areas, a disadvantage of this technique is the need for advanced equipment (cleanroom, reactive-ion etching facility). Furthermore, the patterned surfaces are not truly 2D: as a consequence of the etch process used to create the TiO_2/SiO_2 chemical contrast, an etch step of typically 10–20 nm is introduced between pattern and background.

In the MAPL process, developed by Falconnet *et al.* [243], a lithographically produced photoresist pattern is transferred into the desired biochemical pattern by means of spontaneous adsorption of bioactive polymers or proteins, followed by a photoresist lift-off in an organic solvent, and finally adsorption of a nonfouling polymer in the background (Figure 28.13a). This process has the advantage that no costly, complex equipment or infrastructure is needed, and, in this respect, is more comparable to soft lithography. Further advantages of MAPL are the ability to control the ligand density within the patterned areas and the true 2D nature of the patterned surface. The MAPL technique has also been extended to the nanoscale using either nanoimprint lithography [247] or extreme-ultraviolet interference lithography (EUV-IL) to prepattern the photoresist [57]. An application example is shown in Figure 28.13b: Biotin-presenting patterns of 50 nm diameter were functionalized with streptavidin and biotinylated single-stranded DNA to which gold nanoparticles, carrying the complementary ss-DNA, were immobilized through spontaneous hybridization.

Surface-based patterning approaches have been used to gain a better understanding of carbohydrate-mediated bacterial adhesion by creating, for instance, simplified models of the carbohydrate presentation found on eukaryotic cell membranes or to perform basic studies of host–pathogen interactions [248–250]. Figure 28.13c shows micropatterns of PLL-*g*-PEG/PEG-mannoside, created by the MAPL technique, to study the adhesion and further colonization of *E. coli* bacteria on mannosylated surfaces. After passivating the background, bacterial adhesion could be confined strictly to the mannoside presenting areas. When incubated with *E. Coli* in aqueous solution, high-contrast bacteria patterns of living bacteria were found to be of different size and shape, corresponding to the underlying pattern geometry with dimensions ranging from a few to hundreds of micrometers. It was found that the number or density of strongly adhering bacteria could be coregulated by many structural and physical parameters such as type and density of mannoside epitopes (reflecting the different biomolecular binding affinities) as well as by the flow rate used in the bacterial assay. Higher numbers of bacteria attached to the branched trimannose Man(α1–3)(Man(α1–6))Man compared to monomannose or larger oligomannoses exposing Man(α1–2)Man at their reducing ends, which have a lower binding capacity and a lower tendency to develop into biofilms [224].

Cell cultures are traditionally performed on two-dimensional hard substrates such as glass or tissue-culture plastic, that is, environmental conditions far from the ones found *in vivo*. There is an increasing interest to study cell development and function on culture platforms that more closely mimic microenvironmental conditions of cells in a tissue context, with dimensionality, niche size, biomolecular

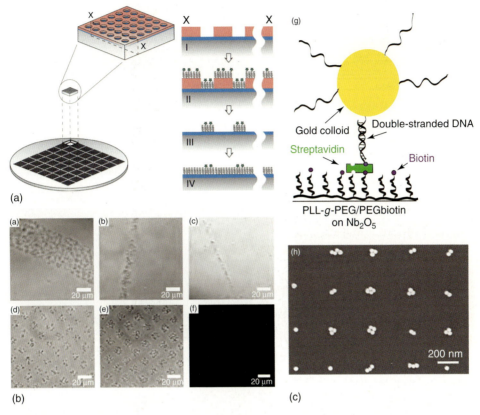

Figure 28.13 Molecular assembly patterning by lift-off (MAPL). (A) Niobia-coated glass (or silicon wafer) is patterned with photoresist by UV illumination through a mask and developed prior to dicing into 1 cm × 1 cm chips [243]. (B) *E. coli* adhering after 30 min incubation in a static assay (10⁹ cfu ml-1) to different mono- and tri-mannoside micropatterns: (a–c) Line patterns of widths 50, 10, and 2 μm, respectively. (d) and (e) 60 × 60 μm² patterns with lower and higher bacteria density reflecting lower and higher affinity to mono- and trimannoside, respectively. (f) Fluorescence image of live/dead stained *E. coli* adhering to 60 × 60 μm² patterned squares presenting mono-mannose (live bacteria appear green, dead bacteria red) [248]. (C) MAPL nanopattern produced by extreme-UV interference lithography and converted through specific DNA-DNA hybridization (g) to an array of immobilized gold nanoparticles (h) [251].

signals, and material stiffness being crucial parameters for cell function [252]. Methods reported in the literature for 3D cell culture range from matrices of fibers to various types of gels and multicellular spheroid systems [253–255]. Systems for single-cell confinement in 3D are also being explored [133, 256, 257]. We have developed a platform based on engineered microwells that enables the confinement of single cells with 3D-shape control, coupled with the ability to control the surface-chemical cues presented to the confined cells, and substrate rigidity [258, 259].

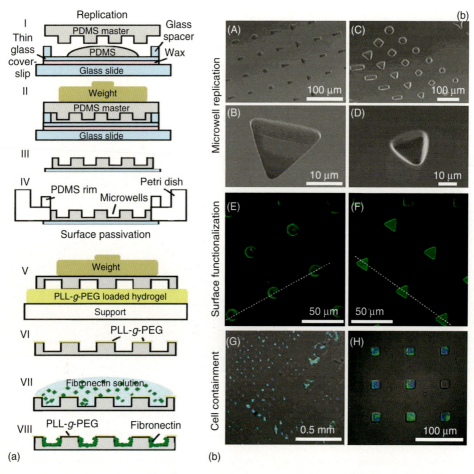

Figure 28.14 (a) Production by replication and surface modification of the microwell substrates for cell culture. Microstructured silicon master (positive structure) replicated into PDMS (negative structure) for the subsequent replication step. Replication into a thin PDMS supported on a glass cover slip. The microwell plateau surface was next passivated by inverted microcontact printing of PLL-g-PEG using a polyacrylamide stamp. After passivation of the plateau, backfill of walls and bottom of wells with cell-adhesive fibronectin (Fn). (b) Scanning electron microscopy images are shown of PDMS replicas of the microfabricated wells of various shapes in hard (A, B; 1.3 MPa) and soft PDMS (C, D; circa 10 kPa). Maximum intensity projections and cross-sections of confocal laser scanning microscopic z-stack images of microwells coated with labeled fibronectin (E) and labeled bilayers (F) are shown, the latter produced by spontaneous and area-selective liposome fusion on wall and floor of the oxidized PDMS culture substrate. Overlays of differential interference contrast (DIC) and fluorescent confocal images are shown (G, H) of human umbilical vein endothelial cells (HUVEC) cells in microwells stained with Alexa 488-phalloidin (actin; green) and ethidium homodimer (nuclei; blue) after 24 h of culture [259].

Arrays of microwells with different geometries (circles, squares, triangles, spindles, etc.) and variable lateral dimensions (80–900 μm^2) with a depth of typically 10 μm, providing a total volume of the order of a single cell were first produced in silicon using standard photolithography techniques and then replicated into standard polydimethylsiloxane (PDMS). These PDMS masters were then used to replicate thin PDMS films of variable stiffness (ranging from 1 MPa to $\leq$ 10 kPa) on glass cover slips compatible with high-resolution fluorescence microscopy (Figure 28.14a,b, A–D). To passivate the plateaus of the surface, an inverted microcontact printing technique was developed to transfer PLL-g-PEG to this area by stamping using a flat polymeric stamp. Walls and bottom of the wells could be backfilled with cell-adhesive proteins such as fibronectin (Figure 28.14b, E,F). Single endothelial cells attached and accommodated in the wells thanks to the contrast of adhesive walls inside the wells and nonfouling chemistry on the upper plateau areas. This cell culturing, microstructured platform is an example to demonstrate the applicability and usefulness of selective molecular assembly of functional polymers to quasi-3D surfaces (Figure 28.14b, G,H).

28.4
Summary and Outlook

Techniques to modify substrate surfaces are critical to many industrial and research applications ranging from tribology to biocompatibility of materials and biomedical devices to biosensors and drug-delivery systems. The toolbox of methods available today covers a large number of approaches, each with their specific pros and cons. Examples include plasma polymerization, "grafting-from" and "grafting-to" polymeric systems, and chemical and physical vapor deposition, among others. Surfaces made from soft polymeric matter have unique properties that are directly related to their macromolecular structure, and offer a myriad of opportunities to tailor their properties to the needs of the envisaged application. Molecularly designed soft surfaces and coatings are particularly attractive for applications where the mastering of biomolecular interactions is key to the functionality and performance of the components, for example, in fields such as packaging of expensive drugs/proteins (preventing loss of material); biosensors for sensing interactions with DNA, proteins, and sugars; drug-delivery containers and medical contrast agents to target specific cells and tissue; or consumer goods that resist attachment and colonization of microbes. Although the fundamental aspects of interactions between biological entities and (soft) surfaces are still only partly understood, there has been substantial progress in the last decade or so in developing empirical design criteria for polymeric surfaces *vis-à-vis* the envisaged properties, partly supported by progress in modeling polymer conformation and biomolecular processes at interfaces.

The present chapter covers only a small, but increasingly attractive sector of the field. *Spontaneous self-assembly and organization of designed molecules and polymers*, incorporating adhesion and additional (bio)functions, constitute cost-effective, easy-to-upscale (dip and rinse) techniques, applicable to a great range of surfaces and devices, from micro- and nanofluidic systems to large areas and three-dimensional devices of complex shape. Furthermore, these polymer coatings exhibit conformity (for thin films) with micro- and nanoscale surface topologies, while requiring only minimal amounts of materials. Three additional benefits of self-assembly approaches are: (i) the possibility, through coassembly of functionalized and nonfunctionalized polymers, to vary and tailor the surface density of (bio)ligands, important in the many biointeractions where affinity (avidity) relies on multivalency; (ii) the quantitatively controlled preparation of surfaces that present two or more bioligands, by coassembly of two or more functional polymers; (iii) the combination with lithography at the micrometer or nanoscale to produce patterns with distinct (bio)chemical heterogeneity. Examples given in this chapter include assembly systems that provide adhesiveness for eucaryotic cells, but repel bacteria, and the combination with μCP to produce biological contrast in 3D-like microstructured cell-culture platforms. Other potential examples are surface presentation of two or more antimicrobial molecules for reducing risks of infections with medical devices, or the combination of multiple release drug systems allowing for temporal release control of two or more drugs.

While the concept of one-step self-assembly of multifunctional polymers that carry both surface-anchorage groups and presented bioligands is simple, it has clear limitations, particularly if the bioligands have higher molecular weight, often competing with the anchorage chemistry for the substrate interface and resulting in poorly defined interfacial architecture. For such cases, self-assembly-based multistep functionalization schemes are useful. Classical examples are the biotin–(strept)avidin–biotin and the NTA-Ni^{2+}–Histag systems covered in this chapter. Another frequent limitation of ultrathin polymeric self-assembled systems is chemical and mechanical stability. Through recent progress, for example, with biomimetic anchorage strategies in combination with multivalent anchors (e.g., catechols derived from MAPs), chemical stability, and longer-term performance of polymer films can be substantially improved. The relatively poor mechanical stability of ultrathin polymer films, however, remains a general issue, limiting the applications to areas where wear resistance is not an important criterion.

In addition to ultrathin, monomolecular films covered in this chapter (typically 1–10 nm layer thickness), multilayers – produced, for example, by the layer-by-layer deposition technique of charged macromolecules or via specific linkage schemes – offer additional opportunities: to tailor the mechanical properties of the surface (e.g., stiffness, through controlled crosslinking), which is highly important in cell–surface interactions; or to provide, for example, a reservoir for growth factors or drugs for tissue engineering or regenerative medicine applications; or to construct complex constructs consisting of artificial material and native constituents such as cells. Another area not covered in this chapter are responsive polymer films that respond to external stimuli, which are of great interest to applications where switching between different surface states (e.g., adhesive/nonadhesive) provides additional functionality.

Acknowledgments

Financial support by ETH Zurich (Grant THG-38/04-2), EU FP7 (Grant ASMENA) and Competence Centre for Materials Science and Technology (CCMX, Grant MatLife) are gratefully acknowledged.

References

1. Janecek, J. and Netz, R.R. (2007) Interfacial water at hydrophobic and hydrophilic surfaces: depletion versus adsorption. *Langmuir*, **23**, 8417–8429.

2. Kasemo, B. (2002) Biological surface science. *Surf. Sci.*, **500**, 656–677.

3. Tirrell, M., Kokkoli, E., and Biesalski, M. (2002) The role of surface science in bioengineered materials. *Surf. Sci.*, **500**, 61–83.

4. Schuler, M., Owen, G.R., Hamilton, D.W., De Wilde, M., Textor, M., Brunette, D.M., and Tosatti, S.G.P. (2006) Biomimetic modification of titanium dental implant model surfaces using the RGDSP-peptide sequence: a cell morphology study. *Biomaterials*, **27**, 4003–4015.

5. Dalsin, J.L., Lin, L.J., Tosatti, S., Vörös, J., Textor, M., and Messersmith, P.B. (2005) Protein resistance of titanium oxide surfaces modified by biologically inspired mPEG-DOPA. *Langmuir*, **21**, 640–646.

6. Cooper, S.L. and Peppas, N.A. (eds) (1982) *Biomaterials: Interfacial Phenomena and Applications*, American Chemical Society, Washington, DC.

7. Gong, P. and Grainger, D.W. (2007) Nonfouling surfaces, in *Microarrays, Synthesis Methods*, vol. 1, 2nd edn, (ed. J.B. Rampal), Humana Press Inc., Totowa, NJ, Vol. 381, pp. 59–92.

8. Ladd, J., Boozer, C., Yu, Q.M., Chen, S.F., Homola, J., and Jiang, S. (2004) DNA-directed protein immobilization on mixed self-assembled monolayers via a Streptavidin bridge. *Langmuir*, **20**, 8090–8095.

9. Huang, N.P., Vörös, J., De Paul, S.M., Textor, M., and Spencer, N.D. (2002) Biotin-derivatized poly(L-lysine)-g-poly(ethylene glycol): a novel polymeric interface for bioaffinity sensing. *Langmuir*, **18**, 220–230.

10. Yu, Q. and Golden, G. (2007) Probing the protein orientation on charged self-assembled monolayers on gold nanohole arrays by SERS. *Langmuir*, **23**, 8659–8662.

11. Vallieres, K., Chevallier, P., Sarra-Bournet, C., Turgeon, S., and Laroche, G. (2007) AFM imaging of immobilized fibronectin: does the surface conjugation scheme affect the protein orientation/conformation?. *Langmuir*, **23**, 9745–9751.

12. Gorton, L. (2005) *Biosensors and Modern Biospecific Analytical Techniques*, Elsevier, Amsterdam, Boston.

13. Harbers, G.M., Emoto, K., Greef, C., Metzger, S.W., Woodward, H.N., Mascali, J.J., Grainger, D.W., and Lochhead, M.J. (2007) Functionalized poly(ethylene glycol)-based bioassay surface Chemistry that facilitates bio-immobilization and inhibits nonspecific protein, bacterial, and mammalian cell adhesion. *Chem. Mater.*, **19**, 4405–4414.

14. Zauscher, S. and Chilkoti, A. (2009) Biological applications of polymer brushes. *Biointerphases*, **4**, FA1–FA2.

15. Vickerman, J.C. and Briggs, D. (2001) *ToF-SIMS: Surface Analysis by Mass Spectrometry*, IM Publication and Surface Spectra Limited, Chichester.

16. Höök, F., Vörös, J., Rodahl, M., Kurrat, R., Böni, P., Ramsden, J.J., Textor, M., Spencer, N.D., Tengvall, P., Gold, J., and Kasemo, B. (2002) A comparative study of protein adsorption on titanium oxide surfaces using in situ ellipsometry, optical waveguide lightmode spectroscopy, and quartz crystal microbalance/dissipation. *Colloids Surf. B*, **24**, 155–170.

17. Vörös, J., Ramsden, J.J., Csucs, G., Szendro, I., De Paul, S.M., Textor, M., and Spencer, N.D. (2002) Optical grating coupler biosensors. *Biomaterials*, **23**, 3699–3710.

18. Michel, R. and Castner, D.G. (2006) Advances in time-of-flight secondary ion mass spectrometry analysis of protein films. *Surf. Interface. Anal.*, **38**, 1386–1392.

19. Grieshaber, D., MacKenzie, R., Vörös, J., and Reimhult, E. (2008) Electrochemical biosensors-sensor principles and architectures. *Sensors*, **8**, 1400–1458.

20. Höök, F., Kasemo, B., Grunze, M., and Zauscher, S. (2009) Quantitative biological surface science: challenges and recent advances. *ACS Nano*, **2**, 2428–2436.

21. Sabbatini, L. and Zambonin, P.G. (1996) XPS and SIMS surface chemical analysis of some important classes of polymeric biomaterials. *J. Electron. Spectrosc. Relat. Phenom.*, **81**, 285–301.

22. Binnig, G., Quate, C.F., and Gerber, C. (1986) Atomic force microscopy. *Phys. Rev. Lett.*, **56**, 930–933.

23. Liu, H.-B., Venkataraman, N.V., Bauert, T.E., Textor, M., and Xiao, S.-J. (2008) Multiple transmission-reflection infrared spectroscopy for high-sensitivity measurement of molecular monolayers on silicon surfaces. *J. Phys. Chem. A*, **112**, 12372–12377.

24. Konradi, R., Textor, M., and Reimhult, E. (2011) Using complementary techniques for quantitative monitoring and understanding of biomolecular processes at interfaces in situ, in *Surface Analysis and Techniques in Biology* (ed. V.S. Smentkowski), Springer-Verlag, (in press).

25. Hadjichristidis, N., Pispas, S., and Floudas, G. (2003) *Block Copolymers Synthetic Strategies, Physical Properties, and Applications*, Wiley-Interscience, Hoboken, NJ, pp. 232–254.

26. Bearinger, J.P., Terretaz, S., Michel, R., Tirreli, N., Vogel, H., Textor, M., and Hubbell, J.A. (2003) Chemisorbed poly(propylene sulphide)-based copolymers resist biomolecular interactions. *Nature Mater.*, **2**, 259–264.

27. Feller, L., Bearinger, J.P., Wu, L., Hubbell, J.A., Textor, M., and Tosatti, S. (2008) Micropatterning of gold substrates based on poly(propylene sulfide-bl-ethylene glycol), (PPS–PEG) background passivation and the molecular-assembly patterning by lift-off (MAPL) technique. *Surf. Sci.*, **602**, 2305–2310.

28. Decher, G., Hong, J.D., and Schmitt, J. (1992) Buildup of ultrathin multilayer films by a self-assembly process: III. Consecutively alternating adsorption of anionic and cationic polyelectrolytes on charged surfaces. *Thin Solid Films*, **210–211**, 831–835.

29. Wiegel, F.W. (1977) Adsorption of a macromolecule to a charged surface. *J. Phys. A: Math. Gen.*, **10**, 299–303.

30. Decher, G. (1997) Fuzzy nanoassemblies: toward layered polymeric multicomposites. *Science*, **277**, 1232–1237.

31. Muthukumar, M. (1987) Adsorption of a polyelectrolyte chain to a charged surface. *J. Chem. Phys.*, **86**, 7230–7235.

32. Baumgartner, A. and Muthukumar, M. (1991) Effects of surface-roughness on adsorbed polymers. *J. Chem. Phys.*, **94**, 4062–4070.

33. Kong, C.Y. and Muthukumar, M. (1998) Monte Carlo study of adsorption of a polyelectrolyte onto charged surfaces. *J. Chem. Phys.*, **109**, 1522–1527.

34. Vagharchakian, L., Desbat, B., and Henon, S. (2004) Adsorption of weak polyelectrolytes to an oppositely charged Langmuir film: change in the conformation of the adsorbed molecules and saturation of the ionization degree. *Macromolecules*, **37**, 8715–8720.

35. Burke, S.E. and Barrett, C.J. (2003) Acid-base equilibria of weak polyelectrolytes in multilayer thin. *Langmuir*, **19**, 3297–3303.

36. Erol, M., Du, H., and Sukhishvili, S. (2006) Control of specific attachment of proteins by adsorption of polymer layers. *Langmuir*, **22**, 11329–11336.

37. Menchaca, J.-L., Jachimska, B., Cuisinier, F., and Perez, E. (2003) In situ surface structure study of polyelectrolyte multilayers by liquid-cell

AFM. *Colloids Surf. A: Physicochem. Eng. Aspects*, **222**, 185–194.

38. Shiratori, S.S. and Rubner, M.F. (2000) pH-dependent thickness behavior of sequentially adsorbed layers of weak polyelectrolytes. *Macromolecules*, **33**, 4213–4219.

39. von Goeler, F. and Muthukumar, M. (1996) Stretch-collapse transition of polyelectrolyte brushes in a poor solvent. *J. Chem. Phys.*, **105**, 11335–11346.

40. Steitz, R., Leiner, V., Siebrecht, R., and von Klitzing, R. (2000) Influence of the ionic strength on the structure of polyelectrolyte films at the solid/liquid interface. *Colloids Surf. A: Physicochem. Eng. Aspects*, **163**, 63–70.

41. Kumar, N.A. and Seidel, C. (2005) Polyelectrolyte brushes with added salt. *Macromolecules*, **38**, 9341–9350.

42. Dubas, S.T. and Schlenoff, J.B. (1999) Factors controlling the growth of polyelectrolyte multilayers. *Macromolecules*, **32**, 8153–8160.

43. Fery, A., Scholer, B., Cassagneau, T., and Caruso, F. (2001) Nanoporous thin films formed by salt-induced structural changes in multilayers of poly(acrylic acid) and poly(allylamine). *Langmuir*, **17**, 3779–3783.

44. Tannenbaum, R., King, S., Lecy, J., Tirrell, M., and Potts, L. (2004) Infrared study of the kinetics and mechanism of adsorption of acrylic polymers on alumina surfaces. *Langmuir*, **20**, 4507–4514.

45. Salomaki, M., Laiho, T., and Kankare, J. (2004) Counteranion-controlled properties of polyelectrolyte multilayers. *Macromolecules*, **37**, 9585–9590.

46. Adamczyk, Z., Bratek, A., Jachimska, B., Jasinski, T., and Warszynski, P. (2006) Structure of poly(acrylic acid) in electrolyte solutions determined from simulations and viscosity measurements. *J. Phys. Chem. B*, **110**, 22426–22435.

47. Jonsson, B., Broukhno, A., Forsman, J., and Akesson, T. (2003) Depletion and structural forces in confined polyelectrolyte solutions. *Langmuir*, **19**, 9914–9922.

48. Bullard, J.W. and Cima, M.J. (2006) Orientation dependence of the isoelectric point of TiO$_2$ (rutile) surfaces. *Langmuir*, **22**, 10264–10271.

49. Malmsten, M., Emoto, K., and Van Alstine, J.M. (1998) Effect of chain density on inhibition of protein adsorption by poly(ethylene glycol) based coatings. *J. Colloid Interface Sci.*, **202**, 507–517.

50. Brink, C., Osterberg, E., Holmberg, K., and Tiberg, F. (1992) Using poly(Ethylene Imine) to graft poly(ethylene glycol) or polysaccharide to polystyrene. *Colloids Surf. A: Physicochem. Eng. Aspects*, **66**, 149–156.

51. Kenausis, G.L., Vörös, J., Elbert, D.L., Huang, N., Hofer, R., Ruiz-Taylor, L., Textor, M., Hubbell, J.A., and Spencer, N.D. (2000) Poly(L-lysine)-g-poly(ethylene glycol) layers on metal oxide surfaces: attachment mechanism and effects of polymer architecture on resistance to protein adsorption. *J. Phys. Chem. B*, **104**, 3298–3309.

52. Fawcett, W.R. and Smagala, T.G. (2006) New developments in the theory of the diffuse double layer. *Langmuir*, **22**, 10635–10642.

53. Salomaki, M., Vinokurov, I.A., and Kankare, J. (2005) Effect of temperature on the buildup of polyelectrolyte multilayers. *Langmuir*, **21**, 11232–11240.

54. Kharlampieva, E. and Sukhishvili, S.A. (2003) Ionization and pH stability of multilayers formed by self-assembly of weak polyelectrolytes. *Langmuir*, **19**, 1235–1243.

55. Farhat, T.R. and Schlenoff, J.B. (2001) Ion transport and equilibria in polyelectrolyte multilayers. *Langmuir*, **17**, 1184–1192.

56. Schlenoff, J.B. and Dubas, S.T. (2001) Mechanism of polyelectrolyte multilayer growth: charge overcompensation and distribution. *Macromolecules*, **34**, 592–598.

57. Blättler, T., Huwiler, C., Ochsner, M., Stadler, B., Solak, H., Vörös, J., and Grandin, H.M. (2006) Nanopatterns with biological functions. *J. Nanosci. Nanotechnol.*, **6**, 2237–2264.

58. Alcantar, N.A., Aydil, E.S., and Israelachvili, J.N. (2000) Polyethylene

glycol-coated biocompatible surfaces. *J. Biomed. Mater. Res.*, **51**, 343–351.

59. Nie, F.Q., Xu, Z.K., Huang, X.J., Ye, P., and Wu, J. (2003) Acrylonitrile-based copolymer membranes containing reactive groups: Surface modification by the immobilization of poly(ethylene glycol) for improving antifouling property and biocompatibility. *Langmuir*, **19**, 9889–9895.

60. Zhang, F., Kang, E.T., Neoh, K.G., and Huang, W. (2001) Modification of gold surface by grafting of poly(ethylene glycol) for reduction in protein adsorption and platelet adhesion. *J. Biomater. Sci. Polym. Ed.*, **12**, 515–531.

61. Ulman, A. (1996) Formation and structure of self-assembled monolayers. *Chem. Rev.*, **96**, 1533–1554.

62. Sagiv, J. (1980) Organized monolayers by adsorption. 1. Formation and structure of oleophobic mixed monolayers on solid-surfaces. *J. Am. Chem. Soc.*, **102**, 92–98.

63. Porter, M.D., Bright, T.B., Allara, D.L., and Chidsey, C.E.D. (1987) Spontaneously organized molecular assemblies. 4. Structural characterization of normal-alkyl thiol monolayers on gold by optical ellipsometry, infrared-spectroscopy, and electrochemistry. *J. Am. Chem. Soc.*, **109**, 3559–3568.

64. Bain, C.D., Troughton, E.B., Tao, Y.T., Evall, J., Whitesides, G.M., and Nuzzo, R.G. (1989) Formation of monolayer films by the spontaneous assembly of organic thiols from solution onto gold. *J. Am. Chem. Soc.*, **111**, 321–335.

65. Prime, K.L. and Whitesides, G.M. (1993) Adsorption of proteins onto surfaces containing end-attached oligo(Ethylene Oxide) – a model system using self-assembled monolayers. *J. Am. Chem. Soc.*, **115**, 10714–10721.

66. Ulman, A. (1991) *An Introduction to Ultrathin Organic Thin Films: From Langmuir Blodgett to Self-Assembly*, Academic Press, San Diego, CA.

67. Hanson, E.L., Schwartz, J., Nickel, B., Koch, N., and Danisman, M.F. (2003) Bonding self-assembled, compact organophosphonate monolayers to the native oxide surface of silicon. *J. Am. Chem. Soc.*, **125**, 16074–16080.

68. Lewington, T.A., Alexander, M.R., Thompson, G.E., and McAlpine, E. (2002) Bodycote Prize Paper Characterisation of alkyl phosphonic acid monolayers self assembled on hydrated surface of aluminium. *Surf. Eng.*, **18**, 228–232.

69. Scheres, L., Arafat, A., and Zuilhof, H. (2007) Self-assembly of high-quality covalently bound organic monolayers onto silicon. *Langmuir*, **23**, 8343–8346.

70. Marcinko, S. and Fadeev, A.Y. (2004) Hydrolytic stability of organic monolayers supported on TiO_2 and ZrO_2. *Langmuir*, **20**, 2270–2273.

71. Van Alsten, J.G. (1999) Self-assembled monolayers on engineering metals: Structure, derivatization, and utility. *Langmuir*, **15**, 7605–7614.

72. Gawalt, E.S., Avaltroni, M.J., Koch, N., and Schwartz, J. (2001) Self-assembly and bonding of alkanephosphonic acids on the native oxide surface of titanium. *Langmuir*, **17**, 5736–5738.

73. Tao, F. and Bernasek, S.L. (2007) Understanding odd-even effects in organic self-assembled monolayers. *Chem. Rev.*, **107**, 1408–1453.

74. Gao, W., Dickinson, L., Grozinger, C., Morin, F.G., and Reven, L. (1996) Self-assembled monolayers of alkylphosphonic acids on metal oxides. *Langmuir*, **12**, 6429–6435.

75. Steiner, G., Sablinskas, V., Savchuk, O., Bariseviciute, R., Jahne, E., Adler, H.J., and Salzer, R. (2003) Characterization of self assembly layers of octadecanephosphonic acid by polarisation modulation FT-IRRA spectroscopy mapping. *J. Mol. Struct.*, **661**, 429–435.

76. Dordi, B., Schonherr, H., and Vancso, G.J. (2003) Reactivity in the confinement of self-assembled monolayers: chain length effects on the hydrolysis of N-hydroxysuccinimide ester disulfides on gold. *Langmuir*, **19**, 5780–5786.

77. Silien, C., Dreesen, L., Cecchet, F., Thiry, P.A., and Peremans, A. (2007) Orientation and order of self-assembled p-benzenedimethanethiol films on

Pt(111) obtained by direct adsorption and via alkanethiol displacement. *J. Phys. Chem. C*, **111**, 6357–6364.

78. Adden, N., Gamble, L.J., Castner, D.G., Hoffmann, A., Gross, G., and Menzel, H. (2006) Phosphonic acid monolayers for binding of bioactive molecules to titanium surfaces. *Langmuir*, **22**, 8197–8204.

79. Chapman, R.G., Ostuni, E., Yan, L., and Whitesides, G.M. (2000) Preparation of mixed self-assembled monolayers (SAMs) that resist adsorption of proteins using the reaction of amines with a SAM that presents interchain carboxylic anhydride groups. *Langmuir*, **16**, 6927–6936.

80. Vyklicky, L., Afzali-Ardakani, A., and Kagan, C.R. (2005) Self-assembly and oligomerization of alkyne-terminated molecules on metal and oxide surfaces. *Langmuir*, **21**, 11574–11577.

81. Kitaev, V., Seo, M., McGovern, M.E., Huang, Y., and Kumacheva, E. (2001) Mixed monolayers self-assembled on mica surface. *Langmuir*, **17**, 4274–4281.

82. Wang, R.L.C., Kreuzer, H.J., and Grunze, M. (1997) Molecular conformation and solvation of oligo(ethylene glycol)-terminated self-assembled monolayers and their resistance to protein adsorption. *J. Phys. Chem. B*, **101**, 9767–9773.

83. Harder, P., Grunze, M., Dahint, R., Whitesides, G.M., and Laibinis, P.E. (1998) Molecular conformation in oligo(ethylene glycol)-terminated self-assembled monolayers on gold and silver surfaces determines their ability to resist protein adsorption. *J. Phys. Chem. B*, **102**, 426–436.

84. Kingshott, P., Thissen, H., and Griesser, H.J. (2002) Effects of cloud-point grafting, chain length, and density of PEG layers on competitive adsorption of ocular proteins. *Biomaterials*, **23**, 2043–2056.

85. Dannenberger, O., Weiss, K., Himmel, H.J., Jager, B., Buck, M., and Woll, C. (1997) An orientation analysis of differently endgroup-functionalised alkanethiols adsorbed on Au substrates. *Thin Solid Films*, **307**, 183–191.

86. Sushko, M.L. and Shluger, A.L. (2007) Dipole-dipole interactions and the structure of self-assembled monolayers. *J. Phys. Chem. B*, **111**, 4019–4025.

87. Zwahlen, M., Tosatti, S., Textor, M., and Haehner, G. (2002) Orientation in methyl- and hydroxyl-terminated self-assembled alkanephosphate monolayers on titanium oxide surfaces investigated with soft X-ray absorption. *Langmuir*, **18**, 3957–3962.

88. Malmsten, M. (1998) *Biopolymers at Interfaces: (ed. M. Malmsten)*, M. Dekker, New York.

89. Yan, L., Marzolin, C., Terfort, A., and Whitesides, G.M. (1997) Formation and reaction of interchain carboxylic anhydride groups on self-assembled monolayers on gold. *Langmuir*, **13**, 6704–6712.

90. Flynn, N.T., Tran, T.N.T., Cima, M.J., and Langer, R. (2003) Long-term stability of self-assembled monolayers in biological media. *Langmuir*, **19**, 10909–10915.

91. Husseman, M., Malmström, E.E., McNamara, M., Mate, M., Mecerreyes, D., Benoit, D.G., Hedrick, J.L., Mansky, P., Huang, E., Russell, T.P., and Hawker, C.J. (1999) Controlled synthesis of polymer brushes by "Living" free radical polymerization techniques. *Macromolecules*, **32**, 1424–1431.

92. Beyer, D., Knoll, W., Ringsdorf, H., Wang, J.H., Timmons, R.B., and Sluka, P. (1997) Reduced protein adsorption on plastics via direct plasma deposition of triethylene glycol monoallyl ether. *J. Biomed. Mater. Res.*, **36**, 181–189.

93. Bretagnol, F., Lejeune, M., Papadopoulou-Bouraoui, A., Hasiwa, M., Rauscher, H., Ceccone, G., Colpo, P., and Rossi, F. (2006) Fouling and non-fouling surfaces produced by plasma polymerization of ethylene oxide monomer. *Acta Biomater.*, **2**, 165–172.

94. Shen, M., Martinson, L., Wagner, M.S., Caster, D.G., Ratner, B.D., and Horbett, T.A. (2002) PEO-like plasma polymerized tetraglyme surface interactions with leukocytes and proteins: in

vitro and in vivo studies. *J. Biomater. Sci. Polym. Ed.*, **13**, 367–390.

95. Wu, Y.L.J., Timmons, R.B., Jen, J.S., and Molock, F.E. (2000) Non-fouling surfaces produced by gas phase pulsed plasma polymerization of an ultra low molecular weight ethylene oxide containing monomer. *Colloids Surf. B-Biointerfaces*, **18**, 235–248.

96. Advincula, R. (2006) Polymer brushes by anionic and cationic surface-initiated polymerization (SIP), in *Surface-Initiated Polymerization I*, Advances in Polymer Science, vol. 197 (ed. R. Jordan), Springer-Verlag, Berlin, pp. 107–136.

97. Jordan, R., West, N., Ulman, A., Chou, Y.M., and Nuyken, O. (2001) Nanocomposites by surface-initiated living cationic polymerization of 2-oxazolines on functionalized gold nanoparticles. *Macromolecules*, **34**, 1606–1611.

98. Prucker, O. and Ruhe, J. (1998) Mechanism of radical chain polymerizations initiated by azo compounds covalently bound to the surface of spherical particles. *Macromolecules*, **31**, 602–613.

99. Ma, H.W., Li, D.J., Sheng, X., Zhao, B., and Chilkoti, A. (2006) Protein-resistant polymer coatings on silicon oxide by surface-initiated atom transfer radical polymerization. *Langmuir*, **22**, 3751–3756.

100. Matyjaszewski, K., Miller, P.J., Shukla, N., Immaraporn, B., Gelman, A., Luokala, B.B., Siclovan, T.M., Kickelbick, G., Vallant, T., Hoffmann, H., and Pakula, T. (1999) Polymers at interfaces: Using atom transfer radical polymerization in the controlled growth of homopolymers and block copolymers from silicon surfaces in the absence of untethered sacrificial initiator. *Macromolecules*, **32**, 8716–8724.

101. Matyjaszewski, K. and Xia, J. (2001) Atom transfer radical polymerization. *Chem. Rev.*, **101**, 2921–2990.

102. Zhao, B. and Brittain, W.J. (2000) Polymer brushes: surface-immobilized macromolecules. *Prog. Polym. Sci.*, **25**, 677–710.

103. Currie, E.P.K., Norde, W., and Stuart, M.A.C. (2003) Tethered polymer chains: surface chemistry and their impact on colloidal and surface properties. *Adv. Colloid Interface Sci.*, **100**, 205–265.

104. Luzinov, I., Minko, S., and Tsukruk, V.V. (2004) Adaptive and responsive surfaces through controlled reorganization of interfacial polymer layers. *Prog. Polym. Sci.*, **29**, 635–698.

105. Edmondson, S., Osborne, V.L., and Huck, W.T.S. (2004) Polymer brushes via surface-initiated polymerizations. *Chem. Soc. Rev.*, **33**, 14–22.

106. Advincula, R.C., Brittain, W.J., Caster, K.C., and Rühe, J. (2004) *Polymer Brushes: Synthesis, Characterization, Applications*, Wiley-VCH Verlag GmbH & Co. KGaA, Weinheim, Germany.

107. Jones, D.M. and Huck, W.T.S. (2001) Controlled surface-initiated polymerizations in aqueous media. *Adv. Mater.*, **13**, 1256–1259.

108. Huang, W., Kim, J.-B., Bruening, M.L., and Backer, G.L. (2002) Functionalization of surfaces by water-accelerated atom-transfer radical polymerization of hydroethyl methacrylate and subsequent derivatization. *Macromolecules*, **35**, 1175–1179.

109. Li, L., Zhu, Y., and Gao, C. (2008) Fabrication of thermoresponsive polymer gradients for study of cell adhesion and detachment. *Langmuir*, **24**, 13632–13639.

110. Mendez, S., Ista, L.K., and Lopez, G.P. (2003) Use of stimuli responsive polymers grafted on mixed self-assembled monolayers to tune transitions in surface energy. *Langmuir*, **19**, 8115–8116.

111. Hucknall, A., Rangarajan, S., and Chilkoti, A. (2009) In pursuit of zero: polymer brushes that resist the adsorption of proteins. *Adv. Mater.*, **21**, 2441–2446.

112. Ma, H., Wells, M., Beebe, T.P., and Chilkoti, A. (2006) Surface-initiated atom transfer radical polymerization of oligo(ethylene glycol) methyl methacrylate from a mixed self-assembled monolayer on gold. *Adv. Funct. Mater.*, **16**, 640–648.

113. Chow, D.C., Lee, W.K., Zauscher, S., and Chilkoti, A. (2005) Enzymatic fabrication of DNA nanostructures:

extension of a self-assembled oligonucleotide monolayer on gold arrays. *J. Am. Chem. Soc.*, **127**, 14122–14123.

114. Brough, B., Christman, K.L., Wong, T.S., Kolodziej, C.M., Forbes, J.G., Wang, K., Maynard, H.D., and Ho, C.M. (2007) Surface initiated actin polymerization from top-down manufactured nanopatterns. *Soft Matter*, **3**, 541–546.

115. Raynor, J.E., Petrie, T.A., Garcia, A.J., and Collard, D.M. (2007) Controlling cell adhesion to titanium: functionalization of poly(Oligo(ethylene glycol) Methacrylate) brushes with cell adhesive peptides. *Adv. Mater.*, **19**, 1724–1728.

116. Petrie, T.A., Raynor, J.E., Reyes, C.D., Burns, K.L., Collard, D.M., and Garcia, A.J. (2008) The effect of integrin-specific bioactive coatings on tissue healing and implant osseointegration. *Biomaterials*, **29**, 2849–2857.

117. Halperin, A., Tirrell, M., and Lodge, T.P. (1992) Tethered chains in polymer microstructures. *Adv. Polym. Sci.*, **100**, 31.

118. Brittain, W.J. and Minko, S. (2007) A structural definition of polymer brushes. *J. Polym. Sci. A: Polym. Chem.*, **45**, 3505–3512.

119. Zdyrko, B., Klep, V., and Luzinov, I. (2003) Synthesis and surface morphology of high- density poly(ethylene glycol) grafted layers. *Langmuir*, **19**, 10179.

120. Albertorio, F., Diaz, A.J., Yang, T., Chapa, V.A., Kataoka, S., Castellana, E.T., Cremer, P.S., and Paul, S. (2005) Fluid and air-stable lipopolymer membranes for biosensor applications. *Langmuir*, **21**, 7476–7482.

121. Daniel, S., Albertorio, F., and Cremer, P.S. (2006) Making lipid membranes rough, tough, and ready to hit the road. *MRS Bull.*, **31**, 536–540.

122. Kaufmann, S., Papastavrou, G., Karthik, K., Textor, M., and Reimhult, E. (2009) A detailed investigation of the formation kinetics and layer structure of poly(ethylene glycol) tether supported lipid bilayers. *Soft Matter*, **5**, 2804–2814.

123. Kuhl, T.L., Leckband, D.E., Lasic, D.D., and Israelachvili, J.N. (1994) Modulation of interaction forces between bilayers exposing short-chained ethylene oxide headgroups. *Biophys. J.*, **66**, 1479–1488.

124. Richter, R.P., Hock, K.K., Burkhartsmeyer, J., Boehm, H., Bingen, P., Wang, G., Steinmetz, N.F., Evans, D.J., and Spatz, J.P. (2007) Membrane-grafted hyaluronan films: a well-defined model system of glycoconjugate cell coats. *J. Am. Chem. Soc.*, **129**, 5306–5307.

125. Albertorio, F., Danicl, S., and Cremer, P.S. (2006) Supported lipopolymer membranes as nanoscale filters: simultaneous protein recognition and size-selection assays. *J. Am. Chem. Soc.*, **128**, 7168–7169.

126. Petrash, S., Cregger, T., Zhao, B., Pokidysheva, E., Foster, M.D., Brittain, W.J., Sevastianov, V., and Majkrzak, C.F. (2001) Changes in protein adsorption on self-assembled monolayers with monolayer order: comparison of human serum albumin and human gamma globulin. *Langmuir*, **17**, 7645–7651.

127. Efremova, N.V., Bondurant, B., O'Brien, D.F., and Leckband, D.E. (2000) Measurements of interbilayer forces and protein adsorption on uncharged lipid bilayers displaying poly(ethylene glycol) chains. *Biochemistry*, **39**, 3441–3451.

128. Yang, Z.H., Galloway, J.A., and Yu, H.U. (1999) Protein interactions with poly(ethylene glycol) self-assembled monolayers on glass substrates: Diffusion and adsorption. *Langmuir*, **15**, 8405–8411.

129. Halperin, A. (1999) Polymer brushes that resist adsorption of model proteins: design parameters. *Langmuir*, **15**, 252–2533.

130. Pasche, S., Textor, M., Mcagher, L., Spencer, N.D., and Griesser, H.J. (2005) Relationship between interfacial forces measured by colloid-probe atomic force microscopy and protein resistance of poly(ethylene

glycol)-grafted poly(L-lysine) adlayers on niobia surfaces. *Langmuir*, **21**, 6508–6520.

131. Pasche, S., Vörös, J., Griesser, H.J., Spencer, N.D., and Textor, M. (2005) Effects of ionic strength and surface charge on protein adsorption at PEGylated surfaces. *J. Phys. Chem. B*, **109**, 17545–17552.

132. Chapman, R.G., Ostuni, E., Liang, M.N., Meluleni, G., Kim, E., Yan, L., Pier, G., Warren, H.S., and Whitesides, G.M. (2001) Surveying for surfaces that resist the adsorption of proteins. *Langmuir*, **17**, 1225–1233.

133. Ostuni, E., Chapman, R.G., Holmlin, R.E., Takayama, S., and Whitesides, G.M. (2001) A survey of structure-property relationships of surfaces that resist the adsorption of protein. *Langmuir*, **17**, 5605–5620.

134. Martins, M.C.L., Wang, D., Ji, J., Feng, L., and Barbosa, M.A. (2003) Albumin and fibrinogen adsorption on PU-PHEMA surfaces. *Biomaterials*, **24**, 2067–2076.

135. Yoshikawa, C., Goto, A., Tsujii, Y., Ishizuka, N., Nakanishi, K., and Fukuda, T. (2006) Protein repellency of well-defined concentrated poly(2-hydroxyethylmethacrylate) brushes by size exclusion effect. *Macromolecules*, **39**, 2284–2290.

136. Chen, S.F., Liu, L.Y., and Hang, S.Y. (2006) Strong resistance of oligo(phosphorylcholine) self assembled monolayers to protein adsorption. *Langmuir*, **22**, 2418–2421.

137. Golander, C.G., Jonsson, S., Vladkova, T., Stenius, P., and Eriksson, J.C. (1986) Preparation and protein adsorption properties of photopolymerized hydrophilic films containing N-vinylpyrolidone (NVP), acrylic-acid (Aa) or ethyleneoxide (EO) units as studied by ESCA. *Colloids Surf.*, **21**, 149–165.

138. Konradi, R., Pidhatika, B., Mühlebach, A., and Textor, M. (2007) Poly-2-methyl-2-oxazoline: a peptide-like polymer for protein-repellent surfaces. *Langmuir*, **24**, 613–616.

139. Woodle, M.C., Engbers, C.M., and Zalipsky, S. (1994) New amphipatic polymer lipid conjugates forming long-circulating reticuloendothelial system-evading liposomes. *Bioconjug. Chem.*, **5**, 493–496.

140. Bosker, W.T.E., Patzsch, K., Cohen Stuart, M.A., and Norde, W. (2007) Sweet brushes and dirty proteins. *Soft Matter*, **3**, 754–762.

141. Perrino, C., Lee, S., Choi, S.W., Maruyama, A., and Spencer, N.D. (2008) A biomimetic alternative to poly(ethylene glycol) as an antifouling coating: resistance to nonspecific protein adsorption of poly(L-lysine)-graft-dextran. *Langmuir*, **24**, 8850–8856.

142. Harris, J.M. and Zalipsky, S. (1997) *Poly(ethylene Glycol) Chemistry: Chemistry and Biological Applications*, American Chemical Society, Washington, DC.

143. Merrill, E.W. (1992) Poly(ethylene oxide) and blood contact: A chronicle of one laboratory, in *Glycol Chemistry: Biotechnological and Biomedical Applications* (ed. J.M. Harris), Plenum, New York, pp. 199–220.

144. Hoffman, A.S. (1999) Nonfouling surface technologies. *J. Biomater. Sci., Polym. Ed.*, **10**, 1011–1014.

145. Leckband, D., Sheth, S., and Halperin, A. (1999) Grafted poly(ethylene oxide) brushes as nonfouling surface coatings. *J. Biomater. Sci., Polym. Ed.*, **10**, 1125–1147.

146. Statz, R., Kuang, J., Ren, C., Barron, A.E., Szleifer, I., and Messersmith, P.B. (2009) Experimental and theoretical investigation of chain length and surface coverage on fouling of surface grafted polypeptoids. *Biointerphases*, **4**, FA22–FA32.

147. Pasche, S., DePaul, S.M., Vörös, J., Spencer, N.D., and Textor, M. (2003) Poly(L-lysine)-graft-poly(ethylene glycol) assembled monolayers on niobium oxide surfaces: a quantitative study of the influence of polymer interfacial architecture on resistance to protein adsorption by ToF-SIMS and in situ OWLS. *Langmuir*, **19**, 9216–9225.

148. Szleifer, I. (1997) Protein adsorption on surfaces with grafted polymers: A theoretical approach. *Biophys. J.*, **72**, 595–612.

149. Feller, L.M., Cerritelli, S., Textor, M., Hubbell, J.A., and Tosatti, S.G.P. (2005) Influence of poly(propylene sulfide-block-ethylene glycol) di- and triblock copolymer architecture on the formation of molecular adlayers on gold surfaces and their effect on protein resistance: a candidate for surface modification in biosensor research. *Macromolecules*, **38**, 10503–10510.

150. Malmsten, M. (1998) Formation of adsorbed protein layers. *J. Colloid Interface Sci.*, **207**, 186–199.

151. McPherson, T., Kidane, A., Szleifer, I., and Park, K. (1998) Prevention of protein adsorption by tethered poly(ethylene oxide) layers: experiments and single-chain mean-field analysis. *Langmuir*, **14**, 176–186.

152. Stolnik, S., Daudali, B., Arien, A., Whetstone, J., Heald, C.R., Garnett, M.C., Davis, S.S., and Illum, L. (2001) The effect of surface coverage and conformation of poly(ethylene oxide) (PEO) chains of poloxamer 407 on the biological fate of model colloidal drug carriers. *Biochim. Biophys. Acta Biomembr.*, **1514**, 261–279.

153. Unsworth, L.D., Sheardown, H., and Brash, J.L. (2005) Protein resistance of surfaces prepared by sorption of end-thiolated poly(ethylene glycol) to gold: Effect of surface chain density. *Langmuir*, **21**, 1036–1041.

154. Kenausis, G.L., Vörös, J., Elbert, D.L., Huang, N.P., Hofer, R., Ruiz-Taylor, L., Textor, M., Hubbell, J.A., and Spencer, N.D. (2000) Poly(L-lysine)-g-poly(ethylene glycol) layers on metal oxide surfaces: attachment mechanism and effects of polymer architecture on resistance to protein adsorption. *J. Phys. Chem. B*, **104**, 3298–3309.

155. Huang, N.P., Michel, R., Vörös, J., Textor, M., Hofer, R., Rossi, A., Elbert, D.L., Hubbell, J.A., and Spencer, N.D. (2001) Poly(L-lysine)-g-poly(ethylene glycol) layers on metal oxide surfaces: surface-analytical characterization and resistance to serum and fibrinogen adsorption. *Langmuir*, **17**, 489–498.

156. Wagner, M.S., Pasche, S., Castner, D.G., and Textor, M. (2004) Characterization of poly(L-lysine)-graft-poly(ethylene glycol) assembled monolayers on niobium pentoxide substrates using time-of-flight secondary ion mass spectrometry and multivariate analysis. *Anal. Chem.*, **76**, 1483–1492.

157. Feuz, L., Leermakers, F.A.M., Textor, M., and Borisov, O. (2008) Adsorption of molecular brushes with polyelectrolyte backbones onto oppositely charged surfaces: a self-consistent field theory. *Langmuir*, **24**, 7232–7244.

158. VandeVondele, S., Vörös, J., and Hubbell, J.A. (2003) RGD-grafted poly-l-lysine-graft-(polyethylene glycol) copolymers block non-specific protein adsorption while promoting cell adhesion. *Biotechnol. Bioeng.*, **82**, 784–790.

159. Wattendorf, U., Kreft, O., Textor, M., Sukhorukov, G., and Merkle, H.P. (2008) Stable stealth function for hollow polyelectrolyte microcapsules through a polyethylene glycol grafted polyelectrolyte adlayer. *Biomacromolecules*, **9**, 100–108.

160. Heuberger, M., Drobek, T., and Vörös, J. (2004) About the role of water in surface-grafted poly(ethylene glycol) layers. *Langmuir*, **20**, 9445–9448.

161. Feuz, L. (2006) On the conformation of graft-copolymers with polyelectrolyte backbone in solution and adsorbed on surfaces. Dissertation, Eidgenössische Technische Hochschule ETH Zürich, Nr 16644, 2006.

162. Wattendorf, U., Koch, M.C., Walter, E., Vörös, J., Textor, M., and Merkle, H.P. (2006) Phagocytosis of poly(L-lysine)-graft-poly(ethylene glycol) coated microspheres by antigen presenting cells: Impact of grafting ratio and polyethylene glycol chain length on cellular recognition. *Biointerphases*, **1**, 123–133.

163. Harris, L.G., Tosatti, S., Wieland, M., Textor, M., and Richards, R.G. (2004) Staphylococcus aureus adhesion to titanium oxide surfaces coated with non-functionalized and peptide-functionalized

poly(L-lysine)-graft- poly(ethylene glycol) copolymers. *Biomaterials*, **25**, 4135–4148.

164. Branch, D.W., Wheeler, B.C., Brewer, G.J., and Leckband, D.E. (2001) Long-term stability of microstamped substrates of polylysine and grafted polyethylene glycol in cell culture conditions. *Biomaterials*, **22**, 1035–1047.

165. Roosjen, A., De Vries, J., Van der Mei, H.C., Norde, W., and Busscher, H.J. (2005) Stability and effectiveness against bacterial adhesion of poly(ethylene oxide) coatings in biological fluids. *J. Biomed. Mater. Res. Appl. Biomater.*, **73**, 347–354.

166. Statz, A.R., Meagher, R.J., Barron, A.E., and Messersmith, P.B. (2005) New peptidomimetic polymers for antifouling surfaces. *J. Am. Chem. Soc.*, **127**, 7972–7973.

167. Adams, N. and Schubert, U.S. (2007) Poly(2-oxazolines) in biological and biomedical application contexts. *Adv. Drug Deliv. Rev.*, **59**, 1504–1520.

168. Zalipsky, S., Hansen, C.B., Oaks, J.M., and Allen, T.M. (1996) Evaluation of blood clearance rates and biodistribution of poly(2-oxazoline)-grafted liposomes. *J. Pharm. Sci.*, **85**, 133–137.

169. Mecke, A., Dittrich, C., and Meier, W. (2006) Biomimetic membranes designed from amphiphilic block copolymers. *Soft Matter*, **2**, 751–759.

170. Broz, P., Benito, S.M., Saw, C.L., Burger, P., Heider, H., Pfisterer, M., Marsch, S., Meier, W., and Hunziker, P. (2005) Cell targeting by a generic receptor-targeted polymer nanocontainer platform. *J. Control. Release*, **102**, 475–488.

171. Miller, S.M., Simon, R.J., Zuckermann, R.N., Kerr, J.M., and Moos, W.H. (1995) Comparison of the proteolytic susceptibilities of homologous L-Amino Acid, D-amino Acid and N-substituted glycine peptide and peptoid oligomers. *Drug Dev. Res.*, **35**, 20–32.

172. Pidhatika, B., Möller, J., Vogel, V., and Konradi, R. (2008) Nonfouling surface coatings based on poly(2-methyl-2-oxazoline). *Chimia*, **62**, 264–269.

173. Holland, N.B., Qiu, Y., Ruegsegger, M., and Marchant, R.E. (1998) Biomimetic engineering of non-. adhesive glycocalyx-like surfaces using oligosaccharide surfactant polymers. *Nature*, **392**, 799–801.

174. McArthur, S.L., McLean, K.M., Kingshott, P., St John, H.A.W., Chatelier, R.C., and Griesser, H.J. (2000) Effect of polysaccharide structure on protein adsorption. *Colloids Surf., B*, **17**, 37–48.

175. Osterberg, E., Bergstrom, K., Holmberg, K., Schuman, T.P., Riggs, J.A., Burns, N.L., Van Alstine, J.M., and Harris, J.M. (1995) Protein-rejecting ability of surface-bound dextran in end-on and side-on configurations: comparison to PEG. *J. Biomed. Mater. Res.*, **29**, 741–747.

176. Zhu, J., and Marchant, R.E. (2006) Dendritic saccharide surfactant polymers as antifouling interface materials to reduce platelet adhesion. *Biomacromolecules*, **7**, 1036–1041.

177. Glasmästar, K., Larsson, C., Höök, F., and Kasemo, B. (2002) Protein adsorption on supported phospholipid bilayers. *J. Colloid Interface Sci.*, **246**, 40–47.

178. Vroman, L. and Adams, A.L. (1969) Identification of rapid changes at plasma-solid interfaces. *J. Biomed. Mater. Res.*, **3**, 43–67.

179. Love, J.C., Estroff, L.A., Kriebel, J.K., Nuzzo, R.G., and Whitesides, G.M. (2005) Self-assembled monolayers of thiolates on metals as a form of nanotechnology. *Chem. Rev.*, **105**, 1103–1170.

180. Dalsin, J.L. and Messersmith, P.B. (2005) Bioinspired antifouling polymers. *Mater. Today*, **8**, 38–46.

181. Lee, H., Dellatore, S.M., Miller, W.M., and Messersmith, P.B. (2007) Mussel-inspired surface chemistry for multifunctional coatings. *Science*, **318**, 426–430.

182. Fan, X.W., Lin, L.J., and Messersmith, P.B. (2006) Cell fouling resistance of polymer brushes grafted from Ti substrates by surface-initiated polymerization: effect of ethylene glycol

side chain length. *Biomacromolecules*, **7**, 2443–2448.

183. Dalsin, J.L., Hu, B.H., Bruce, P.L., and Messersmith, P.B. (2003) Mussel adhesive protein mimetic polymers for the preparation of nonfouling surfaces. *J. Am. Chem. Soc.*, **125**, 4253–4258.

184. Lee, H., Scherer, N.F., and Messersmith, P.B. (2006) Single-molecule mechanics of mussel adhesion. *Proc. Natl. Acad. Sci.*, **103**, 12999–13003.

185. Zürcher, S., Wackerlin, D., Bethuel, Y., Malisova, B., Textor, M., Tosatti, S., and Gademann, K. (2006) Biomimetic surface modifications based on the cyanobacterial iron chelator anachelin. *J. Am. Chem. Soc.*, **128**, 1064–1065.

186. Amstad, E., Gillich, T., Bilecka, I., Textor, M., and Reimhult, E. (2009) Ultrastable iron oxide nanoparticle colloidal suspensions using dispersants with catechol-derived anchor groups. *Nano Lett.*, **9**, 4042–4048.

187. Amstad, E., Zürcher, S., Mashaghi, A., Wong, J.Y., Textor, M., and Reimhult, E. (2009) Surface functionalization of single superparamagnetic iron oxide nanoparticles for targeted magnetic resonance imaging. *Small*, **5**, 1334–1342.

188. Saxer, S.S., Portmann, C., Tosatti, S., Gademann, K., Zürcher, S., and Textor, M. (2010) Surface assembly of catechol-functionalized poly(L-lysine)-graft-poly(ethylene glycol) copolymer on titanium exploiting combined electrostatically driven self-organization and biomimetic strong adhesion. *Macromolecules*, **43**, 1050–1060.

189. Uchida, K., Otsuka, H., Kaneko, M., Kataoka, K., and Nagasaki, Y. (2005) A reactive poly(ethylene glycol) layer to achieve specific surface plasmon resonance sensing with a high S/N ratio: the substantial role of a short underbrushed PEG layer in minimizing nonspecific adsorption. *Anal. Chem.*, **77**, 1075–1080.

190. Otsuka, H., Nagasaki, Y., and Kataoka, K. (2004) Characterization of aldehyde-PEG tethered surfaces: influence of PEG chain length on the specific biorecognition. *Langmuir*, **20**, 11285–11287.

191. Ruiz-Taylor, L.A., Martin, T.L., and Wagner, P. (2001) X-ray photoelectron spectroscopy and radiometry studies of biotin-derivatized poly(L-lysine)-grafted-poly(ethylene glycol) monolayers on metal oxides. *Langmuir*, **17**, 7313–7322.

192. Zhen, G.L., Zobeley, E., Eggli, V., Vörös, J., Zammaretti, P., Glockshuber, R., and Textor, M. (2004) Immobilization of the enzyme β-lactamase on biotin-derivatized poly(L-lysine)-g-poly(ethylene glycol) coated sensor chips: a study on oriented attachment and surface activity by enzyme kinetics and in situ optical sensing. *Langmuir*, **20**, 10464–10473.

193. Heuberger, R., Sukhorukov, G., Vörös, J., Textor, M., and Möhwald, H. (2005) Biofunctional polyelectrolyte multilayers and microcapsules: control of nonspecific and bio-specific protein adsorption. *Adv. Funct. Mater.*, **15**, 357–366.

194. Porath, J., Carlsson, J., Olsson, I., and Belfrage, G. (1975) Metal chelate affinity chromatography, a new approach to protein fractionation. *Nature*, **258**, 598–599.

195. Hochuli, E., Döbeli, H., and Schacher, A. (1987) New metal chelate adsorbent selective for proteins and peptides containing neighbouring histidine residues. *J. Chromatogr. A*, **411**, 177–184.

196. Wingren, C. and Borrebaeck, C.A.K. (2004) High-throughput proteomics using antibody microarrays. *Expert Rev. Proteomics*, **1**, 355–364.

197. Ueda, E.K.M., Gout, P.W., and Morganti, L. (2003) Current and prospective applications of metal ion–protein binding. *J. Chromatogr. A*, **988**, 1–23.

198. Nieba, L., Nieba-Axmann, S.E., Persson, A., Hämäläinen, M., Edebratt, F., Hansson, A., Lidholm, J., Magnusson, K., and Frostell Karlsson, A. (1997) BIACORE analysis of histidine-tagged proteins using a chelating NTA sensor chip. *Anal. Biochem.*, **252**, 217–228.

199. Gershon, P.D. and Khilko, S. (1995) Stable chelating linkage for reversible immobilization of oligohistidine tagged proteins in the BIAcore surface plasmon resonance detector. *J. Immunol. Methods*, **183**, 65–76.

200. Kröger, D., Liley, M., Schiweck, W., Skerra, A., and Vogel, H. (1999) Immobilization of histidine-tagged proteins on gold surfaces using chelator thioalkanes. *Biosens. Bioelectron.*, **14**, 155–161.

201. Sigal, G.B., Bamdad, C., Barberis, A., Strominger, J., and Whitesides, G.M. (1996) A self-assembled monolayer for the binding and study of histidine-tagged proteins by surface plasmon resonance. *Anal. Chem.*, **68**, 490–497.

202. Altin, J.G., White, F.A.J., and Easton, C.J. (2001) Synthesis of the novel chelator lipid nitrilotriacetic acid (NTA-DTDA) and its use with the IAsys biosensor to study receptor–ligand interactions on model membranes. *Biochim. Biophys. Acta*, **1513**, 131–148.

203. Schmitt, L., Dietrich, C., and Tampe, R. (1994) Synthesis and characterization of chelator-lipids for reversible immobilization of engineered proteins at self-assembled lipid interfaces. *J. Am. Chem. Soc.*, **116**, 8485–8491.

204. Gizeli, E. and Glad, J. (2004) Single-step formation of a biorecognition layer for assaying histidine-tagged proteins. *Anal. Chem.*, **76**, 3995–4001.

205. Cha, T., Guo, A., Jun, Y., Pei, D.Q., and Zhu, X.Y. (2004) Immobilization of oriented protein molecules on poly(ethylene glycol)-coated Si(111). *Proteomics*, **4**, 1965–1976.

206. Lata, S. and Piehler, J. (2005) Stable and functional immobilization of histidine-tagged proteins via multivalent chelator headgroups on a molecular poly(ethylene glycol) brush. *Anal. Chem.*, **77**, 1096–1105.

207. Zhen, G., Falconnet, D., Kuennemann, E., Vörös, J., Spencer, N.D., Textor, M., and Zürcher, S. (2006) Nitrilotriacetic acid functionalized graft copolymers: a polymeric interface for selective and reversible binding of histidine-tagged proteins. *Adv. Funct. Mater.*, **16**, 243–251.

208. Xu, F., Zhen, G.L., Textor, M., and Knoll, W. (2006) Surface plasmon optical detection of beta-lactamase binding to different interfacial matrices combined with fiber optic absorbance spectroscopy for enzymatic active assays. *Biointerphases*, **1**, 73–81.

209. Pierschbacher, M.D. and Ruoslahti, E. (1984) Cell attachment activity of fibronectin can be duplicated by small synthetic fragments of the molecule. *Nature*, **309**, 30–33.

210. Ruoslahti, E. (1996) RGD and other recognition sequences for integrins. *Annu. Rev. Cell Dev. Biology*, **12**, 697–715.

211. Boyan, B.D., Lohmann, C.H., Dean, D.D., Sylvia, V.L., Cochran, D.L., and Schwartz, Z. (2001) Mechanisms involved in osteoblast response to implant surface morphology. *Annu. Rev. Mater. Res.*, **31**, 357–371.

212. Schuler, M., Kunzler, T.P., de Wild, M., Sprecher, C.M., Trentin, D., Brunette, D.M., Textor, M., and Tosatti, S.G.P. (2009) Fabrication of TiO$_2$-coated epoxy replicas with identical dual-type surface topographies used in cell culture assays. *J. Biomed. Mater. Res. A*, **88A**, 12–22.

213. Maddikeri, R.R., Tosatti, S., Schuler, M., Chessari, S., Textor, M., Richards, R.G., and Harris, L.G. (2008) Reduced medical infection related bacterial strains adhesion on bioactive RGD modified titanium surfaces: a first step toward cell selective surfaces. *J. Biomed. Mater. Res. A*, **84A**, 425–435.

214. Geesey, G.G. (2001) Bacterial behavior at surfaces. *Curr. Opin. Microbiol.*, **4**, 296–300.

215. Lejeune, P. (2003) Contamination of abiotic surfaces: what a colonizing bacterium sees and how to blur it. *Trends Microbiol.*, **11**, 179–184.

216. Vincent, J.L. (2003) Nosocomial infections in adult intensive-care units. *Lancet*, **361**, 2068–2077.

217. von Eiff, C., Peters, G., and Heilmann, C. (2002) Pathogenesis of infections due to coagulase-negative staphylococci. *Lancet Infect. Dis.*, **2**, 677–685.

218. Gristina, A.G., Naylor, P., and Myrvik, Q. (1998) Infections from biomaterials and implants: a race for the surface. *Med. Prog. Technol.*, **14**, 205–224.

219. Costerton, J.W., Stewart, P.S., and Greenberg, E.P. (1999) Bacterial biofilms: a common cause of persistent infections. *Science*, **284**, 1318–1322.

220. Schuler, M., Trentin, D., Textor, M., and Tosatti, S.G.P. (2006) Biomedical interfaces: titanium surface technology for implants and cell carriers. *Nanomedicine*, **1**, 449–463.

221. Bertozzi C.R., Kiessling L.L. (2001) Chemical Glycobiology. *Science*, **291**, 2357–2364.

222. Seeberger, P. and Werz, D.B. (2007) Synthesis and medical applications of oligosaccharides. *Nature*, **446**, 1046–1051.

223. Sharon, N. (2006) Carbohydrates as future anti-adhesion drugs for infectious diseases. *Biochim. Biophys. Acta*, **1760**, 527–537.

224. Barth, K., Coullerez, G., Nilsson, L.M., Riccardo, C., Seeberger, P.H., Vogel, V., and Textor, M. (2008) An engineered Mannoside presenting platform: Escherichia coli adhesion under static and dynamic conditions. *Adv. Funct. Mater.*, **18**, 1459–1469.

225. Coullerez, G., Seeberger, P.H., and Textor, M. (2006) Merging organic and polymer chemistries to create glycomaterials for glycomics applications. *Macromol. Biosci.*, **6**, 634–647.

226. Adams, E.W., Ratner, D.M., Bokesch, H.R., McMahon, J.B., O'Keefe, B.R., and Seeberger, P.H. (2004) Oligosaccharide and glycoprotein microarrays as tools in HIV glycobiology: glycan-dependent gp120/protein interactions. *Chem. Biol.*, **11**, 875–881.

227. Van Kooyk, Y. and Geitjenbeek, T.B.H. (2003) DC-SIGN: escape mechanism for pathogens. *Nature Rev. Immunol.*, **3**, 697–709.

228. Barth, K. (2009) Carbohydrate-functionalized surfaces for glycomics applications. PhD Thesis, ETH Zürich.

229. Wattendorf, U., Coullerez, G., Vörös, J., Textor, M., and Merkle, H.P. (2008) Mannose-based molecular patterns on stealth microspheres for receptor-specific targeting of human antigen-presenting cells. *Langmuir*, **24**, 11790–11802.

230. Chen, C.S., Mrksich, M., Huang, S., Whitesides, G.M., and Ingber, D.E. (1997) Geometric control of cell life and death. *Science*, **276**, 1425–1428.

231. Engler, A.J., Sen, S., Sweeney, H.L., and Discher, D.E. (2006) Matrix elasticity directs stem cell lineage specification. *Cell*, **126**, 677–689.

232. Georges, P.C. and Janmey, P.A. (2005) Cell type-specific response to growth on soft materials. *J. Appl. Physiol.*, **98**, 1547–1553.

233. McBeath, R., Pirone, D.M., Nelson, C.M., Bhadriraju, K., and Chen, C.S. (2004) Cell shape, cytoskeletal tension, and RhoA regulate stem cell lineage commitment. *Dev. Cell*, **6**, 483–495.

234. Thery, M. and Bornens, M. (2006) Cell shape and cell division. *Curr. Opin. Cell Biol.*, **18**, 648–657.

235. Kumar, A., Biebuyck, H.A., and Whitesides, G.M. (1994) Patterning self-assembled monolayers – applications in materials science. *Langmuir*, **10**, 1498–1511.

236. Mrksich, M. and Whitesides, G.M. (1995) Patterning self-assembled monolayers using microcontact printing – a new technology for biosensors. *Trends Biotechnol.*, **13**, 228–235.

237. Mrksich, M. and Whitesides, G.M. (1996) Using self-assembled monolayers to understand the interactions of man-made surfaces with proteins and cells. *Annu. Rev. Biophys. Biomol. Struct.*, **25**, 55–78.

238. Bernard, A., Delamarche, E., Schmid, H., Michel, B., Bosshard, H.R., and Biebuyck, H. (1998) Printing patterns of proteins. *Langmuir*, **14**, 2225–2229.

239. Csucs, G., Michel, R., Lussi, J.W., Textor, M., and Danuser, G. (2003) Microcontact printing of novel co-polymers in combination with proteins for cell-biological applications. *Biomaterials*, **24**, 1713–1720.

240. Scholl, M., Sprossler, C., Denyer, M., Krause, M., Nakajima, K., Maelicke, A., Knoll, W., and Offenhausser, A. (2000) Ordered networks of rat hippocampal neurons attached to silicon

oxide surfaces. *J. Neurosci. Methods*, **104**, 65–75.

241. Lange, S.A., Benes, V., Kern, D.P., Horber, J.K.H., and Bernard, A. (2004) Microcontact printing of DNA molecules. *Anal. Chem.*, **76**, 1641–1647.

242. Michel, R., Lussi, J.W., Csucs, G., Reviakine, I., Danuser, G., Ketterer, B., Hubbell, J.A., Textor, M., and Spencer, N.D. (2002) Selective molecular assembly patterning: A new approach to micro- and nanochemical patterning of surfaces for biological applications. *Langmuir*, **18**, 3281–3287.

243. Falconnet, D., Koenig, A., Assi, T., and Textor, M. (2004) A combined photolithographic and molecular-assembly approach to produce functional micropatterns for applications in the biosciences. *Adv. Funct. Mater.*, **14**, 749–756.

244. Lussi, J.W., Michel, R., Reviakine, I., Falconnet, D., Goessl, A., Csucs, G., Hubbell, J.A., and Textor, M. (2004) A novel generic platform for chemical patterning of surfaces. *Prog. Surf. Sci.*, **76**, 55–69.

245. Michel, R., Reviakine, I., Sutherland, D., Fokas, C., Csucs, G., Danuser, G., Spencer, N.D., and Textor, M. (2002) A novel approach to produce biologically relevant chemical patterns at the nanometer scale: selective molecular assembly patterning combined with colloidal lithography. *Langmuir*, **18**, 8580–8586.

246. Lussi, J.W., Tang, C., Kuenzi, P.A., Staufer, U., Csucs, G., Vörös, J., Danuser, G., Hubbell, J.A., and Textor, M. (2005) Selective molecular assembly patterning at the nanoscale: a novel platform for producing protein patterns by electron-beam lithography on SiO_2/indium tin oxide-coated glass substrates. *Nanotechnology*, **16**, 1781–1786.

247. Falconnet, D., Pasqui, D., Park, S., Eckert, R., Schift, H., Gobrecht, J., Barbucci, R., and Textor, M. (2004) A novel approach to produce protein nanopatterns by combining nanoimprint lithography and molecular self-assembly. *Nano Lett.*, **4**, 1909–1914.

248. Disney, M. and Seeberger, P. (2004) The use of carbohydrate microarrays to study carbohydrate-cell interactions and to detect pathogens. *Chem. Biol.*, **11**, 1701.

249. Liang, M.N., Smith, S.P., Metallo, S.J., Choi, I.S., Prentiss, M., and Whitesides, G.M. (2000) Arrays of Self-assembled monolayers for studying inhibition of bacterial adhesion. *Proc. Natl. Acad. Sci. USA*, **97**, 13092–13096.

250. Qian, X., Metallo, S.J., Choi, I.S., Wu, H., Liang, M.N., and Whitesides, G.M. (2002) Arrays of self-assembled monolayers for studying inhibition of bacterial adhesion. *Anal. Chem.*, **74**, 1805–18010.

251. Stadler, B., Solak, H.H., Frerker, S., Bonroy, K., Frederix, F., Vörös, J., and Grandin, H.M. (2007) Nanopatterning of gold colloids for label-free biosensing. *Nanotechnology*, **18**, 155306.

252. Dusseiller, M.R., Smith, M.L., Vogel, V., and Textor, M. (2006) Microfabricated three-dimensional environments for single cell studies. *Biointerphases*, **1**, P1–P4.

253. Kunz-Schughart, L.A., Freyer, J.P., Hofstaedter, F., and Ebner, R. (2004) The use of 3-D cultures for high-throughput screening: the multicellular spheroid model. *J. Biomol. Screen.*, **9**, 273–285.

254. Underhill, G.H. and Bhatia, S.N. (2007) High-throughput analysis of signals regulating stem cell fate and function. *Curr. Opin. Chem. Biol.*, **11**, 357–366.

255. Zhang, S., Gelain, F., and Zhao, X. (2005) Designer self-assembling peptide nanofiber scaffolds for 3D tissue cell cultures. *Semin. Cancer Biol.*, **15**, 413–420.

256. Di Carlo, D. and Lee, L.P. (2006) Dynamic single-cell analysis for quantitative biology. *Anal. Chem.*, **78**, 7918–7925.

257. Lovchik, R., von Arx, C., Viviani, A., and Delamarche, E. (2008) Cellular microarrays for use with capillary-driven

microfluidics. *Anal. Bioanal. Chem.*, **390**, 801–808.

258. Dusseiller, M.R., Schlaepfer, D., Koch, M., Kroschewski, R., and Textor, M. (2005) An inverted microcontact printing method on topographically structured polystyrene chips for arrayed micro-3-D culturing of single cells. *Biomaterials*, **26**, 5917–5925.

259. Ochsner, M., Dusseiller, M.R., Grandin, H.M., Luna-Morris, S., Textor, M., Vogel, V., and Smith, M.L. (2007) Micro-well arrays for 3D shape control and high resolution analysis of single cells. *Lab Chip*, **7**, 1074–1077.

29
Fabrication, Properties, and Biomedical Applications of Nanosheets

Toshinori Fujie, Yosuke Okamura, and Shinji Takeoka

29.1
Introduction

Several important clinical technologies have emerged as a direct result of advances in nanobioscience and engineering, such as novel drug-delivery systems (DDSs) [1] and regenerative medicine. DDS has been developed as a new pharmacological approach to improve the efficacy and safety of drugs. One key technological advance in the medicinal field is tissue engineering, which fabricates tailor-made transplantable biological tissues outside the body of the patient by using an artificial cellular matrix. However, such technology has some major drawbacks such as the possibility of infection due to long periods of cell culture, high levels of expenditure, and nonapplicability for emergency care. Therefore, innovative biomaterials derived from noninfectious, mass-produced, inexpensive materials are urgently required for minimally invasive repair of surgical defects, particularly in the field of critical-care medicine.

The recent developments in material science has led us to discover a method of obtaining versatile freestanding nanosheets by utilizing molecular assembly (Table 29.1). Micro-/nanomechanical studies have shown that these nanosheets possess unexpectedly compliant and robust properties even though their aspect ratio is 10–100 times greater than that of conventional films with micrometer thicknesses. Given the wide-ranging applications of conventional static nanosheets on solid substrates (e.g., sensing, controlled release, and optical devices), it is imperative to fully assess the unique properties of the freestanding nanosheets and to demonstrate their practical applications.

In this review, we focus on the sheet-shaped material as a novel DDS carrier and a dressing material by utilizing their unique properties such as a large contact area, high levels of flexibility, noncovalent adhesion, minimum mass introduction and heterofunctionality (Figure 29.1). Such nanosheets are expected to provide suitable matrices or scaffolds for platelet adhesion (Figure 29.2a). Provided the overall size of the nanosheet is sufficiently large to cover the targeted wound it can serve as a sealing biomaterial (Figure 29.2b). We introduce two types of novel sheet-shaped biomaterials that are biocompatible and bioabsorbable. There are no

Functional Polymer Films, First Edition. Edited by Wolfgang Knoll and Rigoberto C. Advincula.
© 2011 Wiley-VCH Verlag GmbH & Co. KGaA. Published 2011 by Wiley-VCH Verlag GmbH & Co. KGaA.

Table 29.1   Series of freestanding polymeric nanosheets.

Film procedure	Materials	Exfoliation	References
Organic			
Spin coating	Conventional polymers	Solvent-assisted transfer	Forrest [2]
Layer-by-layer (LbL)	Polyelectrolytes	Sacrificial layer	Kotov [3]
Crosslinked LbL	Polyelectrolytes	Supporting film	Whitesides [4]
Spin-coating assisted LbL	Polyelectrolytes	Sacrificial layer	Tsukruk [5]
Self-assembled monolayer	Biphenylthiol	Electroetching	Eck [6]
Direct surfactant-assembly	Surfactants	Water evaporation	Ichinose [7]
LbL alternating spray	Polyelectrolytes	Sacrificial layer	Decher [8]
Langmuir–Blodgett	Amphiphilic polymers	Sacrificial layer	Miyashita [9]
Crosslinked resins	Thermal crosslinkable resins	Sacrificial layer	Kunitake [10]
Filtration of nanofibers	Nanofibers	Solvent-assisted transfer	Ichinose [11]
Inorganic			
Electropolishing	Tantalum oxide	Solvent-assisted transfer	Kruse [12]
Silicon electrodeposition	Silicon	Electroetching	Striemer [13]
Hybrid			
Sol/gel interpenetrating network	Organic/inorganic polymers	Sacrificial layer	Kunitake [14]

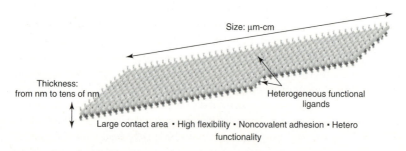

Figure 29.1   General features of a freestanding nanosheet.

previous studies on the biomedical applications of nanosheets, as far as we are aware. With clinical applications such as DDS for internal medical applications and wound-dressing materials for surgical treatments, we propose two practical approaches for the development of nanosheets. One type is an injectable nanosheet with a sheet shape similar to that of a native platelet, as a conceptual platelet substitute. The second type is a giant nanosheet with a huge aspect ratio, in excess of 10^6, which can be transferred onto human skin or organs to serve as a "nanoadhesive plaster." We report the biomedical applications of these nanosheets, such as tissue-defect repair, and touch on further functionalization of the surfaces.

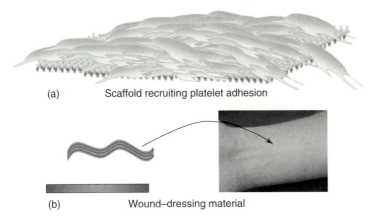

(a)   Scaffold recruiting platelet adhesion

(b)   Wound–dressing material

Figure 29.2 Biomedical application of nanosheets: (a) platelet substitutes as a scaffold recruiting platelet adhesion and (b) an adhesive plaster for wound dressing.

29.2
Nanosheets of Micrometer Size as Platelet Substitutes

29.2.1
Fabrication of Freestanding Nanosheets Derived from Proteins and Nanoparticles

Phospholipid vesicles, micelles, emulsions, and biodegradable nanoparticles have been extensively studied as carriers for DDS (e.g., for delivering drugs, recognition proteins, enzymes, genes). There are two concepts for the development of DDS; namely passive- and active-targeting systems. In the latter case, recognition proteins such as antibodies and various ligands are conjugated to the surface of the carriers for targeting tissue epitopes or specific cells. We have also developed biocompatible and bioabsorbable nanoparticles such as albumin-based nanoparticles [15–18] and vesicles [19, 20] carrying recombinant fragments of platelet membrane proteins [16, 17, 19, 21] and a dodecapeptide of fibrinogen as a recognition site for activated platelets [15, 18, 20, 22]. These nanoparticles specifically recognize the site of bleeding injury or activated platelets. However, sheet-shaped carriers such as nanosheets are expected to have several advantages over spherical-shaped carriers because they have a larger contact area for targeting, bilateral surfaces with obverse and reverse surfaces leading to heterofunctionality by surface modification, and unique dynamics caused by their inherent flexibility. The construction methods of the sheet-shaped materials are based on conventional surface technologies such as cast films [23], layer-by-layer (LbL) assemblies of polyelectrolyte multilayers [3, 24–27], crosslinked amphiphilic Langmuir–Blodgett films [28], self-assembled monolayers (SAMs) [6, 29], and assemblies of triblock copolymers [30]. In particular, both the SAMs and LbL methods rely on simple macromolecular assembling procedures for the fabrication of freestanding nanostructures.

SAMs have been widely used to control the physical and chemical properties of the surfaces of glass, quartz, SiO_2/Si wafers or silica particles [31]. Furthermore, SAMs are an excellent tool to study the immobilization of proteins such as redox proteins [32], enzymes [33], or immunoglobulins using covalent or noncovalent bonds (e.g., electrostatic interactions, hydrogen bonding, van der Waals forces, or hydrophobic interactions). Generally, it is easy to construct patterned SAMs with uniform sizes and shapes on SiO_2 or gold substrates using conventional photolithographic processes [34]. Two-dimensional patterns with steady repeatability in the particle array have also been achieved by site-selective deposition of particles using chemical bonding or electrostatic interactions [35], an electrophotography method [36], a micromold method and gravity [37], a micromold method and a lateral capillary force [38], a patterned Au film and a drying process of a colloidal solution onto the patterned Au film [39]. Furthermore, a novel method for fabrication of a close-packed particle monolayer onto a patterned hydrophilic SAM was also recently reported using a liquid mold under a drying process [40]. However, there are no reports on the preparation of freestanding protein- or nanoparticle-based nanosheets having a uniform micrometer size, nanometer thickness and heterogeneous surfaces, for use as sheet-shaped carriers.

Using patterned SAMs as a template for the fabrication of the nanosheet, we proposed a novel method to fabricate a freestanding nanosheet having heterosurfaces by a combination of four processes; (i) specific adsorption and two-dimensionally crosslinking of proteins or nanoparticles, which were deposited onto a patterned octadecyltrimethoxysilane self-assembled monolayer region (ODS-SAM), (ii) surface

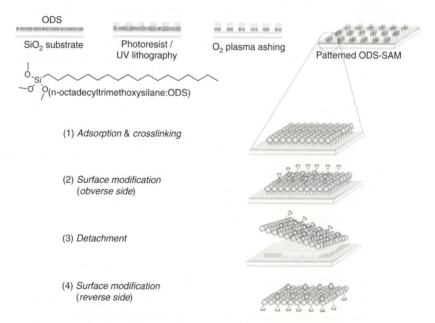

Figure 29.3 Fabrication processes of freestanding nanosheets having heterosurfaces.

modification (obverse side) of the resulting nanosheet, (iii) detachment from the ODS-SAM using a surfactant or a water-soluble sacrificial layer, and (iv) surface modification of the reverse side (Figure 29.3).

29.2.2
Freestanding Albumin Nanosheets

In order to crosslink recombinant human serum albumin (rHSA) on the patterned ODS-SAM, we introduced thiol groups to rHSA molecules (thiolated rHSA (SH-rHSA), 7.4 ± 1.2 SH groups par rHSA molecule) by N-succinimidyl 3-(2-pyridyldithio) propionate under a reductive condition using dithiothreitol. When the substrate of the rectangle patterned ODS-SAM was immersed in an acetate buffer solution (pH 5.0) of the TRITC-labeled SH-rHSA at a concentration of $1 \mu g/ml$, the rectangular patterns ($10 \mu m \times 30 \mu m$) were completely and selectively stained, as shown in Figure 29.4a [41].

After removing the nonadsorbed SH-rHSA by washing with an acetate buffer solution, the SH-rHSA molecules adsorbed on the patterned ODS-SAM were crosslinked at pH 5.0 in the presence of $1 \mu M$ copper (II) ion, and then immersed in a 1% deca(oxyethylene) dodecyl ether ($C_{12}E_{10}$) solution to detach the rectangles from the substrate. By atomic force microscopy (AFM), the rectangular patterns ($10 \mu m \times 30 \mu m$) were vividly embossed by rHSA, and nonspecific adsorption of rHSA was scarcely observed on the SiO_2 regions. From the AFM cross-sectional image, the thickness of the rHSA nanosheets was estimated to be $4.5 \pm 1.0 \, nm$, which agrees with the dimensions of rHSA.

In order to modify the surface of the resulting nanosheet, the NBD-labeled latex beads were conjugated onto the obverse side of the TRITC-labeled rHSA nanosheets, which had been adsorbed on the ODS-SAM. This procedure demonstrated the preparation of nanosheets having heterosurfaces. As shown in the image of the confocal laser scanning microscopy, there were many rHSA nanosheets in various conformations of rectangles, including bent forms. Surprisingly, however, no broken sheets were detected, suggesting that the rHSA nanosheet labeled possess significant shape stability and flexibility. The bent form of the nanosheets generated some interesting optical effects. When the rhodamine-labeled sheets were excited, the entire sheet turned red (Figure 29.4b). However, 7-chloro-4-nitrobenzo-2-oxa-1,3-diazole (NBD) on the obverse surface of the sheet was detected as yellow, and the reverse side (bent site) was significantly quenched. Judging from the thickness of the rHSA nanosheet, the NBD emission from the bent side of the nanosheet was quenched by a fluorescent resonance energy transfer (FRET) effect from NBD to rhodamine as illustrated in Figure 29.4c. These results also indicate that the NBD-latex beads are attached only to the obverse side of the nanosheet. We also observed the latex beads-conjugated rHSA nanosheets on ODS-SAM using an scanning electron microscope (SEM). Many latex beads were specifically conjugated to the obverse side of the rHSA nanosheet. The contrast of the nanosheet was clear and uniform, as shown in Figure 29.4d, suggesting that the rHSA nanosheets were thin and flat.

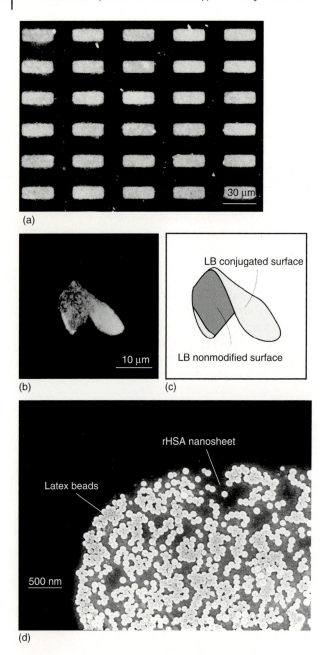

(a)

(b)

(c)

(d)

Figure 29.4 Fluorescent microscopic images of (a) SH-rHSA adsorbed onto the patterned ODS-SAM. (b) Confocal laser scanning microscopic image, (c) schematic image, and (d) SEM image of rhodamine-labeled rHSA nanosheets, of which the obverse sides were modified with NBD-labeled latex beads.

29.2.3
Freestanding Nanoparticle-Fused Nanosheets

Latex beads with a diameter of 100 nm as a model particle were coated with rHSA molecules to stabilize their dispersion state, to avoid nonspecific binding of the latex beads to the substrate, and to be able to conjugate various molecules to the amino groups of rHSA [42]. We adopted a conventional dry-patterning process for the specified adsorption of the rHSA-latex beads onto the patterned hydrophobic ODS-SAM as follows. After dropping the dispersion of latex beads (1.0×10^{11}/ml, pH 5.0) onto the substrate, any suspension remaining on the substrate was slowly blown off with a horizontal stream of nitrogen gas. In this way, the latex beads were spread and arranged as a monolayer on the entire substrate, regardless of hydrophobic and hydrophilic regions. It is possible that the assembling of latex beads involves nucleation initiated by a capillary force and growth driven by laminar flow due to evaporation of the water. Particles are then forced to arrange in the form of a monolayer [43]. Next, the immediate and repeated washing of the substrate with an acetate buffer (pH 5.0) detached the rHSA-latex beads from the hydrophilic SiO_2 region, resulting in the remaining beads forming rectangular patterns with the hydrophobic octadecyl region (5 μm × 10 μm). The rHSA-latex beads were firmly adsorbed onto the patterned ODS-SAM by hydrophobic interactions because the net charge of the rHSA-latex beads at pH 5.0, near the isoelectric point of rHSA, would be approximately zero. Thus, the ζ-potential of the SiO_2 region was extremely negative (circa −50 mV) [44]. After repeating the adsorption and washing of the latex beads on the substrate, the rHSA-latex beads were found to adopt a closely packed monolayer pattern. In order to thermally fuse the rHSA-latex beads adsorbed on the patterned ODS-SAM, the latex bead-adsorbed substrate was heated at 110 °C (above the T_g of the latex beads: 109.9 °C) for 60 s on a hotplate after drying with nitrogen gas. Though the surface of the resulting sheet maintained the spherical configuration of the constituent latex beads, the neighboring latex beads were sufficiently fused.

The poly(acrylic acid) (PAA) solution was cast on the substrate where the nanosheets had been adsorbed. It was easy to peel off the resulting transparent PAA film from the substrate, as shown in Figure 29.5a. Observation of the PAA film using SEM confirmed that the latex bead-fused nanosheet was completely transferred to the PAA film as shown in Figure 29.5b. In agreement with the report by Stroock *et al.* [4], the transfer mechanism has been successfully reproduced in the patterned nanosheets. Furthermore, it was easy to dissolve the resulting PAA film to release the latex bead-fused nanosheets in a phosphate buffer at pH 7.4, and the freestanding nanosheets were collected on the membrane filter as shown in Figure 29.5c. It was confirmed that these nanosheets maintained their rectangular shape, of which the obverse surface was rough and the reverse surface smooth.

The PAA film supporting the latex bead-fused nanosheets was confirmed to emerge on the reverse smooth side of the nanosheet onto the surface of the resulting PAA film (data not shown). Therefore, either side of the nanosheet could be selectively modified by this method. As an example of the

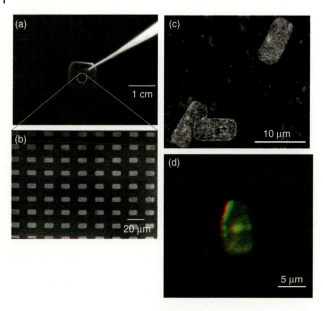

Figure 29.5 (a) Photo and (b) SEM image of latex bead-fused nanosheets transferred from the ODS-SAM to the PAA film. (c) SEM image of freestanding latex bead-fused sheets after dissolution of the resulting PAA film. (d) Images of TRITC and FITC-labeled latex bead-fused sheets using confocal laser fluorescence microscopy.

heteromodification of the nanosheet, two kinds of water-soluble fluorescent probes, tetramethylrhodamine-5-(and-6)-isothiocyanate (TRITC) and fluorescein-isothiocyanate (FITC), were used. An excess of TRITC (5 μM) was firstly added to the amino groups of the rHSA-adsorbing latex beads on the patterned ODS-SAM. The PAA solution was cast on the substrate and the dried PAA film peeled off. Under conditions where the PAA film is insoluble (pH 4.0 and saturated sodium chloride), an excess of FITC (5 μM) was added to the reverse side of the PAA film and then dissolved in phosphate buffer at pH 7.4 to allow collection of the latex bead-fused nanosheets. Using a confocal laser scanning microscope, the abundance of rectangular nanosheets was confirmed, on one or other of the two surfaces where either fluorescent probe (TRITC or FITC) was localized as shown in Figure 29.5d.

The biodegradable poly(DL-lactide-co-glycolide) (PLGA) nanosheets were fabricated onto 3-μm disk-shape patterned hydrophobic octadecyl regions on a SiO$_2$ substrate by depositing PLGA nanoparticles (160 nm) coated with chitosan instead of HSA [45]. Dodecapeptide (H12) bearing polyethylene glycol (PEG) chains were conjugated onto the surface of the PLGA nanosheet with PEG chains. The resulting H12-PLGA nanosheets specifically interacted with the activated platelets adhered on the collagen surface at twice the rate of the H12-PLGA microparticles (2 μm diameter) under the same flow conditions, and showed platelet thrombus formation in a two-dimensional spreading manner. We are currently studying the *in vivo* efficacy of injectable nanosheets as a candidate for a novel platelet substitute. These results also raise the possibility of injectable nanosheets as carriers of recognition molecules and drugs.

29.3
Giant Nanosheets as Nanoadhesive Plasters

29.3.1
Quasi-Two-Dimensional Freestanding Nanosheets

Recent developments in nanotechnology have resulted in a series of step changes in the methodology used for the fabrication of giant nanosheets with a size in excess of a centimeter and a huge aspect ratio [3, 8–14, 45, 46]. One novel methodology for the fabrication of nanosheets involving a wide variety of macromolecules is a LbL technique [47]. The LbL method involves alternative deposition of oppositely charged polyelectrolytes by noncovalent bonding such as electrostatic interactions, hydrogen bonding, or hydrophobic interactions [47–51]. Application of LbL-based nanomaterials has been explored in several fields such as electrochemical devices, chemical sensors, nanomechanical sensors, nanoscale chemical/biological reactors and in DDSs [52].

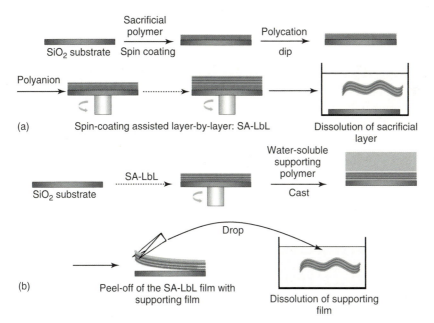

Figure 29.6 Preparative scheme and methodology for exfoliation of the nanosheets: (a) the sacrificial layer method and (b) the supporting film method.

We have adopted a spin-coating-assisted layer-by-layer (SA-LbL) method to fabricate a freestanding nanosheet [5, 53, 54] (Figures 29.6a,b). Spin coating of each polyelectrolyte was preformed alternatively on a substrate covered with a sacrificial layer, followed by subsequent dissolution of the sacrificial layer to release the freestanding ultrathin membrane within several minutes (unlike a conventional dipping LbL process that requires many hours). Because the membrane is composed of polymers with a sheet-like structure of nanometer thickness, they possess a huge aspect ratio ($\geq 10^6$ depending on the size of the substrate).

These structures have been referred as *"polymer nanosheets"* by Miyashita and coworkers [46]. The freestanding nanosheets detached from the substrate were fabricated not only by the SA-LbL method but also by a Langmuir–Blodgett method followed by crosslinking amphiphilic copolymers or by a sol/gel method with organic/inorganic interpenetrating networks [14, 46]. These freestanding nanosheets have been reported to be well-organized, compliant and robust materials for micro/nanomechanical studies [55–57]. Moreover, SA-LbL is the simple way of generating nanosheets without using a chemical crosslinking agent, making them ideal for biomedical applications.

Freestanding nanosheets could be obtained from the solid substrate by a "sacrificial layer method" (Figure 29.6a) or a "(water-soluble) supporting film method" (Figure 29.6b). In the former method, the sacrificial polymer film is dissolved by certain organic solvents such as acetone or ethanol that do not dissolve the SA-LbL film. In the latter method, the water-soluble supporting film, such as a poly(vinyl alcohol) (PVA) film, is prepared on the surface of the SA-LbL film and allows the convenient collection of the freestanding nanosheet by peeling the complex film from a solid substrate, followed by dissolution of the PVA film. It is noted that in the complex film, which consists of the nanosheet and the PVA film, the interaction between these layers should be greater than that between the nanosheet and the SiO_2 substrate. This transfer method is an efficient means of moving the nanosheet from the solid substrate to any surface, including human skin or organs. Moreover, this water-mediated release method should be suitable for biomaterials because the procedure does not require any organic solvents.

29.3.2
Freestanding Polysaccharide Nanosheets

In order to fabricate the nanosheet, we use polysaccharide electrolytes such as chitosan and sodium alginate (Na alginate), which have amino and carboxylic groups as cationic and anionic groups at ambient pH. These polysaccharides are used in biomedical fields, such as wound dressing and artificial skin, because of their high biocompatibility and bioabsorbability. By using the sacrificial-layer method, we obtained a freestanding polysaccharide nanosheet in acetone after each polysaccharide was assembled by the SA-LbL method. The polysaccharide nanosheet floating in acetone maintained the size of the original SiO_2 substrate (Figure 29.7a) and was transferred onto a fresh silicon wafer. Large-scale (90 μm × 90 μm) topographic images by AFM revealed that the surface of the polysaccharide

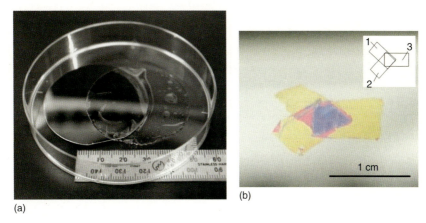

(b)

(a)

Figure 29.7 General properties of the polysaccharide nanosheet: (a) the bilayered film, consisting of a 75-nm polysaccharide nanosheet and a 70-mm supporting PVA film and (b) triple-adsorbed polysaccharide nanosheets (inset: a schematic illustration of the adsorption order) on a SiO_2 substrate.

nanosheet was as smooth and flat as the silicon wafer surface without any corrugations and wrinkles. From the cross-sectional analysis of the nanosheet edge, the thickness of the nanosheet with 10.5 layer pairs was estimated to be 30.2 ± 4.3 nm, corresponding to the ellipsometric thickness (30.7 ± 4.5 nm) of the nanosheet on the SiO_2 substrate. The smooth and flat surface was obtained due to the high-speed horizontal diffusion of the polymers during spin coating. The nanosheets adsorbed on the SiO_2 substrate showed a brilliant color change depending on the adsorbed number of nanosheets (Figure 29.7b) [58].

A microscratch test [59] was carried out for evaluation of macroscopic adhesive properties of ultrathin films such as nanosheets. The microscratch tester employs a diamond stylus that oscillates parallel to the surface of the nanosheet on the SiO_2 substrate. The adhesive failure of the nanosheet with the stylus was detected as the "critical load" of the nanosheet, which is a relative value of the adhesive force. Interestingly, the critical load of the polysaccharide nanosheets drastically increased as their thickness decreased below 200 nm; the critical load of a 39-nm thick nanosheet was approximately 7.5 times greater than that with a thickness of 1482-nm (Figure 29.8a). Moreover, microscopic observation revealed different trail marks after scratching depending on the thickness of the nanosheet, such as "cut-off (1482 nm)" and "drawn (77 nm)" like trails (Figure 29.8b,c). This observation suggested that the elasticity of the nanosheet is critically reduced at a thickness of less than 200 nm.

The bulge test has been frequently used for evaluation of the mechanical strength of ultrathin films [60–62]. Three types of polysaccharide nanosheets with different thicknesses of 35, 75, and 114 nm, were prepared and transferred onto steel plates with either a 1- or 6-mm diameter circular hole in the center (Figure 29.9)

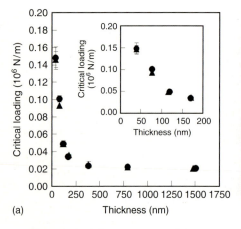

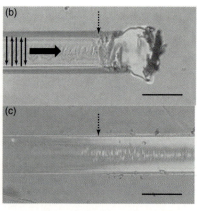

Figure 29.8 Adhesive property of the polysaccharide nanosheet. (a) The relationship between the critical load against the different thickness of the polysaccharide nanosheet between before (circle) and after (triangle) detachment of the nanosheet (Inset: magnification of the graph under the 200 nm thickness). (b) and (c) Microscopic observation of the polysaccharide nanosheets after microscratch testing. Two different thicknesses of nanosheet were used, 1482 and 77 nm. Black arrows show the direction of the stylus on the nanosheet; dashed arrows indicate the detached points from the SiO_2 substrate. Scale bars show 100 μm.

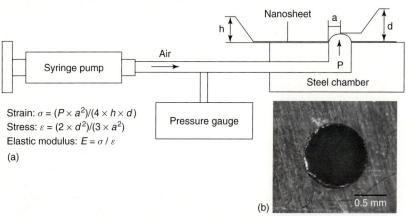

Strain: $\sigma = (P \times a^2)/(4 \times h \times d)$
Stress: $\varepsilon = (2 \times d^2)/(3 \times a^2)$
Elastic modulus: $E = \sigma / \varepsilon$

(a)

Figure 29.9 Mechanical properties of the polysaccharide nanosheets: (a) schematic representation of the bulge test apparatus. (b) Top-view of the circular hole in the steel plate covered with the 75-nm polysaccharide nanosheet.

and kept under ambient conditions (temperature: $25 \pm 1\,°C$, humidity: $37 \pm 3\%$). It is noteworthy that the nanosheet easily adhered onto the steel plate without using chemical adhesion. As pressure was applied to the polysaccharide nanosheet through the circular hole, deflection of the nanosheet was monitored from a side view of the plates until distortion occurred. The relationship between pressure and deflection was nonlinear, suggesting that the elasticity of the polysaccharide nanosheet was dependent on the total film thickness. From the initial elasticity of the stress–strain curve, the ultimate tensile strength (σ_{max}), elongation (ε_{max}), and elastic modulus (E) were calculated for the different thickness of the nanosheets.

The E of the 35-nm polysaccharide nanosheet was 1.1 ± 0.4 GPa, which is a similar value to that (1.5 GPa) of the nanosheet from poly(allylamine hydrochloride) and poly(sodium styrenesulfonate) with the same thickness (35 nm) reported by the Tsukruk group [1] and considerably less than that of a cellulose film ($E = 15$ GPa) with a thickness of over $1\,\mu$m. This result suggested that a nanosheet tens of nm thick was quite flexible due to the low elastic modulus. As the thickness of the polysaccharide nanosheet was increased, the elastic modulus increased to approach the bulk value (75 nm: 8.1 ± 2.5 GPa, 114 nm: 11.0 ± 1.6 GPa). Generally, nanosheets show a glass-transition temperature lower than the corresponding bulk [62]. This would reflect a specific interfacial property of the nanosheet such as unrestricted macromolecular mobility, and is probably the reason why such nanosheets have low elastic moduli.

29.3.3
Freestanding Poly(L-Lactic Acid) Nanosheets

We also succeeded in the fabrication of a freestanding poly(L-lactic acid) (PLLA) nanosheet as shown in Figure 29.10a. A 5 mg/mL solution of PLLA was dropped on a SiO_2 substrate and then spin coated at 4000 rpm for 20 s, followed by drying. The

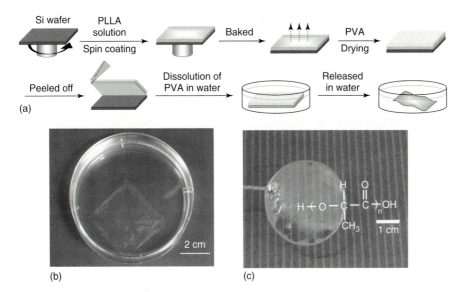

Figure 29.10 Fabrication of a freestanding PLLA nanosheet:
(a) preparative procedure of a freestanding PLLA nanosheet.
(b) Macroscopic image of a PLLA nanosheet with a thickness of 23 nm floating on water. (c) Macroscopic image of a PLLA nanosheet scooped up using a wire loop.

surface of the substrate was light red in color, and the thickness was measured to be 23 ± 5 nm and the root-mean-square roughness (RMS) was as low as 3.6 ± 0.5 nm. The PVA solution was cast on the substrate previously spin coated with the PLLA solution and dried at 70 °C to prepare a water-soluble supporting film. The PVA film was easily peeled from the substrate and the resulting film was tough and transparent. A freestanding PLLA nanosheet was thus prepared by dissolution of the resulting PVA film in distilled water (Figure 29.10b). The nanosheet with an extremely high size–aspect ratio of greater than 10^6 maintained the same shape and size as the SiO$_2$ substrate (4×4 cm^2). Observation of the PLLA nanosheet on an anodisc membrane using SEM confirmed that the nanosheet had a flat and uniform structure without cracks. The transparent PLLA nanosheet could be scooped up and held in air with a wire frame (Figure 29.10c) and was stable without bursting for at least one year. Furthermore, the thickness of the nanosheet could be easily controlled because the thickness was proportional to the PLLA concentration used for spin coating; the minimal thickness of the freestanding PLLA nanosheet was estimated to be 23 ± 5 nm when the concentration of the PLLA solution was 5 mg/ml [63].

The mechanical properties of the resulting PLLA nanosheet were analyzed by a bulge test. The PLLA nanosheet with thicknesses of 23 ± 5 nm swelled gradually and gave an almost semicircular deflection until a pressure of approximately 4 kPa was reached. From the slope of the initial elastic region of the stress–strain curve, the elastic modulus of the nanosheets was calculated to be 1.7 ± 0.1 GPa. The values were quite low in comparison with that of bulk PLLA (7–10 GPa) as reported by Eling *et al.* [64]. As described for the polysaccharide nanosheet, the difference indicates that the PLLA nanosheet was softer than bulk PLLA. In fact, the nanosheet was exquisitely flexible, suggesting that the glass transition temperature (T_g) of the PLLA nanosheet would be lower than that of bulk PLLA (T_g: 60 °C).

Mattsson *et al.* [65] showed that T_g of an ultrathin poly(styrene) film (thickness: 22 nm) was significantly decreased to 37 °C from that of bulk poly(styrene) (T_g: 109 °C). This phenomenon was explained in terms of reduced interaction of the polymer chains in a thin film. Unfortunately, the elastic moduli of the PLLA nanosheet with thickness greater than 100 nm could not be measured by the bulge method because the nanosheet was easily detached from the steel plate, resulting in an air leak.

29.3.4
Nanoadhesive Plaster Including Ubiquitous Transference of the Nanosheets

Freestanding nanosheets in liquid tended to be overwhelmed in air due to their large aspect ratio. In order to transfer the nanosheet from the surface of one substrate to another without distorting its overall shape we used a water-soluble sacrificial film such as PVA. Due to its flexibility, silicon rubber was sometimes chosen as a substrate for the PVA film. We prepared the three-layered structure of a nanosheet (i.e. 60 nm)/PVA layer (1.2 μm)/silicon rubber layer (0.7 mm). Here, we named the layered composite film a "nanoadhesive plaster" because the nanosheet could attach to skin or organ when the water-soluble membrane was finally dissolved. To test the practical applications of the nanoadhesive plaster, we attached the three-layered structure on the wet skin of a human arm. The contour on the nanosheet surface was observed due to differences in reflectivity. The nanosheet was then detached from the PVA-silicone rubber substrate within a few seconds by the dissolution of the PVA layer with a drop of 500 μl D.I. water. The nanosheet on the skin was barely visible from the top view under visible light (Figure 29.11a), perhaps because the surface relief of the skin perfectly matched that of the flexible sheet at the nanometer scale. When the polysaccharide nanosheet was stained with a luminescent dye, luminescent emission from the modified nanosheet confirmed that the shape and size of the nanosheet were well preserved on the skin (Figure 29.11b). Moreover, the nanosheet stably remained on the skin for at least 24 h, and was rinsed away by washing with soap [66].

We also succeeded in attaching the nanosheet onto the cecum of a living rat by utilizing the nanoadhesive plaster consisting of the polysaccharide nanosheet and the water-soluble PVA film as a supporting layer. The PVA film was subsequently dissolved with a few drops of saline from a syringe. The overall shape of the nanosheet faithfully replicated the surface relief of the organ due to its considerable flexibility (Figure 29.11c). Our results demonstrate the potential biomedical applications of a freestanding nanosheet as a unique cosmetic or surgical material.

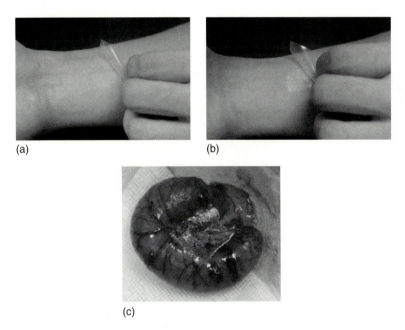

(a) (b)

(c)

Figure 29.11 A nanoadhesive plaster on human skin: (a) after release from the silicone rubber or (b) captured in the dark. (c) Nanoadhesive plaster on rat cecum.

29.4
Surgical Applications of Nanoadhesive Plasters

29.4.1
Freestanding Polysaccharide Nanosheet Integrated for Tissue-Defect Repair

We have developed the concept of a "nanoadhesive plaster" for wound repair in surgery. Surgical repair for tissue defects is achieved by three fundamental methods; suture, plication, and overlapping [67]. Despite their high reliabilities for wound repair, the conventional repair of a pleural defect by suture and plication usually reduces the volume of pulmonary tissue, thereby decreasing respiratory functions. Pulmonary air leakage due to visceral pleural injury is one of the most common postoperative complications after thoracic surgery. Such complications might be caused by the prolonged placement of a drainage tube and/or longer hospitalization, and may even lead to thoracic empyema. Therefore, tight and firm repair of a pleural injury/defect is critically important in order to prevent air leakage [68, 69]. Nevertheless, it is sometimes difficult to suture or plicate a large defect or fragile tissue of the emphysematous lung. Overlapping is therefore seen as an ideal means of repairing a pleural defect, because it simply seals the injured surface without reducing the tissue volume of the injured lung. As a conventional sealant, fibrin glue (sheet) composed of fibrin-glue-coated collagen fleece, a typical adhesive material, is effective for the repair of a visceral pleural defect [70]. However, this material sometimes causes severe pleural adhesion. In the case of high-risk patients with respiratory failure such a severe pleural adhesion might further deteriorate pulmonary dysfunctions, a serious complication for compromised patients. Therefore, a novel tissue sealant that does not cause tissue adhesion is required.

We demonstrated the practical application of a 75-nm polysaccharide nanosheet for an *in vivo* visceral pleural defect model in beagle dogs. The polysaccharide nanosheet, or a fibrin sheet used as a positive control, was placed onto a pleural defect area prepared by a 3.2-cm^2 aorta punch on the right anterior, middle, and posterior lobes. The 75-nm polysaccharide nanosheet was placed on a supporting PVA film (70 μm in thickness) for handling. As the PVA film dissolved with a PBS solution, the underlying nanosheet fitted on the curvature of the remaining tissue fully overlapping the pleural defect without any chemical adhesive reagents. After drying for a few minutes, the nanosheet was completely assimilated to the tissue surface. The airway pressure at which the air leakage occurred was measured after repair using a manometer and termed *"bursting pressure."* The highest airway pressure applied was 60 cm H_2O because air leakage could occur from the intact pulmonary hilum at pressures above 60 cm H_2O. At 5 min after repair, the nanosheet showed a bursting pressure (31.7 ± 10.3 cm H_2O) lower than the fibrin sheet (45.0 ± 5.5 cm H_2O). The bursting pressure of the nanosheet was slightly lower than that found in the bulge test (circa 45 cm H_2O for the 6 mm diameter hole prepared on the steel substrate). At 3 h after repair, the outline of the square shaped nanosheet assimilating to the tissue surface could be faintly seen, and the

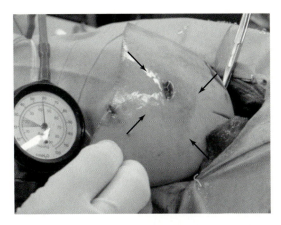

Figure 29.12 Polysaccharide nanosheet with a thickness of 75 nm secured the repaired defect when pressurized to over 50 cm H_2O pressure for 3 h after repair. The region indicated by arrows shows the nanosheet sealed area.

bursting pressure of the nanosheet reached 56.7 ± 6.1 cm H_2O, which is the same level as that of the fibrin sheet (Figure 29.12).

From histological examinations, it is noteworthy that the wound healing after treatment with the nanosheet was quite distinct from that with the fibrin sheet. Although it was difficult to observe the nanosheet overlap on the pleural defect, the formation of flat-shaped blood clots localized along the nanosheet was clearly observed in the region of the defect at 3 h after repair without significant inflammatory response (Figure 29.13a). This finding suggested that blood cells initially deposited under the nanosheet were subsequently transformed to stable blood clots. At three days after repair, fibroblasts had grown around the blood clots, replacing the preformed clots. At seven days after repair, angiogenesis was observed where the blood clots had originally formed under the nanosheet. Importantly, the sequence of the wound-healing process never occurred on the outside of the polysaccharide nanosheet (Figure 29.13b). Hence, no incidence of postsurgical adhesive lesion in the thoracic cavity was observed. At 30 days after repair, the original tissue-defect site was no longer confirmed.

In contrast to the polysaccharide nanosheet, repair of the pleural defect by the fibrin sheet exhibited large vacant air spaces at 3 h because the thick fibrin sheet was too firm to densely overlap the defect site (Figure 29.13c). This would cause the haphazard retention of blood components in the overlapped area.

At three days, the random growth of fibroblasts was observed as well as the induction of inflammatory tissue reaction, such as the emergence of macrophages. Furthermore, it is a critically important clinical issue that the fibrin sheet also strongly adheres to the chest wall (Figure 29.13d). Severe pleural adhesions might reduce respiratory function and may cause a reoccurrence of pneumothorax.

The polysaccharide nanosheet has very desirable properties: physical adhesiveness due to flexibility and sufficient mechanical strength without chemical and

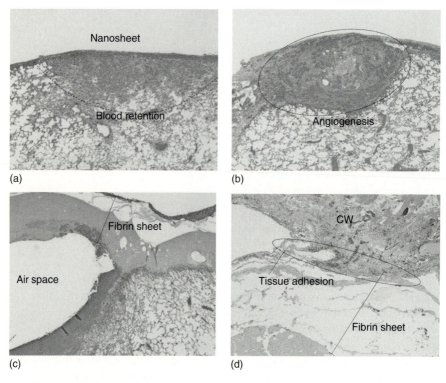

Figure 29.13 Representative histological findings at different time points after repair (hematoxylin-eosin staining, magnification 4×) using the polysaccharide nanosheet or fibrin sheet. Upper and lower panels correspond, respectively, to the polysaccharide nanosheet and the fibrin sheet, at 3 h (a) and (c) and seven days (b) and (d) after repair. At seven days, angiogenesis occurred in the polysaccharide nanosheet groups while strong adhesion between the fibrin sheet and the chest wall (CW) was observed.

biological adhesives such as a fibrin sheet. Thus, repair by overlapping a tissue defect with a nanosheet has significant advantages in maintaining the function of the remaining lung against sustained ventilation and the pressure from respiration and bleeding.

29.4.2
Freestanding PLLA Nanosheet for Sealing Operations

In surgical operations, tissue repair has classically been performed using suture and ligation. Though these procedures are highly reliable and effective, when the repaired tissue is edematous and/or fragile due to severe tissue damage or inflammation, repair by suture and ligation may become technically difficult and unreliable. Furthermore, severe sepsis resulting from anastomotic breakdown is considered to be one of the most critical complications after gastrointestinal surgery because of its associated high mortality rate. Here, we demonstrate a potential biomedical application of the nanosheet as a novel wound dressing in repairing a gastric incision by sealing, instead of suturing [63].

An incision of approximately 1 cm in length was made in the anterior wall of the stomach in mice using a surgical knife. A supporting suture was stitched (without ligation) at the middle of incision line to invert the reflected mucosa. Thereafter, the PLLA nanosheet supported with a PVA supporting film (typically 1.5×1.0 cm) was placed over the incision site. Immediately after covering, the supporting suture (no ligation) was pulled out. The PVA supporting film was then dissolved with saline. At seven days after surgical intervention (PLLA nanosheet- and suture-treated), the stomachs were removed from the mice. Sealing treatment with the nanosheet did not cause tissue adhesion, and surprisingly, few postoperative cicatrices remained on the surface of the stomach (Figure 29.14a). In contrast, tissue adhesion was observed in several of the suture-treated mice with apparent cicatrization in the stomach, causing severe deformity and shrinkage (Figure 29.14b). Histological observation also showed noticeably different wound healing between nanosheet sealing and suture/ligation. In the nanosheet-sealed mice, the gastric mucosa at the incision site was loosely bent because the PLLA nanosheet just sealed the surface of the gastric serosa. Fibroblasts regenerating in response to wound healing grew normally and smoothly sealed the incision site; the thickness of fibroblasts was equal to that of serosa around the incision site. However, in the conventional suture/ligation-treated mice, gastric mucosa was tightly stitched by suturing. Regenerating fibroblasts were markedly increased in number at the incision site, as a part of the normal wound-healing process following the conventional suturing treatment. Because PLLA cannot induce cell adhesion, its surface is commonly modified with laminin, fibronectin (Fn), and Fn-related arginine-glycine-aspartic acid (RGD) tripeptide for improvement of cell adhesion [71, 72]. Our results suggest that the PLLA nanosheet functions as a biointerface for balancing conflicting phenomena involved in tissue repair and resistance to tissue adhesion. Specifically, when the surface of the PLLA nanosheet adheres directly to the stomach (obverse surface) it is exposed to blood and tissue fluid containing various growth factors. Fibroblasts grow normally on the surface of the PLLA nanosheet in the presence of growth factors. However, it is intrinsically difficult to adhere various cells to the outer surface of the PLLA nanosheet (reverse surface).

(a)   (b)

Figure 29.14 Sealing with the PLLA-nanosheet (a). Conventional suture/ligation (b).

Thus, PLLA nanosheets display a therapeutic sealing effect for a gastric incision procedure, which is restricted in conventional suturing surgery. In addition, using a nanosheet the incision was repaired without cicatrices.

29.5
Nanosheets for the Next Generation

29.5.1
Nanosheets Bearing Stimuli-Responsive Functional Surfaces

The freestanding nanosheet bearing thermoresponsive polymer brushes was prepared by atom-transfer radical polymerization (ATRP), [73, 74] of poly(N-isopropylacrylamide). Morphological transformation was macroscopically observed through reversible structural color changes on the air/water interface. It was demonstrated that the highly flexible freestanding nanosheet is capable of acting as a unique platform for inducing stimuli-responsive behavior in nanomaterials. Several methods have already been developed using temperature-tuned materials such as ultrathin films or gels with poly(N-isopropylacrylamide) (pNIPAM) [75–77]. However, we need to maintain the freestanding ultrathin structure in a substrate-free condition in order to utilize specific properties of the polysaccharide nanosheets. To address this issue, we reasoned that the contact area between the stimuli-responsive surface of the nanosheet and water surface should be as large as possible in order to enhance mobility on the nanosheets bearing pNIPAM brushes (pNIPAM-nanosheets).

The freestanding pNIPAM-nanosheet was colored pale blue at 25 °C on the air/water interface and was slightly flexible with a creased surface (Figure 29.15a). As the water temperature was increased to 40 °C, the pNIPAM-nanosheet turned yellow (Figure 29.15b). Furthermore, the nanosheet lost its flexibility and solidified at the air/water interface. Interestingly, this morphological transformation is

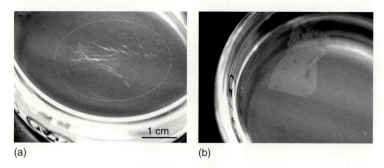

(a)   (b)

Figure 29.15 *In-situ* morphological transformation of a free-standing pNIPAM-nanosheet (1 cm × 2 cm) in water being warmed from (a) 25 °C to (b) 40 °C. A dotted circle in (a) shows the location of the freestanding pNIPAM-nanosheet in water.

reversible: the sheet is transparent pale blue with a flexible surface at 25 °C, and cloudy yellow with a solid surface at 40 °C.

From a previous study [78], we know the color of the polysaccharide nanosheets and showed that structural color is principally a consequence of thickness. Therefore, the color changes in the nanosheet can be seen as an indicator of its varying thickness. The morphological transformation of the freestanding nanosheets can be explained as follows. pNIPAM at 25 °C adopts an extended structure that is hydrophilic. Therefore, the surface of the pNIPAM nanosheet is flexible and transparent, bearing water molecules on the surface. However, pNIPAM at 40 °C is

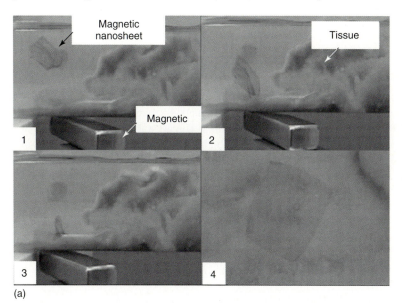

(a)

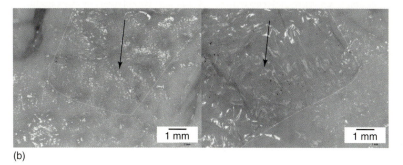

(b)

Figure 29.16 (a) Freestanding magnetic nanosheet driving in water (PLLA 10 mg/ml with φ 40 nm magnetic nanoparticles) manipulated by moving the permanent magnet at a typical distance of 4 cm. The nanosheet was deposited on the tissue surface by holding the magnet underneath the tissue. Macroscopic images of magnetic nanosheets landing on the mucosa where (b) φ 40 nm and (c) φ 200 nm magnetic nanoparticles were embedded inside of the PLLA nanosheet, respectively.

collapsed and hydrophobic, resulting in a surface that is solid and cloudy, with hydrophobically organized pNIPAM brushes.

29.5.2
Remote Controllable Nanosheets Driven by a Magnetic Field

We are investigating the possibility of taking advantage of the unusual properties of nanosheets for two specific applications: (i) as "nanoadhesive plasters" for closing incisions made during surgical procedures and (ii) as drug-loaded patches on the inner organ walls or on ulcers. In particular, nanosheets are an attractive material for endoluminal procedures.

Freestanding magnetic nanosheets can be fabricated by incorporation of magnetic nanoparticles (ϕ: 40 or 200 nm) inside the PLLA nanosheet. The resulting nanosheet, can be manipulated in a physiological solution (Figure 29.16a), using a permanent magnet or a micropipette. After removing the liquid, they can thus be attached on gastric mucosa. Furthermore, optical microscopy imaging highlighted that the particle size embedded inside the magnetic nanosheet influences its adhesion to the tissue. An intimate continuous contact was obtained using particles with a diameter of 40 nm (Figure 29.16b), but not with those of diameter 200 nm (Figure 29.16c). In particular, the grooves of the mucosa appeared completely covered and it was not possible to distinguish the presence of the magnetic nanosheet when we used the 40-nm magnetic nanoparticles. These observations suggest that the particle diameter is an important parameter that determines the mechanical properties of the freestanding nanosheets.

29.6
Concluding Remarks

In conclusion, we have succeeded in preparing two types of freestanding nanosheets that are suitable for biomedical applications. The first is an injectable nanosheet that can act as a platelet substitute. The sheet-shaped material is similar to native platelets, which can be constructed on a micropatterned substrate. The second is a giant nanosheet constructed *via* a LbL assembling method or conventional spin-coating method with a huge aspect ratio of over 10^6. Such material can be used as nanoadhesive plasters that ubiquitously transfer the polysaccharide nanosheet onto human skin or organ. We have also investigated the biomedical applicability of nanosheets for repairing surgical trauma. Biomedical application of nanosheets will radically diversify the range of new minimally invasive treatments in clinical practice.

Acknowledgments

This work was supported by "High-Tech Research Center" Project for Waseda University: matching fund subsidy from MEXT, Japan. Grant-in-Aid for Scientific

Research (B) (No. 21300181) from MEXT, Japan. The authors also acknowledged to Dr. Saitoh Daizo and Dr. Manabu Kinoshita at National Defense Medical College, Dr. Rigoberto C. Advincula at University of Houston, and Dr. Paolo Dario at Scuola Superiore Sant'Anna.

References

1. Tomii, Y. (2002) *Curr. Pharm. Des.*, **8**, 467–474.
2. Forrest, J.A., Dalnoki-Veress, K., Stevens, J.R., and Dutcher, J.R. (1996) *Phys. Rev. Lett.*, **77**, 2002–2005.
3. Mamedov, A.A. and Kotov, N.A. (2000) *Langmuir*, **16**, 5530–5533.
4. Stroock, A.D., Kane, R.S., Weck, M., Metallo, S.J., and Whitesides, G.M. (2003) *Langmuir*, **19**, 2466–2472.
5. Jiang, C., Markutsya, S., and Tsukruk, V.V. (2004) *Adv. Mater.*, **16**, 157–161.
6. Eck, W., Küller, A., Grunze, M., Völkel, B., and Gölzhäuser, A. (2005) *Adv. Mater.*, **17**, 2583–2587.
7. Jin, J., Huang, J., and Ichinose, I. (2005) *Angew. Chem. Int. Edit.*, **44**, 4532–4535.
8. Ono, S.S., and Decher, G. (2006) *Nano Lett.*, **6**, 592–598.
9. Endo, H., Kado, Y., Mitsuishi, M., and Miyashita, T. (2006) *Macromolecules*, **39**, 5559–5563.
10. Watanabe, H., Ohzono, T., and Kunitake, T. (2007) *Macromolecules*, **40**, 1369–1391.
11. Peng, X., Jin, J., Ericsson, E.M., and Ichinose, I. (2007) *J. Am. Chem. Soc.*, **129**, 8625–8633.
12. Singh, S., Greiner, M.T., and Kruse, P. (2007) *Nano Lett.*, **7**, 2583–2676.
13. Striemer, C.C., Gaborski, T.R., McGrath, J.L., and Fauchet, P.M. (2007) *Nature*, **445**, 749–753.
14. Vendamme, R., Onoue, S., Nakao, A., and Kunitake, T. (2006) *Nature Mater.*, **5**, 494–501.
15. Takeoka, S., Teramura, Y., Okamura, Y., Handa, M., Ikeda, Y., and Tsuchida, E. (2001) *Biomacromolecules*, **2**, 1192–1197.
16. Takeoka, S., Teramura, Y., Ohkawa, H., Ikeda, Y., and Tsuchida, E. (2000) *Biomacromolecules*, **1**, 290–295.
17. Teramura, Y., Okamura, Y., Takeoka, S., Tsuchiyama, H., Narumi, H., Kainoh, M., Handa, M., Ikeda, Y., and Tsuchida, E. (2003) *Biochem. Biophys. Res. Commun.*, **306**, 256–260.
18. Okamura, Y., Takeoka, S., Teramura, Y., Maruyama, Y., Tsuchida, E., Handa, M., and Ikeda, Y. (2005) *Transfusion*, **45**, 1221–1228.
19. Takeoka, S., Teramura, Y., Okamura, Y., Tsuchida, E., Handa, M., and Ikeda, Y. (2002) *Biochem. Biophys. Res. Commun.*, **296**, 765–770.
20. Okamura, Y., Maekawa, I., Teramura, Y., Maruyama, Y., Tsuchida, E., Handa, M., Ikeda, Y., and Takeoka, S. (2005) *Bioconjugate Chem.*, **16**, 1589–1596.
21. Okamura, Y., Handa, M., Suzuki, H., Ikeda, Y., and Takeoka, S. (2006) *J. Artif. Organs*, **9**, 251–258.
22. Takeoka, S., Okamura, Y., Teramura, Y., Watanabe, N., Suzuki, H., Tsuchida, E., Handa, M., and Ikeda, Y. (2003) *Biochem. Biophys. Res. Commun.*, **312**, 773–779.
23. Mattson, J., Forrest, J.A., and Borjesson, L. (2000) *Phys. Rev. E*, **62**, 5187–5200.
24. Tang, Z., Kotov, N.A., Magonov, S., and Ozturk, B. (2003) *Nature Mater.*, **2**, 413–418.
25. Mallwitz, F. and Laschewsky, A. (2005) *Adv. Mater.*, **17**, 1296–1299.
26. Mamedov, A., Kotov, N.A., Prato, M., Guldi, D.M., Wicksted, J.P., and Hirsch, A. (2002) *Nature Mater.*, **1**, 190–194.
27. Huck, W.T., Stroock, A.D., and Whitesides, G.M. (2000) *Angew. Chem., Int. Ed. Engl.*, **39**, 1058–1061.
28. Mallwitz, F., and Goedel, W.A. (2001) *Angew. Chem., Int. Ed. Engl.*, **40**, 2645–2647.
29. Xu, H. and Goedel, W.A. (2002) *Langmuir*, **18**, 2363–2367.

30. Nardin, C., Winterhalter, M., and Meier, W. (2000) *Langmuir*, **16**, 7708–7712.

31. Ulman, A. (1991) *An Introduction to Ultrathin Organic Films from Langmuir-Blodgett to Self-Assembly*, Academic Press, San Diego, CA.

32. Khoshtariya, D.E., Wei, J., Liu, H., Yue, H., and Waldeck, D.H. (2003) *J. Am. Chem. Soc.*, **125**, 7704–7714.

33. Ferapontova, E.E., Shipovskov, S., and Gorton, L. (2007) *Biosens. Bioelectron.*, **22**, 2508–2515.

34. Niwa, D., Yamada, Y., Homma, T., and Osaka, T. (2004) *J. Phys. Chem. B*, **108**, 3240–3245.

35. Masuda, Y., Seo, W.S., and Koumoto, K. (2000) *Jpn. J. Appl. Phys.*, **39**, 4596–4600.

36. Fudouzi, H., Kobayashi, M., Egashira, M., and Shinya, N. (1997) *Adv. Powder Technol.*, **8**, 251–262.

37. Wen, W.J., Wang, N., Zheng, D.W., Chen, C., and Tu, K.N. (1999) *J. Mater. Res.*, **14**, 1186–1189.

38. Sun, Y. and Walker, G.C. (2002) *J. Phys. Chem. B*, **106**, 2217–2223.

39. Guo, Q., Arnoux, C., and Palmer, R.E. (2001) *Langmuir*, **17**, 7150–7155.

40. Masuda, Y., Tomimoto, K., and Koumoto, K. (2003) *Langmuir*, **19**, 5179–5183.

41. Okamura, Y., Goto, T., Niwa, D., Fukui, Y., Otsuka, M., Motohashi, N., Osaka, T., and Takeoka, S. (2009) *J. Biomed. Mater. Res. A.*, **89**, 233–241.

42. Okamura, Y., Utsunomiya, S., Suzuki, H., Niwa, D., Osaka, T., and Takeoka, S. (2008) *Colloids Surf., A*, **318**, 184–190.

43. Denkov, N.D., Velev, O.D., Kralchevsky, P.A., Ivanov, I.B., Yoshimura, H., and Nagayama, K. (1992) *Langmuir*, **8**, 3183–3190.

44. Hozumi, A., Asakura, S., Fuwa, A., Shirahata, N., and Kameyama, T. (2005) *Langmuir*, **21**, 8234–8242.

45. Jiang, C., Markutsya, S., Pikus, Y., and Tsukruk, V.V. (2004) *Nature Mater.*, **3**, 721–728.

46. Kado, Y., Mitsuishi, M., and Miyashita, T. (2005) *Adv. Mater.*, **17**, 1857–1861.

47. Lvov, Y., Decher, G., and Mohwald, H. (1993) *Langmuir*, **9**, 481–486.

48. Lvov, Y., Ariga, K., Ichinose, I., and Kunitake, T. (1995) *J. Am. Chem. Soc.*, **117**, 6117–6123.

49. Decher, G., Lvov, Y., and Schmitt, J. (1994) *Thin Solid Films*, **244**, 772–777.

50. Tsukruk, V.V., Bliznyuk, V.N., Visser, D., Campbell, A.L., Buning, T.J., and Adams, W.W. (1997) *Macromolecules*, **30**, 6615–6625.

51. Decher, G. (1997) *Science*, **277**, 1232–1237.

52. Decher, G. and Schlenoff, J.B. (2003) *Multilayer Thin Films*, Wiley-VCH Verlag GmbH, Weinheim.

53. Cho, J., Char, K., Hong, J., and Lee, K. (2001) *Adv. Mater.*, **13**, 1076–1078.

54. Cho, J. and Char, K. (2004) *Langmuir*, **20**, 4011–4016.

55. Markutuya, S., Jiang, C., Pikus, Y., and Tsukruk, V.V. (2005) *Adv. Funct. Mater.*, **15**, 771–780.

56. Jiang, C., Singamaneni, S., Merrick, E., and Tsukruk, V.V. (2006) *Nano Lett.*, **6**, 2254–2259.

57. Jiang, C., McConney, M.E., Singamaneni, S., Merrick, E., Chen, Y., Zhao, J., Zhang, L., and Tsukruk, V.V. (2006) *Chem. Mater.*, **18**, 2632–2634.

58. Fujie, T., Okamura, Y., and Takeoka, S. (2007) *Adv. Mater.*, **19**, 3549–3553.

59. Baba, S., Midorikawa, T., and Nakano, T. (1999) *Appl. Surf. Sci.*, **144**, 344.

60. Vlassak, J.J. and Nix, W.D. (1992) *J. Mater. Res.*, **7**, 3242.

61. Markutsya, S., Jiang, C., Pikus, Y., and Tsukruk, V.V. (2005) *Adv. Funct. Mater.*, **15**, 771.

62. Watanabe, H., Ohzono, T., and Kunitake, T. (2002) *Macromolecules*, **40**, 1369.

63. Okamura, Y., Kabata, K., Kinoshita, M., Saitoh, D., and Takeoka, S. (2009) *Adv. Mater.*, **21**, 4388.

64. Eling, B., Gogolewski, S., and Pennings, A.J. (1982) *Polymer*, **23**, 1587.

65. Mattsson, J., Forrest, J.A., and Börjesson, L. (2000) *Phys. Rev. E*, **62**, 5187.

66. Takeoka, S., Okamura, Y., Fujie, T., and Fukui, Y. (2008) *Pure Appl. Chem.*, **80**, 2259.

67. Fujie, T., Matsutani, N., Kinoshita, M., Okamura, Y., Saito, A., and Takeoka, S. (2009) *Adv. Funct. Mater.*, **19**, 2560.

68. Porte, H.L., Jany, T., Akkad, R., Conti, M., Gillet, P.A., Guidat, A., and Wurtz, A.J. (2001) *Ann. Thorac. Surg.*, **71**, 1618.

69. Kawamura, M., Gika, M., Izumi, Y., Horinouchi, H., Shinya, N., Mukai, M., and Kobayashi, K. (2005) *Eur. J. Cardiothorac. Surg.*, **28**, 39.

70. Gika, M., Kawamura, M., Izumi, Y., and Kobayashi, K. (2007) *Interact. Cardiovasc. Thorac. Surg.*, **6**, 12.

71. Ohya, Y., Matsunami, H., Yamabe, E., and Ouchi, T. (2003) *J. Biomed. Mater. Res.*, **65A**, 79.

72. Eid, K., Chen, E., Griffith, L., and Glowacki, J. (2001) *J. Biomed. Mater. Res.*, **57**, 224.

73. Fujie, T., Park, J.-Y., Murata, A., Estillore, N.C., Tria, M.C.R., Takeoka, S., and Advincula, R.C. (2009) *ACS. Appl. Mater. Interfaces*, **1**, 1404.

74. Matyjaszewski, K., and Xia, J. (2001) *Chem. Rev.*, **101**, 2921.

75. Mizutani, A., Kikuchi, A., Yamato, M., Kanazawa, H., and Okano, T. (2008) *Biomaterials*, **29**, 2073.

76. Quinn, J.F. and Caruso, F. (2004) *Langmuir*, **20**, 20.

77. Li, M.-H., Keller, P., Yang, J., and Albouy, P.-A. (2004) *Adv. Mater.*, **16**, 1922.

78. Fujie, T., Okamura, Y., and Takeoka, S. (2009) *Colloids Surf., A*, **334**, 28.

30
Hybrid Multilayer Films Containing Nano-Objects

Yeongseon Jang, Bongjun Yeom, and Kookheon Char

30.1
Introduction

Nanotechnology has grown to be a research area with remarkable scientific and economic potentials. Various types of nano-objects (i.e. inorganic, organic, polymeric, and biological materials) can be synthesized with precise control of their nanoscopic features such as size, shape, and compositions as well as in their corresponding optical, electrical, magnetic, and biological functions. Although great achievements have been obtained in the synthesis and the characterization of nanoscopic functional materials, their collective properties within the macroscopic structure have not been sufficiently investigated, which are inevitably required to pursue the fundamental understanding and practical applications of *noble hybrid systems based on multinanocomponents.*

The *layer-by-layer (LbL) deposition* methods have been regarded as one of the most efficient and practical techniques to build up nanoscopic functional materials with the desired architectures because of their: (i) simplicity in the fabrication steps and (ii) versatility in the material and substrate choices. The deposition is an adaptable method to prepare *hybrid multilayer films* on various kinds of substrates with nanometer-level control of film thickness and structures. This method can also be used to modify or functionalize surfaces as coatings with facile incorporation of functional nano-objects at desired positions into multilayer films.

Functional hybrid multilayer thin films prepared by LbL deposition method have been developed with a rapid pace. The fundamental understanding of functional nanometer-scale building blocks and the new technologies derived from them are continually being developed. As devices, hybrid multilayer films based on LbL assembly are being developed as drug-delivery systems, fuel cells, batteries, solar cells, actuators, and light-emitting diodes, to name a few. Quite a number of functional hybrid multilayer films have been reported by LbL deposition methods.

This chapter serves to give a background on the current preparation methods for hybrid multilayer films prepared by the LbL assembly (i.e. solution dipping, spin coating, and spraying methods) and the building-block materials for

Functional Polymer Films, First Edition. Edited by Wolfgang Knoll and Rigoberto C. Advincula.

hybrid multilayer films divided into classes of inorganic, organic, and biological nano-objects in detail. Furthermore, various applications of functional hybrid multilayer films composed of various nano-objects will be introduced in the each chapter.

30.2
Preparation Methods for Hybrid Multilayer Films

30.2.1
Layer-by-Layer Assembly

The LbL deposition technique is a simple, versatile, and reproducible method for the preparation of multilayer thin films on various kinds of substrates. The concept of multilayer films was first suggested by Iler [1] who first reported multilayer films using positively and negatively charged colloid particles. Several decades later, Decher and coworkers [2, 3] realized and established the formation of multilayer films from oppositely charged amphiphiles and polyelectrolytes. Afterwards, a great number of papers concerning LbL multilayer films have been published and many research groups have reported a wide range of applications for the LbL deposition method. More specific information is referred to in recent review articles [4–12].

The LbL deposition method can be tailored to allow multimaterial assembly of several functional nano-objects without special chemical modifications. This gives access to multilayer films with complex functionality tailored toward specific surface interactions and the fabrication of thin film devices.

The driving force of LbL deposition is not restricted to electrostatic interactions [13]. Additionally, two complementary materials with intermolecular interactions such as hydrogen bonding [11, 14], covalent bonding [15], donor–acceptor interaction (charge-transfer interaction) [16], $\pi-\pi$ interactions [17], and hydrophobic interactions [18] have been demonstrated to prepare the multilayer films. Newly synthesized or commercially available charged polymers have been the focus for LbL multilayer films in the early stage. It has now been expanded to include polymeric micelles [19–21], dendrimers [22], dyes [23–25], nanoparticles [26, 27], nanotubes [28, 29], and nanowires (NWs) [30], nanosheets [31–33], and biomolecules (i.e. DNA [34], proteins [35], and viruses [36, 37], etc.). These various kinds of materials can be adsorbed on the planar substrate such as silicon wafer, quartz, glass, indium tin oxide (ITO), or even Teflon. This includes nonplanar structures including nano- or microsized particles [38], porous structures [39], stents [40], and enzymes [41], and so on. The other important feature is the precise control of layer thickness at the nanometer level. This control can be achieved by varying the number of bilayers or pairs and the adsorption conditions such as concentration of polyelectrolytes, ionic strength [42], temperature [43], pH condition [44], solvent polarity [45], and so on. In addition, surface property and morphology can also be modulated by these parameters.

Furthermore, the technical deposition methods play an important role with respect to the final film characteristics. In particular, surface and interface properties of the multilayer films can be tuned by deposition conditions. Usually LbL-assembled multilayer films are constructed by dipping a substrate into an aqueous solution (see Section 30.2.2). This has also been extended to alternate spin-assisted assembly (see Section 30.2.3) and spraying method (see Section 30.2.4). Spin coating and spraying-based LbL deposition extend the parameter space of LbL deposition even further. However, it is to be expected that these methods will contribute to the general acceptance of the technology. A detailed introduction of the preparation methods for hybrid multilayer film formation will be presented in the next section.

30.2.2
Solution-Dipping Method

Multilayer film structures composed of polyelectrolytes or other charged molecular or colloidal objects are fabricated using solution-dipping methods, as schematically outlined in Figure 30.1. The method was described in 1992 by Decher and coworkers [2, 3] and has received considerable attention since then. The solution dipping LbL process consists of bringing a given substrate alternatively into contact with anionic and cationic polyelectrolyte solution (see Figure 30.1). The substrate is first exposed to a solution of oppositely charged polyelectrolyte to the substrate, which then begins to adsorb the first layer. This adsorption is limited to one monolayer due to electrostatic repulsion between same-charged molecules, and results in the neutralization of charges on the substrate as well as a reversal of charge at the film surface. The film is rinsed to remove weakly bonded and entangled polymer chains after adsorption, and then exposed to another polyelectrolyte solution of the opposite charge to the previous adsorbed polymer, where the surface force and charge are once again reversed due to adsorption of countercharged polyelectrolyte chains. This process can be repeated as many times as desired, in order to prepare films up to any desired thickness on many types of substrates with various shapes and sizes [39, 40]. A large number of papers concerning LbL multilayer films based on the solution-dipping method have been published by many research groups, and a wide range of advantages and applications of dip-assisted LbL films are reported in review articles [4–12]. Other methods (i.e. the spin-coating method or spraying processing) result in LbL assembled multilayer films similar to those assembled by the solution-dipping method, but with some differences in film properties such as thickness depending on the exact parameters of film assembly or processing time.

30.2.3
Spin-Coating Method

The spin-assisted LbL assembly (spin-coating) method is a powerful tool to fabricate well-defined multilayer films in a very short processing time [46–48]. The method was introduced in 2001 by Char and coworkers [46] as a combination of the conventional LbL growth with the spin-coating routine. Polyelectrolyte solutions are dropped onto the substrate and then the substrate is rotated with a spin coater

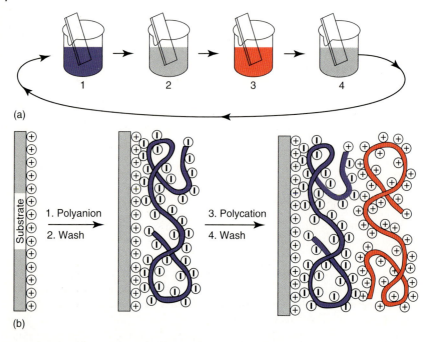

(a)

(b)

Figure 30.1 (a) Schematic illustration of the layer-by-layer (LbL)-assembled multilayer film deposition process via solution dipping method. (b) Simplified molecular picture of the first two adsorption steps, depicting film deposition starting with a positively charged substrate.

at fixed spinning rate. After deposition of the polyelectrolyte layer, the substrate is washed out with deionized water to remove weakly bonded polyelectrolyte chains during spinning process. This process is repeated until the desired number of layers and highly ordered multilayer films are prepared. This spinning process can induce strong centrifugal forces, viscous forces, and air-shear forces. The viscous force is caused by fast solvent drainage of the polymer solutions. The other is the strong adsorption driving forces as well as intermolecular interactions. On the other hand, the centrifugal force and air-shear force also work as desorption driving forces for weakly bonded polyelectrolyte chains. Besides this, the air-shear force directed outward in the radial direction at the interface between air and polymer chains also plays an important role in the planarization of multilayer films. The fabrication process of multilayer films via the spin-coating method is summarized in Figure 30.2a.

The spin-coating method has features of low surface roughness, low chain interpenetration of each layer, and high surface coverage (i.e. dense packing). In addition, the spin-coating method has the advantage of very time-efficient processing as compared with the conventional solution-dipping method in which equilibrium adsorption is achieved after 10–20 min. The internal structure of multilayer films can be characterized by X-ray and neutron reflectivity.

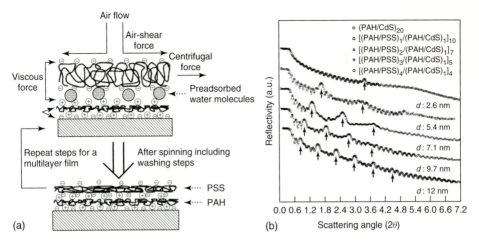

Figure 30.2 (a) Schematic illustration of the spin-coating-assisted LbL deposition method. (b) X-ray reflectivity curves of $[(PAH/PSS)_n/(PAH/CdS)_1]_m$ ($[n+1] \times m =$ 20 or 21) films prepared by the spin coating method. The increase of bilayer number of (PAH/PSS) from zero to four causes the increase of the d spacing between polyelectrolyte and nanoparticle from 2.6 to 12 nm. The arrow symbols in the figure indicate the Bragg peaks of such an internal structure.

As shown in Figure 30.2b, organic/inorganic [poly(allyl amine hydrochloride) $(PAH/PSS)_n(PAH/CdS)_1]_m$ hybrid multilayer films clearly show Bragg-reflection peaks originating from the highly ordered internal structure. Due to low degrees of interdigitation between adjacent layers, many kinds of applications using the spin assembly method have been reported, such as electroluminescent devices [48], surface patterns [49, 50], phosphorescent emitting layers [51], freestanding films [52], tunable LbL platforms for comparative cell assays [53], and so on. Moreover, a desktop system for the automated production of LbL multilayer films via the spin-coating method was designed and verified recently. The utility of this system was demonstrated by fabricating polymer/clay hybrid multilayer films [54].

30.2.4
Spraying Method

Preparation of functional hybrid multilayer films can be simplified and sped up enormously by applying the polyelectrolyte or other functional objects contained in liquid by spray. The spraying deposition method was introduced in 2000 by Schlenoff and coworkers [55]. Spraying method is convenient, fast, and more generally applicable for the coating of large surfaces. As shown in Figure 30.3, the polyelectrolyte solutions and rinsing solutions are supplied by enforced spraying. The procedure requires two spraying episodes per single layer with waiting time in between. This waiting time is to allow for solution drainage. The second supply of the same polyelectrolyte refills the loss of material removed by drainage. Spraying with rinsing solution is needed after a complete layer is formed. Then, after rinsing,

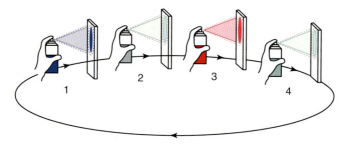

Figure 30.3 Schematic illustration of alternating adsorption cycle in spraying method: (1) polycation solution, (2) rinsing solution, (3) polyanion solution, and (4) rinsing solution.

there is a drainage time after which the layer of oppositely charged polyelectrolyte solution can be sprayed. The main advantage of spraying is a meaningful reduction of time needed for multilayer formation. The time needed for preparation using spraying is reduced by a factor of 64 compared to dipping methods [56].

A multilayer film deposited by the spraying method has a composition and thickness similar to those of a multilayer film constructed by the conventional deposition method in solution [56, 57]. However, it has become clear that the spraying method turns out to have more general advantages in fabrication time and processing, and therefore can be easily exploited for its potential. For example, by simply varying the flow rate of charged species passing through an electrospun material during spray-assisted LbL deposition, individual fibers within the matrix can be conformally functionalized for ultrahigh surface area catalysis, or bridged to form a networked sublayer with complimentary properties. This a powerful and economical technique that has been utilized for developing multiple coatings of different morphologies and functions within a single textile membrane, enabling scientists to engineer the properties of a material from nanoscopic level in commercially viable quantities [58]. Furthermore, Facca and coworkers [59] suggested three-dimensional active biological gels for tissue-engineering applications built by spray-assisted LbL deposition with a large number of different components, including various cell types, polyelectrolytes, drugs, proteins, peptides, or DNA. This approach is based on the simple and progressive spraying of such elements in order to form a highly functionalized and structured platform. Recently, the spraying method has been applied to produce high-performance fuel-cell-membrane electrodes. The advantage is the fast fabrication process with nanoscale multilayer structures that allows specific tuning of their function [60].

30.3
Building-Block Materials for Hybrid Multilayer Films

The LbL deposition technique has quickly become one of the most popular and well-established methods for the preparation of multifunctional thin films not only due to its simplicity, but also robustness and versatility. What makes the LbL deposition method particularly versatile for the build-up of functional hybrid multilayer films is that any dimensional organic or inorganic building-block materials can be incorporated. Primary studies of LbL deposition have focused on the fundamentals of LbL films made up of synthetic and charged polymers with molecularly linear chains. The fundamentals of LbL films with typical strong polyelectrolytes and weak polyelectrolytes have been extensively studied by many research groups. Their physicochemical properties have been finely tuned based on the thorough examination of intrinsic properties of each polyelectrolyte LbL film for practical applications such as drug delivery and functional optical coatings [42, 44, 61–63]. Aside from the use of ordinary polyelectrolytes in LbL films, nano-objects have also been adapted in LbL studies as advanced building blocks to exploit their unique functionalities.

Practically, organic and inorganic hybrid multilayer films have further enriched the functionality and applicability of LbL deposition method. In principle, any type of molecular species, including *inorganic nano-objects* (nanoparticles, nanorods, and nanowires (NWs), nanosheets, etc.), *organic nano-objects* (polymeric micelles, dyes, carbon nanotubes (CNTs), etc.), and *biological nano-objects* (nucleic acids, proteins, viruses, and so on) can be successfully used as LbL assembly building-block materials. The LbL deposition method has also been enlarged to an unlimited number of structural and functional combinations of colloids and macromolecules. In this section, the state-of-the-art in synthesis and properties of hybrid multilayer films will be introduced in detail on the basis of inorganic, organic, and biological nano-objects.

30.3.1
Inorganic Nano-Objects

30.3.1.1 Nanoparticles

LbL assembly has been extensively utilized to create functional hybrid multilayer thin films containing inorganic nanoparticles. Analogous to multilayers comprising oppositely charged polyelectrolytes, charged nanoparticles can be assembled into multilayers paired with an oppositely charged polyelectrolyte. Kotov and coworkers [64, 65] and Lvov *et al.* [66] have shown that a variety of nanoparticles such as CdS, PbS, TiO_2, and SiO_2 with unique optical, magnetic, catalytic, and electronic properties can be readily assembled into thin films using the LbL technique. Also, by controlling the concentration of nanoparticle suspension, the size of nanoparticles, and the ionic strength and pH of nanoparticle suspensions, the structure and properties of nanoparticle hybrid multilayers could be precisely tuned. The LbL method has also been extended to create inorganic nanoparticle hybrid multilayer films on colloidal particles that are not readily achievable by other techniques [38, 67, 68]. Lee *et al.* [27] demonstrated that all-nanoparticle-based multilayer films (*inorganic nanoparticle hybrid multilayer films*) comprising oppositely charged nanoparticles can be created using the LbL deposition method. They used positively charged TiO_2 nanoparticles and negatively charged SiO_2 nanoparticles to create the inorganic nanoparticle hybrid multilayer films as a model system. These all-nanoparticle-based thin-film coatings have a number of advantages in antireflection, antifogging, and self-cleaning properties.

Hybrid structures of quantum dots (QDs) can be also constructed by the LbL deposition method having unique optical and electronic properties [26, 69, 70]. The colors and emissions of QD hybrid multilayer films are tunable by controlling the deposition order and number of bilayers. In addition, combining the LbL assembly with photolithographic techniques has been developed to fabricate patterned QD hybrid microstructures with different emission colors, and such multicolor microstructures that have great potential for use in light-emitting devices (LEDs) and biosensors [26]. Recently, Cho and coworkers [70] developed hybrid multilayer films consisting of polymer and inorganic nanoparticles (CdSe@ZnS, Au, and Pt), and demonstrated that the hybrid film exhibits a variety of interesting physical properties, such as photoluminescent durability, facile color tuning, and the ability to prepare functional freestanding films, as shown in Figure 30.4.

In summary, a number of useful functional multilayer thin films can be generated based on the LbL assembly of inorganic nanoparticles. By taking advantage of the unique properties of nanoparticles, thin-film devices for LED [26], photovoltaic [71], and biosensor [72] applications have been investigated. Most recently, surfaces exhibiting extreme wetting behaviors were prepared by creating nanoporous multilayers containing silica nanoparticles and titanium oxide nanoparticles [73, 74].

30.3.1.2 Nanorods and Nanowires (NWs)

One-dimensional nanomaterials such as inorganic nanorods and NWs have been studied as a candidate for next-generation devices. The LbL-assembly approach was also extended to NWs by alternate adsorption of polymers and inorganic wires. Liz-Marzan and coworkers [75] reported LbL films composed of typical polyelectrolyte, poly(diallyl dimetyl ammonium chloride) (PDDA)/PSS (poly(styrene sulfonate)), with Au NWs. The density of NW in LbL films was controlled by varying deposition time and the number of polymer layers inserted between Au NW layers. The longitudinal surface plasmon band was extensively red shifted and broadened, and this plasmon coupling was finely tuned by the degree of distance between each NW along the transverse axis of NWs. Aroca *et al.* [30] prepared dendrimer/silver NW LbL films as substrates for surface-enhanced Raman scattering. One hundred nanometer diameter Ag NWs are transferred onto a single fifth-generation poly(propylene imine) dendrimers containing amines at the periphery (DAB-Am) dendrimer layer on a glass slide. The dendrimer effectively immobilized the Ag NWs via shielding of electrostatic repulsions between NWs, resulting in significant enhancement of Raman signals throughout the visible range for various analytes deposited using both casting and Langmuir–Blodgett monolayer techniques. In addition, Te NWs LbL films were introduced via transformation of CdTe NWs using the Cd-complexing agent EDTA, as shown in Figure 30.5 [76]. The electrical resistance of PDDA/NW (poly(diallyl-dimethylammonium chloride)) multilayer films was measured in the dark and under light, to examine their photoconducting

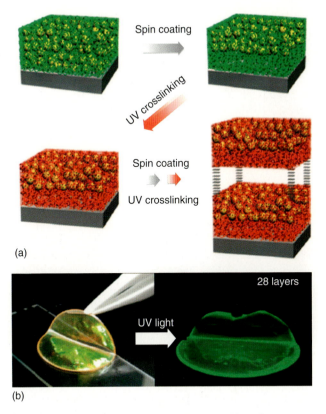

(a)

(b)

Figure 30.4 (a) Schematic illustration for the buildup of $(PS\text{-}N_3\text{-}SH\text{-}QDs:PS\text{-}N_3)_n$ multilayers and (b) photographic images of highly flexible freestanding $(PS\text{-}N_3\text{-}SH\text{-}stabilized}$ $QDs:PS\text{-}N_3)_n$ film containing green emissive QDs ($n = 28$, film thickness $\sim 6 \pm 1$ μm).

properties. The "light-on–light-off" cycle was shown at room temperature in stable form, which could serve as light detectors for various optoelectronic applications.

30.3.1.3 Nanosheets

Ultrathin inorganic nanosheets have received growing attention as building blocks for LbL-assembled multilayer films. Alternating inorganic/organic or inorganic/inorganic multistacking two-dimensional nanosheets can be successfully fabricated using a LbL electrostatic assembly technique, with a view toward atomic-scale-layered thin films [77, 78]. A variety of nanostructured hybrid multilayer films have been fabricated using a number of functional inorganic nanosheets such as $TiO_{1-\delta}O_2$, MnO_2, NbO_{17}.

For example, Manga et al. [31] demonstrated the fabrication of LbL-assembled multilayer thin films consisting of alternating titania ($Ti_{0.91}O_2$) and graphene oxide (GO) nanosheets. Schematic illustration of the formation of hybrid multilayer films

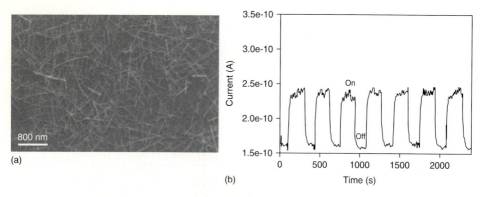

Figure 30.5 (a) Scanning electron microscope (SEM) image of (PDDA/Nanowire)$_2$. (b) Photoresponse of the Te nanowire hybrid LbL thin films for repetitive switching of the He:Ne laser between "on" and "off" states.

consisting of Ti$_{0.91}$O$_2$ nanosheets and GO, followed by subsequent UV-assisted photocatalytic reduction of GO into G, and high-resolution transmission electron microscopy (HR-TEM) image of the Ti$_{0.91}$O$_2$/G multilayer film are shown in Figure 30.6. The photocurrents increased linearly with the thickness of the Ti$_{0.91}$O$_2$/G bilayers, which provide evidence for efficient cross-surface charge percolation, and such hybrid multilayer films could be used to achieve functional separation of charge transport and storage.

Inorganic nanosheet hybrid multilayer films which have the potential application in selective oxidation and photosynthesis as well as in the photodegradation of dyes by the different wavelength light irradiation can be also fabricated by the LbL deposition method. For instance, Bi$_2$WO$_6$ nanosheets with high visible-light photoactivity ($\lambda > 420$ nm) hybrid multilayer films have the spectral selectivity of the photocatalytic degradation of Rhodamine B (RhB). Under the wavelength greater than 300 nm, the RhB molecules tend to be transformed to rhodamine over Bi$_2$WO$_6$ films. However, under the shorter-wavelength ($\lambda = 254$ nm) light irradiation, the RhB molecules can be photodegraded completely.

Therefore, using inorganic nanosheets, a lot of *functional hybrid multilayer films* that have useful potential in photocatalysis, capacitors, and sensors can be constructed via the LbL deposition method. And, in general, the activity of hybrid multilayer films can be modified easily by changing the deposition cycles.

30.3.2
Organic Nano-Objects

30.3.2.1 Polymeric Micelles
Polymeric micelles have been investigated for a long time due to their many advantages in their character as functional nanocontainers. Block-copolymer micelles (BCMs) especially have a capacity to load functional hydrophobic molecules such as nanoparticles, dyes, and drugs into a hydrophobic core, enabling these functional

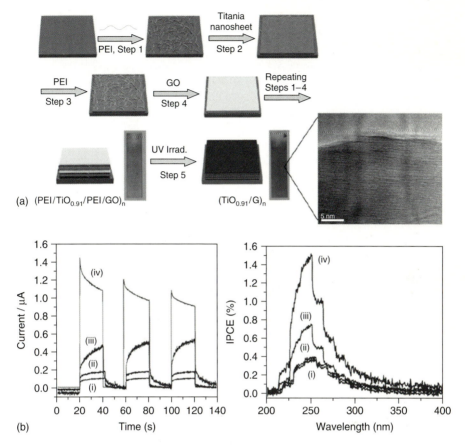

Figure 30.6 (a) Schematic illustration of the formation of the $Ti_{0.91}O_2/GO$ multilayer films, followed by subsequent UV-assisted photocatalytic reduction of GO into G and HR-TEM image of the $Ti_{0.91}O_2/G$ multilayer films (inserted figure). (b) Photoelectrochemical responses of (i) multilayered $(PEI/Ti_{0.91}O_2)_{10}$ films; multilayered $(PEI/Ti_{0.91}O_2/PEI/GO)_{10}$ films (ii) before UV and (iii) after UV irradiation for 24 h; and (iv) multilayered $(PEI/Ti_{0.91}O_2/PEI/GO)_{10}$ films after annealing at 400 °C for 1 h and corresponding photocurrent action spectrum.

molecules to be soluble in aqueous solution. In addition, BCM can protect the hydrophobic functional molecules from environmental conditions owing to the corona region and ligands substituted at the hydrophilic shell part, and it can offer various opportunity for further applications [79, 80].

Using the LbL deposition method, polymeric micelles can be easily incorporated into multilayer films interacting with general linear types of polymers by electrostatic bonding [81] or hydrogen bonding [19, 82, 83] as well as maintaining their functions. In detail, Park and coworkers [81] demonstrated the immobilization of heparin on the metal surface via subsequent build-up of a therapeutic LbL multilayer composed of paclitaxel (PTX; drug) encapsulated poly(lactic-co-glycolic acid) grafted hyaluronic acid (HA-g-PLGA) micelles, heparin, and poly-L-lysine (PLL) for drug eluting stent applications. Zhang and coworkers [84] described the use of BCMs as building blocks for incorporation of water-insoluble dyes and then fabricated multilayer films by alternating deposition of the micelles of poly(styrene-b-acrylic acid) (P(S-b-AA)) and PDDA. They also demonstrated a small organic molecule, pyrene incorporated into the micelle can be released from the multilayer films by immersing the films into solutions of different ionic strength. In addition, Char and coworkers [19] investigated the effect of LbL deposition methods (solution-dipping *vs.* spin-coating methods) on the surface morphology and wetting behavior of multilayer films based on hydrogen-bonding interaction between poly(acrylic acid) (PAA) and hydrophobically modified poly(ethylene oxide) (HM-PEO) micelles. They confirmed a three-dimensional surface structure in the dip-assisted multilayer films as it appeared above a critical number of layer pairs due to the formation of micelles of HM-PEO in its aqueous dipping solution. On the other hand, in the case of spin-assisted HM-PEO/PAA multilayer films, no such surface morphology was observed, regardless of the layer-pair number, owing to the limited rearrangement and aggregation of HM-PEO micelles during spin deposition. These results indicate that hybrid multilayer films composed of functional polymeric micelles can be easily constructed via LbL assembly and that control of LbL film surface property is determined by the preparation methods.

Furthermore, the multilayer films can be constructed without employing linear polymers but by using BCMs only exclusively. For example, protonated polystyrene-block-poly(4-vinylpyridine) (PS-b-P4VP) and anionic polystyrene-block-poly(acrylic acid) (PS-b-PAA) BCMs were used as building blocks for the LbL assembled micelle/micelle multilayer films onto flat [20] and colloidal substrates [21]. Char and coworkers [21] demonstrated that multifunctional films with tunable superhydrophobicity and multicolor rendering can be realized via LbL deposition of these micelles. Besides, the incorporation of hydrophobic fluorophores such as CdSe@ZnS QDs and an organic dyes into the PS cores of BCMs leads to the combination of multicolor emission with superhydrophobicity as shown in Figure 30.7. Therefore, polymeric micelles, as building block for hybrid multilayer films, have the potential for controlled assembly and release of water-insoluble functional nano-objects. The physical and chemical properties of polymeric micelle hybrid multilayer films can be easily controlled with the advantages of the LbL deposition process.

30.3.2.2 Organic Dyes
The LbL deposition method has been extended from linear polyelectrolytes to other charged small molecules such as organic dyes. Electrostatic alternate adsorption was successfully employed for low molecular weight dyes, leading to a large variety of dye and polyelectrolyte LbL deposition. Cooper *et al.* [23] used LbL assembly via electrostatic interactions between charged polypeptides and dyes, and they demonstrated the feasibility of preparing ordered multilayer films composed of charged macromolecules and dye bilayers back in 1995. Several years after this

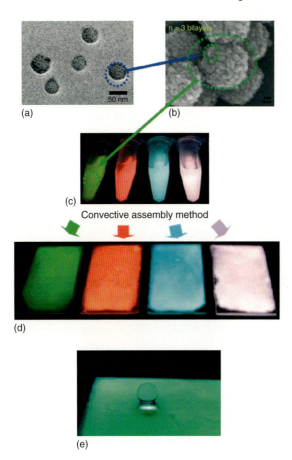

(a)

(b)

(c)

Convective assembly method

(d)

(e)

Figure 30.7 Polymeric micelle hybrid multilayer films: (a) TEM image of $PS_{49.5k}$-b-$P4VP_{16.5k}$ crew-cut micelles with the PS cores loaded with hydrophobic CdSe@ZnSQDs. (b) SEM image of silica colloids coated with ($PS_{49.5k}$-b-$P4VP_{16.5k}$ loaded with CdSe@ZnS/PS_{16k}-b-PAA_{4k})$_3$. (c) Colloidal suspensions and (d) films coated with {$PS_{49.5k}$-b-$P4VP_{16.5k}$ loaded with CdSe@ZnS(5 nm)/PS_{16k}-b-PAA_{4k}}$_3$, {$PS_{49.5k}$-b-$P4VP_{16.5k}$ loaded with CdSe@ZnS(5.4 nm)/PS_{16k}-b-PAA_{4k}}$_3$, {$PS_{49.5k}$-b-$P4VP_{16.5k}$ loaded with CdSe@ZnS(5 nm)/PS_{16k}-b-PAA_{4k} with Coumarin 30}$_3$, and {$PS_{49.5k}$-b-$P4VP_{16.5k}$ loaded with CdSe@ZnS(5.4 nm)/PS_{16k}-b-PAA_{4k} with Coumarin 30}$_3$ multilayers (from left to right). (e) The shape of a water droplet in contact with a colloidal silica film coated with {PS49.5k-b-($PS_{49.5k}$-b-$P4VP_{16.5k}$ loaded with CdSe@ZnS(5 nm)/PS_{16k}-b-PAA_{4k})}$_3$ at pH 4/4. The measured water contact angle is greater than $170°$.

report, Ariga and coworkers [24] investigated the assembling process of individual layers of Congo Red (CR; dye) and PDDA (polyelectrolyte) with quartz crystal microbalance (QCM) measurement. They demonstrated that the dye adsorption occurred at a rate similar to that of conventional polyelectrolyte adsorption by *in-situ* QCM measurements. The formation of well-packed monomolecular dye layers affected by size, number of charge, and spatial orientation of dye molecules was suggested by comparison of the film thickness as estimated from the QCM frequency shift with the molecular dimension of individual dyes. These extensions to incorporate small functional molecules into multilayer films enhance the utility of the LbL assembly, as a means of preparing functional hybrid multilayer films with ordered molecular arrangement and unique properties.

30.3.2.3 Carbon Nanotubes

CNTs have been considered as possible building blocks for LbL films adopting their unique mechanical, electrical, and optical properties, which are varied by the rolling direction type and the number of walls: single-walled carbon nanotubes and multiwalled carbon nanotubes (SWNTs, MWNTs) [85]. Among these unique properties, CNTs have been found to have exceptional mechanical properties with Young's modulus of 0.5–5 TPa and tensile strength of 15–50 GPa, making them one of the strongest candidates to substitute conventional reinforcing fillers for high-strength composites [86]. Electrically, each CNT can be sorted as either semiconducting or metallic, which is dependent on the chiral vector indices (n,m) of the graphene layer, and the type distribution of CNTs in composites affects collectively electrical properties of the macroscale performance. Recently, the LbL assembly method has been introduced to conducting CNT composites for advanced usages for high-performance devices, which could not be achieved by conventional coating technologies.

A variety of adsorption interaction can be utilized in LbL assembly of CNTs and polymers, such as electrostatic force, hydrogen bonds, van der Waals interactions and covalent bonds. The early investigation of *CNT hybrid multilayer film via LbL assembly* was achieved by Mamedov *et al.* [28] reporting the uniform dispersion of SWNTs in nanolayered structures with poly(ethylene imine) (PEI) (Figure 30.8a). This SWNT hybrid film shows distinguished mechanical properties even containing weak polymers, PEI, originating from high loading of CNTs dispersed with exceptional uniformity and functionally activated interfacial bonding between CNTs and polyelectrolyte matrix. Following experimental studies, it has been reported that dipping-assisted LbL assembly usually induces linear growth with constant slope and the surface roughness rely on the growth rate of LbL assembly and thickness of each layer [87, 88]. The loading amount of CNT in LbL-assembled films can be controlled by many variables such as kinds of polymer, stabilizing method of CNTs, and the building process of composite LbL films that affect the degree of exfoliation of CNTs and their dispersion in polymer composites [89].

The main advantage of CNT hybrid multilayer films includes the unique electro-optical functionality of CNT that can be fully exploited toward various applications by selective matching of complementary polymers. Conductive polymers such as poly(aniline) (PAN) and poly(3,4-ethylenedioxythiphene)/poly(styrene sulfonate) (PEDOT/PSS) multilayer were used for p-n heterojunction diodes as shown in Figure 30.8b and transparent conductors [90, 91]. CNTs are wrapped with Nafion and Pt catalysts for highly conductive and chemically durable electrodes in fuel cells [92]. Moreover, bioactive and biocompatible polymers such as PLGA [93], antimicrobial lysozyme (Lys) [94], antisense oligodeoxynucleotide (ASODN)

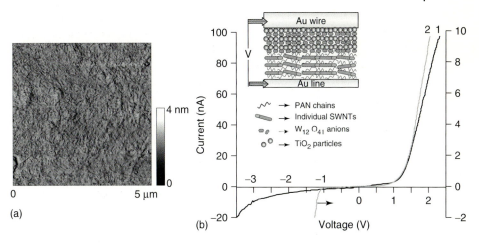

Figure 30.8 (a) Tapping-mode atomic force microscopy (AFM) image of a silicon wafer bearing (PEI/PAA)(PEI/SWNT)$_5$. (b) Schematic diagram of (TiO$_2$/W$_{12}$O$_{41}$)$_3$/(PAN/SWNTox)$_3$PAN multilayer film structure and electrode positioning for electrical measurements (inset) and I–V curves of Au(TiO$_2$/W$_{12}$O$_{41}$)$_8$/(PAN/SWNTox)$_{19}$PAN/AU (1), Au(TiO$_2$/W$_{12}$O$_{41}$)$_7$/TiO$_2$/Au (2) thin-film devices.

[95] enzymes are useful for biomedical applications such as prosthetic devices and stem-cell growth platforms. Also, electrical-property change of CNTs by physical and chemical stimulus enables CNT composite LbL films to be use in sensing applications, such as detection of biological materials and toxic chemicals as well as the observation of collapse or corrosion of construction materials [96–99].

30.3.2.4 Graphene Oxide (GO)

Graphene and its derivatives have attracted considerable attention in recent years because of their novel electronic, mechanical, and thermal properties [100, 101]. For practical use, homogeneous dispersion of graphite into individual plates of graphene is essential, followed by systematic integration of the graphene hybrids over large areas. Since graphite cannot be dispersed in water and it forms micrometer-sized irregular aggregates in organic solvents, graphene oxide exfoliated from graphite has been introduced to build thin films of graphene oxide hybrids, including the subsequent reduction process toward graphenes. Below, we summarize the investigations on the development of the LbL assembly of graphene oxide for electrochemical applications.

Fendler and coworkers [102, 103] preliminary reported that ultrathin films composed of self-assembled GO platelets and polyelectrolytes (P) and the subsequent *in situ* reduction of GO to more conductive graphene state (G) by chemical and electrochemical methods is possible (Figure 30.9). The conductivity of the self-assembled P/G films is of the same order of magnitude or greater than that found for conductive organic multilayers after doping. Self-assembled P/G films are likely to demonstrate high planar electron mobility similar to

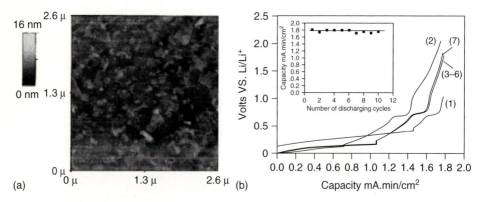

Figure 30.9 (a) Atomic force microscopy image of GO self-assembled film and (b) voltage–capacity curves for the first seven intercalation–deintercalation cycles (0.05 mA/s) of S-(PDDA/GO/PEO)₁₀. Inset shows the cycle life behavior for 10 cycles.

that observed for graphite. Furthermore, a high-density rechargeable lithium-ion battery was constructed using self-assembled a graphite oxide (GO) nanoplatelets LbL films with poly(ethylene oxide) (PEO) and PDDA as cathode. The concept of graphene oxide LbL films was revisited in detail by Kovtyukhova *et al.* [104] *via* examination on the quality and morphology of GO LbL films as well as on the nature of the surface-priming layer and the pH of the adsorbing solutions, with regard to the electronic characteristics. To promote the homogeneous integration of GO LbL films, various trials have been reported. Graphene was modified with polyelectrolytes by covalent bonding to introduce charges on the nanoplates [105], and vacuum-assisted deposition was utilized for the LbL assembly of gold-nanoparticle-decorated graphene thin films [106]. It is believed that the GO LbL films could be used for the transparent platforms that have high potential for use in a variety of applications, such as biosensors, advanced optical, and electronic devices.

30.3.3
Biological Nano-Objects

30.3.3.1 Nucleic Acids

The concept of DNA/polymer hybrid multilayer films prepared by LbL assembly was suggested by Decher and coworkers [34] in 1993. They reported on the LbL assembly procedure for the fabrication of multilayer thin films with alternation of DNA and PAH layers using the solution-dipping method. Since after this report, many research groups have fabricated *DNA hybrid LbL films* by alternative assembly of anionic charged DNA strands and cationic polyelectrolytes such as PAH, PEI, and studied the DNA structure within the films by circular dichroism (CD) spectroscopy. The driving force on the formation of DNA hybrid multilayer films originates from the electrostatic interaction between the negatively charged phosphate groups of DNA strands and the positively charged groups of polyelectrolyte chains [107, 108].

Moreover, the DNA and polymer hybrid microcapsules can be obtained by the sequential assembly of DNA-grafted micelles. Caruso and coworkers [109] reported the synthesis and characterization of DNA-grafted poly(N-isopropylacrylamide) (PNIPAM) micelles, and obtained DNA/polymer hybrid microcapsules by the alternate deposition of PNIPAM-A$_{30}$ and PNIPNAM-T$_{30}$ (DNA-grafted micelles containing the cDNA sequences polyA$_{30}$ and PolyT$_{30}$) onto silica particles followed by removal of the template core (see Figure 30.10). This DNA/polymer hybrid capsule system, offered by the combination of functional nanodomains

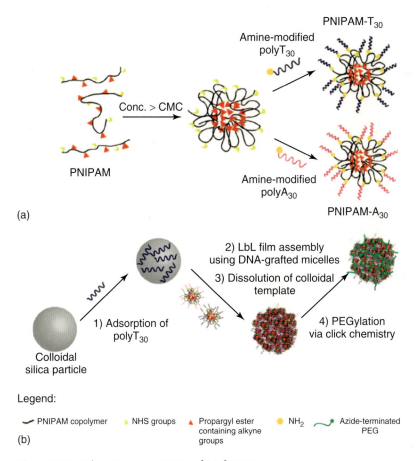

(a)

(b)

Legend:

⌒ PNIPAM copolymer ▸ NHS groups ▲ Propargyl ester containing alkyne groups ● NH$_2$ ⌇✳ Azide-terminated PEG

Figure 30.10 Schematic representation of (a) formation of PNIPAM-A30 and PNIPAMT30 micelles and (b) multilayer assembly of PNIPAM-A30 and PNIPAM-T30 micelles onto silica particles, removal of the template particle, and functionalization via "click" chemistry to afford PEGylated DNA-PNIPAM capsules.

(i.e. hydrophobic and reactive "click" micelles) and degradability of DNA, is envisioned to find applications in the controlled delivery of therapeutics.

Not only DNA, but also RNA can be used as building blocks of LbL-assembled hybrid multilayer films. Very recently, small interfering RNAs (siRNAs) and cationic polymers, branched or linear PEI, was assembled into multilayer films for application in substrate-mediated electroporation with high efficiency. Electroporation microarrays, developed for the high-throughput of expression constructs and siRNA into living mammalian cells, have the potential to provide a platform for cell-based analysis of gene functions. The LbL films of siRNA and PEI facilitated increase in the surface density of loaded siRNA, thus, the transfection efficiency electroporated into cells was improved in the siRNA loaded hybrid film on an electrode [110].

The nucleic acid hybrid multilayer films have many advantages due to good bioactivity and binding ability with different DNA or RNA-intercalated molecules, including antitumor drugs [7]. Furthermore, multilayer films fabricated from nucleic acids can be used to promote localized and surface-mediated cell transfection [111].

30.3.3.2 Proteins

Multilayer films that contain ordered layers of more than one protein species were assembled by alternate electrostatic adsorption mostly with charged polyelectrolytes. Lvov *et al.* [35] used cytochrome c (Cyt c), myoglobin (Mb), lysozyme (Lys), histone f3, hemoglobin (Hb), glucoamylase (GA), and glucose oxidase (GOD) as water-soluble proteins that interact with linear polymers to prepare *protein hybrid multilayer films*. They demonstrated that charged protein layers formed multilayers with linear polymers acting as glue or filler, and the assembly process was monitored by QCM measurements and UV spectroscopy. They also constructed the hybrid multilayer film consisting of alternating montmorillonite, PEI, and GOD layers, and suggested the biomolecular architecture and hybrid protein multilayer films that can carry out complex enzymatic reactions.

The functional characteristics of the proteins onto/or within the multilayer films have been studied for many biological applications such as immunosensing or activating of enzyme reactions. Caruso *et al.* [112] suggested *anti-Immunoglobulin (Anti-IgG) hybrid multilayer films* by the alternate LbL deposition with negatively charged polyelectrolyte (PSS). The utility of the anti-IgG hybrid multilayer film for immunosensing has been investigated via their subsequent interaction with IgG. This anti-IgG hybrid multilayer film increases the binding layer capacity (i.e. sensitivity) with respect to IgG. The films have promise in that the sensitivity can be tuned by fabricating the desired number of protein layers, and the selectivity can be modified by selecting the desired biospecific biomolecules. One example of the practical use of the LbL films containing enzymes is shown in Figure 30.11 [113], where two enzymes, GA and GOD, were assembled in the same film [114]. Additionally, enhanced stability of enzymes against temperature and pH changes within the LbL films has also been demonstrated [115]. This soft fixing of enzyme structures within the hybrid LbL films, multienzyme reactor of LbL film on ultrafilter, helps to suppress protein denaturation.

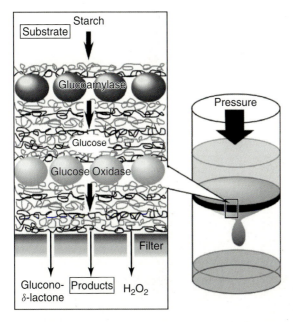

Figure 30.11 Multienzyme reactor of LbL multilayer film on ultrafilter.

In summary, LbL assembly has important features of simplicity and mildness for various biological nano-objects due to preparation conditions without harsh chemical treatment and in aqueous media [116]. Furthermore, biologically active multilayer films containing functional proteins are extremely interesting not only because of their increased capacity to bind incoming antigens with specific recognition but also because the distance between active layers may be controlled.

30.3.3.3 Viruses

Viruses, infecting agents to replicate only within living cells, can be regarded as biological nanoparticles composed of a small number of different biopolymers such as nucleic acids and proteins, from the viewpoint of material science. Using the LbL deposition method, densely packed virus layers that are sandwiched in multilayer films (*virus hybrid multilayer films*) can be constructed with various intermolecular interactions. Young and coworkers [117] prepared three types of multilayer films incorporating viral protein cages using ionic interactions or complementary biological interactions (i.e. streptavidin and biotin). These results demonstrate a number of different ways for fabricating and controlling the composition of multilayer films incorporating viral protein cages such as cowpea chlorotic mottle virus (CCMV) using the LbL deposition technique. The viral protein cages incorporated into a multilayer film can be used to impart individual functions because it can be viewed as scaffolds supporting independently addressable residues. The interior surfaces of the protein cages have been modified

Figure 30.12 (a) An experimental procedure for monolayer assembly of M13 virus on the polyelectrolyte multilayer of LPEI/PAA. (b) AFM image of the randomly stacked and aggregated structure of M13 viruses on 5.5 bilayers of strong polyelectrolyte multilayer (PEM) of PAH/PSS. (c) AFM image of a closely packed monolayer of M13 virus on weak PEM of LPEI/PAA.

to promote the size-constrained formation of a variety of inorganic materials and used to encapsulate functional nano-objects such as drug or inorganic catalysts.

Yoo and coworkers [37] presented the spontaneous assembly behavior of M13 viruses driven by competitive electrostatic binding on multilayered polymer surfaces. They demonstrated that the steric constraints inherent to the competitive charge binding between M13 viruses and two oppositely charged weak polyelectrolytes leads to interdiffusion and the virtual floating of viruses to the film surface (Figure 30.12). The linear poly(ethylene imine) (LPEI)/PAA polyelectrolyte pair is a strong candidate for the interdiffusion behavior. It has revealed that reversible interdiffusion can take place within LbL-assembled multilayer films consisting of certain pairs of weak polyelectrolytes, which typically show the characteristic of superlinear growth [118, 119]. Using this interdiffusion behavior of a weak polyelectrolyte pair, they embodied the spontaneous formation of a two-dimensional monolayer structure of viruses atop a cohesive polyelectrolyte multilayer. This system represents a functional interface that provides a general platform for the systematic incorporation and assembly of organic, biological, and inorganic nano-objects within or onto the hybrid multilayer films. In addition, an advantage of this method is that it can provide a tool for the construction of functionally controllable biomolecular surfaces on the flexible polymer film. In detail, Nam *et al.* [120] used viruses to synthesize and assemble NWs of cobalt oxide at room temperature for lithium ion battery applications. Combining virus-templated synthesis at the peptide level and methods for controlling two-dimensional assembly of viruses on polyelectrolyte multilayers provides a systematic platform for integrating these nanomaterials to form thin, flexible lithium ion batteries.

30.4
Conclusion

Functional hybrid multilayer films can be constructed by the LbL deposition method that virtually has no limitations in selecting building-block materials on the nanometer scale and can utilize any type of intermolecular interactions. Solution-dipping, spin-coating, and spraying methods are introduced as the representative LbL deposition techniques for the preparation of hybrid multilayer films. Each method can be selectively utilized to achieve target structure or function of the hybrid multilayer film according to different adsorption driving forces, processing time, experimental environment, and so on. Moreover, many functional nano-objects are promising building-block materials for hybrid multilayer films using various intermolecular interactions. Inorganic nano-objects from zero-dimensional nanoparticles to two-dimensional nanosheets are assembled into multilayer films with diverse functionality. Organic nano-objects such as polymeric micelles, dyes, and CNTs also can be used as building blocks for functional hybrid multilayer films with their characteristic physical and chemical properties. In addition, biological nano-objects like DNA, RNA, proteins, or viruses can be incorporated into LbL assembled multilayer films for many biological applications. Therefore, the LbL deposition method will certainly leads the way, in near future, how ultrathin films with hybrid composition and nanostructure are designed to have diverse unique functions and potential applications.

References

1. Iler, R.K. (1966) Multilayers of colloidal particles. *J. Colloid Interface Sci.*, **21**, 569–594.
2. Decher, G. and Hong, J.D. (1991) Buildup of ultrathin multilayer films by a self-assembly process. 1. Consecutive adsorption of anionic and cationic bipolar amphiphiles on charged surfaces. *Makromol. Chem.-Makromol. Symp.*, **46**, 321–327.
3. Decher, G. (1997) Fuzzy nanoassemblies: toward layered polymeric multicomposites. *Science*, **277**, 1232–1237.
4. Hammond, P.T. (1999) Recent explorations in electrostatic multilayer thin film assembly. *Curr. Opin. Colloid Interface Sci.*, **4**, 430–442.
5. Hammond, P.T. (2004) Form and function in multilayer assembly: New applications at the nanoscale. *Adv. Mater.*, **16**, 1271–1293.
6. Jaber, J.A. and Schlenoff, J.B. (2006) Recent developments in the properties

and applications of polyelectrolyte multilayers. *Curr. Opin. Colloid Interface Sci.*, **11**, 324–329.

7. Tang, Z., Wang, Y., Podsiadlo, P., and Kotov, N.A. (2006) Biomedical applications of layer-by-layer assembly: from biomimetics to tissue engineering. *Adv. Mater.*, **18**, 3203–3224.

8. Jiang, C.Y. and Tsukruk, V.V. (2006) Freestanding nanostructures via layer-by-layer assembly. *Adv. Mater.*, **18**, 829–840.

9. Quinn, A., Such, G.K., Quinn, J.F., and Caruso, F. (2008) Polyelectrolyte blend multilayers: a versatile route to engineering interfaces and films. *Adv. Funct. Mater.*, **18**, 17–26.

10. Wang, Y., Angelatos, A.S., and Caruso, F. (2008) Template synthesis of nanostructured materials via layer-by-layer assembly. *Chem. Mater.*, **20**, 848–858.

11. Kharlampieva, E., Kozlovskaya, V., and Sukhishvili, S.A. (2009) Layer-by-layer hydrogen-bonded polymer films: from fundamentals to applications. *Adv. Mater.*, **21**, 3053–3065.

12. Schlenoff, J.B. (2009) Retrospective on the future of polyelectrolyte multilayers. *Langmuir*, **25**, 14007–14010.

13. Quinn, J.F., Johnston, A.P.R., Such, G.K., Zelikin, A.N., and Caruso, F. (2007) Next generation, sequentially assembled ultrathin films: beyond electrostatics. *Chem. Soc. Rev.*, **36**, 707–718.

14. Kharlampieva, E. and Sukhishvili, S.A. (2006) Hydrogen-bonded layer-by-layer polymer films. *Polym. Rev.*, **46**, 377–395.

15. Such, G.K., Quinn, J.F., Quinn, A., Tjipto, E., and Caruso, F. (2006) Assembly of ultrathin polymer multilayer films by click chemistry. *J. Am. Chem. Soc.*, **128**, 9318–9319.

16. Shimazaki, Y., Mitsuishi, M., Ito, S., and Yamamoto, M. (1997) Preparation of the layer-by-layer deposited ultrathin film based on the charge-transfer interaction. *Langmuir*, **13**, 1385–1387.

17. Tang, T.J., Qu, J.Q., Mullen, K., and Webber, S.E. (2006) Molecular layer-by-layer self-assembly of water-soluble perylene diimides

through pi-pi and electrostatic interactions. *Langmuir*, **22**, 26–28.

18. Serizawa, T., Hashiguchi, S., and Akashi, M. (1999) Stepwise assembly of ultrathin poly(vinyl alcohol) films on a gold substrate by repetitive adsorption/drying processes. *Langmuir*, **15**, 5363–5368.

19. Seo, J., Lutkenhaus, J.L., Kim, J., Hammond, P.T., and Char, K. (2008) Effect of the layer-by-layer (LbL) deposition method on the surface morphology and wetting behavior of hydrophobically modified PEO and PAA LbL films. *Langmuir*, **24**, 7995–8000.

20. Cho, J.H., Hong, J.K., Char, K., and Caruso, F. (2006) Nanoporous block copolymer micelle/micelle multilayer films with dual optical properties. *J. Am. Chem. Soc.*, **128**, 9935–9942.

21. Hong, J., Bae, W.K., Lee, H., Oh, S., Char, K., Caruso, F., and Cho, J. (2007) Tunable superhydrophobic and optical properties of colloidal films coated with block-copolymer-micelles/micelle-multilayers. *Adv. Mater.*, **19**, 4364–4369.

22. He, J.A., Valluzzi, R., Yang, K., Dolukhanyan, T., Sung, C.M., Kumar, J., Tripathy, S.K., Samuelson, L., Balogh, L., and Tomalia, D.A. (1999) Electrostatic multilayer deposition of a gold-dendrimer nanocomposite. *Chem. Mater.*, **11**, 3268–3274.

23. Cooper, T.M., Campbell, A.L., and Crane, R.L. (1995) Formation of polypeptide-dye multilayers by an electrostatic self-assembly technique. *Langmuir*, **11**, 2713–2718.

24. Ariga, K., Lvov, Y., and Kunitake, T. (1997) Assembling alternate dye-polyion molecular films by electrostatic layer-by-layer adsorption. *J. Am. Chem. Soc.*, **119**, 2224–2231.

25. Baussard, J.F., Habib-Jiwan, J.L., and Laschewsky, A. (2003) Enhanced Forster resonance energy transfer in electrostatically self-assembled multilayer films made from new fluorescently labeled polycations. *Langmuir*, **19**, 7963–7969.

26. Lin, Y.W., Tseng, W.L., and Chang, H.T. (2006) Using a layer-by-layer

assembly technique to fabricate multicolored-light-emitting films of CdSe@CdS and CdTe quantum dots. *Adv. Mater.*, **18**, 1381–1138.

27. Lee, D., Rubner, M.F., and Cohen, R.E. (2006) All-nanoparticle thin-film coatings. *Nano. Lett.*, **6**, 2305–2312.

28. Mamedov, A.A., Kotov, N.A., Prato, M., Guldi, D.M., Wicksted, J.P., and Hirsch, A. (2002) Molecular design of strong single-wall carbon nanotube/polyelectrolyte multilayer composites. *Nature Mater.*, **1**, 190–194.

29. Jiang, C.Y., Ko, H.Y., and Tsukruk, V.V. (2005) Strain-sensitive Raman modes of carbon nanotubes in deflecting freely suspended nanomembranes. *Adv. Mater.*, **17**, 2127–2212.

30. Aroca, R.F., Goulet, P.J.G., dos Santos, D.S., Alvarez-Puebla, R.A., and Oliveira, O.N. (2005) Silver nanowire layer-by-layer films as substrates for surface-enhanced Raman scattering. *Anal. Chem.*, **77**, 378–382.

31. Manga, K.K., Zhou, Y., Yan, Y.L., and Loh, K.P. (2009) Multilayer hybrid films consisting of alternating graphene and titania nanosheets with ultrafast electron transfer and photoconversion properties. *Adv. Funct. Mater.*, **19**, 3638–3643.

32. Zhang, S., Shen, J., Fu, H., Dong, W., Zheng, Z., and Shi, L. (2007) Bi_2WO_6 photocatalytic films fabricated by layer-by-layer technique from Bi_2WO_6 nanoplates and its spectral selectivity. *J. Solid State Chem.*, **180**, 1456–1463.

33. Izawa, K., Yamada, T., Unal, U., Ida, S., Altuntasoglu, O., Koinuma, M., and Matsumoto, Y. (2006) Photoelectrochemical oxidation of methanol on oxide nanosheets. *J. Phys. Chem. B*, **110**, 4645–4650.

34. Lvov, Y., Decher, G., and Sukhorukov, G. (1993) Assembly of thin-films by means of successive deposition of alternate layers of DNA and poly(Allylamine). *Macromolecules*, **26**, 5396–5399.

35. Lvov, Y., Ariga, K., Ichinose, I., and Kunitake, T. (1995) Assembly of multicomponent protein films by means of electrostatic layer-by-layer adsorption. *J. Am. Chem. Soc.*, **117**, 6117–6123.

36. Lvov, Y., Haas, H., Decher, G., Mohwald, H., Mikhailov, A., Mtchedlishvily, B., Morgunova, E., and Vainshtein, B. (1994) Successive deposition of alternate layers of polyelectrolytes and a charged virus. *Langmuir*, **10**, 4232–4236.

37. Yoo, P.J., Nam, K.T., Qi, J.F., Lee, S.K., Park, J., Belcher, A.M., and Hammond, P.T. (2006) Spontaneous assembly of viruses on multilayered polymer surfaces. *Nature Mater.*, **5**, 234–240.

38. Caruso, F., Caruso, R.A., and Mohwald, H. (1998) Nanoengineering of inorganic and hybrid hollow spheres by colloidal templating. *Science*, **282**, 1111–1114.

39. Lee, D., Nolte, A.J., Kunz, A.L., Rubner, M.F., and Cohen, R.E. (2006) pH-induced hysteretic gating of track-etched polycarbonate membranes: swelling/deswelling behavior of polyelectrolyte multilayers in confined geometry. *J. Am. Chem. Soc.*, **128**, 8521–8529.

40. Jewell, C.M., Zhang, J.T., Fredin, N.J., Wolff, M.R., Hacker, T.A., and Lynn, D.M. (2006) Release of plasmid DNA from intravascular stents coated with ultrathin multilayered polyelectrolyte films. *Biomacromolecules*, **7**, 2483–2491.

41. Caruso, F., Trau, D., Mohwald, H., and Renneberg, R. (2000) Enzyme encapsulation in layer-by-layer engineered polymer multilayer capsules. *Langmuir*, **16**, 1485–1488.

42. Dubas, S.T. and Schlenoff, J.B. (1999) Factors controlling the growth of polyelectrolyte multilayers. *Macromolecules*, **32**, 8153–8160.

43. Tan, H.L., McMurdo, M.J., Pan, G.Q., and Van Patten, P.G. (2003) Temperature dependence of polyelectrolyte multilayer assembly. *Langmuir*, **19**, 9311–9314.

44. Shiratori, S.S. and Rubner, M.F. (2000) pH-dependent thickness behavior of sequentially adsorbed layers of weak polyelectrolytes. *Macromolecules*, **33**, 4213–4219.

45. Poptoshev, E., Schoeler, B., and Caruso, F. (2004) Influence of solvent

quality on the growth of polyelectrolyte multilayers. *Langmuir*, **20**, 829–834.

46. Cho, J., Char, K., Hong, J.D., and Lee, K.B. (2001) Fabrication of highly ordered multilayer films using a spin self-assembly method. *Adv. Mater.*, **13**, 1076–1078.

47. Cho, J. and Char, K. (2004) Effect of layer integrity of spin self-assembled multilayer films on surface wettability. *Langmuir*, **20**, 4011–4016.

48. Kim, H., Cho, J., Kim, D.Y., and Char, K. (2007) Electroluminescent characteristics of spin-assembled multilayer films with confined layer structure. *Thin Solid Films*, **516**, 78–83.

49. Cho, J., Jang, H., Yeom, B., Kim, H., Kim, R., Kim, S., Char, K., and Caruso, F. (2006) Modulating the pattern quality of micropatterned multilayer films prepared by layer-by-layer self-assembly. *Langmuir*, **22**, 1356–1364.

50. Jang, H., Kim, S., and Char, K. (2003) Multilayer line micropatterning using convective self-assembly in microfluidic channels. *Langmuir*, **19**, 3094–3097.

51. Kharlampieva, E., Kozlovskaya, V., Zavgorodnya, O., Lilly, G.D., Kotov, N.A., and Tsukruk, V.V. (2010) pH-responsive photoluminescent LbL hydrogels with confined quantum dots. *Soft Matter*, **6**, 800–807.

52. Jiang, C.Y., Markutsya, S., and Tsukruk, V.V. (2004) Compliant, robust, and truly nanoscale free-standing multilayer films fabricated using spin-assisted layer-by-layer assembly. *Adv. Mater.*, **16**, 157–161.

53. Seo, J., Lee, H., Jeon, J., Jang, Y., Kim, R., Char, K., and Nam, J.M. (2009) Tunable layer-by-layer polyelectrolyte platforms for comparative cell assays. *Biomacromolecules*, **10**, 2254–2260.

54. Vozar, S., Poh, Y.C., Serbowicz, T., Bachner, M., Podsiadlo, P., Qin, M., Verploegen, E., Kotov, N., and Hart, A.J. (2009) Automated spin-assisted layer-by-layer assembly of nanocomposites. *Rev. Sci. Instrum.*, **80**, 023903.

55. Schlenoff, J.B., Dubas, S.T., and Farhat, T. (2000) Sprayed polyelectrolyte multilayers. *Langmuir*, **16**, 9968–9969.

56. Kolasinska, M., Krastev, R., Gutberlet, T., and Warszynski, P. (2009) Layer-by-layer deposition of polyelectrolytes. Dipping versus spraying. *Langmuir*, **25**, 1224–1232.

57. Izquierdo, A., Ono, S.S., Voegel, J.C., Schaaf, P., and Decher, G. (2005) Dipping versus spraying: exploring the deposition conditions for speeding up layer-by-layer assembly. *Langmuir*, **21**, 7558–7567.

58. Krogman, K.C., Lowery, J.L., Zacharia, N.S., Rutledge, G.C., and Hammond, P.T. (2009) Spraying asymmetry into functional membranes layer-by-layer. *Nature Mater.*, **8**, 512–518.

59. Facca, P.G.S., Stoltz, J.-F., Netter, P., Mainard, D., Voegel, J.-C., and Benkirane-Jessel, N. (2008) Three-dimensional sprayed active biological gels and cells for tissue engineering application. *Bio-Med. Mater. Eng.*, **18**, 231–235.

60. Michel, M., Ettingshausen, F., Scheiba, F., Wolz, A., and Roth, C. (2008) Using layer-by-layer assembly of polyaniline fibers in the fast preparation of high performance fuel cell nanostructured membrane electrodes. *Phys. Chem. Chem. Phys.*, **10**, 3796–3801.

61. Schlenoff, J.B., Ly, H., and Li, M. (1998) Charge and mass balance in polyelectrolyte multilayers. *J. Am. Chem. Soc.*, **120**, 7626–7634.

62. Hiller, J.A., Mendelsohn, J.D., and Rubner, M.F. (2002) Reversibly erasable nanoporous anti-reflection coatings from polyelectrolyte multilayers. *Nature Mater.*, **1**, 59–63.

63. Sukhishvili, S.A. and Granick, S. (2000) Layered, erasable, ultrathin polymer films. *J. Am. Chem. Soc.*, **122**, 9550–9551.

64. Kotov, N.A., Dekany, I., and Fendler, J.H. (1995) Layer-by-layer self-assembly of polyelectrolyte-semiconductor nanoparticle composite films. *J. Phys. Chem.-Us*, **99**, 13065–13069.

65. Ostrander, J.W., Mamedov, A.A., and Kotov, N.A. (2001) Two modes of linear layer-by-layer growth of nanoparticle-polyelectrolyte multilayers and different interactions in the

layer-by-layer deposition. *J. Am. Chem. Soc.*, **123**, 1101–1110.

66. Lvov, Y., Ariga, K., Onda, M., Ichinose, I., and Kunitake, T. (1997) Alternate assembly of ordered multilayers of SiO_2 and other nanoparticles and polyions. *Langmuir*, **13**, 6195–6203.

67. Caruso, R.A., Susha, A., and Caruso, F. (2001) Multilayered titania, silica, and Laponite nanoparticle coatings on polystyrene colloidal templates and resulting inorganic hollow spheres. *Chem. Mater.*, **13**, 400–409.

68. Caruso, F., Spasova, M., Susha, A., Giersig, M., and Caruso, R.A. (2001) Magnetic nanocomposite particles and hollow spheres constructed by a sequential layering approach. *Chem. Mater.*, **13**, 109–116.

69. Lee, B., Kim, Y., Lee, S., Kim, Y.S., Wang, D.Y., and Cho, J. (2010) Layer-by-layer growth of polymer/quantum dot composite multilayers by nucleophilic substitution in organic media. *Angew. Chem. Int. Ed.*, **49**, 359–363.

70. Lee, S., Lee, B., Kim, B.J., Park, J., Yoo, M., Bae, W.K., Char, K., Hawker, C.J., Bang, J., and Cho, J.H. (2009) Free-standing nanocomposite multilayers with various length scales, adjustable internal structures, and functionalities. *J. Am. Chem. Soc.*, **131**, 2579–2587.

71. Guldi, D.M., Zilberman, I., Anderson, G., Kotov, N.A., Tagmatarchis, N., and Prato, M. (2005) Nanosized inorganic/organic composites for solar energy conversion. *J. Mater. Chem.*, **15**, 114–118.

72. Liang, Z.Q., Dzienis, K.L., Xu, J., and Wang, Q. (2006) Covalent layer-by-layer assembly of conjugated polymers and CdSe nanoparticles: multilayer structure and photovoltaic properties. *Adv. Funct. Mater.*, **16**, 542–548.

73. Kommireddy, D.S., Patel, A.A., Shutava, T.G., Mills, D.K., and Lvov, Y.M. (2005) Layer-by-layer assembly of TiO_2 nanoparticles for stable hydrophilic biocompatible coatings. *J. Nanosci. Nanotechnol.*, **5**, 1081–1087.

74. Cebeci, F.C., Wu, Z.Z., Zhai, L., Cohen, R.E., and Rubner, M.F. (2006) Nanoporosity-driven superhydrophilicity: a means to create multifunctional antifogging coatings. *Langmuir*, **22**, 2856–2862.

75. Vial, S., Pastoriza-Santos, I., Perez-Juste, J., and Liz-Marzan, L.M. (2007) Plasmon coupling in layer-by-layer assembled gold nanorod films. *Langmuir*, **23**, 4606–4611.

76. Wang, Y., Tang, Z.Y., Podsiadlo, P., Elkasabi, Y., Lahann, J., and Kotov, N.A. (2006) Mirror-like photoconductive layer-by-layer thin films of Te nanowires: the fusion of semiconductor, metal, and insulator properties. *Adv. Mater.*, **18**, 518–522.

77. Kleinfeld, E.R. and Ferguson, G.S. (1994) Stepwise formation of multilayered nanostructural films from macromolecular precursors. *Science*, **265**, 370–373.

78. Keller, S.W., Kim, H.N., and Mallouk, T.E. (1994) Layer-by-layer assembly of intercalation compounds and heterostructures on surfaces - toward molecular beaker epitaxy. *J. Am. Chem. Soc.*, **116**, 8817–8818.

79. Park, J.H., Lee, S., Kim, J.H., Park, K., Kim, K., and Kwon, I.C. (2008) Polymeric nanomedicine for cancer therapy. *Prog. Polym. Sci.*, **33**, 113–137.

80. Peer, D., Karp, J.M., Hong, S., Farokhzad, O.C., Margalit, R., and Langer, R. (2007) Nanocarriers as an emerging platform for cancer therapy. *Nature Nano.*, **2**, 751–760.

81. Kim, T.G., Lee, H., Jang, Y., and Park, T.G. (2009) Controlled release of paclitaxel from heparinized metal stent fabricated by layer-by-layer assembly of polylysine and hyaluronic acid-g-poly(lactic-co-glycolic acid) micelles encapsulating paclitaxel. *Biomacromolecules*, **10**, 1532–1539.

82. Zhao, Y., Bertrand, J., Tong, X., and Zhao, Y. (2009) Photo-cross-linkable polymer micelles in hydrogen-bonding-built layer-by-layer films. *Langmuir*, **25**, 13151–13157.

83. Seo, J., Lutkenhaus, J.L., Kim, J., Hammond, P.T., and Char, K. (2007) Development of surface morphology in multilayered films prepared

by layer-by-layer deposition using poly(acrylic acid) and hydrophobically modified poly(ethylene oxide). *Macromolecules*, **40**, 4028–4036.

84. Ma, N., Zhang, H., Song, B., Wang, Z., and Zhang, X. (2005) Polymer micelles as building blocks for layer-by-layer assembly: an approach for incorporation and controlled release of water-insoluble dyes. *Chem. Mater.*, **17**, 5065–5069.

85. Dresselhaus, M.S., Dresselhaus, G., and Jorio, A. (2004) Unusual properties and structure of carbon nanotubes. *Annu. Rev. Mater. Res.*, **34**, 247–278.

86. Yu, M.F., Lourie, O., Dyer, M.J., Moloni, K., Kelly, T.F., and Ruoff, R.S. (2000) Strength and breaking mechanism of multiwalled carbon nanotubes under tensile load. *Science*, **287**, 637–640.

87. Shi, J.H., Qin, Y.J., Luo, H.X., Guo, Z.X., Woo, H.S., and Park, D.K. (2007) Covalently attached multilayer self-assemblies of single-walled carbon nanotubols and diazoresins. *Nanotechnology*, **18**, 365704.

88. Paloniemi, H., Lukkarinen, M., Aaritalo, T., Areva, S., Leiro, J., Heinonen, M., Haapakka, K., and Lukkari, J. (2006) Layer-by-layer electrostatic self-assembly of single-wall carbon nanotube polyelectrolytes. *Langmuir*, **22**, 74–83.

89. Xue, W. and Cui, T.H. (2007) Characterization of layer-by-layer self-assembled carbon nanotube multilayer thin films. *Nanotechnology*, **18**, 365704.

90. Kovtyukhova, N.L. and Mallouk, T.E. (2005) Nanowire p-n heterojunction diodes made by templated assembly of multilayer carbon-nanotube/polymer/semiconductor-particle shells around metal nanowires. *Adv. Mater.*, **17**, 187–192.

91. Ham, H.T., Choi, Y.S., Chee, M.G., Cha, M.H., and Chung, I.J. (2008) PEDOT-PSS/single-wall carbon nanotubes composites. *Polym. Eng. Sci.*, **48**, 1–10.

92. Michel, M., Taylor, A., Sekol, R., Podsiadlo, P., Ho, P., Kotov, N., and Thompson, L. (2007) High-performance nanostructured membrane electrode assemblies for fuel cells made by layer-by-layer assembly of carbon nanocolloids. *Adv. Mater.*, **19**, 3859–3864.

93. Koh, L.B., Rodriguez, I., and Zhou, J.J. (2008) Platelet adhesion studies on nanostructured poly(lactic-co-glycolic-acid)-carbon nanotube composite. *J. Biomed. Mater. Res. A*, **86A**, 394–401.

94. Nepal, D., Balasubramanian, S., Simonian, A.L., and Davis, V.A. (2008) Strong antimicrobial coatings: single-walled carbon nanotubes armored with biopolymers. *Nano. Lett.*, **8**, 1896–1901.

95. Jia, N.Q., Lian, Q., Shen, H.B., Wang, C., Li, X.Y., and Yang, Z.N. (2007) Intracellular delivery of quantum dots tagged antisense oligodeoxynucleotides by functionalized multiwalled carbon nanotubes. *Nano. Lett.*, **7**, 2976–2980.

96. Loh, K.J., Kim, J., Lynch, J.P., Kam, N.W.S., and Kotov, N.A. (2007) Multifunctional layer-by-layer carbon nanotube-polyelectrolyte thin films for strain and corrosion sensing. *Smart Mater. Struct.*, **16**, 429–438.

97. Korkut, S., Keskinler, B., and Erhan, E. (2008) An amperometric biosensor based on multiwalled carbon nanotube-poly(pyrrole)-horseradish peroxidase nanobiocomposite film for determination of phenol derivatives. *Talanta*, **76**, 1147–1152.

98. Loh, K.J., Lynch, J.P., Shim, B.S., and Kotov, N.A. (2008) Tailoring piezoresistive sensitivity of multilayer carbon nanotube composite strain sensors. *J. Intell. Mater. Syst. Struct.*, **19**, 747–764.

99. Ma, H.Y., Zhang, L.P., Pan, Y., Zhang, K.Y., and Zhang, Y.Z. (2008) A novel electrochemical DNA biosensor fabricated with layer-by-layer covalent attachment of multiwalled carbon nanotubes and gold nanoparticles. *Electroanalysis*, **20**, 1220–1226.

100. Geim, A.K. and Novoselov, K.S. (2007) The rise of graphene. *Nature Mater.*, **6**, 183–191.

101. Ruoff, R. (2008) Graphene: calling all chemists. *Nature Nano.*, **3**, 10–11.

102. Kotov, N.A., Dekany, I., and Fendler, J.H. (1996) Ultrathin graphite oxide-polyelectrolyte composites prepared by self-assembly: transition between conductive and non-conductive states. *Adv. Mater.*, **8**, 637–641.

103. Cassagneau, T. and Fendler, J.H. (1998) High density rechargeable lithium-ion batteries self-assembled from graphite oxide nanoplatelets and polyelectrolytes. *Adv. Mater.*, **10**, 877–881.

104. Kovtyukhova, N.I., Ollivier, P.J., Martin, B.R., Mallouk, T.E., Chizhik, S.A., Buzaneva, E.V., and Gorchinskiy, A.D. (1999) Layer-by-layer assembly of ultrathin composite films from micron-sized graphite oxide sheets and polycations. *Chem. Mater.*, **11**, 771–778.

105. Shen, J.F., Hu, Y.Z., Li, C., Qin, C., Shi, M., and Ye, M.X. (2009) Layer-by-layer self-assembly of graphene nanoplatelets. *Langmuir*, **25**, 6122–6128.

106. Kong, B.S., Geng, J.X., and Jung, H.T. (2009) Layer-by-layer assembly of graphene and gold nanoparticles by vacuum filtration and spontaneous reduction of gold ions. *Chem. Commun.*, 2174–2176.

107. Sukhorukov, G.B., Montrel, M.M., Petrov, A.I., Shabarchina, L.I., and Sukhorukov, B.I. (1996) Multilayer films containing immobilized nucleic acids. Their structure and possibilities in biosensor applications. *Biosens. Bioelectron.*, **11**, 913–922.

108. Montrel, M.M., Sukhorukov, G.B., Petrov, A.I., Shabarchina, L.I., and Sukhorukov, B.I. (1997) Spectroscopic study of thin multilayer films of the complexes of nucleic acids with cationic amphiphiles and polycations: their possible use as sensor elements. *Sens. Actuators B-Chem.*, **42**, 225–231.

109. Cavalieri, F., Postma, A., Lee, L., and Caruso, F. (2009) Assembly and functionalization of DNA-polymer microcapsules. *ACS Nano*, **3**, 234–240.

110. Fujimoto, H., Kato, K., and Iwata, K. (2010) Layer-by-layer assembly of small interfering RNA and poly(ethyleneimine) for substrate-mediated electroporation with high efficiency. *Anal. Bioanal. Chem.*, **397**, 571–578.

111. Jewell, C.M. and Lynn, D.M. (2008) Multilayered polyelectrolyte assemblies as platforms for the delivery of DNA and other nucleic acid-based therapeutics. *Adv. Drug Deliv. Rev.*, **60**, 979–999.

112. Caruso, F., Niikura, K., Furlong, D.N., and Okahata, Y. (1997) Assembly of alternating polyelectrolyte and protein multilayer films for immunosensing. 2. *Langmuir*, **13**, 3427–3433.

113. Ariga, K., Hill, J.P., and Ji, Q.M. (2007) Layer-by-layer assembly as a versatile bottom-up nanofabrication technique for exploratory research and realistic application. *Phys. Chem. Chem. Phys.*, **9**, 2319–2340.

114. Hill, J.P., Wakayama, Y., Schmitt, W., Tsuruoka, T., Nakanishi, T., Zandler, M.L., McCarty, A.L., D'Souza, F., Milgrom, L.R., and Ariga, K. (2006) Regulating the stability of 2D crystal structures using an oxidation state-dependent molecular conformation. *Chem. Commun.*, 2320–2322.

115. Hill, J.P., Wakayama, Y., and Ariga, K. (2006) How molecules accommodate a 2D crystal lattice mismatch: an unusual 'mixed' conformation of tetraphenylporphyrin. *Phys. Chem. Chem. Phys.*, **8**, 5034–5037.

116. Ariga, K., Hill, J.P., and Ji, Q. (2008) Biomaterials and biofunctionality in layered macromolecular assemblies. *Macromol. Biosci.*, **8**, 981–990.

117. Suci, P.A., Klem, M.T., Arce, F.T., Douglas, T., and Young, M. (2006) Assembly of multilayer films incorporating a viral protein cage architecture. *Langmuir*, **22**, 8891–8896.

118. Picart, C., Mutterer, J., Richert, L., Luo, Y., Prestwich, G.D., Schaaf, P., Voegel, J.C., and Lavalle, P. (2002) Molecular basis for the explanation of the exponential growth of polyelectrolyte multilayers. *Proc. Natl. Acad. Sci. USA*, **99**, 12531–12535.

119. Lavalle, P., Gergely, C., Cuisinier, F.J.G., Decher, G., Schaaf, P., Voegel,

J.C., and Picart, C. (2002) Comparison of the structure of polyelectrolyte multilayer films exhibiting a linear and an exponential growth regime: an in situ atomic force microscopy study. *Macromolecules*, **35**, 4458–4465.

120. Nam, K.T. (2008) Virus-enabled synthesis and assembly of nanowires for lithium ion battery electrodes. *Science*, **322**, 44; (2006) **312**, 885.

31
Light-Directed Smart Responses in Azobenzene-Containing Liquid-Crystalline Polymer Thin Films

Takahiro Seki

31.1
Introduction

Light is of particular use in thin-film technology. Photoresists and photocurable polymer materials are typical widely known examples that are now indispensable and of great demand in industries. Such processes, however, only utilize light excitation to allow a photoreaction without effectively gaining directional information, namely, polarization, propagation direction, coherency, and so on, on the material. Furthermore, the processes are of once use. On the other hand, a recent stream of research attempts to adopt reversible photoreactions (photochromic reactions) that further recognize the directional information into materials. In the studies of photochromism, it has long been studied for fabrication of photon-mode optical memories and switching of various performances including physical, chemical, and biological functions [1–3]. The versatility of photochromism is rapidly expanding to alter various physical properties and to trigger morphological and shape changes of materials. Photocrosslinkable units are also of significance in addition to photochromic units.

It is worth pointing out that light can access various types of thin films possessing various hierarchical size features [4–8]. The electronic interaction is first achieved between light and a molecule, and this nanoscale event can successively lead to larger events of the surroundings in ranges of mesoscale (10–100 nm, defined as in this chapter), and micro(meter)-scale larger than 1000 nm. When liquid-crystalline materials are employed, large amplification effects can emerge because of their strong cooperativity in terms of molecular orientation. Typical examples are the surface photoalignment of liquid crystals and photomechanical responses in liquid-crystalline polymer films [9–12].

This chapter is arranged in three parts on the photoresponsive properties from molecular scale to larger ones as follow. (i) Nanoscale (1–10 nm) control. Structural anisotropy possessed on a surface can be transferred to preferred alignment of liquid-crystalline molecules. (ii) Mesoscale (10–100 nm) control. The dimension of macromolecules or supramolecular assemblies covers this range. Some examples on the photoalignment of mesoscale phase separation

Functional Polymer Films, First Edition. Edited by Wolfgang Knoll and Rigoberto C. Advincula.
© 2011 Wiley-VCH Verlag GmbH & Co. KGaA. Published 2011 by Wiley-VCH Verlag GmbH & Co. KGaA.

structures of block-copolymers are introduced. (iii) Microscale (>1000 nm) control. As exemplified here, light can be used as powerful tools for structural and orientational controls of various size features, and should play important roles in thin-film technologies in the future.

31.2
Photoalignment of Molecular Aggregates (Nanoscale Regions)

A photochromic and photocrosslinkable polymer films control the alignment of liquid crystalline (LC) materials contacting with the surface. Pioneering work was reported by Ichimura, Seki and coworkers in 1988 [9, 13, 14]. They showed that the *trans–cis* photoisomerization reaction of an azobenzene (Az) monolayer can reversibly switch the alignment of nematic liquid crystals, and proposed a concept of command surface effect (Figure 31.1a). When a photoreactive polymer film is exposed to linearly polarized light (LPL) or oblique irradiation, all the liquid-crystal molecules included in a micrometer-thickness cell align according to the angular selective photoreaction (Figure 31.1b). Photoalignment processes are now widely known, therefore, this chapter does not touch on then further. It is sufficient to mention that, after 20 years since the discovery, the photoalignment technique is now eventually adopted in the industry to produce large screens of liquid-crystal TV display as the replacement for a rubbing alignment process [15].

On the other hand, efforts for photoalignment of other types of organized liquid-crystalline materials have been made over several years. They include lyotropic liquid crystals, a certain type of polysilane [16], and even anisotropic structures of inorganic/organic hybrid materials [17, 18]. The inplane control

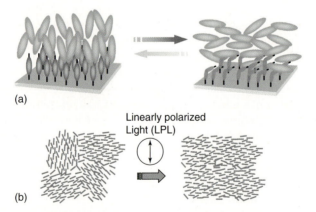

(a)

Linearly polarized
Light (LPL)

(b)

Figure 31.1 (a) Schematic illustration of the command surface for orientational switching of liquid-crystal molecules on a photoresponsive surface of azobenzene monolayer, and (b) a scheme of photoinduced orientation of LC molecular assemblies by linearly polarized light (LPL).

utilizes the angular selective reaction of photoreactive units by irradiation with LPL followed by the ordering due to the cooperative self-assembly [4–6]. In this section, some attempts to align organic/silica nanohybrids by using a photoaligning polymer film are introduced.

31.2.1
Surfactant Aggregate/Silica Nanohybrids

Defined-sized mesoporous metal-oxide materials are synthesized by templating organic molecular assemblies. To date, a vast number of studies have been undertaken to fabricate mesoporous silica materials [19–23] using various surfactant-templated nanostructures. To attain new functions in optics and molecular electronics, and so on, macroscopic uniaxial alignment of the hexagonal columnar mesostructured silica is of importance. In this context, many efforts have been made to align macroscopically the nanohybrids and resulting mesoporous materials by applying external electric [24] and magnetic fields [25], depositions on rubbed polyimide films [26, 27], Langmuir–Blodgett (LB) films [28], and incorporation into a porous alumina membrane [29]. Thus, the immobilization of fluid-ordered lyotropic liquid-crystalline states by the sol-gel condensation processes constitute a central strategy in constructing desired nanostructures.

The photoalignment of surfactant/silica nanohybrid by an Az monolayer is first demonstrated by two-step transfer via photoalignment of polysilane thin film [17, 18, 30, 31]. An alternate simpler way is to use a photocrosslinkable liquid-crystalline polymer film containing a cinnamate unit and a mesogen in the side chain (chemical structure, **1** in Figure 31.2) [32]. The photocrosslinkable polymer as the alignment layer possesses higher stability against heating and solvent exposure and is more favorable for the fixation of the lyotropic liquid-crystalline state and the

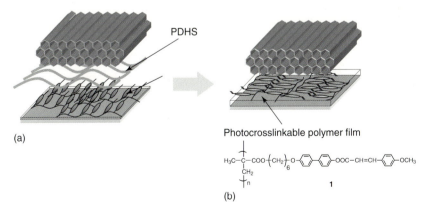

(a)

(b)

Figure 31.2 Photoalignment of silica nanochannels by photoreactive polymer layers based on the photoisomerization of azobenzene monolayer (a) and photodimerization of cinnamoyl units of polymer **1** (b).

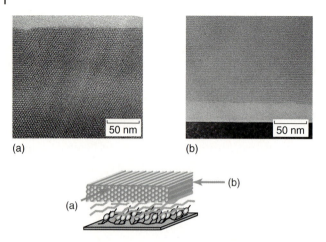

Figure 31.3 Transmission electron microscopic images of photoaligned silica nanochannels. (a) and (b) indicate the directions of observation. (Reproduced from Ref. [30] with permission of The Royal Society of Chemistry.)

siloxane condensation [18, 31]. X-ray diffraction measurements and transmission electron microscopic (TEM) observations (Figure 31.3) both indicate that the surface anisotropy imposed by LPL is transferred to the orientation of the rod-like micelle template, therefore, photoaligned mesochannels silica is obtained after removal of the organic component. Interestingly, even when the dip coating is applied, the channel orientation is predominantly controlled by the direction of the photoaligning polymer layer and not by the lifting direction [18]. Thus, the orienting power from the photoaligned polymer film is stronger than that of the flow orientation during the lifting. In a micropatterning experiment, a resolution of circa 10 μm is obtained [18].

31.2.2
Chromonic Dye Aggregate/Silica Nanohybrids

Another type of lyotropic liquid crystal is chromonic dye aggregates systems [33, 34]. The chromonic liquid crystals are composed of disc-like or plank-like aromatic dye molecules with hydrophilic units at the peripheries, and they can spontaneously self-assemble via π–π stacking interaction and stack face to face to form columnar structures in aqueous solutions. In this section, the first demonstration of the immobilization of columnar structure of the chromonic lyotropic liquid crystal by formation of silica networks is presented for C.I. direct blue 67 (B67) [35–37] (chemical structure shown in Figure 31.4). The resulting hybrid films are found to possess macroscopic inplane alignment of the chromonic structure when the dipping deposition is adopted.

When the silica network is formed simply from tetraethoxysilane (TEOS), the columnar structure B67 aggregates is transformed to a lamella phase. This

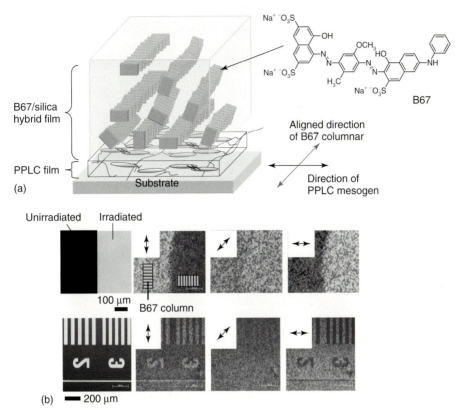

(a)

(b) ▬ 200 μm

Figure 31.4 Schematic illustration of photoaligned nanocolumns of chromonic liquid crystal of B67 embedded in silica (a). In (b), optical microscopic images of photopatterned film observed through a polarizer are indicated. (Reproduced from Ref. [38] with permission of The Royal Society of Chemistry.)

can be ascribed to a destabilization of change repulsion between the anionic nature of both B67 and silica surface (silanol groups). The immobilization of the columnar structure is successfully attained when an appropriate mediating molecule, 2-(2-aminoethoxy)ethanol (AEE), is present [39]. This molecule neutralizes the surface anionic charge of the B67 column and adapts the column into the silica network. The significant role of such mediator molecules in the structural stabilization of lyotropic liquid crystalline (LLC) materials is realized.

A macroscopic uniform inplane alignment of B67 is attained when the dipping method is applied to prepare the hybrid film. The column orientation is parallel to the lifting (dipping) direction, namely, the dye plane is orthogonal to it [39]. When a photocrosslinkable liquid-crystalline polymer film (chemical structure, **1** in Figure 31.2) is employed as the photoalignment layer, the macroscopic photoalignment, and photopatterning of this columnar nanohybrid can be attained. Irradiation with LPL at 313 nm to the film of **1** followed by an appropriate annealing leads to a strong inplane molecular orientation in the film (Figure 31.4). Onto this aligned film, the precursor sol solution containing B67 and TEOS in the presence of an anionic surfactant and AEE is replaced by the static deposition or dipping method. After drying, the chromonic columns hybridized with silica are macroscopically oriented orthogonal to the LPL direction, namely, the orientation of dye molecule being parallel to it. In the dip-coating process, a combination of the photoalignment and flow-induced orientation allowed a clear micropatterning of the column orientations, which can be readily discerned when observed with a polarizing film (images displayed lower in Figure 31.4) [38].

31.2.3
Azobenzene (Az)-Containing Liquid-Crystalline Grafted Polymer Films

The photoalignment of Az mesogens in side-chain polymer in thin films including spin-cast films and LB layers are widely investigated. In these films, the smectic layers are oriented parallel to the substrate plane and the Az mesogens in the normal direction. This section introduces a system in which the orientation is just reversed by anchoring one end of the polymer to the substrate. In recent years, polymer brushes have become fascinating subjects in polymer research. They are defined as dense layers of end-grafted polymer chains confined to a solid surface or interface [40, 41]. Recently, high graft density, and well-defined polymer-brush structures have been developed using controlled radical polymerization, such as atom-transfer radical polymerization (ATRP) [42, 43]. Since ATRP can be performed using various types of monomers, many functional polymer brushes can be designed.

Uekusa *et al.* [44] have synthesized LC polymer brushes bearing an Az mesogenic group in side chains by adopting surface-initiated ATRP (chemical structure **2** in Figure 31.5). UV-visible spectroscopic and X-ray diffraction (XRD) measurements reveal that the periodic layer structure of smectic liquid-crystal phase orients perpendicular to the substrate plane with Az mesogens being aligned parallel to the substrate. Such orientations are observed for a first time only by tethering one terminal to the substrate. The diffraction peak is observed at $2\theta = 2.48°$, which corresponds to a layer spacing (d) of 3.56 nm. In contrast, a diffraction pattern on the spin-cast film of the homopolymer (chemical structure, **3**) is observed only in the out-of-plane direction at $2\theta = 2.73°$ ($d = 3.26$ nm), which is derived from the smectic LC phase.

Since the photoresponsive mesogens are oriented parallel to the substrate plane, this orientation feature enhances the probability of light absorption, leading to the highly ordered inplane photoalignment of Az mesogens when exposed to LPL [45]. In the scheme below in Figure 31.5 displays a schematic illustration of the photoinduced alignment of Az mesogens and smectic layers after irradiation with LPL. The inplane dichroism is most effectively generated at 60 °C, a slightly higher temperature above the glass-transition temperature (T_g) of the polymer. This level of high orientational order is not attained for the spin-cast film with the smectic layers orienting parallel to the substrate. It is expected that the surface-grafted polymer films with highly inplane LPL-responsive mesogenic groups will provide new types

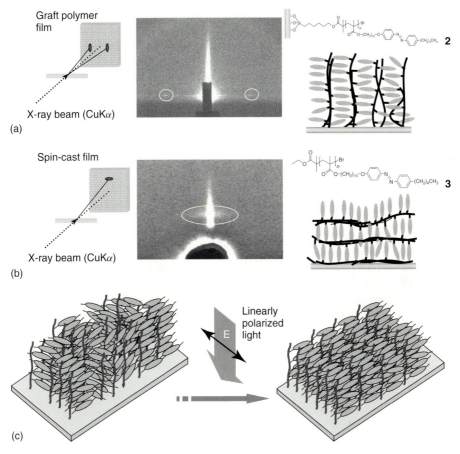

Figure 31.5 The mesogenic group and smectic layer orientations of grafted and spin-cast thin films of a liquid-crystalline Az polymer. Surface-grafted polymer **2** and the corresponding free polymer **3** have the identical chemical structure. In (a) and (b), grazing-angle incidence X-ray diffraction patterns are shown. Scheme (c) indicates the illustration of LPL induced inplane reorientation of the Az mesogenic groups. Due to the parallel orientation of Az, the orientational order in the grafted chain becomes significantly higher than that of the spin-cast film of **3**. (Reproduced from Refs. [44] and [45] with permission of The American Chemical Society.)

of optically functional surfaces, alignment layers for liquid crystals, and smart surfaces that show anisotropic friction properties.

31.3
Block-Copolymer Microphase Separation (MPS) Structure (Mesoscale Region)

Microphase separation (MPS, regular mesostructure formation) of block-copolymers in thin films has been the subject of intensive study from the viewpoint of self-assembly in confined states [46, 47]. From practical viewpoints, such mesostructures have also been receiving considerable attention for fabrication of ever smaller feature sizes than those obtained by the photolithography process. They have potential applications for high-density data-storage media [48, 49], ultrafine filters or membranes [50, 51], templates for metal nanowires [52], and so on. The MPS morphology is generally altered by changing the block length ratio of the different polymer components. When the morphological structure and alignment of MPS is altered by the irradiation procedure, new possibilities in the block-copolymer technology will be developed. This section introduces some attempts to alter the MPS structure by light irradiation.

31.3.1
Photocontrolled Morphological Change of MPS

Photomechanical response is observed in monolayers of Az-containing amphiphilic polymers on a water [53–55] or mica [56, 57] surface. UV and visible light irradiation induces reversible expansion and contraction, respectively, of the monolayer. These motions can be applied to alter the MPS change, when one of the blocks is composed of a photoresponsive segment (area-variable component), the change in the area fraction of the blocks may alter the MPS mesostructures formed in the monolayer. Such light-induced modulations in the monolayer have been actually attained for a triblock-copolymer (chemical structure, **4** in Figure 31.6) [58]. The *trans* and *cis* isomeric states of Az result in mesoscaled dot and strip morphologies, respectively,

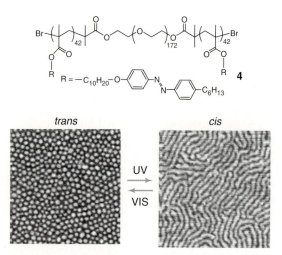

Figure 31.6 Photocontrolled MPS structure of block-copolymer **4** on water. $3 \times 3 \ \mu m$ topological atomic force microscopy (AFM) images are indicated. *Trans*- and *cis*- Az isomers provide the dot and stripe mesostructure, respectively. The morphological change is essentially reversible. (Reproduced from Ref. [58] with permission of The American Chemical Society.)

of the Az-containing domain. The light-induced morphology changes are essentially reversible (Figure 31.6).

This fully tunable behavior of the morphology also provides an important aspect on the chain conformation on water. The middle block of the poly(ethylene oxide) (PEO) chain should adopts predominantly loop conformations rather than bridge ones. This fact coincides with the tendency of segregation rather than interpenetration of polymer chains in the 2D state [59]. This morphological change is also observed on a solid substrate of mica surface when a high humidity condition is provided [60].

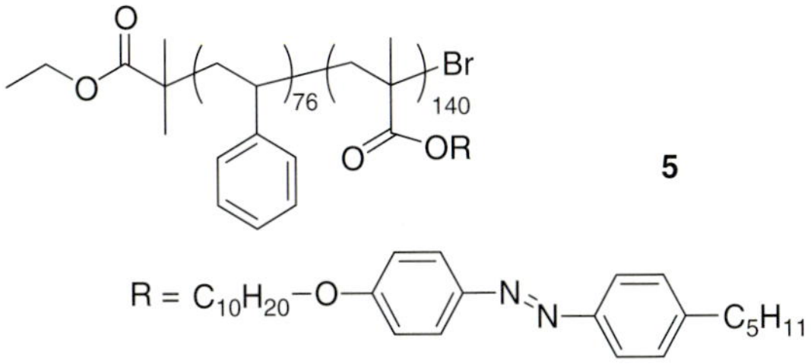

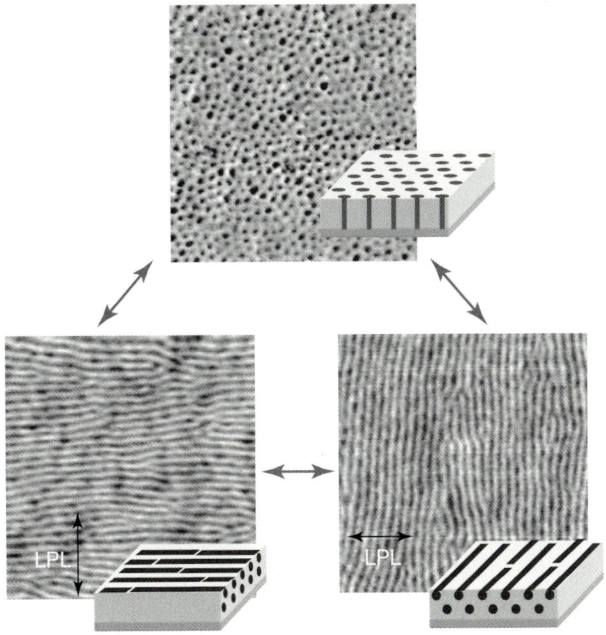

Figure 31.7 Light-directed orientation changes of MPS nanocylinders of polystyrene in a thin film of block-copolymer **5**. The figures indicate 3 × 3 μm phase mode AFM images. The out-of-plane and inplane directions are convertible by photoirradiation with each other. (Reproduced from Ref. [64] with permission of The American Chemical Society.)

31.3.2
Photoalignment of MPS Structure

Many efforts have been made to induce macroscopic alignment of the Materials Research Society (MRS) patterns from the viewpoints of new nanotechnologies [61]. When block-copolymers with LC nature are employed, the regular MPS structure is formed over large areas by the cooperative effect. The importance of liquid crystallinity for large-scale alignment of MPS structure is demonstrated by Iyoda and coworkers [62] by using a series of PEO-*block*-Az liquid-crystalline polymers.

The obvious effect of LPL irradiation to align MPS in block-copolymer films in the inplane direction is demonstrated for a soft PEO-based Az-containing block-copolymer film [63] and a polystyrene (PS)-based block-copolymers [64] possessing higher T_g (chemical structure, **5** in Figure 31.7). The MPS structure of nanocylinders of the light-inert blocks aligned orthogonal to the direction of the electric field vector of the irradiated LPL. The normally oriented nanocylinder in the initial state can be changed to an inplane direction orthogonal to LPL direction. This direction can be further altered to another inplane direction corresponding to the subsequent illumination of LPL. Irradiation with a nonpolarized light in the normal incidence leads to the out-of-plane orientation of the cylinders again. These alignments are convertible with each other (Figure 31.7). For a rigid segment polymer of **5**, an adjustment of annealing temperature is an important factor for the successful alignment. The macroscopically aligned MPS structure evolves at an annealing temperature slightly above the T_g of the PS block and below the smectic to isotropic transition of the LC Az polymer block. It is to be noted that the MPS alignment can be successively altered on-demand in both the inplane and out-of-plane directions [64]. The requirement of illumination above the order–disorder transition temperature of the block-copolymer implies that the erasure of the MPS structure is needed for subsequent evolution of the MPS structure directing another direction. The successful convertible alignment between the inplane and out-of-plane directions strongly suggests this assumption.

31.4
Surface-Relief Formation (Microscale Regions)

31.4.1
Features of Liquid-Crystalline Polymer Materials

Surface-relief gratings (SRGs, regular topological surface modification) formed via irradiation with an interference pattern of coherent argon ion laser beam (488 or 512 nm) were found in 1995 [65, 66] and is perhaps the most interesting target in the current research in Az polymers. A great deal of data has been accumulated rapidly due to its basic phenomenological interest and its technological applications [67, 68]. Another type of photoactivated mass migration systems has been proposed by Seki and coworkers [69] using soft liquid-crystalline polymer systems such as

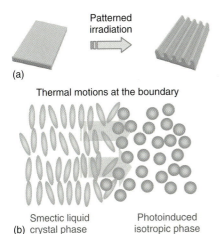

Patterned
irradiation

(a)

Thermal motions at the boundary

Smectic liquid
(b) crystal phase

Photoinduced
isotropic phase

Figure 31.8 Schematic illustrations of photoinduced mass transport to form a surface-relief structure (a) and the transport motions at the boundary between the liquid-crystalline and isotropic phases (b). The transport can be driven by the disparity of the surface tension and fluidity of the two phases. This mechanism can be regarded as phase transition (PT)-type transport motions.

binary component materials and random copolymers containing an oligo(ethylene oxide) (EO) segment [70, 71]. The hybrid films were irradiated with nonpolarized UV light in advance to attain a *cis*-rich photoequilibrated state (UV light treatment). Starting from this state, an argon ion laser beam (488 nm) or a 436-line from a mercury lamp, which induces the isomerization to the *trans* form, is irradiated to the film. The mass migration is completed at surprisingly low dose levels ($< 100 \, \text{mJ cm}^{-2}$), which is 3 orders of magnitude smaller than those required for the conventional amorphous polymer systems. Recent explorations have revealed that the photochemical phase change of the smectic LC to isotropic phase is essential for the migration (Figure 31.8) [72, 73]. This mass-transfer process indicates negligible dependence of polarization of the irradiated light, and further the triplet sensitization from a longer-wavelength absorbing dye can induce the relief formation. Thus, this class of materials can be dubbed as the phase transition (PT)-type [73]. The mass transfer occurs at the boundary between the smectic phase (*trans*-Az rich) and isotropic one (*cis*-Az rich). The motions direct from the smectic to isotropic regions. The driving factor is the thermal process in essence, and seems to be initiated from the disparity in the surface tension and fluidity of the two phases.

31.4.2
Hierarchical Structure Formation in Block-Copolymer Systems

The phototriggered mass migrations have been studied mostly for homopolymers and random copolymers. Recently, a block-copolymer containing PEO and Az polymer (chemical structure, **6** in Figure 31.9) has been subjected to the mass-migration

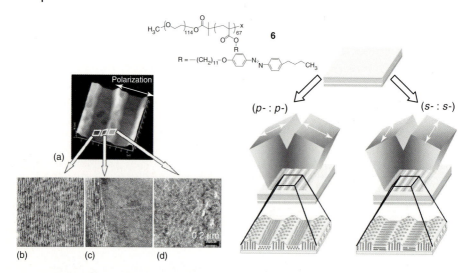

Figure 31.9 Hierarchical structures involved in photo-generated SRG of a block-copolymer. Topographical image (a), phase images at a trough (b), boundary (c), and crest (d) regions. Hierarchical structures of SRG of the block-copolymer generated in different polarization modes. (Reproduced from Ref. [74] with permission of Wiley-VCH.)

process [74]. After the inscription of the relief structure with an interfering laser beam followed by an appropriate annealing procedure, hierarchical structures of MPS, and molecular orientations are involved in the resulting film (Figure 31.9). Two types of holographic irradiation ((*p*-:*p*-) and (*s*-:*s*-)) were performed. When the thickness after the migration is adjusted so as to possess thicknesses above and below 70 nm in the thick and thin regions, respectively, the out-of-plane control is achieved, namely, the cylinders align normal (thicker area) or parallel (thinner area) to the substrate plane depending on the relief thickness [74]. The difference in the holographic irradiation mode of (*p*-:*p*-) and (*s*-:*s*-) leads to contrasting inplane orientations of the laid cylinders in the thin areas. The (*p*-:*p*-) and (*s*-:*s*-) mode interferences provide cylinders oriented parallel and orthogonal to the relief undulations, respectively (Figure 31.9). It is stressed here that a single irradiation leads to the formation of three different hierarchy levels, namely, the molecular orientation (nanometer level), MPS structure (several tens of nanometers), and SRG (micrometers).

31.4.3
Supramolecular Strategy

The Az unit is essential for the photoinduced mass migration, but after the relief formation the existence of this strongly light-absorbing chromophore can be a drawback for optical applications such as a waveguide coupler and a liquid-crystal alignment layer. To overcome this issue, the supramolecular strategy employing hydrogen bonding proposed by Kato and coworkers [75, 76] should be of promising approach.

Zettsu *et al.* [77] synthesized a supramolecular polymer in which an Az-imidazole base is introduced via hydrogen bonding (chemical structure, **7** in Figure 31.10). The complexation is attained with a random copolymer consisting of poly(acrylate) with 4-oxybenzoic acid moiety as the hydrogen-bonding donor part and poly (methacrylate) with a flexible oligo-EO chain as a plasticizing and crosslinking part. The selective extraction of the Azo-imidazole from the film is performed after the chemical crosslinking of the polymer under mild conditions in a vapor phase [70, 71]. The Az unit attached via hydrogen bonding can be readily extracted selectively by ethanol or tetrahydrofuran from the film with retension of the periodic relief feature. For the purpose of crosslinking, the photochemical procedure employing

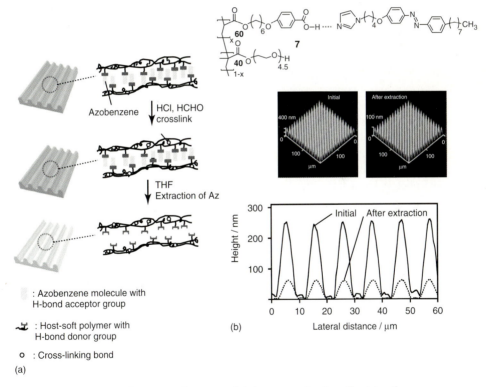

: Azobenzene molecule with
  H-bond acceptor group

: Host-soft polymer with
  H-bond donor group

o : Cross-linking bond

(a)

(b)

Figure 31.10 Schematic illustration of the SRG formation in a hydrogen-bonded supramolecular liquid-crystalline polymer film **7** and the strategy for the selective de-tachment of azobenzene unit (a). Selective extraction of Az unit from the SRG film is shown in (b). Topographical AFM images and their cross-sectional profiles taken from the film after the crosslinking procedure (denoted as initial) and subsequent rinsing with tetrahydrofuran (after extraction) are indicated. (Reproduced from Ref. [77] with permission of Wiley-VCH.)

[2 + 2] dimerization of cinnamate units is an alternative approach [78]. After the photocrosslinking, the relief structure is maintained up to 300 °C.

The strategy of noncovalent attachment of Az side chain has also been reported by Klikovska *et al.* [79]. They adopted the ion-complexation between an Az compound having carboxylic acid and a cationic polyelectrolyte. This approach is appealing from a practical view since both components are commercially available and inexpensive, and therefore, one need not start from tedious synthetic procedures. Furthermore, the processing can be achieved with water, an environmentally friendly solvent.

Quite recently, the same group proposed another system for detaching Az units after SRG formation [80]. The film is prepared by simply spin coating an isocyanate oligomer with 4-aminoazobenzene in the presence of a catalyst, which leads to the formation of a urea unit between the oligomer and the Az unit. After the SRG is formed, the decoloration (detachment of Az unit) is readily achieved by simply heating at 270 °C. The structure remains unchanged at temperatures up to 300 °C.

31.4.4
Organic–Inorganic Hybrid Materials

Organic–inorganic nanohybrids synthesized via the sol-gel reaction containing organic components have been the subject of particular significance in material nanotechnologies, and an explosive number of investigations have been undertaken [19, 81, 82]. The organic–inorganic hybrid materials have superior features possessing both flexibility and functionality of organic components, and chemical, thermal, and mechanical stability of inorganic ones. Despite this large stream of research, however, little effort has been made to fabricate hybrid systems that show photoinduced mass migration. In this respect, Darracq *et al.* [83] reported a pioneering work about 10 years ago using Az-containing silica-based sol-gel system. They succeeded in fabricating the SRGs on the hybrid film by holographic irradiation of an argon ion laser beam, but in this amorphous-type hybrid a vast amount of recording energy as $60–140 \, J \, cm^{-2}$ ($0.6–1.4 \, J \, mm^{-2}$ in the paper) is required [22]. More recently, Kulikovska *et al.* [84] reported a supramolecular sol-gel material system based on the ionic interaction between an oppositely charged Az unit and silica for optical generation of surface-relief structures. The modulation depths of the recorded gratings are large, but also in this type hybrid material, a large amount of recording energy as $30–900 \, J \, cm^{-2}$ is required. The mass transfer based on the PT-type mechanism mentioned above is expected to lead to substantial reduction of light dose required.

Nishizawa *et al.* [85] developed a new type of Az-containing hybrid material that shows liquid-crystalline nature. In this case, titanium oxide (titania) is adopted as the inorganic component that is anticipated to be converted to photofunctional materials such as photocatalyst via removal of the organic component. A liquid-crystalline titania-Az hybrid material (chemical structure, **8** in Figure 31.11) shows a smectic phase (probably smectic A) in the *trans*-Az form, in which organic Az and titania layers are alternatively piled up in the thin film at a period of circa

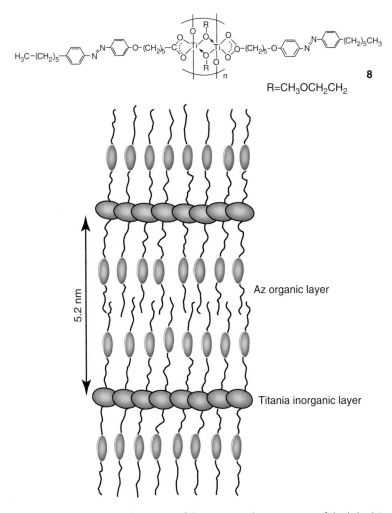

$R=CH_3OCH_2CH_2$

8

Figure 31.11 Schematic illustration of the titania/Az layer structure of the hybrid **8**.

5 nm (scheme in Figure 31.11). The irradiation with 365-nm light brings about the phase transition to the isotropic, just as observed for polymer materials. The mass transfer can be performed by patterned UV irradiation through a photomask of very low energy doses (circa 10^3-10^5 times less than hitherto known sol-gel materials) at temperatures around 130 °C. Since no transfer occurs for the nonliquid-crystalline analog without an appropriate spacer, the importance of the cooperative nature of liquid crystallinity is suggested.

The topological pattern of the hybrid can be converted to pure titanium oxide (Figure 31.12) [86]. The process for removal of organics involves two-step procedures. At the initial step of decomposition of organic components, the relief film is by exposure to UV light at 185 nm of a low-pressure Hg lamp for 20 min

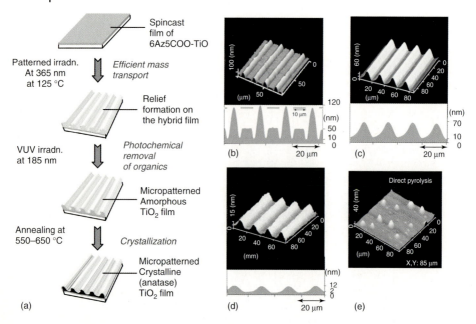

Figure 31.12 A flow chart of the process for removal of Az from the relief film of **8** (a). Topographical AFM images and cross-sectional height profiles are shown for the photoinscribed **8** film at 125 °C through a 10-μm line-and-space photomask (b), the photochemically decomposed film using a low-pressure Hg lamp for 20 min (c), the pyrolized film at 650 °C (d), and directly pyrolyzed film at 550 °C without going through the photodecomposition (e). (Reproduced from Ref. [86] with permission of The Royal Society of Chemistry.)

(Figure 31.12b,c). The photochemical decomposition process modified the morphological characteristics from the two-level-height relief to a simply undulated one with circa 70 nm modulation depth (Figure 31.12c). Next, this film is annealed at 550 °C for 1 h and subsequently at 650 °C for 1 h to achieve full pyrolysis. The periodical feature of the relief structure is essentially retained after the annealing. An attempt at the direct pyrolysis procedure for film B without the photochemical treatment was also made. After heating film B at 550 °C, the periodical relief structure is fully collapsed to give an irregular dot-like morphology (Figure 31.12e). X-ray analysis and TEM observation reveal that the anatase-type crystal grains are obtained. Because large surface areas are provided, the obtained film may be favorably applied to photofunctional devices such as solar cells.

31.5
Summary

Due to the strong cooperativity in liquid-crystalline materials, the introduction of a photoresponsive unit (typically an Az moiety) into molecular or polymer systems gives rise to large amplification photoresponses. Such effects can be exemplified at a wide range of size features, including the molecular nanoscales (photoresponsive surface alignment of liquid-crystalline materials, effective molecular photoalignment in a surface-grafted polymer film), meso (10–100 nm)-scale (photoalignment of MPS structures of block-copolymers), and micrometer scale (mass migrations).

Nanoscale. The surface photoalignment method using "command surface" is now extending to a branch of organic–inorganic nanohybrid systems. Thus, surface-mediated processes provide new opportunities to exert micropatterning of simple components to various composite systems. Such approaches would be of particular use in fabrication of optical devices. Surface-grafted photoresponsive polymer film can be a new target of liquid-crystalline film research. The orientation of the Az mesogenic group strongly influences the photoresponse behavior and sensitivity. Since the planar-arranged Az mesogens should be more sensitive to LPL, fabrications of highly sensitive photorecording media may be realized. Furthermore, smart photoresponse functions such as photoswitchable anisotropic friction properties may be anticipated.

Mesoscale. With respect to the alignment control, the marked research advance in the block-copolymer systems is to be emphasized. Here, the photoalignment is not limited at the molecular level but amplified to regulate the MPS structure of the larger feature size. Understanding the interplay between different-size hierarchies with retention of dynamic features should be a significant issue to create future soft materials. The nanopatterned surface of block-copolymer films can be applied as a template surface to fabricate various functional materials, including biological macromolecules, conjugated polymers, inorganic materials, metals, and so forth. When such patterns are tuned by light, more sophisticated structures that will be suited for device fabrication would be realized.

Microscale. In the mass-migration systems introduced in this chapter, the motions are basically driven by thermal self-assembly processes of the soft materials. The relief formation occurring at the boundary of the liquid-crystalline/isotropic phases is most likely to be the consequence the disparity of surface tension and fluidity. This situation is in marked contrast with the migration mechanism in the conventional amorphous materials. As shown here, the detatchment of Az chromophore (bleaching the color) will be a promising strategy to expand optical applications of the photogenerated relief structure. In the organic–inorganic system, the detachment of Az is readily achieved by photodecomposition followed by pyrolysis. The annealing provides the formation of anatase crystal involved in the relief structure. This can be a new proposal of the photodirected processing of functional inorganic materials.

The topics dealt with in this chapter are mostly in 2D systems, but the ultimate goal would be to create sophisticated dynamic 3D systems similar to the biological ones. It is anticipated that accumulations on the various types light-directed motions at various scales of photoresponsive films in 2D states will give proper clues to future directions of materials processing and applications.

References

1. Irie, M. (ed.) (2000) Special issue of photochromism: memories and switches. *Chem. Rev.*, **100**, 1685–1890.
2. Sekkat, Z. and Knoll, W. (eds) (2002) *Photoreactive Organic Thin Films*, Academic Press, San Diego.
3. Zhao, Y. and Ikeda, T. (2009) *Smart Light-Responsive Materials – Azobenzene-Containing Polymers and Liquid Crystals*, John Wiley & Sons, Inc., Hoboken.
4. Natansohn, A. and Rochon, P. (2002) Photoinduced motions in azo-containing polymers. *Chem. Rev.*, **102**, 4139–4175.
5. Seki, T. (2004) Dynamic photoresponsive functions in organized layer systems comprised of azobenzene-containing polymers. *Polym. J.*, **36**, 435–454.
6. Seki, T. (2007) Smart photoresponsive polymer systems organized in two dimensions. *Bull. Chem. Soc. Jpn.*, **80**, 2084–2109. (Award Account).
7. Ikeda, T. (2003) Photomodulation of liquid crystal orientations for photonic applications. *J. Mater. Chem.*, **13**, 2037–2057.
8. Ikeda, T., Mamiya, J., and Yu, Y. (2007) Photomechanics of liquid-crystalline elastomers and other polymers. *Angew. Chem. Int. Ed.*, **46**, 506–528.
9. Ichimura, K. (2000) Photoalignment of liquid crystal systems. *Chem. Rev.*, **100**, 1847–1873.
10. Schadt, M., Schmitt, K., Kozinkov, V., and Chigrinov, V. (1992) Surface-induced parallel alignment of liquid crystals by linearly polymerized photopolymer. *Jpn. J. Appl. Phys.*, **31**, 2155–2164.
11. Gibbons, W.M., Shannon, P.J., Sun, S.-T., and Swetlin, B.J. (1991) Surface-mediated alignment of nematic liquid crystals with polarized laser light. *Nature*, **351**, 49–50.
12. Zebger, I., Rutloh, M., Hoffmann, U., Stumpe, J., Siesler, H.W., and Hvilsted, S. (2003) Photoorientation of a liquid-crystalline polyester with azobenzene side groups: effects of irradiation with linearly polarized red light after photochemical pretreatment. *Macromolecules*, **36**, 9373–9382.
13. Ichimura, K., Suzuki, Y., Seki, T., Hosoki, A., and Aoki, K. (1988) Reversible change in alignment mode of nematic liquid crystals regulated photochemically by "command surfaces" modified with an azobenzene monolayer. *Langmuir*, **4**, 1214–1216.
14. Seki, T., Sakuragi, M., Kawanishi, Y., Tamaki, T., Fukuda, R., and Ichimura, K. (1993) "Command surfaces" of Langmuir-Blodgett films. Photoregulation of liquid crystal alignment by molecularly tailored surface azobenzene layers. *Langmuir*, **9**, 211–218.
15. Otani, T. (2009) Sharp commercializes photo-alignment technology. Nikkei Electronics Asia.
16. Seki, T., Fukuda, K., and Ichimura, K. (1999) Photocontrol of polymer chain organization using a photochromic monolayer. *Langmuir*, **15**, 5098–5101.
17. Kawashima, Y., Nakagawa, M., Seki, T., and Ichimura, K. (2002) Photoorientation of mesostructured silica via hierarchical multiple transfer. *Chem. Mater.*, **14**, 2842–2844.
18. Fukumoto, H., Nagano, S., Kawatsuki, N., and Seki, T. (2006) Photoalignment behavior of mesoporous silica thin films synthesized on a photo-crosslinkable polymer film. *Chem. Mater.*, **18**, 1226–1234.
19. Lu, G.Q. and Zhao, X.S. (eds) (2004) *Nanoporous Materials: Science and Engineering*, vol. 4, Imperial College Press, Singapore.
20. Kresge, C.T., Leonowicz, M.E., Roth, W.J., Vartuli, J.C., and Beck, J.S. (1992) Ordered mesoporous molecular sieves synthesized by a liquid-crystal template mechanism. *Nature*, **359**, 710–712.
21. Yanagisawa, T., Shimizu, T., Kuroda, K., and Kato, C. (1990) The preparation of alkyltriinethylaininonium– kaneinite complexes and their conversion to microporous materials. *Bull. Chem. Soc. Jpn.*, **63**, 988–992.
22. Huo, Q., Leon, R., Petroff, P.M., and Stucky, G.D. (1995) Mesostructure design with gemini surfactants: supercage formation in a three-dimensional hexagonal array. *Science*, **268**, 1324–1327.

23. Bagshaw, S.A., Prouzet, E., and Pinnavaia, T.J. (1995) Templating of mesoporous molecular sieves by non-ionic polyethylene oxide surfactants. *Science*, **269**, 1242–1244.

24. Kuraoka, K., Tanaka, Y., Yamashita, M., and Yazawa, T. (2004) Preparation of a membrane with aligned nanopores using an organic– inorganic hybrid technique. *Chem. Commun.*, 1198–1199.

25. Yamauchi, Y., Sawada, M., Sugiyama, A., Osaka, T., Sakka, Y., and Kuroda, K. (2006) Magnetically induced orientation of mesochannels in 2D-hexagonal mesoporous silica films. *J. Mater. Chem.*, **16**, 3693–3700.

26. Miyata, H. and Kuroda, K. (1999) Preferred alignment of mesochannels in a mesoporous silica film grown on a silicon (110) surface. *J. Am. Chem. Soc.*, **121**, 7618–7624.

27. Miyata, H., Kawashima, Y., Itoh, M., and Watanabe, M. (2005) Preparation of a mesoporous silica film with a strictly aligned porous structure through a sol-gel process. *Chem. Mater.*, **17**, 5323–5327.

28. Miyata, H. and Kuroda, K. (1999) Alignment of mesostructured silica on a Langmuir-Blodgett film. *Adv. Mater.*, **11**, 1448–1452.

29. Yamaguchi, A., Uejo, F., Yoda, T., Uchida, T., Tanamura, Y., Yamashita, T., and Teramae, N. (2004) Self-assembly of a silica–surfactant nanocomposite in a porous alumina membrane. *Nature Mater.*, **3**, 337–341.

30. Kawashima, Y., Nakagawa, M., Ichimura, K., and Seki, T. (2004) Photo-orientation of mesoporous silica materials via transfer from azobenzene-containing polymer monolayer. *J. Mater. Chem.*, **14**, 328–335.

31. Fukumoto, H., Nagano, S., Kawatsuki, N., and Seki, T. (2005) Photo-orientation of mesoporous silica thin films on photo-crosslinkable polymer film. *Adv. Mater.*, **17**, 1035–1039.

32. Kawatsuki, N., Goto, K., Kawakami, T., and Yamamoto, T. (2002) Reversion of alignment direction in the thermally enhanced photoorientation of photo-cross-linkable polymer liquid crystal films. *Macromolecules*, **35**, 706–713.

33. Vasilevskaya, A.S., Generalova, E.V., and Sonin, A.S. (1989) Chromonic mesophases. *Russ. Chem. Rev.*, **58**, 904–916.

34. Lydon, J.E. (2004) Chromonic mesophases. *Curr. Opin. Colloid Interface Sci.*, **8**, 480–490.

35. Ichimura, K., Momose, M., Kudo, K., Akiyama, H., and Ishizuki, N. (1995) Surface-assisted photolithography to form anisotropic dye layers as a new horizon of command surfaces. *Langmuir*, **11**, 2341–2343.

36. Ichimura, K., Fujiwara, T., Momose, M., and Matsunaga, D. (2002) Surface-assisted photoalignment control of lyotropic liquid crystals. Part 1. Characterisation and photoalignment of aqueous solutions of a water-soluble dye as lyotropic liquid crystals. *J. Mater. Chem.*, **12**, 3380–3386.

37. Ruslim, C., Matsunaga, D., Hashimoto, M., Tamaki, T., and Ichimura, K. (2003) Structural characteristics of the chromonic mesophase of C.I. direct blue 67. *Langmuir*, **19**, 3686–3691.

38. Hara, M., Nagano, S., Kawatsuki, N., and Seki, T. (2008) Photoalignment and patterning of chromonic/silica nanohybrid on photocrosslinkable polymer thin film. *J. Mater. Chem.*, **18**, 3259–3263.

39. Hara, M., Nagano, S., Mizoshita, N., and Seki, T. (2007) Chromonic/silica nanohybrids. synthesis and macroscopic alignment. *Langmuir*, **23**, 12350–12355.

40. Advincula, R.C., Brittain, W.J., Caster, K.C., and Rühe, J. (eds) (2004) *Polymer Brushes*, Wiley-VCH Verlag GmbH.

41. Milner, S.T. (1991) Polymer brushes. *Science*, **251**, 905–914.

42. Tsujii, Y., Ohno, K., Yamamoto, S., Goto, A., and Fukuda, T. (2006) Structure and properties of high-density polymer brushes prepared by surface-initiated living radical polymerization. *Adv. Polym. Sci.*, **197**, 1–45.

43. Edmondoson, S., Osborne, V.L., and Huck, W.T.S. (2004) Polymer brushes via surface initiated polymerizations. *Chem. Soc. Rev.*, **33**, 14.

44. Uekusa, T., Nagano, S., and Seki, T. (2007) Unique molecular orientation in a smectic liquid crystalline film attained by surface-initiated graft polymerization. *Langmuir*, **23**, 4642–4645.

45. Uekusa, T., Nagano, S., and Seki, T. (2009) Highly ordered in-plane photoalignment attained by the brush architecture of liquid crystalline azobenzene polymer. *Macromolecules*, **42**, 312–318.

46. Green, P.F. and Limary, R. (2001) Block copolymer thin films: pattern formation and phase behavior. *Adv. Colloid Interface Sci.*, **94**, 53–81.

47. Krausch, G. (1995) Surface induced self-assembly in thin polymer films. *Mater. Sci. Eng.*, **R14**, 1–94.

48. Park, C., Yoon, J., and Thomas, E.L. (2003) Enabling nanotechnology with self-assembled block copolymer patterns. *Polymer*, **44**, 6725–6760.

49. Park, M., Harrison, C., Chaikin, P.M., Register, R.A., and Adamson, D.H. (1997) Block copolymer lithography: periodic arras of ~10^{11} holes in 1 square centimeter. *Science*, **276**, 1401–1404.

50. Widawski, G., Rawiso, M., and François, B. (1994) Self-organized honeycomb morphology of star-polymer polystyrene films. *Nature*, **369**, 387–389.

51. Shidorenko, A., Tokaref, I., Minko, S., and Stamm, M. (2003) Ordered reactive nanomembranes/nanotemplates from thin films of block copolymer supramolecular assembly. *J. Am. Chem. Soc.*, **125**, 12211–12216.

52. Turn-Albrecht, T., Schotter, J., Kästle, G.A., Emley, N., Shibauchi, T., Krusin-Elbaum, L., Guarini, K., Black, C.T., Tuominen, M.T., and Russell, T.P. (2000) Ultrahigh-density nanowire arrays grown in self-assembled diblock copolymer templates. *Science*, **290**, 2126–2129.

53. Seki, T., Sekizawa, H., Morino, S., and Ichimura, K. (1998) Inherent and cooperative photomechanical motions in monolayers of an azobenzene containing polymer at the air-water interface. *J. Phys. Chem. B*, **102**, 5313–5321.

54. Kago, K., Fuerst, M., Matsuoka, H., Yamaoka, H., and Seki, T. (1999) Direct observation of photoisomerization of polymer monolayer on water surface by X-ray reflectometry. *Langmuir*, **15**, 2237–2240.

55. Ohe, C., Kamijyo, H., Arai, M., Adachi, M., Miyazawa, H., Ito, K., and Seki, T. (2008) Sum frequency generation spectroscopic study on photoinduced isomerization of a poly(vinyl alcohol) containing azobenene side chain at the air water interface. *J. Phys. Chem. C*, **112**, 172–178.

56. Seki, T., Kojima, J., and Ichimura, K. (2000) Multifarious photoinduced morphologies in monomolecular films of azobenzene side chain polymer on mica. *Macromolecules*, **33**, 2709–2717.

57. Seki, T., Kojima, J., and Ichimura, K. (1999) Light-driven dot films consisting of single polymer chain. *J. Phys. Chem. B*, **103**, 10338–10340.

58. Kadota, S., Aoki, K., Nagano, S., and Seki, T. (2005) Photocontrolled microphase separation of a block copolymer in two dimensions. *J. Am. Chem. Soc.*, **127**, 8266–8267.

59. Sato, N., Ito, S., and Yamamoto, M. (1998) Molecular weight dependence of shear viscosity of a polymer monolayer: evidence for the lack of chain entanglement in the two-dimensional plane. *Macromolecules*, **31**, 2673–2675.

60. Kadota, S., Aoki, K., Nagano, S., and Seki, T. (2006) Morphological conversions of nanostructures in monolayers of an ABA triblock copolymer having azobenzene moiety. *Colloids Surf. A*, **284/285**, 535–541.

61. Lazzari, M. and De Rosa, C. (2006) *Block Copolymers in Nanoscience*, Chapter 9, Wiley-VCH Verlag GmbH, pp. 191–231.

62. Tian, Y., Watanabe, K., Kong, X., Abe, J., and Iyoda, T. (2002) Synthesis, nanostructures, and functionality of amphiphilic liquid crystalline block copolymers with azobenzene moieties. *Macromolecules*, **35**, 3739–3747.

63. Yu, H.F., Iyoda, T., and Ikeda, T. (2006) Photoinduced alignment of nanocylinders by supramolecular cooperative motions. *J. Am. Chem. Soc.*, **128**, 11010–11011.

64. Morikawa, Y., Kondo, T., Nagano, S., and Seki, T. (2007) 3D photoalignment and patterning of microphase separated nanostructure in polystyrene-based block copolymer. *Chem. Mater.*, **19**, 1540–1542.

65. Rochon, P., Batalla, E., and Natansohn, A. (1995) Optically induced surface gratings on azoaromatic polymer films. *Appl. Phys. Lett.*, **66**, 136–138.

66. Kim, D.Y., Tripathy, S.K., Li, L., and Kumar, J. (1995) Laser-induced holographic surface relief gratings on nonlinear optical polymer films. *Appl. Phys. Lett.*, **66**, 1166–1168.

67. Viswanathan, N.K., Kim, D.K., Bian, S., Williams, J., Liu, W., Li, L., Samuelson, L., Kumar, J., and Tripathy, S.K. (1999) Surface relief structures on azo polymer films. *J. Mater. Chem.*, **9**, 1941–1955.

68. Yager, K.J. and Barrett, C.J. (2001) All-optical patterning of azo polymer films. *Curr. Opin. Solid State Mater. Sci.*, **5**, 487–494.

69. Ubukata, T., Seki, T., and Ichimura, K. (2000) Surface relief gratings in host-guest supramolecular materials. *Adv. Mater.*, **12**, 1675–1678.

70. Zettsu, N., Ubukata, T., Seki, T., and Ichimura, K. (2001) Soft crosslinkable azo polymer for rapid surface relief formation and persistent fixation. *Adv. Mater.*, **13**, 1693–1697.

71. Zettsu, N. and Seki, T. (2004) Highly efficient photogeneration of surface relief structure and its immobilization in cross-linkable liquid crystalline azobenzene polymers. *Macromolecules*, **37**, 8692–8698.

72. Zettsu, N., Ogasawara, T., Arakawa, R., Nagano, S., Ubukata, T., and Seki, T. (2007) Highly photosensitive surface relief gratings formation in a liquid crystalline azobenzene polymer: new implications for the migration process. *Macromolecules*, **40**, 4607–4613.

73. Seki, T. (2006) Photoresponsive self-assembly motions in polymer thin films. *Curr. Opin. Solid State Mater. Sci.*, **10**, 241–248.

74. Morikawa, Y., Nagano, S., Watanabe, K., Kamata, K., Iyoda, T., and Seki, T. (2006) Optical alignment and patterning of nanoscale microdomains in a block copolymer thin film. *Adv. Mater.*, **18**, 883–886.

75. Kato, T. and Frechet, J.M.J. (1989) New approach to mesophase stabilization through hydrogen-bonding molecular interactions in binary mixtures. *J. Am. Chem. Soc.*, **111**, 8533–8534.

76. Kato, T. (2002) Self-assembly of phase-segregated liquid crystal structures. *Science*, **295**, 2414–2418.

77. Zettsu, N., Ogasawara, T., Mizoshita, N., Nagano, S., Ubukata, T., and Seki, T. (2008) Photo-triggered surface relief gratings in supramolecular liquid crystalline polymer system with detachable azobenzene unit. *Adv. Mater.*, **20**, 516–531.

78. Li, W., Nagano, S., and Seki, T. (2009) Photo-crosslinkable liquid-crystalline azo-polymer for surface relief gratings and persistent fixation. *New J. Chem.*, **33**, 1343–1348.

79. Klikovska, O., Goldenberg, L.M., and Stumpe, J. (2007) Supramolecular azobenzene-based materials for optical generation of microstructures. *Chem. Mater.*, **19**, 3343–3348.

80. Goldenberg, L.M., Kulikovsky, L., Kulikovska, O., and Stumpe, J. (2009) New materials with detachable azobenzene: effective, colourless and extremely stable surface relief gratings. *J. Mater. Chem.*, **19**, 8068–8071.

81. Hoffmann, F., Cornelius, M., Morell, J., and Froba, M. (2006) Silica-based mesoporous organic-inorganic hybrid materials. *Angew. Chem., Int. Ed.*, **45**, 3216–3251.

82. Fujita, S. and Inagaki, S. (2008) Self-organization of organosilica solids with molecular-scale and mesoscale periodicities. *Chem. Mater.*, **20**, 891–908.

83. Darracq, B., Chaput, F., Lahlil, K., Levy, Y., and Boilot, J.P. (1998) Photoinscription of surface relief gratings on azo-hybrid gels. *Adv. Mater.*, **10**, 1133–1136.

84. Kulikovska, O., Goldenberg, L.M., Kulikovsky, L., and Stumpe, J. (2008) Smart ionic sol-gel-based azobenzene materials for optical generation of microstructures. *Chem. Mater.*, **20**, 3528–3534.

85. Nishizawa, K., Nagano, S., and Seki, T. (2009) Novel liquid crystalline organic-inorganic hybrid for highly sensitive photoinscriptions. *Chem. Mater.*, **21**, 2624–2631.

86. Nishizawa, K., Nagano, S., and Seki, T. (2009) Micropatterning of titanium oxide via phototactic motions. *J. Mater. Chem.*, **19**, 7191–7194.

32
Thin-Film Applications of Electroactive Polymers

Jennifer A. Irvin and Katie Winkel

32.1
Introduction

32.1.1
Background

Electroactive polymers (EAPs) have been the subject of considerable research since Heeger, MacDiarmid, and Shirakawa won the Nobel Prize in Chemistry in 2000 for their work with those materials. Of principle interest is the potential of EAPs for their use in a wide variety of applications, including light-emitting diodes, sensors, electrochromics (ECs), charge storage, photovoltaics (PVs), corrosion inhibition, electromechanical actuators, and static dissipation. While some of these applications may involve bulk EAPs, thin films of EAPs are by far more common.

EAPs are conjugated polymers possessing π bonds that can be delocalized along the polymer backbone. This class of polymers includes intrinsically/inherently conducting polymers (ICPs), which are polymers that conduct electricity without the need for conductive fillers such as carbonaceous materials or metals. EAPs also include semiconducting polymers such as poly(p-phenylene vinylene) (PPV) [1]. Conducting polymers not included in the EAPs are filled polymers (such as polypropylene filled with carbon black) and strictly ionically conducting polymers (such as poly(ethylene oxide) and Nafion®); these classes of polymers will not be addressed in this chapter.

EAPs can generally be switched between two or more stable oxidation and reduction (redox) states (Figure 32.1), giving rise to changes in properties including conductivity, color, and volume [2]. The reversible changes in these properties have led to applications including electrochromics, charge storage, actuators, and sensors.

In their neutral form, EAPs are either insulating or semiconducting. The neutral polymers are converted to the doped, electrically conducting form by addition of electrons (reduction to the n-doped state) or removal of electrons (oxidation to the p-doped state.) During p-doping, anions from the electrolyte compensate for the positive charges formed along the polymer chain. Conversely, during n-doping,

Functional Polymer Films, First Edition. Edited by Wolfgang Knoll and Rigoberto C. Advincula.
© 2011 Wiley-VCH Verlag GmbH & Co. KGaA. Published 2011 by Wiley-VCH Verlag GmbH & Co. KGaA.

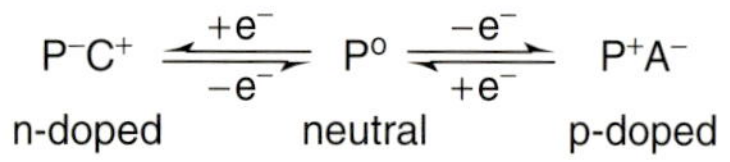

$$P^-C^+ \underset{-e^-}{\overset{+e^-}{\rightleftharpoons}} P^0 \underset{+e^-}{\overset{-e^-}{\rightleftharpoons}} P^+A^-$$

n-doped   neutral   p-doped

Figure 32.1 Electroactive polymers (indicated here by P) can be reversibly switched between their doped state(s) and their neutral state; some polymers both p-dope and n-dope, while others readily undergo only one reversible doping process.

cations from the electrolyte compensate for the negative charges formed along the polymer chain.

32.1.2
Common EAPs

The field of EAPs began with the discovery that polyacetylene (PA) (Figure 32.2a) is highly conductive in the doped state [3]. PA has so far been found to be of little practical use due to its poor environmental stability and poor processability [4]. EAPs with much better stabilities have since been synthesized; it is these polymers that are potentially useful in a wide variety of applications. Polyaniline (PANI, Figure 32.2b), poly(p-phenylene) (PPP, Figure 32.2c), PPV (Figure 32.2d), poly(9,9-dialkylfluorenes) (PF, Figure 32.2e), and the polyheterocycles polypyrrole (PPy, Figure 32.2f), polythiophene (PT, Figure 32.2g), poly(3,4-ethylenedioxythiophene) (PEDOT, Figure 32.2h), and poly(3,4-propylenedioxythiophene) (PProDOT, Figure 32.2i) have all been extensively studied.

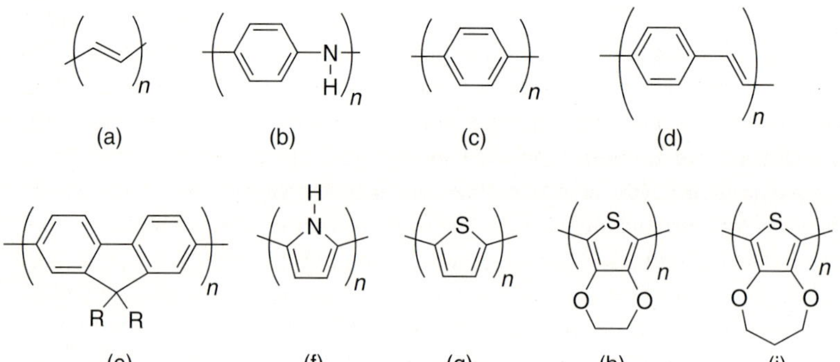

Figure 32.2 Structures of some common EAPs: poly-acetylene (PA, a), polyaniline (PANI, b), poly(p-phenylene) (PPP, c), poly(p-phenylenevinylene) (PPV, d), poly(9,9-dialkylfluorenes) (PF, e), polypyrrole (PPy, f), poly-thiophene (PT, g), poly(3,4-ethylenedioxythiophene) (PEDOT, h), and poly(3,4-propylenedioxythiophene) (PProDOT, i).

32.1.3
Polymer Solubility and Processing

Due to the conjugated, rigid-rod nature of the EAPs shown in Figure 32.2, solubilities and processabilities tend to be quite poor. All of these polymers have been modified to incorporate substituents that impart solubility and/or change electronic properties. In many cases, ease of processing is the one characteristic that makes EAPs more desirable than their inorganic counterparts. When soluble EAPs can be formed, the polymers can be fabricated into devices using solution-based techniques such as spin casting, spray casting, gravure coating (in which ink is spread on a template image and transferred to substrate), and inkjet printing (examples are given below).

Incorporation of alkyl or alkyl ether substituents is commonly employed to impart solubility. With linear alkyl substituents greater than four carbons long, solution disorder is increased (solubility is increased), but solid-state order is increased (alignment of the side chains causes conjugation lengths to increase, improving electronic properties) [5]. However, sterically bulky side chains (such as branched alkyl groups) cause steric twisting of the backbone, reducing conjugation lengths [6]. There is often a slight impact of this functionalization on electronic properties of the resultant polymers, with derivatized polymers exhibiting minor decreases in conductivity [7].

32.2
Applications

32.2.1
Field Effect Transistors

32.2.1.1   Background
A field effect transistor (FET) behaves as a capacitor with a conducting channel between a source and a drain electrode [8]. The amount of electric field is applied to the gate electrode controls the amount of charge carriers flowing through the system. Carriers can be either holes (in the case of p-type semiconductors) or electrons (in the case of n-type semiconductors). FETs are used to modulate current in electrical circuits, and can be applied to devices such as flexible displays, radio-frequency identification tags, memories, and electrochemical sensors [9] (use in sensors will be discussed in more detail later).

While silicon-based FETs dominate the field, hole mobilities of many p-type organic semiconductors are higher than that of silicon ($1.0 \, cm^2/V \, s$) [10]. EAPs also have other advantages over traditional semiconductors for use in FETs; they can be fabricated using simple processes at low cost, they are able to cover large areas, and they are lightweight and flexible [9, 11–14]. The main limitation imposed by the use of EAPs as the active layer is that EAPs typically require higher voltages for switching than inorganic devices; however, the effect can be minimized by careful selection

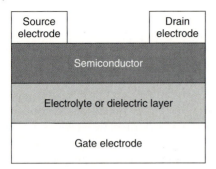

Source electrode		Drain electrode
Semiconductor		
Electrolyte or dielectric layer		
Gate electrode		

Figure 32.3 Schematic of a typical field effect transistor (FET).

of the conducting layer and/or the dielectric [12]. While the semiconducting nature of many neutral EAPs makes them promising FET materials [15], the use of doped, conducting EAPs as the active layer has also been explored [16].

Although there are variations in the way organic field effect transistors (OFETs) are fabricated, there is always a dielectric (active) layer as well as source, drain, and gate electrodes (see Figure 32.3) [11, 12, 17]. The source electrode is where the current to be modulated enters the device. If current is allowed to pass through, it will leave the FET at the drain. The gate electrode acts as part of a gate or switch to determine how much current will flow between the source and drain. The gate and electrolyte layer are components of another circuit, which can be described as an ionic circuit [11]. A change in current in the gate electrode results in a rearrangement of charges in the electrolyte layer, creating a potential gradient. The concentration of mobile charges at the interface with the semiconductor allows current to flow through the semiconductor and completes the electrical circuit between the source and drain [11, 17]. Metrics of principal interest when evaluating OFET design are operating voltage (the lower the better), carrier mobility (the higher the better), and on/off current ratio (the larger the better).

32.2.1.2 Materials for FETs

The combination of high carrier mobilities and ease of fabrication makes OFETs promising alternatives to inorganic FETs. Carrier mobilities of many EAPs have been investigated for use in OFETs. An excellent review of organic materials for OFETs is available, with some attention being given to electroactive oligomers and polymers [10].

The majority of EAPs used as semiconductors in OFETs are as p-type, hole carriers because of their high highest occupied molecular orbital (HOMO) levels and electron-donating ability. PPy, PT, PEDOT:PSS (where PSS is poly(styrene sulfonate)), PANI and their derivatives all demonstrate a field effect and can be used as p-type semiconductor layers in OFETs [15]. An all-polymer FET has been fabricated using PEDOT:PSS electrodes (circa 600 nm thick) and a PPy layer (no thickness given) as the semiconductor [13]. This FET allowed current to pass through when no voltage was applied and gradually turned off the current as the voltage applied to the gate electrode increased.

In an attempt to improve the mobility of the EAP active layer, vinylene bridges have been incorporated into regioregular PTs [9]. The resulting OFETs provided high performance in terms of carrier mobility and switching compared to those with low regioregularity and no conjugated bridges. PEDOT:PSS was used as the active layer in an OFET to study the effects of the thickness of the dielectric layer as well as the environment [18]; it was found that both water and oxygen are required for optimal device performance. Monolayer films (circa 20 Å) of poly(3-hexylthiophene) (P3HT) have been used as the active layer in OFETs to study the effect of film thickness and orientation on field effect mobility [19]; thicker films (circa 200 Å) exhibit 10–100 times greater field effect mobility than those of the monolayers.

Stability has been an issue with n-type EAPs for use as electron carriers in OFETs. The reaction of organic anions with water and oxygen is the main reason for instability in air [20]. Electron-accepting molecules with low HOMO levels are desirable for n-type semiconductor materials. A recent review summarizes attempts to prepare n-type organic semiconductors with high electron mobilities [21] and lists important considerations for design of good n-type semiconductors. First, a large electron affinity is needed to stabilize the materials against reaction with water and oxygen. Incorporation of strongly electron withdrawing groups (fluoro, cyano, alkanoyl, etc.) raises electron affinity. Next, intermolecular overlap should be maximized for optimal electron mobility; extended conjugation groups such as perylene- and naphthylene-bis(dicarboximide) provide excellent overlap. Finally, solution processability is key for commercialization; incorporation of fluoroalkyl, cyano, and alkyl groups improves solubility. Performance can also be strongly influenced by processing parameters; very high purity materials and continuous, oriented films are required.

The n-type semiconducting polymer with the highest electron mobility (up to $0.85 \, cm^2/Vs$) reported to date is a naphthylene-bis(dicarboximide)/bithiophene copolymer bearing solubilizing substituents [22]. The solution-processable nature of this polymer allowed the researchers to fabricate a variety of devices using spin casting, gravure printing, and inkjet printing. The devices were stable in air, showing little performance degradation over 10 weeks. The authors attribute the excellent mobilities to the very lowest unoccupied molecular orbital (LUMO) ($-4.0 \, eV$), electron-poor nature of this polymer.

One group found that low electron mobilities are typically due to electron trapping by residual silanol hydroxyl groups on the commonly used SiO_2 dielectric layer [23]. By replacing SiO_2 with an organic, hydroxyl-free dielectric layer, they were able to improve electron mobilities of polyfluorene derivatives and PPV derivatives. Electron mobilities were improved (to between 10^{-3} and $10^{-2} \, cm^2/Vs$) relative to the same semiconducting polymers on SiO_2 dielectrics, but mobilities are still much lower than hole mobilities in corresponding p-type semiconducting polymers.

32.2.1.3 Processing Considerations for OFETs

The best incentive for use of EAPs in FETs is related to their ease of manufacture. Liu *et al.* [13] first described the use of inkjet printing of all components of an all-polymer FET. All materials chosen were water soluble, and careful choice of the dielectric allowed for low-voltage operation of the finished device. Reverse gravure-coating processes have also been used to apply the semiconductor and dielectric layers in conjunction with inkjet printing of the electrodes [12]. The five-step process is compatible with mass fabrication in an ambient atmosphere and produces low-voltage OFETs. One n-type semiconducting polymer was incorporated into OFETs using a variety of processing techniques [22]; device performance was analyzed as a function of deposition method, with spin-coated devices providing somewhat better performance.

32.2.2
Polymer Light-Emitting Devices

32.2.2.1 Background
In polymer light-emitting devices (PLEDs), light is produced in the polymer via rapid decay of excited molecular states, the color of which depends upon the energy difference between the ground state and the excited state [24]. PLEDs have potential for use in displays in cell phones, digital cameras, televisions, and computers, and also as two-dimensional light sources [25–28]. PLEDs offer many advantages compared to traditional display technology, such as cathode ray tubes or liquid-crystal displays. They have potential to use less power, produce higher color purity, enhance brightness, and offer larger viewing angles [29, 30]. In addition, polymeric materials allow these devices to be lightweight, thin, and flexible [31–33]. Many conducting polymers are solution processable, allowing for the possibility of low-cost bulk manufacturing of PLED devices [34–36].

Most known EAPs are p-type and have much higher hole mobility than electron mobility, so multilayer designs such as the one shown in Figure 32.4 are most often employed [37]. The structure of PLEDs includes an anode (typically indium tin oxide, ITO) on one side and a cathode (typically a metal like aluminum or silver, often accompanied by a layer of another metal like calcium) on the opposite side [26, 29]. Sandwiched between these outer layers are an optional hole-transport layer (HTL), an emissive layer (EML), and an optional electron-transport layer (ETL).

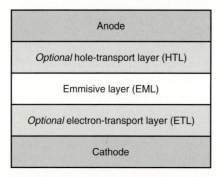

Figure 32.4 Schematic of a typical multilayer PLED.

Incorporation of the HTL and ETL improves efficiency of injection/transport of holes and electrons, respectively, resulting in greatly improved overall device efficiencies [38]. PEDOT-PSS or electron-rich amine-containing materials are the most commonly reported HTLs, while a broad range of electron-deficient nitrogen-containing heterocycles have been studied for use as ETLs [26, 38].

Most EAP-based research in PLEDs is for use as EMLs; in this application, thin (circa 100 nm) films of the polymers are deposited in the neutral (insulating or semiconducting) state. When an electric field is applied, holes are injected into the EML HOMO at the anode/ (or HTL/ EML interface, and electrons are injected into the EML LUMO at the cathode/ (or ETL/) EML interface [39]. The holes and electrons migrate through the EML, and if they meet, they combine to form excitons. Some of the excitons that decay to the ground state within the emitting layer emit photons of light. When charge transport or injection into the EML is not balanced, the result is poor device performance [40], because efficiency is determined by the ratio of light emitted to the current injected [38, 41]. Unbalanced charge can result from a number of factors, including the work function of the electrodes, the structure and resulting conductivity of the materials chosen as transport layers, and film morphology [42–44]. Each of these factors must be taken into account to produce the most efficient PLED.

32.2.2.2 PLED Emissive Layer Materials

The EML can be carefully selected and tuned to produce a specific frequency of light that corresponds to the desired color. Strategies applied for color selection include synthesis of new monomers [30, 33], copolymerization of two or more monomers with target properties [37, 40], blending polymers with desired characteristics [45], and incorporating nonpolymeric additives [25, 28]. The active layers used in PLEDs are commonly made up of polyfluorenes, PPPs, PPVs, and PTs. For example, to produce red light, poly(2-methoxy-5-(2-ethylhexyloxy)-1,4-phenylenevinylene) (MEH-PPV) or a cyano-derivatized PPV can be used [39, 45]. MEH-PPV has the added advantage of greatly improved solubility, facilitating the use of solution-based deposition methods in device fabrication. A single fluoro-substituent on the phenyl ring of PPV (F-PPV) shifts the emission to yellow [46]. Green PLEDs can be manufactured from other PPV derivatives such as phenyl-substituted PPV, or from polyfluorenes [34, 39]. Several representative electroluminescent polymers covering the visible spectrum are shown in Table 32.1 [45, 47].

Table 32.1 Light-emitting polymers across the visible spectrum.

Polymer	Color (λ_{max} (nm))	Reference
CN-PPV	Red (710)	[45]
MEH-PPV	Red-orange (610)	[47]
F-PPV	Yellow (410)	[47]
PPV	Green (550)	[47]
PPP	Blue (459)	[47]

In order to achieve full-color displays through the mixing of colors, EMLs are needed that provide all three primary colors of light: red, blue, and green [48, 49]. All three colors of emitters have been developed, but lifetimes (especially for blue-emitting polymers) need improvement. Another obstacle in producing commercial full color devices is that blue light-emitting materials have not yet met the efficiency and brightness requirements for commercialization [50, 51].

A vast quantity of research has been conducted in an effort to find better blue emitters. Many blue emitters have high HOMO energy levels, making hole injection a challenge [34]. Polyfluorenes and ladder-type poly(p-phenylenes) (LPPPs) have received much attention for their potential as blue emitters [33, 49]. In order to eliminate the longer-wavelength shift and excimers, bulky side chains and copolymerization have been explored [37, 49, 50]. Applying the former approach, Wu *et al.* [49] developed a series of hyperbranched LPPP films that exhibited blue emission; the three-dimensional nature of these polymers makes them soluble in common organic solvents, facilitating formation of good, amorphous films. Chen and coworkers [42] presented a series of copolymers of cyanophenyl-substituted spirobifluorene and tricarbazole-triphenylamines that emitted deep blue when blended with 30% [5-biphenyl-2-(4-tert-butyl)phenyl-1,3,4-oxadiazole)].

White light-emitting devices are also highly desired, although mainly for use in lighting applications rather than displays [25, 52]. There are several approaches to realizing white emission. In one approach, a blue-emitting device is altered by addition of phosphors [25, 28, 52]. The phosphors absorb some of the high-energy blue emissions and re-emit at lower energy (orange); the additive effect results in white light. Another approach utilizes multicolored dyes as additives or a secondary layer to produce white light [25, 28]. A third approach combines red-, blue-, and green-emitting polymers; the combined emission is white light [53].

32.2.2.3 Processing Advances for PLEDs

In order for these display or lighting technologies to be commercially feasible, the potential for low-cost roll-to-roll fabrication must be realized. Solution processing methods, including spin casting, inkjet printing, screen printing, photopatterning, and thermal patterning, offer these benefits and can be used with flexible substrates for lightweight devices [29, 31, 32, 35, 43]. Each of these processes has challenges that must be overcome.

The use of inkjet printing allows selective deposition of materials in defined areas with high accuracy, without requiring masking, or etching. Inkjet printing of a single layer of a multilayered PLED device has been demonstrated [35]. The partially inkjet-printed device required higher voltage to reach its optical threshold, most likely related to the variable thickness of the HTL caused by joined drops and the drying process and impurities. Pretreatment with H_2SO_4/H_2O_2 solution enhances the inkjet-fabricated device performance through a soft etching of the electrode to remove impurities and surface irregularities [36].

Screen printing allows patterning simultaneously with coating; this method has been used to deposit a paste of PEDOT:PSS as a flexible electrode [32]. Although the resulting electrode was comparable to others, all of the other layers were deposited in a thermal evaporator, so the fabrication method used for the total device is not suitable for commercialization. A problem encountered in spin casting is that addition of a subsequent layer may cause dissolution of the previous layer. Mixing of the layers may occur, causing device degradation. In order to make the earlier layer insoluble, it is possible to embed the HTL into an inert crosslinked polymer that has been cured prior to deposition of the follow-on layer [29].

A multilayered, full-color PLED device has been produced using spin casting with dry photopatterning [34]. Treatment of red-emitting MEH-PPV and green-emitting phenyl-substituted PPV (P-PPV) films with ultraviolet light and heat caused the films to become crosslinked and therefore insoluble. When the P-PPV layer is crosslinked, the emission shifts from green to blue. By depositing the layers separately and selectively crosslinking, three-color pixelation was achieved.

32.2.3
Photovoltaics

32.2.3.1 Background

Clean, renewable energy sources are being sought for a broad range of applications. Although harnessing solar energy through the use of PV devices is promising, it is currently prohibitively expensive [54]. Commercial PV devices are now silicon-based; device efficiencies as high as 41% have been achieved [55], with most commercially available units offering external efficiencies between 18 and 20%. Organic polymer photovoltaic (OPV) devices using lower-cost materials could make solar energy more affordable, if the efficiency of these devices could be improved [56–61]. Polymers offer additional advantages of being lightweight and compatible with flexible substrates due to solution processing; ultimately these properties would allow roll-to-roll manufacture of large-area devices [57, 58].

PV devices operate in a process that is the inverse of that used in light-emitting devices [62]. When light of a compatible frequency strikes the active layer of a PV device, photons are absorbed. The photons induce the formation of excitons, or electron–hole pairs. The excitons have a very short decay time, but if the decay occurs at the interface between an electron donor and acceptor, the electrons will move into the electron-acceptor layer and holes will form in the donor layer. The separate charges are then transported to the electrodes to be collected. The electrons collected at the cathode can be used to provide direct current, before being reintroduced at the anode to complete the circuit.

The simplest structure for a PV device is a single layer of active material sandwiched between the electrodes. Because few polymers have both high electron and hole mobility, and because recombination occurs, this architecture is only used to study specific device properties [56]. If a second layer is applied to separate charge transport, a bilayer device results. Although a bilayer device is an improvement over a single-layer device, the interface between donor and acceptor materials is relatively small and can be too far from the location of exciton generation to be very efficient.

Excitons can travel only about 10 nm before they decay, so they should be formed within 10 nm of a disassociation site in order for a device to be efficient [56].

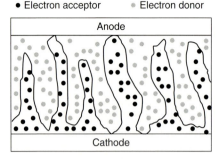

Figure 32.5 Schematic of a bulk heterojunction device.

There are several approaches to building devices that allow for shorter distances between donor/acceptor interfaces. The most common solution has been the bulk heterojunction (BHJ) concept (Figure 32.5). In a BHJ device, the light absorber is usually an EAP, which is blended with an electron acceptor such as a fullerene derivative prior to film deposition [56, 62]. The phase separation of the polymer and electron acceptor can be adjusted using different techniques, with the goal of creating interfacial boundaries with large surface area within close proximity of each other; thickness of the active layer ranges from 30 to 100 nm. BHG devices have reported power-conversion efficiencies (PCEs) of up to 6.1% [63].

32.2.3.2 Materials for Photovoltaics

The most efficient all-polymer bilayer device reported to date had an average efficiency of 1.5% [56]. The device was fabricated with poly[3-(4-n-octyl)-phenylthiophene] (POPT) as the active layer with poly[2-methoxy-5-(2′-ethylhexyl-oxy)-1,4-(1-cyanovinylene)phenylene] (CNPPV) as the electron acceptor.

The most common EAPs used in BHJ devices are P3HT or PPV, but other polymer/fullerene blends have shown promising efficiencies. A quinoxaline-based donor polymer, poly[2,7-(9,9-dioctylfluorene)-alt-5,5-(5′, 8′-di-2-thienyl-2′, 3′-dip-henylquinoxaline)] (N-P7), has been used in a BHJ device with PCE of 5.5% [57]. Liang and coworkers [64] synthesized a series of polymers based on alternating thieno[3, 4-b]thiophene and benzodithiophene units for use in PV devices. A PCE of over 6% was reportedly obtained from a fluorinated composite film BHJ device.

32.2.3.3 Novel Approaches for Photovoltaic Device Fabrication

A BHG active layer can be used in the development of solar power wires [58]. The core of the device is stainless steel wire, coated with a TiOx ETL. Next is the active layer, a P3HT:fullerene blend. PEDOT:PSS is used as a hole-transport (electron-blocking) layer. A very fine, silver-coated wire is wrapped around the polymer layers to serve as the secondary electrode, but light absorption is maintained because coils are separated by 1.25 cm. The entire device is jacketed in a transparent, photocurable epoxide for protection. The outer diameter of the finished wire is 1/100th of an inch. Samples of the solar-power wires had an average efficiency of 3%.

A second approach to constructing efficient PV devices is that of the interpenetrating polymer networks (IPNs) structure [65]. A 1 : 1 mixture of a carbazole derivative and diacrylate perylene was used to cast a thin film. The film was then subjected to photopolymerization to form the crosslinked perylene network for light absorption and electron transport [59, 65]. An electrochemical oxidative process is then used to crosslink the carbazole derivatives used as electron donors. Although the IPN film shows both p- and n-dopable properties, final device efficiencies have not been reported. Layer-by-layer deposition was used to form an interpenetrating network of a rhenium-containing hyperbranched polymer and poly[2-(3-thienyl)ethoxy-4-butylsulfonate] (PTEBS). The resulting devices had relatively low efficiencies, but the method provides a versatile approach to the manufacture of PV devices [66].

In order to facilitate reproducible manufacturing techniques, dyad-based devices can be used [60]. Dyads are covalently bonded donor and acceptor molecules, which prevent the phase-separation problems encountered with BHJs and ensures the close proximity of the donor and acceptor material. The electron-donating oligo(p-phenylene-vinylene) was attached to a fullerene group through a long, flexible spacer. The resulting solar cell had a PCE of only 1.28%, but it was the highest efficiency reported for a dyad-based solar cell.

32.2.4
Electrochromics

32.2.4.1 Background
EC materials are those that exhibit a change in color resulting from a change in absorbance, reflectance, or transmission of electromagnetic waves arising from an electrochemical oxidation or reduction [67]. One main class of EC materials is comprised of nonpolymeric organic materials such as the molecular dye Prussian blue or viologens. These EC materials can exhibit excellent color contrast and can change colors very rapidly, but they lack the robust mechanical properties of polymer-based electrochromics. Another class of EC materials is comprised of metal-oxide films derived from tungsten, nickel, or molybdenum. These materials are widely used in commercial applications including autodimming rearview mirrors in automobiles, but they frequently suffer from slow switching rates [68]. The third main class of EC materials is the EAPs; in these polymers, EC properties result from conjugation changes as a function of electrochemical potential.

EAPs have the prerequisite characteristics for use in EC devices including fast switching, high EC contrast, high coloration efficiencies, and low voltage requirements [68, 69]. EAPs offer several advantages over other EC materials. The basic polymer structures can be manipulated through substitution or copolymerization to obtain specific colors. Polymers may exhibit stability at several electronic states, resulting in multichromism. In addition, EAPs can be processed using electrochemical deposition, spin or spray coating, drop casting, and even inkjet printing [67, 70].

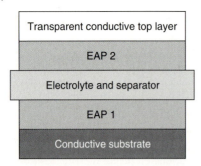

Figure 32.6 Schematic of a typical two-polymer electrochromic device; the conductive substrate may be either transparent or opaque, depending upon the desired application.

A typical EAP-based EC device design is shown in Figure 32.6. The top layer is typically a transparent conductive electrode substrate top layer, such as ITO coated glass or polyester. Adhered to this layer is an EC EAP. The EAP is coated with a layer of electrolyte to compensate for the charges formed during doping. Next, an insulating, porous separator is used to keep the top EAP layer from the bottom EAP layer, which is adhered to another conductive electrode substrate. The bottom substrate may be transparent or opaque, depending on the desired application. The two EAP layers are chosen so that the colors of each achieved as the device is switched are complementary. Film thickness is particularly important in EC devices; thinner films (as thin as 50 nm) are desirable for faster switching, while thicker films (2000 nm or more) produce more intense colors [71]. Film thickness and roughness have also been shown to significantly impact coloration of EAP films [72].

EC properties of EAPs are characterized by their bandgap (E_g), or the energy difference between the HOMO and the LUMO. The optical absorbance of the polymer is related to E_g and frequently falls within the visible range of the spectrum [73]. When EAPs are oxidized, single electrons are removed from the polymer backbone, resulting in formation of delocalized radical cations (polarons) along the polymer chain (Figure 32.7). The polaronic state is electronically different from the neutral state, with new electronic transitions at lower energies resulting in changes in absorption. The color of the polaronic state is shifted toward longer wavelengths relative to the neutral state. The polaronic state can be further oxidized by removal of unpaired electrons to form delocalized dications (bipolarons), resulting in a second optical shift to even longer wavelengths. This process is generally reversible through reduction to return to the polaronic state and then to the neutral state with the originally associated optical properties.

32.2.4.2 Materials for Electrochromics
Many different EAPs have been used in EC devices. The EC properties of several polymers families have been established; these are summarized briefly below and

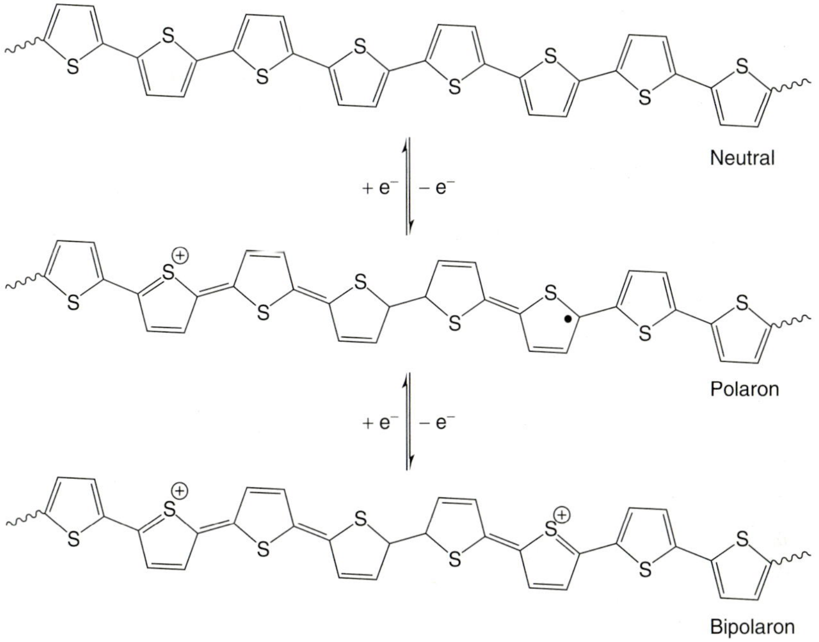

Figure 32.7 Reversible oxidative doping of EAPs such as polythiophene yield polarons (radical cations), which are further oxidized to bipolarons (dications).

more thoroughly elsewhere [67]. However, the basic polymers can be tailored through structural modifications to achieve a variety of colors; incredibly complex polymer structures have been synthesized in order to fine tune EC properties and produce multichromism (more than two stable colors), as can be seen in a very thorough recent review [74].

The PT family of polymers has been extensively studied for use in EC devices. Garnier and coworkers [75, 76] reported in 1983 that PT films exhibit a red color in the neutral state, a blue oxidized state, and a black-green reduced state. When an ethylenedioxy bridge is added to thiophene, the resulting PEDOT has a lower E_g and exhibits different coloration; PEDOT switches between a transmissive sky blue in its oxidized form and a deep absorptive blue neutral state [77]. A wide range of PEDOT derivatives have been synthesized for control of electronic and optical properties [67, 74, 77].

Another family of polymers used for EC devices is the polypyrroles. Unsubstituted polypyrrole is yellow when neutral but oxidizes to black [78]. Alkyl substitutions at the N-position on the pyrrole do not change the color of the films but do reduce conductivity [79]. Substitution at the 3-position of pyrrole does not cause a decrease in conductivity and can produce different colorations, but these polymers are difficult to synthesize and are air sensitive [67].

PANI films show multiple oxidation states with associated hues. Neutral films of PANI are transmissive yellow, but upon oxidation they turn green [80]. Further oxidation yields a deeper green, followed by blue, and finally violet. The neutral and fully oxidized states are insulating, but the intermediate oxidation states are conducting. PANI is pH sensitive; at pH higher than 4 films appear black and do not switch [81].

32.2.4.3 Electrochromic Device Designs

Adaptive camouflage makes use of the variety of colors available using EC conducting polymers. For other applications, such as smart windows or sunglasses, the active layer need only switch between transmissive and colored. A prototype pair of smart sunglasses has been constructed utilizing poly[3, 3-dimethyl-3, 4-dihydro-2H-thieno [3, 4-b] [1, 4]dioxepine] (PProDOT-Me$_2$) [82]. The lenses of the sunglasses can be switched between a transmittance of 57% in the transparent state to only 12% in the colored state. The transition takes less than 1 s in either direction, and the device will retain its coloration state without an applied current for 30 days. The sunglasses can be switched thousands of times using one commercial 1.5 V battery, and the polymer does not degrade significantly over 100 000 cycles.

Individually addressable red, green, and blue elements are the basis for the standard three-color pixel display. By controlling intensities of each of the three colors, the full spectrum of colors is attainable. Sonmez and coworkers [83] demonstrated that all three colors are possible using EAPs; poly(3-alkylthiophenes) were used to impart red color, PEDOT provided blue color, and a complex PT derivative was constructed to provide the green color. The oxidation states of the three polymers were independently controlled, allowing color mixing to attain a broad spectrum of colors. While the authors prepared their EC devices using electrochemical deposition, they propose inkjet deposition for rapid, low-cost mass production of displays.

A milestone in polymeric EC materials was recently reached when a dual-polymer EC device was made to switch from black to transmissive [84]. The pseudo-three-electrode EC device was made from a PProDOT ester derivative, which switches from purple to transmissive, and from another PProDOT derivative that switches from green to transmissive. The combination of the green and purple can display colored images, or be used as a reflective or mirrored surface depending on the backplane.

Although much research to date has revolved around color changes in the visible spectrum, changes that occur in the infrared and even microwave regions are now being explored [70, 85–87]. Chandrasekhar and coworkers [88] have developed infrared-active devices to be employed as emittance control for use on spacecraft. Spacecraft thermal control is needed to conserve heat while the craft is on the darkside and reject heat while on the sunside.

Optical changes that occur outside of the visible region can also be applied to devices such as electrochromic variable optical attenuators (EC-VOAs) which are used as components of fiber-optic communication networks. A substituted poly(3,4-propylenedioxythiophene) (PProDOT-(CH$_2$OEtHx)$_2$) has been used to create a reflective, flexible, near-infrared (NIR) device [70]. This device could be applied as an EC-VOA because of attractive characteristics such as low operating voltage, low optical loss, and fast switching [70]. Another NIR range EC polymer, a nickel dithiolene-containing pendant indole, can function as a three-way optical switch [87]. As such it could also be used as an EC-VOA for use in the telecommunications field.

In order for NIR devices to be applied in EC-VOAs, electrodes used in such devices must be transmissive at the longer wavelengths. Indium tin oxide on glass (ITO/glass) electrodes are typically used as transparent electrodes, but their transmissivity declines quickly beyond 1200 nm. Instead, a thin continuous film of single-walled carbon nanotubes (SWNTs) has been used as a transparent electrode in a NIR device utilizing poly[5,8-bix-(3-dihydro-thieno [3,4-b][1,4]dioxin-5-yl)-2,3-diphenyl-pyrido[3,4-b]pyrazine] [86].

32.2.5
EAP Battery Electrodes

32.2.5.1 Background
The reduction and oxidation (redox) processes in EAPs make it possible to use these materials to store energy, either as battery or capacitor electrodes. While metals, metal oxides, and carbonaceous materials are traditionally used in these devices, the potential for reduced cost, weight, and environmental impact of EAP electrodes makes these polymers attractive alternatives. EAPs can also be structurally tailored to control conductivity, voltage window, reversibility, storage capacity, and chemical/electrochemical/environmental stability.

Performance of batteries and electrochemical capacitors (Section 32.2.6) is characterized by the energy content delivered and the rate of discharge. Energy content is usually described using specific energy (Wh/kg, also known as *gravimetric energy density*), while the discharge rate is usually described using specific power (W/kg, also known as *gravimetric power density*). Also of interest is the open-circuit voltage (OCV) (the voltage across the device when no external current flows).

EAPs electrodes in batteries may be used either as thin films or in bulk, depending on the desired application. Ultimately, for large-scale energy storage, bulk materials will probably be needed, while thin-film electrodes will find use in microbatteries such as those needed to power microelectromechanical devices (MEMs).

32.2.5.2 EAPs as Battery Materials
Most of the classes of EAPs have been studied for use as battery cathodes and/or anodes at one time or another. While PA received considerable attention in the 1980s and 1990s [89–92], stability and processing issues led researchers to move on to other types of EAPs. A summary of properties of battery materials is given in Table 32.2.

PANI and its derivatives have been used extensively as battery cathode materials. In aqueous cells, zinc is typically used as the anode material, OCVs up to 1.5 V are common [93–95]. Nonaqueous cells commonly use lithium or magnesium as the anode material, and much higher open-circuit voltages (up to 4 V) are regularly

Table 32.2 Characteristics of common EAP battery cathodes.

EAP	OCV (V)	Charge density (Ah/kg)	Energy density (Wh/kg)	References
PANI aqueous	1.0–1.5	50–150	100–350	[93–95]
PANI nonaqueous	3.0–4.0	50–150	100–350	[96–99]
PPy nonaqueous	3.0–4.0	50–170	50–350	[100–102]
PT nonaqueous	3.2–4.2	25–100	50–326	[93, 103]
PPP nonaqueous	3.2–4.5	20–140	300	[104–106]

achieved [96–99]. Charge densities of PANI-based cells are usually on the order of 50–150 Ah/kg, and energy densities usually range from 100 to 350 Wh/kg; variations in these numbers are due to device-design differences.

Cathode properties of polypyrrole (PPy) and PPy derivatives are similar to those of PANI [100–102]: OCVs range from 3.0 to 4.0 V in nonaqueous Li anode cells with charge densities between 50 and 170 Ah/kg and energy densities of 50–350 Wh/kg. Composites of polypyrroles with a variety of materials may also be used as cathodes; modest improvements in properties relative to PPy alone are seen [107, 108].

PT and a wide range of PT derivatives have been used as battery cathodes. Cathodes from PTs typically exhibit charge densities between 25 and 100 Ah/kg with energy densities between 50 and 325 Wh/kg [109], with OCVs of nonaqueous cells ranging from 3.2 to 4.2 V [103]. As with the polypyrroles, composites of PTs result in modest improvements to cell performance [110, 111]. Because derivatized PTs can be good n-dopants, many PTs have been investigated for use as battery anodes. The most common of these are poly(3-phenylthiophene) [112], poly(3-(4-fluorophenyl)thiophene) [113], and polydithieno[3,2-b:2′, 3′-d]thiophene [114]. Reported charge densities (18–77 Wh/kg) tend to be lower than for PT cathodes.

PPP is not as highly redox active as the previously mentioned EAP battery materials, but it is a good cathode material due in part to its crystallinity and its resemblance to graphite [115]. Typical OCVs range from 3.2 to 4.5 V [104], higher than any other EAPs mentioned herein. Charge densities are 20–140 Ah/kg [105, 106], and energy densities as high as 300 Wh/kg have been reported.

32.2.6
EAP-Based Electrochemical Capacitors

32.2.6.1 Background
There are many different types of capacitors. While EAPs have found some use in electrolytic capacitors [116–118], the bulk of EAP use in capacitors is in the area of electrochemical capacitors. Electrochemical capacitors are a specific subset of capacitors in which charge is stored and released via electrochemical (redox) processes (Figure 32.8) [119]. Most EAP-based electrochemical capacitors (EPECs)

require thin films of EAPs due to the need for rapid charging/discharging cycles; charging and discharging rates are dependent upon film thickness [120].

There are four possible configurations for EPECs (Figure 32.9) [109, 120]. Type I is a symmetrical EPEC having one p-dopable EAP used for both electrodes. In

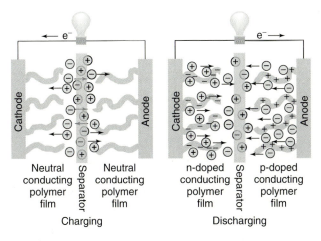

Figure 32.8 Schematic representation of charging and discharging processes in EAP-based Type-IV EPEC. Reproduced by permission of Sigma-Aldrich.

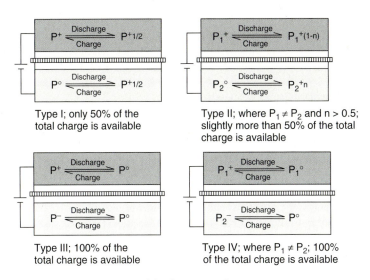

Figure 32.9 Representation of the four types of supercapacitors, changing from the charged state (on the right side of each diagram) to the discharged state (on the left side of each diagram); polymer electrodes are separated by electrolyte (white) and battery separator paper (striped).

the charged state, one polymer layer is fully oxidized and the other is neutral. In the discharged state, each layer is 50% oxidized, meaning that only 50% of the total charge capacity is available. Type II is an asymmetric EPEC in which two different p-dopable EAPs are selected, based on the differences in the potential ranges over which they can be p-doped. In the charged state, the polymer with the higher oxidation potential is completely oxidized, while the polymer with the lower oxidation potential is neutral. In the discharged state, the polymer with the higher oxidation potential is less than 50% oxidized, and the polymer with the lower oxidation potential is more than 50% oxidized. In Type-III EPECs, a single polymer that is both p-dopable and n-dopable is used for both electrodes. In the charged state, one polymer layer is fully p-doped while the other is fully n-doped, and in the discharged state, both polymer layers are neutral. Type-IV EPECs are fabricated with an n-dopable polymer on one electrode and a different p-dopable polymer on the other. In the charged state, one polymer is fully p-doped while the other is fully n-doped, and in the discharged state, both polymers are neutral. Type-III and Type-IV EPECs are of greatest promise in terms of achievable energy density and have advantages over the Type-I and Type-II devices in terms of available conducting states and greater instantaneous power density [113, 120, 121]. For a thorough discussion of EPECs, please see a recent review chapter [109].

In addition to charge density and energy density (Section 32.2.5), capacitors are also characterized by their charge-storage ability. Both capacity (Coulombs per gram) and capacitance (capacity per volt, farad per gram, or farad per cubic centimeter)[1] are used to describe charge-storage ability [122]. Coulombic efficiency and voltage efficiency are usually expressed in terms of percentage of total charge or voltage available. Charge density is the mass of active polymer required per unit charge (Ah/kg), while total device charge density includes the mass of active polymer, electrolyte, solvent, charge collection/distribution electrodes, and any solid supports or encapsulation materials.

32.2.6.2 Electrochemical Capacitor Materials

All the classes of EAPs displayed in Figure 32.2 have been studied for use in EPECs with varied degrees of success. Most of the recent research has focused on structural modification of these polymers to improve specific capacitance (thus maximizing energy storage), reduce electrical resistance (thus maximizing power density), and improve stability. EAPs have significant advantages over the metal oxides commonly employed in electrochemical capacitors: charge can be stored throughout the volume of the polymeric materials, rather than just at or near the surface of the metal oxides, resulting in increased charge storage capacities. EAP oxidation and reduction processes can be very fast, and, most EAPs are more cost effective than metal oxides or carbonaceous materials. However, the performance of EPECs is currently inferior to that of metal-oxide-based systems; specific capacitances of EPECs

---

1) Strictly speaking, *capacitance* and *specific capacitance* refer to charge per volt accumulated in the double layer, while *pseudocapacitance* refers to charge storage per volt in association with a redox reaction. For a thorough treatment of pseudocapacitance, see the discussion by Conway *et al.* [122].

are currently lower than those of hydrous ruthenium-oxide-based electrochemical capacitors, which exhibit a specific capacitance of 720 F/g [123, 124]. The best EPEC specific capacitances are currently only 250 F/g [125], although theoretical values are higher than those of ruthenium oxides. On the other hand, EPEC specific capacities are 10–15 Ah/kg, which is circa five times better than those of high surface area carbonaceous electrodes [126].

Most EPEC research has focused on Type-I devices. Typical PPy Type-I EPECs exhibit specific capacitances from 40 to 200 F/g [121]. One study [127] investigated the effects of PPy film thickness on device performance; specific capacitance remained constant from 40 to 1700 nm. The best specific capacitance of an EPEC (250 F/g) was observed in both a PANI-based device [125] and a PT-based device [128]. Despite this excellent specific capacitance, PT-based supercapacitors are uncommon due to life-cycle issues. However, PT derivatives are prevalent in Type-II, -III, and -IV supercapacitors (see below).

Relatively little research is available on Type-II EPECs, perhaps because the slight improvements gained by the inclusion of two p-doping polymers are offset by the added difficulties associated with device fabrication and operation. Specific capacitances are similar to or poorer than the corresponding Type-I EPECs [121, 129]. A PEDOT/PProDOT Type-II EPEC was used to study effects of electrolyte composition [130]. Both switching speed and life cycle were greatly improved by switching to an ionic liquid-based electrolyte. An additional advantage of ionic liquid electrolytes for use in EAP-based devices is that the devices operate over a much wider temperature range than do solutions of traditional electrolytes [119, 131].

Due to the promising theoretical performance of Type-III and Type-IV EPECs, a great deal of research has focused on synthesis and characterization of n-doping polymers. Unfortunately, most n-doping polymers are unstable because they contain highly reactive carbanions. PA, PPPs, and PT are all too unstable for practical device use [132]. Researchers have attempted to improve stabilities by incorporating groups that transfer electrons away from n-doped polythiophene. Toward that end, poly-3-phenylthiophenes substituted with fluoro- or cyano-substituents have been extensively studied [120]. Poly-3-(4-fluorophenyl)thiophene (PFPT) undergoes stable n-doping processes, and reaches high charge densities than its nonfluorinated analog. While PFPT has been extensively studied by Ferraris and others. Ferraris [133] reports that analogous difluoro- and cyano-substituted polymers are more stable than PFPT and may yield better EPECs. Another approach [134] tries to minimize the bandgap (see Section 32.2.4), thereby increasing intrinsic charge-carrier densities and producing more stable n-doping polymers. A third approach [135] uses self n-doping polymers in the hope of alleviating cation-transport problems and enhancing the n-doping process. While more stable than unsubstituted PT, all of the derivatized polymers are relatively unstable, with even trace amounts of water, oxygen, or acid resulting in rapid decay of device performance. Research continues in hopes of improving stability of n-doping polymers, but additional efforts have focused on attempts to prevent exposure of the polymers to water, oxygen, and acid.

Problems associated with instability of n-doping EAPs may be avoided through the use of hybrid devices, in which a p-doping EAP anode is combined with a carbonaceous cathode. In addition to improved stabilities, specific capacitances may be significantly improved; a PANI/activated carbon hybrid device exhibited specific capacitance of 380 F/g and was stable over 4000 cycles [136]. Hybrid devices from poly(3-methylthiophene) and activated carbon are claimed to outperform double-layer activated-carbon capacitors [137]. On the other hand, PFPT/activated-carbon hybrid capacitors only yielded 68 F/g [138].

32.2.7
Sensors

32.2.7.1 Background
Electrochemical sensors based on EAPs can be used to detect a variety of solid, liquid, and gaseous analytes. There are two mechanisms used in EAP-based sensors. In one type of sensor, the analyte chemically binds to the EAP, thus changing its doping level and electronic properties [139]. The other mechanism is set in motion when a gas molecule that oxidizes or reduces adsorbs onto the surface of the polymer. A charge transfer may occur between the charged gas and the film, thus changing the polymer's conductivity [139].

Conducting polymer-based sensors are the topic of much research due to the advantages they provide over traditional sensing technologies. Electrical-resistance-type sensors are very sensitive and have short response times [139–141]. Polymeric materials provide simple fabrication at a low cost with a lightweight and flexible device as the end result [140, 141]. For these reasons, EAP-based sensors are being explored for diverse applications such as biosensors, chiral sensors, or to detect pH, gases, radiation, or humidity.

32.2.7.2 Materials for Sensors
The structure of polymers can be manipulated to yield desired characteristics, and other functionalities can be immobilized on the thin films as receptors [142, 143]. A wide variety of EAPs has been exploited for use in sensors, with the bulk of the polymers utilizing common EAPs (PANI, PT, PPy, PPV) as the functionalizable polymer unit. Among the factors considered when choosing an EAP substrate are compatibility with the receptors and/or analytes, ease of functionalization, sensitivity, availability, cost, and stability.

32.2.7.3 Sensor Designs
The design of biosensors is often based on a thin (typically <5 μm) polymer film that has specific molecule systems, such as antibodies or natural protein receptors, attached to the surface by adsorption or covalent bonding. The ideal receptors are sensitive to the specific analyte to be detected [139]. A biosensor with cholesterol oxidase (ChOx) covalently coupled to a PANI-based polymer has been used to determine the level of free cholesterol in serum [141]. From the amount of free

cholesterol found, the total cholesterol can be extrapolated assuming that free cholesterol is approximately 30% of the total.

Electrochemical DNA sensors using conducting polymers have a wide spectrum of potential applications, including forensics as well as diagnosis of infectious diseases (such as HIV or hepatitis) and genetic mutations (such as cystic fibrosis) [144–146]. The most commonly used polymers in DNA sensing devices are PPy, PANI, PT, and their derivatives. An excellent review of the application of EAPs to DNA sensing is available [144].

Polypyrrole films with covalently attached oligonucleotides have been used as a label-free DNA sensor. The hybridization of the oligonucleotides with the target DNA cause a change in the ability of ions to pass into the polymer film, altering its doping level and therefore its conductivity [146]. An amperometric nanostructured PANI device, fabricated for the detection of N. Gonorrhoeae, is currently in clinical trials. The device is highly specific for the detection of the analyte at 0.5×10^{-15} M and takes only 60 s [145].

EAP-based biosensors may be built as transistors. For example, Shim et al. [147] constructed an all-plastic glucose sensor using PEDOT:PSS as a channel and gate electrode. Glucose oxidase and ferrocene were deposited into a well across the channel and gate electrode. The interaction of glucose with glucose oxidase starts a reaction cycle in which ferrocene is a mediator, wherein PEDOT is doped and shows an increase in conductivity. The device can detect glucose in the micromolar range, making it compatible with levels present in human saliva. The ability to test glucose using saliva rather than blood makes this device noninvasive and highly attractive.

Another EAP-based sensor using a transistor mechanism is a chiral sensor presented by Torsi and colleagues [148]. Two different conjugated alkoxyphenylene-thiophene derivative oligomers were used in bilayer devices. One of the oligomers in each device is achiral, while the second becomes chiral when derivatized with either l-phenylalanine or β-D-glucosidic substituents. Discrimination between β-citronellol enantiomers using the β-D-glucosidic oligomers was shown, as was discrimination between carvone enantiomers using the l-phenylalanine-substituted oligomers.

pH sensors using EAPs are another analytical tool under development; pH sensors based on PANI or a PANI and o-anthranilic acid (AA) copolymer have been reported [143]. The pH of a solution can be obtained using the UV-Vis absorption spectrum of the polymer, which changes upon doping. The frequency increases with the pH of the analyte. The PANI film exhibits higher sensitivity to changes in pH than does the copolymer film, but the accuracy of PANI between pHs of 6 and 9 is low due to rapid changes in absorbance near pKa. The copolymer film could be used over a wider pH range of 5–12. Ultramicroelectrodes have been modified with PANI and modified-PANI films to detect pH based on potentiometric response [149]. A linear dependence on pH in the range of 3–10 was observed in preliminary experiments. The small size of the devices, between 25 and 100 μm diameter, allows them to be used for very small volumes.

A number of sensors for the detection of harmful gases have been developed using polymers such as PANI, PT, and PPy. The electrode of a quartz crystal microbalance has been coated with PANI to detect dichloromethane, 1,2-dichloroethane, chloroform, and carbon tetrachloride [150]. The resultant sensor showed satisfactory linear correlation in concentrations in the parts per million range and could be used to discriminate between the analytes using diffusion coefficients. EAP-based sensors have been produced to detect gases including nitrogen dioxide, ammonia, and carbon monoxide [139, 142, 151].

Humidity control is important for many industries and technologies, as well as for comfort. Su and Huang [152] report a composite film of PPy and TiO_2 nanoparticles used in a humidity detector that has good sensitivity, linearity, and long-term stability of at least 54 days. A blended film of PANI and polystyrene was used in a humidity sensor by Matsuguchi *et al.* [140]. This sensor was stable for a month and allowed DC operation.

Arrays of sensors can be combined for monitoring more complex mixtures of different chemicals in the air; these arrays have become known as *electronic noses*. While much of the work in this area utilizes carbon-black-based sensors [153], sensors based on PEDOT-PSS have been found to be more sensitive than carbon black in discrimination between nonpolar analytes [154]. Arrays of individual EAP-based sensors have also been used as electronic noses, where different concentrations of volatile organic compounds allow discrimination between wood, paper, plastic, cotton, and cigarette smoke [155]. Similar sensor arrays have also been developed to detect the volatiles generated by bacteria metabolites in wounds [156]. Early detection and identification of bacteria allow for rapid and targeted treatment of specific bacterial infections. A FET has been prepared with human olfactory receptors covalently attached to the surface of polypyrrole nanotubes; this so-called bioelectronic nose is able to discriminate between a variety of odors [157].

A final example of an EAP-based sensor is the gamma-radiation sensor prepared from poly[2-methoxy-5-(2′-ethyl-hexyloxy)-1,4-phenylene vinylene] (MEH-PPV) thin film to detect up to 25 kGy doses of gamma radiation [158]. The UV-Vis absorbance spectrum shows a blue shift as the gamma radiation increases. This device is intended for use in monitoring the radiation doses used in the sterilization of health-care products.

32.2.8
Miscellaneous Applications

32.2.8.1 Antistatic Coatings
Perhaps the biggest commercial application of thin films of EAPs is in antistatic coatings. Static charging in dry air caused by friction generates charges up to several thousand volts, which is discharged when connection is made to ground. The discharge is accompanied by a flash of light, which destroys photographic film. Maintaining a surface resistance below 10^9 Ω/cm prevents static charging [159]. Thin films of neutral (semiconducting) PEDOT-PSS reduce surface resistance of polyethylene terephthalate to 10^6 Ω/cm [160]; more than 100 million square meters

of photographic film are coated with PEDOT-PSS every year. Antistatic treatment of other plastics, glass, carpet, and electronics are also possible using a variety of EAPs. Along similar lines, coating explosives with thin films of EAPs prevents undesirable electrostatic detonation [161].

32.2.8.2 Transparent Polymeric Electrodes

Neutral EAPs that are transparent in the doped, conductive state, such as PEDOT-PSS, can be used as transparent, conductive electrodes. The resulting films serve as a flexible, solution-castable, inexpensive alternative to metal-based conductive thin-film electrodes, such as ITO or gold, which are brittle, expensive materials deposited using expensive, time-consuming thermal processes. PEDOT-PSS electrodes 150 nm thick are 80% transmissive with very low surface resistance (circa 100 Ω/cm; conductivities up to 2500 S/cm); these values are as good as or better than those of ITO films [162]. PEDOT-PSS transparent electrodes have been used as the top electrode in a prototype electronic paper device [163] and as alternatives to gold source/drain electrodes in FETs [164]. While the ability to paint on PEDOT-PSS electrodes from solution offers significant advantages over thermally deposited gold electrodes, the lower conductivity and higher contact resistance of PEDOT-PSS result in higher voltage requirements and lower on/off ratios.

32.2.8.3 Corrosion Inhibition

Chromium-based coatings have been used to prevent corrosion of metals for many years, but mandates controlling occupational and environmental exposure to chromium (VI) have essentially resulted in a need to eliminate chromium compounds from corrosion-inhibiting coatings. EAPs are one of the most promising alternatives to chromium compounds [165]. While the exact mechanism of corrosion inhibition by EAPs is unknown, several polymers have been shown to be effective at inhibiting saltwater-based corrosion of metals [166, 167]. Initial efforts focused on PANI, but PANI proved to be ineffective in marine environments [2], so other structures are under consideration. One promising material is an alkylamino-substituted PPV derivative [47], which has been shown to adhere to aluminum alloys in simulated seawater (pH $\sim$ 8) and to retard corrosion.

32.3
Conclusions and Future Outlook

The ability to reversibly change oxidation states of EAPs allows researchers to modulate properties between insulating and conducting, opaque and transparent, and absorptive and emissive. This has led to their incorporation in a wide range of applications. The ability to tailor these properties using synthetic modifications greatly improves the utility of these materials. Interestingly, several promising applications of these materials require them to remain in either the conductive or the neutral state, while others capitalize on the ability to rapidly change properties.

Using the EAPs in the form of thin films is advantageous, because response times and sensitivities tend to be better for thin films. While many p-doping and p-type semiconducting polymers exhibit excellent stability and utility, it is clear that future advances in EAPs will be in the area of n-doping and n-type semiconducting polymers. Attempts to increase stability are crucial to the development of EAP-based devices for use in commercial applications.

Acknowledgments

The authors would like to thank Dr. David Irvin for his helpful assistance with this manuscript.

References

1. Reynolds, J.R., Karasz, F.E., Chien, J.C.W., Gourley, K.D., and Lillya, C.P. (1983) Electrically conducting aromatic polymers. *J. Phys. Colloq.*, **44**, 693–696.
2. Zarras, P. and Irvin, J.A. (2004) Electrically active polymers, in *Encyclopedia of Polymer Science and Technology*, John Wiley & Sons, Inc., New York.
3. Shirakawa, H., Louis, E.J., MacDiarmid, A.G., Chiang, C.K., and Heeger, A.J. (1977) Synthesis of electrically conducting organic polymers: halogen derivatives of polyacetylene, $(CH)_x$. *J. Chem. Soc. Chem. Commun.*, 578–580.
4. Shirakawa, H. (1998) Metal-insulator transition in doped conducting polymers, in *Handbook of Conducting Polymers*, 2nd edn, Chapter 2 (eds T.A. Skotheim, R.L. Elsenbaumber, and J.R. Reynolds), Marcel Dekker, New York, pp. 27–84.
5. McCullough, R.D. and Jayaraman, M. (1995) The tuning of conjugation by recipe: the synthesis and properties of random head-to-tail poly(3-alkylthiophene) copolymers. *J. Chem. Soc. Chem. Commun.*, 135–138.
6. Goedel, W.A., Somanathan, N.S., Enkelmann, V., and Wegner, G. (1992) Steric effects in 3-substituted polythiophenes: comparing band gap, nonlinear optical susceptibility and conductivity of poly(3-cyclohexylthiophene)

and poly(3-hexylthiophene). *Makromol. Chem.*, **193**, 1195–1206.
7. Kumar, A., Welsh, D.M., Morvant, M.C., Piroux, F., Abboud, K.A., and Reynolds, J.R. (1998) Conducting poly(3,4-alkylenedioxythiophene) derivatives as fast electrochromics with high contrast ratios. *Chem. Mater.*, **10**, 896–902.
8. Lilienfeld, J.E. (1930) Method and apparatus for controlling electric currents. US Patent 1,745,175, and issued Jan. 28, 1930.
9. Lu, K., Sun, X., Liu, Y., Di, C., Xi, H., Yu, G., Gao, X., and Du, C. (2009) Linking polythiophene chains with vinylene-bridges: a way to improve charge transport in polymer field-effect transistors. *J. Polym. Sci. Pol. Chem.*, **47**, 1381–1392.
10. Yamashita, Y. (2009) Organic semiconductors for organic field-effect transistors. *Sci. Technol. Adv. Mater.*, **10**, 024313.
11. Bernards, D.A. and Malliaras, G.G. (2007) Steady-state and transient behavior of organic electrochemical transistors. *Adv. Funct. Mater.*, **17**, 3538–3544.
12. Tobjörk, D., Kaihovirta, N.J., Mäkelä, T., Pettersson, F.S., and Österbacka, R. (2008) All-printed low-voltage organic transistors. *Org. Electron.*, **9**, 931–935.
13. Liu, Y., Varahramyan, K., and Ciu, T. (2005) Low-voltage all-polymer

field-effect transistor fabricated using and inkjet printing technique. *Macromol. Rapid Commun.*, **26**, 1955–1959.

14. Pal, B.N., Dhar, B.M., See, K.C., and Katz, H.E. (2009) Solution-deposited sodium beta-alumina gate dielectrics for low-voltage and transparent field-effect transistors. *Nature Mater.*, **8**, 898–902.

15. Prigodin, V.N., Hsu, F.C., Park, J.H., Waldmann, O., and Epstein, A.J. (2008) Electron-ion interaction in doped conducting polymers. *Phys. Rev. B.*, **78**, 035203-1–035203-9.

16. Lu, J., Pinto, N.J., and MacDiarmid, A.G. (2002) Apparent dependence of conductivity of a conducting polymer on an electric field in a field effect transistor configuration. *J. Appl. Phys.*, **92**, 6033–6038.

17. Katz, H.E. and Huang, J. (2009) Thin-film organic electronic devices. *Annu. Rev. Mater. Res.*, **39**, 71–92.

18. Stricker, J.T., Gudmundsdóttir, A.D., Smith, A.P., Taylor, B.E., and Durstock, M.F. (2007) Fabrication of organic thin-film transistors using layer-by-layer assembly. *J. Phys. Chem. B.*, **111**, 6322–6326.

19. Sandberg, H.G.O., Frey, G.L., Shkunov, M.N., Sirringhaus, H., Friend, R.H., Nielsen, M.M., and Kumpf, C. (2002) Ultrathin regioregular poly(3-hexyl thiophene) field-effect transistors. *Langmuir*, **18**, 10176–10182.

20. de Leew, D.M., Simenon, M.M.J., Brown, A.R., and Einerhand, R.E.F. (1997) Stability of n-type doped conducting polymers and consequences for polymeric microelectronic devices. *Synth. Met.*, **87**, 53–59.

21. Wen, Y. and Liu, Y. (2010) Recent progress in n-channel organic thin-film transistors. *Adv. Mater.*, published online January 14, 2010. doi: 10.1002/adma.200901454.

22. Yan, H., Chen, Z., Zhen, Y., Newman, C., Quinn, J.R., Dotz, F., Kastler, M., and Facchetti, A. (2009) A high-mobility electron-transporting polymer for printed transistors. *Nature*, **457**, 679–686.

23. Chua, L.-L., Zaumseil, J., Chang, J.-F., Ou, E.C.-W., Ho, P.K.-H.,

Sirringhaus, H., and Friend, R.H. (2005) General observation of n-type field-effect behaviour in organic semiconductors. *Nature*, **434**, 194–199.

24. Bernius, M.T., Inbasekaran, M., O'Brien, J., and Wu, W. (2000) Progress with light-emitting polymers. *Adv. Mater.*, **12**, 1737–1750.

25. D'Andrade, B.W. and Forrest, S.R. (2004) White organic light-emitting devices for solid-state lighting. *Adv. Mater.*, **16**, 1585–1595.

26. Kulkarni, A.P., Tonzola, C.J., Babel, A., and Jenekhe, S.A. (2004) Electron transport materials for organic light-emitting diodes. *Chem. Mater.*, **16**, 4556–4573.

27. Park, J.H., Kim, C., and Kim, Y.C. (2009) Dual functions of a new n-type conjugated dendrimer: light-emitting material and additive for polymer electroluminescent devices. *J. Phys. D: Appl. Phys.*, **42**, 035101-1–035101-6.

28. So, F., Krummacher, B., Mathai, M.K., Poplavskyy, D., Choulis, S.A., and Choong, V. (2007) Recent progress in solution processable organic light emitting devices. *J. Appl. Phys.*, **102**, 091101-1–09110.21.

29. Zhou, Z., Sheng, X., Nauka, K., Zhao, L., Gibson, G., Lam, S., Yang, C.C., Brug, J., and Elder, R. (2010) Multilayer structured polymer light emitting diodes with cross-linked polymer matrices. *Appl. Phys. Lett.*, **96**, 013504-1–013504-3.

30. Kim, J., Jin, Y., Kim, J., Jung, J., Kim, S.H., Lee, K., and Suh, H. (2009) PCPP Derivatives containing carbazole pendant as hole transporting moiety for efficient blue electroluminescence. *J. Polym. Sci. A.*, **47**, 1327–1342.

31. Gordon, T.J., Yu, J., Yang, C., and Holdcroft, S. (2007) Direct thermal patterning of a π-conjugated polymer. *Chem. Mater.*, **19**, 2155–2161.

32. Huh, J.W., Kim, Y.M., Park, Y.W., Choi, J.H., Lee, J.W., Lee, J.W., Yang, J.W., Ju, S.H., Paek, K.K., and Ju, B.K. (2008) Characteristics of organic light-emitting diodes with conducting polymer anodes on plastic substrates. *J. Appl. Phys.*, **103**, 044502-1–044502-6.

33. Saleh, M., Park, Y., Baumgarten, M., Kim, J., and Müllen, K. (2009) Conjugated triphenylene polymers for blue OLED devices. *Macromol. Rapid Commun.*, **30**, 1279–1283.

34. Deng, X. and Wong, K.Y. (2009) Cross-linked conjugated polymers for achieving patterned three-color and blue polymer light-emitting diodes with multi-layer structures. *Macromol. Rapid Commun.*, **30**, 1570–1576.

35. Villani, F., Vacca, P., Miscioscia, R., Nenna, G., Burrasca, G., Fasolino, T., Minarini, C., and della Sala, D. (2009) OLED with hole-transporting layer fabricated by ink-jet printing. *Macromol. Symp.*, **286**, 101–106.

36. Villani, F., Vacca, P., Nenna, G., Valentino, O., Burrasca, G., Fasolino, T., Minarini, C., and della Sala, D. (2009) Inkjet printed polymer layer on flexible substrate for OLED applications. *J. Phys. Chem. C.*, **113**, 13398–13402.

37. Lin, Y., Chen, Z., Ye, T., Dai, Y., Ma, D., Ma, Z., Liu, Q., and Chen, Y. (2010) Conjugated copolymers comprised cyanophenyl-substituted spirobifluorene and tricarbazole-triphenylamine repeat units for blue-light-emitting diodes. *J. Polym. Sci. Pol. Chem.*, **48**, 292–301.

38. Huang, F., Zhang, Y., Liu, M.S., and Jen, A.K. (2009) Electron-rich alcohol-soluble neutral conjugated polymers as highly efficient electron-injecting materials for polymer light-emitting diodes. *Adv. Funct. Mater.*, **19**, 2457–2466.

39. Christian-Pandya, H., Vaidyanathan, S., and Galvin, M. (2007) Polymers for use in polymeric light-emitting diodes: structure-property relationships, in *Handbook of Conducting Polymers*, 3rd edn, Chapter 5 (eds T.A. Skotheim and J.R. Reynolds), CRC Press, Boca Raton, FL, pp. 5-3–5-35.

40. Tseng, S.R., Chen, Y.S., Meng, H.F., Lai, H.C., Yeh, C.H., Horng, S.F., Liao, H.H., and Hsu, C.S. (2008) Electron transport and electroluminescent efficiency of conjugated polymers. *Synth. Met.*, **159**, 137–141.

41. Forrest, S.R., Bradley, D.D.C., and Thompson, M.E. (2003) Measuring the efficiency of organic light-emitting devices. *Adv. Mater.*, **15**, 1043–1048.

42. Chen, M., Hung, W., Su, A., Chen, S., and Chen, S. (2009) Nanoscale ordered structure distribution in thin solid film of conjugated polymers: its significance in charge transport across the film and in performance of electroluminescent device. *J. Phys. Chem. B.*, **113**, 11124–11133.

43. Li, S., Zhao, P., Huang, Y., Li, T., Tang, C., Zhu, R., Zhao, L., Fan, Q., Huang, S., Xu, Z., and Huang, W. (2009) Poly(p-phenylene vinylenes) with pendent 2,4-difluorophenyl and fluorenyl moieties: synthesis, characterization, and device performance. *J. Polym. Sci. Pol. Chem.*, **47**, 2500–2508.

44. Tonzola, C.J., Alam, M.M., Bean, B.A., and Jenekhe, S.A. (2004) New soluble n-type conjugated polymers for use as electron transport materials in light-emitting diodes. *Macromolecules*, **37**, 3554–3563.

45. Carter, J.C., Grizzi, I., Heeks, S.K., Lacey, D.J., Latham, S.G., May, P.J., de los Panos, O.R., Pichler, K., Towns, C.R., and Wittmann, H.F. (1997) Operating stability of light-emitting polymer diodes based on poly(p-phenylene vinylene). *Appl. Phys. Lett.*, **71**, 34–36.

46. Kang, I.-N. and Shim, H.-K. (1997) Yellow-light-emitting fluorine-substituted PPV derivative. *Chem. Mater.*, **9**, 746–749.

47. Zarras, P., Prokopuk, N., Anderson, N., and Stenger-Smith, J.D. (2007) Investigation of electroactive polymers and other pretreatments as replacements for chromate conversion coatings: a neutral salt fog and electrochemical impedance spectroscopy study, in *New Developments in Coatings Technology*, ACS Symposium Series, Vol. 962, Chapter 4 (eds P. Zarras, T. Wood, B. Richey, and B.C. Benicewicz), American Chemical Society, pp. 40–53.

48. Wang, P., Jin, H., Yang, Q., Liu, W., Shen, Z., Chen, X., Fan, X., Zou, D., and Zhou, Q. (2009) Synthesis, characterization, and electroluminescence of

novel copolyfluorenes and their applications in white light emission. *J. Polym. Sci. Pol. Chem.*, **47**, 4555–4565.

49. Wu, Y., Hao, X., Wu, J., Jin, J., and Ba, X. (2010) Pure blue-light-emitting materials: hyperbranched ladder-type poly (*p*-phenylene)s containing truxene units. *Macromolecules*, **43**, 731–738.

50. Tang, W., Ke, L., Tan, L., Lin, T., Kietzke, T., and Chen, Z. (2007) Conjugated copolymers based on fluorene-thieno[3,2-*b*]thiophene for light-emitting diodes and photovoltaic cells. *Macromolecules*, **40**, 6164–6171.

51. Mo, Y., Deng, X., Jiang, X., and Cui, Q. (2009) Blue electroluminescence from 3,6-silafluorene-based copolymers. *J. Polym Sci. Pol. Chem.*, **47**, 3286–3295.

52. McNeill, C.R. and Greenham, N.C. (2009) Conjugated-polymer blends for optoelectronics. *Adv. Mater.*, **21**, 3840–3850.

53. Hwang, D.-H., Lee, J.-H., Lee, J.-I., Lee, C.-H., and Kim, Y.-B. (2003) White electroluminescent devices using polymer blends. *Mol. Cryst. Liq. Cryst.*, **405**, 127–135.

54. Bernède, J.C. (2008) Organic photovoltaic cells: history, principle and techniques. *J. Chil. Chem. Soc.*, **53**, 1549–1564.

55. Boeing spectrolab terrestrial solar cell surpasses 40 percent efficiency. Dec. 6, 2006. *http://www.boeing.com/news/releases/2006/q4/061206b_nr.html* (accessed Feb. 1, 2011).

56. Bernéde, J.C., Holcombe, T.W., Woo, C.H., Kavulak, D.F.J., Thompson, B.C., and Fréchet, J.M.J. (2009) All-polymer photovoltaic devices of poly(3-(4-n-octyl)-phenylthiophene) from Grignard metathesis (GRIM) polymerization. *J. Am. Chem. Soc.*, **131**, 14160–14161.

57. Kitazawa, D., Watanabe, N., Yamamoto, S., and Tsukamoto, J. (2009) Quinoxaline-based π-conjugated donor polymer for highly efficient organic thin-film solar cells. *Appl. Phys. Lett.*, **95**, 053701-1–053701-3.

58. Lee, M.R., Eckhart, R.D., Forberich, K., Dennler, G., Brabec, C.J., and Gaudiana, R.A. (2009) Solar power wires based on organic photovoltaic materials. *Science*, **324**, 232–235.

59. Lav, T.-X., Tran-Van, F., Bonnet, J.-P., Chevrot, C., Peralta, S., Teyssié, D., and Grazulevicius, J.V. (2007) P and n dopable semi-interpenetrating polymer networks. *J. Solid State Electrochem.*, **11**, 859–866.

60. Nishizawa, T., Lim, H.K., Tajima, K., and Hashimoto, K. (2009) Efficient dyad-based organic solar cells with a highly crystalline donor group. *Chem. Commun.*, **18**, 2469–2471.

61. Mei, J., Ogawa, K., Kim, Y.-G., Heston, N.C., Arenas, D.J., Nasrollahi, Z., McCarley, T.D., Tanner, D.B., Reynolds, J.R., and Schanze, K.S. (2009) Low-band-gap platinum acetylide polymers as active materials for organic solar cells. *ACS Appl. Mater. Interfaces*, **1**, 150–161.

62. De Boer, B. and Facchetti, A. (2008) Semiconducting polymeric materials. *Polym. Rev.*, **48**, 423–431.

63. Kim, K., Liu, J., Namboothiry, M.A.G., and Carroll, D.L. (2007) Roles of donor and acceptor nanodomains in 6% efficient thermally annealed polymer photovoltaics. *Appl. Phys. Lett.*, **90**, 163511-1–163511-3.

64. Liang, Y., Feng, D., Wu, Y., Tsai, S.-T., Li, G., Ray, C., and Yu, L. (2009) Highly efficient solar cell polymers developed via fine-tuning of structural and electronic properties. *J. Am. Chem. Soc.*, **131**, 7792–7799.

65. Lav, T.X., Tran-Van, F., Vidal, F., Péralta, S., Chevrot, C., Teyssié, D., Grazulevicius, J.V., Getautis, V., Derbal, H., and Nunzi, J.-M. (2008) Synthesis and characterization of p and n dopable interpenetrating polymer networks for organic photovoltaic devices. *Thin Solid Films*, **516**, 7223–7229.

66. Tse, C.W., Man, K.Y.K., Cheng, K.W., Mak, C.S.K., Chan, W.K., Yip, C.T., Liu, Z.T., and Djurišić, A.B. (2007) Layer-by-layer deposition of rhenium-containing hyperbranched polymers and fabrication of photovoltaic cells. *Chem. Eur. J.*, **13**, 328–335.

67. Dyer, A.L. and Reynolds, J.R. (2008) Electrochromism of conjugated conducting polymers, in *Handbook of Conducting Polymers*, 3rd edn, Chapter 20 (eds T. Skotheim and J. Reynolds), CRC Press, Boca Raton, FL, pp. 20-1–20-63.

68. Sapp, S.A., Sotzing, G.A., and Reynolds, J.R. (1998) High contrast ratio and fast-switching dual polymer electrochromic devices. *Chem. Mater.*, **10**, 2101–2108.

69. Beaupre, S., Breton, A.C., Dumas, J., and Leclerc, M. (2009) Multicolored electrochromic cells based on poly(2,7-carbazole) derivatives for adaptive camouflage. *Chem. Mater.*, **21**, 1504–1513.

70. Dyer, A.L., Grenier, C.R.G., and Reynolds, J.R. (2007) A poly(3,4-alkylenedioxythiophene) electrochromic variable optical attenuator with near-infrared reflectivity tuned independently of the visible region. *Adv. Funct. Mater.*, **17**, 1480–1486.

71. DeLongchamp, D.M., Kastantin, M., and Hammond, P.T. (2003) High-contrast electrochromism from layer-by-layer polymer films. *Chem. Mater.*, **15**, 1575–1586.

72. Mortimer, R.J., Graham, K.R., Grenier, C.R.G., and Reynolds, J.R. (2009) Influence of the film thickness and morphology on the colorimetric properties of spray-coated electrochromic disubstituted 3,4-propylenedioxythiophene polymers. *ACS Appl. Mater. Interface*, **1**, 2269–2276.

73. Rasmussen, S.C. and Pomerantz, M. (2008) Low bandgap conducting polymers, in *Handbook of Conducting Polymers*, 3rd edn, Chapter 12 (eds T. Skotheim and J. Reynolds), CRC Press, Boca Raton, FL, pp. 12-1–12-42.

74. Beaujuge, P.M. and Reynolds, J.R. (2010) Color control in π-conjugated organic polymers for use in electrochromic devices. *Chem. Rev.*, **110**, 268–320.

75. Garnier, F., Tourillon, G., Gazard, M., and DuBois, J.C. (1983) Organic conducting polymers derived from substituted thiophenes as electrochromic

material. *J. Electroanal. Chem.*, **148**, 299–303.

76. Gazard, M., DuBois, J.C., Champagne, M., Garnier, F., and Tourillon, G. (1983) Electrooptical properties of thin films of polyheteroclcyles. *J. Phys. C.*, **3**, 537–542.

77. Groenendaal, L.B., Jonas, F., Freitag, D., Pielartzik, H., and Reynolds, J.R. (2000) Poly(3,4-ethylenedioxythiophene) and its derivatives: past, present, and future. *Adv. Mater.*, **12** (7), 481–494.

78. Kanazawa, K.K., Diaz, A.F., Geiss, R.H., Gill, W.D., Kwak, J.F., Logan, J.A., Rabolt, J.F., and Street, G.B. (1979) 'Organic metals': polypyrrole, a stable synthetic 'metallic' polymer. *J. Chem. Soc. Chem. Commun.*, **19**, 854–855.

79. Diaz, A.F., Castillo, J., Kanazawa, K.K., and Logan, J.A. (1981) Conducting poly-*N*-alkylpyrrole polymer films. *J. Electroanal. Chem.*, **129**, 115–132.

80. MacDiarmid, A.G., Yang, L.S., Huang, W.S., and Humphrey, B.D. (1987) Polyaniline: electrochemistry and application to rechargeable batteries. *Synth. Met.*, **18**, 393–398.

81. Prakash, R. (2002) Electrochemistry of polyaniline: study of the pH effect and electrochromism. *J. Appl. Polym. Sci.*, **83**, 378–385.

82. Ma, C., Taya, M., and Xu, C. (2008) Smart sunglasses based on electrochromic polymers. *Polym. Eng. Sci.*, **48** (11), 2224–2228.

83. Sonmez, G., Sonmez, H.B., Shen, C.K.F., and Wudl, F. (2004) Red, green, and blue colors in polymeric electrochromics. *Adv. Mater.*, **16**, 1905–1908.

84. Unur, E., Beaujuge, P.M., Ellinger, S., Jung, J.H., and Reynolds, J.R. (2009) Black to transmissive switching in a pseudo three-electrode electrochromic device. *Chem. Mater.*, **21**, 5145–5153.

85. Chandrasekhar, P., Zay, B.J., McQueeney, T., Birur, G.C., Sitaram, V., Menon, R., Coviello, M., and Elsenbaumer, R.L. (2005) Physical, chemical, theoretical aspects of conducting polymer electrochromics in the visible, IR and microwave regions. *Synth. Met.*, **155**, 623–627.

86. Nikolou, M., Dyer, A.L., Steckler, T.T., Donoghue, E.P., Wu, Z., Heston, N.C., Rinzler, A.G., Tanner, D.B., and Reynolds, J.R. (2009) Dual *n*- and *p*-type electrochromic devices employing transparent carbon nanotube electrodes. *Chem. Mater.*, **21**, 5539–5547.

87. Dalgleish, S. and Robertson, N. (2009) A stable near IR switchable electrochromic polymer based on an indole-substituted nickel dithiolene. *Chem. Commun.*, 5826–5828.

88. Chandrasekhar, P., Zay, B.J., McQueeney, T., Scara, A., Ross, D., Birur, G.C., Haapenen, S., Kauder, L., Swanson, T., and Douglas, D. (2003) Conducting polymer (CP), infrared electrochromics in spacecraft thermal control and military applications. *Synth. Met.*, **135–136**, 23–24.

89. MacDiarmid, A.G., Mammone, R.J., Somasiri, N.L.D., and Krawczyk, J.R. (1984) Fuel cells and batteries employing polyacetylene electrodes in aqueous electrolytes. *Energy Tech.*, **11**, 577–593.

90. Scrosati, B., Panero, S., Prosperi, P., Corradini, A., and Mastragostino, M. (1987) Kinetics of semiconducting polymer electrodes in lithium cells. *J. Power Sources*, **19**, 27–36.

91. Naegele, D. and Bittihn, R. (1988) Electrically conductive polymers as rechargeable battery electrodes. *Solid State Ionics*, **28–30**, 983–989.

92. Ibrisagic, Z. (1993) Recent advances in the electrochemical formation of redox polymer films for rechargeable high energy battery applications. I. Positive electrode materials. *Chem. Biochem. Eng. Q.*, **7**, 191–198.

93. Kitani, A., Kaya, M., and Sasaki, K. (1986) Performance study of aqueous polyaniline batteries. *J. Electrochem. Soc.*, **133**, 1069–1073.

94. Somasiri, N.L.D. and MacDiarmid, A.G. (1988) Polyaniline: characterization as a cathode active material in rechargeable batteries in aqueous electrolytes. *J. Appl. Electrochem.*, **18**, 92–95.

95. Trinidad, F., Montemayor, M.C., and Fatas, E. (1991) Performance study of Zn/ZnCl$_2$, NH$_4$Cl/polyaniline/carbon

battery. *J. Electrochem. Soc.*, **138**, 3186–3189.

96. MacDiarmid, A.G., Yang, L.S., Huang, W.S., and Humphrey, B.D. (1987) Polyaniline: electrochemistry and application to rechargeable batteries. *Synth. Met.*, **18**, 393–398.

97. Yamamoto, K., Yamada, M., and Nishiumi, T. (2000) Doping reaction of redox-active dopants into polyaniline. *Polym. Adv. Technol.*, **11**, 710–715.

98. Ryu, K.S., Hong, Y.-S., Park, Y.J., Wu, X., Kim, K.M., Lee, Y.-G., Chang, S.H., and Lee, S.J. (2004) Polyaniline doped with dimethylsulfate as a polymer electrode for all solid-state power source system. *Solid State Ionics*, **175**, 759–763.

99. Cai, Z., Geng, M., and Tang, Z. (2004) Novel battery using conducting polymers: Polyindole and polyaniline as active materials. *J. Mater. Sci.*, **39**, 4001–4003.

100. Chattaraj, A. and Basumallick, I.N. (1993) Improved conducting polymer cathodes for lithium batteries. *J. Power Sources*, **45**, 237–242.

101. Moon, D.-K., Padias, A.B., Hall, H.K., Huntoon, T., and Calvert, P.D. Jr. (1995) Electroactive polymeric materials for battery electrodes: copolymers of pyrrole and pyrrole derivatives with oligo(ethyleneoxy) chains at the 3-position. *Macromolecules*, **28**, 6205–6210.

102. Otero, T.F. and Cantero, I. (1999) Conducting polymers as positive electrodes in rechargeable lithium-ion batteries. *J. Power Sources*, **81–82**, 838–841.

103. Lee, C., Lee, M.H., Kuang, Y.K., Moon, B.S., and Rhee, S.B. (1993) The preparation of polypyrrole and polythiophene in the presence of ferrocene derivatives. *Synth. Met.*, **55**, 1119–1122.

104. Satoh, M., Tabata, M., Kaneto, K., and Yoshino, K. (1985) A highly conducting poly (*p*-phenylene) film. *J. Electroanal. Chem. Interfacial. Electrochem.*, **195**, 203–206.

105. Shacklette, L.W., Toth, J.E., Murthy, N.S., and Baughman, R.H. (1985) Polyacetylene and polyphenylene as anode materials for nonaqueous secondary

batteries. *J. Electrochem. Soc.*, **132**, 1529–1535.

106. Satoh, M., Tabata, M., Kaneto, K., and Yoshino, K. (1986) Characteristics of rechargeable battery using conducting poly(p-phenylene) film. *Jpn. J. Appl. Phys., Part 2 Lett.*, **25**, L73–L74.

107. Levine, K. and Iroh, J.O. (2001) Electrochemical behavior of a composite of polyimide and polypyrrole. *J. Mater. Chem.*, **11**, 2248–2252.

108. Ramasamy, R.P., Veeraraghavan, B., Haran, B., and Popov, B.N. (2003) Electrochemical characterization of a polypyrrole/Co$_{0.2}$CrO$_x$ composite as a cathode material for lithium ion batteries. *J. Power Sources*, **124**, 197–203.

109. Irvin, J.A., Irvin, D.J., and Stenger-Smith, J.D. (2008) Electroactive polymers for batteries and supercapacitors, in *Handbook of Conducting Polymers*, 3rd edn, Chapter 9 (eds T. Skotheim and J. Reynolds), CRC Press, Boca Raton, FL, pp. 9-1–9-29.

110. Arbizzani, C., Balducci, A., Mastragostino, M., Rossi, M., and Soavi, F. (2003) Li$_{1.01}$Mn$_{1.97}$O$_4$ surface modification by poly(3,4-ethylenedioxythiophene). *J. Power Sources*, **119**, 695–700.

111. Du Pasquier, A., Laforgue, A., and Simon, P. (2004) Li$_4$Ti$_5$O$_{12}$/poly(methyl)thiophene asymmetric hybrid electrochemical device. *J. Power Sources*, **125**, 95–102.

112. Sato, M., Tanaka, S., and Kaeriyama, K. (1987) Electrochemical preparation of highly anode-active poly(3-phenylthiophene). *J. Chem. Soc. Chem. Commun.*, 1725–1726.

113. Rudge, A., Raistrick, I., Gottesfeld, S., and Ferraris, J.P. (1994) A study of the electrochemical properties of conducting polymers for application in electrochemical capacitors. *Electrochim. Acta*, **39**, 273–287.

114. Arbizzani, C., Catellani, M., Mastragostino, M., and Mingazzini, C. (1995) n- and p-Doped polydithieno[3,4-b: 3′,4′-d] thiophene: a narrow band gap polymer for redox supercapacitors. *Electrochim. Acta*, **40**, 1871–1876.

115. Pruss, A. and Beck, F. (1984) Reversible electrochemical insertion of anions in poly-p-phenylene from aqueous electrolytes. *J. Electroanal. Chem.*, **172**, 281–288.

116. Nogami, K., Sakamoto, K., Hayakawa, T., and Kakimoto, M. (2007) The effects of hyperbranched poly(siloxysilane)s on conductive polymer aluminum solid electrolytic capacitors. *J. Power Sources*, **166**, 584–589.

117. Toita, S. and Inoue, K. (2010) Improved electrical and thermal properties of polypyrrole prepared by the repetition of a combination of chemical polymerization and 2-naphthalenesulfonic acid treatment. *Synth. Met.*, **160**, 516–518.

118. Mondal, S.K., Barai, K., and Munichandraiah, N. (2007) High capacitance properties of polyaniline by electrochemical deposition on a porous carbon substrate. *Electrochim. Acta*, **52**, 3258–3264.

119. Stenger-Smith, J.D. and Irvin, J.A. (2009) Ionic liquids for energy storage applications. *Mater. Matters*, **4**, 103–105.

120. Rudge, A., Davey, J., Raistrick, I., Gottesfeld, S., and Ferraris, J.P. (1994) Conducting polymers as active materials in electrochemical supercapacitors. *J. Power Sources*, **47**, 89–107.

121. Arbizzani, C., Mastragostino, M., Meneghello, L., and Paraventi, R. (1996) Electronically conducting polymers and activated carbon: electrode materials in supercapacitor technology. *Adv. Mater.*, **8**, 331–334.

122. Conway, B.E., Birss, V., and Wojtowicz, J. (1997) The role and utilization of pseudocapacitance for energy storage by supercapacitors. *J. Power Sources*, **66**, 1–14.

123. Sarangapani, S., Tilak, B.V., and Chen, C.P. (1996) Materials for electrochemical capacitors: theoretical and experimental constraints. *J. Electrochem. Soc.*, **143**, 3791–3799.

124. Wohlfahrt-Mehrens, M., Schenk, J., Wilde, P.M., Abdelmula, E., Axmann, P., and Garche, J. (2002) New materials

for supercapacitors. *J. Power Sources*, **105**, 182–188.

125. Prasad, K.R. and Munichandraiah, N. (2002) Electrochemical studies of polyaniline in a gel polymer electrolyte – high energy and high power characteristics of a solid-state redox supercapacitor. *Electrochem. Solid State Lett.*, **5**, A271–A274.

126. Suematsu, S. and Naoi, K. (2001) Quinone-introduced oligomeric supramolecule for supercapacitor. *J. Power Sources*, **97–98**, 816–818.

127. Garcia-Belmonte, G. and Bisquert, J. (2002) Impedance analysis of galvanostatically synthesized polypyrrole films. Correlation of ionic diffusion and capacitance parameters with the electrode morphology. *Electrochim. Acta*, **47**, 4263–4272.

128. Laforgue, A., Sarrazin, C., and Fauvarque, J.-F. (1999) Polythiophene-based supercapacitors. *J. Power Sources*, **80**, 142–148.

129. Clemente, A., Panero, S., Spila, E., and Scrosati, B. (1996) Solid-state, polymer-based, redox capacitors. *Solid State Ion.*, **85**, 273–277.

130. Stenger-Smith, J.D., Webber, C.K., Anderson, N., Chafin, A.P., Zong, K., and Reynolds, J.R. (2002) Poly(3,4-alkylenedioxythiophene)-based supercapacitors using ionic liquids as supporting electrolytes. *J. Electrochem. Soc.*, **149**, A973–A977.

131. Stenger-Smith, J.D., Guenthner, A., Cash, J., Irvin, J.A., and Irvin, D.J. (2010) Poly(propylenedioxy)thiophene-based supercapacitors operating at low temperatures. *J. Electrochem. Soc.*, **157**, A298–A304.

132. Borjas, R. and Buttry, D.A. (1991) EQCM studies of film growth, redox cycling, and charge trapping of n-doped and p-doped poly(thiophene). *Chem. Mater.*, **3**, 872–878.

133. Ferraris, J.P., Eissa, M.M., Brotherston, I.D., and Loveday, D.C. (1998) Performance evaluation of poly 3-(phenylthiophene) derivatives as active materials for electrochemical capacitor applications. *Chem. Mater.*, **10**, 3528–3535.

134. Sonmez, G., Meng, H., and Wudl, F. (2003) Very stable low band gap polymer for charge storage purposes and near-infrared applications. *Chem. Mater.*, **15**, 4923–4929.

135. Loveday, D.C., Hmyene, M., and Ferraris, J.P. (1997) Synthesis and characterization of p- and n-dopable polymers. Electrochromic properties of poly-3-(p-trimethylammoniumphenyl) bithiophene. *Synth. Met.*, **84**, 245–246.

136. Park, J.H. and Park, O.O. (2002) Hybrid electrochemical capacitors based on polyaniline and activated carbon electrodes. *J. Power Sources*, **111**, 185–190.

137. Mastragostino, M., Arbizzani, C., and Soavi, F. (2002) Conducting polymers as electrode materials in supercapacitors. *Solid State Ion.*, **148**, 493–498.

138. Laforgue, A., Simon, P., Favarque, J.F., Sarrau, J.F., and Laillier, P. (2001) Hybrid supercapacitors based on activated carbons and conducting polymers. *J. Electrochem. Soc.*, **148**, A1130–A1134.

139. Soloducho, J., Cabaj, J., and Swist, A. (2009) Structure and sensor properties of thin ordered solid films. *Sensors*, **9**, 7733–7752.

140. Matsuguchi, M., Yamanaka, T., Yoshida, M., Kojima, S., and Okumura, S. (2009) Long-term stability of humidity sensor using polyaniline blend films upon dc operation. *J. Electrochem. Soc.*, **156**, J299–J302.

141. Khan, R., Solanki, P.R., Kaushik, A., Singh, S.P., Ahmad, S., and Malhotra, B.D. (2009) Cholesterol biosensor based on electrochemically prepared polyaniline conducting polymer film in presence of a nonionic surfactant. *J. Polym. Res.*, **16**, 363–373.

142. Deshpande, N.G., Gudage, Y.G., Ma, Y.R., Lee, Y.P., and Sharma, R. (2009) Room-temperature gas sensing studies of polyaniline thin films deposited on different substrates. *Smart Mater. Struct.*, **18**, 1–6.

143. Ayad, M.M., Salahuddin, N.A., Abou-Seif, A.K., and Alghaysh, M.O. (2008) pH Sensor based on polyaniline and aniline-anthranilic acid copolymer films using quartz crystal microbalance

and electronic absorption spectroscopy. *Polym. Adv. Technol.*, **19**, 1143–1148.

144. Peng, H., Zhang, L., Soeller, C., and Travas-Sejdic, J. (2008) Conducting polymers for electrochemical DNA sensing. *Biomaterials*, **30**, 2132–2148.

145. Singh, R., Prasad, R., Sumana, G., Arora, K., Sood, S., Gupta, R.K., and Malhotra, B.D. (2009) STD Sensor based on nucleic acid functionalized nanostructured polyaniline. *Biosens. Bioelectron.*, **24**, 2232–2238.

146. Riccardi, C.D., Yamanaka, H., Josowicz, M., Kowalik, J., Mizaikoff, B., and Kranz, C. (2006) Label-free DNA detection based on modified conducting polypyrrole films at microelectrodes. *Anal. Chem.*, **78**, 1139–1145.

147. Shim, N.Y., Bernards, D.A., Macaya, D.J., DeFranco, J.A., Nikolou, M., Owens, R.M., and Malliaras, G.G. (2009) All-plastic electrochemical transistor for glucose sensing using a ferrocene mediator. *Sensors*, **9**, 9896–9902.

148. Torsi, L., Farinola, G.M., Marinelli, F., Tanese, M.C., Omar, O.H., Valli, L., Babudri, F., Palmisano, F., Zambonin, P.G., and Naso, F. (2008) A sensitivity-enhanced field-effect chiral sensor. *Nature Mater.*, **7**, 412–417.

149. Slim, C., Ktari, N., Cakara, D., Kanoufi, F., and Combellas, C. (2008) Polyaniline films based ultramicroelectrodes sensitive to pH. *J. Electroanal. Chem.*, **612**, 53–62.

150. Ayad, M.M., El-Hefnawey, G., and Torad, N.L. (2008) Quartz crystal microbalance sensor coated with polyaniline emeralidine base for determination of chlorinated aliphatic hydrocarbons. *Sens. Actuators, B.*, **134**, 887–894.

151. Paul, S., Chavan, N.N., and Radhakrishnan, S. (2009) Polypyrrole functionalized with ferrocenyl derivative as a rapid carbon monoxide sensor. *Synth. Met.*, **159**, 415–418.

152. Su, P.G. and Huang, L.N. (2007) Humidity sensors based on TiO_2 nanoparticles/polypyrrole composite thin films. *Sens. Actuators, B*, **123**, 501–507.

153. Woodka, M.D., Brunschwig, B.S., and Lewis, N.S. (2007) Use of spatiotemporal response information from sorption-based sensor arrays to identify and quantify the composition of analyte mixtures. *Langmuir*, **23**, 13232–13241.

154. Sotzing, G.A., Briglin, S.M., Grubbs, R.H., and Lewis, N.S. (2000) Preparation and properties of vapor detector arrays formed from poly(3,4-ethylenedioxy)thiophene-poly(styrene sulfonate)/insulating polymer composites. *Anal. Chem.*, **72**, 3181–3190.

155. Scorsone, E., Pisanelli, A.M., and Persaud, K.C. (2006) Development of an electronic nose for fire detection. *Sens. Actuators B*, **116**, 55–61.

156. Bailey, A.L.P.S., Pisanelli, A.M., and Persaud, K.C. (2008) Development of conducting polymer sensor arrays for wound monitoring. *Sens. Actuators B*, **131**, 5–9.

157. Yoon, H., Lee, S.H., Kwon, O.S., Song, H.S., Oh, E.H., Park, T.H., and Jang, J. (2009) Polypyrrole nanotubes conjugated with human oftactory receptors: high-performance transducers for FET-type bioelectronic noses. *Angew. Chem. Int. Ed.*, **48**, 2755–2758.

158. Bazani, D.L.M., Lima, J.P.H., and de Andrade, A.M. (2009) MEH-PPV Thin films for radiation sensor applications. *IEEE Sens. J.*, **9**, 748–751.

159. Groenendaal, L., Jonas, F., Freitag, D., Pielartzik, H., and Reynolds, J.R. (2000) Poly(3,4-ethylenedioxythiophene) and its derivatives: past, present, and future. *Adv. Mater.*, **12**, 481–494.

160. Jonas, F., Krafft, W., and Muys, B. (1995) Poly(3,4-ethylenedioxythiophene): conductive coatings, technical applications, and properties. *Macromol. Symp.*, **100**, 169–173.

161. Fallis, S. and Irvin, J.A. (2006) Electrostatic charge dissipation in energetic materials. US Patent 7,108,758, and issued Sept. 19, 2006.

162. Kim, J.-Y., Kim, T.-W., Lee, J.-H., Kwon, S.-J., Jung, W.-G., and Ju, S.-H. (2009) Highly conductive and transparent poly(3,4-ethylenedioxythiophene):p-toluenesulfonate films as a flexible

organic electrode. *Jpn. J. Appl. Phys.*, **48**, 091501–091506.

163. Nishii, M., Sakurai, R., Sugie, K., and Masuda, Y. (2009) The use of transparent conductive polymer for electrode materials in flexible electronic paper. Society for Information Display 2009 Digest, pp. 768–771.

164. Pal, B.N., Dhar, B.M., See, K.C., and Katz, H.E. (2009) Solution-deposited sodium beta-alumina gate dielectrics for low-voltage and transparent field-effect transistors. *Nature Mater.*, **8**, 898–903.

165. Zarras, P., Anderson, N., Webber, C., Irvin, D.J., Irvin, J.A., Guenthner, A., and Stenger-Smith, J.D. (2003) Progress in using conductive polymers as corrosion-inhibiting coatings. *Radiat. Phys. Chem.*, **68**, 387–394.

166. Tallman, D.E., Spinks, G., Dominis, A., and Wallace, G.G. (2002) Electroactive conducting polymers for corrosion control. *J. Solid State Electrochem.*, **6**, 76–84.

167. Michalik, A. and Rohwerder, M. (2005) Conducting polymers for corrosion protection: a critical view. *Z. Phys. Chem.*, **219**, 1547–1559.

33

Hybrid Nanomaterials in Ultrathin Films: the Sol-Gel Method and Π-Conjugated Polymers

Antonio Francesco Frau and Rigoberto C. Advincula

33.1
Why Hybrid Nanomaterials and Thin Films?

Together with colloidal nanoparticles and self-assemblies, thin (thickness below $1\,\mu m$) and ultrathin (thickness in the nanometer range) films are one of the cogent proofs that matter manipulation at the nanoscale can be precise, systematic, and rigorous. Moreover, when technological needs trigger the scientific interest, research efforts can be directed towards novel properties, implementing unusual components into a pre-established matrix, "mixing and mingling" dissimilar moieties confined within very small dimensions. The work behind hybrid materials that bridge research and technological innovation points toward advances in materials by focusing at the nanoscale (Figure 33.1).

Such nanomaterials are an exigency. Theoretical and practical implications of nanoscience have made their way into the scientific community at an accelerated pace in the last decade. Designing molecular materials having optical, electrical, magnetic, and biomedical applications has been a hot topic. Universal protocols that can ensure reliable, environment-friendly, and high-quality results are of paramount importance in the fabrication of such nanostructures. In an interview [1], Gero Decher – a codiscoverer of the "layer-by-layer" (LbL) film deposition, stated that "In advertising for the LbL assembly technique, I have always used phrases like 'multifunctional coatings for surfaces of almost any kind and any shape'." Smaller and smarter devices that can be fabricated more efficiently with precise molecular control have become a *de facto* objective for most applied materials investigations. Multifunctionality and/or structural flexibility are advantageous when heterogeneous materials are put together. The ultimate goal is to mimic the nanotechnology of Nature: (i) some of the most efficient optoelectronic nanosystems are the photosynthetic proteins in the cell membranes; (ii) the highest turnover frequency is displayed by simple prokaryotes that fix molecular N_2; and (iii) lastly, the finest chemical sensors are the receptor cells of many mammalians [2].

Fabricating thin and ultrathin materials with superior performance is challenging. The assembly of nanostructured molecular species requires deposition of specific amounts of material and careful removal of unwanted excess.

Functional Polymer Films, First Edition. Edited by Wolfgang Knoll and Rigoberto C. Advincula.
© 2011 Wiley-VCH Verlag GmbH & Co. KGaA. Published 2011 by Wiley-VCH Verlag GmbH & Co. KGaA.

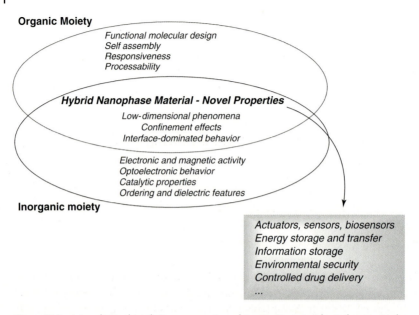

Figure 33.1 Interrelationships between organic and inorganic materials at the nanoscale.

Mutual, unpredictable interactions between dissimilar components (e.g., discrete molecules, polymers, crystalline compounds) may lead to heterogeneous film properties. The goal is to obtain properties greater than a simple sum of the parts, that is, synergistic. This last point is most intriguing, because the assembly of dissimilar organic and inorganic phases within the same material affords the so-called hybrid materials. The LbL-driven hybrid systems, for example, have shown structural flexibility for the formation of freestanding films filled with nanoparticles [3] and the fine tuning of their optoelectronic properties [4]. Manipulation of the film down to the molecular level, pH-dependant thickness control [5], and unique structure–property relationships arise from insertion of dissimilar components. This has found applications from microelectronics all the way to biomolecule-sensing electrodes [6].

33.2
How to Fabricate Hybrid, Layered, Thin-Film Nanomaterials

Among all the fabrication techniques of nanostructured, textured materials (self-assembled monolayers (SAMs), molecular beam epitaxy, laser ablation, chemical vapor deposition, chemical vapor transport, Langmuir–Blodgett (LB) method, etc.) the LbL method has been most popular and widely applied. In a *Science* article, Decher [7] discussed how an old idea by Iler [8] about the alternate sorption of silica and alumina colloidal particles onto flat substrates was revived and applied to *ad hoc* synthesized, anionic and cationic molecules [9]. Decher and coworkers

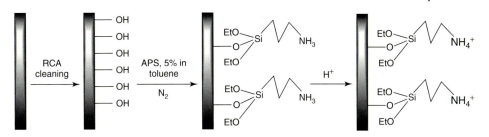

Figure 33.2 Preparation and functionalization of an inorganic surface by APS.

first demonstrated the validity of the LbL approach with amphiphilic molecules having rigid biphenyl cores and low charge density. This initially did not facilitate a stepwise, straightforward LbL deposition. Nonetheless, LbL has reached a vast popularity that is witnessed by the steady exponential growth of scientific publications through the years. In its simplest depiction, the LbL assembly is the repeated sequence of adsorption/rinse of alternately charged, polyions on an opportunely prepared (uniformly functionalized) substrate. By functionalization, an inorganic surface is treated with specific functional groups that render it chemically reactive toward other species. A typical example is the treatment of inorganic surfaces (Si, SiO$_2$, soda-lime glass, etc.) with 3-ethoxy(aminopropyl)silane (APS) (Figure 33.2) [10]. Functionalization of the clean, as-activated surface is subsequently carried out by nucleophilic displacement [11] of the ethoxy groups. Hence, the amino head group sticks out in a densely packed fashion as a monolayer. Protonation results in a uniformly charged surface available for adsorption. Surface functionalization can be carried out in other ways that are mostly SAM-related protocols. Such ideas can be traced back to Zisman, Kuhn, and Nuzzo's work on surfactants, silanes, and alkyl sulfides or thiols [12]. Functionalization is done in such a way that the first polyion can be chemisorbed and bound tightly to the aforementioned APS-monolayer [13].

The entire LbL process is pictorially represented in Figure 33.3: the functionalized substrate (i) is immersed in the polyion solution, (ii) to bring oppositely charged species into contact. If the surface is functionalized with protonated APS, then the first chemisorbed layer will be the negatively charged polyion. Obviously, the opposite process (absorption of cationic polyelectrolyte on a negatively charged surface) is also possible. This affords charge quenching at first and then charge reversal. This phenomenon is termed as *charge overcompensation*. Subsequent washing of the same substrate (iii) is required for removal of excess, unbound material whose anchoring is loose. The next absorption of a cationic polyion (iv) switches the surface charge. Again, a rinsing step is required afterwards (v). Polyions, whose alternate deposition builds up the film, are cast mostly from dilute aqucous solutions (mg/ml scale). Such experimental conditions enable attainment of the adsorption isotherm very easily [14], according to a typical Brunauer, Emmett, and Teller (BET) model. Note that the process can be easily automated.

This straightforward setup displays a plethora of chemical *raisons d'être*: the interlayer attractions are *either* strong (ion–dipole attractions, circa 15 kJ mol^{-1}) *or*

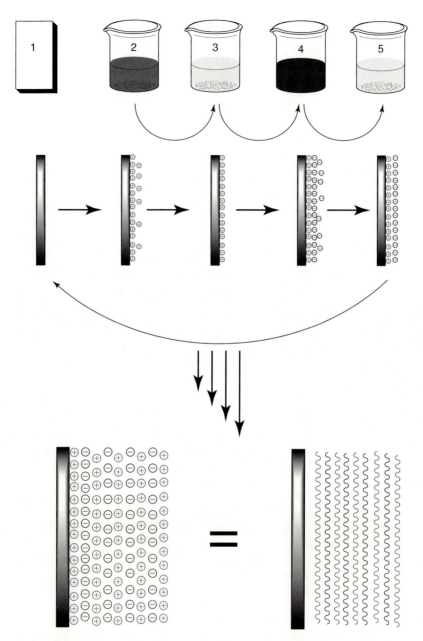

Figure 33.3 General scheme of layer-by-layer (LbL) assembly of polyions (2, 4) and their excess rinse (3, 5) on a functionalized support (1).

very strong (ionic bonds, circa $250 \, \text{kJ} \, \text{mol}^{-1}$) [15]. These interactions are stronger than the van der Waals/London dispersion forces normally utilized in a LB assembly. Because of their high charge density, polycharged species – rather than small, singly charged molecules – can be optimally adsorbed and uniformly cover the solid support, indeed layer after layer. Washing steps are also essential, favoring the removal of nonadsorbed species that may negatively influence successive linear adsorptions [16]. A parallel development in recent years is the LbL electrostatic assembly of core-shell and hollow-shell particles and nanoparticles systems that has been well reported. Although a field of equally high interest, this review will not cover this topic [17].

The iteration of the LbL process *per se* allows fabrication of films in which the thickness and periodicity can be probed by X-rays [18]. Kleinfeld and Ferguson [18] first reported the structural order in poly(diallyldimethylammonium) chloride (PDDA)-hectorite clay multilayered films exhibiting Bragg peaks corresponding to layer spacings and thickness consistent with an ABABAB alternation of silicate sheets (0.96 nm thick) and polydiallyldimethyl am-monium chloride (PDDA) layers (0.4–0.5 nm thick). Further proof of structural organization in LbL films can also come from X-ray reflectivity [19]: Decher *et al.* demonstrated that poly(styrenesulfonate) sodium salt (PSS), poly(allylamine) hydrochloride (PAH), and poly-(1-(4-(3-carboxy-4-hydrophenylazo)-benzenesulfonamido)-1,2-ethanediyl) sodium salt (PAZO) could be assembled as (PSS/PAH/PAZO/PAH)$_n$ multilayered films and give rise to Bragg peaks and quantifiable Kiessig fringes. The spacing that they obtained (93.4 Å) corresponds to the (PSS/PAH/PAZO/PAH)$_n$ repeating unit in a superlattice.

Other than X-ray methods, a wide variety of surface-sensitive, spectroscopic, and microscopic techniques can be utilized for investigating these films. If the macromolecular species bear chromophore groups, UV-Vis/IR analyses are a valid measurement that can confirm linear film growth. A myoglobin/PSS thin film was monitored by Kunitake and coworkers [20] by UV absorption together with quartz crystal microbalance (QCM) measurements. Absorbance increase per layer of the Soret band associated to the protein at 409 nm was observed. Ram *et al.* [21] fabricated LbL films of PSS and protonated (doped) poly(aniline) (PANI). The absorbances of the benzenoid ring ($\pi - \pi^*$ transition, 340 nm) and the polaronic band (450 nm) of PANI gained intensity per layer, according to the Lambert–Beer law. Kunitake and coworkers [22, 23] demonstrated the usefulness of QCM measurements for similar systems [20] and inorganic oxide-based thin films.

Spectroscopic ellipsometry [24] and microscopic imaging such as transmission electron microscopy (TEM) or atomic force microscopy (AFM) [25] are potent tools for structural analysis of organic and inorganic films. AFM or surface probe microscopy (SPM) in general is capable of very high lateral resolution (more than the optical diffraction limit). Moreover, AFM can provide 3D surface morphological profiles and is applicable from single atoms to biological macromolecules [26]. Topographic imaging of LbL surfaces is of high interest especially when features become technologically relevant [27]. With ellipsometry, films can be probed with respect to their subnanometer scale surface structure/roughness, composition, and multilayer architectures [28] based on interaction with elliptically polarized light.

Thus, LbL self-assembly is very accessible and easily adaptable for high turnover fabrication protocols. The as-fabricated films may seemingly appear as "fuzzy" systems but positional order of the single slabs is possible. Whether these materials are polyelectrolytes, colloids, nanoparticles, or proteins, the outcome is "the design of smart materials that can have much higher complexity" [1]. The quest for more complex, LbL-based nanostructures that exploit other interactions to assemble dissimilar molecular moieties is a possibility. H-bonded polymer films [29] (key point for stimuli-responsive, biological applications), inclusion of electron-pool π-systems in a supramolecular fashion [30], through-metal coordination of inorganic sheets (such as α-ZrP/Ti$_2$NbO$_7{}^-$) to polycations [31], Pt(II) ions to poly(amidoamine) dendrimers [32] or Zr-based coordination polymers [33], mutual interaction between donor polymers and acceptor polymers by charge transfer [34], are some of the possibilities. Clark and Hammond [35] demonstrated LbL growth toward specific regions of a surface based on steric repulsions. In their model, a series of polyamines absorbed onto patterned SAM surfaces displayed preferential deposition according to pH (degree of ionization), hydrophobic/hydrophilic interactions depending on the polyamine backbone and tuning of the polyion–surface interaction.

Assembling dissimilar molecular moieties is an ulterior expansion of the LbL field of applicability. It can deal with a wide variety of heterogeneous "components" in the same nanostructure to achieve novel, "smart" materials that exhibit unusual features and properties. Some fundamental device functions are: sensors (nanomaterials capable of recognizing, measuring a specific analyte into a detectable signal), processors (nanomaterials that allow unidirectional or mutual conversion), and actuators (nanodevices that, given a specific amount of energy, convert it into work) [36]. For materials choices, other molecular and macromolecular species within an LbL framework are possible. Examples include the assembly of dendrimers [37], quantum dots [38], biological macromolecules such as nucleic acids [39], polysaccharides [40], nanowires [41].

33.3
Hybrid Nanomaterials: More than Just "Clay" and "Plastic"

The incorporation of "hard," water-insoluble, or colloidal materials such as metal-oxide layers and particles into complex nanostructured architectures is intriguing. The fabrication of such nanocomposites is expected to be a technological advancement, that is, also widely applicable to a variety of functions including: biomolecular recognition [42], hybrid solar cells (a polymer serves as light absorber/hole transporter and the metal oxide works as electron transporter) [43], nonspecific protein adsorption [44], modulation of electrical resistance within the film depending upon the degree of organic-inorganic blend [45], and improved room-temperature conductivity [46].

Different classifications of such nanocomposites have been proposed based on the type of association between organic and inorganic components. Gangopadhyay and De [47] proposed a distinction between organic-in-inorganic (the organic polymer is confined into an inorganic template) and inorganic-in-organic (the inorganic nano-object is embedded in an organic matrix) materials, with a special focus on conducting polymers (CPs). Novak [48] proposed a classification based on phase-separation issues: in a hybrid, textured structure, the organic–inorganic synergism is governed by interfacial properties. Thus, host–guest frictions, aggregation phenomena, and subsequent precipitation/isolation of a compound in the matrix may be an unwanted outcome [49]. A "tandem" solar cell is a specific example, where an inorganic semiconductor (TiO$_2$, mostly) is paired up to functionalized fullerenes blended with CPs [50]. Aging of such a heterogeneous material leads to phase separation and photoinduced degradation. The degree of phase separation can vary, but nevertheless is detrimental. Based upon these experimental findings, Novak described "modified inorganic glasses" and "modified organic polymers," with evident stress on the two aforementioned compound classes.

Metal-oxide-conducting polymer (MO-CP) nanocomposites are of high interest for potential low-cost optoelectronic devices [51]. The combination of high power efficiency – typical for inorganic semiconductors such as TiO$_2$, with ease of processability over large areas (typical for polymers), is of high interest [52]. The two materials – TiO$_2$ and CP, are very dissimilar and yet embed intriguing, potentially complementary characteristics. The inorganic oxide – as a white pigment, n-type semiconductor (E_g circa 3.2 eV *vs.* NHE at pH $= 0$) [53], is also a photocatalyst and has high electron-transporting capabilities suitable for efficient solar-energy converters [54]. In crystalline TiO$_2$, the electrons occupy energy bands as a consequence of the extended bonding network and the bandgap determines the optical-absorption wavelength. When TiO$_2$ is irradiated at 388 nm (the wavelength corresponding to the E_g of the anatase phase) or lower, the valence-band electrons move up to the conduction band and leave positive holes behind. This mechanism is known as *charge separation* and its efficiency is of dramatic importance in photocatalytic reactions and photocurrent generation [43b].

33.4
Sol-Gel Chemistry for Nanostructuring

33.4.1
Basics

In general, the deposition of metal oxides as a thin film is in high demand. In particular, titanium-based materials have found an exceptionally vast array of applications. As crystalline/nanocrystalline phase materials, they are known for their optical (the refractive index of TiO$_2$ is higher than 2.5) [55] and electronic properties. As *amorphous* TiO$_2$ materials, they have remarkable dielectric characteristics [56]. A recent study by Busani and Devine [57] has shed some light on the so-called "semiconductor roadmap." This involved the search for binary or ternary oxides whose permittivity affords superior performances compared to Si-based amorphous oxides. From their viewpoint, an ideal candidate to replace amorphous SiO_2 on crystalline Si films in the electronics industry should have large $\varepsilon (\geq 25)$ value, possess a wide E_g, be thermodynamically stable at high temperatures and have reasonably big band offset at the heterojunction. They fabricated TiO_2 films by plasma-enhanced chemical vapor deposition (PECVD) and, although its bandgap is relatively small, their report showed that such films have surprisingly high dielectric constants. Moreover, the deposition technique *per se* afforded crystalline TiO_2 (both anatase and rutile), which may be detrimental due to possible electrical leakage between semiconductor and dielectric coating at the interface.

New ways of depositing inexpensive dielectric TiO_2 films/coatings are important. To this effect, the *sol-gel* techniques have been most useful. Conceived in the mid-1800s as a scientific curiosity [58], sol-gel chemistry nowadays enables fabrication of materials (mostly metal oxides), and coatings on any scale [59]. Extensive studies in the 1950s and the 1960 [60], allowed for the synthesis of a large number of novel ceramic oxides – which could not otherwise be obtained by classical "shake & bake" methods. The order of magnitude of sol-gel-related publications, as of 2007, has been 10^4 per year.[1] The range of applicability of such a technique is vast: sol-gel-based materials can function as fluorescence-based sensors [61], selective catalysts [62], efficient drug carriers [63], starting materials for nanocluster research [64], advanced electrochromic devices [65], thermal insulators [66], solar cells [67], and the list can go on.

Although the process was originally intended for the synthesis of glasses and siliceous materials [68], its applicability to a host of inorganic and hybrid materials has been demonstrated: transition-metal (TM) oxides (to cite a few, TiO_2, ZrO_2, WO_3, V_2O_5) [69], polyelectrolyte-siliceous hybrids [70], TM complexes–inorganic oxides [71]. In this last example, O'Regan and Grätzel [71] reported on a high surface area TiO_2 film deposited by sol-gel methods and sensitized with the trimeric $RuL_2[\mu\text{-}(CN)Ru(CN)L_2]_2$ ruthenium complex (L and L' are 2,2'-bipyridine-4,4'-dicarboxylic acid and 2,2'-bipyridine, respectively). Such a dye-sensitized solar cell could harvest solar energy and convert it to electrical current efficiently (current density up to $12\,\text{mA cm}^{-2}$, 5 million turnovers with no decomposition).

Sol-gel chemistry is a *chemie douce* method [72] for fabricating oxide materials from molecular precursors (typically, metal alkoxides or chlorides) under mild reaction conditions. An oxide network can be grown by progressive polycondensation reactions of molecular precursors in a liquid medium. The main steps of the process are:

1) hydrolysis of the molecular precursor(s);
2) polycondensation;
3) sol-gel transition (also known as *"gelation"*).

1) Rough estimation based on SciFinder data.

Further steps such as aging and drying of the as-formed materials are optional. The steps *per se* are the same no matter what the starting materials. For the sake of simplicity, silicon-oxide-containing materials will be primarily discussed and should be synonymous to other metal-oxide materials.

33.4.2
Hydrolysis versus Condensation

The *leitmotiv* behind the growth of the Si–O network is the transformation of Si–OR and Si–OH containing species in siloxanes. Formally, the chemistry comprises a hydrolysis step,

$$\overset{|}{\underset{|}{-\text{Si}}}\text{/////OR} \quad + \quad H_2O \quad \longrightarrow \quad \overset{|}{\underset{|}{-\text{Si}}}\text{/////OH} \quad + \quad ROH$$

and either alcohol-producing or water-producing condensation mechanisms:

$$\overset{|}{\underset{|}{-\text{Si}}}\text{/////OR} \quad + \quad \overset{|}{\underset{|}{-\text{Si}}}\text{/////OR} \quad \longrightarrow \quad \overset{|}{\underset{|}{-\text{Si}}}\text{/////O}-\overset{|}{\underset{|}{\text{Si}}}\text{///// } \quad + \quad ROH$$

$$\overset{|}{\underset{|}{-\text{Si}}}\text{/////OR} \quad + \quad \overset{|}{\underset{|}{-\text{Si}}}\text{/////OR} \quad \longrightarrow \quad \overset{|}{\underset{|}{-\text{Si}}}\text{/////O}-\overset{|}{\underset{|}{\text{Si}}}\text{///// } \quad + \quad H_2O$$

The most common molecular precursors are hydrolyzable metal alkoxides $M(OR)_n$ or chlorides MCl_n of a polymeric network-forming cation. Such alkoxides include those of the metals of Group III, IV as well as TMs [73].

The choice of the molecular precursor is critical in terms of metal center and ligands. As the electronegativity of the former decreases, its reactivity toward hydrolysis increases and so does the polarity of the M–O bond [74]. The hydrolysis and polycondensation reactions initiate at numerous sites as mixing occurs. When sufficient interconnected M–O–M bonds are formed in a region, they participate cooperatively as colloidal submicrometer particles or films. The size of the particles and their crosslinking depend on the pH and the so-called alkoxide-to-water ratio, $R_W = [M(OR)_n]/[H_2O]$. The overall reaction for sol-gel processing of, say, silicon tetra-alkoxides is

$$Si(OR)_4 + 2H_2O \rightarrow SiO_2 + 4ROH$$

$R_W \gg 2$ favors condensations whilst $R_W \leq 2$ favors hydrolysis. This is also true for titanium-based sol-gel processes [75]. Hydrolysis and condensation reactions compete with each other [76] and this happens for *any* $M(OR)_n$ system [77].

The reactivity of TM alkoxides in these conditions is much higher than that of silicon alkoxides, owing to the lower electronegativity of M compared to Si.

Moreover, the coordination number of TMs is higher than their valency [78]. It turns out that metal alkoxides are Lewis acids. Often, bidentate ligands (BLs) are employed for slowing down their hydrolysis [79] in such a manner that a fraction of the alkoxide ligands is substituted by BL [80]. The reactivity of the new molecular precursor, $M(OR)_{n-y}(BL)_y$, is lower than the original $M(OR)_n$ due to its behavior as a metal complex chelated by multidentate BL ligands [81]. The chelate effect exerted on M affords a stronger M–L bond and therefore a lower tendency of L to hydrolysis.

This is an important feature of sol-gel molecular precursors (and their modifications): upon controlling the reactivity of the $M(OR)_n$ species, appropriate organic functionalities can be introduced into the material. For instance, methacrylic acid can be used instead of acetylacetonate [82] to modify TM alkoxides and tether polymerizable double bonds to M *via* carboxylate ligands.

The sol particles can be interconnected by covalent bonds, van der Waals forces or hydrogen bonds, depending on the nature of the M–L surface groups and their charges [83–85]. Gel formation can be attained by several means: rapid evaporation of the solvent, flocculation due to entanglement of the polymer chains, pH- or ionic-strength-induced decrease of the surface potential. A polymer oxide network can be attained in such a way. Depending on the further use of such materials, thermal processes are implemented to obtain dense ceramic films, monolithic objects, or glasses.

33.4.3
Sol-Gel Process and Materials Science: the State-of-the-Art

Nowadays, sol-gel-driven methods are ubiquitous. Advantages include low temperature processing and shaping, sample homogeneity and purity, easy tailoring of composition and properties, possibility of incorporating specific functions. The sol-gel or sol-gel-like techniques provide materials chemists with a most versatile means for the synthesis and fabrication of unique objects. Sol-gel methods can afford either extremely light or extremely tough oxide materials [86]. Optical applications of sol-gel materials that include interesting developments are transparent ceramics [87] and/or solid-state lasers [88]. Dip coating of suitable substrates in $In(NO_3)_3/SnCl_4$ molecular precursors allowed Reidinger *et al.* [89] to obtain transparent conductive indium tin-doped oxide (ITO) coatings (Figure 33.4):

Sol-gel glasses, with their porous structures, enable entrapment of a variety of guest molecules. When such species are capable of emissions from singlet or triplet excited states, fluorescent or phosphorescent hybrid materials can be fabricated [90]. Waveguides [91] are of interest along with organically modified silicates (ORMOSILs) [92] which may act as dye hosts. Other applications of sol-gel-cast systems recently explored include the synthesis of electro-optical materials for telecommunication industries. Nonlinear optical (NLO) materials have found their way into such applications as self-frequency doubling lasers [93]. These optoelectronic sol-gel materials have structure–property relationships within the material itself and not necessarily in the guest molecules. For instance, sol-gel

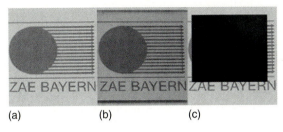

(a)   (b)   (c)

Figure 33.4 ZAE Bayern's logo without coating (a), with optimized ITO coating (b) and with heavily doped, nontransparent ITO coating (c). Image reproduced from Ref. [88].

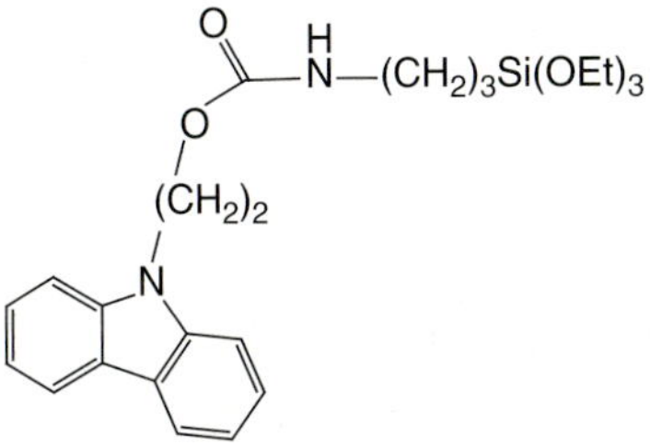

Figure 33.5 Carbazole-modified triethoxysilane, Ref. [95].

processing can afford photorefractive composites [94]: Choi *et al.* [95] reacted a chromophore photosensitizer, carbazole, with a 3-isocyanatopropyltriethoxysilane to afford a carbazole-modified silane (Figure 33.5):

The electro-optical features of such a modified silane could be modulated by simply varying the composition of the two starting materials, rather than fabricating a customary sol-gel framework and dope it with sensitizers.

One of the most important electro-optical applications of the sol-gel method, however, is the fabrication of titanium-based solar cells [71]. Such photovoltaic devices have demonstrated to be a more processable alternative to conventional devices, especially when used in conjunction with liquid electrolytes [96]. Electron transfer at the organic–inorganic heterojunction can be optimized and afford a photon-to-current yield as high as 33%, when a sol-gel cast TiO_2 mesoporous film is sensitized with 2,2′,7,7′- tetrakis(N,N-di-p-methoxyphenyl-amine)9,9′-spirobifluorene [97].

Hybrid polymer/TiO_2 solar cells have improved [98] interfacial areas (a key feature for efficient charge separation and generation of photocurrent) between the dissimilar moieties. Attempts have been reported by different groups [99]. It has been recognized [100] that a device's overall performance is limited by poor penetration of the polymer into the porous film. Since then, novel strategies have been implemented involving the fabrication of a TiO_2/polymer bulk heterojunction by *in situ* conversion of Ti(^{i}OPr)$_4$ codeposited from solution with a π-conjugated polymer [101].

Other than Ti-based materials: TMs such as V, W, Zr are quite popular for optoelectronic applications. Schweiss *et al.* [102] recently reported on the LbL fabrication of a sol-gel-based V_2O_5-poly(aniline) nanocomposite. Electrochemical surface plasmon resonance spectroscopy (EC-SPR) demonstrated that the molecularly ordered assembly generated an electrochemical response governed by the vanadium xerogel and an optical response governed by the π-conjugated polymer. The same system studied by Huguein *et al.* [103] showed that the material could be a good candidate for lithium batteries.

Composite TiO_2-WO_3 sol-gel systems, pioneered by Bechinger *et al.* [104], have found practical applications as photochemical materials. Their activity for water photo-oxidation was recently assessed by Luca *et al.* [105]. NiO and NiO/TiO$_2$ sol-gel systems are employed as important electrochromic devices [106], Ag-doped TiO_2 multilayered films have proved bactericidal activity [107], M-doped TiO_2/CeO_2 ultrathin films (M = alkali, alkali earth, transition, rare earth ions) [108], and mixed SiO_2/ZrO_2 dielectrics are valuable as substrates for organic transistors [109], M_xO_y (M = Al, Ti, Zr, Nb, Ta) or $M'O$-TiO_2 (M' = Pb, Ba, Sr) composites showed ordered mesoporous structures [110].

An important, common point so far is the nanoscale contact between active materials. Whether the interest is on exciton generation or bandgap modulation [111], the surface contact between active "ingredients" must be maximized. According to Luca *et al.*, several possible contacts are feasible down to the nanoscale level (Figure 33.6): (a) multilayered, LbL-like films; (b) spherical, shell-core architectures; (c) smaller nano-objects agglomerated onto bigger nano-objects; and (d) "muffin-like" nano-objects incorporated into a mesoporous matrix.

From these examples, it is clear that sol-gel cast, metal-oxide materials are nearly ubiquitous in materials science owing to their dielectric [112], semiconducting

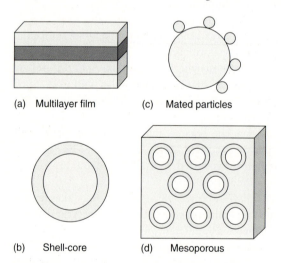

(a) Multilayer film

(c) Mated particles

(b) Shell-core

(d) Mesoporous

Figure 33.6 Nanoscale contacts according to Ref. [106]: (a) multilayered, LbL-like films, (b) spherical core-shell architectures, (c) smaller nano-objects agglomerated onto bigger nano-objects, and (d) nano-objects incorporated onto a mesoporous matrix.

[113], photocatalytic [114], gas-sensing [115], and protecting [116] properties. These facts render sol-gel chemistry of metal oxides a versatile method for synthesizing oxides and hybrid entities. Recently, an even more exciting aspect of the process was brought up: hydrolysis-based methods can be carried out *without* hydrolysis (other functional groups can be used, that is) [117]. As pointed out by Vioux and Mutin [118], the last development is intriguing because it opens up a plethora of new synthetic routes with unusual solvents [119].

33.5
Conducting Polymers for Nanostructuring

33.5.1
Basics

CPs or π-conjugated polymers have been of high interest over the last three decades after the first reports on resistance lowering in I_2-doped poly(pyrrole) [120] and the discovery of the tunable conductivity of poly(acetylene) [121].

CPs (the most common ones are depicted in Figure 33.7) possess an extended, delocalized π-system that generally runs along the polymer backbone. This fact, *per se*, renders CPs semiconductors in their pristine state. However, when charge carriers are brought into their HOMO–LUMO energy levels, the electrical conductivity increases dramatically: p-doped (oxidized) poly(acetylene) shows values as high as 10^5 S cm^{-1} – much like metallic Ag [122]. Metals possess high electronic conductivity because their electrons are free to move through their band structure. In a similar manner, CPs must have charge carriers *and* accessible molecular orbitals, that is, a conjugated, continuous overlapping of π-orbitals along the backbone meets the latter requirement. However, most CPs do not possess intrinsic charge carriers: therefore, doping is necessary. Through such a process, charged defects (polarons, bipolarons, and/or solitons) can be introduced and become available as charge carriers. The doping, namely the introduction of controlled, small (usually

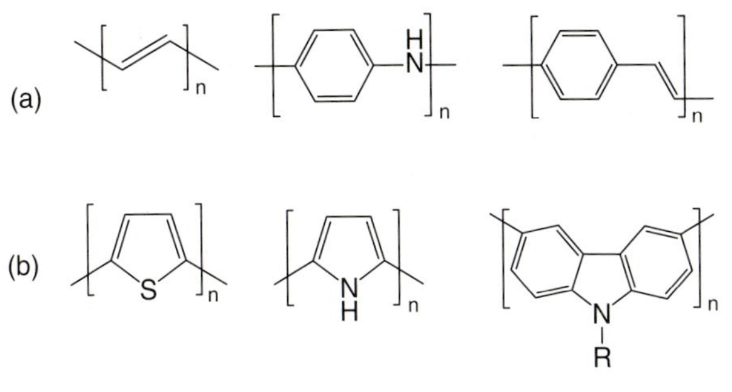

(a)

(b)

Figure 33.7 Neutral (undoped) forms of common CPs. Left to right, (a) trans-poly(acetylene), poly(aniline), poly(p-phenylene vinylene). Left to right, (b) poly(thiophene), poly(pyrrole), poly(carbazole).

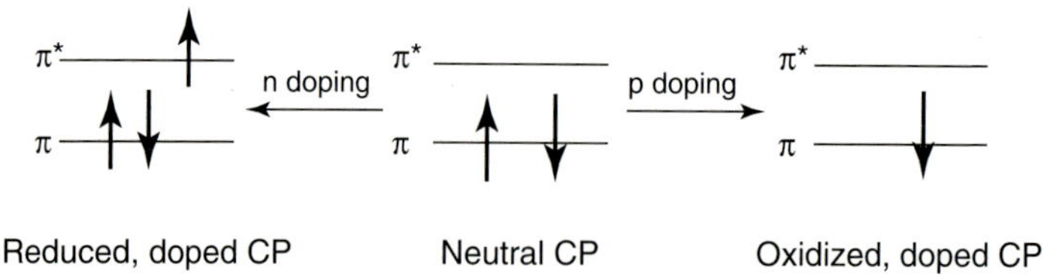

Figure 33.8 Schematic depiction of p- and n-doping in a CP in terms of π and π^* MOs.

less than 0.1% in weight) amounts of impurities in an otherwise "pure" or intrinsic semiconductor, can be done by chemical or electrochemical means. Chemical doping (introduction of a dopant ion in the polymer matrix) or electrochemical doping (application of a voltage so as to force a specific anion to enter the polymer matrix) produces a p-doped or n-doped CP, with an associated increase of the conductivity. In the former case, since electrons are knocked off the delocalized π-backbone, p-doping can be termed as a *partial oxidation*; conversely, n-doping can be regarded as the partial reduction of such a backbone within the CP. Figure 33.8 summarizes this idea.

Examples are numerous: poly(acetylene) can be *chemically* p-doped by oxidation with I_2 or Br_2 and displays tunable metal-to-insulator transition as the dopant concentration (the I_3^- ion) varies [123]:

$$trans - [CH]_x + (3/2)xy I_2 \rightarrow [(CH)^{+y}(I_3)_y^-]_x \quad (y \leq 0.07)$$

On the other hand, *electrochemical* p-doping of poly(acetylene) with $LiClO_4$ is usually carried out by anodic oxidation in propylene carbonate [124]:

$$trans - [CH]_x + xy ClO_4^- \rightarrow [(CH)^{y+}(ClO_4)_y^-] + xy e^-$$

Chemical and electrochemical n-doping of CPs is possible, although uncommon – it usually involves reduction via alkali metals or application of a cathodic voltage to a CP solution. In all the above cases, peculiar spectroscopic signatures of the doping states – polarons, bipolarons, even excitons – can be readily tracked [125].

Further generalizations of the concept of doping extend to processes in which there are neither dopant ions (photodoping, charge-injection doping) nor redox reactions. The latter is termed as *nonredox*, in that no electrons are injected on/knocked off the π-backbone. Only the electronic levels of the CP are rearranged. Poly(anilines), PANIs, are a typical class of CPs whose conductivity can be increased up to 10^3 S cm^{-1} by protonation with strong acids (HCl, H_2SO_4, H_3PO_4, HBF_4) [126]. Current efforts aim at tuning such a remarkable property within wider ranges of pH [127] because PANI shows sensing and electrochromic capabilities.

33.5.2
Electro-Optical Thin-Film Materials

Recently, the journal *Chemical Communications* [128] marked the 25th anniversary of the seminal discovery of the first CP, a halogen derivative of poly(acetylene), by Shirakawa *et al.* [129]. Before then, polymers had always been regarded as electrically insulating materials. That finding opened up the road toward electronic applications of polymers [130] culminating in a Nobel Prize in Chemistry to Shirakawa, MacDiarmid, and Heeger in 2000 [131]. Nowadays, materials science relies on CPs for a most diverse host of applications: light-emitting diodes [132], solar cells and electrochromic devices [133], sensitivity for chemo- and biosensors [125], anticorrosion coatings [134].

A polymer light-emitting diode (PLED) is essentially an electronic light source whose emitting material is a polymer film [135]. This film is sandwiched between a transparent anode and a low work function cathode. When a voltage is applied across the device it causes the injection of electrons from the cathode to the emissive layer. This, in turn, causes an electron flow to travel through the device (cathode → anode) and a hole to flow to travel in the opposite direction (anode → cathode). Electrons and holes eventually recombine in the proximity of the emissive layer. In organic semiconductors, holes are more mobile than electrons and this causes radiative emission whose wavelength is in the visible region. Hence, a PLED is a device that converts an electrical impulse in light. Following Burroughes *et al.*'s work at Cambridge University in 1990 [136], it was demonstrated that [137] poly(2-methoxy, 5-(2′-ethylhexoxy)-1,4-phenylene vinylene) (MEH-PPV), was an excellent candidate as active material. MEH-PPV films (~120 nm thick) were spin coated on patterned ITO and topped with calcium or indium cathodes. A forward bias of 4 V was sufficient to observe electroluminescence in the device. An energy-level diagram of a typical LED (ITO/MEH-PPV/M, wherein M = Ca, In, Ag, Al, Cu, Au) is given in Figure 33.9. In the forward (cathode → anode) regime, electrons and holes are injected from the cathode and ITO, respectively, and propagate by drift and/or diffusion until they capture one another to give excitons.

Subsequent radiative decay of singlet excitons results in light emission. In the previous example, holes from ITO are injected into the HOMO level of MEH-PPV, which is higher in energy by about 0.2 eV; on the other hand, electrons are injected from the Ca cathode (the most efficient material) into the LUMO level of MEH-PPV, which is lower by about the same amount. Both barriers are relatively small and match each other. As a result, the ITO/MEH-PPV/Ca PLED exhibits high

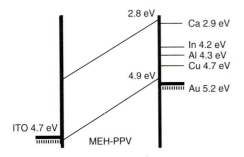

Figure 33.9 Energy diagram of a PLED in forward bias with several cathode materials.

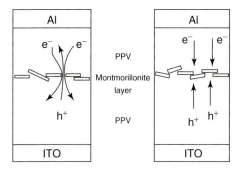

Figure 33.10 Charge transport in the device containing an "isolating" clay layer. Here, e⁻ and h⁺ represent electrons and holes, respectively. From Ref. [138b].

efficiency and low driving voltage. Inorganic LEDs are based, however, on *doped* semiconductors and several junctions within the device. Most PLEDs are based on pristine luminescent CPs whose doping is absent. In other words, the emissive layer in a PLED is basically an insulator that must be kept thin so as to avoid series resistance and maximize the induced electrical field under bias. An obvious circumvention of such issues involves the use of doped polymers in PLEDs [138]. Eckle and Decher [138b] reported on the control of the current–voltage profiles in a PLED by intercalation of an isolated layer of montmorillonite, a clay. In an LbL architecture, buildup of [poly(p-phenylene vinylene)/poly(methylacrylic acid)] (PPV/PMA) layers lowered the current density and increased the light output in a simple manner (Figure 33.10).

Photovoltaic devices are a hot topic because of the increasing demand for energy and the quest for environmentally clean alternative energy resources [139]. Indeed, solar radiation is ideal but necessitates valid ideas to harvest incident photons with higher efficiency. Polymer-based photovoltaic devices offer a lightweight, processable, and cost-effective solution [140]. A photovoltaic device can be thought of as an "upside-down" LED, because it converts light in electricity by the photovoltaic effect. Once an electromagnetic radiation strikes an opportune material, the generated electrons are transferred resulting in the buildup of a voltage difference between two electrodes. A solar cell consists of two materials, a p-type and an n-type semiconductor, sandwiched together to form the so-called p-n junction. When solar photons strike the semiconductor, an electron–hole pair is generated. The electrons are knocked off and can therefore migrate through the conduction band. By connecting the device to an external circuit, the electrons are able to flow and generate photocurrent.

Clearly, an efficient semiconducting CP is appealing as the active component of the donor–acceptor heterojunction [141]. Crucial features of such CP should include:

1) exciton diffusion;
2) charge-carrier generation;

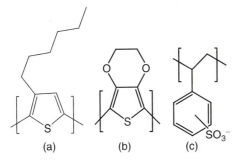

Figure 33.11 P3HT (a), PEDOT (b), and PSS (c).

3) charge-carrier transport;
4) open-circuit voltage.

Strict control of the nanostructure must be attained by carefully choosing the monomer unit. It is no surprise that the LbL approach [142] is a popular method for fabricating photovoltaic devices. In a typical cell, solar-energy conversion occurs by dissociation of photoinduced excitons at the interface of a charge-transporting electrolyte and an n-type inorganic semiconductor (e.g., TiO_2). CPs can replace the electrolyte as hole-transporting materials [143]. An example of a good hole conductor is a heteroaromatic compound, poly(3-hexylthiophene) (P3HT) (Figure 33.11a):

Its use [144], along with nanocrystalline TiO_2, afforded external quantum efficiencies as high as ~10% according to Coakley and McGehee [145]. The most widespread conjugated polymer for photovoltaic purposes is actually a polymer mixture of two ionomers, poly(3,4-ethylenedioxythiophene) and poly(styrenesulfonate), known as PEDOT:PSS (Figure 33.11). Ravirajan *et al.* [146]. proposed some hypotheses about the role of PEDOT:PSS in solar cells – chemical doping of the other polymer (either PSS, MEH-PPV, or P3HT) used in the device fabrication that reduces the contact resistance along the heterojunction resulting in overall improvement of the photon collection. A typical example of such photovoltaic cells can be found in a recent report by Kim *et al.* (Figure 33.12) [147].

In this case, a layered device is fabricated with a simple, all-solution processing of water-soluble PEDOT:PSS and chlorobenzene-soluble P3HT. The distinct moieties are kept apart by a sol-gel cast, intercalated TiO_x slab. The device fabrication is in the metal–insulator–metal configuration. Impinging light strikes the dispositive from the glass side, beyond which a transparent bilayer electrode (P3HT/PEDOT:PSS, hole collectors) and an Al electrode (electron collector) can be found. The charge recombination along the heterojunction can be circumvented by facing electrodes (here, the conjugated polymers and the metal) whose work functions are dissimilar (Figure 33.12b). This breaks the symmetry of the device, thereby providing a driving force for the photogenerated electrons and holes to head toward the respective electrodes.

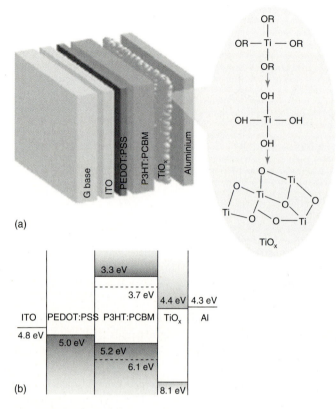

(a)

(b)

Figure 33.12 Schematic representation of the device structure (a) from Ref. [147] with associated energy levels (b). Note the sketched flow-chart with the sol-gel process that affords the TiO$_x$-based spacer.

33.5.3
Anticorrosion Coatings

An emblematic, further example of the wide applicability of CPs to everyday life is their use as anticorrosion materials [148]. Since DeBerry's work in 1985 [149], many more examples of CPs as valid anticorrosion coatings on active (i.e. engineering) metals have been reported [150]. Being electroactive materials, their conductivity can be finely tuned, along with their redox behavior. Moreover, the most common CPs – poly(aniline), poly(thiophene), poly(pyrrole) – are oxidants toward any active metal (i.e. possess higher E_g) and they can act as an efficient anodic protection when coated on, for example, steel coupons. Table 33.1 summarizes this, by comparing the most common engineering metals to the most common CPs. Note that such polymers are oxidized (or p-doped) and, thus, contain counter anions for overall charge neutrality. Tallman *et al.* [151]. proposed a classification of CPs for such purposes into three general groups (Table 33.1), according to the nature

Table 33.1 Reduction potentials for corrosion-relevant redox couples.

Redox couple	$E\pi$, V (*vs.* SHE)
Mg/Mg^{2+}	-2.54
Al/Al_2O_3	-1.93
Zn/Zn^{2+}	-0.94
Fe/Fe^{2+}	-0.62
$Cr_2O_7^{2-}/Cr_2O_3$	$+0.45$
Poly(pyrrole)	-0.1 to $+0.3$
Poly(aniline)	$+0.4$ to $+1.0$
Poly(thiophene)	$+0.8$ to $+1.2$

Source: Ref. [151].
SHE, standard hydrogen electrode.

of the oxidation process: protonic/electronic doping involving *both* H^+ *and* anion incorporation into the polymer matrix (e.g., poly(aniline), PANI) – type 1; electronic doping with anion incorporation (e.g., poly(pyrrole), pPy, or poly(thiophene), pTh) type 2; electronic doping with cation expulsion, either from covalently attached acid group (e.g., sulfonated PANI) or from a sufficiently large, physically entrapped acid or salt (e.g., polyelectrolyte) – type 3.

Figures 33.13–33.15 represent such doping processes for these polymers. Upon redox cycling (e.g., cyclic voltammetry, CV), both cation and anion movement can be observed. Moreover, from Figure 33.13 it can be inferred that the partially oxidized and conducting form of PANI, the emeraldine salt (ES-PANI) can be dedoped without concomitant electron transfer, that is, by base treatment, to afford the nonconducting yet still oxidized emeraldine base PANI, EB PANI.

As noted above, the equilibrium potentials for CPs are positive relative to most metals, for example, Fe and Al. Thus, anodic protection can be a likely mechanism that alters the corrosion behavior of such metals when coated with CPs. However, other factors may also be important: metal complexation [152], interface stabilization, and counter anion release [153].

33.6
Hybrid Inorganic-Oxide–Polymer Materials: the State-of-the-Art

33.6.1
Survey

Nanotechnology emerged in the 1990s as a research field at the nanometer scale (1–100 nm) [154]. The interest in nanostructured materials arises from the possibility of tailoring specific features with nanometer precision. In a bottom-up

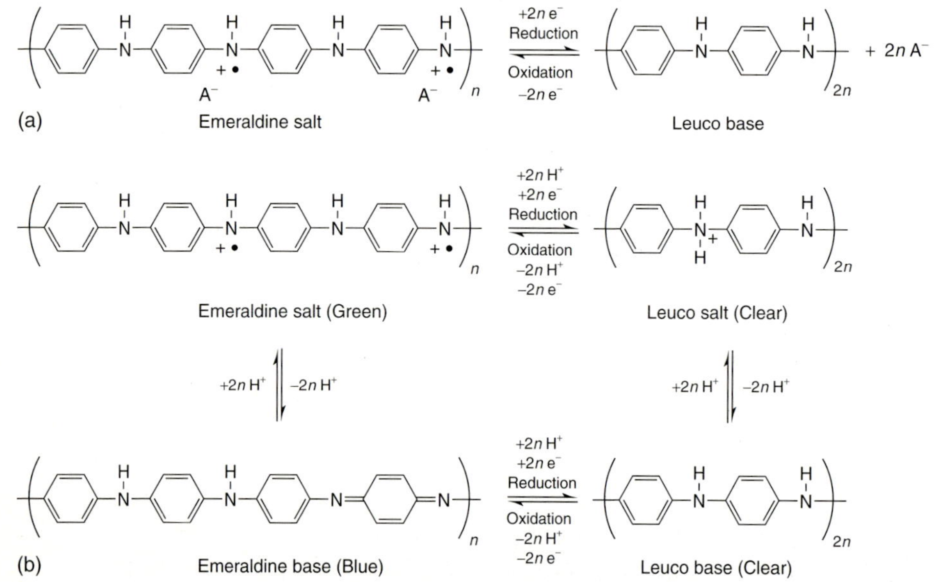
+2n e⁻
Reduction
Oxidation
−2n e⁻
A⁻
A⁻
+ 2n A⁻
(a)
Emeraldine salt
Leuco base
+2n H⁺
+2n e⁻
Reduction
Oxidation
−2n H⁺
−2n e⁻
Emeraldine salt (Green)
Leuco salt (Clear)
+2n H⁺
−2n H⁺
(b)
Emeraldine base (Blue)
Leuco base (Clear)

Figure 33.13 Redox scheme for type-1 conducting polymers (here, PANI). In scheme (a), anion expulsion upon reduction is displayed. In scheme (b), H^+ and electron transfer are shown. A^- is a generic anion (not shown in b for clarity) arising from acidic treatment with HA.

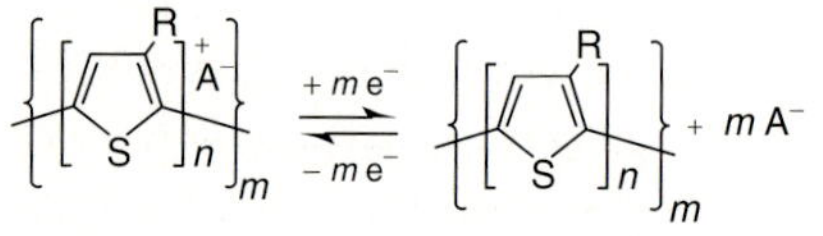
+ me⁻
− me⁻
+ mA⁻

Figure 33.14 Redox scheme for type 2 conducting polymer (here, pTh) showing anion (A^-) expulsion upon reduction. Here, R = H and *n* symbolizes the extension of the polaronic unit, meaning a positive charge every 2–4 monomer units ($n = 2$–4). Adapted from Ref. [151].

approach, nanomatter manipulation involves the use of nanobuilding blocks. When the building blocks can be categorized as "inorganic" and "organic," the outcome is a hybrid nanomaterial. To date, molecular architectures have been synthesized mainly by covalent bonds [155], self-assembly of supramolecular constituents [156], metal-organic frameworks [157], and coordination polymers [158]. Sol-gel chemistry fits into this scheme in that nanoscale processing of molecular precursors can nowadays afford nanocrystalline [159] or amorphous [160] materials. A summary of the *status quo* will be furnished with special highlights on hybrid nanomaterials and sol-gel processes (Section 33.6.2), hybrid nanomaterials for OLEDs/photovoltaic devices (Section 33.6.3), and hybrid nanomaterials for anticorrosion purposes (Section 33.6.4).

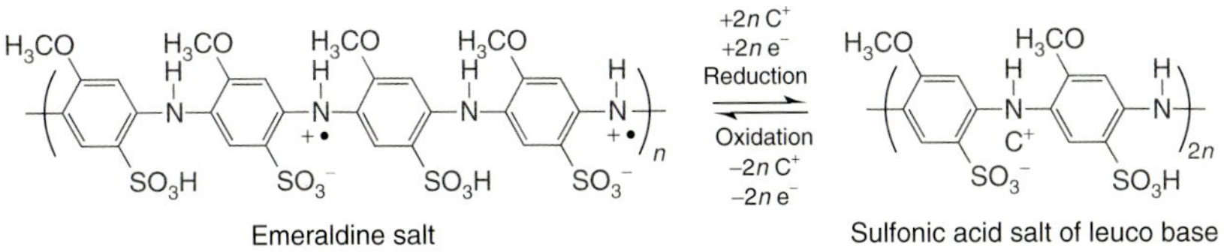

Figure 33.15 Redox scheme for Type 3 conducting polymer (here, sulfonated PANI or SPANI) showing cation uptake (C^+) upon reduction. Left: oxidized form, ES-SPANI. Right: reduced form, leucoemeraldine base or LB-SPANI. Adapted from Ref. [151].

33.6.2
Hybrid Nanomaterials and Sol-Gel Process

Reaction conditions in sol-gel chemistry (pH, temperature, postgelation treatments, catalysts, etc.) may induce a well-defined, nanosized order. A nanostructured sol-gel material exhibits porosity in the nanometer range: a proof of this can be found in Shan's work [161], in which nanoporous silica films were obtained using $Si(OEt)_4$ as molecular precursor. The as-formed silica "flakes" showed superior sequestering capabilities toward biomolecules, owing to the ordered nanoporous structure. The organic entity may be placed within such an inorganic matrix by either mixing it into the sol-gel solution or impregnating it into the dry, porous gel. When the former route is chosen, the organic moiety can be chemically grafted onto the oxo-hydroxo polymer. In the latter case, it is assumed to adsorb on the pore surface [162].

Nowadays, these approaches are termed as *type-I hybrids* (no organic–inorganic bond) and *type-II hybrids* (organic–inorganic bonds) [163]. Examples of the first type are interpenetrating networks, where simultaneous formation of inorganic gel and organic gel is possible: Ogoshi *et al.* [164] demonstrated that the crosslinking reaction of bipyridyl-containing polymer was carried out together with the sol-gel reaction of $Si(OMe)_4$. Type-I hybrids can also be gels whose pores are filled with polymers, rather than mutually interpenetrating networks. A recent patent [165] demonstrated how $Al(OBu)_3$ could be used as a molecular precursor, along with methylmethacrylate, to afford an $AlO(OH)$ nanoparticle network imbibed in polymerized PMMA.

Type-II hybrids, on the other hand, have shown that further functionalization of sol-gel precursors can render interesting properties at the nanoscale. A typical reaction within this class [166] is between $Si(OEt)_4$ and poly(dimethylsiloxane), (PDMS), where organic–inorganic crosslinking takes place. Structure–property relationships and potential applications of such materials can be displayed by carefully bonding one or more pendant groups to the silanol backbone [167]. Examples of such Type-II materials are displayed in Figure 33.16. It is evident that different functionalities grafted onto the silanol moiety will impart mechanical or electronic features to the overall material: a hydrophobic head (Figure 33.16a), a further crosslinkable functional group (Figure 33.16b), an elastomeric–rubbery backbone

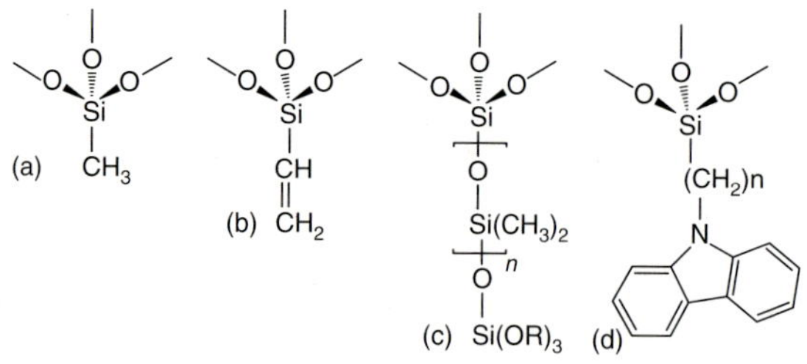

Figure 33.16 Different organic units that may be grafted onto oxide backbones in type-II hybrids. (a–d) Aliphatic, vinyl, poly(dimethylsiloxane), carbazole.

(Figure 33.16c) or even an optoelectronic pendant unit. This is represented by the: (a) methyl, (b) vinyl, (c) siloxane, and (d) carbazole groups (Figure 33.16d).

33.6.3
Hybrid Nanomaterials for Optoelectronic Devices

Organic–inorganic objects at the nanoscale have appealing features very different from their bulk counterparts. The building-block approach in a "bottom-up" fashion [168] has shown promising results for optoelectronic devices and nanopatterning [169]. Combining organic [170] and inorganic [171] nanoscale effects has dramatic relevance in the overall performance of an optoelectronic device. A third requirement is a winning strategy for assembling, manipulating, and depositing dissimilar moieties [172]. Optical coatings are a typical example of this: Biswas *et al.* [173]. demonstrated how poly(methylsilsesquioxane) films with specific porosity could be prepared with PMMA within the inorganic matrix. Such thin (10 nm up to ~1 μm) films showed excellent performances as antireflection coatings. Rubner and Cohen's groups [174] reported on a hybrid SPS/PDAC/PDMS coating (sulfonated poly(styrene) (SPS), poly(diallyldimethylammonium) chloride (PDAC), poly(dimethylsiloxane) or PDMS) coating whose antireflection capabilities could be retained even after several cycles of mechanical deformations (Figure 33.17):

If antireflection coatings can be developed based on silicon-based hybrid materials, hybrid solar cells are the realm of nanocrystalline, TiO_2-based materials. Apart from its well-known photocatalytic activity under UV irradiation (the so-called Honda–Fujishima effect) [175], titanium oxide (as anatase) is utilized as wide E_g, n-type material for solar energy conversion along with dyes [70], CPs [176], carbon nanotubes [177], or quantum dots [178]. There is a quest for the best p-type semiconductor, because nanocrystalline anatase has by far shown superior performances as an electron collector, excellent charge carrier, and optical spacer – three paradigmatic features for an efficient solar cell [179]. Such combinations of materials are required because, apart from the obvious p-n junction within the photovoltaic device, nanocrystalline TiO_2 is a strong UV-absorber (anatase and rutile band edges are at 388 and 413 nm, respectively) and necessitates sensitization to widen its absorption to the visible region [180]. Such an improvement in solar cells may

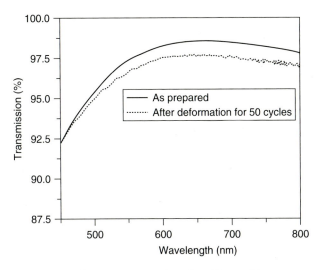

Figure 33.17 Change in transmission of an AR-coated lens after deformation to a strain level of 2.66% for 50 cycles. From Ref. [174].

come from incorporating poly(2,7-carbazole) materials. According to Blouin *et al.* [181], there is a good match between theoretical and experimental values of HOMO and LUMO levels for a series of carbazole-based copolymers. This may open up an easy route to intelligent p-type materials suitable for photovoltaic applications.

Electrodeposition of hybrid, nanostructured materials for optoelectronic applications has an advantage in terms of high quality-to-cost ratio in films. Indeed, electrodeposition of hybrid thin films is now regarded [182] as an alternative to gas-phase methods such as chemical vapor deposition (CVD). In spite of difficulties to fully control the film structure, studies about electrodeposition can now deal with a wide variety of materials and also prove their high quality and functionality [183]. Electrochemical self-assembly of ZnO/dye hybrid thin films has been reported as a valid alternative to TiO_2/dye solar cells due to comparable E_g values and similar semiconducting behavior [184].

33.6.4
Hybrid Nanomaterials for Anticorrosion Coatings

Protecting engineering metals by coating with organic coatings containing a binder, a pigment, an additive (such as a hardening agent) or all of the above in one single step is ideal. The coating matrix will determine the basic physical and chemical property according to its predominant component (i.e. organic or inorganic), in the case of a hybrid coating. It used to be customary [185] to protect an active metal with another, less noble, metal (say, Zn): this is the so-called *anodic protection*. Nowadays, hybrid materials can be nanostructured so that efficient anticorrosion mechanisms and performances can be attained.

The inorganic moiety is, in most cases, a sol-gel matrix [186]. Ease of preparation, versatile composition ranges, and practical applicability on almost any surface make sol-gel an efficient primary protection coating [187]. An ulterior implementation to the coating is its loading or pretreatment with corrosion inhibitors. Hybrid, silane-based thin layers were investigated as prospective pretreatments for many metallic substrates [188]. However, interaction of the inhibitors with the sol-gel layer very often led to significant shortcomings in the stability of the coating itself and/or inhibitor activity. Voevodin *et al.*, for example [189], found that sol-gel films with $NaVO_3$ and Na_2MoO_4 did not provide adequate corrosion protection due to decrease of the sol-gel network stability in the presence of additives. Among the inorganic inhibitors, cerium-based additives are popular. They work better below a certain critical concentration (0.2–0.6 wt%), because above this threshold the formation of defects occurs in sol-gel coatings [188a].

Van Ooij and coworkers [190] introduced triazole-based organic inhibitors into a hybrid sol-gel matrix. Such inhibitors are thought of as local insulators whose action is triggered by pH changes in pits, crevices, and scratches. The main idea is about the development of nanounits that can be sensitive to the external (e.g., mechanical damage) or internal (e.g., pH changes) corrosion triggers. When the local environment undergoes changes or if the corrosion process is started on the metal surface, the nanocontainers release encapsulated inhibitors *directly* into the damaged area, thus preventing undesirable leakage of the inhibitor and reduction of the barrier properties of the sol-gel coating. The rate of release is strongly dependent on the isoelectric points and ζ-potential of the sol-gel matrix and inhibitor [191]. Hence, in many cases the release of the organic inhibitor from the sol-gel film is impossible or very slow. It is no surprise that the state-of-the-art in anticorrosion coatings deals with nanoreservoirs and LbL approaches [187a, 192]. Shchukin *et al.* [193] recently reported on an intelligent modification on silica nanocontainers (diameter $\sim$70 nm): coating such nanoparticles with poly(ethylene imine)/poly(styrene sulfonate) (PEI/PSS) LbL layers afforded an ordered structure that was loaded up with the corrosion inhibitor (benzotriazole) and embedded in a hybrid epoxy-functionalized ZrO_2/SiO_2 sol-gel matrix (Figure 33.18).

Impedance analysis [150] revealed that the corrosion-protection performance of the hybrid material reached an optimum for specific (0.13 wt%) inhibitor concentrations. The idea of nanocontainers has been further expanded and improved by the same research group [194]. Halloysite nanotubes were utilized in another, LbL-driven anticorrosion study. The primer coating was, once again, a silica-zirconia-based film and the nanocontainers were loaded with 2-mercaptobenzothiazole (Figure 33.19).

The ideas behind the use of halloysite, a 2D aluminosilicate, are economical and practical: the possibility of employing such inexpensive, naturally occurring nanotubes (Figure 33.20a) as prospective nanocontainers for anticorrosion coatings with an active inhibitor present (2-mercaptobenzothiazole) has been demonstrated [194]. Nanotubes were loaded with the inhibitor and then incorporated into the zirconia-silica sol-gel coating. To prevent undesirable leakage of the loaded inhibitor from the halloysite interior, the outer surface of the

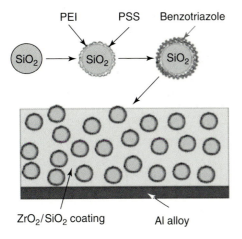

Figure 33.18 Fabrication of the nanocomposite coating loaded with benzotriazole-impregnated silica nanoreservoirs. From Ref. [193].

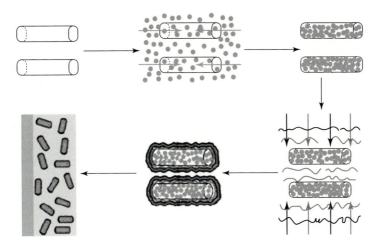

Figure 33.19 Fabrication of the nanocomposite coating, Ref. [194]. From top left to bottom left corners are shown: initial halloysites, inhibitor loading, washing, polyelectrolyte LbL assembly, incorporation into coating.

nanotubes was LbL-modified with poly(allylamine hydrochloride)/poly(styrene sulfonate) (PAH/PSS) (Figure 33.20b).

Such nanotubes acted as efficient reservoirs for 2-mercaptobenzothiazole, preventing their direct interaction with the sol-gel coating and harmful leakage. Polyelectrolyte multilayers around the nanotubes provided storage and on-demand release of the inhibitor in the damaged zones. Local pH changes in corrosion areas triggered such a release due to the sensitivity of polyelectrolyte multilayers to changes in the surrounding environment.

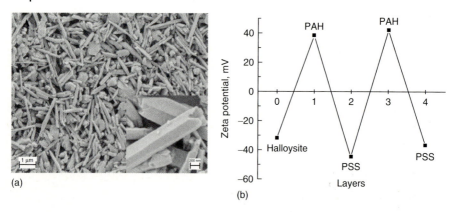

Figure 33.20 (a) SEM image of the halloysite nanocontainers. (b) ζ-potential for LbL assembly of PAH and PSS on halloysite nanotubes, pH 7.5. From Ref. [194].

33.7
Summary

The fabrication of hybrid materials at the nanoscale can take advantage of straightforward (LbL) and inexpensive (sol-gel chemistry) protocols. Implementing a combination of organic (CPs) and inorganic (silicon and titanium oxides) moieties within a layered structure holds much promise in practical materials science. It should be clear that hybrid materials holds promise in terms of combining the synthetic ability of two major classes of materials enabling solid-state chemistry and polymer synthesis to coexist synergistically in the design and function of new materials. This area of research is yet to be fully exploited and there is a need to have materials chemists trained with the best of both backgrounds to truly innovate this field.

Acknowledgement

The authors acknowledge the contribution of the past and present members of the Advincula Research Group. The authors would like to acknowledge funding from NSF DMR-0602896, DMR-1006776, CBET-0854979, CHE 10-41300, Robert A Welch Foundation, E-1551, and Texas NHARP 01846. We would also like to thank KSV Instruments (Biolin), Viscotek (Malvern Instruments), Agilent Technologies, and Optrel for their technical support.

References

1. ESI Special Topic (2002) Molecular Self-Assembly, published April 2002.

2. Ariga, K., Nakanishi, T., and Michinobu, T. (2006) *J. Nanosci. Nanotechnol.*, **6**, 2278–2301.

3. Mamedov, A.A. and Kotov, N.A. (2000) *Langmuir*, **16**, 5530–5533.

4. Lynn, D.M. (2007) *Adv. Mater.*, **19**, 4118–4130.

5. Rogach, A.L., Kotysh, D.S., and Harrison, M. (2000) *Chem. Mater.*, **12**, 1526–1528.

6. Kim, E., Kim, K., Yang, H., Kim, Y.T., and Kwak, J. (2003) *Anal. Chem.*, **75**, 5665–5672.

7. Decher, G. (1997) *Science*, **277**, 1232–1237.

8. Iler, R.K. (1966) *J. Colloid Interface Sci.*, **21**, 569–594.

9. (a) Decher, G. and Hong, J.-D. (1991) *Chem. Macrom. Symp.*, **46**, 321–327; (b) Decher, G. and Hong, J.-D. (1991) *Ber. Bunsenges. Phys. Chem.*, **95**, 1430–1434.

10. (a) Leyden, D.E. (ed.) (1985) *Silanes, Surfaces and Interfaces*, Gordon and Breach, New York; (b) Leyden, D.E. (1978) Immobilization of organic functional groups on silica surfaces, in *Silylated Surfaces* (ed. W.T. Collins), Gordon and Breach, New York, pp. 55–72.

11. Kern, W. (1984) *Semicond. Int.*, **4**, 94–99.

12. (a) Bigelow, W.C., Pickett, D.L., and Zisman, W.Z. (1946) *J. Colloid Interface Sci.*, **1**, 513–538; (b) Nuzzo, R.G. and Allara, D.L. (1983) *J. Am. Chem. Soc.*, **105**, 4481–4483; (c) Kuhn, H. and Ulman, A. (1995) Supramolecular assemblies: vision and strategy, in *Thin Films*, vol. 20 (ed. A. Ulman), Academic Press, New York.

13. Robel, I., Subramanian, V., Kuno, M., and Kamat, P. (2006) *J. Am. Chem. Soc.*, **128**, 2385–2393, and references therein.

14. (a) Langmuir, I. (1915) *Phys. Rev.*, **6**, 79–80; (b) Langmuir, I. (1916) *J. Am. Chem. Soc.*, **38**, 2221–2224; (c) Langmuir, I. (1921) *Trans. Farad. Soc.*, **17**, 4–8; (d) Brunauer, S., Emmett, P.H., and Teller, E. (1938) *J. Am. Chem. Soc.*, **60**, 309–319.

15. (a) Huheey, J.E., Keiter, E.A., and Keiter, R.L. (1993) *Inorganic Chemistry: Principles of Structure and Reactivity*, 4th edn, Harper Collins College Publishers; (b) Atkins, P.W. (1994) *Physical Chemistry*, 5th edn, Oxford University Press.

16. Wu, A., Yoo, D., Lee, J.-K., and Rubner, M.F. (1999) *J. Am. Chem. Soc.*, **121**, 4883–4891; (b) Mendelsohn, J.D., Barrett, C.J., Chan, V.V., Pal, A.J., Mayers, A.M., and Rubner, M.F. (2000) *Langmuir*, **16**, 5017–5023.

17. Srivastava, S. and Kotov, N.A. (2008) *Acc. Chem. Res.*, **41**, 1831–1841.

18. Kleinfeld, E.R. and Ferguson, G.S. (1994) *Science*, **265**, 370–373.

19. Decher, G., Lvov, Y., and Schmitt, J. (1994) *Thin Solid Films*, **244**, 772–777.

20. Lvov, Y., Ariga, K., Ichinose, I., and Kunitake, T. (1995) *J. Am. Chem. Soc.*, **117**, 6117–6123.

21. Ram, M.K., Salerno, M., Adami, M., Faraci, P., and Nicolini, C. (1999) *Langmuir*, **15**, 1252–1259.

22. Ichinose, I., Senzu, H., and Kunitake, T. (1996) *Chem. Lett.*, **25**, 831–832.

23. Buttry, D.A. and Ward, M.D. (1992) *Chem. Rev.*, **92**, 1335–1379.

24. Truijen, I., Bael, M.K., Rul, H., D'Haen, J., and Mullens, J. (2007) *J. Sol-Gel Sci. Technol.*, **43**, 291–297.

25. Cheung, J.H., Fou, A.F., and Rubner, M.F. (1994) *Thin Solid Films*, **244**, 985–989.

26. Sugimoto, Y., Pou, P., Abe, M., Jelinek, P., Pérez, R., Morita, S., and Custance, O. (2007) *Nature*, **446**, 64–67.

27. Fulghum, T.M., Estillore, N.C., Vo, C.-D., Armes, S.P., and Advincula, R.C. (2008) *Macromolecules*, **41**, 429–435.

28. Tompkins, H. G. and Irene, E.A. (eds) (2005) *Handbook of Ellipsometry*, W. Andrew Publishing and Springer, New York.

29. (a) Lutkenhaus, J.L., Hrabak, K.D., McEnnis, K., and Hammond, P.T. (2005) *J. Am. Chem. Soc.*, **127**, 17228–17234; (b) Kharlampieva, E. and Sukhishvili, S.A. (2006) *J. Macromol. Sci.-Pol. R.*, **46**, 377–395.

30. Ikeda, A., Hatano, T., Shinkai, S., Akiyama, T., and Yamada, S. (2001) *J. Am. Chem. Soc.*, **123**, 4855–4856.

31. Keller, S., Kim, H.-N., and Mallouk, T. (1994) *J. Am. Chem. Soc.*, **116**, 8817–8818.

32. Watanabe, S. and Regen, S. (1994) *J. Am. Chem. Soc.*, **116**, 8855–8856.

33. Byrd, H., Holloway, C.E., Pogue, J., Kircus, S., Advincula, R.C., and Knoll, W. (2000) *Langmuir*, **16**, 10322–10328.

34. Shimazaki, Y., Mitsuishi, M., Ito, S., and Yamamoto, M. (1997) *Langmuir*, **13**, 1385–1387.

35. Clark, S.L. and Hammond, P.T. (2000) *Langmuir*, **16**, 10206–10214.

36. Takagi, T. (1990) *J. Intel. Mater. Syst. Structure*, **1**, 149–156.

37. He, J.A., Valluzzi, R., Yang, K., Dolukhanyan, T., Sung, C., Kumar, J., Tripathy, S.K., Samuelson, L., Balogh, L., and Tomalia, D.A. (1999) *Chem. Mater.*, **11**, 3268–3274.

38. (a) Kotov, N.A., Dekany, I., and Fendler, J.H. (1995) *J. Phys. Chem.*, **99**, 13065–13069; (b) Liang, Z., Dzienis, K.L., Xu, J., and Wang, Q. (2006) *Adv. Funct. Mater.*, **16**, 542–548.

39. Lvov, Y., Decher, G., and Sukhorukov, G. (1993) *Macromolecules*, **26**, 5396–5399.

40. Richert, L., Lavalle, P., Vautier, D., Senger, B., Stoltz, J.F., Schaaf, P., Voegel, J.C., and Picart, C. (2002) *Biomacromolecules*, **3**, 1170–1178.

41. Mamedov, A.A., Kotov, N.A., Prato, M., Guldi, D.M., Wicksted, J.P., and Hirsch, A. (2002) *Nature Mater.*, **1**, 190–194.

42. (a) Dickey, F.H. (1955) *J. Phys. Chem.*, **59**, 695–707; (b) Lee, S.-W., Ichinose, I., and Kunitake, T. (1998) *Langmuir*, **14**, 2857–2863.

43. (a) Peiró, A.M., Ravirajan, P., Govender, K., Boyle, D.S., O'Brien, P., Bradley, D.D.C., Nelson, J., and Durrant, J.R. (2006) *J. Mater. Chem*, **16**, 2088–2096; (b) Itoh, E., Takamizawa, Y., and Miyairi, K. (2008) *Jpn. J. Appl. Phys.*, **47**, 509–512.

44. Advincula, M., Fan, X., Lemons, J., and Advincula, R. (2005) *Colloid Surf. B*, **42**, 29–43.

45. Vassilion, J.K., Ziebarth, R.P., and Disalvo, F.J. (1990) *Chem. Mater.*, **2**, 738–741.

46. Nazar, L.F., Zhang, Z., and Zinkweg, D. (1992) *J. Am. Chem. Soc.*, **114**, 6239–6240.

47. Gangopadhyay, R. and De, A. (2000) *Chem. Mater.*, **12**, 608–622.

48. Novak, B.M. (1993) *Adv. Mater.*, **5**, 422–433.

49. Manson, J.A. and Sperling, L.H. (1976) *Polymer Blends and Composites*, Plenum.

50. Ravirajan, P., Bradley, D.D.C., Nelson, J., Haque, S.A., Durrant, J.R., Smit, H.J.P., and Kroon, J.M. (2005) *Appl. Phys. Lett.*, **86**, 143101.

51. Kim, J.Y., Lee, K., Coates, N.E., Moses, D., Nguyen, T.-Q., Dante, M., and Heeger, A.J. (2007) *Science*, **317**, 222–225.

52. (a) Yu, G., Gao, J., Hummelen, J.C., Wudl, F., and Heeger, A.J. (1995) *Science*, **270**, 1789–1791; (b) Padlinger, F., Rittberger, R., and Sariciftci, N.S. (2003) *Adv. Funct. Mater.*, **13**, 85–90.

53. Mills, A., Davies, R.H., and Worsley, D. (1993) *Chem. Soc. Rev.*, **22**, 417–425.

54. Bach, U., Lupo, D., Compte, P., Moser, J.E., Weissörtel, F., Salbeck, J., Spreitzer, H., and Grätzel, M. (1998) *Nature*, **395**, 583–585.

55. Pedrotti, F.L., Pedrotti, L.M., and Pedrotti, L.S. (2007) *Introduction to Optics*, 3rd edn, Addison-Wesley.

56. Samsonov, G.V. (1973) *The Oxide Handbook*, Plenum, New York.

57. Busani, T. and Devine, R.A.B. (2005) *Semicond. Sci. Technol.*, **20**, 870–875.

58. (a) Ebelmen, M. (1846) *Ann. Chim. Phys.*, **16**, 129; (b) Ebelmen, M. (1847) *C. R. Acad. Sci.*, **25**, 854; (c) Graham, T. (1864) *J. Chem. Soc.*, **17**, 318–327.

59. Brinker, C.J. and Scherer, G.W. (1990) *Sol-Gel Science: The Physics and Chemistry of Sol-Gel Processing*, Academic Press.

60. (a) Roy, D.M. and Roy, R. (1954) *Am. Mineral.*, **39**, 957–975; (b) Roy, R. (1956) *J. Am. Ceram. Soc.*, **39**, 145–148; (c) Roy, R. (1969) *J. Am. Ceram. Soc.*, **52**, 344–353.

61. MacCraith, B. and McDonagh, C. (2002) *J. Fluoresc.*, **12**, 333–342.

62. Coles, M.P., Lugmair, C.G., Terry, K.W., and Tilley, T.D. (2000) *Chem. Mater.*, **12**, 122–131.

63. Jain, T.K., Roy, I., De, T.K., and Maitra, A. (1998) *J. Am. Chem. Soc.*, **120**, 11092–11095.

64. Morris, C.A., Anderson, M.L., Merzbacher, C.I., and Rolison, D.R. (1999) *Science*, **284**, 622–624.

65. Rosseinsky, D.R. and Mortimer, R.J. (2001) *Adv. Mater.*, **13**, 783–793.

66. Fricke, J. (1992) *J. Non-Cryst. Solids*, **147**, 356–362.

67. Chen, D.G. (2001) *Sol. Eng. Mater. Sol. Cells*, **68**, 313–336.

68. (a) Iler, R.K. (1955) *The Chemistry of Silica*, John Wiley & Sons, Inc., New York; (b) Stober, W., Fink, A., and Bohn, E. (1968) *J. Colloid Interface Sci.*, **26**, 62–69.

69. (a) Livage, J. (1991) *Chem. Mater.*, **3**, 578–593; (b) Gavrilov, V.Y. (1997) *Kinet. Catal.*, **38**, 697–702.

70. Shi, Y. and Seliskar, C.J. (1997) *Chem. Mater.*, **9**, 821–829.

71. O'Regan, B. and Grätzel, M. (1991) *Nature*, **353**, 737–740.

72. Hench, L.L. and West, J.K. (1990) *Chem. Rev.*, **90**, 33–72.

73. Pope, M.T. (1983) *Heteropoly and Isopoly Oxometalates*, Springer-Verlag, New York.

74. Holleman, A.F. and Wiberg, E. (2001) *Inorganic Chemistry*, Academy Press.

75. Handy, B.E., Maciejewski, M., Baiker, A., and Wokaun, A. (1992) *J. Mater. Chem.*, **2**, 833–840.

76. Kosmulski, M. (2001) *Chemical Properties of Material Surfaces*, Marcel Dekker Inc.

77. Yoldas, B.E. (1984) *J. Non-Cryst. Solids*, **63**, 145–154.

78. Schubert, U. (2005) *J. Mater. Chem.*, **15**, 3701–3715.

79. Sanchez, C., Livage, J., Henry, M., and Babonneau, F. (1988) *J. Non-Cryst. Solids*, **100**, 65–76.

80. Livage, J., Henry, C., and Sanchez, M. (1988) *Prog. Solid State Chem.*, **18**, 259–341.

81. Schwarzenbach, G. (1952) *Helv. Chim. Acta*, **35**, 2344–2359.

82. Schubert, U., Hüsing, N., and Lorenz, A. (1995) *Chem. Mater.*, **7**, 2010–2027.

83. Hunter, R.J. (1989) *Foundations of Colloid Science*, Oxford University Press.

84. Sahimi, M. (1994) *Applications of Percolation Theory*, Taylor & Francis.

85. Meakin, P. and Family, F. (1988) *Phys. Rev. A*, **38**, 2110–2123.

86. Onoda, G.Y. Jr. (1979) in *Ceramic Processing Before Firing* (ed. L.L. Hench), John Wiley & Sons, Inc., New York City.

87. Yanes, A.C., Velazquez, J.J., Del-Castillo, J., Mendez-Ramos, J., and Rodriguez, V.D. (2009) *J. Sol-Gel Sci. Technol.*, **51**, 4–9.

88. Del-Castillo, J., Mendez-Ramos, J., Yanes, A.C., Velazquez, J.J., and Rodriguez, V.D. (2008) *J. Non-Cryst. Solids*, **354**, 2000–2003.

89. Reidinger, M., Rydzek, M., Scherdel, C., Arduini-Schuster, M., and Manara, J. (2009) *Thin Solid Films*, **517**, 3096–3099.

90. Levy, D. and Esquivias, L. (1995) *Adv. Mater.*, **7**, 120–129.

91. Enami, Y., Fukuda, T., and Suye, S. (2007) *Appl. Phys. Lett.*, **91**, 203507-1–203507-3.

92. Laranjo, M.T., Stefani, V., Benvenutti, E.V., Costa, T.M.H., Ramminger, Gde., O., and Gallas, M.R. (2007) *J. Non-Cryst. Solids*, **353**, 24–30.

93. Sanchez, C., Lebeau, B., Chaput, F., and Boilot, J.-P. (2003) *Adv. Mater.*, **15**, 1969–1994.

94. Gunter, P. and Huignard, J.P. (1988) *Photorefractive Materials and their Applications*, Springer, Berlin.

95. Choi, D.H., Jun, W., Oh, K.Y., and Kim, J.H. (2002) *Polym. Bull.*, **49**, 173–180.

96. Nazeeruddin, M.K., Kay, A., Rodicio, I., Humphry-Baker, R., Müller, E., Liska, P., Vlachopoulos, N., and Grätzel, M. (1993) *J. Am. Chem. Soc.*, **115**, 6382–6390.

97. Salbeck, J., Yu, N., Bauer, J., Weissörte, F., and Bestgen, H. (1997) *Synth. Met.*, **91**, 209–215.

98. Arango, A.C., Johnson, L.R., Bliznyuk, V.N., Schlesinger, Z., Carter, S.A., and Horhold, H.H. (2000) *Adv. Mater.*, **12**, 1689–1692.

99. (a) Breeze, A.J., Schlcsinger, Z., Carter, S.A., and Brock, P.J. (2001) *Phys. Rev. B*, **64**, 125205–125201; (b) Coakley, K.M. and McGehee, M.D. (2003) *Appl. Phys. Lett.*, **83**, 3380–3382; (c) Ravirajan, P., Haque, S.A., Durrant,

J.R., Poplavskyy, D., Bradley, D.D.C., and Nelson, J. (2004) *J. Appl. Phys.*, **95**, 1473–1480.

100. Ioannidis, A., Facci, J.S., and Abkowitz, M.A. (1998) *J. Appl. Phys.*, **84**, 1439–1444.

101. Van Hal, P.A., Wienk, M.M., Kroon, J.M., Verhees, W.J.H., Slooff, L.H., van Gennip, W.J.H., Jonkheijm, P., and Janssen, R.A.J. (2003) *Adv. Mater.*, **15**, 118–121.

102. Schweiss, R., Zhang, N., and Knoll, W. (2007) *J. Sol-Gel Sci. Technol.*, **44**, 1–5.

103. Huguenin, F., Ferreira, M., Zucolotto, V., Nart, F.C., Torresi, R.M. and Oliveira, O.N. Jr. (2004) *Chem. Mater.*, **16**, 2293–2299.

104. Bechinger, C., Ferrere, S., Zabane, A., Sprague, J., and Gregg, B.A. (1996) *Nature*, **383**, 608–609.

105. Luca, V., Blackford, M.G., Finnie, K.S., Evans, P.J., James, M., Lindsay, M.J., Skyllas-Kazacos, M., and Barnes, P.R.F. (2007) *J. Phys. Chem. C*, **111**, 18479–18492.

106. Al-Kalhout, A., Heusing, S., and Aegerter, M.A. (2006) *J. Sol-Gel Sci. Technol.*, **39**, 195–206.

107. Yuan, W., Ji, J., Fu, J., and Shen, J. (2008) *J. Biomed. Mater. Res. B*, **85**, 556–563.

108. (a) Nagashima, K., Kokusen, H., Ueno, N., Matsuyoshi, A., Kosaka, T., Hasegawa, M., Hoshi, T., Ebitani, K., Kaneda, K., Aritani, H., and Hasegawa, S. (2000) *Chem. Lett.*, 264–265; (b) He, J., Ichinose, I., Fujikawa, S., Kunitake, T., and Nakao, A. (2002) *Chem. Mater.*, **14**, 3493–3500.

109. Jeong, S., Lee, S., Kim, D., Shin, H., and Moon, J. (2007) *J. Phys. Chem. C*, **111**, 16083–16087.

110. Fan, J., Boettcher, S.W., and Stucky, G.D. (2006) *Chem. Mater.*, **18**, 6391–6396.

111. (a) Henglein, A. (1989) *Chem. Rev.*, **89**, 1861–1873; (b) Kavan, L., Stoto, T., Graetzel, M., Fitzmaurice, D., and Shklover, V. (1993) *J. Phys. Chem.*, **97**, 9493–9499.

112. Hou, Y.-Q., Zhuang, D.-M., Zhang, G., Zhao, M., and Wu, M.-S. (2003) *Appl. Surf. Sci.*, **218**, 97–105.

113. Yildiz, A., Lisesivdin, S.B., Kasap, M., and Mardare, D. (2008) *J. Non-Cryst. Solids*, **354**, 4944–4947.

114. Langlet, M., Permpoon, S., Riassetto, D., Berthome, G., Pernot, E., and Joud, J.C. (2006) *J. Photochem. Photobiol. A*, **181**, 203–214.

115. Mohammadi, M.R., Fray, D.J., and Ghorbani, M. (2008) *Solid State Sci.*, **10**, 884–893.

116. Shanaghi, A., Sabour Rouhaghdam, A., Shahrabi, T., and Aliofkhazraei, M. (2008) *Mater. Sci.*, **44**, 233–247.

117. Mutin, P.H. and Vioux, A. (2009) *Chem. Mater.*, **21**, 582–596.

118. Vioux, A. and Mutin, P.H. (2005) *Handbook of Sol-Gel Science and Technology*, vol. 1, Kluwer Academic Publishers, Norwell, MA.

119. Sui, R., Rizkalla, A.S., and Charpentier, P.A. (2006) *Langmuir*, **22**, 4390–4396.

120. Bolto, B.A., McNeill, R., and Weiss, D.E. (1963) *Aust. J. Chem.*, **16**, 1090–1103.

121. Chiang, C.K., Druy, M.A., Gau, S.C., Heeger, A.J., Louis, E.J., MacDiarmid, A.G., Park, Y.W., and Shirakawa, H. (1978) *J. Am. Chem. Soc.*, **100**, 1013–1015.

122. MacDiarmid, A.G. (2001) *Rev. Mod. Phys.*, **73**, 701–712.

123. Chiang, C.K., Fincher, C.R. Jr., Park, Y.W., Heeger, A.J., Shirakawa, H., Louis, E.J., Gau, S.C., and MacDiarmid, A.G. (1977) *Phys. Rev. Lett.*, **39**, 1098–1101.

124. Nigrey, P.J., MacDiarmid, A.G., and Heeger, A.J. (1979) *J. Chem. Soc. Chem. Commun.*, 594–595.

125. Skotheim, T.A., Elsenbaumer, R.L., and Reynolds, J.R. (eds) (1998) *Handbook of Conducting Polymers*, Marcel Dekker, Inc., New York.

126. MacDiarmid, A.G. and Epstein, A.J. (1989) *Faraday Discuss. Chem. Soc.*, **88**, 317–332.

127. (a) Hatchett, D.W., Josowicz, M., and Janata, J. (1999) *J. Phys. Chem. B*, **103**, 10992–10998; (b) Luo, J., Wang, X., Li, J., Zhao, X., and Wang, F. (2007) *Electrochem. Commun.*, **9**, 1175–1179; (c) Ge, C., Armstrong, N.R., and Saavedra, S.S. (2007) *Anal. Chem.*, **79**, 1401–1410.

128. Hall, N. (2003) *Chem. Commun.*, 1–4.
129. Shirakawa, H., Louis, E.J., MacDiarmid, A.G., Chiang, C.K., and Heeger, A.J. (1977) *J. Chem. Soc. Chem. Commun.*, 578–580.
130. Miller, J.S. (1993) *Adv. Mater.*, **5**, 671–676.
131. MacDiarmid, A.G. (2001) *Angew. Chem., Int. Ed.*, **40**, 2581–2590 (Nobel Lecture).
132. Becker, H., Spreitzer, H., Kreuder, W., Kluge, E., Vestweber, H., Schenk, H., and Treacher, K. (2001) *Synth. Met.*, **122**, 105–110.
133. Deibel, C. and Dyakonov, V. (2008) *Phys. J.*, **7**, 51–54.
134. Biallozor, S. and Kupniewska, A. (2005) *Synth. Met.*, **155**, 443–449.
135. Davids, P.S., Saxena, A., and Smith, D.L. (1995) *J. Appl. Phys.*, **78**, 4244–4252.
136. Burroughes, J.H., Bradley, D.D.C., Brown, A.R., Marks, R.N., Mackay, K., Friend, R.H., Burn, P.L., and Holmes, A.B. (1990) *Nature*, **347**, 539–541.
137. Kalinowski, D. (1999) *J. Phys. D: Appl. Phys.*, **32**, R179–R249.
138. (a) Karg, S., Scott, J.C., Salem, J.R., and Angelopoulos, M. (1996) *Synth. Met.*, **80**, 111–117; (b) Eckle, M. and Decher, G. (2001) *Nano Lett.*, **1**, 45–49.
139. (a) Dresselhaus, M.S. and Thomas, I.L. (2001) *Nature*, **414**, 332–337; (b) Crabtree, G.W., Dresselhaus, M.S., and Buchanan, M.V. (2004) *Phys. Today*, **57**, 39–44.
140. Brabec, J.C. (2004) *Sol. Energy Mater. Sol. Cells*, **83**, 273–292.
141. Brabec, J.C., Dyakonov, V., Parisi, J., and Sariciftci, N.S. (eds) (2003) *Organic Photovoltaics: Concepts and Realization*, Springer, Berlin.
142. Chan, W.K., Man, K.Y., Cheng, K.W., and Tse, C.W. (2008) Fabrication of photovoltaic devices by layer-by-layer polyelectrolyte deposition method, in *Nanoscale Phenomena* (eds Z. Tang and P. Sheng), Springer, New York, pp. 185–190.
143. (a) Beek, W.J.E., Wienke, M.M., and Janssen, R.A.J. (2004) *Adv. Mater.* **16**, 1009–1012; (b) Beek, W.J.E., Wienk, M.M., Kemerink, M., Yang, X., and Janssen, R.A.J. (2005) *J. Phys. Chem. B*, **109**, 9505–9516.
144. McQuade, D.T., Pullen, A.E., and Swager, T.M. (2000) *Chem. Rev.*, **100**, 2537–2574.
145. Coakley, K.M. and McGehee, M.D. (2003) *Appl. Phys. Lett.*, **83**, 3380–3382.
146. Ravirajan, P., Bradley, D.D.C., Nelson, J., Haque, S.A., Durrant, J.R., Smit, H.J.P., and Kroon, J.M. (2005) *Appl. Phys. Lett.*, **86**, 143101.
147. Kim, J.Y., Kim, S.H., Lee, H.-H., Lee, K., Ma, W., Gong, X., and Heeger, A.J. (2006) *Adv. Mater.*, **18**, 572–576.
148. MacDiarmid, A.G. (1985) *Short Course on Conductive Polymers*, SUNY, New Platz, NY.
149. DeBerry, D.W. (1985) *J. Electrochem. Soc.*, **132**, 1022–1026.
150. See, for instance (a) Grundmeiter, G., Schmidt, W., and Stratmann, M. (2000) *Electrochim. Acta*, **45**, 2515–2533; (b) Rammelt, U., Nguyen, P.T., and Plieth, W. (2003) *Electrochim. Acta*, **48**, 1257–1262.
151. Tallman, D.E., Spinks, G., Dominis, A., and Wallace, G.G. (2002) *J. Solid State Electrochem.*, **6**, 73–84.
152. Ferreira, C.A., Aeiyach, S., Aaron, J.J., and Lacaze, P.C. (1996) *Electrochim. Acta*, **41**, 1801–1809.
153. Fraoua, K., Aeiyach, S., Aubard, J., Delamar, M., Lacaze, P.C., and Ferreira, C.A. (1999) *J. Adhes. Sci. Technol.*, **13**, 517–522.
154. Drexler, K.E. (1986) *Engines of Creation*, Anchor Books/Doubleday.
155. (a) Grimsdale, A.C. and Müllen, K. (2005) *Angew. Chem.*, **117**, 5732–5772; (b) Grimsdale, A.C. and Müllen, K. (2005) *Angew. Chem. Int. Ed.*, **44**, 5592–5692.
156. Haupt, K. (2003) *Chem. Commun.*, 171–178.
157. Kaskel, S., Schüth, F., and Stöker, M. (eds) (2004) Metal organic open frameworks, in *Microporous and Mesoporous Materials*, Vol. 73, pp. 1–104.
158. Kitagawa, S., Kitakura, R., and Noro, S. (2004) *Angew. Chem.*, **116**, 2388–2430.
159. Julian, B., Planelles, J., Cordoncillo, E., Escribano, P., Aschehoug, P., Sanchez, C., Viana, B., and Pelle, F. (2006) *J. Mater. Chem.*, **16**, 4612–4618.

160. Yao, N., Xiong, G., Sheng, S., He, M., and Yeung, K.L. (2002) *Stud. Surf. Sci. Catal.*, **143**, 715–722.

161. Shan, W. (2005) *Chem. Commun.*, **14**, 1877–1879.

162. Sanchez, C. and Ribot, F. (1994) *New J. Chem.*, **18**, 1007–1047.

163. Mackenzie, J.D. and Berscher, E.P. (2007) *Acc. Chem. Res.*, **40**, 810–818.

164. Ogoshi, T., Itoh, H., Kim, K.-M., and Chujo, Y. (2002) *Macromolecules*, **35**, 334–338.

165. Wohlrab, S., Friedrich, J., and Meyer-Plath, A. (2008) US Patent 2008148619.

166. Teowee, G., McCarthy, K.C., Baertlein, C.D., Boulton, J.M., Motaker, S., Bukovski, T.J., Alexander, T.P., and Uhlmann, D.R. (1996) *Mater. Res. Soc. Symp. Proc.*, **435**, 559–564.

167. Mackenzie, J.D. and Beschler, E.P. (1998) *J. Sol-Gel Sci. Technol.*, **13**, 371–377.

168. Ozin, G.A. (1992) *Adv. Mater.*, **4**, 612–649.

169. Donthu, S., Pan, Z., Myers, B., Shekhawat, G., Wu, N., and Dravid, V. (2005) *Nano Lett.*, **5**, 1710–1715.

170. Tekin, E., Egbe, D.A.M., Kranenburg, J.M., Ulbricht, C., Rathgeber, S., Birckner, E., Rehmann, N., Meerholz, K., and Schubert, U.S. (2008) *Chem. Mater.*, **20**, 2727–2735.

171. Yang, C.-H., Bhongale, C.J., Chou, C.-H., Yang, S.-H., Lo, C.-N., Chen, T.-M., and Hsu, C.-S. (2007) *Polymer*, **48**, 116–128.

172. Advincula, R.C., Inaoka, S., Roitman, D., Frank, C., Knoll, W., Baba, A., and Kaneko, F. (2000) *Mater. Res. Soc. Symp. Proc.*, **558**, 415–420.

173. Biswas, K., Gangopadhyay, S., Kim, H.-C., and Miller, R.D. (2006) *Thin Solid Films*, **514**, 350–354.

174. Wu, Z., Walish, J., Nolte, A., Zhai, L., Cohen, R.E., and Rubner, M.F. (2006) *Adv. Mater.*, **18**, 2699–2702.

175. Fujishima, A. and Honda, K. (1972) *Nature*, **238**, 37–38.

176. Al-Dmour, H., Taylor, D.M., and Cambridge, J.A. (2007) *J. Phys. D: Appl. Phys.*, **40**, 5034–5038.

177. Hu, G., Meng, X., Feng, X., Ding, Y., Zhang, S., and Yang, M. (2007) *J. Mater. Sci.*, **42**, 7162–7170.

178. Robel, I., Kuno, M., and Kamat, P.V. (2007) *J. Am. Chem. Soc.*, **129**, 4136–4137.

179. Kaune, G., Wang, W., Metwalli, E., Ruderer, M., Roßner, R., Roth, S.V., and Müller-Buschbaum, P. (2008) *Eur. Phys. J. E.*, **26**, 73–79.

180. Ikeda, N. and Miyasaka, T. (2005) *Chem. Commun.*, **14**, 1886–1888.

181. Blouin, N., Michaud, A., Gendron, D., Wakim, S., Blair, E., Neagu-Plesu, R., Belletête, M., Durocher, G., Tao, Y., and Leclerc, M. (2008) *J. Am. Chem. Soc.*, **130**, 732–742.

182. Yoshida, T., Zhang, J., Komatsu, D., Sawatani, S., Minoura, H., Pauporté, T., Lincot, D., Oekermann, T., Schlettwein, D., Tada, H., Wöhrle, D., Funabiki, K., Matsui, M., Miura, H., and Yanagi, H. (2009) *Adv. Funct. Mater.*, **19**, 17–43.

183. Hodes, G. (1995) in *Physical Electrochemistry: Principles, Methods, Applications* (ed. I. Rubinstein), Marcel Dekker, New York, p. 515.

184. Dhere, N.G. (2006) *Sol. Energy Mater. Sol. Cells*, **90**, 2181–2190.

185. Jones, D.A. (1996) *Principles and Prevention of Corrosion*, Prentice Hall, New York.

186. Kasten, L.S., Grant, J.T., Grebasch, N., Voevodin, N., Arnold, F.E., and Donley, M.S. (2001) *Surf. Coat. Technol.*, **140**, 11–15.

187. (a) Khramov, A.N., Voevodin, N.N., Balbyshev, V.N., and Donley, M.S. (2004) *Thin Solid Films*, **447**, 549–557; (b) Zheludkevich, M.L., Salvado, I.M., and Ferreira, M.G.S. (2005) *J. Mater. Chem.*, **15**, 5099–5103.

188. (a) García-Heras, M., Jimenez-Morales, A., Casal, B., Galvan, J.C., Radzki, S., and Villegas, M.A. (2004) *J. Alloys Compd.*, **380**, 219–224; (b) Pepe, A., Aparicio, M., Cere, S., and Duran, A. (2004) *J. Non-Cryst. Solids*, **348**, 162–171.

189. Voevodin, N.N., Grebasch, N.T., Soto, W.S., Arnold, F.E., and Donley, M.S. (2001) *Surf. Coat. Technol.*, **140**, 24–28.

190. (a) Palanivel, V., Huang, Y., and van Ooij, W.J. (2005) *Prog. Org. Coat.*, **53**, 153–168; (b) van Ooij, W.J., Zhu, D., Stacy, M., Seth, A., Mugada, T., Gandhi, J., and Puomi, P. (2005) *Tsinghua Sci. Technol.*, **10**, 639–664.

191. Vreugdenhil, A.J. and Woods, M.E. (2005) *Prog. Org. Coat.*, **53**, 119–125.

192. Zheludkevich, M., Shchukin, D.G., Yasakau, K.A., Möhwald, H., and Ferreira, M.G.S. (2007) *Chem. Mater.*, **19**, 402–411.

193. Shchukin, D.G., Zheludkevich, M., Yasakau, K., Lamaka, S., Ferreira, M.G.S., and Möhwald, H. (2006) *Adv. Mater.*, **18**, 1672–1678.

194. Shchukin, D.G., Lamaka, S.V., Yasakau, K.A., Zheludkevich, M.L., Ferreira, M.G.S., and Möhwald, H. (2008) *J. Phys. Chem. C*, **112**, 958–964.

Index

Functional Polymer Films, First Edition. Edited by Wolfgang Knoll and Rigoberto C. Advincula.
© 2011 Wiley-VCH Verlag GmbH & Co. KGaA. Published 2011 by Wiley-VCH Verlag GmbH & Co. KGaA.